Logic Functions and Equations

Bernd Steinbach • Christian Posthoff

Logic Functions and Equations

Fundamentals and Applications using the XBOOLE-Monitor

Third Edition

Bernd Steinbach
Computer Science
TU Bergakademie Freiberg (retired)
Chemnitz, Sachsen, Germany

Christian Posthoff
Computing and Information Technology
University of the West Indies (retired)
Chemnitz, Sachsen, Germany

ISBN 978-3-030-88947-0   ISBN 978-3-030-88945-6   (eBook)
https://doi.org/10.1007/978-3-030-88945-6

This Springer imprint is published by the registered company Springer Nature Switzerland AG
The registered company address is: Gewerbestrasse 11, 6330 Cham, Switzerland

Foreword

Logic functions and equations are foundational to many areas of Mathematics, Computer Science, and Computer Engineering. A thorough understanding of the fundamentals and how they are applied is critical to solving problems and developing applications in Information Technology. This textbook provides students and practitioners with a well-structured coverage of the theory and application of logic functions and equations.

Part I introduces the fundamentals underpinning the practical application of the programming system XBOOLE and the XBOOLE-monitor XBM 2, which is a powerful interactive tool developed and made freely available by the authors. Provision of this software will be of tremendous assistance to readers in exploring the concepts covered in this book and in undertaking their subsequent work in the area.

The conceptual background of XBOOLE is of importance in its own right and knowledge of that background will provide the reader with a very practical understanding of an effective approach to computation involving logic functions and equations. For this book, XBOOLE and the XBOOLE-monitor XBM 2 are of particular importance, as they are used in solving exercises emphasizing the ideas introduced in each chapter. Indeed, it is a major strength of this textbook that the authors have provided exercises with sample solutions and challenging supplementary exercises throughout the book.

Part II provides a comprehensive progressive introduction of the fundamentals of logic functions and equations. The order of topics is very well planned.

Basic algebraic structures are introduced, leading to the detailed introduction of logic functions and logic equations. The coverage of these areas is very well done and together with the exercises provided will give the reader a firm foundation for the topics to follow.

The more advanced fundamentals: Boolean differential calculus, sets, lattices, and classes of logic functions are also well covered. A good understanding of the material in Part II will serve the reader well in considering both theoretical and practical applications of logic functions and equations.

Part III covers a variety of application areas based on the fundamentals introduced in Part II using XBOOLE and the XBOOLE-monitor XBM 2 introduced in Part I. The breadth of application areas introduced is a particular strength of this textbook.

Part III begins with discussions of logic and arithmetic applications that are traditionally covered in books on logic functions and equations. However, here we also find coverage of special functions and, in particular, bent functions that are fundamental to cryptography applications and cybersecurity, two particularly critical areas in modern IT. The solved exercises in this area are particularly informative and would be very difficult to undertake without the support of XBOOLE.

The chapters on SAT-problems and extremely complex problems are very significant in that they illustrate the complexities of various problems as well as methods for addressing those complexities employing logic functions and equations. Study of these chapters will provide the reader with new ways to visualize, formulate, and solve complex problems they may encounter in future work. The presentation of this material is again greatly facilitated by the use of XBOOLE.

Applications related to combinational and sequential circuits are covered in many books. The strength here is that the material preceding those topics sets the discussion of circuit issues in a clear computational and complexity context. The reader has been given both the in-depth fundamentals and a strong understanding of approaches to complex problems to thoroughly understand the problems and potential solutions.

I strongly recommend this textbook to students and others seeking a comprehensive knowledge of the theory and practical application of logic functions and equations. The authors have addressed a broad spectrum of issues in a clear and consistent manner.

Victoria, Canada,   D. Michael Miller
January 2021

Preface

This textbook provides comprehensive knowledge about logic functions and equations. The basics of this knowledge are used, at least intuitively, by everyone. These basics are successively extended by details and approaches that belong to the state of the art of applications of logic functions and equations and reach topics belonging to recent areas of research. Hence, no previous knowledge is needed to read this book, and the study of this book increases the level of knowledge to get degrees from universities and even beyond.

The common ground of logic functions and equations are variables that can store only two different values; hence, these variables are the simplest variables of all. Depending on the application, the two different values can arbitrarily be chosen. The values True and False are used for logic application, the values On and Off can be used to determine the position of a switch, but very often the digits 1 and 0 are used as values of such a logic variable, synonymously also denoted by Boolean or a binary variable.

Due to the use of these simplest variables, *K. Zuse* successfully built the first computer, and the strong progress in the recent age of digitization is still based on these variables. The drawback of these simplest variables is that many of them are usually needed to express the required different states and the number of different binary vectors of logic variables exponentially increases with the number of logic values in such a vector.

The manipulation of a large number of logic values is both time consuming and error prone for human beings; computer solves such elementary tasks much faster and provides the correct results. However, the computer must know in which manner the logic values must be combined. Logic functions and solution-sets of logic equations are convenient objects in algorithms that solve tasks for a wide range of logic problems. However, programming languages provide only operations to compute single logic values or vectors of such values, but due to the already mentioned complexity problem, such languages do not directly provide operations for logic functions or solution-sets of logic equations. To close this gap, we developed the programming system XBOOLE.

Learning a comprehensive and ready-to-use body of knowledge requires not only the study of the theory but also the application and consolidation of the insights. Therefore, we extended this third edition of this textbook by exercises. Provided solutions for a number of these exercises support the readers when verifying their own solutions. All exercises can be solved using the XBOOLE-monitor XBM 2 that provides easy access to the functions of the XBOOLE-library and can be used by everyone free of charge. The benefit of the XBOOLE-monitor XBM 2 is that the logic tasks are solved very efficiently, skills about programming languages are not needed, and the huge overhead of elementary manipulations of bits will be avoided. We noticed that the exploration of problems and the creation of XBOOLE-problem-programs for their solution significantly increased the level of education of our students. We hope that this textbook contributes to spreading this welcome observation to all readers.

The two chapters of Part I introduce in a compact manner the concepts of XBOOLE and the use of the XBOOLE-monitor XBM 2 so that exercises based on this monitor can be solved after reading these two chapters. Part I can later be used as a helpful reference. More details about the fundamentals of XBOOLE are provided in Part II of this textbook.

The five chapters of Part II explain the fundamentals needed to solve problems that can be specified using the already mentioned logic variables. Logic functions will be defined based on the sets $\mathbb{B}$ and $\mathbb{B}^n$. Boolean Algebras facilitate useful different views of logic functions and equations. The strong interrelation between logic functions and logic equations becomes obvious with the possibility to use ternary vector lists (TVLs), the main data structure of XBOOLE, to represent different forms of logic functions as well as solution-sets of logic equations. We explain this data structure in detail and show that a single ternary vector that contains d dash $(-)$ elements expresses 2^d binary vectors; hence, TVLs are able to reduce the required memory as well as the required computation time exponentially.

Digital systems are specified not only by the values of certain logic variables but also by the changes of these values. The *Boolean Differential Calculus* expands the static concepts of logic functions and equations by operations that specify several types of changes. A very welcome property of a subset of derivative operations is that their results depend on fewer variables than the evaluated logic functions. Efficient algorithms that compute the results of derivative operations of the Boolean Differential Calculus have been implemented in the XBOOLE-library and can be used in the XBOOLE-monitor XBM 2.

There are applications in which not only a single logic function but a set of such functions occur. In many cases, a set of logic functions satisfies the conditions of a lattice. We show that not all functions of such a lattice must be computed separately, that two mark-functions suffice to describe all logic functions of the lattice, and that the computation can be restricted to these mark-functions. Even all derivative operations of the Boolean Differential calculus can be computed for lattices of logic functions using both mark-functions; generalized lattices result can be used in subsequent computations.

The solution of logic equations with regard to variables is a very fundamental concept. Several beneficial applications of this concept will be demonstrated in the third part of this textbook. Derivative operations of the Boolean differential calculus are used to check whether a given logic equation can be solved with regard to a certain subset of variables, to decide whether the solution is a single logic function or a lattice of logic functions, and to compute the single solution-function or the mark-functions of the lattice.

Boolean differential equations extend the field of applications. Different to a logic equation that has a set of binary vectors as its solution, a Boolean differential equation has a set of logic functions as its solution. For certain Boolean differential equations, the solution consists of a set of equivalence classes of logic functions. We show how Boolean differential equations can be solved using the functions of the XBOOLE-monitor XBM 2.

The five chapters of Part III demonstrate the use of logic functions and equations within an important subset of fields of applications. We encourage the readers to utilize the provided fundamentals of Part II and computation power of the XBOOLE-monitor XBM 2 also for tasks in other fields.

Logic functions influence our everyday life: we use the propositional logic for our decisions, arithmetic operations are executed inside of the computers using the binary arithmetic, or coding helps to avoid the errors of transmitted data or increase the security of secret information. The specific normal forms or most complex functions extend our knowledge about logic functions, and bent functions are used in the field of cryptography.

Many practical problems can be modeled as satisfiability (SAT)-problem that is a restrictive logic equation, where the logic function on the left-hand side has a conjunctive form. SAT-problems belong to the set of most complex problems. We demonstrate how ternary vector lists can be used to solve such problems and describe approaches to shorten the required solution time. Several examples are used to show how a given problem can be modeled and solved as a SAT-problem using the XBOOLE-monitor XBM 2.

It is very hard to solve a problem that requires a large number n of logic variables for its specification because 2^n potential solution patterns must be considered. We demonstrate how we solved such an extremely complex problem for $n = 648$, which means we were faced with the gigantic set of more than 10^{195} potential solutions. The key to solving such an extremely complex problem consists of the detection of all hidden properties of the given problem and their utilization with the appropriate approaches of logic functions and equations.

All digital systems contain combinational circuits. There is a strong relation between logic functions and combinational circuits. Both the logic gates of combinational circuits and the behavior of the outputs of combinational circuits can be described by logic functions. New results in analysis, synthesis, and test of combinational circuits contribute to the technological progress of digital systems and vice versa; the requirements of such systems force the development of innovative approaches for improved combinational circuits. In addition to well-known covering methods, we explain powerful decomposition methods that utilize the properties of lattices of logic functions to synthesize completely testable combinational circuits with short delays, small areas, and low power consumption.

Sequential circuits realize finite state machines. These circuits combine combinational circuits and memory-elements. Sequential circuits must be distinguished between asynchronous and synchronous circuits. For these two types of sequential circuits, we explain the methods of analysis and synthesis. The solution of logic equations with regard to variables facilitates the common utilization of properties determined by the required behavior and the lattices that specify the behavior of the selected flip-flops.

In this textbook, we combine the fundamentals of logic functions and equations that represent the state of the art with the results of our own research in this field over more than four decades. Some more or less independent areas existed when we started this cooperative research. One part is very related to Mathematics or can even be considered as a fundamental tool for the construction of *axiomatic Mathematics* altogether. This part goes back to *G. Boole* and other famous mathematicians.

The second field is the use of the binary number system that has its roots in the papers of *G. F. Leibniz* and was *electrically implemented* by *C. E. Shannon* for relay circuits. It is nowadays a giant part of *Information Technology* and one of the fundamental concepts of circuit design.

The third intention, which only started at that time, had the aim to contribute to the development of *Artificial Intelligence* by using rule-based systems. Implication has played a major role in this field. We saw only later that our concepts will also cover this part. The developed algorithms are very efficient; it was possible to solve many problems in a very instructive way.

At the beginning of our cooperative research, only some roots of *Boolean Differential Calculus* were known. This encouraged us to extend this calculus and apply it to find successful solutions for previously open problems.

The most important stimulus for this third edition of this textbook was the possibility to present the research work of about 40 years to others who can benefit from these results, theoretically as well as practically. The combination of the presentation of the theoretical basis with exercises to enlarge the knowledge learned supports the teaching starting from zero to reaching a level as high as necessary or possible. Logic functions, logic equations, and the Boolean differential calculus are now an inseparable theoretical basis of algorithms to solve problems belonging to the Boolean domain. The XBOOLE-system supplements the theoretical basis with software tools for practical solutions. We see this textbook as both the summit and completion of our successful collaboration.

Many people, colleagues, and students contributed to this book by comments and discussions. We are grateful to all of them. We are also very grateful to Springer, who gave us the excellent possibility to transform the research work of many years into a textbook that is available for everyone and helpful for all the readers.

Chemnitz, Germany   Bernd Steinbach

Chemnitz, Germany   Christian Posthoff
January 2022

Contents

Acronyms

AF	Antivalence form
AP	Antivalence polynomial
ASCII	American Standard Code for Information Interchange
ANF	Antivalence normal form
BCD	Binary-coded decimal
BDD	Binary decision diagram
CF	Conjunctive form
CMOS	Complementary Metal Oxide Semiconductor
CNF	Conjunctive normal form
DEO	Derivative operation
DC	Don't-care
DF	Disjunctive form
DNF	Disjunctive normal form
EF	Equivalence form
EP	Equivalence polynomial
ESOP	Exclusive-OR sum of products
EXOR	Exclusive-OR
ENF	Equivalence normal form
FSM	Finite state machine
GCD	Greatest Common Divisor
IS	Intermediate solution
ISF	Incompletely specified function
LCA	Logic cell array
LCM	Least Common Multiple
NTV	Number of ternary vectors
MCF	Most complex function
MOSFET	Metal Oxide Semiconductor Field Effect Transistor
ML	Memory list
ODA	Orthogonal disjunctive form
OKE	Orthogonal conjunctive form
PLA	Programmable logic array
PLD	Programmable logic device
RAM	Random-access memory
RGB	Red–green–blue
ROBDD	Reduced ordered binary decision diagram
ROM	Read-only memory
RT	Remaining task

SA0-error	Stuck-at-zero error
SA1-error	Stuck-at-one error
SAT-error	Stuck-at-T error
SAT	Satisfiability
SD	Space definition
SL	Space list
SNF	Special (or specific) normal form
TM	Ternary matrix
TVL	Ternary vector list
UCP	Unate covering problem
VL	Variable list
VT	Variable tuple
VV	Vector of existing variables

The Concepts of XBOOLE

1

Abstract

A logic function that depends on n variables specifies 2^n function values. The maximal number of solutions of a logic equation of n variables is also equal to 2^n. This exponential increasing number of elements limits the number of logic variables in tasks that can be successfully solved by human beings; hence, computers must be used to solve tasks beyond a certain limit of logic variables. Programming languages of computers can store and manipulate single logic variables or bounded vectors of logic values. The creation of programs that solve challenging logic problems requires comprehensive skills in programming and is time consuming due to the low level of available elements of the programming language. To close this gap, we developed the XBOOLE-system that can efficiently store and manipulate logic functions and solution-sets of logic equations. In this chapter, we explain the concepts of XBOOLE in a compact manner. This background knowledge about XBOOLE supports the readers of this textbook to solve the exercises provided in the following chapters.

Supplementary Information The online version of this chapter (https://doi.org/10.1007/978-3-030-88945-6_1 contains supplementary material which is available for authorized users. Please, follow the link belonging to the version of the XBOOLE-monitor XBM 2 that fits best for your operating system. This XBOOLE-monitor is needed to solve all tasks provided in all subsequent chapters. Instructions for starting the downloaded XBOOLE-monitor XBM 2 are given at the end of this chapter.

XBOOLE-monitor XBM 2 for Windows 10
32 bits
https://doi.org/10.1007/978-3-030-88945-6_1_MOESM1_ESM.zip (15,091 KB)

64 bits
https://doi.org/10.1007/978-3-030-88945-6_1_MOESM2_ESM.zip (14,973 KB)

XBOOLE-monitor XBM 2 for Linux Ubuntu
32 bits
https://doi.org/10.1007/978-3-030-88945-6_1_MOESM3_ESM.zip (29,522 KB)

64 bits
https://doi.org/10.1007/978-3-030-88945-6_1_MOESM4_ESM.zip (28,422 KB)

1.1   Motivation and Aims

We are living in the digital age. More and more digital procedures and devices replace their previously used analog counterparts, and very much so far unsolved problems become both technically and economically realizable.

A unique property of such digital problems is that they can be expressed using a *finite number* of states or items. Due to this finite number, an encoding of all states or items by means of logic (binary, Boolean) variables is possible.

The use of logic variables is a primum mobile for digital devices and applications. A logic variable can have only two different values: *False* and *True*, which are often expressed by the values 0 and 1. Technically, it is very easy to express these two values by two different voltages and even in the case of a perturbation non-overlapping intervals of voltages facilitate to distinguish these two values.

Programming languages of computers usually provide both the data type `Boolean` to store such logic values and the associated logic operations. Hence, the basic requirement to solve any digital problem is given. However, the encoding of the different states requires many logic variables well-ordered in a binary vector. Such vectors and associated logic operations are also available in programming languages of computers, but they are restricted to the width of the registers of the used computer; usually to 32, 64, or 128 bits.

The states or items, which express a certain property of a digital device, require sets of binary vectors of a length that can be greater than the width of the registers of the available computer, and programming languages do not provide operations to manipulate sets of large binary vectors. Of course, the logic operations mentioned above can be used to realize the required logic operations for given sets of binary vectors of an arbitrary length. Such basic software is needed for almost all applications of digital systems; hence, repeated developments of such software waste development efforts and increase the prices of the final digital products.

There is one more critical issue. The maximal count of binary vectors of n bits is equal to 2^n. This exponential growth is a strong challenge for the development of a basic software that is able to operate large sets of binary vectors. Hence, a basic software that is able to compute large sets of binary vectors and that satisfies the following aims is desirable and very useful for the digital system:

- unlimited number of logic variables;
- small memory space to store the needed sets of binary vectors;
- fast computation of sets of binary vectors;
- an appropriate, but small set of operations that facilitates us to solve almost all digital problems;
- the basic software should be both efficiently usable for commercial applications and clear for teaching.

There are several software tools for logic calculations. We developed and optimized the software system *XBOOLE* that satisfies the aims listed above. In this book we use the tool that is denoted by *XBOOLE-monitor XBM 2*. It provides enough logic operations that can easily be combined to solve very many logic problems. The benefits of such a "Boolean pocket calculator" are:

- fast executions of the operations (possible errors caused by human beings are avoided);
- alternative and easy-to-use options for the input of the problem;
- several well-visible representations of the results;
- possibility to solve digital problems of a very wide range.

The knowledge of the basic concepts of XBOOLE that satisfies the aims stated above is very helpful for the use of the XBOOLE-monitor XBM 2. Therefore, we give in the next sections a brief introduction into these concepts.

1.2   Ternary Matrix (TM)

Matrices of binary vectors seem to be the self-evident data structure to store and compute sets of binary vectors. The drawback of a binary matrix of n columns is that this matrix can require up to 2^n rows. This exponential requirement of memory leads to a computation time that also exponentially increases depending on the width n of the binary vectors.

XBOOLE extenuates this drawback by the use of ternary vectors instead of binary vectors. The basic idea of this concept is as follows:

- if two binary vectors differ only in a single position, then these two binary vectors are summarized into a single ternary vector that contains a dash $(-)$ in this position and the same values 0 or 1 as in the given binary vectors in the other positions;
- if two ternary vectors differ only in a single position with a combination of a value 0 in one vector and a value 1 in the other vector, then these two ternary vectors are summarized into a single ternary vector that contains a dash $(-)$ in this position and the same values 0, 1, or $-$ as in the given ternary vectors in the other positions.

Pairs of binary or ternary vectors can be repeatedly combined as long as the conditions mentioned above are satisfied. In this way, a ternary vector with d dashes can be built; such a single ternary vector represents 2^d binary vectors. Hence, the concept of ternary vectors has the effect that the needed memory space of a set of binary vectors *decreases* exponentially. A strongly decreased number of vectors decrease also the time needed for their computation. The great benefit of the ternary representation becomes obvious when all 2^n binary vectors of n logic variables are represented by a single ternary vector consisting of n dashes. Chapters 4 and 5 provide more details about this fundamental concept.

The ternary vectors that represent a given set of binary vectors are stored as rows of a *ternary matrix* (TM). A ternary matrix is the most important abstract data type of XBOOLE.

1.3   Space Concept

The space concept of XBOOLE is characterized as follows:

- the user can define an arbitrary number of Boolean spaces;
- each such *space definition* (SD) is an *object* within the XBOOLE-system;
- the user defines a Boolean space by specifying the maximal number of logic variables (denoted by *VMAX*) that can be assigned to this space as well as a unique number of this space;
- based on the maximal number of variables, the required number of machine words to store a ternary vector is computed (denoted by *TYPE*) and stored in the XBOOLE-object of this Boolean space;
- the XBOOLE-object of each Boolean space contains furthermore a list of references to the logic variables used in this space;
- a Boolean space cannot be deleted or recreated within one XBOOLE-system, and only its list of references to Boolean variables can be extended until the maximal number of logic variables is reached;
- the XBOOLE-object *space list* (SL) stores the references to the defined spaces using the unique number of the space as index; hence, the user of XBOOLE can define an unlimited number of Boolean spaces;

- each *ternary matrix* (TM) belongs to exactly one Boolean space; hence, the ternary vectors of all TMs of this space have the same length, and each position in a selected machine word uniquely determines a logic variable for all TMs associated to the same space.

This space concept contributes to satisfy the aims of a small memory space to store the ternary matrices as well as their fast computation.

At first glance, the aims of an unlimited number of variables, an exponential growth of the number of binary vectors depending on the number of variables, and a small memory space to store the ternary matrices seem to be a contradiction. These aims would be really a contradiction when all logic variables are inserted in each binary vector used to solve a given problem and a large number of variables are needed. XBOOLE resolves this contradiction in the distribution of the logic variables to several Boolean spaces.

The users of XBOOLE can specify the maximal number of variables of each Boolean space as large as needed to solve the problem, but they can decrease the computation effort when they choose these numbers as small as possible. In this way, the maximal size of a TM is restricted. An unlimited number of Boolean spaces facilitates that the count of all logic variables remains unlimited, but this number is also restricted in each Boolean space according to the requirements of the problem determined by the user.

The position where a logic variable is stored in the given TMs is a key issue for their fast computation. Basically, logic variables can be assigned in an arbitrary order to the columns of a TM and the set of variables of two TMs can be identical, partially overlapped, or even disjoint (no variable of the first TM appears in the second TM). The fixed unique positions of the variables in the stored ternary vectors of one Boolean space avoid time-consuming selections of the values for one variable from different positions.

1.4   Unlimited Number of Logic Variables

The count of logic variables usable in one Boolean space is determined as a fixed finite number by the user. The possibility to define an unlimited number of Boolean spaces implicitly provides an unlimited number of logic variables within one XBOOLE-system.

It remains the question how the names of the logic variables are administrated within XBOOLE. The names of these variables are needed for the input and output, but they hamper the internal computation; hence, XBOOLE stores the names of logic variable within a *variable list* (VL).

Each name of a used logic variable is stored exactly once in the VL so that the associated index uniquely indicates the name of the variable. The unique index of a logic variable in the VL is used as reference to this variable in the definition of a Boolean space (SD) and determines therefore the position (column) of this variable in all TMs of this space. In this way, the names of variable must not be considered during the computation of TMs so that their computation becomes faster.

1.5   Memory Concept

Usually the needed memory space to store a TM is not known before its computation. The user of XBOOLE is not faced with the problem to specify the memory space required to store a TM; this task is implicitly solved by the *box system* of XBOOLE that assigns as much boxes of memory as needed to a TM. Such a box has a fixed size so that boxes of deleted TMs can be reused without a fragmentation of the memory.

All *objects* of XBOOLE are stored in these boxes. The size of a box is determined by a constant in the XBOOLE software. This size has been chosen as tradeoff between the needed number of boxes for one TM and the unused memory space in the last box of a TM. One rule of XBOOLE is that a ternary vector does not need more memory space than the size of a box. For extremely long ternary vectors, the constant of the box size can be increased.

The boxes of XBOOLE are chained in two directions in a so-called list of empty boxes. Boxes needed for a XBOOLE-*object* are taken from the head of this list. Boxes of a deleted XBOOLE-object are append at the tail of the list of empty boxes so that these boxes can be reused for other XBOOLE-objects.

XBOOLE dynamically requests memory space needed to create a certain number of boxes from the operating system. XBOOLE requests more memory space from the operating system and configures this memory into boxes that are used to extend the list of empty boxes in the case that all boxes of the list of empty boxes are assigned to XBOOLE-objects.

Each XBOOLE-object is stored as a chain of boxes and a pointer refers to its head. A *fix code* in the boxes is used to verify that the used pointer refers to the head of a box and a *variable code* specifies the type of the stored object.

All needed manipulations with these boxes are internally done by the XBOOLE-system. Users of the XBOOLE-library get access to the XBOOLE-objects by means of pointers associated to the head box of the object, and users of the XBOOLE-monitor XBM 2 refer to the TMs (relevant XBOOLE-objects) simply by numbers.

1.6 Form Predicate of a Ternary Vector List

A ternary vector can be used to represent either a conjunction or a disjunction of Boolean literals. A negated Boolean variable is represented by an element 0 and a non-negated variable by an element 1, respectively. Variables of a logic function that does not occur in the represented conjunction (disjunction) are expressed by dashes in the corresponding positions. Using this mapping, all four basic forms of logic functions:

- the *disjunctive form* (D) is a disjunction of conjunctions;
- the *antivalence form* (A) is an antivalence of conjunctions;
- the *conjunctive form* (K) is a conjunction of disjunctions;
- the *equivalence form* (E) is an equivalence of disjunctions

are represented in XBOOLE using a *ternary vector list* (TVL) and a stored *form predicate*. A TVL is a subtype of a TM.

If all pairs of ternary vectors of a TVL are disjoint, this list satisfies the *orthogonality* condition. In this case the TVL of a logic function in D-form is equivalent to this function in A-form (stored as predicate ODA) and the logic function in K-form can be expressed as function in E-form (stored as predicate OKE). Orthogonal TVLs are preferred in XBOOLE.

The *code* stored in an object of XBOOLE specifies an abstract ternary matrix (TM) as a concrete ternary vector list (TVL). Each TVL has additionally a *form predicate* to distinguish between the six forms. More details about logic functions in the forms mentioned above and their associated TVLs are provided in Chap. 4.

1.7 Sets and Ordered Sequences of Logic Variables

XBOOLE provides operations that compute a certain TVL controlled by a set or ordered sequences of logic variables. Both a set of logic variables and an ordered sequence of logic variables can be the result of a XBOOLE-operation.

A set of logic variables is a subset of variables of a Boolean space, and each ternary matrix (TM) specifies such a subset of variables. XBOOLE stores this predicate (the set of logic variables) as a *vector of existing variables* (VV), a hidden vector on top of each TM; hence, any TM can be used to specify a set of logic variables. The ternary vectors of a TM are not evaluated when the set of its variables is needed; only its VV determines this set.

A set of logic variables that has been computed by XBOOLE is represented by an empty TVL in ODA-form; the information about this set is stored as VV of this TVL.

The Boolean space, to which the TM belongs, assures that each variable can occur only once in the vector of existing variables VV. A logic variable can (but must not) occur in an ordered sequence of logic variables more than once.

XBOOLE represents such an ordered sequence of logic variables also as ternary matrix where each ternary vector contains besides of dashes only a single value 1 in the position of a variable of this sequence. The order of such ternary vectors in a TM determines the order of the Boolean variables of this ordered sequence. The property of this fixed order of special ternary vectors is stored as special *code* that specifies the abstract ternary matrix into the concrete *tuple of variables* (VT).

1.8 The XBOOLE-System: Management of TVLs and VTs

XBOOLE can compute and store an unlimited number of ternary matrices (TMs). Each TM is either a *ternary vector list* (TVL) or a *tuple of variables* (VT). References (object addresses *oa*) to these two types of objects are stored in a *memory list* (ML). The index in the ML to a TVL or a VT is used in the XBOOLE-monitor XBM 2 to refer to these objects.

Figure 1.1 shows the structure of the XBOOLE-system. The three head objects *variable list* (VL), *space list* (SL), and *memory list* (ML) provide the access to all other XBOOLE-*objects*. References to the TVLs or VTs are stored in the ML. Each TVL or VT has a reference to the associated *space definition* (SD). Each SD stores the TYPE (number of machine words needed to store one ternary vector), the maximal count of variables (VMAX—specified by the user), and an ordered list of numbers that indicate the names of the associated variables in the *variable list* (VL).

Using the three head objects, a whole XBOOLE-system can be stored as a file with the file type .sdt and re-established from such a file.

1.9 Parallel Computation of the Data Stored in Ternary Vectors

Due to the fixed order of logic variables within a Boolean space of XBOOLE, all ternary elements of vectors with a number of variables up to the word width of the computer are computed completely in parallel by means of logic operations on word level provided by programming languages and realized in the CPU. This leads to a significant speedup in comparison with computation of the truth values of single variables.

Sequential evaluations of machine words are needed for such ternary vectors that contain more logic variables than the number of bits of a machine word. Most frequently the operation to check the orthogonality of ternary vectors must be executed in XBOOLE. Very often, the sequential evaluation

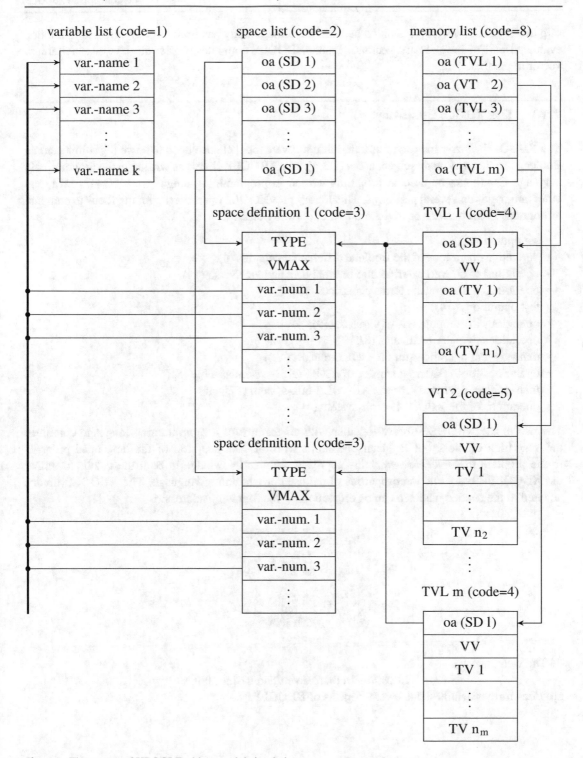

Fig. 1.1 The system of XBOOLE-objects and their relations

of parts of ternary vectors is avoided because the orthogonality has been already detected in the first evaluated part of long ternary vectors. Hence, XBOOLE computes almost all data of each ternary vector in parallel.

1.10   The XBOOLE-Library

The XBOOLE-*library* xb_port is a portable software tool [2] developed to solve high-dimensional Boolean problems of many subject areas [1,3,4]. The XBOOLE-library is written in the programming language C and can be used in programs written in programming languages C, C++, Java, and other languages on several platforms. This library provides 108 operations with the focus to compute orthogonal TVLs that can be divided into the following classes:

- set operations (6);
- derivative operations of the Boolean differential calculus (6);
- operations to convert external into internal data and vice versa (14);
- operations to manipulate ternary matrices (13);
- test operations (14);
- operations to compute sets of variables (9);
- operations to query predicates (8);
- management operations for XBOOLE-objects (12);
- operations for the external storage of a XBOOLE-system as a file (5);
- management operations for the XBOOLE-box-system (17); and
- operations of the XBOOLE-error-system (4).

The use of the XBOOLE-library together with all features of a programming language combines the very high efficiency of Boolean operations with the best adaption of the developed program to the problem to solve. However, this approach requires knowledge in both the details to utilize the XBOOLE-library and the properties of the used programming language. The XBOOLE-library adapted to the needed platform can be ordered using the following address:

> Steinbeis-TZ Logische Systeme
> Nelkentor 7
> D-09126 Chemnitz
> Germany
> FAX: +49 371 53 81 929
> Email: stz158@stw.de.

The web page:

https://tu-freiberg.de/en/fakult1/inf/xboole

provides further details about several aspects of XBOOLE.

1.11   The XBOOLE-Monitor XBM 2

The XBOOLE-monitor XBM 2 is a graphical user interface that provides the use of a large subset of XBOOLE-operations without the need to know a programming language like C or C++. Therefore, the XBOOLE-monitor XBM 2 can be used like a pocket calculator to solve many different Boolean problems. The XBOOLE-monitor XBM 2 allows us to combine the XBOOLE-operations with higher algorithms that solve the given problems. These properties make the XBOOLE-monitor XBM 2 to be an excellent tool to support the education in the Boolean domain. For that reason, we use in this textbook the newest version XBM 2 of several implementations of XBOOLE-monitors to solve exercises related to logic functions and equations.

All subsequent chapters of this book provide in Sections "Exercises" and "Supplementary Exercises" more than 400 tasks that can be solved using the XBOOLE-monitor XBM 2. These tasks are related to both the theory and a wide field of applications of logic functions and equations. Easy tasks help you to become familiar to the use of the XBOOLE-monitor XBM 2. Challenging tasks stimulate you to think about learned relations of logic functions, and solving these tasks contributes to increasing your knowledge in the Boolean domain and more generally in finding approaches to solve real-world problems. Section "Solutions" in the subsequent chapters contain the solutions of all tasks determined in Section "Exercises"; hence, you have the possibility to verify your found solution or get suggestions for possible steps of the solution algorithm. The XBOOLE-monitor XBM 2 works very efficiently and computes the solution of almost all solutions without a noticeable delay. On the high end, we provide some tasks that require very large numbers of logic variables (more than seven hundred) or extremely large numbers of operations to execute, caused by nested loops; the XBOOLE-monitor XBM 2 needs only few minutes to solve even such extreme problems.

The XBOOLE-monitor XBM 2 has been developed for the operating systems Windows 10 and LINUX—Ubuntu. The XBOOLE-monitor XBM 2 exists for these operating systems in the versions of 32 bits and 64 bits. Note that the XBOOLE-monitor XBM 2 for the operating systems Windows 10 can also be used for subsequent Windows systems. The XBOOLE-monitor XBM 2 supports the user with a comprehensive help system. The language of this XBOOLE-monitor and the associated help system can be changed at run time between German and English (may be even more languages in the future).

Each of the mentioned four versions of the XBOOLE-monitor XBM 2 can be *downloaded by everyone free of charge* from the online version of the first chapter of this book.

Authorized users of the online version of this chapter (https://doi.org/10.1007/978-3-030-88945-6_1) can download the XBOOLE-monitor XBM 2 directly from the web page

https://link.springer.com/chapter/10.1007/978-3-030-88945-6_1

where the links for the download of the XBOOLE-monitor XBM 2 are located in the part "Supplementary Information" (below the part "Abstract"). The headline above such a link indicates the associated zip-file of the XBOOLE-monitor XBM 2. The sizes of the zip-files have been provided behind the links and can be used to verify the download. A click on the link of the wanted version of the XBOOLE-monitor XBM 2 starts the download.

Readers of the hardcopy of this book get access to the XBOOLE-monitor XBM 2 using the URL

https://link.springer.com/chapter/10.1007/978-3-030-88945-6_1

to download the first two pages of the first chapter. After this download, the same procedure as the authorized users of the online version of a chapter can be used to download the wanted version of the XBOOLE-monitor XBM 2.

The zip-files of the four versions of XBOOLE-monitor XBM 2 are the same in all chapters; hence, a repeated download is not needed.

The XBOOLE-monitor XBM 2 must not be installed, but must be unzipped into an arbitrary directory of your computer. A convenient tool for unzipping the downloaded zip-file is usually available as part of the operating system or can be downloaded from the Internet.

The executable file of the two versions (32 or 64 bits) for Windows 10 of the XBOOLE-monitor XBM 2 is XBM2.exe; the other files in the expanded directory must remain unchanged. A double-click on the executable file XBM2.exe within the Explorer of Windows starts the XBOOLE-monitor XBM 2.

The unzipped folder of the XBOOLE-monitor XBM 2 contains in the case of the operating system LINUX—Ubuntu only the executable file XBM2-i386.AppImage for the version of 32 bits or XBM2-x86_64.AppImage for the version of 64 bits of the XBOOLE-monitor XBM 2. A double-click on the created AppImage-file within the file manager of LINUX—Ubuntu starts the XBOOLE-monitor XBM 2. Alternatively, you can move this AppImage-file by drag-and-drop to your desktop. A click on the created item also starts the XBOOLE-monitor XBM 2.

All exercises of this book (including the supplementary exercises) have been solved using the 64-bit version of the XBOOLE-monitor XBM 2 that can be downloaded from the web pages in the Springer domain mentioned above. Hence, solutions in exactly the same order of identical ternary vectors as provided in this book will be computed when this XBOOLE-monitor XBM 2 is used.

Alternatively, the web page:

https://tu-freiberg.de/en/fakult1/inf/xboole/download

can be used for the free download of each of the four versions of the XBOOLE-monitor XBM 2. However, extended versions (additional languages, additional commands, further optimized algorithms) will be provided on the web page in the domain of the Freiberg University of Mining and Technology. Hence, the use of such a changed version of the XBOOLE-monitor XBM 2 can cause different representations of the same solutions.

References

1. D. Bochmann, B. Steinbach, *Logic Design with XBOOLE (in German: Logikentwurf mit XBOOLE)* in German (Verlag Technik GmbH, Berlin, 1991). ISBN: 3-341-01006-8
2. F. Dresig et al., Programming with XBOOLE (in German: Programmieren mit XBOOLE), in *Series of Scientific Publications of the Chemnitz University of Technology (in German: Wissenschaftliche Schriftenreihe der Technischen Universität Chemnitz)* in German (1992), pp. 1–119. ISSN: 0863-0755
3. B. Steinbach, XBOOLE—a toolbox for modelling, simulation, and analysis of large digital systems. Syst. Anal. Model. Simul. **9**(4), 297–312 (1992). ISSN: 0232-9298
4. B. Steinbach, C. Posthoff, *EAGLE Start-up Aid – Efficient Computations with XBOOLE (in German: EAGLE Starthilfe – Effiziente Berechnungen mit XBOOLE)* (Edition am Gutenbergplatz, Leipzig, 2015). ISBN: 978-3-95922-081-1

The XBOOLE-Monitor XBM 2

2

Abstract

The XBOOLE-monitor XBM 2 is a software tool for efficient Boolean calculation. It wraps most of the functions of the XBOOLE-library within a graphical user interface; hence, the XBOOLE-monitor XBM 2 is convenient for calculations using logic functions so that sometimes the term "Boolean pocket calculator" is used for this tool. The XBOOLE-monitor XBM 2 uses the set operations of the XBOOLE-library to solve logic equations or even systems of such equations. The XBOOLE-monitor XBM 2 allows us to combine XBOOLE-operations and control operations to higher algorithms, which can be stored and reused as the so-called problem-programs (PRP) to solve the problems of many areas. An unlimited number of Boolean spaces can be used to solve the given problems. The XBOOLE-monitor XBM 2 is equipped with a comprehensive help system. All these properties make the XBOOLE-monitor XBM 2 an excellent tool to support the education in the Boolean domain. The XBOOLE-monitor XBM 2 has been developed for the operating systems Windows 10 and LINUX—Ubuntu. The XBOOLE-monitor XBM 2 exists for these operating systems in the versions of 32 bits and 64 bits and can be downloaded free of charge from the Internet and used by everyone.

Supplementary Information The online version of this chapter (https://doi.org/10.1007/978-3-030-88945-6_2) contains supplementary material which is available for authorized users. Please, follow the link belonging to the version of the XBOOLE-monitor XBM 2 that fits best for your operating system. This XBOOLE-monitor is needed to solve all tasks of this chapter. Instructions for starting the downloaded XBOOLE-monitor XBM 2 are given at the beginning of Section 'Exercises' in this chapter.

XBOOLE-monitor XBM 2 for Windows 10
32 bits
https://doi.org/10.1007/978-3-030-88945-6_2_MOESM1_ESM.zip (15,091 KB)

64 bits
https://doi.org/10.1007/978-3-030-88945-6_2_MOESM2_ESM.zip (14,973 KB)

XBOOLE-monitor XBM 2 for Linux Ubuntu
32 bits
https://doi.org/10.1007/978-3-030-88945-6_2_MOESM3_ESM.zip (29,522 KB)

64 bits
https://doi.org/10.1007/978-3-030-88945-6_2_MOESM4_ESM.zip (28,422 KB)

2.1   The Program Window of XBM 2

The window of the XBOOLE-monitor XBM 2 is shown in Fig. 2.1; it comprises several parts. In the following, the main purpose of these parts will be explained.

The *headline* of the XBOOLE-monitor XBM 2 shows the icon, the title, and the version of the program "XBM 2 (64 Bit)." When the recent XBOOLE-system is stored as an sdt-file, its name is shown on the right-hand side of the program title. Such an sdt-file allows to interrupt the work with the XBOOLE-monitor XBM 2. After loading a stored sdt-file, the work with the XBOOLE-monitor XBM 2 based on previous data can be continued.

The XBOOLE-monitor XBM 2 provides many actions that can be specified by certain parameters and combined in the wanted order by the user. The same action can be executed in different ways as follows:

- the selection of an item in the *menu* and the specification of the parameters in the corresponding dialog window;

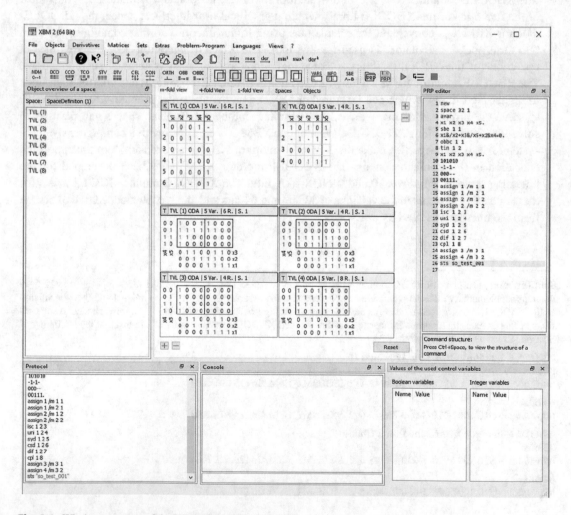

Fig. 2.1 Window structure of the XBOOLE-monitor XBM 2

- the selection of a button in the *toolbar* and the specification of the parameters in the corresponding dialog window;
- typing the command and the needed parameters for the input line of the *console*; or
- executing the completely specified command as part of a *problem-program*. (PRP)

The *menu bar* is located below the headline of the XBOOLE-monitor XBM 2. The menu is structured like a tree and allows us to control a very large part of the behavior of the XBOOLE-monitor XBM 2. Main categories of the menu such as *File*, *Objects*, *Derivatives*, ... guide the use to the needed item to activate the wanted action. The meaning of these actions will be explained later step by step. The menu items will be selected using the mouse. The advantage of the menu is that the user must not know the commands of the XBOOLE-monitor XBM 2.

The window area of the XBOOLE-monitor XBM 2 can be modified so that more space is available for certain parts of this program. Windows with the icon ▣ in their headlines can be moved to a place outside of the area of XBM 2 or docked at another position of XBM 2. Almost all of these windows have also the icon ✕ in their headlines; a click on such an icon closes the corresponding window but preserves the content. Using the appropriate item in the menu *Views* brings these parts back to the XBOOLE-monitor XBM 2. The exception of this feature is the window *Object overview of a space*; it can be moved to another position, but due to its important role for the access to TVLs and VTs, it cannot be deleted.

The *toolbars* are located below the menu bar. The buttons of the tool bars are associated to the most often used items of the menu bar. A single click on such a button starts the respective action (e.g., execute a PRP) or opens the dialog to specify required data. Hence, the buttons of the tool bars shorten the interactions of the user to activate a wanted action.

At the beginning when the XBOOLE-monitor XBM 2 is used without any experience, the *help system* of this monitor supports the user. Later this system is very useful to get any information about the needed details.

A click on the toolbar button ❷ opens the window of the XBM 2 help system as shown in Fig. 2.2. The help system of Fig. 2.2 provides information about all parts and all details of the XBOOLE-monitor XBM 2. The needed information can be found using the tree structure of the rider *Contents*, the lexicographic ordered commands and topics provided by the rider *Index*, or the search for a certain term in the rider *Search*. The selected content is shown in the right part of this help window. Links in the help pages support the access to the wanted information furthermore.

The toolbar button of button ⮷? provides a context sensitive access to the needed information. A click on this button changes the cursor as shown in Fig. 2.3, and the consecutive click on an item in the menu, a button of the toolbar, or any area of the XBOOLE-monitor XBM 2 displays the associated help information and returns to the normal use of the cursor.

Almost all actions of the XBOOLE-monitor XBM 2 can be executed by means of commands; exceptions are very few actions that change the graphical user interface of the XBOOLE-monitor XBM 2 (e.g., the temporary remove certain views or the change between the display of a TVL and the associated Karnaugh-map). Each command and the associated parameters can be typed in the command line of the console. The *Console* is located in the middle at the bottom of the XBOOLE-monitor XBM 2. The console provides above of the command line an unchangeable text field where the executed commands of the console are shown. These commands can be restored into the command line of the console by means of the cursor keys UP and DOWN. In this way, any previous command of the console can be executed again either directly or after some changes using the editor of the command line. All commands are explained in the help system of the XBOOLE-monitor XBM 2. An overview about all commands is given in Sect. 2.4 of this chapter.

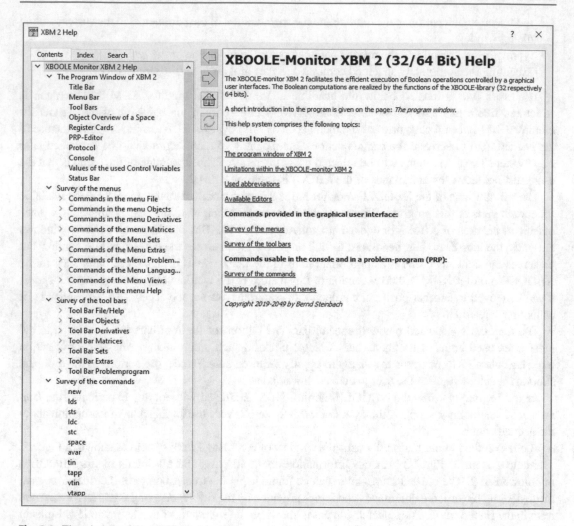

Fig. 2.2 The window of the XBM 2 help system

Fig. 2.3 Cursor in the
case of the activated
context help

A *protocol* of all actions of the XBOOLE-monitor XBM 2 is shown in the view *Protocol* at the bottom left. Each executed action of the XBOOLE-monitor XBM 2 is stored as a command in this protocol independent of whether this action has been initiated by means of a menu item, a toolbar button, as command in the console, or a command of a problem-program (PRP). The protocol shows in each of these cases the executed complete command. In this way, the protocol helps to learn the command language of the XBOOLE-monitor XBM 2. The protocol gives also hints in the case that an error occurs during the execution of a command.

The protocol cannot be changed, but any selected area of the protocol can be stored by means of CTRL-C into the clipboard and thereafter using CTRL-V into the PRP-editor or the command line of the console. This feature facilitates the repeated execution of sequences of commands as well as their storing into a prp-file.

The *PRP-editor* of the XBOOLE-monitor XBM 2 is located on its right-hand side. Using this editor, sequences of commands can be typed and then executed step by step, continuously until the next break point, or completely until the end of the PRP. The possibility to store and load such PRPs into files makes PRPs to be a very useful approach for the development of algorithms and to provide a solution for a wide variety of tasks. The implemented highlighting of both pairs of parentheses and the syntax and the command structure supports the user to edit a PRP.

The results of the most XBOOLE-operations are ternary vector lists (TVLs) or tuples of variables (VTs); however, there are also XBOOLE-operations with an integer result, e.g., the number of ternary vectors (NTV) of a TVL, or a Boolean result, e.g., the truth value whether the result of a set operation is empty. Such results can be stored and manipulated in control variables of the XBOOLE-monitor XBM 2. Both the value and the type (integer vs. Boolean) of these variables can be changed. Control variables are defined and changed simply by their use. The actual values of these variables are shown in the view *Values of the used control variables* at the bottom right of XBOOLE-monitor XBM 2.

Both given and computed TVLs and VTs can be displayed in the viewports in the center of the XBOOLE-monitor XBM 2. There is a single viewport in the *one-fold View*, four viewports are displayed in two rows and two columns in the *four-fold View*, and $m = r * c$ viewports are arranged in r rows and c columns in the *m-fold View*. Using buttons $+$ or $-$, the number of rows and columns of the m-fold view can be changed. A click on the *Reset*-button at the bottom right of the m-fold view reduces this view to a single viewport. A click on the rider of these views brings the selected view in the foreground. Figure 2.1 shows an example of a m-fold view with six viewports arranged in three rows and two columns. Each TVL has a form predicate that is displayed in the headline of the viewport; hence, each TVL represents a logic function. Such a function with up to ten variables can be alternatively shown as a Karnaugh-map in a viewport. A button on the left-hand side of the headline in the viewport labeled by K switches this viewport to the Karnaugh-map and the changed button labeled by T switches the Karnaugh-map back to the TVL. Both a TVL and a Karnaugh-map can be edited in such a viewport.

All TVLs and VTs of a selectable space are listed at the left-hand side of the viewports in the window *Object overview of a space*. These objects can be assigned to a viewport by drag&drop (press the left mouse button to a TVL or VT in the mentioned list, move the mouse with the pressed left button to the wanted viewport, and release the left mouse button). Alternatively, the command `assign` can be used for this action. Scroll bars appear in viewport when the displayed object needs a larger area than available; using these scroll bars, all parts of a TVL, Karnaugh-map, or VT can be shown.

The view of the rider *Space* displays the details of all defined Boolean spaces, and the view of the rider *Objects* shows a list of existing XBOOLE-objects (TVLs and VTs) together with their properties.

2.2 Basic Steps to Use XBM 2

We use a very simple example to become familiar with the use of the XBOOLE-monitor XBM 2:

- a set of tokens is given, where each token:
    - is encoded by four Boolean variables x_1 to x_4;
    - has either the color red or blue;
    - has either the shape of a circle or a square;
- three subsets of tokens are known and specified in Fig. 2.4 by TVLs;

$$\text{ODA}(f_1) = \begin{array}{ccc} x_1 & x_2 & x_4 \\ \hline - & 1 & - \\ 1 & 0 & 1 \end{array} \qquad \text{ODA}(f_2) = \begin{array}{ccc} x_2 & x_3 & x_4 \\ \hline - & 1 & 1 \\ 1 & 1 & 0 \end{array} \qquad \text{ODA}(f_3) = \begin{array}{cccc} x_1 & x_2 & x_3 & x_4 \\ \hline 0 & 0 & 1 & 0 \\ 1 & 0 & 0 & 0 \end{array}$$

  a   **b**   **c**

Fig. 2.4 TVLs of the given subsets of tokens: (**a**) red tokens, (**b**) tokens with square shapes, and **c** blue tokens with circled shapes

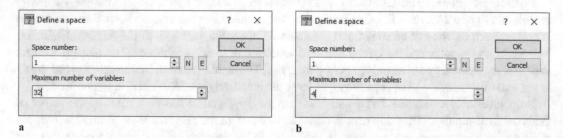

a   b

Fig. 2.5 Dialog to define a Boolean space: (**a**) suggested maximal 32 variables, (**b**) changed to maximal four variables

- the missing two subsets of tokens of one property (all blue tokens, all tokens with the shape of a circle) must be computed;
- the missing three subsets of tokens determining both properties (color and shape) must be computed;
- both the given and the already computed subsets can be used to compute the other subsets of tokens.

Besides the steps to solve this task, we also show some properties of the XBOOLE-monitor XBM 2 and derive preferred methods to use this program. We suggest that the reader executes all steps described in this section to become quickly familiar with the use of the XBOOLE-monitor XBM 2.

As explained in Sect. 1 of this chapter, a Boolean space must be defined first. This can be done in the started program XBM 2 by a click on the menu *Objects* and there by a second click on the item *Define a space*. These two clicks open the dialog of Fig. 2.5a, where the Boolean space with the number 1 and a maximal number of variables 32 is suggested. We know that we need only 4 Boolean variables in this example (see Fig. 2.4); hence, we change the maximal number of variables to 4 (as shown in Fig. 2.5b) and press the OK button.

The associated command `space 4 1` has been included in the protocol view; hence, we learn that a Boolean space can be defined by a command `space` followed by the maximal permitted number of variables and thereafter the number of the space. The properties of this space can be seen in the view *Spaces* in the center of XBM 2 (see Fig. 2.6). The value 2 of the property *type* means that two machine words are used to store a single ternary vector of this space; this is the smallest possible value because each ternary element is encoded by two bits that are stored in separate machine words for fast computations. Furthermore, we see in the view *Spaces* our defined maximal number of variables 4 and to the right that no variable is assigned until now to this space.

The assignment of the variables to a Boolean space of the XBOOLE-monitor XBM 2 can be realized explicitly by the user or implicitly by the program. We study first the implicit assignment of the variables by the program. The benefit of this approach is that all variables used in a TVL, a VT, or a Boolean equation are assigned to the space automatically; the drawback is that the order of the variables is determined by their first occurrences, which can be different to a wanted natural order. We demonstrate this effect by the input of the first two TVLs of our example.

Space name	Space number	Type	Maximum number of variables	Assigned variables
SpaceDefinition (1)	1	2	4	0

Fig. 2.6 Initial properties of the defined space of four variables

Fig. 2.7 Completed
dialog to define the empty
TVL of the red tokens

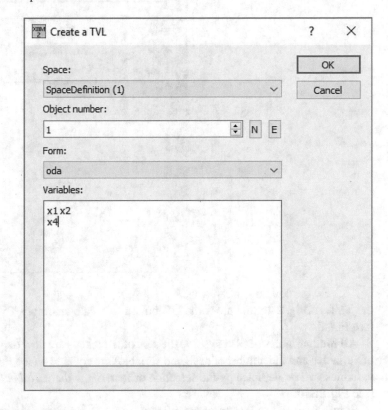

The input of a TVL is split into two steps when the menu or the toolbar is used:

1. the definition of the variables (columns) of the TVL without specification of any row; and
2. the rows of the TVL are appended to the empty TVL created in the first step.

For step 1 of the input of the TVL of the red tokens (see Fig. 2.4a), we click in the menu on *Objects* and there on the item *Create a TVL*; this opens a dialog window where we must not change neither the predefined space number 1, the object number 1, nor the ODA-form, but we type the names of the variables separated by a space or a line break. Figure 2.7 shows this dialog completed with the variables needed to define the red tokens; a click on the OK button closes this dialog window and creates the empty TVL 1.

One click on the button $\overset{+}{\text{TVL}}$ in the toolbar leads to the same dialog window as two clicks in the menu used before; hence, the toolbar provides most of the action of the menu with less interactions. Figure 2.8 shows that automatically the object number 2 has been suggested, which is convenient for the creation of our second TVL. The up and down buttons in this input-field can be used to decrease or increase this number by a mouse click. The button N selects the *next* unused object number in the memory list and the button E the object number at the *end* of this list. We type the three variables of

Fig. 2.8 Completed
dialog to define the empty
TVL of the tokens with a
shape of a square

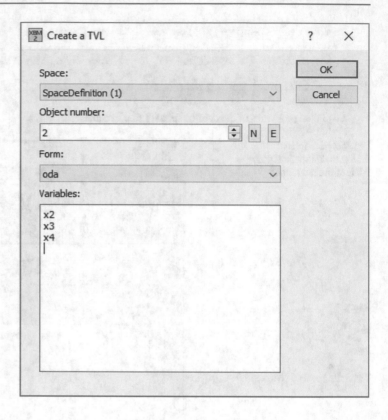

the TVL of Fig. 2.4b and press the OK button to create an empty TVL for the set of square tokens
(see Fig. 2.8).

All four variables needed to solve the task of the tokens are now defined; both the maximal number
of variables and the number of assigned variables are equal to 4 (see Fig. 2.9a); the list of the assigned
variables will be depicted by the selection of item *Show the variables of a space* in the menu *Extras*
(see Fig. 2.9b).

Figure 2.9b shows that an order different to usual lexicographic order of the variables can result
from the implicit assignment of the variables and the use of the variables in TVLs 1 and 2. This order
determines the order of the ternary elements to complete the TVLs in the second step of the input
procedure. The defined order of the variables of the TVLs can be seen when the TVLs are assigned
to a viewport by drag&drop. Figure 2.10 shows the created empty TVLs assigned to the upper two
viewports in the four-fold view. The exchange of the variables x_3 and x_4 must be considered in case
of inputs and all evaluations. This demonstrates the possible drawback of the implicit assignment of
the variables.

We discard all the previous work (the complete XBOOLE-system) using the item *New* in the menu
File or simply the button ⬜ of the toolbar to avoid the drawback of implicitly unordered sequence
of the variables. The solution of our task with regard to the tokens requires a space of at least four
variables but can also be computed with a larger space. To demonstrate another useful aspect, we
define a Boolean space of 32 variables by two mouse clicks; the first one to the toolbar button ⬚,
which leads like the click on the menu *Objects* and there on the item *Create a TVL* to the dialog of
Fig. 2.5a. Pressing the OK button without any change creates the space number 1 with maximal 32
variables.

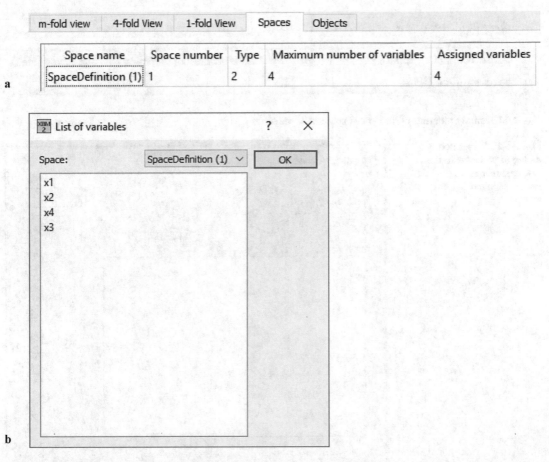

| m-fold view | 4-fold View | 1-fold View | Spaces | Objects |

Space name	Space number	Type	Maximum number of variables	Assigned variables
SpaceDefinition (1)	1	2	4	4

a

List of variables   ?   ✕

Space: SpaceDefinition (1) ⌄   OK

x1
x2
x4
x3

b

Fig. 2.9 Implicitly defined variables of the Boolean space 1: (**a**) parameters of the space 1, (**b**) list of the defined variables

| m-fold view | 4-fold View | 1-fold View | Spaces | Objects |

| K TVL (1) ODA │ 3 Var. │ 0 R. │ S. 1 | K TVL (2) ODA │ 3 Var. │ 0 R. │ S. 1 |

Fig. 2.10 Created empty TVLs 1 and 2 assigned to the upper two viewports in the four-fold view

The properties of this space can again be seen in the view *Spaces* in the center of XBM 2 (see Fig. 2.11). The value 2 of the property *type* confirms that each ternary vector of 32 variables needs the same number of two machine words as the space of only four variables; hence, the definition of Boolean spaces where the maximal number of variables is a multiple of 32 give, on the one hand, more freedom for additional variables and save, on the other hand, memory space.

m-fold view	4-fold View	1-fold View	Spaces	Objects

Space name	Space number	Type	Maximum number of variables	Assigned variables
SpaceDefinition (1)	1	2	32	0

Fig. 2.11 Initial properties of the defined space of 32 variables

Fig. 2.12 Completed
dialog to variables to the
Boolean space 1 in
lexicographic order

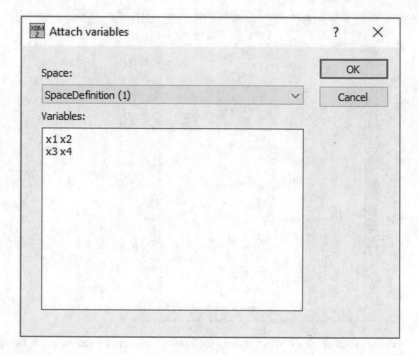

Now we explicitly attach the variables to the space in order to get the lexicographic order of
the variables. This can be done by a click on the menu *Objects* and there by a second click on the
item *Attach variables*. These two clicks open the dialog of Fig. 2.12 where the Boolean space with
the number 1 is suggested. We type the names of the variables in the wanted lexicographic order
separated by a space or a line break as shown in Fig. 2.12; a click on the OK button closes this dialog
window and attaches the variables to the space in the wanted order. The protocol window shows the
executed command for this action; it consists of two lines—avar 1 in the first line declares that
variables have to be attached to the space number 1 and "x1 x2 x3 x4." in the next line defines
the variables in the order to use. The variables in such a list are separated by spaces, can be written
over several lines, and must be determined by a dot.

All four variables needed to solve the task of the tokens are now assigned to the space 1
(see Fig. 2.13a); the list of the assigned variables confirms the specified lexicographic order (see
Fig. 2.13b).

We demonstrate now three different methods for the input of a TVL into the XBOOLE-monitor
XBM 2. The first method uses only the menu and the associated dialog windows.

The empty TVL 1 for the red tokens (see Fig. 2.4a) can be created as shown above using the item
Create a TVL in the menu *Objects* and the input of the three variables as shown in Fig. 2.7. Next, we
append two ternary vectors to this TVL using the item *Append ternary vector(s)* in the menu *Objects*;

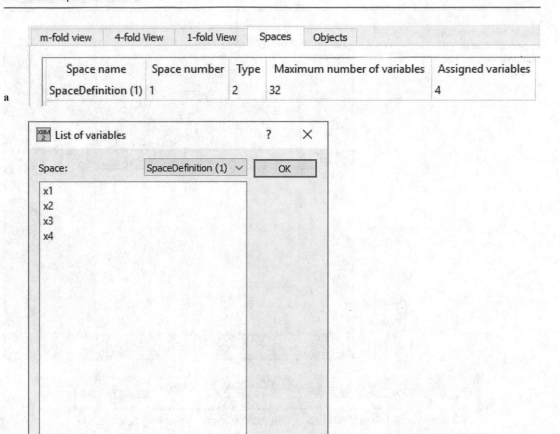

| m-fold view | 4-fold View | 1-fold View | Spaces | Objects |

a

Space name	Space number	Type	Maximum number of variables	Assigned variables
SpaceDefinition (1)	1	2	32	4

b

Fig. 2.13 Explicitly defined variables of the Boolean space 1: (**a**) parameters of the space 1, (**b**) list of the defined variables

in the appearing dialog window of Fig. 2.14, we type the two ternary vectors of Fig. 2.4a and close the dialog by a click on the OK button.

Appending a ternary vector to a TVL can result in a non-orthogonal TVL; therefore, the form predicate of TVL 1 has been changed from ODA to D as can be seen in Fig. 2.15. We come back to the needed ODA-form of the TVL 1 using the item *ORTH—Orthogonalize a TVL* in the menu *Matrices*; in the appearing dialog window, we click on the button G (like given) to change the suggested number 2 of the output-object to the number 1 of the given TVL and close this dialog using the OK button.

In our example we know three sets and have to compute five sets. The associated eight TVLs can commonly be shown in the *m*-fold view when we add two columns of viewports by clicking twice the + button on the right and also two rows by clicking twice the + button at the bottom of this view. We assign the created orthogonal TVL 1 to the viewport in the upper row and left column by drag&drop of TVL (1) of the window *Object overview of a space*. Figure 2.16 shows the *m*-fold view after this assignment. The information in the headline of viewport (1,1) declare that the TVL number 1 has an ODA-form, depends on three variables, consists of two rows, and is defined in the space number 1.

The second method of the input of a TVL uses two buttons of the toolbar and the possibility of XBOOLE-monitor XBM 2 that a TVL can be edited when it is shown in a viewport. We use this

Fig. 2.14 Append the ternary vectors to TVL 1 that describes the red tokens

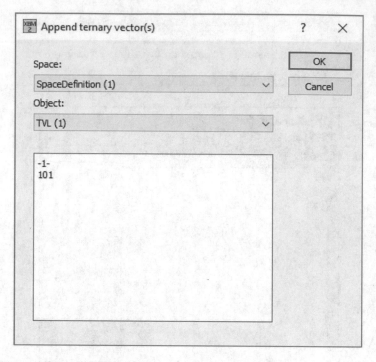

Fig. 2.15 Parameters of TVL 1 after appending two ternary vectors

method for the input of the second TVL of Fig. 2.4b, which represents the tokens with the square shapes.

A click on the button $\overset{+}{\text{TvL}}$ in the toolbar opens the dialog window of Fig. 2.8 where we type the three names of the variables that TVL 2 is depending on and press the OK button. Next, we assign this TVL by drag&drop to any viewport; we use viewport (1,2) in the first row and the second column of the *m*-fold view. We see that the order of the variables remains unchanged due to the predefined variables in the Boolean space 1.

All details that can be used to edit a TVL are provided in the help system of the XBOOLE-monitor XBM 2 using the links *Available Editors* and *TVL-editor*. In the given stage, we have an empty TVL 2; hence, no ternary element can be edited. A double-click with the left mouse button, while the control-key (Strg, respectively, Ctrl) is pressed, appends a row of dashes to an empty TVL, so that we get ternary elements that can be edited. In this first row we need values 1 for the columns x3 and x4. A click with the left mouse button on the dash in the column x3 of the created row selects this element. We type the needed value 1 and press the tabulator-key to reach the next element to the right. Here we type again the needed value 1 and press the tabulator-key; we see that a new row is created and the left element of this new row is prepared for the input of the next ternary element because we have edited the right element in the last row. In the second row, we type the needed values 1, 1, and 0, each confirmed by the tabulator button (or the enter-key).

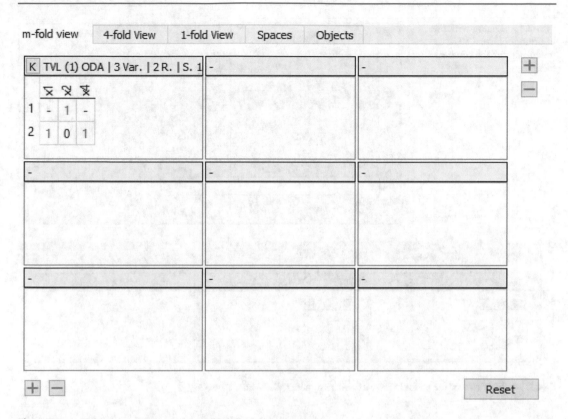

Fig. 2.16 *m*-Fold view with the assigned TVL 1 of the red tokens

The needed two rows of TVL 2 have now the right values, and we can break the edit procedure by pressing the escape-key. Appending a new row when the right element of the last row has been edited is very useful but creates an additional row of dashes when the last edited element is the element on the right column in the last row. A double-click with the left mouse button, while the control-key is pressed, removes this unneeded last row of dashes.

The orthogonality of a TVL can be lost by the change of any element of a TVL or the appending of a row; for that reason, the form of the TVL has been changed to a D-form. A click on the button $\overset{\text{ORTH}}{\perp}$ opens the dialog window in which we select the TVL 2 for both the input- and output-object of the orthogonalization (see Fig. 2.17) and click on the OK button.

Figure 2.18 shows the *m*-fold view after the finished input of TVL 2. The information in the headline of viewport (1,2) declare that the TVL number 2 has an ODA-form, depends on three variables, consists of two rows, and is defined in the space number 1.

The three buttons G, N, and E in the dialog of Fig. 2.17 (and also in several similar dialogs) on the right-hand side of field of the output-object can be used to determine the number of the output-object as follows:

- button G: the number of the *g*iven object specified as input-object is used;
- button N: the number of the *n*ext free object in the memory list is used; and
- button E: the number of the *e*nd of the memory list is used.

The third method of the input of a TVL uses a single command of the XBOOLE-monitor XBM 2 and creates the needed TVL in ODA-form without any further actions. We use this method for the input of the third TVL of Fig. 2.4c, which represents the blue tokens with a shape of a circle.

Fig. 2.17 Dialog window to orthogonalize TVL 2 that describes the square tokens

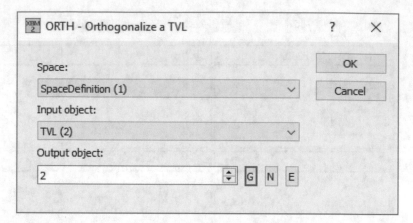

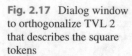

Fig. 2.18 First row of the *m*-fold view with the assigned TVLs 1 and 2

The format of the command can be seen in the protocol because it has been implicitly used to create the first two TVLs. It consists of three parts. In the first line we write the name of the command `tin` (short for TVL-input) followed by the number of the space to use (1 in our case), the number of the TVL to create (3 in our case), and the specification of the form (/oda in our case). We write this command in the command line of the console and press the enter-key. This command is now moved into the protocol part of the console, and the input of the second part of this command is expected.

The second part of the command `tin` consists of a list of variables of the TVL that are separated by spaces (or line breaks) and terminated by a dot. The dot (.) can be written directly behind the last variable or on the next line. Any wanted order of the variables can be used; the XBOOLE-monitor XBM 2 implicitly adjusts the variable to the correct columns of the space. We use the given lexicographic order of the variables and write "x1 x2 x3 x4." into the command line of the console and press the enter-key. This second part of the command is now moved into the protocol part of the console, and the input of the third part of this command is expected.

The third part of the command `tin` consists of a list of ternary vectors of the TVL written in the order as before specified by the list of variables; each ternary vector must be written in the command line without spaces. Pressing the enter-key moves the typed ternary vector in the protocol part of the console and expects the next ternary vector. For the input of the TVL of the blue tokens with the shape of a circle specified in Fig. 2.4c, we type 0010 and press the enter-key and thereafter 1000 and press again the enter-key. The input of a dot (.) confirmed by the enter-key terminates the input of this command and creates the complete TVL in the wanted form. No further actions are needed because the orthogonalization will be executed implicitly due to the requested ODA-form in the first line of the command. The complete command is now appended to the protocol of the XBOOLE-monitor XBM 2, and TVL 3 is added to the list in the window *Object overview of a space*.

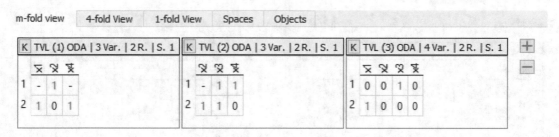

Fig. 2.19 First row of the *m*-fold view with the assigned three given TVLs

We assign the created TVL 3 to viewport (1,3) by drag&drop. Figure 2.19 shows the first row of the *m*-fold view after the additional assignment of TVL 3. The information in the headline of viewport (1,3) declare that the TVL number 3 has an ODA-form, depends on four variables, consists of two rows, and is defined in the Boolean space number 1.

The computation of the five wanted sets is now quite easy:

- the set of red tokens with the shape of a square is the result of the intersection of the red tokens (TVL 1) and tokens with the shape of a square (TVL 2); using the menu, we activate the item *ISC—Intersection* in the menu *Sets* (alternatively, the button ⊞ in the toolbar can be used), select TVLs 1 and 2 as input-objects, and use the number 4 for the resulting TVL (output-object);
- we get the set of the red tokens with the shape of a circle when we compute the difference between the set of the red tokens (TVL 1) and the set of tokens with the shape of a square (TVL 2); using the menu, we activate the item *DIF—Difference* in the menu *Sets* (alternatively, the button ⊞ in the toolbar can be used), select TVLs 1 and 2 as input-objects, and use the number 5 for the resulting TVL (output-object);
- the difference of these sets where the minuend and the subtrahend are exchanged results in the set of blue tokens with the shape of a square; now, we alternatively use the command and write in the command line of the console `dif 2 1 6` confirmed by the enter-key to compute the difference between the set of tokens with the shape of a square (TVL 2) and the set of red tokens (TVL 2) and store the result (the set of blue tokens with the shape of a square) as TVL number 6;
- the set of the tokens with the shape of a circle is the union between the given set of blue tokens with the shape of a circle (TVL 3) and the computed set of red tokens with the shape of a circle (TVL 5); using the menu, we activate the item *UNI—Union* in the menu *Sets* (alternatively, the button ⊞ in the toolbar can be used), select TVLs 3 and 5 as input-objects, and use the number 7 for the resulting TVL (output-object); and finally
- the set of the blue tokens is the union between the given set of blue tokens with the shape of a circle (TVL 3) and the computed set of blue tokens with the shape of a square (TVL 6); here, we alternatively use the command and write in the command line of the console `uni 3 6 8` confirmed by the enter-key to compute the union of the set of blue tokens with the shape of a circle (TVL 3) and the set of blue tokens with the shape of a square (TVL 6) and store the result (the set of blue tokens) as TVL number 8.

It is possible that certain ternary vectors of the computed TVL can be merged into a single ternary vector. We are interested in short TVLs and execute, therefore, the orthogonal block-building for all computed TVLs without the change of the sets of binary vectors. This can be done using the item *OBB—Orthogonal block-building* in the menu *Matrices*; in the opened dialog window, we choose the input-TVL (4 for our first computed TVL), press the button G to use the same object number for minimized TVL, and close the dialog using OK button. The same dialog window will be opened by a

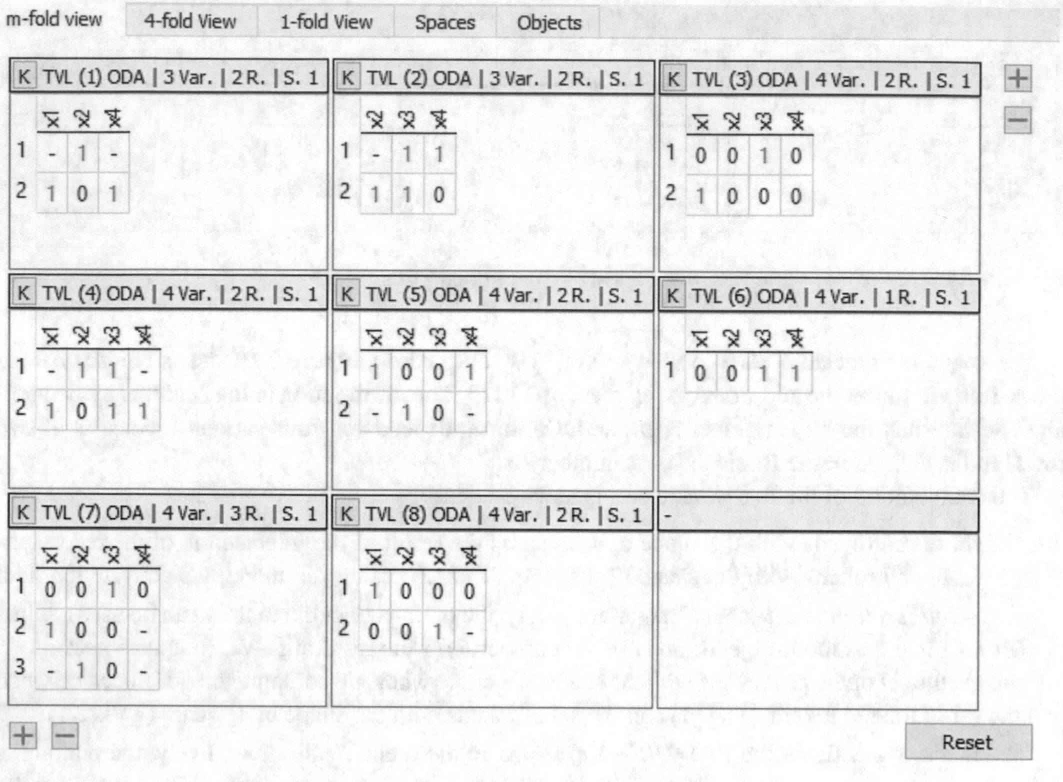

| m-fold view | 4-fold View | 1-fold View | Spaces | Objects |

K | TVL (1) ODA | 3 Var. | 2 R. | S. 1

	x_1	x_2	x_3
1	-	1	-
2	1	0	1

K | TVL (2) ODA | 3 Var. | 2 R. | S. 1

	x_1	x_2	x_3
1	-	1	1
2	1	1	0

K | TVL (3) ODA | 4 Var. | 2 R. | S. 1

	x_1	x_2	x_3	x_4
1	0	0	1	0
2	1	0	0	0

K | TVL (4) ODA | 4 Var. | 2 R. | S. 1

	x_1	x_2	x_3	x_4
1	-	1	1	-
2	1	0	1	1

K | TVL (5) ODA | 4 Var. | 2 R. | S. 1

	x_1	x_2	x_3	x_4
1	1	0	0	1
2	-	1	0	-

K | TVL (6) ODA | 4 Var. | 1 R. | S. 1

	x_1	x_2	x_3	x_4
1	0	0	1	1

K | TVL (7) ODA | 4 Var. | 3 R. | S. 1

	x_1	x_2	x_3	x_4
1	0	0	1	0
2	1	0	0	-
3	-	1	0	-

K | TVL (8) ODA | 4 Var. | 2 R. | S. 1

	x_1	x_2	x_3	x_4
1	1	0	0	0
2	0	0	1	-

-

Reset

Fig. 2.20 Given and computed TVLs of the tokens in the m-fold view

single click on the button ⸬ of the toolbar. The command for the orthogonal block-building is obb followed by the numbers of the input and the output-TVL separated by spaces; the default value of the output-TVL is the specified value of the input-TVL; hence, when we write in the command line of the console obb 7, the completed command obb 7 7 is executed to compute the minimized orthogonal TVL 7. After the assignment of all computed sets (TVLs 4 to 8), we get the result of our task as shown in Fig. 2.20.

Alternatively to the TVL, a Karnaugh-map can be displayed in the viewport for a TVL with up to ten variables. When a TVL is shown in a viewport, a click on the button K in its headline switches to the representation of the associated Karnaugh-map. Vice versa, when Karnaugh-map is shown in a viewport, a click on the button T in its headline switches to the representation of the associated TVL. Figure 2.21 shows the representation of all sets of tokens as Karnaugh-maps.

Figure 2.22 shows a real-world representation of the evaluated sets of tokens. This figure can be used to verify all steps of the solution-process.

2.3　Problem-Programs

A *problem-program* (PRP) is a sequence of commands of the XBOOLE-monitor XBM 2. All commands and their parameters are explained in detail in the help system of the XBOOLE-monitor XBM 2. A PRP can be edited with an arbitrary text-editor and must be stored using the file extension prp. The benefit of a PRP is that it can be reloaded into the XBOOLE-monitor XBM 2 and

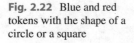

| m-fold view | 4-fold View | 1-fold View | Spaces | Objects |

T TVL (1) ODA | 3 Var. | 2 R. | S. 1

```
0   0 1 1 0
1   0 1 1 1
x4  0 1 1 0 x2
    0 0 1 1 x1
```

T TVL (2) ODA | 3 Var. | 2 R. | S. 1

```
0   0 0 1 0
1   0 1 1 0
x4  0 1 1 0 x3
    0 0 1 1 x2
```

T TVL (3) ODA | 4 Var. | 2 R. | S. 1

```
00   0 0 0 1
01   0 0 0 0
11   0 0 0 0
10   1 0 0 0
x3 x4  0 1 1 0 x2
       0 0 1 1 x1
```

T TVL (4) ODA | 4 Var. | 2 R. | S. 1

```
00   0 0 0 0
01   0 0 0 0
11   0 1 1 1
10   0 1 1 0
x3 x4  0 1 1 0 x2
       0 0 1 1 x1
```

T TVL (5) ODA | 4 Var. | 2 R. | S. 1

```
00   0 1 1 0
01   0 1 1 1
11   0 0 0 0
10   0 0 0 0
x3 x4  0 1 1 0 x2
       0 0 1 1 x1
```

T TVL (6) ODA | 4 Var. | 1 R. | S. 1

```
00   0 0 0 0
01   0 0 0 0
11   1 0 0 0
10   0 0 0 0
x3 x4  0 1 1 0 x2
       0 0 1 1 x1
```

T TVL (7) ODA | 4 Var. | 3 R. | S. 1

```
00   0 1 1 1
01   0 1 1 1
11   0 0 0 0
10   1 0 0 0
x3 x4  0 1 1 0 x2
       0 0 1 1 x1
```

T TVL (8) ODA | 4 Var. | 2 R. | S. 1

```
00   0 0 0 1
01   0 0 0 0
11   1 0 0 0
10   1 0 0 0
x3 x4  0 1 1 0 x2
       0 0 1 1 x1
```

-

Fig. 2.21 Given and computed sets of the tokens shown as Karnaugh-maps in the m-fold view

Fig. 2.22 Blue and red tokens with the shape of a circle or a square

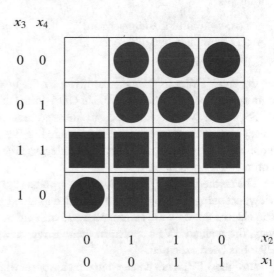

repeatedly executed. The execution of a PRP can be done command by command, automatically until a specified break point and then step by step, or completely until the end of the PRP is reached. All executed commands are appended to the protocol where also possible errors are indicated. Using the error message, the associated command can be directly corrected in the PRP-editor of the XBOOLE-monitor XBM 2. Further benefits of this special PRP-editor are an auto-completion feature, a component that shows the command structure, and the highlighting of pairs of parentheses.

```
1    new                  12   110.              23   (
2    space 32 1           13   tin  1  3         24   obb  $i
3    avar                 14   x1  x2  x3  x4.    25   )
4    x1 x2 x3 x4.         15   0010              26   assign  1  /m 1  1
5    tin 1 1 /oda         16   1000.             27   assign  2  /m 1  2
6    x1 x2 x4.           17   isc  1  2  4       28   assign  3  /m 1  3
7    -1-                 18   dif  1  2  5       29   assign  4  /m 2  1
8    101.                19   dif  2  1  6       30   assign  5  /m 2  2
9    tin 1 2             20   uni  3  5  7       31   assign  6  /m 2  3
10   x2 x3 x4.           21   uni  3  6  8       32   assign  7  /m 3  1
11   -11                 22   for  $i  1  8      33   assign  8  /m 3  2
```

Fig. 2.23 Problem-program that solves the task of colored tokens with different shapes specified and explored in Sect. 2.2

Figure 2.23 shows a PRP that solves the task of sets of tokens explored in the previous section; hence, the reader can compare the solution-steps used above with the commands in this PRP.

The command new in line 1 removes a possible old XBOOLE-system; in this way, conflicts with the new defined space number 1 with 32 variables in line 2 of this PRP are avoided. The command avar (attach variables) in lines 3 and 4 assures that the four variables are lexicographically ordered in the Boolean space 1. The input of the three TVLs of the given subsets of tokens of Fig. 2.4 is realized in lines 5 to 16 using three commands tin (TVL-input).

The set operations in lines 17 to 21 compute the five wanted subsets of tokens as before. The command:

- isc computes the intersection;
- dif computes the difference; and
- uni computes the union

of the sets (TVLs in ODA-form) given by the next two object numbers; the third number in these commands specifies the set (TVL in ODA-form) of the result.

As before, we are interested in the representation of orthogonal TVLs with a small number of rows; hence, we must execute the command obb (orthogonal block-building) for all eight TVLs. The for-loop in lines 22 to 25 counts the integer variable $i from 1 to 8 so that we must only write one command obb in line 24.

The created three TVLs 1, 2, and 3 as well as the five computed TVLs 4 to 8 can be assigned to any viewport using the drag&drop approach explained above. It is very convenient that these assignments can also be done by the command assign as shown in lines 26 to 33 of the PRP in Fig. 2.23; in this way, the wanted TVLs appear in the existing viewports without additional manipulations when the PRP has been executed.

Assigned TVLs are changed in the viewports whenever these TVLs are changed by any XBOOLE-command. The time to prepare the representation in the viewport slows down the solution-procedure; hence, only the important TVLs should be assigned to a viewport and not all intermediate TVLs.

We prepared the PRP of Fig. 2.23 in the PRP-editor of the XBOOLE-monitor XBM 2 and stored this PRP using the item *Save the PRP with a new file name* in the menu *Problem-Program* with the name

br_cs_tokens.prp.

The name of a stored (or loaded) PRP is shown in the headline of the PRP-editor. The execution of the PRP of the PRP-editor can be done command by command using the button ⁵≡ in the toolbar or completely using the button ▷. A click with the left mouse button on the number of a line in the PRP-editor toggles a break point (represented by a red dot) in this line. The continuous execution stops when a break point is reached; this allows us to study intermediate results. The execution of the PRP after a reached break point can be continued command by command or continuously as described before. Pressing the button ■ in the toolbar aborts the execution of a partially executed PRP.

We prepared the *m*-fold view with nine viewports structured in three rows and three columns and executed the PRP listed in Fig. 2.23. Without any further actions, this *m*-fold view shows the same result of Fig. 2.20 as created using the menu, the toolbar, and the console in the previous section.

2.4 An Overview About the Commands of XBM 2

The XBOOLE-monitor XBM 2 provides so much features that we cannot explain all of them in detail in this chapter. We will do this in the following chapters when we utilize these features to solve the tasks from different areas. However, we give here a compact overview about the available commands and their main concepts. This overview can help the readers when they use the XBOOLE-monitor XBM 2 to solve their own Boolean problems.

Table 2.1 gives an overview about commands usable for administrative tasks in lexicographic order. The short remarks about their effects are sufficient to understand their application in most cases.

An sdt-file stores the variables, spaces, TVLs, and VTs of a XBOOLE-system in a compact format based on the internal data structures of XBOOLE. These files can be used to continue the work on a certain problem with the XBOOLE-monitor XBM 2 at a later point in time. An sdt-file can be used either for the 32-bit or for the 64-bit version of XBOOLE because internal data structures are used.

The XBOOLE-monitor XBM 2 implements for the first time the commands stc and ldc, which store or load a cxb-file. A cxb-file contains the same user data as an sdt-file, but a cxb-file can be read in both the 32- and 64-bit versions of the XBOOLE-monitor XBM 2. A cxb-file is larger than the equivalent sdt-file; hence, cxb-files should be used as a bridge between both versions of XBOOLE.

The assignment of a TVL or VT to a viewport establishes a dynamic link between the unique XBOOLE-object and its representation in one or several viewports. That means the change of a TVL causes the execution of the function to convert it into its representation of an external TVL or a Karnaugh-map. To avoid useless conversions, assigned TVLs or VTs can be removed from a viewport using the command clear.

A lattice of logic functions can be represented by its ON-set (function $q(bx)$) and OFF-set (function $r(\mathbf{x})$). The command assign_qr creates a common Karnaugh-map (up to ten variables) of such two mark-functions within a single viewport. This supports applications of lattices of logic functions in the XBOOLE-monitor XBM 2.

Results created with the XBOOLE-monitor XBM 2 are often the topic of a publication. The three export commands support the work of the authors by creating files of a TVL, a Karnaugh-map of a function, or a Karnaugh-map of a lattice function; these files can directly be included in a publication with LaTeX. The help system of the XBOOLE-monitor XBM 2 provides all details for use of the exported `tex`-files.

The solution of a Boolean equation is not a function of the XBOOLE-library but uses several conversion functions to create elementary TVLs as well as set operations of XBOOLE to realize the logic operations. The possibility to solve both a single Boolean equation or a system of several Boolean equations extends the field of application of the XBOOLE-monitor XBM 2 significantly.

The derivative operations of the Boolean Differential Calculus (see Table 2.2) are implemented in the XBOOLE-library and directly usable in the XBOOLE-monitor XBM 2. The theory of the Boolean Differential Calculus is described in [2, 3], Chapter 4 of [1], and Chap. 6 of this book. The k-fold and vectorial derivative operations extend the fields of applications of the XBOOLE-monitor XBM 2.

Even all derivative operations for lattices can be computed with the provided derivative operations for logic functions. The single derivative operations are special cases of both the k-fold and the vectorial derivative operations where a single variable is used; hence, no extra commands are provided

Table 2.1 Administration of XBOOLE-objects and to solve Boolean equations

Command	Effect
assign	Assigns a TVL or VT to a viewport
assign_qr	Assigns a lattice $\langle q, r \rangle$ to a viewport to show a Karnaugh-map
avar	Attach variables to a Boolean space
clear	Remove the assigned XBOOLE-object from a viewport
copy	Copies a TVL or VT
ctin	Creates a TVL of the constant 0 or 1
del	Deletes a TVL or VT
export_km	Creates a file of a Karnaugh-map usable in LaTeX
export_kmqr	Creates a file of a Karnaugh-map of lattice usable in LaTeX
export_tvl	Creates a file of a TVL usable in LaTeX
ldc	Loads a XBOOLE-system (cxb-file) compatible for 32 or 64 bits
lds	Loads a XBOOLE-system (sdt-file) either 32 or 64 bits
new	Creates a new XBOOLE-system
sbe	Solves a Boolean equation
sform	Specifies the form of a TVL
space	Defines a Boolean space
space_trans	Transfers a TVL or VT between Boolean spaces
stc	Stores a XBOOLE-system (cxb-file) compatible for 32 or 64 bits
sts	Stores a XBOOLE-system (sdt-file) either 32 or 64 bits
tapp	Appends ternary vectors to a TVL
tin	Creates a complete TVL
vtapp	Appends variables to a VT
vtin	Creates a complete VT

Table 2.2 Derivative operations of the Boolean Differential Calculus

Command	Effect
derk	k-fold derivative
derv	Vectorial derivative
maxk	k-fold maximum
maxv	Vectorial maximum
mink	k-fold minimum
minv	Vectorial minimum

Table 2.3 Matrix operations

Command	Effect
cco	Change columns of a TVL
cel	Exchange elements of selected columns
con	Concatenate two TVLs or two VTs
dco	Delete columns of a TVL
dtv	Delete a ternary vector of a TVL
ndm	Negation according to De Morgan
obb	Orthogonal block-building
obbc	Orthogonal block-building with changes
orth	Orthogonalize a TVL
stv	Select a ternary vector out of a TVL or VT
tco	Transfer columns of a TVL

Table 2.4 Set operations

Command	Effect
cpl	Complement of a TVL
csd	Complement of the symmetric difference of two TVLs
dif	Difference between two TVLs
isc	Intersection of two TVLs
syd	Symmetric difference of two TVLs
uni	Union of two TVLs

for single derivative operations. We suggest for single derivative operations to use the k-fold derivative operations with a single variable because this variable is removed in the resulting TVL.

Most of the matrix operations of Table 2.3 are related to columns or rows of a TVL or VT, which is handled like a matrix. Two commands are provided to minimize the number of rows of an orthogonal TVL:

- obb merges pairs of ternary vectors that have a zero–one combination in a single column and identical values in all other columns; and
- obbc executes several sweeps of obb followed by the change to an alternative representation of the ternary vectors; this leads usually to even shorter TVLs but requires more time.

The command orth realizes the orthogonalization of all non-orthogonal TVLs depending on their given form into the associated orthogonal TVL.

The set operations of Table 2.4 create always an orthogonal TVL but require in most cases that orthogonal TVLs are given. These operations are also used to compute logic operations for functions represented by TVLs. For TVLs in ODA-form, these operations compute the negation (cpl), the equivalence (csd), the difference (dif), the conjunction (isc), the antivalence (syd), or the disjunction (uni).

Besides the computation of sets or functions represented by TVLs, Boolean problems require also manipulations with regard to the Boolean variables. The variables, which occur in a TVL or a VT, are stored in these XBOOLE-objects as a set (vector) of existing variables (VV). In this way, a TVL or VT also represents a set of variables. The operations for sets of variables of Table 2.5 take the VVs of TVLs or VTs as arguments and compute an empty TVL in ODA-form in which the VV represents the computed set of variables.

The command sv_get creates a TVL in ODA-form in which the VV is stored as a vector of values 1. The command sv_make makes a set of variables that contains a single variable determined by the given set ordered with regard to the Boolean space and an index in this set.

Table 2.5 Operations for sets of variables

Command	Effect
sv_cpl	Complement of a set of variables
sv_csd	Complement of the symmetric difference of two sets of variables
sv_dif	Difference between two sets of variables
sv_get	Represents the set of variables of a TVL by a single vector of values 1
sv_isc	Intersection of two sets of variables
sv_make	Makes a set of variables that contains a single selected variable
sv_next	Selects the next variable of an ordered set of variables
sv_syd	Symmetric difference of two sets of variables
sv_uni	Union of two sets of variables

Table 2.6 Commands with direct specification of variables

Command	Effect
_cco	Change columns of a TVL
_cel	Exchange elements of selected columns
_dco	Delete columns of a TVL
_derk	k-fold derivative
_derv	Vectorial derivative
_maxk	k-fold maximum
_maxv	Vectorial maximum
_mink	k-fold minimum
_minv	Vectorial minimum
_tco	Transfer columns of a TVL

There are several tasks that require to evaluate certain properties iteratively for each variable of a given set. The command sv_next supports this requirement. It creates a set of variables containing a single variable that is the most left variable for a given empty set of variables or the next variable for a given set of a single variable. Additionally, the command sv_next has a Boolean value a result that can be used to control a loop in which each variable of a set is selected once.

A set of variables is needed to specify the variables for which a derivative operation must be computed. In case of the commands cco and tco, even ordered sets of variables (VTs) are needed to specify the pairs of columns that must be manipulated. Therefore, the commands in Table 2.2 use a TVL or VT to control the operation, and the commands in Table 2.3, which are related to columns, require VTs to determine the ordered pairs of variables (when the order is not needed, a TVL can be used to specify the set of variables). All commands in Table 2.6 indicate by an underline (_) as first character that each required (ordered) set of variables can be specified by a list of variables separated by spaces and enclosed by angled bracket. This example

$$\_cco \ 1 \ < a \ b \ c > < x \ y \ z > \ 2$$

uses two such ordered sets of variables. The commands in Table 2.6 are comfortable when such a set of variables is only needed once.

PRPs of the previous XBOOLE-monitor were restricted to sequences of command; hence, not all algorithms could be realized. The XBOOLE-monitor XBM 2 eliminates this restriction by the command if for the conditional execution of commands as well as the commands for and while for loops (see Table 2.7). These commands need either an integer value (for) or a Boolean value (if and while) for their control. For that reason, we implemented control variables for these two types. Control variables are indicated by a character dollar ($) as first character of the name and are

Table 2.7 Commands to control the PRPs

Command	Effect
for	For-loop
if	Conditional execution of commands
print	Show the value of a control variable
set	Set a control variable
while	While-loop

Table 2.8 Commands with a Boolean return value

Command	Effect
and	Logic *and* of two Boolean values
eq	True when integer 1 is equal to integer 2
ge	True when integer 1 is greater than or equal to integer 2
gt	True when integer 1 is greater than integer 2
le	True when integer 1 is less than or equal to integer 2
lt	True when integer 1 is less than integer 2
ne	True when integer 1 is not equal to integer 2
not	Logic *negation* of a Boolean value
or	Logic *or* of two Boolean values
sv_next	True when a variable has been selected
te	Test whether a TVL is empty
te_orth	Test whether the result of an orthogonalization is empty
te_cpl	Test whether the result of a complement is empty
te_csd	Test whether the result of the command csd is empty
te_dif	Test whether the result of a difference is empty
te_isc	Test whether the result of an intersection is empty
te_syd	Test whether the result of a symmetric difference is empty
te_uni	Test whether the result of a union is empty
te_derk	Test whether the result of a k-fold derivative is empty
te_derv	Test whether the result of a vectorial derivative is empty
te_maxk	Test whether the result of a k-fold maximum is empty
te_maxv	Test whether the result of a vectorial maximum is empty
te_mink	Test whether the result of a k-fold minimum is empty
te_minv	Test whether the result of a vectorial minimum is empty
xor	Logic *xor* of two Boolean values

implicitly defined by their use. Both the type and the value of a control variable can be changed by the assignment of a value, which can be done using the command set or implicitly by means of the command for. The command print displays the current value of a control variable in the protocol.

A special feature of the loop commands is that they are interrupted after a certain number of sweeps, and the user can decide at this point in time whether the execution of the loop has to proceed further or must be interrupted. The number of sweeps until this check can be determined as a parameter in the command; the default value is 1000. This feature is helpful to avoid infinite sweeps of a loop and facilitates the study of intermediate results after a certain number of sweeps of a loop.

Each command of Table 2.8 returns a Boolean value that:

- can be evaluated to decide about the execution of the body of a command if;
- can be evaluated to terminate a while-loop; or
- can be assigned to a Boolean control variable using the command set.

Table 2.9 Commands
with an integer return value

Command	Effect
add	Adds two integer values to compute the sum
div	Divides two integer values to compute the quotient
mod	Divides two integer values to compute the remainder (modulo)
mul	Multiplies two integer values to compute the product
ntv	Computes the number of ternary vectors of a ternary matrix
sub	Subtracts two integer values to compute the difference
sv_size	Computes the number of variables of a ternary matrix

The four commands of logic operations (not, and, or, and xor) facilitate the computation of a new Boolean value using given Boolean values. The six comparison commands (eq, ge, gt, le, lt, and ne) determine a Boolean value based on two integer values. The command sv_next is the only command with two results, a set that contains a single variable as well as a Boolean value (it has already been explained above); therefore, we included this command in both Tables 2.5 and 2.8.

Most of the commands in Table 2.5 begin with the letters te (test empty). When an underline (_) follows behind these two letters, the appended command determines the TVL that is tested for the property *empty*. However, these combined commands realize not simply a sequence of two commands; these commands are implemented in the XBOOLE-library such that the computation of the appended operation is immediately stopped when it is known that the resulting TVL of the associated operation is not empty. The time to solve a problem can be reduced when these compounded test commands are used.

The arithmetic commands in Table 2.9 compute an integer value using two given integer values.

The internal representation of both a TVL and a VT is a ternary matrix (TM). The command ntv returns the number of rows of a TM (each repeated variable of a VT is counted). A set of variables is determined by the variables that occur in a TM (the number of columns). The command sv_size determines the number of variables (columns) of a TM (repeated variables of a VT are counted only once).

2.5   Exercises

It is the aim of the exercises of this section that the readers become familiar with the possibilities to use the XBOOLE-monitor XBM 2. We suggest that you solve all exercises first by yourself and use the solutions provided in the next section to verify your solutions. The solutions of all exercises of this book require that the XBOOLE-monitor XBM 2 has been downloaded, unzipped, and started on the available computer.

If you have not yet prepared the XBOOLE-monitor XBM 2 on your computer, you can get this XBOOLE-monitor free of charge by means of the following three steps:

1. **Download**:
   There are four versions of the XBOOLE-monitor XBM 2, two for Windows 10 or subsequent Windows systems (32 or 64 bits) and two for LINUX—Ubuntu (also 32 or 64 bits); you must download the version of the XBOOLE-monitor XBM 2 that fits to your operating system.
   Authorized users of the online version of this chapter (https://doi.org/10.1007/978-3-030-88945-6_2) can download the XBOOLE-monitor XBM 2 directly from the web page

https://link.springer.com/chapter/10.1007/978-3-030-88945-6_2

where the links for the download of the XBOOLE-monitor XBM 2 are located in the part "Supplementary Information" (below the part "Abstract"). The headline above such a link indicates the associated zip-file of the XBOOLE-monitor XBM 2. The sizes of the zip-files have been provided behind the links and can be used to verify the download. A click on the link of the wanted version of the XBOOLE-monitor XBM 2 starts the download.

Readers of the hardcopy of this book get access to the XBOOLE-monitor XBM 2 using the URL

https://link.springer.com/chapter/10.1007/978-3-030-88945-6_2

to download the first two pages of this chapter. After this download, the same procedure as the authorized users of the online version of a chapter can be used to download the wanted version of the XBOOLE-monitor XBM 2.

2. **Unzip**: The XBOOLE-monitor XBM 2 must not be installed but must be unzipped into an arbitrary directory of your computer. A convenient tool for unzipping the downloaded zip-file is usually available as part of the operating system or can be downloaded from the Internet.

3. **Execute**:

   - Windows:
     The executable file of the two versions (32 or 64 bits) for Windows 10 (or subsequent Windows systems) of the XBOOLE-monitor XBM 2 is XBM2.exe; the other files in the expanded directory must remain unchanged. A double-click on the executable file XBM2.exe within the Explorer of Windows starts the XBOOLE-monitor XBM 2.
   - LINUX—Ubuntu:
     The unzipped folder of the XBOOLE-monitor XBM 2 contains for this operating system only the executable file XBM2-i386.AppImage for the version of 32 bits or XBM2-x86_64.AppImage for the version of 64 bits of the XBOOLE-monitor XBM 2. A double-click on the created AppImage-file within the file manager of LINUX—Ubuntu starts the XBOOLE-monitor XBM 2.

   Dialog windows of the XBOOLE-monitor XBM 2 are used to specify the parameters for an operation. Closing such a dialog window using the OK button starts the specified computation. We do not explicitly repeat this needed standard action for each of such dialog window in the provided solutions.

Exercise 2.1 (Boolean Spaces) Define three Boolean spaces with 10, 100, and 1000 logic variables and use the numbers 1, 2, and 3 for these spaces.

(a) Use the menu to define the space number 1 with maximal 10 logic variables.
(b) Use the toolbar to define the space number 2 with maximal 100 logic variables.
(c) Learn the command to define a Boolean space from the protocol view and define the space number 3 with maximal 1000 logic variables using the console.
(d) How many machine words are needed to store one ternary vector within these Boolean spaces? Use the view "Spaces" to answer this question.
(e) Verify that the maximal number of logic variables of a defined Boolean space cannot be changed.
(f) Prepare a PRP that defines the three Boolean spaces of this exercise. Use the command new in the first row of this PRP to avoid conflicts with already defined Boolean spaces. Execute this PRP and verify the results in the view "Spaces."

Exercise 2.2 (Input of Ternary Vector Lists (TVLs)) Define the three TVLs as shown in Fig. 2.24. Remember that a sufficiently large Boolean space must be defined before the input of a TVL. A lexicographic order of the variables is requested.

(a) Define the largest Boolean space that requires the smallest number of machine words to store one ternary vector for all TVLs of Fig. 2.24.
(b) Attach the eight variables x1, ..., x8 to this space using the menu.
(c) Define the empty TVL 1 in A-form (antivalence form) with the variables of $f_1(\mathbf{x})$ using the menu.
(d) Append the four ternary vectors to TVL 1 using the menu.
(e) Show TVL 1 in top-left viewport of the four-fold view.
(f) Define the empty TVL 2 in ODA-form with the variables of $f_2(\mathbf{x})$ using the toolbar.
(g) Show TVL 2 in top-right viewport of the four-fold view and insert the three ternary vectors of $f_2(\mathbf{x})$ using the edit feature of a TVL in a viewport.
(h) Notice that the form of TVL 2 is changed to the D-form. Use the toolbar to orthogonalize this TVL into the ODA-form.
(i) Use the command tin in the console for the input of TVL 3 of $f_3(\mathbf{x})$ in K-form.
(j) Show TVL 3 in bottom-left viewport of the four-fold view.
(k) Prepare a PRP that creates the three TVLs of this exercise. Use at the end of this PRP three commands assign to show the three TVLs as before in this exercise.

Exercise 2.3 (Input of Tuples of Variables (VTs)) Define the VTs VT 1 $= \langle x_1, x_2, x_3 \rangle$, VT 2 $= \langle x_3, x_2 \rangle$, VT 3 $= \langle x_3, x_2, x_4, x_2 \rangle$, and VT 4 $= \langle x_3, x_2, x_1, x_2 \rangle$ within a sufficiently large Boolean space using several methods and show these VTs in the four-fold view.

(a) Define a Boolean space with 32 variables using a command in the console.
(b) Create VT 1 using the menu.
(c) Create VT 2 using the toolbar.
(d) Copy VT 2 to VT 3 and append the two missing variables using the menu.
(e) Concatenate VT 2 and VT 1 (in this order) and store the result as VT 4 using the menu; thereafter, remove the last variable (that is, the fifth variable x3 of the created tuple) using the toolbar.
(f) Show all four VTs in the four-fold view.
(g) Prepare a PRP that creates and shows the four VTs of this exercise as before.

Exercise 2.4 (Set Operations) Two sets of binary vectors are specified by values 1 in the Karnaugh-maps of Fig. 2.25.

Compute all set operations for the two given sets S_1 and S_2 using several methods and show the solution-sets as Karnaugh-maps in the m-fold view with four rows and three columns. Use the help system of XBM 2 to learn how the set operations can be executed using a menu item, a toolbar button, or a command.

(a) Define a Boolean space of 32 variables and two empty TVLs with the variables x1, x2, x3, x4, and x5 in the created space using the object numbers 1 and 2.

$$
\begin{array}{c}
\begin{array}{cccc}
x_2 & x_4 & x_6 & x_8 \\
\hline
1 & - & - & - \\
- & 1 & - & - \\
- & - & 1 & - \\
- & - & - & 1 \\
\end{array}
\end{array}
\qquad
\begin{array}{c}
\begin{array}{ccccc}
x_4 & x_5 & x_6 & x_7 & x_8 \\
\hline
0 & 0 & - & - & 1 \\
1 & - & - & 1 & 1 \\
- & - & 1 & 0 & 0 \\
\end{array}
\end{array}
\qquad
\begin{array}{c}
\begin{array}{cccccc}
x_1 & x_2 & x_3 & x_4 & x_5 & x_6 \\
\hline
0 & 0 & 0 & 0 & 0 & 0 \\
1 & 1 & 1 & 1 & 1 & 1 \\
1 & - & 0 & - & 1 & - \\
- & 1 & - & 0 & - & 1 \\
\end{array}
\end{array}
$$

$A(f_1) = \qquad\qquad\qquad\quad ODA(f_2) = \qquad\qquad\quad K(f_3) =$

Fig. 2.24 Given TVLs used in Exercise 2.2

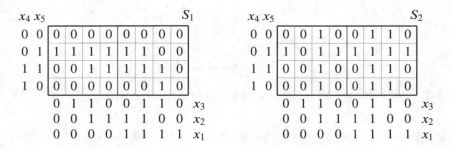

Fig. 2.25 Sets S_1 and S_2 specified by values 1 in the associated Karnaugh-maps

(b) TVL 1 has been created to express the set S_1. The change of the empty TVL 1 to a TVL that represents the set S_1 can be easily done using a Karnaugh-map. Move TVL 1 into the top-left viewport of the prepared m-fold view and click on the button K to show the Karnaugh-map. Edit this Karnaugh-map to the set S_1 as shown in Fig. 2.25. The Karnaugh-map-editor is explained in the help system.

(c) TVL 2 has been created to express the set S_2. Move TVL 2 into the middle viewport of the first row of the prepared m-fold view and click on the button K to show the Karnaugh-map. Edit this Karnaugh-map to the set S_2 as shown in Fig. 2.25.

(d) Compute the complement of S_1 as object 3 using the menu and show the result as Karnaugh-map in viewport (2,1) of the m-fold view.

(e) Compute the complement of S_2 as object 4 using the toolbar and show the result as Karnaugh-map in viewport (2,2) of the m-fold view.

(f) Compute the difference of $S_1 \setminus S_2$ as object 5 using the toolbar and show the result as Karnaugh-map in viewport (3,1) of the m-fold view.

(g) Compute the difference of $S_2 \setminus S_1$ as object 6 using a command in the console and show the result as Karnaugh-map in viewport (3,2) of the m-fold view.

(h) Compute the symmetric difference of $S_1 \Delta S_2$ as object 7 using the menu and show the result as Karnaugh-map in viewport (3,3) of the m-fold view.

(i) Compute the union $S_1 \cup S_2$ as object 8 using the toolbar and show the result as Karnaugh-map in viewport (4,1) of the m-fold view.

(j) Compute the intersection $S_1 \cap S_2$ as object 9 using a command in the console and show the result as Karnaugh-map in viewport (4,2) of the m-fold view.

(k) Compute the complement of the symmetric difference of $S_1 \overline{\Delta} S_2$ as object 10 using the menu and show the result as Karnaugh-map in viewport (4,3) of the m-fold view.

(l) Compare the given and computed sets represented as Karnaugh-maps and explore the properties of these sets.

(m) Prepare a PRP that creates TVLs of S_1 and S_2, compute all set operations as before, and show the ten TVLs in the m-fold view as Karnaugh-maps using the same viewports as before.

Exercise 2.5 (Operations Related to Columns) The three provided operations to manipulate columns:

- dco: delete columns;
- cco: exchange columns; and
- tco: transfer columns

are explored in this exercise using the TVL of Fig. 2.26 and three tuples of variables.

$$\text{ODA}(f_1) = \begin{array}{ccccc} x_1 & x_2 & x_3 & x_4 & x_5 \\ \hline - & 0 & 1 & 0 & 1 \\ 1 & - & 0 & 0 & 1 \\ 0 & 1 & - & 0 & 1 \end{array}$$

Fig. 2.26 TVL 1 used in Exercise 2.5 to explore possible manipulations of their columns

(a) Define a Boolean space of 32 variables, the TVL of Fig. 2.26, and the VTs VT 2 $= \langle x_2, x_3 \rangle$, VT 3 $= \langle x_4, x_5 \rangle$, and VT 4 $= \langle x_6, x_7 \rangle$ using commands in the console. Prepare the m-fold view with three rows and three columns and show these XBOOLE-objects in viewports (1,1), (1,2), (1,3), and (2,1), respectively.

(b) Delete the columns x2 and x3 using the toolbar and show the result in viewport (3,1).

(c) Exchange the columns x2 with x4 and x3 with x5 using the menu and show the result in viewport (2,2).

(d) Transfer the column x2 to x4 and the column x3 to x5 using the toolbar and show the result in viewport (2,3).

(e) Exchange the columns x2 with x6 and x3 with x7 using a command in the console and show the result in viewport (3,2).

(f) Transfer the column x2 to x6 and the column x3 to x7 using a command in the console and show the result in viewport (3,3).

(g) Prepare a PRP that solves all tasks of this exercise and verify that its execution leads to the same result as before.

Exercise 2.6 (Set Operations for Sets of Variables) Sometimes a set of variables is needed that can be created by set operations of given sets of variables. The XBOOLE-monitor XBM 2 provides commands for such set operations usable in the console or a PRP; however, these commands are not available in the menu and toolbar. A set of variables is implicitly specified by the variables that occur in a TVL or VT.

Two sets of variables $SV_1 = \{x_1, x_2, x_3, x_4, x_5\}$ and $SV_2 = \{x_3, x_4, x_5, x_6, x_7\}$ are defined within a Boolean space of nine variables x_1 to x_9. Prepare a PRP that computes all set operations for the given two sets of variables SV_1 and SV_2 using the commands mentioned above. Use the help system of XBM 2 to learn the details about commands that realize set operations for sets of variables.

(a) Prepare the m-fold view with four rows and three columns.

(b) Prepare a PRP that defines a Boolean space of 9 variables, attaches the variables x1 to x9 to this space, define two empty TVLs as objects 1 and 2 for the sets of variables SV_1 and SV_2, and show these TVLs in viewports (1,1) and (1,2).

(c) Extend the PRP such that the complement of the set of variables SV_1 with regard to all variables of the Boolean space number 1 will be computed, stored as object number 3, and displayed in viewport (2,1) of the m-fold view.

(d) Extend the PRP such that the complement of the set of variables SV_2 with regard to all variables of the Boolean space number 1 will be computed, stored as object number 4, and displayed in viewport (2,2) of the m-fold view.

(e) Extend the PRP such that the difference of the sets of variables $SV_1 \setminus SV_2$ will be computed, stored as object number 5, and displayed in viewport (3,1) of the m-fold view.

(f) Extend the PRP such that the difference of the sets of variables $SV_2 \setminus SV_1$ will be computed, stored as object number 6, and displayed in viewport (3,2) of the m-fold view.

(g) Extend the PRP such that the symmetric difference of the sets of variables $SV_1 \triangle SV_2$ will be computed, stored as object number 7, and displayed in viewport (3,3) of the m-fold view.

Fig. 2.27 TVL 1 used in Exercise 2.7 to explore control variables and control commands of XBM 2

$$ODA(f_1) = \begin{array}{cccccc} x_1 & x_2 & x_3 & x_4 & x_5 & x_6 \\ \hline 1 & 1 & 1 & 0 & 0 & 0 \\ 0 & 0 & - & - & 1 & 1 \\ 1 & 0 & 1 & - & - & - \\ - & - & 0 & 1 & 0 & - \end{array}$$

(h) Extend the PRP such that the union of the sets of variables $SV_1 \cup SV_2$ will be computed, stored as object number 8, and displayed in viewport (4,1) of the m-fold view.

(i) Extend the PRP such that the intersection of the sets of variables $SV_1 \cap SV_2$ will be computed, stored as object number 9, and displayed in viewport (4,2) of the m-fold view.

(j) Extend the PRP such that the complement of the symmetric difference of the sets of variables $SV_1 \,\overline{\Delta}\, SV_2$ will be computed, stored as object number 10, and displayed in viewport (4,3) of the m-fold view.

(k) Execute the prepared PRP and explore the properties of these sets of variables.

Exercise 2.7 (Control Variables and Control Commands) Sometimes it is necessary to manipulate each single ternary vector of the TVL separately. Both variables and control commands of the XBOOLE-monitor XBM 2 are used in this exercise to distribute the ternary vectors of a given TVL to several TVLs. We assume that the given TVL as well as the created TVLs must be shown in the first two rows and three columns of the m-fold view. Therefore, only the first five ternary vectors can be selected. A Boolean variable $complete shall be used to provide the information whether all ternary vectors have been selected and are shown. Use the TVL of Fig. 2.27 as example for the explained distribution of ternary vectors.

(a) Prepare the m-fold view with two rows and three columns.

(b) Prepare a PRP that defines a Boolean space of 32 variables and the TVL of Fig. 2.27 and show this TVL in the top left viewport of the m-fold view.

(c) Extend the PRP with commands that initialize the Boolean variable $complete and the integer variable $last (the number of the last ternary vector to be selected) for the worst case that the given TVL consists of more than five ternary vectors.

(d) The integer variable $r is used to indicate the row and the integer variable $c is used to indicate the column of a viewport. Extend the PRP with commands that initialize the integer variables $r and $c such that they determine second viewport of the first row.

(e) Extend the PRP such that the number of rows of TVL 1 is stored as integer variable $n.

(f) Extend the PRP such that the values of the variables $complete and $last are changed to the right values in the case that TVL 1 does not contain more than five ternary vectors.

(g) Use a command for at the end of the PRP that iterates from the first ternary vector of TVL 1 to the ternary vector determined by the index $last, selects in each sweep one of these ternary vectors, stores the created TVLs using the subsequent object numbers behind TVL 1, and assigns the created TVLs line-by-line to viewports of the m-fold view starting at viewport (1,2).

(h) Execute the created PRP and verify the results in the m-fold view and the view "Values of the used control variables."

2.6   Solutions

Solution 2.1 (Boolean Spaces) There are several possibilities to define Boolean spaces in the XBOOLE-monitor XBM 2.

Space name	Space number	Type	Maximum number of variables	Assigned variables
SpaceDefinition (1)	1	2	10	0
SpaceDefinition (2)	2	8	100	0
SpaceDefinition (3)	3	64	1000	0

(above the table: m-fold view | 4-fold View | 1-fold View | Spaces | Objects)

Fig. 2.28 The view "Spaces" that shows the three defined Boolean spaces

```
1   new                                3   space  100  2
2   space  10  1                       4   space  1000  3
```

Fig. 2.29 Problem-program that defines three Boolean spaces with 10, 100, and 1000 variables

(a) A click on the item *Define a space* in the menu *Objects* opens a dialog window where the space number 1 and the maximal number of 10 variables can be specified.

(b) A click on the toolbar button $\stackrel{+}{\square}$ opens a dialog window where the space number 2 and the maximal number of 100 variables can be specified.

(c) The command `space 1000 3` defines the space number 3 and the maximal number of 1000 variables.

(d) Figure 2.28 shows the properties of the defined spaces. The numbers in the column "Type" are the answers to this question: ternary vectors of 10, 100, or 1000 variables are stored in two, eight, or 64 machine words.

(e) The message "invalid index" indicates the error when you try to define a Boolean space that already exists. This can be seen, e.g., when you use the command `space 200 2` after the execution of the command `space 100 2`. The properties of the defined spaces as shown in Fig. 2.28 remain unchanged.

(f) Figure 2.29 shows the PRP that defines the three Boolean spaces of 10, 100, and 1000 variables. The execution of this PRP leads to the same result as shown in Fig. 2.28.

Solution 2.2 (Input of Ternary Vector Lists)

(a) The Boolean space of 32 variables can be defined using one of the methods applied in Exercise 2.1.

(b) A click on the item "Attach variables" of the menu "Objects" opens a dialog window where the names of the eight variables $x1, \ldots, x8$ must be typed in lexicographic order separated by a space or a line break.

(c) A click on the item "Create a TVL" of the menu "Objects" opens a dialog window where the suggested "Object number" 1 remains unchanged, the form "a" (antivalence form) must be selected from the drop-down list, and the names of the four variables $x2$, $x4$, $x6$, and $x8$ must be typed.

(d) A click on the item "Append ternary vector(s)" of the menu "Objects" opens a dialog window where each of the four ternary vectors of the TVL $A(f_1)$ as shown in Fig. 2.24 must be typed in one line. A click on the OK button closes this dialog window and appends the four ternary vectors to TVL 1.

(e) The name "TVL (1)" of create TVL is shown in the view "Object overview of a space"; this item must be moved by drag&drop to the top-left viewport of the four-fold view.

Fig. 2.30 Six lines of the
command `tin` used in the
console to define TVL 3

```
tin 1 3 /k
x1 x2 x3 x4 x5 x6.
000000
111111
1- 0- 1-
- 1- 0- 1.
```

(f) A click on the button $\overset{+}{\text{TVL}}$ in the toolbar opens a dialog window where both the suggested "Object number" 2 and the form "oda" (orthogonal disjunctive or antivalence form) remain unchanged, and the names of the five variables x4, x5, x6, x7, and x8 must be typed.

(g) The name "TVL (2)" of the created TVL is shown in the view "Object overview of a space"; this item must be moved by drag&drop to the top-right viewport of the four-fold view. Hints to edit this TVL are provided in the help system of the XBOOLE-monitor XBM 2: "Contents— Available Editors—TVL-Editor." Insert the three ternary vectors as shown in Fig. 2.24 for the TVL ODA(f_2).

(h) A click on the button $\overset{\text{ORTH}}{\perp}$ in the toolbar opens a dialog window where TVL (2) must be selected in the drop-down list "Input object" and a click on the button G (like given) copies the number 2 into the field "Output object"; a click on the OK button closes this dialog window and realizes the orthogonalization of TVL 2.

(i) The six lines of Fig. 2.30 must be typed into the command line of the console, each of them confirmed by pressing the ENTER-key.

(j) The name "TVL (3)" of create TVL is shown in the view "Object overview of a space"; this item must be moved by drag&drop to the bottom-left viewport of the four-fold view. Figure 2.31 shows the three created TVLs in the four-fold view.

(k) Figure 2.32 shows the PRP that defines the three TVLs of Fig. 2.24 and assigns these TVLs to the viewports of the four-fold view as requested in Exercise 2.2. The execution of this PRP leads to the same result as shown in Fig. 2.31.

Solution 2.3 (Input of Tuples of Variables)

(a) The command `space` without a parameter creates a Boolean space of 32 variables due to the default values 32 for vmax and the next free space number for sn; read the hints for this command in the help system. The variables appear in this exercise in lexicographic order; hence, the assignment of the variables to the space can be avoided.

(b) A click on the item "Create a TV" of the menu "Objects" opens a dialog window where the suggested "Object number" 1 remains unchanged and the names of the three variables x1, x2, and x3 must be typed separated by a space or a line break. No further actions are needed after closing the dialog window using the OK button.

(c) A click on the button $\overset{+}{\text{VT}}$ opens a dialog window where the suggested "Object number" 2 remains unchanged and the names of the two variables x3 and x2 must be typed as requested for VT 2 in this order.

(d) The first two variables of VT 3 are the same as in VT 2; hence, VT 2 can be copied to VT 3. This can be done by a click on the item "Copy object" of the menu "Objects"; a dialog window opens where "VT (2)" must be chosen in the list "Input object" and the suggested object number 3 for "Output object" remains unchanged. Thereafter, a click on the item "Append Variable(s) to a VT"

| m-fold view | 4-fold View | 1-fold View | Spaces | Objects |

| K TVL (1) A | 4 Var. | 4 R. | S. 1 |

	$x2$	$x4$	$x6$	$x8$
1	1	-	-	-
2	-	1	-	-
3	-	-	1	-
4	-	-	-	1

| K TVL (2) ODA | 5 Var. | 3 R. | S. 1 |

	$x4$	$x5$	$x6$	$x7$	$x8$
1	0	0	-	-	1
2	1	-	-	1	1
3	-	-	1	0	0

| K TVL (3) K | 6 Var. | 4 R. | S. 1 |

	$x1$	$x2$	$x3$	$x4$	$x5$	$x6$
1	0	0	0	0	0	0
2	1	1	1	1	1	1
3	1	-	0	-	1	-
4	-	1	-	0	-	1

Fig. 2.31 Three TVLs created as Exercise 2.2

```
1   new                              13   00--1
2   space 32 1                       14   1--11
3   avar 1                           15   --100.
4   x1 x2 x3 x4 x5 x6 x7 x8.         16   tin 1 3 /k
5   tin 1 1 /a                       17   x1 x2 x3 x4 x5 x6.
6   x2 x4 x6 x8.                     18   000000
7   1---                             19   111111
8   -1--                             20   1-0-1-
9   --1-                             21   -1-0-1.
10  ---1.                            22   assign 1 /4 1 1
11  tin 1 2 /oda                     23   assign 2 /4 1 2
12  x4 x5 x6 x7 x8.                  24   assign 3 /4 2 1
```

Fig. 2.32 Problem-program that defines the three TVLs that are specified in Fig. 2.24

of the menu "Objects" opens a dialog window where "VT (3)" must be chosen as "Object" and the needed variables $x4$ and $x2$ must be typed in this order separated by a space or a line break.

(e) The concatenation of VT 2 and VT 1 creates a VT that is equal to VT 4 when the last variable $x3$ has been deleted. Hence, two actions without the input of a variable create VT 4 as follows:

　　a. a click on the item "Concatenate ternary matrices" of the menu "Matrices" opens a dialog window where "VT (2)" must be chosen in the list "Input object 1"; the suggested object number 1 for "Input object 2" as well as the suggested number 4 for "Output object" remains unchanged.

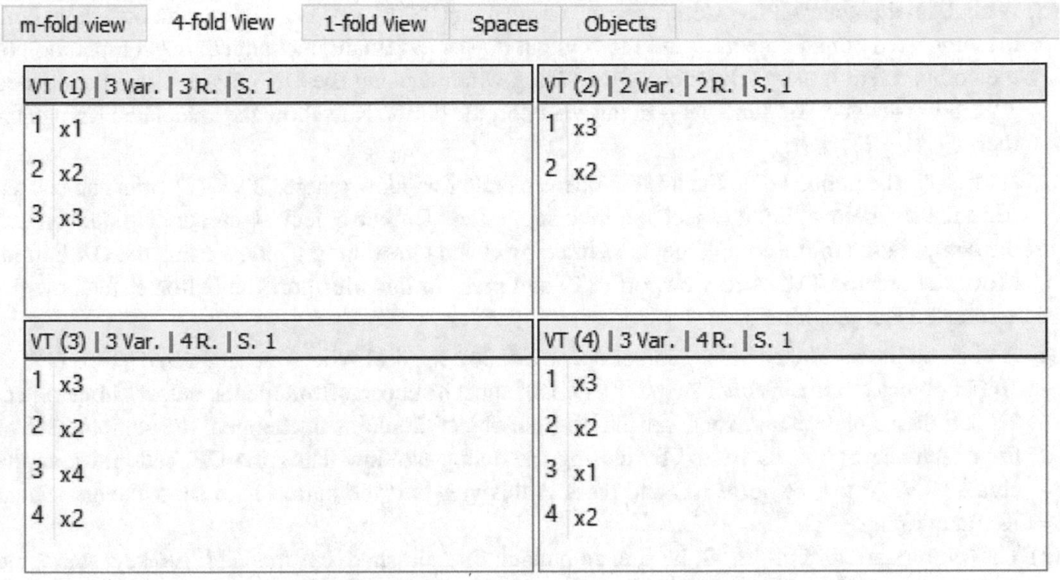

| m-fold view | 4-fold View | 1-fold View | Spaces | Objects |

VT (1) | 3 Var. | 3 R. | S. 1	VT (2) | 2 Var. | 2 R. | S. 1
1 x1 2 x2 3 x3	1 x3 2 x2

VT (3) | 3 Var. | 4 R. | S. 1	VT (4) | 3 Var. | 4 R. | S. 1
1 x3 2 x2 3 x4 4 x2	1 x3 2 x2 3 x1 4 x2

Fig. 2.33 Four VTs created as Exercise 2.3

1	new	6	x3 x2 .	11	dtv 4 5 4
2	space 32 1	7	copy 2 3	12	assign 1 /4 1 1
3	vtin 1 1	8	vtapp 3	13	assign 2 /4 1 2
4	x1 x2 x3 .	9	x4 x2 .	14	assign 3 /4 2 1
5	vtin 1 2	10	con 2 1 4	15	assign 4 /4 2 2

Fig. 2.34 Problem-program that defines the four VTs that are specified in Exercise 2.3

b. a click on the button $\overset{\text{DTV}}{\equiv}$ opens a dialog window where "VT (4)" in the list "Input object" and the number 5 in the input-field "Number of the vector" must be chosen; the number of the "Output object" can be specified to 4 by pressing the button G (like given).

(f) The four created VTs can be moved by drag&drop from the view "Object overview of a space" in the four viewports of the four-fold view. Figure 2.33 shows these four created VTs.

(g) Figure 2.34 shows the PRP that defines the four VTs and assigns these VTs to the viewports of the four-fold view as requested in Exercise 2.3.

Solution 2.4 (Set Operations)

(a) The Boolean space of 32 variables can be defined using one of the methods applied in Exercise 2.1. A click on the button $\overset{+}{\text{TVL}}$ in the toolbar opens a dialog window where both the suggested "Object number" 1 and the form "oda" (orthogonal disjunctive or antivalence form) remain unchanged, and the names of the five variables x1, x2, x3, x4, and x5 must be typed. Typing the command copy 1 2 in the command line of the console copies the empty TVL 1 of five variables to TVL 2.

(b) Move the created TVL 1 to viewport (1,1), press in this viewport the button K to show the associated Karnaugh-map, and insert the values 1 of S_1 as shown in Fig. 2.25.

(c) Move the created TVL 2 to viewport (1,2), press in this viewport the button K to show the associated Karnaugh-map, and insert the values 1 of S_2 as shown in Fig. 2.25.

(d) A click on the item "CPL—Complement" of the menu "Sets" opens a dialog window where both the suggested "Input object" 1 and the "Output object" 3 remain unchanged; the computation of the complement is started by closing the dialog window using the OK button. Move the created TVL 3 to viewport (2,1) and press in this viewport the button K to show the associated Karnaugh-map.

(e) A click on the button ⊡ in the toolbar opens a dialog window where "TVL (2)" must be chosen from the list below "Input object" and the suggested "Output object" 4 can remain unchanged; the computation of the complement is started by closing the dialog window using the OK button. Move the created TVL 4 to viewport (2,2) and press in this viewport the button K to show the associated Karnaugh-map.

(f) A click on the button ⊡ in the toolbar opens a dialog window where the suggested "TVL (1)" as "Input object 1" remains unchanged, "TVL (2)" must be chosen from the list below "Input object 2," and the suggested number 5 for the "Output object" remains unchanged; the computation of the difference $S_1 \setminus S_2$ is started by closing the dialog window using the OK button. Move the created TVL 5 to viewport (3,1) and press in this viewport the button K to show the associated Karnaugh-map.

(g) The command `dif 2 1 6` in the command line finished by the ENTER-key starts the computation of the difference $S_2 \setminus S_1$ and stores the result as TVL number 6. Move the created TVL 6 to viewport (3,2) and press in this viewport the button K to show the associated Karnaugh-map.

(h) A click on the item "SYD—Symmetric difference" of the menu "Sets" opens a dialog window where the suggested "TVL (1)" as "Input object 1" remains unchanged, "TVL (2)" must be chosen from the list below "Input object 2," and the suggested number 7 for the "Output object" remains unchanged; the computation of the symmetric difference $S_1 \triangle S_2$ is started by closing the dialog window using the OK button. Move the created TVL 7 to viewport (3,3) and press in this viewport the button K to show the associated Karnaugh-map.

(i) A click on the button ⊡ in the toolbar opens a dialog window where the suggested "TVL (1)" as "Input object 1" remains unchanged, "TVL (2)" must be chosen from the list below "Input object 2," and the suggested number 8 for the "Output object" remains unchanged; the computation of the union $S_1 \cup S_2$ is started by closing the dialog window using the OK button. Move the created TVL 8 to viewport (4,1) and press in this viewport the button K to show the associated Karnaugh-map.

(j) The command `isc 1 2 9` in the command line finished by the ENTER-key starts the computation of the intersection $S_1 \cap S_2$ and stores the result as TVL number 9. Move the created TVL 9 to viewport (4,2) and press in this viewport the button K to show the associated Karnaugh-map.

(k) A click on the item "CSD—Complement of the symmetric difference" of the menu "Sets" opens a dialog window where the suggested "TVL (1)" as "Input object 1" remains unchanged, "TVL (2)" must be chosen from the list below "Input object 2," and the suggested number 10 for the "Output object" remains unchanged; the computation of the complement of the symmetric difference $S_1 \overline{\triangle} S_2$ is started by closing the dialog window using the OK button. Move the created TVL 10 to viewport (4,3) and press in this viewport the button K to show the associated Karnaugh-map.

(l) Figure 2.35 shows the Karnaugh-maps of the two created and all computed sets. The change of values in a Karnaugh-map leads implicitly to the representation of this set as TVL in ODA-form. This is a welcome property because the most of the set operations require orthogonal TVLs as input.

m-fold view | 4-fold View | 1-fold View | Spaces | Objects

TVL (1) ODA | 5 Var. | 3 R. | S. 1

```
0 0   0 0 0 0 0 0 0 0
0 1   1 1 1 1 1 1 0 0
1 1   0 0 1 1 1 1 1 0
1 0   0 0 0 0 0 0 1 0
x4 x5 0 1 1 0 0 1 1 0 x3
      0 0 1 1 1 1 0 0 x2
      0 0 0 0 1 1 1 1 x1
```

TVL (2) ODA | 5 Var. | 3 R. | S. 1

```
0 0   0 0 1 0 0 1 1 0
0 1   1 0 1 1 1 1 1 1
1 1   0 0 1 0 0 1 1 0
1 0   0 0 1 0 0 1 1 0
x4 x5 0 1 1 0 0 1 1 0 x3
      0 0 1 1 1 1 0 0 x2
      0 0 0 0 1 1 1 1 x1
```

-

TVL (3) ODA | 5 Var. | 7 R. | S. 1

```
0 0   1 1 1 1 1 1 1 1
0 1   0 0 0 0 0 0 1 1
1 1   1 1 0 0 0 0 0 1
1 0   1 1 1 1 1 1 0 1
x4 x5 0 1 1 0 0 1 1 0 x3
      0 0 1 1 1 1 0 0 x2
      0 0 0 0 1 1 1 1 x1
```

TVL (4) ODA | 5 Var. | 3 R. | S. 1

```
0 0   1 1 0 1 1 0 0 1
0 1   0 1 0 0 0 0 0 0
1 1   1 1 0 1 1 0 0 1
1 0   1 1 0 1 1 0 0 1
x4 x5 0 1 1 0 0 1 1 0 x3
      0 0 1 1 1 1 0 0 x2
      0 0 0 0 1 1 1 1 x1
```

-

TVL (5) ODA | 5 Var. | 2 R. | S. 1

```
0 0   0 0 0 0 0 0 0 0
0 1   0 1 0 0 0 0 0 0
1 1   0 0 0 1 1 0 0 0
1 0   0 0 0 0 0 0 0 0
x4 x5 0 1 1 0 0 1 1 0 x3
      0 0 1 1 1 1 0 0 x2
      0 0 0 0 1 1 1 1 x1
```

TVL (6) ODA | 5 Var. | 3 R. | S. 1

```
0 0   0 0 1 0 0 1 1 0
0 1   0 0 0 0 0 0 1 1
1 1   0 0 0 0 0 0 0 0
1 0   0 0 1 0 0 1 0 0
x4 x5 0 1 1 0 0 1 1 0 x3
      0 0 1 1 1 1 0 0 x2
      0 0 0 0 1 1 1 1 x1
```

TVL (7) ODA | 5 Var. | 5 R. | S. 1

```
0 0   0 0 1 0 0 1 1 0
0 1   0 1 0 0 0 0 1 1
1 1   0 0 0 1 1 0 0 0
1 0   0 0 1 0 0 1 0 0
x4 x5 0 1 1 0 0 1 1 0 x3
      0 0 1 1 1 1 0 0 x2
      0 0 0 0 1 1 1 1 x1
```

TVL (8) ODA | 5 Var. | 5 R. | S. 1

```
0 0   0 0 1 0 0 1 1 0
0 1   1 1 1 1 1 1 1 1
1 1   0 0 1 1 1 1 1 0
1 0   0 0 1 0 0 1 1 0
x4 x5 0 1 1 0 0 1 1 0 x3
      0 0 1 1 1 1 0 0 x2
      0 0 0 0 1 1 1 1 x1
```

TVL (9) ODA | 5 Var. | 4 R. | S. 1

```
0 0   0 0 0 0 0 0 0 0
0 1   1 0 1 1 1 1 0 0
1 1   0 0 1 0 0 1 1 0
1 0   0 0 0 0 0 0 1 0
x4 x5 0 1 1 0 0 1 1 0 x3
      0 0 1 1 1 1 0 0 x2
      0 0 0 0 1 1 1 1 x1
```

TVL (10) ODA | 5 Var. | 8 R. | S. 1

```
0 0   1 1 0 1 1 0 0 1
0 1   1 0 1 1 1 1 0 0
1 1   1 1 1 0 0 1 1 1
1 0   1 1 0 1 1 0 1 1
x4 x5 0 1 1 0 0 1 1 0 x3
      0 0 1 1 1 1 0 0 x2
      0 0 0 0 1 1 1 1 x1
```

Reset

Fig. 2.35 Two given sets S_1 (TVL 1) and S_2 (TVL 2) and the results of set operations for these sets

The comparison of the Karnaugh-maps of the upper two rows confirms that the complement of a set of binary vectors consists of all these elements of the universe (here, the set of all binary vectors of 5 variables x1 to x5) that do not belong to the given set; hence, values 1 in the of upper row are changed to values 0 of second row of m-fold view and vice versa.

The Karnaugh-maps in viewports (3,1) and (3,2) confirm that the difference of two sets is not a commutative operation. The set $S_1 \setminus S_2$ in viewport (3,1) contains all elements of the set S_1 (shown as values 1) that do not belong to the set S_2. Vice versa, the set $S_2 \setminus S_1$ in viewport (3,2) contains all elements of the set S_2 that do not belong to the set S_1.

The symmetric difference $S_1 \triangle S_2$ in viewport (3,3) is equal to the union of the two differences $S_1 \setminus S_2$ and $S_2 \setminus S_1$ shown in viewports (3,1) and (3,2), respectively. The comparison of the symmetric difference $S_1 \triangle S_2$ in viewport (3,3) with the given two sets S_1 in viewport (1,1) and

```
 1   new                    11   -11--              21   assign  1  /m  1  1
 2   space  32  1           12   --001.             22   assign  2  /m  1  2
 3   tin  1  1              13   cpl  1  3           23   assign  3  /m  2  1
 4   x1  x2  x3  x4  x5.    14   cpl  2  4           24   assign  4  /m  2  2
 5   -1--1                  15   dif  1  2  5        25   assign  5  /m  3  1
 6   1011-                  16   dif  2  1  6        26   assign  6  /m  3  2
 7   00-01.                 17   syd  1  2  7        27   assign  7  /m  3  3
 8   tin  1  2              18   uni  1  2  8        28   assign  8  /m  4  1
 9   x1  x2  x3  x4  x5.    19   isc  1  2  9        29   assign  9  /m  4  2
10   101--                 20   csd  1  2  10       30   assign  10 /m  4  3
```

Fig. 2.36 Problem-program that defines the sets S_1 (TVL 1) and S_2 (TVL 2), computes the set operations as specified in Exercise 2.4, and shows the results in the m-fold view

Fig. 2.37 Commands that
define the Boolean space,
TVL 1, and three VTs used
in Exercise 2.5

```
space  32  1               vtin  1  2
tin  1  1  /oda            x2  x3.
x1  x2  x3  x4  x5.        vtin  1  3
-0101                      x4  x5.
1-001                      vtin  1  4
01-01.                     x6  x7.
```

S_2 in viewport (1,2) confirms that an element belongs to the symmetric difference if it belongs either to S_1 or to S_2, but not to both of them.

The union $S_1 \cup S_2$ in viewport (4,1) contains all elements of these two sets. The intersection $S_1 \cap S_2$ in viewport (4,2) contains all these elements that belong to both the set S_1 and the set S_2. The comparison of the complement of the symmetric difference $S_1 \overline{\Delta} S_2$ in viewport (4,3) with the symmetric difference $S_1 \Delta S_2$ in viewport (3,3) confirms that these two sets are complements to each other. The comparison of the complement of the symmetric difference $S_1 \overline{\Delta} S_2$ in viewport (4,3) with the given two sets S_1 in viewport (1,1) and S_2 in viewport (1,2) shows that an element belongs to $S_1 \overline{\Delta} S_2$ if either it belongs to both sets S_1 and S_2 or it belongs neither to S_1 nor to S_2.

(m) Figure 2.36 shows the PRP that defines the Boolean space of 32 variables, defines the two sets S_1 and S_2 as TVLs 1 and 2, computes the set operations as requested in Exercise 2.4, and shows these sets as Karnaugh-maps.

The TVLs of the sets S_1 and S_2 can easily be defined in the PRP based on the implicitly generated TVLs as a result of the changed Karnaugh-maps shown in the upper row of the m-fold view; pressing the button T (for show TVL), the TVL belonging to the Karnaugh-map will be displayed. Note that the m-fold view with at least four rows and three columns must be prepared before the PRP of Fig. 2.36 is executed; otherwise, not all TVLs will be displayed. The change between the TVL and the Karnaugh-map in a viewport is only possible when a TVL has been assigned; hence, the change to the view of a Karnaugh-map must be done after the execution of the PRP. Each viewport remembers the property whether a TVL or a Karnaugh-map has to be displayed.

Solution 2.5 (Operations Related to Columns)

(a) Assuming that no Boolean space has been defined before, the 12 lines of Fig. 2.37 must be typed into the command line, each of them confirmed by pressing the ENTER-key.

The created TVL 1 and the three VTs with the object numbers 2, 3, and 4 can be moved by drag&drop from the view "Object overview of a space" to the requested viewports in the upper row of the m-fold view and to viewport (2,1), respectively. Figure 2.38 shows in the mentioned viewports these XBOOLE-objects.

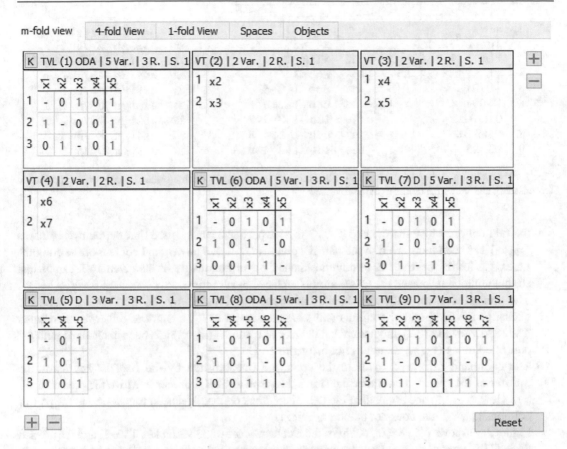

Fig. 2.38 The defined TVL 1, three VTs, and the results of the operations that change the columns of this TVL.

(b) a click on the button opens a dialog window where "TVL (1)" remains unchanged as "Input object 1," "VT (2)" must be chosen in the list "Input object 2," and the number 5 of the "Output object" remains also unchanged. The operation to delete the columns $x2$ and $x3$ is started by closing this dialog window using the OK button. The created TVL 5 can be moved by drag&drop to viewport (3,1) as shown in Fig. 2.38, and the displayed TVL 5 confirms that the columns $x2$ and $x3$ are deleted. The form is set to a D-form because the orthogonality is lost by deleting these columns.

(c) A click on the item "CCO—Exchange columns" of the menu "Matrices" opens a dialog window where both the suggested "TVL (1)" as "Input object 1" and "VT (2)" as "Input object 2" remain unchanged, "VT (3)" must be chosen from the list below "Input object 3," and the suggested number 6 for the "Output object" remains unchanged; the exchange of the specified columns is started by closing the dialog window using the OK button. Move the created TVL 6 to viewport (2,2) and notice that the pair of columns $\langle x_2, x_3 \rangle$ has been exchanged with the pair of columns $\langle x_4, x_5 \rangle$. No column is lost when columns are exchanged; hence, the orthogonality of TVL 1 remains unchanged in the created TVL 6.

(d) A click on the button opens a dialog window where both the suggested "TVL (1)" as "Input object 1" and "VT (2)" as "Input object 2" remain unchanged, "VT (3)" must be chosen from the list below "Input object 3," and the suggested number 7 for the "Output object" remains unchanged; the transfer of the specified columns is started by closing the dialog window using

1	new	10	vtin 1 3
2	space 32 1	11	x4 x5.
3	tin 1 1	12	vtin 1 4
4	x1 x2 x3 x4 x5.	13	x6 x7.
5	−0101	14	dco 1 2 5
6	1−001	15	cco 1 2 3 6
7	01−01.	16	tco 1 2 3 7
8	vtin 1 2	17	cco 1 2 4 8
9	x2 x3.	18	tco 1 2 4 9

19	assign 1 /m 1 1
20	assign 2 /m 1 2
21	assign 3 /m 1 3
22	assign 4 /m 2 1
23	assign 5 /m 3 1
24	assign 6 /m 2 2
25	assign 7 /m 2 3
26	assign 8 /m 3 2
27	assign 9 /m 3 3

Fig. 2.39 Problem-program that defines TVL 1 and three VTs, computes operations related to columns as specified in Exercise 2.5, and shows the results in the m-fold view

the OK button. Move the created TVL 7 to viewport (2,3) and notice that the pair of columns $\langle x_2, x_3 \rangle$ has been transferred to the pair of columns $\langle x_4, x_5 \rangle$. The original content of the columns $\langle x_4, x_5 \rangle$ is lost by the transfer of columns; hence, the orthogonality of the given TVL can be lost so that without any check the ODA-form is replaced by a D-form.

(e) The command cco 1 2 4 8 in the command line finished by the ENTER-key starts the exchange of the specified columns and stores the result as TVL number 8. Move the created TVL 8 to viewport (3,2) and notice that the pair of columns $\langle x_2, x_3 \rangle$ has been exchanged with the pair of columns $\langle x_6, x_7 \rangle$ that does not belong to TVL 1.

(f) The command tco 1 2 4 9 in the command line finished by the ENTER-key starts the transfer of the specified columns and stores the result as TVL number 9. Move the created TVL 9 to viewport (3,3) and notice that the pair of columns $\langle x_2, x_3 \rangle$ has been transferred to the pair of columns $\langle x_6, x_7 \rangle$ that does not belong to TVL 1.

(g) Figure 2.39 shows the PRP that defines the Boolean space of 32 variables, TVL 1, and the needed three VTs, computes thereafter the operations related to columns as requested in Exercise 2.5, and shows all used XBOOLE-objects in nine viewports of the m-fold view (see Fig. 2.37).

Solution 2.6 (Set Operations for Sets of Variables)

(a) Four rows and three columns of viewports are prepared using the buttons in the m-fold view.

(b) Figure 2.40 shows the complete PRP that prepares an empty XBOOLE-system in line 1, defines the requested Boolean space of 9 variables in line 2, attaches the nine variables in lines 3 and 4, creates TVL 1 of SV_1 in lines 5–7, assigns this TVL to the top-left viewport of the m-fold view in line 8, creates TVL 2 of SV_2 in lines 9–11, and assigns this TVL to viewport (1,2) of the m-fold view in line 12.

(c) The command sv_cpl 1 3 in line 13 computes the complement of SV_1 with regard to the Boolean space 1, and the command assign in line 14 assigns the computed set of variables to viewport (2,1).

(d) The command sv_cpl 2 4 in line 15 computes the complement of SV_2 with regard to the Boolean space 1, and the command assign in line 16 assigns the computed set of variables to viewport (2,2).

(e) The command sv_dif 1 25 in line 17 computes the difference of the set of variables $SV_1 \setminus SV_2$, and the command assign in line 18 assigns the computed set of variables to viewport (3,1).

(f) The command sv_dif 2 1 6 in line 19 computes the difference of the set of variables $SV_2 \setminus SV_1$ using the reverse order of the operands, and the command assign in line 20 assigns the computed set of variables to viewport (3,2).

```
 1   new                                15   sv_cpl  2  4
 2   space  9  1                        16   assign  4  /m  2  2
 3   avar  1                            17   sv_dif  1  2  5
 4   x1  x2  x3  x4  x5  x6  x7  x8  x9. 18   assign  5  /m  3  1
 5   tin  1  1                          19   sv_dif  2  1  6
 6   x1  x2  x3  x4  x5.                 20   assign  6  /m  3  2
 7   .                                  21   sv_syd  1  2  7
 8   assign  1  /m  1  1                 22   assign  7  /m  3  3
 9   tin  1  2                          23   sv_uni  1  2  8
10   x3  x4  x5  x6  x7.                 24   assign  8  /m  4  1
11   .                                  25   sv_isc  1  2  9
12   assign  2  /m  1  2                 26   assign  9  /m  4  2
13   sv_cpl  1  3                        27   sv_csd  1  2  10
14   assign  3  /m  2  1                 28   assign  10  /m  4  3
```

Fig. 2.40 Problem-program defines two sets of variables SV_1 and SV_2 using TVLs 1 and 2 within a Boolean space of nine variables, computes all set operations of these sets of variables, and shows all created objects

(g) The command sv_syd 1 2 6 in line 21 computes the symmetric difference of the set of variables $SV_1 \Delta SV_2$, and the command assign in line 22 assigns the computed set of variables to viewport (3,3).

(h) The command sv_uni 1 2 8 in line 23 computes the union of the set of variables $SV_1 \cup SV_2$, and the command assign in line 24 assigns the computed set of variables to viewport (4,1).

(i) The command sv_isc 1 2 9 in line 25 computes the intersection of the set of variables $SV_1 \cap SV_2$, and the command assign in line 26 assigns the computed set of variables to viewport (4,2).

(j) The command sv_csd 1 2 10 in line 27 computes the complement of the symmetric difference of the set of variables $SV_1 \overline{\Delta} SV_2$ with regard to the Boolean space 1, and the command assign in line 28 assigns the computed set of variables to viewport (4,3).

(k) Figure 2.41 shows the two given sets of variables SV_1 and SV_2 in the first row of the m-fold view. The comparison of the sets of variables of the upper two rows confirms that the complement of a set of variables consists of all these variables of the universe (here, the set of all variables x1 to x9 of the Boolean space 1) that do not belong to the given set of variables.

The sets of variables in viewports (3,1) and (3,2) confirm that the difference of two sets of variables is not a commutative operation. The set of variables $SV_1 \setminus SV_2$ in viewport (3,1) contains all variables of the set SV_1 that do not belong to the set of variables SV_2. Vice versa, the set of variables $SV_2 \setminus SV_1$ in viewport (3,2) contains all variables of the set SV_2 that do not belong to the set of variables SV_1.

The symmetric difference of sets of variables $SV_1 \Delta SV_2$ in viewport (3,3) is equal to the union of the two differences $SV_1 \setminus SV_2$ and $SV_2 \setminus SV_1$ shown in viewports (3,1) and (3,2), respectively. The comparison of the symmetric difference of sets of variables $SV_1 \Delta SV_2$ in viewport (3,3) with the given two sets of variables SV_1 in viewport (1,1) and SV_2 in viewport (1,2) confirms that a variable belongs to the symmetric difference of these sets if it belongs either to SV_1 or to SV_2, but not to both of them.

The union of sets of variables $SV_1 \cup SV_2$ in viewport (4,1) contains all variables of these two sets. The intersection of sets of variables $SV_1 \cap SV_2$ in viewport (4,2) contains all those variables that belong to both the sets SV_1 and SV_2.

The comparison of the complement of the symmetric difference of sets of variables $SV_1 \overline{\Delta} SV_2$ in viewport (4,3) with the symmetric difference of sets of variables $SV_1 \Delta SV_2$ in viewport (3,3)

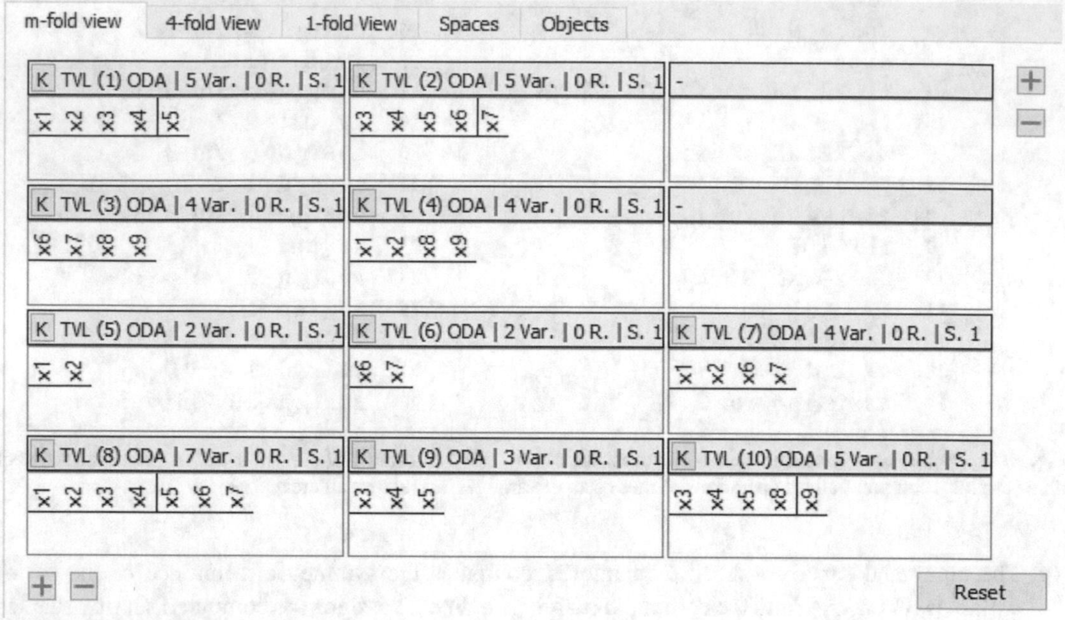

| m-fold view | 4-fold View | 1-fold View | Spaces | Objects |

K | TVL (1) ODA | 5 Var. | 0 R. | S. 1
x1 x2 x3 x4 x5

K | TVL (2) ODA | 5 Var. | 0 R. | S. 1
x3 x4 x5 x6 x7

K | TVL (3) ODA | 4 Var. | 0 R. | S. 1
x6 x7 x8 x9

K | TVL (4) ODA | 4 Var. | 0 R. | S. 1
x1 x2 x8 x9

K | TVL (5) ODA | 2 Var. | 0 R. | S. 1
x1 x2

K | TVL (6) ODA | 2 Var. | 0 R. | S. 1
x6 x7

K | TVL (7) ODA | 4 Var. | 0 R. | S. 1
x1 x2 x6 x7

K | TVL (8) ODA | 7 Var. | 0 R. | S. 1
x1 x2 x3 x4 x5 x6 x7

K | TVL (9) ODA | 3 Var. | 0 R. | S. 1
x3 x4 x5

K | TVL (10) ODA | 5 Var. | 0 R. | S. 1
x3 x4 x5 x8 x9

Reset

Fig. 2.41 Two given sets of variables SV_1 (TVL 1) and SV_2 (TVL 2) and the results of the set operations for these sets of variables

confirms that these two sets are complements to each other using the Boolean space 1 as universe. The comparison of the complement of the symmetric difference of the sets of variables $SV_1 \, \overline{\Delta} \, SV_2$ in viewport (4,3) with the given two sets of variables SV_1 in viewport (1,1) and SV_2 in viewport (1,2) shows that a variable belongs to $SV_1 \, \overline{\Delta} \, SV_2$ if either it belongs to both sets of variables SV_1 and SV_2 (the subset $\{x_3, x_4, x_5\}$) or it belongs neither to SV_1 nor to SV_2 (the subset $\{x_8, x_9\}$).

Solution 2.7 (Control Variables and Control Commands)

(a) Two rows and three columns of viewports are prepared using the buttons in the *m*-fold view.
(b) Figure 2.42 shows the complete PRP that prepares an empty XBOOLE-system in line 1, defines the requested Boolean space of 32 variables in line 2, creates TVL 1 of Fig. 2.27 in lines 3–8, and assigns this TVL to the top-left viewport of the *m*-fold view in line 9.
(c) The value `false` has been assigned to the Boolean control variable `$complete` in line 10, and the integer control variable `$last` is initialized with the value 5 in line 11 because additionally to the given TVL only five other TVLs can be shown in the six viewports of two rows and three columns of the *m*-fold view.
(d) Two commands `set` are used to specify the row (`$r`) in line 12 and the column (`$c`) in line 13 of a viewport. The assigned values determine viewport (1,2).
(e) The command `ntv` in line 14 returns the number of ternary vectors of the specified TVL 1, and the subsequent command `set` assigns this value to the integer variable `$n` in the same line.
(f) The command `if` in lines 15–19 checks in line 15 whether the number of ternary vectors of TVL 1 stored in the variable `$n` is less than or equal to 5; if this condition is satisfied, the commands enclosed in the subsequent pair of parentheses are executed. These two commands change the value of `$complete` to `true` and the value of `$last` to `$n` because up to five ternary vectors can be displayed in the prepared viewports.

```
 1   new                               16  (
 2   space 32 1                        17  set $complete true
 3   tin 1 1                           18  set $last $n
 4   x1 x2 x3 x4 x5 x6.                19  )
 5   111000                            20  for $i 1 $last
 6   00--11                            21  (
 7   101---                            22  set $result (add $i 1)
 8   --010-.                           23  stv 1 $i $result
 9   assign 1 /m 1 1                   24  assign $result /m $r $c
10   set $complete false               25  set $c (add $c 1)
11   set $last 5                       26  if (ge $c 4)
12   set $r 1                          27  (
13   set $c 2                          28  set $c 1
14   set $n (ntv 1)                    29  set $r (add $r 1)
15   if (le $n 5)                      30  ))
```

Fig. 2.42 Problem-program that defines TVL 1 and uses both control variables and control commands to select the ternary vectors as separate TVLs

(g) The command for in lines 20–30 initializes the variable $i and increments this value after each sweep by 1 until the last sweep of the embedded command list is executed for the value of $i equal to the value of $last.

The command list of this command for consists of five commands:

    a. the created TVL of the first ternary vector must be stored as TVL 2; hence, the control variable $result is used as index of the result and must be equal to $i + 1 (see line 22);

    b. the command stv in line 23 selects the ternary vector with the index $i from TVL 1 and stores the created TVL using the index $result;

    c. the TVL created in the previous command is assigned in line 24 to the viewport of the row $r and the column $c of the m-fold view;

    d. the next TVL must be assigned line-by-line to next free viewport; hence, the column $c is incremented in line 25;

    e. when $c is greater than or equal to 4, the first viewport of the next row must be chosen; the command if in lines 26–30 checks this condition, and the commands set determine left viewport (line 28) of the next row (line 29).

(h) Figure 2.43 confirms that all four ternary vectors of TVL 1 have been selected, stored as TVLs 2 to 5, and assigned line-by-line behind TVL 1 in the subsequent viewports.

Figure 2.44 shows the values of the control variables after the execution of the PRP. It can be seen that the value of the variable $complete is equal to true (that means that all ternary vectors have been selected from TVL 1), the value of the variable $n is equal to 4 (there are four ternary vectors in TVL 1), and the value of the variable $result is equal to 5 (the last created resulting TVL carries the object number 5).

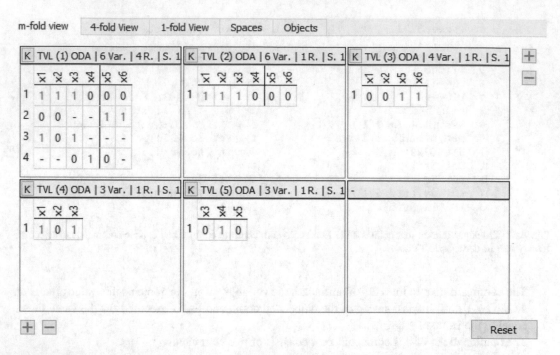

Fig. 2.43 Given TVL 1 and TVLs of all ternary vectors selected from this TVL

Fig. 2.44 The values stored in the defined control variables after the execution of the PRP shown in Fig. 2.42

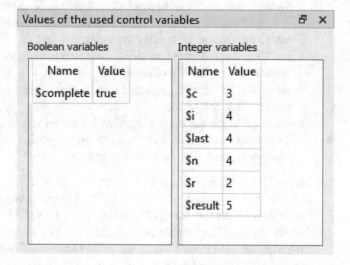

2.7   Supplementary Exercises

Exercise 2.8 (Change the Form and the Elements of TVLs)

Four similar functions are given:

$$f_1 = x_1 \vee x_2 \vee x_3 \,,$$

$$f_2 = x_1 \oplus x_2 \oplus x_3 \,,$$

$$f_3 = \overline{x}_1 \wedge \overline{x}_2 \wedge \overline{x}_3 \, ,$$

$$f_4 = \overline{x}_1 \odot \overline{x}_2 \odot \overline{x}_3 \, .$$

Prepare a PRP that:

(a) creates a Boolean space;
(b) defines TVL 1 of f_1 using the command tin;
(c) computes TVL 2 of f_2 based on TVL 1 using the command sform;
(d) computes TVL 3 of f_3 based on TVL 1 using the commands cel and sform;
(e) computes TVL 4 of f_4 based on TVL 3 using the command sform; and
(f) shows all four TVLs in the first column of the m-fold view using the command assign within the body of a command for.

Use the help system of XBM 2 to learn the details about the commands sform and cel. Execute the PRP and verify the displayed results.

Exercise 2.9 (Orthogonalization of TVLs)
    Both the form and the content of an orthogonalized TVL depend on both the form and the content of the given TVL. Extend the PRP of Exercise 2.8 such that all four TVLs 1 to 4 are orthogonalized using the command orth and stored as TVLs 5 to 8. The command orth utilizes both the form and the content of the given TVL; hence, all four tasks can be solved using the same command orth, and only different object numbers must be used. Show the computed TVLs in the second column of the m-fold view and verify the results.

Exercise 2.10 (Minimization of Orthogonal TVLs)
    Figure 2.45 shows a TVL of nine rows depending on six variables.
    Prepare a PRP that:

(a) creates a Boolean space;
(b) defines the orthogonal TVL 1 of f_1 shown in Fig. 2.45 using the command tin;
(c) computes the orthogonal TVL 2 that expresses the same function f_1 with less rows using TVL 1 and the command obb;
(d) tries to reduce the number of rows of the TVL 2 even more using the command obbc and store the result as TVL 3; and
(e) shows all three TVLs in the four-fold view using commands assign with applicable parameters.

Use the help system of XBM 2 to learn the details about the commands obb and obbc. Execute the PRP and verify the displayed results. The display of the TVLs reveals the minimization and the display of the associated Karnaugh-maps confirms that the represented sets remain unchanged.

Fig. 2.45 TVL 1 used in Exercise 2.10 to explore the commands to minimize the number of rows of an orthogonal TVL

$$\mathrm{ODA}(f_1) = $$

x_1	x_2	x_3	x_4	x_5	x_6
1	1	–	0	1	0
0	–	1	0	1	1
1	1	1	1	1	0
0	0	1	0	0	1
0	1	1	–	1	0
1	1	0	1	1	0
1	0	–	0	1	0
0	–	1	1	1	1
0	0	1	1	0	1

Fig. 2.46 Given TVLs used in Exercise 2.11

$$
\begin{array}{ccc}
x_1 & x_2 & x_3 \\
\hline
1 & 1 & 0 \\
\mathrm{ODA}(f_1) = \quad 0 & - & 1 \\
0 & 0 & 0
\end{array}
\qquad
\begin{array}{ccc}
x_3 & x_4 & x_5 \\
\hline
0 & 0 & 1 \\
\mathrm{ODA}(f_2) = \quad - & 1 & 0 \\
1 & - & 1
\end{array}
$$

Fig. 2.47 TVL 1 used in Exercise 2.12 to explore the direct specification of VTs in commands

$$
\mathrm{ODA}(f_1) =
\begin{array}{ccccc}
x_1 & x_2 & x_3 & x_4 & x_5 \\
\hline
1 & 1 & 0 & - & 0 \\
1 & - & 1 & 1 & 0 \\
1 & 0 & 0 & 1 & 0 \\
1 & 0 & - & 0 & 0
\end{array}
$$

Exercise 2.11 (Concatenation of TMs)

The command con can be used to concatenate either two TVLs or two VTs. Figure 2.46 shows two TVLs used to demonstrate their concatenation. Furthermore, the VTs VT 3 $= \langle x_3, x_1, x_2 \rangle$ and VT 4 $= \langle x_3, x_4, x_5, x_6 \rangle$ are given.

Prepare a PRP that:

(a) creates a Boolean space of 32 variables;
(b) attaches six variables x1 to x6 in lexicographic order to the defined space;
(c) defines the orthogonal TVL 1 of f_1 shown in Fig. 2.46 using the command tin;
(d) defines the orthogonal TVL 2 of f_2 shown in Fig. 2.46 using the command tin;
(e) defines VT 3 as specified in Exercise 2.11 using the command vtin;
(f) defines VT 4 as specified in Exercise 2.11 using the command vtin;
(g) concatenates TVLs 1 and 2 and stores the result as TVL 5 using the command con;
(h) concatenates VTs 3 and 4 and stores the result as VT 6 using the command con;
(i) shows the three TVLs in the first row and the three VTs in the second row of the m-fold view using commands assign.

Use the help system of XBM 2 to learn the details about the command con. Prepare the m-fold view with two rows and three columns and execute the PRP. Verify the displayed results.

Exercise 2.12 (Direct Specifications of Tuples of Variables in a Command) There are several commands that need VTs to determine the columns that must be manipulated. The explicit definition of such VTs can be avoided when alternative commands are used, which directly determine the needed VTs within the command. The first letter of all these commands is an underline (_).

The aim of this exercise is that based on TVL 1 shown in Fig. 2.47 two new TVLs 2 and 3 must be computed, where in TVL 2 the columns of TVL 1 are rotated by one column to the left and in TVL 3 the columns of TVL 1 are rotated by one column to the right.

The rotation of all columns to the left means that the columns of x2 to x5 are moved one column to the left (to x1 to x4), and the left column x1 fills thereafter the column x5. This rotation can be realized using a command cco that swaps neighboring columns. Similarly, the rotation of all columns to the right can be executed.

Prepare a PRP that:

(a) creates a Boolean space of 32 variables;
(b) defines the orthogonal TVL 1 of f_1 shown in Fig. 2.47 using the command tin;
(c) realizes the rotation of all columns of TVL 1 by one column to the left using a command _cco and stores the result as TVL 2;
(d) realizes the rotation of all columns of TVL 1 by one column to the right using a command _cco and stores the result as TVL 3; and

Fig. 2.48 Structure of a chained system

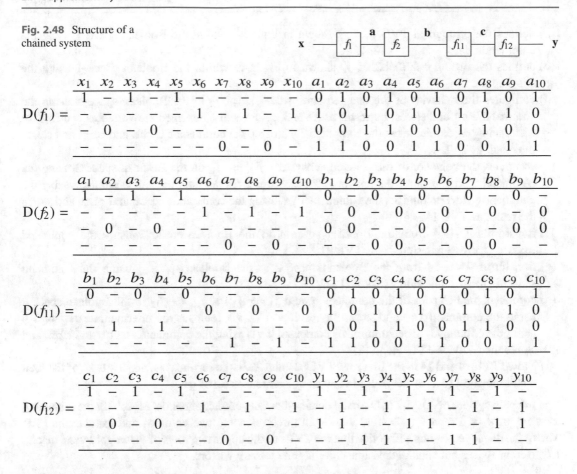

Fig. 2.49 Given TVLs used in Exercise 2.13 to specify four chained subsystems

$D(f_1) =$

x_1	x_2	x_3	x_4	x_5	x_6	x_7	x_8	x_9	x_{10}	a_1	a_2	a_3	a_4	a_5	a_6	a_7	a_8	a_9	a_{10}
1	–	1	–	1	–	–	–	–	–	0	1	0	1	0	1	0	1	0	1
–	–	–	–	–	1	–	1	–	1	1	0	1	0	1	0	1	0	1	0
–	0	–	0	–	–	–	–	–	–	0	0	1	1	0	0	1	1	0	0
–	–	–	–	–	–	0	–	0	–	1	1	0	0	1	1	0	0	1	1

$D(f_2) =$

a_1	a_2	a_3	a_4	a_5	a_6	a_7	a_8	a_9	a_{10}	b_1	b_2	b_3	b_4	b_5	b_6	b_7	b_8	b_9	b_{10}
1	–	1	–	1	–	–	–	–	–	0	–	0	–	0	–	0	–	0	–
–	–	–	–	–	1	–	1	–	1	–	0	–	0	–	0	–	0	–	0
–	0	–	0	–	–	–	–	–	–	0	0	–	–	0	0	–	–	0	0
–	–	–	–	–	–	0	–	0	–	–	–	0	0	–	–	0	0	–	–

$D(f_{11}) =$

b_1	b_2	b_3	b_4	b_5	b_6	b_7	b_8	b_9	b_{10}	c_1	c_2	c_3	c_4	c_5	c_6	c_7	c_8	c_9	c_{10}
0	–	0	–	0	–	–	–	–	–	0	1	0	1	0	1	0	1	0	1
–	–	–	–	–	0	–	0	–	0	1	0	1	0	1	0	1	0	1	0
–	1	–	1	–	–	–	–	–	–	0	0	1	1	0	0	1	1	0	0
–	–	–	–	–	–	1	–	1	–	1	1	0	0	1	1	0	0	1	1

$D(f_{12}) =$

c_1	c_2	c_3	c_4	c_5	c_6	c_7	c_8	c_9	c_{10}	y_1	y_2	y_3	y_4	y_5	y_6	y_7	y_8	y_9	y_{10}
1	–	1	–	1	–	–	–	–	–	1	–	1	–	1	–	1	–	1	–
–	–	–	–	–	1	–	1	–	1	–	1	–	1	–	1	–	1	–	1
–	0	–	0	–	–	–	–	–	–	1	1	–	–	1	1	–	–	1	1
–	–	–	–	–	–	0	–	0	–	–	–	1	1	–	–	1	1	–	–

(e) shows the three TVLs in the first row of the m-fold view using commands `assign`.

Use the help system of XBM 2 to learn the details about the command `_cco`. Prepare the m-fold view with one row and three columns and execute the PRP. Verify the displayed results.

Exercise 2.13 (Transformation of TVLs Between Boolean Spaces) The aim of this exercise is the utilization of the command `space_trans` in order to use several small spaces instead of one large space.

Figure 2.48 shows the structure of a chained system that consists of four subsystems. Each subsystem has ten inputs and ten outputs. The four TVLs in D-form shown in Fig. 2.49 describe the behavior of these subsystems.

The global behavior can be computed as intersection of these TVLs (after their orthogonalization). This computation needs a Boolean space of 50 variables where each ternary vector needs four machine words. One ternary vector of 32 variables can be stored in two machine words; hence, the input–output-behavior of the chain of four subsystems $f_1 \Rightarrow f_2 \Rightarrow f_{11} \Rightarrow f_{12}$ can be computed using three Boolean spaces of 32 variables.

Prepare a PRP that:

(a) creates three Boolean spaces of 32 variables;

(b) defines the orthogonal TVL 1 of f_1 shown in Fig. 2.49 within the Boolean space 1 using the command `tin`;

(c) defines the orthogonal TVL2 of f_2 shown in Fig. 2.49 within the Boolean space 1 using the command `tin`;

(d) computes the behavior of the chained subsystem $f_1 \Rightarrow f_2$ in the Boolean space 1 using the command `isc` and stores the result as TVL 3, removes the intermediate variables **a** using the command `_dco`, minimizes the computed TVL using the commands `orth` and `obb`, and stores the result as TVL 4;

(e) computes the behavior of the chained subsystem $f_{11} \Rightarrow f_{12}$ in the Boolean space 2 using the command `isc` and stores the results as TVL 13, removes the intermediate variables **c** using the command `_dco`, minimizes the computed TVL using the commands `orth` and `obb`, and stores the result as TVL 14;

(f) transforms TVL 4 from the Boolean space 1 to the Boolean space 3 using the command `space_trans` and stores it as TVL 21;

(g) transforms TVL 14 from the Boolean space 2 to the Boolean space 3 using the command `space_trans` and stores it as TVL 22;

(h) computes the behavior of the complete system $f_1 \Rightarrow f_2 \Rightarrow f_{11} \Rightarrow f_{12}$ in the Boolean space 3 using the command `isc` and stores the result as TVL 23, removes the intermediate variables **b** using the command `_dco`, minimizes computed TVL using the commands `orth` and `obb`, and stores the result as TVL 24; and

(i) shows TVLs 4 and 14 in the first row, TVLs 21 and 22 in the second row, and TVL 24 of the final result viewport (3,1) of the m-fold view using commands `assign`.

Use the help system of XBM 2 to learn the details about the used commands, especially the command `space_trans`. Prepare the m-fold view with three rows and two columns and execute the PRP. Verify the displayed results and notice that the TVLs of the upper two rows of the m-fold view belong to different spaces but contain the same columns and ternary vectors.

Exercise 2.14 (Save and Load a XBOOLE-System)

An `sdt`-file can be used to save the data of a complete XBOOLE-system at the end of one PRP and to load these data at the beginning of another PRP. In this way, the execution of an algorithm implemented in one PRP can be reused for different data prepared by several other PRPs. This exercise demonstrates this approach based on the chained Exercises 2.8 and 2.9:

(a) Extend the PRP of Exercises 2.8 such that all four created non-orthogonal TVLs are stored as file `four_no_tvls.sdt` and execute this new PRP.

(b) Implement a PRP the loads first the stored file `four_no_tvls.sdt` and realizes thereafter the orthogonalization and displays the results as requested in Exercises 2.9.

Use the help system of XBM 2 to learn the details about the commands `sts` and `lds` that either save or load a complete XBOOLE-system.

References

1. C. Posthoff, B. Steinbach, *Logic Functions and Equations – Binary Models for Computer Science*, 2nd edn. (Springer, Cham, 2019). ISBN: 978-3-030-02419-2. https://doi.org/10.1007/978-3-030-02420-8
2. B. Steinbach, C. Posthoff, Boolean Differential Calculus – Theory and Applications. J. Comput. Theor. Nanosci. **7**(6), 933–981 (2010). ISSN: 1546-1955. https://doi.org/10.1166/jctn.2010.1441
3. B. Steinbach, C. Posthoff, *Boolean Differential Calculus* (Morgan & Claypool Publishers, San Rafael, 2017). ISBN: 978-1-6270-5922-0. https://doi.org/10.2200/S00766ED1V01Y201704DCS052

Part II
Fundamentals

Basic Algebraic Structures

3

Abstract

We start our considerations with the basic concepts of *binary algebraic structures*. These concepts go back to the outstanding scientists Gottfried Wilhelm Leibniz who invented the Binary Mathematics, George Boole who created the foundations of propositional logic, and Claude E. Shannon who used this binary calculus for the description of the behavior of relay circuits. The set $\mathbb{B} = \{0, 1\}$ contains exactly distinct elements and is the root of all theories and applications in the Boolean domain. Relations between elements of this set are subsets of the cross-product $\mathbb{B} \times \mathbb{B}$ and will be explored with regard to their properties. Changing the relational point of view into the operational one allows us to introduce the basic logic operations conjunction, disjunction, antivalence, equivalence, and negation. The combination of the set $\mathbb{B}$, selected logic operations, and neutral elements determine two dual Boolean Algebras. Similarly, two dual Boolean Rings are introduced. The knowledge about these algebraic structures and their properties supports the application of the huge amount of transformation laws existing in the Boolean domain. The set $\mathbb{B}^n$ consists of all 2^n binary vectors. The algebraicvadjust structures, introduced for Boolean space

Supplementary Information The online version of this chapter (https://doi.org/10.1007/978-3-030-88945-6_3) contains supplementary material which is available for authorized users. Please, follow the link belonging to the version of the XBOOLE-monitor XBM 2 that fits best for your operating system. This XBOOLE-monitor is needed to solve all tasks of this chapter. Instructions for starting the downloaded XBOOLE-monitor XBM 2 are given at the beginning of Section 'Exercises' in this chapter.

XBOOLE-monitor XBM 2 for Windows 10
32 bits
https://doi.org/10.1007/978-3-030-88945-6_3_MOESM1_ESM.zip (15,091 KB)

64 bits
https://doi.org/10.1007/978-3-030-88945-6_3_MOESM2_ESM.zip (14,973 KB)

XBOOLE-monitor XBM 2 for Linux Ubuntu
32 bits
https://doi.org/10.1007/978-3-030-88945-6_3_MOESM3_ESM.zip (29,522 KB)

64 bits
https://doi.org/10.1007/978-3-030-88945-6_3_MOESM4_ESM.zip (28,422 KB)

$\mathbb{B}$, will be generalized for the set $\mathbb{B}^n$ and due to the concept of isomorphism to power sets and set operations. Further concepts are introduced, and all explained concepts support the understanding in the further chapters.

3.1   The Roots of Logic Concepts

Logic is nowadays a fundamental field that can be traced back to the ancient Greek mathematicians and philosophers. The modern development relates strongly to the name of *George Boole* (1815–1864) who created the foundations of *propositional logic* in the nineteenth century.

Very surprisingly, this extremely specialized field became one of the most important parts of Engineering and Computing Sciences. This development started in the thirties of the last century; however, the main breakthrough resulted from publications of *Claude E. Shannon* (1916–2001) who used the binary calculus of propositional logic for the description of the behavior of relay circuits. And this is at present a certain dilemma for the education in different disciplines.

Logic itself and the application of logic as a fundamental scientific discipline for all the sciences with even an important meaning in philosophy are mostly taught in higher courses of the advanced level at universities, not even at too many places. The application of the same binary concepts and theories in all fields of Computer Science and Computer Science applications, however, are to be understood and taught "nearly everywhere"; these concepts are used in all the programming languages, and *logic programming* is a whole area of programming built upon logic. The design of hardware, combinational and sequential circuits, coding theory, artificial neural networks, machine learning, image processing, artificial intelligence, relational databases, and many more have a direct relation to these theories.

Very often, a student meets these concepts in different courses, mostly under different names using different symbols and concepts, and it is not easy to build an ordered understanding of this area, often due to the existing time pressure or other reasons.

The explanations and denominations in various publications are also not unique, different names are used for the same structures (like truth function, switching function, logic function, Boolean function, binary function, etc.), but the opposite situation can also be met, and the same name can have quite different meanings in different contexts.

Hence, one main goal of this book is the presentation of the fundamental concepts such that a student has to read (to understand, to learn) these concepts only once and will then be able to apply all of them in the different areas of applications.

Consequently, we use the binary approach exclusively, and $\mathbb{B} = \{0, 1\}$ is the basic set to be considered; all the different concepts will be built upon this set. We start in all areas with the basic and fundamental concepts in order to ensure that the understanding is fundamental. We want to transfer an exact and precise presentation of these foundations. We restrict our explanations to the basic concepts because further steps might leave the introductory level, or other very good and comprehensive books or articles are available that can be used quite well after the basic understanding has been achieved or when more specialized courses have to be taken, for theoretical foundations of logic functions and mathematical theories we mention, for instance [3, 9], and for decision trees or decision diagrams [2, 6, 18].

The presentation of numerical methods based on the parallel processing of ternary vectors is an essential point. Many *classical* methods are outdated and can be replaced.

Fig. 3.1 Gottfried Wilhelm Leibniz (https://www.mathematik.ch/mathematiker/leibniz.php)

We emphasize the numerical approach, based on a good understanding of the fundamentals, because it can be seen that for many problems in this area, good algorithms and efficient program implementations are available that allow the solution of many binary problems, even for large numbers of variables. Hence, some of the "awfully looking" formulas can be applied quite easily and extend quite considerably the range of problems that can be solved. And this is another "hot topic"—to teach mathematical methods with the background that the solution-mechanisms will be built into software packages and visible to the outside only in principle. The user must be able to transform his problems into this environment, he must be able to use the theoretical formalism for the modeling of his special applications or problems, the solution-process itself does not need special attention, but thereafter, the correct interpretation and application of the results is again the users' responsibility. And this is the only possibility to solve real-world problems.

XBOOLE is such a software package and publicly available at

https://tu-freiberg.de/en/fakult1/inf/xboole,

and anybody can download and use the XBOOLE-monitor XBM 2 for her/his purposes, together with many more examples, prototype solutions, tutorials, etc. [12, 14]. In this way, the problem-solving process in a professional environment reaches a much higher level, as can be seen by the exercises and solutions provided in this book.

This book can be used as a textbook for a course of two or even three semesters, starting preferably in the first semester of all areas of Computer Science or Computer Engineering programs. Some basic knowledge in Discrete Mathematics (sets, relations, functions) is recommendable. We did not try to give a full bibliography for the areas of this book (which might be quite impossible), and we simply included and stated those books and papers that we held in our hands.

We find three starting points for our considerations.

Gottfried Wilhelm Leibniz (Fig. 3.1) "invented" the *Binary Mathematics*, i.e., the way of building the Arithmetic on two digits and using the number 2 as the foundation of this system, in contrast to the *decimal* or other number systems.

Based on large tables of dual representations of natural numbers (see Fig. 3.2), he discovered, for instance, periodicities and symmetries in the columns of these representations and introduced algebraic operations that used these periodicities as objects. Other intended applications in geometry

TABLE
DES
NOMBRES.

Pour *l'Addition*
par exemple.

Pour la *Sou-
ſtraction.*

Pour la *Mul-
tiplication.*

Pour la *Diviſion.*

Fig. 3.2 Using binary numbers by G. W. Leibniz (G. W. Leibniz: "Explication de l'Arithmétique Binaire", 1703)

Fig. 3.3 George Boole
(1815–1864) (January 1,
1860)

and number theory and even intentions to relate the dual numbers to topics of Chinese philosophy
and theology are more of philosophical and historical interest.

George Boole (Fig. 3.3) built the binary structures to be used as a foundation of Logic and
Mathematics. He had the goal to find a calculus that decides whether mathematical theorems or
combined statements given in natural language are true or false. His basic objects were classes
of elements (nowadays sets) defined by predicates. Then operations with these classes had to be
considered. He introduced the famous laws:

- $x(1-x) = 0$ *Law of Contradiction*;
- $x + (1-x) = 1$ *Law of Excluded Middle*.

He stated that a statement that has been used twice is as true as the statement itself. Hence, the
equation $x \times x = x$ must hold. The resulting equation $x^2 = x$ (he used multiplication for the
simultaneous consideration of two statements) or $x^2 - x = 0$ has the two solutions $x = 1$ and $x = 0$,

Fig. 3.4 The propositional calculus by George Boole

THE MATHEMATICAL ANALYSIS

OF LOGIC,

BEING AN ESSAY TOWARDS A CALCULUS
OF DEDUCTIVE REASONING.

BY GEORGE BOOLE.

Ἐπικοινωνοῦσι δὲ πᾶσαι αἱ ἐπιστῆμαι ἀλλήλαις κατὰ τὰ κοινά. Κοινὰ δὲ
λέγω, οἷς χρῶνται ὡς ἐκ τούτων ἀποδεικνύντες· ἀλλ᾽ οὐ περὶ ὧν δεικνύουσιν
οὐδὲ ὃ δεικνύουσι.

ARISTOTLE, *Anal. Post.*, lib. I. cap. XI

CAMBRIDGE:
MACMILLAN, BARCLAY, & MACMILLAN;
LONDON: GEORGE BELL.

———

1847

and he identified 1 with *true* and 0 with *false* and concluded that any calculus that deals with the logic values of statements will allow only two values. For these two values, $1 - x$ converts $x = 0$ into $x = 1$ and vice versa, i.e., this defines the *complement* (*negation*).

The Law of Contradiction expresses the fact that a statement and its complement cannot be true at the same time. The multiplication will later be seen as the *conjunction* since $0 \times 0 = 0 \times 1 = 1 \times 0 = 0$ and $1 \times 1 = 1$. $x + y$ existed only for an empty intersection; hence, he defined $0 + 0 = 0$, $1 + 0 = 0 + 1 = 1$ and used $1 + 1 = 0$, which means that he used the *antivalence* (the *symmetric difference* in set theory) as the basic additive operation. This fact is not very well-known, and the *disjunction* in the sense of $\vee$ has been introduced later.

He took the stated arithmetic-based equations (laws) or properties as correct (since arithmetics had a long and well established history that he considered as valid and applicable without any doubt) and developed the foundations of the *Propositional Logic*, see Fig. 3.4, recently reprinted [1].

It is not well-known that he always used the understanding of expressions built on these operations and variables in the sense of *equations*. He wrote, for instance, an equation such as $ax \oplus b\overline{x} = 1$ and tried to find a solution $x(a, b)$ that transforms the original equation into an identity $0 = 0$ or $1 = 1$.

A completely other field is the use of logic functions in the analysis and synthesis of digital systems. This goes back to the famous paper "Symbolic Analysis of Relay and Switching Circuits" of *Claude Shannon* (Fig. 3.5) who used these *logic functions* to describe the behavior of relay circuits.

As shown in Fig. 3.6: if a circuit contains two relays sequentially in one line, then both of them must be closed in order to get a current that flows on this line. This requirement can be expressed by *and*: the first contact must be closed, *and* the second contact must be closed. If the two relays are

Fig. 3.5 Claude E.
Shannon (1916–2001) (ca.
1963, Oberwolfach Photo
Collection)

arranged in parallel, then one can be closed *or* the other *or* even both of them, and in all three cases
the current is flowing. Only if both relay contacts are open, then there will be no current on the line.
And from there logic functions are now the main tool for the synthesis of digital systems with an
overwhelming importance for many problems. In this regard very often the term "switching function"
can be found.

Based on the findings of Leibniz, Boole, and Shannon, and especially the applications in analysis
and synthesis of digital circuits, a very intensive research started. A huge number of publications have
been published about the minimization or decomposition of logic functions for use in circuits, for test
of such circuits, or for logic functions with special properties, like a high degree of nonlinearity for,
e.g., applications in cryptography.

We do not emphasize one of these approaches or intentions—we build the calculus just from the
beginning and apply it to various fields and problems, without any special preferences. Additionally,
many new application areas appeared, based on origins in Computer Science, such as complexity
of logic functions (SAT-problems, Artificial Intelligence), binary coding, binary neural networks,
machine learning, and many more. The reader will be able to understand all these problems after
studying this book carefully and to use his knowledge for further studies in many different areas.

The application of the Boolean Differential Calculus is now a huge area and can be studied later
based on the introductory knowledge that will be presented here (see, for instance, [13, 15]).

3.2   The Set $\mathbb{B}$

Let $\mathbb{B}$ be the set $\{0, 1\}$ with two distinct elements 0 and 1; this set is also denoted by *binary space*
or *Boolean space*. In technical applications, 0 and 1 will be used all the time, and in logic and in
programming languages, we also find **true** or **t** being used instead of 1, and **false** or **f** instead of 0.

As a first step, we introduce a *reflexive* and an *irreflexive order relation* (designated by $\leq$ and $<$,
resp.):

$$\leq \quad \equiv \quad \{(0, 0), (0, 1), (1, 1)\}\,, \tag{3.1}$$

$$< \quad \equiv \quad \{(0, 1)\}\,. \tag{3.2}$$

II. Series-Parallel Two-Terminal Circuits

Fundamental Definitions and Postulates

We shall limit our treatment of circuits containing only relay contacts and switches, and therefore at any given time the circuit between any two terminals must be either open (infinite impedance) or closed (zero impedance). Let us associate a symbol X_{ab} or more simply X, with the terminals a and b. This variable, a function of time, will be called the hindrance of the two-terminal circuit $a - b$. The symbol 0 (zero) will be used to represent the hindrance of a closed circuit, and the symbol 1 (unity) to represent the hindrance of an open circuit. Thus when the circuit $a - b$ is open $X_{ab} = 1$ and when closed $X_{ab} = 0$. Two hindrances X_{ab} and X_{cd} will be said to be equal if whenever the circuit $a - b$ is open, the circuit $c - d$ is open, and whenever $a - b$ is closed, $c - d$ is closed. Now let the symbol + (plus) be defined to mean the series connection of the two-terminal circuits whose hindrances are added together. Thus $X_{ab} + X_{cd}$ is the hindrance of the circuit $a - d$ when b and c are connected together. Similarly the product of two hindrances $X_{ab} \cdot X_{cd}$ or more briefly $X_{ab} X_{cd}$ will be defined to mean the hindrance of the circuit formed by connecting the circuits $a - b$ and $c - d$ in parallel. A relay contact or switch will be represented in a circuit by the symbol in Figure 1, the letter being the corresponding hindrance function. Figure 2 shows the interpretation of the plus sign and Figure 3 the multiplication sign. This choice of symbols makes the manipulation of hindrances very similar to ordinary numerical algebra.

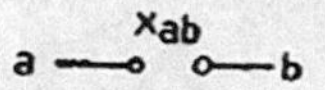

Figure 1 (left). Symbol for hindrance function

Figure 2 (right). Interpretation of addition

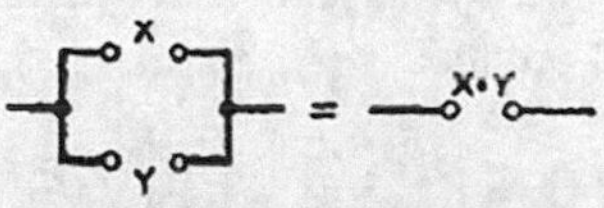

Figure 3 (middle). Interpretation of multiplication

It is evident that with the above definitions the following postulates will hold:

Postulates

1. a. $0 \cdot 0 = 0$ — A closed circuit in parallel with a closed circuit is a closed circuit.

   b. $1 + 1 = 1$ — An open circuit in series with an open circuit is an open circuit.

2. a. $1 + 0 = 0 + 1 = 1$ — An open circuit in series with a closed circuit in either order (i.e., whether the open circuit is to the right or left of the closed circuit) is an open circuit.

   b. $0 \cdot 1 = 1 \cdot 0 = 0$ — A closed circuit in parallel with an open circuit in either order is a closed circuit.

3. a. $0 + 0 = 0$ — A closed circuit in series with a closed circuit is a closed circuit.

   b. $1 \cdot 1 = 1$ — An open circuit in parallel with an open circuit is an open circuit.

4. At any given time either $X = 0$ or $X = 1$.

Fig. 3.6 Page of the pioneering paper C. E. Shannon [10]

Table 3.1 Transitivity of
the $\leq$ - relation

x	y	z	$x \leq y$	$y \leq z$	$x \leq y$ and $y \leq z$	$x \leq z$	Theorem
0	0	0	**True**	**True**	**True**	**True**	True
0	0	1	**True**	**True**	**True**	**True**	True
0	1	0	True	False	False	True	True
0	1	1	**True**	**True**	**True**	**True**	True
1	0	0	False	True	False	False	True
1	0	1	False	True	False	True	True
1	1	0	True	False	False	False	True
1	1	1	**True**	**True**	**True**	**True**	True

The pairs are to be understood in such a way that the relation sign ($\leq$ or $<$) connects the first and the second element:

$$0 \leq 0 \,, \; 0 \leq 1 \,, \; 1 \leq 1 \,, \quad \text{or} \quad 0 < 1 \,. \tag{3.3}$$

In order to show that the relation $\leq$ is a reflexive order, the following properties must be verified:

$\forall x \in \mathbb{B}:$	$x \leq x$	*Reflexivity* ;
$\forall x \, \forall y \in \mathbb{B}:$	$x \leq y$ or $y \leq x$	*Linearity* ;
$\forall x \, \forall y \, \forall z \in \mathbb{B}:$	if $x \leq y$ and $y \leq z$, then $x \leq z$	*Transitivity* ;
$\forall x \, \forall y \in \mathbb{B}:$	if $x \leq y$ and $y \leq x$, then $x = y$	*Antisymmetry* .

Remark 3.1 The symbols $\forall x$ and $\exists x$ will be used as abbreviations: *for all x ...* and *there is an x ...*, respectively.

Note 3.1 A relation R (like $\leq$, $<$, or others) has pairs as its elements. Hence, a relation R can always be considered as a subset of the cross-product

$$\mathbb{B} \times \mathbb{B} = \{(x, y) \mid x \in \mathbb{B}, y \in \mathbb{B}\} \,.$$

Since the set $\mathbb{B}$ has only two elements, it is not difficult to verify these four properties. We check all possibilities—a method that will be used very often.

The *reflexivity* holds because of $0 \leq 0$ and $1 \leq 1$.

To check the *linearity*, the four cases $\{x = 0, y = 0\}$, $\{x = 0, y = 1\}$, $\{x = 1, y = 0\}$, and $\{x = 1, y = 1\}$ have to be explored. By using $0 \leq 0$ for the first case, $0 \leq 1$ for the second and third case, and $1 \leq 1$ for the last, we can confirm this property. It should be noted that in the second case we use $x \leq y$, whereas in the third case we use $y \leq x$.

Transitivity requires the inclusion of three variables x, y, z. All possible combinations of values are represented in Table 3.1.

Remark 3.2 In Table 3.1 we have already used concepts of logic that will be explained later. The combination of two statements by *and* will be true if and only if both parts are true; otherwise, it will be false. A theorem of the format *if ... then ...* is true by definition of the implication when the assumption (the expression behind *if*) is false; furthermore, it is also true when the assumption as well as the conclusion (the expression behind *then*) is true. This means that only those lines have to be checked where the two initial conditions are true (marked in bold). In all four lines the conclusion is also true; hence, the whole theorem is true.

Fig. 3.7 The (**a**) reflexive
and (**b**) irreflexive order
relations of $\mathbb{B}$

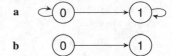

The *antisymmetry* requires the consideration of two cases $0 \leq 0$ and $1 \leq 1$, resp., which leads to $0 = 0$ and $1 = 1$. The relation

$$id = \{(0, 0), (1, 1)\} \tag{3.4}$$

is denoted by *identity relation* and describes the identity of each element with itself. It can be used for an easy transition between reflexive and irreflexive orders. If we eliminate, for instance, all pairs belonging to the identity from $\leq$ (consisting of $\{(0, 0), (0, 1), (1, 1)\}$), then we obtain the irreflexive order $\{(0, 1)\}$. Hence, given one of these order relations, the transition to the other one is possible conveniently. In general, the irreflexive order $<$ in $\mathbb{B}$ is characterized by the following conditions:

$\neg \exists x \in \mathbb{B}:$	$x < x$	*Irreflexivity* ;
$\forall x \forall y \forall z \in \mathbb{B}:$	if $x < y$ and $y < z$, then $x < z$	*Transitivity* ;
$\forall x \forall y \in \mathbb{B}:$	if $x < y$, then $y \not< x$	*Asymmetry* .

Remark 3.3 $\neg \exists x \dots$ has to be read as *there is no* $x \dots$

The irreflexive order relation $<$ does not satisfy the linearity because, e.g., for $x = 0$ and $y = 0$ neither $(x = 0) < (y = 0)$ nor $(y = 0) < (x = 0)$ holds. The verification of the three conditions of the order relation $<$ is recommended as an exercise. The evaluation of all combinations as shown in Table 3.1 can be used to solve this task.

It is very often useful to represent a relation by the *graph of the relation*. We use the elements of $\mathbb{B}$ as *nodes*, and there is a directed edge from node x to node y if and only if (x, y) is an element of the relation. The reflexive and irreflexive order relations in $\mathbb{B}$ are represented in Fig. 3.7.

As we have seen, an order relation gives the possibility of comparing any two elements. This makes the following definition meaningful.

Definition 3.1 For any two elements $x, y \in \mathbb{B}$,

$$min\{x, y\} = \begin{cases} x & \text{if } x \leq y \\ y & \text{otherwise} \end{cases} \tag{3.5}$$

is the *minimum* of x and y, and

$$max\{x, y\} = \begin{cases} y & \text{if } x \leq y \\ x & \text{otherwise} \end{cases} \tag{3.6}$$

is the *maximum* of x and y.

When we consider the four possible combinations of x and y, we get Table 3.2.

The set $\mathbb{B}$ has a *smallest element* 0 and a *largest element* 1. For these elements, it holds that

$\forall x$	(if $x \in \mathbb{B}$, then $0 \leq x$)	*minimum* ,
$\forall x$	(if $x \in \mathbb{B}$, then $x \leq 1$)	*maximum* .

Table 3.2 Minimum and maximum in $\mathbb{B}$

x	y	$min\{x, y\}$	$max\{x, y\}$
0	0	0	0
0	1	0	1
1	0	0	1
1	1	1	1

Table 3.3 Conjunction and disjunction in $\mathbb{B}$

x	y	$x \wedge y$	$x \vee y$
0	0	0	0
0	1	0	1
1	0	0	1
1	1	1	1

Table 3.4 Antivalence (exclusive-or) and equivalence (exclusive-and) in $\mathbb{B}$

x	y	$x \oplus y$	$x \odot y$
0	0	0	1
0	1	1	0
1	0	1	0
1	1	0	1

Table 3.5 Negation in $\mathbb{B}$

x	$\overline{x}$
0	1
1	0

Instead of the relational point of view, it is also possible to use a *functional* or *operational* understanding. $min\{x, y\}$ and $max\{x, y\}$ assign the minimum or the maximum to each pair (x, y), respectively. The minimum will be denoted by *conjunction*, the maximum will be denoted by *disjunction*, and we will use the symbols $\wedge$ and $\vee$, respectively (Table 3.3).

Remark 3.4 The symbol $\wedge$ will be very often omitted (like the multiplication dot). Hence, we understand ab as $a \wedge b$, $(a \wedge b) \vee (c \wedge d)$ can be written as $ab \vee cd$, etc.

The application of these operations in logic gave $\wedge$ the name *and*: if one part of a proposition is true *and* a second part is true, then the whole proposition will be true. $\vee$ is denoted by *or*, and in order to get a true statement, the first proposition *or* the second proposition must be true. Since the overall proposition is also true when both parts are true, this function is the *inclusive-or*. Subsequently, the names of the functions have also applied in technical applications.

Two other functions are very instructive for indicating the *inequality* and *equality* of the arguments. The *antivalence function*, which is very often denoted by *exclusive-or*, shows the first property and is written as $\oplus$ (see Table 3.4). In order to be very precise, it can be replaced by *either ... or*. The second property is shown by the *equivalence function* (designated by $\odot$); the special case that the equivalence is equal to 1 (if and only if, iff) is expressed by the identity sign $\equiv$.

It is visible that the equivalence function is equal to 1 when x and y have the same value. The value of the antivalence is equal to 1 when the values of x and y are different.

It is remarkable that one column on the right-hand side of Table 3.4 can be transformed into the other by replacing 0 by 1 and vice versa. This operation has the name *negation*. It is most frequently expressed by an overline, $\overline{x}$, sometimes also by $\neg x$ (Table 3.5).

The close connection between conjunction and minimum as well as disjunction and maximum allows us to replace inequalities by equalities.

Table 3.6 The connection between $\leq$ and $=$ in $\mathbb{B}$

x	y	$\overline{x}$	$\overline{y}$	$x \wedge \overline{y}$	$\overline{x} \vee y$	$x \leq y$	$x \wedge \overline{y} = 0$	$\overline{x} \vee y = 1$
0	0	1	1	0	1	True	True	True
0	1	1	0	0	1	True	True	True
1	0	0	1	1	0	False	False	False
1	1	0	0	0	1	True	True	True

Table 3.7 The connection between $<$ and $=$ in $\mathbb{B}$

x	y	$\overline{x}$	$\overline{y}$	$\overline{x} \wedge y$	$x \vee \overline{y}$	$x < y$	$\overline{x} \wedge y = 1$	$x \vee \overline{y} = 0$
0	0	1	1	0	1	False	False	False
0	1	1	0	1	0	True	True	True
1	0	0	1	0	1	False	False	False
1	1	0	0	0	1	False	False	False

Theorem 3.1

$$\forall x \, \forall y \in \mathbb{B}: \quad x \leq y \quad \equiv \quad x \wedge \overline{y} = 0 \quad \equiv \quad \overline{x} \vee y = 1 \,, \tag{3.7}$$

$$\forall x \, \forall y \in \mathbb{B}: \quad x < y \quad \equiv \quad \overline{x} \wedge y = 1 \quad \equiv \quad x \vee \overline{y} = 0 \,. \tag{3.8}$$

Proof Tables 3.6 and 3.7 show the validity of this theorem.

The symbol $\equiv$ is already an application of the identity in a logic context. The expression on the left-hand side of $=$ must be identical to the expression on the right-hand side: both expressions result for all combinations of x and y in the same value; either both of them are true or both of them are false (see Table 3.6). The proof of (3.8) follows the same idea (see Table 3.7). □

Hence, any pairs of elements that satisfy the relations $\leq$ or $<$ in $\mathbb{B}$ also satisfy the respective equations and vice versa.

3.3 Boolean Algebras

Now we will put the set $\mathbb{B}$ and the functions that have been defined so far into a broader context. In order to do this, we need the algebraic structure of a *Boolean Algebra*.

Definition 3.2 Let be given a set B, the two functions $+$ and $\cdot$ from $B \times B$ into B, one function $\overline{x}$ from B into B, and two special elements 0 and 1 with $0 \neq 1$. Then the sixtuple $(B, +, \cdot, \overline{}, 0, 1)$ is a *Boolean Algebra* if the following axioms are satisfied:

Commutative Laws

$$\forall x \, \forall y \in B: \qquad\qquad\qquad\qquad x + y = y + x \,, \tag{3.9}$$

$$\forall x \, \forall y \in B: \qquad\qquad\qquad\qquad x \cdot y = y \cdot x \,, \tag{3.10}$$

Distributive Laws

$$\forall x \, \forall y \, \forall z \in B: \qquad\qquad x + (y \cdot z) = (x + y) \cdot (x + z) \,, \tag{3.11}$$

$$\forall x \, \forall y \, \forall z \in B: \qquad\qquad x \cdot (y + z) = (x \cdot y) + (x \cdot z) \,, \tag{3.12}$$

Neutral Elements

$$\forall x \in B : \qquad\qquad\qquad\qquad 0 + x = x \,, \qquad\qquad (3.13)$$

$$\forall x \in B : \qquad\qquad\qquad\qquad 1 \cdot x = x \,, \qquad\qquad (3.14)$$

Complement

$$\forall x \in B \; \exists \overline{x} \in B : \qquad\qquad\qquad x + \overline{x} = 1 \,, \qquad\qquad (3.15)$$

$$\forall x \in B \; \exists \overline{x} \in B : \qquad\qquad\qquad x \cdot \overline{x} = 0 \,. \qquad\qquad (3.16)$$

Note 3.2 B is the *carrier set*, a set of elements, $+$ and $\cdot$ are operations that take two elements of B and determine another element of B; they are denoted by *disjunction* and *conjunction*, respectively. The *complement* is defined for each element separately. It is always assumed that the result of an operation (like $x + y$, $x \cdot y$, $\overline{x}$) is again an element of B. The set B must be *closed* under these operations. The functions must be defined for all elements of $B \times B$ (for $+$ and $\cdot$) and for all elements of B (for $\overline{x}$).

Note 3.3   Very often the operations in a Boolean Algebra are indicated immediately by $\vee$ instead of $+$ and by $\wedge$ instead of $\cdot$. In the rest of the book, we will follow this line. Here $+$ and $\cdot$ have been used to emphasize the independent, very general algebraic point of view.

The relations $\leq, \geq, <,$ and $>$ can be introduced in each Boolean Algebra as follows.

Definition 3.3   It holds for two elements $x, y \in B$:

$$x \leq y \quad \equiv \quad \overline{x} + y = 1 \,,$$
$$x \leq y \quad \equiv \quad x \cdot \overline{y} = 0 \,,$$
$$y \geq x \quad \equiv \quad x \leq y \,,$$
$$x < y \quad \equiv \quad x \leq y \text{ and } x \neq y \,,$$
$$y > x \quad \equiv \quad x < y \,.$$

It can be shown (in the general theory of Boolean Algebras) that these relations $\leq$ and $<$ are a *reflexive* or an *irreflexive partial order*; hence, they show the following properties:

$\forall x \in B :$	$x \leq x$	*Reflexivity*;
$\forall x \forall y \forall z \in B :$	if $x \leq y$ and $y \leq z$, then $x \leq z$	*Transitivity*;
$\forall x \forall y \in B :$	if $x \leq y$ and $y \leq x$, then $x = y$	*Antisymmetry*;
$\neg \exists x \in B :$	$x < x$	*Irreflexivity*;
$\forall x \forall y \forall z \in B :$	if $x < y$ and $y < z$, then $x < z$	*Transitivity*;
$\forall x \forall y \in B :$	if $x < y$, then $y \not< x$	*Asymmetry*.

Such a system of axioms will be used in *Abstract Algebra* to build up the theory that follows from these axioms, i.e., all theorems, which can be proven, appropriate definitions and new concepts, etc. This theory is afterward available for all kinds of applications in the following sense: in order to apply

	x	y	z	$(y \wedge z)$	$x \vee (y \wedge z)$	$x \vee y$	$x \vee z$	$(x \vee y) \wedge (x \vee z)$
Table 3.8 The first distributive law	0	0	0	0	0	0	0	0
	0	0	1	0	0	0	1	0
	0	1	0	0	0	1	0	0
	0	1	1	1	1	1	1	1
	1	0	0	0	1	1	1	1
	1	0	1	0	1	1	1	1
	1	1	0	0	1	1	1	1
	1	1	1	1	1	1	1	1

this theory, we must define the *carrier set B*, two functions (operations) that can be used for $+$ and $\cdot$ as well as two special elements *0* and *1* that satisfy altogether the axioms. If this can be done, then all the theorems are valid automatically, and all of them are already proven. This is very efficient because often there are multiple *models* of a theory.

Hence, we can look, for instance, at the following:

Theorem 3.2 $(\mathbb{B}, \vee, \wedge, ^{-}, 0, 1)$ *is a Boolean Algebra.*

Proof We identify successively B and $\mathbb{B}$, $+$ and $\vee$, $\cdot$ and $\wedge$, $\bar{x}$ in B with $\bar{x}$ in $\mathbb{B}$, *0* and 0, *1* and 1 and show that the four groups of axioms are satisfied.

First, we consider the *complement*. There arc two possibilities $x = 0, \bar{x} = 1$ and $x = 1, \bar{x} = 0$. In these two cases the disjunction is equal to 1 and the conjunction is equal to 0.

The *neutral elements* can also be seen without problems: $0 \vee 0 = 0, 0 \vee 1 = 1, 1 \wedge 0 = 0, 1 \wedge 1 = 1$.

The *commutative laws* are just as easy as the others. We check the four possibilities and find always the correct equality.

For the *first distributive law*, a full consideration of all possibilities will show that this axiom is satisfied (see Table 3.8).

The two columns to be investigated have the same values in the corresponding cells. The proof of the sense, these two rings are even *fields*. To extend the second distributive law as well as the associative laws follows the same procedure. □

Another well-known example of a Boolean Algebra is the power set of a given finite set.

Theorem 3.3 *Let* $X = \{x_1, \ldots, x_n\}$ *be a finite set,* $P(X)$ *be the power set (the set of all subsets) of* X, $Y \in P(X)$. *Then* $(P(X), \cup, \cap, ^{-}, \emptyset, X)$ *is a Boolean Algebra.*

For a given set X, the *intersection* ($\cap$) and the *union* ($\cup$) of two subsets are again subsets of X. The same applies to the *complement* $\bar{Y}$ of a given subset Y. The *empty set* $\emptyset$ and the set X itself are the neutral elements. The subset relations $\subseteq$ or $\subset$ play the role of the partial orders. The validity of all the axioms is proven in set theory.

As a last example, it will be indicated that there are also very "unusual" Boolean Algebras. We use the set $N = \{1, 2, 3, 5, 6, 10, 15, 30\}$ as the carrier, 1 as the element *0*, 30 as *1*, the *Least Common Multiple*, LCM, as $+$, and the *Greatest Common Divisor*, GCD, as $\cdot$. The complement needs some consideration. Let us determine $\bar{2}$. By checking all the available numbers, it can be seen that 15 is the only number n with $\text{LCM}(2, n) = 30, \text{GCD}(2, n) = 1$; hence, $\bar{2} = 15$ and $\overline{15} = 2$. Generally, the complement of the number x is the quotient $30 : x$. It should not be too difficult for the reader to verify that all axioms are satisfied.

This example can be generalized in the following way. Let $P = \{p_1, \ldots, p_n\}$ be a set of different prime numbers, $N = p_1 * p_2 * \cdots * p_n$. Then the set D of all divisors of N (including 1 and N) is a Boolean Algebra when LCM and GCD are used for the disjunction and conjunction, respectively, and the complement is defined appropriately, as above. The exploration of this result gives an idea of *isomorphic structures*. We identify the empty set $\emptyset$ with 1, the set P with N, and we can assign each divisor of N a subset of P, the set of all prime numbers that are used for this divisor (each divisor can be built up by some of the prime numbers—6 is a divisor of 30, $6 = 2 * 3$, $\{2, 3\}$ is a subset of $\{2, 3, 5\}$). The intersection corresponds to the GCD, the union to the LCM.

Very remarkable and important is the next property. When we exchange the operation symbols $+$ and $\cdot$ simultaneously everywhere in all the axioms as well as 0 and 1, then we get another set of *dual* axioms. This property is denoted by *duality*.

That means, among others, that any theorem that can be derived and that includes $\vee, \wedge, 0$, and 1 can be transformed into a second theorem just by implementing the exchange mentioned above. Intuitively spoken, we can say that the theorems of a Boolean Algebra come "in pairs."

We can see the theorems of De Morgan as an example. We assume that the first law already has been proven:

$$\overline{x_1 \wedge x_2} = \overline{x}_1 \vee \overline{x}_2 . \tag{3.17}$$

The exchange of $\wedge$ and $\vee$ on the left-hand side as well as $\vee$ and $\wedge$ on the right-hand side results immediately in the second theorem:

$$\overline{x_1 \vee x_2} = \overline{x}_1 \wedge \overline{x}_2 . \tag{3.18}$$

Based on this property, the following theorem is obvious.

Theorem 3.4 $(\mathbb{B}, \wedge, \vee, \bar{}, 1, 0)$ *is a Boolean Algebra.*

The operations are essentially the same; however, the previous maximum element is now the new minimum element and vice versa. The new order relation reverses the arrow from 0 to 1 into an arrow from 1 to 0. This difference is in many applications not very important; hence, this explanatory note should be sufficient.

It is not very well-known and has been omitted in many textbooks that the algebraic concepts of *rings* and *fields* can also be applied in the Boolean domain. In order to introduce these notions, we need the axioms of rings and fields. If the reader is not too familiar with these concepts, then she/he should think of the set of integers together with addition and multiplication, which is also a ring.

Definition 3.4 A quintuple $(R, +, \cdot, 0, 1)$ is a *commutative ring* if the following axioms are satisfied:

> *Additive Laws*

$\forall x \; \forall y \; \forall z \in R :$	$(x + y) + z = x + (y + z)$	*Associativity*;
$\forall x \; \forall y \in R :$	$x + y = y + x$	*Commutativity*;
$\forall x \in R :$	$x + 0 = x$	*Zero Element*;
$\forall x \in R \; \exists (-x) \in R :$	$x + (-x) = 0$	*Inverse Element*;

Multiplicative Laws

$$\forall x \ \forall y \ \forall z \in R: \qquad (x \cdot y) \cdot z = x \cdot (y \cdot z) \qquad\qquad Associativity;$$

$$\forall x \ \forall y \in R: \qquad\qquad x \cdot y = y \cdot x \qquad\qquad\qquad Commutativity;$$

$$\forall y \in R: \qquad\qquad\qquad 1 \cdot y = y \qquad\qquad\qquad\quad Unit\ Element;$$

Distributive Law

$$\forall x \ \forall y \ \forall z \in R: \qquad x \cdot (y + z) = (x \cdot y) + (x \cdot z) \qquad Distributivity.$$

It should be noted that the "addition" in any ring is always assumed to be commutative; however, for the "multiplication" this would be a special requirement expressed by the naming of *commutative rings*.

For rings, the operations will be very often denoted by "addition" and "multiplication" even then when we are not talking about numbers.

Theorem 3.5 $(\mathbb{B}, \oplus, \wedge, 0, 1)$ *and* $(\mathbb{B}, \odot, \vee, 1, 0)$ *are commutative rings.*

Proof This proof uses again the principle of complete enumeration, which has already been used several times, so that the reader should be able to use it without any difficulties. For Boolean Rings, it is a special requirement that the multiplications (here $\wedge$ and $\vee$) are associative operations. This is not the case for a Boolean Algebra—there it can be shown that these operations are always associative.   □

Note 3.4 In a strict algebraic sense, these two rings are even *fields*. To extend the structure of a ring to a field, the following property must hold:

$$\text{if } a \cdot b = a \cdot c \text{ and } a \neq 0, \text{ then } b = c.$$

Intuitively speaking, in a field, the equation $a \cdot b = a \cdot c$ can be "divided by" a (if $a \neq 0$).

In terms of the given rings $(\mathbb{B}, \oplus, \wedge, 0, 1)$ and $(\mathbb{B}, \odot, \vee, 1, 0)$, we have

$$a \wedge b = a \wedge c, a \neq 0 \qquad \Rightarrow \qquad a = 1,$$
$$1 \wedge b = 1 \wedge c \qquad \Rightarrow \qquad b = c,$$
$$a \vee b = a \vee c, a \neq 1 \qquad \Rightarrow \qquad a = 0,$$
$$0 \vee b = 0 \vee c \qquad \Rightarrow \qquad b = c.$$

The symbol $\Rightarrow$ indicates the implication *if ... then*.

Very important are the following relations.

Theorem 3.6 *De Morgan's Laws in* $\mathbb{B}$:

$$\forall x \ \forall y \in \mathbb{B}: \qquad\qquad \overline{x \wedge y} = \overline{x} \vee \overline{y}, \qquad\qquad (3.19)$$

$$\forall x \ \forall y \in \mathbb{B}: \qquad\qquad \overline{x \vee y} = \overline{x} \wedge \overline{y}. \qquad\qquad (3.20)$$

These rules will be used for the simplification of expressions. The negation of a conjunction or disjunction of variables can be transferred into the negation of single variables by changing the operation to a disjunction or conjunction, respectively.

Theorem 3.7 *Idempotence:*

$$\forall x \in \mathbb{B}: \qquad x \vee x = x \,, \tag{3.21}$$

$$\forall x \in \mathbb{B}: \qquad x \wedge x = x \,. \tag{3.22}$$

Note 3.5 George Boole wrote $x \times x = x^2 = x$. He concluded that 0 and 1 are the only integer solutions of this equation ($0^2 = 0$, $1^2 = 1$). Hence, a calculus that is appropriate for logic must be built upon these two values only. He used the laws of arithmetic as given and true and added the argument that a repeated statement is as true as the statement itself.

A commutative ring with the property of idempotence will be denoted by *Boolean Ring*. Hence, the commutative rings specified in Theorem 3.5 are also Boolean Rings.

The difference between *Boolean Algebra* and *Boolean Ring* is for the set $\mathbb{B}$ not very important because the used functions and operations can be defined directly by tables, and thereafter, it can be shown that the axioms are satisfied. All of them can be used at the same time. However, it can be helpful for simplifications of Boolean expressions to restrict to one of these algebraic structures. Transformations from a *Boolean Ring* to a *Boolean Algebra* and vice versa can be executed in the following manner:

1. Given the Boolean Ring $(\mathbb{B}, \oplus, \wedge, 0, 1)$, then, by using the identities

$$x \oplus 1 = \overline{x} \,, \tag{3.23}$$

$$x \oplus y = (x \wedge \overline{y}) \vee (\overline{x} \wedge y) \,, \tag{3.24}$$

the negation can be introduced and the antivalence eliminated. We get the Boolean Algebra $(\mathbb{B}, \vee, \wedge, \overline{}, 0, 1)$.

2. Given the Boolean Ring $(\mathbb{B}, \odot, \vee, 1, 0)$, then, by using the identities

$$x \odot 0 = \overline{x} \,, \tag{3.25}$$

$$x \odot y = (x \wedge y) \vee (\overline{x} \wedge \overline{y}) \,, \tag{3.26}$$

the equivalence will be eliminated. The negation can be introduced, and we get Boolean Algebra $(\mathbb{B}, \wedge, \vee, \overline{}, 1, 0)$.

3. Given the Boolean Algebra $(\mathbb{B}, \vee, \wedge, \overline{}, 0, 1)$. Using

$$x \vee y = x \oplus y \oplus (x \wedge y) \,, \tag{3.27}$$

$$\overline{x} = 1 \oplus x \,, \tag{3.28}$$

we can eliminate the operation $\vee$ and the negation by means of the antivalence and get the Boolean Ring $(\mathbb{B}, \oplus, \wedge, 0, 1)$.

4. Given the Boolean Algebra $(\mathbb{B}, \wedge, \vee, \overline{}, 1, 0)$. Using

$$x \wedge y = x \odot y \odot (x \vee y), \tag{3.29}$$

$$\overline{x} = 0 \odot x, \tag{3.30}$$

we can eliminate the operation $\wedge$ and the negation using the equivalence and get the Boolean Ring $(\mathbb{B}, \odot, \vee, 1, 0)$.

Once again, there is hardly a need to emphasize the algebraic background, and all the operations are available all the time and can be used according to our convenience. However, sometimes the utilization of a fitting algebraic structure can facilitate the solution-steps of a given problem. Furthermore, the knowledge of the two dual Boolean Algebras as well as the two dual Boolean Rings can be used to map a known valid rule from one of these structures to the dual one and restricts in this way the number of rules we have to remember.

We will end this section with a comprehensive summary of all the operations defined and used so far. The user should ensure that she/he understands all these rules and is able to use them properly. They are valid in any Boolean Algebra or Boolean Ring, and the most important applications are the set $\mathbb{B}$ (which has already been considered) and the set $\mathbb{B}^n$ (to be introduced next).

Commutative Laws

$$\forall x\ \forall y \in \mathbb{B}: \qquad\qquad x \vee y = y \vee x$$

$$\forall x\ \forall y \in \mathbb{B}: \qquad\qquad x \wedge y = y \wedge x$$

$$\forall x\ \forall y \in \mathbb{B}: \qquad\qquad x \oplus y = y \oplus x$$

$$\forall x\ \forall y \in \mathbb{B}: \qquad\qquad x \odot y = y \odot x$$

Associative Laws

$$\forall x\ \forall y\ \forall z \in \mathbb{B}: \qquad (x \wedge y) \wedge z = x \wedge (y \wedge z) = x \wedge y \wedge z$$

$$\forall x\ \forall y\ \forall z \in \mathbb{B}: \qquad (x \vee y) \vee z = x \vee (y \vee z) = x \vee y \vee z$$

$$\forall x\ \forall y\ \forall z \in \mathbb{B}: \qquad (x \oplus y) \oplus z = x \oplus (y \oplus z) = x \oplus y \oplus z$$

$$\forall x\ \forall y\ \forall z \in \mathbb{B}: \qquad (x \odot y) \odot z = x \odot (y \odot z) = x \odot y \odot z$$

Distributive Laws

$$\forall x\ \forall y\ \forall z \in \mathbb{B}: \qquad x \wedge (y \vee z) = (x \wedge y) \vee (x \wedge z)$$

$$\forall x\ \forall y\ \forall z \in \mathbb{B}: \qquad x \vee (y \wedge z) = (x \vee y) \wedge (x \vee z)$$

$$\forall x\ \forall y\ \forall z \in \mathbb{B}: \qquad x \wedge (y \oplus z) = (x \wedge y) \oplus (x \wedge z)$$

$$\forall x\ \forall y\ \forall z \in \mathbb{B}: \qquad x \vee (y \odot z) = (x \vee y) \odot (x \vee z)$$

Negation (Complement)

$$\forall x\ \forall y \in \mathbb{B}: \qquad\qquad \overline{x \wedge y} = \overline{x} \vee \overline{y}$$

$$\forall x\ \forall y \in \mathbb{B}: \qquad\qquad \overline{x \vee y} = \overline{x} \wedge \overline{y}$$

$$\forall x\ \forall y \in \mathbb{B}: \qquad\qquad \overline{x \oplus y} = \overline{x} \oplus y = x \oplus \overline{y}$$

$$\forall x \; \forall y \in \mathbb{B} : \qquad \overline{x \oplus y} = 1 \oplus x \oplus y$$

$$\forall x \; \forall y \in \mathbb{B} : \qquad \overline{x \odot y} = \overline{x} \odot y = x \odot \overline{y}$$

$$\forall x \; \forall y \in \mathbb{B} : \qquad \overline{x \odot y} = 0 \odot x \odot y$$

$$\overline{0} = 1, \; \overline{1} = 0$$

Identities

$$\forall x \in \mathbb{B} : \qquad\qquad\qquad x \wedge 0 = 0$$

$$\forall x \in \mathbb{B} : \qquad\qquad\qquad x \vee 0 = x$$

$$\forall x \in \mathbb{B} : \qquad\qquad\qquad x \oplus 0 = x$$

$$\forall x \in \mathbb{B} : \qquad\qquad\qquad x \odot 0 = \overline{x}$$

$$\forall x \in \mathbb{B} : \qquad\qquad\qquad x \wedge 1 = x$$

$$\forall x \in \mathbb{B} : \qquad\qquad\qquad x \vee 1 = 1$$

$$\forall x \in \mathbb{B} : \qquad\qquad\qquad x \oplus 1 = \overline{x}$$

$$\forall x \in \mathbb{B} : \qquad\qquad\qquad x \odot 1 = x$$

$$\forall x \in \mathbb{B} : \qquad\qquad\qquad x \wedge \overline{x} = 0$$

$$\forall x \in \mathbb{B} : \qquad\qquad\qquad x \vee \overline{x} = 1$$

$$\forall x \in \mathbb{B} : \qquad\qquad\qquad x \oplus \overline{x} = 1$$

$$\forall x \in \mathbb{B} : \qquad\qquad\qquad x \odot \overline{x} = 0$$

$$\forall x \; \forall y \in \mathbb{B} : \qquad\qquad x \oplus y = (x \wedge \overline{y}) \vee (\overline{x} \wedge y)$$

$$\forall x \; \forall y \in \mathbb{B} : \qquad\qquad x \odot y = (x \wedge y) \vee (\overline{x} \wedge \overline{y})$$

$$\forall x \; \forall y \in \mathbb{B} : \qquad\qquad x \vee y = x \oplus y \oplus (x \wedge y)$$

$$\forall x \; \forall y \in \mathbb{B} : \qquad\qquad x \wedge y = x \odot y \odot (x \vee y)$$

$$\forall x \; \forall y \in \mathbb{B} : \qquad\qquad x \oplus y = \overline{x} \oplus \overline{y}$$

$$\forall x \; \forall y \in \mathbb{B} : \qquad\qquad x \odot y = \overline{x} \odot \overline{y}$$

$$\forall x \; \forall y \in \mathbb{B} : \qquad x \oplus y = 0 \qquad \equiv \qquad x = y$$

$$\forall x \; \forall y \in \mathbb{B} : \qquad x \odot y = 1 \qquad \equiv \qquad x = y$$

$$\forall x \; \forall y \; \forall z \in \mathbb{B} : \qquad x \oplus z = y \oplus z \qquad \equiv \qquad x = y$$

$$\forall x \; \forall y \; \forall z \in \mathbb{B} : \qquad x \odot z = y \odot z \qquad \equiv \qquad x = y$$

$$\forall x \; \forall y \; \forall z \in \mathbb{B} : \qquad x \oplus y = z \qquad \equiv \qquad x = y \oplus z$$

$$\forall x \; \forall y \; \forall z \in \mathbb{B} : \qquad x \odot y = z \qquad \equiv \qquad x = y \odot z$$

Idempotence

$\forall x \in \mathbb{B}:$ $\qquad\qquad\qquad x \wedge x = x$

$\forall x \in \mathbb{B}:$ $\qquad\qquad\qquad x \vee x = x$

$\forall x \in \mathbb{B}:$ $\qquad\qquad\qquad x \oplus x = 0$

$\forall x \in \mathbb{B}:$ $\qquad\qquad\qquad x \odot x = 1$

Absorption

$\forall x \, \forall y \in \mathbb{B}:$ $\qquad\qquad\qquad x \vee (x \wedge y) = x$

$\forall x \, \forall y \in \mathbb{B}:$ $\qquad\qquad\qquad x \wedge (x \vee y) = x$

$\forall x \, \forall y \in \mathbb{B}:$ $\qquad\qquad\qquad x \vee (\overline{x} \wedge y) = x \vee y$

$\forall x \, \forall y \in \mathbb{B}:$ $\qquad\qquad\qquad x \wedge (\overline{x} \vee y) = x \wedge y$

Involution

$\forall x \in \mathbb{B}:$ $\qquad\qquad\qquad \overline{\overline{x}} = x$

Consensus

$\forall x \, \forall y \, \forall z \in \mathbb{B}:$ $\qquad (x \wedge y) \vee (\overline{x} \wedge z) \vee (y \wedge z) = (x \wedge y) \vee (\overline{x} \wedge z)$

$\forall x \, \forall y \, \forall z \in \mathbb{B}:$ $\qquad (x \vee y) \wedge (\overline{x} \vee z) \wedge (y \vee z) = (x \vee y) \wedge (\overline{x} \vee z)$

Partial Order

$\forall x \in \mathbb{B}:$ $\qquad\qquad\qquad x \leq x$

$\forall x \in \mathbb{B}:$ $\qquad\qquad\qquad 0 \leq x \leq 1$

$\forall x \, \forall y \in \mathbb{B}:$ $\qquad\qquad\qquad (x \wedge y) \leq x \leq (x \vee y)$

$\forall x \, \forall y \in \mathbb{B}:$ $\qquad x \leq y \ \text{ and } \ y \leq z \ \Rightarrow \ x \leq z$

$\forall x \, \forall y \in \mathbb{B}:$ $\qquad x \leq y \ \text{ and } \ y \leq x \ \Rightarrow \ x = y$

Equality and Inequality

$\forall x \, \forall y \in \mathbb{B}:$ $\qquad\qquad\qquad x \leq y \quad \equiv \quad x \wedge \overline{y} = 0$

$\forall x \, \forall y \in \mathbb{B}:$ $\qquad\qquad\qquad x \leq y \quad \equiv \quad \overline{x} \vee y = 1$

$\forall x \, \forall y \in \mathbb{B}:$ $\qquad\qquad\qquad r \leq y \quad = \quad \overline{y} \leq \overline{x}$

$\forall x \, \forall y \in \mathbb{B}:$ $\qquad\qquad\qquad x \leq y \quad \equiv \quad x \vee y = y$

$\forall x \, \forall y \in \mathbb{B}:$ $\qquad\qquad\qquad x \leq y \quad \equiv \quad x \wedge y = x$

$\forall x \, \forall y \, \forall z \in \mathbb{B}:$ $\qquad x \leq z, \ y \leq z \ \Rightarrow \ x \vee y \leq z$

$\forall x \, \forall y \, \forall z \in \mathbb{B}:$ $\qquad x \leq y, \ x \leq z \ \Rightarrow \ x \leq y \wedge z$

$\forall x \, \forall y \, \forall z \in \mathbb{B}:$ $\qquad\qquad x \leq y \ \Rightarrow \ x \vee z \leq y \vee z$

$\forall x \, \forall y \, \forall z \in \mathbb{B}:$ $\qquad\qquad x \leq y \ \Rightarrow \ x \wedge z \leq y \wedge z$

$$\forall x\ \forall y\ \forall z \in \mathbb{B}: \qquad\qquad (x \vee y) \oplus (x \vee z) \ \leq\ (y \oplus z)$$

$$\forall x\ \forall y\ \forall z \in \mathbb{B}: \qquad\qquad (x \wedge y) \oplus (x \wedge z) \ \leq\ (y \oplus z)$$

$$\forall x\ \forall y \in \mathbb{B}: \qquad\qquad\qquad\qquad x \ \leq\ y \vee (x \oplus y)$$

$$\forall x\ \forall y\ \forall z \in \mathbb{B}: \qquad\qquad\qquad x \oplus y \ \leq\ z \vee (x \oplus y)$$

$$\forall x\ \forall y\ \forall u\ \forall w \in \mathbb{B}: \qquad (x \vee y) \oplus (u \vee w) \ \leq\ (x \oplus u) \vee (y \oplus w)$$

$$\forall x\ \forall y\ \forall u\ \forall w \in \mathbb{B}: \qquad (x \wedge y) \oplus (u \wedge w) \ \leq\ (x \oplus u) \vee (y \oplus w).$$

Note 3.6 It is a considerable simplification and improvement of the understanding of expressions when the $\wedge$ is omitted (like the multiplication sign in arithmetic). The first three distributive laws, for instance, would appear in the following format:

$$x(y \vee z) = xy \vee xz\,,$$

$$x \vee yz = (x \vee y)(x \vee z)\,,$$

$$x(y \oplus z) = xy \oplus xz\,.$$

Proof All these statements follow from the axioms of a Boolean Algebra or a Boolean Ring and can be derived one after the other, based on these axioms. We will give only some examples for these proofs. As a first example, we will prove the idempotence of $\vee$: $x \vee x = x$. We get successively:

$x = 0 \vee x$	*Neutral Element*
$= x \vee 0$	*Commutativity*
$= x \vee (x \wedge \overline{x})$	*Complement*
$= (x \vee x) \wedge (x \vee \overline{x})$	*Distributivity*
$= (x \vee x) \wedge 1$	*Complement*
$= (x \vee x)\,.$	*Neutral Element*

In this example, only axioms have been used. The respective axiom for each step is stated at the right-hand side in the line of its use. The axioms can be used from the left to the right as well as from the right to the left. Now this law is available for other proofs.

As a next example, the first part of the *consensus* will be proven. We use the properties $x \vee \overline{x} = 1$ and $x \wedge 1 = x$. The expression A of the left-hand side can be transformed as follows:

$$A = xy \vee \overline{x}z \vee yz$$

$$= xy1 \vee \overline{x}1z \vee 1yz$$

$$= xy(z \vee \overline{z}) \vee \overline{x}(y \vee \overline{y})z \vee (x \vee \overline{x})yz\,.$$

The application of the distributive law results in

$$A = xyz \vee xy\overline{z} \vee \overline{x}yz \vee \overline{x}\,\overline{y}z \vee xyz \vee \overline{x}yz\,.$$

Now the idempotence can be used to avoid duplications of the same terms ($x \vee x = x$). Hence,

$$A = xyz \vee xy\overline{z} \vee \overline{x}yz \vee \overline{x}\,\overline{y}z \,.$$

Using the same means for the right-hand side $B = xy \vee \overline{x}z$ gives

$$B = xyz \vee xy\overline{z} \vee \overline{x}yz \vee \overline{x}\,\overline{y}z \,.$$

Since only identities have been used to get identical expressions for both A and B, the proof is complete.

The enumeration of all possible combinations for x, y, z would be a second option to prove this identity.

The method that has been used here is a very characteristic procedure. Every conjunction is extended in such a way that all variables appear in each conjunction:

$$xy = xy1$$
$$= xy(z \vee \overline{z})$$
$$= xyz \vee xy\overline{z} \,.$$

After the elimination of duplications, there is a characteristic expression, the disjunction of conjunctions where each conjunction contains all the variables (negated or non-negated). It will be seen later on that the concept of a *disjunctive normal form* has been used.

The axioms of a Boolean Algebra state that

$$0 \vee x = x \quad \text{and} \quad 1 \wedge x = x \,.$$

Therefore, it must be desirable or necessary to prove

$$1 \vee x = 1 \quad \text{and} \quad 0 \wedge x = 0 \,.$$

We prove $1 \vee x = 1$ as follows:

$$1 \vee x = (1 \vee x) \wedge 1$$
$$= (1 \vee x) \wedge (x \vee \overline{x})$$
$$= (x \vee 1) \wedge (x \vee \overline{x})$$
$$= x \vee (1 \wedge \overline{x})$$
$$= x \vee (\overline{x} \wedge 1)$$
$$= x \vee \overline{x}$$
$$= 1 \,.$$

The reader can fill in some occasional simplifications that are based on the two axioms mentioned above.

In order to show that $\overline{x \vee y} = \overline{x} \wedge \overline{y}$, it has to be shown that

$$(x \vee y) \wedge (\overline{x} \wedge \overline{y}) = 0 \,,$$

$$(x \lor y) \lor (\overline{x} \land \overline{y}) = 1 \, .$$

By omitting the $\land$, we get step by step

$$
\begin{aligned}
(x \lor y)(\overline{x} \, \overline{y}) &= (\overline{x} \, \overline{y})(x \lor y) \\
&= (\overline{x} \, \overline{y}) \, x \lor (\overline{x} \, \overline{y}) \, y \\
&= x \, (\overline{x} \, \overline{y}) \lor (\overline{x} \, \overline{y}) \, y \\
&= 0 \, \overline{y} \lor \overline{x} \, 0 \\
&= 0 \lor 0 \\
&= 0 \, .
\end{aligned}
$$

This is the proof of the first identity, and the other part follows analogously.   □

All the other properties can be confirmed in the same way, either by calculations (taking the axioms as starting point) or, for the set $\mathbb{B}$, by inserting all possible combinations of 0 and 1 into the formulas.

3.4   The Set $\mathbb{B}^n$

As next we will increase the number of elements of the Boolean Algebras under consideration. In order to do this, the set of all binary vectors of a given length n will be considered.

Definition 3.5 Let $\mathbb{B} = \{0, 1\}$, and then the set

$$\mathbb{B}^n = \{\mathbf{x} \mid \mathbf{x} = (x_1, x_2, \ldots, x_{n-1}, x_n), x_i \in \mathbb{B} \; \forall i = 1, \ldots, n\} \tag{3.31}$$

is denoted by *binary space* or *Boolean space* $\mathbb{B}^n$.

Informally, $\mathbb{B}^n$ is the set of all *binary vectors* with n components. $\mathbb{B}^n$ can be understood as the *cross-product* $\mathbb{B} \times \mathbb{B} \times \cdots \times \mathbb{B}$, which is using the set $\mathbb{B}$ n times. Very often, the commas between the components will be omitted in order to simplify the representation. As examples, we can see, for instance,

$$
\begin{aligned}
\mathbb{B}^2 =&\{(00), (01), (10), (11)\} \, , \\
\mathbb{B}^4 =&\{(0000), (0001), (0010), (0011), (0100), (0101), (0110), (0111), \\
&(1000), (1001), (1010), (1011), (1100), (1101), (1110), (1111)\} \, .
\end{aligned}
$$

It is easy to understand and to prove the following theorem:

Theorem 3.8 *The set $\mathbb{B}^n$ has 2^n elements.*

Proof The proof uses the induction principle. Let $n = 1$, and then $\mathbb{B}^1 = \mathbb{B}$ has two elements. If $\mathbb{B}^n$ contains 2^n binary vectors, then each vector can be extended either by 0 or by 1, which results in two new vectors, and this gives $2 * 2^n = 2^{n+1}$ vectors with $n + 1$ components in $\mathbb{B}^{n+1}$.   □

Intuitively, there are n positions, and for each position, two different values are possible, independent of the other positions. This results in $2 \times 2 \times \cdots \times 2 = 2^n$ different binary vectors.

The relations $\leq$ and $<$, respectively, will be introduced as follows:

Definition 3.6 Let $\mathbf{x} = (x_1, x_2, \ldots, x_n)$ and $\mathbf{y} = (y_1, y_2, \ldots, y_n)$. Then

$$\mathbf{x} \leq \mathbf{y} \quad \equiv \quad x_1 \leq y_1, \ldots, x_n \leq y_n \,, \tag{3.32}$$

$$\mathbf{x} < \mathbf{y} \quad \equiv \quad \mathbf{x} \leq \mathbf{y} \text{ and } \mathbf{x} \neq \mathbf{y} \,. \tag{3.33}$$

This means that the two vectors have to be compared component by component, and if the inequality holds for each component, then it also holds for the two vectors.

Example 3.1

1. Let $\mathbf{x} = (0101)$, $\mathbf{y} = (1101)$, $\mathbf{z} = (0001) \in \mathbb{B}^4$. Then it holds that

$$\mathbf{x} \leq \mathbf{y} \,, \qquad \mathbf{z} \leq \mathbf{x} \,, \qquad \mathbf{z} \leq \mathbf{y} \,,$$

   or, expressed by one term,

$$\mathbf{z} \leq \mathbf{x} \leq \mathbf{y} \,.$$

2. There are incomparable elements. For $\mathbf{x} = (0100)$ and $\mathbf{y} = (0001)$ neither $\mathbf{x} \leq \mathbf{y}$ nor $\mathbf{y} \leq \mathbf{x}$ holds. The *linearity* of $\leq$ is lost for $\mathbb{B}^n$ with $n > 1$.

Remark 3.5 In graphs for partial orders, the transitive edges are mostly omitted to make the graphs more understandable. If there is an edge from node $\mathbf{a}$ to node $\mathbf{b}$ and from node $\mathbf{b}$ to node $\mathbf{c}$, then the transitive edge from $\mathbf{a}$ to $\mathbf{c}$ is not represented.

Figure 3.8 shows the partial orders for $\mathbb{B}^3$. This kind of presentation is often denoted by *Hasse* diagram. It has a very characteristic shape: the vector $\mathbf{0}$ (with no values of 1) is on level 0 (the lowest left corner), and all the vectors with exactly one value of 1 are on the first level—one edge is necessary to reach such a node from $\mathbf{0}$; all the vectors with exactly two values of 1 are on the second level—two edges are necessary to reach nodes from $\mathbf{0}, \ldots$, finally, on the level $n - 1$ all vectors with $n - 1$ values of 1 (i.e., one value of 0) can be found, and the vector $\mathbf{1}$ is on the last and highest level. As already mentioned, the transitive arrows are not included in order to make the diagrams more readable.

Fig. 3.8 Partial orders of $\mathbb{B}^3$, transitive edges omitted: (a) the reflexive partial order $\leq$ of $\mathbb{B}^3$ and (b) the irreflexive partial order $<$ of $\mathbb{B}^3$

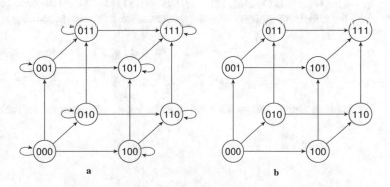

Theorem 3.9 *For $n > 1$, the relations $(\mathbb{B}^n, \leq)$ and $(\mathbb{B}^n, <)$ are a reflexive and an irreflexive partial order, respectively.*

Proof The vector $\mathbf{0} = (0, 0, \ldots, 0)$ can be compared with all vectors $\mathbf{x} = (x_1, \ldots, x_n)$, and it holds that $\mathbf{0} \leq \mathbf{x}$ (or $\mathbf{0} < \mathbf{x}$ for $\mathbf{x} \neq \mathbf{0}$). The vector $\mathbf{1} = (1, 1, \ldots, 1)$ can be compared with all vectors $\mathbf{x} = (x_1, \ldots, x_n)$, and it holds that $\mathbf{x} \leq \mathbf{1}$ (or $\mathbf{x} < \mathbf{1}$ for $\mathbf{x} \neq \mathbf{1}$). $\qquad\square$

The operations $^{-}$, $\wedge$, $\vee$, $\oplus$, $\odot$ are introduced for $\mathbb{B}^n$ by applying them to every component of the vector(s). For the vectors $\mathbf{x}, \mathbf{y} \in \mathbb{B}^n$, we get

$$\bar{\mathbf{x}} = (\bar{x}_1, \ldots, \bar{x}_n) \,, \tag{3.34}$$

$$\mathbf{x} \wedge \mathbf{y} = (x_1 \wedge y_1, \ldots, x_n \wedge y_n) \,, \tag{3.35}$$

$$\mathbf{x} \vee \mathbf{y} = (x_1 \vee y_1, \ldots, x_n \vee y_n) \,, \tag{3.36}$$

$$\mathbf{x} \oplus \mathbf{y} = (x_1 \oplus y_1, \ldots, x_n \oplus y_n) \,, \tag{3.37}$$

$$\mathbf{x} \odot \mathbf{y} = (x_1 \odot y_1, \ldots, x_n \odot y_n) \,. \tag{3.38}$$

All the properties of a *Boolean Algebra* and a *Boolean Ring*, which have been established for $\mathbb{B}$, are also valid for $\mathbb{B}^n$. The properties of a *field*, however, will be lost. Take, for instance, $\mathbf{a} = (0011)$, $\mathbf{b} = (1101)$, $\mathbf{c} = (1001)$, then

$$\mathbf{a} \neq \mathbf{0} \,, \quad \mathbf{a} \wedge \mathbf{b} = \mathbf{a} \wedge \mathbf{c} \,, \quad \text{but} \quad \mathbf{b} \neq \mathbf{c} \,.$$

Theorem 3.10 *The following algebraic structures can be found:*

- $(\mathbb{B}^n, \vee, \wedge, ^{-}, \mathbf{0}, \mathbf{1})$ *is* Boolean Algebra.
- $(\mathbb{B}^n, \wedge, \vee, ^{-}, \mathbf{1}, \mathbf{0})$ *is a* Boolean Algebra.
- $(\mathbb{B}^n, \oplus, \wedge, \mathbf{0}, \mathbf{1})$ *is a* commutative Boolean Ring.
- $(\mathbb{B}^n, \odot, \vee, \mathbf{1}, \mathbf{0})$ *is a* commutative Boolean Ring.

It has already been mentioned that the power set of a given finite set $X = \{x_1, \ldots, x_n\}$ together with union, intersection, complement and the elements $\emptyset$ and X is a *Boolean Algebra*. It will now be shown that there is a very characteristic relation between the power set of a finite set with n elements and the set $\mathbb{B}^n$.

Example 3.2 Let $n = 3$, $X = \{x_1, x_2, x_3\}$. Then the power set $P(X)$ has the elements $\emptyset, \{x_1\}, \{x_2\}, \{x_3\}, \{x_1, x_2\}, \{x_1, x_3\}, \{x_2, x_3\}, \{x_1, x_2, x_3\}$. For each element Y of this power set, we now construct a binary *characteristic vector* $\mathbf{b}(Y) = (b_1, b_2, b_3) \in \mathbb{B}^3$ in a unique way:

$$b_i = 1 \quad \equiv \quad (x_i \in Y) \,, \qquad i = 1, 2, 3 \,,$$

$$b_i = 0 \quad \equiv \quad (x_i \notin Y) \,, \qquad i = 1, 2, 3 \,.$$

For the sets above, we get successively the vectors represented in Table 3.9.

Table 3.9 The relation between power set and binary vectors

Set	b_1	b_2	b_3
$\emptyset$	0	0	0
$\{x_1\}$	1	0	0
$\{x_2\}$	0	1	0
$\{x_3\}$	0	0	1
$\{x_1, x_2\}$	1	1	0
$\{x_1, x_3\}$	1	0	1
$\{x_2, x_3\}$	0	1	1
$\{x_1, x_2, x_3\}$	1	1	1

There is a unique correspondence between binary vectors and sets. Different binary vectors define different sets and vice versa. The notation $\mathbf{b}(Y)$ is used for the vector that corresponds to the subset Y. Now it can be seen how the operations $\overline{}$, $\wedge$, $\vee$, $\oplus$, $\odot$ are reflected by set-theoretic operations:

$$\overline{\mathbf{b}(Y_1)} = \mathbf{b}(\overline{Y_1}) \,,$$

$$\mathbf{b}(Y_1) \wedge \mathbf{b}(Y_2) = \mathbf{b}(Y_1 \cap Y_2) \,,$$

$$\mathbf{b}(Y_1) \vee \mathbf{b}(Y_2) = \mathbf{b}(Y_1 \cup Y_2) \,,$$

$$\mathbf{b}(Y_1) \oplus \mathbf{b}(Y_2) = \mathbf{b}(Y_1 \bigtriangleup Y_2) \,,$$

$$\mathbf{b}(Y_1) \odot \mathbf{b}(Y_2) = \mathbf{b}(Y_1 \overline{\bigtriangleup} Y_2) \,.$$

The $\bigtriangleup$ is the *symmetric difference* of two sets:

$$X \bigtriangleup Y = (X \cap \overline{Y}) \cup (\overline{X} \cap Y) \,.$$

Any element must be in one given set, but not in the other, in order to belong to the set specified by $X \bigtriangleup Y$. This directly reflects the *antivalence*. $\overline{\bigtriangleup}$ is the *complement of the symmetric difference* of two sets:

$$X \overline{\bigtriangleup} Y = (X \cap Y) \cup (\overline{X} \cap \overline{Y}) \,.$$

Any element must be in both given sets or in none of them (as expressed by the *equivalence*) in order to belong to the set specified by $X \overline{\bigtriangleup} Y$.

This is a very characteristic relation between the two different algebraic structures. $\mathbf{b}(Y_1 \cap Y_2)$ uses first the intersection of the two sets and in a second step the characteristic vector, $\mathbf{b}(Y_1) \wedge \mathbf{b}(Y_2)$, takes the characteristic vectors of Y_1 and Y_2 and combines these two vectors using the *conjunction*. The result is the same.

That means that we can work in the power set and transfer the results into bit vectors, or we work in $\mathbb{B}^n$ and transfer the results into the power set, according to our convenience. The two algebraic structures are *isomorphic*. There is a one-to-one correspondence between the elements of the two sets and the operations, respectively:

Table 3.10 The Greatest Common Divisor as Boolean $\wedge$ and the Lowest Common Multiple as Boolean $\vee$

GCD	1	2	3	5	6	10	15	30	LCM	1	2	3	5	6	10	15	30
1	1	1	1	1	1	1	1	**1**	1	1	2	3	5	6	10	15	**30**
2	1	2	1	1	2	2	**1**	2	2	2	2	6	10	6	10	**30**	30
3	1	1	3	1	3	**1**	3	3	3	3	6	3	15	6	**30**	15	30
5	1	1	1	5	**1**	5	5	5	5	5	10	15	5	**30**	10	15	30
6	1	2	3	**1**	6	2	3	6	6	6	6	6	**30**	6	30	30	30
10	1	2	**1**	5	2	10	5	10	10	10	10	**30**	10	30	10	30	30
15	1	**1**	3	5	3	5	15	15	15	15	**30**	15	15	30	30	15	30
30	**1**	2	3	5	6	10	15	30	30	**30**	30	30	30	30	30	30	30

Theorem 3.11

- *Each power set $P(X)$ of a finite set $X = \{x_1, \ldots, x_n\}$ with the operations $\cup$, $\cap$, $^-$ and the elements $\emptyset$ and X is isomorphic to $(\mathbb{B}^n, \vee, \wedge, ^-, \mathbf{0}, \mathbf{1})$.*
- *Each power set $P(X)$ of a finite set $X = \{x_1, \ldots, x_n\}$ with the operations $\cap$, $\cup$, $^-$ and the elements X and $\emptyset$ is isomorphic to $(\mathbb{B}^n, \wedge, \vee, ^-, \mathbf{1}, \mathbf{0})$.*
- *Each power set $P(X)$ of a finite set $X = \{x_1, \ldots, x_n\}$ with the operations $\triangle$, $\cap$ and the elements $\emptyset$ and X is isomorphic to $(\mathbb{B}^n, \oplus, \wedge, \mathbf{0}, \mathbf{1})$.*
- *Each power set $P(X)$ of a finite set $X = \{x_1, \ldots, x_n\}$ with the operations $\overline{\triangle}$, $\cup$ and the elements X and $\emptyset$ is isomorphic to $(\mathbb{B}^n, \odot, \vee, \mathbf{1}, \mathbf{0})$.*

In set theory, there is shown a further extension that states that there are no other *finite Boolean Algebras* at all.

Theorem 3.12 (Stone's Theorem) *Every finite Boolean Algebra is isomorphic to the power set of a finite set with the operations $\cup$, $\cap$, $^-$ and the elements $\emptyset$ and X, and, hence, to a Boolean Algebra $(\mathbb{B}^n, \vee, \wedge, ^-, \mathbf{0}, \mathbf{1})$.*

When we study the structures

$$(\mathbb{B}^n, \vee, \wedge, ^-, \mathbf{0}, \mathbf{1}) \quad \text{or} \quad (\mathbb{B}^n, \oplus, \wedge, \mathbf{0}, \mathbf{1}),$$

then we study all possible finite Boolean Algebras or Boolean Rings.

This theorem was extremely important to create numerical algorithms for the solution of Boolean equations. Any such equation will be transformed into sets, and the respective operations have been implemented as set operations and included into the XBOOLE-library [11, 14, 16, 17].

Note 3.7 George Boole himself studied sets together with the intersection and the symmetric difference, i.e., he was working with Boolean Rings. Only his successors "forgot" about the symmetric difference (or the antivalence) and used "and" and "or" (or the intersection and the union). A comprehensive summary of his work can be found in [4, 5].

Example 3.3 Let $X = \{1, 2, 3, 5, 6, 10, 15, 30\}$, i.e., the set of all divisors of $n = 30$ together with the *Lowest Common Multiple* LCM and the *Greatest Common Divisor* GCD (Table 3.10).

We mentioned already that these two operations LCM and GCD define a Boolean Algebra if:

- 1 is taken as the element 0.
- 30 is taken as the element 1.

Table 3.11 The isomorphism between different structures

Number	Vector	Subset
1	(000)	$\emptyset$
2	(100)	$\{x_1\}$
3	(010)	$\{x_2\}$
5	(001)	$\{x_3\}$
6	(110)	$\{x_1, x_2\}$
10	(101)	$\{x_1, x_3\}$
15	(011)	$\{x_2, x_3\}$
30	(111)	$\{x_1, x_2, x_3\}$

- The bold elements in the main diagonal characterize the complement:
  $x \wedge \overline{x} = 0$, $x \vee \overline{x} = 1$, e.g., $\overline{6} = 5$, $\overline{5} = 6$, etc.
- It has been checked that all the axioms have been satisfied.

Since this Boolean Algebra has eight elements, $\mathbb{B}^3$ or a set $Y = \{y_1, y_2, y_3\}$ with three elements can be used for the creation of isomorphic structures. The assignments shown in Table 3.11 establish the isomorphism.

It was already mentioned that the partial orders $\leq$ or $<$ in $\mathbb{B}^n$ are not linear, i.e., there are incomparable elements (e.g., $(001) \not\leq (100)$, and also $(100) \not\leq (001)$). This can be seen very characteristically in the graph of the partial order: there is no *path* (a sequence of directed edges) from (001) to (100) or vice versa (see Fig. 3.8).

Sometimes it is useful to use the concept of an *interval*.

Definition 3.7 For $\mathbf{x}, \mathbf{y} \in \mathbb{B}^n$, $\mathbf{x} \leq \mathbf{y}$, $\mathbf{x} \neq \mathbf{y}$, the subset $I_{\mathbf{xy}} \subseteq \mathbb{B}^n$ with

$$I_{\mathbf{xy}} = \{\mathbf{z} | \mathbf{x} < \mathbf{z} < \mathbf{y}\} \tag{3.39}$$

is denoted by *interval $I_{\mathbf{xy}}$*.

Intuitively speaking, the interval $I_{\mathbf{xy}}$ contains all the elements of $\mathbb{B}^n$ *between* $\mathbf{x}$ and $\mathbf{y}$, including $\mathbf{x}$ and $\mathbf{y}$. Every interval has at least two elements. If $\mathbf{x} = \mathbf{0}$ and $\mathbf{y} = \mathbf{1}$, then $I_{01} = \mathbb{B}^n$. An analogous definition can be given for the irreflexive partial order $<$:

$$I'_{\mathbf{xy}} = \{\mathbf{z} | \mathbf{x} < \mathbf{z} < \mathbf{y}\} . \tag{3.40}$$

Each interval $I_{\mathbf{xy}}$ of the reflexive partial order can be transformed into an interval $I'_{\mathbf{xy}}$ of the irreflexive order by deleting $\mathbf{x}$ and $\mathbf{y}$ in $I_{\mathbf{xy}}$. $I'_{\mathbf{xy}} = \emptyset$ if $\mathbf{x}$ and $\mathbf{y}$ are the only elements in $I_{\mathbf{xy}}$. In this case, $\mathbf{x}$ and $\mathbf{y}$ are directly connected by one edge. The largest interval is I'_{01} containing all elements of $\mathbb{B}^n$ except $\mathbf{0}$ and $\mathbf{1}$.

Note 3.8 The concept of an interval can naturally be defined for any Boolean Algebra. We are using this restricted definition simply because we mainly use $\mathbb{B}^n$.

Example 3.4 We consider the two vectors $\mathbf{x} \neq \mathbf{0}$, $\mathbf{y} \neq \mathbf{1}$. Let $\mathbf{x} = (0010)$, $\mathbf{y} = (0111)$. By comparing these two vectors, it can be seen that the first and third components are constant, and the second and fourth change the value from 0 to 1. We build a vector $\mathbf{t}$ where we use the constant values 0 and 1 as given and mark the two positions with a changing value by "$-$": $\mathbf{t} = (0 - 1-)$. Now all the vectors

Table 3.12 The elements
of $\mathbb{B}^n$ as binary numbers
and the decimal equivalent
$dec(\mathbf{x})$

x_3	x_2	x_1	x_0	$dec(\mathbf{x})$	x_3	x_2	x_1	x_0	$dec(\mathbf{x})$
0	0	0	0	0	1	0	0	0	8
0	0	0	1	1	1	0	0	1	9
0	0	1	0	2	1	0	1	0	10
0	0	1	1	3	1	0	1	1	11
0	1	0	0	4	1	1	0	0	12
0	1	0	1	5	1	1	0	1	13
0	1	1	0	6	1	1	1	0	14
0	1	1	1	7	1	1	1	1	15

of the interval can be found quite easily by replacing the marked positions (the positions with a $-$) by all possible binary vectors of length 2, i.e., by (00), (01), (10), (11), and we receive the interval $I = \{(0010), (0011), (0110), (0111)\}$. The *ternary vector* $\mathbf{t}$ is used as a comfortable abbreviation for the interval I.

The changeable components (two in our case) define a *subalgebra* of $\mathbb{B}^4$ that is isomorphic to $\mathbb{B}^2$. A subalgebra is characterized by the fact that the carrier is a subset of $\mathbb{B}^n$, and all the axioms of a Boolean Algebra are satisfied.

Generally, for n components and k changeable components, the interval of the reflexive order is isomorphic to $\mathbb{B}^{n-k}$. The changeable positions can easily be found by using the antivalence. The vector $\mathbf{x} \oplus \mathbf{y}$ is equal to 0 in all constant positions and equal to 1 in all changeable positions.

An important order relation in $\mathbb{B}^n$ can be defined in a constructive way (see Table 3.12):

1. Define a table with n columns and 2^n rows.
2. In the rightmost column, the sequence 01 is used 2^{n-1} times.
3. Now the elements are "doubled" (which results in 0011), and this sequence is used in the last but one column 2^{n-2} times.
4. Then we use 00001111 in the rightmost empty column.
5. Finally, we use 2^{n-1} values of 0 and 2^{n-1} values of 1 in the first column.

This construction results in 2^n different vectors, i.e., in all the vectors of $\mathbb{B}^n$. We consider the elements as ordered in the sequence as they appear in the table. The "smallest" (first) element is the vector (0000) in the first row, and the "largest" (last) element is the vector (1111) in the last row. Please note that the table has been split into two halves (behind the element (0111) follows the element (1000)):

$$(0000) \prec (0001) \prec (0010) \prec \ldots \prec (1110) \prec (1111) \, .$$

All the elements are now in a sequence. This construction defines an order relation. The graph of this relation is a *chain*: each element $\mathbf{x} \neq \mathbf{0}$, $\mathbf{x} \neq \mathbf{1}$ has exactly one predecessor and exactly one successor. $\mathbf{0}$ has exactly one successor and no predecessor, and $\mathbf{1}$ has exactly one predecessor and no successor.

The vectors are checked from the left to the right. All vectors with 0 in the left position ($i = n - 1$) are smaller than the vectors with 1 in the left position. When the values in the left position are equal to each other, then the next position to the right will be checked, etc.

This principle is used to sort the words in a dictionary (based on

$$a \prec b \prec \cdots \prec x \prec y \prec z$$

and appropriate consideration of the different lengths of words, which is not necessary for $\mathbb{B}^n$); hence, the name is *lexicographic order*.

Table 3.13 Largest decimal numbers that can be represented with n bits	n	$2^n - 1$	n	$2^n - 1$	n	$2^n - 1$
	1	1	11	2.047	21	2.097.151
	2	3	12	4.095	22	4.194.303
	3	7	13	8.191	23	8.388.607
	4	15	14	16.383	24	16.777.215
	5	31	15	32.767	25	33.554.431
	6	63	16	65.535	26	67.108.863
	7	127	17	131.071	27	134.217.727
	8	255	18	262.143	28	268.435.455
	9	511	19	524.287	29	536.870.911
	10	1.023	20	1.048.575	30	1.073.741.823

A second understanding of this order relation can be found when we use the order of the natural numbers $\{0, 1, \ldots, 2^n - 1\}$:

$$0 < 1 < 2 < \ldots < 2^n - 2 < 2^n - 1 \,.$$

For each binary vector $\mathbf{x} = (x_{n-1}, \ldots, x_1, x_0)$, we define the decimal equivalent $\mathrm{dec}(\mathbf{x})$ by

$$\mathrm{dec}(\mathbf{x}) = x_{n-1} * 2^{n-1} + x_{n-2} * 2^{n-2} + \ldots + x_1 * 2^1 + x_0 * 2^0$$

$$= \sum_{i=0}^{n-1} x_i * 2^i \,. \tag{3.41}$$

This means that the vector $\mathbf{x} = (x_{n-1}, \ldots, x_1, x_0)$ will be understood as a *binary number* (the representation of a natural number using the base 2) and transformed into the usual decimal representation. This system of binary numbers goes back to *G. W. Leibniz*.

The order relation $\prec$ can be defined in the following way:

$$\mathbf{x} \prec \mathbf{y} \quad \equiv \quad \mathrm{dec}(\mathbf{x}) < \mathrm{dec}(\mathbf{y}) \,. \tag{3.42}$$

In a strict algebraic way, this approach needs some attention because in $\mathbb{B} = \{0, 1\}$ nothing is said about digits and multiplication. However, it is quite feasible "to forget" for a moment the algebraic structure of $\mathbb{B}$ and use the values 0 and 1 as digits in the sense of arithmetic. Hence, in the example, the vectors $\mathbf{x} = (0000), \ldots, (1111)$ correspond to the numbers $0, \ldots, 15$ and are ordered according to their decimal equivalent $\mathrm{dec}(\mathbf{x})$, see Table 3.12. n positions (*bits*) facilitate the representation of the numbers between 0 and $2^n - 1$. If a number k to be represented is larger than $2^n - 1$ (see Table 3.13), then more bits have to be used.

We can find the necessary condition for the number of bits to be used (see Table 3.14):

$$k \le 2^n - 1 \,,$$

$$(k + 1) \le 2^n \,,$$

$$\mathrm{ld}(k + 1) \le n \,, \qquad\qquad \text{ld is the logarithm at base 2}\,,$$

$$3.322 \cdot \lg(k + 1) \le n \,, \qquad\qquad \text{lg is the logarithm at base 10}\,.$$

It will be seen later that the interpretation of binary vectors as representations of numbers is one of the most important applications of these vectors.

Table 3.14 Number of bits required

$k+1$	$lg(k+1)$	$ld(k+1)$	Number of bits
10	1	3.322	4
10^2	2	6.644	7
10^3	3	9.966	10
10^4	4	13.288	14
10^5	5	16.610	17
10^6	6	19.932	20
10^7	7	23.253	24
10^8	8	26.275	27
10^9	9	29.897	30
10^{10}	10	33.219	34

Table 3.15 The elements of $\mathbb{B}^n$ enumerated by the Gray code

x_3	x_2	x_1	x_0	$\text{dec}(\mathbf{x})$	x_3	x_2	x_1	x_0	$\text{dec}(\mathbf{x})$
0	0	0	0	0	1	1	0	0	12
0	0	0	1	1	1	1	0	1	13
0	0	1	1	3	1	1	1	1	15
0	0	1	0	2	1	1	1	0	14
0	1	1	0	6	1	0	1	0	10
0	1	1	1	7	1	0	1	1	11
0	1	0	1	5	1	0	0	1	9
0	1	0	0	4	1	0	0	0	8

The second order relation over $\mathbb{B}^n$ is known as the *Gray code*. The same table with n columns and 2^n rows will be used, see Table 3.15.

We start in the rightmost column with 0 in the first and 1 in the second row, for the next two rows; however, the mirrored sequence 10 is used, i.e., we use 0110 down to the end of the table 2^{n-2} times. The other columns follow the same principle after the duplication of the elements:

$$0011 \mid 1100, 00001111 \mid 11110000,$$

etc., as long as necessary. This results again in 2^n different binary vectors, and the order of these vectors is defined according to their appearance in Table 3.15.

This enumeration of the elements of $\mathbb{B}^n$ has a very remarkable property:

> In order to get the next vector, exactly the value of *one bit* of the vector has to change. The last vector (1000), in the example, will be transformed into the first also by one bit change.

The transformation can also be done in the following way (using an n-bit vector of the binary code): let $(b_{n-1}, \ldots, b_1, b_0)$ be a vector in the binary code and $(g_{n-1}, \ldots, g_1, g_0)$ a vector in the *Gray code*; then

$$g_{n-1} = b_{n-1}, \quad g_i = b_{i+1} \oplus b_i, \quad i = 0, \ldots, n-2. \tag{3.43}$$

The leftmost bit is the same in the two representations. For the next vector components, two neighbored components of the dual vector are combined by antivalence in order to get the next bit of the *Gray code* representation, and this has to be done for the right $n-1$ bits (Table 3.16).

The change of only one bit indicates that there is a very short "distance" from one vector to the other. In order to formalize this understanding, we introduce the concept of a *metric space*.

Table 3.16 Ordered binary vectors: **a** binary code and **b** Gray code

	b_2	b_1	b_0		g_2	g_1	g_0
	0	0	0		0	0	0
	0	0	1		0	0	1
	0	1	0		0	1	1
a	0	1	1	**b**	0	1	0
	1	0	0		1	1	0
	1	0	1		1	1	1
	1	1	0		1	0	1
	1	1	1		1	0	0

Definition 3.8 A function d from a set $S \times S$ into the set R of real numbers is a *metric* if the following axioms are satisfied $(x, y, z \in S)$:

1. $d(x, y) \geq 0$, $d(x, x) = 0$;   if $d(x, y) = 0$, then $x = y$.
2. $d(x, y) = d(y, x)$.
3. $d(x, z) \leq d(x, y) + d(y, z)$.

Note 3.9 The last axiom is the *triangular inequality*: the "direct way" between x and z is supposed to be the shortest connection, and any other point in between extends the distance. When these axioms are satisfied, then (S, d) is a *metric space*. In order to define $\mathbb{B}^n$ as a metric space, we introduce the *norm* of $\mathbf{x}$ as the number of values of 1 in the vector.

Definition 3.9 For $\mathbf{x} \in \mathbb{B}^n$, $\|\mathbf{x}\| = \sum_{i=1}^{n} x_i$ is the *norm* of the vector $\mathbf{x}$.

Again we use the elements of $\mathbb{B}$ as numbers. By counting the components, we determine the number of components with a value of 1. It follows immediately that, for any $\mathbf{x} \in \mathbb{B}^n$,

$$0 \leq \|\mathbf{x}\| \leq n . \tag{3.44}$$

There are

$$\binom{n}{0} \qquad \text{vectors with} \qquad \|\mathbf{x}\| = 0 ,$$

$$\binom{n}{1} \qquad \text{vectors with} \qquad \|\mathbf{x}\| = 1 ,$$

$$\binom{n}{2} \qquad \text{vectors with} \qquad \|\mathbf{x}\| = 2 ,$$

..., and finally,

$$\binom{n}{n} \qquad \text{vectors with} \qquad \|\mathbf{x}\| = n .$$

$$\binom{n}{k} = \frac{n!}{k!(n-k)!}$$

is the number of possibilities to select k positions for the value 1, $n - k$ positions for the value 0, with $n! = 1 * 2 * \ldots * (n-1) * n$. Since every vector must have a given norm between 0 and n, the following identity holds:

Table 3.17 The triangular inequality for one component

x_i	y_i	z_i	$x_i \oplus y_i$	$y_i \oplus z_i$	$x_i \oplus z_i$	$(x_i \oplus y_i) + (y_i \oplus z_i)$
0	0	0	0	0	0	0
0	0	1	0	1	1	1
0	1	0	1	1	0	2
0	1	1	1	0	1	1
1	0	0	1	0	1	1
1	0	1	1	1	0	2
1	1	0	0	1	1	1
1	1	1	0	0	0	0

Table 3.18 Arrangement of vectors according to the Hamming metric $h(\mathbf{x}, \mathbf{y})$

$h = 0$		$h = 1$		$h = 2$		$h = 3$		$h = 4$
				(0011)				
		(0001)		(0101)		(0111)		
(0000)	$\Rightarrow$	(0010)	$\Rightarrow$	(1001)	$\Rightarrow$	(1011)	$\Rightarrow$	(1111)
		(0100)		(0110)		(1101)		
		(1000)		(1010)		(1110)		
				(1100)				
$\binom{4}{0} = 1$		$\binom{4}{1} = 4$		$\binom{4}{2} = 6$		$\binom{4}{3} = 4$		$\binom{4}{4} = 1$

$$\binom{n}{0} + \binom{n}{1} + \cdots + \binom{n}{n} = 2^n . \tag{3.45}$$

The number of vectors with a "small" or "large" norm is small, and most of the vectors are "in the middle." This norm now allows the introduction of a metric in $\mathbb{B}^n$ (mostly known as *Hamming* metric).

Theorem 3.13 *The function* $h(\mathbf{x}, \mathbf{x}_0) = \|\mathbf{x} \oplus \mathbf{x}_0\|$ *is a metric in* $\mathbb{B}^n$.

Proof This function counts the number of positions where $\mathbf{x}$ differs from $\mathbf{x}_0$. The first two axioms are satisfied immediately by using the appropriate properties of the antivalence. The most interesting case is the *triangular inequality*. In order to prove this inequality, we show that the inequality already holds for each component, and the summation of the component values will give the final result (see Table 3.17).   □

We use $n = 4$ as an example and show in Table 3.18 the distribution of binary vectors according to their distance from a given vector.

The existence of a metric offers numerous possibilities to introduce concepts that are well-known in geometry. We explain some of them as an illustration and define others when they are used in some applications.

Definition 3.10 For $\mathbf{x}, \mathbf{x}_0 \in \mathbb{B}^n$ and the metric $h(\mathbf{x}, \mathbf{x}_0)$:

1. $\mathbf{x}$ is a *neighbor* of $\mathbf{x}_0$, and vice versa, if $h(\mathbf{x}, \mathbf{x}_0) = 1$.
2. $\mathbf{x}$ is *opposite* to $\mathbf{x}_0$, and vice versa, if $h(\mathbf{x}, \mathbf{x}_0) = n$. Then $\bar{\mathbf{x}} = \mathbf{x}_0$, $\bar{\mathbf{x}}_0 = \mathbf{x}$.

The function $h(\mathbf{x}, \mathbf{x}_0)$ is not the only metric in $\mathbb{B}^n$; another metric considers a "directed equality" of two vectors $\mathbf{x}$ and $\mathbf{x}_0$ and has been explored in [7, 8].

The following inequalities hold for $h(\mathbf{x}, \mathbf{x}_0)$:

$$0 \leq h(\mathbf{x}, \mathbf{x}_0) \leq n \; ; \qquad (3.46)$$

hence, $h(\mathbf{x}, \mathbf{y})$ will only have the values $0, 1, 2, \ldots, n$. Taking these values as radius r, we can define two types of spheres:

Definition 3.11 For a metric space with the metric function $h(\mathbf{x}, \mathbf{x}_0)$, the set

$$K_r^o = \{\mathbf{x} | h(\mathbf{x}, \mathbf{x}_0) < r\} \qquad (3.47)$$

is an *open sphere*, and

$$K_r^c = \{h(\mathbf{x}, \mathbf{x}_0) \leq r\} \qquad (3.48)$$

is a *closed sphere*, both with radius r and center $\mathbf{x}_0$.

We use $\mathbb{B}^3$ and find for the *Hamming metric* $h(\mathbf{x}, \mathbf{x}_0)$ and the center $\mathbf{x}_0 = (000)$:

$$K_0^o = \emptyset \, ,$$
$$K_1^o = \{(000)\} \qquad\qquad = \quad K_0^c \, ,$$
$$K_2^o = \{(001), (010), (100)\} \cup K_1^o \quad = \quad K_1^c \, ,$$
$$K_3^o = \{(011), (101), (110)\} \cup K_2^o \quad = \quad K_2^c \, ,$$
$$K_4^o = \{(111)\} \cup K_3^o \qquad\qquad = \quad K_3^c \quad = \quad \mathbb{B}^3 \, .$$

It is easy to see that for any n:

$$K_0^o = \emptyset \, ,$$
$$K_{r+1}^o = K_r^o \cup \{\mathbf{x} \mid h(\mathbf{x}, \mathbf{x}_0) = r\}, \; r = 0, \ldots, n \, ,$$
$$K_{r+1}^o = K_r^c, \; r = 0, \ldots, n \, .$$

The Hamming metric $h(\mathbf{x}, \mathbf{x}_0)$ specifies for a fixed radius r a shell:

Definition 3.12 The set

$$S_r = \{\mathbf{x} | h(\mathbf{x}, \mathbf{x}_0) = r\} \qquad (3.49)$$

is the *shell* of the closed sphere with the radius r and contains $\binom{n}{r}$ vectors; the number of components of $\mathbf{x}$ that differ from $\mathbf{x}_0$ is equal to r, $0 \leq r \leq n$ for $\mathbb{B}^n$.

Most of the elements of $\mathbb{B}^n$ belong to the shells in the middle ($i \approx \frac{n}{2}$), the first shell S_0 has only the center point $\mathbf{x}_0$ as an element, and S_n contains only $\mathbf{x} = \bar{\mathbf{x}}_0$ (see Table 3.18) because all the components of $\mathbf{x}$ must be different from the components of $\mathbf{x}_0$.

Due to the relations $<$ in (3.47), $\leq$ in (3.48), and $=$ in (3.49), we get for $\mathbb{B}^3$ and the center point $\mathbf{x}_0 = (000)$:

- $S_0 = K_0^c \setminus K_0^o = \{(000)\}$;
  there is no difference.
- $S_1 = K_1^c \setminus K_1^o = \{(001), (010), (100)\}$;
  there is a difference in one position in comparison to the used center point.
- $S_2 = K_2^c \setminus K_2^o = \{(011), (101), (110)\}$;
  there is a difference in two positions in comparison to the used center point.

- $S_3 = K_3^c \setminus K_3^o = \{(111)\}$;
  the shell S_3 differs in all three positions of the center point $\mathbf{x}_0 = (000)$.

It can be seen that the shell S_r extends the open sphere K_r^o to the closed sphere K_r^c. Generalized for an arbitrary Boolean space $\mathbb{B}^n$, it holds:

$$K_0^c = \{\mathbf{x}_0\} = S_0 \,,$$

$$K_r^c = K_r^o \cup \{\mathbf{x} \mid h(\mathbf{x}, \mathbf{x}_0) = r\} = K_r^o \cup S_r \,, \quad r = 0, \ldots, n \,.$$

3.5   Exercises

Prepare for each task of each exercise of this chapter a PRP for the XBOOLE-monitor XBM 2 using a convenient order of the variables. Use the help system of the XBOOLE-monitor XBM 2 to learn the details of the needed commands. Execute the created PRPs and verify the computed results.

If you have not yet prepared the XBOOLE-monitor XBM 2 on your computer, you can get this XBOOLE-monitor free of charge by means of the following three steps:

1. **Download**:
   There are four versions of the XBOOLE-monitor XBM 2, two for Windows 10 or subsequent Windows systems (32 or 64 bits) and two for LINUX - Ubuntu (also 32 or 64 bits); you must download the version of the XBOOLE-monitor XBM 2 that fits to your operating system. Authorized users of the online version of this chapter (https://doi.org/10.1007/978-3-030-88945-6_3) can download the XBOOLE-monitor XBM 2 directly from the web page

   https://link.springer.com/chapter/10.1007/978-3-030-88945-6_3

   where the links for the download of the XBOOLE-monitor XBM 2 are located in the part "Supplementary Information" (below the part "Abstract"). The headline above such a link indicates the associated zip-file of the XBOOLE-monitor XBM 2. The sizes of the zip-files have been provided behind the links and can be used to verify the download. A click on the link of the wanted version of the XBOOLE-monitor XBM 2 starts the download.

   Readers of the hardcopy of this book get access to the XBOOLE-monitor XBM 2 using the URL

   https://link.springer.com/chapter/10.1007/978-3-030-88945-6_3

   to download the first two pages of this chapter. After this download, the same procedure as the authorized users of the online version of a chapter can be used to download the wanted version of the XBOOLE-monitor XBM 2.

2. **Unzip**: The XBOOLE-monitor XBM 2 must not be installed but unzipped into an arbitrary directory of your computer. A convenient tool for unzip the downloaded zip-file is usually available as part of the operating system or can be downloaded from the Internet.

3. **Execute**:

   - Windows:
     The executable file of the two versions (32 or 64 bits) for Windows 10 (or subsequent Windows systems) of the XBOOLE-monitor XBM 2 is XBM2.exe; the other files in the expanded directory must remain unchanged. A double-click on the executable file XBM2.exe within the Explorer of Windows starts the XBOOLE-monitor XBM 2.

- LINUX - Ubuntu:
  The unzipped folder of the XBOOLE-monitor XBM 2 contains for this operating system only the executable file `XBM2-i386.AppImage` for the version of 32 bits or `XBM2-x86_64.AppImage` for the version of 64 bits of the XBOOLE-monitor XBM 2. A double-click on the created AppImage-file within the file manager of LINUX - Ubuntu starts the XBOOLE-monitor XBM 2.

Exercise 3.1 (Relations in $\mathbb{B}$) The definitions of the relation $\leq$ and the operations $\wedge$, $\vee$, and $\Rightarrow$ provide the needed information to solve the following problems using appropriate tables. These tables can be created in PRPs as TVLs, where dashes are not needed in this simple example. The intersection of the two relation tables results in a new relation table, where the two given relations are satisfied; hence, we can use the command `isc` to combine the TVLs of the relation tables.

Prepare for each of the following tasks a PRP for the XBOOLE-monitor XBM 2 that specifies the elementary relations and operations as TVLs, computes the relation to verify, and shows the table of the complete relation. Verify the computed relation tables:

(a) Verify that $xy \leq x$.
(b) Verify if $x_1 \leq y_1$ and $x_2 \leq y_2$, then $x_1 x_2 \leq y_1 y_2$.
(c) Verify if $x \leq y$ or $x \leq z$, then $x \leq (y \vee z)$.
(d) Verify that $x \leq yz$ is equivalent to $(x \leq y)$ and $(x \leq z)$.

Exercise 3.2 (Binary Vectors)
There are several approaches to solve the following four tasks using the XBOOLE-monitor XBM 2. Both a single ternary vector and a set of binary vectors can be represented by a TVL; control variables of the type integer are convenient to represent the decimal equivalent or the number of binary vectors of a set. Use the help system of the XBOOLE-monitor XBM 2 to find the details about the appropriate commands. Besides the commands for control variables and the input of TVLs and VTs, the commands `cel`, `cco`, `dif`, and `te_isc` can be usefully utilized.

Prepare for each of the following tasks a PRP for the XBOOLE-monitor XBM 2. Execute the created PRPs and verify the computed results:

(a) Find the decimal equivalent $dec(\mathbf{x})$ for the vector $(110101) \in \mathbb{B}^6$.
(b) Find the vector $\mathbf{x} \in \mathbb{B}^5$ with $dec(\mathbf{x}) = 23$.
(c) Find the binary vectors $\mathbf{x}$ with $2^{n-1} \leq dec(\mathbf{x}) < 2^n$ in $\mathbb{B}^n$ for $n = 10$. How many such binary vectors exist?
(d) Let be given the vector $\mathbf{x} = (10010101) \in \mathbb{B}^8$.

  - Find all $\mathbf{y} \in \mathbb{B}^8$ with $\mathbf{y} \leq \mathbf{x}$.
  - Find all $\mathbf{z} \in \mathbb{B}^8$ with $\mathbf{z} \geq \mathbf{x}$.

Represent the sets of binary vectors $\{\mathbf{y}\}$ and $\{\mathbf{z}\}$ as ternary vectors.

Exercise 3.3 (Binary Vectors and Sets of Variables)
Each ternary matrix (TVL or VT) contains the implicitly stored existence vector (VV) that describes the set of variables determined by the variables that occur in this TVL or VT; hence, each ternary matrix can be used as argument of operations of sets of variables. The result of a set operation of sets of variables is an empty TVL (no row) where the columns (determined by VV) describe the computed set of variables.

The command `stv` selects a ternary vector of a TVL and stores this vector such that in the created TVL only values 0 and 1 occur; hence, the commands `stv` and `cel` can together be used to determine subsets of variables specified by either the values 0 or the values 1 of a binary vector.

The command `sv_get` maps the existence vector of a given ternary matrix (TM) to the single row of the resulting TVL such that 1-elements of this row indicate the set of variables of the given TM; hence, this command can be used to construct a binary vector based on a given set of variables.

The command `sv_size` computes the number of variables of the set of variables of a given TM; hence, the norm of a binary vector (Definition 3.9) can be computed using the set of variables corresponding to the values 1 of the binary vector.

Prepare for each of the following tasks a PRP for the XBOOLE-monitor XBM 2 using a lexicographic order of the variables x_0 to x_9. Use the help system of the XBOOLE-monitor XBM 2 to learn the details of the needed commands. Execute the created PRPs and verify the computed results:

(a) Compute the set of variables, represented as an empty TVL, for which $x_i = 1$ in the given binary vector $(x_0, x_1, x_2, x_3, x_4, x_5, x_6, x_7, x_8, x_9) = (1100101101)$.

(b) Create the TVL of one binary vector whose values 1 are determined by the set of variables $\{x_2, x_3, x_4, x_6, x_7, x_9\}$ and whose values 0 are determined by the set of variables $\{x_0, x_1, x_5, x_8\}$.

(c) Verify that $\|\bar{\mathbf{x}}\| = \|\mathbf{x} \vee \bar{\mathbf{x}}\| - \|\mathbf{x}\|$ for $\mathbf{x} = (x_0, x_1, x_2, x_3, x_4, x_5, x_6, x_7, x_8, x_9) = (1101101001)$.

(d) Verify that $\|\mathbf{x}_1 \vee \mathbf{x}_2\| = \|\mathbf{x}_1\| + \|\mathbf{x}_2\| - \|\mathbf{x}_1\mathbf{x}_2\|$, where $\mathbf{x}_1$ is the binary vector of values 1 for $x_i \in \{x_2, x_3, x_4, x_5, x_7, x_9\}$ and $\mathbf{x}_2$ is the binary vector of values 1 for $x_i \in \{x_0, x_1, x_4, x_5\}$; both $\mathbf{x}_1$ and $\mathbf{x}_2$ depend on ten variables x_0 to x_9.

Exercise 3.4 (Shells and Spheres)

Each Boolean space $\mathbb{B}^n$ can be structured into shells $S_r(\mathbf{x}, \mathbf{x}_0)$ and closed spheres $K_r^c(\mathbf{x}, \mathbf{x}_0) = \bigcup_{i=0}^r S_r(\mathbf{x}, \mathbf{x}_0)$, where both $\mathbf{x} \in \mathbb{B}^n$ and $\mathbf{x}_0 \in \mathbb{B}^n$.

Prepare for each of the following tasks a PRP for the XBOOLE-monitor XBM 2 using a lexicographic order of the variables x_0 to x_8. Use the help system of the XBOOLE-monitor XBM 2 to learn the details of the needed commands. Execute the created PRPs and verify the computed results:

(a) Compute all shells $S_r(\mathbf{x}, \mathbf{x}_0 = \mathbf{0}) = \{\mathbf{x} | h(\mathbf{x}, \mathbf{x}_0) = r\}$ for the Boolean spaces $\mathbb{B}^n$ with $n = 1, \dots, 9$ using a recursive procedure and the initial shells $S_0^1(x_0, 0) = \{(0)\}$ and $S_1^1(x_0, 0) = \{(1)\}$ of $\mathbb{B}^1$. Store each computed shell S_r using the object number $k = 10 \cdot n + r$ and assign it to viewport (r, n) of the m-fold view. Save the XBOOLE-system using the name "shell_1_9.sdt" for later use. How much binary vectors belong to the shell $S_i(\mathbf{x}, \mathbf{x}_0 = \mathbf{0})$?

(b) Load the `sdt`-file "shell_1_9.sdt," saved at the end of Exercise 3.4 (a), and compute all closed spheres $K_r^c(\mathbf{x}, \mathbf{x}_0 = \mathbf{0})$ for Boolean spaces $\mathbb{B}^n$ with $n = 1, \dots, 9$. Minimize the computed closed spheres using the command `obbc`. Store the computed closed spheres $K_r^c(\mathbf{x}, \mathbf{x}_0 = \mathbf{0})$ using the object numbers $k = 10 \cdot n + r$ so that the shells $S_r(\mathbf{x}, \mathbf{x}_0 = \mathbf{0})$ are replaced by the closed spheres $K_r^c(\mathbf{x}, \mathbf{x}_0 = \mathbf{0})$ and assign them to viewports (r, n) of the m-fold view.

(c) Reuse the `sdt`-file "shell_1_9.sdt," saved at the end of Exercise 3.4 (a), and compute all shells $S_r(\mathbf{x}, \mathbf{x}_0 = (101100110))$ for Boolean spaces $\mathbb{B}^n$ with $n = 1, \dots, 9$. Store each computed shell $S_r(\mathbf{x}, \mathbf{x}_0 = (101100110))$ with the changed center using the object number $k = 10 \cdot n + r$ and assign it to viewport (r, n) of the m-fold view. Does the changed center of the shells change the numbers of binary vectors belonging to the shells?

(d) Load the `sdt`-file "shell_1_9.sdt," saved at the end of Exercise 3.4 (a) because certain shells are useful to split the set of binary vectors computed in this task. Compute the neighbors of $\mathbf{x} = (x_0, x_1, x_2, x_3, x_4, x_5, x_6, x_7, x_8) = (101101010)$ and split these neighbors according to their belonging to a shell $S_r(\mathbf{x}, \mathbf{x}_0 = \mathbf{0})$. Use the commands `while` and `sv_next` to select successively each variable of $\mathbf{x}$ needed to compute one of the neighbors. Show the computed vectors with a smaller, respectively, larger, norm besides the given vector in the m-fold view. What is the relation between the values 0 and 1 of the given binary vector and the numbers of binary vectors in the solution-sets?

1	new	7	100	13	100
2	space 32 1	8	111.	14	111.
3	tin 1 1	9	tin 1 2	15	isc 1 2 3
4	x y axy.	10	axy x axylex.	16	assign 1 /m 1 1
5	000	11	001	17	assign 2 /m 1 2
6	010	12	011	18	assign 3 /m 1 3

Fig. 3.9 Problem-program that computes the relation $xy \leq x$

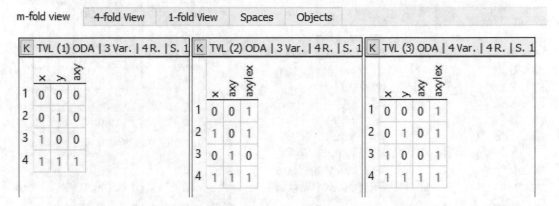

Fig. 3.10 Elementary and complete relation tables of the relation $xy \leq x$

3.6 Solutions

Solution 3.1 (Relations in $\mathbb{B}$)

(a) Figure 3.9 shows the PRP that computes the complete relation table for the relation $xy \leq x$. Figure 3.10 shows the defined relation tables of the conjunction (TVL 1) and the relation $\leq$ (TVL 2) in the left two viewports of the m-fold view. The right viewport of this figure shows the computed complete relation table for this task, where in the column $axylex$ occur only values 1; this confirms that the evaluated relation is valid for all pairs of variables (x, y). This figure shows that the relation $x \leq axy$ (shown in the viewport in the middle) is not satisfied for $x = 0$ and $axy = 1$; however, such a combination does not occur in the operation $axy = x \wedge y$ (shown in the left viewport).

(b) Figure 3.11 shows on the left-hand side the PRP that computes the complete relation table for the relation $if((x_1 \leq y_1) \wedge (x_2 \leq y_2)), then(x_1x_2 \leq y_1y_2)$ and on the right-hand side the computed relation table. Elementary relation tables are prepared for the three relations $\leq$ (TVLs 1, 2, and 6), the three operations $\wedge$ (TVLs 3, 4, and 5), and the implication (TVL 7). The intersections of these TVLs can be executed in an arbitrary order; the intermediate results of the assumptions are stored as TVLs 10 and 11 and of the conclusions as TVLs 20 and 21 in the PPR of Fig. 3.11. TVL 30 combines the assumption and the conclusion, and the final TVL 31 applies the implication between these relations. In the column $result$ of the computed relation table occur only values 1; this confirms that the evaluated relation is valid for all 16 combinations of the values 0 and 1 of the variables (x_1, x_2, y_1, y_2).

(c) Figure 3.12 shows both the PRP that computes the complete relation table for the relation $((x \leq y) \vee (x \leq z)) \Rightarrow (x \leq (y \vee z))$ and the computed relation table. Elementary relation tables are prepared for the three relations $\leq$ (TVLs 1, 2, and 5), the two operations $\vee$ (TVLs 3 and 4), and the implication (TVL 6). The intersections in lines 49–53 compute the relation table (TVL 31)

#	new / program		program
1	new	32	010
2	space 32 1	33	100
3	avar 1	34	111.
4	x1 x2 y1 y2	35	tin 1 5
5	x1ley1 x2ley2	36	y1 y2 ay1y2.
6	ax1x2 ay1y2	37	000
7	assumption	38	010
8	conclusion	39	100
9	result.	40	111.
10	tin 1 1	41	tin 1 6
11	x1 y1 x1ley1.	42	ax1x2 ay1y2
12	001	43	conclusion.
13	011	44	001
14	100	45	011
15	111.	46	100
16	tin 1 2	47	111.
17	x2 y2 x2ley2.	48	tin 1 7
18	001	49	assumption
19	011	50	conclusion
20	100	51	result.
21	111.	52	001
22	tin 1 3	53	011
23	x1ley1 x2ley2	54	100
24	assumption.	55	111.
25	000	56	isc 1 2 10
26	010	57	isc 10 3 11
27	100	58	isc 4 5 20
28	111.	59	isc 20 6 21
29	tin 1 4	60	isc 11 21 30
30	x1 x2 ax1x2.	61	isc 30 7 31
31	000	62	assign 31 /1

TVL (31) ODA | 11 Var. | 16 R. | S. 1

	$x1$	$x2$	$y1$	$y2$	x1ley1	x2ley2	ax1x2	ay1y2	assumption	conclusion	result
1	0	0	0	0	1	1	0	0	1	1	1
2	0	0	0	1	1	1	0	0	1	1	1
3	0	1	0	0	1	0	0	0	0	1	1
4	0	1	0	1	1	1	0	0	1	1	1
5	0	0	1	0	1	1	0	0	1	1	1
6	0	0	1	1	1	1	0	1	1	1	1
7	0	1	1	0	1	0	0	0	0	1	1
8	0	1	1	1	1	1	0	1	1	1	1
9	1	0	0	0	0	1	0	0	0	1	1
10	1	0	0	1	0	1	0	0	0	1	1
11	1	1	0	0	0	0	1	0	0	0	1
12	1	1	0	1	0	1	1	0	0	0	1
13	1	0	1	0	1	1	0	0	1	1	1
14	1	0	1	1	1	1	0	1	1	1	1
15	1	1	1	0	1	0	1	0	0	0	1
16	1	1	1	1	1	1	1	1	1	1	1

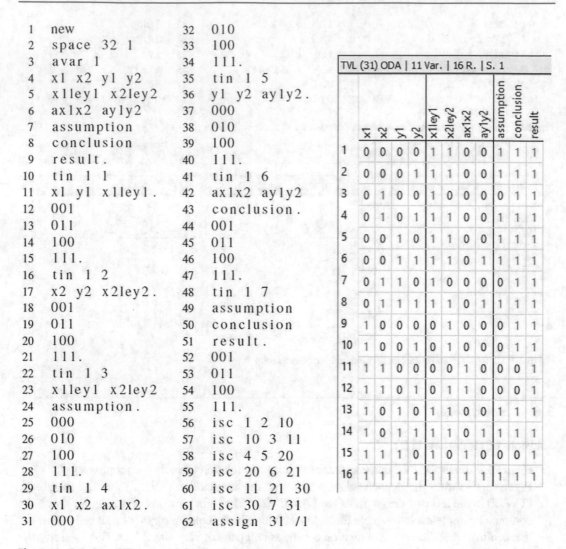

Fig. 3.11 Solution of Exercise 3.1 (b): Problem-program (on the left) and the computed complete relation table (on the right)

of the complete relation of Exercise 3.1 (c) that is shown down to the right in Fig. 3.12. In the column *result* of this computed relation table occur only values 1; this confirms that the evaluated relation is valid for all eight combinations of the values 0 and 1 of the variables (x, y, z).

(d) Figure 3.13 shows the PRP that computes the complete relation table for the relation $(x \leq yz) \equiv ((x \leq y) \wedge (x \leq z))$. Elementary relation tables are prepared for the three relations $\leq$ (TVLs 2, 3, and 4), the two operations $\wedge$ (TVLs 1 and 5), and the equivalence (TVL 6). The relation 1 of the left-hand side of the equivalence (TVL 10) is the intersection of the TVLs 1 and 4. The relation 2 of the right-hand side of the equivalence (TVL 11) is the intersection of TVLs 2, 3, and 5. The intersection of TVLs 10 and 11 merges these two relations into TVL 12, and the final intersection of TVL 12 with TVL 6 of the equivalence computes the relation table of the complete relation specified in Exercise 3.1 (d). Figure 3.14 shows the relation tables of both sides of the equivalence in the left two viewports and the complete relation table in the viewport of the right-hand side. In the column *solution* in viewport (1,3) of the m-fold view in Fig. 3.14 occur only values 1; this

1	new	25	011	49	isc 1 2 10
2	space 32 1	26	101	50	isc 10 3 11
3	avar 1	27	111.	51	isc 4 5 20
4	x y z xley xlez	28	tin 1 4	52	isc 11 20 30
5	oyz xleoyz	29	y z oyz.	53	isc 30 6 31
6	assumption	30	000	54	assign 31 /1
7	conclusion	31	011		
8	result.	32	101		
9	tin 1 1	33	111.		
10	x y xley.	34	tin 1 5		
11	001	35	x oyz		
12	011	36	conclusion.		
13	100	37	001		
14	111.	38	011		
15	tin 1 2	39	100		
16	x z xlez.	40	111.		
17	001	41	tin 1 6		
18	011	42	assumption		
19	100	43	conclusion		
20	111.	44	result.		
21	tin 1 3	45	001		
22	xley xlez	46	011		
23	assumption.	47	100		
24	000	48	111.		

K	\multicolumn TVL (31) ODA | 9 Var. | 8 R. | S. 1								
	x	y	z	xley	xlez	oyz	assumption	conclusion	result
1	0	0	0	1	1	0	1	1	1
2	0	0	1	1	1	1	1	1	1
3	0	1	0	1	1	1	1	1	1
4	0	1	1	1	1	1	1	1	1
5	1	0	0	0	0	0	0	0	1
6	1	0	1	0	1	1	1	1	1
7	1	1	0	1	0	1	1	1	1
8	1	1	1	1	1	1	1	1	1

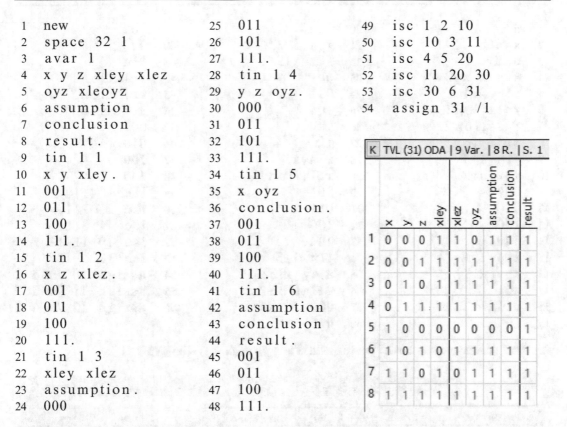

Fig. 3.12 Solution of Exercise 3.1 (c): Problem-program (on the left) and the computed complete relation table (down on the right)

confirms that the evaluated relation is valid for all eight combinations of the values 0 and 1 of the variables (x, y, z).

Solution 3.2 (Binary Vectors)

(a) Figure 3.15 shows the PRP that computes the decimal equivalent dec($\mathbf{x}$) for the vector $(110101) \in \mathbb{B}^6$. This PRP uses Definition (3.41) to compute the decimal equivalent. The command set in line 13 initializes the result value $dec with the value 0 and the command set in line 14 the value $v of the most right bit with $2^0 = 1$. Each sweep of the for-loop in lines 15–23 evaluates one bit of the given binary vector (TVL 1 defined in lines 3–5) starting with x_0 defined as TVL 2 in lines 6–8. The condition of the command if in line 17 is satisfied when the evaluated bit of the binary vector is equal to 1; in this case the appropriate value $v is added to the result variable $dec in line 19. The command cco in line 21 rotates, controlled by VTs 3 and 4, the columns of TVL 1 one position to the right and the commands in line 22 doubles the value of $v; hence, in the next sweep the next higher bit of the binary vector can be evaluated using the unchanged TVL 2.

Figure 3.16 shows the view "Values of the control variables" after the execution of the PRP of Fig. 3.15. Associated to the control variable $dec, the computed decimal equivalent 53 belonging

1	new	20	1 1 1 .	39	1 0 0
2	space 32 1	21	tin 1 3	40	1 1 1 .
3	avar 1	22	x z xlez .	41	tin 1 6
4	x y z ayz	23	0 0 1	42	relation1
5	xley xlez	24	0 1 1	43	relation2
6	relation1	25	1 0 0	44	solution .
7	relation2	26	1 1 1 .	45	0 0 1
8	solution .	27	tin 1 4	46	0 1 0
9	tin 1 1	28	x ayz	47	1 0 0
10	y z ayz .	29	relation1 .	48	1 1 1 .
11	0 0 0	30	0 0 1	49	isc 1 4 10
12	0 1 0	31	0 1 1	50	isc 2 3 11
13	1 0 0	32	1 0 0	51	isc 11 5 11
14	1 1 1 .	33	1 1 1 .	52	isc 10 11 12
15	tin 1 2	34	tin 1 5	53	isc 12 6 12
16	x y xley .	35	xley xlez	54	assign 10 /m 1 1
17	0 0 1	36	relation2 .	55	assign 11 /m 1 2
18	0 1 1	37	0 0 0	56	assign 12 /m 1 3
19	1 0 0	38	0 1 0		

Fig. 3.13 Problem-program that computes the relation $(x \leq yz) \equiv ((x \leq y) \land (x \leq z))$

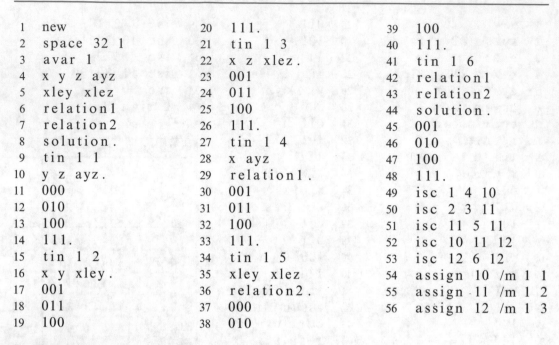

Fig. 3.14 Relation tables of the relations 1 and 2 that are equivalent to each and the complete relation table of the relation $(x \leq yz) \equiv ((x \leq y) \land (x \leq z))$

to the evaluated binary vector (110101) is shown. The values of the control variables $i and $v have been already adjusted at the end of the for-loop for the next (not yet executed) sweep.

(b) Figure 3.17 shows both the PRP that computes the binary vector $\mathbf{x} \in \mathbb{B}^5$ of the given decimal equivalent $\text{dec}(\mathbf{x}) = 23$ and the computed binary vector $\mathbf{x} = (10111)$ as TVL 1. The method

```
1    new                              13   set $dec 0
2    space 32 1                       14   set $v 1
3    tin 1 1                          15   for $i 0 5
4    x5 x4 x3 x2 x1 x0.               16   (
5    110101.                          17   if (not (te_isc 1 2))
6    tin 1 2                          18   (
7    x0.                              19   set $dec (add $dec $v)
8    1.                               20   )
9    vtin 1 3                         21   cco 1 3 4 1
10   x1 x2 x3 x4 x5.                  22   set $v (mul $v 2)
11   vtin 1 4                         23   )
12   x0 x1 x2 x3 x4.
```

Fig. 3.15 Problem-program that computes the decimal equivalent dec(110101)

Fig. 3.16 Computed
decimal equivalent $dec
equal to 53 of the given
binary vector (110101)

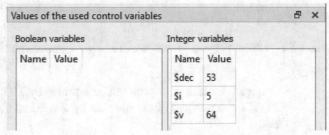

Fig. 3.17 Solution of
Exercise 3.2 (b):
Problem-program and the
computed binary vector

```
1    new                              20   if (eq 0 (mod $dec 2))
2    space 32 1                       21   (
3    set $dec 23                      22   cel 1 2 /-0
4    tin 1 1                          23   )
5    x4 x3 x2 x1 x0.                  24   set $dec (div $dec 2)
6    -----.                           25   cco 2 3 4 2
7    tin 1 2                          26   )
8    x0.                              27   assign 1 /4 1 1
9    1.
10   vtin 1 3
11   x4 x3 x2 x1.
12   vtin 1 4
13   x3 x2 x1 x0.
14   for $i 0 4
15   (
16   if (eq 1 (mod $dec 2))
17   (
18   cel 1 2 /-1
19   )
```

used to convert the given integer into a binary vector consists in changing the five dashes of the
prepared TVL 1 of $\mathbb{B}^5$ into the binary values 0 or 1 depending on the value of the remainder of
the division of the given integer by 2. The for-loop in lines 14–26:

- Iterates over the five columns from x_0 to x_4.
- Changes the dash of the column determined by TVL 2 to the value 1 (command cel in line
  18) if the remainder of the division by 2 is equal to 1 (checked in the command if in line
  16).

```
 1   new                            19   x1 x2 x3 x4 x5 x6 x7 x8 x9.
 2   space 32 1                     20   vtin 1 7
 3   tin 1 1                        21   x0 x1 x2 x3 x4 x5 x6 x7 x8.
 4   x9 x8 x7 x6 x5                 22   set $nbv 1
 5   x4 x3 x2 x1 x0.                23   for $i 0 9
 6   -----------.                   24   (
 7   tin 1 2                        25   set $b0 (not (te_isc 3 4))
 8   x9 x8 x7 x6 x5                 26   set $b1 (not (te_isc 3 5))
 9   x4 x3 x2 x1 x0.                27   if (and $b0 $b1)
10   0----------.                   28   (
11   dif 1 2 3                      29   set $nbv (mul $nbv 2)
12   tin 1 4                        30   )
13   x0.                            31   cco 3 6 7 3
14   0.                             32   )
15   tin 1 5                        33   assign 1 /m 1 1
16   x0.                            34   assign 2 /m 2 1
17   1.                             35   assign 3 /m 3 1
18   vtin 1 6
```

Fig. 3.18 Problem-program that computes the binary vectors $\mathbf{x}$ with $2^9 \le \text{dec}(\mathbf{x}) < 2^{10}$ and their number

- Changes the dash of the column determined by TVL 2 to the value 0 (command cel in line 22) if the remainder of the division by 2 is equal to 0 (checked in the command if in line 20).
- Prepares the evaluation of the next higher bit using the division of given integer by 2 (commands set and div in line 24).
- Rotates the column of TVL 2 by one position to the left over the five columns of TVL 1 (command cco in line 25 controlled by the VTs 3 and 4) so that the next higher bit will be selected in the next sweep.

The computed binary vector $\mathbf{x} = (10111)$ of the given integer 23 is shown down on the right in Fig. 3.17.

(c) Figure 3.18 shows the PRP that computes first the binary vectors $\mathbf{x}$ with $2^{n-1} \le \text{dec}(\mathbf{x}) < 2^{10}$ and counts thereafter the number of these binary vectors.

All binary vectors $\mathbf{x} \in \mathbb{B}^{10}$ with $\text{dec}(\mathbf{x}) < 2^{10}$ are represented by a ternary vector of ten dashes (TVL 1). The binary vector of 2^9 is (1000000000); hence, all binary vectors $\mathbf{x} \in \mathbb{B}^{10}$ with $\text{dec}(\mathbf{x}) < 2^9$ can be represented by a ternary vector of a single 0 followed by nine dashes (TVL 2). The set difference of TVL 1 minus TVL 2 using the command dif in line 11 computes the wanted set of binary vectors $\mathbf{x}$ with $2^9 \le \text{dec}(\mathbf{x}) < 2^{10}$ (TVL 3).

The procedure in lines 22–32 evaluates the ternary solution-vector such that the initial number 1 is multiplied by 2 for each dash. The commands in lines 25–27 verify whether the column, determined by TVLs 4 and 5, contains a dash. The command mul in line 29 multiplies in this case the value $nbv with 2. The command cco in line 31 rotates the columns of TVL 3 so that each ternary element is evaluated once in lines 25 and 26.

Figure 3.19 shows on the left-hand side in the m-fold view the prepared and computed sets of binary vectors, each presented by a single ternary vector. The computed number of binary vectors of the ternary vector of TVL 3 is shown on the right-hand side of Fig. 3.19 as value $nbv = 512$; hence, the single ternary vector of TVL 3 represents 512 binary vectors.

(d) Figure 3.20 shows on the left-hand side the PRP that computes all $\mathbf{y} \in \mathbb{B}^8$ with $\mathbf{y} \le \mathbf{x}$ and all $\mathbf{z} \in \mathbb{B}^8$ with $\mathbf{z} \ge \mathbf{x}$ for $\mathbf{x} = (10010101)$; and on the right-hand side TVL 1 of the given binary vector, the computed set of binary vectors $\mathbf{y}$ as TVL 2, and the computed set of binary vectors $\mathbf{z}$ as TVL 3.

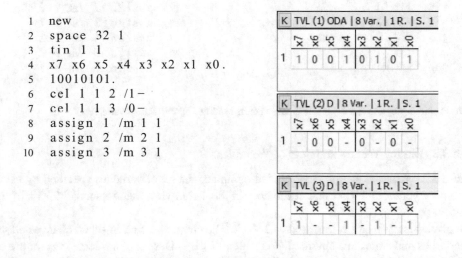

Fig. 3.19 Sets of binary vectors $\mathbf{x} \in \mathbb{B}^{10}$ with $\text{dec}(\mathbf{x}) < 2^{10}$, $\text{dec}(\mathbf{x}) < 2^9$, and $2^9 \leq \text{dec}(\mathbf{x}) < 2^{10}$ as well as the computed number \$nbv $= 512$ of vectors of the last set (TVL 3)

```
1   new
2   space 32 1
3   tin 1 1
4   x7 x6 x5 x4 x3 x2 x1 x0.
5   10010101.
6   cel 1 1 2 /1-
7   cel 1 1 3 /0-
8   assign 1 /m 1 1
9   assign 2 /m 2 1
10  assign 3 /m 3 1
```

Fig. 3.20 Solution of Exercise 3.2 (d): Problem-program and the computed sets of binary vectors

All vectors $\mathbf{y}$ that are smaller or equal to $\mathbf{x} = (10010101)$ can be built by replacing the values 1 of the binary vector $\mathbf{x}$ by dashes. Vise versa, all vectors $\mathbf{z}$ that are larger or equal to $\mathbf{x} = (10010101)$ can be built by replacing the values 0 of the binary vector $\mathbf{x}$ by dashes. The XBOOLE-monitor XBM 2 provides the command cel for the change of elements; this command is used in lines 6 and 7 to solve these tasks. The given binary vector $\mathbf{x}$ (TVL 1) is shown in viewport (1,1), and the two sets of binary vectors $\{\mathbf{y}\}$ (TVL 2) and $\{\mathbf{z}\}$ (TVL 3) are depicted in viewports (2,1) and (3,1) of the m-fold view.

```
1   new                                    7   stv  2 1 3
2   space 32 1                             8   dtv  3 1 4
3   tin 1 1                                9   assign 1 /4 1 1
4   x0 x1 x2 x3 x4 x5 x6 x7 x8 x9 .        10  assign 2 /4 2 1
5   1100101101 .                           11  assign 3 /4 1 2
6   cel 1 1 2 /0 -                         12  assign 4 /4 2 2
```

K	TVL (1) ODA	10 Var.	1 R.	S. 1

	x_0	x_1	x_2	x_3	x_4	x_5	x_6	x_7	x_8	x_9
1	1	1	0	0	1	0	1	1	0	1

K	TVL (3) ODA	6 Var.	1 R.	S. 1

	x_0	x_1	x_4	x_6	x_7	x_9
1	1	1	1	1	1	1

K	TVL (2) D	10 Var.	1 R.	S. 1

	x_0	x_1	x_2	x_3	x_4	x_5	x_6	x_7	x_8	x_9
1	1	1	-	-	1	-	1	1	-	1

K	TVL (4) ODA	6 Var.	0 R.	S. 1

	x_0	x_1	x_4	x_6	x_7	x_9

Fig. 3.21 Solution of Exercise 3.3 (a): Problem-program and the computed sets of variables

```
1   new                                    12  sv_get 1 3
2   space 32 1                             13  sv_get 2 4
3   avar 1                                 14  cel 4 4 5 /10 /01
4   x0 x1 x2 x3 x4                         15  isc 3 5 6
5   x5 x6 x7 x8 x9 .                       16  assign 1 /m 1 1
6   tin 1 1                                17  assign 2 /m 1 2
7   x2 x3 x4 x6 x7 x9 .                    18  assign 3 /m 2 1
8   .                                      19  assign 4 /m 2 2
9   tin 1 2                                20  assign 5 /m 3 2
10  x0 x1 x5 x8 .                          21  assign 6 /m 3 1
11  .
```

Fig. 3.22 Problem-program that computes binary vectors specified by two sets of variables

Solution 3.3 (Binary Vectors and Sets of Variables)

(a) Figure 3.21 shows at the top the PRP that computes the set of variables specified by values 1 in the binary vector of TVL 1 and at the bottom the 4-fold view that depicts all TVLs of the used solution-procedure.

The given binary vector is prepared as TVL 1. The command cel in line 6 changes all values 0 into dashes and stores this intermediate result as TVL 2. Dashes of a selected vector are avoided in the result of the command stv in line 7. TVL 3 specifies already the wanted set of variables. The command dtv in line 8 deletes the single binary vector of TVL 3. The comparison of TVLs 1 and 4 in the 4-fold view confirms that TVL 4 describes the set of variables specified by values 1 in the binary vector of TVL 1. The comparison of TVLs 3 and 4 confirms that the set of variable is not changed by the command dtv.

(b) Figure 3.22 shows the PRP that computes the binary vector $\mathbf{x}$ (TVL 6) specified by the set of variables for which $x_i = 1$ (TVL 1) and the set of variables for which $x_i = 0$ (TVL 2).

The lexicographic order of the ten variables is prepared for space 1 using the command avar in lines 3–5. The set of variables for which the wanted binary vector must be equal to 1 is specified as TVL 1 in lines 6–8. The set of variables for which the wanted binary vector must be equal to 0 is specified as TVL 2 in lines 9–11. Binary vectors of values 1 for the variables of these sets are created by the commands sv_get in lines 12 and 13. The values 1 of TVL 3 can directly

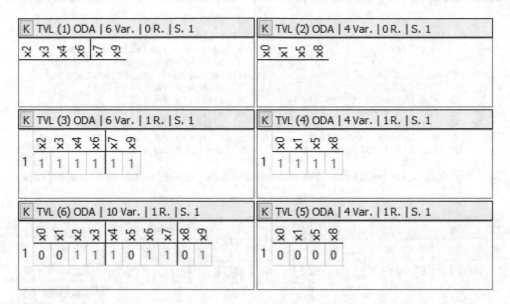

Fig. 3.23 Given sets of variables, intermediate partial binary vectors, and associated merged binary vector

be used to construct the wanted binary vector, but the values 1 of TVL 4 must be negated. The command `cel` in line 14 exchanged the values 0 and 1 in all columns of TVL 4; due to the use of the two parameters /01 and /10, the orthogonality of TVL 4 is preserved in TVL 5. Finally, the two ternary vectors of ones for the set of TVL 1, respectively, zeros for the set of TVL 2, are merged into the wanted binary vector (TVL 6) using the command `isc` in line 15; this is possible because the set operations of XBOOLE compute implicitly the union of the sets of variables of the given TVLs and fill not determined elements of given TVLs by dashes.

Figure 3.23 shows in the upper two viewports the given sets of variables and below in the second row of viewports the associated binary vector created by the commands `sv_get`. The values 1 of the binary vector in TVL 4 have been changed into values 0 in TVL 5 (viewport (3,2)). TVL 6 in viewport (3,1) shows the wanted binary vector, as merging result of TVL 3 (viewport (2,1)) and TVL 5 (viewport (3,2)) by means of the command `isc`. The comparison of TVL 1, 2, and 6 confirms that the computed binary vector (TVL 6) contains elements 1 for all variables of TVL 1 and elements 0 for all variables of TVL 2, respectively.

(c) Figure 3.24 shows the PRP that computes the norm $\|\mathbf{x} \vee \overline{\mathbf{x}}\|$ ($normxnx: the number of all variables of $\mathbf{x}$), the norm $\|\mathbf{x}\|$ ($normx: the number of variables of $\mathbf{x}$ with $x_i = 1$), and $\|\overline{\mathbf{x}}\|$ ($normnx: the number of variables of $\mathbf{x}$ with $x_i = 0$) for the binary vector $\mathbf{x} = (1101101001)$ and verifies the given relation between these integers.

The norm $\|\mathbf{x} \vee \overline{\mathbf{x}}\|$ is equal to the number of all variables of the binary vector $\mathbf{x}$ prepared in lines 3–6; hence, this norm is the result of the command `sv_size` in line 7. The command `cel` in line 8 changes the zeros of $\mathbf{x}$ into dashes so that the command `stv` in line 9 can create TVL 3 that describes the set of values 1 of $\mathbf{x}$; hence, the command `sv_size` in line 10 computes the norm $\|\mathbf{x}\|$. The command `cel` in line 11 computes the complement of $\mathbf{x}$ so that the commands in lines 12–14 compute the norm $\|\overline{\mathbf{x}}\|$ using the same procedure as before. The command `sub` in line 15 computes the difference $\|\mathbf{x} \vee \overline{\mathbf{x}}\| - \|\mathbf{x}\|$, and the command `eq` in line 16 checks whether both sides of the given equation carry the same value.

Figure 3.25 shows:

```
 1   new                              12   cel 4 4 5 /0-
 2   space 32 1                       13   stv 5 1 6
 3   tin 1 1                          14   set $normnx (sv_size 6)
 4   x0 x1 x2 x3 x4                   15   set $d (sub $normxnx $normx)
 5   x5 x6 x7 x8 x9.                  16   set $equal (eq $d $normnx)
 6   1101101001.                      17   assign 1 /m 1 1
 7   set $normxnx (sv_size 1)         18   assign 4 /m 1 2
 8   cel 1 1 2 /0-                    19   assign 2 /m 2 1
 9   stv 2 1 3                        20   assign 3 /m 2 2
10   set $normx (sv_size 3)          21   assign 5 /m 3 1
11   cel 1 1 4 /01 /10               22   assign 6 /m 3 2
```

Fig. 3.24 Problem-program that computes the norm of three binary vectors and verifies the relation $\|\overline{\mathbf{x}}\| = \|\mathbf{x} \vee \overline{\mathbf{x}}\| - \|\mathbf{x}\|$

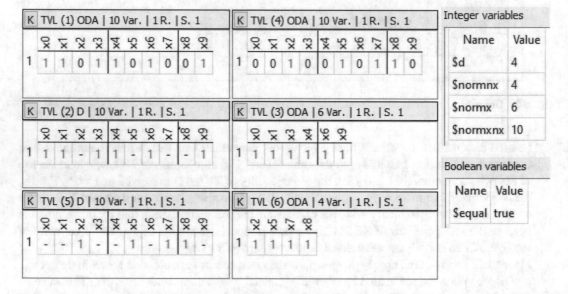

Fig. 3.25 Given binary vector and intermediate results of Exercise 3.3 (c) and the values of the computed control variables

- the given binary vector **x** (TVL 1),
- the associate complement $\overline{\mathbf{x}}$ (TVL 4),
- the intermediate TVL 2 and TVL 3 that describe the set of variables determined by values 1 in TVL 1, and
- the intermediate TVL 5 and TVL 6 that describe the set of variables determined by values 0 in TVL 1.

Figure 3.25 shows furthermore on the right-hand side the computed norms: $\|\overline{\mathbf{x}}\| = 4$ ($normnx), $\|\mathbf{x}\| = 6$ ($normx), and $\|\mathbf{x} \vee \overline{\mathbf{x}}\| = 10$ ($normxnx) as well as the Boolean variable $equal = true that confirms that the given equation is satisfied for the evaluated binary vector **x**.

(d) Figure 3.26 shows on the left-hand side the PRP that computes the four norms and verifies the equation $\|\mathbf{x}_1 \vee \mathbf{x}_2\| = \|\mathbf{x}_1\| + \|\mathbf{x}_2\| - \|\mathbf{x}_1 \mathbf{x}_2\|$ for the binary vectors $\mathbf{x}_1$ and $\mathbf{x}_2$ of ten variables and on the right-hand side the given and computed sets of variables and thereunder the computed control variables.

The command avar in lines 3 and 4 specifies the ten variables x_1 to x_9 in lexicographic order. The given sets of variables are prepared as empty TVL 1 for $\mathbf{x}_1$ and empty TVL 2 for $\mathbf{x}_2$ in lines

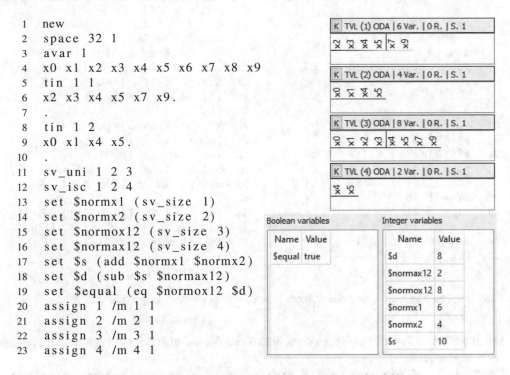

```
1   new
2   space 32 1
3   avar 1
4   x0 x1 x2 x3 x4 x5 x6 x7 x8 x9
5   tin 1 1
6   x2 x3 x4 x5 x7 x9.
7   .
8   tin 1 2
9   x0 x1 x4 x5.
10  .
11  sv_uni 1 2 3
12  sv_isc 1 2 4
13  set $normx1 (sv_size 1)
14  set $normx2 (sv_size 2)
15  set $normox12 (sv_size 3)
16  set $normax12 (sv_size 4)
17  set $s (add $normx1 $normx2)
18  set $d (sub $s $normax12)
19  set $equal (eq $normox12 $d)
20  assign 1 /m 1 1
21  assign 2 /m 2 1
22  assign 3 /m 3 1
23  assign 4 /m 4 1
```

K TVL (1) ODA | 6 Var. | 0 R. | S. 1
K TVL (2) ODA | 4 Var. | 0 R. | S. 1
K TVL (3) ODA | 8 Var. | 0 R. | S. 1
K TVL (4) ODA | 2 Var. | 0 R. | S. 1

Boolean variables

Name	Value
$equal	true

Integer variables

Name	Value
$d	8
$normax12	2
$normox12	8
$normx1	6
$normx2	4
$s	10

Fig. 3.26 Solution of Exercise 3.3 (d): Problem-program and the computed sets of variables

5–10. The command sv_uni in lines 11 computes union of the given sets of variables and the command sv_isc in lines 12 the intersection of them. The norms of the four TVLs 1–4 are computed using the commands sv_size in lines 13–16. The commands add and sub in lines 17 and 18 compute the value $d of the right-hand side of the equation to verify, and the command eq in line 19 checks whether $\|\mathbf{x}_1 \vee \mathbf{x}_2\| = \|\mathbf{x}_1\| + \|\mathbf{x}_2\| - \|\mathbf{x}_1\mathbf{x}_2\|$ is satisfied.

The computed sets of variables as well as the four norms:

- $normx1: $\|\mathbf{x}_1\| = 6$
- $normx2: $\|\mathbf{x}_2\| = 4$
- $normox12: $\|\mathbf{x}_1 \vee \mathbf{x}_2\| = 8$
- $normax12: $\|\mathbf{x}_1\mathbf{x}_2\| = 2$

are shown on the right-hand side of Fig. 3.26. The value true of the control variable $equal confirms that the given equation is satisfied for the evaluated binary vectors $\mathbf{x}_1$ and $\mathbf{x}_2$.

Solution 3.4 (Shells and Spheres)

(a) Figure 3.27 shows the PRP that recursively computes all shells $S_i(\mathbf{x}, \mathbf{x}_0 = \mathbf{0})$ of the Boolean spaces $\mathbb{B}^n, n = 1, \ldots, 9$.

All shells $S_i(\mathbf{x}, \mathbf{x}_0 = \mathbf{0})$ are computed in the Boolean space 1, defined in line 2. The variables of the large shell are attached to this space in lexicographic order in lines 3 and 4.

The recursive procedure extends in each sweep the shells of the variables $x_0, \ldots, x_{i-1}$ by a column of the of variables x_i, where both values 0 and 1 are needed; hence, the initial TVL 1 for $x_0 = 0$ is created in lines 5–7 and the associated complement (TVL 2) is computed using the command cpl in line 8. The rotation of TVLs 1 and 2 by one column to the right is needed to

```
1    new                                19    cco  1  3  4  1
2    space 32 1                         20    cco  2  3  4  2
3    avar 1                             21    set $j (mul $n 10)
4    x0 x1 x2 x3 x4 x5 x6 x7 x8.        22    isc (sub $j 10) 1 $j
5    tin 1 1                            23    assign $j /m 1 $n
6    x0.                                24    for $i 1 (sub $n 1)
7    0.                                 25    (
8    cpl 1 2                            26    set $k (add $j $i)
9    vtin 1 3                           27    isc (sub $k 11) 2 $k
10   x7 x6 x5 x4 x3 x2 x1 x0.           28    isc (sub $k 10) 1 5
11   vtin 1 4                           29    uni $k 5 $k
12   x8 x7 x6 x5 x4 x3 x2 x1.           30    assign $k /m (add $i 1) $n
13   copy 1 10                          31    )
14   copy 2 11                          32    isc (sub $k 10) 2 (add $k 1)
15   assign 10 /m 1 1                   33    set $r (add $i 2)
16   assign 11 /m 2 1                   34    assign (add $k 1) /m $r $n
17   for $n 2 9                         35    )
18   (                                  36    sts shell_1_9
```

Fig. 3.27 Problem-program that computes all shells $S_i(\mathbf{x}, \mathbf{x}_0 = \mathbf{0})$ of the Boolean spaces $\mathbb{B}^1, \ldots, \mathbb{B}^9$

select the next variable in the next sweep; VTs 3 and 4 (created in lines 9–12) have been prepared for this purpose.

The TVLs 1 and 2 are already the shells $S_0(x_0, 0)$ and $S_1(x_0, 0)$ of $\mathbb{B}^1$; hence, these TVLs are copied to TVLs 10 and 11 in lines 13 and 14 and assigned to the requested viewports in lines 15 and 16.

The computation of all shells of the Boolean spaces $\mathbb{B}^2, \ldots, \mathbb{B}^9$ is realized in the `for`-loop in lines 17–35. The commands `cco` in lines 19 and 20 rotate TVLs 1 and 2 by one column to the right. The integer $j = n \cdot 10$ in line 21 has been defined to store the shells of neighbored space $\mathbb{B}^{n-1}$ and $\mathbb{B}^n$ as requested with a distance of 10. The shell S_0 of $\mathbb{B}^n$ results from S_0 of $\mathbb{B}^{n-1}$ extended by a 0 in the next column; the command `isc` in line 22 solves this task and the command `assign` in line 23 assigns this shell in the viewport of the first row and the column $n.

The shells S_1 of $\mathbb{B}^n$ to S_{n-1} of $\mathbb{B}^n$ can be computed using the same rule; the inner `for`-loop in lines 24–31 realizes these computations. The integer $k = j + i$ in line 26 has been defined to store the shells of the space $\mathbb{B}^n$ as requested in neighbored TVLs. The rule

$$S_i(x_0, \ldots, x_n) = x_n S_{i-1}(x_0, \ldots, x_{n-1}) \vee \overline{x}_n S_i(x_0, \ldots, x_{n-1})$$

is realized by two command `isc` and the command `uni` in lines 27–29 to compute this shell using the intermediate TVL 5 and store it as TVL with the index k. The command `assign` in line 30 assigns this shell to viewport $(i + 1, n)$ of the m-fold view.

It remains the computation of the shell S_n of $\mathbb{B}^n$ that results from S_{n-1} of $\mathbb{B}^{n-1}$ extended by a 1 in the next column; the command `isc` in line 32 solves this task and the command `assign` in line 34 assigns this shell to viewport $(i + 1, n)$ of the m-fold view.

The command `sts` in line 36 saves the complete XBOOLE-system using the name "shell_1_9"; the default extension of the file name "sdt" has been avoided.

Figure 3.28 shows due to the restricted space in the upper part the computed shells only for the Boolean spaces $\mathbb{B}^1$, $\mathbb{B}^2$, and $\mathbb{B}^3$. The used principle to construct the shells of $\mathbb{B}^n$ recursively based

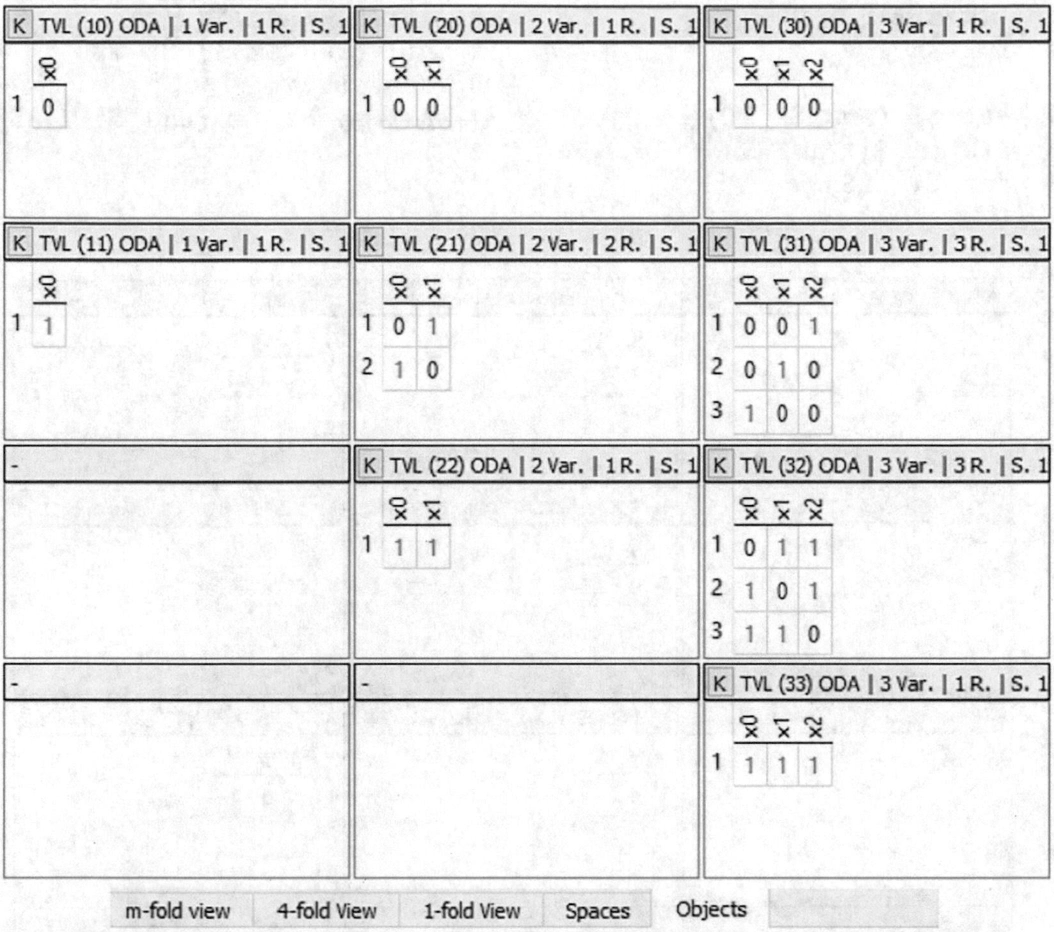

K TVL (10) ODA | 1 Var. | 1 R. | S. 1

	x_0
1	0

K TVL (20) ODA | 2 Var. | 1 R. | S. 1

	x_0	x_1
1	0	0

K TVL (30) ODA | 3 Var. | 1 R. | S. 1

	x_0	x_1	x_2
1	0	0	0

K TVL (11) ODA | 1 Var. | 1 R. | S. 1

	x_0
1	1

K TVL (21) ODA | 2 Var. | 2 R. | S. 1

	x_0	x_1
1	0	1
2	1	0

K TVL (31) ODA | 3 Var. | 3 R. | S. 1

	x_0	x_1	x_2
1	0	0	1
2	0	1	0
3	1	0	0

-

K TVL (22) ODA | 2 Var. | 1 R. | S. 1

	x_0	x_1
1	1	1

K TVL (32) ODA | 3 Var. | 3 R. | S. 1

	x_0	x_1	x_2
1	0	1	1
2	1	0	1
3	1	1	0

-

-

K TVL (33) ODA | 3 Var. | 1 R. | S. 1

	x_0	x_1	x_2
1	1	1	1

m-fold view   4-fold View   1-fold View   Spaces   Objects

Object name	Object number	Code/Form	Space	Variables	Rows	Boxes
TVL (87)	87	4 (ODA)	1	8	8	1
TVL (88)	88	4 (ODA)	1	8	1	1
TVL (90)	90	4 (ODA)	1	9	1	1
TVL (91)	91	4 (ODA)	1	9	9	1
TVL (92)	92	4 (ODA)	1	9	36	1
TVL (93)	93	4 (ODA)	1	9	84	2
TVL (94)	94	4 (ODA)	1	9	126	3
TVL (95)	95	4 (ODA)	1	9	126	3
TVL (96)	96	4 (ODA)	1	9	92	2
TVL (97)	97	4 (ODA)	1	9	36	1
TVL (98)	98	4 (ODA)	1	9	9	1
TVL (99)	99	4 (ODA)	1	9	1	1

Fig. 3.28 Computed shells S_i of the Boolean space $\mathbb{B}^1$, $\mathbb{B}^2$, and $\mathbb{B}^3$ in the m-fold view and details of the computed shells of $\mathbb{B}^9$ in the view "Objects"

```
1   lds shell_1_9              8    set $k (add $j $i)
2   for $n 1 9                 9    uni (sub $k 1) $k $k
3   (                          10   obbc $k $k
4   set $j (mul $n 10)         11   assign $k /m (add $i 1) $n
5   assign $j /m 1 $n          12   )
6   for $i 1 $n                13   )
7   (
```

K | TVL (10) ODA | 1 Var. | 1 R. | S. 1

	x_0
1	0

K | TVL (20) ODA | 2 Var. | 1 R. | S. 1

	x_0	x_1
1	0	0

K | TVL (30) ODA | 3 Var. | 1 R. | S. 1

	x_0	x_1	x_2
1	0	0	0

K | TVL (11) ODA | 1 Var. | 1 R. | S. 1

	x_0
1	-

K | TVL (21) ODA | 2 Var. | 2 R. | S. 1

	x_0	x_1
1	0	1
2	-	0

K | TVL (31) ODA | 3 Var. | 3 R. | S. 1

	x_0	x_1	x_2
1	0	0	1
2	0	1	0
3	-	0	0

K | TVL (22) ODA | 2 Var. | 1 R. | S. 1

	x_0	x_1
1	-	-

K | TVL (32) ODA | 3 Var. | 3 R. | S. 1

	x_0	x_1	x_2
1	1	0	1
2	-	-	0
3	0	-	1

K | TVL (33) ODA | 3 Var. | 1 R. | S. 1

	x_0	x_1	x_2
1	-	-	-

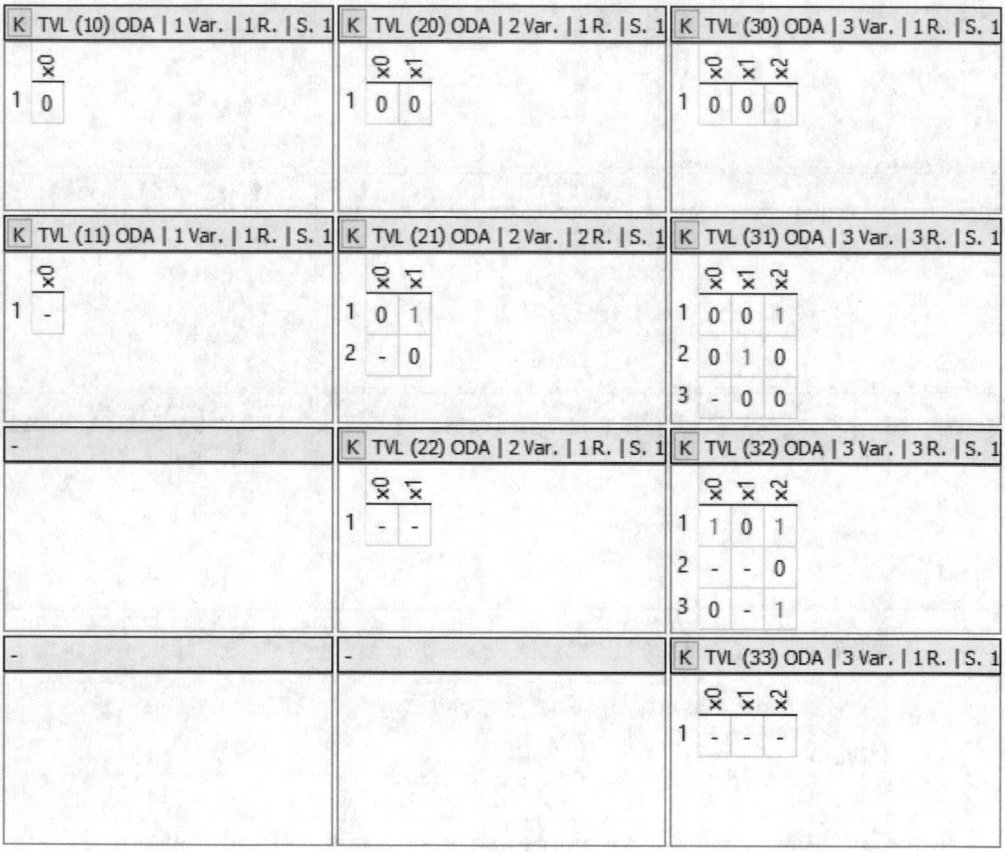

Fig. 3.29 Solution of Exercise 3.4 (b): Problem-program and the computed spheres of the Boolean space $\mathbb{B}^1$, $\mathbb{B}^2$, and $\mathbb{B}^3$

on the shells of $\mathbb{B}^{n-1}$ can be recognized. On a larger screen all computed shells for $\mathbb{B}^1$ to $\mathbb{B}^9$ are shown in the m-fold view prepared for ten rows and nine columns. Details about the computed shells of $\mathbb{B}^9$ are shown in the view "Objects" at the bottom of Fig. 3.28.

Pairs of binary vectors with a distance of 1 belong to different shells; hence, the binary vectors of the shells cannot be merged into ternary vectors and the number of rows of each shown TVL is equal to the number of binary vectors of the represented shell. The evaluation of the computed TVLs shown in the m-fold view, successively in the 1-fold view or in the view "Objects," confirms that the shell $S_i(\mathbf{x}, \mathbf{x}_0)$ of $\mathbb{B}^n$ contains $\binom{n}{i}$ binary vectors.

(b) Figure 3.29 shows at the top the PRP that computes the closed spheres of the Boolean space $\mathbb{B}^1$ to $\mathbb{B}^9$ and thereunder the results of the first three spaces.

The command lds in line 1 loads the XBOOLE-system saved in Exercise 3.4 (a) so that shells S_i of $\mathbb{B}^1$ to $\mathbb{B}^9$ can directly be used to compute the associated closed spheres.

The computation of the closed spheres is realized in two nested loops; the outer for-loop in lines 2–13 iterates over the Boolean spaces $\mathbb{B}^1$ to $\mathbb{B}^9$ and the inner for-loop in lines 6–12 over the closed spheres K_1^c to K_n^c. The sphere K_0^c is equal to S_0 for each Boolean space so that no computation is needed in this case.

The PRP of Fig. 3.29 utilizes the property

$$K_i^c(\mathbf{x}, \mathbf{x}_0 = \mathbf{0}) = K_{i-1}^c(\mathbf{x}, \mathbf{x}_0 = \mathbf{0}) \cup S_i(\mathbf{x}, \mathbf{x}_0 = \mathbf{0}) \; ;$$

the command uni in line 9 realizes this computation. The union of neighbored shells in the closed spheres increases on the one hand the number of binary vectors but facilitates on the other hand their merge into ternary vectors; hence, the command obbc is used in line 10 to represent the computed spheres with a small number of ternary vectors as can be seen in the m-fold view of Fig. 3.29.

The closed spheres $K_n^c(\mathbf{x}, \mathbf{x}_0 = \mathbf{0})$ of $\mathbb{B}^n$ are stored as TVLs with the index $k = 10 \cdot n + n$. A check of the column "Row" in the view "Objects" confirms that that all these closed spheres are stored by a single ternary vector that represents all 2^n binary vectors of these spaces.

(c) Figure 3.30 shows at the top the PRP that computes the shells of the Boolean space $\mathbb{B}^1$ to $\mathbb{B}^9$ based on the appropriate part of the binary vector $\mathbf{x}_0 = (101100110)$ as center and thereunder the results of the first three Boolean spaces.

The center point $\mathbf{x}_0 = \mathbf{0}$ has been used to compute the shells $S_i(\mathbf{x}, \mathbf{x}_0 = \mathbf{0})$ in Exercise 3.4 (a). Due to the rule $S_i = \{\mathbf{x}|h(\mathbf{x}, \mathbf{x}_0) = i\}$ the shells with the changed center can be computed based on the shells $S_i(\mathbf{x}, \mathbf{x}_0 = \mathbf{0})$ by exchanging the values 0 and 1 in the columns where $\mathbf{x}_0$ is equal to 1; hence, the command lds in line 1 is again used to load the XBOOLE-system saved in Exercise 3.4 (a). The binary vector of the requested center is defined as TVL 6 in lines 2–4.

The computation of the modified shells is again realized in two nested loops; the outer for-loop in lines 5–18 iterates over the Boolean spaces $\mathbb{B}^1$ to $\mathbb{B}^9$ and the inner for-loop in lines 12–17 over the shells S_0 to S_n for the selected Boolean space.

The given vector of the center must be restricted to n bits to compute the shells of $\mathbb{B}^n$, and a set of variables is needed that contain the variables where these n bits are equal to 1. This task is solved in lines 8–11 of the outer for-loop. The command sv_dif in line 8 computes the set of variables that must be avoided (TVL 7), and the command dco deletes these columns in TVL 6 and stores the intermediate result as TVL 8. The command cel in line 10 replaces the value 0 by dashes in TVL 8 so that the command stv in line 11 computes TVL 9 of the required set of variables.

The main operation to change the center point of the shells is realized by the command cel in line 15; this command exchanged the values 0 and 1 in the columns (TVL 9) where the center vector is equal to 1 in the appropriate part for $\mathbb{B}^n$. The command assign in line 16 assigns the changed shell to the requested viewport of the m-fold view.

The comparison of Figs. 3.28 and 3.30 confirms that values 0 and 1 have been changed in the columns where the center point $\mathbf{x}_0$ is equal to 1. The number of binary vectors of the shells $S_i = \{\mathbf{x}|h(\mathbf{x}, \mathbf{x}_0) = i\}$ does not depend on the used center vector $\mathbf{x}_0$.

(d) Figure 3.31 shows at the top the PRP that computes the neighbors of the binary vector (101101010) and splits this set of binary vectors into subsets of the shells $S_4(\mathbf{x}, \mathbf{x}_0 = \mathbf{0})$ and $S_6(\mathbf{x}, \mathbf{x}_0 = \mathbf{0})$.

The command lds in line 1 loads the XBOOLE-system saved in Exercise 3.4 (a) so that all shells S_i of $\mathbb{B}^9$ can directly be used to split the computed neighbors with regard to the expected shells.

```
 1  lds shell_1_9                          10  cel 8 8 8 /0-
 2  tin 1 6                                11  stv 8 1 9
 3  x0 x1 x2 x3 x4 x5 x6 x7 x8 .           12  for $i 0 $n
 4  101100110 .                            13  (
 5  for $n 1 9                             14  set $k (add $j $i)
 6  (                                      15  cel $k 9 $k /01 /10
 7  set $j (mul $n 10)                     16  assign $k /m (add $i 1) $n
 8  sv_dif 6 $j 7                          17  )
 9  dco 6 7 8                              18  )
```

K | TVL (10) ODA | 1 Var. | 1 R. | S. 1

	x0
1	1

K | TVL (20) ODA | 2 Var. | 1 R. | S. 1

	x0	x1
1	1	0

K | TVL (30) ODA | 3 Var. | 1 R. | S. 1

	x0	x1	x2
1	1	0	1

K | TVL (11) ODA | 1 Var. | 1 R. | S. 1

	x0
1	0

K | TVL (21) ODA | 2 Var. | 2 R. | S. 1

	x0	x1
1	1	1
2	0	0

K | TVL (31) ODA | 3 Var. | 3 R. | S. 1

	x0	x1	x2
1	1	0	0
2	1	1	1
3	0	0	1

K | TVL (22) ODA | 2 Var. | 1 R. | S. 1

	x0	x1
1	0	1

K | TVL (32) ODA | 3 Var. | 3 R. | S. 1

	x0	x1	x2
1	1	1	0
2	0	0	0
3	0	1	1

K | TVL (33) ODA | 3 Var. | 1 R. | S. 1

	x0	x1	x2
1	0	1	0

Fig. 3.30 Solution of Exercise 3.4 (c): Problem-program and the computed shell of the Boolean space $\mathbb{B}^1$, $\mathbb{B}^2$, and $\mathbb{B}^3$ using the centers (1), (10), and (101)

The given binary vector has been defined as TVL 6 in lines 2–5. The empty TVL 100 is created based on TVL 6 of the given binary vector using the command dtv in line 5.

The while-loop in lines 6–10 is controlled by the command sv_next so that each variable of TVL 6 can be used in one sweep of this loop as the set of variables stored as TVL 7. The command cel in line 8 exchanges in the given vector (TVL 6) in the position of the selected variable (TVL 7) the value 0 into 1 or vice versa so that a neighbor vector is built (TVL 8). The command uni includes the computed neighbor vector into the set of so far computed neighbors (TVL 100).

The given binary vector has five values 1; hence, this vector belongs to the shell $S_5(\mathbf{x}, \mathbf{x}_0 = \mathbf{0})$. The neighbors have a distance of 1 so that each neighbor vector must belong either to the shell

```
1   lds shell_1_9                     9   uni 100 8 100
2   tin 1 6                           10  )
3   x0 x1 x2 x3 x4 x5 x6 x7 x8.       11  isc 100 94 101
4   101101010.                        12  isc 100 96 102
5   dtv 6 1 100                       13  assign 6 /m 1 2
6   while (sv_next 6 7 7)             14  assign 101 /m 1 1
7   (                                 15  assign 102 /m 1 3
8   cel 6 7 8 /01 /10
```

K	TVL (101) ODA	9 Var.	5 R.	S. 1						
	x0	x1	x2	x3	x4	x5	x6	x7	x8	
1	0	0	1	1	0	1	0	1	0	
2	1	0	0	1	0	1	0	1	0	
3	1	0	1	0	0	1	0	1	0	
4	1	0	1	1	0	0	0	1	0	
5	1	0	1	1	0	1	0	0	0	

K	TVL (6) ODA	9 Var.	1 R.	S. 1						
	x0	x1	x2	x3	x4	x5	x6	x7	x8	
1	1	0	1	1	0	1	0	1	0	

K	TVL (102) ODA	9 Var.	4 R.	S. 1						
	x0	x1	x2	x3	x4	x5	x6	x7	x8	
1	1	1	1	1	0	1	0	1	0	
2	1	0	1	1	1	1	0	1	0	
3	1	0	1	1	0	1	1	1	0	
4	1	0	1	1	0	1	0	1	1	

Fig. 3.31 Solution of Exercise 3.4 (d): Problem-program and the computed smaller (TVL 101) and larger neighbors (TVL 102) of the binary vector in the middle

$S_4(\mathbf{x}, \mathbf{x}_0 = \mathbf{0})$ or to the shell $S_6(\mathbf{x}, \mathbf{x}_0 = \mathbf{0})$. The commands isc in lines 11 and 12 split the set of all neighbors (TVL 100) into the sets of smaller (TVL 101) and larger vectors (TVL 102).

The viewports in Fig. 3.31 show from the left to the right the smaller neighbors, the given binary vector, and the larger neighbors. The given vector has four values 0; hence, the change of each of these values into a value 1 results in a larger neighbor. Vice versa, the given vector has five values 1; hence, the change of each of these values into a value 0 results in a smaller neighbor. Figure 3.31 confirms this property.

3.7 Supplementary Exercises

Solve all tasks of the supplementary exercises by means of a PRP.

Exercise 3.5 (Relations in $\mathbb{B}$)

Verify the following relations using the approach of Exercise 3.1:

(a) $x \leq (x \vee y)$.
(b) If $x_1 \leq y_1$ and $x_2 \leq y_2$, then $(x_1 \vee x_2) \leq (y_1 \vee y_2)$.
(c) $(x \oplus y)$ is equivalent to $((x \wedge \overline{y}) \vee (\overline{x} \wedge y))$.
(d) $(x \vee y) \oplus (x \vee z) \leq (y \oplus z)$.

Exercise 3.6 (Binary Vectors)

(a) Find the decimal equivalent $\text{dec}(\mathbf{x})$ for the vector $(1100010101) \in \mathbb{B}^{10}$.
(b) Find the vector $\mathbf{x} \in \mathbb{B}^{10}$ with $\text{dec}(\mathbf{x}) = 567$.
(c) Let be given two vectors $\mathbf{x}, \mathbf{y} \in \mathbb{B}^6$ with $\mathbf{x} \leq \mathbf{y}$: $\mathbf{x} = (010101)$, $\mathbf{y} = (110111)$. Find all vectors $\mathbf{z}$ with $\mathbf{x} \leq \mathbf{z} \leq \mathbf{y}$.

(d) Let be given two vectors $\mathbf{x}, \mathbf{y} \in \mathbb{B}^8$: $\mathbf{x} = (00101101)$, $\mathbf{y} = (10110001)$. Check whether these vectors can specify an interval that is not empty?

Exercise 3.7 (Ternary Vectors and Sets of Variables)

(a) Compute for the given ternary vector $\mathbf{x} = (1 - -010 - 11-)$ three sets of variables, represented by empty TVLs, for which either:

   - $x_i = 1$.
   - $x_i = 0$.
   - $x_i = -$.

(b) Create the TVL of one ternary vector, where the set of variables $\{x_0, x_2, x_4, x_7, x_9\}$ describes values 1, the set of variables $\{x_1, x_5, x_8\}$ describes values 0, and the set of variables $\{x_3, x_6\}$ describes the dash elements.

(c) Compute the number of binary vectors represented by the ternary vector $\mathbf{x} = (0 - -1 - 1 - -0 - -1)$ utilizing the command `sv_size` to count the number of dashes.

(d) How many pairs of binary vectors are specified by the pair of ternary vectors $\mathbf{x}_1 = (0 - -11 - 1001)$ and $\mathbf{x}_2 = (01 - 0 - -1 - 10)$? Compute the minimal and the maximal distance (metric (Theorem 3.13)) for all pairs of binary vectors represented by the given two ternary vectors directly based on these two ternary vectors.

Exercise 3.8 (Shells and Spheres)

(a) Create all shells $S_r(\mathbf{x}, \mathbf{x}_0 = \mathbf{0})$ of $\mathbb{B}^5$ using a split of $(- - - - -)$ into binary vectors and their assignment to the shell it belongs to and save the XBOOLE-system using the name "shell_5.sdt" for later use.

(b) Reuse the `sdt`-file "shell_5.sdt," saved at the end of Exercise 3.8 (a), and compute all $S_r(\mathbf{x}, \mathbf{x}_0 = \mathbf{1})$ of $\mathbb{B}^5$. Verify that $S_r(\mathbf{x}, \mathbf{x}_0 = \mathbf{0}) = S_{n-r}(\mathbf{x}, \mathbf{x}_0 = \mathbf{1})$ for the explored Boolean space $\mathbb{B}^n$ with $n = 5$.

(c) Create the closed sphere $K_2^c(\mathbf{x}, \mathbf{x}_0 = \mathbf{0})$ of $\mathbb{B}^5$ using the approach of Exercise 3.8 (a).

(d) Extend the PRP of 3.8 (c) such that additionally the open sphere $K_3^o(\mathbf{x}, \mathbf{x}_0 = \mathbf{1})$ of $\mathbb{B}^5$ is generated, and verify that $K_3^o(\mathbf{x}, \mathbf{x}_0 = \mathbf{1}) = \overline{K_2^c(\mathbf{x}, \mathbf{x}_0 = \mathbf{0})}$ in $\mathbb{B}^5$.

References

1. G. Boole, *The Mathematical Analysis of Logic: Being an Essay Towards a Calculus of Deductive Reasoning (Classic Reprint)* (Forgotten Books, London, 2016). ISBN: 978-1-4400-6642-9
2. R.E. Bryant, Graph-based algorithms for Boolean function manipulation, in *IEEE Transaction on Computers* C-35.8 (1986), pp. 677–691. ISSN: 0018-9340. https://doi.org/10.1109/TC.1986.1676819
3. P. Clote, E. Krankis, *Boolean Functions and Computation Models* (Springer, Berlin, 2002). ISBN: 978-3-662-04943-3. https://doi.org/10.1007/978-3-662-04943-3
4. I. Grattan-Guinness, G. Bornet, eds., *George Boole: Selected Manuscripts on Logic and its Philosophy*. Science Networks. Historical Studies 20 (Birkhäuser, Berlin, 1997). ISBN: 978-3-7643-5456-5
5. D. MacHale, *The Life and Work of George Boole: A Prelude to the Digital Age* (Cork University Press, Cork, 2014). ISBN: 978-1-78205-004-9
6. C. Meinel, T. Theobald, *Algorithms and Data Structures in VLSI Design* (Springer, Berlin, 1998). ISBN: 978-3-540-64486-6. https://doi.org/10.1007/978-3-642-58940-9
7. C. Posthoff, B. Steinbach, *Logic Functions and Equations—Binary Models for Computer Science* (Springer, The Netherlands, 2004). ISBN: 978-1-44195-261-5. https://doi.org/10.1007/978-1-4020-2938-7
8. C. Posthoff, B. Steinbach, *Logic Functions and Equations—Binary Models for Computer Science, Second Edition* (Springer, Cham, 2019). ISBN: 978-3-030-02419-2. https://doi.org/10.1007/978-3-030-02420-8

9. S. Rudeanu, *Lattice Functions and Equations*. Discrete Mathematics and Theoretical Computer Science (Springer, London, 2001). ISBN: 978-1-85233-266-2. https://doi.org/10.1007/978-1-4471-0241-0

10. C.E. Shannon, A symbolic analysis of relay and switching circuits, in *Transactions of the American Institute of Electrical Engineers*, vol. 57(12) (1938), pp. 713–723. ISSN: 0096-3860. https://doi.org/10.1109/T-AIEE.1938.5057767

11. B. Steinbach, XBOOLE—A Toolbox for Modelling, Simulation, and Analysis of Large Digital Systems, in *Systems Analysis and Modeling Simulation*, vol. 9(4) (1992), pp. 297–312. ISSN: 0232-9298

12. B. Steinbach, C. Posthoff. *Logic Functions and Equations—Examples and Exercises* (Springer, Berlin, 2009). ISBN: 978-1-4020-9594-8. https://doi.org/10.1007/978-1-4020-9595-5

13. B. Steinbach, C. Posthoff, *Boolean Differential Equations* (Morgan and Claypool Publishers, San Rafael, 2013). ISBN: 978-1-6270-5241-2. https://doi.org/10.2200/S00511ED1V01Y201305DCS042

14. B. Steinbach, C. Posthoff, *EAGLE Start-up Aid—Efficient Computations with XBOOLE (in German: EAGLE Starthilfe—Effiziente Berechnungen mit XBOOLE)* (Edition am Gutenbergplatz, Leipzig, 2015). ISBN: 978-3-95922-081-1

15. B. Steinbach, C. Posthoff, *Boolean Differential Calculus* (Morgan and Claypool Publishers, San Rafael, 2017). ISBN: 978-1-6270-5922-0. https://doi.org/10.2200/S00766ED1V01Y201704DCS052

16. B. Steinbach, M. Werner, XBOOLE-CUDA—fast Boolean operations on the GPU, in *Boolean Problems, Proceedings of the 11th International Workshops on Boolean Problems* ed. by B. Steinbach. IWSBP 11 (Freiberg University of Mining and Technology, Freiberg, 2014), pp. 75–84. ISBN: 978-3-86012-488-8

17. B. Steinbach, M. Werner, XBOOLE-CUDA—Fast Calculations on the GPU, in *Problems and New Solutions in the Boolean Domain*, ed. by B. Steinbach (Cambridge Scholars Publishing, Cambridge, 2016), pp. 117–149. ISBN: 978-1-4438-8947-6

18. I. Wegener. *Branching Programs and Binary Decision Diagrams: Theory and Applications*. Discrete Mathematics and Applications (Society for Industrial and Applied Mathematics (SIAM), Philadelphia, 2000). ISBN: 978-0-89871-458-6. https://doi.org/10.1137/1.9780898719789

Logic Functions

4

Abstract

Logic functions belong to the core concepts in the Boolean domain. A logic function of n variables is a unique mapping from $\mathbb{B}^n$ into $\mathbb{B}$. The wide field of applications of logic functions becomes visible by other names, which are synonymously used, like switching functions, binary functions, Boolean functions, constraints, or truth functions. Certain logic functions determine the logic operation introduced in the previous chapter. Several possibilities to present a logic function are explained: function table, several normal forms, simplified forms, ternary vector lists (TVL) in certain forms, polynomials, decision tree, reduced ordered binary decision diagrams (ROBDD), or Karnaugh-maps. The defined logic operations allow the use of formulas and expressions to represent a logic function. Different formulas can be used to express the same logic function; hence, the minimization of logic expressions is an important task and basic approaches to solve this minimization task are introduced. Logic functions can satisfy certain properties that are explained and utilized to determine complete systems of logic functions.

Supplementary Information The online version of this chapter (https://doi.org/10.1007/978-3-030-88945-6_4) contains supplementary material which is available for authorized users. Please, follow the link belonging to the version of the XBOOLE-monitor XBM 2 that fits best for your operating system. This XBOOLE-monitor is needed to solve all tasks of this chapter. Instructions for starting the downloaded XBOOLE-monitor XBM 2 are given at the beginning of Section 'Exercises' in this chapter.

XBOOLE-monitor XBM 2 for Windows 10
32 bits
https://doi.org/10.1007/978-3-030-88945-6_4_MOESM1_ESM.zip (15,091 KB)

64 bits
https://doi.org/10.1007/978-3-030-88945-6_4_MOESM2_ESM.zip (14,973 KB)

XBOOLE-monitor XBM 2 for Linux Ubuntu
32 bits
https://doi.org/10.1007/978-3-030-88945-6_4_MOESM3_ESM.zip (29,522 KB)

64 bits
https://doi.org/10.1007/978-3-030-88945-6_4_MOESM4_ESM.zip (28,422 KB)

4.1   Logic Functions

We will now introduce one of the core concepts—*logic functions*. We want to draw the attention of the reader to the fact that many other names are used synonymously, like *switching functions*, *binary functions*, *Boolean functions*, *constraints,*, *truth functions*, etc. It is always important to check the given definitions and to find out which names are used in a given context.

Definition 4.1 A unique mapping from $\mathbb{B}^n$ into $\mathbb{B}$ is a *logic function* of n variables, expressed by $f(x_1, \ldots, x_n)$.

This definition means that a value of $\mathbb{B}$, i.e., 0 or 1, is assigned to each vector $\mathbf{x} = (x_1, \ldots, x_n) \in \mathbb{B}^n$. n is the *arity* of the function f. The first most basic method to represent a logic function is the *function table* (the term *truth table* is also used very often). This table comprises all elements of $\mathbb{B}^n$ together with the function value. Therefore, each vector of 2^n components represents a function of n variables if a fixed order of the arguments is given.

Theorem 4.1 *There are 2^{2^n} different functions of n variables.*

Proof $\mathbb{B}^n$ has 2^n vectors; for each vector, there are two possible values 0 or 1; hence, there are $\underbrace{2 \times \ldots \times 2}_{2^n \text{ factors } 2}$ possible combinations of function values.   □

Table 4.1 shows the $2^{2^1} = 4$ functions of one variable.
These functions of one variable can be explained as follows:

$$
\begin{aligned}
f_0(x) \text{ is identically equal to } 0: && f_0(x) &= 0(x)\,, \\
f_1(x) \text{ uses the values of } x \text{ directly:} && f_1(x) &= x\,, \\
f_2(x) \text{ uses the negated value of } x: && f_2(x) &= \overline{x}\,, \\
f_3(x) \text{ is identically equal to } 1: && f_3(x) &= 1(x)\,.
\end{aligned}
$$

There are $2^{2^2} = 2^4 = 16$ functions of two variables (see Table 4.2).
The enumeration has been selected in such a way that the binary coding of the index of the function is equal to the vector of the function values (when the first element in a column is taken as the leftmost element of the coding vector).
The complement of the logic function $f_i(x, y)$ is $f_{15-i}(x, y)$, $i = 0, 1, \ldots, 15$. $f_0(x, y)$ and $f_{15}(x, y)$ are identically equal to 0 and 1, respectively. Some other functions are already known as *conjunction*, *disjunction*, *antivalence*, and *equivalence*:

$$
\begin{aligned}
f_1(x, y) &= x \wedge y\,, & (4.1) \\
f_7(x, y) &= x \vee y\,, & (4.2)
\end{aligned}
$$

Table 4.1 Logic functions of one variable

x	$f_0(x)$	$f_1(x)$	$f_2(x)$	$f_3(x)$
0	0	0	1	1
1	0	1	0	1

Table 4.2 Logic functions of two variables

x	y	f_0	f_1	f_2	f_3	f_4	f_5	f_6	f_7	f_8	f_9	f_{10}	f_{11}	f_{12}	f_{13}	f_{14}	f_{15}
0	0	0	0	0	0	0	0	0	0	1	1	1	1	1	1	1	1
0	1	0	0	0	0	1	1	1	1	0	0	0	0	1	1	1	1
1	0	0	0	1	1	0	0	1	1	0	0	1	1	0	0	1	1
1	1	0	1	0	1	0	1	0	1	0	1	0	1	0	1	0	1

$$f_6(x, y) = x \oplus y, \tag{4.3}$$

$$f_9(x, y) = x \odot y. \tag{4.4}$$

When we look at the vector of function values, then we can see that $f_{14}(x, y)$ is the complement of $f_1(x, y)$, and in the same way that $f_8(x, y)$ is the complement of $f_7(x, y)$; hence, these two functions will be designated by

$$\text{NAND}(x, y) = \overline{x \wedge y} = f_{14}(x, y) \qquad \text{Sheffer-function,}$$

$$\text{NOR}(x, y) = \overline{x \vee y} = f_8(x, y) \qquad \text{Peirce-function.}$$

Note 4.1 Sometimes the following function symbols will be used:

$$\text{NAND}(x, y) = x \mid y,$$

$$\text{NOR}(x, y) = x \downarrow y.$$

The function $f_{13}(x, y)$ will be denoted by *implication*, designated by an arrow:

$$f_{13}(x, y) = (x \rightarrow y). \tag{4.5}$$

It will play a leading role in the treatment of logic problems (propositional calculus).

For two sets $A \subseteq M, B \subseteq M$ the difference $A \setminus B = \{x \mid x \in A, x \notin B\}$ is used rather often. We use for elements $m_i \in M$ the logic variables x and y by

$$x \in A \quad \equiv \quad x = 1,$$

$$x \notin A \quad \equiv \quad x = 0,$$

and the same for the set B:

$$y \in B \quad \equiv \quad y = 1,$$

$$y \notin B \quad \equiv \quad y = 0.$$

Then the function $f_2(x, y) = x \wedge \overline{y}$, the *complement* of the implication, reflects this understanding: the function value 1 of $f_2(x, y) = x \wedge \overline{y}$, i.e., $x = 1, y = 0$, expresses the fact that $x \in A$, but $y \notin B$.

The number of logic functions of n variables is growing extremely fast, as can be seen in Table 4.3. We will use $\mathcal{F}_n$ to indicate the set of all logic functions of n variables.

Ten is still a "very small" number; however, there are already $2^{10} = 1024$ different elements in $\mathbb{B}^{10}$; hence, functions of ten variables will have 1024 function values and the number of different

Table 4.3 The number of
logic functions of n
variables

n	2^{2^n}
1	$2^2 = 4$
2	$2^4 = 16$
3	$2^8 = 256$
4	$2^{16} = 65,536$
5	$2^{32} = 4,294,967,296$
6	$2^{64} \approx 1.84467 * 10^{19}$
7	$2^{128} \approx 3.40282 * 10^{38}$
8	$2^{256} \approx 1.15792 * 10^{77}$
9	$2^{512} \approx 1.34078 * 10^{154}$
10	$2^{1024} \approx 1.79769 * 10^{308}$

functions is equal to $2^{2^{10}} = 2^{1024}$. That means that each problem that depends on functions of 10 variables has, at least in principle, a *search space* of 2^{1024} possible solutions.

Hence, the function table can be used only for small numbers of variables (n at most equal to 5, 6, or 7).

Theorem 4.2 *The set of all logic functions of n variables is a Boolean Algebra* $(\mathbb{B}^{2^n}, \vee, \wedge, \bar{\ }, \mathbf{0}, \mathbf{1})$.

Proof The idea of this proof is quite simple. Each logic function of n variables can be represented by a binary vector $\mathbf{x}$ of length 2^n. For each function vector, the different argument vectors $\mathbf{x} \in \mathbb{B}^n$ must have the same order. The operations $\vee, \wedge, \bar{\ }$ can now be applied to these vectors as before. That means that the Boolean Algebras $(\mathbb{B}^{2^n}, \vee, \wedge, \bar{\ }, \mathbf{0}, \mathbf{1})$ are special cases of the general concept that has already been introduced before. $\square$

It is important to understand the meaning of an expression (a formula) like $x \wedge y, x \vee y, \ldots$ The function itself is given by the function table (which shows the function value for each binary vector). The formula represents (describes) the function by computational rules that use the operations defined for $\mathbb{B}$—the replacement of the variables by the values of a binary vector allows the calculation of the function value (according to the definition of the operation(s) to be used). Operations with two operands (like $\wedge, \vee, \oplus, \odot, \rightarrow$) are just using the functions of Table 4.2 and the operation symbols assigned in (4.1), $\ldots$, (4.5).

Sometimes it is useful to understand an expression as a *representation* of a function: the calculation according to the definition of the respective operations results for each vector $\mathbf{x}$ in the function value $f(\mathbf{x})$ (see also Sect. 4.2). The transition between a table of a logic function and an expression representing this function and vice versa will be used so often and in such a very natural way that these principal remarks should be sufficient. If special emphasis is required, then this will be taken into consideration according to the respective needs.

In order to have more compact representations of logic functions, we introduce step by step the concept of *logic formulas*.

We consider

$$f(x_1, \ldots, x_n) = x_1 \wedge x_2 \wedge \ldots \wedge x_n = x_1 x_2 \ldots x_n \,.$$

Since the conjunction is associative, no brackets are required; by an inductive generalization of $x_1 \wedge x_2$, it is easy to see that $x_1 \wedge x_2 \wedge \ldots \wedge x_n$ is equal to 1 for $x_1 = 1, \ldots, x_n = 1$, and for all the other vectors it is equal to 0. If negated variables appear (e.g., $\bar{x}_1 x_2 \bar{x}_3 \ldots x_n$), then the value 0 must be taken for the negated variables, the value 1 for the non-negated variables, and the conjunction again will be equal to 1.

Table 4.4 The correspondence between binary vectors and elementary conjunctions and disjunctions

$\mathbf{x} = (x_1, x_2, x_3)$	Elementary conjunction that is equal to 1	Elementary disjunction that is equal to 0
(000)	$\overline{x}_1\,\overline{x}_2\,\overline{x}_3$	$x_1 \vee x_2 \vee x_3$
(001)	$\overline{x}_1\,\overline{x}_2\,x_3$	$x_1 \vee x_2 \vee \overline{x}_3$
(010)	$\overline{x}_1\,x_2\,\overline{x}_3$	$x_1 \vee \overline{x}_2 \vee x_3$
(011)	$\overline{x}_1\,x_2\,x_3$	$x_1 \vee \overline{x}_2 \vee \overline{x}_3$
(100)	$x_1\,\overline{x}_2\,\overline{x}_3$	$\overline{x}_1 \vee x_2 \vee x_3$
(101)	$x_1\,\overline{x}_2\,x_3$	$\overline{x}_1 \vee x_2 \vee \overline{x}_3$
(110)	$x_1\,x_2\,\overline{x}_3$	$\overline{x}_1 \vee \overline{x}_2 \vee x_3$
(111)	$x_1\,x_2\,x_3$	$\overline{x}_1 \vee \overline{x}_2 \vee \overline{x}_3$

Using the same construction, we can also build disjunctions that will be equal to 0 for exactly one corresponding binary vector and equal to 1 for all the other binary vectors. In general, the relation between binary vectors $\mathbf{b} = (b_1, \ldots, b_n) \in \mathbb{B}^n$ and *elementary conjunctions* and *disjunctions* (see the next definition) can be expressed in the following way:

$$(x_1 \oplus \overline{b}_1) \wedge \ldots \wedge (x_n \oplus \overline{b}_n) = 1,$$
$$(x_1 \oplus b_1) \vee \ldots \vee (x_n \oplus b_n) = 0,$$
$$(x_1 \odot b_1) \wedge \ldots \wedge (x_n \odot b_n) = 1,$$
$$(x_1 \odot \overline{b}_1) \vee \ldots \vee (x_n \odot \overline{b}_n) = 0,$$
$$\text{for} \quad (x_1, \ldots, x_n) = (b_1, \ldots, b_n),$$

$$(x_1 \oplus \overline{b}_1) \wedge \ldots \wedge (x_n \oplus \overline{b}_n) = 0,$$
$$(x_1 \oplus b_1) \vee \ldots \vee (x_n \oplus b_n) = 1,$$
$$(x_1 \odot b_1) \wedge \ldots \wedge (x_n \odot b_n) = 0,$$
$$(x_1 \odot \overline{b}_1) \vee \ldots \vee (x_n \odot \overline{b}_n) = 1,$$
$$\text{for} \quad (x_1, \ldots, x_n) \neq (b_1, \ldots, b_n).$$

Definition 4.2 A *literal* is a negated or a non-negated variable. A conjunction C_n of n literals of n different variables is an *elementary conjunction*, and a disjunction D_n of n literals of n different variables is an *elementary disjunction*.

In an elementary conjunction or disjunction, each variable appears precisely once (and only once), either negated or non-negated.

Example 4.1 Let $n = 3$; then we have the unique correspondence between binary vectors, conjunctions, and disjunctions that is shown in Table 4.4. For every vector $\mathbf{x} = (x_1, x_2, x_3)$ of the first column, the conjunction in the second column is equal to 1, and the disjunction in the third column is equal to 0.

Definition 4.3 (Disjunctive Normal Form) The disjunction of different elementary conjunctions of n variables x_i represents a logic function $f(x_1, \ldots, x_n)$ and is denoted by *disjunctive normal form* (DNF).

Note 4.2 The function $0(\mathbf{x})$ cannot be represented by means of elementary conjunctions. However, $0(\mathbf{x}) = 0$ can be used as its DNF.

Definition 4.4 (Conjunctive Normal Form) The conjunction of different elementary disjunctions of n variables x_i represents a logic function $f(x_1, \ldots, x_n)$ and is denoted by *conjunctive normal form* (CNF).

Note 4.3 The function $1(\mathbf{x})$ cannot be represented by means of elementary disjunctions. However, $1(\mathbf{x}) = 1$ can be used as its CNF.

Note 4.4 These definitions require that each conjunction or disjunction used in a normal form contains all the variables of the given context. Very often these normal forms are denoted by *canonical normal forms*.

Theorem 4.3 *Each logic function* $f(x_1, \ldots, x_n)$ *can be uniquely represented by a* disjunctive normal form (DNF).

Proof Let a function $f(x_1, \ldots, x_n)$ be given by its function table. For each vector $\mathbf{x}$ with $f(\mathbf{x}) = 1$ we build the corresponding elementary conjunction and combine all these conjunctions by $\vee$. The resulting DNF represents the given function. The expression includes all elementary conjunctions corresponding to binary vectors $\mathbf{b}$ where the function has the value $f(\mathbf{b}) = 1$. The function $f(\mathbf{x}) = 0(\mathbf{x})$ is equal to 0 for all $\mathbf{x}$. Hence, no elementary conjunction is applicable for this function, and the DNF $0(\mathbf{x}) = 0$ will be used.   □

Theorem 4.4 *Each logic function* $f(x_1, \ldots, x_n)$ *can be uniquely represented by a* conjunctive normal form (CNF).

Proof Let a function $f(x_1, \ldots, x_n)$ be given by its function table. For each vector $\mathbf{x}$ with $f(\mathbf{x}) = 0$ we build the corresponding elementary disjunction and combine all these disjunctions by $\wedge$. The resulting CNF represents the given function. The expression contains all elementary disjunctions corresponding to binary vectors $\mathbf{b}$ where the function has the value $f(\mathbf{b}) = 0$. The $f(\mathbf{x}) = 1(\mathbf{x})$ is equal to 1 for all $\mathbf{x}$. Hence, no elementary disjunction is applicable for this function, and the CNF $1(\mathbf{x}) = 1$ will be used.   □

Note 4.5 It should be understood that different normal forms comprising the same set of conjunctions or disjunctions, only in a different order, are not considered as being different. The unique representation relates only to the conjunctions or disjunctions that are used, not to the sequence of their notation. The commutative law for $\vee$ could be used to establish the same order of conjunctions in two DNFs, and the commutative law for $\wedge$ facilitates an analog reordering of disjunctions in two CNFs.

This allows still another understanding of the normal forms. For n variables we consider the set of all elementary conjunctions with 2^n elements. Each subset defines exactly one DNF. In this case the empty set $\emptyset$ corresponds to the function $f(\mathbf{x}) = 0(\mathbf{x})$. The same approach can be used for the

Table 4.5 Logic function $f(x_1, x_2, x_3)$ used to construct a DNF and CNF

x	Conjunction	Disjunction	$f(\mathbf{x})$
(000)	$\overline{x}_1\,\overline{x}_2\,\overline{x}_3$	$x_1 \vee x_2 \vee x_3$	0
(001)	$\overline{x}_1\,\overline{x}_2\,x_3$	$x_1 \vee x_2 \vee \overline{x}_3$	1
(010)	$\overline{x}_1\,x_2\,\overline{x}_3$	$x_1 \vee \overline{x}_2 \vee x_3$	1
(011)	$\overline{x}_1\,x_2\,x_3$	$x_1 \vee \overline{x}_2 \vee \overline{x}_3$	0
(100)	$x_1\,\overline{x}_2\,\overline{x}_3$	$\overline{x}_1 \vee x_2 \vee x_3$	1
(101)	$x_1\,\overline{x}_2\,x_3$	$\overline{x}_1 \vee x_2 \vee \overline{x}_3$	0
(110)	$x_1\,x_2\,\overline{x}_3$	$\overline{x}_1 \vee \overline{x}_2 \vee x_3$	1
(111)	$x_1\,x_2\,x_3$	$\overline{x}_1 \vee \overline{x}_2 \vee \overline{x}_3$	0

set of all elementary disjunctions and its subsets. Here the empty set $\emptyset$ corresponds to the function $f(\mathbf{x}) = 1(\mathbf{x})$.

Example 4.2

$$f(\mathbf{x}) = \overline{x}_1\overline{x}_2 x_3 \vee \overline{x}_1 x_2 \overline{x}_3 \vee x_1\overline{x}_2\overline{x}_3 \vee x_1 x_2 \overline{x}_3 \,,$$

$$f(\mathbf{x}) = (x_1 \vee x_2 \vee x_3)(x_1 \vee \overline{x}_2 \vee \overline{x}_3)(\overline{x}_1 \vee x_2 \vee \overline{x}_3)(\overline{x}_1 \vee \overline{x}_2 \vee \overline{x}_3)$$

are the DNF and CNF of the function given in Table 4.5.

Naturally, this representation is not very useful, since the elementary conjunctions and disjunctions have the same length like the binary vectors. However, by applying the rules of a Boolean Algebra, we can find shorter expressions for $f(\mathbf{x})$:

$$f(\mathbf{x}) = \overline{x}_1\overline{x}_2 x_3 \vee \overline{x}_1 x_2 \overline{x}_3 \vee x_1\overline{x}_2\overline{x}_3 \vee x_1 x_2 \overline{x}_3$$
$$= \overline{x}_1\overline{x}_2 x_3 \vee x_1\overline{x}_2\overline{x}_3 \vee (\overline{x}_1 \vee x_1)x_2\overline{x}_3$$
$$= x_1 x_2 x_3 \vee x_1\overline{x}_2\overline{x}_3 \vee x_2\overline{x}_3 \,,$$

using the identities: $\overline{x}_1 \vee x_1 = 1$, $1x_2\overline{x}_3 = x_2\overline{x}_3$.

Sometimes these simplifications are a bit "tricky" and require some skills and training:

$$f(\mathbf{x}) = \overline{x}_1\overline{x}_2 x_3 \vee \overline{x}_1 x_2 \overline{x}_3 \vee x_1\overline{x}_2\overline{x}_3 \vee x_1 x_2 \overline{x}_3$$
$$= \overline{x}_1\overline{x}_2 x_3 \vee \overline{x}_1 x_2 \overline{x}_3 \vee x_1\overline{x}_2\overline{x}_3 \vee x_1 x_2 \overline{x}_3 \vee \underline{x_1 x_2 \overline{x}_3}$$
$$= \overline{x}_1\overline{x}_2 x_3 \vee (\overline{x}_1 \vee x_1)x_2\overline{x}_3 \vee x_1(x_2 \vee \overline{x}_2)\overline{x}_3$$
$$= \overline{x}_1\overline{x}_2 x_3 \vee x_2\overline{x}_3 \vee x_1\overline{x}_3 \,.$$

The underlined conjunction was "artificially" added (which is possible because of $x = x \vee x$) and could be used for a second simplification. Now each conjunction contributes to the values 1 of $f(\mathbf{x})$:

$$\overline{x}_1\overline{x}_2 x_3 \quad \Longrightarrow \quad \{(001)\}\,,$$
$$x_1\overline{x}_3 \quad \Longrightarrow \quad \{(100), (110)\}\,,$$
$$x_2\overline{x}_3 \quad \Longrightarrow \quad \{(010), (110)\}\,.$$

The vector (110) still appears twice; however, no further simplification is possible. Sometimes the following point of view is useful:

- $x_1\overline{x}_3$ is created by "omitting" x_2 or $\overline{x}_2$ in some conjunctions; the number of literals in some conjunctions is reduced.
- In $x_1\overline{x}_3$, no variable can be omitted without changing $f(\mathbf{x})$; hence, $x_1\overline{x}_3$ is denoted by *minterm* of $f(\mathbf{x})$.

In the same way, we can transform

$$f(\mathbf{x}) = (x_1 \vee x_2 \vee x_3)(x_1 \vee \overline{x}_2 \vee \overline{x}_3)(\overline{x}_1 \vee x_2 \vee \overline{x}_3)(\overline{x}_1 \vee \overline{x}_2 \vee \overline{x}_3)$$

$$= (x_1 \vee x_2 \vee x_3)(x_1 \vee \overline{x}_2 \vee \overline{x}_3)(x_2\overline{x}_2 \vee (\overline{x}_1 \vee \overline{x}_3))$$

$$= (x_1 \vee x_2 \vee x_3)(x_1 \vee \overline{x}_2 \vee \overline{x}_3)(\overline{x}_1 \vee \overline{x}_3) \,,$$

using the identities: $\overline{x}_2 \wedge x_2 = 0, 0 \vee (\overline{x}_1 \vee \overline{x}_3) = \overline{x}_1 \vee \overline{x}_3$.

A shorter representation would be given by

$$f(\mathbf{x}) = (\dot{x}_1 \vee x_2 \vee x_3)(\overline{x}_1 \vee \overline{x}_3)(\overline{x}_2 \vee \overline{x}_3) \,.$$

No shorter disjunctions can be used for the representation of $f(\mathbf{x})$; this expression consists of *maxterms*.

Definition 4.5 (Disjunctive Form) Any disjunction of conjunctions that represents a logic function $f(x_1, \ldots, x_n)$ is a *disjunctive form* (DF) of $f(x_1, \ldots, x_n)$. The DF of $f(\mathbf{x}) = 0(\mathbf{x})$ is $0(\mathbf{x}) = 0$.

Definition 4.6 (Conjunctive Form) Any conjunction of disjunctions that represents a logic function $f(x_1, \ldots, x_n)$ is a *conjunctive form* (CF) of $f(x_1, \ldots, x_n)$. The CF of $f(\mathbf{x}) = 1(\mathbf{x})$ is $1(\mathbf{x}) = 1$.

Note 4.6 The function $f(\mathbf{x})$ will depend on those variables $\{x_1, \ldots, x_n\}$ that appear in the conjunctions or disjunctions. Sometimes subsets or supersets of this set have to be considered as well.

Generally, there are many different possibilities of simplifying a conjunctive or disjunctive normal form, among these *disjunctive* and *conjunctive forms*; it can now be tried to find optimal forms, for instance, with the smallest number of conjunctions or variables, etc. There is a huge number of publications devoted to these *minimization problems*; we will provide some algorithmic solutions at a later point in time.

Different elementary conjunctions and disjunctions have one remarkable property: they are *orthogonal* to each other.

Theorem 4.5 *It holds for two different elementary conjunctions C_1, C_2 that*

$$C_1 \wedge C_2 = 0$$

and for two different elementary disjunctions D_1, D_2 that

$$D_1 \vee D_2 = 1 \,.$$

Proof Two different elementary conjunctions must differ in at least one literal, i.e., there is x_i in one conjunction, $\overline{x}_i$ in the other, and $x_i\overline{x}_i = 0$; hence, $C_1 \wedge C_2 = 0$. The same argument applies to two elementary disjunctions where you will find $x_i \vee \overline{x}_i = 1$, so that $D_1 \vee D_2 = 1$. □

Elementary conjunctions and disjunctions are "automatically" mutually orthogonal, but this concept can be defined for any two different conjunctions or disjunctions.

Definition 4.7 Two conjunctions C_1, C_2 are mutually orthogonal if

$$C_1 \wedge C_2 = 0 .$$

Two disjunctions D_1, D_2 are mutually orthogonal if

$$D_1 \vee D_2 = 1 .$$

Theorem 4.6 *It holds for two different conjunctions C_1, C_2 or different disjunctions D_1, D_2 that*

if	$C_1 \wedge C_2 = 0 ,$	*then*	$C_1 \vee C_2 = C_1 \oplus C_2 ,$	(4.6)
if	$D_1 \vee D_2 = 0 ,$	*then*	$D_1 \wedge D_2 = D_1 \odot D_2 .$	(4.7)

Proof Rule (4.6) follows from the identities $x \vee y = x \oplus y \oplus (x \wedge y)$ (3.27) and $x \oplus 0 = x$, where conjunction C_1 is substituted for x and C_2 for y. Alternatively, Table 4.2 can be used to proof rule (4.6) in the following manner:

- The premise $C_1 \wedge C_2 = 0$ is satisfied in the first three rows of Table 4.2 where the conjunction $f_1 = 0$.
- The truth of the conclusion $C_1 \vee C_2 = C_1 \oplus C_2$ follows from identical function values of the disjunction f_7 and the antivalence f_6 in these three rows of Table 4.2.

Rule (4.7) expresses the analog property for the dual algebraic structures. That means, rule (4.7) follows from the identities $x \wedge y = x \odot y \odot (x \vee y)$ (3.29) and $x \odot 1 = x$, where disjunction D_1 is substituted for x and D_2 for y. Alternatively, Table 4.2 can be used to proof rule (4.7) in the following manner:

- The premise $D_1 \vee D_2 = 1$ is satisfied in the last three rows of Table 4.2 where the disjunction $f_7 = 1$.
- The truth of the conclusion $D_1 \wedge D_2 = D_1 \odot D_2$ follows from identical function values of the conjunction f_1 and the equivalence f_9 in these three rows of Table 4.2. □

After we have seen that each function given by a table can be represented by conjunctive and disjunctive normal forms in a unique way and possibly by several conjunctive and disjunctive forms we will now take the opposite track:

> Each conjunction comprising at least one and at most n different literals defines a logic function $f(x_1, \ldots, x_n)$.

Example 4.3 Let $C = \overline{x}_1 x_3 x_4, n = 4$. $C = 1$ holds for $x_1 = 0, x_3 = 1, x_4 = 1$. In all table rows where these values will be found, we set $f(\mathbf{x}) = 1$, and in all the other rows $f(\mathbf{x})$ will be equal to 0 (see Table 4.6).

By stating $n = 4$, it is assumed that four variables have to be considered. If there are more than one conjunction connected by $\vee$, the same procedure takes place for each conjunction (using the same

Table 4.6 A function defined by one conjunction

x_1	x_2	x_3	x_4	$f(\mathbf{x})$
0	0	0	0	0
0	0	0	1	0
0	0	1	0	0
0	0	1	1	1
0	1	0	0	0
0	1	0	1	0
0	1	1	0	0
0	1	1	1	1

x_1	x_2	x_3	x_4	$f(\mathbf{x})$
1	0	0	0	0
1	0	0	1	0
1	0	1	0	0
1	0	1	1	0
1	1	0	0	0
1	1	0	1	0
1	1	1	0	0
1	1	1	1	0

Table 4.7 The function $f(\mathbf{x}) = x_1 x_3 \vee \overline{x}_2 x_4 \vee x_1 x_3 x_4$ defined by a disjunctive form

x_1	x_2	x_3	x_4	$x_1 x_3$	$\overline{x}_2 x_4$	$x_1 x_3 x_4$	$f(\mathbf{x})$
0	0	0	0	0	0	0	0
0	0	0	1	0	1	0	1
0	0	1	0	0	0	0	0
0	0	1	1	0	1	0	1
0	1	0	0	0	0	0	0
0	1	0	1	0	0	0	0
0	1	1	0	0	0	0	0
0	1	1	1	0	0	0	0
1	0	0	0	0	0	0	0
1	0	0	1	0	1	0	1
1	0	1	0	1	0	0	1
1	0	1	1	1	1	1	1
1	1	0	0	0	0	0	0
1	1	0	1	0	0	0	0
1	1	1	0	1	0	0	1
1	1	1	1	1	0	1	1

table). In principle, for each conjunction a separate column could be used, and the different columns are finally combined by $\vee$, but this can also be done immediately, and after the value 1 appeared once, it will stay up to the end of the procedure (because of $1 \vee x = 1$).

Example 4.4 The disjunctive form $f(\mathbf{x}) = x_1 x_3 \vee \overline{x}_2 x_4 \vee x_1 x_3 x_4$ results in the function given by Table 4.7.

The value of $f(\mathbf{x})$ is equal to 1 if at least one conjunction C_i is equal to 1. We see that the expressions are not uniquely defined for a function $f(\mathbf{x})$: in this example, the conjunction $x_1 x_3 x_4$ can be simply omitted because of the absorption law: $x_1 x_3 \vee x_1 x_3 x_4 = x_1 x_3$. The values 1 in the column of $x_1 x_3 x_4$ can already be found in the column of $x_1 x_3$.

Hence, we can state the following.

Theorem 4.7 *Each disjunctive form defines one and only one logic function of n variables.*

It could be seen as well (by using the conjunction $\overline{x}_1 x_3 x_4$) that the variables to be considered must be clearly stated (here x_1, x_2, x_3, x_4). If nothing is explicitly said, only those variables will be used that appear explicitly in the form. It will be a natural intention to simplify the expressions as much as possible; hence, shorter forms will always be preferable. Furthermore, it is very easy to include other variables into consideration:

Fig. 4.1 The introduction of a dummy variable

x_2	x_1	$f(x_1,x_2)=x_1 \vee x_2$
0	0	0
0	1	1
1	0	1
1	1	1

x_3	x_2	x_1	$f^*(x_1,x_2,x_3)=\overline{x}_3 f(x_1,x_2) \vee x_3 f(x_1,x_2)$
0	0	0	0
0	0	1	1
0	1	0	1
0	1	1	1
1	0	0	0
1	0	1	1
1	1	0	1
1	1	1	1

A function that depends on n variables can always be understood as a function of $n+1$ variables.

Let, for instance, $f(\mathbf{x})$ be given by $f(\mathbf{x}) = x_1 \vee x_2$; then we can use the following transformation:

$$f(\mathbf{x}) = x_1 \vee x_2$$
$$= (x_1 \vee x_2) \wedge 1$$
$$= (x_1 \vee x_2)(x_i \vee \overline{x}_i)$$
$$= x_1 x_i \vee x_2 x_i \vee x_1 \overline{x}_i \vee x_2 \overline{x}_i \ .$$

Such a newly introduced variable will be denoted by *dummy variable*. The mechanism of introducing such a variable can easily be understood by the following.

Example 4.5 Let $f(x_1, x_2) = x_1 \vee x_2$. The introduction of x_3 as a dummy variable results in $f^*(x_1, x_2, x_3) = x_1 \overline{x}_3 \vee x_2 \overline{x}_3 \vee x_1 x_3 \vee x_2 x_3$. The relation between $f(x_1, x_2)$ and $f^*(x_1, x_2, x_3)$ is given as follows:

$$f^*(x_1, x_2, x_3 = 0) = x_1 \vee x_2 \ ,$$
$$f^*(x_1, x_2, x_3 = 1) = x_1 \vee x_2 \ ,$$
$$f^*(x_1, x_2, x_3) = \overline{x}_3 f(x_1, x_2) \vee x_3 f(x_1, x_2) \ .$$

Naturally, $f(x_1, x_2)$ and $f^*(x_1, x_2, x_3)$ are two different functions because $f(x_1, x_2) : \mathbb{B}^2 \Rightarrow \mathbb{B}$ is defined depending on two variables and $f^*(x_1, x_2, x_3) : \mathbb{B}^3 \Rightarrow \mathbb{B}$ on three variables. However, since the disjunctive form of $f^*(x_1, x_2, x_3)$ can be transformed into a disjunctive form for $f(x_1, x_2)$, we can very often identify such functions $f(x_1, x_2)$ and $f^*(x_1, x_2, x_3)$ with each other and prefer the shorter representation.

Figure 4.1 shows that the values of the function table of $f(x_1, x_2)$ are used twice in $f^*(x_1, x_2, x_3)$, for $x_3 = 0$ and for $x_3 = 1$. The used order of the binary vectors x_3, x_2, x_1 and the gray shaded ranges make this quite visible.

Considerations like this make it useful to introduce the general concept of a *subfunction*.

Definition 4.8 Let $\mathbf{x} = \{x_1, \ldots, x_n\}$, $f(\mathbf{x}) : \mathbb{B}^n \Rightarrow \mathbb{B}$, $\mathbf{x}_0 = \{x_{i_1}, \ldots, x_{i_k}\}$ be a non-empty subset of $\mathbf{x}$, $\mathbf{x}_1 = \mathbf{x} \setminus \mathbf{x}_0$, $\mathbf{c} = (c_{i_1}, \ldots, c_{i_k}) \in \mathbb{B}^k$; then the function

$$f(x_{i_1} = c_{i_1}, \ldots, x_{i_k} = c_{i_k}, \mathbf{x_1})$$

is denoted by *subfunction* $f(\mathbf{x_0} = \mathbf{c}, \mathbf{x_1})$.

That means that such a subfunction will be given by fixing the values of some of the variables; we split the set $\mathbf{x} = \{x_1, \ldots, x_n\}$ into two parts:

- The set $\mathbf{x_0} = \{x_{i_1}, \ldots, x_{i_k}\}$
- The set of the remaining variables $\mathbf{x_1} = \{x_1, \ldots, x_n\} \setminus \{x_{i_1}, \ldots, x_{i_k}\}$

and fix the values of the first set $\mathbf{x_0}$. At least one variable will be fixed, and half of the function table will be selected, or at most all the variables are fixed, and one single value of the table will be selected as a constant.

Using $f^*(x_1, x_2, x_3)$ from the above, we find, for instance,

$$f^*(x_1, x_2, x_3 = 0) = x_1 \vee x_2 , \qquad\qquad f^*(x_1, x_2, x_3 = 1) = x_1 \vee x_2 ,$$

$$f^*(x_1 = 1, x_2, x_3) = 1(x_2, x_3) , \qquad\qquad f^*(x_1 = 0, x_2, x_3 = 0) = x_2 .$$

Note 4.7 Very often only the variables with fixed values will be stated, and the dependency on the other variables mostly follows from the context.

Theorem 4.8 *A given function* $f(x_1, \ldots, x_n)$ *has* $3^n - 1$ *subfunctions.*

Proof The proof is quite easy. For each variable, there are three possibilities:

- The variable is set to 0.
- The variable is set to 1.
- The variable is not set at all.

This results in 3^n different possibilities. If none of the variables is set to a given value, then $f(\mathbf{x})$ remains unchanged. In all the other cases, we receive a subfunction.   □

The possibility to find an expression for a function $f(\mathbf{x})$ that does not contain a given variable x_i leads to the following.

Definition 4.9 The function $f(\mathbf{x})$ is *independent of* x_i if

$$f(x_i = 0) = f(x_i = 1) .$$

Note 4.8 Sometimes it is also said that the function $f(\mathbf{x})$ does *not depend essentially* on the variable x_i.

We will emphasize again that this daily use of the language is a bit imprecise because a function depends on all of its variables. It expresses precisely the fact that it is possible to find a disjunctive form that does not show a literal for x_i.

The case that a function $f(\mathbf{x}) : \mathbb{B}^n \Rightarrow \mathbb{B}$ is independent of x_i can also be seen from another point of view. Due to the independence of the variables x_i, the given function $f(\mathbf{x})$ can be transformed into the function $f^*(\mathbf{x}) : \mathbb{B}^{n-1} \Rightarrow \mathbb{B}$. Both the functions $f(\mathbf{x})$ and $f^*(\mathbf{x})$ can be expressed by the same disjunctive form. For that reason, the unique name $f(\mathbf{x})$ (instead of $f^*(\mathbf{x})$) is usually used for such a function.

Due to independent variables, a disjunctive form specifies logic functions $f(\mathbf{x}) : \mathbb{B}^n \Rightarrow \mathbb{B}$ for each Boolean space $\mathbb{B}^n$ where n is greater than or equal to the number of different variables in the given expression. For a fixed acceptable number n of variables, each disjunctive form defines *exactly one* function $f(\mathbf{x}) : \mathbb{B}^n \Rightarrow \mathbb{B}$. However, it can be shown that some of these disjunctive forms must define the same function because there are much more disjunctive forms with n variables than functions of n variables.

Theorem 4.9 *There are 2^{3^n} different disjunctive forms of n variables that contain each conjunction at most once.*

Proof We consider a conjunction to be built from n variables; each variable can be used negated, non-negated, or not at all, which results in 3^n different conjunctions (see Theorem 4.8 regarding the number of subfunctions). Each of these conjunctions can be used at most once or cannot be used in a disjunctive form that gives 2^{3^n} different disjunctive forms. The special case where all the variables are omitted can be identified with the constant 0. □

This underlines the requirement that, for a given function $f(\mathbf{x})$, disjunctive forms with special properties can or have to be found, which opens a large field of optimization problems.

Based on these considerations, another compact representation of a logic function $f(\mathbf{x})$ can be introduced. We assume that n variables $x_1, \ldots, x_n$ have to be considered, and the ternary vectors to be constructed have consequently the components $t_1, \ldots, t_n$.

Definition 4.10 For any conjunction C of literals, the *ternary vector* $\mathbf{t}(C)$ is given by

$$t_i(C) = \begin{cases} 0 & \text{for} \quad \overline{x}_i \text{ in } C, \\ 1 & \text{for} \quad x_i \text{ in } C, \\ - & \text{for} \quad x_i \text{ not in } C. \end{cases}$$

Theorem 4.10 *The function $f(\mathbf{x})$ can be represented by lists (matrices) $D(f)$ of ternary vectors $\mathbf{t}_k, k = 1, \ldots, m$, where each ternary vector represents a conjunction of a disjunctive form.*

Proof The construction from the above shows this theorem quite easily. $f(\mathbf{x})$ will (can) be represented by (at least) one disjunctive form (DF). Each conjunction C_l is represented by its ternary vector, and the *list of ternary vectors* (TVL) represents a DF of the function $f(\mathbf{x})$. □

Example 4.6 The disjunctive form of the function

$$f(\mathbf{x}) = x_1 \overline{x}_3 x_5 \vee x_2 x_4 x_6 \vee x_1 \overline{x}_3 \overline{x}_6 \vee x_3 x_4 x_5 \overline{x}_6 \vee x_1 x_3 x_6$$

can be expressed by the *ternary vector list* (TVL)

$$D(f) = \begin{array}{c} \begin{array}{cccccc} x_1 & x_2 & x_3 & x_4 & x_5 & x_6 \end{array} \\ \hline \begin{array}{cccccc} 1 & - & 0 & - & 1 & - \\ - & 1 & - & 1 & - & 1 \\ 1 & - & 0 & - & - & 0 \\ - & - & 1 & 1 & 1 & 0 \\ 1 & - & 1 & - & - & 1 \end{array} \\ \hline \end{array},$$

where $\mathrm{D}(f)$ indicates that this TVL represents a disjunctive form of the function $f(\mathbf{x})$.

Note 4.9 This is a very interesting point in the use of ternary vectors. The vector has been introduced as the coding of a conjunction, the list as the coding of a disjunctive form. All $-$ elements can be replaced in each ternary vector by all combinations of 0 and 1. The use of only 0 gives for a selected ternary vector the smallest element $\mathbf{x}$ of an interval $I_{\mathbf{xy}}$ (3.39), and the use of only 1 gives the largest element $\mathbf{y}$ of this interval. In fact, for all binary vectors $\mathbf{z} \in I_{\mathbf{xy}}$, we have $C_1(\mathbf{z}) = 1$. Hence, the ternary vector is at the same time the description of the set of all binary vectors with a given property (here $C_1(\mathbf{z}) = 1$). We will come back to these considerations when logic equations are considered.

See, for instance, the conjunction $C = x_3 x_4 x_5 \overline{x}_6$ and $\mathbf{t}(C) = (- - 1\ 1\ 1\ 0)$. $C = 1$ for the binary vectors $(00|1110)$, $(01|1110)$, $(10|1110)$, $(11|1110)$. The values for x_3, x_4, x_5, x_6 remain unchanged, and for x_1, x_2, all possible values can be used. That means that the ternary vector $\mathbf{t}(C)$ describes the *interval*

$$(001110) \leq \mathbf{z} \leq (111110) \,,$$

i.e., all binary vectors $\mathbf{z}$ with $C(\mathbf{z}) = 1$. The whole list of ternary vectors represents all binary vectors with $f(\mathbf{x}) = 1$ by a *set of intervals*.

Now it is also very consistent that the "conjunction" where all variables have been omitted (i.e., the ternary vector consisting only of $-$ elements) represents the function $1(\mathbf{x})$ since the replacement of all $-$ results in the whole Boolean space $\mathbb{B}^n$. For instance, the function $1(x_1, x_2, \ldots, x_{10})$ can be represented by the TVL in DF

$$\mathrm{D}(1(x_1, x_2, \ldots, x_{10})) = \frac{x_1\ x_2\ x_3\ x_4\ x_5\ x_6\ x_7\ x_8\ x_9\ x_{10}}{-\ -\ -\ -\ -\ -\ -\ -\ -\ -}\ .$$

There is no conjunction and consequently no ternary vector that represents the function $0(\mathbf{x})$. This property does not restrict the use of a TVL as representation of a logic function because the function $0(\mathbf{x})$ can be represented as an empty TVL. For the function $0(x_1, x_2, \ldots, x_{10})$ we get the TVL in DF

$$\mathrm{D}(0(x_1, x_2, \ldots, x_{10})) = \frac{x_1\ x_2\ x_3\ x_4\ x_5\ x_6\ x_7\ x_8\ x_9\ x_{10}}{}\ .$$

Unfortunately, however, the sets of ternary vectors representing a function $f(\mathbf{x})$ depend on the initial disjunctive form, the intervals can overlap, and the minimization problem still occurs.

On the other side, lists like this can be used in computers with hundreds of columns and thousands of rows and will be the base of many algorithms. In the sequel, we will use *orthogonal lists of ternary vectors* as the main data structure. We already defined the orthogonality of two conjunctions C_1 and C_2. This property can be seen at ternary vectors when it holds for at least one component t_i:

- $t_i(C_1) = 0$ and $t_i(C_2) = 1$
- $t_i(C_1) = 1$ and $t_i(C_2) = 0$

because in these cases one of the associated conjunctions contains $\overline{x}_i$ and the other x_i. If there are two different conjunctions in a disjunctive form that are not mutually orthogonal, then it is possible to transform these conjunctions in such a way that they still represent the same function but are orthogonal to each other.

Because of the importance of this concept that will be the foundation for many algorithms in this area, we will start with a simple.

Example 4.7 Let $f(x_1, x_2) = x_1 \vee x_2$. This is a very simple disjunctive form. The corresponding list of ternary vectors (TVL) can easily be found as

$$\mathrm{D}(f) = \begin{array}{cc} x_1 & x_2 \\ \hline 1 & - \\ - & 1 \end{array} .$$

Now the following transformation can be taken into consideration:

$$f(x_1, x_2) = x_1 \vee x_2 = x_1 \vee (x_1 \vee \overline{x}_1)x_2 = x_1 \vee x_1 x_2 \vee \overline{x}_1 x_2 = x_1 \vee \overline{x}_1 x_2$$

using the absorption law. The two conjunctions $C_1 = x_1$ and $C_2 = \overline{x}_1 x_2$ are orthogonal to each other (due to the existence of x_1 in the first and of $\overline{x}_1$ in the second conjunction). Using the rule (4.6) this orthogonal disjunctive form can be transformed into an orthogonal antivalence form simply by replacing the operation sign $\vee$ by $\oplus$:

$$f(x_1, x_2) = x_1 \vee \overline{x}_1 x_2 = x_1 \oplus \overline{x}_1 x_2 .$$

Due to the same set of conjunctions in these two forms a single TVL can be used to represent both of them. We indicate these alternative forms by the specification of the TVL as ODA (orthogonal disjunctive or antivalence) form:

$$\mathrm{ODA}(f) = \begin{array}{cc} x_1 & x_2 \\ \hline 1 & - \\ 0 & 1 \end{array} .$$

This representation has more variables than $x_1 \vee x_2$ but is still shorter than the disjunctive normal form $f(x_1, x_2) = x_1 \overline{x}_2 \vee \overline{x}_1 x_2 \vee x_1 x_2$. Hence, ODA-forms can be "a little bit" longer than the shortest representations, but they have the very important property that each binary vector $\mathbf{b}$ with $f(\mathbf{b}) = 1$ is generated by one and only one conjunction.

Let us see a second

Example 4.8 $f(\mathbf{x}) = x_1 x_3 \vee x_2 x_4$. We find the corresponding TVL as

$$\mathrm{D}(f) = \begin{array}{cccc} x_1 & x_2 & x_3 & x_4 \\ \hline 1 & - & 1 & - \\ - & 1 & - & 1 \end{array} .$$

Now the orthogonal representation will be found by the following steps:

1. We take the first vector without any changes: $(1 - 1 -)$.
2. In the second vector, find the first position where a $-$ in the second vector meets a value (0 or 1) in the first vector (here immediately the position of x_1). Replace the second vector $(-1 - 1)$ by two vectors $(0\ 1 - 1)$ and $(1\ 1 - 1)$ such that the first of these two vectors is orthogonal to the vector $(1 - 1 -)$, which results in

$$D(f) = \begin{array}{cccc} x_1 & x_2 & x_3 & x_4 \\ \hline 1 & - & 1 & - \\ 0 & 1 & - & 1 \\ 1 & 1 & - & 1 \\ \hline \end{array} \; .$$

The second vector is now orthogonal to the first and to the third, and the third will be transformed in the same way because it is not yet orthogonal to the first vector.

3. Processing the third vector:

$$D(f) = \begin{array}{cccc} x_1 & x_2 & x_3 & x_4 \\ \hline 1 & - & 1 & - \\ 0 & 1 & - & 1 \\ 1 & 1 & - & 1 \\ \hline \end{array} = \begin{array}{cccc} x_1 & x_2 & x_3 & x_4 \\ \hline 1 & - & 1 & - \\ 0 & 1 & - & 1 \\ 1 & 1 & 0 & 1 \\ 1 & 1 & 1 & 1 \\ \hline \end{array} \; .$$

The new third vector is orthogonal to the first and the second, and the last one (understood as a conjunction) is absorbed by the first, or the set of binary vectors covered by the first vector contains (1111); hence, this vector can be omitted, and we finally get

$$ODA(f) = \begin{array}{cccc} x_1 & x_2 & x_3 & x_4 \\ \hline 1 & - & 1 & - \\ 0 & 1 & - & 1 \\ 1 & 1 & 0 & 1 \\ \hline \end{array}$$

as an equivalent orthogonal representation of the same function:

$$f(\mathbf{x}) = x_1 x_3 \vee \overline{x}_1 x_2 x_4 \vee x_1 x_2 \overline{x}_3 x_4 = x_1 x_3 \oplus \overline{x}_1 x_2 x_4 \oplus x_1 x_2 \overline{x}_3 x_4 \; .$$

It is easy to see the introduction of dummy variables and appropriate simplifications in the background.

The understanding

$$x_1 \vee x_2 = x_1 \vee \overline{x}_1 x_2$$

can also be used directly:

$$\begin{aligned} f(\mathbf{x}) &= x_1 x_3 \vee x_2 x_4 \\ &= x_1 x_3 \vee (\overline{x_1 x_3}) \, x_2 x_4 \\ &= x_1 x_3 \vee (\overline{x}_1 \vee \overline{x}_3) \, x_2 x_4 \\ &= x_1 x_3 \vee \overline{x}_1 x_2 x_4 \vee x_2 \overline{x}_3 x_4 \\ &= x_1 x_3 \vee \overline{x}_1 x_2 x_4 \vee (\overline{\overline{x}_1 x_2 x_4}) x_2 \overline{x}_3 x_4 \\ &= x_1 x_3 \vee \overline{x}_1 x_2 x_4 \vee (x_1 \vee \overline{x}_2 \vee \overline{x}_4) x_2 \overline{x}_3 x_4 \\ &= x_1 x_3 \vee \overline{x}_1 x_2 x_4 \vee x_1 x_2 \overline{x}_3 x_4 \; . \end{aligned}$$

Now the three conjunctions are mutually orthogonal to each other. The previously given algorithm implements just this idea.

When we understand ternary vectors as sets or intervals of binary vectors, then we have an orthogonal representation of a function by ternary vectors if and only if the intersection of the sets of binary vectors represented by the ternary vectors is empty:

$$(1 - 1 -) \qquad \Rightarrow \qquad S_1 = \{(1010), (1011), (1110), (1111)\} \,,$$

$$(0\ 1 - 1) \qquad \Rightarrow \qquad S_3 = \{(0101), (0111)\} \,,$$

and the intersection $S_1 \cap S_3$ of these two sets is empty.

In order to orthogonalize $S_2 = (- \ 1 - \ 1)$ with regard to $S_1 = (1 - 1 -)$, where the components are labeled as before by (x_1, x_2, x_3, x_4), we use this idea in the following way:

- Select one vector that will not change. Normally this should be the vector that describes the larger interval (representing the shorter conjunction), i.e., the vector with the largest number of components equal to $-$. Here we select $S_1 = (1 - 1 -)$.
- The interval represented by the vector $S_2 = (- \ 1 - \ 1)$ will not be taken as a whole; only those parts of this interval are of interest that belong to the complement of the first vector. Hence, we find the complement of the vector $S_1 = (1 - 1 -)$ as

$$\overline{S}_1 = \begin{array}{cccc} x_1 & x_2 & x_3 & x_4 \\ 0 & - & - & - \\ 1 & - & 0 & - \end{array} .$$

The vectors of the complement $\overline{S}_1$ must show a difference in the first components in comparison with ternary vector of S_2, or, if the first components are equal to each other, there must be a difference in the third components. Other components are not available.

- The intersection

$$\overline{S}_1 \cap S_2 = \begin{array}{cccc} x_1 & x_2 & x_3 & x_4 \\ 0 & - & - & - \\ 1 & - & 0 & - \end{array} \cap \begin{array}{cccc} x_1 & x_2 & x_3 & x_4 \\ - & 1 & - & 1 \end{array} = \begin{array}{cccc} x_1 & x_2 & x_3 & x_4 \\ 0 & 1 & - & 1 \\ 1 & 1 & 0 & 1 \end{array}$$

results in the required orthogonal representation of the second vector, and we get for the union of S_1 and S_2 the same result as before:

$$S_1 \cup S_2 = S_1 \cup (\overline{S}_1 \cap S_2) = \begin{array}{cccc} x_1 & x_2 & x_3 & x_4 \\ 1 & - & 1 & - \\ 0 & 1 & - & 1 \\ 1 & 1 & 0 & 1 \end{array} .$$

In principle, any order of the vectors to be orthogonalized can be used and the results can be very different. The *orthogonal minimization* of disjunctive, conjunctive, . . . forms is another interesting and difficult problem. The selection of the other vector would result in

$$S_2 \cup S_1 = S_2 \cup (\overline{S}_2 \cap S_1) = \begin{array}{cccc} x_1 & x_2 & x_3 & x_4 \\ - & 1 & - & 1 \\ 1 & 0 & 1 & - \\ 1 & 1 & 1 & 1 \end{array} .$$

Fig. 4.2 The representation of a function $f(\mathbf{x})$ by means of a Karnaugh-map

$x_3\,x_4$					$f(\mathbf{x})$
0 0	0	1	0	0	
0 1	1	1	0	0	
1 1	1	1	1	0	
1 0	0	1	1	0	
	0	1	1	0	x_2
	0	0	1	1	x_1

This approach is very elegant because the vectors of the complement are automatically orthogonal to each other.

The generalization of these steps will result in an algorithm that transforms any disjunctive form into an orthogonal disjunctive form. The border case that can be reached is the disjunctive normal form that shows the mutual orthogonality of all pairs of vectors for sure, but very often this property can be achieved with a much smaller number of conjunctions (in the example above the disjunctive normal form would have seven conjunctions, and $f(\mathbf{x}) = 1$ holds for seven different elementary conjunctions).

Note 4.10 It is very important to indicate that the list of ternary vectors $\mathrm{D}(f)$ describes the vectors for $f(\mathbf{x}) = 1$, i.e., by $\mathrm{D}(f) = \{\, \mathbf{x} \mid f(\mathbf{x}) = 1 \}$, because we will see later that the vectors with $f(\mathbf{x}) = 0$ derived from conjunctive forms can be used in the same way.

A next representation of logic functions that is used in many textbooks is useful for smaller numbers of variables (up to 6 or 7). It is denoted by *Karnaugh-map* or *Karnaugh–Veitch diagram* and uses a matrix-type representation together with the *Gray code* for the arguments.

Example 4.9 Let $f(x_1, x_2, x_3, x_4) = \overline{x}_1 x_4 \vee x_2 x_3 \vee \overline{x}_1 x_2$. Figure 4.2 shows the representation of this function using the Karnaugh-map. The coding of the variables x_1, x_2, and x_3, x_4, resp., follows the Gray code.

The Gray code below the Karnaugh-map determines the arguments of the represented function for the columns, and the Gray code to the left of the Karnaugh-map specifies the rows. The assignment of the variables to the two Gray codes determines the function value of each field of a Karnaugh-map. Different assignments of the variables to the two Gray codes lead to different patterns of the same function within the Karnaugh-map. All permutations of these assignments are acceptable. We prefer the assignment of the variables as shown in Fig. 4.2 of Example 4.9.

Based on the given order of the variables of the function, we assign the first variable to the lowest row of the horizontal Gray code and succeed with the next variables upward. The remaining variables are assigned from the left to the right of the vertical Gray code. This assignment has the useful property that the values of the variables in the given order specify connected intervals of columns or rows that are cut in half intervals for each next variable until the selected column and row are determined.

The Gray code has the nice property that the values 1 described by one conjunction (like $x_2 x_3$) appear in "neighbored" cells. With some training it is not too difficult to determine for a given function $f(\mathbf{x})$ a "good (short)" representation of $f(\mathbf{x})$ by a disjunctive form. Since the Karnaugh-map can be used only for small numbers of variables, this representation will be very often replaced or extended by ternary vectors and appropriate computer programs.

Table 4.8 Logic function used to express as a conjunctive normal form

x_1	x_2	x_3	$f(\mathbf{x})$	x_1	x_2	x_3	$f(\mathbf{x})$
0	0	0	1	1	0	0	0
0	0	1	0	1	0	1	1
0	1	0	0	1	1	0	1
0	1	1	1	1	1	1	0

Because of the easy transition between disjunctive forms and logic functions, we will use a rather "relaxed" language, at least in most cases, like in other fields of mathematics, and make no difference between the logic expressions and the function that is defined by these expressions. If necessary, however, the difference will be emphasized and taken into consideration.

The next important representation of logic functions can be derived in many different ways. We simply go back to the function table and describe the values 0 by elementary disjunctions. Then, the following conclusions are obvious:

- Every 0 of $f(x_1, \ldots, x_n)$ can be described (generated) by a full disjunction of literals (*elementary disjunction*).
- Combining all elementary disjunctions of $f(\mathbf{x})$ by $\wedge$ gives the *conjunctive normal form* (CNF) of f (uniquely defined).
- "Shorter" conjunctive forms (CF) are possible. We have the same minimization problem as for disjunctive forms.
- The function $1(\mathbf{x})$ contains no 0 value so that no elementary disjunction for this function exists. We define the expression 1 for both the CNF and CF of the function $1(\mathbf{x})$.

Example 4.10 The function $f(x_1, x_2, x_3)$ defined in Table 4.8 can be represented by the conjunctive normal form:

$$f(\mathbf{x}) = (x_1 \vee x_2 \vee \overline{x}_3)(x_1 \vee \overline{x}_2 \vee x_3)(\overline{x}_1 \vee x_2 \vee x_3)(\overline{x}_1 \vee \overline{x}_2 \vee \overline{x}_3) \, .$$

Note 4.11 The brackets of conjunctive forms cannot be omitted.

Another approach would be to use the disjunctive normal form or any disjunctive form and apply the distributive law, together with all kinds of possible simplifications. Let us take, for instance, $f(\mathbf{x}) = \overline{x}_1\overline{x}_2x_3 \vee x_1\overline{x}_3 \vee x_2\overline{x}_3$. Then, the application of the distributive law and some simplifications give the following expression:

$$f(\mathbf{x}) = (\overline{x}_1 \vee x_1 \vee x_2)(\overline{x}_1 \vee x_1 \vee \overline{x}_3)(\overline{x}_1 \vee \overline{x}_3 \vee x_2)(\overline{x}_1 \vee \overline{x}_3 \vee \overline{x}_3)$$

$$(\overline{x}_2 \vee x_1 \vee x_2)(\overline{x}_2 \vee x_1 \vee \overline{x}_3)(\overline{x}_2 \vee \overline{x}_3 \vee x_2)(\overline{x}_2 \vee \overline{x}_3 \vee \overline{x}_3)$$

$$(x_3 \vee x_1 \vee x_2)(x_3 \vee x_1 \vee \overline{x}_3)(x_3 \vee \overline{x}_3 \vee x_2)(x_3 \vee \overline{x}_3 \vee \overline{x}_3)$$

$$= (x_1 \vee x_2 \vee x_3)(\overline{x}_1 \vee \overline{x}_3)(\overline{x}_2 \vee \overline{x}_3) \, .$$

The understanding of conjunctions and disjunctions as special functions explains the names *minterm* and *maxterm* quite easily. We go back to

$$f(\mathbf{x}) = \overline{x}_1\overline{x}_2x_3 \vee x_1\overline{x}_3 \vee x_2\overline{x}_3$$

$$= (x_1 \vee x_2 \vee x_3)(\overline{x}_1 \vee \overline{x}_3)(\overline{x}_2 \vee \overline{x}_3)$$

Table 4.9 The concepts of minterm and maxterm

x_1	x_2	x_3	$x_1\overline{x}_3$	$f(\mathbf{x})$	$\overline{x}_2 \vee \overline{x}_3$
0	0	0	0	0	1
0	0	1	0	1	1
0	1	0	0	1	1
0	1	1	0	0	0
1	0	0	1	1	1
1	0	1	0	0	1
1	1	0	1	1	1
1	1	1	0	0	0

and represent $f(\mathbf{x})$, the minterm $x_1\overline{x}_3$, and the maxterm $(\overline{x}_2 \vee \overline{x}_3)$ in Table 4.9.

In the sense of the partial order $\leq$ in $\mathbb{B}^8$ (the function $f(\mathbf{x})$ has eight values), it can be seen immediately that

$$x_1\overline{x}_3 \leq f(\mathbf{x}) \leq \overline{x}_2 \vee \overline{x}_3 .$$

The minterms are limiting $f(\mathbf{x})$ from below by values 1 ($f(\mathbf{x})$ must have the values 1 of the minterm), and the maxterms are limiting $f(\mathbf{x})$ from above by values 0 ($f(\mathbf{x})$ must have the values 0 of the maxterm). The values of $x_1\overline{x}_3$ and $\overline{x}_2 \vee \overline{x}_3$ determine the value of $f(\mathbf{x})$ for four vectors (row 4, 5, 7, 8). These constraints finally (together with all the other minterms or maxterms of $f(\mathbf{x})$) result in a unique definition of $f(\mathbf{x})$.

For each disjunction, we will define a ternary vector that describes the generated values 0 implicitly.

Definition 4.11 For any disjunction D of literals, the ternary vector $\mathbf{t}(D)$ is given by the component-wise definition

$$t_i(D) = \begin{cases} 0 & \text{for} & \overline{x}_i \in D , \\ 1 & \text{for} & x_i \in D , \\ - & \text{for} & x_i \notin D , \end{cases}$$

$$1 \leq i \leq n .$$

The ternary vector $\mathbf{t}(D)$ with d − elements describes a set of 2^d binary vectors $\mathbf{b} = (b_1, b_2, \ldots, b_n)$ with $f(\overline{b}_1, \overline{b}_2, \ldots, \overline{b}_n) = 0$. Note, the components $t_i(D)$ and $t_i(C)$ use the same encoding; however, the literals associated to one ternary vector must be connected by $\vee$ in case of $\mathrm{K}(f)$, but by $\wedge$ in case of $\mathrm{D}(f)$. The concept of orthogonality also applies to conjunctive forms as well as to the associated lists of ternary vectors.

For $f(\mathbf{x}) = (x_1 \vee x_2 \vee x_3)(\overline{x}_1 \vee \overline{x}_3)(\overline{x}_2 \vee \overline{x}_3)$, we get the three ternary vectors of the list

$$\mathrm{K}(f) = \begin{array}{ccc} x_1 & x_2 & x_3 \\ \hline 1 & 1 & 1 \\ 0 & - & 0 \\ - & 0 & 0 \\ \hline \end{array} ,$$

where $\mathrm{K}(f)$ indicates that a conjunctive form (CF) of $f(\mathbf{x})$ is represented in the TVL. We reuse here the form specification of XBOOLE [7–9], where the letter K has been selected for the German word *Konjunktion* as translation of conjunction. After orthogonalization of the second and third vectors, we get

$$\text{OKE}(f) = \begin{array}{|ccc}
x_1 & x_2 & x_3 \\
\hline
1 & 1 & 1 \\
0 & - & 0 \\
1 & 0 & 0
\end{array} \ .$$

The form specification $\text{OKE}(f)$ indicates that this TVL can directly be mapped into an orthogonal conjunctive form $\text{K}(f)$ or an orthogonal equivalence form $\text{E}(f)$ that contains the same disjunctions:

$$f(\mathbf{x}) = (x_1 \vee x_2 \vee x_3) \wedge (\overline{x}_1 \vee \overline{x}_3) \wedge (x_1 \vee \overline{x}_2 \vee \overline{x}_3) \,,$$

$$f(\mathbf{x}) = (x_1 \vee x_2 \vee x_3) \odot (\overline{x}_1 \vee \overline{x}_3) \odot (x_1 \vee \overline{x}_2 \vee \overline{x}_3) \,.$$

For easy denotations, we introduce the following symbols:

Definition 4.12

$$F^1(\mathbf{x}) = \{\mathbf{x} \mid \mathbf{x} \in \mathbb{B}^n, f(\mathbf{x}) = 1\} \,,$$

$$F^0(\mathbf{x}) = \{\mathbf{x} \mid \mathbf{x} \in \mathbb{B}^n, f(\mathbf{x}) = 0\} \,.$$

Summarizing the constructions, it can explicitly be stated:

- Disjunctive (normal) forms and the associated ternary vectors describe explicitly the set $F^1(\mathbf{x})$. The set $F^0(\mathbf{x})$ can be calculated as the complement $\mathbb{B}^n \setminus F^1(\mathbf{x})$.
- The application of the negation with regard to De Morgan to conjunctive (normal) forms $\text{K}(f)$ results in disjunctive (normal) forms of the negated function $\text{D}(\overline{f})$, and the ternary vectors associated to $\text{D}(\overline{f})$ describe explicitly the set $F^0(\mathbf{x})$. The set $F^1(\mathbf{x})$ is given as the complement $\mathbb{B}^n \setminus F^0(\mathbf{x})$.

If the transition from a conjunctive form to a disjunctive form of $f(\mathbf{x})$ is required, then not necessarily the function table has to be used. The comprehensive use of the *distributive laws* and all the simplification rules is also possible.

Example 4.11

$$f(\mathbf{x}) = (x_1 \vee x_2 \vee x_3)(\overline{x}_1 \vee x_2 \vee x_3)$$

$$= x_1\overline{x}_1 \vee x_1x_2 \vee x_1x_3 \vee x_2\overline{x}_1 \vee x_2x_2 \vee x_2x_3 \vee x_3\overline{x}_1 \vee x_3x_2 \vee x_3x_3$$

$$= 0 \vee x_1x_2 \vee x_1x_3 \vee \overline{x}_1x_2 \vee x_2 \vee x_2x_3 \vee \overline{x}_1x_3 \vee x_2x_3 \vee x_3$$

$$= x_2 \vee x_3 \,.$$

The orthogonality of disjunctions can be used for the definition of antivalence normal forms. We leave the algebraic structure of a *Boolean Algebra* and step over to a *Boolean Ring*. The identity $a \vee b = a \oplus b \oplus ab$ allows to replace $\vee$ in any disjunctive form; if a is orthogonal to b, then this step simplifies to

$$a \vee b = a \oplus b \,.$$

Hence, any disjunctive normal form and any orthogonal disjunctive form can be transformed simply by replacing all $\vee$'s by $\oplus$'s. The transformed disjunctive normal form will be called *antivalence normal form* (ANF), and any transformed orthogonal disjunctive form results in an orthogonal *antivalence form* (AF).

Example 4.12

$$f(\mathbf{x}) = \overline{x}_1 x_2 x_3 \vee x_1 \overline{x}_2 x_3 \vee x_1 x_2 x_3$$
$$= \overline{x}_1 x_2 x_3 \oplus x_1 \overline{x}_2 x_3 \oplus x_1 x_2 x_3 \, .$$

The complement can be eliminated by using the identity

$$\overline{x} = (1 \oplus x)$$

followed by the application of the distributive law.

$$f(\mathbf{x}) = \overline{x}_1 x_2 x_3 \oplus x_1 \overline{x}_2 x_3 \oplus x_1 x_2 x_3$$
$$= (1 \oplus x_1) x_2 x_3 \oplus x_1 (1 \oplus x_2) x_3 \oplus x_1 x_2 x_3$$
$$= x_2 x_3 \oplus x_1 x_2 x_3 \oplus x_1 x_3 \oplus x_1 x_2 x_3 \oplus x_1 x_2 x_3$$
$$= x_2 x_3 \oplus x_1 x_3 \oplus x_1 x_2 x_3 \, .$$

The final representation of Example 4.12 is denoted by *antivalence polynomial* (AP), due to the restriction to non-negated variables (positive polarity *Reed–Muller polynomial*, or *Shegalkin polynomial*). Pairs of equal conjunctions can be eliminated because of $a \oplus a = 0$ and $b \oplus 0 = b$. Since the disjunctive normal form is unique, the antivalence normal form as well as the antivalence polynomial are also uniquely defined. It holds the following.

Theorem 4.11

- *Each function $f(\mathbf{x})$ can be uniquely represented by an antivalence normal form.*
- *Each function $f(\mathbf{x})$ can be uniquely represented by an antivalence polynomial.*

At this point in time it should be quite clear that $0(\mathbf{x})$ plays a special role. The simple expression 0 can be used as a representation of $0(\mathbf{x})$ for both the antivalence normal form ANF and the antivalence polynomial AP.

Sometimes it might be possible (or necessary or desirable) to eliminate the complement only for some variables and keep it for other variables. Hence, a large set of *mixed antivalence forms* is possible as well.

Example 4.13 We take the function $f(\mathbf{x})$ of Example 4.12, eliminate, for instance, only the complement of x_1, and obtain

$$f(\mathbf{x}) = \overline{x}_1 x_2 x_3 \oplus x_1 \overline{x}_2 x_3 \oplus x_1 x_2 x_3$$
$$= (1 \oplus x_1) x_2 x_3 \oplus x_1 \overline{x}_2 x_3 \oplus x_1 x_2 x_3$$
$$= x_2 x_3 \oplus x_1 x_2 x_3 \oplus x_1 \overline{x}_2 x_3 \oplus x_1 x_2 x_3$$
$$= x_2 x_3 \oplus x_1 \overline{x}_2 x_3 \, .$$

Another possibility would be to require that some of the variables appear only negated; then we have to replace the variable x_i by $1 \oplus \overline{x}_i$ and use again the distributive law and possible simplifications.

Example 4.14 The final expression $x_2 x_3 \oplus x_1 \overline{x}_2 x_3$ of Example 4.13 would be transformed with regard to only negated variables x_2 as follows:

$$f(\mathbf{x}) = x_2 x_3 \oplus x_1 \overline{x}_2 x_3$$
$$= (1 \oplus \overline{x}_2) x_3 \oplus x_1 \overline{x}_2 x_3$$
$$= x_3 \oplus \overline{x}_2 x_3 \oplus x_1 \overline{x}_2 x_3$$
$$= x_3 \oplus \overline{x}_1 \overline{x}_2 x_3 \ .$$

Many other antivalence forms can similarly be built.

As shown in Example 4.14, an antivalence form can be transformed such that each variable appears either only negated or only non-negated. Polynomials with this property are denoted by *fixed polarity Reed–Muller polynomials* that are uniquely specified for each chosen polarity of the variables.

We will not systematically explore all these possible mixed forms but use them without further considerations if necessary. It is quite understandable that many optimization options exist that result in difficult algorithmic problems.

The construction of *equivalence normal forms* (ENF) follows the same principles using Theorems 4.5 and 4.6 based on a given conjunctive normal form:

$$f(\mathbf{x}) = (x_1 \vee x_2 \vee \overline{x}_3)(\overline{x}_1 \vee x_2 \vee x_3)(\overline{x}_1 \vee \overline{x}_2 \vee \overline{x}_3)(x_1 \vee \overline{x}_2 \vee x_3)$$
$$= (x_1 \vee x_2 \vee \overline{x}_3) \odot (\overline{x}_1 \vee x_2 \vee x_3) \odot (\overline{x}_1 \vee \overline{x}_2 \vee \overline{x}_3) \odot (x_1 \vee \overline{x}_2 \vee x_3) \ .$$

The *equivalence polynomial* results from $\overline{x} = (0 \odot x)$ and the corresponding simplifications; this transformation needs some effort but should be seen and understood at least once:

$$f(\mathbf{x}) = (x_1 \vee x_2 \vee \overline{x}_3)(\overline{x}_1 \vee x_2 \vee x_3)(\overline{x}_1 \vee \overline{x}_2 \vee \overline{x}_3)(x_1 \vee \overline{x}_2 \vee x_3)$$
$$= (x_1 \vee x_2 \vee (0 \odot x_3)) \odot ((0 \odot x_1) \vee x_2 \vee x_3)$$
$$\quad \odot ((0 \odot x_1) \vee (0 \odot x_2) \vee (0 \odot x_3)) \odot (x_1 \vee (0 \odot x_2) \vee x_3)$$
$$= (x_1 \vee x_2 \vee 0) \odot (x_1 \vee x_2 \vee x_3) \odot (0 \vee x_2 \vee x_3) \odot (x_1 \vee x_2 \vee x_3)$$
$$\quad \odot (0 \vee 0 \vee 0) \odot (0 \vee 0 \vee x_3) \odot (0 \vee x_2 \vee 0) \odot (0 \vee x_2 \vee x_3)$$
$$\quad \odot (x_1 \vee 0 \vee 0) \odot (x_1 \vee 0 \vee x_3) \odot (x_1 \vee x_2 \vee 0) \odot (x_1 \vee x_2 \vee x_3)$$
$$\quad \odot (x_1 \vee 0 \vee x_3) \odot (x_1 \vee x_2 \vee x_3)$$
$$= x_1 \odot x_2 \odot x_3 \odot 0 \ .$$

The transformations use the distributive law, $(a \odot a) = 1$, $(a \odot 0) = \overline{a}$, and $(a \odot 1) = a$.

Since the conjunctive normal form is unique, the equivalence normal form as well as the equivalence polynomial is also uniquely defined. It holds the following.

Theorem 4.12

- *Each function $f(\mathbf{x})$ can be uniquely represented by an equivalence normal form.*
- *Each function $f(\mathbf{x})$ can be uniquely represented by an equivalence polynomial.*

Here the function $1(\mathbf{x})$ carries a special role. The simple expression 1 can be used as a representation of the function $1(\mathbf{x})$ for both the equivalence normal form ENF and the equivalence polynomial EP. Mixed forms can be considered in the same way as for the antivalence forms.

It is not necessary to spend too much time onto the equivalence forms because $\odot$ can always be replaced by $\oplus$:

$$x \odot y = \overline{x \oplus y}, \qquad \overline{x \odot y} = x \oplus y \tag{4.8}$$

followed by the respective simplifications.

The following *Expansion Theorem* is very often related to the name of *Claude E. Shannon* [6].

Theorem 4.13 (Expansion Theorem, Shannon Expansion)   *Each function* $f(\mathbf{x})$ *can be represented by*

$$f(\mathbf{x}) = \overline{x}_i f(x_i = 0) \vee x_i f(x_i = 1), \tag{4.9}$$

$$= \overline{x}_i f(x_i = 0) \oplus x_i f(x_i = 1) \tag{4.10}$$

for any $x_i \in \{x_1, \ldots, x_n\}$.

The occurrence of x_i and $\overline{x}_i$ in the expression ensures the orthogonality; hence, both $\vee$ and $\oplus$ can be used for this expansion.

This theorem can be used in two ways: the function $f(\mathbf{x})$ can be built from two *subfunctions* $f(x_i = 0)$ and $f(x_i = 1)$—both of them must not be dependent on x_i, or the given function $f(\mathbf{x})$ can be split into two parts $f(x_i = 0)$ and $f(x_i = 1)$ that can be dealt with separately.

When the expansion is continued with regard to one of the remaining variables, then the original function $f(\mathbf{x})$ is split into four parts:

$$\begin{aligned} f(\mathbf{x}) = \quad & \overline{x}_i \overline{x}_j f(x_i = 0, x_j = 0) \oplus \overline{x}_i x_j f(x_i = 0, x_j = 1) \\ & \oplus x_i \overline{x}_j f(x_i = 1, x_j = 0) \oplus x_i x_j f(x_i = 1, x_j = 1). \end{aligned}$$

If we are using all the variables, then we finally reach the antivalence or disjunctive normal form:

$$\begin{aligned} f(\mathbf{x}) = \quad & \overline{x}_1 \overline{x}_2 \ldots \overline{x}_n f(x_1 = 0, x_2 = 0, \ldots, x_n = 0) \\ & \oplus \overline{x}_1 \overline{x}_2 \ldots x_n f(x_1 = 0, x_2 = 0, \ldots, x_n = 1) \\ & \vdots \\ & \oplus x_1 x_2 \ldots x_n f(x_1 = 1, x_2 = 1, \ldots, x_n = 1). \end{aligned}$$

Analogously,

$$\begin{aligned} f(\mathbf{x}) = \quad & (x_i \vee f(x_i = 0)) \wedge (\overline{x}_i \vee f(x_i = 1)) \\ & = (x_i \vee f(x_i = 0)) \odot (\overline{x}_i \vee f(x_i = 1)), \end{aligned}$$

$$\begin{aligned} f(\mathbf{x}) = \quad & (x_i \vee x_j \vee f(x_i = 0, x_j = 0)) \wedge (x_i \vee \overline{x}_j \vee f(x_i = 0, x_j = 1)) \\ & \wedge (\overline{x}_i \vee x_j \vee f(x_i = 1, x_j = 0)) \wedge (\overline{x}_i \vee \overline{x}_j \vee f(x_i = 1, x_j = 1)), \end{aligned}$$

etc.

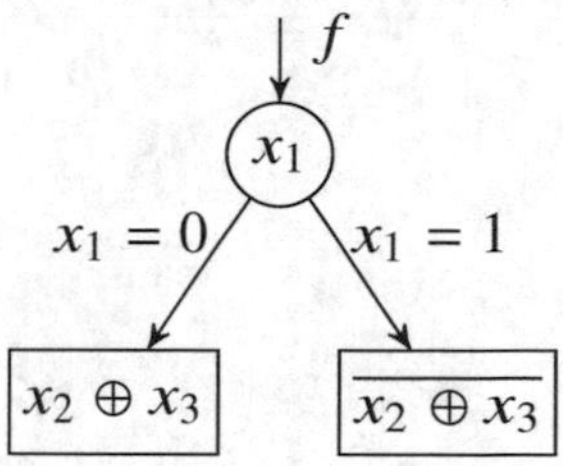

Fig. 4.3 The root of the decision tree and the first construction step

Fig. 4.4 Step 2 of the construction of the decision tree

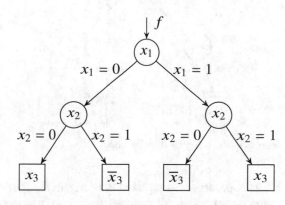

Sometimes it is very illustrative to use *binary decision trees* for the representation of logic functions. Trees are an efficient data structure; many interesting algorithms can be built by using trees. They use the Shannon expansion as long as the function values have not yet been found, based on a given order of the variables.

Example 4.15 As a first example the function $f(x_1, x_2, x_3) = x_1 \oplus x_2 \oplus x_3$ and the natural order $x_1 \Rightarrow x_2 \Rightarrow x_3$ will be used. Figure 4.3 shows the start of the construction of the tree with the *root node* labeled by x_1. The edge labeled by f is directed to this root node. Two edges are leaving the root node. The labels are $x_1 = 0$ at the left edge and $x_1 = 1$ at the right edge; for short the labels 0 and 1 can be used due to the label x_1 in the root node.

The two successor nodes represent the subfunctions $f(x_1 = 0) = x_2 \oplus x_3$ and $f(x_1 = 1) = \overline{x_2 \oplus x_3}$. In this example we draw the edges with the label 0 to the left and with the label 1 to the right.

According to the given order, the nodes of the next level are labeled by x_2, and the four edges leaving these nodes get the labels $x_2 = 0$ and $x_2 = 1$, resp. The four subfunctions represented are (see Fig. 4.4)

$$f(x_1 = 0, x_2 = 0) = x_3 \, ,$$

$$f(x_1 = 0, x_2 = 1) = \overline{x}_3 \, ,$$

$$f(x_1 = 1, x_2 = 0) = \overline{x}_3 \, ,$$

$$f(x_1 = 1, x_2 = 1) = x_3 \, .$$

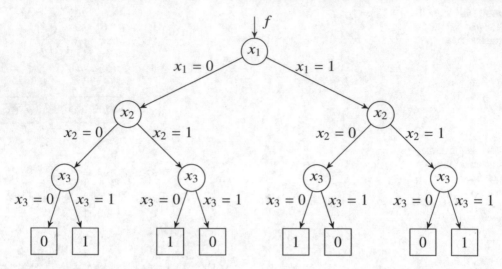

Fig. 4.5 The complete decision tree

This procedure is repeated once again: the four nodes are labeled by x_3, with eight edges leaving these nodes, $x_3 = 0$ to the left and $x_3 = 1$ to the right. This results in eight subfunctions that are constants equal to 0 or 1 (see Fig. 4.5).

$$f(x_1 = 0, x_2 = 0, x_3 = 0) = 0 ,$$
$$f(x_1 = 0, x_2 = 0, x_3 = 1) = 1 ,$$
$$f(x_1 = 0, x_2 = 1, x_3 = 0) = 1 ,$$
$$f(x_1 = 0, x_2 = 1, x_3 = 1) = 0 ,$$
$$f(x_1 = 1, x_2 = 0, x_3 = 0) = 1 ,$$
$$f(x_1 = 1, x_2 = 0, x_3 = 1) = 0 ,$$
$$f(x_1 = 1, x_2 = 1, x_3 = 0) = 0 ,$$
$$f(x_1 = 1, x_2 = 1, x_3 = 1) = 1 .$$

The *terminal nodes* labeled by 0 or 1 in Fig. 4.5 cannot be expanded anymore. The eight subfunctions supply the same information like the function table. So this is another means to represent the function $f(\mathbf{x})$ (and all the subfunctions).

Now the construction (the point of view) can be reversed. Let be given the complete tree of Fig. 4.5. We start with the end nodes and construct the subfunctions for the level x_3 (from the left to the right):

$$0\,\overline{x}_3 \oplus 1\,x_3 = x_3 ,$$
$$1\,\overline{x}_3 \oplus 0\,x_3 = \overline{x}_3 ,$$
$$1\,\overline{x}_3 \oplus 0\,x_3 = \overline{x}_3 ,$$
$$0\,\overline{x}_3 \oplus 1\,x_3 = x_3 .$$

The edge with $x_3 = 0$ shows the value (subfunction) to be taken for $\overline{x}_3$, and the edge with $x_3 = 1$ shows the value (subfunction) to be taken for x_3 (see again the Shannon expansion).

Now the procedure is repeated for x_2:

$$x_3\overline{x}_2 \oplus \overline{x}_3 x_2 \, ,$$

$$\overline{x}_3\overline{x}_2 \oplus x_3 x_2 \, ;$$

these are two subfunctions on the level of x_2, and finally, we get

$$\overline{x}_1(x_3\overline{x}_2 \oplus \overline{x}_3 x_2) \oplus x_1(\overline{x}_3\overline{x}_2 \oplus x_3 x_2)$$

$$= \overline{x}_1\overline{x}_2 x_3 \oplus \overline{x}_1 x_2\overline{x}_3 \oplus x_1\overline{x}_2\overline{x}_3 \oplus x_1 x_2 x_3,$$

which is the antivalence normal form of $f(\mathbf{x}) = x_1 \oplus x_2 \oplus x_3$.

It has to be emphasized that this procedure is working strictly sequentially. One variable will be processed after the other, upward as well as downward.

A welcome property is that a complete expanded decision tree with a fixed order of the variables uniquely represents the associated function $f(\mathbf{x})$. This property follows from the direct mapping between this tree and the ANF or DNF of $f(\mathbf{x})$. A drawback of such a complete expanded decision tree is that it needs $2^n - 1$ decision nodes and 2^n terminal nodes to represent a function of n variables. However, for many functions, a more compact, but also unique, representation can be constructed as an acyclic directed graph. Randal E. Bryant suggested in his pioneering paper [1] simple reduction rules to construct a *Reduced Ordered Binary Decision Diagram* (ROBDD) that remains the property of a unique representation of the function for a fixed order of the variables; for short, such a graph is usually denoted by *Binary Decision Diagram* (BDD).

The reduction of a complete expanded decision tree to an ROBDD can be done using the following rules:

1. Identify all terminal nodes 0 with a single terminal node 0 as well as all terminal nodes 1 with a single terminal node 1 (in this way the given tree is transformed into a directed acyclic graph).
2. Reuse one representation of identical subgraphs, remove all other copies of them, and redirect the related edges to the shared subgraph (these subgraphs describe identical functions).
3. Remove all nodes with outgoing edges pointing to the same node and redirect the incoming edge to the common child node.
4. Repeat the reduction rules 2 and 3 until no further simplifications are possible.

It is not necessary to repeat the name of the variable in the label of an edge because it is the same as the label of the node the edge is coming from. Hence, in the following, we avoid the name of the variable in the labels of the edges. Additionally, we increase the clearness using dashed lines for edges labeled by 0.

Example 4.16 We use the complete decision tree of Fig. 4.5 as an example to create an ROBDD of the function $f(x_1, x_2, x_3) = x_1 \oplus x_2 \oplus x_3$. Figure 4.6a shows this complete tree using the simplified representation.

Due to reduction rule 1, one terminal node 0 and one terminal node 1 are used in Fig. 4.6b as the first step to create the ROBDD.

The subgraphs of the leftmost and rightmost x_3-nodes of Fig. 4.6b are identical. Hence, due to reduction rule 2, we reuse the leftmost x_3-node, redirect the 1-edge of the right x_2-node to this shared

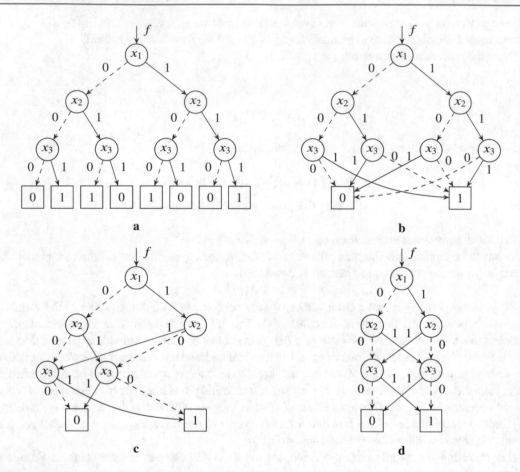

Fig. 4.6 Creation of the ROBDD of the function $f = x_1 \oplus x_2 \oplus x_3$: (**a**) the given complete decision tree using the simplified notation, (**b**) directed acyclic graph after applying reduction rule 1, (**c**) ROBDD after applying reduction rule 2, and (**d**) ROBDD after adjusting the positions of the nodes

node, and remove the rightmost x_3-node. Similarly, the outgoing edges of the two x_3-nodes in the middle of Fig. 4.6b end at the same terminal nodes. Hence, we reuse the left of these x_3-nodes, redirect the 0-edge of the right x_2-node to this shared node, and remove the third x_3-node (counting from the left in Fig. 4.6b). Figure 4.6c shows the result after this second step to create the ROBDD.

The graph of Fig. 4.6c does not contain any node the outgoing edges of which point to the same node. Hence, in this example reduction rule 3 does not result in any simplification. The outgoing edges of the two x_2-nodes do not point to the same x_3-nodes. Hence, no further simplifications are possible, and the graph of Fig. 4.6c is already the unique ROBDD representation of the function $f(x_1, x_2, x_3) = x_1 \oplus x_2 \oplus x_3$ for the natural order of the variables $x_1 \Rightarrow x_2 \Rightarrow x_3$. Figure 4.6d depicts this ROBDD after adjusting the positions of the x_3-nodes.

As mentioned above, it is a benefit that an ROBDD uniquely represents a function for a fixed order of variables. However, it is a drawback that the number of nodes of an ROBDD depends in general on the used order of the variables. There are functions for which the number of needed nodes is linear in the number of variables for the best order, but exponential for the worst order of the variables. Extensive research has been done to find a good order of the variables within an acceptable period of time and to deal with slightly different types of decision diagrams, see e.g., [5]. A very extensive description of decision diagram techniques for micro- and nano-electronic design has been published in the book [10].

We return to the tree representation and notice the same situation that is already known for disjunctive forms. It is possible that shorter conjunctions can be used, and not all the expansions have to be considered. This can be seen in the following.

Example 4.17 Let the order be the natural order $x_1 \Rightarrow x_2 \Rightarrow x_3 \Rightarrow x_4$ and $f(x_1, x_2, x_3, x_4) = \overline{x}_1 x_3 x_4 \vee x_1 x_2 \overline{x}_4 \vee x_1 x_3 \vee x_2 \overline{x}_3 x_4$; then we find step by step the expansion of Fig. 4.7a and the associated tree of Fig. 4.7b.

The terminal nodes occur on different levels, and the numbers of nodes and edges in this tree are smaller; it represents a disjunctive form.

Both the Shannon expansion of level 2 and the tree of Fig. 4.7b show a possibility for further simplification. The subfunction $f(x_1 = 0, x_2 = 1)$ does not depend on x_3, but only on x_4. Hence, the expansion with regard to x_3 is not necessary for this subfunction. The tree of Fig. 4.7b shows this property by identical subtrees of the second x_3-node from the left. Hence, the tree of $f(\mathbf{x})$ can be simplified such that the middle x_3-node and the complete subtree below this x_3-node are removed, and the 1-edge of the left x_2-node is redirected to the remaining rightmost x_4-node.

However, there is now an additional difficult problem, the tree and its parameters (the number of nodes and edges) strongly depend on the selected order of the variables as can be seen by the following.

Example 4.18 Let $f(\mathbf{x}) = \overline{x}_1 \overline{x}_2 \vee x_2 \overline{x}_3 \vee \overline{x}_2 x_3$. The complete tree in natural order $x_1 \Rightarrow x_2 \Rightarrow x_3$ results in the tree of Fig. 4.8a, and in case of the inverse order $x_3 \Rightarrow x_2 \Rightarrow x_1$ in the tree of Fig. 4.8b.

The order of the variables determines the order of function values in the terminal nodes. Identical subgraphs can be merged. Figure 4.8c shows that in the case of the natural order only the leftmost x_3-node can be replaced by the terminal node 1 because bath outgoing edges of this x_3-node point to terminal nodes with the same function value 1. In case of the reverse order of the variables there are three x_1-nodes where the outgoing edges point to terminal nodes with identical function values. Hence, as can be seen in Fig. 4.8d, the reduced tree of the function $f(\mathbf{x})$ needs two non-terminal nodes less. The finding of the "smallest" tree is again a difficult optimization problem. In principle, there are $n!$ different orders to be considered.

The same order of the variables has to be used in each branch of the tree, which is expressed by the name *ordered decision tree*. In principle, it would be possible to use a different order in each branch. However, this would result in very difficult structures that could hardly be handled in a reasonable way. Hence, we restrict our considerations to ordered decision trees.

Another model that can be understood as a graphical representation of a logic function $f(\mathbf{c})$ is the *circuit model*. It will be explored when the implementation of logic functions by means of circuits is discussed (see Chap. 11).

Level 1:
$$f(x_1 = 0) = x_3 x_4 \vee x_2 \overline{x}_3 x_4 \,,$$
$$f(x_1 = 1) = x_2 \overline{x}_4 \vee x_3 \vee x_2 \overline{x}_3 x_4 \,.$$

Level 2:
$$f(x_1 = 0, x_2 = 0) = x_3 x_4 \,,$$
$$f(x_1 = 0, x_2 = 1) = x_3 x_4 \vee \overline{x}_3 x_4 = x_4 \,,$$
$$f(x_1 = 1, x_2 = 0) = x_3 \,,$$
$$f(x_1 = 1, x_2 = 1) = \overline{x}_4 \vee x_3 \vee \overline{x}_3 x_4 = 1 \,.$$

Level 3:
$$f(x_1 = 0, x_2 = 0, x_3 = 0) = 0 \,,$$
$$f(x_1 = 0, x_2 = 0, x_3 = 1) = x_4 \,,$$
$$f(x_1 = 0, x_2 = 1, x_3 = 0) = x_4 \,,$$
$$f(x_1 = 0, x_2 = 1, x_3 = 1) = x_4 \,,$$
$$f(x_1 = 1, x_2 = 0, x_3 = 0) = 0 \,,$$
$$f(x_1 = 1, x_2 = 0, x_3 = 1) = 1 \,.$$

Level 4:
$$f(x_1 = 0, x_2 = 0, x_3 = 1, x_4 = 0) = 0 \,,$$
$$f(x_1 = 0, x_2 = 0, x_3 = 1, x_4 = 1) = 1 \,,$$
$$f(x_1 = 0, x_2 = 1, x_3 = 0, x_4 = 0) = 0 \,,$$
$$f(x_1 = 0, x_2 = 1, x_3 = 0, x_4 = 1) = 1 \,,$$
$$f(x_1 = 0, x_2 = 1, x_3 = 1, x_4 = 0) = 0 \,,$$
$$f(x_1 = 0, x_2 = 1, x_3 = 1, x_4 = 1) = 1 \,.$$

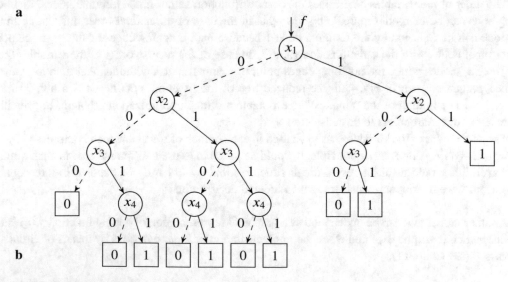

Fig. 4.7 The decision tree with terminal nodes on different levels: (**a**) levels of the expansion, (**b**) tree

4.2   Formulas and Expressions

Up to this point in time, we used *formulas* in a rather intuitive way: the function was given by the function table, and the symbols like $\wedge$, $\vee$, etc., have been used as names of the functions whose values are also specified in tables. Mostly, we used the *infix* notation for the functions, and the symbol of the function has been written between the variables. Only for the complement, the *prefix* notation has been used.

However, very often another approach is used. The formulas are the starting point, and the functions are introduced based on (meaningful) formulas. A formula will be considered, in this sense, as a set or sequence of computational rules by means of which one, some, or all functional values can be calculated. Many problems can be found because the meaning of a formula is not very clear: it cannot be seen whether the formula is a constant (equal to $0(\mathbf{x})$ or $1(\mathbf{x})$), how long it will take to calculate all the values, whether two different formulas express the same function, etc. In order to solve these problems appropriately, we need a precise definition of formulas that can be given in an inductive way.

Definition 4.13 Let $x_1, \ldots, x_n$ be n symbols for variables, and then:

1. $x_1, \ldots, x_n$ as well as the constant values 0 and 1 are formulas.
2. If F is a formula, then also $\overline{F}$.
3. If F_1, F_2 are formulas, then also

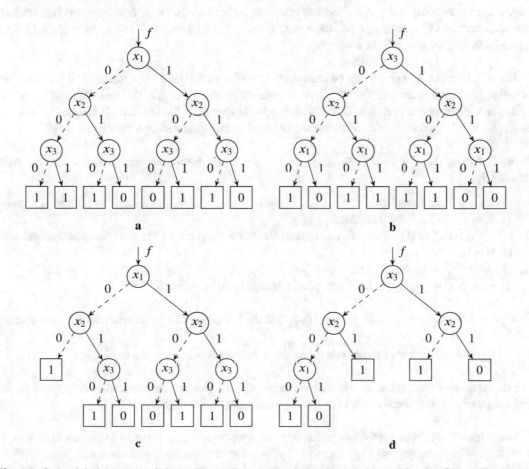

Fig. 4.8 Ordered decision trees of $f = \overline{x}_1\overline{x}_2 \lor x_2\overline{x}_3 \lor \overline{x}_2x_3$: (**a**) complete tree in natural order, (**b**) complete tree in inverse order, (**c**) reduced tree in natural order, (**d**) reduced tree in inverse order of the variables

Table 4.10 Function table for $f(\mathbf{x}) = \overline{x}_1 \vee (1 \oplus x_2)(x_2 \oplus x_3)$

x_1	x_2	x_3	$f(\mathbf{x})$
0	0	0	0
0	0	1	0
0	1	0	0
0	1	1	0
1	0	0	1
1	0	1	0
1	1	0	1
1	1	1	1

$$(F_1 \wedge F_2), (F_1 \vee F_2), (F_1 \oplus F_2), (F_1 \odot F_2), (F_1 \rightarrow F_2), (F_1 \mid F_2), (F_1 \downarrow F_2).$$

4. Any formula F can be constructed by a finite sequence of steps $1, \ldots, 3$, starting with the given variables $x_1, \ldots, x_n$ and optionally the constant values 0 and 1.

Note 4.12 Rules for omitting brackets and operation symbols can be included and used as before. Only those expressions are *correct* formulas that can be built in this way. Sometimes it is necessary that each formula is a sequence of characters. Then, the complement $\overline{F}$ has to be replaced by an appropriate symbol like $\neg F$ (or similarly).

The definition of a function can be built upon formulas using the *interpretation* of a formula. We used this concept already before; it consists in assigning values 0 or 1 to the variables and using the definitions of the operations (i.e., of the functions) that appear in the formula. If this has been done for each $\mathbf{x} \in \mathbb{B}^n$, then the function has been computed and represented as a vector or a table.

Example 4.19 $f(\mathbf{x}) = \overline{x}_1 \vee (1 \oplus x_2)(x_2 \oplus x_3)$. It can be seen that this formula can be built inductively:

1. $x_1, x_2, x_3, 1$ are formulas.
2. $(1 \oplus x_2), (x_2 \oplus x_3)$ are formulas.
3. $((1 \oplus x_2)(x_2 \oplus x_3)), \overline{x}_1$ are formulas, and the outer brackets of the first formula are omitted; $(1 \oplus x_2)(x_2 \oplus x_3)$ is a formula.
4. $(\overline{x}_1 \vee (1 \oplus x_2)(x_2 \oplus x_3))$ is a formula.
5. $\overline{x}_1 \vee (1 \oplus x_2)(x_2 \oplus x_3)$ is a formula; outer brackets can be omitted.

Interpretation Let $\mathbf{x} = (x_1, x_2, x_3) = (101)$; then the following computations have to be performed:

$$\overline{1} \vee (1 \oplus 0)(0 \oplus 1) \Rightarrow \overline{1} \vee 1 \wedge 1 \Rightarrow \overline{0 \vee 1} \Rightarrow \overline{1} \Rightarrow 0.$$

The calculated function values for all assigned patterns can be summarized in a function table; Table 4.10 shows the function values of the explored formula for each $\mathbf{x} \in \mathbb{B}^3$.

Sometimes it might be useful to restrict the used operations of the formula to certain functions. It is even possible to express all logic functions by formulas that use the single operation, e.g., $F_1 \mid F_2$.

We have already seen that different formulas can describe the same function. In fact, all the identities stated above give different formulas for the same function.

Definition 4.14 Two formulas F_1 and F_2 are equivalent if $f(F_1) = f(F_2)$.

$f(F_1)$ means the function represented by the formula F_1. This definition implies that the same set of variables has to be considered for the two formulas (or functions). It can be shown that this relation between formulas is an equivalence relation; hence, the set of all formulas using a given fixed set of variables will be partitioned into equivalence classes. Every class contains the set of all expressions that represent the same function. Again, the use of dummy variables requires some attention, as above.

A formula F that represents the function $1(\mathbf{x})$ for a given set of variables $\mathbf{x}$ is denoted by *tautology* or a *logical law*.

Sometimes it is useful or desirable to have formulas without brackets. The so-called Reverse Polish Notation is usually built upon the structure of trees in a recursive way:

- Each sign of an operation is the root of a tree; this tree has one successor node when the operation is the negation, and for all other 2-ary operations, the two outgoing edges point to a *left* and a *right* subtree.
- Each constant and each variable is a leaf of the tree.
- The "Reverse Polish Notation" is constructed using a post-order procedure by starting with the highest operator and writing expressions of the format

for nodes labeled by negation:	single subtree - root,
for nodes of 2-ary operations:	left subtree - right subtree - root.

- The subtrees are handled in the same way as long as the leaves have not yet been reached.

Example 4.20 Let be given

$$[(x \wedge y) \oplus (y \odot z)] \rightarrow [(x \vee y) \odot \overline{(y \wedge z)}] .$$

Then, it can be seen that $\rightarrow$ is the highest operation (indicated by the bracket structure). Hence, we get

$$\underbrace{(x \wedge y) \oplus (y \odot z)}_{\text{left subtree}} \quad \underbrace{(x \vee y) \odot \overline{(y \wedge z)}}_{\text{right subtree}} \quad \underbrace{\rightarrow}_{\text{root}} .$$

The left and right subtrees are now dealt with in the same way, and we get

$$\underbrace{(x \wedge y)}_{\text{left}} \ \underbrace{(y \odot z)}_{\text{right}} \ \oplus \ \underbrace{(x \vee y)}_{\text{left}} \ \underbrace{\overline{(y \wedge z)}}_{\text{right}} \ \odot \ \rightarrow \ .$$

Next the negation (expressed by $\neg$) is moved behind $(y \wedge z)$, then all the remaining four brackets are transformed in the same way, and we finally get

$$xy \wedge yz \odot \oplus xy \vee yz \wedge \neg \odot \rightarrow \ .$$

Brackets are no longer necessary. Always, starting from the left, the first operation has to be found. The last two values on the left-hand side of the operator are combined according to the definition of the operator. The resulting value replaces the operator and the two arguments. This continues until only one single value remains back, which is the result of the evaluation.

The value of the expression can now be calculated from the left to the right. The negation is applied to the single left argument and all other 2-ary operations to the left two arguments. We calculate the value of the created expression in the "Reverse Polish Notation" for the assignment $x = 1$, $y = 0$, $z = 1$:

x	y	$\wedge$	y	z	$\odot$	$\oplus$	x	y	$\vee$	y	z	$\wedge$	$\neg$	$\odot$	$\rightarrow$,
1	0	$\wedge$	0	1	$\odot$	$\oplus$	1	0	$\vee$	0	1	$\wedge$	$\neg$	$\odot$	$\rightarrow$,
		0	0	1	$\odot$	$\oplus$	1	0	$\vee$	0	1	$\wedge$	$\neg$	$\odot$	$\rightarrow$,
				0	0	$\oplus$	1	0	$\vee$	0	1	$\wedge$	$\neg$	$\odot$	$\rightarrow$,
						0	1	0	$\vee$	0	1	$\wedge$	$\neg$	$\odot$	$\rightarrow$,
								0	1	0	1	$\wedge$	$\neg$	$\odot$	$\rightarrow$,
										0	1	0	$\neg$	$\odot$	$\rightarrow$,
											0	1	1	$\odot$	$\rightarrow$,
													0	1	$\rightarrow$,
															1 .

4.3 Special Logic Functions

There are several functions and classes (sets) of functions with special interesting properties. This section will give an overview. We begin with the following.

Definition 4.15 Let be given two vectors

$$\mathbf{x} = (x_1, \ldots, x_n) \quad \text{and} \quad \mathbf{x}_0 = (x_1, \ldots, x_{i-1}, x_{i+1}, \ldots, x_n) .$$

Then, the function $f(\mathbf{x})$ is

1.	conjunctively degenerated in x_i if	$f(\mathbf{x}) = x_i \wedge f^*(\mathbf{x}_0)$;
2.	disjunctively degenerated in x_i if	$f(\mathbf{x}) = x_i \vee f^*(\mathbf{x}_0)$;
3.	linearly degenerated in x_i if	$f(\mathbf{x}) = x_i \oplus f^*(\mathbf{x}_0)$;
4.	linearly degenerated in x_i if	$f(\mathbf{x}) = x_i \odot f^*(\mathbf{x}_0)$.

This means that the variable x_i appears only on one place and reduces the complexity of the considerations. The search space for $\mathbf{x}_0$ has "only" 2^{n-1} elements, whereas 2^n elements have to be considered for $\mathbf{x}$.

A conjunctively degenerated function has only function values $f(\mathbf{x}) = 1$ for $x_i = 1$. All function values of a conjunctively degenerated function are equal to 0 for $x_i = 0$; hence, we have

$$f(\mathbf{x}_0, x_i = 0) = 0 \quad \text{and} \quad f(\mathbf{x}_0, x_i = 1) = f^*(\mathbf{x}_0) .$$

A disjunctively degenerated function has only function values $f(\mathbf{x}) = 0$ for $x_i = 0$. All function values of a disjunctively degenerated function are equal to 1 for $x_i = 1$; hence, we have

$$f(\mathbf{x}_0, x_i = 0) = f^*(\mathbf{x}_0) \quad \text{and} \quad f(\mathbf{x}_0, x_i = 1) = 1 .$$

Table 4.11 The dual
function $f^*(\mathbf{x})$ of the
function $f(\mathbf{x})$

x_1	x_2	x_3	$f(\mathbf{x})$	$f^*(\mathbf{x})$
0	0	0	0	1
0	0	1	1	0
0	1	0	1	0
0	1	1	0	0
1	0	0	1	1
1	0	1	1	0
1	1	0	1	0
1	1	1	0	1

The function values of a linearly degenerated function $f(\mathbf{x}) = x_i \oplus f^*(\mathbf{x}_0)$ are equal to the function values of the subfunction $f^*(\mathbf{x}_0)$ in the case that $x_i = 0$ or equal to the function values of the complement of the subfunction $f^*(\mathbf{x}_0)$ for $x_i = 1$; hence, we have

$$f(\mathbf{x}_0, x_i = 0) = f^*(\mathbf{x}_0) \quad \text{and} \quad f(\mathbf{x}_0, x_i = 1) = \overline{f^*(\mathbf{x}_0)} \,.$$

The linearly degenerated function $f(\mathbf{x}) = x_i \odot f^*(\mathbf{x}_0)$ with the equivalence can be transformed into a linearly degenerated function that uses $\overline{x}_i$ and the antivalence

$$f(\mathbf{x}) = x_i \odot f^*(\mathbf{x}_0) = \overline{x}_i \oplus f^*(\mathbf{x}_0) \,,$$

and we get

$$f(\mathbf{x}_0, x_i = 0) = \overline{f^*(\mathbf{x}_0)} \quad \text{and} \quad f(\mathbf{x}_0, x_i = 1) = f^*(\mathbf{x}_0) \,.$$

Definition 4.16 The function $f^*(x_1, \ldots, x_n) = \overline{f(\overline{x}_1, \ldots, \overline{x}_n)}$ is *dual* to the function $f(x_1, \ldots, x_n)$.

Example 4.21 Table 4.11 shows the function values of the dual function $f^*(\mathbf{x})$ generated for the given function $f(\mathbf{x})$. It can be seen, for instance, that

$$f^*(0, 0, 0) = \overline{f(\overline{0}, \overline{0}, \overline{0})} = \overline{f(1, 1, 1)} \,,$$

$$f^*(1, 0, 1) = \overline{f(\overline{1}, \overline{0}, \overline{1})} = \overline{f(0, 1, 0)} \,.$$

In general, the order of the vector of function values of $f(\mathbf{x})$ is first reversed, and thereafter, each element is negated.

$f(\mathbf{x})$	:	0	1	1	0	1	1	1	0
$f_R(\mathbf{x})$	:	0	1	1	1	0	1	1	0
$f^*(\mathbf{x})$	:	1	0	0	0	1	0	0	1

The construction of $f^*(\mathbf{x})$ shows immediately that each function has a dual function, and this dual function is uniquely defined.

The following example shows that some functions are dual to themselves:

$$f = f^* \,.$$

These functions are denoted by *self-dual functions*.

$f(\mathbf{x})$	:	0	1	0	1	0	1	0	1
$f_R(\mathbf{x})$	:	1	0	1	0	1	0	1	0
$f^*(\mathbf{x})$	:	0	1	0	1	0	1	0	1

Furthermore, it holds that

$$(f^*(\mathbf{x}))^* = f(\mathbf{x}) \, .$$

When the construction of the dual function is applied twice, then the old function will be re-established.

Identical function values of the upper and lower functions in each column of Table 4.12 shows that these functions are dual to each other; this will be useful for the application of the next theorem.

Theorem 4.14 *If $f(x_1, \ldots, x_n)$ is represented by a formula using the operations (functions) $f_1, \ldots, f_k$, then $f^*(x_1, \ldots, x_n)$ is described by a formula where each $f_1, \ldots, f_k$ has been replaced by $f_1^*, \ldots, f_k^*$.*

Example 4.22

$$f(\mathbf{x}) = x_1 \oplus x_2 \qquad\qquad = x_1 \overline{x}_2 \vee \overline{x}_1 x_2 \qquad\qquad = (x_1 \vee x_2)(\overline{x}_1 \vee \overline{x}_2) \, ,$$
$$f^*(\mathbf{x}) = x_1 \odot x_2 \qquad\qquad = (x_1 \vee \overline{x}_2)(\overline{x}_1 \vee x_2) \qquad\qquad = x_1 x_2 \vee \overline{x}_1 \overline{x}_2 \, .$$

This example and the theorem indicate the following special case: when F is a formula for the function $f(\mathbf{x})$ including $0, 1, {}^{-}, \wedge, \vee$, then we get the formula F^* for the function $f^*(\mathbf{x})$ by exchanging 0 and 1, $\wedge$ and $\vee$, as well as $\oplus$ and $\odot$:

$$F_1 = x_1 \wedge 0 = 0 \, , \qquad\qquad\qquad F_1^* = x_1 \vee 1 = 1 \, ,$$
$$F_2 = x_1 \vee \overline{x}_1 x_2 = x_1 \vee x_2 \, , \qquad\qquad F_2^* = x_1 \wedge (\overline{x}_1 \vee x_2) = x_1 \wedge x_2 \, ,$$
$$F_3 = x_1 \oplus x_2 = x_1 \overline{x}_2 \vee \overline{x}_1 x_2 \, , \qquad F_3^* = x_1 \odot x_2 = (x_1 \vee \overline{x}_2) \wedge (\overline{x}_1 \vee x_2) \, ,$$

etc.

Theorem 4.15 *When there is an identity (a formula)*

$$F(x_1, \ldots, x_n) \equiv G(x_1, \ldots, x_n) \, ,$$

then the identity

$$F^*(x_1, \ldots, x_n) \equiv G^*(x_1, \ldots, x_n)$$

holds as well.

Due to the dual operations (functions) shown in Table 4.12 and Theorem 4.15, the two Boolean Algebras $(\mathbb{B}^n, \vee, \wedge, {}^{-}, \mathbf{0}, \mathbf{1})$ and $(\mathbb{B}^n, \wedge, \vee, {}^{-}, \mathbf{1}, \mathbf{0})$ are dual to each other and the same statement is true for the two Boolean Rings $(\mathbb{B}^n, \oplus, \wedge, \mathbf{0}, \mathbf{1})$ and $(\mathbb{B}^n, \odot, \vee, \mathbf{1}, \mathbf{0})$. Hence, the knowledge of one of these algebraic structures can be reused for the associated other structure.

Table 4.12 Dual functions indicated by identical function values

x	y	0	1	$\overline{x}$	$\overline{y}$	$x \wedge y$	$x \vee y$	$x \oplus y$	$x \odot y$
0	0	0	1	1	1	0	0	0	1
0	1	0	1	1	0	0	1	1	0
1	0	0	1	0	1	0	1	1	0
1	1	0	1	0	0	1	1	0	1

x	y	1^*	0^*	$\overline{x}^*$	$\overline{y}^*$	$(x \vee y)^*$	$(x \wedge y)^*$	$(x \odot y)^*$	$(x \oplus y)^*$
0	0	0	1	1	1	0	0	0	1
0	1	0	1	1	0	0	1	1	0
1	0	0	1	0	1	0	1	1	0
1	1	0	1	0	0	1	1	0	1

Example 4.23

$$\overline{x_1 \wedge x_2 \wedge \ldots \wedge x_n} \equiv \overline{x}_1 \vee \overline{x}_2 \vee \ldots \vee \overline{x}_n$$

results in

$$\overline{x_1 \vee x_2 \vee \ldots \vee x_n} \equiv \overline{x}_1 \wedge \overline{x}_2 \wedge \ldots \wedge \overline{x}_n \, .$$

Note 4.13 Sometimes we will use expressions like this as a generalized NAND and NOR:

$$\mathrm{NAND}(x_1, \ldots, x_n) = \overline{x_1 \wedge \ldots \wedge x_n} \, ,$$

$$\mathrm{NOR}(x_1, \ldots, x_n) = \overline{x_1 \vee \ldots \vee x_n} \, .$$

Definition 4.17 The function $f(\mathbf{x})$ is a *linear* function if $f(\mathbf{x})$ can be written as

$$f(\mathbf{x}) = a_0 \oplus a_1 x_1 \oplus \cdots \oplus a_n x_n \, , \tag{4.11}$$

with $a_0, \ldots, a_n \in \mathbb{B}$ given constants.

Note 4.14 Due to the duality of the two Boolean Rings, the notion of linear functions could also be defined by means of $\odot$ and $\vee$; however, the results achieved in one structure can be directly transferred into the other structure; hence, we limit the considerations to the given definition.

Note 4.15 Sometimes the function (4.11) is denoted by *affine* function, and the term *linear* function is only used for such functions where $a_0 = 0$.

Theorem 4.16 *There are 2^{n+1} linear functions of n variables*

This follows easily from the fact that each vector $\mathbf{a} = (a_0, a_1, \ldots, a_n)$ defines one linear function. For a given vector $(a_1, \ldots, a_n)$, the component a_0 can be equal to 0 or equal to 1, and because of $1 \oplus f(\mathbf{x}) = \overline{f(\mathbf{x})}$, it can be seen that with each linear function $f(\mathbf{x})$ also $\overline{f(\mathbf{x})}$ is a linear function. We use $\mathcal{L}_n$ for the set of all linear functions of n variables.

Example 4.24 Table 4.13 shows all linear functions for $n = 3$ variables; there are only $2^4 = 16$ linear functions of $2^{2^3} = 2^8 = 256$ functions altogether. Only 2^n functions are essentially different

Table 4.13 Linear functions of 3 variables

a	$f(\mathbf{x})$	a	$f(\mathbf{x})$
0000	$0(\mathbf{x})$	1000	$1(\mathbf{x})$
0001	x_3	1001	$1 \oplus x_3$
0010	x_2	1010	$1 \oplus x_2$
0011	$x_2 \oplus x_3$	1011	$1 \oplus x_2 \oplus x_3$
0100	x_1	1100	$1 \oplus x_1$
0101	$x_1 \oplus x_3$	1101	$1 \oplus x_1 \oplus x_3$
0110	$x_1 \oplus x_2$	1110	$1 \oplus x_1 \oplus x_2$
0111	$x_1 \oplus x_2 \oplus x_3$	1111	$1 \oplus x_1 \oplus x_2 \oplus x_3$

(these are the linear functions in the stronger sense, see Note 4.15), and the other functions are their complements.

For $n = 1$, there is $\mathcal{L}_1 = \{0(x), x, 1 \oplus x, 1(x)\} = \mathcal{F}_1$; all the functions depending on one variable are linear functions.

Definition 4.18 A function $f(\mathbf{x})$ preserves 0 if $f(\mathbf{0}) = 0$. The function $f(\mathbf{x})$ preserves 1 if $f(\mathbf{1}) = 1$.

$\mathcal{P}_n^0$ and $\mathcal{P}_n^1$ are the sets of all 0-preserving and 1-preserving functions. There are 2^{2^n-1} elements in $\mathcal{P}_n^0$ and $\mathcal{P}_n^1$. The requirement of the definition determines the value of $f(\mathbf{x})$ in one point ($\mathbf{0}$ or $\mathbf{1}$), and $2^n - 1$ function values can be chosen from the set $\{0, 1\}$.

Example 4.25 The functions $f_1(\mathbf{x}) = x_1 \vee x_2 \vee x_3$ and $f_2(\mathbf{x}) = x_1 \wedge x_2 \wedge x_3$ are 0- and 1-preserving. Furthermore, $0(x_1, x_2, x_3) \in \mathcal{P}_3^0$ and $1(x_1, x_2, x_3) \in \mathcal{P}_3^1$.

A very interesting class is the class $\mathcal{M}_n$ of all monotonous functions.

Definition 4.19 The function $f(\mathbf{x})$ is *monotonously increasing* if

$$\mathbf{x} \leq \mathbf{y} \quad \Rightarrow \quad f(\mathbf{x}) \leq f(\mathbf{y})$$

or *monotonously decreasing* if

$$\mathbf{x} \leq \mathbf{y} \quad \Rightarrow \quad f(\mathbf{x}) \geq f(\mathbf{y}) .$$

This definition is based on the partial order

$$\mathbf{x} \leq \mathbf{y} \quad \equiv \quad x_i \leq y_i, \ i = 1, \ldots, n .$$

The border cases are the functions $0(\mathbf{x})$ and $1(\mathbf{x})$ satisfying this definition because they are constant. For all the other monotonously increasing functions, it must hold that $f(\mathbf{0}) = 0$ and $f(\mathbf{1}) = 1$. For monotonously increasing functions, $f(\mathbf{0}) = 1$ just results in $1(\mathbf{x})$ and $f(\mathbf{1}) = 0$ results in $0(\mathbf{x})$.

Inversely, monotonously decreasing functions, which are not constant, must hold that $f(\mathbf{0}) = 1$ and $f(\mathbf{1}) = 0$.

Example 4.26 We explore two monotonous functions, the monotonously increasing function $f_1(x_1, x_2, x_3) = x_1 x_2 \vee x_3$ and the monotonously decreasing function $f_2(x_1, x_2, x_3) = \overline{x}_1 \overline{x}_2 \vee \overline{x}_2 \overline{x}_3$.

Fig. 4.9 The graphs of two monotonous functions; reflexive loops are omitted: (**a**) the monotonously increasing function $f_1(\mathbf{x}) = x_1 x_2 \vee x_3$, (**b**) the monotonously decreasing function $f_2(\mathbf{x}) = \overline{x}_1 \overline{x}_2 \vee \overline{x}_2 \overline{x}_3$

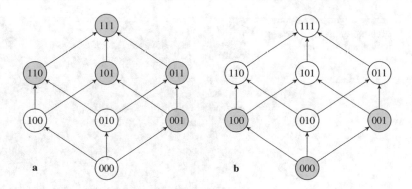

Figure 4.9 shows the graphs of these functions where the labels in the nodes are the values of (x_1, x_2, x_3). Gray shaded nodes indicate function values 1.

The function $f_1(x_1, x_2, x_3)$ can be described by the nodes (001) and (110) for which we have $f_1(x_1, x_2, x_3) = 1$. All larger nodes (in the sense of $\leq$) also have the value $f_1(x_1, x_2, x_3) = 1$, i.e., $f_1(011) = 1$, $f_1(101) = 1$, and $f_1(111) = 1$ as well. The function $f_2(x_1, x_2, x_3)$ can be described by the nodes (001) and (100) for which we have $f_2(x_1, x_2, x_3) = 1$. The smaller node (in the sense of $\leq$) also have the value $f_2(x_1, x_2, x_3) = 1$, i.e., $f_2(000) = 1$.

When the graph of the partial order $\leq$ is considered, together with a monotonous function, then the following interesting properties can be seen:

- On each path from **0** to **1**, there is one and only one edge e_{01} with $f(\mathbf{x}) = 0$ in the initial point and $f(\mathbf{x}) = 1$ in the final point of this edge.
- All the edges on the path before the edge e_{01} have $f(\mathbf{x}) = 0$ in both points of the edge.
- All the edges on the path behind the edge e_{01} have $f(\mathbf{x}) = 1$ in both points of the edge.

This means that this set of edges, or the set of all the initial points of these edges, or the set of all final points of these edges is completely sufficient for the definition of a monotonously increasing function. The set of all monotonous functions of n variables will be indicated by $\mathcal{M}_n$. It is very difficult to determine the number of monotonous functions (i.e., the cardinality of $\mathcal{M}_n$) in dependence of n. Table 4.14 shows all monotonously increasing functions of $\mathbb{B}^1$, $\mathbb{B}^2$, and $\mathbb{B}^3$. The same number of monotonously decreasing functions exists.

It is easy to see an interesting relation between these sets of functions. When any function is taken and $\wedge$ is replaced by $\vee$ and vice versa, then one of these subsets of monotonously increasing functions is transformed into another one. $x_1 x_2$ is transformed into $x_1 \vee x_2$, $x_1 x_2 \vee x_1 x_3$ is transformed into $(x_1 \vee x_2)(x_1 \vee x_3) = x_1 \vee x_2 x_3$, etc.

Another relation exists between monotonously increasing and decreasing functions; the complement transforms a monotonously increasing function into a monotonously decreasing function and vice versa:

- The sum of function values 1 of an arbitrary monotonously increasing or decreasing function and its complement is always equal to 2 for $n = 1$. Both the monotonously increasing function $f_{mi}(x_1) = x_1$ and the monotonously decreasing function $f_{md}(x_1) = \overline{f_{mi}(x_1)} = \overline{x}_1$ have one function value 1; hence, the sum of these two values 1 is equal to 2. The functions $0(x_1)$ and $1(x_1)$ are both monotonously increasing and monotonously decreasing functions. The function $0(x_1)$ has no function value 1, and the function $1(x_1)$ has two ones so that the sum of function values 1 of these two functions is also equal to 2.

Table 4.14 Enumeration of monotonously increasing function

Number of Variables	Function values 1	Functions	Monotonously increasing functions
$n = 1$	0	1	$0(x_1)$
	1	1	x_1
	2	1	$1(x_1)$
$n = 2$	0	1	$0(x_1, x_2)$
	1	1	$x_1 x_2$
	2	2	$x_1, \ x_2$
	3	1	$x_1 \lor x_2$
	4	1	$1(x_1, x_2)$
$n = 3$	0	1	$0(x_1, x_2, x_3)$
	1	1	$x_1 x_2 x_3$
	2	3	$x_1 x_2, \ x_1 x_3, \ x_2 x_3$
	3	3	$x_1 x_2 \lor x_1 x_3, \ x_1 x_2 \lor x_2 x_3, \ x_1 x_3 \lor x_2 x_3$
	4	4	$x_1, \ x_2, \ x_3, \ x_1 x_2 \lor x_1 x_3 \lor x_2 x_3$
	5	3	$x_1 \lor x_2 x_3, \ x_2 \lor x_1 x_3, \ x_3 \lor x_1 x_2$
	6	3	$x_1 \lor x_2, \ x_1 \lor x_3, \ x_2 \lor x_3$
	7	1	$x_1 \lor x_2 \lor x_3$
	8	1	$1(x_1, x_2, x_3)$

- For $n = 2$, the sum of values 1 in a monotonous function and its complement is always equal to 4.
- For $n = 3$, the sum of values 1 in a monotonous function and its complement is always equal to 8, etc.

These two relations can be taken as indication of the following.

Theorem 4.17 *Each monotonously increasing function $f(\mathbf{x})$ can be represented as a disjunctive form without negated variables.*

This theorem directly follows from the concept of a "monotonously increasing function." If only $f(1, 1, \ldots, 1) = 1$, then $f(\mathbf{x}) = x_1 x_2 \ldots x_n$. If there is, however, a value 0 in the vector $\mathbf{x}_0$ with $f(\mathbf{x}_0) = 1$, say $x_i = 0$, then we have $f(x_1, \ldots, x_{i-1}, 0, x_{i+1}, \ldots, x_n) = 1$; because of

$$\mathbf{x}_0 = (x_1, \ldots, x_{i-1}, 0, x_{i+1}, \ldots, x_n) \leq \mathbf{x}_1 = (x_1, \ldots, x_{i-1}, 1, x_{i+1}, \ldots, x_n)$$

and the definition of a monotonously increasing function, the value of $f(\mathbf{x}_1)$ must also be equal to 1. The two conjunctions appearing in the disjunctive normal form or in any disjunctive form can be combined into one conjunction without the variable x_i. And this procedure can be applied to any conjunction with a negated variable.

There is also a weaker form of monotony: the definition can start with the consideration of one variable, and then it can include several variables and finally take all the variables into consideration. We use the following understanding:

the function $f(\mathbf{x})$ is monotonously increasing with regard to x_i if

$$f(x_i = 0) \leq f(x_i = 1) \,.$$

This means, for instance, for two variables and $x_i = x_1$ that

$$f(0,0) \le f(1,0) , \qquad f(0,1) \le f(1,1) .$$

The following functions of two variables would be monotonously increasing in x_1:

$$\{\mathbf{0}, \; x_1 x_2, \; x_1 \overline{x}_2, \; x_1, \; x_2, \; \overline{x}_2, \; x_1 \vee x_2, \; x_1 \vee \overline{x}_2, \; \mathbf{1}\}.$$

These functions are either independent on x_1, or $\overline{x}_1$ does not appear in these formulas. Similarly, we have

the function $f(\mathbf{x})$ is monotonously decreasing with regard to x_i if

$$f(x_i = 0) \ge f(x_i = 1) .$$

Sometimes it will be necessary (or desirable) to compare the number of 0s and the number of 1s in a binary vector.

Definition 4.20

$$\mathrm{MAJ}^n(x_1, \ldots, x_n) = \begin{cases} 1 , & \text{if } \|x\| \ge \frac{n}{2} \\ 0 , & \text{otherwise} \end{cases}$$

is the *majority function* of n variables.

This function counts the numbers of 1s in the argument vector, and if this number is larger than or equal to $\frac{n}{2}$, then the function value is equal to 1. For an odd number n, the value of $\frac{n}{2}$ is not an integer, and then the next larger integer will be taken for the number of components.

Example 4.27 Let $n = 3$, and then two or three components must have the value 1. This is appropriately expressed by

$$\mathrm{MAJ}^3(x_1, x_2, x_3) = x_1 x_2 \vee x_1 x_3 \vee x_2 x_3 .$$

For $n = 4$, the situation is similar; two or more components have to be equal to 1, and we get

$$\mathrm{MAJ}^4(x_1, x_2, x_3, x_4) = x_1 x_2 \vee x_1 x_3 \vee x_1 x_4 \vee x_2 x_3 \vee x_2 x_4 \vee x_3 x_4 .$$

All the majority functions are monotonously increasing functions. They take the value 1 for a given number of components equal to 1, adding more components with the value 1 (i.e., increasing the vector in the sense of $\le$) leaves the value of the function at 1, and it never goes back to 0.

Instead of $\frac{n}{2}$, any value k between 0 and n can be used to count the number of components equal to 1. These functions are denoted by threshold functions and designated by TH^k.

Definition 4.21

$$\mathrm{TH}^k(x_1, \ldots, x_n) = \begin{cases} 1 , & \text{if } \|x\| \ge k \\ 0 , & \text{otherwise} \end{cases}$$

is the *threshold function* of n variables with threshold k.

For each argument vector the number of 1s is counted, and if this number is larger than or equal to k, the function value is set equal to 1. We find successively

$$\|x\| \geq 0 \Rightarrow \mathrm{TH}^0 = 1(\mathbf{x}) \,,$$

$$\|x\| \geq 1 \Rightarrow \text{one or more values are equal to 1, } \mathrm{TH}^1(\mathbf{x}) = x_1 \vee \ldots \vee x_n \,,$$

$$\vdots$$

$$\|x\| \geq n \Rightarrow \text{all values are equal to 1, } \mathrm{TH}^n(\mathbf{x}) = x_1 \wedge \ldots \wedge x_n \,.$$

Again, it is not difficult to see from the definition that all threshold functions are monotonous functions. The majority function is a special case of the threshold functions for $k = \frac{n}{2}$.

Symmetric functions are another interesting class of logic functions.

Definition 4.22 The function $f(\mathbf{x})$ is *symmetric* with regard to the two variables $\{x_i, x_j\}$ if

$$f(x_i, x_j) = f(x_j, x_i) \,.$$

This definition is motivated by functions like

$$f_1(x, y) = x \vee y \,,$$
$$f_2(x, y) = x \wedge y \,,$$
$$f_3(x, y) = x \oplus y \,,$$
$$f_4(x, y, z) = x \vee y \vee z \,,$$
$$f_5(x, y, z) = xy \vee xz \vee yz \,.$$

$f_1(x, y)$, $f_2(x, y)$, and $f_3(x, y)$ are symmetric with regard to $\{x, y\}$; $f_4(x, y, z)$ and $f_5(x, y, z)$ are symmetric with regard to $\{x, y\}$, $\{x, z\}$, and $\{y, z\}$. Any exchange of the variables of a given pair still results in the same function.

In order to get some insight into the consequences of this property, we use the representation

$$f(x_1, x_2) = \overline{x}_1 \, \overline{x}_2 \, f(0, 0) \vee \overline{x}_1 \, x_2 \, f(0, 1) \vee x_1 \, \overline{x}_2 \, f(1, 0) \vee x_1 \, x_2 \, f(1, 1) \,;$$

the exchange of x_1 and x_2 results in

$$f^*(x_2, x_1) = \overline{x}_2 \, \overline{x}_1 \, f(0, 0) \vee \overline{x}_2 \, x_1 \, f(0, 1) \vee x_2 \, \overline{x}_1 \, f(1, 0) \vee x_2 \, x_1 \, f(1, 1)$$
$$= \overline{x}_1 \, \overline{x}_2 \, f(0, 0) \vee x_1 \, \overline{x}_2 \, f(0, 1) \vee \overline{x}_1 \, x_2 \, f(1, 0) \vee x_1 \, x_2 \, f(1, 1) \,.$$

In order to establish the required equality $f(x_1, x_2) = f^*(x_2, x_1)$, the condition

$$f(0, 1) = f(1, 0)$$

must be satisfied, and this has the consequence that we no longer have four independent values $f(0, 0)$, $f(0, 1)$, $f(1, 0)$, and $f(1, 1)$, but only three values $f(0, 0)$, $f(0, 1) = f(1, 0)$, and $f(1, 1)$. Hence, the following eight functions are symmetric in $\{x_1, x_2\}$:

$$f_1(x_1, x_2) = 0 \qquad = (0000) \,,$$

$$f_2(x_1, x_2) = \overline{x}_1 \overline{x}_2 \quad = (1000) \,,$$

$$f_3(x_1, x_2) = x_1 \oplus x_2 = (0110) \,,$$

$$f_4(x_1, x_2) = \overline{x}_1 \vee \overline{x}_2 = (1110) \,,$$

$$f_5(x_1, x_2) = x_1 x_2 \quad = (0001) \,,$$

$$f_6(x_1, x_2) = x_1 \odot x_2 = (1001) \,,$$

$$f_7(x_1, x_2) = x_1 \vee x_2 = (0111) \,,$$

$$f_8(x_1, x_2) = 1 \quad\quad = (1111) \,.$$

The symmetry can easily be seen for all these functions from the formulas, and it follows directly from the commutativity of the functions. The vectors of the function values have the properties in which always the second component is equal to the third component.

When the function $f(\mathbf{x})$ depends on more than two variables, then the required equality applies to the subfunctions:

$$f(x_1 = 0, x_2 = 1, x_3, \ldots, x_n) \equiv f(x_1 = 1, x_2 = 0, x_3, \ldots, x_n) \,.$$

Example 4.28 For $n = 3$ the condition of the symmetry of $f(x_1, x_2, x_3)$ with regard to the variables $\{x_1, x_2\}$ results in two requirements:

$$f(0, 1, 0) \equiv f(1, 0, 0) \quad \text{and} \quad f(0, 1, 1) \equiv f(1, 0, 1) \,.$$

The number of independent values decreases from eight to six.

$$f(x_1, x_2, x_3) = \overline{x}_1 \overline{x}_2 \vee \overline{x}_1 x_3 \vee \overline{x}_2 x_3 \vee x_1 x_2 \overline{x}_3$$

is one function that is symmetric with regard to the variables $\{x_1, x_2\}$.

An interesting property is the *transitivity* of this relation.

Theorem 4.18 *If the function $f(x_i, x_j, x_k)$ is symmetric in $\{x_i, x_j\}$ and symmetric in $\{x_j, x_k\}$, then it is also symmetric in $\{x_i, x_k\}$.*

Proof We exchange successively $\{x_i, x_j\}$, $\{x_j, x_k\}$, and again $\{x_i, x_j\}$, in this order, and get

$$f(x_i, x_j, x_k) \equiv f(x_j, x_i, x_k) \equiv f(x_k, x_i, x_j) \equiv f(x_k, x_j, x_i) \,.$$

Any permutation of the three components must show the same function value; this can be expressed by the conditions

$$f(0, 0, 1) \equiv f(0, 1, 0) \equiv f(1, 0, 0) \,,$$

$$f(0, 1, 1) \equiv f(1, 0, 1) \equiv f(1, 1, 0) \,.$$

$f(0, 0, 0)$ and $f(1, 1, 1)$ are still independent. This property results in only four independent values. Vectors with the same number of 1s (or 0s) have the same function values. The functions

Fig. 4.10 Symmetric
functions of two variables

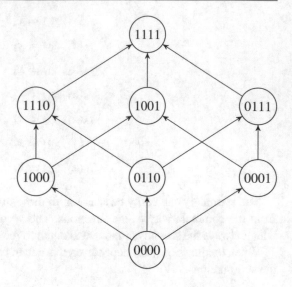

$$f_1(x_1, x_2, x_3) = x_1 \odot x_2 \odot x_3 \,,$$
$$f_2(x_1, x_2, x_3) = x_1 \oplus x_2 \oplus x_3 \,,$$
$$f_3(x_1, x_2, x_3) = x_1 \vee x_2 \vee x_3 \,,$$
$$f_4(x_1, x_2, x_3) = x_1 \wedge x_2 \wedge x_3$$

can be taken as examples.   □

The next theorem can be prepared by Fig. 4.10. We represented the binary vectors of the symmetric functions of two variables (a subset of $\mathbb{B}^{2^2} = \mathbb{B}^4$) according to the partial order $\leq$ in $\mathbb{B}^4$, and it can be seen that we get a structure that is isomorphic to $\mathbb{B}^3$. Hence, the subset of the symmetric functions is already a Boolean Algebra. This will be summarized by the following.

Theorem 4.19 *The subset $\mathcal{S}^n$ of all symmetric functions of n variables is a Boolean Algebra.*

Proof The proof uses the fact that the set $\mathcal{S}^n$ is closed under $\wedge$, $\vee$, and the complement, and that $0(\mathbf{x})$ and $1(\mathbf{x})$ are always symmetric functions. When $f(0, 1) \equiv f(1, 0)$, then $\overline{f(0, 1)} \equiv \overline{f(1, 0)}$. For two functions $f(x, y)$ and $g(x, y)$ with $f(0, 1) \equiv f(1, 0)$ and $g(0, 1) \equiv g(1, 0)$, we can build

$$h_1(x, y) \equiv f(x, y) \wedge g(x, y) \text{ and}$$
$$h_2(x, y) \equiv f(x, y) \vee g(x, y) \,,$$

and $h_1(0, 1) \equiv h_1(1, 0)$ and $h_2(0, 1) \equiv h_2(1, 0)$ follow directly from

$$f(0, 1) \wedge g(0, 1) \equiv f(1, 0) \wedge g(1, 0) \text{ and}$$
$$f(0, 1) \vee g(0, 1) \equiv f(1, 0) \vee g(1, 0) \,.$$

Thus, the property of symmetry reduces the dimension of the Boolean Algebra.   □

4.4 Minimization Problems

A function $f(\mathbf{x})$ given by its function table can be represented in different ways, by means of different expressions. Up to now, the following means are available:

1. The *Disjunctive Normal Form* DNF: an enumeration of all elementary conjunctions C_i with $f(C_i) = 1$, connected by $\vee$, *uniquely defined*; $0(\mathbf{x})$ is represented by the constant value 0.
2. The *Conjunctive Normal Form* CNF: an enumeration of all elementary disjunctions D_i with $f(\overline{D}_i) = 0$, connected by $\wedge$, *uniquely defined*; $1(\mathbf{x})$ is represented by the constant value 1.
3. The *Antivalence Normal Form* ANF follows from the DNF by replacing the $\vee$ by $\oplus$; this is always possible because of the orthogonality of the elementary disjunctions.
4. The *Equivalence Normal Form* ENF follows from the CNF by replacing the $\wedge$ by $\odot$; this is always possible because of the orthogonality of the elementary conjunctions.
5. The *Antivalence Polynomial* AP follows from the ANF by replacing all negated variables $\overline{x}_i$ by $1 \oplus x_i$ and applying the distributive law and possible simplifications.
6. The *Equivalence Polynomial* EP follows from the ENF by replacing all negated variables $\overline{x}_i$ by $0 \odot x_i$ and applying the distributive law and possible simplifications.

These six possible representations are *unique*. Remember: the order of conjunctions or disjunctions is not taken into consideration. When now disjunctive forms (orthogonalized or non-orthogonalized) are considered, then there must be several forms for one function. We already have seen that there are 2^{2^n} functions $f(x_1, \ldots, x_n)$ of n variables, but 2^{3^n} different *disjunctive forms* DF (the DF of $0(\mathbf{x})$ is represented by 0). The number of disjunctive forms is therefore (much) larger than the number of functions, many forms will represent the same function, and this is the reason for different *minimization problems*. Among those different representations, we can look for representations with the smallest number of conjunctions, with the smallest number of literals in the formula, etc. This problem has been thoroughly explored over the last decades. We will present some of the key ideas and problems.

Let $f(\mathbf{x})$ be given by the function table (or equivalently, as a set of elementary conjunctions). It can be seen immediately (by counting) whether $f(\mathbf{x}) = 1(\mathbf{x})$. In this case, nothing has to be done, and the constant 1 is the shortest representation of $1(\mathbf{x})$. The following two functions $f_1(\mathbf{x})$ and $f_2(\mathbf{x})$ will be used (see Table 4.15).

The idea of the procedure consists in finding a number of conjunctions that generate the values 1 of the given function. The conjunctions should be as short as possible, and the number of conjunctions as small as possible. This idea can be based on *maximal intervals*, also denoted by *prime implicants* or *prime conjunctions*. We consider $f_1(\mathbf{x})$ and the conjunctions $x_1\overline{x}_4$, x_1 and $\overline{x}_4$ (see Table 4.16).

Table 4.15 Two functions to be minimized

$f_1(\mathbf{x})$	1	1	0	1	0	0	0	1	1	0	1	0	1	0	1	0
$f_2(\mathbf{x})$	0	0	0	1	0	0	1	0	0	1	0	1	0	0	1	0
x_4	0	1	0	1	0	1	0	1	0	1	0	1	0	1	0	1
x_3	0	0	1	1	0	0	1	1	0	0	1	1	0	0	1	1
x_2	0	0	0	0	1	1	1	1	0	0	0	0	1	1	1	1
x_1	0	0	0	0	0	0	0	0	1	1	1	1	1	1	1	1

Table 4.16 The function $f_1(\mathbf{x})$ to be minimized and three selected conjunctions

$f_1(\mathbf{x})$	1	1	0	1	0	0	0	1	1	0	1	0	1	0	1	0
$x_1\overline{x}_4$	0	0	0	0	0	0	0	0	1	0	1	0	1	0	1	0
x_1	0	0	0	0	0	0	0	0	1	1	1	1	1	1	1	1
$\overline{x}_4$	1	0	1	0	1	0	1	0	1	0	1	0	1	0	1	0

It can be seen that $x_1\overline{x}_4 \leq f_1(\mathbf{x})$; however, $x_1 \not\leq f_1(\mathbf{x}), \overline{x}_4 \not\leq f_1(\mathbf{x})$. Hence, the interval described by $x_1\overline{x}_4$ is a maximal interval, and the conjunction $x_1\overline{x}_4$ itself is a shortest conjunction with regard to the function $f_1(\mathbf{x})$. None of the variables can be omitted, and the interval cannot be extended to a larger interval.

The *consensus-method* can be used to find all prime conjunctions of a given function $f(\mathbf{x})$. The following rule uses the inverse of the consensus and is the main step of the procedure:

$$xC_1 \vee \overline{x}C_2 \equiv xC_1 \vee \overline{x}C_2 \vee C_1C_2 \ . \tag{4.12}$$

If there are two conjunctions that match the left part of this identity, then these two conjunctions can be replaced by the three conjunctions of the right-hand side, i.e., the conjunction C_1C_2 will be added. Additionally, the absorption law will be applied as often as possible, once in the following.

Example 4.29

$$\overline{x}_1\overline{x}_3x_4 \vee \overline{x}_1\overline{x}_2\overline{x}_3\overline{x}_4 \equiv x_4C_1 \vee \overline{x}_4C_2$$

$$\equiv \overline{x}_1\overline{x}_3x_4 \vee \overline{x}_1\overline{x}_2\overline{x}_3\overline{x}_4 \vee C_1C_2$$

$$\equiv \overline{x}_1\overline{x}_3x_4 \vee \overline{x}_1\overline{x}_2\overline{x}_3\overline{x}_4 \vee \overline{x}_1\overline{x}_2\overline{x}_3$$

$$\equiv \overline{x}_1\overline{x}_3x_4 \vee \overline{x}_1\overline{x}_2\overline{x}_3 \ .$$

It is a benefit that the consensus rule also covers the reverses of the distributive law: $a = a(b \vee \overline{b}) = ab \vee a\overline{b}$. This can be seen in the following example.

Example 4.30

$$\overline{x}_1\overline{x}_2\overline{x}_3x_4 \vee \overline{x}_1\overline{x}_2\overline{x}_3\overline{x}_4 \equiv x_4C_1 \vee \overline{x}_4C_2$$

$$\equiv \overline{x}_1\overline{x}_2\overline{x}_3x_4 \vee \overline{x}_1\overline{x}_2\overline{x}_3\overline{x}_4 \vee C_1C_2$$

$$\equiv \overline{x}_1\overline{x}_2\overline{x}_3x_4 \vee \overline{x}_1\overline{x}_2\overline{x}_3\overline{x}_4 \vee \overline{x}_1\overline{x}_2\overline{x}_3$$

$$\equiv \overline{x}_1\overline{x}_2\overline{x}_3 \ .$$

Examples 4.29 and 4.30 show the different cases where the consensus rule can be used.

Now we apply the consensus rule to the functions specified in Table 4.15. We start with $f_1(\mathbf{x})$ and the given elementary conjunctions. Table 4.17 enumerates in the left part the given elementary conjunctions of $f_1(\mathbf{x})$ and shows in the right part the results of all possible applications of the consensus to these conjunctions.

These eight conjunctions absorb all elementary conjunctions, and the process can continue. There are two more possibilities with the same result:

$$(5. + 7.) + (6. + 8.) \quad \Rightarrow \quad x_1\overline{x}_4 \ ,$$

$$(5. + 6.) + (7. + 8.) \quad \Rightarrow \quad x_1\overline{x}_4 \ .$$

The last four conjunctions can be replaced by $x_1\overline{x}_4$, and five conjunctions have to be considered in the next step. Further changes by means of the consensus rule cannot be achieved. Hence, we know the five prime conjunctions:

Table 4.17 Application of the consensus rule to the elementary conjunctions of $f_1(\mathbf{x})$

Elementary conjunctions of $f_1(\mathbf{x})$			Results of the consensus rule		
1.	(0000)	$= \bar{x}_1\bar{x}_2\bar{x}_3\bar{x}_4$	1. + 2.	$\Rightarrow$	$\bar{x}_1\bar{x}_2\bar{x}_3$
2.	(0001)	$= \bar{x}_1\bar{x}_2\bar{x}_3 x_4$	1. + 5.	$\Rightarrow$	$\bar{x}_2\bar{x}_3\bar{x}_4$
3.	(0011)	$= \bar{x}_1\bar{x}_2 x_3 x_4$	2. + 3.	$\Rightarrow$	$\bar{x}_1\bar{x}_2 x_4$
4.	(0111)	$= \bar{x}_1 x_2 x_3 x_4$	3. + 4.	$\Rightarrow$	$\bar{x}_1 x_3 x_4$
5.	(1000)	$= x_1\bar{x}_2\bar{x}_3\bar{x}_4$	5. + 6.	$\Rightarrow$	$x_1\bar{x}_2\bar{x}_4$
6.	(1010)	$= x_1\bar{x}_2 x_3\bar{x}_4$	5. + 7.	$\Rightarrow$	$x_1\bar{x}_3\bar{x}_4$
7.	(1100)	$= x_1 x_2\bar{x}_3\bar{x}_4$	6. + 8.	$\Rightarrow$	$x_1 x_3\bar{x}_4$
8.	(1110)	$= x_1 x_2 x_3\bar{x}_4$	7. + 8.	$\Rightarrow$	$x_1 x_2\bar{x}_4$

Table 4.18 Cover of the function $f_1(\mathbf{x})$

$f_1(0000)$	$f_1(0001)$	$f_1(0011)$	$f_1(0111)$	$f_1(1000)$	$f_1(1010)$	$f_1(1100)$	$f_1(1110)$
$C_2 \vee C_3$	$C_2 \vee C_5$	$C_4 \vee C_5$	C_4	$C_1 \vee C_3$	C_1	C_1	C_1

$$C_1 = x_1\bar{x}_4 \,, \qquad C_2 = \bar{x}_1\bar{x}_2\bar{x}_3 \,, \qquad C_3 = \bar{x}_2\bar{x}_3\bar{x}_4 \,,$$
$$C_4 = \bar{x}_1 x_3 x_4 \,, \qquad C_5 = \bar{x}_1\bar{x}_2 x_4 \,.$$

Now there is the problem that not all of these conjunctions are necessary for a disjunctive form. This problem will be solved by Quine's algorithm sometimes also denoted by *Algorithm of Quine–McCluskey* [2–4]:

1. For each value 1 in the vector $f_1(\mathbf{x})$, it is determined which conjunctions generate (cover) this value 1. Table 4.18 shows for each value 1 in the vector $f_1(\mathbf{x})$ which conjunctions generate (cover) this value 1. If there are more than one conjunction, these conjunctions are combined by $\vee$.
2. The disjunctions of the lower row in Table 4.18 are connected by $\wedge$, simplified according to the rules of logic functions, and used as a *selecting expression*. Each final conjunction of conjunction symbols is one irredundant representation of $f_1(\mathbf{x})$ as disjunctive form. Among those irredundant expressions, the expression with the smallest number of literals and with the smallest number of conjunctions can be found:

$$\begin{aligned} S \;&=\; (C_2 \vee C_3)(C_2 \vee C_5)(C_4 \vee C_5)C_4(C_1 \vee C_3)C_1 C_1 C_1 \\ &=\; C_1 C_2 C_4 \vee C_1 C_3 C_4 C_5 \,. \end{aligned}$$

There are two irredundant possibilities, and the use of C_1, C_2, C_4 has the minimal number of conjunctions and literals:

$$f_1(\mathbf{x}) = x_1\bar{x}_4 \vee \bar{x}_1\bar{x}_2\bar{x}_3 \vee \bar{x}_1 x_3 x_4 \,,$$
$$f_1(\mathbf{x}) = x_1\bar{x}_4 \vee \bar{x}_2\bar{x}_3\bar{x}_4 \vee \bar{x}_1 x_3 x_4 \vee \bar{x}_1\bar{x}_2 x_4 \,.$$

For $f_2(\mathbf{x})$ of Table 4.15, there is only one solution:

$$f_2(\mathbf{x}) = x_1\bar{x}_2 x_4 \vee x_2 x_3\bar{x}_4 \vee \bar{x}_2 x_3 x_4 \,.$$

In the case of $f_1(\mathbf{x})$, even single variables would have been possible because each single literal generates eight values 1. In the second case $f_2(\mathbf{x})$ has only five values of 1, the procedure can stop with conjunctions of length 2, and further simplifications are not possible.

A big number of publications and algorithms exist—all of them deal with refinements and improvements of the presented algorithm, but in principle the basic ideas remain unchanged.

A minimal conjunctive form of $f(\mathbf{x})$ can be constructed using the following algorithm:

1. Create an arbitrary disjunctive form of $g(\mathbf{x}) = \overline{f(\mathbf{x})}$.
2. Construct the minimal disjunctive form of $g(\mathbf{x})$ using the procedure introduced above.
3. Transform the minimal disjunctive form of $g(\mathbf{x})$ into the minimal conjunctive form of $f(\mathbf{x})$ using the negation with regard to De Morgan's laws.

We will come back to this problem when we explore the design of combinational circuits and show some more detailed examples.

A very difficult problem is the minimization of antivalence forms. It was already mentioned that the antivalence polynomial (all variables are non-negated) is unique. The problem is difficult when negated variables are possible.

Example 4.31 Minimal antivalence forms will be created using transformations starting with the minimal disjunctive form of $f_3(\mathbf{x})$:

$$
\begin{aligned}
f_3(\mathbf{x}) &= x_1\overline{x}_2 \vee x_1\overline{x}_3 \vee x_1\overline{x}_4 \vee \overline{x}_1x_2x_3x_4 \\
&= x_1\overline{x}_2 \vee x_1x_2\overline{x}_3 \vee x_1x_2\overline{x}_4 \vee \overline{x}_1x_2x_3x_4 \\
&= x_1\overline{x}_2 \vee x_1x_2\overline{x}_3 \vee x_1x_2x_3\overline{x}_4 \vee \overline{x}_1x_2x_3x_4 \\
&= x_1\overline{x}_2 \oplus x_1x_2\overline{x}_3 \oplus x_1x_2x_3\overline{x}_4 \oplus \overline{x}_1x_2x_3x_4 \\
&= x_1\overline{x}_2 \oplus x_1x_2\overline{x}_3 \oplus x_1x_2x_3\overline{x}_4 \oplus \overline{x}_1x_2x_3x_4 \oplus x_1x_2x_3x_4 \oplus x_1x_2x_3x_4 \\
&= x_1\overline{x}_2 \oplus x_1x_2\overline{x}_3 \oplus x_1x_2x_3 \oplus x_2x_3x_4 \\
&= x_1\overline{x}_2 \oplus x_1x_2 \oplus x_2x_3x_4 \\
&= x_1 \oplus x_2x_3x_4 \ .
\end{aligned}
$$

We changed in these steps of transformations first the minimal disjunctive form into an orthogonal disjunctive form. In this orthogonal form we replaced all $\vee$ by $\oplus$ using Theorem 4.6. The additional $\oplus$ with two times the conjunction $x_1x_2x_3x_4$ does not change the function but facilitates a sequence of simplifications until the minimal antivalence form of only four literals is reached.

4.5   Complete Systems of Logic Functions

In the first section of this chapter we already saw many different expressions for the representation of functions and many identities. We remember that the *conjunction*, the *disjunction*, and the *negation* have been sufficient for the representation of all functions of n variables. Two special cases had to be considered as well, the constant functions $0(\mathbf{x})$ and $1(\mathbf{x})$. The use of ring structures has also been possible: the *conjunction* and the *antivalence* or the *disjunction* and the *equivalence* with the *negation* or the *constants* 0 and 1 had to be used.

Theorem 4.20

- *Each function $f(\mathbf{x})$ can be represented by means of the generalized* NAND.
- *Each function $f(\mathbf{x})$ can be represented by means of the generalized* NOR.

Theorem 4.20 shows that the generalized NAND and NOR can be used as a single operation.

The construction of this representation is quite easy, and the function $f(\mathbf{x}) \neq 0(\mathbf{x})$ will be represented as a disjunctive (normal) form, i.e., as a disjunction of conjunctions:

$$f(\mathbf{x}) = C_1 \vee C_2 \vee \ldots \vee C_k$$

for some conjunctions $C_1, \ldots, C_k$. A two-fold negation does not change $f(\mathbf{x})$:

$$f(\mathbf{x}) = \overline{\overline{C_1 \vee C_2 \vee \ldots \vee C_k}},$$

and the application of De Morgan's laws results in the desired representation; the single conjunctions are negated, and then they are combined by conjunction and negation again:

$$f(\mathbf{x}) = \overline{\overline{C_1} \wedge \overline{C_2} \wedge \ldots \wedge \overline{C_k}}.$$

The function $0(\mathbf{x})$, $x_i \in \mathbf{x}$ can also be presented using only NANDs:

$$0(\mathbf{x}) = \overline{\overline{\overline{x_i x_i} \wedge x_i} \wedge \overline{\overline{x_i x_i} \wedge x_i}}.$$

The representation of the function by means of NOR follows the same procedure starting with the conjunctive (normal) form of $f(\mathbf{x})$. The fundamental understanding of completeness can be achieved by means of the following theorem (*Stone*).

Theorem 4.21 *A set $\mathcal{C} = \{f_1(\mathbf{x}), \ldots, f_k(\mathbf{x})\}$ of logic functions is a complete set of functions if and only if it contains at least one function that is not self-dual, one that is not monotonously increasing, one that is not linear, one that is not 0-preserving, and one that is not 1-preserving.*

Not necessarily all these functions have to be different from each other. It could even be that one function has all these properties (which is already indicated by the theorem with regard to NAND and NOR). Let us consider the NAND-function. The NAND-function:

- is not 0-preserving: $\mathrm{NAND}(0, 0) = 1$;
- is not 1-preserving: $\mathrm{NAND}(1, 1) = 0$;
- is not monotonously increasing: $(0, 0) \leq (1, 1)$,

                                       $\mathrm{NAND}(0, 0) = 1 > \mathrm{NAND}(1, 1) = 0$;
- is not linear: $\mathrm{NAND}(x, y) = 1 \oplus xy$;
- is not self dual: $\overline{\mathrm{NAND}(\overline{x}, \overline{y})} = \overline{\overline{\overline{x}\,\overline{y}}} = \overline{x}\,\overline{y} = \overline{x \vee y}$

                                       $\overline{x \vee y} = \mathrm{NOR}(x, y) \neq \mathrm{NAND}(x, y)$.

A general proof of this theorem takes into consideration that we already know that $0(\mathbf{x})$, $1(\mathbf{x})$, the *conjunction*, and the *negation* are sufficient for the representation of any logic function $f(\mathbf{x})$. The idea of the proof is based on the construction of these functions by means of functions with the stated properties. We will only show how this idea applies to NAND:

- $\text{NAND}(x, x) = \overline{x \wedge x} = \overline{x}$. When we identify the first and second arguments of this function, then the negation of a variable can be generated.
- $\text{NAND}(x, \overline{x}) = 1$. The negation of a variable x has been generated before; the use of a variable and the negated variable results in 1, and another complement results in 0.
- Finally, the negation of $\overline{x \wedge y}$, i.e., $\overline{\overline{x \wedge y}} = x \wedge y$, generates the conjunction, and another negation would even result in the disjunction.

The easiest way of transforming any logic function into this format is the method used above: take a disjunctive form of the function and use the two-fold negation.

The construction using NAND can also be applied to NOR, repeating the considerations from the above step by step.

4.6   Exercises

Prepare for each task of each exercise of this chapter a PRP for the XBOOLE-monitor XBM 2 using a lexicographic order of the variables x_i. Use the help system of the XBOOLE-monitor XBM 2 to learn the details of the needed commands. Execute the created PRPs and verify the computed results.

If you have not yet prepared the XBOOLE-monitor XBM 2 on your computer, you can get this XBOOLE-monitor free of charge by means of the following three steps:

1. **Download**:
   There are four versions of the XBOOLE-monitor XBM 2, two for Windows 10 or subsequent Windows systems (32 or 64 bits) and two for LINUX–Ubuntu (also 32 or 64 bits); you must download the version of the XBOOLE-monitor XBM 2 that fits to your operating system.
   Authorized users of the online version of this chapter (https://doi.org/10.1007/978-3-030-88945-6_4) can download the XBOOLE-monitor XBM 2 directly from the web page

   https://link.springer.com/chapter/10.1007/978-3-030-88945-6_4

   where the links for the download of the XBOOLE-monitor XBM 2 are located in the part "Supplementary Information" (below the part "Abstract"). The headline above such a link indicates the associated zip-file of the XBOOLE-monitor XBM 2. The sizes of the zip-files have been provided behind the links and can be used to verify the download. A click on the link of the wanted version of the XBOOLE-monitor XBM 2 starts the download.

   Readers of the hardcopy of this book get access to the XBOOLE-monitor XBM 2 using the URL

   https://link.springer.com/chapter/10.1007/978-3-030-88945-6_4

   to download the first two pages of this chapter. After this download, the same procedure as the authorized users of the online version of a chapter can be used to download the wanted version of the XBOOLE-monitor XBM 2.

2. **Unzip**: The XBOOLE-monitor XBM 2 must not be installed but unzipped into an arbitrary directory of your computer. A convenient tool for unziping the downloaded zip-file is usually available as part of the operating system or can be downloaded from the Internet.

3. **Execute**:
   - Windows:
     The executable file of the two versions (32 or 64 bits) for Windows 10 (or subsequent Windows systems) of the XBOOLE-monitor XBM 2 is `XBM2.exe`; the other files in

the expanded directory must remain unchanged. A double-click on the executable file XBM2.exe within the Explorer of Windows starts the XBOOLE-monitor XBM 2.

- LINUX–Ubuntu:
  The unzipped folder of the XBOOLE-monitor XBM 2 contains for this operating system only the executable file XBM2-i386.AppImage for the version of 32 bits or XBM2-x86_64.AppImage for the version of 64 bits of the XBOOLE-monitor XBM 2. A double-click on the created AppImage-file within the file manager of LINUX–Ubuntu starts the XBOOLE-monitor XBM 2.

Exercise 4.1 (Methods to Define Logic Functions)

In this exercise, we explore possibilities to define the following elementary logic functions:

$$f_1(x_1) = \overline{x}_1 \,,$$

$$f_2(x_1, x_2) = x_1 \wedge x_2 \,,$$

$$f_3(x_1, x_2) = x_1 \vee x_2 \,,$$

$$f_4(x_1, x_2) = x_1 \oplus x_2 = x_1\overline{x}_2 \wedge \overline{x}_1 x_2 \,,$$

$$f_5(x_1, x_2) = x_1 \odot x_2 = \overline{x}_1\overline{x}_2 \wedge x_1 x_2 \,,$$

$$f_6(x_1, x_2) = x_1 \rightarrow x_2 = \overline{x}_1 \vee x_2 \,,$$

$$f_7(x_1, x_2) = x_1 \mid x_2 = (\overline{x_1 \wedge x_2}) = \overline{x}_1 \vee \overline{x}_2 \,,$$

$$f_8(x_1, x_2) = x_1 \downarrow x_2 = (\overline{x_1 \vee x_2}) = \overline{x}_1 \wedge \overline{x}_2 \,.$$

The aim is the representation of these functions as TVL in ODA-form, i.e., the wanted TVL contains all binary vectors for which the represented function is equal to 1.

(a) Define the function-TVLs of the functions $f_1(x_1)$ to $f_8(x_1, x_2)$ using commands tin. Utilize that this command implicitly realizes the orthogonalization of a given D-form into the requested ODA-form. Show all created TVLs in the m-fold view organized in two rows and four columns.

(b) Define the function-TVLs of the functions $f_1(x_1)$ to $f_8(x_1, x_2)$ using created TVLs of the trivial functions $f_{11}(x_1) = x_1$ and $f_{12}(x_2) = x_2$ and appropriate set operations that realize the logic operations:

$$\mathrm{ODA}(\overline{f}_i) = \overline{\mathrm{ODA}(f_i)} \,, \qquad\qquad \text{command: cpl} \,,$$

$$\mathrm{ODA}(f_i \wedge f_j) = \mathrm{ODA}(f_i) \cap \mathrm{ODA}(f_j) \,, \qquad\qquad \text{command: isc} \,,$$

$$\mathrm{ODA}(f_i \vee f_j) = \mathrm{ODA}(f_i) \cup \mathrm{ODA}(f_j) \,, \qquad\qquad \text{command: uni} \,,$$

$$\mathrm{ODA}(f_i \oplus f_j) = \mathrm{ODA}(f_i) \, \Delta \, \mathrm{ODA}(f_i) \,, \qquad\qquad \text{command: syd} \,,$$

$$\mathrm{ODA}(f_i \odot f_j) = \mathrm{ODA}(f_i) \, \overline{\Delta} \, \mathrm{ODA}(f_j) \,, \qquad\qquad \text{command: csd} \,.$$

Show all created TVLs in the m-fold view organized in two rows and four columns.

(c) Define the function-TVLs of the functions $f_1(x_1)$ to $f_8(x_1, x_2)$ using commands sbe. This command creates a TVL in ODA-form of the specified expression. The sings of operations expected by the command sbe can be seen (and even changed) in the dialog "Options" of the dialog "Solve Boolean Equation" in the XBOOLE-monitor XBM 2. We usually use the following signs to express the operation:

$$\overline{x}_i : \qquad /\text{xi},$$

$$x_i \wedge x_j : \qquad \text{xi\&xj},$$

$$x_i \vee x_j : \qquad \text{xi+xj},$$

$$x_i \oplus x_j : \qquad \text{xi\#xj},$$

$$x_i \odot x_j : \qquad \text{xi}=\text{xj},$$

$$x_i \to x_j : \qquad \text{xi}>\text{xj}.$$

Show all created TVLs in the m-fold view organized in two rows and four columns.

(d) Define the function-TVLs of the functions $f_1(x_1)$ to $f_8(x_1, x_2)$ using the possibility that function values can be changed in a shown Karnaugh-maps. This approach needs changes in prepared Karnaugh-map; hence, prepare the PRP such that functions $f_1(x_1) = 0$ and $f_i(x_1, x_2) = 0, i = 2, \ldots, 8$, are created and shown in the m-fold view twice (for comparison between the Karnaugh-map and the TVL) in columns 1 and 2. Change after the execution of this PRP the left viewports of the m-fold view to Karnaugh-maps and change in these Karnaugh-maps the function values such that the requested functions $f_1(x_1)$ to $f_8(x_1, x_2)$ are defined; the TVL to the right will show the associated ODA-form to the functions $f_1(x_1)$ to $f_8(x_1, x_2)$.

Exercise 4.2 (Normal Forms of Logic Functions)

The function $f(\mathbf{x})$ is given by the following expression:

$$f(x_1, x_2, x_3, x_4) = (x_1 \oplus x_3)(x_2 \oplus x_4) \vee x_1\overline{x}_4 .$$

(a) Compute the disjunctive normal form (DNF) of the given function $f(\mathbf{x})$ and show the TVL in ODA-form of this function and the TVL of the DNF (created as D-form) in the first column of the m-fold view. You can use an arbitrary order of the elementary conjunctions of the DNF. Store the created XBOOLE-system as file "dnf.sdt" for later use. How many elementary conjunctions describe the given function $f(\mathbf{x})$? Use the computed TVL of the DNF as source to write down the DNF of $f(x_1, x_2, x_3, x_4)$ as an expression.

(b) Compute the conjunctive normal form (CNF) of the given function $f(\mathbf{x})$ and show the TVL in ODA-form of this function and the TVL of the CNF (created as K-form) in the first column of the m-fold view. You can use an arbitrary order of the elementary disjunctions of the CNF. Store the created XBOOLE-system as file "cnf.sdt" for later use. How many elementary disjunctions describe the given function $f(\mathbf{x})$? Use the computed TVL of the CNF as source to write down the CNF of $f(x_1, x_2, x_3, x_4)$ as an expression.

(c) Load the file "dnf.sdt," compute disjunctive normal form (DNF) of the function

$$g(x_1, x_2, x_3, x_4) = (x_1 \vee x_2x_3) \oplus ((x_1x_2 \vee x_3)x_4) ,$$

and verify based on the DFNs whether the functions $f(x_1, x_2, x_3, x_4)$ and $g(x_1, x_2, x_3, x_4)$ are equal to each other. Show the TVL in ODA-form of $g(x_1, x_2, x_3, x_4)$ and the TVLs of the DNFs of $f(\mathbf{x})$ and $g(\mathbf{x})$ in the first row of the m-fold view. Prepare a list of equal pairs of elementary conjunctions and in the case that $f(\mathbf{x})$ and $g(\mathbf{x})$ are different functions two sets of elementary conjunctions that describe the differences between these functions.

(d) Load the file "cnf.sdt," compute conjunctive normal form (CNF) of the function

$$h(x_1, x_2, x_3, x_4) = (x_1 \lor x_2 x_3) \oplus (x_2 x_4),$$

and verify based on the CFNs whether the functions $f(x_1, x_2, x_3, x_4)$ and $h(x_1, x_2, x_3, x_4)$ are equal to each other. Show the TVL in ODA-form of $h(x_1, x_2, x_3, x_4)$ and the TVLs of the CNFs of $f(\mathbf{x})$ and $h(\mathbf{x})$ in the first row of the m-fold view. Prepare a list of equal pairs of elementary disjunctions and in the case that $f(\mathbf{x})$ and $h(\mathbf{x})$ are different functions two sets of elementary disjunctions that describe the differences between these functions.

Exercise 4.3 (Function Tables of Logic Functions)

(a) Create a function table of the logic function $f(x_1, x_2, x_3, x_4)$ given as disjunctive form:

$$f(x_1, x_2, x_3, x_4) = x_1 \overline{x}_3 \lor x_2 x_3 \lor x_1 x_4.$$

The binary vectors of (x_1, x_2, x_3, x_4) must be ordered top down from (0000) to (1111) using the binary code. Use the approach that generates first a function table in the requested order for initial function $f(\mathbf{x}) = 0$ and change thereafter the function values based on the given three conjunctions. Show in one row of the m-fold view the function-TVL of $f(\mathbf{x})$, the function table of the initial function $f(\mathbf{x}) = 0$, and the function table of $f(\mathbf{x})$.

(b) Create a function table of the logic function $f(x_1, x_2, x_3, x_4)$ given as conjunctive form:

$$f(x_1, x_2, x_3, x_4) = (x_1 \lor \overline{x}_3)(x_2 \lor x_3)(x_1 \lor x_4).$$

The binary vectors of (x_1, x_2, x_3, x_4) must be ordered top down from (0000) to (1111) using the binary code. Use the approach that generates first a function table in the requested order for the initial function $f(\mathbf{x}) = 1$ and change thereafter the function values based on the given three disjunctions. Show in one row of the m-fold view the function-TVL of $f(\mathbf{x})$, the function table of initial function $f(\mathbf{x}) - 1$, and the function table of $f(\mathbf{x})$.

(c) Create a function table of the logic function $f(x_1, x_2, x_3, x_4)$ given by the following expression:

$$f(x_1, x_2, x_3, x_4) = (x_1 \lor \overline{x}_3) \oplus x_2 x_4 \oplus (x_2 \lor x_3).$$

The binary vectors of (x_1, x_2, x_3, x_4) must be ordered top down from (0000) to (1111) using the binary code. Use the approach that generates first the ordered set of the binary vectors of (x_1, x_2, x_3, x_4), computes thereafter a TVL of the function $F(x_1, x_2, x_3, x_4, f) = f(x_1, x_2, x_3, x_4) \odot f$ that contains for each vector (x_1, x_2, x_3, x_4) the associated function value f of the logic function $f(\mathbf{x})$, and extends finally the ordered set of the binary vectors to the function table using the TVL $F(\mathbf{x}, f)$. Show in one row of the m-fold view the function-TVL of $f(\mathbf{x})$, the ordered set of the binary vectors of (x_1, x_2, x_3, x_4), and the function table of $f(\mathbf{x})$.

(d) Figure 4.11 shows the function table of $f(x_1, x_2, x_3)$. Compute the function-TVL of $f(\mathbf{x})$ as minimized ODA-form and show in one row of the m-fold view both the given function table and the computed function-TVL of $f(\mathbf{x})$.

Exercise 4.4 (Basic Forms of Logic Functions)

A given expression of a logic function can be transformed into each of the four basic forms: the disjunctive form, the conjunctive form, the antivalence form, and the equivalence form. Compute the three other basic forms of the function given in one of these forms and show all four forms in the first row of the m-fold view as TVLs and for comparison in the second row as Karnaugh-maps. Use for all four tasks the assignment of the forms to the columns in the order of their enumeration given above. Utilize the commands orth, cpl, ndm, and sform to solve these tasks for the following functions:

(a) Disjunctive form: $f_1(\mathbf{x}) = x_1\overline{x}_4 \vee x_2x_3 \vee x_3x_4$.
(b) Conjunctive form: $f_2(\mathbf{x}) = (x_1 \vee \overline{x}_4)(x_2 \vee x_3)(x_3 \vee x_4)$.
(c) Antivalence form: $f_3(\mathbf{x}) = x_1\overline{x}_4 \oplus x_2x_3 \oplus x_3x_4$.
(d) Equivalence form: $f_4(\mathbf{x}) = (x_1 \vee \overline{x}_4) \odot (x_2 \vee x_3) \odot (x_3 \vee x_4)$.

Exercise 4.5 (Properties of Logic Functions with Regard to a Variable)

The subfunctions $f(x_i = 0)$ and $f(x_i = 1)$ of the Shannon expansion can be used to verify whether a logic function satisfies certain properties. Utilize this possibility to check whether the following functions are monotonously increasing, monotonously decreasing, or linear with regard to x_3, or even independent of x_3:

(a) $f_1(\mathbf{x}) = (x_1x_2 \oplus x_3x_4) \vee (x_1\overline{x}_4 \oplus \overline{x}_2\overline{x}_3x_4 \oplus x_2x_3x_4)$.
(b) $f_2(\mathbf{x}) = \overline{x}_2\overline{x}_4 \vee (x_1 \oplus x_2)x_4 \vee x_1x_3x_4 \vee x_2\overline{x}_3x_4$.
(c) $f_3(\mathbf{x}) = x_1\overline{x}_2x_4 \vee x_1x_4 \vee \overline{x}_2\overline{x}_3\overline{x}_4 \vee \overline{x}_1\overline{x}_4$.
(d) $f_4(\mathbf{x}) = \overline{x}_2\overline{x}_3(x_1 \oplus x_4) \vee (x_1 \vee x_2)x_3x_4 \vee \overline{x}_1\overline{x}_2x_3\overline{x}_4 \vee x_2\overline{x}_3\overline{x}_4$.

4.7　Solutions

Solution 4.1 (Methods to Define Logic Functions)

(a) Figure 4.12 shows in the upper part the PRP that defines the functions $f_1(x_1)$ to $f_8(x_1, x_2)$ using commands tin and below the created TVLs in ODA-form.

The first parameter of a command tin is the amount of the used Boolean space that has been defined in line 2, and the second parameter specifies the number of the TVLs created by this command. The third parameter of a command tin determines the form of the created TVL; this parameter has been omitted in all commands tin in lines 3–31 so that the default option /oda is used. The option /oda determines that an orthogonal TVL usable as disjunctive or antivalence form must be created; however, the XBOOLE-monitor XBM 2 cannot be sure that the TVL provided by the user behind the command line tin is really an orthogonal TVL. Therefore, the command tin with the (default) option /oda takes the given TVL as D-form and realizes

Fig. 4.11 Function table of the logic function $f(x_1, x_2, x_3)$

x_1	x_2	x_3	$f(\mathbf{x})$
0	0	0	1
0	0	1	0
0	1	0	0
0	1	1	1
1	0	0	1
1	0	1	1
1	1	0	0
1	1	1	0

Fig. 4.12
Problem-program that defines the functions $f_1(x_1)$ to $f_8(x_1, x_2)$ using commands tin and the created TVLs in ODA-form

```
1    new                13   tin  1 4          25   tin  1 7
2    space 32 1         14   x1 x2.            26   x1 x2.
3    tin  1 1           15   10                27   0-
4    x1.                16   01.               28   -0.
5    0.                 17   tin  1 5          29   tin  1 8
6    tin  1 2           18   x1 x2.            30   x1 x2.
7    x1 x2.             19   00                31   00.
8    11.                20   11.               32   for $i 1 4
9    tin  1 3           21   tin  1 6          33   (
10   x1 x2.             22   x1 x2.            34   assign $i /m 1 $i
11   1-                 23   0-                35   set $j (add $i 4)
12   -1.                24   -1.               36   assign $j /m 2 $i
                                               37   )
```

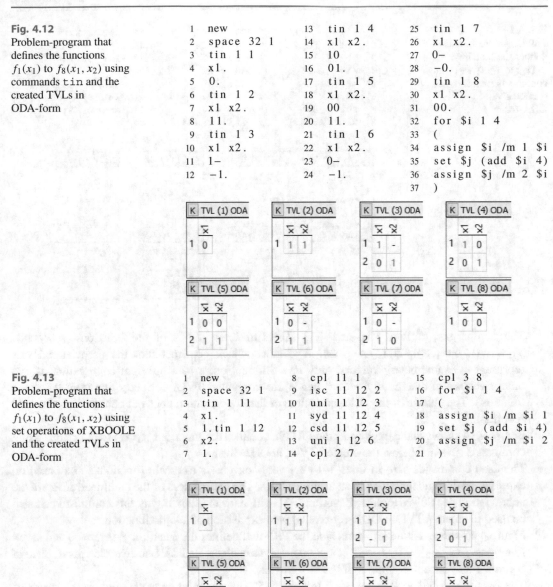

Fig. 4.13
Problem-program that defines the functions $f_1(x_1)$ to $f_8(x_1, x_2)$ using set operations of XBOOLE and the created TVLs in ODA-form

```
1    new            8    cpl 11 1          15   cpl 3 8
2    space 32 1     9    isc 11 12 2       16   for $i 1 4
3    tin 1 11       10   uni 11 12 3       17   (
4    x1.            11   syd 11 12 4       18   assign $i /m $i 1
5    1. tin 1 12    12   csd 11 12 5       19   set $j (add $i 4)
6    x2.            13   uni 1 12 6        20   assign $j /m $i 2
7    1.             14   cpl 2 7           21   )
```

the orthogonalization implicitly. Hence, each ternary vector behind the used commands tin describes one conjunction of the given disjunctive forms.

(b) Figure 4.13 shows in the upper part the PRP that defines the functions $f_1(x_1)$ to $f_8(x_1, x_2)$ using set operations of XBOOLE and below the created TVLs in ODA-form. The TVLs in ODA-form of the functions $f_1(x_1)$ to $f_5(x_1, x_2)$ are directly created in lines 8–12 using one set operation and the TVLs of the defined elementary functions $f_{11}(x_1) = x_1$ (TVL 11) and $f_{12}(x_1) = x_2$ (TVL 12). The TVLs of $f_1(x_1)$ to $f_3(x_1, x_2)$ are reused to compute the TVLs in ODA-form of the functions $f_6(x_1, x_2)$ to $f_8(x_1, x_2)$ in lines 13–15.

Fig. 4.14
Problem-program that
defines the functions
$f_1(x_1)$ to $f_8(x_1, x_2)$ using
commands sbe and the
created TVLs in
ODA-form

```
 1   new                  9   sbe  1 4          17   sbe  1 8
 2   space 32 1          10   x1 # x2.          18   /( x1 + x2).
 3   sbe  1 1            11   sbe  1 5          19   for $i  1 4
 4   /x1.                12   x1 = x2.          20   (
 5   sbe  1 2            13   sbe  1 6          21   assign  $i  /m 1  $i
 6   x1 & x2.            14   x1 > x2.          22   set  $j  (add  $i  4)
 7   sbe  1 3            15   sbe  1 7          23   assign  $j  /m 2  $i
 8   x1 + x2.            16   /( x1 & x2).      24   )
```

K	TVL (1) ODA
	$\bar{x}_1$
1	0

K	TVL (2) ODA	
	$\bar{x}_1$	$\bar{x}_2$
1	1	1

K	TVL (3) ODA	
	$\bar{x}_1$	$\bar{x}_2$
1	0	1
2	1	-

K	TVL (4) ODA	
	$\bar{x}_1$	$\bar{x}_2$
1	0	1
2	1	0

K	TVL (5) ODA	
	$\bar{x}_1$	$\bar{x}_2$
1	1	1
2	0	0

K	TVL (6) ODA	
	$\bar{x}_1$	$\bar{x}_2$
1	0	0
2	-	1

K	TVL (7) ODA	
	$\bar{x}_1$	$\bar{x}_2$
1	0	1
2	-	0

K	TVL (8) ODA	
	$\bar{x}_1$	$\bar{x}_2$
1	0	0

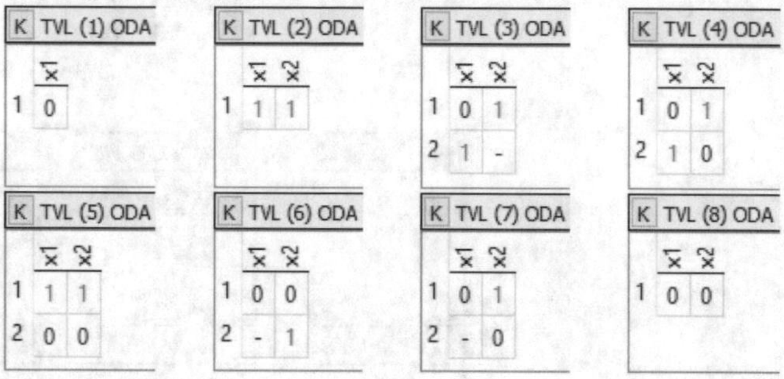

The comparison of Figs. 4.12 and 4.13 shows different TVLs of the functions $f_3(x_1, x_2)$, $f_5(x_1, x_2)$, $f_6(x_1, x_2)$, and $f_7(x_1, x_2)$. A detailed check confirms that the associated TVLs represent the same binary vectors, only in a different order or a different aggregation of two binary vectors to one ternary vector; hence, the computed TVLs correctly represent the given functions. The change of the order of vectors or their aggregation does not change a represented function.

(c) Figure 4.14 shows in the upper part the PRP that defines the functions $f_1(x_1)$ to $f_8(x_1, x_2)$ using commands sbe and below the created TVLs in ODA-form.

The used commands sbe in lines 3–18 are very convenient to create function-TVLs based on expressions of logic functions in an arbitrary form. The three letters of the command sbe are the abbreviation of "Solve Boolean Equation." We will learn in Chap. 5 why this command can also be used to create a TVL in ODA-form of a given expression of a logic function.

(d) Figure 4.15 shows on the left-hand side the PRP that defines the function $f_1(x_1) = 0$ and seven functions $f_i(x_1, x_2) = 0$, $i = 2, \ldots, 8$ and assigns each created TVL to two viewports, once in column 1 and once in column 2 of the m-fold view.

The function $f_1(x_1) = 0$ is specified in lines 3–5. This function depends only on the variables x_1 that are determined in line 4. The function value 0 is defined because no ternary vector occurs before the dot in line 5. The commands assign in lines 6 and 7 assign the created TVL of $f_1(x_1)$ to the viewports in the columns 1 and 2 of the row 1 in the m-fold view.

The other seven initially constant 0 functions uniquely depend on the variables x_1 and x_2; hence, these TVLs can analogously be created and assigned to the requested viewports in the for-loop in lines 8–15.

After the execution of this PRP, the left viewports must be switched to display Karnaugh-maps by pressing the buttons K. Now the Karnaugh-maps can be edited to get the requested functions; click on the position where the function value 1 is needed and press the key 1. All details that can be used to edit a Karnaugh-map are described in the help system of the XBOOLE-monitor XBM 2. The corresponding TVL in ODA-form appears immediately in the viewport in the same row and column 2 because the same TVL has been assigned to neighbored viewports of a row

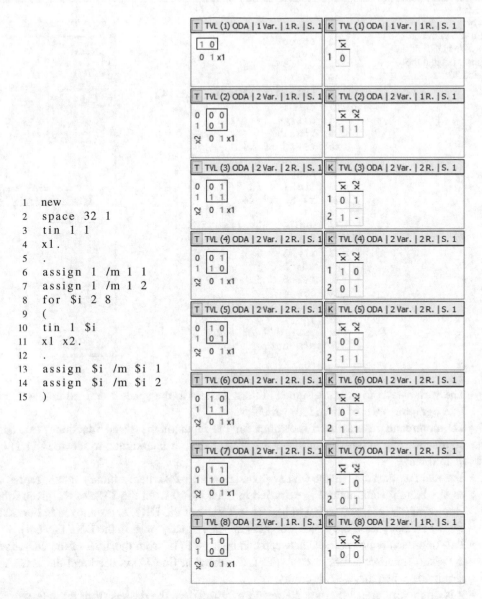

```
1   new
2   space 32 1
3   tin 1 1
4   x1.
5   .
6   assign 1 /m 1 1
7   assign 1 /m 1 2
8   for $i 2 8
9   (
10  tin 1 $i
11  x1 x2.
12  .
13  assign $i /m $i 1
14  assign $i /m $i 2
15  )
```

Fig. 4.15 Problem-program that defines the functions $f_1(x_1) = 0$ to $f_8(x_1, x_2) = 0$ that have been edited thereafter within the shown Karnaugh-maps and their additional representations as TVLs in ODA-form

and the Karnaugh-map is only another view to this TVL. A click in the region of the values of the variable removes the gray displayed selection of a function value.

Solution 4.2 (Normal Forms of Logic Functions)

(a) Figure 4.16 shows on the left-hand side the PRP that computes the TVL of the DNF belonging to the given function $f(\mathbf{x})$ and on the right-hand side both the function-TVL of $f(\mathbf{x})$ (TVL 1) and the TVL of the DNF of this function (TVL 4).
The used method to compute the requested DNF is as follows:

Fig. 4.16
Problem-program that computes the DNF of the functions $f(\mathbf{x})$ and the associated TVLs

```
1   new
2   space  32  1
3   avar  1
4   x1  x2  x3  x4.
5   sbe  1  1
6   (x1#x3)&(x2#x4)+(x1&/x4).
7   obb  1  1
8   assign  1  /m  1  1
9   tin  1  2
10  x1  x2  x3  x4.
11  .
12  copy  1  3
13  tin  1  4  /d
14  x1  x2  x3  x4.
15  .
16  while  (not  (te  3))
17  (
18  stv  3  1  5
19  uni  5  2  5
20  cel  5  5  5  /-0
21  con  4  5  4
22  dif  3  5  3
23  )
24  assign  4  /m  1  2
25  sts  dnf
```

K	TVL (1) ODA	4 Var.	5 R.	S. 1	
	x_1	x_2	x_3	x_4	
1	1	-	1	0	
2	0	0	1	1	
3	1	0	0	-	
4	0	1	1	0	
5	1	1	0	0	

K	TVL (4) D	4 Var.	7 R.	S. 1	
	x_1	x_2	x_3	x_4	
1	1	0	1	0	
2	1	1	1	0	
3	1	1	0	0	
4	0	1	1	0	
5	1	0	0	0	
6	1	0	0	1	
7	0	0	1	1	

- The variables x_1 to x_4 are assigned in lines 3 and 4 to the space 1 defined in line 2 to get a lexicographic order of these four variables.
- The command sbe is used thereafter for the input of the given function $f(\mathbf{x})$, and the command obb in line 7 minimizes this TVL before it is assigned to viewport (1,1) of the m-fold view.
- The main work is done in the while-loop in lines 16–23 where a binary vector (representing an elementary conjunction) is extracted in lines 18–20 from the TVL of the given function (copied as TVL 3), connected in line 21 to TVL 4 of the DNF, and removed in line 22 from TVL 3 so that each binary vector of $f(\mathbf{x})$ appears exactly once in the DNF (TVL 4).
- This approach removes in each sweep one binary vector from the TVL of the ODA-form of the given function; hence, a copy (TVL 3 created in line 12) is used and the while-loop terminates when this TVL is empty.
- It is often very useful that the command stv removes the dashes from the selected row of the TVL; however, here we need the complete ternary vector including the dashes; therefore, the empty TVL 2 has be created in lines 9–11 and used in the command uni in line 19 to get the complete ternary vector.
- The command cel in line 20 replaces all dashes of the selected ternary vector by values 0 so that one binary vector (elementary conjunction) of $f(\mathbf{x})$ is prepared.
- The empty TVL 4 in D-form, created in lines 13–15, collects the separated binary vectors in the wanted DNF.
- Finally, TVL 4 of the DNF is assigned to viewport (2,1) of the m-fold view in line 24, and the created XBOOLE-system is stored as file "dnf.sdt" in line 25.

The given function $f(x_1, x_2, x_3, x_4)$ is described by 7 elementary conjunctions as shown in the following DNF:

Fig. 4.17
Problem-program that computes the CNF of the functions $f(\mathbf{x})$ and the associated TVLs

```
 1  new
 2  space 32 1
 3  avar 1
 4  x1 x2 x3 x4.
 5  sbe 1 1
 6  (x1#x3)&(x2#x4)+(x1&/x4).
 7  obb 1 1
 8  assign 1 /m 1 1
 9  tin 1 2 /oke
10  x1 x2 x3 x4.
11  .
12  ndm 1 3
13  cpl 3 3
14  tin 1 4 /k
15  x1 x2 x3 x4.
16  .
17  while (not (te 3))
18  stv 3 1 5
19  uni 5 2 5
20  cel 5 5 5 /-0
21  con 4 5 4
22  dif 3 5 3
23  )
24  assign 4 /m 2 1
25  sts cnf
```

K	TVL (1) ODA | 4 Var. | 5 R. | S. 1			
	$\overline{x}_1$	$\overline{x}_2$	$\overline{x}_3$	$\overline{x}_4$
1	1	-	1	0
2	0	0	1	1
3	1	0	0	-
4	0	1	1	0
5	1	1	0	0

K	TVL (4) K | 4 Var. | 9 R. | S. 1			
	$\overline{x}_1$	$\overline{x}_2$	$\overline{x}_3$	$\overline{x}_4$
1	1	1	0	1
2	0	0	1	0
3	1	0	1	0
4	1	0	1	1
5	0	0	0	0
6	1	0	0	0
7	1	1	1	0
8	1	1	1	1
9	0	1	0	0

$$f(x_1, x_2, x_3, x_4) = \; x_1\overline{x}_2x_3\overline{x}_4 \vee x_1x_2x_3\overline{x}_4 \vee x_1x_2\overline{x}_3\overline{x}_4 \vee \overline{x}_1x_2x_3\overline{x}_4$$

$$\vee \; x_1\overline{x}_2\overline{x}_3\overline{x}_4 \vee x_1\overline{x}_2\overline{x}_3x_4 \vee \overline{x}_1\overline{x}_2x_3x_4 \,.$$

(b) Figure 4.17 shows on the left-hand side the PRP that computes the TVL of the CNF belonging to the given function $f(\mathbf{x})$ and on the right-hand side both the function-TVL of $f(\mathbf{x})$ (TVL 1) and the TVL of the CNF of this function (TVL 4).

The used method to compute the requested CNF is similar to the method to compute the DNF but requires the following changes:

- The main work is done in the while-loop in lines 17–23 where a binary vector (representing an elementary disjunction) is extracted in lines 18–20 from TVL 3 of OKE-form (prepared in lines 12 and 13) of the given function, connected in line 21 to TVL 4 of the CNF and removed in line 22 from TVL 3 so that each binary vector of $f(\mathbf{x})$ appears exactly once in the CNF (TVL 4).
- This approach removes in each sweep one binary vector from the TVL of the OKE-form of the given function; hence, the command ndm in line 12 transforms the ODA-form of $f(\mathbf{x})$ (TVL 1) into the OKE-form of $\overline{f(\mathbf{x})}$; the subsequent command cpl computes the complement of this intermediate result so that TVL 3 represents the function $f(\mathbf{x})$ in OKE-form and can be used to terminate the while-loop when this TVL is empty.
- The command stv in line 18 selects a vector in OKE-form; hence, the empty TVL 2 must be created in lines 9–11 in OKE-form so that the command uni in line 19 can compute the complete ternary vector.
- The empty TVL 4 in K-form, created in lines 14–16, collects the separated binary vectors in the wanted CNF.

• Finally, TVL 4 of the CNF is assigned to viewport (2,1) of the m-fold view in line 24, and the created XBOOLE-system is stored as file "cnf.sdt" in line 25.

The given function $f(x_1, x_2, x_3, x_4)$ is described by 9 elementary disjunctions as shown in the following CNF:

$$f(\mathbf{x}) = (x_1 \vee x_2 \vee \overline{x}_3 \vee x_4)(\overline{x}_1 \vee \overline{x}_2 \vee x_3 \vee \overline{x}_4)(x_1 \vee \overline{x}_2 \vee x_3 \vee \overline{x}_4)$$

$$(x_1 \vee \overline{x}_2 \vee x_3 \vee x_4)(\overline{x}_1 \vee \overline{x}_2 \vee \overline{x}_3 \vee \overline{x}_4)(x_1 \vee \overline{x}_2 \vee \overline{x}_3 \vee \overline{x}_4)$$

$$(x_1 \vee x_2 \vee x_3 \vee \overline{x}_4)(x_1 \vee x_2 \vee x_3 \vee x_4)(\overline{x}_1 \vee x_2 \vee \overline{x}_3 \vee \overline{x}_4) \ .$$

Each elementary disjunction of the CNF covers a value 0 of the logic function $f(\mathbf{x})$ and each elementary conjunction (computed in the previous task) a value 1 of the logic function $f(\mathbf{x})$; $9 + 7 = 16 = 2^4$ that confirms that all function values are covered by the CNF and the DNF.

(c) Figure 4.18 shows in the upper part the PRP that computes the DNF of the function $g(\mathbf{x})$ and shows below the TVL in ODA-form of $g(\mathbf{x})$ and the TVLs of the DNFs of $f(\mathbf{x})$ and $g(\mathbf{x})$.

The PRP of Fig. 4.18 loads the XBOOLE-system "dnf.sdt" that has been stored at the end of the PRP shown in Fig. 4.16 so that the TVL of the DNF of $f(\mathbf{x})$ is available as TVL 4. The computation of the DNF of the function $g(\mathbf{x})$ is realized in the same manner as before, only the object numbers are increased by 10.

The DNF of $g(\mathbf{x})$ (TVL 14 shown in viewport (1,3) of the m-fold view) has seven binary vectors (elementary conjunctions) as the DNF of $f(\mathbf{x})$ (TVL 4 shown in viewport (1,2) of the m-fold view); hence, $f(\mathbf{x})$ and $g(\mathbf{x})$ can be equal functions. Equal pairs of elementary conjunctions of $(f(\mathbf{x}), g(\mathbf{x}))$ are:

$$\{(1, 4), (2, 5), (3, 3), (4, 7), (5, 2), (6, 6), (7, 1)\} \ .$$

Each number appears in these pairs once in the first (indicating the associated elementary conjunction of $f(\mathbf{x})$) and once in the second position (indicating the associated elementary conjunction of $g(\mathbf{x})$); hence, the different expressions of $f(\mathbf{x})$ and $g(\mathbf{x})$ describe the same logic function.

(d) Figure 4.19 shows in the upper part the PRP that computes the CNF of the function $h(\mathbf{x})$ and shows below the TVL in ODA-form of $h(x_1, x_2, x_3, x_4)$ and the TVLs of the CNFs of $f(\mathbf{x})$ and $h(\mathbf{x})$.

The PRP of Fig. 4.19 loads the XBOOLE-system "cnf.sdt" that has been stored at the end of the PRP shown in Fig. 4.17 so that the TVL of the CNF of $f(\mathbf{x})$ is available as TVL 4. The computation of the CNF of the function $h(\mathbf{x})$ is realized in the same manner as before, and only the object numbers are increased by 10.

The CNF of $h(\mathbf{x})$ (TVL 14 shown in viewport (1,3) of the m-fold view) has eight binary vectors (elementary disjunctions), but the CNF of $f(\mathbf{x})$ (TVL 4 shown in viewport (1,2) of the m-fold view) has nine binary vectors (elementary disjunctions); hence, $f(\mathbf{x})$ and $g(\mathbf{x})$ cannot be equal functions. Equal pairs of elementary disjunctions of $(f(\mathbf{x}), h(\mathbf{x}))$ are

$$\{(1, 8), (2, 1), (4, 4), (5, 2), (6, 3), (7, 6), (8, 5)\} \ ;$$

hence, these elementary disjunctions determine equal function values 0 of $f(\mathbf{x})$ and $h(\mathbf{x})$. The elementary disjunctions $\{3, 9\}$ of the CNF of $f(\mathbf{x})$ determine functions values 0 of $f(\mathbf{x})$ for which $h(\mathbf{x}) = 1$. Vice versa, the elementary disjunction number 7 of the CNF of $h(\mathbf{x})$ determines function values 0 of $h(\mathbf{x})$ for which $f(\mathbf{x}) = 1$. At all, the functions $f(\mathbf{x})$ and $h(\mathbf{x})$ coincide only in 13 of 16 patterns of $\mathbf{x}$ so that $f(\mathbf{x})$ and $h(\mathbf{x})$ are different logic functions.

Fig. 4.18
Problem-program that computes the DNF of the functions $g(\mathbf{x})$ and the computed TVLs of $g(\mathbf{x})$ as well as the DNFs of $f(\mathbf{x})$ and $g(\mathbf{x})$

```
 1  lds dnf                 11  .
 2  sbe 1 11                12  while (not (te 13))
 3  (x1+x2&x3) #            13  (
 4  ((x1&x2+x3)&x4).        14  stv 13 1 15
 5  obb 11 11               15  uni 15 2 15
 6  assign 11 /m 1 1        16  cel 15 15 15 /-0
 7  assign 4 /m 1 2         17  con 14 15 14
 8  copy 11 13              18  dif 13 15 13
 9  tin 1 14                19  )
10  x1 x2 x3 x4.            20  assign 14 /m 1 3
```

K	TVL (11) ODA	4 Var.	4 R.	S. 1
	$\bar{x}_1$	$\bar{x}_2$	$\bar{x}_3$	$\bar{x}_4$
1	0	0	1	1
2	0	1	1	0
3	1	0	0	1
4	1	-	-	0

K	TVL (4) D	4 Var.	7 R.	S. 1
	$\bar{x}_1$	$\bar{x}_2$	$\bar{x}_3$	$\bar{x}_4$
1	1	0	1	0
2	1	1	1	0
3	1	1	0	0
4	0	1	1	0
5	1	0	0	0
6	1	0	0	1
7	0	0	1	1

K	TVL (14) D	4 Var.	7 R.	S. 1
	$\bar{x}_1$	$\bar{x}_2$	$\bar{x}_3$	$\bar{x}_4$
1	0	0	1	1
2	1	0	0	0
3	1	1	0	0
4	1	0	1	0
5	1	1	1	0
6	1	0	0	1
7	0	1	1	0

Fig. 4.19
Problem-program that computes the CNF of the functions $h(\mathbf{x})$ and the computed TVLs of $h(\mathbf{x})$ as well as the CNFs of $f(\mathbf{x})$ and $h(\mathbf{x})$

```
 1  lds cnf                 11  .
 2  sbe 1 11                12  while (not (te 13))
 3  (x1+x2&x3)#(x2&x4).     13  (
 4  obb 11 11               14  stv 13 1 15
 5  assign 11 /m 1 1        15  uni 15 2 15
 6  assign 4 /m 1 2         16  cel 15 15 15 /-0
 7  ndm 11 13               17  con 14 15 14
 8  cpl 13 13               18  dif 13 15 13
 9  tin 1 14 /k             19  )
10  x1 x2 x3 x4.            20  assign 14 /m 1 3
```

K	TVL (11) ODA	4 Var.	4 R.	S. 1
	$\bar{x}_1$	$\bar{x}_2$	$\bar{x}_3$	$\bar{x}_4$
1	0	1	0	1
2	0	1	1	0
3	1	-	-	0
4	1	0	-	1

K	TVL (4) K	4 Var.	9 R.	S. 1
	$\bar{x}_1$	$\bar{x}_2$	$\bar{x}_3$	$\bar{x}_4$
1	1	1	0	1
2	0	0	1	0
3	1	0	1	0
4	1	0	1	1
5	0	0	0	0
6	1	0	0	0
7	1	1	1	0
8	1	1	1	1
9	0	1	0	0

K	TVL (14) K	4 Var.	8 R.	S. 1
	$\bar{x}_1$	$\bar{x}_2$	$\bar{x}_3$	$\bar{x}_4$
1	0	0	1	0
2	0	0	0	0
3	1	0	0	0
4	1	0	1	1
5	1	1	1	1
6	1	1	1	0
7	1	1	0	0
8	1	1	0	1

Solution 4.3 (Function Tables of Logic Functions)

(a) Figure 4.20 shows on the left-hand side the PRP that computes the function table of the function $f(\mathbf{x})$ given in D-form and on the right-hand side the function-TVL of $f(\mathbf{x})$ (TVL 1), the generated initial function table of $f(\mathbf{x}) = 0$ (TVL 2), and the computed function table (TVL 8).

The initial function table of a constant function equal to 0 (TVL 2) has been prepared in lines 9–25 based on the initial ternary vector defined in line 11. The principle of construction is as follows:

- The command cel in line 21 changes the values of TVL 2 in the column determined by the TVL 3 form 0 to 1 and stores the result as intermediate TVL 6; in the first sweep of the for-loop in lines 19–24, we get from (00000) (TVL 2) the single vector (00010).

```
1   new
2   space 32 1
3   tin 1 1 /d
4   x1 x2 x3 x4 .
5   1-0-
6   -11-
7   1--1.
8   assign 1 /m 1 1
9   tin 1 2
10  x1 x2 x3 x4 f .
11  00000.
12  tin 1 3
13  x4 .
14  .
15  vtin 1 4
16  x1 x2 x3 .
17  vtin 1 5
18  x2 x3 x4 .
19  for $i 1 4
20  (
21  cel 2 3 6 /01 /10
22  uni 2 6 2
23  cco 3 4 5 3
24  )
25  assign 2 /m 1 2
26  vtin 1 7
27  f .
28  ctin 1 8 /0
29  for $i 1 (ntv 2)
30  (
31  stv 2 $i 9
32  for $j 1 (ntv 1)
33  (
34  stv 1 $j 10
35  if (not(te_isc 9 10))
36  (
37  cel 9 7 9 /01
38  ))
39  con 8 9 8
40  )
41  assign 8 /m 1 3
```

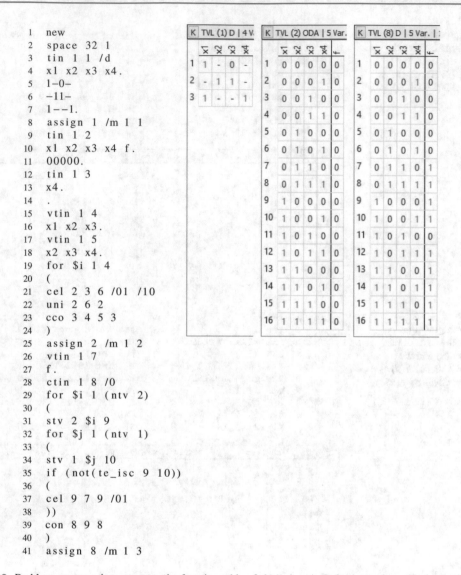

Fig. 4.20 Problem-program that computes the function table of $f(\mathbf{x})$ given in D-form together with the results

- The command uni in line 22 concatenates TVL 6 at the end of TVL 2 and checks the already given orthogonality of the result that consists of the first two rows of the initial function table after the first execution of this command.
- The command cco in line 23 rotates the columns of TVL 3 due to the controlling VTs 4 and 5 (defined in lines 15–18) so that in the second, third, and fourth sweep of the for-loop the columns x_3, x_2, and x_1 are determined by TVL 3.
- The for-loop in lines 19–24 iterates over the four variables, and the command uni in line 22 extends TVL 2 to four, eight, and 16 rows in the remaining sweeps.

The actual function values are inserted in the initial function table in two nested for-loops; the outer for-loop in lines 29–40 iterates over the 16 rows of the function table (TVL 2), and the inner for-loop in line 32–38 over the three conjunctions of the function $f(\mathbf{x})$ (TVL 1). The

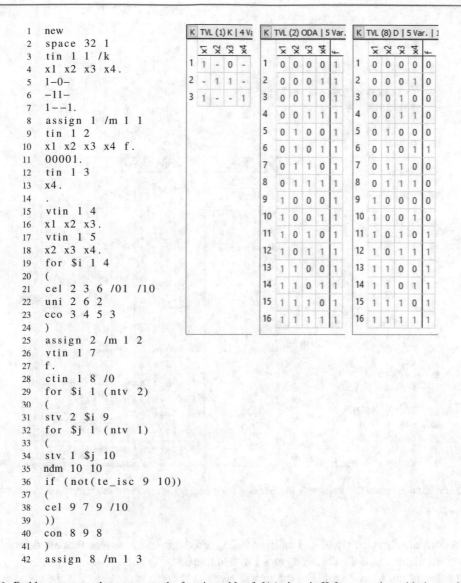

```
1   new
2   space 32 1
3   tin 1 1 /k
4   x1 x2 x3 x4.
5   1-0-
6   -11-
7   1--1.
8   assign 1 /m 1 1
9   tin 1 2
10  x1 x2 x3 x4 f.
11  00001.
12  tin 1 3
13  x4.
14  .
15  vtin 1 4
16  x1 x2 x3.
17  vtin 1 5
18  x2 x3 x4.
19  for $i 1 4
20  (
21  cel 2 3 6 /01 /10
22  uni 2 6 2
23  cco 3 4 5 3
24  )
25  assign 2 /m 1 2
26  vtin 1 7
27  f.
28  ctin 1 8 /0
29  for $i 1 (ntv 2)
30  (
31  stv 2 $i 9
32  for $j 1 (ntv 1)
33  (
34  stv 1 $j 10
35  ndm 10 10
36  if (not(te_isc 9 10))
37  (
38  cel 9 7 9 /10
39  ))
40  con 8 9 8
41  )
42  assign 8 /m 1 3
```

TVL (1) K | 4 Var.

	x_1	x_2	x_3	x_4
1	1	-	0	-
2	-	1	1	-
3	1	-	-	1

TVL (2) ODA | 5 Var.

	x_1	x_2	x_3	x_4	f
1	0	0	0	0	1
2	0	0	0	1	1
3	0	0	1	0	1
4	0	0	1	1	1
5	0	1	0	0	1
6	0	1	0	1	1
7	0	1	1	0	1
8	0	1	1	1	1
9	1	0	0	0	1
10	1	0	0	1	1
11	1	0	1	0	1
12	1	0	1	1	1
13	1	1	0	0	1
14	1	1	0	1	1
15	1	1	1	0	1
16	1	1	1	1	1

TVL (8) D | 5 Var. | 1

	x_1	x_2	x_3	x_4	f
1	0	0	0	0	0
2	0	0	0	1	0
3	0	0	1	0	0
4	0	0	1	1	0
5	0	1	0	0	0
6	0	1	0	1	1
7	0	1	1	0	0
8	0	1	1	1	0
9	1	0	0	0	0
10	1	0	0	1	0
11	1	0	1	0	1
12	1	0	1	1	1
13	1	1	0	0	1
14	1	1	0	1	1
15	1	1	1	0	1
16	1	1	1	1	1

Fig. 4.21 Problem-program that computes the function table of $f(\mathbf{x})$ given in K-form together with the results

function value belonging to the binary vector selected in line 34 is changed from 0 to 1 in line 37 when the condition in line 35 is satisfied.

Each binary vector with the right function values is appended in line 39 to the function table (TVL 8) that has been crated as empty TVL in line 28.

(b) Figure 4.21 shows on the left-hand side the PRP that computes the function table of the function $f(\mathbf{x})$ given in K-form and on the right-hand side the function-TVL of $f(\mathbf{x})$ (TVL 1), the generated initial function table of $f(\mathbf{x}) = 1$ (TVL 2), and the computed function table (TVL 8).

The function table of the function $f(\mathbf{x})$ given in K-form can be computed similar to the function table of a D-form; however, the different forms require some changes. Each disjunction of a conjunctive form determines a function value 0; hence, the initial function table has been created

```
1   new
2   space 32 1
3   avar 1
4   x1 x2 x3 x4 .
5   sbe 1 1
6   (x1+/x3)#x2&x4#(x2+x3) .
7   assign 1 /m 1 1
8   tin 1 2
9   x1 x2 x3 x4 .
10  0000 .
11  tin 1 3
12  x4 .
13  .
14  vtin 1 4
15  x1 x2 x3 .
16  vtin 1 5
17  x2 x3 x4 .
18  for $i 1 4
19  (
20  cel 2 3 6 /01 /10
21  uni 2 6 2
22  cco 3 4 5 3
23  )
24  assign 2 /m 1 2
25  sbe 1 7
26  f .
27  csd 1 7 7
28  ctin 1 8 /0
29  for $i 1 (ntv 2)
30  (
31  stv 2 $i 9
32  isc 9 7 9
33  con 8 9 8
34  )
35  assign 8 /m 1 3
```

TVL (1) ODA

K	x_1	x_2	x_3	x_4
1	1	1	1	1
2	0	0	1	-
3	0	1	1	0
4	-	1	0	1
5	-	0	0	-

TVL (2) ODA | 4 Var.

K	x_1	x_2	x_3	x_4
1	0	0	0	0
2	0	0	0	1
3	0	0	1	0
4	0	0	1	1
5	0	1	0	0
6	0	1	0	1
7	0	1	1	0
8	0	1	1	1
9	1	0	0	0
10	1	0	0	1
11	1	0	1	0
12	1	0	1	1
13	1	1	0	0
14	1	1	0	1
15	1	1	1	0
16	1	1	1	1

TVL (8) D | 5 Var. |

K	x_1	x_2	x_3	x_4	f
1	0	0	0	0	1
2	0	0	0	1	1
3	0	0	1	0	1
4	0	0	1	1	1
5	0	1	0	0	0
6	0	1	0	1	1
7	0	1	1	0	1
8	0	1	1	1	0
9	1	0	0	0	1
10	1	0	0	1	1
11	1	0	1	0	0
12	1	0	1	1	0
13	1	1	0	0	0
14	1	1	0	1	1
15	1	1	1	0	0
16	1	1	1	1	1

Fig. 4.22 Problem-program that computes the function table of $f(\mathbf{x})$ given in an arbitrary form together with the results

for a constant function equal to 1 in lines 9–25, and the function values that are determined by the given disjunctions are changed from 1 to 0 in line 38.

The value in the function table of a function given in K-form is equal to 0 if the complement of one of the disjunctions is equal to 1; hence, the disjunction selected in line 34 must be negated using the command ndm in line 35 before it can be checked whether the condition in line 36 is satisfied.

(c) Figure 4.22 shows on the left-hand side the PRP that computes the function table of the function $f(\mathbf{x})$ given as expression in arbitrary form and on the right-hand side the function-TVL of $f(\mathbf{x})$ (TVL 1), the generated ordered list of binary vectors (x_1, x_2, x_3, x_4) (TVL 2), and the computed function table (TVL 8).

The main steps of the used approach to compute the function table of a given expression are:

- Lines 1–4: define a Boolean space within a new XBOOLE-system and attach the variables $x_1, \ldots, x_4$.
- Lines 5–7: compute and show the function-TVL of the given expression using the command sbe.
- Lines 8–24: generate the ordered set of binary vectors (x_1, x_2, x_3, x_4).

Fig. 4.23
Problem-program that computes the function table of f given in an arbitrary form together with the results

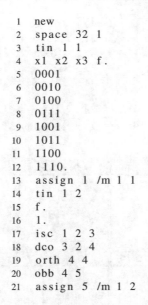

```
 1   new
 2   space 32 1
 3   tin 1 1
 4   x1 x2 x3 f.
 5   0001
 6   0010
 7   0100
 8   0111
 9   1001
10   1011
11   1100
12   1110.
13   assign 1 /m 1 1
14   tin 1 2
15   f.
16   1.
17   isc 1 2 3
18   dco 3 2 4
19   orth 4 4
20   obb 4 5
21   assign 5 /m 1 2
```

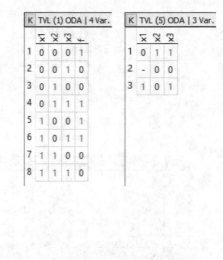

K	TVL (1) ODA | 4 Var.			
	x_1	x_2	x_3	f
1	0	0	0	1
2	0	0	1	0
3	0	1	0	0
4	0	1	1	1
5	1	0	0	1
6	1	0	1	1
7	1	1	0	0
8	1	1	1	0

K	TVL (5) ODA | 3 Var.		
	x_1	x_2	x_3
1	0	1	1
2	-	0	0
3	1	0	1

- Lines 25–27: compute the TVL of the function $F(\mathbf{x}, f) = f(\mathbf{x}) \odot f$ using the command sbe with the simple expression f and the command csd—the ternary vectors of the computed TVL 7 of the function $F(\mathbf{x}, f)$ contain for each binary vector (x_1, x_2, x_3, x_4) the associated function value f.
- Lines 28–35: transform TVL 7 of $F(\mathbf{x}, f)$ into the wanted function table (TVL 8) using the ordered set of binary vectors (TVL 2) and show the result.

(d) Figure 4.23 shows on the left-hand side the PRP that computes the minimized function-TVL of $f(\mathbf{x})$ in ODA-form using a given function table and on the right-hand side the function table (TVL 1) and the computed function-TVL (TVL 5).

The command tin in lines 3–12 has been used for the input of the given function table, and the command assign in line 13 shows this function table in viewport (1,1) of the m-fold view.

TVL 2 (created in lines 14–16) specifies the function value $f = 1$ so that the intersection in line 17 selects the rows with $f(\mathbf{x}) = 1$ of the given function table (TVL 1) and stores the intermediate result as TVL 3.

The function-TVL depends only on the variables $\mathbf{x}$; hence, the column f can be deleted using the command dco in line 18.

The orthogonality of a TVL can be lost when a column is deleted; the command dco changes therefore the ODA-form into a D-form even if the orthogonality remains. The command orth re-establishes the ODA-form in line 19, and the command obb in line 20 minimizes the function-TVL before it is assigned to viewport (1,2) of the m-fold view.

Solution 4.4 (Basic Forms of Logic Functions)

(a) Figure 4.24 shows in the upper part the PRP that transforms the function $f_1(\mathbf{x})$ from the given D-form into the associated K-, A-, and E-form and below the created TVLs in all four basic forms and the associated Karnaugh-maps.

```
1   new                      10   cpl 5 6
2   space 32 1               11   ndm 6 6
3   tin 1 1 /d               12   sform 6 /k 2
4   x1 x2 x3 x4.             13   sform 6 /e 4
5   1--0                     14   for $i 1 4
6   -11-                     15   (
7   --11.                    16     assign $i /m 1 $i
8   orth 1 5                 17     assign $i /m 2 $i
9   sform 5 /a 3             18   )
```

K	TVL (1) D | 4 Var. | 3 Z. | R. 1			
	x̄1	x̄2	x̄3	x̄4
1	1	-	-	0
2	-	1	1	-
3	-	-	1	1

K	TVL (2) K | 4 Var. | 3 Z. | R. 1			
	x̄1	x̄2	x̄3	x̄4
1	1	1	0	1
2	-	-	1	0
3	1	-	1	1

K	TVL (3) A | 4 Var. | 3 Z. | R. 1			
	x̄1	x̄2	x̄3	x̄4
1	1	-	-	0
2	0	1	1	0
3	-	-	1	1

K	TVL (4) E | 4 Var. | 3 Z. | R. 1			
	x̄1	x̄2	x̄3	x̄4
1	1	1	0	1
2	-	-	1	0
3	1	-	1	1

```
T  TVL (1) D | 4 Var. | 3 Z. | R. 1
0 0    0 0 1 1
0 1    0 0 0 0
1 1    1 1 1 1
1 0    0 1 1 1
x3 x4  0 1 1 0 x2
       0 0 1 1 x1
```

```
T  TVL (2) K | 4 Var. | 3 Z. | R. 1
0 0    0 0 1 1
0 1    0 0 0 0
1 1    1 1 1 1
1 0    0 1 1 1
x3 x4  0 1 1 0 x2
       0 0 1 1 x1
```

```
T  TVL (3) A | 4 Var. | 3 Z. | R. 1
0 0    0 0 1 1
0 1    0 0 0 0
1 1    1 1 1 1
1 0    0 1 1 1
x3 x4  0 1 1 0 x2
       0 0 1 1 x1
```

```
T  TVL (4) E | 4 Var. | 3 Z. | R. 1
0 0    0 0 1 1
0 1    0 0 0 0
1 1    1 1 1 1
1 0    0 1 1 1
x3 x4  0 1 1 0 x2
       0 0 1 1 x1
```

Fig. 4.24 Problem-program that transforms the function $f_1(\mathbf{x})$ from the given D-form into the associated K-, A-, and E-form and shows the TVLs and Karnaugh-maps of $f_1(\mathbf{x})$ in all four basic forms

The command orth in line 8 considers the given D-form and creates TVL 5 of $f_1(\mathbf{x})$ in ODA-form. A TVL in ODA-form describes the expression of both the D-form and the A-form of the given function; hence, the command sform in line 9 creates the wanted TVL in A-form.

The two subsequent commands cpl and ndm in lines 10 and 11 transform the TVL of $f_1(\mathbf{x})$ from the ODA-form (TVL 5) to the OKE-form (TVL 6) of the same function. A TVL in OKE-form describes the expression of both the K-form and the E-form of the given function; hence, the commands sform in lines 12 and 13 create the wanted TVLs in K-form and E-form.

The two commands assign in the for-loop in lines 14–18 realize the assignment of the four basic forms to four viewport of the first two rows of the m-fold view. Clicks to the buttons K in the second row of the m-fold view switch these viewports to the display of the wanted Karnaugh-maps. The four Karnaugh-maps confirm that the different TVLs 1 to 4 describe the same function $f_1(\mathbf{x})$.

(b) Figure 4.25 shows in the upper part the PRP that transforms the function $f_2(\mathbf{x})$ from the given K-form into the associated D-, A-, and E-form and below the created TVLs in all four basic forms and the associated Karnaugh-maps.

The PRP of Fig. 4.25 uses an analog approach. The command orth in line 8 considers the given K-form and creates TVL 6 of $f_2(\mathbf{x})$ in OKE-form. A TVL in OKE-form describes the expression of both the K-form and the E-form of the given function; hence, the command sform in line 9 creates the wanted TVL in E-form.

The two subsequent commands cpl and ndm in lines 10 and 11 transform the TVL of $f_2(\mathbf{x})$ from the OKE-form (TVL 6) to the ODA-form (TVL 5) of the same function. A TVL in ODA-form describes the expression of both the D-form and the A-form of the given function; hence, the commands sform in lines 12 and 13 create the wanted TVLs in D-form and A-form.

```
 1   new                        10   cpl  6 5
 2   space 32 1                 11   ndm  5 5
 3   tin 1 2 /k                 12   sform 5 /d 1
 4   x1 x2 x3 x4 .              13   sform 5 /a 3
 5   1--0                       14   for $i 1 4
 6   -11-                       15   (
 7   --11 .                     16   assign $i /m 1 $i
 8   orth 2 6                   17   assign $i /m 2 $i
 9   sform 6 /e 4               18   )
```

K	TVL (1) D | 4 Var. | 3R. | S. 1			
	$\tilde{x}_1$	$\tilde{x}_2$	$\tilde{x}_3$	$\tilde{x}_4$
1	1	1	0	1
2	-	-	1	0
3	1	-	1	1

K	TVL (2) K | 4 Var. | 3R. | S. 1			
	$\tilde{x}_1$	$\tilde{x}_2$	$\tilde{x}_3$	$\tilde{x}_4$
1	1	-	-	0
2	-	1	1	-
3	-	-	1	1

K	TVL (3) A | 4 Var. | 3R. | S. 1			
	$\tilde{x}_1$	$\tilde{x}_2$	$\tilde{x}_3$	$\tilde{x}_4$
1	1	1	0	1
2	-	-	1	0
3	1	-	1	1

K	TVL (4) E | 4 Var. | 3R. | S. 1			
	$\tilde{x}_1$	$\tilde{x}_2$	$\tilde{x}_3$	$\tilde{x}_4$
1	1	-	-	0
2	0	1	1	0
3	-	-	1	1

T	TVL (1) D | 4 Var. | 3R. | S. 1			
0 0	0	0	0	0
0 1	0	0	1	0
1 1	0	0	1	1
1 0	1	1	1	1
$x_3 x_4$	0	1	1	0 x2
	0	0	1	1 x1

T	TVL (2) K | 4 Var. | 3R. | S. 1			
0 0	0	0	0	0
0 1	0	0	1	0
1 1	0	0	1	1
1 0	1	1	1	1
$x_3 x_4$	0	1	1	0 x2
	0	0	1	1 x1

T	TVL (3) A | 4 Var. | 3R. | S. 1			
0 0	0	0	0	0
0 1	0	0	1	0
1 1	0	0	1	1
1 0	1	1	1	1
$x_3 x_4$	0	1	1	0 x2
	0	0	1	1 x1

T	TVL (4) E | 4 Var. | 3R. | S. 1			
0 0	0	0	0	0
0 1	0	0	1	0
1 1	0	0	1	1
1 0	1	1	1	1
$x_3 x_4$	0	1	1	0 x2
	0	0	1	1 x1

Fig. 4.25 Problem-program that transforms the function $f_2(\mathbf{x})$ from the given K-form into the associated D-, A-, and E-form and shows the TVLs and Karnaugh-maps of $f_2(\mathbf{x})$ in all four basic forms

The assignment to the TVLs of the four forms is again done by two commands `assign` in the `for`-loop at the end of the PRP. The four Karnaugh-maps confirm that the different TVLs 1 to 4 describe the same function $f_2(\mathbf{x})$; however, despite the same three ternary vectors of $f_2(\mathbf{x})$ in K-form (in comparison to $f_1(\mathbf{x})$ in D-form), these two functions are different (even in the number of function values 1).

(c) Figure 4.26 shows in the upper part the PRP that transforms the function $f_3(\mathbf{x})$ from the given A-form into the associated D-, K-, and E-form and below the created TVLs in all four basic forms and the associated Karnaugh-maps.

The command `orth` in line 8 considers the given A-form and creates TVL 5 of $f_3(\mathbf{x})$ in ODA-form so that the command `sform` in line 9 can be used to create the wanted TVL in D-form.

The two subsequent commands `cpl` and `ndm` in lines 10 and 11 transform the TVL of $f_3(\mathbf{x})$ from the ODA-form (TVL 5) into the OKE-form (TVL 6) of the same function so that the commands `sform` in lines 12 and 13 can be used to create the wanted TVLs in K- and E-form.

The assignment to the TVLs of the four forms is again done by two commands `assign` in the `for` loop at the end of the PRP. The four Karnaugh-maps confirm that the different TVLs 1 to 4 describe the same function $f_3(\mathbf{x})$; however, despite the same three ternary vectors of $f_3(\mathbf{x})$ in K-form (in comparison to $f_1(\mathbf{x})$ in D-form and $f_2(\mathbf{x})$ in K-form), these three functions are different (even in the number of function values 1).

(d) Figure 4.27 shows in the upper part the PRP that transforms the function $f_4(\mathbf{x})$ from the given E-form into the associated D-, K-, and A-form and below the created TVLs in all four basic forms and the associated Karnaugh-maps.

The command `orth` in line 8 can also be used to compute the OKE-form of the given TVL in E-form; hence, the command `sform` in line 9 creates the wanted TVL in K-form.

```
1   new                        10   cpl 5 6
2   space 32 1                 11   ndm 6 6
3   tin 1 3 /a                 12   sform 6 /k 2
4   x1 x2 x3 x4.               13   sform 6 /e 4
5   1--0                       14   for $i 1 4
6   -11-                       15   (
7   --11.                      16   assign $i /m 1 $i
8   orth 3 5                   17   assign $i /m 2 $i
9   sform 5 /d 1               18   )
```

K	TVL (1) D | 4 Var. | 4 R. | S. 1
	$\bar{x}_1$ $\bar{x}_2$ $\bar{x}_3$ $\bar{x}_4$
1	1 0 1 0
2	0 1 1 0
3	- 0 1 1
4	1 - 0 0

K	TVL (2) K | 4 Var. | 5 R. | S. 1
	$\bar{x}_1$ $\bar{x}_2$ $\bar{x}_3$ $\bar{x}_4$
1	1 1 0 1
2	- 0 0 0
3	1 - 1 1
4	0 0 0 1
5	- - 1 0

K	TVL (3) A | 4 Var. | 3 R. | S. 1
	$\bar{x}_1$ $\bar{x}_2$ $\bar{x}_3$ $\bar{x}_4$
1	1 - - 0
2	- 1 1 -
3	- - 1 1

K	TVL (4) E | 4 Var. | 5 R. | S. 1
	$\bar{x}_1$ $\bar{x}_2$ $\bar{x}_3$ $\bar{x}_4$
1	1 1 0 1
2	- 0 0 0
3	1 - 1 1
4	0 0 0 1
5	- - 1 0

T	TVL (1) D | 4 Var. | 4 R. | S. 1
0 0	0 0 1 1
0 1	0 0 0 0
1 1	1 0 0 1
1 0	0 1 0 1
$x_3 x_4$	0 1 1 0 x2
	0 0 1 1 x1

T	TVL (2) K | 4 Var. | 5 R. | S. 1
0 0	0 0 1 1
0 1	0 0 0 0
1 1	1 0 0 1
1 0	0 1 0 1
$x_3 x_4$	0 1 1 0 x2
	0 0 1 1 x1

T	TVL (3) A | 4 Var. | 3 R. | S. 1
0 0	0 0 1 1
0 1	0 0 0 0
1 1	1 0 0 1
1 0	0 1 0 1
$x_3 x_4$	0 1 1 0 x2
	0 0 1 1 x1

T	TVL (4) E | 4 Var. | 5 R. | S. 1
0 0	0 0 1 1
0 1	0 0 0 0
1 1	1 0 0 1
1 0	0 1 0 1
$x_3 x_4$	0 1 1 0 x2
	0 0 1 1 x1

Fig. 4.26 Problem-program that transforms the function $f_3(\mathbf{x})$ from the given A-form into the associated D-, K-, and E-form and shows the TVLs and Karnaugh-maps of $f_3(\mathbf{x})$ in all four basic forms

The two subsequent commands cpl and ndm in lines 10 and 11 transform the TVL of $f_4(\mathbf{x})$ from the OKE-form (TVL 6) into the ODA-form (TVL 5) of the same function so that the commands sform in lines 12 and 13 can be used to create the wanted TVLs in D- and A-form.

The assignment to the TVLs of the four forms is again done by two commands assign in the for-loop at the end of the PRP. The four Karnaugh-maps confirm that the different TVLs 1 to 4 describe the same function $f_4(\mathbf{x})$; however, despite the same three ternary vectors of $f_4(\mathbf{x})$ in E-form (in comparison to the three previous functions) all four functions are different (even in the number of function values 1).

We learn from this exercise:

• The command orth can be used to get either the ODA-form (from a D- or A-form) or the OKE-form (from a K- or E-form) for all four basic forms without the change of the represented function.

• The command sform can be used to specify a D-form or an A-form of an ODA-form as well as a K-form or an E-form of an OKE-form without the change of the represented function.

• The command ndm changes the TVL in ODA-form of the function $f(\mathbf{x})$ into the TVL in OKE-form of the complement function $\overline{f(\mathbf{x})}$ or vice versa.

• The command cpl changes the orthogonal TVL (ODA- or OKE-form) of a function $f(\mathbf{x})$ into the orthogonal TVL complement function $\overline{f(\mathbf{x})}$ without changing the form.

• The two subsequent commands ndm and cpl transform a TVL of an ODA-form into an OKE-form (or vice versa) without the change of the function; the changed order of these commands (first cpl and thereafter ndm) leads to exactly the same result.

1	new	10	cpl 6 5
2	space 32 1	11	ndm 5 5
3	tin 1 4 /e	12	sform 5 /d 1
4	x1 x2 x3 x4.	13	sform 5 /a 3
5	1--0	14	for $i 1 4
6	-11-	15	(
7	--11.	16	assign $i /m 1 $i
8	orth 4 6	17	assign $i /m 2 $i
9	sform 6 /k 2	18	)

K | TVL (1) D | 4 Var. | 5 Z. | R. 1

	x_1	x_2	x_3	x_4
1	1	1	0	1
2	-	0	0	0
3	1	-	1	1
4	0	0	0	1
5	-	-	1	0

K | TVL (2) K | 4 Var. | 4 Z. | R. 1

	x_1	x_2	x_3	x_4
1	1	0	1	0
2	0	1	1	0
3	-	0	1	1
4	1	-	0	0

K | TVL (3) A | 4 Var. | 5 Z. | R. 1

	x_1	x_2	x_3	x_4
1	1	1	0	1
2	-	0	0	0
3	1	-	1	1
4	0	0	0	1
5	-	-	1	0

K | TVL (4) E | 4 Var. | 3 Z. | R. 1

	x_1	x_2	x_3	x_4
1	1	-	-	0
2	-	1	1	-
3	-	-	1	1

T | TVL (1) D | 4 Var. | 5 Z. | R. 1

```
0 0  1 0 0 1
0 1  1 0 1 0
1 1  0 0 1 1
1 0  1 1 1 1
x3 x4  0 1 1 0 x2
       0 0 1 1 x1
```

T | TVL (2) K | 4 Var. | 4 Z. | R. 1

```
0 0  1 0 0 1
0 1  1 0 1 0
1 1  0 0 1 1
1 0  1 1 1 1
x3 x4  0 1 1 0 x2
       0 0 1 1 x1
```

T | TVL (3) A | 4 Var. | 5 Z. | R. 1

```
0 0  1 0 0 1
0 1  1 0 1 0
1 1  0 0 1 1
1 0  1 1 1 1
x3 x4  0 1 1 0 x2
       0 0 1 1 x1
```

T | TVL (4) E | 4 Var. | 3 Z. | R. 1

```
0 0  1 0 0 1
0 1  1 0 1 0
1 1  0 0 1 1
1 0  1 1 1 1
x3 x4  0 1 1 0 x2
       0 0 1 1 x1
```

Fig. 4.27 Problem-program that transforms the function $f_4(\mathbf{x})$ from the given E-form into the associated D-, K-, and A-form and shows the TVLs and Karnaugh-maps of $f_4(\mathbf{x})$ in all four basic forms

Solution 4.5 (Properties of Logic Functions with Regard to a Variable)

All four tasks of this exercise check whether a given function is monotonously increasing, monotonously decreasing, or linear with regard to the specified variable x_3, or even independent of x_3. These properties are defined by relations between subfunctions $f(x_3 = 0)$ and $f(x_3 = 1)$:

- Monotonously increasing with regard to x_3: $f(x_3 = 0) \le f(x_3 = 1)$
- Monotonously decreasing with regard to x_3: $f(x_3 = 0) \ge f(x_3 = 1)$
- Linear with regard to x_3: $f(x_3 = 0) \equiv \overline{f(x_3 = 1)}$
- Independent of x_3: $f(x_3 = 0) \equiv f(x_3 = 1)$

The relations can be transformed such that a certain function that is equal to 0 indicates that one of these properties is satisfied:

- Monotonously increasing with regard to x_3: $f_{minc} = f(x_3 = 0) \wedge \overline{f(x_3 = 1)} = 0$.
- Monotonously decreasing with regard to x_3: $f_{mdec} = f(x_3 = 1) \wedge \overline{f(x_3 = 0)} = 0$.
- Linear with regard to x_3 $f_{lin} = f(x_3 = 0) \odot f(x_3 = 1) = 0$.
- Independent of x_3: $f_{indep} = f(x_3 = 0) \oplus f(x_3 = 1) = 0$.

The commands te_dif, te_csd, or te_syd can be used to check whether a function $f(\mathbf{x})$ satisfies the associated property based on the subfunction given $f(x_3 = 0)$ and $f(x_3 = 1)$ represented as TVLs in ODA-form.

(a) Figure 4.28 shows on the left-hand side the PRP that verifies the properties of the function $f_1(\mathbf{x})$ and on the right-hand side the computed results.

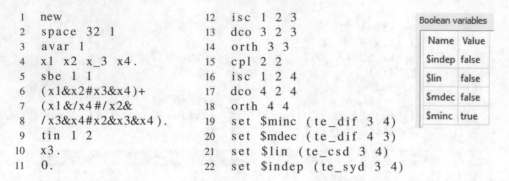

```
1    new                    12   isc  1 2 3
2    space  32  1           13   dco  3 2 3
3    avar  1                14   orth  3 3
4    x1  x2  x_3  x4 .       15   cpl  2 2
5    sbe  1 1               16   isc  1 2 4
6    ( x1&x2#x3&x4 )+        17   dco  4 2 4
7    ( x1&/x4#/x2&           18   orth  4 4
8    / x3&x4#x2&x3&x4 ).     19   set  $minc  ( te_dif  3  4 )
9    tin  1 2               20   set  $mdec  ( te_dif  4  3 )
10   x3 .                   21   set  $lin  ( te_csd  3  4 )
11   0 .                    22   set  $indep  ( te_syd  3  4 )
```

Boolean variables	
Name	Value
$indep	false
$lin	false
$mdec	false
$minc	true

Fig. 4.28 Problem-program that verifies four properties of the function $f_1(\mathbf{x})$ with regard to x_3 together with the computed solutions

The PRP of Fig. 4.28 prepares the function-TVL of $f_1(\mathbf{x})$ using the command sbe and TVL 2 of $x_3 = 0$ within a Boolean space with lexicographically ordered variables $x_1, \ldots, x_4$ in lines 1–11. The subfunction $f_1(x_3 = 0)$ is computed in lines 12–14 as TVL in ODA-form. The command isc in line 12 uses TVL 2 to restrict TVL 1 to $f_1(x_3 = 0)$ and stores the result as TVL 3. The command dco in line 13 deletes the column of x_3 in TVL 3, and the command orth in line 14 re-establishes the orthogonality of TVL 3.

Similarly, the subfunction $f_1(x_3 = 1)$ is computed in lines 15–18 as TVL in ODA-form. The command cpl in line 15 computes the complement of TVL 2 so that the TVL 2 describes now $x_3 = 1$. The command isc in line 16 uses this changed TVL 2 to restrict TVL 1 to $f_1(x_3 = 1)$ and stores the result as TVL 4. The command dco in line 17 deletes the column of x_3 in TVL 4, and the command orth in line 18 re-establishes the orthogonality of TVL 4.

Based on the two subfunctions $f_1(x_3 = 0)$ (TVL 3) and $f_1(x_3 = 1)$ (TVL 4), it can easily be verified whether the given function $f_1(\mathbf{x})$ satisfies one of the four properties:

- The Boolean variable $minc (monotonously increasing with regard to x_3) is set to the value true when the difference TVL 3 minus TVL 4 is empty, and otherwise to the value false.
- The Boolean variable $mdec (monotonously decreasing with regard to x_3) is set to the value true when the difference of TVL 4 minus TVL 3 is empty (note the exchanged order of the TVLs), and otherwise to the value false.
- The Boolean variable $lin (linear in x_3) is set to the value true when the complement of the symmetric difference of TVL 3 and TVL 4 is empty, and otherwise to the value false.
- The Boolean variable $indep (independent of x_3) is set to the value true when the symmetric difference of TVL 3 and TVL 4 is empty, and otherwise to the value false.

The value true of the variable $minc in the table of Boolean variables on the right-hand side of Fig. 4.28 confirms that the function $f_1(\mathbf{x})$ is monotonously increasing with regard to x_3.

(b) Figure 4.29 shows on the left-hand side the PRP that verifies the properties of the function $f_2(\mathbf{x})$ and on the right-hand side the computed results.

The PRP of Fig. 4.29 prepares the function-TVL of $f_2(\mathbf{x})$ using the command sbe and reuses the last 14 lines of the PRP of Fig. 4.28 without changes.

The value true of the variable $indep in the table of Boolean variables on the right-hand side of Fig. 4.29 confirms that the function $f_2(\mathbf{x})$ is independent of x_3. Due to the independence of x_3, we have: $f_2(x_3 = 0) \equiv f_2(x_3 = 1)$; hence, both the relations $f_2(x_3 = 0) \le f_2(x_3 = 1)$ and $f_2(x_3 = 0) \ge f_2(x_3 = 1)$ are satisfied so that the function $f_2(\mathbf{x})$ is also monotonously increasing and decreasing with regard to x_3 indicated by $minc=true and $mdec=true.

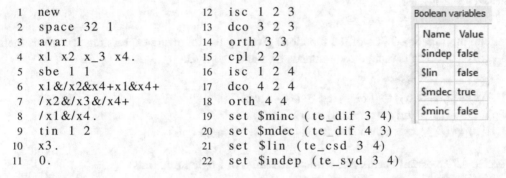

```
 1   new                    12   isc  1  2  3
 2   space  32  1           13   dco  3  2  3
 3   avar  1                14   orth  3  3
 4   x1  x2  x_3  x4.       15   cpl  2  2
 5   sbe  1  1              16   isc  1  2  4
 6   /x2&/x4+               17   dco  4  2  4
 7   (x1#x2)&x4+            18   orth  4  4
 8   x1&x3&x4+x2&/x3&x4.    19   set  $minc  (te_dif  3  4)
 9   tin  1  2              20   set  $mdec  (te_dif  4  3)
10   x3.                    21   set  $lin  (te_csd  3  4)
11   0.                     22   set  $indep  (te_syd  3  4)
```

Boolean variables	
Name	Value
$indep	true
$lin	false
$mdec	true
$minc	true

Fig. 4.29 Problem-program that verifies four properties of the function $f_2(\mathbf{x})$ with regard to x_3 together with the computed solutions

```
 1   new                    12   isc  1  2  3
 2   space  32  1           13   dco  3  2  3
 3   avar  1                14   orth  3  3
 4   x1  x2  x_3  x4.       15   cpl  2  2
 5   sbe  1  1              16   isc  1  2  4
 6   x1&/x2&x4+x1&x4+       17   dco  4  2  4
 7   /x2&/x3&/x4+           18   orth  4  4
 8   /x1&/x4.               19   set  $minc  (te_dif  3  4)
 9   tin  1  2              20   set  $mdec  (te_dif  4  3)
10   x3.                    21   set  $lin  (te_csd  3  4)
11   0.                     22   set  $indep  (te_syd  3  4)
```

Boolean variables	
Name	Value
$indep	false
$lin	false
$mdec	true
$minc	false

Fig. 4.30 Problem-program that verifies four properties of the function $f_3(\mathbf{x})$ with regard to x_3 together with the computed solutions

(c) Figure 4.30 shows on the left-hand side the PRP that verifies the properties of the function $f_3(\mathbf{x})$ and on the right-hand side the computed results.
The PRP of Fig. 4.30 prepares the function-TVL of $f_2(\mathbf{x})$ using the command sbe and reuses the last 14 lines of the PRP of Fig. 4.28 without changes.
The value true of the variable $mdec in the table of Boolean variables on the right-hand side of Fig. 4.30 confirms that the function $f_3(\mathbf{x})$ is monotonously decreasing with regard to x_3.
(d) Figure 4.31 shows on the left-hand side the PRP that verifies the properties of the function $f_4(\mathbf{x})$ and on the right-hand side the computed results.
The PRP of Fig. 4.31 prepares the function-TVL of $f_4(\mathbf{x})$ using the command sbe and reuses the last 14 lines of the PRP of Fig. 4.28 without changes.
The value true of the variable $lin in the table of Boolean variables on the right-hand side of Fig. 4.31 confirms that the function $f_4(\mathbf{x})$ is linear with regard to x_3.

4.8 Supplementary Exercises

Exercise 4.6 (Logic Functions Computed by Set Operations)

The function-TVL of a logic function that is specified by an arbitrary expression can be computed based on the TVLs of trivial functions $f_i(x_i) = x_i$ and appropriate set operations (see Exercise 4.1).

```
 1   new                        13   isc 1 2 3
 2   space 32 1                 14   dco 3 2 3
 3   avar 1                     15   orth 3 3
 4   x1 x2 x_3 x4.              16   cpl 2 2
 5   sbe 1 1                    17   isc 1 2 4
 6   /x2&/x3&(x1#x4)+           18   dco 4 2 4
 7   (x1+x2)&x3&x4+             19   orth 4 4
 8   /x1&/x2&x3&/x4+            20   set $minc (te_dif 3 4)
 9   x2&/x3&/x4.               21   set $mdec (te_dif 4 3)
10   tin 1 2                    22   set $lin (te_csd 3 4)
11   x3.                        23   set $indep (te_syd 3 4)
12   0.
```

Boolean variables	
Name	Value
$indep	false
$lin	true
$md	false
$mi	false

Fig. 4.31 Problem-program that verifies four properties of the function $f_4(\mathbf{x})$ with regard to x_3 together with the computed solutions

Compute the function-TVLs of the functions given by the following expressions using this method and verify your results using the commands sbe and te_syd:

(a) $f_1(\mathbf{x}) = x_1 x_2 \vee (x_3 \overline{x}_4 \oplus \overline{x}_5 x_6)$.
(b) $f_2(\mathbf{x}) = (x_1 \to x_2) \oplus ((x_2 \to x_3) \odot (x_3 \to x_1))$.
(c) $f_3(\mathbf{x}) = ((x_1 \vee x_2) \oplus (x_1 \vee x_3)) \wedge (x_2 \odot x_3)$.
(d) $f_4(\mathbf{x}) = (x_1 \vee x_2 \vee x_3 \vee x_4) \oplus (x_5 \wedge (x_6 \odot (x_7 \to x_8)))$.

Exercise 4.7 (Normal Forms and Function Tables of Logic Functions)

(a) Create the antivalence normal form (ANF) of the logic function $f(x_1, x_2, x_3, x_4)$ given as antivalence form:

$$f(x_1, x_2, x_3, x_4) = x_1 \overline{x}_3 \oplus x_2 x_3 \oplus x_1 x_4$$

using the Shannon expansion of the conjunctions followed by removing of pairs of elementary conjunctions and verify the result.

(b) Create the equivalence normal form (ENF) of the logic function $f(x_1, x_2, x_3, x_4)$ given as equivalence form:

$$f(x_1, x_2, x_3, x_4) = (x_1 \vee \overline{x}_3) \odot (x_2 \vee x_3) \odot (x_1 \vee x_4)$$

using the Shannon expansion of the disjunctions followed by removing of pairs of elementary disjunctions and verify the result.

(c) Create a function table of the logic function $f(x_1, x_2, x_3, x_4)$ given as antivalence form:

$$f(x_1, x_2, x_3, x_4) = x_1 \overline{x}_3 \oplus x_2 x_3 \oplus x_1 x_4 .$$

The binary vectors of (x_1, x_2, x_3, x_4) the must be order top down from (0000) to (1111) using the binary code. Use the approach that generates first a function table in the requested order for initial function $f(\mathbf{x}) = 0$ and change thereafter the function values based on the given three conjunctions taking into account that the given conjunctions can overlap each other and verify the result.

(d) Create a function table of the logic function $f(x_1, x_2, x_3, x_4)$ given as equivalence form:

$$f(x_1, x_2, x_3, x_4) = (x_1 \vee \overline{x}_3) \odot (x_2 \vee x_3) \odot (x_1 \vee x_4) .$$

The binary vectors of (x_1, x_2, x_3, x_4) the must be order top down from (0000) to (1111) using the binary code. Use the approach that generates first a function table in the requested order for initial function $f(\mathbf{x}) = 1$ and change thereafter the function values based on the given three disjunctions taking into account that the given disjunctions can overlap each other and verify the result.

Exercise 4.8 (Polynomials of Logic Functions)

Compute for the logic function:

$$f(\mathbf{x}) = \overline{x}_1\overline{x}_2\overline{x}_4 \vee x_1\overline{x}_3x_4 \vee x_1x_2 \vee x_2x_3.$$

(a) The antivalence polynomial also known as positive polarity Reed–Muller polynomial, or Shegalkin polynomial
(b) The equivalence polynomial
(c) Fixed polarity Reed–Muller polynomial where the variables x_2 and x_3 occur only negated
(d) The equivalence polynomial where the variables x_2 and x_4 occur only negated

in which the conjunctions (or disjunctions) are well ordered and verify your results. Utilize convenient ordered function tables and the duality between antivalence and equivalence polynomials.

Exercise 4.9 (Selected Function Values of Logic Functions)

A logic function that depends on 20 variables determines $2^{20} = 1,048,576$ function values. Compute the function values of

$$\mathbf{x}_1 = (x_1, x_2, \ldots, x_{20}) = (0000\ 0000\ 0000\ 0000\ 0000),$$

$$\mathbf{x}_2 = (x_1, x_2, \ldots, x_{20}) = (1010\ 1010\ 1010\ 1010\ 1010),$$

$$\mathbf{x}_3 = (x_1, x_2, \ldots, x_{20}) = (0011\ 1101\ 0011\ 1100\ 0000),$$

$$\mathbf{x}_4 = (x_1, x_2, \ldots, x_{20}) = (1100\ 1010\ 0111\ 0010\ 1101),\text{ and}$$

$$\mathbf{x}_5 = (x_1, x_2, \ldots, x_{20}) = (1111\ 1111\ 1111\ 1111\ 1111)$$

for the functions:

(a)
$$f_1(\mathbf{x}) = x_1\overline{x}_2x_3\overline{x}_4x_5 \vee \overline{x}_6\overline{x}_7\overline{x}_8x_9x_{10} \vee \overline{x}_{11}x_{12}\overline{x}_{13}x_{14}\overline{x}_{15}$$
$$\vee x_{16}x_{17}x_{18}x_{19}\overline{x}_{20} \vee \overline{x}_1\overline{x}_2x_8x_{14}\overline{x}_{19} \vee x_3x_4\overline{x}_{10}\overline{x}_{16}.$$

(b)
$$f_2(\mathbf{x}) = x_1\overline{x}_2x_3\overline{x}_4x_5 \oplus \overline{x}_6\overline{x}_7\overline{x}_8x_9x_{10} \oplus \overline{x}_{11}x_{12}\overline{x}_{13}x_{14}\overline{x}_{15}$$
$$\oplus x_{16}x_{17}x_{18}x_{19}\overline{x}_{20} \oplus \overline{x}_1\overline{x}_2x_8x_{14}\overline{x}_{19} \oplus x_3x_4\overline{x}_{10}\overline{x}_{16}.$$

(c)
$$f_3(\mathbf{x}) = ((x_1x_2 \vee x_3\overline{x}_4x_5) \oplus (x_6x_7 \vee x_8\overline{x}_9x_{10}))$$
$$\rightarrow ((x_{11}x_{12} \vee x_{13}\overline{x}_{14}x_{15}) \oplus (x_{16}x_{17} \vee x_{18}\overline{x}_{19}x_{20})).$$

(d)
$$f_4(\mathbf{x}) = \quad ((x_{11}x_{12} \vee x_{13}\overline{x}_{14}x_{15}) \oplus (x_{16}x_{17} \vee x_{18}\overline{x}_{19}x_{20}))$$
$$\rightarrow ((x_1x_2 \vee x_3\overline{x}_4x_5) \oplus (x_6x_7 \vee x_8\overline{x}_9x_{10})) \,.$$

Specify all four functions for one of the given five binary vectors the same function value?

Exercise 4.10 (Global Properties of Logic Functions)

Check whether the following functions are self-dual, linear, monotonously increasing, or monotonously decreasing:

(a)
$$f_1(\mathbf{x}) = \quad x_1\overline{x}_2x_4\overline{x}_6 \vee x_1\overline{x}_4x_5x_6 \vee x_1x_2x_3\overline{x}_4\overline{x}_6 \vee x_4x_6$$
$$\vee x_1\overline{x}_2x_3\overline{x}_4x_5\overline{x}_6 \vee x_1x_2\overline{x}_4\overline{x}_5x_6 \vee x_2x_4\overline{x}_6 \,.$$

(b)
$$f_2(\mathbf{x}) = \quad \overline{x}_1\overline{x}_4\overline{x}_5x_6 \vee x_1x_3\overline{x}_4 \vee x_1x_4x_6 \vee x_1\overline{x}_3x_4\overline{x}_5$$
$$\vee x_1x_2\overline{x}_4x_6 \vee \overline{x}_3\overline{x}_4x_5x_6 \vee \overline{x}_1x_2x_3x_4x_5 \vee \overline{x}_1x_3x_4x_6 \,.$$

(c)
$$f_3(\mathbf{x}) = \quad x_1\overline{x}_2\overline{x}_4\overline{x}_5 \vee \overline{x}_1\overline{x}_2x_4\overline{x}_6 \vee \overline{x}_1\overline{x}_4 \vee \overline{x}_2\overline{x}_3\overline{x}_4\overline{x}_6 \,.$$

(d)
$$f_4(\mathbf{x}) = \quad \overline{x}_1x_2\overline{x}_3x_4x_6 \vee x_1x_2\overline{x}_4x_6 \vee \overline{x}_1x_2x_3x_4x_6 \vee x_1\overline{x}_2x_4\overline{x}_5x_6$$
$$\vee \overline{x}_1\overline{x}_2\overline{x}_4x_6 \vee x_1\overline{x}_2x_4x_5x_6 \vee x_1x_2x_4\overline{x}_6 \vee \overline{x}_1\overline{x}_2x_3x_4x_5\overline{x}_6$$
$$\vee \overline{x}_1x_2\overline{x}_4\overline{x}_6 \vee \overline{x}_1\overline{x}_2x_3x_4\overline{x}_6 \vee x_1\overline{x}_2x_4\overline{x}_6 \vee \overline{x}_1\overline{x}_2x_3x_4\overline{x}_6 \,.$$

Utilize the possibility that some of these global properties can be verified by the corresponding properties with regard to each of their variables.

References

1. R.E. Bryant, Graph-based algorithms for Boolean function manipulation. IEEE Trans. Comput. **C-35**(8), 677–691 (Aug. 1986). ISSN:0018-9340. https://doi.org/10.1109/TC.1986.1676819
2. E.J. McCluskey, Minimization of Boolean functions. Bell Syst. Tech. J. **35**(6), 1417–1444 (Nov. 1956). ISSN:0005-8580. https://doi.org/10.1002/j.1538-7305.1956.tb03835.x
3. W.V. Quine, The problem of simplifying truth functions. Am. Math. Monthly **59**(8), 521–531 (Oct. 1952). https://doi.org/10.2307/2308219
4. W.V. Quine, A way to simplify truth functions. Am. Math. Monthly **62**(9), 627–631 (Nov. 1955). https://doi.org/10.2307/2307285
5. T. Sasao, M. Fujita (eds.), *Representations of Discrete Functions* (Kluwer Academic Publishers, Boston, London, Dordrecht, 1996). ISBN:0-7923-9720-7. https://doi.org/10.1007/978-1-4613-1385-4
6. C.E. Shannon, A symbolic analysis of relay and switching circuits. Trans. Am. Inst. Electric. Eng. **57**(12), 713–723 (Dec. 1938). ISSN:0096-3860. https://doi.org/10.1109/T-AIEE.1938.5057767
7. B. Steinbach, XBOOLE–A toolbox for modelling, simulation, and analysis of large digital systems. Syst. Anal. Model. Simul. **9**(4), 297–312 (Sept. 1992). ISSN:0232-9298

8. B. Steinbach, M. Werner, XBOOLE-CUDA - Fast Boolean operations on the GPU, in *Proceedings of the 11th International Workshops on Boolean Problems*, eds. by B. Steinbach. IWSBP 11 (Freiberg University of Mining and Technology, Freiberg, Germany, Sept. 2014), pp. 75–84. Isbn:978-3-86012-488-8

9. B. Steinbach, M. Werner, XBOOLE-CUDA - Fast calculations on the GPU, in *Problems and New Solutions in the Boolean Domain*, ed. by B. Steinbach (Cambridge Scholars Publishing, Newcastle upon Tyne, UK, Apr. 2016), pp. 117–149. ISBN:978-1-4438-8947-6

10. S.N. Yanushkevich et al. (eds.), *Decision Diagram Techniques for Micro- and Nanoelectronic Design - Handbook* (CRC Press, Taylor & Francis Group, Boca Raton, London, New York, 2006). ISBN:978-0-8493-3424-5. https://doi.org/10.1201/9781420037586

Logic Equations

5

Abstract

Logic equations are strongly related to logic functions; hence, logic equations also belong to the core concepts in the Boolean domain. A logic equation consists of two logic functions, which are connected by an equal sign. The solution of a logic equation is a set of binary vectors for which the two functions are either both equal to 0 or both equal to 1. Vice versa, the solution-set of a logic equation can be used to determine a characteristic logic function. Homogenous equations are special logic equations where one of the two functions is constant either equal to 1 or equal to 0. The value 1 determines that the other function characterizes certain properties; hence, we use in this case the name homogenous characteristic equation. The other case determines the value 0 that the specification of the other function is forbidden, which leads to the name homogenous restrictive equation. Each logic inequality can be transformed into logic equation so that no special solution-procedures are needed. Several logic equations determine a system of logic equations. Any system of logic equations and inequalities can be transformed into a single logic equation with the same solution-set. Lists of ternary vectors are used as the main data structure to solve logic equations based on set operations. Alternative solution-procedures are provided for special equation-systems.

Supplementary Information The online version of this chapter (https://doi.org/10.1007/978-3-030-88945-6_5) contains supplementary material which is available for authorized users. Please, follow the link belonging to the version of the XBOOLE-monitor XBM 2 that fits best for your operating system. This XBOOLE-monitor is needed to solve all tasks of this chapter. Instructions for starting the downloaded XBOOLE-monitor XBM 2 are given at the beginning of Section 'Exercises' in this chapter.

XBOOLE-monitor XBM 2 for Windows 10
32 bits
https://doi.org/10.1007/978-3-030-88945-6_5_MOESM1_ESM.zip (15,091 KB)

64 bits
https://doi.org/10.1007/978-3-030-88945-6_5_MOESM2_ESM.zip (14,973 KB)

XBOOLE-monitor XBM 2 for Linux Ubuntu
32 bits
https://doi.org/10.1007/978-3-030-88945-6_5_MOESM3_ESM.zip (29,522 KB)

64 bits
https://doi.org/10.1007/978-3-030-88945-6_5_MOESM4_ESM.zip (28,422 KB)

5.1   Basic Problems and Definitions

For an easy access to the concepts in question, we will go back to the basic understanding of an equation, as it is known from elementary mathematics. An equation

$$ax + b = c \qquad \text{or} \qquad ax^2 + bx + c = d$$

is a constraint for the values of x, and a solution x_1 allows the transformation of the equations in the identity $c \equiv c$ by calculating $ax_1 + b$, which results in c or the identity $d \equiv d$ by calculating $ax_1^2 + bx_1 + c$ that has to result in d. It is well-known that these equations are equivalent to the equations $ax + (b - c) = 0$ or $ax^2 + bx + (c - d) = 0$ (i.e., these equations have the same solutions as the original equations). Further transformations result in

$$x + \frac{b - c}{a} = 0 \qquad \text{or} \quad x^2 + \frac{b}{a}x + \frac{c - d}{a} = 0$$

which can be changed to

$$x + x_0 = 0 \qquad \text{or} \qquad x^2 + px + q = 0 \,,$$

and the solutions are

$$x = -x_0 \quad \text{or} \qquad x_{1,2} = -\frac{p}{2} \pm \sqrt{\frac{p^2}{4} - q} \,,$$

under consideration of several conditions, like $a \neq 0$, $\frac{p^2}{4} - q \geq 0$, etc.

In order to transform these ideas into the area of logic equations, we start with a problem that seems to look a bit strange. Let be given the equation

$$x_1 \vee x_2 = a \wedge b \,.$$

Since we are dealing with logic functions, the identities can only have the format $0 \equiv 0$ or $1 \equiv 1$. $x_1 \vee x_2 = 0$ holds only for $x_1 = 0$ and $x_2 = 0$. $a \wedge b = 0$ holds for $a = 0$, $b = 0$; $a = 0$, $b = 1$; and $a = 1$, $b = 0$; hence, we have the following set of solution-vectors with the components (x_1, x_2, a, b)

$$\{(0000), (0001), (0010)\} \,.$$

Now the identity $1 \equiv 1$ has to be explored, and, according to the definition of $\vee$ and $\wedge$ we get the following solutions:

$$\{(0111), (1011), (1111)\} \,.$$

Altogether, this equation has six solution-vectors. The strange character of this equation comes from the fact that there are different variables on the left-hand side and the right-hand side of the equation. In order to equalize the variables on both sides, we use dummy variables:

$$(x_1 \vee x_2)(a \vee \overline{a})(b \vee \overline{b}) = ab(x_1 \vee \overline{x}_1)(x_2 \vee \overline{x}_2) \,,$$

and this finally results in the equation

$$x_1ab \vee x_2ab \vee x_1a\overline{b} \vee x_2a\overline{b} \vee x_1\overline{a}b \vee x_2\overline{a}b \vee x_1\overline{a}\overline{b} \vee x_2\overline{a}\overline{b} =$$

$$abx_1x_2 \vee abx_1\overline{x}_2 \vee ab\overline{x}_1x_2 \vee ab\overline{x}_1\overline{x}_2 \ .$$

By checking all 16 vectors of $\mathbb{B}^4$, it can be seen that this equation has the same solutions as the original equation $x_1 \vee x_2 = ab$. Hence, we can assume that in equation $f(\mathbf{x}) = g(\mathbf{x})$ the functions $f(\mathbf{x})$ and $g(\mathbf{x})$ depend on the same set of variables.

Definition 5.1 Let $\mathbf{x} = (x_1, \ldots, x_n)$, $f(\mathbf{x})$, $g(\mathbf{x})$ be two logic functions, then

$$f(\mathbf{x}) = g(\mathbf{x})$$

is a *logic equation* of n variables. The binary vector $\mathbf{b} = (b_1, \ldots, b_n)$ is a *solution* of this equation if $f(\mathbf{b}) \equiv g(\mathbf{b})$, i.e., either $f(\mathbf{b}) \equiv g(\mathbf{b}) \equiv 0$ or $f(\mathbf{b}) \equiv g(\mathbf{b}) \equiv 1$.

Note 5.1 The term *Boolean equation* can synonymously be used to express the term *logic equation*.

It is important to understand that the functions on the left-hand side and on the right-hand side are given by formulas. If the functions would be given by function tables or vectors, then the problem is easy—it would be necessary only to compare the two vectors or tables element by element. The argument vectors of such rows of a function table belong to the set of solution-vectors for which both functions have the same value, i.e., $0 = 0$ or $1 = 1$. Similarly, the argument vectors for equal values in the two function vectors must be determined and collected in the solution-set.

An interesting question is to identify which changes of a given logic equation do not change the associated solution-set. The following theorem answers this question.

Theorem 5.1 *The solution-set of a logic equation $f(\mathbf{x}) = g(\mathbf{x})$ remains unchanged when:*

- *the expressions of $f(\mathbf{x})$ and/or $g(\mathbf{x})$ are equivalently transformed so that these two functions remain unchanged;*
- *the functions $f(\mathbf{x})$ and $g(\mathbf{x})$ are replaced by their complement so that the equation $\overline{f(\mathbf{x})} = \overline{g(\mathbf{x})}$ is solved;*
- *an arbitrary function $h(\mathbf{x})$ is added on both sides using antivalence so that the equation $f(\mathbf{x}) \oplus h(\mathbf{x}) = g(\mathbf{x}) \oplus h(\mathbf{x})$ is solved;*
- *an arbitrary function $h(\mathbf{x})$ is added on both sides using equivalence so that the equation $f(\mathbf{x}) \odot h(\mathbf{x}) = g(\mathbf{x}) \odot h(\mathbf{x})$ is solved.*

Proof Equivalent transformations change the expression, but not the associated function. Only the functions $f(\mathbf{x})$ and $g(\mathbf{x})$ determine the solution-set of the equation $f(\mathbf{x}) = g(\mathbf{x})$.

The evaluation of the logic equation leads for solution-vectors to the tautologies $0 \equiv 0$ or $1 \equiv 1$. The complement of both $f(\mathbf{x})$ and $g(\mathbf{x})$ changes $0 \equiv 0$ into $1 \equiv 1$ or vice versa. Hence, the equations

$$\overline{f(\mathbf{x})} = \overline{g(\mathbf{x})} \quad \text{and} \quad f(\mathbf{x}) = g(\mathbf{x})$$

have the same solution-set.

There are two cases for the logic equation $f(\mathbf{x}) \oplus h(\mathbf{x}) = g(\mathbf{x}) \oplus h(\mathbf{x})$. For binary vectors $\mathbf{b}$ with $h(\mathbf{b}) = 0$ we get the original equation $f(\mathbf{x}) = g(\mathbf{x})$ and for $h(\mathbf{b}) = 1$ the equation $\overline{f(\mathbf{x})} = \overline{g(\mathbf{x})}$. These two equations have the same solutions; hence, the equations

$$f(\mathbf{x}) \oplus h(\mathbf{x}) = g(\mathbf{x}) \oplus h(\mathbf{x}) \quad \text{and} \quad f(\mathbf{x}) = g(\mathbf{x})$$

have the same solution-set.

In the case of the equation $f(\mathbf{x}) \odot h(\mathbf{x}) = g(\mathbf{x}) \odot h(\mathbf{x})$ the same arguments hold with exchanged values of $h(\mathbf{b})$. Hence the equations

$$f(\mathbf{x}) \odot h(\mathbf{x}) = g(\mathbf{x}) \odot h(\mathbf{x}) \quad \text{and} \quad f(\mathbf{x}) = g(\mathbf{x})$$

have the same solution-set.   □

In the same way as above, it is possible to reduce the considerations to homogenous equations.

Theorem 5.2 *The equation $f(\mathbf{x}) = g(\mathbf{x})$ is equivalent to the*

homogenousrestrictiveequation	$f(\mathbf{x}) \oplus g(\mathbf{x}) = 0\,,$	(5.1)
homogenouscharacteristicequation	$f(\mathbf{x}) \odot g(\mathbf{x}) = 1\,.$	(5.2)

Proof This theorem follows directly from the definition of $\oplus$ and $\odot$.   □

The logic function on the left-hand side of a *homogenous restrictive equation* is equal to 1 for all argument vectors of forbidden cases. Vice versa, the logic function on the left-hand side of a *homogenous characteristic equation* is equal to 1 for all argument vectors of valid cases. The intentions of *forbidden* and *valid* are naturally related to a given context.

5.2   Systems of Equations and Inequalities

In contrast to other areas of mathematics, systems of logic equations can be reduced to one single equation.

Theorem 5.3 *Let*

$$f_1(\mathbf{x}) = g_1(\mathbf{x})$$
$$f_2(\mathbf{x}) = g_2(\mathbf{x})$$
$$\vdots \qquad \vdots$$
$$f_m(\mathbf{x}) = g_m(\mathbf{x}) \tag{5.3}$$

be a system of m logic equations of n variables. Then, this system of equations is equivalent to the equations

$$(f_1(\mathbf{x}) \oplus g_1(\mathbf{x})) \vee \ldots \vee (f_m(\mathbf{x}) \oplus g_m(\mathbf{x})) = 0 \tag{5.4}$$

or

$$(f_1(\mathbf{x}) \odot g_1(\mathbf{x})) \wedge \ldots \wedge (f_m(\mathbf{x}) \odot g_m(\mathbf{x})) = 1\,. \tag{5.5}$$

Proof This theorem follows directly from the definition of $\oplus$, $\odot$, $\wedge$, $\vee$.

In a first step, all the equations are transformed into the homogenous form (either the value 0 or 1 is used on the right-hand side); these homogenous equations are then transformed into one equation according to

$$a = 0 , \quad b = 0 \quad \equiv \quad a \vee b = 0 , \tag{5.6}$$

$$a = 1 , \quad b = 1 \quad \equiv \quad a \wedge b = 1 . \tag{5.7}$$

$\square$

This means that, at least in principle, only single homogenous logic equations of n variables have to be considered. The practical way will very often have the reverse direction: a very complicate logic equation will be split into "smaller" parts, and the solution-sets of these parts will be found and thereafter combined into the solution-set of the given logic equation. This approach will be explained in detail in the next section.

It is also not very difficult to deal with *inequalities*; this already had been indicated in the previous chapter. Two cases have to be considered (according to the definition of $\leq$ or $<$ in $\mathbb{B}$):

$$f(\mathbf{x}) \leq g(\mathbf{x}) , \tag{5.8}$$

$$f(\mathbf{x}) < g(\mathbf{x}) . \tag{5.9}$$

Inequality (5.8) has the solutions $(f(\mathbf{x}), g(\mathbf{x})) = \{(00), (01), (11)\}$. This solution-set can be translated into a logic equation with the same solution-set; there are two such equivalent equations:

$$\overline{f(\mathbf{x})} \vee g(\mathbf{x}) = 1 \quad \equiv \quad f(\mathbf{x})\overline{g(\mathbf{x})} = 0 .$$

The solution-set of the second inequality (5.9) contains such vectors $\mathbf{x}$ that satisfy both $f(\mathbf{x}) = 0$ and $g(\mathbf{x}) = 1$; this property can be expressed by the equation(s)

$$f(\mathbf{x}) \vee \overline{g(\mathbf{x})} = 0 \quad \equiv \quad \overline{f(\mathbf{x})}g(\mathbf{x}) = 1 .$$

This is already a proof of the following.

Theorem 5.4 *Each inequality $f(\mathbf{x}) \leq g(\mathbf{x})$ and $f(\mathbf{x}) < g(\mathbf{x})$ can be transformed into equivalent homogenous equations:*

$$f(\mathbf{x}) \leq g(\mathbf{x}) \quad \equiv \quad \overline{f(\mathbf{x})} \vee g(\mathbf{x}) = 1 \quad \equiv \quad f(\mathbf{x})\overline{g(\mathbf{x})} = 0 , \tag{5.10}$$

$$f(\mathbf{x}) < g(\mathbf{x}) \quad \equiv \quad f(\mathbf{x}) \vee \overline{g(\mathbf{x})} = 0 \quad \equiv \quad \overline{f(\mathbf{x})}g(\mathbf{x}) = 1 . \tag{5.11}$$

Hence, homogenous equations also cover inequalities between logic functions as well as systems of inequalities and all kinds of "mixed" systems.

Similarly, to logic equations, equivalent transformations within the expressions of the left-hand side and/or right-hand side do not change the solution-set of an inequality. However, equivalent transformations of an inequality that change the functions $f(\mathbf{x})$ and $g(\mathbf{x})$ are restricted to the negation and require the change between $\leq$ and $\geq$ or $<$ and $>$:

$$f(\mathbf{x}) \leq g(\mathbf{x}) \quad \equiv \quad \overline{f(\mathbf{x})} \geq \overline{g(\mathbf{x})} ,$$

$$f(\mathbf{x}) < g(\mathbf{x}) \quad \equiv \quad \overline{f(\mathbf{x})} > \overline{g(\mathbf{x})} .$$

5.3 Ternary Vectors as the Main Data Structure

The most important approach for the solution of logic equations is the implementation of appropriate methods, algorithms, and programs. Traditional systems of symbolic Mathematics hardly consider logic functions and equations. However, there are many specialized systems available, mainly as systems for the logic synthesis of digital systems or as systems for the SAT-problem (SAT-solvers). The attention for the SAT-problem increased considerably over the last 30 years, even international competitions took and take place.

In this chapter we will consider the solution of all kinds of *logic equations* and *inequalities* without their transformation into homogenous equations. The comprehensive use of solution-sets and their representation by *ternary vectors* will be the base for the implementation of parallel algorithms.

The basic theoretical ideas use the isomorphism between Boolean Algebras for these algorithms and follow a set of theoretically well-founded principles:

- a complex problem (equation) will be split into smaller problems that are "easier to solve";
- the solution-sets of these smaller problems will be represented by (orthogonal) lists of ternary vectors (TVL) that represent a set of binary vectors;
- the logic operations like $\vee, \wedge, \overline{}, \ldots$ are mapped to set operations such as $\cup, \cap, \overline{}, \ldots$;
- in fact, we use the isomorphism between the two Boolean Algebras

$$(\mathbb{B}^n, \vee, \wedge, \overline{}, \mathbf{0}, \mathbf{1}) \quad \text{and} \quad (P(X), \cap, \cup, \overline{}, \emptyset, X)$$

as well as the two Boolean Rings

$$(\mathbb{B}^n, \oplus, \wedge, \mathbf{0}, \mathbf{1}) \quad \text{and} \quad (P(X), \Delta, \cup, \emptyset, X) \, ;$$

- the bitwise coding of the ternary vectors allows the parallel implementation of many parts of the algorithms and reduces the complexity as much as possible;
- all these steps have been included into the system XBOOLE; see
$$\text{https://tu-freiberg.de/en/fakult1/inf/xboole}$$
and [1–4].

It is already known by previous considerations that a ternary vector

$$\mathbf{t} = (t_1, \ldots, t_n)$$

with the values "0," "1," and "−" for the components describes a *set of binary vectors S* (an *interval*). The symbol "−" can be replaced arbitrarily with the values "0" or "1." The replacement of all "−" by "0" gives the minimum of the interval, the replacement of all "−" by "1" gives the maximum of the interval, any combinations of "0" and "1" give all the vectors of the interval.

Definition 5.2 Let $\mathbf{t} = (t_1, \ldots, t_n), t_i \in \{0, 1, -\}, i = 1, \ldots, n$, be a ternary vector. Then, $S(\mathbf{t})$ is the set of all vectors $\mathbf{x} \in \mathbb{B}^n$ generated by the ternary vector $\mathbf{t}$.

$S(\mathbf{t})$ indicates the meaning *solution-set* of the ternary vector $\mathbf{t}$.

Example 5.1 Let $\mathbf{t} = (0 - 1 - 0)$, then

$$S(\mathbf{t}) = \{(00100), (00110), (01100), (01110)\} \, .$$

Table 5.1 Intersection of two ternary vectors	t_1	0	0	0	1	1	1	–	–	–
	t_2	0	1	–	0	1	–	0	1	–
	$t_1 \cap t_2$	0	∅	0	∅	1	1	0	1	–

As a border case, the vector $\mathbf{t} = (- - \ldots -)$ generates $\mathbb{B}^n$. A vector $\mathbf{x}$ without even one "−" will be considered as the singleton $\{\mathbf{x}\}$.

The use of several ternary vectors will be possible by *lists of ternary vectors* (synonymously *ternary matrices* (TMs) and *ternary vector lists* (TVLs)).

Definition 5.3 Let $L = (\mathbf{t}_1, \mathbf{t}_2, \ldots, \mathbf{t}_k)$ be a list of ternary vectors of the same length, then

$$S(L) = S(\mathbf{t}_1) \cup S(\mathbf{t}_2) \cup \cdots \cup S(\mathbf{t}_k) \, .$$

Hence, when a list of ternary vectors is given, then we generate the set of binary vectors for each ternary vector of the list, the final set is the union of these sets.

Unfortunately, the union of sets can meet one vector several times; it is very difficult to see how many vectors are in the set $S(L)$ as long as it is not fully computed. We already saw in Chap. 4 that the *orthogonality* of ternary vectors avoids this duplication of binary vectors and that it is always possible to orthogonalize a list of ternary vectors in such a way that all pairs of ternary vectors of this list are mutually orthogonal. Because of the importance of this fundamental concept, we summarize the previous knowledge.

Definition 5.4 Two vectors $\mathbf{t}_1$ and $\mathbf{t}_2$ are *mutually orthogonal* ($\mathbf{t}_1 \perp \mathbf{t}_2$) if $S(\mathbf{t}_1) \cap S(\mathbf{t}_2) = \emptyset$.

Theorem 5.5 *For two ternary vectors $\mathbf{t}_1$ and $\mathbf{t}_2$, the property $\mathbf{t}_1 \perp \mathbf{t}_2$ holds if and only if for at least one component t_i the combination $\mathbf{t}_{1i} = 0, \mathbf{t}_{2i} = 1$ or $\mathbf{t}_{2i} = 0, \mathbf{t}_{1i} = 1$ appears.*

This property is very easy to test and very useful because now the sets $S(\mathbf{t}_1)$ and $S(\mathbf{t}_2)$ have an *empty intersection*. Assuming $\mathbf{t}_1 \perp \mathbf{t}_2$, then any $\mathbf{x} \in \mathbb{B}^n$ is (if at all) either in $S(\mathbf{t}_1)$ or in $S(\mathbf{t}_2)$, but not in both of them.

The orthogonality is one extreme property. The two sets are completely separated. The other extreme case occurs when one set is a subset of the other set, e.g.,

$$S(\mathbf{t}_1) \subseteq S(\mathbf{t}_2) \, .$$

This means that each binary vector generated by $\mathbf{t}_1$ is also generated by $\mathbf{t}_2$. In this case the vector $\mathbf{t}_1$ can simply be omitted in the list L, and it does not contribute to the set $S(L)$.

In order to test this property, we keep in mind that $A \subseteq B$ if and only if $A \cap B = A$. The intersection $S(\mathbf{t}_1) \cap S(\mathbf{t}_2)$ of $S(\mathbf{t}_1)$ and $S(\mathbf{t}_2)$ can be computed using Table 5.1 for each component of the two vectors.

This follows from the fact that "−" always can be replaced by "0" and "1." The empty intersection $\emptyset$ occurs for the combinations (0 1) or (1 0), resp.; hence, (0 −) and (− 0) result in 0, (1 −) and (− 1) result in 1, etc.

Definition 5.5 $\mathbf{t}_1$ will be *absorbed* (or *covered*) by $\mathbf{t}_2$ if

$$S(\mathbf{t}_1) \cap S(\mathbf{t}_2) = S(\mathbf{t}_1) \, , \tag{5.12}$$

$$\mathbf{t}_1 \cap \mathbf{t}_2 = \mathbf{t}_1 \,. \tag{5.13}$$

Once again this test is very easy to implement, and the TVL can be reduced by omitting those vectors that are covered by other vectors.

Example 5.2 Let $\mathbf{t}_1 = (1\ 0\ 0\ 1 - 0)$, $\mathbf{t}_2 = (1\ 0 - 1 - 0)$. We calculate the intersection

t_1	1	0	0	1	–	0
t_2	1	0	–	1	–	0
$t_1 \cap t_2$	1	0	0	1	–	0

and get $\mathbf{t}_1 \cap \mathbf{t}_2 = \mathbf{t}_1$. Hence, $\mathbf{t}_1$ is covered by $\mathbf{t}_2$. It can easily be seen that in this case $\mathbf{t}_1$ can be created by replacing the left "−" in $\mathbf{t}_2$ with "0."

Now we can start to build the solution-set of "small" and "simple" equations.

5.4   The Solution of Simple Equations

The expression of the given equation is split into simple parts. These parts can be used to create homogenous equations. The solution-sets of such simple equations are afterward combined to the solution-set of the whole given equation.

Characteristic Equation of a Conjunction of Literals: $C = 1$
Let $C = x_1 \bar{x}_3 x_5$. In order to have $C = 1$, every literal must be equal to 1, i.e., $x_1 = 1$, $x_3 = 0$, $x_5 = 1$. By the given context, we assume that x_2 and x_4 also play a given role; hence,

$$\mathbf{t} = (1 - 0 - 1)$$

is the solution-set of the logic equation $C = 1$.

The coding of this solution-set will follow a very simple rule and has a very special meaning. In order to find the ternary vector for a single conjunction equal to 1, a variable will be encoded by "1," a negated variable will be encoded by "0," and a variable that is missing in the given conjunction is encoded by "−." The resulting ternary vector defines the interval for which the given conjunction is equal to 1. Normally the variables to be considered should be known at the beginning. It is, however, easy to extend the set of variables in a later stage (if necessary): missing variables are simply added by means of a "−" at the appropriate position.

Characteristic Equation of a Disjunction of Literals: $D = 1$
Let $D = x_1 \vee \bar{x}_3 \vee x_5$. In order to have $D = 1$, at least one literal must be equal to 1, and three vectors must be used to represent the solution-set:

$$\begin{aligned}
\mathbf{t}_1 &= (1 - - - -)\,, \\
\mathbf{t}_2 &= (- - 0 - -)\,, \\
\mathbf{t}_3 &= (- - - - 1)\,.
\end{aligned}$$

The solution-set for $D = 1$ is equal to

$$S = \mathbf{t}_1 \cup \mathbf{t}_2 \cup \mathbf{t}_3 .$$

It can be seen that now the problem of *double solutions* exists; $(1 - 0 - -)$ can be generated by $\mathbf{t}_1$, but also by $\mathbf{t}_2$:

$$(1 - - - -) = \begin{cases} (1 - 0 - -) \\ (1 - 1 - -) \end{cases} ,$$

$$(- - 0 - -) = \begin{cases} (1 - 0 - -) \\ (0 - 0 - -) \end{cases} .$$

This problem can be solved or avoided by a "tricky coding." We are coding the solution-set for $D = x_1 \vee \overline{x}_3 \vee x_5$ directly in an orthogonal manner:

$$\mathbf{t}_1 = (\ 1 - - - -\ ) ,$$
$$\mathbf{t}_2 = (\ 0 - 0 - -\ ) ,$$
$$\mathbf{t}_3 = (\ 0 - 1 - 1\ ) .$$

The first vector contains all vectors with $x_1 = 1$. Next, by considering $\overline{x}_3 = 1$ we get $x_3 = 0$, and it can be additionally assumed that $x_1 = 0$ (the complement of the value selected for $\mathbf{t}_1$). In the same way, $x_5 = 1$ can assume $x_1 = 0$ and $x_3 = 1$, and all the other vectors are already covered by $\mathbf{t}_1$ or $\mathbf{t}_2$. The solution vectors are now mutually orthogonal:

$$\mathbf{t}_1 \perp \mathbf{t}_2 , \quad \mathbf{t}_1 \perp \mathbf{t}_3 , \quad \mathbf{t}_2 \perp \mathbf{t}_3 .$$

Note 5.2 Any other order of the variables will also be possible.

For the order $x_3 \ \rangle \ x_5 \rightarrow x_1$, for instance, we get

$$\mathbf{t}_1 = (\ - - 0 - -\ ) ,$$
$$\mathbf{t}_2 = (\ - - 1 - 1\ ) ,$$
$$\mathbf{t}_3 = (\ 1 - 1 - 0\ ) .$$

Restrictive Equation of a Conjunction of Literals: $C = 0$
Let $C = x_1 \overline{x}_3 x_5$. In order to have $C = 0$, it is sufficient that one literal is equal to 0. Hence, we need again three vectors to represent the solution-set:

$$\mathbf{t}_1 = (\ 0 - - - -\ ) ,$$
$$\mathbf{t}_2 = (\ - - 1 - -\ ) ,$$
$$\mathbf{t}_3 = (\ - - - - 0\ ) .$$

This situation is the same as for $D = 1$. The existing overlap of the solutions can be eliminated as before by transferring the matrix into the orthogonalized format:

$$\mathbf{t}_1 = (\ 0 - - - -\ ) ,$$
$$\mathbf{t}_2 = (\ 1 - 1 - -\ ) ,$$
$$\mathbf{t}_3 = (\ 1 - 0 - 0\ ) .$$

Restrictive Equation of a Disjunction of Literals: $D = 0$

Let $D = x_1 \vee \overline{x}_3 \vee x_5$. In order to have $D = 0$, every literal of the disjunction must be equal to 0. The vector $\mathbf{t} = (0 - 1 - 0)$ will describe the solution-set.

In the following we use the two vectors

$$\mathbf{x} = (x_1, \ldots, x_n) \quad \text{and} \quad \mathbf{x}_1 = (x_1, \ldots, x_{i-1}, x_{i+1}, \ldots, x_n)$$

and consider some special cases where it is rather easy to find solutions.

The Variable x_i Is Linearly Separable
In this case the function $f(\mathbf{x})$ can be represented in the format

$$f(\mathbf{x}) = x_i \oplus g(\mathbf{x}_1) \,.$$

In the case of $f(\mathbf{x}) = 0$ we get $x_i \oplus g(\mathbf{x}_1) = 0$; hence, the solution-set consists of two subsets:

- $x_i = 0$ and the solutions of $g(\mathbf{x}_1) = 0$;
- $x_i = 1$ and the solutions of $g(\mathbf{x}_1) = 1$.

This requires mainly the solution of one homogenous equation with the function $g(\mathbf{x}_1)$, and the second solution-set is the complement of the first.

The equation $x_i \oplus g(\mathbf{x}_1) = 1$ holds for

- $x_i = 1$ and the solutions of $g(\mathbf{x}_1) = 0$;
- $x_i = 0$ and the solutions of $g(\mathbf{x}_1) = 1$;

hence, only one equation with $g(\mathbf{x}_1)$ on the left-hand side and 0 or 1 on the right-hand side has to be considered.

The Variable x_i Is Conjunctively Separable
The function can be represented in the format $f(\mathbf{x}) = x_i \wedge g(\mathbf{x}_1)$. Two subsets have to be considered for the construction of the solution-set of the logic equation $f(\mathbf{x}) = x_i \wedge g(\mathbf{x}_1) = 0$:

- $x_i = 0$ together with the whole $\mathbb{B}^{n-1}$; and
- $x_i = 1$ together with the solutions of $g(\mathbf{x}_1) = 0$.

The equation $x_i \wedge g(\mathbf{x}_1) = 1$ holds for $x_i = 1$ and $g(\mathbf{x}_1) = 1$.

The Variable x_i Is Disjunctively Separable
The function can be represented in the format $f(\mathbf{x}) = x_i \vee g(\mathbf{x}_1)$. The solution of $f(\mathbf{x}) = x_i \vee g(\mathbf{x}_1) = 0$ requires $x_i = 0$ and the solution of $g(\mathbf{x}_1) = 0$.

The equation $f(\mathbf{x}) = x_i \vee g(\mathbf{x}_1) = 1$ has the solutions:

- $x_i = 1$ together with the whole $\mathbb{B}^{n-1}$; and
- $x_i = 0$ together with the solutions of $g(\mathbf{x}_1) = 1$.

In all three cases the complexity of the problem is reduced to the half; since only 2^{n-1} elements of $\mathbb{B}^{n-1}$ have to be considered, the vector $\mathbf{x}_1$ has only $n - 1$ variables. The variable $\overline{x}_i$ can also be used for the separation, and the considerations remain the same.

5.5   Operations with Solution-Sets

The Union of Solution-Sets
Based on the following procedure, it can be seen that each TVL can be orthogonalized: we select one vector, normally the vector with the largest number of "−"-elements, and orthogonalize (as shown above) all the other vectors with regard to this one, omitting all the vectors absorbed (covered).

Thereafter a next vector is selected leaving the first one untouched, and the rest of the TVL will be orthogonalized with regard to the second one, etc.

This means that we have now an algorithm for the *union* of solution-sets. The result is free of double solutions. We combine the two TVLs of the solution-sets into one TVL and orthogonalize the resulting vector list. Since we can assume that the two lists are already orthogonalized, only the second list must be orthogonalized with regard to the first one.

If there are more than two TVLs, then we process the lists one after the other.

Note 5.3 The union itself is already achieved when TVLs have been written below each other. The orthogonalization eliminates vectors that appear in several lists and gives back every solution-vector exactly once. Very often the intermediate lists can remain in non-orthogonal formats for further processing, and the orthogonalization will take place only at the end of a computational process in order to define the final result.

Example 5.3 Let be given

$$
S_1 = \begin{array}{ccccc} x_1 & x_2 & x_3 & x_4 & x_5 \\ \hline 1 & - & - & - & - \\ 0 & 0 & - & - & - \\ 0 & 1 & 1 & - & - \end{array}, \quad
S_2 = \begin{array}{ccccc} x_1 & x_2 & x_3 & x_4 & x_5 \\ \hline - & 1 & 0 & 0 & - \\ - & 1 & - & 1 & 1 \\ - & - & 1 & - & 0 \end{array},
$$

then we have

$$
S_1 \cup S_2 = \begin{array}{ccccc} x_1 & x_2 & x_3 & x_4 & x_5 \\ \hline 1 & - & - & - & - \\ 0 & 0 & - & - & - \\ 0 & 1 & 1 & - & - \\ - & 1 & 0 & 0 & - \\ - & 1 & - & 1 & 1 \\ - & - & 1 & - & 0 \end{array}.
$$

The orthogonalization of this list results in

$$
S_1 \cup S_2 = \begin{array}{ccccc} x_1 & x_2 & x_3 & x_4 & x_5 \\ \hline 1 & - & - & - & - \\ 0 & 0 & - & - & - \\ 0 & 1 & 1 & - & - \\ 0 & 1 & 0 & 0 & - \\ 0 & 1 & 0 & 1 & 1 \end{array}.
$$

The orthogonal list is even shorter than the original list. It can be seen that the number of values 0 and 1 grows from top to bottom.

The Complement of a Solution-Set
The complement of a given solution-set can be found in a very direct way. The basic relation is

$$\overline{t_1 \cup t_2 \cup \ldots \cup t_k} = \overline{t_1} \cap \overline{t_2} \cap \ldots \cap \overline{t_k} .$$

We build the complement of each ternary vector of a given set and use the intersection to combine these complements (see the next paragraph for details of the intersection). When we start with rather large intervals, then the complements of these vectors are rather small, and the intersections will still reduce these intervals.

Example 5.4 We consider the equation $x_1 \vee \overline{x}_2 \vee x_3 = 1$ that has the solution-set

$$S = \begin{array}{ccc} x_1 & x_2 & x_3 \\ \hline 1 & - & - \\ 0 & 0 & - \\ 0 & 1 & 1 \end{array} .$$

We know that

$$\overline{S} = \begin{array}{ccc} x_1 & x_2 & x_3 \\ \hline 0 & 1 & 0 \end{array}$$

is the solution of the complementary equation

$$x_1 \vee \overline{x}_2 \vee x_3 = 0 .$$

Our algorithm will show that the same result will be obtained very fast:

$$S_1 = \begin{array}{ccc} x_1 & x_2 & x_3 \\ \hline 0 & - & - \end{array}$$

is the complement of the first vector $(1 - -)$ of S;

$$S_2 = \begin{array}{ccc} x_1 & x_2 & x_3 \\ \hline 1 & - & - \\ 0 & 1 & - \end{array}$$

is the complement of the second vector of S. The intersection of these two sets is

$$S_1 \cap S_2 = \begin{array}{ccc} x_1 & x_2 & x_3 \\ \hline 0 & 1 & - \end{array} .$$

$$S_3 = \begin{array}{ccc} x_1 & x_2 & x_3 \\ \hline 1 & - & - \\ 0 & 0 & - \\ 0 & 1 & 0 \end{array}$$

is the complement of the third vector of S, and the intersection of the sets S_i, $i = 1, 2, 3$ results in

$$(S_1 \cap S_2) \cap S_3 = \begin{array}{ccc} x_1 & x_2 & x_3 \\ \hline 0 & 1 & 0 \end{array}.$$

The complement is very useful when we want to change the right-hand side of the equation; instead of solving $f(\mathbf{x}) = 1$, it might be easier first to solve $f(\mathbf{x}) = 0$ and thereafter to find the complement of the solution-set, or vice versa. We will use this approach very often.

The Intersection of Solution-Sets
We already saw that the intersection of two ternary vectors can be calculated according to Table 5.1 on page 199.

When we have two lists of ternary vectors and have to find the intersection of the corresponding solution-sets, then we remember that one list represents the union of the sets associated to the single vectors; hence, we have to find the intersection of two sets that are constructed in the following way:

$$(S(\mathbf{t}_1) \cup S(\mathbf{t}_2) \cup \ldots) \cap (S(\mathbf{u}_1) \cup S(\mathbf{u}_2) \cup \ldots) ,$$

and this results in the union of pairwise intersections

$$(S(\mathbf{t}_1) \cap S(\mathbf{u}_1)) \cup (S(\mathbf{t}_1) \cap S(\mathbf{u}_2)) \cup \ldots \cup (S(\mathbf{t}_2) \cap S(\mathbf{u}_1)) \cup (S(\mathbf{t}_2) \cap S(\mathbf{u}_2)) \cup \ldots$$

This means that each ternary vector of one list has to be combined with each vector of the second list using the definition above. Fortunately, many of these intersections might be empty and omitted. For n_1 vectors in the first list and n_2 vectors in the second list, the intersection produces at most $n_1 * n_2$ vectors.

Example 5.5 Let be given again

$$S_1 = \begin{array}{ccccc} x_1 & x_2 & x_3 & x_4 & x_5 \\ \hline 1 & - & - & - & - \\ 0 & 0 & - & - & - \\ 0 & 1 & 1 & - & - \end{array}, \quad S_2 = \begin{array}{ccccc} x_1 & x_2 & x_3 & x_4 & x_5 \\ \hline - & 1 & 0 & 0 & - \\ - & 1 & - & 1 & 1 \\ - & - & 1 & - & 0 \end{array},$$

then we have

$$S_1 \cap S_2 = \begin{array}{ccccc} x_1 & x_2 & x_3 & x_4 & x_5 \\ \hline 1 & 1 & 0 & 0 & - \\ 1 & 1 & - & 1 & 1 \\ 1 & - & 1 & - & 0 \\ 0 & 0 & 1 & - & 0 \\ 0 & 1 & 1 & 1 & 1 \\ 0 & 1 & 1 & - & 0 \end{array}.$$

This matrix is already orthogonalized due to the used orthogonal TVLs S_1 and S_2; all the vectors of the TVL $S_1 \cap S_2$ are mutually orthogonal.

In order not to be afraid of the $n_1 * n_2$ possible intersections, we keep in mind that

$$S_1 \cap S_2 \subseteq S_1 , \quad S_1 \cap S_2 \subseteq S_2 .$$

The number of *binary* vectors in $S_1 \cap S_2$ is always less than or equal to the number of binary vectors in S_1 and S_2. For a better understanding of this problem, let us consider all ternary vectors of length 3:

$$(000), \ (001), \ (00-), \ \ldots, \ (--0), \ (--1), \ (---) \,.$$

These are 27 ternary vectors altogether. We get as detailed exploration of their orthogonality:

- a binary vector (like (011)) is orthogonal to 19 other ternary vectors, and there are eight different binary vectors;
- a vector with one "$-$" (like (01$-$)) is orthogonal to 15 other ternary vectors, and there are 12 such vectors;
- a vector with "$-$" in two positions (like ($-0-$)) is still orthogonal to nine other ternary vectors, and there are six such vectors;
- the vector ($---$) is not orthogonal to any other vector.

Hence, there are $27 * 27 = 729$ possible intersections, and among those

$$8 * 19 + 12 * 15 + 6 * 9 + 1 * 0 = 386$$

are empty, i.e., 52.95%, more than half of the conjunctions. Possible absorptions will reduce this number further.

The Solution-Set of an Arbitrary Logic Equation
Now all the possible equations (i.e., equations with any of the available operations) can be solved by using the methods stated and explored above. The following summary should be easy to understand:

- $f(\mathbf{x}) \wedge g(\mathbf{x}) = 1$: solve $f(\mathbf{x}) = 1$, $g(\mathbf{x}) = 1$ and calculate $S = S_{f1} \cap S_{g1}$;
- $f(\mathbf{x}) \vee g(\mathbf{x}) = 1$: solve $f(\mathbf{x}) = 1$, $g(\mathbf{x}) = 1$ and calculate $S = S_{f1} \cup S_{g1}$;
- $f(\mathbf{x}) \wedge g(\mathbf{x}) = 0$: solve $f(\mathbf{x}) = 0$, $g(\mathbf{x}) = 0$ and calculate $S = S_{f0} \cup S_{g0}$;
- $f(\mathbf{x}) \vee g(\mathbf{x}) = 0$: solve $f(\mathbf{x}) = 0$, $g(\mathbf{x}) = 0$ and calculate $S = S_{f0} \cap S_{g0}$.

In all these cases both non-orthogonal and orthogonal TVLs can be used.

In the case of antivalence we use the well-known transformations and the corresponding set operations:

$$f(\mathbf{x}) \oplus g(\mathbf{x}) = 1 : \quad \text{transform into } f(\mathbf{x}) \wedge \overline{g(\mathbf{x})} \vee \overline{f(\mathbf{x})} \wedge g(\mathbf{x}) = 1 \,;$$

$$f(\mathbf{x}) \wedge \overline{g(\mathbf{x})} = 1 : \quad \text{solve } f(\mathbf{x}) = 1, \ g(\mathbf{x}) = 0, \quad \text{calculate } S_{10} = S_{f1} \cap S_{g0} \,;$$

$$\overline{f(\mathbf{x})} \wedge g(\mathbf{x}) = 1 : \quad \text{solve } f(\mathbf{x}) = 0, \ g(\mathbf{x}) = 1, \quad \text{calculate } S_{01} = S_{f0} \cap S_{g1} \,;$$

$$f(\mathbf{x}) \oplus g(\mathbf{x}) = 1 \quad \equiv \quad f(\mathbf{x}) \wedge \overline{g(\mathbf{x})} \vee f(\mathbf{x}) \wedge \overline{g(\mathbf{x})} = 1 : \text{calculate } S = S_{10} \cup S_{01} \,.$$

$$f(\mathbf{x}) \oplus g(\mathbf{x}) = 0 : \quad \text{transform into } f(\mathbf{x}) \wedge g(\mathbf{x}) \vee \overline{f(\mathbf{x})} \wedge \overline{g(\mathbf{x})} = 1 \,;$$

$$f(\mathbf{x}) \wedge g(\mathbf{x}) = 1 : \quad \text{solve } f(\mathbf{x}) = 1, \ g(\mathbf{x}) = 1, \quad \text{calculate } S_{11} = S_{f1} \cap S_{g1} \,;$$

$$\overline{f(\mathbf{x})} \wedge \overline{g(\mathbf{x})} = 1 : \quad \text{solve } f(\mathbf{x}) = 0, \ g(\mathbf{x}) = 0, \quad \text{calculate } S_{00} = S_{f0} \cap S_{g0} \,;$$

$$f(\mathbf{x}) \oplus g(\mathbf{x}) = 0 \quad \equiv \quad f(\mathbf{x}) \wedge g(\mathbf{x}) \vee \overline{f(\mathbf{x})} \wedge \overline{g(\mathbf{x})} = 1 : \text{calculate } S = S_{11} \cup S_{00} \,.$$

Homogenous equations of two logic functions connected by the equivalence do not require special consideration because they can be solved by means of $\oplus$:

$$f(\mathbf{x}) \odot g(\mathbf{x}) = 1 \text{ has the same solutions as } f(\mathbf{x}) \oplus g(\mathbf{x}) = 0 \,,$$

$$f(\mathbf{x}) \odot g(\mathbf{x}) = 0 \text{ has the same solutions as } f(\mathbf{x}) \oplus g(\mathbf{x}) = 1 \,.$$

The implication $f(\mathbf{x}) \rightarrow g(\mathbf{x})$ can be expressed using the disjunction and the negation:

$$f(\mathbf{x}) \rightarrow g(\mathbf{x}) \quad \equiv \quad \overline{f(\mathbf{x})} \vee g(\mathbf{x}) \,.$$

Hence,

$$f(\mathbf{x}) \rightarrow g(\mathbf{x}) = 1 \text{ has the same solutions as } \overline{f(\mathbf{x})} \vee g(\mathbf{x}) = 1 \,,$$

$$f(\mathbf{x}) \rightarrow g(\mathbf{x}) = 0 \text{ has the same solutions as } \overline{f(\mathbf{x})} \vee g(\mathbf{x}) = 0 \,,$$

and the equivalent equations can be solved as before.

Example 5.6 In order to give a detailed example for the application of set-theoretic operations using lists (sets) of ternary vectors as the only data structure, we use the function

$$f(\mathbf{x}) = [x_1 x_2 \oplus (x_2 \odot x_3)] \rightarrow [(x_1 \vee x_2) \odot x_2 x_3]$$

and solve the characteristic equation $f(\mathbf{x}) = 1$ step by step.

We start with the left bracket $[x_1 x_2 \oplus (x_2 \odot x_3)]$ and get the lists

$$S_1 = \begin{array}{ccc} x_1 & x_2 & x_3 \\ \hline 1 & 1 & - \end{array} \,, \quad S_2 = \begin{array}{ccc} x_1 & x_2 & x_3 \\ \hline - & 0 & 0 \\ - & 1 & 1 \end{array}$$

for $x_1 x_2 = 1$ and $(x_2 \odot x_3) = 1$. The antivalence can be transferred to the sets by means of the *symmetric difference* $\triangle$:

$$S_1 \triangle S_2 = (\overline{S}_1 \cap S_2) \cup (S_1 \cap \overline{S}_2) \,.$$

Hence, we continue with

$$\overline{S}_1 = \begin{array}{ccc} x_1 & x_2 & x_3 \\ \hline 0 & - & - \\ 1 & 0 & - \end{array} \,, \quad \overline{S}_2 = \begin{array}{ccc} x_1 & x_2 & x_3 \\ \hline - & 1 & 0 \\ - & 0 & 1 \end{array} \,,$$

get the intermediate results:

$$S_3 = \overline{S}_1 \cap S_2 = \begin{array}{ccc} x_1 & x_2 & x_3 \\ \hline 0 & 0 & 0 \\ 0 & 1 & 1 \\ 1 & 0 & 0 \end{array} \,, \quad S_4 = S_1 \cap \overline{S}_2 = \begin{array}{ccc} x_1 & x_2 & x_3 \\ \hline 1 & 1 & 0 \end{array} \,,$$

and finally we have

$$S_5 = S_1 \triangle S_2 = S_3 \cup S_4 = \begin{array}{ccc} x_1 & x_2 & x_3 \\ \hline 0 & 0 & 0 \\ 0 & 1 & 1 \\ 1 & 0 & 0 \\ 1 & 1 & 0 \end{array} = \begin{array}{ccc} x_1 & x_2 & x_3 \\ \hline 0 & 0 & 0 \\ 0 & 1 & 1 \\ 1 & - & 0 \end{array}$$

as the solution of $x_1 x_2 \oplus (x_2 \odot x_3) = 1$.

For the right bracket $[(x_1 \lor x_2) \odot x_2 x_3]$, we find in the same way the TVLs for $x_1 \lor x_2 = 1$ and $x_2 x_3 = 1$:

$$S_6 = \begin{array}{ccc} x_1 & x_2 & x_3 \\ \hline 1 & - & - \\ 0 & 1 & - \end{array} \, , \qquad\qquad S_7 = \begin{array}{ccc} x_1 & x_2 & x_3 \\ \hline - & 1 & 1 \end{array} \, ,$$

$$\overline{S}_6 = \begin{array}{ccc} x_1 & x_2 & x_3 \\ \hline 0 & 0 & - \end{array} \, , \qquad\qquad \overline{S}_7 = \begin{array}{ccc} x_1 & x_2 & x_3 \\ \hline - & 0 & - \\ - & 1 & 0 \end{array} \, .$$

For the equivalence, the *complement of the symmetric difference* $\overline{\triangle}$ will be used:

$$S_6 \,\overline{\triangle}\, S_7 = (S_6 \cap S_7) \cup (\overline{S}_6 \cap \overline{S}_7) \, ,$$

and we get

$$S_8 = S_6 \,\overline{\triangle}\, S_7 = \begin{array}{ccc} x_1 & x_2 & x_3 \\ \hline - & 1 & 1 \end{array} \cup \begin{array}{ccc} x_1 & x_2 & x_3 \\ \hline 0 & 0 & - \end{array} = \begin{array}{ccc} x_1 & x_2 & x_3 \\ \hline - & 1 & 1 \\ 0 & 0 & - \end{array}$$

as solution-set of $[(x_1 \lor x_2) \odot x_2 x_3] = 1$.

For the implication $S_5 \to S_8$, the identity $\overline{S}_5 \cup S_8$ can be used, and we get finally

$$\overline{S}_5 = \begin{array}{ccc} x_1 & x_2 & x_3 \\ \hline 0 & 0 & 1 \\ 0 & 1 & 0 \\ 1 & - & 1 \end{array}$$

$$S = S_5 \to S_8 = \overline{S}_5 \cup S_8 = \begin{array}{ccc} x_1 & x_2 & x_3 \\ \hline 0 & 0 & 1 \\ 0 & 1 & 0 \\ 1 & - & 1 \end{array} \cup \begin{array}{ccc} x_1 & x_2 & x_3 \\ \hline - & 1 & 1 \\ 0 & 0 & - \end{array} = \begin{array}{ccc} x_1 & x_2 & x_3 \\ \hline 0 & 0 & 1 \\ 0 & 1 & 0 \\ 1 & - & 1 \\ 0 & 1 & 1 \\ 0 & 0 & 0 \end{array} = \begin{array}{ccc} x_1 & x_2 & x_3 \\ \hline 1 & - & 1 \\ 0 & - & - \end{array} \, .$$

The TVL S represents the six binary vectors that are the solution of the characteristic equation

$$[x_1 x_2 \oplus (x_2 \odot x_3)] \to [(x_1 \lor x_2) \odot x_2 x_3] = 1 \, .$$

The two solutions of the restrictive equation

$$[x_1 x_2 \oplus (x_2 \odot x_3)] \to [(x_1 \lor x_2) \odot x_2 x_3] = 0$$

are the complement of this solution ($\overline{S}$) and can be expressed by a single ternary vector:

$$\overline{S} = \begin{array}{ccc} x_1 & x_2 & x_3 \\ \hline 1 & - & 0 \end{array} \, .$$

5.6 Special Equation-Systems

In this section two special systems of logic equations will be considered. It could be seen already that each equation-system can be replaced by one equation with the same solution-set. However, these systems will be a good example that the transformation into one single equation can sometimes be much more complicated than a direct numerical approach.

We consider two different formats: the conjunctions of variables $x_1, \ldots, x_n$ and coefficients $a_1, \ldots, a_n$ are combined by $\vee$ or by $\oplus$.

Disjunctive Equations

Let be given the following system of equations:

$$a_{11}x_1 \vee \cdots \vee a_{1n}x_n = b_1$$
$$a_{21}x_1 \vee \cdots \vee a_{2n}x_n = b_2$$
$$\vdots \qquad\qquad \vdots \qquad \vdots$$
$$a_{m1}x_1 \vee \cdots \vee a_{mn}x_n = b_m$$

for $a_{ij}, b_i \in \mathbb{B}, i = 1, \ldots, m, j = 1, \ldots, n$.

The normal understanding of such a system is as follows: the coefficients a_{ij} are given constants; any $\mathbf{x} = (x_1, \ldots, x_n) \in \mathbb{B}^n$ that satisfies all the equations is a solution. An analytical exploration is given in the following example starting necessary assignments due to the first equation:

Example 5.7 Let be given the following system with $n = 5$ variables and $m = 3$ equations:

$$x_1 \qquad \vee x_3 \qquad \vee x_5 = 0$$
$$x_1 \vee x_2 \qquad \vee x_4 \qquad = 1$$
$$x_1 \qquad \vee x_3 \qquad\qquad = 1 .$$

According to the explanations above, we conclude from the first equation immediately

$$x_1 = 0, \quad x_3 = 0, \quad x_5 = 0 .$$

This satisfies the first and simplifies the second equation to $x_2 \vee x_4 = 1$. The third equation, however, results in a contradiction: the left-hand side is necessarily equal to 0, whereas the right-hand side equal to 1; hence, the solution-set of this system of logic equations is empty.

We change the system of equations and consider the following system:

$$x_1 \qquad \vee x_3 \qquad \vee x_5 = 0$$
$$x_1 \vee x_2 \qquad \vee x_4 \qquad = 1$$
$$x_1 \qquad \vee x_4 \qquad = 1 .$$

We get as before $x_1 = 0$, $x_3 = 0$, $x_5 = 0$, and the simplified system:

$$x_2 \vee x_4 = 1$$
$$x_4 = 1$$

gives immediately $x_4 = 1$ and $x_2 \vee 1 = 1$, i.e., $x_2 \in \{0, 1\}$. As a whole, the second system of equations is solved by

$$S = \frac{x_1 \ x_2 \ x_3 \ x_4 \ x_5}{0 \ - \ 0 \ 1 \ 0} \ .$$

Naturally this is only a small example; however, programs for these types of equations are not complicated. Hence, simple programs can solve such systems of equations for (more or less) "any" size.

We can also code the solution-sets of the single equations (because they are single disjunctions) and intersect these sets as before. This is especially useful when not only variables but also negated variables are allowed.

Example 5.8 Let be given the following system with $n = 5$ variables and $m = 3$ equations:

$$
\begin{aligned}
x_1 \vee \overline{x}_2 \qquad\qquad\quad \vee x_5 &= 0 \\
x_1 \qquad \vee x_3 \vee \overline{x}_4 \qquad &= 1 \\
x_2 \vee \overline{x}_3 \qquad\qquad &= 1 \ .
\end{aligned}
$$

Then, we get three intermediate solution-sets to be intersected:

$$
\frac{x_1 \ x_2 \ x_3 \ x_4 \ x_5}{0 \ 1 \ - \ - \ 0} \cap
\frac{x_1 \ x_2 \ x_3 \ x_4 \ x_5}{\begin{array}{ccccc} 1 & - & - & - & - \\ 0 & - & 1 & - & - \\ 0 & - & 0 & 0 & - \end{array}} \cap
\frac{x_1 \ x_2 \ x_3 \ x_4 \ x_5}{\begin{array}{ccccc} - & 1 & - & - & - \\ - & 0 & 0 & - & - \end{array}}
$$

with the final solution:

$$S = \frac{x_1 \ x_2 \ x_3 \ x_4 \ x_5}{\begin{array}{ccccc} 0 & 1 & 1 & - & 0 \\ 0 & 1 & 0 & 0 & 0 \end{array}} \ .$$

An analogous system of inequalities can be handled in the same way. The condition ≤ 0 on the right-hand side requires that all literals on the left-hand side are equal to 0; the condition ≤ 1 is always satisfied, and it is no restriction. Let us see the following example immediately.

Example 5.9 Let be given the following system with $n = 5$ variables and $m = 3$ inequalities:

$$
\begin{aligned}
x_1 \qquad \vee x_3 \qquad \vee x_5 &\leq 1 \\
x_1 \vee x_2 \qquad \vee x_4 \qquad &\leq 1 \\
x_1 \qquad \vee x_3 \qquad &\leq 0 \ .
\end{aligned}
$$

This results in $x_1 = x_3 = 0$ and $x_5 \leq 1$, $x_2 \vee x_4 \leq 1$. x_2, x_4, x_5 can have any values, and the whole solution-set is given by

$$S = \frac{x_1 \ x_2 \ x_3 \ x_4 \ x_5}{0 \ - \ 0 \ - \ -} \ .$$

The transformation $a \leq b \equiv a\overline{b} = 0$ is applicable as well:

$$x_1 \vee x_3 \leq 0 \; \equiv \; (x_1 \vee x_3)\overline{0} = 0 \; \equiv \; (x_1 \vee x_3)1 = 0 \; \equiv \; x_1 \vee x_3 = 0 \,,$$

leading to $x_1 = 0, x_3 = 0$;

$$x_2 \vee x_4 \leq 1 \; \equiv \; (x_2 \vee x_4)\overline{1} = 0 \; \equiv \; (x_2 \vee x_4)0 = 0 \,,$$

this equation is identically satisfied for any values of x_2 and x_4, etc.

Inequalities with $\geq$ can be transformed into equations using:

$$a \geq b \; \equiv \; a \vee \overline{b} = 1 \,.$$

This leads for an inequality with the value 1 of the right-hand side to the transformation:

$$x_1 \vee x_3 \vee x_5 \geq 1$$
$$\equiv \; x_1 \vee x_3 \vee x_5 \vee \overline{1} = 1 \; \equiv \; x_1 \vee x_3 \vee x_5 \vee 0 = 1 \; \equiv \; x_1 \vee x_3 \vee x_5 = 1 \,,$$

so that at least one of the three variables must be equal to 1. In case of a value 0 on the right-hand side we get, for example,

$$x_1 \vee x_3 \geq 0 \; \equiv \; x_1 \vee x_3 \vee \overline{0} = 1 \; \equiv \; x_1 \vee x_3 \vee 1 = 1 \; \equiv \; 1 = 1 \,,$$

so that any assigned value to the variables x_1 and x_3 satisfies the given inequality. A case like this mostly will be excluded in the stage of finding the conditions because "any values ..." does not restrict the set of solutions; hence, the inequality has no special meaning in this context.

Linear Antivalence Equations
These equations can be understood as strictly linear (in the sense of *Boolean Rings*), as follows:

$$
\begin{aligned}
a_{11}x_1 \oplus \cdots \oplus a_{1n}x_n &= b_1 \\
a_{21}x_1 \oplus \cdots \oplus a_{2n}x_n &= b_2 \\
\vdots \qquad\qquad \vdots \qquad \vdots & \\
a_{m1}x_1 \oplus \cdots \oplus a_{mn}x_n &= b_m
\end{aligned}
$$

for $a_{ij}, b_i \in \mathbb{B}, i = 1, \ldots, m, j = 1, \ldots, n$.

This system can be approached with a purely combinatorial methodology:

- for $b_i = 1$, the number of variables equal to 1 on the left-hand side must be an odd number (1, 3, 5, ...);
- for $b_i = 0$, the number of variables equal to 1 on the left-hand side must be an even number (0, 2, 4, ...).

However, this method is rather complex and not easy to handle. Very often the solution consists of single bit vectors, and no solution intervals can be found. It is more elegant to use a method that in the "normal" linear Algebra carries the name of C. F. Gauß.

It is easy to see that if $a = b$ and $c = d$, then $a \oplus c = b \oplus d$. That means that if there are two equations, then the left-hand sides and the right-hand sides can be "added" (i.e., combined by $\oplus$), and the equality still holds. The two systems

$$a = b \qquad \text{and} \qquad a \qquad = b$$
$$c = d \qquad\qquad\qquad a \oplus c = b \oplus d$$

have the same solutions. The second equation has been replaced by the "sum" of the first and the second equation.

Example 5.10 Let be given the system of equations as follows:

$$x_1 \oplus x_2 \oplus x_3 \oplus x_4 = 0$$
$$x_1 \oplus \qquad x_3 \oplus x_4 = 1$$
$$x_2 \oplus x_3 \oplus x_4 = 0$$
$$x_3 \oplus x_4 = 1 .$$

The solution of this system of equations can be found as follows:

- keep the first equation without changes:

$$x_1 \oplus x_2 \oplus x_3 \oplus x_4 = 0 ;$$

- replace the second equation with the $\oplus$ of the first and the second equation:

$$(x_1 \oplus x_2 \oplus x_3 \oplus x_4) \oplus (x_1 \oplus x_3 \oplus x_4) = 0 \oplus 1,$$

which results in $\qquad\qquad\qquad x_2 = 1 ;$
- this value can be substituted in the third equation:

$$1 \oplus x_3 \oplus x_4 = 0 ,$$

which is equivalent to $\qquad\qquad x_3 \oplus x_4 = 1 .$

This equation can be omitted, since it is equal to the fourth equation. Hence, we have the following equivalent system of equations:

$$x_1 \oplus x_2 \oplus x_3 \oplus x_4 = 0$$
$$x_2 \qquad\qquad = 1$$
$$x_3 \oplus x_4 = 1 .$$

This system has now a very characteristic format. There are no variables below the main diagonal, and the system has a triangular shape. We have three equations and four variables, and both $x_4 = 0$ and $x_4 = 1$ are possible. The values for the other variables will then be calculated from the last equation back to the first:

$$x_4 = 0 \;\rightarrow\; x_3 = 1, \; x_2 = 1 \;\rightarrow\; x_1 = 0 ;$$
$$x_4 = 1 \;\rightarrow\; x_3 = 0, \; x_2 = 1 \;\rightarrow\; x_1 = 0 .$$

Linear Equivalence Equations
Logic equations of this type do not need a special consideration, and they can be translated into equivalent antivalence equations:

$$a \odot b = 1 \quad \equiv \quad 1 \oplus a \oplus b = 1 \quad \equiv \quad a \oplus b = 0 .$$

After this transformation, the system of equivalence equations can be solved as above.

5.7 Exercises

Prepare for each task of each exercise of this chapter a PRP for the XBOOLE-monitor XBM 2. Prefer a lexicographic order of the variables x_i. Use the help system of the XBOOLE-monitor XBM 2 to learn the details of the needed commands. Execute the created PRPs, verify the computed results, and explain the observed properties.

If you have not yet prepared the XBOOLE-monitor XBM 2 on your computer, you can get this XBOOLE-monitor free of charge by means of the following three steps:

1. **Download**:
   There are four versions of the XBOOLE-monitor XBM 2, two for Windows 10 or subsequent Windows systems (32 or 64 bits) and two for LINUX–Ubuntu (also 32 or 64 bits); you must download the version of the XBOOLE-monitor XBM 2 that fits to your operating system.
   Authorized users of the online version of this chapter (https://doi.org/10.1007/978-3-030-88945-6_5) can download the XBOOLE-monitor XBM 2 directly from the web page

   https://link.springer.com/chapter/10.1007/978-3-030-88945-6_5

   where the links for the download of the XBOOLE-monitor XBM 2 are located in the part "Supplementary Information" (below the part "Abstract"). The headline above such a link indicates the associated zip-file of the XBOOLE-monitor XBM 2. The sizes of the zip-files have been provided behind the links and can be used to verify the download. A click on the link of the wanted version of the XBOOLE-monitor XBM 2 starts the download.
   Readers of the hardcopy of this book get access to the XBOOLE-monitor XBM 2 using the URL

   https://link.springer.com/chapter/10.1007/978-3-030-88945-6_5

   to download the first two pages of this chapter. After this download, the same procedure as the authorized users of the online version of a chapter can be used to download the wanted version of the XBOOLE-monitor XBM 2.

2. **Unzip**: The XBOOLE-monitor XBM 2 must not be installed but must be unzipped into an arbitrary directory of your computer. A convenient tool for unzipping the downloaded zip-file is usually available as part of the operating system or can be downloaded from the Internet.

3. **Execute**:
   - Windows:
     The executable file of the two versions (32 or 64 bits) for Windows 10 (or subsequent Windows systems) of the XBOOLE-monitor XBM 2 is XBM2.exe; the other files in the expanded directory must remain unchanged. A double-click on the executable file XBM2.exe within the Explorer of Windows starts the XBOOLE-monitor XBM 2.
   - LINUX–Ubuntu:
     The unzipped folder of the XBOOLE-monitor XBM 2 contains for this operating system only the executable file XBM2-i386.AppImage for the version of 32 bits or XBM2_x86_64.AppImage for the version of 64 bits of the XBOOLE-monitor XBM 2. A double-click on the created AppImage-file within the file manager of LINUX–Ubuntu starts the XBOOLE-monitor XBM 2.

Exercise 5.1 (Logic Functions and Equations)

Use in this exercise the logic functions:

$$f(\mathbf{x}) = x_1 x_2 \vee \overline{x}_2 \overline{x}_3 x_4 \qquad \text{and}$$

$$g(\mathbf{x}) = x_1 \overline{x}_4 \vee x_2 \overline{x}_3 x_4$$

to explore the relations between logic functions and equations.

(a) Use the command $\mathtt{sbe}$ for both the input of the logic function $f(\mathbf{x})$ and the solution of the logic equation $f(\mathbf{x}) = 1$. Show the Karnaugh-map of the logic function $f(\mathbf{x})$ and the Karnaugh-map in which the values 1 indicated the binary vectors of the solution-set of the logic equation $f(\mathbf{x}) = 1$ in the first row of the m-fold view. Explain your observations.

(b) Use the command $\mathtt{sbe}$ for both the input of the logic function $\overline{f(\mathbf{x})}$ and the solution of the logic equation $f(\mathbf{x}) = 0$. Show the Karnaugh-map of the logic function $\overline{f(\mathbf{x})}$ and the Karnaugh-map in which the values 1 indicated the binary vectors of the solution-set of the logic equation $f(\mathbf{x}) = 0$ in the first row of the m-fold view. Explain your observations.

(c) Use the command $\mathtt{sbe}$ to compute:

  • the solution-set S of the logic equation $f(\mathbf{x}) = g(\mathbf{x})$;
  • the solution-set S_{f1} of the logic equation $f(\mathbf{x}) = 1$;
  • the solution-set S_{g1} of the logic equation $g(\mathbf{x}) = 1$;
  • the solution-set S_{f0} of the logic equation $f(\mathbf{x}) = 0$; and
  • the solution-set S_{g0} of the logic equation $g(\mathbf{x}) = 0$.

Compute thereafter the set

$$S' = (S_{f1} \cap S_{g1}) \cup (S_{f0} \cap S_{g0})$$

and compare the sets S and S'. Show all computed sets as Karnaugh-maps using a convenient assignment to the viewports of the m-fold view.

(d) Use the command $\mathtt{sbe}$ to compute:

  • the logic function $h(\mathbf{x}) = f(\mathbf{x}) \odot g(\mathbf{x})$ represented as ODA-form and
  • the solution-set S of the logic equation $f(\mathbf{x}) = g(\mathbf{x})$

and compare the Karnaugh-maps of the logic function $h(\mathbf{x})$ and the set of solutions S.

Exercise 5.2 (Equivalent Transformations of Logic Equations)

We use in this exercise the logic functions:

$$f(\mathbf{x}) = x_2 x_3 \vee \overline{x}_1 x_4 \,,$$

$$g(\mathbf{x}) = x_3 \overline{x}_4 \vee x_1 x_2 \,, \qquad \text{and}$$

$$h(\mathbf{x}) = x_1 \oplus (x_2 \vee \overline{x}_3 x_4)$$

to verify that the equivalent transformations of Theorem 5.1 does not change the set of solutions of the logic equation $f(\mathbf{x}) = g(\mathbf{x})$.

(a) Modify the expression of:

$$f(\mathbf{x}) \quad \text{into} \quad f_m(\mathbf{x}) = x_2 x_3 \oplus \overline{x}_1 x_4 \oplus \overline{x}_1 x_2 x_3 x_4 \qquad \text{and}$$

$$g(\mathbf{x}) \quad \text{into} \quad g_m(\mathbf{x}) = x_3 \overline{x}_4 \oplus x_1 x_2 \oplus x_1 x_2 x_3 \overline{x}_4$$

and verify that the logic equations $f(\mathbf{x}) = g(\mathbf{x})$ and $f_m(\mathbf{x}) = g_m(\mathbf{x})$ have the same solution-set.
(b) Verify that the logic equations $f(\mathbf{x}) = g(\mathbf{x})$ and $\overline{f(\mathbf{x})} = \overline{g(\mathbf{x})}$ have the same solution-set.
(c) Verify that the logic equations $f(\mathbf{x}) = g(\mathbf{x})$ and $f(\mathbf{x}) \oplus h(\mathbf{x}) = g(\mathbf{x}) \oplus h(\mathbf{x})$ have the same solution-set.
(d) Verify that the logic equations $f(\mathbf{x}) = g(\mathbf{x})$ and $f(\mathbf{x}) \odot h(\mathbf{x}) = g(\mathbf{x}) \odot h(\mathbf{x})$ have the same solution-set.

Show for each task the Karnaugh-maps in which the values 1 indicated the binary vectors of the solution-sets of the two solved equations in the first row of the m-fold view.

Exercise 5.3 (Systems of Logic Equations)
Explore different approaches to solve the following system of logic equations:

$$\overline{x}_1 \vee \overline{x}_2 = x_3 \oplus x_4$$

$$x_2 \vee x_4 = x_1 \vee \overline{x}_2 \vee x_3$$

$$x_3 \vee x_4 = x_1 x_2 \oplus \overline{x}_3$$

and verify that the differently computed solution-sets contain the same binary vectors. Use the approach of the following task (a) for comparison and show the solution-sets as Karnaugh-maps.

(a) Use three commands sbe to solve each of the logic equations separately and compute the solution-set of the system of equations as intersection of these three solution-sets.
(b) Utilize the possibility that the command sbe solves a system of equations when each pair of consecutive logic equations is separated by a comma.
(c) Transform the given system of logic functions into a homogenous characteristic equation and solve this equation using a single command sbe.
(d) Transform the given system of logic functions into a homogenous restrictive equation and solve this equation using a single command sbe.

5.8 Solutions

Solution 5.1 (Logic Functions and Equations)

(a) Figure 5.1 shows on the left-hand side the PRP that computes the ODA-form of the function $f(\mathbf{x})$ (TVL 1) and solves the equation $f(\mathbf{x}) = 1$ (TVL 2); the Karnaugh-maps of these two TVLs are shown on the right-hand side.
   The command sbe in lines 5 and 6 creates the ODA-form of the given expression of the function $f(\mathbf{x})$. The argument line 8 specifies the logic equation $f(\mathbf{x}) = 1$ that is solved by the second command sbe in line 7.

Fig. 5.1
Problem-program that computes ODA-form of the function $f(\mathbf{x})$, solves the equation $f(\mathbf{x}) = 1$, and compares the results together with the related Karnaugh-maps

```
1   new
2   space  32  1
3   avar  1  1
4   x1  x2  x3  x4 .
5   sbe  1  1
6   x1&x2+/x2&/x3&x4 .
7   sbe  1  2
8   x1&x2+/x2&/x3&x4 = 1 .
9   set $equal (te_syd 1 2)
10  assign 1 /m 1 1
11  assign 2 /m 1 2
```

T	TVL (1) ODA	4 Var.
0 0	0 0 1 0
0 1	1 0 1 1
1 1	0 0 1 0
1 0	0 0 1 0
x4 x3	0 1 1 0 x2
	0 0 1 1 x1

T	TVL (2) ODA	4 Var.
0 0	0 0 1 0
0 1	1 0 1 1
1 1	0 0 1 0
1 0	0 0 1 0
x4 x3	0 1 1 0 x2
	0 0 1 1 x1

Fig. 5.2 Problem-program that computes ODA-form of the function $\overline{f(\mathbf{x})}$, solves the equation $f(\mathbf{x}) = 0$, and compares the results together with the related Karnaugh-maps

```
1   new
2   space 32 1
3   avar 1 1
4   x1 x2 x3 x4.
5   sbe 1 1
6   /(x1&x2+/x2&/x3&x4).
7   sbe 1 2
8   x1&x2+/x2&/x3&x4=0.
9   set $equal (te_syd 1 2)
10  assign 1 /m 1 1
11  assign 2 /m 1 2
```

T	TVL (1) ODA	4 Var.
0 0	1 1 0 1
0 1	0 1 0 0
1 1	1 1 0 1
1 0	1 1 0 1
x3 x4	0 1 1 0 x2
	0 0 1 1 x1

T	TVL (2) ODA	4 Var.
0 0	1 1 0 1
0 1	0 1 0 0
1 1	1 1 0 1
1 0	1 1 0 1
x3 x4	0 1 1 0 x2
	0 0 1 1 x1

Both the result $equal = true of the commands in line 9 and the comparison of the related Karnaugh-maps of Fig. 5.1 confirm that TVLs 1 and 2 are equal to each other; hence, these two TVLs describe the same set of binary vectors.

A conjunction of an expression of a logic function in disjunctive form is represented by a ternary vector in the associated TVL in ODA-form. The substitution of the values 0 and 1 of this ternary vector into the associated conjunction leads to a value 1 of this conjunction; hence, this ternary vector belongs to the solution-set of the equation $f(\mathbf{x}) = 1$ due to the value 1 on the right-hand side of this equation and the disjunctive form of the expression on the left-hand side.

(b) Figure 5.2 shows on the left-hand side the PRP that computes the ODA-form of the function $\overline{f(\mathbf{x})}$ (TVL 1) and solves the equation $f(\mathbf{x}) = 0$ (TVL 2); the Karnaugh-maps of these two TVLs are shown on the right-hand side.

The command sbe in lines 5 and 6 creates the ODA-form of the function $\overline{f(\mathbf{x})}$ using the given expression of the function $f(\mathbf{x})$ surrounded by parentheses and a negation. The argument line 8 specifies the logic equation $f(\mathbf{x}) = 0$ that is solved by the second command sbe in line 7.

Both the result $equal = true of the commands in line 9 and the comparison of the related Karnaugh-maps of Fig. 5.2 confirm that TVLs 1 and 2 are equal to each other; hence, these two TVLs describe the same set of binary vectors.

The negated function $\overline{f(\mathbf{x})}$ is equal to 1 for all binary vectors $\mathbf{x}$ that belong to the solution-set S of the logic equation $f(\mathbf{x}) = 0$. The negation of the function $f(\mathbf{x})$ and the change of the right-hand side of the logic equation from 1 to 0 have the same effect.

(c) Figure 5.3 shows in the upper part the PRP that solves the logic equation $f(\mathbf{x}) = g(\mathbf{x})$ and verifies whether the solution of $S' = (S_{f1} \cap S_{g1}) \cup (S_{f0} \cap S_{g0})$ based on all homogenous equations of the given functions leads to the same result. The Karnaugh-maps in the lower part of Fig. 5.3 show the detailed properties of the solution-sets of the solved equations.

The command sbe in lines 5–7 solves the logic equation $f(\mathbf{x}) = g(\mathbf{x})$ for the given functions $f(\mathbf{x})$ and $g(\mathbf{x})$ and stores the solution-set as TVL 1. The commands sbe in lines 8–15 solve all four homogenous equations of the functions $f(\mathbf{x})$ and $g(\mathbf{x})$ and store the associated solution-sets as follows:

- TVL 2 stores the solution-set S_{f1} of the equation $f(\mathbf{x}) = 1$;
- TVL 3 stores the solution-set S_{g1} of the equation $g(\mathbf{x}) = 1$;
- TVL 4 stores the solution-set S_{f0} of the equation $f(\mathbf{x}) = 0$;
- TVL 5 stores the solution-set S_{g0} of the equation $g(\mathbf{x}) = 0$.

The Karnaugh-maps of these four solution-sets are shown in the viewports of the first two rows and left two columns of the m-fold view.

The command isc in line 16 computes TVL 6 as intersection of the sets S_{f1} and S_{g1}; hence, values 1 in the Karnaugh-map of TVL 6 (see viewport (1,3)) indicate patterns $\mathbf{x}$ for which both the function $f(\mathbf{x})$ and the function $g(\mathbf{x})$ are equal to 1.

```
 1   new                              15   x1&/x4+x2&/x3&x4=0.
 2   space 32 1                       16   isc 2 3 6
 3   avar 1 1                         17   isc 4 5 7
 4   x1 x2 x3 x4.                     18   uni 6 7 8
 5   sbe 1 1                          19   set $equal (te_syd 1 8)
 6   x1&x2+/x2&/x3&x4=                20   assign 2 /m 1 1
 7   x1&/x4+x2&/x3&x4.                21   assign 3 /m 1 2
 8   sbe 1 2                          22   assign 4 /m 2 1
 9   x1&x2+/x2&/x3&x4=1.              23   assign 5 /m 2 2
10   sbe 1 3                          24   assign 6 /m 1 3
11   x1&/x4+x2&/x3&x4=1.              25   assign 7 /m 2 3
12   sbe 1 4                          26   assign 8 /m 3 3
13   x1&x2+/x2&/x3&x4=0.              27   assign 1 /m 3 2
14   sbe 1 5
```

T TVL (2) ODA | 4 Var. | 4 R. | S. 1

```
0 0   0 0 1 0
0 1   1 0 1 1
1 1   0 0 1 0
1 0   0 0 1 0
x3 x4   0 1 1 0 x2
        0 0 1 1 x1
```

T TVL (3) ODA | 4 Var. | 2 R. | S. 1

```
0 0   0 0 1 1
0 1   0 1 1 0
1 1   0 0 0 0
1 0   0 0 1 1
x3 x4   0 1 1 0 x2
        0 0 1 1 x1
```

T TVL (6) ODA | 4 Var. | 2 R. | S. 1

```
0 0   0 0 1 0
0 1   0 0 1 0
1 1   0 0 0 0
1 0   0 0 1 0
x3 x4   0 1 1 0 x2
        0 0 1 1 x1
```

T TVL (4) ODA | 4 Var. | 5 R. | S. 1

```
0 0   1 1 0 1
0 1   0 1 0 0
1 1   1 1 0 1
1 0   1 1 0 1
x3 x4   0 1 1 0 x2
        0 0 1 1 x1
```

T TVL (5) ODA | 4 Var. | 3 R. | S. 1

```
0 0   1 1 0 0
0 1   1 0 0 1
1 1   1 1 1 1
1 0   1 1 0 0
x3 x4   0 1 1 0 x2
        0 0 1 1 x1
```

T TVL (7) ODA | 4 Var. | 4 R. | S. 1

```
0 0   1 1 0 0
0 1   0 0 0 0
1 1   1 1 0 1
1 0   1 1 0 0
x3 x4   0 1 1 0 x2
        0 0 1 1 x1
```

T TVL (1) ODA | 4 Var. | 6 R. | S. 1

```
0 0   1 1 1 0
0 1   0 0 1 0
1 1   1 1 0 1
1 0   1 1 1 0
x3 x4   0 1 1 0 x2
        0 0 1 1 x1
```

T TVL (8) ODA | 4 Var. | 6 R. | S. 1

```
0 0   1 1 1 0
0 1   0 0 1 0
1 1   1 1 0 1
1 0   1 1 1 0
x3 x4   0 1 1 0 x2
        0 0 1 1 x1
```

Fig. 5.3 Problem-program and Karnaugh-maps that demonstrate the detailed properties of the solution of the logic equation $f(\mathbf{x}) = g(\mathbf{x})$

The command isc in line 17 computes TVL 7 as intersection of the sets S_{f0} and S_{g0}; hence, values 1 in the Karnaugh-map of TVL 7 (see viewport (2,3)) indicate patterns $\mathbf{x}$ for which both the function $f(\mathbf{x})$ and the function $g(\mathbf{x})$ are equal to 0.

The command uni in line 18 computes TVL 8 (see viewport (3,3)) as union of the two sets, which indicate that either both functions $f(\mathbf{x})$ and $g(\mathbf{x})$ are equal to 1 or these two functions are equal to 0.

The Karnaugh-map of the solution-set of the equation $f(\mathbf{x}) = g(\mathbf{x})$ is shown in viewport (3,2) of the m-fold view for easy comparison. Both the result $equal = true of the commands in line 19 and the comparison of the Karnaugh-maps in the third row of Fig. 5.3 confirm that TVLs 1 and 8 are equal to each other; hence, the solution-set of equation $f(\mathbf{x}) = g(\mathbf{x})$ consists of all

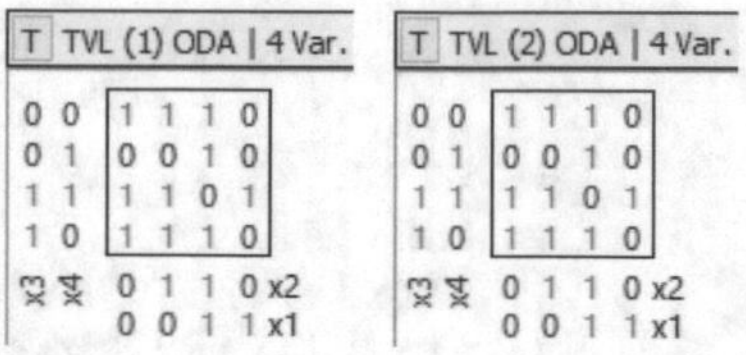

Fig. 5.4
Problem-program that
computes the function
$h(\mathbf{x}) = f(\mathbf{x}) \odot g(\mathbf{x})$
represented as ODA-form
and the solution-set S of
the logic equation
$f(\mathbf{x}) = g(\mathbf{x})$ together with
the associated
Karnaugh-maps

```
1   new
2   space 32 1
3   avar 1 1
4   x1 x2 x3 x4.
5   sbe 1 1
6   (x1&x2+/x2&/x3&x4)=
7   (x1&/x4+x2&/x3&x4).
8   sbe 1 2
9   x1&x2+/x2&/x3&x4=
10  x1&/x4+x2&/x3&x4.
11  set $equal (te_syd 1 2)
12  assign 1 /m 1 1
13  assign 2 /m 1 2
```

T	TVL (1) ODA | 4 Var.
0 0	1 1 1 0
0 1	0 0 1 0
1 1	1 1 0 1
1 0	1 1 1 0
	0 1 1 0 x2
	0 0 1 1 x1

T	TVL (2) ODA | 4 Var.
0 0	1 1 1 0
0 1	0 0 1 0
1 1	1 1 0 1
1 0	1 1 1 0
	0 1 1 0 x2
	0 0 1 1 x1

binary vectors for which either both functions $f(\mathbf{x})$ and $g(\mathbf{x})$ are equal to 1 or these two functions are equal to 0.

(d) Figure 5.4 shows on the left-hand side the PRP that computes the ODA-form of the function $h(\mathbf{x}) = f(\mathbf{x}) \odot g(\mathbf{x})$ (TVL 1) and solves the equation $f(\mathbf{x}) = g(\mathbf{x})$ (TVL 2); the Karnaugh-maps of these two TVLs are shown on the right-hand side.

The command sbe in lines 5–7 creates the ODA-form of the function $h(\mathbf{x}) = f(\mathbf{x}) \odot g(\mathbf{x})$ based on the given expressions of the functions $f(\mathbf{x})$ and $g(\mathbf{x})$; these functions are enclosed in parentheses for clearness. The logic operation $\odot$ is expressed in the command sbe by the sign =. The parentheses can be avoided because the equivalence operation ($\odot$, =) has the lowest priority of all logic operations.

Next, the command sbe in lines 8–10 solves the logic equation $f(\mathbf{x}) = g(\mathbf{x})$ for the given functions $f(\mathbf{x})$ and $g(\mathbf{x})$ and stores the solution-set as TVL 2. The sign = is used in this equation to separate the two functions $f(\mathbf{x})$ and $g(\mathbf{x})$ of the logic equation to solve. Parentheses around these two functions are not needed due to the lowest priority of the operation sign =.

The comparison of the two commands sbe shows that there is no difference (after removing the not needed parentheses) between these two commands; the used sign = has the same effect as logic operation $\odot$ and as the separation of the two functions $f(\mathbf{x})$ and $g(\mathbf{x})$ of the logic equation to solve, and only the context is different.

Both the result $equal = true of the commands in line 11 and the comparison of the two Karnaugh-maps of Fig. 5.4 confirm that TVLs 1 and 2 are equal to each other. The equation $h(\mathbf{x}) = 1$ has the same solution (as demonstrated in task 5.1(a)); hence, adding "=1" at the end of the expression of a command sbe does not change the solution-set. This is the theoretical background why the command sbe can be used to create the TVL in ODA-form of a logic function that is given by an expression of logic variables and operations.

Solution 5.2 (Equivalent Transformations of Logic Equations)

(a) Figure 5.5 shows on the left-hand side the PRP that solves the logic equations $f(\mathbf{x}) = g(\mathbf{x})$ (TVL 1) and $f_m(\mathbf{x}) = g_m(\mathbf{x})$ (TVL 2) and confirms by the Karnaugh-maps of these two TVLs (shown on the right-hand side) and the computed result $equal = true that these two equations have the same solution-set.

(b) Figure 5.6 shows on the left-hand side the PRP that solves the logic equations $f(\mathbf{x}) = g(\mathbf{x})$ (TVL 1) and $\overline{f(\mathbf{x})} = \overline{g(\mathbf{x})}$ (TVL 2) and confirms by the Karnaugh-maps of these two TVLs (shown on the right-hand side) and the computed result $equal = true that these two equations have the same solution-set.

(c) Figure 5.7 shows on the left-hand side the PRP that solves the logic equations $f(\mathbf{x}) = g(\mathbf{x})$ (TVL 1) and $f(\mathbf{x}) \oplus h(\mathbf{x}) = g(\mathbf{x}) \oplus h(\mathbf{x})$ (TVL 2) and confirms by the Karnaugh-maps of these two

Fig. 5.5
Problem-program and
computed equal
solution-sets of the
equations $f(\mathbf{x}) = g(\mathbf{x})$ and
$f_m(\mathbf{x}) = g_m(\mathbf{x})$

Fig. 5.6
Problem-program and
computed equal
solution-sets of the
equations $f(\mathbf{x}) = g(\mathbf{x})$ and
$\overline{f(\mathbf{x})} = \overline{g(\mathbf{x})}$

Fig. 5.7
Problem-program and
computed equal
solution-sets of the
equations $f(\mathbf{x}) = g(\mathbf{x})$ and
$f(\mathbf{x}) \oplus h(\mathbf{x}) = g(\mathbf{x}) \oplus h(\mathbf{x})$

Fig. 5.8
Problem-program and
computed equal
solution-sets of the
equations $f(\mathbf{x}) = g(\mathbf{x})$ and
$f(\mathbf{x}) \odot h(\mathbf{x}) = g(\mathbf{x}) \odot h(\mathbf{x})$

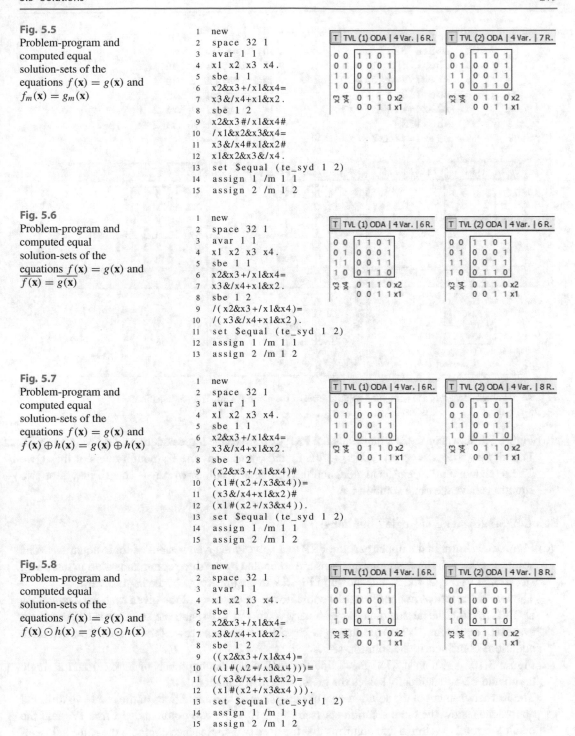

TVLs (shown on the right-hand side) and the computed result $equal = `true` that these two
equations have the same solution-set.

```
1   new                9    isc  1  2  4
2   space  32  1       10   isc  4  3  5
3   sbe  1  1          11   assign  1  /m  1  1
4   /x1+/x2=x3#x4 .     12   assign  2  /m  1  2
5   sbe  1  2          13   assign  3  /m  1  3
6   x2+x4=x1+/x2+x3 .   14   assign  4  /m  2  1
7   sbe  1  3          15   assign  5  /m  2  2
8   x3+x4=x1&x2#/x3 .
```

T TVL (1) ODA | 4 Var. | 6 R. | S. 1	T TVL (2) ODA | 4 Var. | 4 R. | S. 1	T TVL (3) ODA | 4 Var. | 4 R. | S. 1

0 0   0 0 1 0
 0 1   1 1 0 1
 1 1   0 0 1 0
 1 0   1 1 0 1
 	
 x3 x4   0 1 1 0 x2
       0 0 1 1 x1
	0 0 0 0 1 0
	 0 1 1 0 1 1
	 1 1 1 1 1 1
	 1 0 0 1 1 0
	
	 x3 x4 0 1 1 0 x2
	       0 0 1 1 x1
		0 0 0 0 1 0
		 0 1 1 1 0 1
		 1 1 0 0 1 0
		 1 0 0 0 1 0
		
		 x3 x4 0 1 1 0 x2
		       0 0 1 1 x1

T TVL (4) ODA | 4 Var. | 5 R. | S. 1	T TVL (5) ODA | 4 Var. | 4 R. | S. 1	-

0 0   0 0 1 0
 0 1   1 0 0 1
 1 1   0 0 1 0
 1 0   0 1 0 0
 	
 x3 x4   0 1 1 0 x2
       0 0 1 1 x1
	0 0 0 0 1 0
	 0 1 1 0 0 1
	 1 1 0 0 1 0
	 1 0 0 0 0 0
	
	 x3 x4 0 1 1 0 x2
	       0 0 1 1 x1	

Fig. 5.9 Problem-program and computed solution-sets of the given system of three equations

(d) Figure 5.8 shows on the left-hand side the PRP that solves the logic equations $f(\mathbf{x}) = g(\mathbf{x})$ (TVL 1) and $f(\mathbf{x}) \odot h(\mathbf{x}) = g(\mathbf{x}) \odot h(\mathbf{x})$ (TVL 2) and confirms by the Karnaugh-maps of these two TVLs (shown on the right-hand side) and the computed result $equal = true that these two equations have the same solution-set.

Solution 5.3 (Systems of Logic Equations)

(a) Figure 5.9 shows in the upper part the PRP that solves the given system of logic equations. The solution-sets of the three single equations are computed by the three commands sbe in lines 3–8, stored as TVLs 1, 2, and 3, and shown in Fig. 5.9 in the upper row of the m-fold view.

The two intersections in lines 9 and 10 compute the solution-set of the given system of equations as TVL 5. Both the intermediate set of binary vectors (TVL 4) and the solution-set of the given system of equations (TVL 5) are shown in Fig. 5.9 in the second row of the m-fold view. All three equations contribute to the solution-set.

(b) Figure 5.10 shows on the left-hand side the PRP using the approach of task (a) and a single command sbe that directly solves the given system of equations.

The Karnaugh-maps of the solution-sets of the system of logic equations of the used two different approaches show the same solution-set (see Fig. 5.10). The two commands in line 17 yield the result $equal = true that confirms that the used two approaches compute the same solution-set of the given system of logic equations.

(c) Figure 5.11 shows on the left-hand side the PRP using the approach of task (a) and a single command sbe in lines 12–15 that solves the homogenous characteristic equation

$$((\overline{x}_1 \vee \overline{x}_2) \odot (x_3 \oplus x_4)) \wedge ((x_2 \vee x_4) \odot (x_1 \vee \overline{x}_2 \vee x_3)) \wedge ((x_3 \vee x_4) \odot (x_1 x_2 \oplus \overline{x}_3)) = 1$$

Fig. 5.10
Problem-program that uses two approaches to computed solution-sets of the given system of three equations—the second approach utilizes the property of the command sbe to directly solve a system of logic equations

```
1   new
2   space 32  1
3   sbe  1  1
4   /x1+/x2=x3#x4.
5   sbe  1  2
6   x2+x4=x1+/x2+x3.
7   sbe  1  3
8   x3+x4=x1&x2#/x3.
9   isc  1  2  4
10  isc  4  3  5
11  assign  5  /m 1  1
12  sbe  1  10
13  /x1+/x2=x3#x4,
14  x2+x4=x1+/x2+x3,
15  x3+x4=x1&x2#/x3.
16  assign  10  /m 2  1
17  set $equal (te_syd 5 10)
```

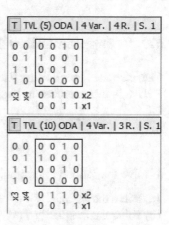

Fig. 5.11
Problem-program that uses two approaches to computed solution-sets of the given system of three equations—the second approach uses the command sbe to solve the homogenous characteristic equation of the given system of equations

```
1   new
2   space 32  1
3   sbe  1  1
4   /x1+/x2=x3#x4.
5   sbe  1  2
6   x2+x4=x1+/x2+x3.
7   sbe  1  3
8   x3+x4=x1&x2#/x3.
9   isc  1  2  4
10  isc  4  3  5
11  assign  5  /m 1  1
12  sbe  1  10
13  ((/x1+/x2)=(x3#x4))&
14  ((x2+x4)=(x1+/x2+x3))&
15  ((x3+x4)=(x1&x2#/x3))=1.
16  assign  10  /m 2  1
17  set $equal (te_syd 5 10)
```

of the given system of equations.

The Karnaugh-maps of the solution-sets of the system of logic equations of the used two different approaches show the same solution-set (see Fig. 5.11). The two commands in line 17 yield the result $equal = true that confirms that the used two approaches compute the same solution-set of the given system of logic equations.

(d) Figure 5.12 shows on the left-hand side the PRP using the approach of task (a) and a single command sbe in lines 12–15 that solves the homogenous restrictive equation

$$((x_1 \vee x_2) \oplus (x_3 \oplus x_4)) \vee ((x_2 \vee x_4) \oplus (x_1 \vee x_2 \vee x_3)) \vee ((x_3 \vee x_4) \oplus (x_1 x_2 \oplus \overline{x}_3)) = 0$$

of the given system of equations.

The Karnaugh-maps of the solution-sets of the system of logic equations of the used two different approaches show the same solution-set (see Fig. 5.12). The two commands in line 17 yield the result $equal = true that confirms that the used two approaches compute the same solution-set of the given system of logic equations.

Fig. 5.12
Problem-program that uses
two approaches to
computed solution-sets of
the given system of three
equations—the second
approach uses the
command sbe to solve the
homogenous restrictive
equation of the given
system of equations

```
 1   new
 2   space  32  1
 3   sbe  1  1
 4   /x1+/x2=x3#x4 .
 5   sbe  1  2
 6   x2+x4=x1+/x2+x3 .
 7   sbe  1  3
 8   x3+x4=x1&x2#/x3 .
 9   isc  1  2  4
10   isc  4  3  5
11   assign  5  /m  1  1
12   sbe  1  10
13   ((/x1+/x2)#(x3#x4))+
14   ((x2+x4)#(x1+/x2+x3))+
15   ((x3+x4)#(x1&x2#/x3))=0 .
16   assign  10  /m  2  1
17   set  $equal  (te_syd  5  10)
```

T	TVL (5) ODA | 4 Var. | 4 R. | S. 1

0 0	0 0 1 0
0 1	1 0 0 1
1 1	0 0 1 0
1 0	0 0 0 0
x3 x4	0 1 1 0 x2
	0 0 1 1 x1

T	TVL (10) ODA | 4 Var. | 3 R. | S. 1

0 0	0 0 1 0
0 1	1 0 0 1
1 1	0 0 1 0
1 0	0 0 0 0
x3 x4	0 1 1 0 x2
	0 0 1 1 x1

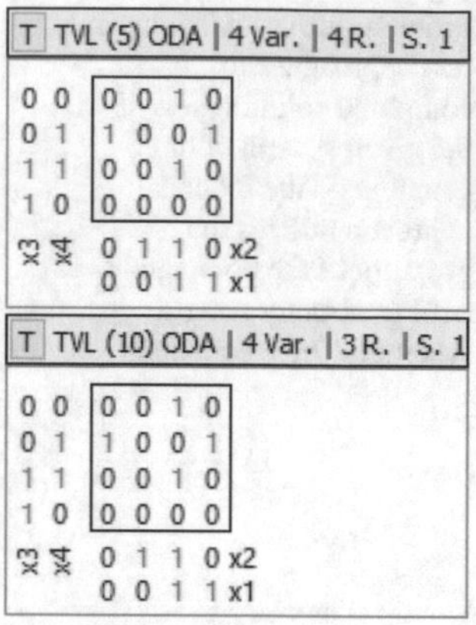

5.9   Supplementary Exercises

Exercise 5.4 (Inequalities Transformed into Homogenous Logic Equations)

Use in this exercise the logic functions:

$$f(\mathbf{x}) = x_1 \bar{x}_3 x_6 \vee \bar{x}_2 \bar{x}_3 x_4 \vee x_4 x_5 x_6 \vee x_1 \bar{x}_4 x_5 \qquad \text{and}$$

$$g(\mathbf{x}) = x_1 x_2 \vee \bar{x}_2 \bar{x}_3 x_4 \vee \bar{x}_5 x_6$$

to solve the following inequalities by means of equivalent homogenous logic equations:

(a) solve the inequality $f(\mathbf{x}) \leq g(\mathbf{x})$ transformed in a characteristic equation;
(b) solve the inequality $f(\mathbf{x}) \leq g(\mathbf{x})$ transformed in a restrictive equation;
(c) solve the inequality $f(\mathbf{x}) < g(\mathbf{x})$ transformed in a characteristic equation;
(d) solve the inequality $f(\mathbf{x}) < g(\mathbf{x})$ transformed in a restrictive equation;

and verify your results by comparing the Karnaugh-maps of $f(\mathbf{x}) = 1$, $g(\mathbf{x}) = 1$, and the solved homogenous equations.

Exercise 5.5 (Systems of Inequalities and Logic Equations)

Use in this exercise the logic functions:

$$f_1(\mathbf{x}) = x_1 \bar{x}_2 x_5 \vee x_3 x_4 \,,$$

$$f_2(\mathbf{x}) = x_2 \bar{x}_3 x_1 \vee x_4 x_5 \,,$$

$$f_3(\mathbf{x}) = x_3 \bar{x}_4 x_2 \vee x_5 x_1 \,, \qquad \text{and}$$

$$f_4(\mathbf{x}) = x_4 \bar{x}_5 x_3 \vee x_1 x_2$$

to solve the following systems of inequalities and logic equations:

(a) transform the inequality of the system:

$$f_1(\mathbf{x}) > f_2(\mathbf{x})$$

$$f_3(\mathbf{x}) = f_4(\mathbf{x})$$

into a characteristic equation and solve the created system of equations using a single command
sbe;
(b) transform the inequality of the system:

$$f_1(\mathbf{x}) = f_2(\mathbf{x})$$
$$f_3(\mathbf{x}) \geq f_4(\mathbf{x})$$

into a characteristic equation and solve the created system of equations using a single command
sbe;
(c) transform the inequality of the system:

$$f_1(\mathbf{x}) < f_2(\mathbf{x})$$
$$f_3(\mathbf{x}) = f_4(\mathbf{x})$$

into a restrictive equation, solve the two equations separately, and compute the solution-set using
the command isc;
(d) solve the equations $f_1(\mathbf{x}) = 1, \ldots, f_4(\mathbf{x}) = 1$ and use these solution-sets and set operations to
compute the solution-set of:

$$f_1(\mathbf{x}) \geq f_2(\mathbf{x})$$
$$f_3(\mathbf{x}) \rightarrow f_4(\mathbf{x}) = 1 .$$

Verify the computed results by comparing the Karnaugh-maps of the solution-sets of the inequality,
respectively, equation with the solution-set of the system.

References

1. D. Bochmann, B. Steinbach, *Logic Design with XBOOLE* (in German: Logikentwurf mit XBOOLE) (Verlag Technik GmbH, Berlin, Germany, 1991). ISBN:3-341-01006-8
2. F. Dresig et al., Programming with XBOOLE (in German: Programmieren mit XBOOLE), in *Series of Scientific Publications of the Chemnitz University of Technology* (in German: Wissenschaftliche Schriftenreihe der Technischen Universität Chemnitz) (1992), pp. 1–119. ISSN:0863-0755
3. B. Steinbach, XBOOLE—A toolbox for modelling, simulation, and analysis of large digital systems. Syst. Anal. Model. Simul. **9**(4), 297–312 (Sept. 1992). ISSN:0232-9298
4. B. Steinbach, C. Posthoff, *EAGLE Start-up Aid - Efficient Computations with XBOOLE* (in German: EAGLE Starthilfe - Effiziente Berechnungen mit XBOOLE) (Edition am Gutenbergplatz, Leipzig, Germany, 2015). ISBN:978-3-95922-081-1

Boolean Differential Calculus

6

Abstract

The *Boolean Differential Calculus* extends the Boolean Algebra by evaluating the change behavior of logic functions. Vectorial derivative operations explore the value change between pairs of function values reached by the change of an arbitrary number of variables. Different function values in these pairs are indicated by the vectorial derivative, unchanged values 1 by the vectorial minimum, and at least one value 1 by the vectorial maximum. A special case of the vectorial derivative operations are the single derivative operations where only a single variable changes its value. Repeated execution of the same type of a single derivative operation with regard to different variables results in k-fold derivative operations. These operations evaluate the function values of certain subspaces of the given function. Derivative operation can be efficiently calculated and has many applications in determining properties of logic functions, in circuit design, test, and others. The differential of a given Boolean variable is also a Boolean variable that is equal to 1 when the given variable changes its value. Logic equations that commonly use Boolean variables and the associated differentials can be used to describe the edges of a graph. Similarly to the differential of Boolean variables, differential operations of logic function are defined. Differential operations summarize associated derivative operations for several directions of change.

Supplementary Information The online version of this chapter (https://doi.org/10.1007/978-3-030-88945-6_6) contains supplementary material which is available for authorized users. Please, follow the link belonging to the version of the XBOOLE-monitor XBM 2 that fits best for your operating system. This XBOOLE-monitor is needed to solve all tasks of this chapter. Instructions for starting the downloaded XBOOLE-monitor XBM 2 are given at the beginning of Section 'Exercises' in this chapter.

XBOOLE-monitor XBM 2 for Windows 10
32 bits
https://doi.org/10.1007/978-3-030-88945-6_6_MOESM1_ESM.zip (15,091 KB)

64 bits
https://doi.org/10.1007/978-3-030-88945-6_6_MOESM2_ESM.zip (14,973 KB)

XBOOLE-monitor XBM 2 for Linux Ubuntu
32 bits
https://doi.org/10.1007/978-3-030-88945-6_6_MOESM3_ESM.zip (29,522 KB)

64 bits
https://doi.org/10.1007/978-3-030-88945-6_6_MOESM4_ESM.zip (28,422 KB)

© Springer Nature Switzerland AG 2022
B. Steinbach, C. Posthoff, *Logic Functions and Equations*,
https://doi.org/10.1007/978-3-030-88945-6_6

6.1   Preliminaries

The *Boolean Differential Calculus* is an interesting and useful extension of the concepts that have been used so far for $\mathbb{B}^n$. A monograph [1] of this calculus was published in German language. An early comprehensive English edition is [7]. Very recent publications with a huge amount of relationships and theorems open a wide area of applications of the Boolean Differential Calculus [4–6].

Always when changes of the values of Boolean variables or logic functions are important in the context of the problem to be solved, the Boolean Differential Calculus can help to find a solution. Another important application area is the test of digital systems where the key topic consists in changes by faults.

There are two basic understandings of the approach to capture the concepts of changing the values of variables and the resulting changes of the values of binary functions.

Traditionally in Mathematics the concepts of derivatives have been based on *limits of difference quotients*. For a real or complex function $y = f(x)$ the derivative is defined as

$$\frac{dy}{dx} = \lim_{\Delta x \to 0} \frac{f(x + \Delta x) - f(x)}{\Delta x} .$$

Here all the values of $f(x + \Delta x)$, $f(x)$, x, Δx and y are *real* or *complex numbers*, and there exists the *division* as well as an appropriate definition of the *limit*.

This way of thinking has been applied to logic variables and functions as well. The derivative of a logic function $y = f(x_i, \mathbf{x}_1)$ with regard to the single variable x_i, for example, has been written in the following way:

$$\frac{\partial f(x_i, \mathbf{x}_1)}{\partial x_i} = f(x_i, \mathbf{x}_1) \oplus f(\overline{x}_i, \mathbf{x}_1) . \tag{6.1}$$

It must be observed, however, that this is only a symbolic expression to emphasize the consideration of changing the values of x_i and $f(x_i, \mathbf{x}_1)$. The fraction bar has no definite meaning.

If we want to emphasize that we work with finite sets and without *division* and *limit*, then we can go back to the structure of a Boolean Ring as background and use the antivalence as addition as well as difference in order to consider the changing of values of variables and functions. Exactly spoken, we use the antivalence to calculate the *difference* of the two values $f(x_i, \mathbf{x}_1)$ and $f(\overline{x}_i, \mathbf{x}_1)$. This difference is:

- Equal to 0 for the pairs of function values $(0, 0)$ and $(1, 1)$
- Equal to 1 for the pairs of function values $(0, 1)$ and $(1, 0)$

Writing

$$\operatorname*{der}_{x_i} f(x_i, \mathbf{x}_1) = f(x_i, \mathbf{x}_1) \oplus f(\overline{x}_i, \mathbf{x}_1) \tag{6.2}$$

we describe the *derivative* of the function $f(x_i, \mathbf{x}_1)$ with regard to the single variable x_i without a fraction bar. However, the notation of (6.2) does not change the meaning of the notation (6.1) that has been used in previous publications like [1–6]. Hence,

$$\operatorname*{der}_{x_i} f(x_i, \mathbf{x}_1) = \frac{\partial f(x_i, \mathbf{x}_1)}{\partial x_i} = f(x_i, \mathbf{x}_1) \oplus f(\overline{x}_i, \mathbf{x}_1), \tag{6.3}$$

and we use in the following uniquely $\text{der}_{x_i} f(x_i, \mathbf{x}_1)$. This notation fits also very well to the already used notation of the minimum $\min_{x_i} f(x_i, \mathbf{x}_1)$ and the maximum $\max_{x_i} f(x_i, \mathbf{x}_1)$, which will be explained later.

6.2 Vectorial Derivative Operations

Vectorial derivative operations explore the value change between pairs of function values. These pairs are determined by the simultaneous value change of an arbitrary but fixed subset of variables $\mathbf{x}_0 \subseteq \mathbf{x}$. The chosen set of variables $\mathbf{x}_0$ decides on the *direction of change* explored by the vectorial derivative operation. There are $2^n - 1$ directions of change for a logic function of n variables.

Figure 6.1a shows the $2^3 - 1 = 7$ directions of change in the Boolean space $\mathbb{B}^3$ for the selected point $(x_1, x_2, x_3) = (100)$. Pairs of function values that are compared in vectorial derivative operations with regard to the simultaneous change of the variables x_1 and x_2 are connected by arrows in Fig. 6.1b.

Definition 6.1 Let $\mathbf{x}_0 = (x_1, x_2, \ldots, x_k)$, $\mathbf{x}_1 = (x_{k+1}, x_{k+2}, \ldots, x_n)$ be two disjoint sets of Boolean variables, and $f(\mathbf{x}_0, \mathbf{x}_1) = f(\mathbf{x})$ a logic function of n variables; then

$$\text{der}_{\mathbf{x}_0} f(\mathbf{x}_0, \mathbf{x}_1) = f(\mathbf{x}_0, \mathbf{x}_1) \oplus f(\overline{\mathbf{x}}_0, \mathbf{x}_1) \tag{6.4}$$

is the *vectorial derivative* of the logic function $f(\mathbf{x}_0, \mathbf{x}_1)$ with regard to the variables of $\mathbf{x}_0$.

Note 6.1 Sometimes the *vectorial derivative* is also denoted by *vectorial difference*.

Note 6.2 The *negated vectorial derivative* can be defined replacing $\oplus$ by $\odot$ in (6.4):

$$\text{nder}_{\mathbf{x}_0} f(\mathbf{x}_0, \mathbf{x}_1) = f(\mathbf{x}_0, \mathbf{x}_1) \odot f(\overline{\mathbf{x}}_0, \mathbf{x}_1) \,. \tag{6.5}$$

This negated vectorial derivative can be expressed by the negation of the already defined vectorial derivative (6.4):

$$\text{nder}_{\mathbf{x}_0} f(\mathbf{x}_0, \mathbf{x}_1) = \overline{\text{der}_{\mathbf{x}_0} f(\mathbf{x}_0, \mathbf{x}_1)} \,.$$

Hence, we avoid the detailed exploration of $\text{nder}_{\mathbf{x}_0} f(\mathbf{x}_0, \mathbf{x}_1)$.

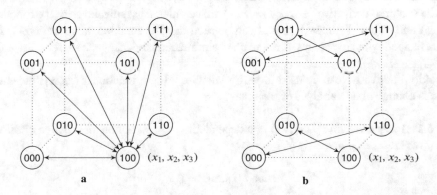

Fig. 6.1 Directions of change in the Boolean space $\mathbb{B}^3$: (**a**) all directions of change shown for the point $(x_1, x_2, x_3) = (100)$, (**b**) pairs of function values for the directions of change (x_1, x_2)

Table 6.1 Calculation of
the vectorial derivative of
the logic function (6.6)
with regard to (a, c)

Pair	a	b	c	$f(a,b,c)$	$\bar{a}$	b	$\bar{c}$	$f(\bar{a},b,\bar{c})$	$\mathrm{der}_{(a,c)} f(a,b,c)$
1	0	0	0	0	1	0	1	1	1
	1	0	1	1	0	0	0	0	1
2	0	0	1	0	1	0	0	0	0
	1	0	0	0	0	0	1	0	0
3	0	1	0	1	1	1	1	1	0
	1	1	1	1	0	1	0	1	0
4	0	1	1	0	1	1	0	0	0
	1	1	0	0	0	1	1	0	0

The vectorial derivative $\mathrm{der}_{\mathbf{x}_0} f(\mathbf{x}_0, \mathbf{x}_1)$ is a logic function that depends generally on all variables of $\mathbf{x} = (\mathbf{x}_0, \mathbf{x}_1)$. A function value 1 of the vectorial derivative $\mathrm{der}_{\mathbf{x}_0} f(\mathbf{x}_0, \mathbf{x}_1)$ indicates that the corresponding value of $f(\mathbf{x}_0, \mathbf{x}_1)$ will change if all variables of $\mathbf{x}_0$ change their values simultaneously.

Example 6.1 Let

$$f(a,b,c) = \bar{a}\,b\,\bar{c} \oplus a\,c\,. \tag{6.6}$$

Based on (6.4) of Definition 6.1, the vectorial derivative $\mathrm{der}_{(a,c)} f(a,b,c)$ can be calculated as follows:

$$
\begin{aligned}
\mathop{\mathrm{der}}_{(a,c)} f(a,b,c) &= f(a,b,c) \oplus f(\bar{a},b,\bar{c}) \\[4pt]
&= (\bar{a}\,b\,\bar{c} \oplus a\,c) \oplus (a\,b\,c \oplus \bar{a}\,\bar{c}) \\[4pt]
&= \bar{a}\,b\,\bar{c} \oplus a\,c \oplus a\,b\,c \oplus \bar{a}\,\bar{c} \\[4pt]
&= \bar{a}\,\bar{c}\,(b \oplus 1) \oplus a\,c\,(b \oplus 1) \\[4pt]
&= \bar{b}\,(\bar{a}\,\bar{c} \oplus a\,c) \\[4pt]
&= \bar{b}\,(\bar{a} \oplus c)\,.
\end{aligned}
$$

This result shows that a vectorial derivative can depend on all the variables of the given function.

Alternatively, a function table can be used to calculate this vectorial derivative. Table 6.1 demonstrates this calculation for the function (6.6). Based on the function table of $f(a,b,c)$, the changed values $(\bar{a}, b, \bar{c})$ and the resulting function $f(\bar{a}, b, \bar{c})$ are appended to the right. Finally, the values of the vectorial derivative $\mathrm{der}_{(a,c)} f(a,b,c)$ are calculated as antivalence of the function values of $f(a,b,c)$ and $f(\bar{a}, b, \bar{c})$ in each row. It can be seen that the function value of $\mathrm{der}_{(a,c)} f(a,b,c)$ remains unchanged when the values of a and c are simultaneously changed.

The vectorial derivative can be used to verify whether a logic function $f(\mathbf{x}_0, \mathbf{x}_1)$ depends on the simultaneous change of all variables of the set $\mathbf{x}_0$.

Theorem 6.1 *A logic function $f(\mathbf{x}_0, \mathbf{x}_1)$ is independent of the simultaneous change of all variables of the set of variables $\mathbf{x}_0$ if and only if*

$$\mathop{\mathrm{der}}_{\mathbf{x}_0} f(\mathbf{x}_0, \mathbf{x}_1) = 0\,. \tag{6.7}$$

Proof The condition that a logic function $f(\mathbf{x}_0, \mathbf{x}_1)$ does not depend on the simultaneous change of all variables of the set $\mathbf{x}_0$ is

$$f(\mathbf{x}_0, \mathbf{x}_1) = f(\overline{\mathbf{x}}_0, \mathbf{x}_1) \,.$$

This equation can be equivalently transformed into

$$f(\mathbf{x}_0, \mathbf{x}_1) \oplus f(\overline{\mathbf{x}}_0, \mathbf{x}_1) = 0$$

$$\operatorname*{der}_{\mathbf{x}_0} f(\mathbf{x}_0, \mathbf{x}_1) = 0 \,,$$

where in the final step (6.4) of Definition 6.1 has been substituted to get (6.7). All steps of this proof can also be executed in the reverse direction.   □

Another property that can easily be verified by means of the vectorial derivative is the self-duality of a logic function.

Theorem 6.2 *A logic function $f(\mathbf{x})$ is self-dual if and only if*

$$\operatorname*{der}_{\mathbf{x}} f(\mathbf{x}) = 1 \,. \tag{6.8}$$

Proof A dual logic function $f^*(\mathbf{x}) = \overline{f(\overline{\mathbf{x}})}$ (see Definition 4.16) is dual to itself if $f^*(\mathbf{x}) = f(\mathbf{x})$. Hence, the condition that a logic function $f(\mathbf{x})$ is self-dual is

$$f(\mathbf{x}) = \overline{f(\overline{\mathbf{x}})} \,.$$

This equation can be equivalently transformed into

$$f(\mathbf{x}) = f(\overline{\mathbf{x}}) \oplus 1$$

$$f(\mathbf{x}) \oplus f(\overline{\mathbf{x}}) = 1$$

$$\operatorname*{der}_{\mathbf{x}} f(\mathbf{x}) = 1 \,,$$

where in the final step (6.4) of Definition 6.1 for $\mathbf{x}_0 = \mathbf{x}$ and $\mathbf{x}_1 = \emptyset$ has been substituted to get (6.8). All steps of this proof can also be executed in the reverse direction.   □

The replacement of $\oplus$ of the vectorial derivative by $\wedge$ leads to the *vectorial minimum*.

Definition 6.2 Let $\mathbf{x}_0 = (x_1, x_2, \ldots, x_k)$, $\mathbf{x}_1 = (x_{k+1}, x_{k+2}, \ldots, x_n)$ be two disjoint sets of Boolean variables, and $f(\mathbf{x}_0, \mathbf{x}_1) = f(x_1, x_2, \ldots, x_n) = f(\mathbf{x})$ a logic function of n variables; then

$$\operatorname*{min}_{\mathbf{x}_0} f(\mathbf{x}_0, \mathbf{x}_1) = f(\mathbf{x}_0, \mathbf{x}_1) \wedge f(\overline{\mathbf{x}}_0, \mathbf{x}_1) \tag{6.9}$$

is the *vectorial minimum* with regard to the variables of $\mathbf{x}_0$.

A function value 1 occurs in the vectorial minimum $\min_{\mathbf{x}_0} f(\mathbf{x}_0, \mathbf{x}_1)$ if the corresponding values of $f(\mathbf{x}_0, \mathbf{x}_1)$ remain equal to 1 when all the variables of $\mathbf{x}_0$ change their values simultaneously. Because of this restriction, the vectorial minimum is less than or equal to the function $f(\mathbf{x}_0, \mathbf{x}_1)$. Therefore, the name *vectorial minimum* was chosen.

Table 6.2 Calculation of the vectorial minimum of the logic function (6.6) with regard to (a, c)

Pair	a	b	c	$f(a,b,c)$	$\overline{a}$	b	$\overline{c}$	$f(\overline{a},b,\overline{c})$	$\min_{(a,c)} f(a,b,c)$
1	0	0	0	0	1	0	1	1	0
	1	0	1	1	0	0	0	0	0
2	0	0	1	0	1	0	0	0	0
	1	0	0	0	0	0	1	0	0
3	0	1	0	1	1	1	1	1	1
	1	1	1	1	0	1	0	1	1
4	0	1	1	0	1	1	0	0	0
	1	1	0	0	0	1	1	0	0

Example 6.2 For an easy comparison with the vectorial derivative, we take the same function (6.6) and calculate the vectorial minimum $\min_{(a,c)} f(a, b, c)$ based on (6.9) of Definition 6.2:

$$\min_{(a,c)} f(a, b, c) = f(a, b, c) \wedge f(\overline{a}, b, \overline{c})$$

$$= (\overline{a}\, b\, \overline{c} \oplus a\, c) \wedge (a\, b\, c \oplus \overline{a}\, \overline{c})$$

$$= (\overline{a}\, b\, \overline{c} \oplus a\, b\, c)$$

$$= b\,(\overline{a}\, \overline{c} \oplus a\, c)$$

$$= b\,(\overline{a} \oplus c)\,.$$

This result shows that a vectorial minimum can depend on all the variables of the given function. Table 6.2 emphasizes the pairs of function values that determine the vectorial minimum of the same function and the same direction of change.

Definition 6.3 Let $\mathbf{x}_0 = (x_1, x_2, \ldots, x_k)$, $\mathbf{x}_1 = (x_{k+1}, x_{k+2}, \ldots, x_n)$ be two disjoint sets of Boolean variables, and $f(\mathbf{x}_0, \mathbf{x}_1) = f(x_1, x_2, \ldots, x_n) = f(\mathbf{x})$ a logic function of n variables; then

$$\max_{\mathbf{x}_0} f(\mathbf{x}_0, \mathbf{x}_1) = f(\mathbf{x}_0, \mathbf{x}_1) \vee f(\overline{\mathbf{x}}_0, \mathbf{x}_1) \tag{6.10}$$

is the *vectorial maximum* with regard to the variables of $\mathbf{x}_0$.

The comparison of (6.9) and (6.10) shows that the same pair of function values of $f(\mathbf{x}_0, \mathbf{x}_1)$ determines the function value of the vectorial maximum. It is sufficient that at least one of them is equal to 1 in order get the function value 1 of $\max_{\mathbf{x}_0} f(\mathbf{x}_0, \mathbf{x}_1)$. Since $f(\mathbf{x}_0, \mathbf{x}_1)$ occurs as part of a $\vee$-operation in (6.10) of Definition 6.3, the vectorial maximum is larger than or equal to the corresponding function $f(\mathbf{x}_0, \mathbf{x}_1)$. This explains the name *vectorial maximum* for this derivative operation.

Example 6.3 For an easy comparison with the other two vectorial derivative operations, we take the same function (6.6) and calculate the vectorial maximum $\max_{(a,c)} f(a, b, c)$ based on (6.10) of Definition 6.3:

$$\max_{(a,c)} f(a, b, c) = f(a, b, c) \vee f(\overline{a}, b, \overline{c})$$

$$= (\overline{a}\, b\, \overline{c} \oplus a\, c) \vee (a\, b\, c \oplus \overline{a}\, \overline{c})$$

$$= (\overline{a}\, b\, \overline{c} \vee a\, c) \vee (a\, b\, c \vee \overline{a}\, \overline{c})$$

$$= a\, c \vee \overline{a}\, \overline{c}$$

$$= \overline{a} \oplus c\,.$$

This result of a vectorial maximum does not depend on all the variables of the given function; it is independent of the variable b. Table 6.3 shows that pairs of function values also determine the vectorial maximum of the same function and the same direction of change.

A vectorial maximum can also depend on all the variables of the given function, as can be seen for the slightly changed function:

$$f(a, b, c) = \overline{a}\, b \oplus a\, c\,.$$

$$\max_{(a,c)} f(a, b, c) = f(a, b, c) \vee f(\overline{a}, b, \overline{c})$$

$$= (\overline{a}\, b \oplus a\, c) \vee (a\, b \oplus \overline{a}\, \overline{c})$$

$$= (\overline{a}\, b \vee a\, c) \vee (a\, b \vee \overline{a}\, \overline{c})$$

$$= b(\overline{a} \vee a) \vee a\, c \vee \overline{a}\, \overline{c}$$

$$= b \vee (\overline{a} \oplus c)\,.$$

This vectorial maximum depends on all variables of the given function.

Relations between the vectorial derivative operations can easily be seen in Fig. 6.2, which summarizes the examples of the vectorial derivative operations. The small arrows in the Karnaugh-map of $f(a, b, c)$ indicate the pairs of function values that cause the calculated results of the three derivative operations with regard to (a, c).

The Karnaugh-maps in the second row of Fig. 6.2 show the results of the derivative operations calculated by (6.9), (6.4), and (6.10), respectively. The left two of these Karnaugh-maps show that the result of vectorial derivative operations generally depends on all variables of the given function. The vectorial maximum in the rightmost Karnaugh-map does not depend on the variable b. A simple example where the vectorial maximum also depends on all variables has been shown at the end of Example 6.3.

The comparison of the outer Karnaugh-maps of the lower row and the Karnaugh-map on top of Fig. 6.2 reveals that the order relation (6.11) holds between the vectorial minimum $\min_{\mathbf{x}_0} f(\mathbf{x}_0, \mathbf{x}_1)$,

Table 6.3 Calculation of	Pair	a	b	c	$f(a, b, c)$	$\overline{a}$	b	$\overline{c}$	$f(\overline{a}, b, \overline{c})$	$\max_{(a,c)} f(a, b, c)$
the vectorial maximum of	1	0	0	0	0	1	0	1	1	1
the logic function (6.6)		1	0	1	1	0	0	0	0	1
with regard to (a, c)	2	0	0	1	0	1	0	0	0	0
		1	0	0	0	0	0	1	0	0
	3	0	1	0	1	1	1	1	1	1
		1	1	1	1	0	1	0	1	1
	4	0	1	1	0	1	1	0	0	0
		1	1	0	0	0	1	1	0	0

the logic function $f(\mathbf{x}_0, \mathbf{x}_1)$, and the vectorial maximum $\max_{\mathbf{x}_0} f(\mathbf{x}_0, \mathbf{x}_1)$:

$$\min_{\mathbf{x}_0} f(\mathbf{x}_0, \mathbf{x}_1) \le f(\mathbf{x}_0, \mathbf{x}_1) \le \max_{\mathbf{x}_0} f(\mathbf{x}_0, \mathbf{x}_1) \,. \tag{6.11}$$

From (6.11) follow the next three equations:

$$\min_{\mathbf{x}_0} f(\mathbf{x}_0, \mathbf{x}_1) \wedge \overline{f(\mathbf{x}_0, \mathbf{x}_1)} = 0 \,, \tag{6.12}$$

$$f(\mathbf{x}_0, \mathbf{x}_1) \wedge \overline{\max_{\mathbf{x}_0} f(\mathbf{x}_0, \mathbf{x}_1)} = 0 \,, \tag{6.13}$$

$$\min_{\mathbf{x}_0} f(\mathbf{x}_0, \mathbf{x}_1) \wedge \overline{\max_{\mathbf{x}_0} f(\mathbf{x}_0, \mathbf{x}_1)} = 0 \,. \tag{6.14}$$

The Karnaugh-maps in the lower row of Fig. 6.2 show furthermore that the antivalence of all three vectorial derivative operations is equal to 0 (6.15), and the vectorial minimum is orthogonal to the vectorial derivative (6.16):

$$\min_{\mathbf{x}_0} f(\mathbf{x}_0, \mathbf{x}_1) \oplus \operatorname*{der}_{\mathbf{x}_0} f(\mathbf{x}_0, \mathbf{x}_1) \oplus \max_{\mathbf{x}_0} f(\mathbf{x}_0, \mathbf{x}_1) = 0 \,, \tag{6.15}$$

$$\min_{\mathbf{x}_0} f(\mathbf{x}_0, \mathbf{x}_1) \wedge \operatorname*{der}_{\mathbf{x}_0} f(\mathbf{x}_0, \mathbf{x}_1) = 0 \,. \tag{6.16}$$

Based on (6.15), each vectorial derivative operation can be calculated using the two other operations:

$$\min_{\mathbf{x}_0} f(\mathbf{x}_0, \mathbf{x}_1) = \operatorname*{der}_{\mathbf{x}_0} f(\mathbf{x}_0, \mathbf{x}_1) \oplus \max_{\mathbf{x}_0} f(\mathbf{x}_0, \mathbf{x}_1) \,,$$

$$\operatorname*{der}_{\mathbf{x}_0} f(\mathbf{x}_0, \mathbf{x}_1) = \min_{\mathbf{x}_0} f(\mathbf{x}_0, \mathbf{x}_1) \oplus \max_{\mathbf{x}_0} f(\mathbf{x}_0, \mathbf{x}_1) \,,$$

$$\max_{\mathbf{x}_0} f(\mathbf{x}_0, \mathbf{x}_1) = \min_{\mathbf{x}_0} f(\mathbf{x}_0, \mathbf{x}_1) \oplus \operatorname*{der}_{\mathbf{x}_0} f(\mathbf{x}_0, \mathbf{x}_1) \,.$$

Alternative rules can be used as well:

$$\min_{\mathbf{x}_0} f(\mathbf{x}_0, \mathbf{x}_1) = \max_{\mathbf{x}_0} f(\mathbf{x}_0, \mathbf{x}_1) \wedge \overline{\operatorname*{der}_{\mathbf{x}_0} f(\mathbf{x}_0, \mathbf{x}_1)} \,,$$

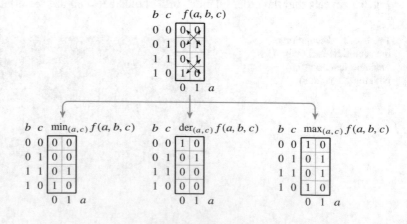

Fig. 6.2 The Karnaugh-maps of $f(a, b, c) = \overline{a}\,b\,\overline{c} \oplus a\,c$ and all vectorial derivative operations with regard to (a, c)

$$\operatorname*{der}_{\mathbf{x}_0} f(\mathbf{x}_0, \mathbf{x}_1) = \operatorname*{max}_{\mathbf{x}_0} f(\mathbf{x}_0, \mathbf{x}_1) \wedge \overline{\operatorname*{min}_{\mathbf{x}_0} f(\mathbf{x}_0, \mathbf{x}_1)},$$

$$\operatorname*{max}_{\mathbf{x}_0} f(\mathbf{x}_0, \mathbf{x}_1) = \operatorname*{min}_{\mathbf{x}_0} f(\mathbf{x}_0, \mathbf{x}_1) \vee \operatorname*{der}_{\mathbf{x}_0} f(\mathbf{x}_0, \mathbf{x}_1).$$

The vectorial derivative $g(\mathbf{x}_0, \mathbf{x}_1) = \operatorname{der}_{\mathbf{x}_0} f(\mathbf{x}_0, \mathbf{x}_1)$ is a logic function that generally depends on all n variables of $\mathbf{x} = (\mathbf{x}_0, \mathbf{x}_1)$ the given function $f(\mathbf{x}_0, \mathbf{x}_1)$ is depending on. However, this vectorial derivative satisfies the following restriction:

$$\begin{aligned}
g(\mathbf{x}_0, \mathbf{x}_1) &= \operatorname*{der}_{\mathbf{x}_0} f(\mathbf{x}_0, \mathbf{x}_1) \\
&= f(\mathbf{x}_0, \mathbf{x}_1) \oplus f(\overline{\mathbf{x}}_0, \mathbf{x}_1) \\
&= f(\overline{\overline{\mathbf{x}}}_0, \mathbf{x}_1) \oplus f(\overline{\mathbf{x}}_0, \mathbf{x}_1) \\
&= f(\overline{\mathbf{x}}_0, \mathbf{x}_1) \oplus f(\overline{\overline{\mathbf{x}}}_0, \mathbf{x}_1) \\
&= g(\overline{\mathbf{x}}_0, \mathbf{x}_1).
\end{aligned} \tag{6.17}$$

Hence, the vectorial derivative $g(\mathbf{x}_0, \mathbf{x}_1) = \operatorname{der}_{\mathbf{x}_0} f(\mathbf{x}_0, \mathbf{x}_1)$ consists of pairs of equal function values $g(\mathbf{c}_0, \mathbf{c}_1) \equiv g(\overline{\mathbf{c}}_0, \mathbf{c}_1)$ so that only $2^{2^{n-1}}$ of all 2^{2^n} functions of n variables are the result of a vectorial derivative. Hence, the vectorial derivative is simpler to the effect that it does not depend on the simultaneous change of the variables $\mathbf{x}_0$ anymore:

$$\begin{aligned}
\operatorname*{der}_{\mathbf{x}_0}(\operatorname*{der}_{\mathbf{x}_0} f(\mathbf{x}_0, \mathbf{x}_1)) &= \operatorname*{der}_{\mathbf{x}_0}(f(\mathbf{x}_0, \mathbf{x}_1) \oplus f(\overline{\mathbf{x}}_0, \mathbf{x}_1)) \\
&= (f(\mathbf{x}_0, \mathbf{x}_1) \oplus f(\overline{\mathbf{x}}_0, \mathbf{x}_1)) \oplus (f(\overline{\mathbf{x}}_0, \mathbf{x}_1) \oplus f(\overline{\overline{\mathbf{x}}}_0, \mathbf{x}_1)) \\
&= (f(\mathbf{x}_0, \mathbf{x}_1) \oplus f(\overline{\mathbf{x}}_0, \mathbf{x}_1)) \oplus (f(\overline{\mathbf{x}}_0, \mathbf{x}_1) \oplus f(\mathbf{x}_0, \mathbf{x}_1)) \\
&= (f(\mathbf{x}_0, \mathbf{x}_1) \oplus f(\overline{\mathbf{x}}_0, \mathbf{x}_1)) \oplus (f(\mathbf{x}_0, \mathbf{x}_1) \oplus f(\overline{\mathbf{x}}_0, \mathbf{x}_1)) \\
&= 0.
\end{aligned}$$

The restriction (6.17) utilizes only the commutativity of the $\oplus$-operation. Both the $\wedge$-operation of the vectorial minimum and the $\vee$-operation of the vectorial maximum also satisfy this property. Hence, all vectorial derivative operations are simpler than arbitrary logic functions of n variables because they do not depend on the simultaneous change of $\mathbf{x}_0$:

$$\operatorname*{der}_{\mathbf{x}_0}(\operatorname*{der}_{\mathbf{x}_0} f(\mathbf{x}_0, \mathbf{x}_1)) = 0,$$

$$\operatorname*{der}_{\mathbf{x}_0}(\operatorname*{min}_{\mathbf{x}_0} f(\mathbf{x}_0, \mathbf{x}_1)) = 0,$$

$$\operatorname*{der}_{\mathbf{x}_0}(\operatorname*{max}_{\mathbf{x}_0} f(\mathbf{x}_0, \mathbf{x}_1)) = 0.$$

It can easily be shown that neither the vectorial minimum nor the vectorial maximum changes the result of any vectorial derivative operation with regard to the same direction of change:

$$\operatorname*{min}_{\mathbf{x}_0}(\operatorname*{der}_{\mathbf{x}_0} f(\mathbf{x}_0, \mathbf{x}_1)) = \operatorname*{der}_{\mathbf{x}_0} f(\mathbf{x}_0, \mathbf{x}_1),$$

$$\operatorname*{min}_{\mathbf{x}_0}(\operatorname*{min}_{\mathbf{x}_0} f(\mathbf{x}_0, \mathbf{x}_1)) = \operatorname*{min}_{\mathbf{x}_0} f(\mathbf{x}_0, \mathbf{x}_1),$$

$$\min_{\mathbf{x}_0}(\max_{\mathbf{x}_0} f(\mathbf{x}_0, \mathbf{x}_1)) = \max_{\mathbf{x}_0} f(\mathbf{x}_0, \mathbf{x}_1) \, ,$$

$$\max_{\mathbf{x}_0}(\text{der}_{\mathbf{x}_0} f(\mathbf{x}_0, \mathbf{x}_1)) = \text{der}_{\mathbf{x}_0} f(\mathbf{x}_0, \mathbf{x}_1) \, ,$$

$$\max_{\mathbf{x}_0}(\min_{\mathbf{x}_0} f(\mathbf{x}_0, \mathbf{x}_1)) = \min_{\mathbf{x}_0} f(\mathbf{x}_0, \mathbf{x}_1) \, ,$$

$$\max_{\mathbf{x}_0}(\max_{\mathbf{x}_0} f(\mathbf{x}_0, \mathbf{x}_1)) = \max_{\mathbf{x}_0} f(\mathbf{x}_0, \mathbf{x}_1) \, .$$

A sequence of vectorial derivative operations with regard to different directions of change simplifies step by step the given function $f(\mathbf{x}_0, \mathbf{x}_1)$ of n variables. There are $2^n - 1$ different directions of change specified by the set of variables $\mathbf{x}_0$. However, only n of these directions are linearly independent of each other. A vectorial derivative with regard to the set of variables $\mathbf{x}_0$ that is linearly dependent on previous applied directions of change results in a function that is equal to 0.

Example 6.4 The set of variables $(\mathbf{x}_{0b}, \mathbf{x}_{0c})$ linearly depends on the sets of variables $(\mathbf{x}_{0a}, \mathbf{x}_{0b})$ and $(\mathbf{x}_{0a}, \mathbf{x}_{0c})$. Hence, the following sequence of vectorial derivatives results for each given function $f(\mathbf{x}_{0a}, \mathbf{x}_{0b}, \mathbf{x}_{0c}, \mathbf{x}_1)$ in the logic function that is constant equal to 0:

$$\text{der}_{(\mathbf{x}_{0b}, \mathbf{x}_{0c})} (\, \text{der}_{(\mathbf{x}_{0a}, \mathbf{x}_{0c})} (\, \text{der}_{(\mathbf{x}_{0a}, \mathbf{x}_{0b})} f(\mathbf{x}_{0a}, \mathbf{x}_{0b}, \mathbf{x}_{0c}, \mathbf{x}_1)))$$

$$= \text{der}_{(\mathbf{x}_{0b}, \mathbf{x}_{0c})} (\, \text{der}_{(\mathbf{x}_{0a}, \mathbf{x}_{0c})} (f(\mathbf{x}_{0a}, \mathbf{x}_{0b}, \mathbf{x}_{0c}, \mathbf{x}_1) \oplus f(\overline{\mathbf{x}}_{0a}, \overline{\mathbf{x}}_{0b}, \mathbf{x}_{0c}, \mathbf{x}_1)))$$

$$= \text{der}_{(\mathbf{x}_{0b}, \mathbf{x}_{0c})} ((f(\mathbf{x}_{0a}, \mathbf{x}_{0b}, \mathbf{x}_{0c}, \mathbf{x}_1) \oplus f(\overline{\mathbf{x}}_{0a}, \overline{\mathbf{x}}_{0b}, \mathbf{x}_{0c}, \mathbf{x}_1))$$

$$\oplus ((f(\overline{\mathbf{x}}_{0a}, \mathbf{x}_{0b}, \overline{\mathbf{x}}_{0c}, \mathbf{x}_1) \oplus f(\overline{\overline{\mathbf{x}}}_{0a}, \overline{\mathbf{x}}_{0b}, \overline{\mathbf{x}}_{0c}, \mathbf{x}_1))))$$

$$= (f(\mathbf{x}_{0a}, \mathbf{x}_{0b}, \mathbf{x}_{0c}, \mathbf{x}_1) \oplus f(\overline{\mathbf{x}}_{0a}, \overline{\mathbf{x}}_{0b}, \mathbf{x}_{0c}, \mathbf{x}_1)$$

$$\oplus f(\overline{\mathbf{x}}_{0a}, \mathbf{x}_{0b}, \overline{\mathbf{x}}_{0c}, \mathbf{x}_1) \oplus f(\overline{\overline{\mathbf{x}}}_{0a}, \overline{\mathbf{x}}_{0b}, \overline{\mathbf{x}}_{0c}, \mathbf{x}_1))$$

$$\oplus (f(\mathbf{x}_{0a}, \overline{\mathbf{x}}_{0b}, \overline{\mathbf{x}}_{0c}, \mathbf{x}_1) \oplus f(\overline{\mathbf{x}}_{0a}, \overline{\overline{\mathbf{x}}}_{0b}, \overline{\mathbf{x}}_{0c}, \mathbf{x}_1)$$

$$\oplus f(\overline{\mathbf{x}}_{0a}, \overline{\mathbf{x}}_{0b}, \overline{\overline{\mathbf{x}}}_{0c}, \mathbf{x}_1) \oplus f(\overline{\overline{\mathbf{x}}}_{0a}, \overline{\overline{\mathbf{x}}}_{0b}, \overline{\overline{\mathbf{x}}}_{0c}, \mathbf{x}_1))$$

$$= f(\mathbf{x}_{0a}, \mathbf{x}_{0b}, \mathbf{x}_{0c}, \mathbf{x}_1) \oplus f(\overline{\mathbf{x}}_{0a}, \overline{\mathbf{x}}_{0b}, \mathbf{x}_{0c}, \mathbf{x}_1)$$

$$\oplus f(\overline{\mathbf{x}}_{0a}, \mathbf{x}_{0b}, \overline{\mathbf{x}}_{0c}, \mathbf{x}_1) \oplus f(\mathbf{x}_{0a}, \overline{\mathbf{x}}_{0b}, \overline{\mathbf{x}}_{0c}, \mathbf{x}_1)$$

$$\oplus f(\mathbf{x}_{0a}, \overline{\mathbf{x}}_{0b}, \overline{\mathbf{x}}_{0c}, \mathbf{x}_1) \oplus f(\overline{\mathbf{x}}_{0a}, \mathbf{x}_{0b}, \overline{\mathbf{x}}_{0c}, \mathbf{x}_1)$$

$$\oplus f(\overline{\mathbf{x}}_{0a}, \overline{\mathbf{x}}_{0b}, \mathbf{x}_{0c}, \mathbf{x}_1) \oplus f(\mathbf{x}_{0a}, \mathbf{x}_{0b}, \mathbf{x}_{0c}, \mathbf{x}_1)$$

$$= 0 \, .$$

Similarly, a vectorial minimum or maximum with regard to a direction of change that linearly depends on previous directions of change does not change the given function of a vectorial derivative operation that has been calculated with regard to the mentioned previous directions of change.

Example 6.5 Let, for example,

$$g(\mathbf{x}_{0a}, \mathbf{x}_{0b}, \mathbf{x}_{0c}, \mathbf{x}_1) = \max_{(\mathbf{x}_{0a}, \mathbf{x}_{0c})} (\, \text{der}_{(\mathbf{x}_{0a}, \mathbf{x}_{0b})} f(\mathbf{x}_{0a}, \mathbf{x}_{0b}, \mathbf{x}_{0c}, \mathbf{x}_1))) \, ,$$

and then

$$\min_{(\mathbf{x}_{0b},\mathbf{x}_{0c})} g(\mathbf{x}_{0a}, \mathbf{x}_{0b}, \mathbf{x}_{0c}, \mathbf{x}_1) = g(\mathbf{x}_{0a}, \mathbf{x}_{0b}, \mathbf{x}_{0c}, \mathbf{x}_1) \; .$$

Similarly, it holds for

$$h(\mathbf{x}_{0a}, \mathbf{x}_{0b}, \mathbf{x}_{0c}, \mathbf{x}_1) = \operatorname*{der}_{(\mathbf{x}_{0a},\mathbf{x}_{0c})} (\min_{(\mathbf{x}_{0a},\mathbf{x}_{0b})} f(\mathbf{x}_{0a}, \mathbf{x}_{0b}, \mathbf{x}_{0c}, \mathbf{x}_1)))$$

that

$$\max_{(\mathbf{x}_{0b},\mathbf{x}_{0c})} h(\mathbf{x}_{0a}, \mathbf{x}_{0b}, \mathbf{x}_{0c}, \mathbf{x}_1) = h(\mathbf{x}_{0a}, \mathbf{x}_{0b}, \mathbf{x}_{0c}, \mathbf{x}_1) \; .$$

The statements of these examples remain true when any other vectorial derivative operation is used with regard to the fixed directions of change $(\mathbf{x}_{0a}, \mathbf{x}_{0b})$ or $(\mathbf{x}_{0a}, \mathbf{x}_{0c})$, respectively.

For transformations of expressions containing vectorial derivative operation, it is useful to know the result of these operations when instead of the function $f(\mathbf{x}_0, \mathbf{x}_1)$ their complement is given:

$$\operatorname*{der}_{\mathbf{x}_0} \overline{f(\mathbf{x}_0, \mathbf{x}_1)} = \operatorname*{der}_{\mathbf{x}_0} f(\mathbf{x}_0, \mathbf{x}_1) \; , \tag{6.18}$$

$$\operatorname*{min}_{\mathbf{x}_0} \overline{f(\mathbf{x}_0, \mathbf{x}_1)} = \overline{\operatorname*{max}_{\mathbf{x}_0} f(\mathbf{x}_0, \mathbf{x}_1)} \; , \tag{6.19}$$

$$\operatorname*{max}_{\mathbf{x}_0} \overline{f(\mathbf{x}_0, \mathbf{x}_1)} = \overline{\operatorname*{min}_{\mathbf{x}_0} f(\mathbf{x}_0, \mathbf{x}_1)} \; . \tag{6.20}$$

These and the following relations directly follow from the definitions of the vectorial derivative operations. If the functions $f(\mathbf{x}_0, \mathbf{x}_1)$ and $g(\mathbf{x}_0, \mathbf{x}_1)$ are connected by the same logic operation as used in the definition of a vectorial derivative operation, a separation is possible due to the commutative laws:

$$\operatorname*{der}_{\mathbf{x}_0}(f(\mathbf{x}_0, \mathbf{x}_1) \oplus g(\mathbf{x}_0, \mathbf{x}_1)) = \operatorname*{der}_{\mathbf{x}_0} f(\mathbf{x}_0, \mathbf{x}_1) \oplus \operatorname*{der}_{\mathbf{x}_0} g(\mathbf{x}_0, \mathbf{x}_1) \; , \tag{6.21}$$

$$\operatorname*{min}_{\mathbf{x}_0}(f(\mathbf{x}_0, \mathbf{x}_1) \wedge g(\mathbf{x}_0, \mathbf{x}_1)) = \operatorname*{min}_{\mathbf{x}_0} f(\mathbf{x}_0, \mathbf{x}_1) \wedge \operatorname*{min}_{\mathbf{x}_0} g(\mathbf{x}_0, \mathbf{x}_1) \; , \tag{6.22}$$

$$\operatorname*{max}_{\mathbf{x}_0}(f(\mathbf{x}_0, \mathbf{x}_1) \vee g(\mathbf{x}_0, \mathbf{x}_1)) = \operatorname*{max}_{\mathbf{x}_0} f(\mathbf{x}_0, \mathbf{x}_1) \vee \operatorname*{max}_{\mathbf{x}_0} g(\mathbf{x}_0, \mathbf{x}_1) \; . \tag{6.23}$$

In the case that the simultaneous change of all variables of the set $\mathbf{x}_0$ does not change any function value of the function $f(\mathbf{x}_0, \mathbf{x}_1)$, it holds

$$\operatorname*{der}_{\mathbf{x}_0} f(\mathbf{x}_0, \mathbf{x}_1) = 0 \; , \tag{6.24}$$

$$\operatorname*{min}_{\mathbf{x}_0} f(\mathbf{x}_0, \mathbf{x}_1) - f(\mathbf{x}_0, \mathbf{x}_1) \; , \tag{6.25}$$

$$\operatorname*{max}_{\mathbf{x}_0} f(\mathbf{x}_0, \mathbf{x}_1) = f(\mathbf{x}_0, \mathbf{x}_1) \; . \tag{6.26}$$

Now we assume that the function $g(\mathbf{x}_0, \mathbf{x}_1)$ is independent of the simultaneous change of all variables of the set $\mathbf{x}_0$, i.e., $g(\mathbf{x}_0, \mathbf{x}_1)$ satisfies

$$\operatorname*{der}_{\mathbf{x}_0} g(\mathbf{x}_0, \mathbf{x}_1) = 0 \; ,$$

and then the following relationships hold for the connection with $\wedge$:

$$\underset{\mathbf{x}_0}{\operatorname{der}}(g(\mathbf{x}_0, \mathbf{x}_1) \wedge f(\mathbf{x}_0, \mathbf{x}_1)) = g(\mathbf{x}_0, \mathbf{x}_1) \wedge \underset{\mathbf{x}_0}{\operatorname{der}} f(\mathbf{x}_0, \mathbf{x}_1),$$

$$\underset{\mathbf{x}_0}{\min}(g(\mathbf{x}_0, \mathbf{x}_1) \wedge f(\mathbf{x}_0, \mathbf{x}_1)) = g(\mathbf{x}_0, \mathbf{x}_1) \wedge \underset{\mathbf{x}_0}{\min} f(\mathbf{x}_0, \mathbf{x}_1),$$

$$\underset{\mathbf{x}_0}{\max}(g(\mathbf{x}_0, \mathbf{x}_1) \wedge f(\mathbf{x}_0, \mathbf{x}_1)) = g(\mathbf{x}_0, \mathbf{x}_1) \wedge \underset{\mathbf{x}_0}{\max} f(\mathbf{x}_0, \mathbf{x}_1),$$

the connection with $\vee$:

$$\underset{\mathbf{x}_0}{\operatorname{der}}(g(\mathbf{x}_0, \mathbf{x}_1) \vee f(\mathbf{x}_0, \mathbf{x}_1)) = \overline{g(\mathbf{x}_0, \mathbf{x}_1)} \wedge \underset{\mathbf{x}_0}{\operatorname{der}} f(\mathbf{x}_0, \mathbf{x}_1),$$

$$\underset{\mathbf{x}_0}{\min}(g(\mathbf{x}_0, \mathbf{x}_1) \vee f(\mathbf{x}_0, \mathbf{x}_1)) = g(\mathbf{x}_0, \mathbf{x}_1) \vee \underset{\mathbf{x}_0}{\min} f(\mathbf{x}_0, \mathbf{x}_1),$$

$$\underset{\mathbf{x}_0}{\max}(g(\mathbf{x}_0, \mathbf{x}_1) \vee f(\mathbf{x}_0, \mathbf{x}_1)) = g(\mathbf{x}_0, \mathbf{x}_1) \vee \underset{\mathbf{x}_0}{\max} f(\mathbf{x}_0, \mathbf{x}_1),$$

and the connection with $\oplus$:

$$\underset{\mathbf{x}_0}{\operatorname{der}}(g(\mathbf{x}_0, \mathbf{x}_1) \oplus f(\mathbf{x}_0, \mathbf{x}_1)) = \underset{\mathbf{x}_0}{\operatorname{der}}(f(\mathbf{x}_0, \mathbf{x}_1)),$$

$$\underset{\mathbf{x}_0}{\min}(g(\mathbf{x}_0, \mathbf{x}_1) \oplus f(\mathbf{x}_0, \mathbf{x}_1)) = \overline{g(\mathbf{x}_0, \mathbf{x}_1)} \wedge \underset{\mathbf{x}_0}{\min} f(\mathbf{x}_0, \mathbf{x}_1)$$

$$\vee \, g(\mathbf{x}_0, \mathbf{x}_1) \wedge \underset{\mathbf{x}_0}{\min} \overline{f(\mathbf{x}_0, \mathbf{x}_1)},$$

$$\underset{\mathbf{x}_0}{\max}(g(\mathbf{x}_0, \mathbf{x}_1) \oplus f(\mathbf{x}_0, \mathbf{x}_1)) = \overline{g(\mathbf{x}_0, \mathbf{x}_1)} \wedge \underset{\mathbf{x}_0}{\max} f(\mathbf{x}_0, \mathbf{x}_1)$$

$$\vee \, g(\mathbf{x}_0, \mathbf{x}_1) \wedge \underset{\mathbf{x}_0}{\max} \overline{f(\mathbf{x}_0, \mathbf{x}_1)}.$$

There are many applications of vectorial derivative operations that will be explained on other places of this book. Here we explore the utilization of vectorial derivatives to detect the symmetry of a logic function with regard to the pair of variables (x_i, x_j) and with regard to all variables.

A logic function $f(x_i, x_j, \mathbf{x}_1)$ is *symmetric* with regard to the pair of variables (x_i, x_j) if it satisfies

$$(x_i \oplus x_j) \wedge \underset{(x_i, x_j)}{\operatorname{der}} f(x_i, x_j, \mathbf{x}_1) = 0. \tag{6.27}$$

This property can be satisfied for several variables $(x_1, \ldots, x_k, \mathbf{x}_1)$:

$$\bigvee_{j=2}^{k} (x_1 \oplus x_j) \wedge \underset{(x_1, x_j)}{\operatorname{der}} f(x_1, \ldots, x_k, \mathbf{x}_1) = 0, \tag{6.28}$$

and for all variables $\mathbf{x} = (x_i, \mathbf{x}_1)$:

$$\bigvee_{x_j \in \mathbf{x}_1} (x_i \oplus x_j) \wedge \underset{(x_i, x_j)}{\operatorname{der}} f(x_i, \mathbf{x}_1) = 0. \tag{6.29}$$

The rule (6.27) is used in (6.28) for $k - 1$ pairs of variables. All other directions of change must not be explicitly evaluated due to the transitivity of the symmetry relation. The evaluation of $n - 1$ pairs of variables is for the same reason sufficient to verify in (6.29) whether a logic function of n variables is totally symmetric.

6.3 Single Derivative Operations

The single derivative operations may be considered as a special case of the vectorial derivative operations where the vector $\mathbf{x}_0$ contains only the single variable x_i. Hence, all rules of Sect. 6.2 can also be used for single derivative operations; the vector $\mathbf{x}_0$ must only be replaced by the single variable x_i.

Figure 6.3 visualizes this property for the Boolean space $\mathbb{B}^3$. Figure 6.3a indicates by solid arrows the three pairs of function values for the point $(x_1, x_2, x_3) = (100)$ where a single variable is changed. Dashed arrows show the remaining $(2^3 - 1) - 3 = 4$ directions of change. Figure 6.3b depicts the four pairs of function values where the single variable x_1 changes its value.

Knowing the definitions of vectorial derivative operations, we define all single derivative operations together by replacing the vector $\mathbf{x}_0$ by the single variable x_i:

Definition 6.4 Let $f(\mathbf{x}) = f(x_1, \ldots, x_i, \ldots, x_n)$ be a logic function of n variables; then

$$\det_{x_i} f(\mathbf{x}) = f(x_1, \ldots, x_i, \ldots, x_n) \oplus f(x_1, \ldots, \overline{x}_i, \ldots, x_n) \qquad (6.30)$$

is the (*single*) *derivative*,

$$\operatorname{nder}_{x_i} f(\mathbf{x}) = f(x_1, \ldots, x_i, \ldots, x_n) \odot f(x_1, \ldots, \overline{x}_i, \ldots, x_n) \qquad (6.31)$$

is the *negated* (*single*) *derivative*,

$$\min_{x_i} f(\mathbf{x}) = f(x_1, \ldots, x_i, \ldots, x_n) \wedge f(x_1, \ldots, \overline{x}_i, \ldots, x_n) \qquad (6.32)$$

is the (*single*) *minimum*,

$$\max_{x_i} f(\mathbf{x}) = f(x_1, \ldots, x_i, \ldots, x_n) \vee f(x_1, \ldots, \overline{x}_i, \ldots, x_n) \qquad (6.33)$$

is the (*single*) *maximum*, of the function $f(\mathbf{x})$ with regard to the variable x_i.

Fig. 6.3 Directions of change in the Boolean space $\mathbb{B}^3$: (**a**) all directions of change shown for the point $(x_1, x_2, x_3) = (100)$ (dashed if more than one variable is changed), (**b**) pairs of function values for the directions of change of the single variable x_1

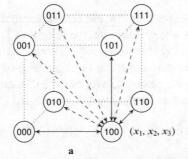
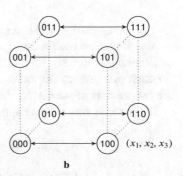

Table 6.4 Calculation of
the derivative of the logic
function $f(a, b, c)$ (6.34)
with regard to c

Pair	a	b	c	$f(a, b, c)$	$\text{der}_c\, f(a, b, c)$
1	0	0	0	1	1
	0	0	1	0	1
2	0	1	0	0	0
	0	1	1	0	0
3	1	0	0	0	0
	1	0	1	0	0
4	1	1	0	1	0
	1	1	1	1	0

Note 6.3 The term *single* emphasizes that the value of a single variable will change. This term is used to distinguish single derivative operations from vectorial derivative operations. In general, the term *single* is omitted for single derivative operations.

Note 6.4 This negated single derivative (6.31) can be expressed by the negation of the single derivative (6.30):

$$\operatorname*{nder}_{x_i} f(x_i, \mathbf{x}_1) = \overline{\operatorname*{der}_{x_i} f(x_i, \mathbf{x}_1)}\,.$$

Hence, we avoid the detailed exploration of $\operatorname{nder}_{x_i} f(x_i, \mathbf{x}_1)$.

It follows from (6.30) that the derivative $\operatorname{der}_{x_i} f(\mathbf{x})$ is a logic function that is equal to 1 if the function value of $f(x_i, \mathbf{x}_1)$ differs from the function value after the change of x_i, expressed by $f(\overline{x}_i, \mathbf{x}_1)$. With other words, the derivative of $f(\mathbf{x})$ with regard to x_i describes whether the change of the variable x_i causes a change of the function value.

Example 6.6 Let

$$f(a, b, c) = \overline{a}\,\overline{b}\,\overline{c} \oplus a\,b\,. \tag{6.34}$$

Based on (6.30) the derivative $\operatorname{der}_c f(a, b, c)$ can be calculated as follows:

$$\operatorname*{der}_c f(a, b, c) = f(a, b, c) \oplus f(a, b, \overline{c})$$

$$= (\overline{a}\,\overline{b}\,\overline{c} \oplus a\,b) \oplus (\overline{a}\,\overline{b}\,c \oplus a\,b)$$

$$= \overline{a}\,\overline{b}\,(\overline{c} \oplus c)$$

$$= \overline{a}\,\overline{b}.$$

The interpretation of this result is easy when we use the representation of Table 6.4.

The variable c has for each pair of arguments either the value 0 or 1, and the values for a and b are constant equal to (00), (01), (10), and (11), respectively. The antivalence of the values of $f(a, b, c)$ for each pair defines the values of the derivative; the function value of the derivative is the same for the two vectors of such a pair due to the commutativity of the antivalence used in (6.30).

In the same way we get the derivatives for the two other variables:

$$\operatorname*{der}_a f(a, b, c) = (\overline{a}\,\overline{b}\,\overline{c} \oplus a\,b) \oplus (a\,\overline{b}\,\overline{c} \oplus \overline{a}\,b) = \overline{b}\,\overline{c} \oplus b = b \vee \overline{c}\,,$$

$$\operatorname*{der}_b f(a, b, c) = (\overline{a}\,\overline{b}\,\overline{c} \oplus a\,b) \oplus (\overline{a}\,b\,\overline{c} \oplus a\,\overline{b}) = \overline{a}\,\overline{c} \oplus a = a \vee \overline{c}\,.$$

The derivative of $f(\mathbf{x})$ with regard to one single variable does not depend on this variable anymore. This property is explicitly observable by the following:

Theorem 6.3 *It holds for* $f(\mathbf{x}) = f(x_i, \mathbf{x}_1)$ *that*

$$\mathop{\mathrm{der}}_{x_i} f(\mathbf{x}) = f(x_i = 0, \mathbf{x}_1) \oplus f(x_i = 1, \mathbf{x}_1) \,. \tag{6.35}$$

This formula is important for numerical calculations since the direct use of the constants can result in a considerable amount of simplifications. On the other side it can be used to determine whether a function $f(\mathbf{x})$ does not depend on the variable x_i:

Theorem 6.4 *A logic function* $f(\mathbf{x}) = f(x_i, \mathbf{x}_1)$ *is* independent *of the variable* x_i *if and only if*

$$\mathop{\mathrm{der}}_{x_i} f(\mathbf{x}) = 0 \,. \tag{6.36}$$

Proof A function $f(x_i, \mathbf{x}_1)$ is independent of the variable x_i

$$f(x_i = 0, \mathbf{x}_1) = f(x_i = 1, \mathbf{x}_1) \,.$$

This equation can be equivalently transformed into

$$f(x_i = 0, \mathbf{x}_1) \oplus f(x_i = 1, \mathbf{x}_1) = 0 \,,$$

$$\mathop{\mathrm{der}}_{x_i} f(x_i, \mathbf{x}_1) = 0 \,.$$

All these steps can be executed in the reverse direction. □

The other extreme case of the derivative of $f(\mathbf{x}) = f(x_i, \mathbf{x}_1)$ with regard to x_i can be used to verify the following property of logic function:

Theorem 6.5 *A logic function* $f(\mathbf{x}) = f(x_i, \mathbf{x}_1)$ *is* linear *in* x_i *if and only if*

$$\mathop{\mathrm{der}}_{x_i} f(x_i, \mathbf{x}_1) = 1 \,. \tag{6.37}$$

Proof A logic function $f(x_i, \mathbf{x}_1)$ is linear in x_i if it can be expressed by

$$f(x_i, \mathbf{x}_1) = x_i \oplus f(\mathbf{x}_1) \,. \tag{6.38}$$

We show first that the linear function (6.38) satisfies the condition (6.37). The substitution of (6.38) into (6.37) leads to

$$\mathop{\mathrm{der}}_{x_i}(x_i \oplus f(\mathbf{x}_1)) = (0 \oplus f(\mathbf{x}_1)) \oplus (1 \oplus f(\mathbf{x}_1))$$

$$= 1 \,.$$

For the proof of the reverse direction we use (6.35) of Theorem 6.3 in (6.37) and get

$$f(x_i = 0, \mathbf{x}_1) \oplus f(x_i = 1, \mathbf{x}_1) = 1,$$
$$f(x_i = 1, \mathbf{x}_1) = 1 \oplus f(x_i = 0, \mathbf{x}_1),$$
$$f(x_i = 1, \mathbf{x}_1) = \overline{f(x_i = 0, \mathbf{x}_1)}. \tag{6.39}$$

The Shannon decomposition of an arbitrary function $f(x_i, \mathbf{x}_1)$ is

$$f(x_i, \mathbf{x}_1) = \overline{x}_i f(x_i = 0, \mathbf{x}_1) \vee x_i f(x_i = 1, \mathbf{x}_1),$$

and the substitution of (6.39) describes functions that satisfy (6.37)

$$f(x_i, \mathbf{x}_1) = \overline{x}_i f(x_i = 0, \mathbf{x}_1) \vee x_i \overline{f(x_i = 0, \mathbf{x}_1)}.$$

Choosing the cofactor $f(x_i = 0, \mathbf{x}_1)$ as function $f(\mathbf{x}_1)$, we get

$$f(x_i, \mathbf{x}_1) = \overline{x}_i f(\mathbf{x}_1) \vee x_i \overline{f(\mathbf{x}_1)}$$
$$= x_i \oplus f(\mathbf{x}_1),$$

which is the wanted logic function that is linear in x_i.   □

The minimum of $f(\mathbf{x})$ is less than or equal to the logic function $f(\mathbf{x})$ because $f(\mathbf{x})$ occurs as part of a $\wedge$-operation in (6.32) of Definition 6.4. Therefore, this derivative operation is denoted by *minimum*.

Example 6.7 We take the function (6.34) from the previous example and calculate the minimum $\min_c f(a, b, c)$:

$$\min_c f(a, b, c) = f(a, b, c) \wedge f(a, b, \overline{c})$$
$$= (\overline{a}\,\overline{b}\,\overline{c} \oplus a b) \wedge (\overline{a}\,\overline{b}\, c \oplus a b)$$
$$= \overline{a}\,\overline{b}\,\overline{c} \wedge \overline{a}\,\overline{b}\, c \oplus \overline{a}\,\overline{b}\,\overline{c} \wedge a b \oplus a b \wedge \overline{a}\,\overline{b}\, c \oplus a b \wedge a b$$
$$= a b.$$

In order to get a good understanding of this operation, we use Table 6.5 with the same structure as above.

The two values of $f(a, b, c)$ within one group have now to be combined by the conjunction $\wedge$. The relation

$$\min_c f(a, b, c) \le f(a, b, c)$$

can be seen immediately. Again the result does not depend on the variable c.

The results for the minimum with regard to the two other variables are not very surprising because the function of the example has many values 0:

$$\min_a f(a, b, c) = (\overline{a}\,\overline{b}\,\overline{c} \oplus a b) \wedge (a\,\overline{b}\,\overline{c} \oplus \overline{a}\, b) = 0,$$

Table 6.5 Calculation of the minimum of the logic function $f(a, b, c)$ (6.34) with regard to c

Pair	a	b	c	$f(a, b, c)$	$\min_c f(a, b, c)$
1	0	0	0	1	0
	0	0	1	0	0
2	0	1	0	0	0
	0	1	1	0	0
3	1	0	0	0	0
	1	0	1	0	0
4	1	1	0	1	1
	1	1	1	1	1

$$\min_b f(a, b, c) = (\overline{a}\,\overline{b}\,\overline{c} \oplus a\,b) \wedge (\overline{a}\,b\,\overline{c} \oplus a\,\overline{b}) = 0 \, .$$

Theorem 6.6 *It holds for* $f(\mathbf{x}) = f(x_i, \mathbf{x}_1)$ *that*

$$\min_{x_i} f(\mathbf{x}) = f(x_i = 0, \mathbf{x}_1) \wedge f(x_i = 1, \mathbf{x}_1) \, . \tag{6.40}$$

When we take the two equations

$$\min_{x_i} f(x_i, \mathbf{x}_1) = 0 \quad \text{and} \quad \min_{x_i} f(x_i, \mathbf{x}_1) = 1$$

under consideration, then we gain the following insights:

- The vectors with $\min_{x_i} f(x_i, \mathbf{x}_1) = 0$ are the vectors for which

$$f(x_i = 0, \mathbf{x}_1) = 0 \quad \text{and} \quad f(x_i = 1, \mathbf{x}_1) = 0 \, ,$$
$$f(x_i = 0, \mathbf{x}_1) = 0 \quad \text{and} \quad f(x_i = 1, \mathbf{x}_1) = 1 \, , \text{ or}$$
$$f(x_i = 0, \mathbf{x}_1) = 1 \quad \text{and} \quad f(x_i = 1, \mathbf{x}_1) = 0 \, .$$

- The vectors with $\min_{x_i} f(\mathbf{x}) = 1$ are the vectors for which the change of the variable x_i leaves the values of $f(x_i, \mathbf{x}_1)$ constant equal to 1:

$$f(x_i = 0, \mathbf{x}_1) = 1 \quad \text{and} \quad f(x_i = 1, \mathbf{x}_1) = 1 \, .$$

Finally, we explore the properties of the single maximum of the logic function $f(x_i, \mathbf{x}_1)$ with regard to x_i. The name *maximum* emphasizes that the result of this derivative operation is larger than or equal to the basic function $f(\mathbf{x})$ because $f(\mathbf{x})$ occurs as part of a $\vee$-operation in (6.33).

The following characteristics of the *single maximum* can be seen:

- The vectors with $\max_{x_i} f(x_i, \mathbf{x}_1) = 0$ are the vectors for which the change of the variable x_i leaves the values of $f(x_i, \mathbf{x}_1)$ constant equal to 0:

$$f(x_i = 0, \mathbf{x}_1) = 0 \quad \text{and} \quad f(x_i = 1, \mathbf{x}_1) = 0 \, .$$

- The vectors with $\max_{x_i} f(x_i, \mathbf{x}_1) = 1$ are the vectors for which

Table 6.6 Calculation of the maximum of the logic function $f(a, b, c)$ (6.34) with regard to c

Pair	a	b	c	$f(a, b, c)$	$\max_c f(a, b, c)$
1	0	0	0	1	1
	0	0	1	0	1
2	0	1	0	0	0
	0	1	1	0	0
3	1	0	0	0	0
	1	0	1	0	0
4	1	1	0	1	1
	1	1	1	1	1

$$f(x_i = 0, \mathbf{x}_1) = 0 \quad \text{and} \quad f(x_i = 1, \mathbf{x}_1) = 1 \,,$$

$$f(x_i = 0, \mathbf{x}_1) = 1 \quad \text{and} \quad f(x_i = 1, \mathbf{x}_1) = 0 \,, \text{ or}$$

$$f(x_i = 0, \mathbf{x}_1) = 1 \quad \text{and} \quad f(x_i = 1, \mathbf{x}_1) = 1 \,.$$

Example 6.8 We take again the function (6.34) from the previous examples, transform it into $f(a, b, c) = \overline{a}\,\overline{b}\,\overline{c} \vee a\,b$, and calculate the single maximum $\max_c f(a, b, c)$:

$$\max_c f(a, b, c) = f(a, b, c) \vee f(a, b, \overline{c})$$

$$= (\overline{a}\,\overline{b}\,\overline{c} \vee a\,b) \vee (\overline{a}\,\overline{b}\,c \vee a\,b)$$

$$= \overline{a}\,\overline{b}\,(\overline{c} \vee c) \vee a\,b$$

$$= \overline{a}\,\overline{b} \vee a\,b.$$

Table 6.6 uses the same structure as above and helps us to get a good understanding of the single maximum.

The two values of $f(a, b, c)$ within one group have now to be combined by the disjunction $\vee$. The relation

$$f(a, b, c) \leq \max_c f(a, b, c)$$

can be seen immediately. Again the result does not depend on the variable c.

The results for the maximum with regard to the two other variables are

$$\max_a f(a, b, c) = (\overline{a}\,\overline{b}\,\overline{c} \oplus a\,b) \vee (a\,\overline{b}\,\overline{c} \oplus \overline{a}\,b) = \overline{b}\,\overline{c} \vee b = b \vee \overline{c} \,,$$

$$\max_b f(a, b, c) = (\overline{a}\,\overline{b}\,\overline{c} \oplus a\,b) \vee (\overline{a}\,b\,\overline{c} \oplus a\,\overline{b}) = \overline{a}\,\overline{c} \vee a = a \vee \overline{c} \,.$$

The next theorem works in the same way as for the previous operations.

Theorem 6.7 *It holds for* $f(\mathbf{x}) = f(x_i, \mathbf{x}_1)$ *that*

$$\max_{x_i} f(\mathbf{x}) = f(x_i = 0, \mathbf{x}_1) \vee f(x_i = 1, \mathbf{x}_1) \,. \tag{6.41}$$

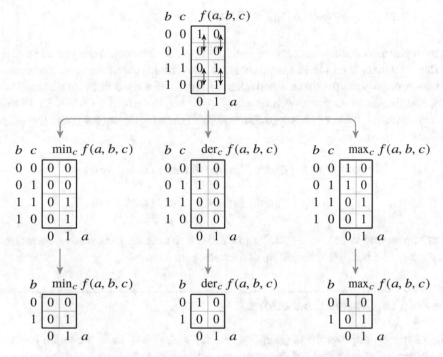

Fig. 6.4 The Karnaugh-maps of $f(a, b, c) = \overline{a}\,\overline{b}\,\overline{c} \oplus a\,b$ and all single derivative operations with regard to c

Figure 6.4 summarizes the results of the previous examples. The small arrows within the Karnaugh-map of $f(a, b, c)$ indicate the pairs of function values that cause the calculated results of the derivative operations with regard to c.

The Karnaugh-maps in the middle row of Fig. 6.4 show the results of the derivative operations calculated by (6.32), (6.30), and (6.33) of Definition 6.4. These Karnaugh-maps confirm that single derivative operations do not depend on the variable that has been selected for the calculation. Hence, single derivative operations satisfy

$$\operatorname*{der}_{x_i}(\operatorname*{der}_{x_i} f(x_i, \mathbf{x}_1)) = 0 \,,$$

$$\operatorname*{der}_{x_i}(\operatorname*{min}_{x_i} f(x_i, \mathbf{x}_1)) = 0 \,,$$

$$\operatorname*{der}_{x_i}(\operatorname*{max}_{x_i} f(x_i, \mathbf{x}_1)) = 0 \,,$$

which confirm the welcome effect of simplification:

> Each single derivative operation of a logic function of n variables with regard to the variable x_i, the given function is depending on, results in a *simpler* logic function of $n - 1$ variables that does not depend on the variable x_i anymore.

The comparison of the left and right Karnaugh-maps in the middle row and the Karnaugh-map of the given function confirms the $\leq$-relation between these three functions. In general, it holds that

$$\min_{x_i} f(x_i, \mathbf{x}_1) \leq f(x_i, \mathbf{x}_1) \leq \max_{x_i} f(x_i, \mathbf{x}_1) \,. \tag{6.42}$$

The Karnaugh-maps in the lowest row of Fig. 6.4 show the reduction when the variable c has been omitted. They originate from the Karnaugh-maps in the middle row of Fig. 6.4. The comparison of these Karnaugh-maps confirms that the antivalence of all three simple derivative operations is equal to 0 (6.43), and the single minimum is orthogonal to the single derivative (6.44). These rules can be easily verified by substitution of (6.30), (6.32), and (6.33) into (6.43) or (6.44) and the simplification of the resulting expression:

$$\min_{x_i} f(x_i, \mathbf{x}_1) \oplus \operatorname{der}_{x_i} f(x_i, \mathbf{x}_1) \oplus \max_{x_i} f(x_i, \mathbf{x}_1) = 0 \,, \tag{6.43}$$

$$\min_{x_i} f(x_i, \mathbf{x}_1) \wedge \operatorname{der}_{x_i} f(x_i, \mathbf{x}_1) = 0 \,. \tag{6.44}$$

These rules correspond to the rules (6.15) and (6.16) of the vectorial derivative operations, and all derived rules can also be applied for single derivative operations.

6.4   *k*-Fold Derivative Operations

Since single derivative operations of $f(x_1, x_2, \ldots, x_n)$ with regard to x_i are again logic functions, further single derivative operations of the same type with regard to another variable can be calculated.

It is known from (6.35), (6.40), and (6.41) that the result of each single derivative operation does not depend on the variable that already has been used for this calculation.

Consequently:

The result of k-fold derivative operations of $f(x_1, x_2, \ldots, x_k, x_{k+1}, \ldots, x_n)$ with regard to $\mathbf{x}_0 = (x_1, x_2, \ldots, x_k)$ depends only on $n - k$ variables $\mathbf{x}_1 = (x_{k+1}, \ldots, x_n)$. Therefore, k-fold derivative operations describe properties of whole subspaces specified by $\mathbf{x}_1 = const$.

Definition 6.5 Let $\mathbf{x}_0 = (x_1, x_2, \ldots, x_k)$, $\mathbf{x}_1 = (x_{k+1}, \ldots, x_n)$ be two disjoint sets of Boolean variables, and $f(\mathbf{x}_0, \mathbf{x}_1) = f(\mathbf{x})$ a logic function of n variables; then

$$\operatorname{der}^k_{\mathbf{x}_0} f(\mathbf{x}_0, \mathbf{x}_1) = \operatorname{der}_{x_k} \left(\ldots \left(\operatorname{der}_{x_2} \left(\operatorname{der}_{x_1}(f(\mathbf{x}_0, \mathbf{x}_1)) \right) \right) \ldots \right) \tag{6.45}$$

is the *k-fold derivative* of the logic function $f(\mathbf{x}_0, \mathbf{x}_1)$ with regard to the subset of variables $\mathbf{x}_0$.

The k-fold derivative is a logic function that is equal to 1 for such subspaces $\mathbf{x}_1 = const.$ where the function $f(\mathbf{x}_0, \mathbf{x}_1 = const.)$ has an odd number of function values 1. The following example shows this property.

Example 6.9 Let $f(a, b, c, d) = b\,d \vee \overline{a}\,(c \vee \overline{d})$. Based on (6.45) of Definition 6.5 and (6.35) of Theorem 6.3, the 2-fold derivative $\operatorname{der}^2_{(c,d)} f(a, b, c, d)$ can be calculated as follows:

$$\operatorname{der}^2_{(c,d)} f(a, b, c, d) = \operatorname{der}_d \left(f(a, b, c = 0, d) \oplus f(a, b, c = 1, d) \right)$$

$$= f(a, b, c = 0, d = 0) \oplus f(a, b, c = 1, d = 0)$$

$$\oplus f(a, b, c = 0, d = 1) \oplus f(a, b, c = 1, d = 1)$$

$$= \overline{a} \oplus \overline{a} \oplus b \oplus (b \vee \overline{a})$$

$$= b \oplus b \oplus \overline{a} \oplus \overline{a} b$$

$$= \overline{a}(1 \oplus b)$$

$$= \overline{a}\,\overline{b}\ .$$

In order to evaluate the properties of the 2-fold derivative, we calculate the four subfunctions of $f(a, b, c, d)$ depending on (c, d),

$$f(a = 0, b = 0, c, d) = c \vee \overline{d}\ ,$$

$$f(a = 0, b = 1, c, d) = d \vee c \vee \overline{d} = 1\ ,$$

$$f(a = 1, b = 0, c, d) = 0\ ,$$

$$f(a = 1, b = 1, c, d) = d\ ,$$

and we observe that only for $a = 0, b = 0$ three function values 1 exist determined by $c \vee \overline{d} = 1$. Because this is the only subspace with an odd number of function values 1, the 2-fold derivative is equal to 1 only for $(a = 0, b = 0)$.

Definition 6.6 Let $\mathbf{x}_0 = (x_1, x_2, \ldots, x_k)$, $\mathbf{x}_1 = (x_{k+1}, \ldots, x_n)$ be two disjoint sets of Boolean variables, and $f(\mathbf{x}_0, \mathbf{x}_1) = f(\mathbf{x})$ a logic function of n variables; then

$$\min_{\mathbf{x}_0}^k f(\mathbf{x}_0, \mathbf{x}_1) = \min_{x_k} \left(\ldots \left(\min_{x_2} \left(\min_{x_1} (f(\mathbf{x}_0, \mathbf{x}_1)) \right) \right) \ldots \right) \tag{6.46}$$

is the *k-fold minimum* of the logic function $f(\mathbf{x}_0, \mathbf{x}_1)$ with regard to the subset of variables $\mathbf{x}_0$.

The k-fold minimum is a logic function that is equal to 1 for such subspaces $\mathbf{x}_1 = const.$ where the function $f(\mathbf{x}_0, \mathbf{x}_1 = const.)$ is constant equal to 1 for all patterns of the remaining variables $\mathbf{x}_0$.

Example 6.10 We take again the function $f(a, b, c, d) = bd \vee \overline{a}(c \vee \overline{d})$. Based on (6.46) of Definition 6.6 and (6.40) of Theorem 6.6, the 2-fold minimum

$$\min_{(c,d)}^2 f(a, b, c, d)$$

can be calculated as follows:

$$\min_{(c,d)}^2 f(a, b, c, d) = \min_d \left(f(a, b, c = 0, d) \wedge f(a, b, c = 1, d) \right)$$

$$= f(a, b, c = 0, d = 0) \wedge f(a, b, c = 1, d = 0)$$

$$\wedge f(a, b, c = 0, d = 1) \wedge f(a, b, c = 1, d = 1)$$

$$= \overline{a} \wedge \overline{a} \wedge b \wedge (b \vee \overline{a})$$

$$= \overline{a}\,b \vee \overline{a}\,b$$

$$= \overline{a}\,b .$$

Only $f(a = 0, b = 1, c, d)$ is constant equal to 1.

It follows from (6.42) and (6.46) that the k-fold minimum of a logic function with regard to a set of variables is less than or equal to a $(k - 1)$-fold minimum of the same function with regard to a subset of this set of variables:

$$\min_{(x_i, \mathbf{x}_0)}{}^k f(x_i, \mathbf{x}_0, \mathbf{x}_1) \leq \min_{\mathbf{x}_0}{}^{k-1} f(x_i, \mathbf{x}_0, \mathbf{x}_1) \leq f(x_i, \mathbf{x}_0, \mathbf{x}_1) . \tag{6.47}$$

Definition 6.7 Let $\mathbf{x}_0 = (x_1, x_2, \ldots, x_k)$, $\mathbf{x}_1 = (x_{k+1}, \ldots, x_n)$ be two disjoint sets of Boolean variables, and $f(\mathbf{x}_0, \mathbf{x}_1) = f(\mathbf{x})$ a logic function of n variables; then

$$\max_{\mathbf{x}_0}{}^k f(\mathbf{x}_0, \mathbf{x}_1) = \max_{x_k} \left(\ldots \left(\max_{x_2} \left(\max_{x_1} (f(\mathbf{x}_0, \mathbf{x}_1)) \right) \right) \ldots \right) \tag{6.48}$$

is the *k-fold maximum* of the logic function $f(\mathbf{x}_0, \mathbf{x}_1)$ with regard to the subset of variables $\mathbf{x}_0$.

The k-fold maximum is a logic function that is equal to 1 for such subspaces $\mathbf{x}_1 = const.$ where at least one function value of the function $f(\mathbf{x}_0, \mathbf{x}_1 = const.)$ is equal to 1.

Example 6.11 We take again the same function $f(a, b, c, d) = b\,d \vee \overline{a}\,(c \vee \overline{d})$ and calculate the 2-fold maximum $\max_{(c,d)}{}^2 f(a, b, c, d)$ based on (6.48) of Definitions 6.7 and (6.41) of Theorem 6.7:

$$\max_{(c,d)}^2 f(a, b, c, d) = \max_d \left(f(a, b, c = 0, d) \vee f(a, b, c = 1, d) \right)$$

$$= f(a, b, c = 0, d = 0) \vee f(a, b, c = 1, d = 0)$$

$$\vee f(a, b, c = 0, d = 1) \vee f(a, b, c = 1, d = 1)$$

$$= \overline{a} \vee \overline{a} \vee b \vee (b \vee \overline{a})$$

$$= \overline{a} \vee b .$$

In three subspaces $(a = 0, b = 0)$, $(a = 0, b = 1)$, and $(a = 1, b = 1)$ exist function values 1 of $f(a, b, c, d)$. Only the function $f(a = 1, b = 0, c, d)$ is constant equal to 0.

It follows from (6.42) and (6.48) of Definition 6.7 that the k-fold maximum of a logic function with regard to a set of variables is larger than or equal to a $(k - 1)$-fold maximum of the same function with regard to a subset of variables:

$$f(x_i, \mathbf{x}_0, \mathbf{x}_1) \leq \max_{\mathbf{x}_0}{}^{k-1} f(x_i, \mathbf{x}_0, \mathbf{x}_1) \leq \max_{(x_i, \mathbf{x}_0)}{}^k f(x_i, \mathbf{x}_0, \mathbf{x}_1) . \tag{6.49}$$

Another k-fold derivative operation can be defined as a useful abbreviation.

Definition 6.8 Let $\mathbf{x}_0 = (x_1, x_2, \ldots, x_k)$, $\mathbf{x}_1 = (x_{k+1}, \ldots, x_n)$ be two disjoint sets of Boolean variables, and $f(\mathbf{x}_0, \mathbf{x}_1) = f(\mathbf{x})$ a logic function of n variables; then

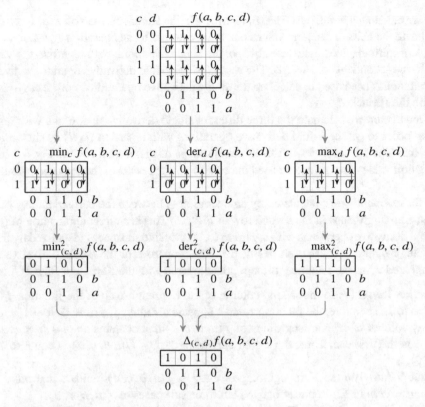

Fig. 6.5 The Karnaugh-maps of the function $f(a, b, c, d) = b\,d \vee \overline{a}\,(c \vee \overline{d})$ and all 2-fold derivative operations with regard to (c, d)

$$\Delta_{\mathbf{x}_0} f(\mathbf{x}_0, \mathbf{x}_1) = \min_{\mathbf{x}_0}^{k} f(\mathbf{x}_0, \mathbf{x}_1) \oplus \max_{\mathbf{x}_0}^{k} f(\mathbf{x}_0, \mathbf{x}_1) \tag{6.50}$$

is the Δ-operation of the logic function $f(\mathbf{x}_0, \mathbf{x}_1)$ with regard to the set of variables $\mathbf{x}_0$.

It follows from (6.50), (6.47), and (6.49) that $\Delta_{\mathbf{x}_0} f(\mathbf{x}_0, \mathbf{x}_1)$ is equal to 1 if $\min_{\mathbf{x}_0}^{k} f(\mathbf{x}_0, \mathbf{x}_1) = 0$ and $\max_{\mathbf{x}_0}^{k} f(\mathbf{x}_0, \mathbf{x}_1) = 1$.

Consequently, the Δ-operation characterizes such subspaces where the function is not constant.

Example 6.12 We reuse the results from the previous examples and calculate the Δ-operation of $f(a, b, c, d) = b\,d \vee \overline{a}\,(c \vee \overline{d})$ with regard to (c, d):

$$\Delta_{(c,d)} f(a, b, c, d) = \min_{(c,d)}^{2} f(a, b, c, d) \oplus \max_{(c,d)}^{2} f(a, b, c, d)$$

$$= (\overline{a}\,b) \oplus (\overline{a} \vee b)$$

$$= \overline{a}\,b \oplus \overline{a} \oplus b \oplus \overline{a}\,b$$

$$= \overline{a} \oplus b\,.$$

Figure 6.5 summarizes the results of all k-fold derivative operations of $f(a, b, c, d)$ with regard to (c, d), calculated in the examples above. Based on (6.46), (6.45), and (6.48), the k-fold derivative operations are calculated iteratively. The small arrows in the Karnaugh-map of $f(a, b, c, d)$ on top of Fig. 6.5 indicate the pairs of function values that are connected by an operation $\wedge$, $\oplus$, or $\vee$, in order to get the single minimum, the single derivative, or the single maximum with regard to the variable d as visualized in the second row of Fig. 6.5. The results of the k-fold derivative operations do not depend on the order that has been used to calculate the single derivative operations. For an easy visualization we start with the variable d.

In the second step we calculate the three different single derivative operations with regard to the variable c in order to get the k-fold derivative operations with regard to (c, d) as shown in the third row of Fig. 6.5. The small arrows in the Karnaugh-maps of the second row of Fig. 6.5 indicate the pairs of function values that have to be used in order to get the results of the required k-fold derivative operations.

Finally, the Δ-operation is calculated by an operation $\oplus$ between the corresponding values of the first and last Karnaugh-maps in the second row of Fig. 6.5. This is a direct application of (6.50).

The 2-fold derivative operations with regard to (c, d) evaluate subspaces visualized by the columns of the Karnaugh-map of $f(a, b, c, d)$ in Fig. 6.5. The properties of the k-fold derivative operations can be recognized when their Karnaugh-maps are compared with the Karnaugh-map of $f(a, b, c, d)$:

- *All function values* ($\forall$) in the second column of the Karnaugh-map of the function $f(a, b, c, d)$ are equal to 1. Therefore, the function value 1 appears for $\min_{(c,d)}^2 f(a = 0, b = 1, c, d)$.
- The only column of the Karnaugh-map of $f(a, b, c, d)$ that contains *an odd number of function values* 1 is the first one. Thus, the 2-fold derivative $\text{der}_{(c,d)}^2 f(a, b, c, d)$ is equal to 1 for $(a = 0, b = 0)$.
- The three values 1 in the Karnaugh-map of $\max_{(c,d)}^2 f(a, b, c, d)$ indicate that *there is at least one function value* 1 ($\exists$) in the associated left three subspaces of $f(a, b, c, d)$.
- *Both function values, 0 and 1, occur* in the first and third columns of the Karnaugh-map of $f(a, b, c, d)$. Therefore, $\Delta_{(c,d)} f(a, b, c, d)$ is equal to 1 for $(a = 0, b = 0)$ and $(a = 1, b = 1)$.

There are many relations between k-fold derivative operations. The general equation (6.51) follows from (6.50) of Definition 6.8:

$$\min_{\mathbf{x}_0}^k f(\mathbf{x}_0, \mathbf{x}_1) \oplus \Delta_{\mathbf{x}_0} f(\mathbf{x}_0, \mathbf{x}_1) \oplus \max_{\mathbf{x}_0}^k f(\mathbf{x}_0, \mathbf{x}_1) = 0 \ . \tag{6.51}$$

Using two of the k-fold derivative operations, the third k-fold derivative operation can be calculated:

$$\min_{\mathbf{x}_0}^k f(\mathbf{x}_0, \mathbf{x}_1) = \Delta_{\mathbf{x}_0} f(\mathbf{x}_0, \mathbf{x}_1) \oplus \max_{\mathbf{x}_0}^k f(\mathbf{x}_0, \mathbf{x}_1) \ ,$$

$$\max_{\mathbf{x}_0}^k f(\mathbf{x}_0, \mathbf{x}_1) = \Delta_{\mathbf{x}_0} f(\mathbf{x}_0, \mathbf{x}_1) \oplus \min_{\mathbf{x}_0}^k f(\mathbf{x}_0, \mathbf{x}_1) \ .$$

Some more relations can be found as before:

$$\min_{\mathbf{x}_0}^k f(\mathbf{x}_0, \mathbf{x}_1) \leq f(\mathbf{x}_0, \mathbf{x}_1) \leq \max_{\mathbf{x}_0}^k f(\mathbf{x}_0, \mathbf{x}_1) \ ,$$

$$\min_{\mathbf{x}_0}^k f(\mathbf{x}_0, \mathbf{x}_1) \wedge \overline{\max_{\mathbf{x}_0}^k f(\mathbf{x}_0, \mathbf{x}_1)} = 0 \ .$$

The Δ-operation is orthogonal to the complement of the k-fold maximum:

$$\Delta_{\mathbf{x}_0} f(\mathbf{x}_0, \mathbf{x}_1) \wedge \overline{\max_{\mathbf{x}_0}^k f(\mathbf{x}_0, \mathbf{x}_1)}$$

$$= (\min_{\mathbf{x}_0}^k f(\mathbf{x}_0, \mathbf{x}_1) \oplus \max_{\mathbf{x}_0}^k f(\mathbf{x}_0, \mathbf{x}_1)) \wedge \overline{\max_{\mathbf{x}_0}^k f(\mathbf{x}_0, \mathbf{x}_1)}$$

$$= \min_{\mathbf{x}_0}^k f(\mathbf{x}_0, \mathbf{x}_1) \wedge \overline{\max_{\mathbf{x}_0}^k f(\mathbf{x}_0, \mathbf{x}_1)}$$

$$= \min_{\mathbf{x}_0}^k f(\mathbf{x}_0, \mathbf{x}_1) \wedge \min_{\mathbf{x}_0}^k \overline{f(\mathbf{x}_0, \mathbf{x}_1)}$$

$$= \min_{\mathbf{x}_0}^k \left(f(\mathbf{x}_0, \mathbf{x}_1) \wedge \overline{f(\mathbf{x}_0, \mathbf{x}_1)} \right)$$

$$= 0 .$$

Furthermore, the orthogonality between the *k*-fold minimum and the Δ-operation follows from (6.50):

$$\min_{\mathbf{x}_0}^k f(\mathbf{x}_0, \mathbf{x}_1) \wedge \Delta_{\mathbf{x}_0} f(\mathbf{x}_0, \mathbf{x}_1)$$

$$= \min_{\mathbf{x}_0}^k f(\mathbf{x}_0, \mathbf{x}_1) \wedge (\min_{\mathbf{x}_0}^k f(\mathbf{x}_0, \mathbf{x}_1) \oplus \max_{\mathbf{x}_0}^k f(\mathbf{x}_0, \mathbf{x}_1))$$

$$= \min_{\mathbf{x}_0}^k f(\mathbf{x}_0, \mathbf{x}_1) \oplus (\min_{\mathbf{x}_0}^k f(\mathbf{x}_0, \mathbf{x}_1) \wedge \max_{\mathbf{x}_0}^k f(\mathbf{x}_0, \mathbf{x}_1))$$

$$= \min_{\mathbf{x}_0}^k f(\mathbf{x}_0, \mathbf{x}_1) \wedge (1 \oplus \max_{\mathbf{x}_0}^k f(\mathbf{x}_0, \mathbf{x}_1))$$

$$= \min_{\mathbf{x}_0}^k f(\mathbf{x}_0, \mathbf{x}_1) \wedge \overline{\max_{\mathbf{x}_0}^k f(\mathbf{x}_0, \mathbf{x}_1)}$$

$$= 0 .$$

Alternative rules for *k*-fold derivative operations use $\wedge$ and $\vee$:

$$\Delta_{\mathbf{x}_0} f(\mathbf{x}_0, \mathbf{x}_1) = \max_{\mathbf{x}_0}^k f(\mathbf{x}_0, \mathbf{x}_1) \wedge \overline{\min_{\mathbf{x}_0}^k f(\mathbf{x}_0, \mathbf{x}_1)} , \tag{6.52}$$

$$\min_{\mathbf{x}_0}^k f(\mathbf{x}_0, \mathbf{x}_1) = \max_{\mathbf{x}_0}^k f(\mathbf{x}_0, \mathbf{x}_1) \wedge \overline{\Delta_{\mathbf{x}_0} f(\mathbf{x}_0, \mathbf{x}_1)} , \tag{6.53}$$

$$\max_{\mathbf{x}_0}^k f(\mathbf{x}_0, \mathbf{x}_1) = \min_{\mathbf{x}_0}^k f(\mathbf{x}_0, \mathbf{x}_1) \vee \Delta_{\mathbf{x}_0} f(\mathbf{x}_0, \mathbf{x}_1) . \tag{6.54}$$

Further properties of *k*-fold derivative operations are explored below. The repeated calculation of the same *k*-fold derivative operations leads to the following results:

$$\operatorname{der}_{\mathbf{x}_0}^k \left(\operatorname{der}_{\mathbf{x}_0}^k f(\mathbf{x}_0, \mathbf{x}_1) \right) = 0 , \tag{6.55}$$

$$\min_{\mathbf{x}_0}^k \left(\min_{\mathbf{x}_0}^k f(\mathbf{x}_0, \mathbf{x}_1) \right) = \min_{\mathbf{x}_0}^k f(\mathbf{x}_0, \mathbf{x}_1) , \tag{6.56}$$

$$\max_{\mathbf{x}_0}^k \left(\max_{\mathbf{x}_0}^k f(\mathbf{x}_0, \mathbf{x}_1) \right) = \max_{\mathbf{x}_0}^k f(\mathbf{x}_0, \mathbf{x}_1) , \tag{6.57}$$

$$\Delta_{\mathbf{x}_0} \left(\Delta_{\mathbf{x}_0} f(\mathbf{x}_0, \mathbf{x}_1) \right) = 0 . \tag{6.58}$$

The results of all *k*-fold derivative operations do not depend on all *k* variables $x_i \in \mathbf{x}_0$:

$$\bigvee_{i=1}^{k} \operatorname*{der}_{x_i} \left(\operatorname*{der}_{\mathbf{x}_0}^{k} f(\mathbf{x}_0, \mathbf{x}_1) \right) = 0 \,, \tag{6.59}$$

$$\bigvee_{i=1}^{k} \operatorname*{der}_{x_i} \left(\operatorname*{min}_{\mathbf{x}_0}^{k} f(\mathbf{x}_0, \mathbf{x}_1) \right) = 0 \,, \tag{6.60}$$

$$\bigvee_{i=1}^{k} \operatorname*{der}_{x_i} \left(\operatorname*{max}_{\mathbf{x}_0}^{k} f(\mathbf{x}_0, \mathbf{x}_1) \right) = 0 \,, \tag{6.61}$$

$$\bigvee_{i=1}^{k} \operatorname*{der}_{x_i} \left(\Delta_{\mathbf{x}_0} f(\mathbf{x}_0, \mathbf{x}_1) \right) = 0 \,. \tag{6.62}$$

The k-fold derivative operations of negated functions can be transformed such that non-negated functions can be used:

$$\operatorname*{der}_{\mathbf{x}_0}^{k} \overline{f(\mathbf{x}_0, \mathbf{x}_1)} = \operatorname*{der}_{\mathbf{x}_0}^{k} f(\mathbf{x}_0, \mathbf{x}_1) \,, \tag{6.63}$$

$$\operatorname*{min}_{\mathbf{x}_0}^{k} \overline{f(\mathbf{x}_0, \mathbf{x}_1)} = \operatorname*{max}_{\mathbf{x}_0}^{k} f(\mathbf{x}_0, \mathbf{x}_1) \,, \tag{6.64}$$

$$\operatorname*{max}_{\mathbf{x}_0}^{k} \overline{f(\mathbf{x}_0, \mathbf{x}_1)} = \operatorname*{min}_{\mathbf{x}_0}^{k} f(\mathbf{x}_0, \mathbf{x}_1) \,, \tag{6.65}$$

$$\Delta_{\mathbf{x}_0} \overline{f(\mathbf{x}_0, \mathbf{x}_1)} = \Delta_{\mathbf{x}_0} f(\mathbf{x}_0, \mathbf{x}_1) \,. \tag{6.66}$$

The k-fold derivative operations of the functions $f(\mathbf{x}_0, \mathbf{x}_1)$ and $g(\mathbf{x}_0, \mathbf{x}_1)$ that are connected by the same operation as used in the definition can be split into two k-fold derivative operations:

$$\operatorname*{der}_{\mathbf{x}_0}^{k} (f(\mathbf{x}_0, \mathbf{x}_1) \oplus g(\mathbf{x}_0, \mathbf{x}_1)) = \operatorname*{der}_{\mathbf{x}_0}^{k} f(\mathbf{x}_0, \mathbf{x}_1) \oplus \operatorname*{der}_{\mathbf{x}_0}^{k} g(\mathbf{x}_0, \mathbf{x}_1) \,,$$

$$\operatorname*{min}_{\mathbf{x}_0}^{k} (f(\mathbf{x}_0, \mathbf{x}_1) \wedge g(\mathbf{x}_0, \mathbf{x}_1)) = \operatorname*{min}_{\mathbf{x}_0}^{k} f(\mathbf{x}_0, \mathbf{x}_1) \wedge \operatorname*{min}_{\mathbf{x}_0}^{k} g(\mathbf{x}_0, \mathbf{x}_1) \,,$$

$$\operatorname*{max}_{\mathbf{x}_0}^{k} (f(\mathbf{x}_0, \mathbf{x}_1) \vee g(\mathbf{x}_0, \mathbf{x}_1)) = \operatorname*{max}_{\mathbf{x}_0}^{k} f(\mathbf{x}_0, \mathbf{x}_1) \vee \operatorname*{max}_{\mathbf{x}_0}^{k} g(\mathbf{x}_0, \mathbf{x}_1) \,.$$

For the following k-fold derivative operations, we assume that the function $g(\mathbf{x}_0, \mathbf{x}_1)$ does not depend on the variables $x_i \in \mathbf{x}_0$:

$$\bigvee_{i=1}^{k} \operatorname*{der}_{x_i} (g(\mathbf{x}_0, \mathbf{x}_1)) = 0 \,,$$

so that $g(\mathbf{x}_0, \mathbf{x}_1)$ can be replaced by $g(\mathbf{x}_1)$.

In k-fold derivative operations of conjunctions of $g(\mathbf{x}_1)$ and $f(\mathbf{x}_0, \mathbf{x}_1)$, the function $g(\mathbf{x}_1)$ operates like a filter; the results can be equal to 1 only for subspaces with $g(\mathbf{x}_1) = 1$:

$$\operatorname*{der}_{\mathbf{x}_0}^{k} (g(\mathbf{x}_1) \wedge f(\mathbf{x}_0, \mathbf{x}_1)) = g(\mathbf{x}_1) \wedge \operatorname*{der}_{\mathbf{x}_0}^{k} f(\mathbf{x}_0, \mathbf{x}_1) \,,$$

$$\operatorname*{min}_{\mathbf{x}_0}^{k} (g(\mathbf{x}_1) \wedge f(\mathbf{x}_0, \mathbf{x}_1)) = g(\mathbf{x}_1) \wedge \operatorname*{min}_{\mathbf{x}_0}^{k} f(\mathbf{x}_0, \mathbf{x}_1) \,,$$

$$\operatorname*{max}_{\mathbf{x}_0}^{k} (g(\mathbf{x}_1) \wedge f(\mathbf{x}_0, \mathbf{x}_1)) = g(\mathbf{x}_1) \wedge \operatorname*{max}_{\mathbf{x}_0}^{k} f(\mathbf{x}_0, \mathbf{x}_1) \,,$$

$$\Delta_{\mathbf{x}_0} (g(\mathbf{x}_1) \wedge f(\mathbf{x}_0, \mathbf{x}_1)) = g(\mathbf{x}_1) \wedge \Delta_{\mathbf{x}_0} f(\mathbf{x}_0, \mathbf{x}_1) \,.$$

Using the disjunctions of $g(\mathbf{x}_1)$ and $f(\mathbf{x}_0, \mathbf{x}_1)$ in k-fold derivative operations, we get the following results:

$$\operatorname{der}_{\mathbf{x}_0}^k(g(\mathbf{x}_1) \vee f(\mathbf{x}_0, \mathbf{x}_1)) = \overline{g(\mathbf{x}_1)} \wedge \operatorname{der}_{\mathbf{x}_0}^k f(\mathbf{x}_0, \mathbf{x}_1) ,$$

$$\min_{\mathbf{x}_0}^k(g(\mathbf{x}_1) \vee f(\mathbf{x}_0, \mathbf{x}_1)) = g(\mathbf{x}_1) \vee \min_{\mathbf{x}_0}^k f(\mathbf{x}_0, \mathbf{x}_1) ,$$

$$\max_{\mathbf{x}_0}^k(g(\mathbf{x}_1) \vee f(\mathbf{x}_0, \mathbf{x}_1)) = g(\mathbf{x}_1) \vee \max_{\mathbf{x}_0}^k f(\mathbf{x}_0, \mathbf{x}_1) ,$$

$$\Delta_{\mathbf{x}_0}(g(\mathbf{x}_1) \vee f(\mathbf{x}_0, \mathbf{x}_1)) = \overline{g(\mathbf{x}_1)} \wedge \Delta_{\mathbf{x}_0} f(\mathbf{x}_0, \mathbf{x}_1) .$$

The antivalence of $g(\mathbf{x}_1)$ and $f(\mathbf{x}_0, \mathbf{x}_1)$ changes the function $f(\mathbf{x}_0, \mathbf{x}_1)$ into $\overline{f(\mathbf{x}_0, \mathbf{x}_1)}$ for subspaces with $g(\mathbf{x}_1) = 1$ so that we get for the k-fold derivative operations the rules:

$$\operatorname{der}_{\mathbf{x}_0}^k(g(\mathbf{x}_1) \oplus f(\mathbf{x}_0, \mathbf{x}_1)) = \operatorname{der}_{\mathbf{x}_0}^k f(\mathbf{x}_0, \mathbf{x}_1) ,$$

$$\min_{\mathbf{x}_0}^k(g(\mathbf{x}_1) \oplus f(\mathbf{x}_0, \mathbf{x}_1)) = g(\mathbf{x}_1) \min_{\mathbf{x}_0}^k \overline{f(\mathbf{x}_0, \mathbf{x}_1)} \vee \overline{g(\mathbf{x}_1)} \min_{\mathbf{x}_0}^k f(\mathbf{x}_0, \mathbf{x}_1) ,$$

$$\max_{\mathbf{x}_0}^k(g(\mathbf{x}_1) \oplus f(\mathbf{x}_0, \mathbf{x}_1)) = g(\mathbf{x}_1) \max_{\mathbf{x}_0}^k \overline{f(\mathbf{x}_0, \mathbf{x}_1)} \vee \overline{g(\mathbf{x}_1)} \max_{\mathbf{x}_0}^k f(\mathbf{x}_0, \mathbf{x}_1) ,$$

$$\Delta_{\mathbf{x}_0}(g(\mathbf{x}_1) \oplus f(\mathbf{x}_0, \mathbf{x}_1)) = \Delta_{\mathbf{x}_0} f(\mathbf{x}_0, \mathbf{x}_1) .$$

The reader might not be afraid of this amount of formulas. It is not necessary to learn all these relations and formulas by heart. The respective applications will show how to use these operations and how to understand the results.

6.5 Differential of a Boolean Variable

At a given point in time, a Boolean variable x_i can have one of the values 0 or 1. This variable will give us the information about the actual value, but no information about the change of this value is available. In order to close this gap, we define the differential of a Boolean variable as follows:

Definition 6.9 Let x_i be a Boolean variable; then

$$\mathrm{d}x_i = \begin{cases} 1, & \text{if the variable } x_i \text{ changes its value} \\ 0, & \text{if the variable } x_i \text{ does not change its value} \end{cases} \tag{6.67}$$

is the *differential of the Boolean variable* x_i.

Note 6.5 The *differential* $\mathrm{d}x_i$ is a Boolean variable too; hence, all concepts considered before remain valid for differentials of Boolean variables.

Depending on the value of the differential, the value of the associated variable will change or will not change. In order to distinguish between the value of x_i before and after a possible change, we introduce a Boolean variable x_i' that shows the value of x_i after the change according to the differential $\mathrm{d}x_i$:

$$x_i' = x_i \oplus \mathrm{d}x_i . \tag{6.68}$$

Equation (6.68) describes how the new value x_i' can be calculated using the given value of the variable x_i and its differential dx_i. All possible changes of a single variable x_i are visualized in Fig. 6.6.

The variable x_i and its differential dx_i specify independently of each other the Boolean values of the variable and its possible change. A conjunction of these two variables (negations may be included) describes the direction of the change in more detail. The fact that the value of x_i changes from 0 to 1 will be expressed by the equation $\overline{x}_i \wedge dx_i = 1$. The equation $x_i \wedge \overline{dx_i} = 1$ describes that the value 1 of the variable x_i does not change. The concept of changing the value of a variable assumes, at least implicitly, an understanding of an ordering (time, direction, etc.) so that *before*, *change*, and *after* make a definite sense.

Boolean differentials can be defined for each Boolean variable. The vector $\mathbf{dx} = (dx_1, dx_2, \ldots, dx_n)$ is the *differential of the vector of Boolean variables* $\mathbf{x} = (x_1, x_2, \ldots, x_n)$ and specifies a *direction of change* in the Boolean space. Formula (6.69) describes which point $\mathbf{x}' = (x_1', x_2', \ldots, x_n')$ in the Boolean space will be reached starting in $\mathbf{x}$ and changing all the values x_i where $dx_i = 1$:

$$\mathbf{x}' = \mathbf{x} \oplus \mathbf{dx} \,. \tag{6.69}$$

The set of all differential vectors $\mathbf{dx}$ defines the Boolean space of all possible changes $d\mathbb{B}^n$. On the one hand this is a Boolean space as defined in the first chapter. On the other hand this Boolean space may be visualized by direction vectors embedded in the Boolean space $\mathbb{B}^n$ of all vectors $\mathbf{x}$. Figure 6.7 shows the second type of visualization of $d\mathbb{B}^2$ embedded in the Boolean space $\mathbb{B}^2$ for the point (node, vector) $(x_1, x_2) = (01)$.

The variable x_i and the associated differential dx_i differ in their meaning, but both are Boolean variables. Hence, we can define logic functions $F(\mathbf{x}, \mathbf{dx})$ as mappings from $\mathbb{B}^n \times d\mathbb{B}^n$ into $\mathbb{B}$ and consider the logic equation

$$f_1(\mathbf{x}, \mathbf{dx}) = f_2(\mathbf{x}, \mathbf{dx}) \,. \tag{6.70}$$

The solution of this equation is a set of binary vectors $(\mathbf{x}, \mathbf{dx})$. Each of these vectors describes the combination of a point $\mathbf{x} \in \mathbb{B}^n$ and a direction vector $\mathbf{dx} \in d\mathbb{B}^n$. Hence, these solution-vectors specify directed edges of a graph, and Eq. (6.70) can be denoted by *graph equation*. Many conceivable manipulations on graphs can be expressed by solutions of certain graph equations.

Example 6.13 Let $n = 3$,

$$f_1(\mathbf{x}, \mathbf{dx}) = \overline{x}_1 \overline{x}_2 \overline{dx}_1 \overline{dx}_3 (x_3 \oplus \overline{dx}_2) \oplus x_2 \overline{dx}_1 dx_3 (\overline{x}_1 \oplus x_3 \overline{dx}_2)$$

$$\oplus \overline{x}_1 x_3 dx_1 dx_2 (x_2 \oplus dx_3) \oplus x_1 \overline{x}_2 x_3 \overline{dx}_1 \overline{dx}_2 dx_3$$

$$\oplus x_1 x_3 dx_1 \oplus x_1 x_2 dx_1 dx_2 \overline{dx}_3 \oplus x_2 \overline{dx}_1 \overline{dx}_2 dx_3$$

and

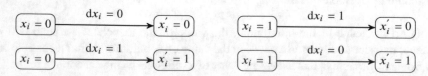

Fig. 6.6 Possible changes of the value of the variable x_i

Fig. 6.7 Direction vectors of $d\mathbb{B}^2$ embedded in $\mathbb{B}^2$

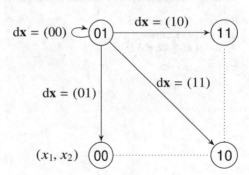

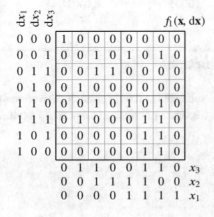

dx_1 dx_2 dx_3		$f_1(\mathbf{x}, \mathbf{dx})$							
0 0 0	1	0	0	0	0	0	0	0
0 0 1	0	0	1	0	1	0	1	0
0 1 1	0	0	1	1	0	0	0	0
0 1 0	0	1	0	0	0	0	0	0
1 1 0	0	0	1	0	1	0	1	0
1 1 1	0	1	0	0	0	1	1	0
1 0 1	0	0	0	0	0	1	1	0
1 0 0	0	0	0	0	0	1	1	0
	0	1	1	0	0	1	1	0	x_3
	0	0	1	1	1	1	0	0	x_2
	0	0	0	0	1	1	1	1	x_1

dx_1 dx_2 dx_3		$f_2(\mathbf{x}, \mathbf{dx})$							
0 0 0	0	0	1	1	1	1	1	1
0 0 1	1	1	0	0	1	1	1	1
0 1 1	1	1	0	0	1	1	1	1
0 1 0	0	0	1	1	1	1	1	1
1 1 0	1	1	1	1	0	0	0	1
1 1 1	1	1	1	1	1	0	0	1
1 0 1	1	1	1	1	1	0	0	1
1 0 0	1	1	1	1	1	0	0	0
	0	1	1	0	0	1	1	0	x_3
	0	0	1	1	1	1	0	0	x_2
	0	0	0	0	1	1	1	1	x_1

Fig. 6.8 The Karnaugh-maps of the functions $f_1(\mathbf{x}, \mathbf{dx})$ and $f_2(\mathbf{x}, \mathbf{dx})$

$$f_2(\mathbf{x}, \mathbf{dx}) = \overline{x}_1\overline{dx}_1(x_2 \oplus dx_3) \oplus x_1\overline{x}_3 dx_1 dx_3 \oplus x_1 \oplus dx_1$$

$$\oplus\, x_1\overline{x}_3 dx_1\overline{dx}_3(x_2 \oplus dx_2)\,.$$

The Karnaugh-maps of these two logic functions are shown in Fig. 6.8.

The solution-set $G(\mathbf{x}, \mathbf{dx})$ of the graph equation (6.70) can easily be found by comparing the values in the two Karnaugh-maps:

$$G(\mathbf{x}, \mathbf{dx}) = \begin{array}{cccccc} x_1 & x_2 & x_3 & dx_1 & dx_2 & dx_3 \\ \hline 0 & 0 & 1 & 1 & 1 & 1 \\ 1 & 0 & 1 & 0 & 0 & 1 \\ 1 & 0 & 0 & 1 & 0 & 0 \\ 0 & 0 & 1 & 0 & 0 & 0 \\ - & 1 & 0 & 0 & 0 & 1 \\ 0 & 0 & 0 & 0 & 1 & 0 \\ - & 1 & 1 & 1 & 1 & 0 \\ \hline \end{array}\;.$$

Each solution-vector $(\mathbf{x}, \mathbf{dx})$ corresponds to an edge of the graph shown in Fig. 6.9.

This graph contains three cycles:

1. There is a loop at the node (001).
2. The cycle $(001) \to (110) \to (111) \to (001)$ is connected to this loop.
3. The cycle $(000) \to (010) \to (011) \to (101) \to (100) \to (000)$ has no common edges or nodes with the two other cycles.

Fig. 6.9 Graph of the solution of Equation $f_1(\mathbf{x}, \mathbf{dx}) = f_2(\mathbf{x}, \mathbf{dx})$

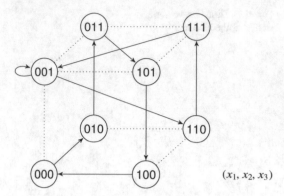

(x_1, x_2, x_3)

6.6   Differential Operations of Logic Functions

The evaluated direction of change or the explored subspace is fixed for each derivative operation. Using the differential dx_i of a Boolean variable x_i, it becomes possible to commonly explore several directions of change.

> Differential operations explore the change of logic functions with regard to several or even all directions of change or several subspaces.

The direction of change is specified by the differentials dx_i as follows:

$$x_i \oplus dx_i = \begin{cases} x_i \,, & \text{if the value of } x_i \text{ is not changed, i.e., } dx_i = 0 \,, \\ \overline{x}_i \,, & \text{if the value of} x_i \text{ is changed, i.e., } dx_i = 1 \,. \end{cases}$$

The substitution of this expression for all variables x_i of the function $f(\mathbf{x})$ creates the function

$$f(\mathbf{x} \oplus \mathbf{dx}) = f(x_1 \oplus dx_1, x_2 \oplus dx_2, \dots, x_n \oplus dx_n) \tag{6.71}$$

that determines the function values by both the selection of a point in the Boolean space $\mathbf{x} = \mathbf{c}$ and the direction of change $\mathbf{dx} = \mathbf{dc}$. Using this function, the complete change behavior of a logic function can be explored.

Definition 6.10 Let $f(\mathbf{x}) = f(x_1, x_2, \dots, x_n)$ be a logic function of n variables; then

$$d_{\mathbf{x}} f(\mathbf{x}) = f(\mathbf{x}) \oplus f(\mathbf{x} \oplus \mathbf{dx}) \tag{6.72}$$

is the *(total) differential*,

$$\text{Min}_{\mathbf{x}} f(\mathbf{x}) = f(\mathbf{x}) \wedge f(\mathbf{x} \oplus \mathbf{dx}) \tag{6.73}$$

is the *(total) differential minimum*,

$$\text{Max}_{\mathbf{x}} f(\mathbf{x}) = f(\mathbf{x}) \vee f(\mathbf{x} \oplus \mathbf{dx}) \tag{6.74}$$

is the *(total) differential maximum*, and

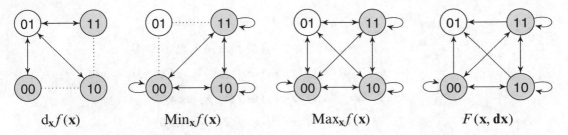

$$d_\mathbf{x} f(\mathbf{x}) \qquad \text{Min}_\mathbf{x} f(\mathbf{x}) \qquad \text{Max}_\mathbf{x} f(\mathbf{x}) \qquad F(\mathbf{x}, \mathbf{dx})$$

Fig. 6.10 Graphs of all total differential operations of the function $f(\mathbf{x}) = x_1 \vee \overline{x}_2$

$$F(\mathbf{x}, \mathbf{dx}) = f(\mathbf{x}) \wedge \bigwedge_{i=1}^{n} (dx_i \vee \overline{dx_i}) \tag{6.75}$$

is the *(total) differential expansion* of $f(\mathbf{x})$ with regard to all variables $\mathbf{x}$.

Note 6.6 The term *total* in Definition 6.10 emphasizes that all possible directions of change are explored. This is the general case for differential operations; hence, this term can be avoided.

From Definition 6.10 directly follows:

$$d_\mathbf{x} f(\mathbf{x}) \oplus \text{Min}_\mathbf{x} f(\mathbf{x}) \oplus \text{Max}_\mathbf{x} f(\mathbf{x}) = 0 , \tag{6.76}$$

$$\text{Min}_\mathbf{x} f(\mathbf{x}) \le F(\mathbf{x}, \mathbf{dx}) \le \text{Max}_\mathbf{x} f(\mathbf{x}) . \tag{6.77}$$

All the total differential operations map $\mathbb{B}^n \times d\mathbb{B}^n$ into $\mathbb{B}$; hence, the results of these operations are functions that depend on the variables $(\mathbf{x}, \mathbf{dx})$. Thus, function values 1 of a differential operation can be visualized by an edge of a graph. The nodes of this graph may be additionally labeled by the function values of the used logic function $f(\mathbf{x})$.

Due to Definition 6.10:

- The result of the total differential describes edges in a graph that connect nodes where the function $f(\mathbf{x})$ has different values.
- The result of the total differential minimum describes edges in a graph that connect nodes where the function values of both $f(\mathbf{x})$ and $f(\mathbf{x} \oplus \mathbf{dx})$ are equal to 1.
- The result of the total differential maximum describes edges in a graph that connect nodes where the function $f(\mathbf{x})$ carries at least once the value 1.
- The result of the total differential expansion describes edges in a graph that connect nodes where the function $f(\mathbf{x})$ is equal to 1, with all nodes of the graph.

Example 6.14 We take the function $f(x_1, x_2) = x_1 \vee \overline{x}_2$ and calculate all four total differential operations. Figure 6.10 shows the graphs where edges indicate that the associated differential operation is equal to 1. The nodes in these graphs are encoded by (x_1, x_2). Gray filled nodes indicate function values 1.

The Shannon decomposition can be used to find the relationship between the total differential operations and the derivative operations. We demonstrate these equivalent transformations for $d_{(x_1, x_2)} f(x_1, x_2)$:

$$d_{(x_1,x_2)} f(x_1, x_2) = f(x_1, x_2) \oplus f(x_1 \oplus dx_1, x_2 \oplus dx_2)$$

$$= \overline{dx}_1 (f(x_1, x_2) \oplus f(x_1 \oplus 0, x_2 \oplus dx_2))$$

$$\oplus dx_1 (f(x_1, x_2) \oplus f(x_1 \oplus 1, x_2 \oplus dx_2))$$

$$= \overline{dx}_1 (f(x_1, x_2) \oplus f(x_1, x_2 \oplus dx_2))$$

$$\oplus dx_1 (f(x_1, x_2) \oplus f(\overline{x}_1, x_2 \oplus dx_2))$$

$$= \overline{dx}_2 (\ \overline{dx}_1 (f(x_1, x_2) \oplus f(x_1, x_2 \oplus 0))$$

$$\oplus dx_1 (f(x_1, x_2) \oplus f(\overline{x}_1, x_2 \oplus 0)))$$

$$\oplus dx_2 (\ \overline{dx}_1 (f(x_1, x_2) \oplus f(x_1, x_2 \oplus 1))$$

$$\oplus dx_1 (f(x_1, x_2) \oplus f(\overline{x}_1, x_2 \oplus 1)))$$

$$= (f(x_1, x_2) \oplus f(x_1, x_2)) \overline{dx}_1 \overline{dx}_2$$

$$\oplus (f(x_1, x_2) \oplus f(\overline{x}_1, x_2)) dx_1 \overline{dx}_2$$

$$\oplus (f(x_1, x_2) \oplus f(x_1, \overline{x}_2)) \overline{dx}_1 dx_2$$

$$\oplus (f(x_1, x_2) \oplus f(\overline{x}_1, \overline{x}_2)) dx_1 dx_2$$

$$= \operatorname*{der}_{x_1} f(x_1, x_2) dx_1 \overline{dx}_2 \oplus \operatorname*{der}_{x_2} f(x_1, x_2) \overline{dx}_1 dx_2$$

$$\oplus \operatorname*{der}_{(x_1,x_2)} f(x_1, x_2) dx_1 dx_2 \ .$$

These equivalent transformations show that the total differential of $f(x_1, x_2)$ summarizes all vectorial derivatives of this function. Similar transformation steps can be executed for the other differential operations and lead to the following relationships:

$$\mathrm{Min}_{(x_1,x_2)} f(x_1, x_2) = f(x_1, x_2) \wedge f(x_1 \oplus dx_1, x_2 \oplus dx_2)$$

$$= f(x_1, x_2) \overline{dx}_1 \overline{dx}_2 \vee \operatorname*{min}_{x_1} f(x_1, x_2) dx_1 \overline{dx}_2$$

$$\vee \operatorname*{min}_{x_2} f(x_1, x_2) \overline{dx}_1 dx_2 \vee \operatorname*{min}_{(x_1,x_2)} f(x_1, x_2) dx_1 dx_2 \ ,$$

$$\mathrm{Max}_{(x_1,x_2)} f(x_1, x_2) = f(x_1, x_2) \vee f(x_1 \oplus dx_1, x_2 \oplus dx_2)$$

$$= f(x_1, x_2) \overline{dx}_1 \overline{dx}_2 \vee \operatorname*{max}_{x_1} f(x_1, x_2) dx_1 \overline{dx}_2$$

$$\vee \operatorname*{max}_{x_2} f(x_1, x_2) \overline{dx}_1 dx_2 \vee \operatorname*{max}_{(x_1,x_2)} f(x_1, x_2) dx_1 dx_2 \ .$$

These formulas can be utilized in two ways:

- Knowing all vectorial derivative operations of a given logic function $f(\mathbf{x})$, the associated differential operation can be computed.
- Knowing a total differential operation of a given logic function $f(\mathbf{x})$, all associated vectorial derivative operations can be separated by the assignment of constant values to the differentials dx_i.

The relations between the total differential operations explained above for functions of two variables can be generalized for function of an arbitrary number of variables. The fact that each total differential operation summarizes all associated vectorial derivative operations is a great benefit, and this is often the reason for the use of the total differential operations. The drawback, however, is that the number of independent variables of a total differential operation is two times the number of variables of the used function $f(\mathbf{x})$. For that reason, partial differential operations are considered for task where the change behavior is not needed for all directions of change.

Definition 6.11 Let $\mathbf{x}_0 = (x_1, x_2, \ldots, x_k)$, $\mathbf{x}_1 = (x_{k+1}, \ldots, x_n)$ be two disjoint sets of Boolean variables, and $f(\mathbf{x}_0, \mathbf{x}_1) = f(\mathbf{x})$ a logic function of n variables; then

$$d_{\mathbf{x}_0} f(\mathbf{x}_0, \mathbf{x}_1) = f(\mathbf{x}_0, \mathbf{x}_1) \oplus f(\mathbf{x}_0 \oplus \mathbf{dx}_0, \mathbf{x}_1) \tag{6.78}$$

is the *partial differential*,

$$\text{Min}_{\mathbf{x}_0} f(\mathbf{x}_0, \mathbf{x}_1) = f(\mathbf{x}_0, \mathbf{x}_1) \wedge f(\mathbf{x}_0 \oplus \mathbf{dx}_0, \mathbf{x}_1) \tag{6.79}$$

is the *partial differential minimum*,

$$\text{Max}_{\mathbf{x}_0} f(\mathbf{x}_0, \mathbf{x}_1) = f(\mathbf{x}_0, \mathbf{x}_1) \vee f(\mathbf{x}_0 \oplus \mathbf{dx}_0, \mathbf{x}_1) \tag{6.80}$$

is the *partial differential maximum*, and

$$F(\mathbf{x}_0, \mathbf{x}_1, \mathbf{dx}_0) = f(\mathbf{x}_0, \mathbf{x}_1) \wedge \bigwedge_{i=1}^{k} (dx_i \vee \overline{dx_i}) \tag{6.81}$$

is the *partial differential expansion* of $f(\mathbf{x}_0, \mathbf{x}_1)$ with regard to the subset of variables $\mathbf{x}_0 = (x_1, x_2, \ldots, x_k)$.

The partial differential operations describe the same change behavior as the total differential operations but take into account only changes in the directions specified by $\mathbf{dx}_0$. The following formulas show for $\mathbf{x}_0 = (x_1, x_2)$ how the vectorial derivative operations are summarized within the partial differential operations:

$$d_{(x_1, x_2)} f(\mathbf{x}_0, \mathbf{x}_1) = \underset{x_1}{\text{der}}\, f(\mathbf{x}_0, \mathbf{x}_1) dx_1 \overline{dx_2} \oplus \underset{x_2}{\text{der}}\, f(\mathbf{x}_0, \mathbf{x}_1) \overline{dx_1} dx_2$$

$$\oplus \underset{(x_1, x_2)}{\text{der}}\, f(\mathbf{x}_0, \mathbf{x}_1) dx_1 dx_2 \,,$$

$$\text{Min}_{(x_1, x_2)} f(\mathbf{x}_0, \mathbf{x}_1) = f(\mathbf{x}_0, \mathbf{x}_1) \overline{dx_1}\,\overline{dx_2} \vee \underset{x_1}{\text{min}}\, f(\mathbf{x}_0, \mathbf{x}_1) dx_1 \overline{dx_2}$$

$$\vee \underset{x_2}{\text{min}}\, f(\mathbf{x}_0, \mathbf{x}_1) \overline{dx_1} dx_2 \vee \underset{(x_1, x_2)}{\text{min}}\, f(\mathbf{x}_0, \mathbf{x}_1) dx_1 dx_2 \,,$$

$$\text{Max}_{(x_1, x_2)} f(\mathbf{x}_0, \mathbf{x}_1) = f(\mathbf{x}_0, \mathbf{x}_1) \overline{dx_1}\,\overline{dx_2} \vee \underset{x_1}{\text{max}}\, f(\mathbf{x}_0, \mathbf{x}_1) dx_1 \overline{dx_2} \vee$$

$$\vee \underset{x_2}{\text{max}}\, f(\mathbf{x}_0, \mathbf{x}_1) \overline{dx_1} dx_2 \vee \underset{(x_1, x_2)}{\text{max}}\, f(\mathbf{x}_0, \mathbf{x}_1) dx_1 dx_2 \,.$$

Fig. 6.11 Graphs of all
parital differential
operations of
$f(\mathbf{x}) = x_1\overline{x}_2\overline{x}_3 \vee \overline{x}_1 x_2$
with regard to
$\mathbf{x}_0 = (x_1, x_2)$

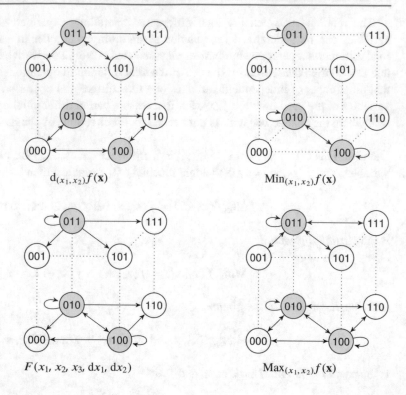

$d_{(x_1, x_2)} f(\mathbf{x})$

$\text{Min}_{(x_1, x_2)} f(\mathbf{x})$

$F(x_1, x_2, x_3, dx_1, dx_2)$

$\text{Max}_{(x_1, x_2)} f(\mathbf{x})$

These formulas can be generalized for an arbitrary number of variables of $\mathbf{x}_0$. The exploration of the changes is restricted to subspaces $\mathbf{x}_1 = \mathbf{c}$.

Example 6.15 We take the function $f(x_1, x_2, x_3) = x_1\overline{x}_2\overline{x}_3 \vee \overline{x}_1 x_2$ and calculate all four partial differential operations. Figure 6.11 shows the graphs where edges indicate that the associated partial differential operation is equal to 1. The nodes in these graphs are encoded by (x_1, x_2, x_3). Gray filled nodes indicate function values 1.

The existence of an edge in a graph of a partial differential operation depends only on the function values at the start node and the end node. This is the same condition as for the total differential operations, and therefore, the adopted properties hold:

$$d_{\mathbf{x}_0} f(\mathbf{x}_0, \mathbf{x}_1) \oplus \text{Min}_{\mathbf{x}_0} f(\mathbf{x}_0, \mathbf{x}_1) \oplus \text{Max}_{\mathbf{x}_0} f(\mathbf{x}_0, \mathbf{x}_1) = 0 \,. \tag{6.82}$$

$$\text{Min}_{\mathbf{x}_0} f(\mathbf{x}_0, \mathbf{x}_1) \leq F(\mathbf{x}_0, \mathbf{x}_1, d\mathbf{x}_0) \leq \text{Max}_{\mathbf{x}_0} f(\mathbf{x}_0, \mathbf{x}_1) \,. \tag{6.83}$$

Figure 6.11 can be used to verify these general relationships between the partial differential operations.

The k-fold differential operations show quite different properties. The formal way to compute such operations will be based on partial differential operations with regard to single variables. A partial differential operation of $f(\mathbf{x})$ with regard to one variable x_i is again a logic function and can be the object of a next partial differential operation of the same type with regard to another variable. The results of k-fold differential operations are independent of the order in which the different variables are taken into consideration.

Definition 6.12 Let $\mathbf{x}_0 = (x_1, x_2, \ldots, x_k)$, $\mathbf{x}_1 = (x_{k+1}, x_{k+2}, \ldots, x_n)$ be two disjoint sets of Boolean variables, and $f(\mathbf{x}_0, \mathbf{x}_1) = f(x_1, x_2, \ldots, x_n) = f(\mathbf{x})$ a logic function of n variables; then

$$d_{\mathbf{x}_0}^k f(\mathbf{x}_0, \mathbf{x}_1) = d_{x_k}(\ldots (d_{x_2}(d_{x_1} f(\mathbf{x}_0, \mathbf{x}_1))) \ldots) \tag{6.84}$$

is the *k-fold differential*,

$$\text{Min}_{\mathbf{x}_0}^k f(\mathbf{x}_0, \mathbf{x}_1) = \text{Min}_{x_k}(\ldots (\text{Min}_{x_2}(\text{Min}_{x_1} f(\mathbf{x}_0, \mathbf{x}_1))) \ldots) \tag{6.85}$$

is the *k-fold differential minimum*,

$$\text{Max}_{\mathbf{x}_0}^k f(\mathbf{x}_0, \mathbf{x}_1) = \text{Max}_{x_k}(\ldots (\text{Max}_{x_2}(\text{Max}_{x_1} f(\mathbf{x}_0, \mathbf{x}_1))) \ldots) \tag{6.86}$$

is the *k-fold differential maximum*, and

$$\vartheta_{\mathbf{x}_0} f(\mathbf{x}_0, \mathbf{x}_1) = \text{Min}_{\mathbf{x}_0}^k f(\mathbf{x}_0, \mathbf{x}_1) \oplus \text{Max}_{\mathbf{x}_0}^k f(\mathbf{x}_0, \mathbf{x}_1) \tag{6.87}$$

is the *ϑ-operation* of the logic function $f(\mathbf{x}_0, \mathbf{x}_1)$ with regard to the set of variables $\mathbf{x}_0 = (x_1, x_2, \ldots, x_k)$.

The k-fold differential has a special property that is different to the other k-fold differential operations. It is not easy to observe this property by means of (6.84) of Definition 6.12. An example will help to understand the k-fold differential:

Example 6.16 Let $f(x_1, x_2, x_3) = \overline{x}_1 \overline{x}_2 \overline{x}_3 \oplus x_1 x_2$. Based on (6.84), the 2-fold differential $d_{(x_2, x_3)}^2 f(x_1, x_2, x_3)$ has to be calculated iteratively.

$$\begin{aligned}
d_{(x_2, x_3)}^2 f(x_1, x_2, x_3) &= d_{x_3}(d_{x_2}(\overline{x}_1 \overline{x}_2 \overline{x}_3 \oplus x_1 x_2)) \\
&= d_{x_3}(\overline{x}_1 \overline{x}_2 \overline{x}_3 \oplus x_1 x_2 \oplus \overline{x}_1 (\overline{x_2 \oplus dx_2}) \overline{x}_3 \oplus x_1 (x_2 \oplus dx_2)) \\
&= d_{x_3}(\overline{x}_1 \overline{x}_2 \overline{x}_3 \oplus x_1 x_2 \oplus \overline{x}_1 \overline{x}_2 \overline{x}_3 \oplus \overline{x}_1 \overline{x}_3 dx_2 \oplus x_1 x_2 \oplus x_1 dx_2) \\
&= d_{x_3}(\overline{x}_1 \overline{x}_3 dx_2 \oplus x_1 dx_2) \\
&= \overline{x}_1 \overline{x}_3 dx_2 \oplus x_1 dx_2 \oplus \overline{x}_1 (\overline{x_3 \oplus dx_3}) dx_2 \oplus x_1 dx_2 \\
&= \overline{x}_1 \overline{x}_3 dx_2 \oplus \overline{x}_1 \overline{x}_3 dx_2 \oplus \overline{x}_1 dx_2 dx_3 \\
&= \overline{x}_1 dx_2 dx_3 \\
&= (\overline{x}_1 \overline{x}_2 \overline{x}_3 \oplus \overline{x}_1 x_2 \overline{x}_3 \oplus \overline{x}_1 \overline{x}_2 x_3 \oplus \overline{x}_1 x_2 x_3) dx_2 dx_3 .
\end{aligned}$$

In this example the variable x_1 is not included into the set $\mathbf{x}_0$. No differential dx_1 occurs in the result of $d_{(x_2, x_3)}^2 f(x_1, x_2, x_3)$ due to the properties of partial differentials in the steps of the calculation. Consequently, the graph of the 2-fold differential $d_{(x_2, x_3)}^2 f(x_1, x_2, x_3)$ (see Fig. 6.12) can only contain edges in the subspaces $x_1 = 0$ and $x_1 = 1$, respectively. The orthogonal expansion in the last line of the calculation shows the start nodes of the edges.

The differentials dx_2 and dx_3 occur in all conjunctions of $d_{(x_2, x_3)}^2 f(x_1, x_2, x_3)$ because the k-fold differential has been calculated with regard to (x_2, x_3). As can be seen in Fig. 6.12, the edges of this k-fold differential begin either on each node or on no node of subspaces $x_1 = const$.

Fig. 6.12 Graph of the 2-fold differential $d^2_{(x_2,x_3)} f(x_1, x_2, x_3)$ for $f(\mathbf{x}) = \overline{x}_1 \overline{x}_2 \overline{x}_3 \oplus x_1 x_2$

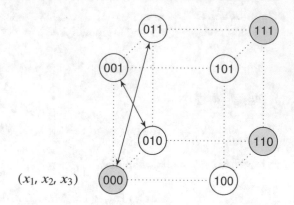

Edges of a k-fold differential occur in a subspace that contains an odd number of nodes with the values $f(\mathbf{x}) = 1$.

This observation can be generalized for the calculation of the k-fold differential $d^k_{\mathbf{x}_0} f(\mathbf{x}_0, \mathbf{x}_1)$. The iterative calculation of partial differentials according to (6.84) as well as the k-fold derivative using (6.45) for the variables of $\mathbf{x}_0$ leads directly to

$$d^k_{\mathbf{x}_0} f(\mathbf{x}_0, \mathbf{x}_1) = \left(\bigoplus_{\mathbf{c} \in B^k} f(\mathbf{x}_0 = \mathbf{c}, \mathbf{x}_1) \right) dx_1 \dots dx_k = \mathop{\mathrm{der}}_{\mathbf{x}_0}{}^k f(\mathbf{x}_0, \mathbf{x}_1) dx_1 \dots dx_k \,.$$

Hence, the k-fold differential evaluates only subspaces of the dimension defined by $\mathbf{x}_0$. This property does not hold for other k-fold differential operations as can be seen by the following example.

Example 6.17 We reuse the function $f(x_1, x_2, x_3) = \overline{x}_1 \overline{x}_2 \overline{x}_3 \oplus x_1 x_2$ of Example 6.16. Figure 6.13 shows the graphs of the 2-fold differential minimum, the 2-fold differential maximum, and the ϑ-operation of this function with regard to (x_2, x_3) where edges indicate that the associated differential operation is equal to 1. The nodes in these graphs are encoded by (x_1, x_2, x_3). Gray filled nodes indicate function values 1.

Figure 6.13 shows that k-fold differential operations do not describe relations between selected pairs of function values, but global properties of subspaces. Non-negated differentials dx_i of an edge indicate the free variables of the associated subspace. The remaining variables are bounded and define the evaluated subspace.

A loop in such a graph indicates a subspace of $2^0 = 1$ node. Loops occur in the graph of $\mathrm{Min}^2_{(x_2,x_3)} f(x_1, x_2, x_3)$ on nodes with $f(\mathbf{x}) = 1$ because no other node with $f(\mathbf{x}) = 0$ belongs to such a minimal subspace. Loops also occur in the graph of $\mathrm{Max}^2_{(x_2,x_3)} f(x_1, x_2, x_3)$ on nodes with $f(\mathbf{x}) = 1$ because at least for one node (the only node of this subspace) the function value of $f(\mathbf{x})$ is equal to 1. No loops occur in the graph of $\vartheta_{(x_2,x_3)} f(x_1, x_2, x_3)$ because such a minimal subspace cannot carry different function values.

An edge with exactly one non-negated differential dx_i indicates a subspace of $2^1 = 2$ nodes. In the graph of $\mathrm{Min}^2_{(x_2,x_3)} f(x_1, x_2, x_3)$ satisfies only one edge this property; this subspace is specified by $(x_1 = 1, x_2 = 1)$, and the function values of all nodes of this subspace are equal to 1. Edges with one non-negated differential dx_i connect the nodes in the graph of $\mathrm{Max}^2_{(x_2,x_3)} f(x_1, x_2, x_3)$ if the associated subspace contains at least one node for which the function value is equal to 1. The condition that $\vartheta_{(x_2,x_3)} f(x_1, x_2, x_3) = 1$ is that the specified subspace contains different function values. The edges with exactly one non-negated differential dx_i in the right graph of Fig. 6.13 satisfy this condition.

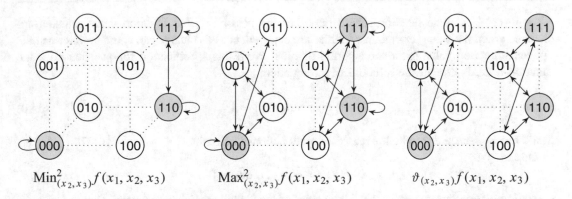

$$\text{Min}^2_{(x_2,x_3)} f(x_1, x_2, x_3) \qquad \text{Max}^2_{(x_2,x_3)} f(x_1, x_2, x_3) \qquad \vartheta_{(x_2,x_3)} f(x_1, x_2, x_3)$$

Fig. 6.13 Graphs of the 2-fold differential minimum, the 2-fold differential maximum, and the ϑ-operation of $f(\mathbf{x}) = \overline{x}_1 \overline{x}_2 \overline{x}_3 \oplus x_1 x_2$ with regard to (x_2, x_3)

An edge $dx_1 dx_2$ connects two nodes of a subspace of $2^2 = 4$ nodes and determines the associated subspace where the remaining variables are constant. Figure 6.13 shows that either no or two such edges occur in a subspace of four nodes that are specified by $x_1 = 0$ or $x_1 = 1$. The graph of $\text{Min}^2_{(x_2,x_3)} f(x_1, x_2, x_3)$ does not contain such edges because neither the subspace $x_1 = 0$ nor $x_1 = 1$ contains nodes where all function values are equal to 1. Both the subspaces $x_1 = 0$ and $x_1 = 1$ contain at least one node with $f(\mathbf{x}) = 1$; hence, the graph of $\text{Max}^2_{(x_2,x_3)} f(x_1, x_2, x_3)$ contains all possible edges $dx_1 dx_2$. All these edges also occur in the graph of $\vartheta_{(x_2,x_3)} f(x_1, x_2, x_3) = 1$ because both the subspaces $x_1 = 0$ and $x_1 = 1$ contain nodes with different function values.

The observations of Example 6.17 can be summarized for an arbitrary function $f(\mathbf{x}_0, \mathbf{x}_1)$ and $\mathbf{x}_0 = (x_1, x_2)$ as follows:

$$\text{Min}^2_{\mathbf{x}_0} f(\mathbf{x}_0, \mathbf{x}_1) = f(\mathbf{x}_0, \mathbf{x}_1) dx_1 \overline{dx}_2 \vee \min_{x_1} f(\mathbf{x}_0, \mathbf{x}_1) dx_1 \overline{dx}_2$$

$$\vee \min_{x_2} f(\mathbf{x}_0, \mathbf{x}_1) \overline{dx}_1 dx_2 \vee \min^2_{(x_1,x_2)} f(\mathbf{x}_0, \mathbf{x}_1) dx_1 dx_2 \,,$$

$$\text{Max}^2_{\mathbf{x}_0} f(\mathbf{x}_0, \mathbf{x}_1) = f(\mathbf{x}_0, \mathbf{x}_1) \overline{dx}_1 \overline{dx}_2 \vee \max_{x_1} f(\mathbf{x}_0, \mathbf{x}_1) dx_1 \overline{dx}_2$$

$$\vee \max_{x_2} f(\mathbf{x}_0, \mathbf{x}_1) \overline{dx}_1 dx_2 \vee \max^2_{(x_1,x_2)} f(\mathbf{x}_0, \mathbf{x}_1) dx_1 dx_2 \,,$$

$$\vartheta_{\mathbf{x}_0} f(\mathbf{x}_0, \mathbf{x}_1) = \vartheta_{x_1} f(\mathbf{x}_0, \mathbf{x}_1) dx_1 \overline{dx}_2 \vee \vartheta_{x_2} f(\mathbf{x}_0, \mathbf{x}_1) \overline{dx}_1 dx_2$$

$$\vee \vartheta_{(x_1,x_2)} f(\mathbf{x}_0, \mathbf{x}_1) dx_1 dx_2 \,.$$

These formulas can be generalized for an arbitrary number of variables of $\mathbf{x}_0$. Subspaces $\mathbf{x}_1 = \mathbf{c}$ and all embedded subspaces are explored.

The k-fold differential operations satisfy the rules:

$$\vartheta_{\mathbf{x}_0} f(\mathbf{x}_0, \mathbf{x}_1) \oplus \text{Min}^k_{\mathbf{x}_0} f(\mathbf{x}_0, \mathbf{x}_1) \oplus \text{Max}^k_{\mathbf{x}_0} f(\mathbf{x}_0, \mathbf{x}_1) = 0 \,, \tag{6.88}$$

$$\text{Min}^k_{\mathbf{x}_0} f(\mathbf{x}_0, \mathbf{x}_1) \le F(\mathbf{x}_0, \mathbf{x}_1, \mathbf{dx}_0) \le \text{Max}^k_{\mathbf{x}_0} f(\mathbf{x}_0, \mathbf{x}_1) \,, \tag{6.89}$$

which can be seen as aggregations of analogue rules of the k-fold derivative operations.

The following example demonstrates the expressive power of differential operations.

Example 6.18 A static function-hazard describes the possibility that function value remains unchanged when the values of certain variables are changed, but this change can cause an intermediate opposite function value for a short period of time. From this specification follows that all static function-hazards $\mathbf{x}\,\mathbf{dx}$ are the solutions of the equation:

$$\overline{\mathrm{d}_{\mathbf{x}}^{k} f(\mathbf{x})} \wedge \vartheta_{\mathbf{x}} f(\mathbf{x}) = 1 \, . \tag{6.90}$$

Static function-hazards can be the reason of malfunctions of circuits.

Due to

$$\vartheta_{\mathbf{x}} f(\mathbf{x}) \wedge \overline{\mathrm{d}x_1}\,\overline{\mathrm{d}x_2} \ldots \overline{\mathrm{d}x_1} \equiv 0,$$

no static function-hazard occurs when the values of the variable remain unchanged. Due to

$$\mathrm{d}_{x_i} f(\mathbf{x}) \equiv \vartheta_{x_i} f(\mathbf{x}),$$

it follows that

$$\overline{\mathrm{d}_{x_i} f(\mathbf{x})} \wedge \vartheta_{x_i} f(\mathbf{x}) \equiv 0 \, ,$$

and consequently, there is no static function-hazard when the value of only one variable is changed. Static function-hazard can occur for all other directions of change and is described by the short formula (6.90).

6.7 Exercises

Prepare for each task of each exercise of this chapter a PRP for the XBOOLE-monitor XBM 2. Prefer a lexicographic order of the variables x_i. Use the help system of the XBOOLE-monitor XBM 2 to learn the details of the needed commands. Execute the created PRPs, verify the computed results, and explain the observed properties.

If you have not yet prepared the XBOOLE-monitor XBM 2 on your computer, you can get this XBOOLE-monitor free of charge by means of the following three steps:

1. **Download**:
   There are four versions of the XBOOLE-monitor XBM 2, two for Windows 10 or subsequent Windows systems (32 or 64 bits) and two for LINUX - Ubuntu (also 32 or 64 bits); you must download the version of the XBOOLE-monitor XBM 2 that fits to your operating system.
   Authorized users of the online version of this chapter (https://doi.org/10.1007/978-3-030-88945-6_6) can download the XBOOLE-monitor XBM 2 directly from the web page

           https://link.springer.com/chapter/10.1007/978-3-030-88945-6_6

   where the links for the download of the XBOOLE-monitor XBM 2 are located in the part "Supplementary Information" (below the part "Abstract"). The headline above such a link indicates the associated zip-file of the XBOOLE-monitor XBM 2. The sizes of the zip-files have been provided behind the links and can be used to verify the download. A click on the link of the wanted version of the XBOOLE-monitor XBM 2 starts the download.
   Readers of the hardcopy of this book get access to the XBOOLE-monitor XBM 2 using the URL

           https://link.springer.com/chapter/10.1007/978-3-030-88945-6_6

   to download the first two pages of this chapter. After this download, the same procedure as the authorized users of the online version of a chapter can be used to download the wanted version of the XBOOLE-monitor XBM 2.

2. **Unzip**: The XBOOLE-monitor XBM 2 must not be installed but unzipped into an arbitrary directory of your computer. A convenient tool for unziping the downloaded zip-file is usually available as part of the operating system or can be downloaded from the Internet.

3. **Execute**:

   - Windows:
     The executable file of the two versions (32 or 64 bits) for Windows 10 (or subsequent Windows systems) of the XBOOLE-monitor XBM 2 is XBM2.exe; the other files in the expanded directory must remain unchanged. A double-click on the executable file XBM2.exe within the Explorer of Windows starts the XBOOLE-monitor XBM 2.
   - LINUX - Ubuntu:
     The unzipped folder of the XBOOLE-monitor XBM 2 contains for this operating system only the executable file XBM2-i386.AppImage for the version of 32 bits or XBM2-x86_64.AppImage for the version of 64 bits of the XBOOLE-monitor XBM 2. A double-click on the created AppImage-file within the file manager of LINUX - Ubuntu starts the XBOOLE-monitor XBM 2.

Exercise 6.1 (Vectorial Derivative Operations) Use in this exercise the logic function:

$$f(\mathbf{x}) = x_1 x_2 \vee \overline{x}_3 \overline{x}_5 \vee x_2 x_4$$

to explore the vectorial derivative operations.

(a) Use Definition (6.4) to compute the vectorial derivative of the function $f(\mathbf{x})$ with regard to $\mathbf{x}_0 = (x_1, x_3, x_4)$. Show the Karnaugh-maps of $f(\mathbf{x}_0, \mathbf{x}_1)$, $f(\overline{\mathbf{x}}_0, \mathbf{x}_1)$, and $\mathrm{der}_{\mathbf{x}_0} f(\mathbf{x}_0, \mathbf{x}_1)$ in one row of the m-fold view. Check whether the result depends on all five variables $x_1, \ldots, x_5$ using the command te_derk.

(b) Use the command derv to compute the vectorial derivative of the function $f(\mathbf{x})$ with regard to $\mathbf{x}_0 = (x_1, x_3, x_4)$. Verify that the result is equal to the use of Definition (6.4). Show the Karnaugh-maps of $f(\mathbf{x}_0, \mathbf{x}_1)$ and the two differently computed vectorial derivatives in one row of the m-fold view.

(c) Use the commands minv and maxv to compute the vectorial minimum and maximum of the function $f(\mathbf{x})$ with regard to $\mathbf{x}_0 = (x_1, x_3, x_4)$. Verify that the computed results satisfy the inequality (6.11) and the derived restrictive equations (6.12), (6.13), and (6.14). Show the Karnaugh-maps of $f(\mathbf{x}_0, \mathbf{x}_1)$, $\min_{\mathbf{x}_0} f(\mathbf{x}_0, \mathbf{x}_1)$, and $\max_{\mathbf{x}_0} f(\mathbf{x}_0, \mathbf{x}_1)$ in one row of the m-fold view.

(d) Verify that the three vectorial derivative operations of the function $f(\mathbf{x})$ with regard to $\mathbf{x}_0 = (x_1, x_3, x_4)$ satisfy Eqs. (6.15) and (6.16). Show the Karnaugh-maps of the three computed derivative operations in one row of the m-fold view.

Exercise 6.2 (Properties of Vectorial Derivative Operations) Use in this exercise the logic function:

$$f(\mathbf{x}_0, \mathbf{x}_1) = x_1 x_3 \overline{x}_4 \vee x_2 \overline{x}_3 \vee x_3 \overline{x}_4 \overline{x}_5$$

and the vectorial derivative operations with regard to $\mathbf{x}_0 = (x_3, x_5)$:

$$f_1(\mathbf{x}_0, \mathbf{x}_1) = \operatorname*{der}_{\mathbf{x}_0} f(\mathbf{x}_0, \mathbf{x}_1)$$

$$f_2(\mathbf{x}_0, \mathbf{x}_1) = \min_{\mathbf{x}_0} f(\mathbf{x}_0, \mathbf{x}_1)$$

$$f_3(\mathbf{x}_0, \mathbf{x}_1) = \max_{\mathbf{x}_0} f(\mathbf{x}_0, \mathbf{x}_1)$$

to explore properties of these vectorial derivative operations.

(a) Verify that all three vectorial derivative operations satisfy

$$f_i(\mathbf{x}_0, \mathbf{x}_1) = f_i(\overline{\mathbf{x}}_0, \mathbf{x}_1), \quad i = 1, \dots, 3,$$

where $\mathbf{x}_0 = (x_3, x_5)$ specifies the direction of change used to compute the vectorial derivative operations $f_i(\mathbf{x}_0, \mathbf{x}_1)$. Show the Karnaugh-maps of $f_i(\mathbf{x}_0, \mathbf{x}_1)$ in the first row and the associated Karnaugh-maps of $f_i(\overline{\mathbf{x}}_0, \mathbf{x}_1)$ in the second row of the m-fold view for visual comparison and compute three Boolean values that express these verifications.

(b) Verify that all three vectorial derivative operations satisfy

$$\mathop{\mathrm{der}}_{\mathbf{x}_0}(f_i(\mathbf{x}_0, \mathbf{x}_1)) = 0, \quad i = 1, \dots, 3,$$

where $\mathbf{x}_0 = (x_3, x_5)$ specifies the direction of change used to compute the vectorial derivative operations $f_i(\mathbf{x}_0, \mathbf{x}_1)$.

(c) Verify that all three vectorial derivative operations satisfy

$$\min_{\mathbf{x}_0}(f_i(\mathbf{x}_0, \mathbf{x}_1)) = f_i(\mathbf{x}_0, \mathbf{x}_1), \quad i = 1, \dots, 3,$$

where $\mathbf{x}_0 = (x_3, x_5)$ specifies the direction of change used to compute the vectorial derivative operations $f_i(\mathbf{x}_0, \mathbf{x}_1)$.

(d) Verify that all three vectorial derivative operations satisfy

$$\max_{\mathbf{x}_0}(f_i(\mathbf{x}_0, \mathbf{x}_1)) = f_i(\mathbf{x}_0, \mathbf{x}_1), \quad i = 1, \dots, 3,$$

where $\mathbf{x}_0 = (x_3, x_5)$ specifies the direction of change used to compute the vectorial derivative operations $f_i(\mathbf{x}_0, \mathbf{x}_1)$.

Exercise 6.3 (Single Derivative Operations)

Use in this exercise the logic function:

$$f(\mathbf{x}) = x_1 x_3 \vee \overline{x}_2 x_4 \vee x_3 x_5$$

to explore the single derivative operations.

(a) Use Eq. (6.30) of Definition 6.4 to compute the (single) derivative of the function $f(\mathbf{x})$ with regard to x_3. Show the Karnaugh-maps of $f(\mathbf{x})$, $f(x_1, x_2, \overline{x}_3, x_4, x_5)$, and $\mathrm{der}_{x_3} f(\mathbf{x})$ in one row of the m-fold view. Use the command obb to minimize the computed derivative and show in the second row of the m-fold view the TVLs of the given function as well as the computed derivative before and after this minimization. Use five commands te_derk to check whether the result depends on each of the five variables.

(b) Compute the (single) derivative of the function $f(\mathbf{x})$ with regard to x_3 based on Eq. (6.35) of Theorem 6.3. Use the command derk to compute the (single) derivative of the function $f(\mathbf{x})$

with regard to x_3. Verify that these two results are equal to each other. Show the Karnaugh-maps of $f(\mathbf{x})$ and the two differently computed derivatives in one row of the m-fold view.

(c) Single derivative operations are special cases of both vectorial derivative operations and k-fold derivative operations. Compute the derivative, the minimum, and the maximum of $f(\mathbf{x})$ with regard to x_1 using first the commands of the vectorial derivative operations and second the commands of the k-fold derivative operations. Verify that the computed single derivative operations represent the same function independent of the used definition although the variable x_1 occurs in the results when the commands of the vectorial derivative operations are executed.

(d) Verify that the three single derivative operations of the functions $f(\mathbf{x})$ with regard to x_1 satisfy Eqs. (6.42), (6.43), and (6.44). Use the commands `minv` and `maxv` to compute the minimum and maximum of $f(\mathbf{x})$ with regard to x_1 for the verification of (6.42) and show the Karnaugh-maps of $\min_{x_1} f(\mathbf{x})$, $f(\mathbf{x})$, and $\max_{x_1} f(\mathbf{x})$ in the first row of the m-fold view. Use the preferred commands of the k-fold derivative operations to compute $\min_{x_1} f(\mathbf{x})$, $\text{der}_{x_1} f(\mathbf{x})$, and $\max_{x_1} f(\mathbf{x})$ and show the associated Karnaugh-maps in the second row of the m-fold view.

Exercise 6.4 (Properties of Logic Functions Explored by Means of Single Derivative Operations)

(a) A logic function $f(x_i, \mathbf{x}_1)$ does not depend on x_i if it satisfies

$$\underset{x_i}{\text{der}}\, f(x_i, \mathbf{x}_1) = 0\,.$$

When this condition is fulfilled, the simplified function $f(\mathbf{x}_1)$ can be computed as follows:

$$f(\mathbf{x}_1) = \underset{x_i}{\max}\, f(x_i, \mathbf{x}_1)\,.$$

Verify whether the function

$$f(\mathbf{x}) = (\overline{x}_2 x_3 x_5 \vee (x_1 \oplus x_3) x_4 \vee x_2 x_3 \overline{x}_4 x_5 \vee x_3 \overline{x}_4 \overline{x}_5 \vee x_1 x_5) \oplus x_1 x_4 \overline{x}_5$$

depends on all variables. Simplify this function such that independent variables do not occur in its representation. From which variables is this function not depending on? Show the Karnaugh-map of the given function and the TVL of the simplified function side by side in the m-fold view. Verify that the given and simplified functions are equal to each other. Create a short expression of the simplified function based on the computed TVL.

(b) A logic function $f(x_i, \mathbf{x}_1)$ is monotonously decreasing with regard to x_i if it satisfies

$$x_i \wedge f(x_i, \mathbf{x}_1) \wedge \overline{\underset{x_i}{\min}\, f(x_i, \mathbf{x}_1)} = 0\,.$$

Verify for all five variables x_i whether the function

$$f(\mathbf{x}) = x_1 x_3 x_5 \vee \overline{x}_3 x_4 \overline{x}_5 \vee x_2 x_5 \vee x_2 \overline{x}_3 \overline{x}_4 x_5 \vee x_2 \overline{x}_4 \overline{x}_5 \vee x_1 (\overline{x}_3 \vee x_5)$$

is monotonously decreasing. Assign the value `true` to the Boolean variable `$md` when the explored function is monotonously decreasing at least with regard to one variable and otherwise the value `false`. Store each variable for which the given function is monotonously decreasing in one row of a TVL in D-form. Show the Karnaugh-map of the given function and the TVL of the detected variables in the first row of the m-fold view.

(c) A logic function $f(x_i, \mathbf{x}_1)$ that is monotonously decreasing with regard to x_i can be expressed by

$$f(x_i, \mathbf{x}_1) = f_0(\mathbf{x}_1) \vee \overline{x}_i \wedge f_1(\mathbf{x}_1) \,,$$

where the two subfunctions $f_0(\mathbf{x}_1)$ and $f_1(\mathbf{x}_1)$ do not depend on x_i. The function $f_0(\mathbf{x}_1)$ is uniquely specified by

$$f_0(\mathbf{x}_1) = \min_{x_i} f(x_i, \mathbf{x}_1) \,.$$

Generally, there are several functions $f_1(\mathbf{x}_1)$; the function $f_1(\mathbf{x}_1)$ with the smallest number of function values 1 can be calculated as follows:

$$f_1(\mathbf{x}_1) = \max_{x_i} \left(\overline{x}_i \wedge f(x_i, \mathbf{x}_1) \wedge \overline{f_0(\mathbf{x}_1)} \right) \,.$$

Compute the functions $f_0(\mathbf{x}_1)$ and $f_1(\mathbf{x}_1)$ for the function $f(\mathbf{x})$ specified in Exercise 6.4(b) with regard to the variable for which this function is monotonously decreasing (see the solution of Exercise 6.4(b)). Compute the function $f(\mathbf{x})$ using the subfunctions $f_0(\mathbf{x}_1)$ and $f_1(\mathbf{x}_1)$ and verify whether this solution-function is equal to the given function. Show the Karnaugh-maps of the given and computed functions $f(\mathbf{x})$ in the left column of the m-fold view and the TVLs of $f_0(\mathbf{x}_1)$ and $f_1(\mathbf{x}_1)$ in the right column of this view. Create a short expression of the function $f(\mathbf{x})$ based on the computed TVLs of $f_0(\mathbf{x}_1)$ and $f_1(\mathbf{x}_1)$.

(d) A logic function $f(x_i, \mathbf{x}_1)$ is linear with regard to x_i if it satisfies

$$\operatorname*{der}_{x_i} f(x_i, \mathbf{x}_1) = 1 \,.$$

When this condition is fulfilled, the function $f(x_i, \mathbf{x}_1)$ can be expressed by

$$f(x_i, \mathbf{x}_1) = x_i \oplus f_1(\mathbf{x}_1) \,,$$

where

$$f_1(\mathbf{x}_1) = \max_{x_i} (x_i \oplus f(x_i, \mathbf{x}_1)) \,.$$

A logic function can be linear with regard to several variables x_i; hence, the linear separation of more than one variable x_i is possible.

Verify for each variable x_i whether the function

$$f(\mathbf{x}) = (x_1 \vee \overline{x}_2)x_3\overline{x}_4\overline{x}_5 \oplus \overline{x}_2(x_1 \vee \overline{x}_3 \vee x_4) \oplus \overline{x}_3 x_5 \oplus x_3 x_4 x_5 \oplus \overline{x}_1 x_2 x_3 \overline{x}_4 x_5$$

is linear with regard to this variable. Create a TVL in A-form of the function

$$f_0(\mathbf{x}_0) = \bigoplus_{x_i \in \mathbf{x}_0} x_i$$

that expresses the antivalence of all variables $x_i \in \mathbf{x}_0$ for which the function $f(\mathbf{x})$ is linear. Compute also a TVL in ODA-form of the function

$$f_1(\mathbf{x}_1) = f(\mathbf{x}) \oplus f_0(\mathbf{x}_0)$$

that depends only on the variables $\mathbf{x}_1 = (\mathbf{x} \setminus \mathbf{x}_0)$ in which the given function is not linear. Show these two TVLs in the right column of the m-fold view.

Verify that the given function $f(\mathbf{x})$ is equal to $f'(\mathbf{x}) = f_0(\mathbf{x}_0) \oplus f_1(\mathbf{x}_1)$. Show the Karnaugh-maps of $f(\mathbf{x})$ and $f'(\mathbf{x})$ in the left column of the m-fold view.

Create a short expression of the simplified function based on the computed TVLs of $f_0(\mathbf{x}_0)$ and $f_1(\mathbf{x}_1)$.

Exercise 6.5 (k-fold Derivative Operations)

Use in this exercise the logic function:

$$f(\mathbf{x}) = x_1(x_2 \oplus x_4) \vee x_1 x_3(\overline{x}_2 \vee \overline{x}_5) \vee x_2 x_5(\overline{x}_3 \vee \overline{x}_4) \vee \overline{x}_2 x_3 \overline{x}_4 \overline{x}_5$$

to explore the k-fold derivative operations.

(a) Compute the k-fold derivative of the function $f(\mathbf{x})$ with regard to $\mathbf{x}_0 = (x_4, x_5)$ using the command `derk`. Show the result as Karnaugh-map to the right of the Karnaugh-map of the given function in the first row of the m-fold view. Which property of the given function in a subspace $\mathbf{x}_1 = (x_1, x_2, x_3) =$ const. originates a value 1 of this k-fold derivative?

Use Definition (6.45) to compute the k-fold derivative of the function $f(\mathbf{x})$ with regard to $\mathbf{x}_0 = (x_4, x_5)$ by two successive single derivatives using command `_derk`. Verify that the successive execution of two single derivatives with regard to these two variables leads to the same result; compute for this verification the two successive single derivatives using the two possible orders of the variables x_4 and x_5. Show Karnaugh-maps of the different intermediate results in the second row of the m-fold view and the Karnaugh-maps computed by the second single derivative in the viewports thereunder.

Utilize the possibility to change the arrangement of the variables of the Karnaugh-maps such that variables $\mathbf{x}_1 = (x_1, x_2, x_3)$ are shown on the lower border, as in the case of the given function. In this way, the evaluation of the function values within the columns of the Karnaugh-maps becomes well visible. A right-click in the region of the Gray code of a Karnaugh-map creates the button

Change the distribution of the variables

and a left-click on this button opens a dialog window. The variable to change (here x_3) can be selected by a left-click on it and moved to the variables of the X-axis by a left-click on the button

<- Move to X

in this dialog window. This dialog must be closed using the OK button.

(b) Compute the k-fold minimum, maximum, and the delta-operation of the function $f(\mathbf{x})$ with regard to $\mathbf{x}_0 = (x_3, x_5)$ using the appropriate commands. Show the Karnaugh-maps of the given function and the computed k-fold derivative operations. Verify the rule (6.51) and substantiate this property of the k-fold derivative operations.

(c) Compute the maximum of $f(\mathbf{x})$ with regard to x_3 and the 2-fold maximum of $f(\mathbf{x})$ with regard to $\mathbf{x}_0 = (x_3, x_5)$ using command `_maxk` in which the names of the variables to be used can directly be specified. Verify the two inequalities of

$$f(\mathbf{x}) \le \max_{x_3} f(\mathbf{x}) \le \max_{(x_3,x_5)}^2 f(\mathbf{x})$$

using command `te_dif`.

The number of variables, the result of a k-fold derivative operation is depending on, decreases for larger values of k. Emphasize this property by showing the Karnaugh-maps of $f(\mathbf{x})$, $\max_{x_3} f(\mathbf{x})$, and $\max_{(x_3,x_5)}^2 f(\mathbf{x})$ in three viewports of the upper row in the m-fold view.

The property that each additional maximum can increase the number of values 1 of the result is better visible when all the basically occurring variables are also used in the Karnaugh-maps of the computed k-fold derivative operations. Find a possibility to map the computed $\max_{x_3} f(\mathbf{x})$ and $\max_{(x_3,x_5)}^2 f(\mathbf{x})$ back to functions defined for all variables $\mathbf{x} = (x_1, x_2, x_3, x_4, x_5)$. Show the Karnaugh-maps of $f(\mathbf{x})$, $\max_{x_3} f(\mathbf{x})$, and $\max_{(x_3,x_5)}^2 f(\mathbf{x})$ represented by all five variables $\mathbf{x} = (x_1, x_2, x_3, x_4, x_5)$ in three viewports of the second row in the m-fold view and verify again the two inequalities explored in this task.

(d) Verify the rules (6.63), (6.64), (6.65), and (6.66) of k-fold derivative operations of negated functions with regard to $\mathbf{x}_0 = (x_3, x_5)$. Show for this purpose the Karnaugh-maps of $f(\mathbf{x})$ and the results of the left-hand sides of these equations in the viewports of the first column and the Karnaugh-maps of $\overline{f(\mathbf{x})}$ and the results of the right-hand sides of these equations in the corresponding viewports of the second column of the m-fold view. Confirm these Karnaugh-maps the explored rules?

Exercise 6.6 (Properties of Logic Functions Explored by Means of k-fold Derivative Operations)

Compute a function $g(\mathbf{x}_1)$ that is equal to 1 for subspaces specified by $\mathbf{x}_1 = (x_1, x_4)$ in which the function

$$f(\mathbf{x}_0, \mathbf{x}_1) = \overline{x}_1 x_2 x_3 \vee \overline{x}_4 \overline{x}_5 \vee \overline{x}_1 x_5 \vee x_2 \overline{x}_4 :$$

(a) Is constant equal to 1
(b) Is constant equal to 0
(c) Has an odd number of function values 1
(d) Has different function values

Specify a rule using a k-fold derivative operation for each of these tasks? Show the Karnaugh-maps of the given function and the computed result in one row of the m-fold view.

Exercise 6.7 (Differentials of Variables, Graphs, and Differential Operations)

(a) Solve the graph equation

$$x_1 dx_2 + x_2 dx_3 + x_3 dx_1 = \overline{(x_3 dx_2 + x_2 dx_1 + x_1 dx_3)} .$$

Show both the TVL of the solution-set and the associated Karnaugh-map in the first row of the m-fold view. Draw the graph that is determined by this equation. What is the longest cycle in this graph when a single differential is equal to 1 for the used edges? What is the longest cycle in this graph when the values of exactly two variables are changed at the same time?

(b) Compute the total differential of $f(\mathbf{x}) = x_1 \vee x_2 \overline{x}_3$ using Definition (6.72) and show both the TVL and the associated Karnaugh-map of this differential of $f(\mathbf{x})$ in the first row of the m-fold view. Draw the graph that is determined by $d_{\mathbf{x}} f(\mathbf{x}) = 1$; emphasize in this graph the nodes where $f(\mathbf{x}) = 1$. Store the XBOOLE-system as `sdt`-file for the use in the next task.

(c) Load the sdt-file of the previous task that contains the function $f(\mathbf{x})$ as TVL 1 and the total differential of this function as TVL 3. Separate all seven vectorial derivatives with regard to one to three variables from $d_{\mathbf{x}} f(\mathbf{x}) = 1$ (TVL 3) and verify that these results are equal to the vectorial derivatives computed by the corresponding command derv.

(d) Compute the 2-fold differential minimum $\mathrm{Min}^2_{(x_1,x_2)} f(\mathbf{x})$ of $f(\mathbf{x}) = x_1 \overline{x}_2 \vee \overline{x}_1 \overline{x}_3$ using all needed k-fold minima of this function and the associated differentials of the variables. Show both the TVL of the given function and the associated Karnaugh-map in the first row and the same representations of the computed 2-fold differential minimum in the second row of the m-fold view. Draw the graph that is determined by $\mathrm{Min}^2_{(x_1,x_2)} f(\mathbf{x}) = 1$.

6.8 Solutions

Solution 6.1 (Vectorial Derivative Operations)

(a) Figure 6.14 shows in the upper part the PRP that computes the vectorial derivative of the function $f(\mathbf{x}_0, \mathbf{x}_1)$ with regard to $\mathbf{x}_0 = (x_1, x_3, x_4)$ using Definition (6.4) and verifies whether this derivative depends on all five variables $x_1, \ldots, x_5$. The created Karnaugh-maps of $f(\mathbf{x}_0, \mathbf{x}_1)$, $f(\overline{\mathbf{x}}_0, \mathbf{x}_1)$, and $\mathrm{der}_{\mathbf{x}_0} f(\mathbf{x}_0, \mathbf{x}_1)$ are shown below this PRP.

The input of the function $f(\mathbf{x}_0, \mathbf{x}_1)$ is realized by the command sbe in lines 5 and 6, and the command vtin in lines 7 and 8 has been used to specify the variables $\mathbf{x}_0 = (x_1, x_3, x_4)$ needed to computed the vectorial derivative.

The command cel in line 9 computes $f(\overline{\mathbf{x}}_0, \mathbf{x}_1)$, and the command syd in line 10 the wanted vectorial derivative $\mathrm{der}_{(x_1,x_3,x_4)} f(\mathbf{x})$ according to Definition (6.4).

The while-loop in lines 12–17 uses the command te_derk in line 14 to check for each variable $x_1, \ldots, x_5$ (selected by the command sv_next) whether $\mathrm{der}_{x_i} (\mathrm{der}_{(x_1,x_3,x_4)} f(\mathbf{x})) \neq 0$. The result $\$depend = true$ confirms that $\mathrm{der}_{(x_1,x_3,x_4)} f(\mathbf{x})$ depends on all five variables $x_1, \ldots, x_5$.

```
 1   new                          11   set $depend true
 2   space 32 1                   12   while (sv_next 4 5 5)
 3   avar 1                       13   (
 4   x1 x2 x3 x4 x5.              14   if (te_derk 4 5)
 5   sbe 1 1                      15   (
 6   x1&x2+/x3&/x5+x2&x4.         16   set $depend false
 7   vtin 1 2                     17   ))
 8   x1 x3 x4.                    18   assign 1 /m 1 1
 9   cel 1 2 3 /01 /10            19   assign 3 /m 1 2
10   syd 1 3 4                    20   assign 4 /m 1 3
```

T	TVL (1) ODA	5 Var.	5 R.	S. 1
0 0	1 0 0 1	1 1 0 1
0 1	0 0 0 0	1 1 0 0
1 1	0 0 1 1	1 1 0 0
1 0	1 0 1 1	1 1 0 1
x_4 x_5	0 1 1 0 0 1 1 0 x3
	0 0 1 1 1 1 0 0 x2
	0 0 0 0 1 1 1 1 x1

T	TVL (3) ODA	5 Var.	5 R.	S. 1
0 0	0 1 1 1	1 1 1 0
0 1	0 0 1 1	1 1 0 0
1 1	0 0 1 1	0 0 0 0
1 0	0 1 1 1	0 1 1 0
x_4 x_5	0 1 1 0 0 1 1 0 x3
	0 0 1 1 1 1 0 0 x2
	0 0 0 0 1 1 1 1 x1

T	TVL (4) ODA	5 Var.	6 R.	S. 1
0 0	1 1 1 0	0 0 1 1
0 1	0 0 1 1	0 0 0 0
1 1	0 0 0 0	1 1 0 0
1 0	1 1 0 0	1 0 1 1
x_4 x_5	0 1 1 0 0 1 1 0 x3
	0 0 1 1 1 1 0 0 x2
	0 0 0 0 1 1 1 1 x1

Fig. 6.14 Problem-program that computes $\mathrm{der}_{(x_1,x_3,x_4)} f(\mathbf{x})$ and verifies that this derivative depends on all five variables $x_1, \ldots, x_5$ together with the related Karnaugh-maps

```
1   new                         9   cel 1 2 3 /01 /10
2   space 32 1                  10  syd 1 3 4
3   avar 1                      11  derv 1 2 5
4   x1 x2 x3 x4 x5.             12  set $equal (te_syd 4 5)
5   sbe 1 1                     13  assign 1 /m 1 1
6   x1&x2+/x3&/x5+x2&x4.        14  assign 4 /m 1 2
7   vtin 1 2                    15  assign 5 /m 1 3
8   x1 x3 x4.
```

T	TVL (1) ODA | 5 Var. | 5 R. | S. 1
0 0	1 0 0 1 | 1 1 0 1
0 1	0 0 0 0 | 1 1 0 0
1 1	0 0 1 1 | 1 1 0 0
1 0	1 0 1 1 | 1 1 0 1
x4 x5	0 1 1 0 0 1 1 0 x3
	0 0 1 1 1 1 0 0 x2
	0 0 0 0 1 1 1 1 x1

T	TVL (4) ODA | 5 Var. | 6 R. | S. 1
0 0	1 1 1 0 | 0 0 1 1
0 1	0 0 1 1 | 0 0 0 0
1 1	0 0 0 0 | 1 1 0 0
1 0	1 1 0 0 | 1 0 1 1
x4 x5	0 1 1 0 0 1 1 0 x3
	0 0 1 1 1 1 0 0 x2
	0 0 0 0 1 1 1 1 x1

T	TVL (5) ODA | 5 Var. | 6 R. | S. 1
0 0	1 1 1 0 | 0 0 1 1
0 1	0 0 1 1 | 0 0 0 0
1 1	0 0 0 0 | 1 1 0 0
1 0	1 1 0 0 | 1 0 1 1
x4 x5	0 1 1 0 0 1 1 0 x3
	0 0 1 1 1 1 0 0 x2
	0 0 0 0 1 1 1 1 x1

Fig. 6.15 Problem-program that computes $\text{der}_{(x_1,x_3,x_4)} f(\mathbf{x})$ using two approaches and compares the results together with the related Karnaugh-maps

(b) Figure 6.15 shows in the upper part the PRP that computes the vectorial derivative of the function $f(\mathbf{x}_0, \mathbf{x}_1)$ with regard to $\mathbf{x}_0 = (x_1, x_3, x_4)$ first based on Definition (6.4) (TVL 4) and thereafter using the command derv. The created Karnaugh-maps of $f(\mathbf{x}_0, \mathbf{x}_1)$, $\text{der}_{(x_1,x_3,x_4)} f(\mathbf{x})$ computed based on Definition (6.4) (TVL 4) and using the command derv (TVL 5) are shown below this PRP.

Both the result $equal = true and the identical Karnaugh-maps of TVLs 4 and 5 confirm that the used two approaches to compute the wanted vectorial derivative result in the same function. Indeed, the command derv uses internally the execution of the commands cel and syd and deletes implicitly the intermediate TVL of $f(\overline{\mathbf{x}}_0, \mathbf{x}_1)$.

(c) Figure 6.16 shows in the upper part the PRP that computes both the vectorial minimum and maximum of the function $f(\mathbf{x}_0, \mathbf{x}_1)$ with regard to $\mathbf{x}_0 = (x_1, x_3, x_4)$ using the commands minv and maxv. Furthermore, the two inequalities of (6.11) and the three restrictive equations (6.12), (6.13), and (6.14) are verified. The Karnaugh-maps of $f(\mathbf{x}_0, \mathbf{x}_1)$ (TVL 1), $\min_{(x_1,x_3,x_4)} f(\mathbf{x})$ (TVL 3), and $\max_{(x_1,x_3,x_4)} f(\mathbf{x})$ (TVL 4) are shown in Fig. 6.16 this PRP.

The results $minvlef = true and $flemaxv = true as well as the comparison of the Karnaugh-maps of $f(\mathbf{x}_0, \mathbf{x}_1)$, $\min_{(x_1,x_3,x_4)} f(\mathbf{x})$, and $\max_{(x_1,x_3,x_4)} f(\mathbf{x})$ confirm that the inequality (6.11) is satisfied.

The complements of $f(\mathbf{x}_0, \mathbf{x}_1)$ and $\max_{(x_1,x_3,x_4)} f(\mathbf{x})$ are needed to verify Eqs. (6.12), (6.13), and (6.14); the command cpl in lines 13 and 14 solves this task. The results

- $minvanf = true
- $fanmaxv = true
- $minvanmaxv = true

confirm that the three restrictive equations (6.12), (6.13), and (6.14) are satisfied for the given function and the computed vectorial minimum and maximum.

(d) Figure 6.17 shows in the upper part the PRP that computes $\min_{(x_1,x_3,x_4)} f(\mathbf{x})$ (TVL 3), $\text{der}_{(x_1,x_3,x_4)} f(\mathbf{x})$ (TVL 4), and $\max_{(x_1,x_3,x_4)} f(\mathbf{x})$ (TVL 5) using the commands minv, derv,

```
 1  new                        11  set $minvlef (te_dif 3 1)
 2  space 32 1                 12  set $flemaxv (te_dif 1 4)
 3  avar 1                     13  cpl 1 5
 4  x1 x2 x3 x4 x5.            14  cpl 4 6
 5  sbe 1 1                    15  set $minvanf (te_isc 3 5)
 6  x1&x2+/x3&/x5+x2&x4.       16  set $fanmaxv (te_isc 1 6)
 7  vtin 1 2                   17  set $minvanmaxv (te_isc 3 6)
 8  x1 x3 x4.                  18  assign 1 /m 1 1
 9  minv 1 2 3                 19  assign 3 /m 1 2
10  maxv 1 2 4                 20  assign 4 /m 1 3
```

```
T TVL (1) ODA | 5 Var. | 5 R. | S. 1     T TVL (3) ODA | 5 Var. | 6 R. | S. 1     T TVL (4) ODA | 5 Var. | 8 R. | S. 1

0 0  1 0 0 1 | 1 1 0 1                   0 0  0 0 0 1 | 1 1 0 0                   0 0  1 1 1 1 | 1 1 1 1
0 1  0 0 0 0 | 1 1 0 0                   0 1  0 0 0 0 | 1 1 0 0                   0 1  0 0 1 1 | 1 1 0 0
1 1  0 0 1 1 | 1 1 0 0                   1 1  0 0 1 1 | 0 0 0 0                   1 1  0 0 1 1 | 1 1 0 0
1 0  1 0 1 1 | 1 1 0 1                   1 0  0 0 1 1 | 0 1 0 0                   1 0  1 1 1 1 | 1 1 1 1

x4                                       x4                                       x4
x5   0 1 1 0 0 1 1 0 x3                   x5   0 1 1 0 0 1 1 0 x3                   x5   0 1 1 0 0 1 1 0 x3
     0 0 1 1 1 1 0 0 x2                        0 0 1 1 1 1 0 0 x2                        0 0 1 1 1 1 0 0 x2
     0 0 0 0 1 1 1 1 x1                        0 0 0 0 1 1 1 1 x1                        0 0 0 0 1 1 1 1 x1
```

Fig. 6.16 Problem-program that computes $\min_{(x_1,x_3,x_4)} f(\mathbf{x})$ and $\max_{(x_1,x_3,x_4)} f(\mathbf{x})$ and verifies the rules (6.11), ..., (6.14) together with the related Karnaugh-maps

```
 1  new                        10  derv 1 2 4
 2  space 32 1                 11  maxv 1 2 5
 3  avar 1                     12  syd 3 4 6
 4  x1 x2 x3 x4 x5.            13  set $xorvdo0 (te_syd 6 5)
 5  sbe 1 1                    14  set $minvaderv0 (te_isc 3 4)
 6  x1&x2+/x3&/x5+x2&x4.       15  assign 3 /m 1 1
 7  vtin 1 2                   16  assign 4 /m 1 2
 8  x1 x3 x4.                  17  assign 5 /m 1 3
 9  minv 1 2 3
```

```
T TVL (3) ODA | 5 Var. | 6 R. | S. 1     T TVL (4) ODA | 5 Var. | 6 R. | S. 1     T TVL (5) ODA | 5 Var. | 8 R. | S. 1

0 0  0 0 0 1 | 1 1 0 0                   0 0  1 1 1 0 | 0 0 1 1                   0 0  1 1 1 1 | 1 1 1 1
0 1  0 0 0 0 | 1 1 0 0                   0 1  0 0 1 1 | 0 0 0 0                   0 1  0 0 1 1 | 1 1 0 0
1 1  0 0 1 1 | 0 0 0 0                   1 1  0 0 0 0 | 1 1 0 0                   1 1  0 0 1 1 | 1 1 0 0
1 0  0 0 1 1 | 0 1 0 0                   1 0  1 1 0 0 | 1 0 1 1                   1 0  1 1 1 1 | 1 1 1 1

x4                                       x4                                       x4
x5   0 1 1 0 0 1 1 0 x3                   x5   0 1 1 0 0 1 1 0 x3                   x5   0 1 1 0 0 1 1 0 x3
     0 0 1 1 1 1 0 0 x2                        0 0 1 1 1 1 0 0 x2                        0 0 1 1 1 1 0 0 x2
     0 0 0 0 1 1 1 1 x1                        0 0 0 0 1 1 1 1 x1                        0 0 0 0 1 1 1 1 x1
```

Fig. 6.17 Problem-program that computes the three vectorial derivative operations of $f(\mathbf{x})$ with regard to (x_1, x_3, x_4) and verifies the rules (6.15) and (6.16) together with the Karnaugh-maps of these three vectorial derivative operations

and maxv. Thereafter the rules (6.15) and (6.16) are verified. The Karnaugh-maps of the three vectorial derivative operations of $f(\mathbf{x})$ with regard to (x_1, x_3, x_4) are shown below this PRP. The result $xorvdo0 = true$ (computed in lines 12 and 13) confirms that (6.15) is satisfied for the computed three vectorial derivative operations. The comparison of the three Karnaugh-maps of Fig. 6.17 leads to the same result; each function value 1 of $\max_{(x_1,x_3,x_4)} f(\mathbf{x})$ (TVL 5) occurs either for $\min_{(x_1,x_3,x_4)} f(\mathbf{x})$ (TVL 3) or for $\mathrm{der}_{(x_1,x_3,x_4)} f(\mathbf{x})$ (TVL 4) so that the antivalence of these three vectorial derivative operations is equal to 0. The orthogonality of $\min_{(x_1,x_3,x_4)} f(\mathbf{x})$ and $\mathrm{der}_{(x_1,x_3,x_4)} f(\mathbf{x})$ is checked in line 14 and confirmed by the result `$minvaderv0 = true`.

Solution 6.2 (Properties of Vectorial Derivative Operations)

(a) Figure 6.18 shows in the upper part the PRP that verifies that all three vectorial derivative operations of the function $f(\mathbf{x}_0, \mathbf{x}_1)$ with regard to $\mathbf{x}_0 = (x_3, x_5)$ satisfy $f_i(\mathbf{x}_0, \mathbf{x}_1) = f_i(\overline{\mathbf{x}}_0, \mathbf{x}_1)$. Below this PRP the created Karnaugh-maps of $f_i(\mathbf{x}_0, \mathbf{x}_1)$ are shown in the first row of the m-fold view and the Karnaugh-maps of the associated functions $f_i(\overline{\mathbf{x}}_0, \mathbf{x}_1)$ thereunder in the second row. The three vectorial derivative operations $f_i(\mathbf{x}_0, \mathbf{x}_1)$ of the function $f(\mathbf{x}_0, \mathbf{x}_1)$ with regard to $\mathbf{x}_0 = (x_3, x_5)$ are computed in lines 10–12. The command `cel` in lines 13, 15, and 17 computes the modified functions $f_i(\overline{\mathbf{x}}_0, \mathbf{x}_1)$. Both the results:

- `$f1eqf1nx0 = true`
- `$f2eqf2nx0 = true`
- `$f3eqf3nx0 = true`

```
 1   new                    13   cel 3 2 6 /01 /10
 2   space 32 1             14   set $f1eqf1nx0 (te_syd 3 6)
 3   avar 1                 15   cel 4 2 7 /01 /10
 4   x1 x2 x3 x4 x5 .       16   set $f2eqf2nx0 (te_syd 4 7)
 5   sbe 1 1                17   cel 5 2 8 /01 /10
 6   x1&x3&/x4+x2&/x3+      18   set $f3eqf3nx0 (te_syd 5 8)
 7   x3&/x4&/x5 .           19   assign 3 /m 1 1
 8   vtin 1 2               20   assign 4 /m 1 2
 9   x3 x5 .                21   assign 5 /m 1 3
10   derv 1 2 3             22   assign 6 /m 2 1
11   minv 1 2 4             23   assign 7 /m 2 2
12   maxv 1 2 5             24   assign 8 /m 2 3
```

```
T  TVL (3) ODA | 5 Var. | 12 R. | S. 1     T  TVL (4) ODA | 5 Var. | 4 R. | S. 1     T  TVL (5) ODA | 5 Var. | 8 R. | S. 1
0 0   0 1 0 1 0 0 1 1                       0 0   0 0 1 0 1 1 0 0                      0 0   0 1 1 1 1 1 1 1
0 1   1 0 1 0 0 0 1 1                       0 1   0 0 0 1 1 1 0 0                      0 1   1 0 1 1 1 1 1 1
1 1   0 0 1 1 1 1 0 0                       1 1   0 0 0 0 0 0 0 0                      1 1   0 0 1 1 1 1 0 0
1 0   0 0 1 1 1 1 0 0                       1 0   0 0 0 0 0 0 0 0                      1 0   0 0 1 1 1 1 0 0
x4
x5    0 1 1 0 0 1 1 0 x3                    x4 x5  0 1 1 0 0 1 1 0 x3                  x4 x5  0 1 1 0 0 1 1 0 x3
      0 0 1 1 1 1 0 0 x2                           0 0 1 1 1 1 0 0 x2                        0 0 1 1 1 1 0 0 x2
      0 0 0 0 1 1 1 1 x1                           0 0 0 0 1 1 1 1 x1                        0 0 0 0 1 1 1 1 x1

T  TVL (6) ODA | 5 Var. | 12 R. | S. 1     T  TVL (7) ODA | 5 Var. | 4 R. | S. 1     T  TVL (8) ODA | 5 Var. | 8 R. | S. 1
0 0   0 1 0 1 0 0 1 1                       0 0   0 0 1 0 1 1 0 0                      0 0   0 1 1 1 1 1 1 1
0 1   1 0 1 0 0 0 1 1                       0 1   0 0 0 1 1 1 0 0                      0 1   1 0 1 1 1 1 1 1
1 1   0 0 1 1 1 1 0 0                       1 1   0 0 0 0 0 0 0 0                      1 1   0 0 1 1 1 1 0 0
1 0   0 0 1 1 1 1 0 0                       1 0   0 0 0 0 0 0 0 0                      1 0   0 0 1 1 1 1 0 0
x4
x5    0 1 1 0 0 1 1 0 x3                    x4 x5  0 1 1 0 0 1 1 0 x3                  x4 x5  0 1 1 0 0 1 1 0 x3
      0 0 1 1 1 1 0 0 x2                           0 0 1 1 1 1 0 0 x2                        0 0 1 1 1 1 0 0 x2
      0 0 0 0 1 1 1 1 x1                           0 0 0 0 1 1 1 1 x1                        0 0 0 0 1 1 1 1 x1
```

Fig. 6.18 Problem-program that verifies that $f_1 = \mathrm{der}_{\mathbf{x}_0} f(\mathbf{x}_0, \mathbf{x}_1)$, $f_2 = \min_{\mathbf{x}_0} f(\mathbf{x}_0, \mathbf{x}_1)$, and $f_3 = \max_{\mathbf{x}_0} f(\mathbf{x}_0, \mathbf{x}_1)$ where $\mathbf{x}_0 = (x_3, x_5)$ satisfy $f_i(\mathbf{x}_0, \mathbf{x}_1) = f_i(\overline{\mathbf{x}}_0, \mathbf{x}_1)$ together with the related Karnaugh-maps

Fig. 6.19
Problem-program that
verifies that the vectorial
derivative with regard to $\mathbf{x}_0$
of each vectorial derivative
operation with regard to
the same set of variables $\mathbf{x}_0$
is equal to 0

```
1   new
2   space 32 1
3   avar 1
4   x1 x2 x3 x4 x5 .
5   sbe 1 1
6   x1&x3&/x4+x2&/x3+x3&/x4&/x5 .
7   vtin 1 2
8   x3 x5 .
9   derv 1 2 3
10  minv 1 2 4
11  maxv 1 2 5
12  set $dervf1x0eq0 (te_derv 3 2)
13  set $dervf2x0eq0 (te_derv 4 2)
14  set $dervf3x0eq0 (te_derv 5 2)
```

Boolean variables

Name	Value
$dervf1x0eq0	true
$dervf2x0eq0	true
$dervf3x0eq0	true

and the comparison of the two Karnaugh-maps in each of the three columns of the m-fold view
confirm that each vectorial derivative operation $f_i(\mathbf{x}_0, \mathbf{x}_1)$ of the function $f(\mathbf{x}_0, \mathbf{x}_1)$ with regard to
$\mathbf{x}_0 = (x_3, x_5)$ satisfies $f_i(\mathbf{x}_0, \mathbf{x}_1) = f_i(\overline{\mathbf{x}}_0, \mathbf{x}_1)$. In subspaces $(x_1, x_2, x_4) = const.$ occur only four
of 16 functions $f(x_3, x_5)$ with expressions 0, $x_3 \oplus x_5$, $\overline{x}_3 \oplus x_5$, and 1; no other function $f(x_3, x_5)$
is possible for the explored vectorial derivative operations $f_i(\mathbf{x}_0, \mathbf{x}_1)$ with regard to $\mathbf{x}_0 = (x_3, x_5)$.

(b) Figure 6.19 shows on the left-hand side the PRP that verifies that

$$\underset{\mathbf{x}_0}{\text{der}}(\underset{\mathbf{x}_0}{\text{der}}\, f(\mathbf{x}_0, \mathbf{x}_1)) = 0 \,,$$

$$\underset{\mathbf{x}_0}{\text{der}}(\underset{\mathbf{x}_0}{\text{min}}\, f(\mathbf{x}_0, \mathbf{x}_1)) = 0 \,, \text{ and}$$

$$\underset{\mathbf{x}_0}{\text{der}}(\underset{\mathbf{x}_0}{\text{max}}\, f(\mathbf{x}_0, \mathbf{x}_1)) = 0 \,.$$

The three computed values `true` (shown on the right-hand side of Fig. 6.19) confirm this
property.

The command `te_derv` in lines 12–14 computes the vectorial derivative of the given function
(indicated by the first parameter) with regard to $\mathbf{x}_0$ (indicated by the second parameter) and
immediately break this computation when the result will not be equal to 0; the result of a
command `te_derv` is equal to `true` if the computed vectorial derivative is equal to 0. These
computed results are assigned to the specified variables in the three command `set` in lines 12–14.

(c) Figure 6.20 shows on the left-hand side the PRP that verifies that

$$\underset{\mathbf{x}_0}{\text{min}}(\underset{\mathbf{x}_0}{\text{der}}\, f(\mathbf{x}_0, \mathbf{x}_1)) = \underset{\mathbf{x}_0}{\text{der}}\, f(\mathbf{x}_0, \mathbf{x}_1) \,,$$

$$\underset{\mathbf{x}_0}{\text{min}}(\underset{\mathbf{x}_0}{\text{min}}\, f(\mathbf{x}_0, \mathbf{x}_1)) = \underset{\mathbf{x}_0}{\text{min}}\, f(\mathbf{x}_0, \mathbf{x}_1) \,, \text{ and}$$

$$\underset{\mathbf{x}_0}{\text{min}}(\underset{\mathbf{x}_0}{\text{max}}\, f(\mathbf{x}_0, \mathbf{x}_1)) = \underset{\mathbf{x}_0}{\text{max}}\, f(\mathbf{x}_0, \mathbf{x}_1) \,.$$

The three computed values `true` (shown on the right-hand side of Fig. 6.20) confirm this
property.

The command `minv` in lines 12, 14, and 16 computes the vectorial minimum with regard
to $\mathbf{x}_0 = (x_3, x_5)$ of $\text{der}_{\mathbf{x}_0}\, f(\mathbf{x}_0, \mathbf{x}_1)$, $\text{min}_{\mathbf{x}_0}\, f(\mathbf{x}_0, \mathbf{x}_1)$, and $\text{max}_{\mathbf{x}_0}\, f(\mathbf{x}_0, \mathbf{x}_1)$, respectively. The
command `te_syd` in the subsequent lines verifies whether these vectorial minimums change
the given derivative operation with regard to the same set of variables $\mathbf{x}_0$ and the command `set`
assigns the computed results to Boolean variables.

Fig. 6.20
Problem-program that
verifies that the vectorial
minimum with regard to $\mathbf{x}_0$
does not change any
vectorial derivative
operation with regard to the
same set of variables $\mathbf{x}_0$

```
1   new
2   space 32 1
3   avar 1
4   x1 x2 x3 x4 x5.
5   sbe 1 1
6   x1&x3&/x4+x2&/x3+x3&/x4&/x5.
7   vtin 1 2
8   x3 x5.
9   derv 1 2 3
10  minv 1 2 4
11  maxv 1 2 5
12  minv 3 2 6
13  set $minvf1x0eqf1 (te_syd 3 6)
14  minv 4 2 7
15  set $minvf2x0eqf2 (te_syd 4 7)
16  minv 5 2 8
17  set $minvf3x0eqf3 (te_syd 5 8)
```

Boolean variables

Name	Value
$minvf1x0eqf1	true
$minvf2x0eqf2	true
$minvf3x0eqf3	true

Fig. 6.21
Problem-program that
verifies that the vectorial
maximum with regard to
$\mathbf{x}_0$ does not change any
vectorial derivative
operation with regard to the
same set of variables $\mathbf{x}_0$

```
1   new
2   space 32 1
3   avar 1
4   x1 x2 x3 x4 x5.
5   sbe 1 1
6   x1&x3&/x4+x2&/x3+x3&/x4&/x5.
7   vtin 1 2
8   x3 x5.
9   derv 1 2 3
10  minv 1 2 4
11  maxv 1 2 5
12  maxv 3 2 6
13  set $maxvf1x0eqf1 (te_syd 3 6)
14  maxv 4 2 7
15  set $maxvf2x0eqf2 (te_syd 4 7)
16  maxv 5 2 8
17  set $maxvf3x0eqf3 (te_syd 5 8)
```

Boolean variables

Name	Value
$maxvf1x0eqf1	true
$maxvf2x0eqf2	true
$maxvf3x0eqf3	true

(d) Figure 6.21 shows on the left-hand side the PRP that verifies that

$$\max_{\mathbf{x}_0}(\operatorname{der}_{\mathbf{x}_0} f(\mathbf{x}_0, \mathbf{x}_1)) = \operatorname{der}_{\mathbf{x}_0} f(\mathbf{x}_0, \mathbf{x}_1) \,,$$

$$\max_{\mathbf{x}_0}(\min_{\mathbf{x}_0} f(\mathbf{x}_0, \mathbf{x}_1)) = \min_{\mathbf{x}_0} f(\mathbf{x}_0, \mathbf{x}_1) \,, \text{ and}$$

$$\max_{\mathbf{x}_0}(\max_{\mathbf{x}_0} f(\mathbf{x}_0, \mathbf{x}_1)) = \max_{\mathbf{x}_0} f(\mathbf{x}_0, \mathbf{x}_1) \,.$$

The three computed values `true` (shown on the right-hand side of Fig. 6.21) confirm this property.

The command `maxv` in lines 12, 14, and 16 computes the vectorial maximum with regard to $\mathbf{x}_0 = (x_3, x_5)$ of $\operatorname{der}_{\mathbf{x}_0} f(\mathbf{x}_0, \mathbf{x}_1)$, $\min_{\mathbf{x}_0} f(\mathbf{x}_0, \mathbf{x}_1)$, and $\max_{\mathbf{x}_0} f(\mathbf{x}_0, \mathbf{x}_1)$, respectively. The checks that a vectorial maximum with regard to $\mathbf{x}_0$ does not change any vectorial derivative operation with regard to the same set of variables $\mathbf{x}_0$ is realized as in the previous task.

1	new
2	space 32 1
3	vtin 1 6
4	x1 x2 x3 x4 x5 .
5	sbe 1 1
6	x1&x3+/x2&x4+x3&x5 .
7	vtin 1 2
8	x3 .
9	cel 1 2 3 /01 /10
10	syd 1 3 4
11	obb 4 5
12	assign 1 /m 1 1
13	assign 3 /m 1 2
14	assign 4 /m 1 3
15	assign 1 /m 2 1
16	assign 4 /m 2 2
17	assign 5 /m 2 3
18	stv 6 1 7
19	set $indepx1 (te_derk 4 7)
20	stv 6 2 7
21	set $indepx2 (te_derk 4 7)
22	stv 6 3 7
23	set $indepx3 (te_derk 4 7)
24	stv 6 4 7
25	set $indepx4 (te_derk 4 7)
26	stv 6 5 7
27	set $indepx5 (te_derk 4 7)

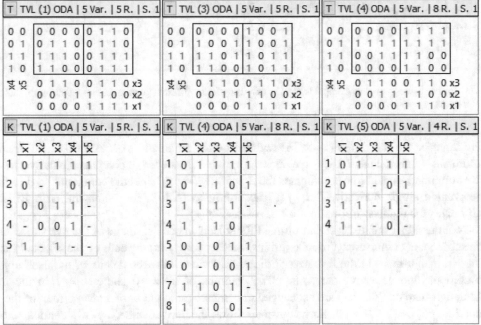

Fig. 6.22 Problem-program that computes $\mathrm{der}_{x_3} f(\mathbf{x})$ and verifies whether this derivative depends on each of the five variables $x_1, \ldots, x_5$ together with the related Karnaugh-maps and TVLs

Solution 6.3 (Single Derivative Operations)

(a) Figure 6.22 shows in the upper part the PRP that computes the (single) derivative of the function $f(x_3, \mathbf{x}_1)$ with regard to x_3 using Eq. (6.30) of Definition 6.4 and verifies whether this derivative depends on each of the five variables $x_1, \ldots, x_5$. Below this PRP the created Karnaugh-maps of $f(x_3, \mathbf{x}_1)$, $f(\overline{x}_3, \mathbf{x}_1)$, and $\mathrm{der}_{x_3} f(x_3, \mathbf{x}_1)$ are shown in the first row of the m-fold view; additionally, the TVLs of the given function as well as the computed derivative before and after the minimization using the command obb are shown in the second row of the m-fold view.

The command vtin in lines 3 and 4 is used at the beginning of this PRP for both the assignment of the five variables $x_1, \ldots, x_5$ to the Boolean space 1 and the possibility to use these variables to check whether $\mathrm{der}_{x_3} f(\mathbf{x})$ depends on each of them.

```
 1   new                         12   orth  4  5
 2   space  32  1                13   cpl  2  6
 3   avar  1                     14   isc  1  6  7
 4   x1  x2  x3  x4  x5.         15   dco  7  6  8
 5   sbe  1  1                   16   orth  8  9
 6   x1&x3+/x2&x4+x3&x5.         17   syd  5  9  10
 7   tin  1  2                   18   derk  1  2  11
 8   x3.                         19   assign  1  /m  1  1
 9   0.                          20   assign  10  /m  1  2
10   isc  1  2  3                21   assign  11  /m  1  3
11   dco  3  2  4
```

```
T  TVL (1) ODA | 5 Var. | 5 R. | S. 1      T  TVL (10) ODA | 4 Var. | 4 R. | S. 1     T  TVL (11) ODA | 4 Var. | 4 R. | S. 1

0 0   0 0 0 0   0 1 1 0                     0 0    0 0 1 1                              0 0    0 0 1 1
0 1   0 1 1 0   0 1 1 0                     0 1    1 1 1 1                              0 1    1 1 1 1
1 1   1 1 1 0   0 1 1 1                     1 1    0 1 1 0                              1 1    0 1 1 0
1 0   1 1 0 0   0 1 1 1                     1 0    0 0 1 0                              1 0    0 0 1 0

x4 x5  0 1 1 0 0 1 1 0  x3                  x4 x5   0 1 1 0  x2                         x4 x5   0 1 1 0  x2
       0 0 1 1 1 1 0 0  x2                          0 0 1 1  x1                                 0 0 1 1  x1
       0 0 0 0 1 1 1 1  x1
```

Fig. 6.23 Problem-program that computes $\text{der}_{x_3} f(\mathbf{x})$ using two different approaches together with the related Karnaugh-maps

The input of the function $f(x_3, \mathbf{x}_1)$ is realized by the command sbe in lines 5 and 6, and the command vtin in lines 7 and 8 specifies the variable x_3 needed to compute the derivative.

The command cel in line 9 computes the function $f(\overline{x}_3, \mathbf{x}_1)$, and the command syd in line 10 the wanted single derivative $\text{der}_{x_3} f(\mathbf{x})$ according to (6.30). The command obb in line 11 minimizes the number of rows in the TVL of $\text{der}_{x_3} f(\mathbf{x})$.

The command stv in the last part of this PRP selects in each case one of the five variables, and the subsequent command te_derk verifies whether $\text{der}_{x_3} f(\mathbf{x})$ depends on the selected variable. The Karnaugh-maps in the first row of the m-fold view show the details of the used approach to compute the derivative $\text{der}_{x_3} f(\mathbf{x})$. The columns of $x_3 = 0$ and $x_3 = 1$ in the middle Karnaugh-map of $f(\overline{x}_3, \mathbf{x}_1)$ are exchanged in comparison to the left Karnaugh-map of the given function $f(x_3, \mathbf{x}_1)$. The right Karnaugh-map shows the derivative $\text{der}_{x_3} f(\mathbf{x})$; identical values in the columns of $x_3 = 0$ and $x_3 = 1$ for fixed values of x_1 and x_2 confirm that $\text{der}_{x_3} f(\mathbf{x})$ does not depend on x_3.

TVL 4 of the derivative $\text{der}_{x_3} f(\mathbf{x})$ contains the same ternary vectors of (x_1, x_2, x_4, x_5) in the upper part for $x_3 = 1$ and below for $x_3 = 0$. This detailed check confirms that $\text{der}_{x_3} f(\mathbf{x})$ does not depend on x_3. In the minimized TVL 5 of the derivative $\text{der}_{x_3} f(\mathbf{x})$ in the right viewport of the second row of the m-fold view contains only dashes in column of x_3; this shows obviously that $\text{der}_{x_3} f(\mathbf{x})$ is independent of x_3.

The value of the computed variable $indepx3 is equal to true, and the other four variables of this check have been set to the value false; this confirms that the computed derivative $\text{der}_{x_3} f(\mathbf{x})$ is independent of x_3 but depends on the other four variables.

(b) Figure 6.23 shows in the upper part the PRP that computes the (single) derivative of the function $f(x_3, \mathbf{x}_1)$ with regard to x_3 first using Eq. (6.35) of Theorem 6.3 and second using the command derk. The created Karnaugh-maps of $f(x_3, \mathbf{x}_1)$ and the differently computed derivative $\text{der}_{x_3} f(x_3, \mathbf{x}_1)$ are shown below this PRP.

```
 1  new                          13  mink  1  2  7
 2  space  32  1                 14  maxk  1  2  8
 3  avar  1                      15  assign  3  /m  1  1
 4  x1  x2  x3  x4  x5 .         16  assign  4  /m  1  2
 5  sbe  1  1                    17  assign  5  /m  1  3
 6  x1&x3 +/ x2&x4+x3&x5 .       18  assign  6  /m  2  1
 7  vtin  1  2                   19  assign  7  /m  2  2
 8  x1 .                         20  assign  8  /m  2  3
 9  derv  1  2  3                21  set  $equalder  (te_syd  3  6)
10  minv  1  2  4                22  set  $equalmin  (te_syd  4  7)
11  maxv  1  2  5                23  set  $equalmax  (te_syd  5  8)
12  derk  1  2  6
```

```
| T TVL (3) ODA | 5 Var. | 4 R. | S. 1 |   | T TVL (4) ODA | 5 Var. | 7 R. | S. 1 |   | T TVL (5) ODA | 5 Var. | 7 R. | S. 1 |

  0 0  | 0 1 1 0 | 0 1 1 0           0 0  | 0 0 0 0 | 0 0 0 0           0 0  | 0 1 1 0 | 0 1 1 0
  0 1  | 0 0 0 0 | 0 0 0 0           0 1  | 0 1 1 0 | 0 1 1 0           0 1  | 0 1 1 0 | 0 1 1 0
  1 1  | 0 0 0 0 | 0 0 0 0           1 1  | 1 1 1 0 | 0 1 1 1           1 1  | 1 1 1 0 | 0 1 1 1
  1 0  | 0 0 1 0 | 0 1 0 0           1 0  | 1 1 0 0 | 0 0 1 1           1 0  | 1 1 1 0 | 0 1 1 1
 x4 x5  0 1 1 0 0 1 1 0 x3          x4 x5  0 1 1 0 0 1 1 0 x3          x4 x5  0 1 1 0 0 1 1 0 x3
        0 0 1 1 1 1 0 0 x2                 0 0 1 1 1 1 0 0 x2                 0 0 1 1 1 1 0 0 x2
        0 0 0 0 1 1 1 1 x1                 0 0 0 0 1 1 1 1 x1                 0 0 0 0 1 1 1 1 x1

| T TVL (6) ODA | 4 Var. | 2 R. | S. 1 |   | T TVL (7) ODA | 4 Var. | 4 R. | S. 1 |   | T TVL (8) ODA | 4 Var. | 6 R. | S. 1 |

  0 0  | 0 1 1 0 |                   0 0  | 0 0 0 0 |                   0 0  | 0 1 1 0 |
  0 1  | 0 0 0 0 |                   0 1  | 0 1 1 0 |                   0 1  | 0 1 1 0 |
  1 1  | 0 0 0 0 |                   1 1  | 1 1 1 0 |                   1 1  | 1 1 1 0 |
  1 0  | 0 0 1 0 |                   1 0  | 1 1 0 0 |                   1 0  | 1 1 1 0 |
 x4 x5  0 1 1 0 x3                  x4 x5  0 1 1 0 x3                  x4 x5  0 1 1 0 x3
        0 0 1 1 x2                         0 0 1 1 x2                         0 0 1 1 x2
```

Fig. 6.24 Problem-program that computes $\mathrm{der}_{x_1} f(\mathbf{x})$, $\min_{x_1} f(\mathbf{x})$, and $\max_{x_1} f(\mathbf{x})$ in two manners, shows the related Karnaugh-maps, and checks the equivalence of the computed results

The cofactor $f(x_3 = 0, \mathbf{x}_1)$ is computed in lines 10–12 using the given function (TVL 1) and the prepared TVL 2 of $x_3 = 0$. Next, the cofactor $f(x_3 = 1, \mathbf{x}_1)$ is computed in lines 13–16 using the given function (TVL 1) and TVL 6 of $x_3 = 1$ created by the command cpl in line 13. The derivative $\mathrm{der}_{x_3} f(x_3, \mathbf{x}_1)$ is now computed using the command syd in line 17.

The same result is directly computed by the single command derk in line 18. The right two Karnaugh-maps confirm that these two methods compute $\mathrm{der}_{x_3} f(x_3, \mathbf{x}_1)$ such that the variable x_3 does not occur in the result.

(c) Figure 6.24 shows in the upper part the PRP that computes all three (single) derivative operations of the function $f(x_1, \mathbf{x}_1)$ with regard to x_1 first using the commands of the vectorial derivative operations in lines 9–11 and second using the commands of the k-fold derivative operations in lines 12–14. Below this PRP the created Karnaugh-maps of $\mathrm{der}_{x_1} f(\mathbf{x})$, $\min_{x_1} f(\mathbf{x})$, and $\max_{x_1} f(\mathbf{x})$ are shown in the first row for the results computed by the commands of the vectorial derivative operations and in the second row for the corresponding results computed by the commands of the k-fold derivative operations.

Due to the used variable x_1 show all three Karnaugh-maps of the first row a mirror symmetry between the left and right parts; this confirms that all three computed derivative operations are independent of x_1. The Karnaugh-maps of the second row are identical with the left part of the corresponding Karnaugh-map of the first row. The benefit of the commands derk, mink, and maxk is that the variable x_1 used for these derivative operations has been removed in the results; hence, these commands should be preferred to compute the (single) derivative operations.

Despite the occurrence of x_1 in the results of the commands derv, minv, and maxv confirm the computed variables $equalder=true, $equalmin=true, and $equalmax=true that both approaches compute the same function for each of the three single derivative operations.

(d) Figure 6.25 shows in the upper part the PRP that verifies the three rules (6.42), (6.43), and (6.44) and thereunder the Karnaugh-maps that easily confirm that these rules are satisfied.

The commands minv and maxv are used in lines 9 and 10 to compute $\min_{x_1} f(\mathbf{x})$ and $\max_{x_1} f(\mathbf{x})$ without removing the variable x_1 from these results. The Karnaugh-maps in the first row of the m-fold view show from the left to the right $\min_{x_1} f(\mathbf{x})$, $f(\mathbf{x})$, and $\max_{x_1} f(\mathbf{x})$; the comparison of these Karnaugh-maps confirms that $\min_{x_1} f(\mathbf{x}) \le f(\mathbf{x})$ and $f(\mathbf{x}) \le \max_{x_1} f(\mathbf{x})$. The values true of the variables $minlef and $flemax computed in lines 11 and 12 also confirm the rule (6.42).

The preferred commands mink, derk, and maxv are used in lines 16–18 to compute $\min_{x_1} f(\mathbf{x})$, $\operatorname{der}_{x_1} f(\mathbf{x})$, and $\max_{x_1} f(\mathbf{x})$ with implicitly removing the variable x_1 from these results. The associated three Karnaugh-maps have been assigned to the viewports of the second row m-fold view. The comparison of these Karnaugh-maps confirms that both the antivalence of $\min_{x_1} f(\mathbf{x})$, $\operatorname{der}_{x_1} f(\mathbf{x})$, and $\max_{x_1} f(\mathbf{x})$ and the conjunction of $\min_{x_1} f(\mathbf{x})$ and $\operatorname{der}_{x_1} f(\mathbf{x})$ are equal to 0. The computed values true of the variables $xorsdo0 and $minader0 additionally confirm that the rules (6.43) and (6.44) are satisfied.

Solution 6.4 (Properties of Logic Functions Explored by Means of Single Derivative Operations)

(a) Figure 6.26 shows in the upper part the PRP that verifies whether a given function depends on each variable that occurs in the given expression and removes the independent variables.

The command sv_next in line 14 selects in each sweep of the while-loop one variable of the given function that must remain unchanged for these selections. Therefore, the command copy in line 12 copies TVL 1 of the given function to TVL 2 that will be simplified as result.

The command te_derk in line 16 checks whether the given function is independent of the selected variable x_i, and the conditionally executed command maxk in line 18 realizes $\max_{x_i} f(\mathbf{x})$ to remove a detected independent variable from the resulting TVL 2. The command obb in line 20 minimizes the simplified orthogonal TVL 2.

The command te_syd in line 22 checks whether the given TVL 1 and the simplified TVL 2 represent the same function; the result $equal=true confirms that these two functions are equal to each other. It is not easy to detect the independence of x_2 and x_4 in the given expression of $f(\mathbf{x})$, but this independence is well visible in the associated Karnaugh-map, shown in viewport (1,1) of the m-fold view, as well as in the resulting TVL 2 in viewport (1,2).

The given function can be expressed by

$$f(\mathbf{x}) = x_1 x_5 \vee x_3 \ ;$$

hence, this function does not depend on the variables x_2 and x_4 that occur in the given expression of this function.

(b) Figure 6.27 shows in the upper part the PRP that verifies for each variable whether the given function is monotonously decreasing with regard to this variable.

```
 1   new                              13   assign  3  /m 1 1
 2   space 32 1                       14   assign  1  /m 1 2
 3   avar 1                           15   assign  4  /m 1 3
 4   x1 x2 x3 x4 x5.                  16   mink 1 2 5
 5   sbe 1 1                          17   derk 1 2 6
 6   x1&x3+/x2&x4+x3&x5.              18   maxk 1 2 7
 7   vtin 1 2                         19   syd 5 6 8
 8   x1.                              20   set $xorsdo0 (te_syd 8 7)
 9   minv 1 2 3                       21   set $minader0 (te_isc 5 6)
10   maxv 1 2 4                       22   assign 5 /m 2 1
11   set $minlef (te_dif 3 1)         23   assign 6 /m 2 2
12   set $flemax (te_dif 1 4)         24   assign 7 /m 2 3
```

T TVL (3) ODA | 5 Var. | 7 R. | S. 1

```
0 0   0 0 0 0   0 0 0 0
0 1   0 1 1 0   0 1 1 0
1 1   1 1 1 0   0 1 1 1
1 0   1 1 0 0   0 0 1 1
x4 x5 0 1 1 0 0 1 1 0 x3
      0 0 1 1 1 1 0 0 x2
      0 0 0 0 1 1 1 1 x1
```

T TVL (1) ODA | 5 Var. | 5 R. | S. 1

```
0 0   0 0 0 0   0 1 1 0
0 1   0 1 1 0   0 1 1 0
1 1   1 1 1 0   0 1 1 1
1 0   1 1 0 0   0 1 1 1
x4 x5 0 1 1 0 0 1 1 0 x3
      0 0 1 1 1 1 0 0 x2
      0 0 0 0 1 1 1 1 x1
```

T TVL (4) ODA | 5 Var. | 7 R. | S. 1

```
0 0   0 1 1 0   0 1 1 0
0 1   0 1 1 0   0 1 1 0
1 1   1 1 1 0   0 1 1 1
1 0   1 1 1 0   0 1 1 1
x4 x5 0 1 1 0 0 1 1 0 x3
      0 0 1 1 1 1 0 0 x2
      0 0 0 0 1 1 1 1 x1
```

T TVL (5) ODA | 4 Var. | 4 R. | S. 1

```
0 0   0 0 0 0
0 1   0 1 1 0
1 1   1 1 1 0
1 0   1 1 0 0
x4 x5 0 1 1 0 x3
      0 0 1 1 x2
```

T TVL (6) ODA | 4 Var. | 2 R. | S. 1

```
0 0   0 1 1 0
0 1   0 0 0 0
1 1   0 0 0 0
1 0   0 0 1 0
x4 x5 0 1 1 0 x3
      0 0 1 1 x2
```

T TVL (7) ODA | 4 Var. | 6 R. | S. 1

```
0 0   0 1 1 0
0 1   0 1 1 0
1 1   1 1 1 0
1 0   1 1 1 0
x4 x5 0 1 1 0 x3
      0 0 1 1 x2
```

Fig. 6.25 Problem-program that computes $\mathrm{der}_{x_1} f(\mathbf{x})$, $\min_{x_1} f(\mathbf{x})$, and $\max_{x_1} f(\mathbf{x})$, shows the related Karnaugh-maps, and verifies the three rules (6.42), (6.43), and (6.44)

The command $\mathtt{ctin}$ in line 10 creates a TVL in D-form that has no row and no column; this TVL has been prepared to store the variables for which the given function is monotonously decreasing. As in the previous task, the command $\mathtt{sv\_next}$ in line 13 selects in each sweep of the $\mathtt{while}$-loop one variable of the given function.

The selected variable x_i is stored as a set of variables in TVL 3 that has one column and no row. This TVL indicates the variable x_i so that $\min_{x_i} f(\mathbf{x})$ can be computed using the command $\mathtt{mink}$ in line 15. The variable x_i is needed in the formula to check whether the given function is monotonously decreasing; the command $\mathtt{sv\_get}$ in line 16 creates TVL 5 that represents this variable. The subsequent command $\mathtt{isc}$ computes the conjunction between the function $f(\mathbf{x})$ (TVL 1) and the variable x_i (TVL 5). The command $\mathtt{te\_dif}$ in line 18 realizes implicitly three required operations: the complement of $\min_{x_i} f(\mathbf{x})$ (TVL 4), the conjunction of this result and the function $(x_i \wedge f(\mathbf{x}))$ (TVL 6), and the check whether the result is equal to 0.

In the case that the function $f(\mathbf{x})$ is monotonously decreasing with regard to x_i, the command $\mathtt{con}$ in line 20 appends this variable to TVL 2 and the subsequent command $\mathtt{set}$ changes the Boolean variable $\mathtt{\$md}$, which was initialized with the value $\mathtt{false}$, to the value $\mathtt{true}$.

Fig. 6.26
Problem-program that
simplifies a function by
removing all such variables
this function is not
depending on together with
the related results

```
 1   new
 2   space  32  1
 3   avar  1
 4   x1  x2  x3  x4  x5.
 5   sbe  1  1
 6   (/ x2&x3&x5 +(x1#x3)&x4
 7   +x2&x3 &/x4&x5
 8   +x3 &/x4&/x5
 9   +x1&x5)
10   #x1&x4 &/x5.
11   assign  1  /m  1  1
12   copy  1  2
13   assign  1  /m  1  1
14   while  (sv_next  1  3  3)
15   (
16   if  (te_derk  1  3)
17   (
18   maxk  2  3  2
19   ))
20   obb  2  2
21   assign  2  /m  1  2
22   set  $equal  (te_syd  1  2)
```

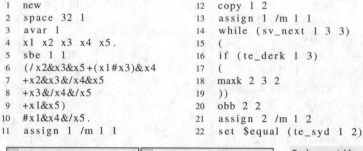

Fig. 6.27
Problem-program that
detects each variable for
that a given function is
monotonously decreasing

```
 1   new
 2   space  32  1
 3   avar  1
 4   x1  x2  x3  x4  x5.
 5   sbe  1  1
 6   x1&x3&x5 +/x3&x4 &/x5
 7   +x2&x5 +x2 &/x3 &/x4&x5
 8   +x2 &/x4 &/x5 +x1 & (/x3+x5 ).
 9   assign  1  /m  1  1
10   ctin  1  2  /0  /d
11   assign  1  /m  1  1
12   set  $md  false
13   while  (sv_next  1  3  3)
14   (
15   mink  1  3  4
16   sv_get  3  5
17   isc  1  5  6
18   if  (te_dif  6  4)
19   (
20   con  2  5  2
21   set  $md  true
22   ))
23   assign  2  /m  1  2
```

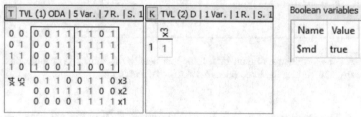

TVL 2 in viewport (1,2) shows that the given function is monotonously decreasing only with regard to the variable x_3. This result of the executed complete analysis can be verified by the comparison of all pairs of function values of the function $f(\mathbf{x})$ shown in the Karnaugh-map of viewport (1,1).

(c) Figure 6.28 shows in the upper part the PRP that computes the subfunctions $f_0(\mathbf{x}_1)$ (TVL 3 in lines 12 and 13) and $f_1(\mathbf{x}_1)$ (TVL 6 in lines 14–16) that are independent of x_3 and can be combined to the given function $f(x_3, \mathbf{x}_1)$ that is monotonously decreasing with regard to x_3.

The variable x_3 is needed in negated form to reconstruct the monotonously deceasing function $f(x_3, \mathbf{x}_1)$; hence, this negated variable is defined as TVL 2 by the command sbe in lines 10 and 11. Note that commands mink in line 12 and maxk in line 16 use only the occurrence of x_3 in TVL 2, but not the negated value of this variable to compute the specified single derivative operation.

Fig. 6.28
Problem-program that
decomposes a
monotonously with regard
to x_3 decreasing function
into the subfunctions
$f_0(\mathbf{x}_1)$ and $f_1(\mathbf{x}_1)$ that are
shown in the right column
of the m-fold view

```
 1   new                        12   mink  1  2  3
 2   space  32  1               13   obb  3  3
 3   avar  1                    14   isc  1  2  4
 4   x1  x2  x3  x4  x5.        15   dif  4  3  5
 5   sbe  1  1                  16   maxk  5  2  6
 6   x1&x3&x5 +/x3&x4&/x5       17   assign  3  /m  1  2
 7   +x2&x5+x2&/x3&/x4&x5       18   assign  6  /m  2  2
 8   +x2&/x4&/x5+x1&(/x3+x5).   19   isc  2  6  7
 9   assign  1  /m  1  1        20   uni  3  7  8
10   sbe  1  2                  21   assign  8  /m  2  1
11   /x3.                       22   set  $equal  (te_syd  1  8)
```

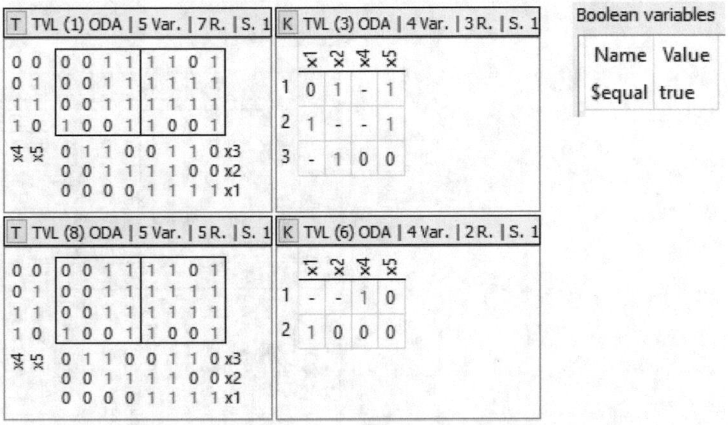

The monotonously with regard to x_3 deceasing function $f(x_3, \mathbf{x}_1)$ (TVL 8) has been reconstructed in lines 19 and 20; the Karnaugh-map of this reconstructed function is shown in viewport (2,1) of the m-fold view.

Both the comparison of the Karnaugh-maps in the left column and the result $\texttt{\$equal=true}$ computed in the last line of the PRP confirm that the given function $f(x_3, \mathbf{x}_1)$ is monotonously decreasing with regard to x_3 and can be expressed by the negated variable $\overline{x}_3$ and the functions $f_0(\mathbf{x}_1)$ and $f_1(\mathbf{x}_1)$.

The TVLs of the functions $f_0(\mathbf{x}_1)$ and $f_1(\mathbf{x}_1)$ are shown in the right column of the m-fold view. Using these two TVLs, we get as short expression of the given function:

$$f(x_3, \mathbf{x}_1) = (x_2 x_5 \vee x_1 x_5 \vee x_2 \overline{x}_4) \vee \overline{x}_3 (x_4 \overline{x}_5 \vee x_1 \overline{x}_2 \overline{x}_5) .$$

This expression shows explicitly that this function is monotonously decreasing with regard to x_3. The variable x_3 occurs only in negated form; hence,

$$f(x_3 = 0, \mathbf{x}_1) \geq f(x_3 = 1, \mathbf{x}_1) .$$

(d) Figure 6.29 shows in the upper part the PRP decomposes a given function into linear and nonlinear subfunctions.

The antivalence operation must be used to connect the linearly separated variables; hence, the empty TVL 2 in A-form is created in line 9 to collect the found variables in which the given function is linear. A copy of the given function is prepared as TVL 3 in the next line; this TVL will be restricted to the nonlinear part of the given function.

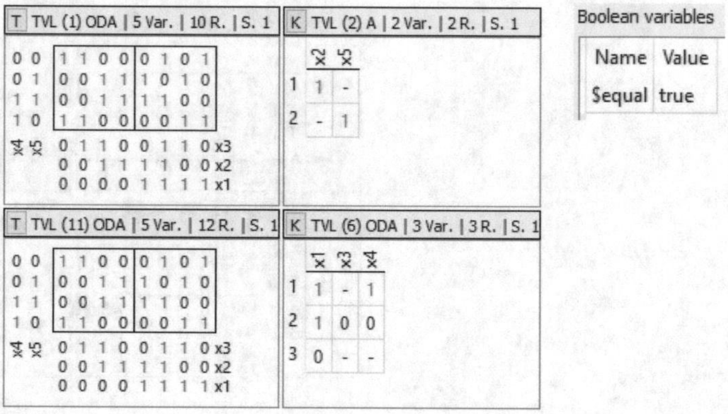

Fig. 6.29
Problem-program that
separates all variables from
a given function in which
this function is linear and
verifies the computed
linear decomposition
together with the
respective results

```
 1   new
 2   space 32 1
 3   avar 1
 4   x1 x2 x3 x4 x5.
 5   sbe 1 1
 6   (x1+/x2)&x3&/x4&/x5
 7   #/x2&(x1+/x3+x4)#/x3&x5
 8   #x3&x4&x5 #/x1&x2&x3&/x4&x5.
 9   ctin 1 2 /0 /a
10   copy 1 3
11   assign 1 /m 1 1
12   while (sv_next 1 4 4)
13   (
14   derk 1 4 5
```

```
15   if (te_cpl 5)
16   (
17   sv_get 4 6
18   con 2 6 2
19   syd 3 6 3
20   maxk 3 6 3
21   ))
22   assign 2 /m 1 2
23   obb 3 3
24   assign 3 /m 2 2
25   orth 2 10
26   syd 10 3 11
27   assign 11 /m 2 1
28   set $equal (te_syd 1 11)
```

T	TVL (1) ODA	5 Var.	10 R.	S. 1
0 0	1 1 0 0	0 1 0 1
0 1	0 0 1 1	1 0 1 0
1 1	0 0 1 1	1 1 0 0
1 0	1 1 0 0	0 0 1 1

```
x̄5 x5   0 1 1 0 0 1 1 0 x3
        0 0 1 1 1 1 0 0 x2
        0 0 0 0 1 1 1 1 x1
```

K	TVL (2) A	2 Var.	2 R.	S. 1
	x̄2 x2
1	1 -
2	- 1

Boolean variables

Name	Value
$equal	true

T	TVL (11) ODA	5 Var.	12 R.	S. 1
0 0	1 1 0 0	0 1 0 1
0 1	0 0 1 1	1 0 1 0
1 1	0 0 1 1	1 1 0 0
1 0	1 1 0 0	0 0 1 1

```
x̄5 x5   0 1 1 0 0 1 1 0 x3
        0 0 1 1 1 1 0 0 x2
        0 0 0 0 1 1 1 1 x1
```

K	TVL (6) ODA	3 Var.	3 R.	S. 1
	x̄1 x̄3 x̄4
1	1 - 1
2	1 0 0
3	0 - -

The separation of all linear variables is realized in the `while`-loop in lines 12–21. In each sweep of this loop the linearity of the function with regard to one variable is checked using the commands `derk` and `te_cpl` in lines 14 and 15.

The TVL in ODA-form of a found linear variable is created by the command `sv_get` in line 17 using the associated set of variables. The subsequent command `con` appends this variable at the end of TVL 2 in A-form. The command `syd` in line 19 changes TVL 3 such that it is independent of this variable and the subsequent command `maxk` removes the associated column from TVL 3 of the nonlinear part. The TVLs of the linear and minimized nonlinear parts are assigned to the column 2 of the m-fold view when all variables have been verified within the `while`-loop.

Using these two TVLs we get as short expression of the given function:

$$f(\mathbf{x}) = (x_2 \oplus x_5) \oplus (\overline{x}_1 \vee \overline{x}_3 \vee x_4) .$$

The function $f(\mathbf{x})$ is reconstructed as TVL 11 in lines 25 and 26. Both the comparison of Karnaugh-maps in the left column of the m-fold view and the result $equal=true (computed in line 28) confirm the computed linear decomposition of the given function.

Solution 6.5 (k-fold Derivative Operations)

(a) Figure 6.30 shows on the left-hand side the PRP that computes the 2-fold derivative of $f(\mathbf{x})$ using several approaches. The Karnaugh-maps on the right-hand side show the computed results.

The command `derk` in line 13 directly computes the 2-fold derivative of $f(\mathbf{x})$ with regard to (x_4, x_5). The Karnaugh-map in viewport $(1, 2)$ confirms that a value 1 of this k-fold derivative indicates that in the associated subspace (a column of the Karnaugh-map to the left) and an odd number of function values 1 occur.

```
 1   new
 2   space  32  1
 3   avar  1
 4   x1  x2  x3  x4  x5 .
 5   sbe  1  1
 6   x1 & (x2 # x4)
 7   +x1 & x3 & (/x2 +/x5)
 8   +x2 & x5 & (/x3 +/x4)
 9   +/x2 & x3 &/x4 &/x5 .
10   assign  1  /m  1  1
11   vtin  1  2
12   x4  x5 .
13   derk  1  2  3
14   assign  3  /m  1  2
15   _derk  1  <x4>  4
16   assign  4  /m  2  1
17   _derk  4  <x5>  5
18   assign  5  /m  3  1
19   _derk  1  <x5>  6
20   assign  6  /m  2  2
21   _derk  6  <x4>  7
22   assign  7  /m  3  2
```

```
T  TVL (1) ODA | 5 Var. | 8 R. | S. 1        T  TVL (3) ODA | 3 Var. | 4 R. | S. 1

  0 0 | 0 1 0 0 | 1 1 1 0                        0 1 1 0 | 1 1 0 0
  0 1 | 0 0 1 1 | 1 1 1 0
  1 1 | 0 0 0 1 | 1 0 1 1                         0 1 1 0 0 1 1 0 x3
  1 0 | 0 0 0 0 | 0 1 1 1                         0 0 1 1 1 1 0 0 x2
x4 x5  0 1 1 0 0 1 1 0 x3                         0 0 0 0 1 1 1 1 x1
       0 0 1 1 1 1 0 0 x2
       0 0 0 0 1 1 1 1 x1


T  TVL (4) ODA | 4 Var. | 5 R. | S. 1        T  TVL (6) ODA | 4 Var. | 5 R. | S. 1

  0 | 0 1 0 0 | 1 0 0 1                         0 | 0 1 1 1 | 0 0 0 0
  1 | 0 0 1 0 | 0 1 0 1                         1 | 0 0 0 1 | 1 1 0 0
x5  0 1 1 0 0 1 1 0 x3                        x4  0 1 1 0 0 1 1 0 x3
    0 0 1 1 1 1 0 0 x2                            0 0 1 1 1 1 0 0 x2
    0 0 0 0 1 1 1 1 x1                            0 0 0 0 1 1 1 1 x1


T  TVL (5) ODA | 3 Var. | 4 R. | S. 1        T  TVL (7) ODA | 3 Var. | 4 R. | S. 1

  0 1 1 0 | 1 1 0 0                             0 1 1 0 | 1 1 0 0

  0 1 1 0 0 1 1 0 x3                            0 1 1 0 0 1 1 0 x3
  0 0 1 1 1 1 0 0 x2                            0 0 1 1 1 1 0 0 x2
  0 0 0 0 1 1 1 1 x1                            0 0 0 0 1 1 1 1 x1
```

Fig. 6.30 Problem-program that computes $\mathrm{der}^2_{(x_4,x_5)} f(\mathbf{x})$ both directly and in two different successive procedures together with the computed results

The command _derk is used in the endmost part of this PRP; this variant of the command allows to specify the names of the variable directly in angled parentheses. The derivative of $f(\mathbf{x})$ with regard to x_4 (TVL 4, see viewport (2,1)) is equal to 1 if the change of x_4 leads to a change of the function value of $f(\mathbf{x})$; that means the pairs of function values either in the upper and lower rows or in the two rows in the middle of the Karnaugh-map of $f(\mathbf{x})$ are compared. The derivative of $f(\mathbf{x})$ with regard to x_5 (TVL 6, see viewport (2,2)) is equal to 1 if the change of x_5 leads to a change of the function value of $f(\mathbf{x})$; the pairs of function values either in the upper two rows or in the lower two rows of the Karnaugh-map of $f(\mathbf{x})$ are compared.

The successive computation of the derivatives of these intermediate results with regard to the not jet used variable leads to the same result:

$$\mathrm{der}^2_{(x_4,x_5)} f(\mathbf{x}) = \underset{x_5}{\mathrm{der}} \left(\underset{x_4}{\mathrm{der}} f(\mathbf{x}) \right) \qquad \text{(TVL 5, see viewport (3,1))} ,$$

$$\mathrm{der}^2_{(x_4,x_5)} f(\mathbf{x}) = \underset{x_4}{\mathrm{der}} \left(\underset{x_5}{\mathrm{der}} f(\mathbf{x}) \right) \qquad \text{(TVL 7, see viewport (3,2))} .$$

(b) Figure 6.31 shows in the upper part the PRP that computes the 2-fold minimum, the 2-fold maximum, and the delta-operation of $f(\mathbf{x})$ with regard to (x_3, x_5) and verifies the rule (6.51). The command mink in line 13 directly computes $\min^2_{(x_3,x_5)} f(\mathbf{x})$; the result is shown in viewport (2, 1). The command maxk in line 15 computes also directly $\max^2_{(x_3,x_5)} f(\mathbf{x})$; the result is shown in viewport (2,2) of the m-fold view. The comparison of these neighbored Karnaugh-maps confirms the property that this 2-fold minimum is smaller than this 2-fold maximum.

```
 1    new                              11    vtin  1  2
 2    space  32  1                     12    x3  x5.
 3    avar  1                          13    mink  1  2  3
 4    x1  x2  x3  x4  x5.              14    assign  3  /m  2  1
 5    sbe  1  1                        15    maxk  1  2  4
 6    x1&(x2#x4)                       16    assign  4  /m  2  2
 7    +x1&x3&(/x2+/x5)                 17    syd  3  4  5
 8    +x2&x5&(/x3+/x4)                 18    assign  5  /m  1  2
 9    +/x2&x3&/x4&/x5.                 19    syd  3  5  6
10    assign  1  /m  1  1             20    set  $xorkdo0  (te_syd  6  4)
```

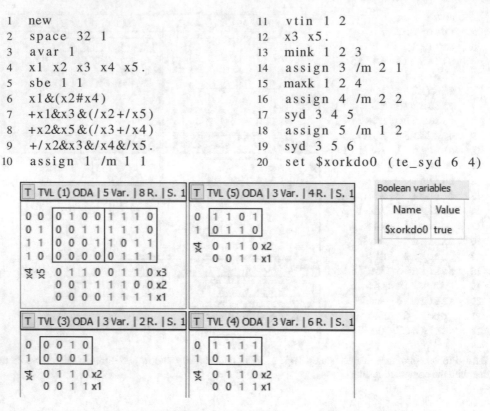

Fig. 6.31 Problem-program that computes $\min^2_{(x_3,x_5)} f(\mathbf{x})$, $\max^2_{(x_3,x_5)} f(\mathbf{x})$, and $\Delta_{(x_3,x_5)} f(\mathbf{x})$, verifies that the antivalence of these three k-fold derivative operation is equal to 0, and shows all computed results

The Δ-operation is defined as antivalence of the k-fold minimum and the k-fold maximum (see (6.50)); hence, the provided commands for these two operations and the command syd in line 17 compute the wanted Δ-operation. The result is shown in viewport (2,1).

The commands in the last two lines verify the rule (6.51) and store the result to the Boolean variable $xorkdo0=true; this result confirms the explored property. The substitution of Definition (6.50) of the Δ-operation into the rule (6.51) leads to an expression in which both the k-fold minimum and the k-fold maximum occur twice and can be replaced by the value 0.

(c) Figure 6.32 shows in the upper part a PRP that computes a single and a 2-fold maximum and verifies that the number of function values 1 increases the more variables are used to compute the k-fold maximum.

The command _maxk in lines 11 and 13 computes the k-fold maxima with regard to the variables specified in the angled parentheses; the results are shown in viewports (1,2) and (1,3). It can be seen that each additional variable used to compute the k-fold maximum reduces the number of variables the associated result is depending on.

The results $le1=true and $le2=true, computed by command te_dif in lines 15 and 16, confirm that the rule (6.49) is satisfied for the computed the k-fold maxima.

There are several possibilities to add independent variables to a function. Here, the constant function $1(x_3, x_5)$ (TVL 5) has been defined in lines 17–19, and the intersection of each k-fold maximum with this function leads to the wanted result. The Karnaugh-maps in the second row

1 new	13 _maxk 1 <x3 x5> 4
2 space 32 1	14 assign 4 /m 1 3
3 avar 1	15 set \$le1 (te_dif 1 3)
4 x1 x2 x3 x4 x5.	16 set \$le2 (te_dif 3 4)
5 sbe 1 1	17 tin 1 5
6 x1&(x2#x4)	18 x3 x5.
7 +x1&x3&(/x2+/x5)	19 --.
8 +x2&x5&(/x3+/x4)	20 isc 3 5 6
9 +/x2&x3&/x4&/x5.	21 isc 4 5 7
10 assign 1 /m 1 1	22 assign 1 /m 2 1
11 _maxk 1 <x3> 3	23 assign 6 /m 2 2
12 assign 3 /m 1 2	24 assign 7 /m 2 3

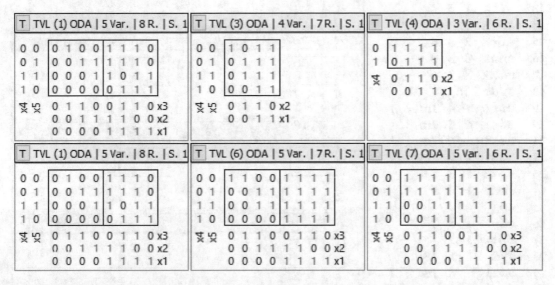

Fig. 6.32 Problem-program that computes $\max_{x_3} f(\mathbf{x})$ and $\max^2_{(x_3,x_5)} f(\mathbf{x})$, shows the related Karnaugh-maps depending either on the needed variables or on all five variables, and verifies the rule (6.49)

confirm that each additional variable used to compute a k-fold maximum increases the number of values 1 in the result.

(d) Figure 6.33 shows on the left-hand side the PRP that verifies the rules (6.63), (6.64), (6.65), and (6.66) of k-fold derivative operations of negated functions with regard to $\mathbf{x}_0 = (x_3, x_5)$.

The upper row of the m-fold view shows the used function and its complement computed by the command cpl in line 11.

The functions of the left-hand sides of the four explored equations are computed in lines 15–18 and shown in the left column of the m-fold view below the given function.

The functions of the right-hand sides of the four explored equations are computed in lines 23–28 and shown in the right column of the m-fold view below the complement of the given function.

Each row from 2 to 5 of the m-fold view shows on the left-hand side the Karnaugh-map of the left-hand side of one of the explored equations in the given order and on the right-hand side the Karnaugh-map of the associated right-hand side of this equation. Equal contents of these Karnaugh-maps in each of these four rows confirm that the rules (6.63), (6.64), (6.65), and (6.66) are satisfied for the k-fold derivative operations of the given function with regard to $\mathbf{x}_0 = (x_3, x_5)$.

```
 1   new
 2   space 32 1
 3   avar 1
 4   x1 x2 x3 x4 x5.
 5   sbe 1 1
 6   x1&(x2#x4)
 7   +x1&x3&(/x2+/x5)
 8   +x2&x5&(/x3+/x4)
 9   +/x2&x3&/x4&/x5.
10   assign 1 /m 1 1
11   cpl 1 2
12   assign 2 /m 1 2
13   vtin 1 3
14   x3 x5.
15   derk 2 3 4
16   mink 2 3 5
17   maxk 2 3 6
18   syd 5 6 7
19   assign 4 /m 2 1
20   assign 5 /m 3 1
21   assign 6 /m 4 1
22   assign 7 /m 5 1
23   derk 1 3 8
24   maxk 1 3 9
25   cpl 9 10
26   mink 1 3 11
27   cpl 11 12
28   syd 9 11 13
29   assign 8 /m 2 2
30   assign 10 /m 3 2
31   assign 12 /m 4 2
32   assign 13 /m 5 2
```

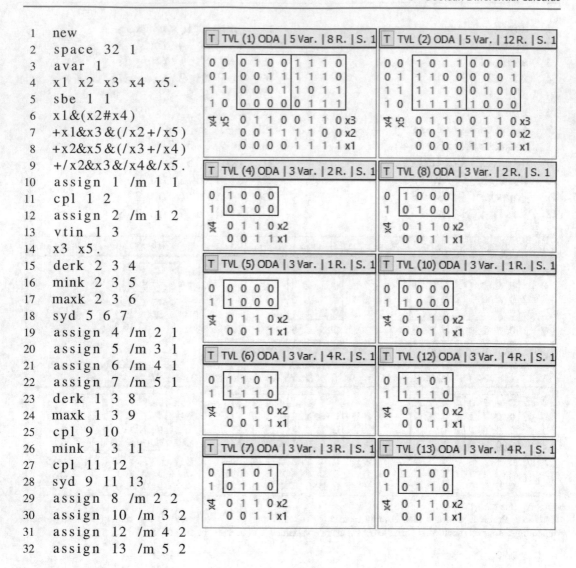

Fig. 6.33 Problem-program that verifies four rules related to k-fold derivative operations of negated functions

Solution 6.6 (Properties of Logic Functions Explored by Means of k-fold Derivative Operations)

(a) The function $f(\mathbf{x}_0, \mathbf{x}_1)$ with $\mathbf{x}_0 = (x_2, x_3, x_5)$ is constant equal to 1 in subspaces defined by $\mathbf{x}_1 = (x_1, x_4)$ for which the function

$$g(\mathbf{x}_1) = \min_{\mathbf{x}_0}^3 f(\mathbf{x}_0, \mathbf{x}_1)$$

is equal to 1. Figure 6.34 shows on the left-hand side the PRP that computes $g(\mathbf{x}_1)$ as TVL 3.

The given function $f(\mathbf{x}_0, \mathbf{x}_1)$ is constant equal to 1 in the subspace specified by $(x_1, x_4) = (0, 0)$.

(b) The function $f(\mathbf{x}_0, \mathbf{x}_1)$ with $\mathbf{x}_0 = (x_2, x_3, x_5)$ is constant equal to 0 in subspaces defined by $\mathbf{x}_1 = (x_1, x_4)$ for which the function

```
1   new
2   space 32 1
3   sbe  1 1
4   / x1&x2&x3 +/ x4&/x5
5   +/x1&x5+x2&/x4 .
6   assign 1 /m 1 1
7   vtin 1 2
8   x2 x3 x5.
9   mink 1 2 3
10  assign 3 /m 1 2
```

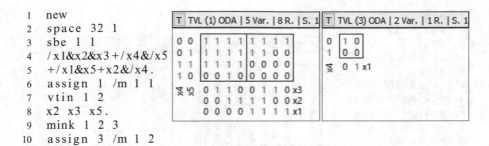

T	TVL (1) ODA | 5 Var. | 8 R. | S. 1		T	TVL (3) ODA | 2 Var. | 1 R. | S. 1

Fig. 6.34 Problem-program that computes $g(\mathbf{x}_1)$ determined by subspaces $\mathbf{x}_1 = (x_1, x_4)$ where the given function $f(\mathbf{x}_0, \mathbf{x}_1)$ is constant equal to 1 together with the computed result

```
1   new
2   space 32 1
3   sbe  1 1
4   / x1&x2&x3 +/ x4&/x5
5   +/x1&x5+x2&/x4 .
6   assign 1 /m 1 1
7   vtin 1 2
8   x2 x3 x5.
9   maxk 1 2 3
10  cpl 3 4
11  assign 4 /m 1 2
```

Fig. 6.35 Problem-program that computes $g(\mathbf{x}_1)$ determined by subspaces $\mathbf{x}_1 = (x_1, x_4)$ where the given function $f(\mathbf{x}_0, \mathbf{x}_1)$ is constant equal to 0 together with the computed result

```
1   new
2   space 32 1
3   sbe  1 1
4   / x1&x2&x3 +/ x4&/x5
5   +/x1&x5+x2&/x4 .
6   assign 1 /m 1 1
7   vtin 1 2
8   x2 x3 x5.
9   derk 1 2 3
10  assign 3 /m 1 2
```

Fig. 6.36 Problem-program that computes $g(\mathbf{x}_1)$ determined by subspaces $\mathbf{x}_1 = (x_1, x_4)$ where the given function $f(\mathbf{x}_0, \mathbf{x}_1)$ has an odd number of function values 1 together with the computed result

$$g(\mathbf{x}_1) = \overline{\max_{\mathbf{x}_0}^3 f(\mathbf{x}_0, \mathbf{x}_1)}$$

is equal to 1. Figure 6.35 shows on the left-hand side the PRP that computes $g(\mathbf{x}_1)$ as TVL 4.

The given function $f(\mathbf{x}_0, \mathbf{x}_1)$ is constant equal to 0 in the subspace specified by $(x_1, x_4) = (1, 1)$.

(c) The function $f(\mathbf{x}_0, \mathbf{x}_1)$ with $\mathbf{x}_0 = (x_2, x_3, x_5)$ has an odd number of function values 1 in subspaces defined by $\mathbf{x}_1 = (x_1, x_4)$ for which the function

$$g(\mathbf{x}_1) = \operatorname*{der}_{\mathbf{x}_0}^3 f(\mathbf{x}_0, \mathbf{x}_1)$$

is equal to 1. Figure 6.36 shows on the left-hand side the PRP that computes $g(\mathbf{x}_1)$ as TVL 3.

```
 1    new
 2    space 32 1
 3    avar 1
 4    x1 x2 x3 x4 x5 .
 5    sbe 1 1
 6    / x1&x2&x3 +/ x4&/ x5
 7    +/ x1&x5 + x2&/ x4 .
 8    assign 1 / m 1 1
 9    vtin 1 2
10    x2 x3 x5 .
11    mink 1 2 3
12    maxk 1 2 4
13    syd 3 4 5
14    assign 5 / m 1 2
```

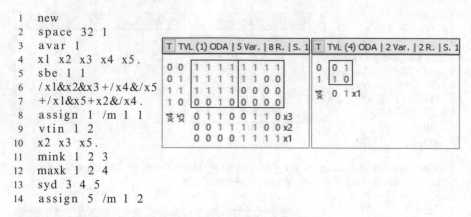

Fig. 6.37 Problem-program that computes $g(\mathbf{x}_1)$ determined by subspaces $\mathbf{x}_1 = (x_1, x_4)$ where the given function $f(\mathbf{x}_0, \mathbf{x}_1)$ has different function values together with the computed result

The given function $f(\mathbf{x}_0, \mathbf{x}_1)$ has an odd number of function values 1 in the subspace specified by $(x_1, x_4) = (0, 1)$.

(d) The function $f(\mathbf{x}_0, \mathbf{x}_1)$ with $\mathbf{x}_0 = (x_2, x_3, x_5)$ has different function values in subspaces defined by $\mathbf{x}_1 = (x_1, x_4)$ for which the function

$$g(\mathbf{x}_1) = \Delta_{\mathbf{x}_0} f(\mathbf{x}_0, \mathbf{x}_1)$$

is equal to 1. Figure 6.37 shows on the left-hand side the PRP that computes $g(\mathbf{x}_1)$ as TVL 5. The given function $f(\mathbf{x}_0, \mathbf{x}_1)$ has different function values in the subspaces where the variables x_1 and x_4 have different values; i.e., $g(\mathbf{x}_1) = x_1 \oplus x_4 = 1$.

Solution 6.7 (Differentials of Variables, Graphs, and Differential Operations)

(a) Figure 6.38 shows on the left-hand side the PRP that solves the given graph equation. Both the TVL and the Karnaugh-map on the right-hand side show the computed solution-set. Each of these representations can be used to construct the graphical representation shown below this PRP.

The command sbe used in lines 6–10 solves the given graph equation; this command computes the differentials of variables in the same manner as all other Boolean variables.

The nodes in the graph are labeled by the values of (x_1, x_2, x_3). Only one value of the variables (x_1, x_2, x_3) is changed when the edge of the graph belongs to the edges of the cube. The longest cycle of edges determined by a single differential equal to 1 is

$$(100) - (110) - (010) - (011) - (001) - (101) - (100) ;$$

hence, six nodes belong to this cycle. Exactly two values of (x_1, x_2, x_3) are changed when the edge of the graph belongs to the diagonals of the side areas of the cube. The longest cycle of edges determined by two differentials equal to 1 is

$$(100) - (010) - (001) - (100) ;$$

hence, three nodes belong to this cycle.

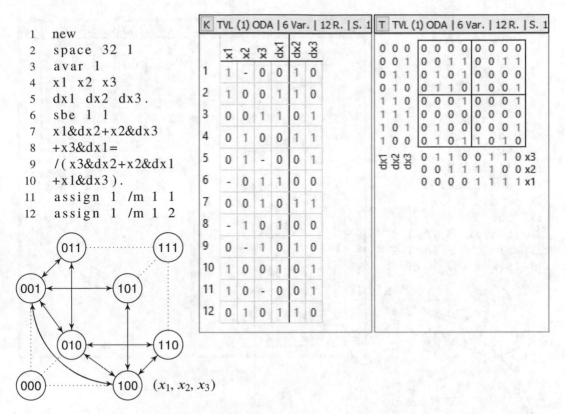

Fig. 6.38 Problem-program that solves a given graph equation together with three representations of the computed result

(b) Figure 6.39 shows on the left-hand side the PRP that computes the total differential $d_{\mathbf{x}} f(\mathbf{x})$ of $f(\mathbf{x}) = x_1 \vee x_2 \overline{x}_3$. The result is shown on the right-hand side both as TVL and as Karnaugh-map. The graph of $df(\mathbf{x}) = 1$ is shown below this PRP. It can be seen that each white node of $f(\mathbf{x}) = 0$ is connected with each gray node of $f(\mathbf{x}) = 1$; hence, an edge of this graph indicates the change of the function value.

Definition (6.72) has been used to compute this differential. The command sbe in lines 6 and 7 realizes the input of the given function $f(\mathbf{x})$. A second command sbe in lines 8–11 prepares the modified function $f(\mathbf{x} \oplus \mathbf{dx})$ in which each variable x_i has been replaced by $x_i \oplus dx_i$. The command in syd in line 12 uses these two functions to compute the differential of the given function.

The command sts in the last line of the PRP stores the XBOOLE-system for later use.

(c) Figure 6.40 shows on the left-hand side the PRP that separates all seven vectorial derivatives with regard to one to three variables from the differential $df(\mathbf{x})$ (TVL 3) and verifies that these results are equal to the vectorial derivatives computed by the corresponding command derv. The result $equal=true shown to the right of the PRP confirms this property.

The command lds in line 1 loads the XBOOLE-system of the previous task that contains the function $f(\mathbf{x})$ as TVL 1 and the total differential $df(\mathbf{x})$ as TVL 3.

```
1   new
2   space 32 1
3   avar 1
4   x1 x2 x3
5   dx1 dx2 dx3 .
6   sbe 1 1
7   x1+x2&/x3 .
8   sbe 1 2
9   (x1#dx1)
10  +(x2#dx2)
11  &/(x3#dx3).
12  syd 1 2 3
13  obb 3
14  assign 3 /m 1 1
15  assign 3 /m 1 2
16  sts e6_7_b_df
```

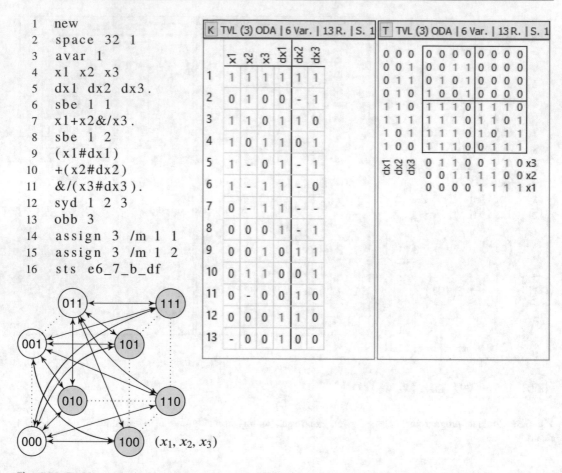

Fig. 6.39 Problem-program that computes the total differential of $f(\mathbf{x}) = x_1 \vee x_2 \overline{x}_3$ together with three representations of the computed result

Fig. 6.40
Problem-program that verifies that the total differential $d_\mathbf{x} f(\mathbf{x})$ of $f(\mathbf{x}) = x_1 \vee x_2 \overline{x}_3$ combines seven vectorial derivatives, separable by the conjunctions of differentials of the three variables

```
1   lds e6_7_b_df        17  1-1
2   tin 1 4 /d           18  -11
3   dx1 dx2 dx3 .        19  111.
4   100                  20  set $equal true
5   010                  21  for $i 1 7
6   110                  22  (
7   001                  23  stv 4 $i 10
8   101                  24  isc 3 10 11
9   011                  25  maxk 11 4 12
10  111.                 26  stv 5 $i 13
11  tin 1 5 /d           27  derv 1 13 14
12  x1 x2 x3 .           28  if (not (te_syd 12 14))
13  1--                  29  (
14  -1-                  30  set $equal false
15  11-                  31  ))
16  --1
```

Boolean variables

Name	Value
$equal	true

1	new	12	--	23	for $i 1 4
2	space 32 1	13	1-	24	(
3	avar 1	14	-1	25	stv 2 $i 5
4	x1 x2 x3	15	11.	26	mink 1 5 6
5	dx1 dx2 .	16	tin 1 3 /d	27	stv 3 $i 7
6	sbe 1 1	17	dx1 dx2 .	28	isc 6 7 8
7	x1&/x2+/x1&/x3 .	18	00	29	uni 4 8 4
8	assign 1 /m 1 1	19	10	30	)
9	assign 1 /m 1 2	20	01	31	obb 4
10	tin 1 2 /d	21	11.	32	assign 4 /m 2 1
11	x1 x2 .	22	ctin 1 4 /0	33	assign 4 /m 2 2

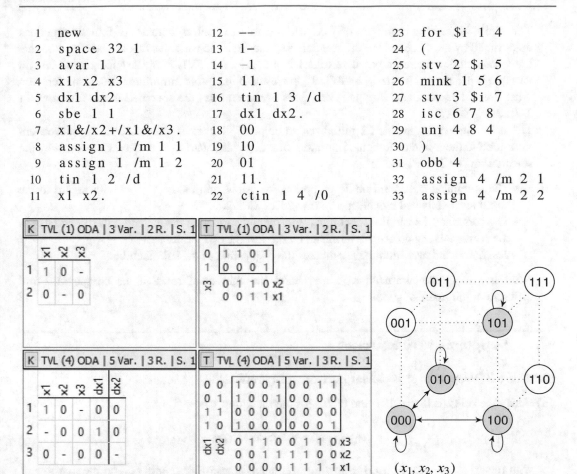

Fig. 6.41 Problem-program that computes the 2-fold differential minimum $\mathrm{Min}^2_{(x_1,x_2)} f(\mathbf{x})$ of $f(\mathbf{x}) = x_1\bar{x}_2 \vee \bar{x}_1\bar{x}_3$ together with three representations of the computed result

The seven needed conjunctions of the differentials dx_1, dx_2, and dx_3 are prepared as TVL 4 to separate the vectorial derivatives from the total differential of $f(\mathbf{x})$. Values 1 in a row of TVL 5 indicate the variables needed to compute one of the seven vectorial derivatives.

The Boolean variable $equal is initialized with the value true.

Each sweep of the for-loop in lines 21–31 uses first the differentials determined in one row of TVL 4 to select one of the seven vectorial derivatives as TVL 12, computes thereafter the corresponding vectorial derivative as TVL 14, and checks their equivalence using the command te_syd in line 28. The value of the variable $equal would be changed to false if one of the seven pairs of differently computed vectorial derivatives specifies different functions; however, such a case does not occur.

(d) Figure 6.41 shows in the upper part the PRP that computes the 2-fold differential minimum $\mathrm{Min}^2_{(x_1,x_2)} f(\mathbf{x})$ of $f(\mathbf{x}) = x_1\bar{x}_2 \vee \bar{x}_1\bar{x}_3$ based on the associated k-fold minima. The TVL and Karnaugh-map of the given function are shown in the first row of the m-fold view, and these representations of computed result are shown thereunder. The graph on the right-hand side visualizes the computed 2-fold differential minimum $\mathrm{Min}^2_{(x_1,x_2)} f(\mathbf{x})$.

The 2-fold differential minimum of $f(\mathbf{x})$ with regard to x_1 and x_2 consists of four subfunctions determined by the values of the differentials dx_1 and dx_2. The variables that are used to compute the k-fold minimum are prepared as values 1 in the rows of TVL 2. In the special case that no variable is determined (first row of TVL 2), the result of the k-fold minimum is the given function $f(\mathbf{x})$ itself. The corresponding pairs of values of dx_1 and dx_2 are specified in the four rows of TVL 3.

The command `ctin` in line 22 initializes an empty TVL that is completed in the subsequent `for`-loop to the 2-fold differential minimum of $f(\mathbf{x})$ with regard to x_1 and x_2. Controlled by the selected row of TVLs 2 and 3:

- The command `mink` in line 26 computes the k-fold minimum of $f(\mathbf{x})$ with regard to the variables determined by values 1 in row `$i` of TVL 2.
- The command `isc` in line 28 computes the conjunction with the differentials dx_1 and dx_2 in the corresponding polarity determined by the row `$i` of TVL 3.
- The command `uni` in line 29 computes the disjunction of the subfunctions.

After this loop, the command `obb` minimizes the number of rows of the computed 2-fold differential minimum of $f(\mathbf{x})$.

6.9   Supplementary Exercises

Exercise 6.8 (Properties of Vectorial Derivative Operations)

(a) Verify the rules (6.18), (6.19), and (6.20) for the function

$$f(\mathbf{x}) = x_1 x_3 \overline{x}_5 \vee \overline{x}_2 \overline{x}_4 x_6 \vee x_1 x_4 \vee \overline{x}_3 \overline{x}_6$$

with regard to $\mathbf{x}_0 = (x_2, x_3, x_4)$. Show the Karnaugh-maps of $f(\mathbf{x})$ and $\overline{f(\mathbf{x})}$ in the first row of the m-fold view and thereunder the Karnaugh-maps of the left-hand sides of these three rules in the first column and the associated right-hand sides in the second column of this view.

(b) Verify the rules (6.21), (6.22), and (6.23) for the functions

$$f(\mathbf{x}) = x_1 \overline{x}_3 x_6 \vee \overline{x}_2 \overline{x}_4 \vee x_4 \overline{x}_5 \,,$$

$$g(\mathbf{x}) = \overline{x}_1 x_3 x_6 \vee x_2 \overline{x}_4 x_5 \vee x_1 x_5 \vee x_3 x_4 \overline{x}_5$$

with regard to $\mathbf{x}_0 = (x_1, x_3)$. Show the Karnaugh-maps of $f(\mathbf{x})$ and $g(\mathbf{x})$ in the first row of the m-fold view and thereunder the Karnaugh-maps of the left-hand sides of these three rules in the first column and the associated right-hand sides in the second column of this view. To study more details, the Karnaugh-maps of the computed derivative operations of the two functions $f(\mathbf{x})$ and $g(\mathbf{x})$ should be shown to the right of the Karnaugh-maps of the verified rule.

(c) Split the function

$$f(\mathbf{x}) = x_1 x_4 \overline{x}_6 \vee \overline{x}_2 \overline{x}_3 x_5 \vee \overline{x}_1 x_2 x_4 \vee x_3 \overline{x}_4$$

into the functions $f_i(\mathbf{x})$ and $f_d(\mathbf{x})$ such that

$$f(\mathbf{x}) = f_i(\mathbf{x}) \vee f_d(\mathbf{x}) \,,$$

$$f_i(\mathbf{x}) \wedge f_d(\mathbf{x}) = 0 \,,$$

$$\operatorname*{der}_{(x_2, x_4, x_6)} f_i(\mathbf{x}) = 0 \, .$$

Verify that the computed functions $f_i(\mathbf{x})$ and $f_d(\mathbf{x})$ satisfy these conditions. Verify furthermore that the function $f_i(\mathbf{x})$ satisfies the rules (6.24), (6.25), and (6.26). Show the Karnaugh-maps of $f(\mathbf{x})$, $f_i(\mathbf{x})$, and $f_d(\mathbf{x})$ in the upper row of the m-fold view and the Karnaugh-maps of $\operatorname{der}_{(x_2, x_4, x_6)} f_i(\mathbf{x})$, $\min_{(x_2, x_4, x_6)} f_i(\mathbf{x})$, and $\max_{(x_2, x_4, x_6)} f_i(\mathbf{x})$ thereunder.

(d) Find all pairs of variables (x_i, x_j) for which the function

$$f(\mathbf{x}) = x_1 \overline{x}_2 x_5 \overline{x}_6 \vee x_1 \overline{x}_2 \overline{x}_4 x_5 \vee x_3 \overline{x}_4 x_6 \vee x_1 x_2 \overline{x}_5 \vee x_1 \overline{x}_2 x_3 x_5 \vee \overline{x}_3 x_4 x_6$$

is symmetric. Store one ternary vector consisting of dashes and two values 1 for each detected pair of symmetrical variables in the solution-TVL. Show the TVLs of the given function and the computed pairs of symmetrical variables in one row of the m-fold view.

Exercise 6.9 (Applications of Single Derivative Operations)

Use for this exercise the function

$$f(\mathbf{x}) = x_1 x_5 \overline{x}_6 \vee \overline{x}_2 \overline{x}_5 x_7 \vee \overline{x}_1 x_3 \vee \overline{x}_4 \overline{x}_7 \overline{x}_8,$$

which depends on eight variables. Use the command obbc to minimize the representation of associated TVL after the input with the command sbe.

(a) It is very convenient to use the maximum with regard to x_i to compute the subfunctions of the Shannon expansion:

$$f(x_i = 0, \mathbf{x}_1) = \max_{x_i} (\overline{x}_i \wedge f(x_i, \mathbf{x}_1)) \, ,$$

$$f(x_i = 1, \mathbf{x}_1) = \max_{x_i} (x_i \wedge f(x_i, \mathbf{x}_1)) \, .$$

Use this approach to compute the subfunctions of the Shannon expansion of the function $f(\mathbf{x})$ with regard to x_2 and verify that the substitution of these subfunctions into (4.9) and (4.10) of Theorem 4.13 restores the given function. Show the TVLs of the given function and the two computed subfunctions in the m-fold view.

(b) The *positive Davio expansion* results from the *Shannon expansion* by the following bijective transformations:

$$f(x_i, \mathbf{x}_1) = \overline{x}_i f(x_i = 0, \mathbf{x}_1) \oplus x_i f(x_i = 1, \mathbf{x}_1) \, ,$$

$$= (x_i \oplus 1) f(x_i = 0, \mathbf{x}_1) \oplus x_i f(x_i = 1, \mathbf{x}_1) \, ,$$

$$= f(x_i = 0, \mathbf{x}_1) \oplus x_i (f(x_i = 0, \mathbf{x}_1) \oplus f(x_i = 1, \mathbf{x}_1)) \, ,$$

$$= f(x_i = 0, \mathbf{x}_1) \oplus x_i \operatorname*{der}_{x_i} f(x_i, \mathbf{x}_1) \, .$$

Compute the subfunctions $f(x_i = 0, \mathbf{x}_1)$ and $\operatorname{der}_{x_i} f(x_i, \mathbf{x}_1)$ of the positive Davio expansion of the function $f(\mathbf{x})$ with regard to $x_i = x_3$ and verify that the substitution of these subfunctions into the formula of this Davio expansion restores the given function. Show the TVLs of the given function and the two computed subfunctions in the m-fold view.

(c) The *negative Davio expansion* results from the *Shannon expansion* by the following bijective transformations:

$$f(x_i, \mathbf{x}_1) = \overline{x}_i f(x_i = 0, \mathbf{x}_1) \oplus x_i f(x_i = 1, \mathbf{x}_1) \,,$$

$$= \overline{x}_i f(x_i = 0, \mathbf{x}_1) \oplus (\overline{x}_i \oplus 1) f(x_i = 1, \mathbf{x}_1) \,,$$

$$= f(x_i = 1, \mathbf{x}_1) \oplus \overline{x}_i \left(f(x_i = 0, \mathbf{x}_1) \oplus f(x_i = 1, \mathbf{x}_1) \right) \,,$$

$$= f(x_i = 1, \mathbf{x}_1) \oplus \overline{x}_i \underset{x_i}{\mathrm{der}}\, f(x_i, \mathbf{x}_1) \,.$$

Compute the subfunctions $f(x_i = 1, \mathbf{x}_1)$ and $\mathrm{der}_{x_i} f(x_i, \mathbf{x}_1)$ of the negative Davio expansion of the function $f(\mathbf{x})$ with regard to $x_i = x_4$ and verify that the substitution of these subfunctions into the formula of this Davio expansion restores the given function. Show the TVLs of the given function and the two computed subfunctions in the m-fold view.

(d) The two types of Davio expansions as well as the Shannon expansion exist for each function and each occurring variable. A decomposition that utilizes that a given function is monotonously increasing is only possible when the function satisfies this property for a selected variable.
A logic function $f(x_i, \mathbf{x}_1)$ that is monotonously increasing with regard to x_i can be expressed by

$$f(x_i, \mathbf{x}_1) = f_0(\mathbf{x}_1) \vee x_i \wedge f_1(\mathbf{x}_1) \,,$$

where the two subfunctions $f_0(\mathbf{x}_1)$ and $f_1(\mathbf{x}_1)$ do not depend on x_i. The function $f_0(\mathbf{x}_1)$ is uniquely specified by

$$f_0(\mathbf{x}_1) = \min_{x_i} f(x_i, \mathbf{x}_1) \,.$$

Generally, there are several functions $f_1(\mathbf{x}_1)$; the function $f_1(\mathbf{x}_1)$ with the smallest number of function values 1 can be calculated as follows:

$$f_1(\mathbf{x}_1) = \max_{x_i} \left(x_i \wedge f(x_i, \mathbf{x}_1) \wedge \overline{f_0(\mathbf{x}_1)} \right) \,.$$

Verify whether the function $f(\mathbf{x})$ is monotonously increasing with regard to one of the eight variables. Break the search for such a variable when the first variable is found. Compute in this case the functions $f_0(\mathbf{x}_1)$ and $f_1(\mathbf{x}_1)$ for the function $f(\mathbf{x})$ with regard to the variable for which this function is monotonously increasing and verify that the substitution of these subfunctions into the formula of decomposition of a monotonously increasing function restores the given function. Show the TVLs of the given function, the found variable, and the two computed subfunctions in the m-fold view.

Exercise 6.10 (Properties of k-fold Derivative Operations)
Use for this exercise the function

$$f(\mathbf{x}) = x_1(x_2 \oplus x_4 x_6) \vee x_1 x_3 (\overline{x}_2 \vee \overline{x}_5) \vee x_2 x_5 (\overline{x}_3 \vee \overline{x}_4) \vee \overline{x}_2 \overline{x}_3 \overline{x}_4 x_5 \vee x_1 x_3 x_6,$$

which depends on six variables. Use the command `obbc` to minimize the representation of the associated TVL after the input with the command `sbe`:

(a) Compute the minimum of $f(\mathbf{x})$ with regard to x_3, the 2-fold minimum of $f(\mathbf{x})$ with regard to $\mathbf{x}_0 = (x_3, x_5)$, and the 3-fold minimum of $f(\mathbf{x})$ with regard to $\mathbf{x}_0 = (x_3, x_5, x_6)$ using command `_maxk` in which the names of the variables to be used can be directly specified. Verify the three inequalities of

$$\min_{(x_3, x_5, x_6)}{}^3 f(\mathbf{x}) \leq \min_{(x_3, x_5)}{}^2 f(\mathbf{x}) \leq \min_{x_3} f(\mathbf{x}) \leq f(\mathbf{x})$$

using command `te_dif`. Show the TVLs of $f(\mathbf{x})$ and all computed k-fold minima in the upper row of the m-fold view in the order of their appearance in this inequality and the associated Karnaugh-maps expanded to all variables of $f(\mathbf{x})$ in the viewports thereunder.

(b) Compute the k-fold minimum, the k-fold maximum, and the Δ-operation of $f(\mathbf{x})$ with regard to $\mathbf{x}_0 = (x_3, x_5, x_6)$. Verify the rules (6.52), (6.53), and (6.54), which use two of these k-fold derivative operations to compute the third one. Show the Karnaugh-maps of $f(\mathbf{x})$ and the basically computed k-fold derivative operations in the first column and the results of the right-hand sides of (6.52), (6.53), and (6.54) in the corresponding viewport of the second column of the m-fold view.

(c) Compute all four k-fold derivative operations of $f(\mathbf{x})$ with regard to $\mathbf{x}_0 = (x_2, x_5)$ and verify the rules (6.55), (6.56), (6.57), and (6.58) in which the same k-fold derivative operation is executed twice. Show the Karnaugh-maps of $f(\mathbf{x})$ and the first computed k-fold derivative operations in the first column and the results of the second execution of the same k-fold derivative operations in the corresponding viewports of the second column of the m-fold view.

(d) A very welcome property of all four k-fold derivative operations is stated in the rules (6.59), (6.60), (6.61), and (6.62). Compute all four k-fold derivative operations of $f(\mathbf{x})$ with regard to $\mathbf{x}_0 = (x_3, x_6)$ and verify this common property.

References

1. D. Bochmann, C. Posthoff, *Binäre Dynamische Systeme* (Akademie, Oldenbourg, 1981). ISBN: 978-3-48625-071-8
2. C. Posthoff, B. Steinbach, *Logic Functions and Equations—Binary Models for Computer Science*. (Springer, The Netherlands, 2004). ISBN: 978-1-44195-261-5. https://doi.org/10.1007/978-1-4020-2938-7
3. B. Steinbach, C. Posthoff, *Logic Functions and Equations—Examples and Exercises* (Springer, Berlin, 2009). ISBN: 978-1-4020-9594-8. https://doi.org/10.1007/978-1-4020-9595-5
4. B. Steinbach, C. Posthoff, Boolean differential calculus—theory and applications. J. Comput. Theor. Nanosci. **7**(6), 933–981 (2010). ISSN: 1546-1955. https://doi.org/10.1166/jctn.2010.1441
5. B. Steinbach, C. Posthoff, *Boolean Differential Equations* (Morgan and Claypool Publishers, San Rafael, 2013). ISBN: 978-1-6270-5241-2. https://doi.org/10.2200/S00511ED1V01Y201305DCS042
6. B. Steinbach, C. Posthoff, *Boolean Differential Calculus* (Morgan and Claypool Publishers, California, 2017). ISBN: 978-1-6270-5922-0. https://doi.org/10.2200/S00766ED1V01Y201704DCS052
7. A. Thayse, *Boolean Calculus of Differences* (Springer, Berlin, 1981). ISBN: 978-3-54010-286-1. https://doi.org/10.1007/3-540-10286-8

Sets, Lattices, and Classes of Logic Functions

7

Abstract

Many applications use sets of logic functions. It is beneficial when not each function of such a set must be computed separately. A strongly simplified computation is possible when a set of functions satisfies certain properties. We explore sets of functions which satisfy the rules of an equivalence relation as well as sets having the structure of a lattice. Very often used are partially defined logic functions which describe a lattice of such functions that is isomorphic to a Boolean Algebra. These lattices can and must be generalized to express the results of derivative operations of a lattice of logic functions. The Boolean Differential Calculus has been extended for all derivative operations of such generalized lattice of logic functions. The solution of a logic equation with regard to variables has a wide field of applications. Both the condition whether a logic equation can be solved with regard to selected variables and the computation of either a single solution function or a lattice of solution functions can be realized using derivative operations of the Boolean Differential Calculus. As special application we explain systems of reversible functions. Sets of functions can also be determined by functional equations. More generally, a Boolean differential equation can be used to describe an arbitrary set of logic functions. We present a basic algorithm that calculates

Supplementary Information The online version of this chapter (https://doi.org/10.1007/978-3-030-88945-6_7) contains supplementary material which is available for authorized users. Please, follow the link belonging to the version of the XBOOLE-monitor XBM 2 that fits best for your operating system. This XBOOLE-monitor is needed to solve all tasks of this chapter. Instructions for starting the downloaded XBOOLE-monitor XBM 2 are given at the beginning of Section 'Exercises' in this chapter.

XBOOLE-monitor XBM 2 for Windows 10
32 bits
https://doi.org/10.1007/978-3-030-88945-6_7_MOESM1_ESM.zip (15,091 KB)

64 bits
https://doi.org/10.1007/978-3-030-88945-6_7_MOESM2_ESM.zip (14,973 KB)

XBOOLE-monitor XBM 2 for Linux Ubuntu
32 bits
https://doi.org/10.1007/978-3-030-88945-6_7_MOESM3_ESM.zip (29,522 KB)

64 bits
https://doi.org/10.1007/978-3-030-88945-6_7_MOESM4_ESM.zip (28,422 KB)

the set of all solution-lattices of a special type of Boolean differential equations and generalize this algorithm for arbitrary sets of logic functions as solution of a Boolean differential equation.

7.1 Partially Defined Functions—Lattices of Functions

Very often it will be the case that a function $f(\mathbf{x}) \in \mathcal{F}_n$ is not defined for all $\mathbf{x} \in \mathbb{B}^n$, but only for a subset $S \subseteq \mathbb{B}^n$. The values are not specified for some $\mathbf{x} \in \mathbb{B}^n \setminus S$ because they are not known (not important, not allowed, etc.). Such not specified values are denoted by *don't-cares* and displayed in a Karnaugh-map by the symbol ϕ or in a function vector by the symbol $-$. A *partially defined function* is sometimes also denoted by *incompletely specified function* (ISF).

Example 7.1 Figure 7.1 shows the Karnaugh-map of a given partially defined function $f(x_1, x_2, x_3, x_4)$. The vectors with three values 1 are excluded.

If the function $f(\mathbf{x})$ is supposed to be represented by an expression, it is important that the expression correctly evaluates only for the vectors $\mathbf{x}$ that have to be considered. $f(\mathbf{x})$ can have *any value* for the vectors that have been excluded. The symbol ϕ indicates in the Karnaugh-map that either the 0 or 1 can be used as function value for such an assignment. The same meaning has the symbol—in a function vector. The use of the symbol—already indicates that such a partially defined function can be understood as the representation of an interval of functions.

$$F = (0\ 0\ 1\ 1\ 0\ 1\ 0 - 1\ 0\ 1 - 1 - - 0)$$

would be the representation as a function vector in the order of the binary code for (x_4, x_3, x_2, x_1); the four elements with a—can be replaced by 16 combinations $(0000), (0001), (0010), \ldots, (1111)$.

It is quite understandable that the minimization problem also exists for these functions. However, the existence of vectors without function values increases the degrees of freedom and, therefore, the complexity of the problem. Instead of one function, 16 functions (in our example) have to be minimized, and then the representation with the smallest number of conjunctions (or other properties) can be found. Figure 7.2 shows three functions that coincide with the partially defined function of Fig. 7.1.

$$f_a(\mathbf{x}) = x_2 \overline{x}_3 \overline{x}_4 \vee \overline{x}_1 \overline{x}_3 x_4 \vee \overline{x}_1 \overline{x}_2 x_4 \vee x_1 \overline{x}_2 x_3 \overline{x}_4 \,,$$

$$f_b(\mathbf{x}) = x_2 \overline{x}_3 \vee \overline{x}_1 x_4 \vee \overline{x}_2 x_3 x_4 \vee x_1 x_3 \overline{x}_4 \,,$$

Fig. 7.1 Karnaugh-map of
a partially defined function

$x_3\ x_4$				$f(\mathbf{x})$
0 0	0	1	1	0
0 1	1	1	ϕ	0
1 1	1	ϕ	0	ϕ
1 0	0	0	ϕ	1
	0	1	1	0 x_2
	0	0	1	1 x_1

$x_3\ x_4$	$f_a(\mathbf{x})$				$x_3\ x_4$	$f_b(\mathbf{x})$				$x_3\ x_4$	$f_c(\mathbf{x})$			
0 0	0	1	1	0	0 0	0	1	1	0	0 0	0	1	1	0
0 1	1	1	0	0	0 1	1	1	1	0	0 1	1	1	1	0
1 1	1	0	0	0	1 1	1	1	0	1	1 1	1	1	0	0
1 0	0	0	0	1	1 0	0	0	1	1	1 0	0	0	1	1
	0	1	1	0 $\,x_2$		0	1	1	0 $\,x_2$		0	1	1	0 $\,x_2$
	0	0	1	1 $\,x_1$		0	0	1	1 $\,x_1$		0	0	1	1 $\,x_1$
a					**b**					**c**				

Fig. 7.2 Three functions belonging to the function interval: (**a**) $f_a(\mathbf{x})$: all ϕ are replaced by 0, (**b**) $f_b(\mathbf{x})$: all ϕ are replaced by 1, (**c**) $f_c(\mathbf{x})$: a compromising replacement

$$f_c(\mathbf{x}) = x_2\overline{x}_3 \vee \overline{x}_1 x_4 \vee x_1 x_3 \overline{x}_4\ .$$

Even this small example gives already quite different solutions. We found one expression with four conjunctions and 13 literals ($f_a(\mathbf{x})$), one expression with four conjunctions and 10 literals ($f_b(\mathbf{x})$), and a third expression with three conjunctions and only seven literals ($f_c(\mathbf{x})$).

In other cases, we will have to deal with sets of logic functions that arise from other considerations. Very often they have the structure of a *lattice*. Even the explored Boolean Algebras and incompletely specified functions determine special lattices. We do not want to go too deep into the *Lattice Algebra*; we restrict our considerations to the basic concepts.

Definition 7.1 (Lattice of Logic Functions)
A set of logic functions that is *closed* with regard to the two operations $\vee$ and $\wedge$ has the structure of a *lattice* $\mathcal{L}$ when the following axioms are satisfied for any functions $f(\mathbf{x}), g(\mathbf{x}), h(\mathbf{x}) \in \mathcal{L}$:

1. Commutative laws:

$$\forall f(\mathbf{x}), \forall g(\mathbf{x}): \qquad f(\mathbf{x}) \vee g(\mathbf{x}) = g(\mathbf{x}) \vee f(\mathbf{x})\,,$$

$$\forall f(\mathbf{x}), \forall g(\mathbf{x}): \qquad f(\mathbf{x}) \wedge g(\mathbf{x}) = g(\mathbf{x}) \wedge f(\mathbf{x})\,,$$

2. Associative laws:

$$\forall f(\mathbf{x}), \forall g(\mathbf{x}): \quad f(\mathbf{x}) \vee (g(\mathbf{x}) \vee h(\mathbf{x})) = (f(\mathbf{x}) \vee g(\mathbf{x})) \vee h(\mathbf{x})\,,$$

$$\forall f(\mathbf{x}), \forall g(\mathbf{x}): \quad f(\mathbf{x}) \wedge (g(\mathbf{x}) \wedge h(\mathbf{x})) = (f(\mathbf{x}) \wedge g(\mathbf{x})) \wedge h(\mathbf{x})\,,$$

3. Absorption laws:

$$\forall f(\mathbf{x}), \forall g(\mathbf{x}): \quad f(\mathbf{x}) \vee (f(\mathbf{x}) \wedge g(\mathbf{x})) = f(\mathbf{x})\,,$$

$$\forall f(\mathbf{x}), \forall g(\mathbf{x}): \quad f(\mathbf{x}) \wedge (f(\mathbf{x}) \vee g(\mathbf{x})) = f(\mathbf{x})\,.$$

A set of functions is *closed with regard to an operation* if the execution of this operation results for any pair of functions of this set again in one function of the given set. The following two identities are also usually regarded as axioms of a lattice, even though they follow from the two absorption laws in the case of $f(\mathbf{x}) \equiv g(\mathbf{x})$.

$$\text{Idempotency laws:} \qquad \forall f(\mathbf{x}): \quad f(\mathbf{x}) \vee f(\mathbf{x}) = f(\mathbf{x})\,,$$

$$\forall f(\mathbf{x}): \quad f(\mathbf{x}) \wedge f(\mathbf{x}) = f(\mathbf{x})\,.$$

It can be seen that Boolean Algebras satisfy these axioms; hence, a Boolean Algebra is a lattice with *additional properties*.

There are subsets of $\mathbb{B}^n$ which satisfy these axioms of a general lattice, but they are not a Boolean Algebra.

Example 7.2 Let us see, for instance, the following set of three vectors which can be understood as three functions of two variables:

$$(0\ 0\ 0\ 0)\ ,$$
$$(0\ 1\ 0\ 1)\ ,$$
$$(1\ 1\ 1\ 1)\ .$$

The $\wedge$ and the $\vee$ between two of these vectors always result in a vector that is again an element of the set, the axioms of a lattice are satisfied, but this set cannot be extended to a Boolean Algebra.

There are two points of view in the case that not all values of a logic function are defined:

1. Such a function is partially defined (incompletely specified), and the values of the don't-cares can arbitrarily be chosen.
2. There are 2^{dc} different binary vectors which can be assigned to the dc don't-cares; hence, we have 2^{dc} functions that satisfy the given specification.

This set of 2^{dc} functions satisfies the laws of a *lattice* $\mathcal{L}$.

We return to Example 7.1; the vector

$$F = (0\ 0\ 1\ 1\ 0\ 1\ 0 - 1\ 0\ 1 - 1 - - 0)$$

of function values contains four dashes, so that a set of $2^4 = 16$ functions can be created.

This set can even be seen as a Boolean Algebra where the usual operations $\wedge$ and $\vee$ are used. The complement must be restricted to the fields of the don't-cares. The needed complement operation can easily be defined based on the following three *mark-functions*.

The function values of an incompletely specified function (ISF) divide the Boolean space into three subspaces. Hence, the three characteristic functions of these subspaces specify both the partially defined function and all 2^{dc} functions of the lattice $\mathcal{L}$:

- $\mathbf{x} \in$ don't-care-set $\Leftrightarrow f_\varphi(x_1, \ldots, x_n) = 1$

    $\Leftrightarrow$ it is allowed to choose the function value of

    $f(\mathbf{x})$ without any restrictions;

- $\mathbf{x} \in$ ON-set $\Leftrightarrow f_q(x_1, \ldots, x_n) = 1$

    $\Leftrightarrow (f_\varphi(x_1, \ldots, x_n) = 0) \wedge (f(x_1, \ldots, x_n) = 1)\ ;$

- $\mathbf{x} \in$ OFF-set $\Leftrightarrow f_r(x_1, \ldots, x_n) = 1$

    $\Leftrightarrow (f_\varphi(x_1, \ldots, x_n) = 0) \wedge (f(x_1, \ldots, x_n) = 0)\ .$

These three mark-functions are mutually disjoint:

$$f_q(\mathbf{x}) \wedge f_r(\mathbf{x}) = 0\ , \tag{7.1}$$

$x_3\,x_4$	$f_q(\mathbf{x})$			
0 0	0	1	1	0
0 1	1	1	0	0
1 1	1	0	0	0
1 0	0	0	0	1
	0	1	1	0 x_2
	0	0	1	1 x_1

a

$x_3\,x_4$	$f_r(\mathbf{x})$			
0 0	1	0	0	1
0 1	0	0	0	1
1 1	0	0	1	0
1 0	1	1	0	0
	0	1	1	0 x_2
	0	0	1	1 x_1

b

$x_3\,x_4$	$f_\varphi(\mathbf{x})$			
0 0	0	0	0	0
0 1	0	0	1	0
1 1	0	1	0	1
1 0	0	0	1	0
	0	1	1	0 x_2
	0	0	1	1 x_1

c

Fig. 7.3 Three mark-functions of the ISF specified in Fig. 7.1: (**a**) $f_q(\mathbf{x})$: ON-set-function, (**b**) $f_r(\mathbf{x})$: OFF-set-function, (**c**) $f_\varphi(\mathbf{x})$: don't-care-function

$$f_q(\mathbf{x}) \wedge f_\varphi(\mathbf{x}) = 0 \,, \tag{7.2}$$

$$f_r(\mathbf{x}) \wedge f_\varphi(\mathbf{x}) = 0 \,, \tag{7.3}$$

and they cover the whole Boolean space:

$$f_q(\mathbf{x}) \vee f_r(\mathbf{x}) \vee f_\varphi(\mathbf{x}) = 1 \,. \tag{7.4}$$

Hence, each of these three mark-functions can be calculated using the other two mark-functions. For many applications the use of $f_q(\mathbf{x})$ and $f_r(\mathbf{x})$ has some advantages. A function $f_i(\mathbf{x})$ belongs to the lattice of logic functions characterized by the mark-functions $f_q(\mathbf{x})$ and $f_r(\mathbf{x})$, in short $f_i(\mathbf{x}) \in \mathcal{L}\langle f_q(\mathbf{x}),\, f_r(\mathbf{x})\rangle$, if

$$f_q(\mathbf{x}) \le f_i(\mathbf{x}) \le \overline{f_r(\mathbf{x})} \,. \tag{7.5}$$

Figure 7.3 shows these three mark-functions for the partially defined function specified in Fig. 7.1.

Using the mark-functions $f_q(\mathbf{x})$ and $f_r(\mathbf{x})$ of the lattice $\mathcal{L}$, a special complement operation for each function $f_i(\mathbf{x}) \in \mathcal{L}\langle f_q(\mathbf{x}),\, f_r(\mathbf{x})\rangle$, indicated by $^{\mathcal{L}}\overline{f_i(\mathbf{x})}$, can be defined:

$$^{\mathcal{L}}\overline{f_i(\mathbf{x})} = \overline{f_i(\mathbf{x})} \wedge \overline{f_r(\mathbf{x})} \vee f_q(\mathbf{x}) \,. \tag{7.6}$$

The smallest function (the *infimum*) of the lattice $\mathcal{L}\langle f_q(\mathbf{x}),\, f_r(\mathbf{x})\rangle$ is the function $f_q(\mathbf{x})$. For each function $f_i(\mathbf{x}) \in \mathcal{L}\langle f_q(\mathbf{x}),\, f_r(\mathbf{x})\rangle$ we have

$$f_i(\mathbf{x}) \wedge ^{\mathcal{L}}\overline{f_i(\mathbf{x})} = f_q(\mathbf{x})$$

$$f_i(\mathbf{x}) \wedge (\overline{f_i(\mathbf{x})} \wedge \overline{f_r(\mathbf{x})} \vee f_q(\mathbf{x})) = f_q(\mathbf{x})$$

$$f_i(\mathbf{x}) \wedge \overline{f_i(\mathbf{x})} \wedge \overline{f_r(\mathbf{x})} \vee f_i(\mathbf{x}) \wedge f_q(\mathbf{x}) = f_q(\mathbf{x})$$

$$f_i(\mathbf{x}) \wedge f_q(\mathbf{x}) = f_q(\mathbf{x})$$

$$f_q(\mathbf{x}) = f_q(\mathbf{x}) \,.$$

The largest function (the *supremum*) of the lattice $\mathcal{L}\langle f_q(\mathbf{x}),\, f_r(\mathbf{x})\rangle$ is the function $\overline{f_r(\mathbf{x})}$. For each function $f_i(\mathbf{x}) \in \mathcal{L}\langle f_q(\mathbf{x}),\, f_r(\mathbf{x})\rangle$ we have

$$f_i(\mathbf{x}) \vee ^{\mathcal{L}}\overline{f_i(\mathbf{x})} = \overline{f_r(\mathbf{x})}$$

$$f_i(\mathbf{x}) \vee (\overline{f_i(\mathbf{x})} \wedge \overline{f_r(\mathbf{x})} \vee f_q(\mathbf{x})) = \overline{f_r(\mathbf{x})}$$

$$f_i(\mathbf{x}) \vee \overline{f_i(\mathbf{x})} \wedge \overline{f_r(\mathbf{x})} \vee f_q(\mathbf{x}) = \overline{f_r(\mathbf{x})}$$

$$f_i(\mathbf{x}) \vee \overline{f_r(\mathbf{x})} \vee f_q(\mathbf{x}) = \overline{f_r(\mathbf{x})}$$

$$\overline{f_r(\mathbf{x})} = \overline{f_r(\mathbf{x})} \ .$$

It can be seen that the operations of $\mathbb{B}^4$ have been transferred to the 16 functions of Example 7.1. Using the functions $f_q(\mathbf{x})$ and $f_r(\mathbf{x})$ of Fig. 7.3, the set of logic functions determined by $\mathcal{L}\langle f_q(\mathbf{x}), f_r(\mathbf{x})\rangle$ and the set $\mathbb{B}^4$ specify *isomorphic Boolean Algebras*.

7.2   Generalized Lattices of Logic Functions

A lattice of logic functions $f_i(\mathbf{x})$ is a set of functions which satisfy several rules (see Sect. 7.1). Such lattices are very useful in circuit design because an arbitrary function of $f_i(\mathbf{x})$ (that is optimal with regard to certain properties) can be chosen out of the lattice $\mathcal{L}\langle f_q(\mathbf{x}), f_r(\mathbf{x})\rangle$ of logic functions. Using the knowledge of Chap. 5, the inequality (7.5) that describes all functions $f_i(\mathbf{x}) \in \mathcal{L}\langle f_q(\mathbf{x}), f_r(\mathbf{x})\rangle$ can be transformed into an equivalent equation:

$$f_q(\mathbf{x}) \wedge \overline{f_i(\mathbf{x})} \vee f_i(\mathbf{x}) \wedge f_r(\mathbf{x}) = 0 \ . \tag{7.7}$$

Derivative operations can be used to detect whether a logic function satisfies a certain property. Basically, it is possible to calculate the needed derivative operation for each function of a lattice; however, this approach is very time consuming. Hence, the derivative operations that commonly calculate the results for all functions $f_i(\mathbf{x}) \in \mathcal{L}\langle f_q(\mathbf{x}), f_r(\mathbf{x})\rangle$ are desirable.

We explore the peculiarity of vectorial derivative operations by the following simple example:

Example 7.3 A lattice $\mathcal{L}\langle f_q(\mathbf{x}), f_r(\mathbf{x})\rangle$ of four logic functions $f_i(a, b, c)$ has been specified by the mark-functions:

$$f_q(a, b, c) = \overline{a}\overline{b} \vee \overline{b}c \vee ab\overline{c} \ ,$$

$$f_r(a, b, c) = \overline{c}(a \oplus b) \ .$$

Our aim is the calculation of the vectorial derivatives $g_i(a, b, c)$ of all functions of this lattice with regard to the simultaneous change of the values of (b, c) and the exploration of the properties of the resulting functions.

Figure 7.4 shows detailed calculation steps using Karnaugh-maps. The function values 1 of the ON-set-function $f_q(a, b, c)$ in the top left Karnaugh-map of Fig. 7.4 determine the values 1 of all functions of the lattice $\mathcal{L}\langle f_q(\mathbf{x}), f_r(\mathbf{x})\rangle$ as can be seen in the Karnaugh-map in the middle of the upper row in Fig. 7.4. The function values 1 of the OFF-set-function $f_r(a, b, c)$ in the top right Karnaugh-map of Fig. 7.4 determine the function values 0 of all functions of this lattice.

The two values ϕ in the Karnaugh-map in the middle of the upper row of Fig. 7.4 indicate that this lattice describes $2^2 = 4$ logic functions which are shown in the second row of Fig. 7.4. The small arrows in these Karnaugh-maps indicate the pairs of function values which determine the vectorial derivatives with regard to (b, c). The Karnaugh-maps in the third row of Fig. 7.4 show the results of the vectorial derivative with regard to (b, c) for each of these four functions.

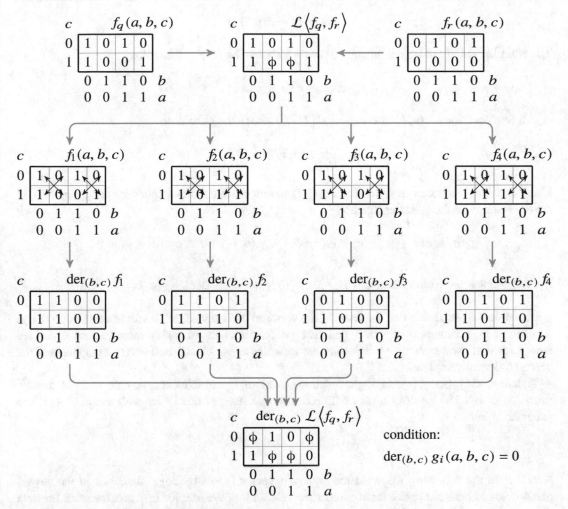

Fig. 7.4 Vectorial derivative of a lattice of logic functions

As final step, we merge the four Karnaugh-maps of the vectorial derivatives (shown in the third row of Fig. 7.4) into a single Karnaugh-map shown at the bottom of Fig. 7.4. The values of this Karnaugh-map have the following meaning:

- Value 1: The value 1 appears in the associated position of all four vectorial derivatives with regard to (b, c);
- Value 0: The value 0 appears in the associated position of all four vectorial derivatives with regard to (b, c);
- Value ϕ: The vectorial derivatives with regard to (b, c) carry either the value 0 or 1 in the associated position.

The result of this merge is a Karnaugh-map that contains four values ϕ; hence, this lattice describes $2^4 = 16$ functions based on the model of lattices introduced in Sect. 7.1.

The evaluation of the four functions of the vectorial derivatives, represented in the Karnaugh-maps in third row of Fig. 7.4, confirms that these functions satisfy the conditions of a lattice and even the conditions of a Boolean Algebra. The smallest element of this Algebra is the ON-set-function

$$g_q(a, b, c) = \overline{a}(b \oplus c) .$$

The complement of the largest function of this Algebra is the OFF-set-function

$$g_r(a, b, c) = a(b \oplus c) .$$

All four functions $g_i(a, b, c) \in \text{der}_{(b,c)} \mathcal{L} \langle f_q, f_r \rangle$ satisfy the condition:

$$\underset{(b,c)}{\text{der}} g_i(a, b, c) = 0 .$$

Using this restriction we can generalize Eq. (7.7) so that exactly the four functions $\text{der}_{(b,c)} g_i(a, b, c)$ are the solution of the lattice equation:

$$g_q(a, b, c) \wedge \overline{g_i(a, b, c)} \vee g_i(a, b, c) \wedge g_r(a, b, c) \vee \underset{(b,c)}{\text{der}} g_i(a, b, c) = 0 .$$

A short characteristic of this lattice is $\mathcal{L} \langle g_q(a, b, c), g_r(a, b, c), \text{der}_{(b,c)} g(a, b, c) \rangle$.

Example 7.3 has shown that the result of the vectorial derivative of all functions of a lattice is also a lattice of logic functions of a more general type. All functions of such a more general lattice are independent of the simultaneous change of the variables $\mathbf{x}_0$ which are used for the calculation of the vectorial derivative operation.

A lattice $\mathcal{L} \langle f_q(\mathbf{x}), f_r(\mathbf{x}) \rangle$ of logic functions $f_i(\mathbf{x})$ of n variables can contain each of the 2^{2^n} functions of $\mathbb{B}^n$. The result of a vectorial derivative operation of this lattice with regard to $\mathbf{x}_0$ is the generalized lattice

$$\mathcal{L} \left\langle g_q(\mathbf{x}), g_r(\mathbf{x}), \underset{\mathbf{x}_0}{\text{der}} g(\mathbf{x}) \right\rangle .$$

Note 7.1 In the following we avoid the argument vector ($\mathbf{x}$) of the logic functions in the angled parentheses of lattices to get a more compact representation. We use, for instance, the short formula $\mathcal{L} \langle g_q, g_r, \text{der}_{\mathbf{x}_0} g \rangle$ instead of the complete formula $\mathcal{L} \langle g_q(\mathbf{x}), g_r(\mathbf{x}), \text{der}_{\mathbf{x}_0} g(\mathbf{x}) \rangle$.

The lattice $\mathcal{L} \langle g_q, g_r, \text{der}_{\mathbf{x}_0} g \rangle$ also contains logic functions $g_i(\mathbf{x})$ of n variables, but these functions belong to a subset of $2^{2^{n-1}}$ functions of $\mathbb{B}^n$ due to the equal values in pairs of function values. In this sense the generalized lattice $\mathcal{L} \langle g_q, g_r, \text{der}_{\mathbf{x}_0} g \rangle$ of a vectorial derivative operation of the lattice $\mathcal{L} \langle f_q, f_r \rangle$ with regard to $\mathbf{x}_0$ is *simpler* than the given lattice.

A subsequent vectorial derivative operation of the given generalized lattice $\mathcal{L} \langle g_q, g_r, \text{der}_{\mathbf{x}_0} g \rangle$ with regard to another direction of change $\mathbf{x}_2$ results in the generalized lattice $\mathcal{L} \langle h_q, h_r, \text{der}_{\mathbf{x}_0} h \vee \text{der}_{\mathbf{x}_2} h \rangle$ that is *even simpler* because the lattice $\mathcal{L} \langle h_q, h_r, \text{der}_{\mathbf{x}_0} h \vee \text{der}_{\mathbf{x}_2} h \rangle$ contains only functions of n variables belonging to a subset of $2^{2^{n-2}}$ functions of $\mathbb{B}^n$ due to the equal values in quadruples of function values.

There are $2^n - 1$ different directions of change in the Boolean space $\mathbb{B}^n$. However, only n of these directions of change are linearly independent of each other. The same result of two subsequent vectorial derivative operations can be calculated using two of three linearly dependent directions of change.

Binary vectors that indicate the direction of change can be used to verify whether certain directions of change are linearly independent of each other.

Definition 7.2 (Binary Vector (BV)) Let $f(\mathbf{x}) = f(x_1, x_2, \ldots, x_n)$ be a logic function and $\mathbf{x}_0 \subseteq \mathbf{x}$ be a subset of variables; then

$$\mathbf{s}_0 = \mathrm{BV}(\mathbf{x}_0) \tag{7.8}$$

is a *binary vector* of n elements where $\mathbf{s}_0[i] = 1$ indicates that $x_i \in \mathbf{x}_0$.

Three directions of change indicated by the subsets $\mathbf{x}_{01}$, $\mathbf{x}_{02}$, and $\mathbf{x}_{03}$ of $\mathbf{x}$ are linearly dependent on each other if:

$$\mathrm{BV}(\mathbf{x}_{01}) \oplus \mathrm{BV}(\mathbf{x}_{02}) \oplus \mathrm{BV}(\mathbf{x}_{03}) = \mathbf{0} \,.$$

The following example uses a simple lattice of four functions to demonstrate the property that two subsequent vectorial derivative operations of the same type with regard to different selections out of three linearly dependent directions of change have the same result.

Example 7.4 The mark-functions of a given lattice $\mathcal{L}\langle f_q, f_r \rangle$ are

$$f_q(a, b, c, d) = \overline{a}\,\overline{b}\,d \vee b\,\overline{c} \vee c\,\overline{d} \,,$$

$$f_r(a, b, c, d) = a\,c\,d \vee \overline{b}\,\overline{c}\,\overline{d} \,.$$

Subsequently the vectorial derivative of the lattice $\mathcal{L}\langle f_q, f_r \rangle$ must be calculated with regard to two of the sets of variables (a, b, c), (b, c, d), and (a, d). These sets are linearly dependent because

$$\mathrm{BV}(a, b, c) \oplus \mathrm{BV}(b, c, d) \oplus \mathrm{BV}(a, d) = \mathbf{0} \,,$$

$$(1110) \qquad \oplus (0111) \qquad \oplus (1001) \quad = \mathbf{0} \,.$$

Figure 7.5 demonstrates that these different calculations result in the same lattice

$$\mathcal{L}\left\langle h_{1q}, h_{1r}, \underset{(a,b,c)}{\mathrm{der}}\ h_1 \vee \underset{(b,c,d)}{\mathrm{der}}\ h_1 \right\rangle \equiv \mathcal{L}\left\langle h_{2q}, h_{2r}, \underset{(b,c,d)}{\mathrm{der}}\ h_2 \vee \underset{(a,d)}{\mathrm{der}}\ h_2 \right\rangle$$

because the sequence of two of these directions of change describes the third direction of change.

The upper row of Fig. 7.5 shows the merge of the mark-function $f_q(a, b, c, d)$ and $f_r(a, b, c, d)$ into the lattice $\mathcal{L}\langle f_q, f_r \rangle$. Due to the two don't-cares this lattice contains $2^2 = 4$ functions.

The Karnaugh-maps in the second row of Fig. 7.5 show the vectorial derivatives of this lattice with regard to (a, b, c) (on the left-hand side) and (b, c, d) (on the right-hand side). The simultaneous change of (a, b, c) reaches one element ϕ from the other one; hence, the lattice $\mathrm{der}_{(a,b,c)} \mathcal{L}\langle f_q, f_r \rangle$ contains only two functions due to the condition $\mathrm{der}_{(a,b,c)} f = 0$. The simultaneous change of (b, c, d) starting from an element ϕ reaches once the value 0 and once the value 1; hence, the lattice $\mathrm{der}_{(b,c,d)} \mathcal{L}\langle f_q, f_r \rangle$ has four elements ϕ and contains furthermore $2^{4/2} = 2^2 = 4$ functions.

The Karnaugh-map on the left-hand side in the lower row of Fig. 7.5 shows the result $\mathcal{L}\langle h_{1q}, h_{1r}, \mathrm{der}_{(a,b,c)}\, h_1 \vee \mathrm{der}_{(b,c,d)}\, h_1 \rangle$ of two subsequently calculated vectorial derivatives of the lattice $\mathcal{L}\langle f_q, f_r \rangle$ with regard to (a, b, c) and (b, c, d). The simultaneous change of (b, c, d) reaches in the lattice $\mathrm{der}_{(a,b,c)} \mathcal{L}\langle f_q, f_r \rangle$ (Karnaugh-map on the left-hand side in the middle row of Fig. 7.5) from both elements ϕ function values 1; hence, we get four elements ϕ in the result of $\mathrm{der}_{(b,c,d)} \left(\mathrm{der}_{(a,b,c)} \mathcal{L}\langle f_q, f_r \rangle \right)$. Due to Condition $\mathrm{der}_{(a,b,c)} h_1 \vee \mathrm{der}_{(b,c,d)} h_1 = 0$ this lattice is structured into quadruples of function values. Identically colored fields in this Karnaugh-map emphasize all function values that belong to such a quadruple. Especially, the four elements ϕ compose one of

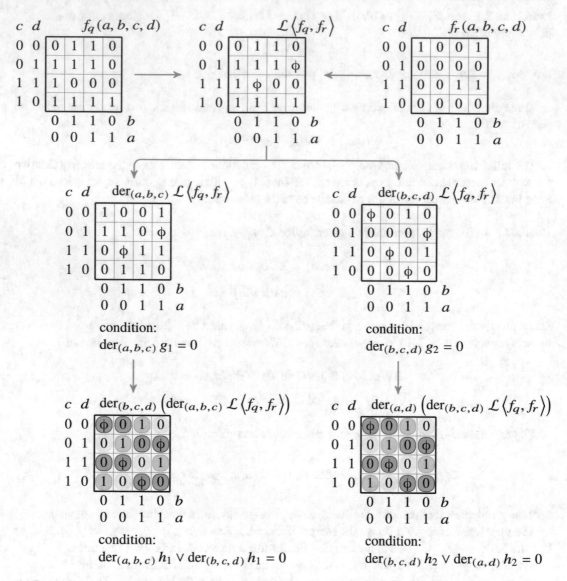

Fig. 7.5 Subsequent vectorial derivatives of a lattice of logic functions with regard to dependent directions of change

these quadruples; hence, all the elements ϕ must have either the value 0 or 1 so that the lattice $\text{der}_{(b,c,d)}\left(\text{der}_{(a,b,c)}\,\mathcal{L}\langle f_q, f_r\rangle\right)$ contains $2^{4/4} = 2^1 = 2$ functions.

The Karnaugh-map on the right-hand side in the lower row of Fig. 7.5 shows the result $\mathcal{L}\langle h_{2q}, h_{2r}, , \text{der}_{(b,c,d)}\,h_2 \vee \text{der}_{(a,d)}\,h_2\rangle$ of two subsequently calculated vectorial derivatives of the lattice $\mathcal{L}\langle f_q, f_r\rangle$ with regard to (b, c, d) and (a, d). The simultaneous change of (a, d) reaches in the lattice $\text{der}_{(b,c,d)}\,\mathcal{L}\langle f_q, f_r\rangle$ (Karnaugh-map on the right-hand side in the middle row of Fig. 7.5) from each element ϕ another element ϕ; hence, the four elements ϕ remain unchanged in the result of $\text{der}_{(a,d)}\left(\text{der}_{(b,c,d)}\,\mathcal{L}\langle f_q, f_r\rangle\right)$. Due to Condition $\text{der}_{(b,c,d)}\,h_2 \vee \text{der}_{(a,d)}\,h_2 = 0$ this lattice is also structured into quadruples of function values. Identically colored fields in this Karnaugh-map emphasize all function values that belong to such a quadruple. Again, the four elements ϕ compose one of these quadruples; hence, all the elements ϕ must have either the value 0 or 1 so that the lattice $\text{der}_{(a,d)} \left(\text{der}_{(b,c,d)} \mathcal{L} \langle f_q, f_r \rangle \right)$ contains $2^{4/4} = 2^1 = 2$ functions.

The comparison of the Karnaugh-maps in the lower row of Fig. 7.5 shows that identical lattices have been calculated as result of two subsequent vectorial derivatives of the lattice $\mathcal{L} \langle f_q, f_r \rangle$ with regard to different but linear dependent directions of change. Six subsequently calculated vectorial derivatives of the lattice $\mathcal{L} \langle f_q, f_r \rangle$ have the same result:

$$
\underset{(b,c,d)}{\text{der}} \left(\underset{(a,b,c)}{\text{der}} \mathcal{L} \langle f_q, f_r \rangle \right) = \underset{(a,d)}{\text{der}} \left(\underset{(a,b,c)}{\text{der}} \mathcal{L} \langle f_q, f_r \rangle \right) = \underset{(a,b,c)}{\text{der}} \left(\underset{(b,c,d)}{\text{der}} \mathcal{L} \langle f_q, f_r \rangle \right)
$$

$$
= \underset{(a,d)}{\text{der}} \left(\underset{(b,c,d)}{\text{der}} \mathcal{L} \langle f_q, f_r \rangle \right) = \underset{(a,b,c)}{\text{der}} \left(\underset{(a,d)}{\text{der}} \mathcal{L} \langle f_q, f_r \rangle \right) = \underset{(b,c,d)}{\text{der}} \left(\underset{(a,d)}{\text{der}} \mathcal{L} \langle f_q, f_r \rangle \right) .
$$

A conclusion of Example 7.4 is that the result of two subsequently calculated vectorial derivatives of the lattice $\mathcal{L} \langle f_q, f_r \rangle$ is not only independent of the used directions of change, but also independent of a third direction of change that is linearly dependent on the used directions of change. A unique specification of all directions of change a lattice of logic functions does not depend on is desirable. The independence matrix satisfies this aim.

Definition 7.3 (Independence Matrix IDM(f))
The *independence matrix* IDM(f) of a logic function $f(x_1, x_2, \ldots, x_n)$ is a matrix of n rows and n columns. The columns of the independence matrix are associated with the n variables of the Boolean space in the fixed order $(x_1, x_2, \ldots, x_n)$. The independence matrix has the shape of an echelon; all elements below the main diagonal are equal to 0. Values 1 of a row of the independence matrix indicate a set of variables for which the vectorial derivative of the function $f(x_1, x_2, \ldots, x_n)$ is equal to 0. The following rules ensure the uniqueness of the independence matrix:

1. Values 1 can only occur to the right of a value 1 in the main diagonal of the independence matrix;
2. All values above a value 1 in the main diagonal of the independence matrix are equal to 0.

The number of independent directions of change of a lattice indicates its simplicity. The independence matrix implicitly comprises this information, and the rank indicates it explicitly.

Definition 7.4 (Rank of an Independence Matrix)
The *rank* of an independence matrix IDM(f) describes the number of linearly independent directions of change of the logic function $f(x_1, x_2, \ldots, x_n)$. The **rank**(IDM(f)) is equal to the number of elements 1 in the main diagonal of the unique echelon shape of IDM(f).

The independence matrix of a lattice describes the directions of change this lattice is independent of. These directions of change can be expressed by a disjunction of vectorial derivatives that is equal to 0. For a short notation of all functions of a lattice we define the independence function $f^{id}(\mathbf{x})$.

Algorithm 1 $s_{min} = \text{MIDC}(\text{IDM}(f), \mathbf{x}_0)$: minimal independent direction of change

---

Input : $\mathbf{x}_0 \subseteq \mathbf{x}$: evaluated subset of variables
Input : $\text{IDM}(f)$: unique independence matrix of n rows and n columns of $f(\mathbf{x})$
Output : s_{min}: minimal direction of change

1: $j \leftarrow 1$
2: $s_{min} \leftarrow \text{BV}(\mathbf{x}_0)$
3: **while** $j \leq n$ **do**
4:     **if** $(s_{min}[j] = 1) \wedge (\text{IDM}(f)[j, j] = 1)$ **then**
5:         $s_{min} \leftarrow s_{min} \oplus \text{IDM}(f)[j]$
6:     **end if**
7:     $j \leftarrow j + 1$
8: **end while**

Definition 7.5 (Independence Function) The *independence function* $f^{id}(\mathbf{x})$ of a logic function $f(\mathbf{x})$ corresponds to the independence matrix $\text{IDM}(f)$ such that

$$f^{id}(\mathbf{x}) = \bigvee_{i=1}^{n} \underset{\mathbf{x}_{0i}}{\text{der}} f(\mathbf{x}) , \tag{7.9}$$

where

$$\underset{\mathbf{x}_{0i}}{\text{der}} f(\mathbf{x}) = 0$$

if all elements of the row i in $\text{IDM}(f)$ are equal to 0, and

$$x_j \in \mathbf{x}_{0i} \text{ if } \text{IDM}(f)[i, j] = 1 .$$

The independence function $f^{id}(\mathbf{x})$ facilitates the definition of more general lattices of logic functions $f_i(\mathbf{x}) \in \mathcal{L}\langle f_q(\mathbf{x}), f_r(\mathbf{x}), f^{id}(\mathbf{x})\rangle$:

$$f_q(\mathbf{x}) \wedge \overline{f_i(\mathbf{x})} \vee f_i(\mathbf{x}) \wedge f_r(\mathbf{x}) \vee f^{id}(\mathbf{x}) = 0 . \tag{7.10}$$

The independence matrix of a lattice must be adjusted after each vectorial derivative operation of the associated lattice. Initially, the independence matrix is a zero matrix. The binary vector $s_0 = \text{BV}(\mathbf{x}_0)$ of the first direction of change $\mathbf{x}_0$ can be stored into the fitting row of the independence matrix.

We know from Example 7.4 that after a subsequently calculated vectorial derivative operation of the lattice with regard to another direction the resulting lattice is independent of two additional directions of change. Our concept to reach the required uniqueness of the independence matrix is that we generate for both directions the binary vectors s_{01} and s_{02} and choose the vector s_{min} with the minimal decimal equivalent.

Algorithm 1 calculates this minimal direction of change based on the set $\mathbf{x}_0$ of the next vectorial derivative operation using the context of the given independence matrix.

The function $\text{BV}(\mathbf{x}_0)$ in line 2 maps the variables of the set $\mathbf{x}_0$ into a *binary vector* based on Definition 7.2. The initial vector s_{min} is modified in line 5 if both the bit $s_{min}[j]$ and the bit in the main diagonal $\text{IDM}(f)[j, j]$ are equal to 1. The result of Algorithm 1 is a uniquely specified vector s_{min}

that must be included into the independence matrix IDM(f) such that the most significant bit belongs to the main diagonal. A result vector $\mathbf{s}_{min} = \mathbf{0}$ of Algorithm 1 for the direction of change $\mathbf{x}_0$ indicates that all functions $f(\mathbf{x})$ belonging to the lattice of the used IDM(f) satisfy

$$\operatorname*{der}_{\mathbf{x}_0} f(\mathbf{x}_0, \mathbf{x}_1) = 0 . \tag{7.11}$$

Algorithm 2 IDM(g) = UM(IDM(f), $\mathbf{x}_0$): unique merge

---

Input : $\mathbf{x}_0 \subseteq \mathbf{x}$: subset of variables that satisfy (7.11) to merge with IDM(f)
Input : IDM(f): unique independence matrix of n rows and n columns of $f(\mathbf{x})$
Output : IDM(g): unique independence matrix of the same size of $g(\mathbf{x})$ with der$_{\mathbf{x}_0}$ $g(\mathbf{x}) = 0$

1: IDM(g) $\leftarrow$ IDM(f)
2: $\mathbf{s}_{min} \leftarrow$ MIDC(IDM(f), $\mathbf{x}_0$)
3: **if** $\mathbf{s}_{min} > \mathbf{0}$ **then**
4:     $j \leftarrow$ IndexOfMostSignificantBit($\mathbf{s}_{min}$)
5:     $i \leftarrow 1$
6:     **while** $i < j$ **do**
7:         **if** IDM(g)[i, j] = 1 **then**
8:             IDM(g)[i] $\leftarrow$ IDM(g)[i] $\oplus \mathbf{s}_{min}$
9:         **end if**
10:         $i \leftarrow i + 1$
11:     **end while**
12:     IDM(g)[j] $\leftarrow \mathbf{s}_{min}$
13: **end if**

Algorithm 2 realizes the *unique merge* of a given direction of change specified by $\mathbf{x}_0$ into the independence matrix IDM(f).

Initial steps of Algorithm 2 copy IDM(f) to IDM(g) and calculate the unique vector $\mathbf{s}_{min}$ for the given set of variables $\mathbf{x}_0$ using Algorithm 1. The copied independence matrix IDM(g) must be changed only if $\mathbf{s}_{min} > \mathbf{0}$. The index of the most significant bit j of $\mathbf{s}_{min}$ indicates within IDM(g) both the column which must be evaluated and the row where $\mathbf{s}_{min}$ must be stored. All rows of IDM(g) must be equal to 0 in the column j except in the main diagonal. The operations within the while-loop in lines 6–11 perform the needed changes by conditional $\oplus$-operations. The new vector $\mathbf{s}_{min}$ is included into the independence matrix IDM(g) in line 12.

Example 7.5 We explore the construction of the independence matrix for the two subsequent vectorial derivatives of the lattice of Example 7.4. Figure 7.6 shows the minimal direction of change $\mathbf{s}_{min}$ calculated by Algorithm 1 (MIDC) and the independence matrix IDM(f) after the unique merge of this binary vector using Algorithm 2 (UM) for three different consecutive vectorial derivatives of the lattice. The given lattice depends on all direction of change; hence, the basic independence matrix IDM(f) is a zero matrix in each of the three explored cases.

$i \backslash j$	1	2	3	4
1	0	0	0	0
2	0	0	0	0
3	0	0	0	0
4	0	0	0	0

$$s_0 = (1\ 1\ 1\ 0)$$
$$s_{min} = (1\ 1\ 1\ 0)$$

$i \backslash j$	1	2	3	4
1	0	0	0	0
2	0	0	0	0
3	0	0	0	0
4	0	0	0	0

$$s_0 = (0\ 1\ 1\ 1)$$
$$s_{min} = (0\ 1\ 1\ 1)$$

$i \backslash j$	1	2	3	4
1	0	0	0	0
2	0	0	0	0
3	0	0	0	0
4	0	0	0	0

$$s_0 = (0\ 1\ 1\ 1)$$
$$s_{min} = (0\ 1\ 1\ 1)$$

$i \backslash j$	1	2	3	4
1	1	1	1	0
2	0	0	0	0
3	0	0	0	0
4	0	0	0	0

$$s_0 = (0\ 1\ 1\ 1)$$
$$s_{min} = (0\ 1\ 1\ 1)$$

$i \backslash j$	1	2	3	4
1	0	0	0	0
2	0	1	1	1
3	0	0	0	0
4	0	0	0	0

$$s_0 = (1\ 0\ 0\ 1)$$
$$s_{min} = (1\ 0\ 0\ 1)$$

$i \backslash j$	1	2	3	4
1	0	0	0	0
2	0	1	1	1
3	0	0	0	0
4	0	0	0	0

$$s_0 = (1\ 1\ 1\ 0)$$
$$s_{min} = (1\ 0\ 0\ 1)$$

a

$i \backslash j$	1	2	3	4
1	1	0	0	1
2	0	1	1	1
3	0	0	0	0
4	0	0	0	0

b

$i \backslash j$	1	2	3	4
1	1	0	0	1
2	0	1	1	1
3	0	0	0	0
4	0	0	0	0

c

$i \backslash j$	1	2	3	4
1	1	0	0	1
2	0	1	1	1
3	0	0	0	0
4	0	0	0	0

Fig. 7.6 Adding several directions of change to the independence matrix IDM(f)

Figure 7.6a shows the independence matrix IDM(f) for all steps of the calculation of

$$\operatorname*{der}_{(b,c,d)} \left(\operatorname*{der}_{(a,b,c)} \mathcal{L} \langle f_q, f_r \rangle \right).$$

The resulting lattices are already shown in the Karnaugh-maps on the left-hand side of Fig. 7.5. The binary vector of the first direction of change (a, b, c) is $s_0 = (1110)$. Due to IDM(f) = 0 Algorithm 1 assigns this vector to $s_{min} = (1110)$, and Algorithm 2 stores s_{min} into the first row of the independence matrix. The binary vector of the second direction of change (b, c, d) is $s_0 = (0111)$. Algorithm 1 determines this vector unchanged as s_{min}; however, Algorithm 2 changes the first row of the independence matrix and stores $s_{min} = (0111)$ into the second row. Hence, the unique pair of directions of change expressed by binary vectors is (1001), (1110).

Figure 7.6b shows the independence matrix IDM(f) for all steps of the calculation:

$$\operatorname*{der}_{(a,d)} \left(\operatorname*{der}_{(b,c,d)} \mathcal{L} \langle f_q, f_r \rangle \right).$$

The resulting lattices are already shown in the Karnaugh-maps on the right-hand side of Fig. 7.5. The binary vector of the first direction of change (b, c, d) is $s_0 = (0111)$ is confirmed by Algorithm 1 as $s_{min} = (0111)$ due to IDM(f) = 0. Algorithm 2 stores this binary vector into the second row due to the leftmost 1-bit in the second position. The binary vector of the second direction of change (a, d) is $s_0 = (1001)$. Algorithm 1 also confirms this vector as $s_{min} = (0111)$, and Algorithm 2 stores it into the first row of the independence matrix without other changes. The same unique independence matrix as in Fig. 7.6a has been generated.

Figure 7.6c shows the independence matrix IDM(f) for all steps of the calculation:

$$\operatorname*{der}_{(a,b,c)} \left(\operatorname*{der}_{(b,c,d)} \mathcal{L} \langle f_q, f_r \rangle \right) .$$

The subsequent vectorial derivatives with regard to these two directions of change result in the same lattice as shown in Example 7.4, but were not explored there in detail. We add this sequence of directions of change to demonstrate the effect of Algorithm 1. The binary vector of the first direction of change (b, c, d) is $\mathbf{s}_0 = (0111)$; it is assigned to $\mathbf{s}_{min} = (0111)$ without changes, and it is stored into the second row of independence matrix as in the case of Fig. 7.6b. The binary vector of the second direction of change (a, b, c) is $\mathbf{s}_0 = (1110)$. Algorithm 1 detects in line 4 that this is not the minimal direction of change and calculates in line 5 $\mathbf{s}_{min} = (1110) \oplus (0111) = (1001)$ where the vector (0111) is taken from the second row of the independence matrix. Algorithm 2 stores $\mathbf{s}_{min} = (1001)$ without other changes into the first row of the independence matrix so that the same unique independence matrix as in Fig. 7.6a and b has been generated.

7.3 Derivative Operations of Lattices of Logic Functions

Using the basic knowledge explored in Sect. 7.2, the mark-functions $g_q(\mathbf{x})$, $g_r(\mathbf{x})$, and $g^{id}(\mathbf{x})$ of the vectorial derivative operations of a generalized lattice $\mathcal{L} \langle f_q(\mathbf{x}), f_r(\mathbf{x}), f^{id}(\mathbf{x}) \rangle$ can be calculated using Theorem 7.1:

Theorem 7.1 (Vectorial Derivative Operations of a Lattice of Logic Functions) *Let $f(\mathbf{x}) = f(\mathbf{x}_0, \mathbf{x}_1) = f(x_1, x_2, \ldots, x_n)$ be a logic function of n variables that belongs to the lattice $\mathcal{L} \langle f_q(\mathbf{x}), f_r(\mathbf{x}), f^{id}(\mathbf{x}) \rangle$ defined by Eq. (7.10) where $f_q(\mathbf{x}) = f_q(\mathbf{x}_0, \mathbf{x}_1)$ and $f_r(\mathbf{x}) = f_r(\mathbf{x}_0, \mathbf{x}_1)$ satisfy (7.1), and $f(\mathbf{x})$ depends on the simultaneous change of the values of all variables of $\mathbf{x}_0$:*

$$\mathrm{MIDC}(\mathrm{IDM}(f), \mathbf{x}_0) > 0 . \tag{7.12}$$

*Then all **vectorial derivatives** of $f(\mathbf{x})$ with regard to $\mathbf{x}_0$*

$$g_1(\mathbf{x}) = \operatorname*{der}_{\mathbf{x}_0} f(\mathbf{x}_0, \mathbf{x}_1)$$

belong to a lattice of logic functions defined by

$$f_q^{\mathrm{der}_{\mathbf{x}_0}}(\mathbf{x}_0, \mathbf{x}_1) \wedge \overline{g_1(\mathbf{x})} \vee g_1(\mathbf{x}) \wedge f_r^{\mathrm{der}_{\mathbf{x}_0}}(\mathbf{x}_0, \mathbf{x}_1) \vee g_1^{id}(\mathbf{x}) = 0$$

with the mark-functions of the vectorial derivative of the lattice with regard to $\mathbf{x}_0$

$$f_q^{\mathrm{der}_{\mathbf{x}_0}}(\mathbf{x}_0, \mathbf{x}_1) = \max_{\mathbf{x}_0} f_q(\mathbf{x}_0, \mathbf{x}_1) \wedge \max_{\mathbf{x}_0} f_r(\mathbf{x}_0, \mathbf{x}_1) , \tag{7.13}$$

$$f_r^{\mathrm{der}_{\mathbf{x}_0}}(\mathbf{x}_0, \mathbf{x}_1) = \min_{\mathbf{x}_0} f_q(\mathbf{x}_0, \mathbf{x}_1) \vee \min_{\mathbf{x}_0} f_r(\mathbf{x}_0, \mathbf{x}_1) , \tag{7.14}$$

and the independence function $g_1^{id}(\mathbf{x})$ associated to

$$\mathrm{IDM}(g_1) = \mathrm{UM}(\mathrm{IDM}(f), \mathbf{x}_0) ;$$

*all **vectorial minima** of $f(\mathbf{x})$ with regard to $\mathbf{x}_0$*

$$g_2(\mathbf{x}) = \min_{\mathbf{x}_0} f(\mathbf{x})$$

belong to a lattice of logic functions defined by

$$f_q^{\min_{\mathbf{x}_0}}(\mathbf{x}_0, \mathbf{x}_1) \wedge \overline{g_2(\mathbf{x})} \vee g_2(\mathbf{x}) \wedge f_r^{\min_{\mathbf{x}_0}}(\mathbf{x}_0, \mathbf{x}_1) \vee g_2^{id}(\mathbf{x}) = 0$$

with the mark-functions of the vectorial minimum of the lattice with regard to $\mathbf{x}_0$

$$f_q^{\min_{\mathbf{x}_0}}(\mathbf{x}_0, \mathbf{x}_1) = \min_{\mathbf{x}_0} f_q(\mathbf{x}_0, \mathbf{x}_1) \,, \tag{7.15}$$

$$f_r^{\min_{\mathbf{x}_0}}(\mathbf{x}_0, \mathbf{x}_1) = \max_{\mathbf{x}_0} f_r(\mathbf{x}_0, \mathbf{x}_1) \,, \tag{7.16}$$

and the independence function $g_2^{id}(\mathbf{x})$ associated to

$$\mathrm{IDM}(g_2) = \mathrm{UM}(\mathrm{IDM}(f), \mathbf{x}_0) \,;$$

*and all **vectorial maxima** of $f(\mathbf{x})$ with regard to $\mathbf{x}_0$*

$$g_3(\mathbf{x}) = \max_{\mathbf{x}_0} f(\mathbf{x})$$

belong to a lattice of logic functions defined by

$$f_q^{\max_{\mathbf{x}_0}}(\mathbf{x}_0, \mathbf{x}_1) \wedge \overline{g_3(\mathbf{x})} \vee g_3(\mathbf{x}) \wedge f_r^{\max_{\mathbf{x}_0}}(\mathbf{x}_0, \mathbf{x}_1) \vee g_3^{id}(\mathbf{x}) = 0$$

with the mark-functions of the vectorial maximum of the lattice with regard to $\mathbf{x}_0$

$$f_q^{\max_{\mathbf{x}_0}}(\mathbf{x}_0, \mathbf{x}_1) = \max_{\mathbf{x}_0} f_q(\mathbf{x}_0, \mathbf{x}_1) \,, \tag{7.17}$$

$$f_r^{\max_{\mathbf{x}_0}}(\mathbf{x}_0, \mathbf{x}_1) = \min_{\mathbf{x}_0} f_r(\mathbf{x}_0, \mathbf{x}_1) \,, \tag{7.18}$$

and the independence function $g_3^{id}(\mathbf{x})$ associated to

$$\mathrm{IDM}(g_3) = \mathrm{UM}(\mathrm{IDM}(f), \mathbf{x}_0) \,.$$

The three independence functions are equal to each other:

$$g_1^{id}(\mathbf{x}) = g_2^{id}(\mathbf{x}) = g_3^{id}(\mathbf{x}) = g^{id}(\mathbf{x}) \,,$$

with

$$\mathrm{IDM}(g) = \mathrm{UM}(\mathrm{IDM}(f), \mathbf{x}_0)$$

and

$$\mathbf{rank}(\mathrm{IDM}(g)) = \mathbf{rank}(\mathrm{IDM}(f)) + 1 \,.$$

The formal proof of Theorem 7.1 is given in [1,4]. Here we explain in a well understandable manner the reasons and effects of these rules. The vectorial derivative operations with regard to $\mathbf{x}_0$ compare correspondent pairs of function values. We get in detail:

- The *vectorial derivative is equal to* 1 if both a function value 0 and 1 occur in such a pair of function values; hence, $f_q^{\text{der}_{\mathbf{x}_0}}(\mathbf{x}_0, \mathbf{x}_1)$ is equal to 1 if both the vectorial maximum of the ON-set-function $f_q(\mathbf{x}_0, \mathbf{x}_1)$ and the OFF-set-function $f_r(\mathbf{x}_0, \mathbf{x}_1)$ are equal to 1.
- The *vectorial derivative is equal to* 0 if the two values of such a pair are either equal to 0 or equal to 1; hence, $f_r^{\text{der}_{\mathbf{x}_0}}(\mathbf{x}_0, \mathbf{x}_1)$ is equal to 1 if the vectorial minimum of either the ON-set-function $f_q(\mathbf{x}_0, \mathbf{x}_1)$ or the OFF-set-function $f_r(\mathbf{x}_0, \mathbf{x}_1)$ are equal to 1.
- The *vectorial minimum is equal to* 1 if both function values of such a pair are equal to 1; hence, $f_q^{\text{min}_{\mathbf{x}_0}}(\mathbf{x}_0, \mathbf{x}_1)$ is equal to 1 if the vectorial minimum of the ON-set-function $f_q(\mathbf{x}_0, \mathbf{x}_1)$ is equal to 1.
- The *vectorial minimum is equal to* 0 if at least one function value of such a pair of function values is equal to 0; hence, $f_r^{\text{min}_{\mathbf{x}_0}}(\mathbf{x}_0, \mathbf{x}_1)$ is equal to 1 if the vectorial maximum of the OFF-set-function $f_r(\mathbf{x}_0, \mathbf{x}_1)$ is equal to 1.
- The *vectorial maximum is equal to* 1 if at least one function value of such a pair is equal to 1; hence, $f_q^{\text{max}_{\mathbf{x}_0}}(\mathbf{x}_0, \mathbf{x}_1)$ is equal to 1 if the vectorial maximum of the ON-set-function $f_q(\mathbf{x}_0, \mathbf{x}_1)$ is equal to 1.
- The *vectorial maximum is equal to* 0 if both function values of such a pair are equal to 0; hence, $f_r^{\text{max}_{\mathbf{x}_0}}(\mathbf{x}_0, \mathbf{x}_1)$ is equal to 1 if the vectorial minimum of the OFF-set-function $f_r(\mathbf{x}_0, \mathbf{x}_1)$ is equal to 1.

In the case that Condition (7.12) of Theorem 7.1 is not satisfied, it holds

$$s_{min} = \text{MIDC}(\text{IDM}(f), \mathbf{x}_0) = \mathbf{0} \qquad (7.19)$$

and all functions of the given lattice do not depend on the simultaneous change of $\mathbf{x}_0 \subseteq \mathbf{x}$. In this case the mark-functions of the vectorial derivative operations of the given lattice $\mathcal{L}\langle f_q(\mathbf{x}), f_r(\mathbf{x}), f^{id}(\mathbf{x})\rangle$ are

$$f_q^{\text{der}_{\mathbf{x}_0}}(\mathbf{x}) = 0, \qquad f_r^{\text{der}_{\mathbf{x}_0}}(\mathbf{x}) = 1, \qquad \text{IDM}(g_1) = I_n , \qquad (7.20)$$

$$f_q^{\text{min}_{\mathbf{x}_0}}(\mathbf{x}) = f_q(\mathbf{x}), \quad f_r^{\text{min}_{\mathbf{x}_0}}(\mathbf{x}) = f_r(\mathbf{x}), \quad \text{IDM}(g_2) = \text{IDM}(f) , \qquad (7.21)$$

$$f_q^{\text{max}_{\mathbf{x}_0}}(\mathbf{x}) = f_q(\mathbf{x}), \quad f_r^{\text{max}_{\mathbf{x}_0}}(\mathbf{x}) = f_r(\mathbf{x}), \quad \text{IDM}(g_3) = \text{IDM}(f) , \qquad (7.22)$$

where I_n is the identity matrix of the size n. From (7.20) follows that the vectorial derivatives with regard to $\mathbf{x}_0$ of all functions $f(\mathbf{x})$ of the given lattice $\mathcal{L}\langle f_q(\mathbf{x}), f_r(\mathbf{x}), f^{id}(\mathbf{x})\rangle$ which do not depend on the simultaneous change of $\mathbf{x}_0$, i.e., $s_{min} = \mathbf{0}$, are equal to the constant function $f(\mathbf{x}) = 0(\mathbf{x})$.

Example 7.6 This example demonstrates the detailed calculation of the vectorial derivative of the lattice given in Example 7.4 with regard to (a, d) using the rules (7.13) and (7.14) of Theorem 7.1. Figure 7.7 shows the Karnaugh-maps of all needed mark-functions as well as the given and the resulting lattice and the associated independence matrices.

The given lattice $\mathcal{L}\langle f_q, f_r, f^{id}\rangle$ depends on all directions of change; hence, the associated independence matrix $\text{IDM}(f)$ is a zero matrix and $f^{id}(a, b, c, d) = 0$.

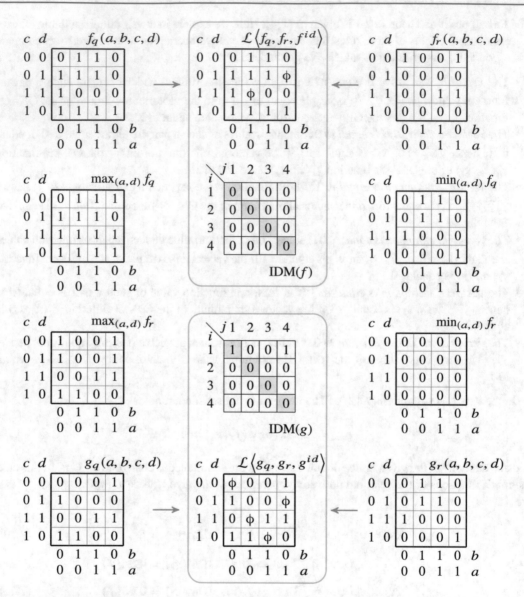

Fig. 7.7 Calculation of a vectorial derivative of a lattice of logic functions using its mark-functions

Using the mark-functions $f_q(a, b, c, d)$ and $f_r(a, b, c, d)$ of this lattice the vectorial maxima with regard to (a, d) have been calculated, and their Karnaugh-maps are shown on the left-hand side of Fig. 7.7. The conjunction of these two functions results in the mark-function $g_q(a, b, c, d)$.

Using again the mark-functions $f_q(a, b, c, d)$ and $f_r(a, b, c, d)$ of the lattice $\mathcal{L}\langle f_q, f_r, f^{id}\rangle$ the vectorial minima with regard to (a, d) have been calculated, and their Karnaugh-maps are shown on the right-hand side of Fig. 7.7. The disjunction of these two functions results in the mark-function $g_r(a, b, c, d)$.

The resulting lattice $\mathcal{L}\langle g_q, g_r, g^{id}\rangle$ of the calculated vectorial derivative with regard to (a, d) is determined by these two mark-function and the independence function

$$g^{id}(a, b, c, d) = \operatorname*{der}_{(a,c)} g(a, b, c, d)$$

belonging to the independence matrix $\text{IDM}(g)$. Due to $\mathbf{rank}(\text{IDM}(g)) = 1$ and the four don't-cares the lattice $\mathcal{L}\langle g_q, g_r, g^{id}\rangle$ contains $2^{4/2^1} = 2^2 = 4$ functions $g_i(a, b, c, d)$ which satisfy

$$g_q(a, b, c, d) \wedge \overline{g_i(a, b, c, d)} \vee g_i(a, b, c, d) \wedge g_r(a, b, c, d) \vee \operatorname*{der}_{(a,c)} g_i(a, b, c, d) = 0 .$$

The restriction of the set of variables $\mathbf{x}_0$ to the single variables x_i leads to Theorem 7.2 that specifies all derivative operations with regard to a single variable for all functions $f(\mathbf{x})$ of the given lattice $\mathcal{L}\langle f_q(\mathbf{x}), f_r(\mathbf{x}), f^{id}(\mathbf{x})\rangle$.

Theorem 7.2 (Single Derivative Operations of a Lattice of Logic Functions) *Let* $f(\mathbf{x}) = f(x_i, \mathbf{x}_1) = f(x_1, x_2, \ldots, x_n)$ *be a logic function of n variables that belongs to the lattice defined by* (7.10) *where* $f_q(\mathbf{x}) = f_q(x_i, \mathbf{x}_1)$ *and* $f_r(\mathbf{x}) = f_r(x_i, \mathbf{x}_1)$ *satisfy* (7.1), *and* $f(\mathbf{x})$ *depends on* x_i:

$$\text{MIDC}(\text{IDM}(f), x_i) > 0 . \tag{7.23}$$

*Then all **derivatives of** $f(\mathbf{x})$ **with regard to** x_i*

$$g_1(\mathbf{x}) = \operatorname*{der}_{x_i} f(x_i, \mathbf{x}_1)$$

belong to a lattice of logic functions defined by

$$f_q^{\mathrm{der}_{x_i}}(\mathbf{x}_1) \wedge \overline{g_1(\mathbf{x})} \vee g_1(\mathbf{x}) \wedge f_r^{\mathrm{der}_{x_i}}(\mathbf{x}_1) \vee g_1^{id}(\mathbf{x}) = 0$$

with the mark-functions of the derivative of the lattice with regard to x_i

$$f_q^{\mathrm{der}_{x_i}}(\mathbf{x}_1) = \max_{x_i} f_q(x_i, \mathbf{x}_1) \wedge \max_{x_i} f_r(x_i, \mathbf{x}_1) , \tag{7.24}$$

$$f_r^{\mathrm{der}_{x_i}}(\mathbf{x}_1) = \min_{x_i} f_q(x_i, \mathbf{x}_1) \vee \min_{x_i} f_r(x_i, \mathbf{x}_1) , \tag{7.25}$$

and the independence function $g_1^{id}(\mathbf{x})$ *associated to*

$$\text{IDM}(g_1) = \text{UM}(\text{IDM}(f), x_i) ;$$

*all **minima of** $f(\mathbf{x})$ **with regard to** x_i*

$$g_2(\mathbf{x}) = \min_{x_i} f(x_i, \mathbf{x}_1)$$

belong to a lattice of logic functions defined by

$$f_q^{\min_{x_i}}(\mathbf{x}_1) \wedge \overline{g_2(\mathbf{x})} \vee g_2(\mathbf{x}) \wedge f_r^{\min_{x_i}}(\mathbf{x}_1) \vee g_2^{id}(\mathbf{x}) = 0$$

with the mark-functions of the minimum of the lattice with regard to x_i

$$f_q^{\min_{x_i}}(\mathbf{x}_1) = \min_{x_i} f_q(x_i, \mathbf{x}_1) \,, \tag{7.26}$$

$$f_r^{\min_{x_i}}(\mathbf{x}_1) = \max_{x_i} f_r(x_i, \mathbf{x}_1) \,, \tag{7.27}$$

and the independence function $g_2^{id}(\mathbf{x})$ associated to

$$\mathrm{IDM}(g_2) = \mathrm{UM}(\mathrm{IDM}(f), x_i) \,;$$

and all **maxima of** $f(\mathbf{x})$ **with regard to** x_i

$$g_3(\mathbf{x}) = \max_{x_i} f(x_i, \mathbf{x}_1)$$

belong to a lattice of logic functions defined by

$$f_q^{\max_{x_i}}(\mathbf{x}_1) \wedge \overline{g_3(\mathbf{x})} \vee g_3(\mathbf{x}) \wedge f_r^{\max_{x_i}}(\mathbf{x}_1) \vee g_3^{id}(\mathbf{x}) = 0$$

with the mark-functions of the maximum of the lattice with regard to x_i

$$f_q^{\max_{x_i}}(\mathbf{x}_1) = \max_{x_i} f_q(x_i, \mathbf{x}_1) \,, \tag{7.28}$$

$$f_r^{\max_{x_i}}(\mathbf{x}_1) = \min_{x_i} f_r(x_i, \mathbf{x}_1) \,, \tag{7.29}$$

and the independence function $g_3^{id}(\mathbf{x})$ associated to

$$\mathrm{IDM}(g_3) = \mathrm{UM}(\mathrm{IDM}(f), x_i) \,.$$

The three independence functions are equal to each other:

$$g_1^{id}(\mathbf{x}) = g_2^{id}(\mathbf{x}) = g_3^{id}(\mathbf{x}) = g^{id}(\mathbf{x}) \,,$$

with

$$\mathrm{IDM}(g) = \mathrm{UM}(\mathrm{IDM}(f), x_i)$$

and

$$\mathbf{rank}(\mathrm{IDM}(g)) = \mathbf{rank}(\mathrm{IDM}(f)) + 1 \,.$$

The derivative operations with regard to a single variable compare pairs of function values which differ in the values of the selected variable x_i. Hence, the rules to calculate the mark-functions of single derivative operations with regard to x_i of a lattice of logic functions $f_i(\mathbf{x}) \in \mathcal{L}\langle f_q(\mathbf{x}), f_r(\mathbf{x}), f^{id}(\mathbf{x})\rangle$ use instead of the vectorial derivative operations with regard to $\mathbf{x}_0$ the associated derivative operations with regard to x_i.

In the case that Condition (7.23) of Theorem 7.2 is not satisfied, it holds that

$$\mathbf{s}_{min} = \mathrm{MIDC}(\mathrm{IDM}(f), x_i) = \mathbf{0} \tag{7.30}$$

and all functions of the given lattice do not depend on the change of x_i. In this case the mark-functions of the derivative operations with regard to x_i of the given lattice $\mathcal{L}\langle f_q(\mathbf{x}), f_r(\mathbf{x}), f^{id}(\mathbf{x})\rangle$ are

$$f_q^{\mathrm{der}_{x_i}}(\mathbf{x}_1) = 0 \,, \qquad f_r^{\mathrm{der}_{x_i}}(\mathbf{x}_1) = 1 \,, \qquad \mathrm{IDM}(g_1) = I_n \,, \tag{7.31}$$

$$f_q^{\min_{x_i}}(\mathbf{x}_1) = f_q(\mathbf{x}_1) \,, \quad f_r^{\min_{x_i}}(\mathbf{x}_1) = f_r(\mathbf{x}_1) \,, \quad \mathrm{IDM}(g_2) = \mathrm{IDM}(f) \,, \tag{7.32}$$

$$f_q^{\max_{x_i}}(\mathbf{x}_1) = f_q(\mathbf{x}_1) \,, \quad f_r^{\max_{x_i}}(\mathbf{x}_1) = f_r(\mathbf{x}_1) \,, \quad \mathrm{IDM}(g_3) = \mathrm{IDM}(f) \,, \tag{7.33}$$

where I_n is the identity matrix of the size n. From (7.31) follows that the derivative with regard to x_i of all functions $f(\mathbf{x})$ of the given lattice $\mathcal{L}\langle f_q(\mathbf{x}), f_r(\mathbf{x}), f^{id}(\mathbf{x})\rangle$ which do not depend on the change of x_i, i.e., $\mathbf{s}_{min} = \mathbf{0}$, is equal to the constant function $f(\mathbf{x}) = 0(\mathbf{x})$.

Repeated derivative operations of the same type with regard to different variables are summarized to k-fold derivative operations. Due to Theorem 7.2 each k-fold derivative operation of a given lattice $\mathcal{L}\langle f_q(\mathbf{x}), f_r(\mathbf{x}), f^{id}(\mathbf{x})\rangle$ results again in a lattice of logic functions.

Theorem 7.3 (*k-fold Derivative Operations of a Lattice of Logic Functions*) *Let $\mathbf{x}_0 = (x_1, x_2, \ldots, x_k)$, $\mathbf{x}_1 = (x_{k+1}, \ldots, x_n)$ be two disjoint sets of variables, and $f(\mathbf{x}) = f(\mathbf{x}_0, \mathbf{x}_1) = f(x_1, x_2, \ldots, x_n)$ be a logic function of n variables that belong to the lattice defined by (7.10) where $f_q(\mathbf{x}) = f_q(\mathbf{x}_0, \mathbf{x}_1)$ and $f_r(\mathbf{x}) = f_r(\mathbf{x}_0, \mathbf{x}_1)$ satisfy (7.1), and $f(\mathbf{x})$ depends at least on one $x_{0i} \in \mathbf{x}_0$:*

$$\exists x_{0i} \in \mathbf{x}_0 : \qquad \mathrm{MIDC}(\mathrm{IDM}(f), x_{0i}) > 0 \,. \tag{7.34}$$

*Then all k-**fold derivatives** of $f(\mathbf{x}) = f(\mathbf{x}_0, \mathbf{x}_1)$ with regard to $\mathbf{x}_0$*

$$g_1(\mathbf{x}_1) = \mathop{\mathrm{der}}_{\mathbf{x}_0}^k f(\mathbf{x}_0, \mathbf{x}_1)$$

belong to a lattice of logic functions defined by

$$f_q^{\mathrm{der}_{\mathbf{x}_0}^k}(\mathbf{x}_1) \wedge \overline{g_1(\mathbf{x}_1)} \vee g_1(\mathbf{x}_1) \wedge f_r^{\mathrm{der}_{\mathbf{x}_0}^k}(\mathbf{x}_1) \vee g_1^{id}(\mathbf{x}) = 0$$

with the mark-functions of the k-fold derivatives with regard to $\mathbf{x}_0$

$$f_q^{\mathrm{der}_{\mathbf{x}_0}^k}(\mathbf{x}_1) = \mathop{\mathrm{der}}_{\mathbf{x}_0}^k f_q(\mathbf{x}_0, \mathbf{x}_1) \wedge \mathop{\min}_{\mathbf{x}_0}^k (f_q(\mathbf{x}_0, \mathbf{x}_1) \vee f_r(\mathbf{x}_0, \mathbf{x}_1)) \,, \tag{7.35}$$

$$f_r^{\mathrm{der}_{\mathbf{x}_0}^k}(\mathbf{x}_1) = \overline{\mathop{\mathrm{der}}_{\mathbf{x}_0}^k f_q(\mathbf{x}_0, \mathbf{x}_1)} \wedge \mathop{\min}_{\mathbf{x}_0}^k (f_q(\mathbf{x}_0, \mathbf{x}_1) \vee f_r(\mathbf{x}_0, \mathbf{x}_1)) \,, \tag{7.36}$$

and the independence function $g_1^{id}(\mathbf{x})$ associated to $\mathrm{IDM}(g_1)$ satisfies

$$\forall x_{0i} \in \mathbf{x}_0 : \qquad \mathrm{MIDC}(\mathrm{IDM}(g_1), x_{0i}) = 0 \,;$$

*all k-**fold minima** of $f(\mathbf{x}) = f(\mathbf{x}_0, \mathbf{x}_1)$ with regard to $\mathbf{x}_0$*

$$g_2(\mathbf{x}_1) = \mathop{\min}_{\mathbf{x}_0}^k f(\mathbf{x}_0, \mathbf{x}_1)$$

belong to a lattice of logic functions defined by

$$f_q^{\min_{\mathbf{x}_0}^k}(\mathbf{x}_1) \wedge \overline{g_2(\mathbf{x}_1)} \vee g_2(\mathbf{x}_1) \wedge f_r^{\min_{\mathbf{x}_0}^k}(\mathbf{x}_1) \vee g_2^{id}(\mathbf{x}) = 0$$

with the mark-functions of the k-fold minima with regard to $\mathbf{x}_0$

$$f_q^{\min_{\mathbf{x}_0}^k}(\mathbf{x}_1) = \min_{\mathbf{x}_0}^k f_q(\mathbf{x}_0, \mathbf{x}_1) \,, \tag{7.37}$$

$$f_r^{\min_{\mathbf{x}_0}^k}(\mathbf{x}_1) = \max_{\mathbf{x}_0}^k f_r(\mathbf{x}_0, \mathbf{x}_1) \,, \tag{7.38}$$

and the independence function $g_2^{id}(\mathbf{x})$ *associated to* $\text{IDM}(g_2)$ *satisfies*

$$\forall x_{0i} \in \mathbf{x}_0 : \quad \text{MIDC}(\text{IDM}(g_2), x_{0i}) = 0 \,;$$

*all k-**fold maxima** of* $f(\mathbf{x}) = f(\mathbf{x}_0, \mathbf{x}_1)$ *with regard to* $\mathbf{x}_0$

$$g_3(\mathbf{x}_1) = \max_{\mathbf{x}_0}^k f(\mathbf{x}_0, \mathbf{x}_1)$$

belong to a lattice of logic functions defined by

$$f_q^{\max_{\mathbf{x}_0}^k}(\mathbf{x}_1) \wedge \overline{g_3(\mathbf{x}_1)} \vee g_3(\mathbf{x}_1) \wedge f_r^{\max_{\mathbf{x}_0}^k}(\mathbf{x}_1) \vee g_3^{id}(\mathbf{x}) = 0$$

with the mark-functions of the k-fold maxima with regard to $\mathbf{x}_0$

$$f_q^{\max_{\mathbf{x}_0}^k}(\mathbf{x}_1) = \max_{\mathbf{x}_0}^k f_q(\mathbf{x}_0, \mathbf{x}_1) \,, \tag{7.39}$$

$$f_r^{\max_{\mathbf{x}_0}^k}(\mathbf{x}_1) = \min_{\mathbf{x}_0}^k f_r(\mathbf{x}_0, \mathbf{x}_1) \,, \tag{7.40}$$

and the independence function $g_3^{id}(\mathbf{x})$ *associated to* $\text{IDM}(g_3)$ *satisfies*

$$\forall x_{0i} \in \mathbf{x}_0 : \quad \text{MIDC}(\text{IDM}(g_3), x_{0i}) = 0 \,;$$

and all **Δ*-operations*** *of* $f(\mathbf{x}) = f(\mathbf{x}_0, \mathbf{x}_1)$ *with regard to* $\mathbf{x}_0$

$$g_4(\mathbf{x}_1) = \Delta_{\mathbf{x}_0} f(\mathbf{x}_0, \mathbf{x}_1)$$

belong to a lattice of logic functions defined by

$$f_q^{\Delta_{\mathbf{x}_0}}(\mathbf{x}_1) \wedge \overline{g_4(\mathbf{x}_1)} \vee g_4(\mathbf{x}_1) \wedge f_r^{\Delta_{\mathbf{x}_0}}(\mathbf{x}_1) \vee g_4^{id}(\mathbf{x}) = 0$$

with the mark-functions of the Δ-operations with regard to $\mathbf{x}_0$

$$f_q^{\Delta \mathbf{x}_0}(\mathbf{x}_1) = \max_{\mathbf{x}_0}^k f_q(\mathbf{x}_0, \mathbf{x}_1) \wedge \max_{\mathbf{x}_0}^k f_r(\mathbf{x}_0, \mathbf{x}_1) \,, \tag{7.41}$$

$$f_r^{\Delta \mathbf{x}_0}(\mathbf{x}_1) = \min_{\mathbf{x}_0}^k f_q(\mathbf{x}_0, \mathbf{x}_1) \vee \min_{\mathbf{x}_0}^k f_r(\mathbf{x}_0, \mathbf{x}_1) \,, \tag{7.42}$$

and the independence function $g_4^{id}(\mathbf{x})$ *associated to* $\mathrm{IDM}(g_4)$ *satisfies*

$$\forall x_{0i} \in \mathbf{x}_0 : \quad \mathrm{MIDC}(\mathrm{IDM}(g_4), x_{0i}) = 0 .$$

The four independence functions are equal to each other:

$$g_1^{id}(\mathbf{x}) = g_2^{id}(\mathbf{x}) = g_3^{id}(\mathbf{x}) = g_4^{id}(\mathbf{x}) = g^{id}(\mathbf{x}) ,$$

with

$$\mathrm{IDM}(g) = \mathrm{UM}(\ldots \mathrm{UM}(\mathrm{IDM}(f), x_1), \ldots, x_k)$$

and

$$\mathbf{rank}(\mathrm{IDM}(f)) + 1 \leq \mathbf{rank}(\mathrm{IDM}(g)) \leq \mathbf{rank}(\mathrm{IDM}(f)) + k .$$

The k-fold derivative operations with regard to $\mathbf{x}_0$ evaluate subspaces of 2^k function values. The rules of Theorem 7.3 can be explained as follows:

- A unique decision whether an odd or an even number of function values occur in a subspace is only possible if all function values belong either to the ON-set or to the OFF-set of the evaluated subspace; this property is satisfied if the k-fold minimum of the disjunction of the ON-set-function $f_q(\mathbf{x}_0, \mathbf{x}_1)$ and the OFF-set-function $f_r(\mathbf{x}_0, \mathbf{x}_1)$ is equal to 1.
- The k-fold derivative is equal to 1 if an odd number of function values 1 occurs in such a subspace; hence, $f_q^{\mathrm{der}_{\mathbf{x}_0}^k}(\mathbf{x}_1)$ is equal to 1 if additionally to the condition of the first item the k-fold derivative of the ON-set-function $f_q(\mathbf{x}_0, \mathbf{x}_1)$ is equal to 1.
- The k-fold derivative is equal to 0 if an even number of function values 1 occurs in such a subspace; hence, $f_r^{\mathrm{der}_{\mathbf{x}_0}^k}(\mathbf{x}_1)$ is equal to 1 if additionally to the condition of the first item the k-fold derivative of the ON-set-function $f_q(\mathbf{x}_0, \mathbf{x}_1)$ is equal to 0.
- The k-fold minimum is equal to 1 if all function values in such a subspace are equal to 1; hence, $f_q^{\mathrm{min}_{\mathbf{x}_0}^k}(\mathbf{x}_1)$ is equal to 1 if the k-fold minimum of the ON-set-function $f_q(\mathbf{x}_0, \mathbf{x}_1)$ is equal to 1.
- The k-fold minimum is equal to 0 if at least one function value in such a subspace is equal to 0; hence, $f_r^{\mathrm{min}_{\mathbf{x}_0}^k}(\mathbf{x}_1)$ is equal to 1 if the k-fold maximum of the OFF-set-function $f_r(\mathbf{x}_0, \mathbf{x}_1)$ is equal to 1.
- The k-fold maximum is equal to 1 if at least one function value in such a subspace is equal to 1; hence, $f_q^{\mathrm{max}_{\mathbf{x}_0}^k}(\mathbf{x}_1)$ is equal to 1 if the k-fold maximum of the ON-set-function $f_q(\mathbf{x}_0, \mathbf{x}_1)$ is equal to 1.
- The k-fold maximum is equal to 0 if all function values in such a subspace are equal to 0; hence, $f_r^{\mathrm{max}_{\mathbf{x}_0}^k}(\mathbf{x}_1)$ is equal to 1 if the k-fold minimum of the OFF-set-function $f_r(\mathbf{x}_0, \mathbf{x}_1)$ is equal to 1.
- The $\wedge$-operation is equal to 1 if both at least one function value 1 and one function value 0 occur in such a subspace; hence, $f_q^{\Delta \mathbf{x}_0}(\mathbf{x}_1)$ is equal to 1 if both the k-fold maximum of the ON-set-function $f_q(\mathbf{x}_0, \mathbf{x}_1)$ and the k-fold maximum of the OFF-set-function $f_r(\mathbf{x}_0, \mathbf{x}_1)$ are equal to 1.
- The Δ-operation is equal to 0 if all function values in such a subspace are either equal to 0 or equal to 1; hence, $f_r^{\Delta \mathbf{x}_0}(\mathbf{x}_1)$ is equal to 1 if either the k-fold minimum of the ON-set-function $f_q(\mathbf{x}_0, \mathbf{x}_1)$ or the k-fold minimum of the OFF-set-function $f_r(\mathbf{x}_0, \mathbf{x}_1)$ is equal to 1.

In the case that Condition (7.34) of Theorem 7.3 is not satisfied, it holds

$$\forall x_{0i} \in \mathbf{x}_0 : \qquad \mathbf{s}_{min} = \text{MIDC}(\text{IDM}(f), x_{0i}) = \mathbf{0} \qquad (7.43)$$

and all functions of the given lattice do not depend on the change of all $x_{0i} \in \mathbf{x}_0$. In this case the mark-functions of the k-fold derivative operations with regard to $\mathbf{x}_0$ of the given lattice $\mathcal{L}\langle f_q(\mathbf{x}), f_r(\mathbf{x}), f^{id}(\mathbf{x})\rangle$ are

$$f_q^{der_{\mathbf{x}_0}^k}(\mathbf{x}_1) = 0 , \qquad f_r^{der_{\mathbf{x}_0}^k}(\mathbf{x}_1) = 1 , \qquad \text{IDM}(g_1) = I_n , \qquad (7.44)$$

$$f_q^{min_{\mathbf{x}_0}^k}(\mathbf{x}_1) = f_q(\mathbf{x}_1) , \quad f_r^{min_{\mathbf{x}_0}^k}(\mathbf{x}_1) = f_r(\mathbf{x}_1) , \quad \text{IDM}(g_2) = \text{IDM}(f) , \qquad (7.45)$$

$$f_q^{max_{\mathbf{x}_0}^k}(\mathbf{x}_1) = f_q(\mathbf{x}_1) , \quad f_r^{max_{\mathbf{x}_0}^k}(\mathbf{x}_1) = f_r(\mathbf{x}_1) , \quad \text{IDM}(g_3) = \text{IDM}(f) , \qquad (7.46)$$

$$f_q^{\Delta \mathbf{x}_0}(\mathbf{x}_1) = 0 , \qquad f_q^{\Delta \mathbf{x}_0}(\mathbf{x}_1) = 1 , \qquad \text{IDM}(g_4) = I_n , \qquad (7.47)$$

where I_n is the identity matrix of the size n. From (7.44) and (7.47) follows that both the k-fold derivative and the Δ-operation with regard to $\mathbf{x}_0$ of all functions $f(\mathbf{x})$ of the given lattice $\mathcal{L}\langle f_q(\mathbf{x}), f_r(\mathbf{x}), f^{id}(\mathbf{x})\rangle$ which do not depend on the change of all $x_{0i} \in \mathbf{x}_0$, i.e., $\mathbf{s}_{min} = \mathbf{0}$ for all these x_{0i}, are equal to the constant function $f(\mathbf{x}) = \mathbf{0}(\mathbf{x})$.

7.4   Solution of Equations with Regard to Variables

Motivation

Another important concept is the solution of logic equations *with regard to variables*. As a motivation, we go back to the equation

$$x^2 + px + q = 0 . \qquad (7.48)$$

Previously we had the understanding that p and q are constant real numbers, and for each pair of these numbers, it can be explored whether the condition

$$\frac{p^2}{4} - q \geq 0$$

is satisfied, and if so, the solutions x_1 and x_2 can be calculated. However, we also can use another point of view. We take the function of three variables

$$f(x, p, q) = x^2 + px + q$$

and ask whether Eq. (7.48) defines, for instance, a function $x(p, q)$ that satisfies the equation, or whether functions $p(x)$ and $q(x)$ can be found satisfying this equation, etc. In order to do this Eq. (7.48) can be used to find all the points (x, p, q) in a three-dimensional space satisfying the given equation. Thereafter, it can be tried to find such functions $x(p, q)$ or pairs $p(x)$, $q(x)$ satisfying the equation.

This problem exists also for logic equations and has many applications. As we have already seen, one equation can be considered as the "summary" of many equations, inequalities, etc. The whole knowledge existing for a given binary problem can be put together, and then it can be tried to find such implicitly given functions that describe the most interesting parts of the "logic model." The applications will show the working of this methodology. Here we introduce the required concepts for the logic equations.

Solution of Restrictive Equations with Regard to Variables

At the beginning let be given the logic equation $f(x_1, \ldots, x_{n+k}) = 0$ and the solution-set S. The set of variables $\{x_1, \ldots, x_{n+k}\}$ can be partitioned into two disjoint subsets $\{x_1, \ldots, x_n\}$ and $\{x_{n+1}, \ldots, x_{n+k}\}$. This partition also partitions each binary vector $\mathbf{b}$ of the solution-set into two parts $(b_1, \ldots, b_n)$ and $(b_{n+1}, \ldots, b_{n+k})$. The equation defines a mapping φ from $\mathbb{B}^n$ into $\mathbb{B}^k$. Each vector $\mathbf{x}$ has been assigned to one, more than one, or even zero vectors of $\mathbb{B}^k$, and this happens for any of the partitions of the set $\{x_1, \ldots, x_{n+k}\}$. In order to have an easy understanding of the role of different variables, we write $f(x_1, \ldots, x_n, y_1, \ldots, y_k)$, but we keep in mind that any partition of $\{x_1, \ldots, x_{n+k}\}$ can be considered. Very often the role of the different variables will be defined by the context of the problem. It can also be said that the equation defines a given relation $R(\mathbf{x}, \mathbf{y})$ between the vectors of $\mathbf{x} \in \mathbb{B}^n$ and $\mathbf{y} \in \mathbb{B}^k$.

Especially interesting are now the cases where the mapping φ is unique and where φ is complete. In the first case every $\mathbf{x} \in \mathbb{B}^n$ has been assigned at most to one element of $\mathbb{B}^k$; however, some $\mathbf{y}$ might not have an assigned element. In the second case each $\mathbf{x} \in \mathbb{B}^n$ will have an assigned element of $\mathbb{B}^k$; however, this is not necessarily defined in a unique way, more than one element is quite possible. These two cases will be considered for the solution with regard to variables.

Definition 7.6 Let $\mathbf{x} = (x_1, \ldots, x_n)$, $\mathbf{y} = (y_1, \ldots, y_k)$, $f(\mathbf{x}, \mathbf{y})$ a logic function of $n + k$ variables. The restrictive equation $f(\mathbf{x}, \mathbf{y}) = 0$ can be solved with regard to the variables $y_1, \ldots, y_k$ if there are functions

$$y_1 = g_1(\mathbf{x})$$
$$y_2 = g_2(\mathbf{x})$$
$$\vdots \quad \vdots$$
$$y_k = g_k(\mathbf{x})$$

that satisfy

$$f(\mathbf{x}, g_1(\mathbf{x}), \ldots, g_k(\mathbf{x})) = 0 \,.$$

The following three examples show different cases that can appear when a logic equation has to be solved with regard to variables.

Example 7.7 Let $f(\mathbf{x}, \mathbf{y}) = \overline{x}_1 y_1 \vee x_1 \overline{y}_1 \vee x_2 y_1 \overline{y}_2 = 0$. Table 7.1 shows the solutions of this equation and the different pairs of functions $(y_1(x_1, x_2), y_2(x_1, x_2))$ that solve this equation.

The mapping from $\mathbb{B}^2$ into $\mathbb{B}^2$ is complete, the domain of the mapping is the whole $\mathbb{B}^2$, but it is not unique, and three of the four $\mathbf{x}$-vectors have two elements assigned to them. The grouping of the solutions follows the values of the $\mathbf{x}$-vectors. Hence, in order to define the functions $g_1(\mathbf{x})$, $g_2(\mathbf{x})$, for $\mathbf{x} = (00)$ two vectors for (y_1, y_2) can be selected, the same occurs for $\mathbf{x} = (01)$ and $\mathbf{x} = (10)$; only

Table 7.1 The solution with regard to variables

					$(g_1(x_1, x_2),\ g_2(x_1, x_2))$							
x_1	x_2	y_1	y_2	$f(\mathbf{x}, \mathbf{y})$	1	2	3	4	5	6	7	8
0	0	0	0	0	00	00	00	00				
0	0	0	1	0					01	01	01	01
0	1	0	0	0	00	00			00	00		
0	1	0	1	0			01	01			01	01
1	0	1	0	0	10		10		10		10	
1	0	1	1	0		11		11		11		11
1	1	1	1	0	11	11	11	11	11	11	11	11

Table 7.2 A unique solution with regard to variables

x_1	x_2	y_1	y_2	$f(\mathbf{x}, \mathbf{y})$	$g_1(\mathbf{x})$	$g_2(\mathbf{x})$
0	0	0	0	0	0	0
0	1	1	1	0	1	1
1	0	1	1	0	1	1
1	1	0	1	0	0	1

the value for $\mathbf{x} = (11)$ is defined in a unique way. The right-hand side of Table 7.1 shows all eight possible combinations of pairs of functions $(g_1(\mathbf{x}), g_2(\mathbf{x}))$.

The left component of these eight columns uniquely specifies the logic function $g_1(\mathbf{x}) = x_1$ by the vector of function values (0011). For $g_2(\mathbf{x})$ we find

$$g_2(\mathbf{x}) = x_1 x_2\,, \qquad g_2(\mathbf{x}) = x_1\,, \qquad g_2(\mathbf{x}) = x_2\,, \qquad g_2(\mathbf{x}) = x_1 \vee x_2\,,$$

$$g_2(\mathbf{x}) = x_1 \odot x_2\,, \quad g_2(\mathbf{x}) = x_1 \vee \overline{x}_2\,, \quad g_2(\mathbf{x}) = \overline{x}_1 \vee x_2\,, \quad g_2(\mathbf{x}) = 1\,,$$

defined by the second component in the eight columns. Let us use, for instance, $y_1 = g_1(\mathbf{x}) = x_1$, $y_2 = g_2(\mathbf{x}) = x_1 \vee \overline{x}_2$, then we get

$$f(\mathbf{x}, \mathbf{y}) = \overline{x}_1 y_1 \vee x_1 \overline{y}_1 \vee x_2 y_1 \overline{y}_2$$

$$= \overline{x}_1 x_1 \vee x_1 \overline{x}_1 \vee x_2 x_1 \overline{x}_1 x_2$$

$$= 0 \vee 0 \vee 0$$

$$= 0\,.$$

As solution of the equation explored in Example 7.7 a single uniquely specified function $y_1 = g_1(\mathbf{x})$ can be combined with one of eight functions $y_2 = g_2(\mathbf{x})$. In the next example both the function $y_1 = g_1(\mathbf{x})$ and $y_2 = g_2(\mathbf{x})$ are uniquely specified by the given equation.

Example 7.8 The logic equation

$$f(\mathbf{x}, \mathbf{y}) = y_1(x_1 x_2 \vee \overline{x}_1 \overline{x}_2) \vee \overline{y}_1(x_1 \overline{x}_2 \vee \overline{x}_1 x_2) \vee y_2 \overline{(x_1 \vee x_2)} \vee \overline{y}_2(x_1 \vee x_2) = 0 \qquad (7.49)$$

has the solution-vectors shown in Table 7.2. The mapping φ is complete and also unique. The pair $y_1 = g_1(\mathbf{x}) = x_1 \oplus x_2$ and $y_2 = g_2(\mathbf{x}) = x_1 \vee x_2$ is the only pair of solution-functions.

Table 7.3 A reversible system of solutions with regard to variables

x_1	x_2	y_1	y_2	$f'(\mathbf{x}, \mathbf{y})$	$g_1(\mathbf{x})$	$g_2(\mathbf{x})$	$h_1(\mathbf{y})$	$h_2(\mathbf{y})$
0	0	0	0	0	0	0	0	0
0	1	1	1	0	1	1	0	1
1	0	1	0	0	1	0	1	0
1	1	0	1	0	0	1	1	1

Now we change to roles of the variables and try to solve Eq. (7.49) with regard to (x_1, x_2), i.e., we are searching functions

$$x_1 = h_1(\mathbf{y}) \,,$$

$$x_2 = h_2(\mathbf{y}) \,.$$

We check Table 7.2 and see that the vector $(y_1, y_2) = (1, 1)$ appears twice, the vector $(y_1, y_2) = (1, 0)$ is missing. Hence, Eq. (7.49) does not specify the needed function values of $x_1 = h_1(\mathbf{y})$ and $x_2 = h_2(\mathbf{y})$ for $(y_1, y_2) = (1, 0)$ so that no solution-functions $h_1(\mathbf{y})$ and $h_2(\mathbf{y})$ exist.

Example 7.9 In order to show that there are equations $f(\mathbf{x}, \mathbf{y}) = 0$ which can be solved with regard to both $\mathbf{y}$ and $\mathbf{x}$ we change Table 7.2 a bit in the third row and get Table 7.3:

In order to find an expression for the function $f'(x_1, x_2, y_1, y_2)$ of Table 7.3 (the function $f'(\mathbf{x}, \mathbf{y})$ is equal to 1 for all patterns $(\mathbf{x}, \mathbf{y})$ not listed in Table 7.3), we use, for instance, the conjunctive normal form:

$$f'(x_1, x_2, y_1, y_2) = (x_1 \lor x_2 \lor y_1 \lor y_2)\,(x_1 \lor \overline{x}_2 \lor \overline{y}_1 \lor \overline{y}_2)$$

$$(\overline{x}_1 \lor x_2 \lor \overline{y}_1 \lor y_2)\,(\overline{x}_1 \lor \overline{x}_2 \lor y_1 \lor \overline{y}_2) \,.$$

Now Table 7.3 can be used in two directions to solve the equation

$$f'(x_1, x_2, y_1, y_2) = 0$$

with regard to both $\mathbf{y}$ and $\mathbf{x}$. We get for the solution with regard to $\mathbf{y}$:

$$y_1 = g_1(\mathbf{x}) = x_1 \oplus x_2 \,,$$

$$y_2 = g_2(\mathbf{x}) = x_2 \,.$$

The values of x_1 and x_2 can be used to calculate the respective values of y_1 and y_2. The substitution of the function $g_1(\mathbf{x})$ and $g_2(\mathbf{x})$ into the equation $f'(x_1, x_2, y_1, y_2) = 0$ confirms this solution:

$$(x_1 \lor x_2 \lor y_1 \lor y_2)\,(x_1 \lor \overline{x}_2 \lor \overline{y}_1 \lor \overline{y}_2)$$

$$\wedge(\overline{x}_1 \lor x_2 \lor \overline{y}_1 \lor y_2)\,(\overline{x}_1 \lor \overline{x}_2 \lor y_1 \lor \overline{y}_2) = 0$$

$$(x_1 \lor x_2 \lor x_1\overline{x}_2 \lor \overline{x}_1 x_2 \lor x_2)\,(x_1 \lor \overline{x}_2 \lor \overline{x}_1\overline{x}_2 \lor x_1 x_2 \lor \overline{x}_2)$$

$$\wedge(\overline{x}_1 \lor x_2 \lor \overline{x}_1\overline{x}_2 \lor x_1 x_2 \lor x_2)\,(\overline{x}_1 \lor \overline{x}_2 \lor x_1\overline{x}_2 \lor \overline{x}_1 x_2 \lor \overline{x}_2) = 0$$

$$(x_1 \lor x_2)\,(x_1 \lor \overline{x}_2)\,(\overline{x}_1 \lor x_2)\,(\overline{x}_1 \lor \overline{x}_2) = 0$$

$$x_1(x_2 \lor \overline{x}_2) \wedge \overline{x}_1(x_2 \lor \overline{x}_2) = 0$$

$$x_1 \wedge \overline{x}_1 = 0$$

$$0 = 0 \,.$$

Table 7.4 Solution of
Eq. (7.50) with regard to
variables under constraints

x_1	x_2	y_1	y_2
0	0	0	0
0	1	1	0
1	0	1	1

As solution of $f'(x_1, x_2, y_1, y_2) = 0$ with regard to **y** we get

$$x_1 = h_1(\mathbf{y}) = y_1 \oplus y_2 \, ,$$

$$x_2 = h_2(\mathbf{y}) = y_2 \, .$$

Here, the values of y_1 and y_2 are used to calculate the values of x_1 and x_2 and the substitution of the function $h_1(\mathbf{y})$ and $h_2(\mathbf{y})$ into the equation $f'(x_1, x_2, y_1, y_2) = 0$ also confirms this solution:

$$(x_1 \vee x_2 \vee y_1 \vee y_2) \, (x_1 \vee \overline{x}_2 \vee \overline{y}_1 \vee \overline{y}_2)$$

$$\wedge (\overline{x}_1 \vee x_2 \vee \overline{y}_1 \vee y_2) \, (\overline{x}_1 \vee \overline{x}_2 \vee y_1 \vee \overline{y}_2) = 0$$

$$(y_1 \overline{y}_2 \vee \overline{y}_1 y_2 \vee y_2 \vee y_1 \vee y_2) \, (y_1 \overline{y}_2 \vee \overline{y}_1 y_2 \vee \overline{y}_2 \vee \overline{y}_1 \vee \overline{y}_2)$$

$$\wedge (\overline{y}_1 \overline{y}_2 \vee y_1 y_2 \vee y_2 \vee \overline{y}_1 \vee y_2) \, (\overline{y}_1 \overline{y}_2 \vee y_1 y_2 \vee \overline{y}_2 \vee y_1 \vee \overline{y}_2) = 0$$

$$(y_1 \vee y_2) \, (\overline{y}_1 \vee \overline{y}_2) \, (\overline{y}_1 \vee y_2) \, (y_1 \vee \overline{y}_2) = 0$$

$$y_1 (y_2 \vee \overline{y}_2) \wedge \overline{y}_1 (y_2 \vee \overline{y}_2) = 0$$

$$y_1 \wedge \overline{y}_1 = 0$$

$$0 = 0 \, .$$

This is an example of a very simple system of *reversible functions* defined by the equation $f'(x_1, x_2, y_1, y_2) = 0$.

We noticed in Example 7.8 that Eq. (7.49) cannot be solved with regard to **x** because no function values of $x_1 = h_1(\mathbf{y})$ and $x_2 = h_2(\mathbf{y})$ are determined for $(y_1, y_2) = (1, 0)$. The following example shows that in such cases a constraint can be defined and solution-function can be determined under the assumption that this constraint is satisfied.

Example 7.10 We consider the logic equation

$$f(\mathbf{x}, \mathbf{y}) = x_1 \overline{y}_2 \vee \overline{x}_1 y_2 \vee x_1 x_2 \vee \overline{y}_1 y_2 \vee x_2 \overline{y}_1 \vee \overline{x}_1 \overline{x}_2 y_1 = 0 \, . \tag{7.50}$$

This equation has only the three solution-vectors shown in Table 7.4. There is no solution with $x_1 = 1$, $x_2 = 1$. The mapping φ is unique; however, it is not complete.

A solution with regard to the variables y_1 and y_2 can be found in such a case when the constraint $x_1 x_2 = 0$ is accepted. This leads to the changed equation

$$f'(\mathbf{x}, \mathbf{y}) = (x_1 \overline{y}_2 \vee \overline{x}_1 y_2 \vee x_1 x_2 \vee \overline{y}_1 y_2 \vee x_2 \overline{y}_1 \vee \overline{x}_1 \overline{x}_2 y_1) \wedge \overline{(x_1 x_2)} = 0 \, . \tag{7.51}$$

for which four pairs of solution-functions ($y_1 = g_1(\mathbf{x})$, $y_2 = g_2(\mathbf{x})$) exist:

$$g_1(\mathbf{x}) = x_1 \oplus x_2 \,, \qquad g_1(\mathbf{x}) = x_1 \oplus x_2 \,, \qquad g_1(\mathbf{x}) = x_1 \vee x_2 \,, \qquad g_1(\mathbf{x}) = x_1 \vee x_2 \,,$$

$$g_2(\mathbf{x}) = x_1 \overline{x}_2 \,, \qquad g_2(\mathbf{x}) = x_1 \,, \qquad g_2(\mathbf{x}) = x_1 \overline{x}_2 \,, \qquad g_2(\mathbf{x}) = x_1 \,.$$

This kind of solution with regard to variables can always be achieved.

In the given examples the role of the $x_1, x_2, \ldots$ and $y_1, y_2, \ldots$ has been defined already at the beginning. In many applications, however, we will solve a logic equation $f(\mathbf{x}) = 0$, and consequently, we try to find a minimal number of variables $x_{i_1}, \ldots, x_{i_k}$, so that the other remaining variables are *functionally dependent* on the *independent variables*. A small number of independent variables is highly desirable in many cases. This approach is very useful especially for diagnostic situations. The following theorem defines the number of independent variables that can be achieved.

Theorem 7.4 *For $\mathbf{x} \in \mathbb{B}^n$ and $\mathbf{y} \in \mathbb{B}^k$ the mapping $\varphi : \mathbb{B}^n \to \mathbb{B}^k$*

- *Can be complete only if the equation $f(\mathbf{x}, \mathbf{y}) = 0$ has at least 2^n solutions;*
- *Can be unique and complete only if the number of solutions of the equation $f(\mathbf{x}, \mathbf{y}) = 0$ is exactly equal to 2^n.*

Proof If the number of solutions of the equation $f(\mathbf{x}, \mathbf{y}) = 0$ is less than 2^n, then not each $\mathbf{x}$ can occur in the solution-set; hence, the mapping φ will not be complete.

When the mapping is supposed to be unique and complete, then each vector $\mathbf{x}$ occurs once and only once in the domain $\mathbb{B}^n$ of φ; hence, the number of solutions must be equal to 2^n.   □

Another interesting possibility is the introduction of independent parameters $t_1, t_2, \ldots$ and the expression of all the solutions by means of these parameters.

Let be given $f(\mathbf{x}, \mathbf{x}) = (x_1 \vee x_2)(y_1 \oplus \overline{y}_2) \vee \overline{x}_1 \overline{x}_2 (y_1 \oplus y_2)$. The equation $f(\mathbf{x}, \mathbf{x}) = 0$ has eight solutions

$$
S =
\begin{array}{cccc}
x_1 & x_2 & y_1 & y_2 \\
\hline
0 & 0 & 0 & 0 \\
0 & 0 & 1 & 1 \\
0 & 1 & 0 & 1 \\
0 & 1 & 1 & 0 \\
1 & 1 & 0 & 1 \\
1 & 1 & 1 & 0 \\
1 & 0 & 0 & 1 \\
1 & 0 & 1 & 0 \\
\end{array}
$$

that can be "encoded" by three parameters t_1, t_2, t_3

$$x_1 = t_1 \,, \qquad x_2 = t_1 \oplus t_2 \,, \qquad y_1 = t_3 \,, \qquad \text{and} \qquad y_2 = (t_1 \vee t_2)\overline{t}_3 \vee \overline{t}_1 \overline{t}_2 t_3$$

define the vectors of the solution-set by means of four functions depending on the independent parameters t_1, t_2, t_3 using the assignment of Table 7.5.

Table 7.5 Encoding of the solution-set by parameters t_1, t_2, t_3

t_1	t_2	t_3	x_1	x_2	y_1	y_2
0	0	0	0	0	0	0
0	0	1	0	0	1	1
0	1	0	0	1	0	1
0	1	1	0	1	1	0
1	0	0	1	1	0	1
1	0	1	1	1	1	0
1	1	0	1	0	0	1
1	1	1	1	0	1	0

The introduction of independent variables is very elegant and efficient. In many applications, the equation (the constraint) is known, and after the solution of the equation, the solution-set can be described (stored) by a number of functions depending directly on a smaller number of parameters. Even the reproduction of the solution-set by means of circuits or programs is quite feasible and can use this approach.

As a next problem, we consider equations of the type

$$f(x_1, \mathbf{y}) = 0, \qquad \mathbf{y} = (y_1, \ldots, y_k).$$

The problem consists in finding functions

$$y_1 = g_1(x_1), \ y_2 = g_2(x_1), \ \ldots, y_k = g_k(x_1)$$

that satisfy the original equation. It was already considered by *George Boole*. He wanted to find out whether it is possible to find an equation where an independent variable (axiom) has been included and other logical statements can be determined by means of this axiom.

Example 7.11 We start with the following equation:

$$x_1\overline{y}_1 \vee x_1 y_2 \vee \overline{x}_1 y_1 \overline{y}_2 = 0.$$

The solution-set is given by

$$S = \begin{array}{ccc} x_1 & y_1 & y_2 \\ \hline 0 & 0 & 0 \\ 0 & 1 & 1 \\ 0 & 0 & 1 \\ \hline 1 & 1 & 0 \end{array},$$

where the three solution-vectors with $x_1 = 0$ are separated from the single solution with $x_1 = 1$.

For $x_1 = 0$ three vectors can be used, and for $x_1 = 1$ only one vector is available. This results in three solutions:

x_1	y_1	y_2
0	0	0
1	1	0

x_1	y_1	y_2
0	1	1
1	1	0

x_1	y_1	y_2
0	0	1
1	1	0

$y_1 = g_1(x_1) = x_1$

$y_2 = g_2(x_1) = 0$

$y_1 = g_1(x_1) = 1$

$y_2 = g_2(x_1) = \overline{x}_1$

$y_1 = g_1(x_1) = x_1$

$y_2 = g_2(x_1) = \overline{x}_1.$

In this way, we have three different possibilities where functions $y_1 = g_1(x_1)$ and $y_2 = g_2(x_1)$ depend on x_1. All the four functions of one variable are used $(0, 1, x_1, \overline{x}_1)$, two of them for $y_1 = g_1(x_1)$, and the other two functions for $y_2 = g_2(x_1)$. However, only three combinations (out of the four possible arrangements) solve the problem.

In order to find a condition for the solution of this problem, we use the representation of the function $f(x_1, y_1, y_2)$ in a Karnaugh-map

$$
\begin{array}{cc}
y_1\,y_2 & f(x_1, y_1, y_2) \\
\begin{array}{cc}
0 & 0 \\
0 & 1 \\
1 & 1 \\
1 & 0
\end{array} &
\begin{array}{|c|c|}
\hline
0 & 1 \\
0 & 1 \\
0 & 1 \\
\hline
1 & 0 \\
\hline
\end{array} \\
& \begin{array}{cc} 0 & 1 \end{array}\ x_1
\end{array}
$$

and see that the functions $y_1 = g_1(x_1)$ and $y_2 = g_2(x_1)$ can be found if there is at least one value 0 in each column. Hence, the equation $f(x_1, y_1, y_2) = 0$ can be solved with regard to the variables y_1 and y_2 if the 2-fold minimum of the function $f(x_1, y_1, y_2)$ with regard to the variables (y_1, y_2) is equal to 0:

$$
\min_{(y_1, y_2)}{}^2 f(x_1, y_1, y_2) = 0 \, .
$$

Example 7.12 In this example we use a function $f(\mathbf{x}, \mathbf{y})$ which depends on five variables

$$
f(\mathbf{x}, \mathbf{y}) = x_1\overline{y}_1 \vee \overline{x}_2 x_3 \overline{y}_1 \vee x_3 \overline{y}_1 \overline{y}_2 \vee x_2 \overline{x}_3 y_1 y_2 \vee \overline{x}_1 x_2 \overline{x}_3 y_1 \vee \overline{x}_1 \overline{x}_2 y_1 \overline{y}_2 \vee x_1 \overline{x}_2 y_2
$$

and solve the equation

$$
f(x_1, x_2, x_3, y_1, y_2) = 0
$$

for this function with regard to the variables y_1 and y_2.

$$
\begin{array}{cl}
y_1\,y_2 & \qquad\qquad f(x_1, x_2, x_3, y_1, y_2) \\
\begin{array}{cc}
0 & 0 \\
0 & 1 \\
1 & 1 \\
1 & 0
\end{array} &
\begin{array}{|c c c c c c c c|}
\hline
0 & 1 & 1 & 0 & 1 & 1 & 1 & 1 \\
0 & 1 & 0 & 0 & 1 & 1 & 1 & 1 \\
0 & 0 & 0 & 1 & 1 & 0 & 1 & 1 \\
1 & 1 & 0 & 1 & 0 & 0 & 0 & 0 \\
\hline
\end{array}
\end{array}
$$

$$
\begin{array}{cccccccc}
0 & 1 & 1 & 0 & 0 & 1 & 1 & 0 \quad x_3 \\
0 & 0 & 1 & 1 & 1 & 1 & 0 & 0 \quad x_2 \\
0 & 0 & 0 & 0 & 1 & 1 & 1 & 1 \quad x_1
\end{array}
$$

The two-fold minimum can be used in order to check whether there is a solution:

$$
\begin{aligned}
\min_{(y_1, y_2)}{}^2 f(x_1, x_2, x_3, y_1, y_2) = \quad & f(x_1, x_2, x_3, 0, 0) \\
\wedge \ & f(x_1, x_2, x_3, 0, 1) \\
\wedge \ & f(x_1, x_2, x_3, 1, 0) \\
\wedge \ & f(x_1, x_2, x_3, 1, 1) \ .
\end{aligned}
$$

Table 7.6 Complete function table and selected pairs of functions

Choice	x_1	x_2	x_3	y_1	y_2	The first 12 pairs of solution-functions		
a1	0	0	0	0	0	Used choice	Function $y_1 = g_1(\mathbf{x})$	Function $y_2 = g_2(\mathbf{x})$
a2	0	0	0	0	1	a1-b1-c1-d1	$y_1 = x_1 \vee \overline{x}_2\, x_3$	$y_2 = \overline{x}_1\, x_3$
a3	0	0	0	1	1	a1-b1-c1-d2	$y_1 = x_1 \vee \overline{x}_2\, x_3$	$y_2 = (\overline{x}_1 \vee x_2)x_3$
	0	0	1	1	1	a1-b1-c2-d1	$y_1 = x_1 \vee x_3$	$y_2 = \overline{x}_1\, \overline{x}_2\, x_3$
b1	0	1	0	0	0	a1-b1-c2-d2	$y_1 = x_1 \vee x_3$	$y_2 = (x_1 \odot x_2)x_3$
b2	0	1	0	0	1	a1-b1-c3-d1	$y_1 = x_1 \vee x_3$	$y_2 = \overline{x}_1\, x_3$
c1	0	1	1	0	1	a1-b1-c3-d2	$y_1 = x_1 \vee x_3$	$y_2 = (\overline{x}_1 \vee x_2)x_3$
c2	0	1	1	1	0	a1-b2-c1-d1	$y_1 = x_1 \vee \overline{x}_2\, x_3$	$y_2 = \overline{x}_1\,(x_2 \vee x_3)$
c3	0	1	1	1	1	a1-b2-c1-d2	$y_1 = x_1 \vee \overline{x}_2\, x_3$	$y_2 = \overline{x}_1\,(x_2 \vee x_3)$
	1	0	0	1	0			$\vee\, x_2\, x_3$
	1	0	1	1	0	a1-b2-c2-d1	$y_1 = x_1 \vee x_3$	$y_2 = \overline{x}_1\,(x_2 \oplus x_3)$
	1	1	0	1	0	a1-b2-c2-d2	$y_1 = x_1 \vee x_3$	$y_2 = \overline{x}_1\,(x_2 \oplus x_3)$
								$\vee\, x_1\, x_2\, x_3$
d1	1	1	1	1	0	a1-b2-c3-d1	$y_1 = x_1 \vee x_3$	$y_2 = \overline{x}_1\,(x_2 \vee x_3)$
d2	1	1	1	1	1	a1-b2-c3-d2	$y_1 = x_1 \vee x_3$	$y_2 = \overline{x}_1\,(x_2 \vee x_3)$
								$\vee\, x_2\, x_3$

Hence, we have to combine the values in each column of the Karnaugh-map by $\wedge$, and we find that

$$\min_{(y_1, y_2)}{}^2 f(x_1, x_2, x_3, y_1, y_2) = 0 \,.$$

Solutions can be found by an ordered grouping of the solution vectors of the equation $f(\mathbf{x}, \mathbf{y}) = 0$, according to the order of the $\mathbf{x}$-vectors, and by selecting one $\mathbf{y}$-vector for each $\mathbf{x}$-vector. This means that all the functions for $y_i = g_i(\mathbf{x})$ will be defined simultaneously. Due to

- Three choices for $(x_1, x_2, x_3) = (000)$
- Three choices for $(x_1, x_2, x_3) = (011)$
- Two choices for $(x_1, x_2, x_3) = (010)$
- Two choices for $(x_1, x_2, x_3) = (111)$

there are at all $3 * 3 * 2 * 2 = 36$ pairs of solution-functions $y_1 = g_1(\mathbf{x})$ and $y_2 = g_2(\mathbf{x})$. These pairs of solution-functions can be identified when all combinations of the function table on the left-hand side of Table 7.6 are used; the first 12 pairs of these solution-functions are enumerated in the right-hand side of Table 7.6.

Table 7.6 already indicates a very general numerical approach to find all (some) solutions of the equation $f(\mathbf{x}, \mathbf{y}) = 0$:

- Use the appropriate grouping of the solutions with regard to the $\mathbf{x}$-vectors.
- For each $\mathbf{x} \in \mathbb{B}^n$, select one vector $\mathbf{y}$ and enter this vector into the associated table for the components of $\mathbf{y}$.
- Determine expressions for the different functions $y_1 = g_1(\mathbf{x}), \ldots, y_k = g_k(\mathbf{x})$.

This method is especially useful when the dependent variables are not given by the problem, but have to be found by the exploration of the solution-set. Step by step it can be checked whether $x_1, \ldots, x_n$ are candidates for independent variables, then any pairs $(x_1, x_2), (x_1, x_3), \ldots, (x_{n-1}, x_n)$ can be considered and so on.

Fig. 7.8 The Karnaugh-maps of the lattices of functions $g_1(\mathbf{x})$ and $g_2(\mathbf{x})$

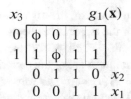
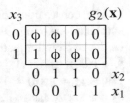

Here it is also understandable that in the case of $f(\mathbf{x}, \mathbf{y}) = 1$ the same approach is possible. The solution-set will be determined and equally arranged. From there the grouping and construction of solution-functions use the same steps.

The following considerations will show all the different functions that play a given role for the solution of the equation with regard to y_1 and y_2. In order to find the algebraic structure of the solution-sets for the different variables, we go back to the grouping of the solutions. In our example, there are eight groups (see Table 7.6):

$$S_0(x_1, x_2, x_3, y_1, y_2) = \{(000\ 00), (000\ 01), (000\ 11)\},$$

$$S_1(x_1, x_2, x_3, y_1, y_2) = \{(001\ 11)\},$$

$$S_2(x_1, x_2, x_3, y_1, y_2) = \{(010\ 00), (010\ 01)\},$$

$$S_3(x_1, x_2, x_3, y_1, y_2) = \{(011\ 01), (011\ 10), (011\ 11)\},$$

$$S_4(x_1, x_2, x_3, y_1, y_2) = \{(100\ 10)\},$$

$$S_5(x_1, x_2, x_3, y_1, y_2) = \{(101\ 10)\},$$

$$S_6(x_1, x_2, x_3, y_1, y_2) = \{(110\ 10)\},$$

$$S_7(x_1, x_2, x_3, y_1, y_2) = \{(111\ 10), (111\ 11)\}.$$

In each group, the **x**-part is constant, only different **y**-vectors occur. The set of possible functions $y_i = g_i(\mathbf{x})$ can now be constructed in the following way, for $\mathbf{x}^l$ being the constant part in $S_l(\mathbf{x}, \mathbf{y})$:

$$g_i(\mathbf{x}^l) = \begin{cases} 0, & \text{if } y_i \text{ is constant equal to 0 in } S_l(\mathbf{x}, \mathbf{y}), \\ 1, & \text{if } y_i \text{ is constant equal to 1 in } S_l(\mathbf{x}, \mathbf{y}), \\ -, & \text{if } y_i \text{ has the value 0 as well as the value 1 in } S_l(\mathbf{x}, \mathbf{y}). \end{cases}$$

Figure 7.8 shows the Karnaugh-maps of $y_1 = g_1(\mathbf{x})$ and $y_2 = g_2(\mathbf{x})$:

When ϕ is replaced by 0 or 1, then one of the possible functions $g_1(\mathbf{x})$ or $g_2(\mathbf{x})$ has been constructed. This procedure will be generalized and summarized in the following theorem.

Theorem 7.5 *If the logic equation $f(\mathbf{x}, \mathbf{y}) = 0$ can be solved with regard to $y_1, \ldots, y_k$, then the set of functions $\{g_{i1}(\mathbf{x}), \ldots, g_{im}(\mathbf{x})\}$ that is possible for the variable y_i is a lattice $\mathcal{L}\langle g_{iq}(\mathbf{x}), g_{ir}(\mathbf{x})\rangle$. Every function of this lattice occurs in at least one vector $(g_1(\mathbf{x}), \ldots, g_k(\mathbf{x}))$ of the solution-functions.*

As usual, the smallest element (the *infimum*) of the lattices $\mathcal{L}\langle g_q(\mathbf{x}), g_r(\mathbf{x})\rangle$ can be found by replacing each ϕ by 0; the largest element (the *supremum*) will be given by replacing each ϕ by 1:

$$\inf \mathcal{L}\langle g_{1q}, g_{1r}\rangle = g_{1q}(\mathbf{x}) = x_1 \vee \overline{x}_2 x_3, \qquad \sup \mathcal{L}\langle g_{1q}, g_{1r}\rangle = \overline{g_{1r}(\mathbf{x})} = x_1 \vee \overline{x}_2 \vee x_3,$$

$$\inf \mathcal{L}\langle g_{2q}, g_{2r}\rangle = g_{2q}(\mathbf{x}) = \overline{x}_1 \overline{x}_2 x_3, \qquad \sup \mathcal{L}\langle g_{2q}, g_{2r}\rangle = \overline{g_{2r}(\mathbf{x})} = \overline{x}_1 \vee x_2 x_3.$$

The construction also shows that for the two lattices $\mathcal{L}\langle g_{1q}, g_{1r}\rangle$ and $\mathcal{L}\langle g_{2q}, g_{2r}\rangle$ not all the combinations of any elements of the two lattices occur; e.g., both the value 0 and 1 occur as solution of the explored equation for both y_1 and y_2 together with $\mathbf{x} = (000)$ in the three combinations $\mathbf{y} = \{(00), (01), (11)\}$, but the solution-set of the equation $f(\mathbf{x}, \mathbf{y}) = 0$ excludes the pair of function values $y_1 = g_1(\mathbf{0}) = 1$ and $y_2 = g_2(\mathbf{0}) = 0$. The values are not independent of each other. In each set S_k one vector has to be selected, and this selection defines the value for all the functions $y_1 = g_1(\mathbf{x})$ and $y_2 = g_2(\mathbf{x})$. In the example above, there are $2^2 * 2^4 = 4 * 16 = 64$ arbitrary combinations of functions from the lattices $\mathcal{L}\langle g_{1q}, g_{1r}\rangle$ and $\mathcal{L}\langle g_{2q}, g_{2r}\rangle$; however, only 36 out of these 64 combinations specify a vector of solution-functions.

The functions $y_1(\mathbf{x}), \dots, y_k(\mathbf{x})$ can be defined in a *unique way* when each column $\mathbf{x}$ of the Karnaugh-map has exactly one position with $f(\mathbf{x}, \mathbf{y}) = 0$. Then exactly one $\mathbf{y}^* \in \mathbb{B}^k$ can be found for each $\mathbf{x}^* \in \mathbb{B}^n$, and the function can be determined by one single choice for each $\mathbf{x}^* \in \mathbb{B}^n$.

Example 7.13 For the choice a1-b2-c3-d1 of Table 7.6 we get the Karnaugh-map of $f(x_1, x_2, x_3, y_1, y_2)$

$$
\begin{array}{cc|cccccccc}
y_1\, y_2 & & & & & f(x_1, x_2, x_3, y_1, y_2) & & & & \\
0\ 0 & & 0 & 1 & 1 & 1 & 1 & 1 & 1 & 1 \\
0\ 1 & & 1 & 1 & 1 & 0 & 1 & 1 & 1 & 1 \\
1\ 1 & & 1 & 0 & 0 & 1 & 1 & 1 & 1 & 1 \\
1\ 0 & & 1 & 1 & 1 & 1 & 0 & 0 & 0 & 0 \\
\end{array}
$$

$$
\begin{array}{cccccccc}
0 & 1 & 1 & 0 & 0 & 1 & 1 & 0 & x_3 \\
0 & 0 & 1 & 1 & 1 & 1 & 0 & 0 & x_2 \\
0 & 0 & 0 & 0 & 1 & 1 & 1 & 1 & x_1 \\
\end{array}
$$

that has exactly one position with $f(\mathbf{x}, \mathbf{y}) = 0$ in each column. Hence, exactly one $\mathbf{y}^* \in \mathbb{B}^2$ exists for each $\mathbf{x}^* \in \mathbb{B}^3$, and the function can be determined by one single choice for each $\mathbf{x}^* \in \mathbb{B}^n$. The equation $f(x_1, x_2, x_3, y_1, y_2) = 0$ has the only solution:

$$y_1 = x_1 \vee x_3 , \qquad\qquad y_2 = \overline{x}_1(x_2 \vee x_3) .$$

Still the 2-fold minimum of $f(\mathbf{x}, \mathbf{y})$ combining the values of each column by $\wedge$ will be equal to 0.

We can see some interesting properties for this type of unique solutions. The minimum $\min_{y_1} f(x_1, x_2, x_3, y_1, y_2)$ for the explored choice a1-b2-c3-d1 can be calculated by

$$f(y_1 = 0) \wedge f(y_1 = 1)$$

and results in the following function:

$$
\begin{array}{c|cccccccc}
y_2 & & & & \min_{y_1} f(x_1, x_2, x_3, y_1, y_2) & & & \\
0 & 0 & 1 & 1 & 1 & 0 & 0 & 0 & 0 \\
1 & 1 & 0 & 0 & 0 & 1 & 1 & 1 & 1 \\
\end{array}
$$

$$
\begin{array}{cccccccc}
0 & 1 & 1 & 0 & 0 & 1 & 1 & 0 & x_3 \\
0 & 0 & 1 & 1 & 1 & 1 & 0 & 0 & x_2 \\
0 & 0 & 0 & 0 & 1 & 1 & 1 & 1 & x_1 \\
\end{array}
$$

which does not depend on y_1 anymore.

There is only one pair $(0, 1)$ for each (x_1, x_2, x_3) due to the single value 0 in each column; the other pair is equal to $(1,1)$ and the result shows only pairs $(0, 1)$, so that we get

$$\min_{y_1} f(x_1, x_2, x_3, y_1, y_2) = f'(x_1, x_2, x_3, y_2)$$

$$= \overline{x}_1(x_2 \vee x_3)\overline{y}_2 \vee (x_1 \vee \overline{x}_2\overline{x}_3)y_2 \ .$$

The computed minimum of $f(\mathbf{x}, \mathbf{y})$ with regard to y_1 reduces solution-set of the given equation to the mapping from $\mathbf{x}$ to y_2. The occurrence of only pairs $(0, 1)$ for each (x_1, x_2, x_3) can be checked by a derivative of $f'(x_1, x_2, x_3, y_2)$ with regard to y_2

$$\operatorname*{der}_{y_2} f'(x_1, x_2, x_3, y_2) = 1(x_1, x_2, x_3) \ ,$$

and determines a unique solution-function $y_2 = g_2(\mathbf{x})$.

It is feasible to combine these two steps to

$$\operatorname*{der}_{y_2} \left(\min_{y_1} f(x_1, x_2, x_3, y_1, y_2) \right) = 1(x_1, x_2, x_3) \ ; \tag{7.52}$$

the explored equation $f(x_1, x_2, x_3, y_1, y_2) = 0$ uniquely specifies the solution-function $y_2 = g_2(\mathbf{x})$ if this differential equation is satisfied for all $\mathbf{x}$.

In order to get an understanding of the size of the solution-sets we go back to the solution of the example and perform the following transformation:

$$y_1 = x_1 \vee x_3 \ , \qquad\qquad\qquad y_2 = \overline{x}_1(x_3 \vee x_2) \ ,$$

$$y_1 \oplus (x_1 \vee x_3) = 0 \ , \qquad\qquad y_2 \oplus (x_1(x_3 \vee x_2)) = 0 \ ,$$

and notice that the unique solution-functions are linear in y_1 and y_2.

Let us consider, for instance, the 16 functions $g_i(x_1, x_2)$ of two variables; each of these functions can be extended to a function $f_i(x_1, x_2, y_1)$ by combining the function $g_i(x_1, x_2)$ and y_1 by $\oplus$, for instance:

$$g_0(\mathbf{x}) = 0 \ , \qquad\qquad\qquad f_0(\mathbf{x}, y_1) = y_1,$$

$$g_1(\mathbf{x}) = x_1 x_2 \ , \qquad\qquad\quad f_1(\mathbf{x}, y_1) = x_1 x_2 \oplus y_1,$$

$$\vdots \qquad\qquad\qquad\qquad\qquad \vdots$$

$$g_{14}(\mathbf{x}) = \overline{x_1 \wedge x_2} \ , \qquad\qquad f_{14}(\mathbf{x}, y_1) = \overline{x_1 \wedge x_2} \oplus y_1 \ ,$$

$$g_{15}(\mathbf{x}) = 1 \ , \qquad\qquad\qquad f_{15}(\mathbf{x}, y_1) = 1 \oplus y_1 \ .$$

There are $2^{2^3} = 2^8 = 256$ functions $f(x_1, x_2, y_1)$ of three variables, for the 16 of them, constructed above, the equation $f(x_1, x_2, y_1) = 0$ has a unique solution $y_1 = g_1(x_1, x_2)$.

This set of functions is determined by means of the differential equation

$$\operatorname*{der}_{y_1} f(x_1, x_2, y_1) = 1 \, ;\tag{7.53}$$

it can be said that this differential equation *characterizes the set of functions* $f(x_1, x_2, y_1)$ for which the equation $f(x_1, x_2, y_1) = 0$ has exactly one solution $y_1 = g_1(x_1, x_2)$. This is a very important point of view for many applications: A given differential equation describes a set of logic functions with a special property. In this case it is easy to see that the set contains all the functions $f(x_1, x_2, y_1)$ that are *linear* in y_1.

The ratio "16 out of 256" also shows that unique solutions are a rare case. For two variables (x_1, x_2) this ratio is

$$2^{2^2}/2^{2^3} = 16/256 = 6.25\%$$

and for three variables (x_1, x_2, x_3) this ratio is already reduced to

$$2^{2^3}/2^{2^4} = 256/65.536 = 0.390625\% \, .$$

Example 7.14 Let

$$f(\mathbf{x}, \mathbf{y}) = (x_1 \vee x_2)\overline{y}_1\overline{y}_2 \vee (x_1 \vee \overline{x}_2)\overline{y}_1 y_2 \vee (\overline{x}_1 \vee \overline{x}_2)y_1 y_2 \vee (\overline{x}_1 \vee x_2)y_1\overline{y}_2 \, .$$

$y_1 y_2$	$f(x_1, x_2, y_1, y_2)$			
0 0	0	1	1	1
0 1	1	0	1	1
1 1	1	1	0	1
1 0	1	1	1	0
	0	1	1	0 x_2
	0	0	1	1 x_1

It can immediately be seen that the equation $f(x_1, x_2, y_1, y_2) = 0$ is uniquely solvable with regard to y_1 and y_2. The single pair of solution-functions is

$$y_1 = g_1(\mathbf{x}) = x_1 \, , \qquad\qquad y_2 = g_2(\mathbf{x}) = x_2 \, .$$

If there are columns without even one location with $f(\mathbf{x}, \mathbf{y}) = 0$, then the functions $y_1 = g_1(\mathbf{x}), \ldots, y_k = g_k(\mathbf{x})$ can be defined only partially. The $\mathbf{x}$-vector of such a column has to be excluded from the domain of the solution-functions.

Example 7.15 We look at the function $f(\mathbf{x}, \mathbf{y})$ given by the next Karnaugh-map. It can be seen that there is one column without a value 0. This function is

$$f(\mathbf{x}, \mathbf{y}) = (x_2 \vee x_3)\overline{y}_1\overline{y}_2 \vee (\overline{x}_2 \vee x_3)\overline{y}_1 y_2 \vee x_1(\overline{y}_1 \vee y_2) \vee \overline{x}_1\overline{x}_3 y_1 \vee \overline{x}_1 y_1\overline{y}_2 \vee x_1 x_2 x_3 \, .$$

Our aim is to solve the equation $f(\mathbf{x}, \mathbf{y}) = 0$ with regard to the variables y_1 and y_2. The condition to solve this task is that the k-fold minimum must be equal to 0. This condition is not satisfied:

$$\operatorname*{min}_{(y_1, y_2)}^2 f(x_1, x_2, x_3, y_1, y_2) = x_1 x_2 x_3 \, .$$

$$
\begin{array}{cc}
y_1\, y_2 & \qquad\qquad f(x_1, x_2, x_3, y_1, y_2)
\end{array}
$$

$y_1\, y_2$								
0 0	0	1	1	1	1	1	1	1
0 1	1	1	1	0	1	1	1	1
1 1	1	0	0	1	1	1	1	1
1 0	1	1	1	1	0	1	0	0
	0	1	1	0	0	1	1	0	x_3
	0	0	1	1	1	1	0	0	x_2
	0	0	0	0	1	1	1	1	x_1

In this case the 2-fold minimum shows the **x**-vector to be excluded.

This means that only those vectors are allowed that satisfy the equation

$$
x_1 x_2 x_3 = 0 \, .
$$

Using this restriction, we get the modified equation:

$$
\big((x_2 \vee x_3)\overline{y}_1\overline{y}_2 \vee (\overline{x}_2 \vee x_3)\overline{y}_1 y_2 \vee x_1(\overline{y}_1 \vee y_2)
$$

$$
\vee\, \overline{x}_1\overline{x}_3 y_1 \vee \overline{x}_1 y_1 \overline{y}_2 \vee x_1 x_2 x_3 \big) \wedge \overline{x_1 x_2 x_3} = 0 \, .
$$

This restricted equation has four pairs of solution-functions:

$$
\begin{aligned}
y_1 = g_1(\mathbf{x}) = x_1 \vee x_3 \, , \qquad\qquad &\qquad y_2 = g_2(\mathbf{x}) = \overline{x}_1(x_2 \vee x_3) \, , \\
y_1 = g_1(\mathbf{x}) = x_1 \vee x_3 \, , \qquad\qquad &\qquad y_2 = g_2(\mathbf{x}) = \overline{x}_1(x_2 \vee x_3) \vee x_2 x_3 \, , \\
y_1 = g_1(\mathbf{x}) = (x_1 \oplus x_3) \vee x_1\overline{x}_2 x_3 \, , \qquad\qquad &\qquad y_2 = g_2(\mathbf{x}) = \overline{x}_1(x_2 \vee x_3) \, , \\
y_1 = g_1(\mathbf{x}) = (x_1 \oplus x_3) \vee x_1\overline{x}_2 x_3 \, , \qquad\qquad &\qquad y_2 = g_2(\mathbf{x}) = \overline{x}_1(x_2 \vee x_3) \vee x_2 x_3 \, .
\end{aligned}
$$

We go back to the equation $f(x_1, \dots, x_n) = 0$ that can be solved with regard to x_n if

$$
\min_{x_n} f(x_1, \dots, x_n) = 0 \, .
$$

This is the case if for any vector $\mathbf{x}^* \in \mathbb{B}^{n-1}$ at least one of the two values $f(\mathbf{x}^*, 0)$ and $f(\mathbf{x}^*, 1)$ is equal to 0, i.e., the equation $f(\mathbf{x}) = 0$ will have at least 2^{n-1} solutions. The condition $\min_{x_n} f(\mathbf{x}) = 0$ ensures that for at least 50% of the **x**-vectors the value of $f(\mathbf{x})$ is equal to 0. For two variables x_n, x_{n-1} at least one 0 has to be found for the vectors $(\mathbf{x}^*, 0, 0), (\mathbf{x}^*, 0, 1), (\mathbf{x}^*, 1, 0), (\mathbf{x}^*, 1, 1), \mathbf{x}^* \in \mathbb{B}^{n-2}$. In this case, only one value out of four has to be equal to 0 (and this is guaranteed automatically when a solution with regard to one variable can be found). Hence, the following theorem is obvious.

Theorem 7.6 *Let* $f(\mathbf{x}) = f(x_1, x_2, \dots, x_{n+k})$. *The equation* $f(\mathbf{x}) = 0$ *can only be solved with regard to* k *variables* $x_{i_1}, \dots, x_{i_k}$ *if there are at least* 2^n *solutions. An equation* $f(\mathbf{x}) = 0$ *with* r *solutions can be solved with regard to at most* $\lfloor \log_2 r \rfloor$ *variables.*[1] *The unique solution with regard to* $x_{i_1}, \dots, x_{i_k}$ *requires exactly* 2^n *solutions of* $f(\mathbf{x}) = 0$.

---

[1] $\lfloor \log_2 r \rfloor$ is the largest integer less or equal to $\log_2 r$ where $\log_2$ is the logarithm to the base 2.

Example 7.16 The equation

$$(\overline{x}_1 \vee \overline{x}_2 \vee \overline{x}_3 \vee x_4)(x_1 \vee \overline{x}_2 \vee x_3 \vee \overline{x}_4) = 0$$

has only the two solutions

$$S = \begin{array}{cccc} x_1 & x_2 & x_3 & x_4 \\ \hline 1 & 1 & 1 & 0 \\ 0 & 1 & 0 & 1 \end{array}.$$

It can be solved with regard to the variables x_2, x_3, x_4 so that the three associated functions only depend on x_1:

$$x_2 = g_2(x_1) = 1, \ x_3 = g_3(x_1) = x_1, \ x_4 = g_4(x_1) = \overline{x}_1 .$$

It cannot be solved for any pair of variables and not at all for any single variable.

The border case would be a single 0 for $f(\mathbf{x})$; then all the functions degenerate to a constant, and it can be stated that $x_1 = c_1, \dots, x_n = c_n$. It should be kept in mind: The smaller the number of solutions of an equation, the larger the number of dependent variables. Only a small number of independent variables *is necessary* to describe the solution-set. For each equation there is a smallest number k for the number of dependent functions $y_1 = g_1(\mathbf{x}), \dots, y_k = g_k(\mathbf{x})$, and this number is uniquely defined. No smaller set of functions can be found. However, the extension to larger sets of dependent functions is always possible.

Theorem 7.7 *If $f(\mathbf{x}, \mathbf{y}) = 0$ can be solved with regard to a variable y_i, which is a component of $\mathbf{y} = (y_1, \dots, y_k)$, then $f(\mathbf{x}, \mathbf{y}) = 0$ can be solved with regard to $\mathbf{y} = (y_1, \dots, y_k)$.*

Proof The proof follows simply from the rule (6.47) which can be specified to

$$\min_{\mathbf{y}}^{k} f(\mathbf{x}, y_1, \dots, y_i, \dots, y_k) \leq \min_{y_i} f(\mathbf{x}, y_i) = 0 .$$

The condition

$$\min_{y_i} f(\mathbf{x}, y_i) = 0$$

generates a *sufficient* number of 0'es.   □

Example 7.17 Let $f(\mathbf{x}) = (x_1 \vee x_2)(x_3 \oplus \overline{x}_4) \vee \overline{x}_1 \overline{x}_2 (x_3 \oplus x_4)$; then

$$\min_{x_3} f(\mathbf{x}) = [(x_1 \vee x_2)\overline{x}_4 \vee \overline{x}_1 \overline{x}_2 x_4] \wedge [(x_1 \vee x_2)x_4 \vee \overline{x}_1 \overline{x}_2 \overline{x}_4] = 0 .$$

Hence, $f(\mathbf{x}) = 0$ can be solved with regard to x_3, as can be seen for

$$x_3 = g_3(x_1, x_2, x_4) = x_4 \oplus (x_1 \vee x_2) .$$

There are also solutions with regard to

$$(x_3, x_1), \ (x_3, x_2), \ (x_3, x_4), \ (x_3, x_1, x_2), \ (x_3, x_1, x_4), \ (x_3, x_2, x_4) .$$

Table 7.7 shows on the left-hand side selected sets of solution-functions and confirms on the right that these solution are correct.

Table 7.7 Sets of solution-functions and their verification

Set of solutions	Verification of the correctness
$x_3 = x_2\overline{x}_4$ $x_1 = \overline{x}_2 x_4$	$f(\mathbf{x}) = (x_1 \vee x_2)(x_3 \oplus \overline{x}_4) \vee \overline{x}_1\overline{x}_2(x_3 \oplus x_4)$ $= ((\overline{x}_2 x_4) \vee x_2)((x_2\overline{x}_4) \oplus \overline{x}_4) \vee \overline{\overline{x}_2 x_4}\,\overline{x}_2((x_2\overline{x}_4) \oplus x_4)$ $= 0$
$x_3 = x_1\overline{x}_4$ $x_2 = x_1 \vee x_4$	$f(\mathbf{x}) = (x_1 \vee x_2)(x_3 \oplus \overline{x}_4) \vee \overline{x}_1\overline{x}_2(x_3 \oplus x_4)$ $= (x_1 \vee (x_1 \vee x_4))((x_1\overline{x}_4) \oplus \overline{x}_4) \vee \overline{x}_1\overline{(x_1 \vee x_4)}((x_1\overline{x}_4) \oplus x_4)$ $= 0$
$x_3 = x_1 \oplus x_2$ $x_4 = x_1 x_2$	$f(\mathbf{x}) = (x_1 \vee x_2)(x_3 \oplus \overline{x}_4) \vee \overline{x}_1\overline{x}_2(x_3 \oplus x_4)$ $= (x_1 \vee x_2)((x_1 \oplus x_2) \oplus \overline{(x_1 x_2)}) \vee \overline{x}_1\overline{x}_2((x_1 \oplus x_2) \oplus (x_1 x_2))$ $= 0$
$x_3 = \overline{x}_4$ $x_1 = \overline{x}_4$ $x_2 = 1$	$f(\mathbf{x}) = (x_1 \vee x_2)(x_3 \oplus \overline{x}_4) \vee \overline{x}_1\overline{x}_2(x_3 \oplus x_4)$ $= (\overline{x}_4 \vee 1)(\overline{x}_4 \oplus \overline{x}_4) \vee \overline{\overline{x}}_4\,\overline{1}(\overline{x}_4 \oplus x_4)$ $= 0$
$x_3 = x_2$ $x_1 = \overline{x}_2$ $x_4 = \overline{x}_2$	$f(\mathbf{x}) = (x_1 \vee x_2)(x_3 \oplus \overline{x}_4) \vee \overline{x}_1\overline{x}_2(x_3 \oplus x_4)$ $= (\overline{x}_2 \vee x_2)(x_2 \oplus \overline{\overline{x}}_2) \vee \overline{\overline{x}}_2\overline{x}_2(x_2 \oplus \overline{x}_2)$ $= 0$
$x_3 = x_1$ $x_2 = \overline{x}_1$ $x_4 = \overline{x}_1$	$f(\mathbf{x}) = (x_1 \vee x_2)(x_3 \oplus \overline{x}_4) \vee \overline{x}_1\overline{x}_2(x_3 \oplus x_4)$ $= (x_1 \vee \overline{x}_1)(x_1 \oplus \overline{\overline{x}}_1) \vee \overline{x}_1\overline{\overline{x}}_1(x_1 \oplus \overline{x}_1)$ $= 0$

The border case would be again the enumeration of all solutions by means of constants:

$$(x_1 = 0, x_2 = 0, x_3 = 0, x_4 = 0), \ldots, (x_1 = 1, x_2 = 1, x_3 = 1, x_4 = 0).$$

As could be seen, the criterion

$$\min_{x_3} f(x_1, x_2, x_3, x_4) = 0$$

ensured for each vector $\mathbf{x} = (x_1, x_2, x_4)$ at least one solution of $f(\mathbf{x}) = 0$. When the criterion

$$\max_{x_3} f(x_1, x_2, x_3, x_4) = 1$$

is added, then for each vector $\mathbf{x}$ at least one value 1 exists, and this makes the relation between $\mathbf{x}$ and x_3 unique.

The following theorem generalizes the previous relations. The starting point not necessarily will be a single component; it can be a whole vector of variables.

Theorem 7.8 *Let* $f(\mathbf{x}, \mathbf{y}) = 0$ *be solvable with regard to the variables of the vector* $\mathbf{y}^* = \{y_{\alpha_1}, y_{\alpha_2}, \ldots, y_{\alpha_k}\}$, *and let* $\mathbf{y}^* \subseteq \mathbf{y} = \{y_{\beta_1}, \ldots, y_{\beta_l}\}$, $l > k$. *Then* $f(\mathbf{x}, \mathbf{y}) = 0$ *can also be solved with regard to the variables of the vector* $\mathbf{y}$.

Proof The proof uses the rule (6.47) that, when the k-fold minimum is equal to 0 for a given set of variables, it is also equal to 0 for each superset:

$$\min_{\mathbf{y}}^{l} f(\mathbf{x}, \mathbf{y}) \le \min_{\mathbf{y}^*}^{k} f(\mathbf{x}, \mathbf{y}) = 0. \qquad \qquad \square$$

Example 7.18 We consider the function $f(x_1, x_2, y_1, y_2, y_3)$ given by the Karnaugh-maps of Fig. 7.9 using two different distributions of the variables $\mathbf{y}$ to illustrate Theorem 7.8.

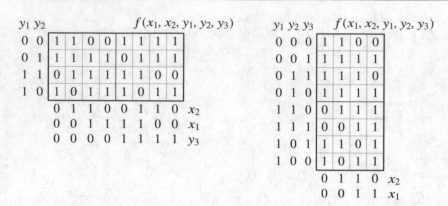

Fig. 7.9 Two Karnaugh-maps of the same function

It can be seen that there is exactly one value 0 in each column of the left Karnaugh-map of Fig. 7.9; hence, there are unique solution-functions $y_1 = g_1(x_1, x_2, y_3)$ and $y_2 = g_2(x_1, x_2, y_3)$. The Karnaugh-map on the right-hand side of Fig. 7.9 contains two values 0 in each column; hence, several solutions with regard to the variables y_1, y_2, and y_3 can be found.

The Inversion of Equation-Systems

Each system of logic equations

$$y_1 = f_1(x_1, \ldots, x_n) \,,$$

$$\vdots \qquad \vdots$$

$$y_k = f_k(x_1, \ldots, x_n)$$

implements a unique mapping F from $\mathbb{B}^n$ into $\mathbb{B}^k$; there is exactly one vector **y** for each vector **x**. We will now explore in which case this mapping can be inverted, i.e., when it is unique and how to find F^{-1}.

Definition 7.7 A system of logic equations

$$y_1 = f_1(x_1, \ldots, x_n) \,,$$

$$\vdots \qquad \vdots$$

$$y_k = f_k(x_1, \ldots, x_n)$$

is *invertible* if and only if the mapping F defined by $f_1(\mathbf{x}), \ldots, f_k(\mathbf{x})$ from $\mathbb{B}^n$ into $\mathbb{B}^k$ is unique, and the inverse mapping can be uniquely defined by functions

$$x_1 = g_1(y_1, \ldots, y_k) \,,$$

$$\vdots \qquad \vdots$$

$$x_n = g_n(y_1, \ldots, y_k) \,.$$

The functions $g_i(\mathbf{x})$ can be unique only if $k = n$. This reduces the problem to the finding of all unique mappings from $\mathbb{B}^n$ onto $\mathbb{B}^n$. Since $\mathbb{B}^n$ is a finite set with 2^n elements, there are $2^n!$ unique mappings to be taken into consideration. The solution of a logic equation with regard to variables can be used to solve this problem. The system

$$y_1 = f_1(\mathbf{x}), \quad \ldots, \quad y_n = f_n(\mathbf{x})$$

will be replaced by the equivalent equation

$$h(\mathbf{x}, \mathbf{y}) = [y_1 \oplus f_1(\mathbf{x})] \vee \cdots \vee [y_n \oplus f_n(\mathbf{x})] = 0 .$$

When the equation can be solved with regard to $x_1, \ldots, x_n$ and all the functions

$$x_1 = g_1(\mathbf{y}), \quad \ldots, \quad x_n = g_n(\mathbf{y})$$

can be defined in a unique way and without constraints, then the system is invertible.

Theorem 7.9 *The set of all functions $h(\mathbf{x}, \mathbf{y})$ that describe invertible systems of equations*

$$h(\mathbf{x}, \mathbf{y}) = [y_1 \oplus f_1(\mathbf{x})] \vee \cdots \vee [y_n \oplus f_n(\mathbf{x})]$$
$$= [x_1 \oplus g_1(\mathbf{y})] \vee \cdots \vee [x_n \oplus g_n(\mathbf{y})] = 0$$

is given as the solution of the system of differential equations

$$\forall x_i \in \mathbf{x}: \quad \operatorname*{der}_{x_i}(\operatorname*{min}_{\mathbf{x}\backslash x_i}^{n-1} h(\mathbf{x}, \mathbf{y})) = 1 , \tag{7.54}$$

$$\forall y_i \in \mathbf{y}: \quad \operatorname*{der}_{y_i}(\operatorname*{min}_{\mathbf{y}\backslash y_i}^{n-1} h(\mathbf{x}, \mathbf{y})) = 1 . \tag{7.55}$$

These two conditions can also be used to check the *invertibility* for a given system of logic equations, after transforming this system into one equation.

We will see now an example where the inverse functions $g_i(\mathbf{x}), i = 1, \ldots, n$, can be given explicitly. We remember the class of *linear functions*:

$$f(x_1, \ldots, x_n) = a_0 \oplus a_1 x_1 \oplus \cdots \oplus a_n x_n ,$$

$$a_i \in \{0, 1\} ,$$

$$a_1 \vee \ldots \vee a_n = 1 .$$

Not all of the coefficients a_i can be equal to 0, at least one x_i must be present. Two linear functions are linearly independent when they differ in the coefficient of at least one variable x_i:

$$f_1(\mathbf{x}) = a_{10} \oplus a_{11} x_1 \oplus \cdots \oplus a_{1n} x_n ,$$

$$f_2(\mathbf{x}) = a_{20} \oplus a_{21} x_1 \oplus \cdots \oplus a_{2n} x_n ,$$

$$(a_{11} \oplus a_{21}) \vee \ldots \vee (a_{1n} \oplus a_{2n}) = 1 .$$

For $n = 2$, all the invertible functions are linear and linearly independent functions.

Theorem 7.10 *The system of equations*

$$y_1 = f_1(x_1, x_2) \,,$$
$$y_2 = f_2(x_1, x_2)$$

can be inverted if and only if $f_1(x_1, x_2)$, $f_2(x_1, x_2)$ are linear and linearly independent functions.

There are eight linear functions of $n = 2$ variables, according to the eight vectors that are possible for a_0, a_1, a_2:

$$0(x_1, x_2), \; x_1, \; x_2, \; x_1 \oplus x_2, \; 1 \oplus x_1, \; 1 \oplus x_2, \; 1 \oplus x_1 \oplus x_2, \; 1(x_1, x_2) \,.$$

The two constant functions cannot be used for $f_1(\mathbf{x})$ and $f_2(\mathbf{x})$, each one of the remaining six functions can be used as a pair with four of these functions; they cannot be used with itself and with its negated function because x_1 and $1 \oplus x_1$, for instance, are not linearly independent. Hence, we find exactly $6 \times 4 = 24$ different pairs, and this is exactly the number of invertible functions (which is equal to 4!). Some of these pairs of invertible equations are

$$y_1 = x_1 \,, \qquad\qquad\qquad y_2 = x_2 \,;$$
$$y_1 = x_1 \oplus x_2 \,, \qquad\qquad y_2 = 1 \oplus x_1 \,;$$
$$y_1 = 1 \oplus x_1 \,, \qquad\qquad y_2 = 1 \oplus x_2 \,.$$

This concept of linear and linearly dependent functions can be generalized.

Theorem 7.11 *Let $f_1(\mathbf{x}), \ldots, f_n(\mathbf{x})$ be linear and pairwise linearly independent functions. Then the system of equations*

$$y_1 = f_1(\mathbf{x}) \,,$$
$$\vdots \quad\;\; \vdots$$
$$y_n = f_n(\mathbf{x})$$

is invertible.

For $n > 2$, however, most of the unique mappings are implemented by nonlinear functions. There are (without consideration of the special role of a_0) 2^{n+1} linear functions which can be combined (without consideration of the linear independency) with each other n times, and this results in $2^{(n+1)n}$ different possibilities, for $n > 2$; however, $2^{(n+1)n} < 2^n!$. For $n = 3$, we have $2^{4 \times 3} = 2^{12} = 4.096$, but $2^3! = 8! = 40.320$, and this difference will be larger and larger when n increases.

Example 7.19 The function $h(x_1, x_2, x_3, y_1, y_2, y_3)$ is given by both the function table and the Karnaugh-map of Fig. 7.10.

This function satisfies the conditions (7.54) and (7.55) for all y_i and x_i, $i = 1, 2, 3$, so that $h(\mathbf{x}, \mathbf{y}) = 0$ is uniquely solvable with regard to both all variable y_i and all variables x_i.

x_1 x_2 x_3	y_1 y_2 y_3	$h(\mathbf{x},\mathbf{y})$
0 0 0	1 0 0	0
0 0 1	1 0 1	0
0 1 0	1 1 1	0
0 1 1	0 0 0	0
1 0 0	0 1 1	0
1 0 1	0 1 0	0
1 1 0	0 0 1	0
1 1 1	1 1 0	0
otherwise		1

$y_1\,y_2\,y_3$ — $h(\mathbf{x},\mathbf{y})$

$y_1\,y_2\,y_3$								
0 0 0	1	1	0	1	1	1	1	1
0 0 1	1	1	1	1	0	1	1	1
0 1 1	1	1	1	1	1	1	1	0
0 1 0	1	1	1	1	1	1	0	1
1 1 0	1	1	1	1	1	0	1	1
1 1 1	1	1	1	0	1	1	1	1
1 0 1	1	0	1	1	1	1	1	1
1 0 0	0	1	1	1	1	1	1	1
	0	1	1	0	0	1	1	0	x_3
	0	0	1	1	1	1	0	0	x_2
	0	0	0	0	1	1	1	1	x_1

Fig. 7.10 Function $h(\mathbf{x},\mathbf{y})$ that describes an invertible system of equations

The function table of Fig. 7.10 shows that all the vectors of $\mathbb{B}^3$ satisfying the equation $h(\mathbf{x},\mathbf{y}) = 0$ occur in the table for (x_1, x_2, x_3) as well as for (y_1, y_2, y_3). In the Karnaugh-map of Fig. 7.10 it can be seen that each row and each column contains exactly one value 0. As solution of $h(\mathbf{x},\mathbf{y}) = 0$ with regard to $\mathbf{y}$ we get the unique solution-functions:

$$y_1 = f_1(\mathbf{x}) = \overline{x}_1 \oplus x_2 x_3 \,,$$

$$y_2 = f_2(\mathbf{x}) = x_2 \oplus x_1 \oplus x_2 x_3 \,,$$

$$y_3 = f_3(\mathbf{x}) = x_3 \oplus x_1 \oplus x_2 \oplus x_1 x_2 \,,$$

and the unique solution-functions of $h(\mathbf{x},\mathbf{y}) = 0$ with regard to $\mathbf{x}$ are

$$x_1 = g_1(\mathbf{y}) = y_2 \oplus y_3 \oplus y_1 y_3 \oplus y_2 y_3 \,,$$

$$x_2 = g_2(\mathbf{y}) = y_2 \oplus \overline{y}_1 \,,$$

$$x_3 = g_3(\mathbf{y}) = y_3 \oplus \overline{y}_1 \oplus y_1 y_2 \,.$$

It is quite easy to remember that equivalence equations can be considered in the same way. The easy transition to antivalence equations makes it superfluous to elaborate these possibilities in detail.

Solution of Characteristic Equations with Regard to Variables

The solution-set of a characteristic equation $f(\mathbf{x},\mathbf{y}) = 1$ comprises all binary vectors for which the function $f(\mathbf{x},\mathbf{y})$ is equal to 1. This is the opposite to a restrictive equation $f(\mathbf{x},\mathbf{y}) = 0$, where the solution-set contains all binary vectors for which the function $f(\mathbf{x},\mathbf{y})$ is equal to 0. Hence, the solution of characteristic equations with regard to variables evaluates the values 1 of the function $f(\mathbf{x},\mathbf{y})$ instead of the values 0 as done in case of the restrictive equations.

Due to this relation between the two special types of equations, which can to be solved with regard to a subset of variables, we restrict ourselves to the definition and most important rules to solve a characteristic equation with regard to variables.

Definition 7.8 Let $\mathbf{x} = (x_1, \ldots, x_n)$, $\mathbf{y} = (y_1, \ldots, y_k)$, $f(\mathbf{x},\mathbf{y})$ be a logic function of $n+k$ variables. The characteristic equation $f(\mathbf{x},\mathbf{y}) = 1$ can be solved with regard to the variables $y_1, \ldots, y_k$ if there are functions

$$y_1 = g_1(\mathbf{x})$$
$$y_2 = g_2(\mathbf{x})$$
$$\vdots \qquad \vdots$$
$$y_k = g_k(\mathbf{x})$$

that satisfy

$$f(\mathbf{x}, g_1(\mathbf{x}), \ldots, g_k(\mathbf{x})) = 1 \,.$$

Transferring our knowledge from the restrictive equations, we get as condition that the characteristic equation $f(\mathbf{x}, \mathbf{y}) = 1$ can be solved with regard to the variables $\mathbf{y}$ that for each $\mathbf{x}$ exists at least one function value 1.

Theorem 7.12 *The characteristic equation $f(\mathbf{x}, \mathbf{y}) = 1$, $\mathbf{x} = (x_1, \ldots, x_n)$, $\mathbf{y} = (y_1, \ldots, y_k)$, can be solved with regard to the k variables $\mathbf{y}$ if and only if*

$$\max_{\mathbf{y}}^k f(\mathbf{x}, \mathbf{y}) = 1 \,. \tag{7.56}$$

Proof The functions $g_1(\mathbf{x}), \ldots, g_k(\mathbf{x})$ of Definition 7.8 define for each $\mathbf{x}$ a function value for the variables $\mathbf{y}$. The function $f(\mathbf{x}, \mathbf{y})$ has the same set of solutions as this uniquely defined system of equations. Hence, the function $f(\mathbf{x}, \mathbf{y})$ has for each assignment of $\mathbf{x}$ exactly one binary vector $\mathbf{y}$ with $f(\mathbf{x}, \mathbf{y}) = 1$ so that Condition (7.56) is satisfied.

Condition (7.56) is satisfied if for each $\mathbf{x}$ at least one binary vector $\mathbf{y}$ with $f(\mathbf{x}, \mathbf{y}) = 1$ exists. Hence, in the case of a single solution of $f(\mathbf{x}, \mathbf{y}) = 1$ for each $\mathbf{x}$ we get the uniquely defined system of equations of Definition 7.8 as solution of the characteristic equation $f(\mathbf{x}, \mathbf{y}) = 1$ with regard to the k variables $\mathbf{y}$; in the case of more than one solution of $f(\mathbf{x}, \mathbf{y}) = 1$ exist for one or several $\mathbf{x}$, at least one solution-function $g_i(\mathbf{x})$ can be chosen out of a lattice of solution-functions. □

In general the solution-function $g_i(\mathbf{x})$ belonging to the variable y_i of the characteristic equation $f(\mathbf{x}, \mathbf{y}) = 1$ can be chosen out of a lattice of functions $\mathcal{L}\langle g_{iq}(\mathbf{x}), g_{ir}(\mathbf{x})\rangle$. The ON-set-function $g_{iq}(\mathbf{x})$ of this lattice (the infimum of $\mathcal{L}$: $\inf \mathcal{L}$) is equal to 1 if there is (independent of the values of the other variables $\mathbf{y} \setminus y_i$) at least one solution of $f(\mathbf{x}, \mathbf{y}) = 1$ for $y_i = 1$, but no solution of this equation for $y_i = 0$:

$$g_{iq}(\mathbf{x}) = \max_{\mathbf{y}}^k (y_i \wedge f(\mathbf{x}, \mathbf{y})) \wedge \overline{\max_{\mathbf{y}}^k (\overline{y}_i \wedge f(\mathbf{x}, \mathbf{y}))} \,.$$

This mark-function must often be calculated so that we try to find an easier formula. With this aim we extend this formula by an expression that is equal to 0 and apply the distributive law and the rules of the Boolean Differential Calculus:

$$g_{iq}(\mathbf{x}) = \left(\max_{\mathbf{y}}^k (y_i \wedge f(\mathbf{x}, \mathbf{y})) \wedge \overline{\max_{\mathbf{y}}^k (\overline{y}_i \wedge f(\mathbf{x}, \mathbf{y}))} \right) \vee 0$$

$$= \left(\max_{\mathbf{y}}^k (y_i \wedge f(\mathbf{x}, \mathbf{y})) \wedge \overline{\max_{\mathbf{y}}^k (\overline{y}_i \wedge f(\mathbf{x}, \mathbf{y}))} \right)$$

$$\vee \left(\max_{\mathbf{y}}^k (\overline{y}_i \wedge f(\mathbf{x}, \mathbf{y})) \wedge \overline{\max_{\mathbf{y}}^k (\overline{y}_i \wedge f(\mathbf{x}, \mathbf{y}))} \right)$$

$$= \overline{\max_{\mathbf{y}}^k \left(\overline{y}_i \wedge f(\mathbf{x}, \mathbf{y}) \right)} \wedge \left(\max_{\mathbf{y}}^k \left(y_i \wedge f(\mathbf{x}, \mathbf{y}) \right) \vee \max_{\mathbf{y}}^k \left(\overline{y}_i \wedge f(\mathbf{x}, \mathbf{y}) \right) \right)$$

$$= \overline{\max_{\mathbf{y}}^k \left(\overline{y}_i \wedge f(\mathbf{x}, \mathbf{y}) \right)} \wedge \left(\max_{\mathbf{y}}^k \left(y_i \wedge f(\mathbf{x}, \mathbf{y}) \vee \overline{y}_i \wedge f(\mathbf{x}, \mathbf{y}) \right) \right)$$

$$= \overline{\max_{\mathbf{y}}^k \left(\overline{y}_i \wedge f(\mathbf{x}, \mathbf{y}) \right)} \wedge \left(\max_{\mathbf{y}}^k \left((y_i \vee \overline{y}_i) \wedge f(\mathbf{x}, \mathbf{y}) \right) \right)$$

$$= \overline{\max_{\mathbf{y}}^k \left(\overline{y}_i \wedge f(\mathbf{x}, \mathbf{y}) \right)} \wedge \left(\max_{\mathbf{y}}^k \left(f(\mathbf{x}, \mathbf{y}) \right) \right) .$$

The right term in the last expression must be equal to 1 due to Condition (7.56); hence, we get the simplified formula:

$$g_{iq}(\mathbf{x}) = \overline{\max_{\mathbf{y}}^k \left(\overline{y}_i \wedge f(\mathbf{x}, \mathbf{y}) \right)} \tag{7.57}$$

to calculate the ON-set-function $g_{iq}(\mathbf{x})$ for the lattice $\mathcal{L}_i \langle g_{iq}(\mathbf{x}), g_{ir}(\mathbf{x}) \rangle$ of the variable y_i.

The OFF-set-function $g_{ir}(\mathbf{x})$ of the lattice $\mathcal{L}_i \langle g_{iq}(\mathbf{x}), g_{ir}(\mathbf{x}) \rangle$ (the complement of the supremum of $\mathcal{L}$: $\overline{\sup \mathcal{L}}$) is equal to 1 if there is (independent of the values of the other variables $\mathbf{y} \setminus y_i$) at least one solution of $f(\mathbf{x}, \mathbf{y}) = 1$ for $y_i = 0$, but no solution of this equation for $y_i = 1$:

$$g_{ir}(\mathbf{x}) = \max_{\mathbf{y}}^k \left(\overline{y}_i \wedge f(\mathbf{x}, \mathbf{y}) \right) \wedge \overline{\max_{\mathbf{y}}^k \left(y_i \wedge f(\mathbf{x}, \mathbf{y}) \right)}$$

that can also be simplified to

$$g_{ir}(\mathbf{x}) = \overline{\max_{\mathbf{y}}^k \left(y_i \wedge f(\mathbf{x}, \mathbf{y}) \right)} . \tag{7.58}$$

It is a benefit that the complement in (7.57) and (7.58) must be calculated for the result of the k-fold maximum because this maximum reduces the number of variables by k. Hence, the alternative formulas

$$g_{iq}(\mathbf{x}) = \min_{\mathbf{y}}^k \left(y_i \vee \overline{f(\mathbf{x}, \mathbf{y})} \right) ,$$

$$g_{ir}(\mathbf{x}) = \min_{\mathbf{y}}^k \left(\overline{y}_i \vee \overline{f(\mathbf{x}, \mathbf{y})} \right)$$

should not be used.

Example 7.20 We use the function

$$f(\mathbf{x}, \mathbf{y}) = x_1 \overline{x}_2 x_3 y_1 \vee \overline{x}_1 \overline{x}_2 x_3 \overline{y}_1 y_2 \vee x_2 \overline{x}_3 y_1 y_2 \vee x_2 x_3 \overline{y}_2 \vee \overline{x}_2 \overline{x}_3 y_1 \overline{y}_2$$

and try to solve the characteristic equation

$$f(x_1, x_2, x_3, y_1, y_2) = 1$$

for this function with regard to the variables y_1 and y_2.

First, we use Condition (7.56) to verify whether this equation is solvable with regard to the variables y_1 and y_2:

$$\max_{\mathbf{y}}^2 \left(x_1\overline{x}_2x_3y_1 \vee \overline{x}_1\overline{x}_2x_3\overline{y}_1y_2 \vee x_2\overline{x}_3y_1y_2 \vee x_2x_3\overline{y}_2 \vee \overline{x}_2\overline{x}_3y_1\overline{y}_2 \right)$$

$$= x_1\overline{x}_2x_3 \vee \overline{x}_1\overline{x}_2x_3 \vee x_2\overline{x}_3 \vee x_2x_3 \vee \overline{x}_2\overline{x}_3$$

$$= 1 \, .$$

The explored function satisfies this condition; hence, the associated characteristic equation can be solved with regard to the variables y_1 and y_2 and we can calculate the mark-function of the lattices $\mathcal{L}\langle g_{1q}(\mathbf{x}), g_{1r}(\mathbf{x})\rangle$ and $\mathcal{L}\langle g_{2q}(\mathbf{x}), g_{2r}(\mathbf{x})\rangle$ using the simplified formulas (7.57) and (7.58) as follows:

$$g_{1q}(\mathbf{x}) = \overline{\max_{\mathbf{y}}^k \left(\overline{x}_1\overline{x}_2x_3\overline{y}_1y_2 \vee x_2x_3\overline{y}_1\overline{y}_2 \right)}$$

$$= \overline{\overline{x}_1\overline{x}_2x_3 \vee x_2x_3}$$

$$= \overline{x}_3 \vee x_1\overline{x}_2 \, ,$$

$$g_{1r}(\mathbf{x}) = \overline{\max_{\mathbf{y}}^k \left(x_1\overline{x}_2x_3y_1 \vee x_2\overline{x}_3y_1y_2 \vee x_2x_3y_1\overline{y}_2 \vee \overline{x}_2\overline{x}_3y_1\overline{y}_2 \right)}$$

$$= \overline{x_1\overline{x}_2x_3 \vee x_2\overline{x}_3 \vee x_2x_3 \vee \overline{x}_2\overline{x}_3}$$

$$= \overline{x}_1\overline{x}_2x_3 \, ,$$

$$g_{2q}(\mathbf{x}) = \overline{\max_{\mathbf{y}}^k \left(x_1\overline{x}_2x_3y_1\overline{y}_2 \vee x_2x_3\overline{y}_2 \vee \overline{x}_2\overline{x}_3y_1\overline{y}_2 \right)}$$

$$= \overline{x_1\overline{x}_2x_3 \vee x_2x_3 \vee \overline{x}_2\overline{x}_3}$$

$$= x_2\overline{x}_3 \vee \overline{x}_1\overline{x}_2x_3 \, ,$$

$$g_{2r}(\mathbf{x}) = \overline{\max_{\mathbf{y}}^k \left(x_1\overline{x}_2x_3y_1y_2 \vee \overline{x}_1\overline{x}_2x_3\overline{y}_1y_2 \vee x_2\overline{x}_3y_1y_2 \right)}$$

$$= \overline{x_1\overline{x}_2x_3 \vee \overline{x}_1\overline{x}_2x_3 \vee x_2\overline{x}_3}$$

$$= x_2 \odot x_3 \, .$$

Figure 7.11 shows on the left-hand side the Karnaugh-maps of $f(\mathbf{x}, \mathbf{y})$ and the calculated mark-functions of the solution-lattices in an arrangement that facilitates a direct comparison between these function; these mark-functions are merged within Karnaugh-maps of the lattices $\mathcal{L}\langle g_{1q}(\mathbf{x}), g_{1r}(\mathbf{x})\rangle$ and $\mathcal{L}\langle g_{2q}(\mathbf{x}), g_{2r}(\mathbf{x})\rangle$ on the right-hand side. Additionally, all functions of these two lattices are given.

The don't-cares of these two lattices do not overlap; hence, each of the eight pairs of solution-function satisfies the explored characteristic equation. We use the simplest solution-functions:

$$y_1 = g_{13}(\mathbf{x}) = x_1 \vee \overline{x}_3 \, ,$$

$$y_2 = g_{22}(\mathbf{x}) = x_2 \oplus x_3$$

to verify this solution:

$$x_1\overline{x}_2x_3y_1 \vee \overline{x}_1\overline{x}_2x_3\overline{y}_1y_2 \vee x_2\overline{x}_3y_1y_2 \vee x_2x_3\overline{y}_2 \vee \overline{x}_2\overline{x}_3y_1\overline{y}_2 = 1$$

Fig. 7.11 Solution of the characteristic equation $f(\mathbf{x}, \mathbf{y}) = 1$ with regard to the variables y_1 and y_2

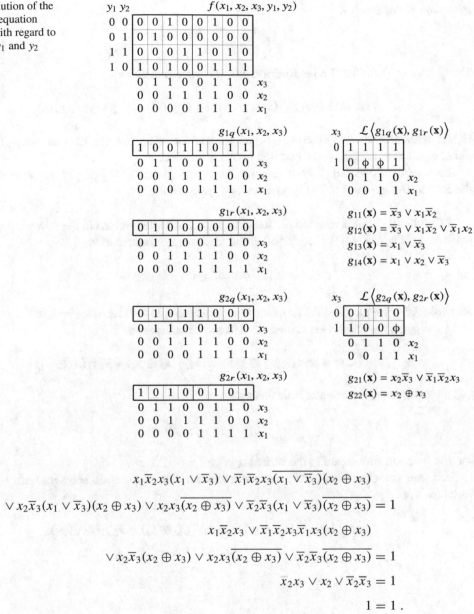

$$x_1\overline{x}_2x_3(x_1 \vee \overline{x}_3) \vee \overline{x}_1\overline{x}_2x_3\overline{(x_1 \vee \overline{x}_3)}(x_2 \oplus x_3)$$

$$\vee x_2\overline{x}_3(x_1 \vee \overline{x}_3)(x_2 \oplus x_3) \vee x_2x_3\overline{(x_2 \oplus x_3)} \vee \overline{x}_2\overline{x}_3(x_1 \vee \overline{x}_3)\overline{(x_2 \oplus x_3)} = 1$$

$$x_1\overline{x}_2x_3 \vee \overline{x}_1\overline{x}_2x_3\overline{x}_1x_3(x_2 \oplus x_3)$$

$$\vee x_2\overline{x}_3(x_2 \oplus x_3) \vee x_2x_3\overline{(x_2 \oplus x_3)} \vee \overline{x}_2\overline{x}_3\overline{(x_2 \oplus x_3)} = 1$$

$$\overline{x}_2x_3 \vee x_2 \vee \overline{x}_2\overline{x}_3 = 1$$

$$1 = 1 \, .$$

It can be verified in the same way that the other seven pairs of solution-functions $y_1 = g_{1i}(\mathbf{x})$ and $y_2 = g_{2i}(\mathbf{x})$ satisfy the explored characteristic equation.

In the case that a characteristic equation $f(\mathbf{x}, \mathbf{y}) = 1$ is solvable with regard to the variables $\mathbf{y}$, one, several, or even all functions can be uniquely specified. The function $y_i = g_i(\mathbf{x})$ is uniquely specified if there is for each $\mathbf{x}$ an assignment to the variables $\mathbf{y} \setminus y_i$ such that the function $f(\mathbf{x}, \mathbf{y})$ is equal to 1 either for $y_i = 1$ or $y_i = 0$.

Theorem 7.13 *Let be given a characteristic equation $f(\mathbf{x}, \mathbf{y}) = 1$, $\mathbf{x} = (x_1, \dots, x_n)$, $\mathbf{y} = (y_1, \dots, y_k)$, that is solvable with regard to the k variables $\mathbf{y}$. Then the function $y_i = g_i(\mathbf{x})$ is uniquely* *determined if and only if*

$$\mathop{\mathrm{der}}_{y_i} \left(\mathop{\max}_{\mathbf{y} \setminus y_i}^{k-1} f(\mathbf{x}, \mathbf{y}) \right) = 1 . \tag{7.59}$$

Proof Due to Definition 7.8 the function $f(\mathbf{x}, \mathbf{y})$ is determined by

$$f(\mathbf{x}, \mathbf{y}) = (y_1 \odot g_1(\mathbf{x})) \wedge \cdots \wedge (y_i \odot g_i(\mathbf{x})) \wedge \cdots \wedge (y_k \odot g_k(\mathbf{x})) .$$

Hence, a uniquely defined function $g_i(\mathbf{x})$ determines for each $\mathbf{x}$ that the function $\max_{\mathbf{y} \setminus y_i}^{k-1} f(\mathbf{x}, \mathbf{y})$ is either equal to y_i or $\overline{y}_i$ so that Condition (7.59) is satisfied.

Reversely, if Condition (7.59) is satisfied, the function $\max_{\mathbf{y} \setminus y_i}^{k-1} f(\mathbf{x}, \mathbf{y})$ is equal to either y_i or $\overline{y}_i$ for each $\mathbf{x}$. Hence, the function $g_i(\mathbf{x})$ is uniquely specified. $\qquad \square$

If Condition (7.59) is satisfied the function $g_i(\mathbf{x})$ can be determined in the following way: $g_i(\mathbf{x})$ is equal to 1 if $(y_i \wedge f(\mathbf{x}, \mathbf{y}))$ is equal to 1 for at least one assignment of $\mathbf{y}$:

$$g_i(\mathbf{x}) = \mathop{\max}_{\mathbf{y}}^k (y_i \wedge f(\mathbf{x}, \mathbf{y})) . \tag{7.60}$$

Example 7.21 We modified the function $f(\mathbf{x}, \mathbf{y})$ of Example 7.20 such that both functions $y_1 = g_1(\mathbf{x})$ and $y_2 = g_2(\mathbf{x})$ are uniquely specified. This leads to the function

$$f(\mathbf{x}, \mathbf{y}) = x_1 x_3 y_1 (x_2 \oplus y_2) \vee \overline{x}_1 x_3 \overline{y}_1 (x_2 \oplus y_2) \vee \overline{x}_3 y_1 (x_2 \odot y_2)$$

and we try to solve the characteristic equation

$$f(x_1, x_2, x_3, y_1, y_2) = 1$$

for this function with regard to the variables y_1 and y_2.

First, we use Condition (7.56) to verify whether this equation is solvable with regard to the variables y_1 and y_2:

$$\mathop{\max}_{\mathbf{y}}^2 \left(x_1 x_3 y_1 (x_2 \oplus y_2) \vee \overline{x}_1 x_3 \overline{y}_1 (x_2 \oplus y_2) \vee \overline{x}_3 y_1 (x_2 \odot y_2) \right)$$

$$= x_1 x_3 \vee \overline{x}_1 x_3 \vee \overline{x}_3$$

$$= 1 .$$

The explored function satisfies this condition; hence, the characteristic equation is solvable with regard to the variables y_1 and y_2 and we verify next whether this equation is even uniquely solvable with regard to the variables y_1 and y_2 using Condition (7.59):

$$\mathop{\mathrm{der}}_{y_1} \left(\mathop{\max}_{y_2} \left(x_1 x_3 y_1 (x_2 \oplus y_2) \vee \overline{x}_1 x_3 \overline{y}_1 (x_2 \oplus y_2) \vee \overline{x}_3 y_1 (x_2 \odot y_2) \right) \right)$$

$$= \mathop{\mathrm{der}}_{y_1} \left(x_1 x_3 y_1 \vee \overline{x}_1 x_3 \overline{y}_1 \vee \overline{x}_3 y_1 \right)$$

$$= (\overline{x}_1 x_3) \oplus (x_1 x_3 \vee \overline{x}_3)$$

$$= 1 ,$$

$$\operatorname*{der}_{y_2} \left(\operatorname*{max}_{y_1} \left(x_1 x_3 y_1 (x_2 \oplus y_2) \vee \overline{x}_1 x_3 \overline{y}_1 (x_2 \oplus y_2) \vee \overline{x}_3 y_1 (x_2 \odot y_2) \right) \right)$$

$$= \operatorname*{der}_{y_2} \left(x_1 x_3 (x_2 \oplus y_2) \vee \overline{x}_1 x_3 (x_2 \oplus y_2) \vee \overline{x}_3 (x_2 \odot y_2) \right)$$

$$= (x_1 x_3 x_2 \vee \overline{x}_1 x_3 x_2 \vee \overline{x}_3 \overline{x}_2) \oplus (x_1 x_3 \overline{x}_2 \vee \overline{x}_1 x_3 \overline{x}_2 \vee \overline{x}_3 x_2)$$

$$= (1 \oplus x_2 \oplus x_3) \oplus (x_2 \oplus x_3)$$

$$= 1 .$$

Hence, the explored characteristic equation $f(\mathbf{x}, \mathbf{y}) = 1$ is uniquely solvable with regard to both y_1 and y_2; the two uniquely determined functions $y_1 = g_1(\mathbf{x})$ and $y_2 = g_2(\mathbf{x})$ can be calculated using (7.60):

$$g_1(\mathbf{x}) = \operatorname*{max}_{\mathbf{y}}{}^2 \left(y_1 \wedge \left[x_1 x_3 y_1 (x_2 \oplus y_2) \vee \overline{x}_1 x_3 \overline{y}_1 (x_2 \oplus y_2) \vee \overline{x}_3 y_1 (x_2 \odot y_2) \right] \right)$$

$$= \operatorname*{max}_{\mathbf{y}}{}^2 \left(x_1 x_3 y_1 (x_2 \oplus y_2) \vee \overline{x}_3 y_1 (x_2 \odot y_2) \right)$$

$$= x_1 x_3 \vee \overline{x}_3$$

$$= x_1 \vee \overline{x}_3 ,$$

$$g_2(\mathbf{x}) = \operatorname*{max}_{\mathbf{y}}{}^2 \left(y_2 \wedge \left[x_1 x_3 y_1 (x_2 \oplus y_2) \vee \overline{x}_1 x_3 \overline{y}_1 (x_2 \oplus y_2) \vee \overline{x}_3 y_1 (x_2 \odot y_2) \right] \right)$$

$$= \operatorname*{max}_{\mathbf{y}}{}^2 \left(x_1 x_3 y_1 \overline{x}_2 y_2 \vee \overline{x}_1 x_3 \overline{y}_1 \overline{x}_2 y_2 \vee \overline{x}_3 y_1 x_2 y_2 \right)$$

$$= x_1 x_3 \overline{x}_2 \vee \overline{x}_1 x_3 \overline{x}_2 \vee \overline{x}_3 x_2$$

$$= x_3 \overline{x}_2 \vee \overline{x}_3 x_2$$

$$= x_2 \oplus x_3 .$$

Figure 7.12 shows on the left-hand side the Karnaugh-maps of $f(\mathbf{x}, \mathbf{y})$ and the calculated unique solution-functions in an arrangement that facilitates a direct comparison between these functions and on the right-hand side the Karnaugh-maps of $g_1(\mathbf{x})$ and $g_2(\mathbf{x})$ in the usual distribution of the variables. Additionally the expressions of the two uniquely functions $y_1 = g_1(\mathbf{x})$ and $y_2 = g_2(\mathbf{x})$ are given.

A single value 1 in each column of the Karnaugh-map of $f(\mathbf{x}, \mathbf{y})$, which uses the assignment of the variables as shown in Fig. 7.12, indicates that the associated characteristic equation $f(\mathbf{x}, \mathbf{y}) = 1$ is uniquely solvable with regard to all variables $\mathbf{y}$.

The substitution of the functions $y_1 = g_1(\mathbf{x})$ and $y_2 = g_2(\mathbf{x})$ into the characteristic equation $f(\mathbf{x}, \mathbf{y}) = 1$ followed by equivalent transformations confirm that these functions satisfy the explored equation:

$$x_1 x_3 y_1 (x_2 \oplus y_2) \vee \overline{x}_1 x_3 \overline{y}_1 (x_2 \oplus y_2) \vee \overline{x}_3 y_1 (x_2 \odot y_2) = 1$$

$$x_1 x_3 (x_1 \vee \overline{x}_3)(x_2 \oplus (x_2 \oplus x_3)) \vee \overline{x}_1 x_3 \overline{(x_1 \vee \overline{x}_3)}(x_2 \oplus (x_2 \oplus x_3))$$

$$\vee \overline{x}_3 (x_1 \vee \overline{x}_3)(x_2 \odot (x_2 \oplus x_3)) = 1$$

$$x_1 x_3 (x_3) \vee \overline{x}_1 x_3 (\overline{x}_1 x_3)(x_3) \vee \overline{x}_3 (1 \oplus x_2 \oplus x_2 \oplus x_3) = 1$$

$$x_1 x_3 \vee \overline{x}_1 x_3 \vee \overline{x}_3 = 1$$

$$1 = 1 .$$

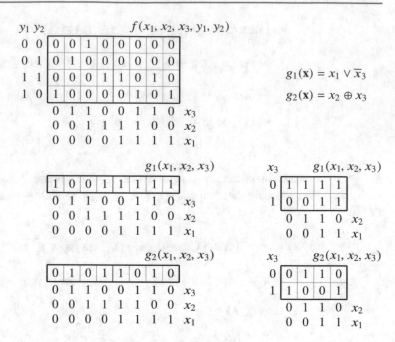

Fig. 7.12 Solution of the characteristic equation $f(\mathbf{x}, \mathbf{y}) = 1$ that is uniquely solvable with regard to the variables y_1 and y_2

7.5   Relations Between Lattices of Solution Functions

The previous sections provided the required information for a good understanding of all the solutions of a single logic equation, of a system of such equations, and a system of equations and inequalities with regard to variables. A broad range of problems is now open for the solution by means of this theory.

The availability of software for the Boolean Differential Calculus as well as the solution of equations allows the user to deal with the numerical solutions directly, but also to use analytical solutions. The reduction of all the problems (systems) to one single equation (at least in principle) defines a unique strategy for all the problems.

In order to consider the relations between the different solution vectors $y_1 = g_1(\mathbf{x}), \ldots, y_k = g_k(\mathbf{x})$, the simplest possibility is the application of the numerical methods that have been mentioned several times:

1. Solve $f(\mathbf{x}, \mathbf{y}) = 0$.
2. Sort the solution-set according to the $\mathbf{x}$-vectors.
3. Select one vector $\mathbf{y}^*$ for each vector $\mathbf{x}^*$ (if possible).
4. Determine the vector of solution-functions $(y_1 = g_1(\mathbf{x}), \ldots, y_k = g_k(\mathbf{x}))$.

We will also give an analytical solution for the problem of determining the relation between the different possibilities for these solution-vectors; however, this can be very often skipped when the explanations given so far are sufficient for the respective problem.

We will use the function $f(x_1, x_2, y_1, y_2)$ defined by the Karnaugh-map of Fig. 7.13a as example to solve the restrictive equation $f(x_1, x_2, y_1, y_2) = 0$ with regard to the variables y_1 and y_2. The four pairs of solution-functions are given by the function table of Fig. 7.13b.

The solution-lattices can be constructed as before; this results in two lattices $\mathcal{L}_1$ and $\mathcal{L}_2$ for y_1 and y_2:

$$\mathcal{L}_1 = \{g_{11}(\mathbf{x}), g_{12}(\mathbf{x})\}, \qquad \mathcal{L}_2 = \{g_{21}(\mathbf{x}), g_{22}(\mathbf{x}), g_{23}(\mathbf{x}), g_{24}(\mathbf{x})\},$$

with

$$g_{11}(x_1, x_2) = \bar{x}_1 x_2, \qquad\qquad g_{21}(x_1, x_2) = x_1 \bar{x}_2,$$
$$g_{12}(x_1, x_2) = 0, \qquad\qquad g_{22}(x_1, x_2) = x_1 \oplus x_2,$$
$$g_{23}(x_1, x_2) = \bar{x}_2,$$
$$g_{24}(x_1, x_2) = \bar{x}_1 \vee \bar{x}_2,$$

and only the combinations

$$(g_{11}(\mathbf{x}), g_{21}(\mathbf{x})), \ (g_{12}(\mathbf{x}), g_{22}(\mathbf{x})), \ (g_{11}(\mathbf{x}), g_{23}(\mathbf{x})), \ (g_{12}(\mathbf{x}), g_{24}(\mathbf{x}))$$

occur in the table of Fig. 7.13b Hence, we have the equivalent equation

$$f(\mathbf{x}, \mathbf{y}) = [(y_1 \oplus g_{11}(\mathbf{x})) \vee (y_2 \oplus g_{21}(\mathbf{x}))] [(y_1 \oplus g_{12}(\mathbf{x})) \vee (y_2 \oplus g_{22}(\mathbf{x}))]$$
$$[(y_1 \oplus g_{11}(\mathbf{x})) \vee (y_2 \oplus g_{23}(\mathbf{x}))] [(y_1 \oplus g_{12}(\mathbf{x})) \vee (y_2 \oplus g_{24}(\mathbf{x}))] = 0$$

and the elimination of the squared brackets results in

$$f(\mathbf{x}, \mathbf{y}) = (y_1 \oplus g_{11}(\mathbf{x}))(y_1 \oplus g_{12}(\mathbf{x}))$$
$$\vee (y_2 \oplus g_{21}(\mathbf{x}))(y_2 \oplus g_{22}(\mathbf{x}))(y_2 \oplus g_{23}(\mathbf{x}))(y_2 \oplus g_{24}(\mathbf{x}))$$
$$\vee (y_1 \oplus g_{11}(\mathbf{x}))(y_2 \oplus g_{22}(\mathbf{x}))(y_2 \oplus g_{24}(\mathbf{x}))$$
$$\vee (y_1 \oplus g_{12}(\mathbf{x}))(y_2 \oplus g_{21}(\mathbf{x}))(y_2 \oplus g_{23}(\mathbf{x})) = 0.$$

The conjunctions in the first two lines of this equation characterize the lattices $\mathcal{L}_1$ and $\mathcal{L}_2$; the combinations of solution-functions $(g_{11}(\mathbf{x}), g_{22}(\mathbf{x}))$, $(g_{11}(\mathbf{x}), g_{24}(\mathbf{x}))$ as well as $(g_{12}(\mathbf{x}), g_{21}(\mathbf{x}))$, $(g_{12}(\mathbf{x}), g_{23}(\mathbf{x}))$ are excluded by the constraints

$$(y_1 \oplus g_{11}(\mathbf{x}))(y_2 \oplus g_{22}(\mathbf{x}))(y_2 \oplus g_{24}(\mathbf{x})) = 0,$$
$$\text{and} \quad (y_1 \oplus g_{12}(\mathbf{x}))(y_2 \oplus g_{21}(\mathbf{x}))(y_2 \oplus g_{23}(\mathbf{x})) = 0.$$

When only the first part of the equation is considered, the result is a function $f^*(\mathbf{x}, \mathbf{y})$ with

$y_1\ y_2$	$f(x_1, x_2, y_1, y_2)$
0 0	0 1 0 1
0 1	0 0 1 0
1 1	1 1 1 1
1 0	1 0 1 1
	0 1 1 0 x_2
a	0 0 1 1 x_1

	solution 1		solution 2		solution 3		solution 4	
$x_1\ x_2$	y_1	y_2	y_1	y_2	y_1	y_2	y_1	y_2
0 0	0	0	0	0	0	1	0	1
0 1	1	0	0	1	1	0	0	1
1 0	0	1	0	1	0	1	0	1
1 1	0	0	0	0	0	0	0	0

Fig. 7.13 Solution of the restrictive equation $f(x_1, x_2, y_1, y_2) = 0$ with regard to the variables y_1 and y_2

$$f^*(\mathbf{x}, \mathbf{y}) = (y_1 \oplus g_{11}(\mathbf{x}))(y_1 \oplus g_{12}(\mathbf{x}))$$

$$\vee \, (y_2 \oplus g_{21}(\mathbf{x}))(y_2 \oplus g_{22}(\mathbf{x}))(y_2 \oplus g_{23}(\mathbf{x}))(y_2 \oplus g_{24}(\mathbf{x})) = 0 \,.$$

Based on the construction, it can be seen that all the solutions of the equation $f(\mathbf{x}, \mathbf{y}) = 0$ with regard to the variables y_1 and y_2 are also solutions of $f^*(\mathbf{x}, \mathbf{y}) = 0$ with regard to these variables; however, some new solutions can be generated by illegal combinations. In our example, for instance, the combination

$$y_1 = g_{11}(\mathbf{x}) = \overline{x}_1 x_2 \,, \quad y_2 = g_{22}(\mathbf{x}) = x_1 \oplus x_2$$

generates the value 0 for $(x_1, x_2, y_1, y_2) = (0111)$. The vectors that change the value from $f(\mathbf{x}, \mathbf{y})$ to $f^*(\mathbf{x}, \mathbf{y})$ can be determined by solving the equation

$$f^*(\mathbf{x}, \mathbf{y}) \oplus f(\mathbf{x}, \mathbf{y}) = 0 \,.$$

In the example, we find two vectors where $f(\mathbf{x}, \mathbf{y}) = 1$ and $f^*(\mathbf{x}, \mathbf{y}) = 0$.

$y_1\,y_2$	$f(\mathbf{x},\mathbf{y}) \oplus f^*(\mathbf{x},\mathbf{y})$			
0 0	0	1	0	0
0 1	0	0	0	0
1 1	0	1	0	0
1 0	0	0	0	0
	0	1	1	0 x_2
	0	0	1	1 x_1

Based on this construction for two variables, the general results can now be formulated.

Theorem 7.14 *Let the equation $f(\mathbf{x}, \mathbf{y}) = 0$ be solvable with regard to $y_1, \ldots, y_k$; let $\mathcal{L}_1, \ldots, \mathcal{L}_k$ be the lattices of solution-functions:*

$$\mathcal{L}_1 = \{g_{11}(\mathbf{x}), \ldots, g_{1i_1}(\mathbf{x})\} \,,$$

$$\vdots$$

$$\mathcal{L}_k = \{g_{k1}(\mathbf{x}), \ldots, g_{ki_k}(\mathbf{x})\} \,;$$

and

$$f^*(\mathbf{x}, \mathbf{y}) = \quad (y_1 \oplus g_{11}(\mathbf{x}))(y_1 \oplus g_{12}(\mathbf{x})) \ldots (y_1 \oplus g_{1i_1}(\mathbf{x}))$$

$$\vee \; \vdots$$

$$\vee \, (y_k \oplus g_{k1}(\mathbf{x}))(y_k \oplus g_{k2}(\mathbf{x})) \ldots (y_k \oplus g_{ki_k}(\mathbf{x})) \,.$$

Then it holds that:

1. *$f^*(\mathbf{x}, \mathbf{y}) \leq f(\mathbf{x}, \mathbf{y})$; the equality holds if and only if all $y_i = g_i(\mathbf{x})$ are unique.*
2. *The equation $f^*(\mathbf{x}, \mathbf{y}) = 0$ has the same lattices of solution-functions as the equation $f(\mathbf{x}, \mathbf{y}) = 0$.*

3. *The set of all functions $h(\mathbf{x}, \mathbf{y})$ with $f^*(\mathbf{x}, \mathbf{y}) \leq h(\mathbf{x}, \mathbf{y}) \leq f(\mathbf{x}, \mathbf{y})$ is a lattice $\mathcal{L}$ of functions with $\inf \mathcal{L} = f^*(\mathbf{x}, \mathbf{y})$ and $\sup \mathcal{L} = f(\mathbf{x}, \mathbf{y})$; each equation $h(\mathbf{x}, \mathbf{y}) = 0$ with a function $h(\mathbf{x}, \mathbf{y})$ of this lattice $\mathcal{L}$ has the same lattices of solution-functions $\mathcal{L}_1, \ldots, \mathcal{L}_k$ like $f(\mathbf{x}, \mathbf{y}) = 0$.*
4. *Only such combinations of functions*

$$g_{1\alpha}(\mathbf{x}) \in \mathcal{L}_1, g_{2\beta}(\mathbf{x}) \in \mathcal{L}_2, \ldots, g_{k\gamma}(\mathbf{x}) \in \mathcal{L}_k$$

can be selected that satisfy

$$f(\mathbf{x}, g_{1\alpha}(\mathbf{x}), \ldots, g_{k\gamma}(\mathbf{x})) \oplus f^*(\mathbf{x}, g_{1\alpha}(\mathbf{x}), \ldots, g_{k\gamma}(\mathbf{x})) = 0 .$$

We go back to our example. The Karnaugh-map for $f(\mathbf{x}, \mathbf{y}) \oplus f^*(\mathbf{x}, \mathbf{y})$ showed already the positions that have been changed from $f(\mathbf{x}, \mathbf{y})$ to $f^*(\mathbf{x}, \mathbf{y})$. We replace now these two positions by ϕ and get the lattice $\mathcal{L}(\mathbf{x}, \mathbf{y})$.

$y_1 y_2$				$\mathcal{L}(\mathbf{x}, \mathbf{y})$
0 0	0	ϕ	0	1
0 1	0	0	1	0
1 1	1	ϕ	1	1
1 0	1	0	1	1
	0	1	1	0 x_2
	0	0	1	1 x_1

The lattice $\mathcal{L}(\mathbf{x}, \mathbf{y})$ contains four different functions. When the two ϕ are replaced by 0, then we get $h_0(\mathbf{x}, \mathbf{y}) = f^*(\mathbf{x}, \mathbf{y})$, when they are replaced by 1, then we get $h_3(\mathbf{x}, \mathbf{y}) = f(\mathbf{x}, \mathbf{y})$, and for the combinations $(1, 0)$ and $(0, 1)$ we get two more functions $h_1(\mathbf{x}, \mathbf{y})$ and $h_2(\mathbf{x}, \mathbf{y})$. All four functions $h_i(\mathbf{x}, \mathbf{y})$ are shown in Fig. 7.14.

We calculate the expression $f(\mathbf{x}, \mathbf{y}) \oplus f^*(\mathbf{x}, \mathbf{y})$ and get

$$f(\mathbf{x}, \mathbf{y}) \oplus f^*(\mathbf{x}, \mathbf{y}) = \overline{x}_1 x_2 (\overline{y}_1 \overline{y}_2 \vee y_1 y_2) .$$

When we insert $y_1 = g_{12}(\mathbf{x}) = 0$, $y_2 = g_{21}(\mathbf{x}) = x_1 \overline{x}_2$ into this equation, then we get

$$f(\mathbf{x}, \mathbf{y}) \oplus f^*(\mathbf{x}, \mathbf{y}) = \overline{x}_1 x_2 (1(\overline{x}_1 \vee x_2) \vee 0 x_1 \overline{x}_2) = \overline{x}_1 x_2 ;$$

this combination of $g_{12}(\mathbf{x})$ and $g_{21}(\mathbf{x})$ will not be permissible.

$y_1 y_2$				$h_0(\mathbf{x}, \mathbf{y})$	$y_1 y_2$				$h_1(\mathbf{x}, \mathbf{y})$	$y_1 y_2$				$h_2(\mathbf{x}, \mathbf{y})$	$y_1 y_2$				$h_3(\mathbf{x}, \mathbf{y})$
0 0	0	0	0	1	0 0	0	0	0	1	0 0	0	1	0	1	0 0	0	1	0	1
0 1	0	0	1	0	0 1	0	0	1	0	0 1	0	0	1	0	0 1	0	0	1	0
1 1	1	0	1	1	1 1	1	1	1	1	1 1	1	0	1	1	1 1	1	1	1	1
1 0	1	0	1	1	1 0	1	0	1	1	1 0	1	0	1	1	1 0	1	0	1	1
	0	1	1	0 x_2		0	1	1	0 x_2		0	1	1	0 x_2		0	1	1	0 x_2
	0	0	1	1 x_1		0	0	1	1 x_1		0	0	1	1 x_1		0	0	1	1 x_1

Fig. 7.14 Karnaugh-maps of the four functions belonging to the lattice $\mathcal{L}(\mathbf{x}, \mathbf{y})$

For $y_1 = g_{11}(\mathbf{x}) = \overline{x}_1 x_2$, $y_2 = g_{23}(\mathbf{x}) = \overline{x}_2$, we get

$$f(\mathbf{x}, \mathbf{y}) \oplus f^*(\mathbf{x}, \mathbf{y}) = \overline{x}_1 x_2((x_1 \vee \overline{x}_2)x_2 \vee \overline{x}_1 x_2 \overline{x}_2) = 0 ;$$

hence, this combination of solution-functions is allowed.

Based on this explanation, it can be seen that a whole lattice of logic functions $h_i(\mathbf{x}, \mathbf{y})$ exists for which the equations $h_i(\mathbf{x}, \mathbf{y}) = 0$ can be solved with regard to y_1 and y_2 and have the same lattices $\mathcal{L}_1, \ldots, \mathcal{L}_k$ of solution-functions as the equation $f(\mathbf{x}, \mathbf{y}) = 0$.

Theorem 7.15 *Let M be the set of all equations $f(\mathbf{x}, \mathbf{y}) = 0$ that can be solved with regard to $y_1, \ldots, y_k$, and let $\mathcal{L}_1, \ldots, \mathcal{L}_k$ be the associated lattices of solution-functions. The relation R over M with $f_1(\mathbf{x}, \mathbf{y})Rf_2(\mathbf{x}, \mathbf{y}) \iff f_1(\mathbf{x}, \mathbf{y})$ and $f_2(\mathbf{x}, \mathbf{y})$ has the same set of solution-lattices for $f_1(\mathbf{x}, \mathbf{y}), f_2(\mathbf{x}, \mathbf{y}) \in M$ is an equivalence relation.*

The reflexivity, transitivity, and symmetry of this relation follow directly from the definition of R. The set M will be partitioned into subsets with the same set of solution-intervals.

Theorem 7.16 *Let be given a lattice $\mathcal{L} = \{g_1(\mathbf{x}), \ldots, g_k(\mathbf{x})\}$ of k logic functions. Then there is a function $f(\mathbf{x}, y_1)$ so that $\mathcal{L}$ is the lattice of solution-functions of $f(\mathbf{x}, y_1) = 0$ with regard to y_1.*

Let M be a given set of lattices of logic functions,

$$M = \{\mathcal{L}_1, \mathcal{L}_2, \ldots, \mathcal{L}_k\} ,$$
$$\mathcal{L}_1 = \{g_{11}(\mathbf{x}), \ldots, g_{1i_1}(\mathbf{x})\} ,$$
$$\vdots$$
$$\mathcal{L}_k = \{g_{k1}(\mathbf{x}), \ldots, g_{ki_k}(\mathbf{x})\} ,$$

then there is a function $h(\mathbf{x}, \mathbf{y})$ such that M is the solution system of the equation $h(\mathbf{x}, \mathbf{y}) = 0$ with regard to $y_1, \ldots, y_k$.

Proof The proof of this theorem can be given by constructing the functions $f(\mathbf{x}, y_1)$ or $h(\mathbf{x}, \mathbf{y})$. The lattice $\mathcal{L}$ is given by $\inf \mathcal{L}$ and $\sup \mathcal{L}$. Each function $g_i(\mathbf{x}) \in \mathcal{L}$ has the value 1 for all $\mathbf{x}$-vectors where $\inf \mathcal{L} = 1$, the value 0 where $\sup \mathcal{L} = 0$, for all the other $\mathbf{x}$-vectors the values 0 and 1 are possible (indicated by ϕ). The function $f(\mathbf{x}, y_1)$ can now be defined as follows:

$$f(\mathbf{x}, y_1 = 1) = 0 \qquad\qquad \Longleftrightarrow \qquad\qquad \inf \mathcal{L} = 1 ,$$
$$f(\mathbf{x}, y_1 = 0) = 0 \qquad\qquad \Longleftrightarrow \qquad\qquad \sup \mathcal{L} = 0 ,$$
$$f(\mathbf{x}, y_1 = 1) = f(\mathbf{x}, y_1 = 0) = 0 \qquad \Longleftrightarrow \qquad \overline{\inf \mathcal{L}} \wedge \sup \mathcal{L} = 1 ,$$
$$f(\mathbf{x}, y_1) = 1 \qquad\qquad\qquad\qquad \text{otherwise.}$$

The function $f(\mathbf{x}, y_1)$ constructed in this way just has the lattice $\mathcal{L}$ as the set of solution-functions of the equation $f(\mathbf{x}, y_1) = 0$ with regard to y_1. If there are several lattices of logic functions $\mathcal{L}_1, \ldots, \mathcal{L}_k$, then a variable y_i is assigned to each lattice, and the values of the resulting functions are defined in the same way.   □

Example 7.22 A given lattice $\mathcal{L}(g_i(\mathbf{x}))$ contains four functions $g_i(\mathbf{x})$:

$\mathcal{L}(g_i(\mathbf{x}))$

0	1	φ	0	1	1	1	φ	
0	1	1	0	0	1	1	0	x_3
0	0	1	1	1	1	0	0	x_2
0	0	0	0	1	1	1	1	x_1

inf $\mathcal{L}(g_i(\mathbf{x}))$

0	1	0	0	1	1	1	0	
0	1	1	0	0	1	1	0	x_3
0	0	1	1	1	1	0	0	x_2
0	0	0	0	1	1	1	1	x_1

sup $\mathcal{L}(g_i(\mathbf{x}))$

0	1	1	0	1	1	1	1	
0	1	1	0	0	1	1	0	x_3
0	0	1	1	1	1	0	0	x_2
0	0	0	0	1	1	1	1	x_1

y_1 $\qquad\qquad f(\mathbf{x}, y_1)$

y_1								
0	0	1	0	0	1	1	1	0
1	1	0	0	1	0	0	0	0
	0	1	1	0	0	1	1	0	x_3
	0	0	1	1	1	1	0	0	x_2
	0	0	0	0	1	1	1	1	x_1

$\overline{\inf \mathcal{L}(g_i(\mathbf{x}))} \wedge \sup \mathcal{L}(g_i(\mathbf{x}))$

0	0	1	0	0	0	0	1	
0	1	1	0	0	1	1	0	x_3
0	0	1	1	1	1	0	0	x_2
0	0	0	0	1	1	1	1	x_1

Fig. 7.15 Construction of the function $f(\mathbf{x}, y_1)$ based on the lattice $\mathcal{L}(g_i(\mathbf{x}))$ that contains the four solution-functions $g_i(\mathbf{x})$ of the equation $f(\mathbf{x}, y_1) = 0$ with regard to the variable y_1

$$g_1(\mathbf{x}) = x_1(x_2 \vee x_3) \vee \overline{x}_2 x_3 , \qquad\qquad g_2(\mathbf{x}) = x_1(x_2 \vee x_3) \vee x_3 ,$$

$$g_3(\mathbf{x}) = x_1 \vee \overline{x}_2 x_3 , \qquad\qquad g_4(\mathbf{x}) = x_1 \vee x_3 ,$$

which are the solutions of an equation $f(\mathbf{x}, y_1) = 0$ with regard to y_1. Figure 7.15 shows the construction of the function $f(\mathbf{x}, y_1)$ based on the given lattice $\mathcal{L}(g_i(\mathbf{x}))$ using the method described in the proof of Theorem 7.16.

Hence, we get the following function $f(\mathbf{x}, y_1)$:

$$f(\mathbf{x}, y_1) = x_1 x_2 \overline{y}_1 \vee \overline{x}_1 \overline{x}_3 y_1 \vee \overline{x}_2 x_3 \overline{y}_1 .$$

The solution of $f(\mathbf{x}, y_1) = 0$ with regard to y_1 just results in the given lattice $\mathcal{L}(g_i(\mathbf{x}))$ of four functions shown in Fig. 7.15.

All the considerations that have been made so far remain valid when there are constraints for the problem. These constraints have to be taken into consideration when the function $f(\mathbf{x}, \mathbf{y})$ has no 0's in some subspaces $\mathbf{x} = \mathbf{x}^*$. These subspaces are solutions of the equation

$$\min_{\mathbf{y}}^k f(\mathbf{x}, \mathbf{y}) = 1 .$$

When these subspaces are excluded from the domain of the functions $y_i = g_i(\mathbf{x})$, all the previous considerations remain valid. The formulas for the lattices $\mathcal{L}_i$ of solution-functions give the value $y_i = g_i(\mathbf{x}^*) = 1$ for the infimum of $\mathcal{L}_i$ and $y_i(\mathbf{x}^*) = 0$ for the complement of the supremum of $\mathcal{L}_i$; since these vectors are excluded, we can generally use $y_i = g_i(\mathbf{x}) = \emptyset$, e.g., for optimization purposes.

Example 7.23 Let $f(\mathbf{x}, \mathbf{y}) = (x_1 \oplus x_2) \vee \overline{y}_2 \vee x_1 y_1$. Then we get

$$\min_{\mathbf{y}}^2 f(\mathbf{x}, \mathbf{y}) = x_1 \oplus x_2$$

Fig. 7.16 Karnaugh-maps of the given function $f(\mathbf{x}, \mathbf{y})$ and the modified function $f^*(\mathbf{x}, \mathbf{y})$ for which the equation $f^*(\mathbf{x}, \mathbf{y}) = 0$ can be solved with regard to y_1 and y_2

$y_1\ y_2$		$f(\mathbf{x}, \mathbf{y})$		
0 0	1	1	1	1
0 1	0	1	0	1
1 1	0	1	1	1
1 0	1	1	1	1

$$\begin{array}{cccc} 0 & 1 & 1 & 0 \quad x_2 \\ 0 & 0 & 1 & 1 \quad x_1 \end{array}$$

$y_1\ y_2$		$f^*(\mathbf{x}, \mathbf{y})$		
0 0	1	0	1	0
0 1	0	0	0	0
1 1	0	0	1	0
1 0	1	0	1	0

$$\begin{array}{cccc} 0 & 1 & 1 & 0 \quad x_2 \\ 0 & 0 & 1 & 1 \quad x_1 \end{array}$$

Fig. 7.17 Karnaugh-maps of the restricted function $f^*(\mathbf{x}, \mathbf{y})$ and the derived lattices $\mathcal{L}_1(g_{1i}(\mathbf{x}))$ and $\mathcal{L}_2(g_{2i}(\mathbf{x}))$

y_1		$\min_{y_2} f^*(\mathbf{x}, \mathbf{y})$		
0	0	0	0	0
1	0	0	1	0

$$\begin{array}{cccc} 0 & 1 & 1 & 0 \quad x_2 \\ 0 & 0 & 1 & 1 \quad x_1 \end{array}$$

y_2		$\min_{y_1} f^*(\mathbf{x}, \mathbf{y})$		
0	1	0	1	0
1	0	0	0	0

$$\begin{array}{cccc} 0 & 1 & 1 & 0 \quad x_2 \\ 0 & 0 & 1 & 1 \quad x_1 \end{array}$$

$$\mathcal{L}_1(g_{1i}(\mathbf{x}))$$

ϕ	ϕ	0	ϕ

$$\begin{array}{cccc} 0 & 1 & 1 & 0 \quad x_2 \\ 0 & 0 & 1 & 1 \quad x_1 \end{array}$$

$$\mathcal{L}_2(g_{2i}(\mathbf{x}))$$

1	ϕ	1	ϕ

$$\begin{array}{cccc} 0 & 1 & 1 & 0 \quad x_2 \\ 0 & 0 & 1 & 1 \quad x_1 \end{array}$$

so that the condition to solve the equation $f(\mathbf{x}, \mathbf{y}) = 0$ with regard to the variables y_1 and y_2 is not satisfied. Using the found restriction we get the modified equation

$$f^*(\mathbf{x}, \mathbf{y}) = ((x_1 \oplus x_2) \vee \overline{y}_2 \vee x_1 y_1) \wedge \overline{(x_1 \oplus x_2)} = 0 \tag{7.61}$$

that can be solved with regard to the variables y_1 and y_2. Figure 7.16 shows both the Karnaugh-map of $f(\mathbf{x}, \mathbf{y})$ and the modified function $f^*(\mathbf{x}, \mathbf{y})$.

Using the method explained above we get the mark-function of the solution-lattices that solve the equation $f^*(\mathbf{x}, \mathbf{y}) = 0$ with regard to y_1 and y_2 as follows:

$$\inf \mathcal{L}_1(g_1(\mathbf{x})) = g_{1q}(\mathbf{x}) = 0 \ ,$$

$$\sup \mathcal{L}_1(g_1(\mathbf{x})) = \overline{g_{1r}(\mathbf{x})} = \overline{x}_1 \vee \overline{x}_2 \ ,$$

$$\inf \mathcal{L}_2(g_2(\mathbf{x})) = g_{2q}(\mathbf{x}) = x_1 \odot x_2 \ ,$$

$$\sup \mathcal{L}_2(g_2(\mathbf{x})) = \overline{g_{2r}(\mathbf{x})} = 1 \ .$$

Figure 7.17 shows in the upper row the Karnaugh-maps of the $f^*(\mathbf{x}, \mathbf{y})$ restricted such that the dependency either only on y_1 or only on y_2 comes visible and below the Karnaugh-maps of the associated lattices $\mathcal{L}_1(g_{1i}(\mathbf{x}))$ and $\mathcal{L}_2(g_{2i}(\mathbf{x}))$ using an encoding of the variable (x_1, x_2) that directly facilitates the comparison with the Karnaugh-maps above and the function $f^*(\mathbf{x}, \mathbf{y})$ of Fig. 7.16. All pairs of functions of these two lattices are solution-functions of Eq. (7.61) with regard to the variables y_1 and y_2.

7.6   Functional Equations

We start with a very simple example that will demonstrate the possible solution-steps.
The Basic Problem

Let be given a function $f(\mathbf{x})$. Find all functions $g(\mathbf{x})$ with $f(\mathbf{x}) \wedge g(\mathbf{x}) = 0(\mathbf{x})$.

The question is now the finding of a set of functions with a given property. In order to emphasize this difference to the solution of equations considered so far, we call this type of equations *Functional Equations*.

The solution-method is quite typical. Since $f(\mathbf{x})$ is given, the function vector of $g(\mathbf{x})$ can be found. If, for a given vector $\mathbf{x}^*$, $f(\mathbf{x}^*) = 0$, then $g(\mathbf{x}^*)$ can take any value. If, however, $f(\mathbf{x}^*) = 1$, then $g(\mathbf{x}^*)$ must be equal to 0. Hence, generally there will be a lattice $\mathcal{L}(g_i(\mathbf{x}))$ of solution-functions $g_1(\mathbf{x})$, $g_2(\mathbf{x}), \ldots$ which satisfy the functional equation.

Note 7.2 Functions $g_i(\mathbf{x})$ that satisfy this equation are *orthogonal* to $f(\mathbf{x})$.

Example 7.24 Let $f(x_1, x_2) = x_1 \oplus x_2$. Table 7.8 demonstrates the approach explained above to solve this functional equation.

The set of solution-functions is a lattice $\mathcal{L}(g_i(x_1, x_2))$ of four functions $g_i(x_1, x_2)$ that can be described by the ternary vector $(-00-)$. The four functions of this lattice are

$$g_1(x_1, x_2) = 0 \,,$$

$$g_2(x_1, x_2) = \overline{x}_1 \overline{x}_2 \,,$$

$$g_3(x_1, x_2) = x_1 x_2 \,,$$

$$g_4(x_1, x_2) = x_1 \odot x_2 \,,$$

where the infimum and supremum of this lattice are the functions:

$$\inf \mathcal{L}(g_i(\mathbf{x})) = g_1(x_1, x_2) \,,$$

$$\sup \mathcal{L}(g_i(\mathbf{x})) = g_4(x_1, x_2) \,.$$

Table 7.8 Solution of a functional equation

x_1	x_2	$f(x_1, x_2)$	$\mathcal{L}(g_i(x_1, x_2))$	$g_1(x_1, x_2)$	$g_2(x_1, x_2)$	$g_3(x_1, x_2)$	$g_4(x_1, x_2)$
0	0	0	ϕ	0	1	0	1
0	1	1	0	0	0	0	0
1	0	1	0	0	0	0	0
1	1	0	ϕ	0	0	1	1

The More General Problem

Find all pairs of functions $(f(\mathbf{x}), g(\mathbf{x}))$ so that $f(\mathbf{x}) \wedge g(\mathbf{x}) = 0$.

The easiest way to solve this problem would be a full enumeration of all possible functions $f(\mathbf{x})$, and for each function $f_i(\mathbf{x})$ the method from above could be applied. As a special case, we can use $f(\mathbf{x}) = 0(\mathbf{x})$; any function $g(\mathbf{x})$ can be used together with $0(\mathbf{x})$. If $f(\mathbf{x}) = 1(\mathbf{x})$, then $g(\mathbf{x}) = 0(\mathbf{x})$ is the only possibility. Another method that is very useful for many of these problems would be the use of *undefined coefficients*. We demonstrate this method for functions depending on two variables. As we have seen before, each function $f(x, y)$ can be represented as

$$f(x_1, x_2) = \overline{x}_1 \overline{x}_2 f_{00} \vee \overline{x}_1 x_2 f_{01} \vee x_1 \overline{x}_2 f_{10} \vee x_1 x_2 f_{11} .$$

We use this form to calculate $f(x_1, x_2) \wedge g(x_1, x_2)$:

$$f(x_1, x_2) \wedge g(x_1, x_2) = \overline{x}_1 \overline{x}_2 f_{00} g_{00} \vee \overline{x}_1 x_2 f_{01} g_{01} \vee x_1 \overline{x}_2 f_{10} g_{10} \vee x_1 x_2 f_{11} g_{11} .$$

The equation $f(x_1, x_2) \wedge g(x_1, x_2) = 0(x_1, x_2)$ will be satisfied if and only if

$$f_{00} \, g_{00} = 0 \, ,$$
$$f_{01} \, g_{01} = 0 \, ,$$
$$f_{10} \, g_{10} = 0 \, ,$$
$$f_{11} \, g_{11} = 0 \, .$$

The solution of this system is not difficult and can be seen in Table 7.9 using a representation of orthogonal ternary vectors.

Table 7.9 Solution coefficients of the general functional equation $f(\mathbf{x}) \wedge g(\mathbf{x}) = 0(\mathbf{x})$

f_{00}	g_{00}	f_{01}	g_{01}	f_{10}	g_{10}	f_{11}	g_{11}
0	–	0	–	0	–	0	–
0	–	0	–	0	–	1	0
0	–	0	–	1	0	0	–
0	–	0	–	1	0	1	0
0	–	1	0	0	–	0	–
0	–	1	0	0	–	1	0
0	–	1	0	1	0	0	–
0	–	1	0	1	0	1	0
1	0	0	–	0	–	0	–
1	0	0	–	0	–	1	0
1	0	0	–	1	0	0	–
1	0	0	–	1	0	1	0
1	0	1	0	0	–	0	–
1	0	1	0	0	–	1	0
1	0	1	0	1	0	0	–
1	0	1	0	1	0	1	0

This method of using undetermined coefficients can be used for all the equations that contain unknown functions. It will be used for the solution of Boolean differential equations as well.

7.7   Boolean Differential Equations

In the following we understand *Boolean Differential Equations* as an extension of *Functional Equations* in a very intuitive way; they contain *unknown functions* as well as *derivatives of these functions* (which have been defined before). Equations of this type are explored since the 1970s; for the first time [2] published a comprehensive theory and solution-algorithms for Boolean differential equations.

All the rules for systems of logic equations and possible transformations are valid for Boolean differential equations as well and will not be repeated. We use some examples in order to get an understanding for the problems that can be solved and the solution-methods. The complete theory and advanced solution-methods can be found, for instance, in [3].

We will explore equations of the following type: On the left-hand side of the equation we have a derivative of a function $f(\mathbf{x})$, on the right-hand side a function $g(\mathbf{x})$, such as

$$\underset{x_i}{\operatorname{der}} f(\mathbf{x}) = g(\mathbf{x}) , \tag{7.62}$$

$$\underset{x_i}{\operatorname{min}} f(\mathbf{x}) = g(\mathbf{x}) , \tag{7.63}$$

$$\underset{x_i}{\operatorname{max}} f(\mathbf{x}) = g(\mathbf{x}) . \tag{7.64}$$

Since the result of a single derivative operation with regard to x_i does not depend on this variable, the Boolean differential equations (7.62), (7.63), and (7.64) have a non-empty set of solution-functions $f(\mathbf{x})$ only if (7.65) holds

$$\underset{x_i}{\operatorname{der}} g(\mathbf{x}) = 0(\mathbf{x}) . \tag{7.65}$$

Example 7.25 We want to find the set of all functions that solve the Boolean differential equation

$$\underset{x_2}{\operatorname{der}} f(x_1, x_2, x_3) = \overline{x}_1 \oplus x_3. \tag{7.66}$$

In a first step Condition (7.65) must be checked:

$$\underset{x_2}{\operatorname{der}} g(x_1, x_3) = \underset{x_2}{\operatorname{der}}(\overline{x}_1 \oplus x_3) = 0(\mathbf{x}) .$$

Solution functions of (7.66) exist because this condition holds. The set of solution-functions can be built as follows:

$$f(x_1, x_2, x_3) = x_2 \,(\overline{x}_1 \oplus x_3) \oplus h(x_1, x_3) .$$

The substitution of this general solution into (7.66) confirms its correctness:

$$\operatorname*{der}_{x_2} f(x_1, x_2, x_3) = \overline{x}_1 \oplus x_3$$

$$\operatorname*{der}_{x_2}(x_2\,(\overline{x}_1 \oplus x_3) \oplus h(x_1, x_3)) = \overline{x}_1 \oplus x_3$$

$$\operatorname*{der}_{x_2}(x_2\,(\overline{x}_1 \oplus x_3)) \oplus \operatorname*{der}_{x_2} h(x_1, x_3) = \overline{x}_1 \oplus x_3$$

$$(\overline{x}_1 \oplus x_3)\,\operatorname*{der}_{x_2}(x_2) = \overline{x}_1 \oplus x_3$$

$$\overline{x}_1 \oplus x_3 = \overline{x}_1 \oplus x_3\;.$$

There are 16 functions $h(x_1, x_3)$ so that the desired set of solution-functions of (7.66) comprises the following functions $f(\mathbf{x}) = f(x_1, x_2, x_3)$:

$$f_0(\mathbf{x}) = x_2\,(\overline{x}_1 \oplus x_3) \oplus 0\;, \qquad\qquad f_8(\mathbf{x}) = x_2\,(\overline{x}_1 \oplus x_3) \oplus (x_3),$$

$$f_1(\mathbf{x}) = x_2\,(\overline{x}_1 \oplus x_3) \oplus (\overline{x}_1\,\overline{x}_3)\;, \qquad\qquad f_9(\mathbf{x}) = x_2\,(\overline{x}_1 \oplus x_3) \oplus (\overline{x}_1 \oplus x_3)\;,$$

$$f_2(\mathbf{x}) = x_2\,(\overline{x}_1 \oplus x_3) \oplus (x_1\,\overline{x}_3)\;, \qquad\qquad f_{10}(\mathbf{x}) = x_2\,(\overline{x}_1 \oplus x_3) \oplus (x_1 \oplus x_3)\;,$$

$$f_3(\mathbf{x}) = x_2\,(\overline{x}_1 \oplus x_3) \oplus (\overline{x}_1\,x_3)\;, \qquad\qquad f_{11}(\mathbf{x}) = x_2\,(\overline{x}_1 \oplus x_3) \oplus (\overline{x}_1 \vee \overline{x}_3)\;,$$

$$f_4(\mathbf{x}) = x_2\,(\overline{x}_1 \oplus x_3) \oplus (x_1\,x_3)\;, \qquad\qquad f_{12}(\mathbf{x}) = x_2\,(\overline{x}_1 \oplus x_3) \oplus (x_1 \vee \overline{x}_3)\;,$$

$$f_5(\mathbf{x}) = x_2\,(\overline{x}_1 \oplus x_3) \oplus (\overline{x}_1)\;, \qquad\qquad f_{13}(\mathbf{x}) = x_2\,(\overline{x}_1 \oplus x_3) \oplus (\overline{x}_1 \vee x_3)\;,$$

$$f_6(\mathbf{x}) = x_2\,(\overline{x}_1 \oplus x_3) \oplus (x_1)\;, \qquad\qquad f_{14}(\mathbf{x}) = x_2\,(\overline{x}_1 \oplus x_3) \oplus (x_1 \vee x_3)\;,$$

$$f_7(\mathbf{x}) = x_2\,(\overline{x}_1 \oplus x_3) \oplus (\overline{x}_3)\;, \qquad\qquad f_{15}(\mathbf{x}) = x_2\,(\overline{x}_1 \oplus x_3) \oplus 1\;.$$

The same condition (7.65) must be satisfied to find a non-empty solution-set of a Boolean differential equation where the single minimum $\min_{x_i} f(\mathbf{x})$ or the single maximum $\max_{x_i} f(\mathbf{x})$ appears on the left-hand side (see (7.63) and (7.64)).

Then the general solution of

$$\min_{x_i} f(\mathbf{x}) = g(\mathbf{x})$$

is equal to

$$f(\mathbf{x}) = g(\mathbf{x}) \vee h(\mathbf{x}) \quad \text{with} \quad \min_{x_i} h(\mathbf{x}) = 0\;,$$

and the general solution of

$$\max_{x_i} f(\mathbf{x}) = g(\mathbf{x})$$

is equal to

$$f(\mathbf{x}) = g(\mathbf{x}) \wedge h(\mathbf{x}) \quad \text{with} \quad \max_{x_i} h(\mathbf{x}) = 1\;. \tag{7.67}$$

Similar general solutions exist when on the left-hand side vectorial derivative operations are used:

$$\operatorname*{der}_{\mathbf{x_0}} f(\mathbf{x}) = g(\mathbf{x})\;, \tag{7.68}$$

$$\min_{\mathbf{x_0}} f(\mathbf{x}) = g(\mathbf{x})\;, \tag{7.69}$$

$$\max_{\mathbf{x_0}} f(\mathbf{x}) = g(\mathbf{x}) . \tag{7.70}$$

For these derivative operations pairs of function values with the distance of the selected change are compared with each other. Therefore, we have the following condition for the solution:

$$\operatorname*{der}_{\mathbf{x_0}} g(\mathbf{x_0}, \mathbf{x_1}) = 0(\mathbf{x_0}, \mathbf{x_1}) . \tag{7.71}$$

Finally, k-fold derivative operations can be used on the left-hand side:

$$\operatorname*{der}_{\mathbf{x_0}}^{k} f(\mathbf{x_0}, \mathbf{x_1}) = g(\mathbf{x_0}, \mathbf{x_1}) , \tag{7.72}$$

$$\operatorname*{min}_{\mathbf{x_0}}^{k} f(\mathbf{x_0}, \mathbf{x_1}) = g(\mathbf{x_0}, \mathbf{x_1}) , \tag{7.73}$$

$$\operatorname*{max}_{\mathbf{x_0}}^{k} f(\mathbf{x_0}, \mathbf{x_1}) = g(\mathbf{x_0}, \mathbf{x_1}) , \tag{7.74}$$

$$\Delta_{\mathbf{x_0}} f(\mathbf{x_0}, \mathbf{x_1}) = g(\mathbf{x_0}, \mathbf{x_1}) . \tag{7.75}$$

The results of all k-fold derivative operations of $f(\mathbf{x_0}, \mathbf{x_1})$ with regard to $\mathbf{x_0}$ are independent of all $x_i \in \mathbf{x_0}$. Thus, in the case of k-fold derivative operations the Boolean differential equations have a non-empty set of solution-functions if (7.76) is satisfied:

$$\Delta_{\mathbf{x_0}} g(\mathbf{x_0}, \mathbf{x_1}) = 0(\mathbf{x_0}, \mathbf{x_1}) . \tag{7.76}$$

The general solution can be constructed similarly to the case of single derivative operations.

As another step to solution-algorithms for more general Boolean differential equations we restrict the elements of this equation to one function $f(\mathbf{x})$ and all possible vectorial derivatives of this function:

$$\mathrm{D}\left(f(\mathbf{x}), \operatorname*{der}_{x_1} f(\mathbf{x}), \operatorname*{der}_{x_2} f(\mathbf{x}), \dots, \operatorname*{der}_{\mathbf{x}} f(\mathbf{x}) \right) = 0 . \tag{7.77}$$

The following definition introduces three concepts which support both the understanding and the practical solution-algorithms of Boolean differential equations.

Definition 7.9 Let $g(\mathbf{x})$ be a solution-function of (7.77). Then

1.
$$\left(g(\mathbf{x}), \operatorname*{der}_{x_1} g(\mathbf{x}), \operatorname*{der}_{x_2} g(\mathbf{x}), \dots, \operatorname*{der}_{\mathbf{x}} g(\mathbf{x}) \right) \Big|_{\mathbf{x}=\mathbf{c}} \tag{7.78}$$

is a *local solution* for $\mathbf{x} = \mathbf{c}$.

2.
$$\mathrm{D}(u_0, u_1, \dots, u_{2^n-1}) = 0 \tag{7.79}$$

is the logic equation, associated to the Boolean differential equation (7.77); and has the *set of local solutions* $SLS(\mathbf{u})$.

3.
$$\nabla g(\mathbf{x}) = \left(g(\mathbf{x}), \operatorname*{der}_{x_1} g(\mathbf{x}), \operatorname*{der}_{x_2} g(\mathbf{x}), \dots, \operatorname*{der}_{\mathbf{x}} g(\mathbf{x}) \right) . \tag{7.80}$$

The local solution (7.78) is the key to solve Boolean differential equations. Assume, the function $f(\mathbf{x}) = g(\mathbf{x})$ is one solution-function of (7.77). The vector of this function and all its vectorial derivatives describes completely the function itself and the changes of its values with respect to each direction. The local solution (7.78) restricts this knowledge to one of the 2^n points of $\mathbb{B}^n$. The local solution is a vector of 2^n values that determine the function value $g(\mathbf{c})$ and the values of all vectorial derivatives of the same function at the same point $\mathbf{x} = \mathbf{c}$. The information of a single local solution is sufficient in order to reconstruct the complete function $g(\mathbf{x})$ because the local function value and its changes in all directions are known.

The associated equation (7.79) helps us to solve the Boolean differential equation (7.77). The elements of (7.77) are the unknown function $f(\mathbf{x})$ and its vectorial derivatives. Since these elements can only appear in non-negated or negated form they can be assigned to logic variables $\mathbf{u}$. The index i of the variable u_i indicates the associated element of (7.77) in such a way that a value 1 in the binary code of the index determines a variable in the change vector of the associated derivative. Using this rule, we map

the function	$f(\mathbf{x})$	to	u_0 ,
the derivative	$\der_{x_1} f(\mathbf{x})$	to	u_1 ,
the derivative	$\der_{x_2} f(\mathbf{x})$	to	u_2 ,
the derivative	$\der_{(x_1,x_2)} f(\mathbf{x})$	to	u_3 ,
$\vdots$			
the derivative	$\der_{\mathbf{x}} f(\mathbf{x})$	to	u_{2^n-1} .

The solution of the logic equation (7.79) is the set $SLS(\mathbf{u})$ of binary vectors. Each of these vectors describes an allowed combination of the function $f(\mathbf{x})$ and all their vectorial derivatives (a zero indicates that the associated element must be negated).

The operator (7.80) helps to shorten expressions of further relations and rules.

After these preparations we study the solution-process of the Boolean differential equation (7.77). In the first step we create the associated equation (7.79), solve it, and get the set of local solutions $SLS(\mathbf{u})$. This solution-set comprises all local solutions (7.78) that are not in conflict with the Boolean differential equation (7.77). As discussed above, a local solution determines a complete function $f(\mathbf{x})$, but the existence of a local solution is only a necessary condition.

The logic function $f(\mathbf{x})$ is a solution-function of the Boolean differential equation (7.77) if the necessary and sufficient condition (7.81) becomes true:

$$\forall \mathbf{c} \in \mathbb{B}^n \quad \nabla f(\mathbf{x})\,|_{\mathbf{x}=\mathbf{c}} \in SLS(\mathbf{u}) . \tag{7.81}$$

The correctness of condition (7.81) follows from the property that a solution-function has local solutions for each point of $\mathbb{B}^n$.

If the expression

$$\exists \mathbf{c}' \in \mathbb{B}^n \quad \nabla f(\mathbf{x})\,|_{\mathbf{x}=\mathbf{c}'} \notin SLS(\mathbf{u}) \tag{7.82}$$

becomes true, the local solution $\nabla f(\mathbf{x})\,|_{\mathbf{x}=\mathbf{c}'}$ does not solve (7.79), and consequently, this $f(\mathbf{x})$ cannot be a solution-function of (7.77).

The conclusion from (7.82) is that 2^n local solutions are necessary so that one function $f(\mathbf{x})$ is part of the solution of (7.77).

Note 7.3 Not all of these local solution must be different.

Next, we analyze how many solution-functions $f(\mathbf{x})$ can be created from these 2^n local solutions.

Theorem 7.17 *If the logic function $f(\mathbf{x})$ is a solution-function of the Boolean differential equation (7.77), then all logic functions*

$$f(x_1, x_2, \ldots, x_n) = f(x_1 \oplus c_1, x_2 \oplus c_2, \ldots, x_n \oplus c_n) \tag{7.83}$$

for $\mathbf{c} = (c_1, \ldots, c_n) \in \mathbb{B}^n$ are also solution-functions of (7.77).

Proof The formula $x_i \oplus c_i$ can be expressed by

$$x_i \oplus c_i = \begin{cases} x_i & : & c_i = 0 \\ \overline{x}_i & : & c_i = 1 \end{cases} \tag{7.84}$$

1. Since

$$f(\mathbf{x}_0, x_i, \mathbf{x}_1) = \overline{x}_i\, f(\mathbf{x}_0, 0, \mathbf{x}_1) \oplus x_i\, f(\mathbf{x}_0, 1, \mathbf{x}_1)\,,$$

$$f(\mathbf{x}_0, \overline{x}_i, \mathbf{x}_1) = \overline{x}_i\, f(\mathbf{x}_0, 1, \mathbf{x}_1) \oplus x_i\, f(\mathbf{x}_0, 0, \mathbf{x}_1)\,,$$

it follows that

$$f(\mathbf{x}_0, x_i, \mathbf{x}_1)|_{(\mathbf{x}_0, x_i, \mathbf{x}_1)=(\mathbf{c}_0, c_i, \mathbf{c}_1)} = f(\mathbf{x}_0, \overline{x}_i, \mathbf{x}_1)|_{(\mathbf{x}_0, x_i, \mathbf{x}_1)=(\mathbf{c}_0, \overline{c}_i, \mathbf{c}_1)}\,,$$

$$\operatorname*{der}_{\mathbf{x}_0} f(\mathbf{x}_0, x_i, \mathbf{x}_1)|_{(\mathbf{x}_0, x_i, \mathbf{x}_1)=(\mathbf{c}_0, c_i, \mathbf{c}_1)} = \operatorname*{der}_{\mathbf{x}_0} f(\mathbf{x}_0, \overline{x}_i, \mathbf{x}_1)|_{(\mathbf{x}_0, x_i, \mathbf{x}_1)=(\mathbf{c}_0, \overline{c}_i, \mathbf{c}_1)}\,,$$

$$\operatorname*{der}_{(\mathbf{x}_0, x_i)} f(\mathbf{x}_0, x_i, \mathbf{x}_1)|_{(\mathbf{x}_0, x_i, \mathbf{x}_1)=(\mathbf{c}_0, c_i, \mathbf{c}_1)} = \operatorname*{der}_{(\mathbf{x}_0, x_i)} f(\mathbf{x}_0, \overline{x}_i, \mathbf{x}_1)|_{(\mathbf{x}_0, x_i, \mathbf{x}_1)=(\mathbf{c}_0, \overline{c}_i, \mathbf{c}_1)}\,,$$

and, therefore (7.83) is true for $\mathbf{c} = (\mathbf{c}_0, c_i, \mathbf{c}_1) = (\mathbf{0}, 1, \mathbf{0})$.
2. All 2^n vectors $\mathbf{c}$ can be created by the antivalence $\oplus$ of vectors with $|\mathbf{c}| = 1$. The theorem is completely proven when the previous item is applied 2^n times for different solution-functions in each iteration. $\qquad\square$

Additionally, it can be proven that (7.83) defines an *equivalence relation*. Consequently, (7.83) divides all logic functions $f(\mathbf{x}) : \mathbb{B}^n \to \mathbb{B}$ into *disjoint classes*, and the solution of the Boolean differential equation (7.77) will only consist of k *complete equivalence classes* of functions defined by (7.83), where $k \geq 0$.

Each vector $\mathbf{u}$ of the set $SLS(\mathbf{u})$ expresses the local solutions by the local function value u_0 and the values of the derivatives $u_i, 1 < i \leq 2^{n-1}$. In order to solve the Boolean differential equation (7.77), it is easier to manipulate function values instead of values of the derivatives. The vector $\mathbf{v}$ of values of the function contains the same information as the vector $\mathbf{u}$ and can be calculated by (7.85). The function $\mathtt{d2v}$ (*derivative to value*) transforms the set $SLS(\mathbf{u})$ into the set $S(\mathbf{v})$:

Table 7.10 Index pairs for the exchange of function values

$i=1$	$i=2$	$i=3$	$i=4$	$i=5$
$0 \Leftrightarrow 1$	$0 \Leftrightarrow 2$	$0 \Leftrightarrow 4$	$0 \Leftrightarrow 8$	$0 \Leftrightarrow 16$
$2 \Leftrightarrow 3$	$1 \Leftrightarrow 3$	$1 \Leftrightarrow 5$	$1 \Leftrightarrow 9$	$1 \Leftrightarrow 17$
$4 \Leftrightarrow 5$	$4 \Leftrightarrow 6$	$2 \Leftrightarrow 6$	$2 \Leftrightarrow 10$	$2 \Leftrightarrow 18$
$6 \Leftrightarrow 7$	$5 \Leftrightarrow 7$	$3 \Leftrightarrow 7$	$3 \Leftrightarrow 11$	$3 \Leftrightarrow 19$
$8 \Leftrightarrow 9$	$8 \Leftrightarrow 10$	$8 \Leftrightarrow 12$	$4 \Leftrightarrow 12$	$4 \Leftrightarrow 20$
$10 \Leftrightarrow 11$	$9 \Leftrightarrow 11$	$9 \Leftrightarrow 13$	$5 \Leftrightarrow 13$	$5 \Leftrightarrow 21$
$12 \Leftrightarrow 13$	$12 \Leftrightarrow 14$	$10 \Leftrightarrow 14$	$6 \Leftrightarrow 14$	$6 \Leftrightarrow 22$
$14 \Leftrightarrow 15$	$13 \Leftrightarrow 15$	$11 \Leftrightarrow 15$	$7 \Leftrightarrow 15$	$7 \Leftrightarrow 23$
$16 \Leftrightarrow 17$	$16 \Leftrightarrow 18$	$16 \Leftrightarrow 20$	$16 \Leftrightarrow 24$	$8 \Leftrightarrow 24$
$18 \Leftrightarrow 19$	$17 \Leftrightarrow 19$	$17 \Leftrightarrow 21$	$17 \Leftrightarrow 25$	$9 \Leftrightarrow 25$
$20 \Leftrightarrow 21$	$20 \Leftrightarrow 22$	$18 \Leftrightarrow 22$	$18 \Leftrightarrow 26$	$10 \Leftrightarrow 26$
$22 \Leftrightarrow 23$	$21 \Leftrightarrow 23$	$19 \Leftrightarrow 23$	$19 \Leftrightarrow 27$	$11 \Leftrightarrow 27$
$24 \Leftrightarrow 25$	$24 \Leftrightarrow 26$	$24 \Leftrightarrow 28$	$20 \Leftrightarrow 28$	$12 \Leftrightarrow 28$
$26 \Leftrightarrow 27$	$25 \Leftrightarrow 27$	$25 \Leftrightarrow 29$	$21 \Leftrightarrow 29$	$13 \Leftrightarrow 29$
$28 \Leftrightarrow 29$	$28 \Leftrightarrow 30$	$26 \Leftrightarrow 30$	$22 \Leftrightarrow 30$	$14 \Leftrightarrow 30$
$30 \Leftrightarrow 31$	$29 \Leftrightarrow 31$	$27 \Leftrightarrow 31$	$23 \Leftrightarrow 31$	$15 \Leftrightarrow 31$

$$v_0 = u_0 \,,$$

$$v_i = u_0 \oplus u_i \,, \quad \text{with} \quad i = 1, 2, \ldots, 2^n - 1 \,. \tag{7.85}$$

It follows from (7.83) that the exchange of x_i and $\overline{x}_i$ does not change the set of solution-functions. This change can be implemented by exchanging pairs of function values v_i (7.86) in the set $S(\mathbf{v})$:

$$v_{(m+2\,k*2^{i-1})} \Longleftrightarrow v_{(m+(2\,k+1)*2^{i-1})} \,, \tag{7.86}$$

$$\text{with} \quad i = 1, 2, \ldots, n \,,$$

$$m = 0, 1, \ldots, 2^{i-1} - 1 \,,$$

$$k = 0, 1, \ldots, 2^{n-i} - 1 \,.$$

The function `epv` (*exchange pairs of values*) exchanges function values of $S(\mathbf{v})$ with regard to a given index i (as defined in formula (7.86)) and returns the set $ST(\mathbf{v})$. Table 7.10 lists in 16 lines explicitly the index pairs defined by (7.86) to solve a Boolean differential equation (7.77) of 5 variables. The value of i indicates which variable x_i of the desired solution-function must be changed.

The set $S(\mathbf{v})$ contains the local solutions. The associated logic functions are possible solution-functions. The actual solution-functions are those functions for which (7.81) becomes true. The other local solutions must be excluded. The following algorithm solves this problem in the steps 4 and 5. The remaining local solutions describe the actual solution-functions of (7.77) by (7.87):

$$
\begin{aligned}
f(x_1, x_2, \ldots, x_n) = \ & \overline{x}_1 \overline{x}_2 \ldots \overline{x}_n \, v_0 \\
& \oplus x_1 \, \overline{x}_2 \ldots \overline{x}_n \, v_1 \\
& \oplus \overline{x}_1 \, x_2 \ldots \overline{x}_n \, v_2 \\
& \oplus \ldots \\
& \oplus x_1 \, x_2 \ldots x_n \, v_{2^n - 1} \,.
\end{aligned}
\tag{7.87}
$$

Algorithm 3 *Separation of Function Classes* solves the Boolean differential equation (7.77) using the solution-procedure explained above.

Algorithm 3 Separation of function classes

Input : Boolean differential equation (7.77) in which the function $f(\mathbf{x})$ depends on n variables
Output : set S of binary vectors $\mathbf{v} = (v_0, v_1, \ldots, v_{2^n-1})$ that describes substituted in (7.87) the set of all solution-functions of the Boolean differential equation (7.77)

1: $SLS(\mathbf{u}) \leftarrow$ solution of the logic equation (7.79) associated to (7.77)
2: $S(\mathbf{v}) \leftarrow \texttt{d2v}(SLS(\mathbf{u}))$
3: **for** $i \leftarrow 1$ to n **do**
4: $\quad ST(\mathbf{v}) \leftarrow \texttt{epv}(S(\mathbf{v}), i)$
5: $\quad S(\mathbf{v}) \leftarrow S(\mathbf{v}) \cap ST(\mathbf{v})$
6: **end for**

Example 7.26 We solve the Boolean differential equation

$$\operatorname*{der}_{x_2} f(\mathbf{x}) \oplus \overline{\operatorname*{der}_{x_1} f(\mathbf{x})} \operatorname*{der}_{(x_1,x_2)} f(\mathbf{x}) \oplus f(\mathbf{x}) \overline{\operatorname*{der}_{x_2} f(\mathbf{x})} \, \overline{\operatorname*{der}_{(x_1,x_2)} f(\mathbf{x})}$$

$$\oplus \overline{f(\mathbf{x})} \, \overline{\operatorname*{der}_{x_1} f(\mathbf{x})} \oplus f(\mathbf{x}) \operatorname*{der}_{x_1} f(\mathbf{x}) \operatorname*{der}_{x_2} f(\mathbf{x}) \operatorname*{der}_{(x_1,x_2)} f(\mathbf{x}) = 0 \tag{7.88}$$

and verify the result.

The logic equation associated to (7.88) is

$$u_2 \oplus \overline{u}_1 u_3 \oplus u_0 \overline{u}_2 \overline{u}_3 \oplus \overline{u}_0 \overline{u}_1 \oplus u_0 u_1 u_2 u_3 = 0 \,. \tag{7.89}$$

- Line 1 of Algorithm 3 *Separation of Function Classes*:
  The set of local solutions of (7.89) is

$$SLS(\mathbf{u}) = \begin{array}{cccc} u_0 & u_1 & u_2 & u_3 \\ \hline 0 & 1 & 0 & - \\ 1 & 1 & 0 & 1 \\ 1 & - & 1 & 1 \\ 0 & 0 & 1 & 0 \\ 0 & 0 & 0 & 1 \end{array} \,.$$

- Line 2 of Algorithm 3 *Separation of Function Classes*:
  The function $\texttt{d2v}$ maps $SLS(\mathbf{u})$ onto the set $S(\mathbf{v})$ of local solutions expressed by function values:

$$S(\mathbf{v}) = \begin{array}{cccc} v_0 & v_1 & v_2 & v_3 \\ \hline 0 & 1 & 0 & - \\ 1 & 0 & 1 & 0 \\ 1 & - & 0 & 0 \\ 0 & 0 & 1 & 0 \\ 0 & 0 & 0 & 1 \end{array} \,.$$

- Line 3 of Algorithm 3 *Separation of Function Classes*:
  First sweep of the `for`-loop $i = 1$.
- Line 4 of Algorithm 3 *Separation of Function Classes*:
  The function $\mathrm{epv}(S(\mathbf{v}), 1)$ exchanges the pairs of values (v_0, v_1) and (v_2, v_3), see Table 7.10, column $i = 1$. From $S(\mathbf{v})$ the transformed set $ST(\mathbf{v})$ is created:

$$
ST(\mathbf{v}) =
\begin{array}{cccc}
v_0 & v_1 & v_2 & v_3 \\
\hline
1 & 0 & - & 0 \\
0 & 1 & 0 & 1 \\
- & 1 & 0 & 0 \\
0 & 0 & 0 & 1 \\
0 & 0 & 1 & 0 \\
\hline
\end{array} .
$$

- Line 5 of Algorithm 3 *Separation of Function Classes*:
  The new set of local solutions $S(\mathbf{v})$ is calculated by the intersection of the previously determined sets $S(\mathbf{v}) \cap ST(\mathbf{v})$. In the example presented here, the intersection does not remove any logic vector:

$$
S(\mathbf{v}) =
\begin{array}{cccc}
v_0 & v_1 & v_2 & v_3 \\
\hline
0 & 1 & 0 & - \\
1 & 0 & 1 & 0 \\
1 & - & 0 & 0 \\
0 & 0 & 1 & 0 \\
0 & 0 & 0 & 1 \\
\hline
\end{array}
\cap
\begin{array}{cccc}
v_0 & v_1 & v_2 & v_3 \\
\hline
1 & 0 & - & 0 \\
0 & 1 & 0 & 1 \\
- & 1 & 0 & 0 \\
0 & 0 & 0 & 1 \\
0 & 0 & 1 & 0 \\
\hline
\end{array}
=
\begin{array}{cccc}
v_0 & v_1 & v_2 & v_3 \\
\hline
1 & 0 & - & 0 \\
0 & - & 0 & 1 \\
- & 1 & 0 & 0 \\
0 & 0 & 1 & 0 \\
\hline
\end{array} .
$$

- Line 3 of Algorithm 3 *Separation of Function Classes*:
  Second sweep of the `for`-loop $i = 2$.
- Line 4 of Algorithm 3 *Separation of Function Classes*:
  The function $\mathrm{epv}(S(\mathbf{v}), 2)$ exchanges pairs of values (v_0, v_2) and (v_1, v_3), see Table 7.10, column $i = 2$. The transformed set $ST(\mathbf{v})$ is created from the last calculated set $S(\mathbf{v})$:

$$
ST(\mathbf{v}) =
\begin{array}{cccc}
v_0 & v_1 & v_2 & v_3 \\
\hline
- & 0 & 1 & 0 \\
0 & 1 & 0 & - \\
0 & 0 & - & 1 \\
1 & 0 & 0 & 0 \\
\hline
\end{array} .
$$

- Line 5 of Algorithm 3 *Separation of Function Classes*:
  The new set of local solutions $S(\mathbf{v})$ is calculated by the intersection of the previously determined sets $S(\mathbf{v}) \cap ST(\mathbf{v})$. In this case, $\mathbf{v} = (1100)$ appears in $S(\mathbf{v})$, but not in $ST(\mathbf{v})$ so that the intersection removes this local solution:

$$
S(\mathbf{v}) =
\begin{array}{cccc}
v_0 & v_1 & v_2 & v_3 \\
\hline
1 & 0 & - & 0 \\
0 & - & 0 & 1 \\
- & 1 & 0 & 0 \\
0 & 0 & 1 & 0 \\
\hline
\end{array}
\cap
\begin{array}{cccc}
v_0 & v_1 & v_2 & v_3 \\
\hline
- & 0 & 1 & 0 \\
0 & 1 & 0 & - \\
0 & 0 & - & 1 \\
1 & 0 & 0 & 0 \\
\hline
\end{array}
=
\begin{array}{cccc}
v_0 & v_1 & v_2 & v_3 \\
\hline
1 & 0 & 1 & 0 \\
0 & 0 & 1 & 0 \\
0 & 1 & 0 & 1 \\
0 & 1 & 0 & 0 \\
0 & 0 & 0 & 1 \\
1 & 0 & 0 & 0 \\
\hline
\end{array} .
$$

- Line 3 of Algorithm 3 *Separation of Function Classes*:
Due to $i = 3 > 2 = n$ the for-loop is finished; hence, each vector of the finally calculated set $S(\mathbf{v})$ describes, after substitution into

$$f_i(x_1, x_2) = \overline{x}_1 \overline{x}_2 v_0 \oplus x_1 \overline{x}_2 v_1 \oplus \overline{x}_1 x_2 v_2 \oplus x_1 x_2 v_3 \tag{7.90}$$

a solution-function of the Boolean differential equation (7.88):

$$f_1(x_1, x_2) = \overline{x}_1 \overline{x}_2 1 \oplus x_1 \overline{x}_2 0 \oplus \overline{x}_1 x_2 1 \oplus x_1 x_2 0 = \overline{x}_1 \,,$$

$$f_2(x_1, x_2) = \overline{x}_1 \overline{x}_2 0 \oplus x_1 \overline{x}_2 0 \oplus \overline{x}_1 x_2 1 \oplus x_1 x_2 0 = \overline{x}_1 x_2 \,,$$

$$f_3(x_1, x_2) = \overline{x}_1 \overline{x}_2 0 \oplus x_1 \overline{x}_2 1 \oplus \overline{x}_1 x_2 0 \oplus x_1 x_2 1 = x_1 \,,$$

$$f_4(x_1, x_2) = \overline{x}_1 \overline{x}_2 0 \oplus x_1 \overline{x}_2 1 \oplus \overline{x}_1 x_2 0 \oplus x_1 x_2 0 = x_1 \overline{x}_2 \,,$$

$$f_5(x_1, x_2) = \overline{x}_1 \overline{x}_2 0 \oplus x_1 \overline{x}_2 0 \oplus \overline{x}_1 x_2 0 \oplus x_1 x_2 1 = x_1 x_2 \,,$$

$$f_6(x_1, x_2) = \overline{x}_1 \overline{x}_2 1 \oplus x_1 \overline{x}_2 0 \oplus \overline{x}_1 x_2 0 \oplus x_1 x_2 0 = \overline{x}_1 \overline{x}_2 \,.$$

There are two classes of solution-functions of the Boolean differential equation (7.88), $\{f_1(\mathbf{x}), f_3(\mathbf{x})\}$ and $\{f_2(\mathbf{x}), f_4(\mathbf{x}), f_5(\mathbf{x}), f_6(\mathbf{x})\}$.

We verify the solution using the function $f_5(x_1, x_2)$ of the second class and calculate first the derivatives:

$$\operatorname*{der}_{x_1} f_5(\mathbf{x}) = x_2 \,, \quad \operatorname*{der}_{x_2} f_5(\mathbf{x}) = x_1 \,, \quad \operatorname*{der}_{(x_1,x_2)} f_5(\mathbf{x}) = x_1 \oplus \overline{x}_2 \,.$$

Now we substitute these functions into (7.88) and simplify this equation:

$$x_1 \oplus \overline{x}_2 (x_1 \oplus \overline{x}_2) \oplus x_1 x_2 \overline{x}_1 \overline{(x_1 \oplus \overline{x}_2)}$$

$$\oplus \overline{(x_1 x_2)} \overline{x}_2 \oplus x_1 x_2 x_2 x_1 (x_1 \oplus \overline{x}_2) =$$

$$x_1 \oplus x_1 \overline{x}_2 \oplus \overline{x}_2 \oplus 0 \oplus \overline{x}_2 \oplus x_1 x_2 =$$

$$x_1(1 \oplus \overline{x}_2) \oplus x_1 x_2 =$$

$$x_1 x_2 \oplus x_1 x_2 = 0 \,.$$

This confirms that $f_5(x_1, x_2)$ is a solution-function of the Boolean differential equation (7.88).

On the contrary we check the function $f_c(x_1, x_2) = \overline{x}_2$. The associated local solution $\mathbf{v} = (1100)$ was removed in the second sweep of the algorithm. The derivatives of $f_c(x_1, x_2)$ are

$$\operatorname*{der}_{x_1} f_c(\mathbf{x}) = 0 \,, \quad \operatorname*{der}_{x_2} f_c(\mathbf{x}) = 1 \,, \quad \operatorname*{der}_{(x_1,x_2)} f_c(\mathbf{x}) = 1 \,.$$

which are substituted into (7.88), and we get a contradiction which confirms that $f_c(x_1, x_2)$ is not a solution-function:

$$1 \oplus \overline{0} \wedge 1 \oplus \overline{x}_2 \wedge \overline{1} \wedge \overline{1} \oplus \overline{x}_2 \wedge \overline{0} \oplus \overline{x}_2 \wedge 0 \wedge 1 \wedge 1 =$$

$$1 \oplus 1 \oplus 0 \oplus x_2 \oplus 0 =$$

$$x_2 \neq 0 \,.$$

We have studied the properties and a solution-method of the Boolean differential equation (7.77) in detail. Now we generalize these results. First we add variables $\mathbf{x}$, differentials $\mathbf{dx}$, and parameters $\mathbf{p}$ to (7.77) and get the Boolean differential equation (7.91):

$$\mathrm{D}\left(f(\mathbf{x}), \underset{x_1}{\operatorname{der}} f(\mathbf{x}), \underset{x_2}{\operatorname{der}} f(\mathbf{x}), \dots, \underset{\mathbf{x}}{\operatorname{der}} f(\mathbf{x}), \mathbf{dx}, \mathbf{p}, \mathbf{x}\right) = 0 . \tag{7.91}$$

The set of local solutions $SLS(\mathbf{u}, \mathbf{dx}, \mathbf{p}, \mathbf{x})$ of the logic equation

$$\mathrm{D}\left(u_0, u_1, \dots, u_{2^n-1}, \mathbf{dx}, \mathbf{p}, \mathbf{x}\right) = 0 \tag{7.92}$$

associated to (7.91) comprises both the logic variables u_i, associated to the derivatives, and the additional variables $\mathbf{dx}, \mathbf{p}, \mathbf{x}$.

A function $f(\mathbf{x})$ is a solution-function of the Boolean differential equation (7.91) if the condition (7.93) becomes true:

$$\forall \mathbf{c}_1 \in \mathbb{B}^n, \forall \mathbf{c}_2 \in \mathrm{d}\mathbb{B}^n : (\nabla f(\mathbf{x}), \mathbf{dx}, \mathbf{p}, \mathbf{x}) \Big|_{\substack{\mathbf{x} = \mathbf{c}_1 \\ \mathbf{dx} = \mathbf{c}_2 \\ \mathbf{p} = \mathbf{p}_0}} \in SLS(\mathbf{u}, \mathbf{dx}, \mathbf{p}, \mathbf{x}) . \tag{7.93}$$

There are three conclusions from (7.93):

1. The Boolean differential equation (7.91) can be solved for each given vector $\mathbf{p}_0$ of the parameters $\mathbf{p}$ separately; vice versa, the parameters make it possible to solve several Boolean differential equations in parallel.
2. The set of local solutions $SLS(\mathbf{u}, \mathbf{dx}, \mathbf{p}, \mathbf{x})$ can be simplified to $SLS(\mathbf{u}, \mathbf{p}, \mathbf{x})$ by

$$SLS(\mathbf{u}, \mathbf{p}, \mathbf{x}) = \min_{\mathbf{dx}}^{|\mathbf{dx}|} SLS(\mathbf{u}, \mathbf{dx}, \mathbf{p}, \mathbf{x}) .$$

3. The local solutions $SLS(\mathbf{u}, \mathbf{p}, \mathbf{x})$ are associated to special points $\mathbf{x}$ of $\mathbb{B}^n$; for that reason not classes of functions are the solutions of the Boolean differential equation (7.91), but arbitrary sets of logic functions.

Algorithm 3 *Separation of Function Classes* has been extended in [2] to Algorithm 4 *Separation of Functions* that solves the Boolean differential equation (7.91). The main difference between these two algorithms is that the formula in step 5 of Algorithm *Separation of Function Classes* is substituted by

$$S(\mathbf{v}, \mathbf{p}, \mathbf{x} \setminus (x_1, \dots, x_i)) \leftarrow S(\mathbf{v}, \mathbf{p}, \mathbf{x})\big|_{x_i=0} \cap \operatorname{epv}(S(\mathbf{v}, \mathbf{p}, \mathbf{x})\big|_{x_i=1}, i) .$$

Algorithm 4 *Separation of Functions* shows all steps to solve a Boolean differential equation (7.91) where the values of the parameter $\mathbf{p}$ are fixed to $\mathbf{p}_0$.

Note 7.4 In each sweep of the `for`-loop in Algorithm 4 one of the variables x_i is removed so that finally $S(\mathbf{v})$ only depends on variables $\mathbf{v}$. Algorithm 4 can also be used for Boolean differential equation (7.91) that contains parameters. These parameters remain in the solution and determine the associated set of solution-functions.

Algorithm 4 Separation of functions

Input : Boolean differential equation (7.91) in which the function $f(\mathbf{x})$ depends on n variables and the parameter $\mathbf{p}$ are fixed to $\mathbf{p}_0$
Output : set S of binary vectors $\mathbf{v} = (v_0, v_1, \ldots, v_{2^n-1})$ that describes substituted in (7.87) the set of all solution-functions of the Boolean differential equation (7.91) for $\mathbf{p} = \mathbf{p}_0$

1: $SLS(\mathbf{u}, \mathbf{dx}, \mathbf{x}) \leftarrow$ solution of the logic equation (7.92) associated to (7.91)
2: $SLS(\mathbf{u}, \mathbf{x}) \leftarrow \min_{\mathbf{dx}}^{|\mathbf{dx}|} SLS(\mathbf{u}, \mathbf{dx}, \mathbf{x})$
3: $S(\mathbf{v}, \mathbf{x}) \leftarrow \mathrm{d2v}(SLS(\mathbf{u}, \mathbf{x}))$
4: **for** $i \leftarrow 1$ to n **do**
5: $\quad S_0(\mathbf{v}, \mathbf{x} \setminus (x_1, \ldots, x_i)) \leftarrow \max_{x_i}[\overline{x}_i \wedge S(\mathbf{v}, \mathbf{x})]$
6: $\quad S_1(\mathbf{v}, \mathbf{x} \setminus (x_1, \ldots, x_i)) \leftarrow \max_{x_i}[x_i \wedge S(\mathbf{v}, \mathbf{x})]$
7: $\quad ST_1(\mathbf{v}, \mathbf{x} \setminus (x_1, \ldots, x_i)) \leftarrow \mathrm{epv}(S_1(\mathbf{v}, \mathbf{x} \setminus (x_1, \ldots, x_i)), i)$
8: $\quad S(\mathbf{v}, \mathbf{x} \setminus (x_1, \ldots, x_i)) \leftarrow S_0(\mathbf{v}, \mathbf{x} \setminus (x_1, \ldots, x_i)) \cap ST_1(\mathbf{v}, \mathbf{x} \setminus (x_1, \ldots, x_i))$
9: **end for**

More general Boolean differential equations that contain arbitrary derivative operations of the function $f(\mathbf{x})$ can be transformed into the unified Boolean differential equation (7.91).

The single minimum and single maximum can be expressed by the function and the single derivative:

$$\min_{x_i} f(\mathbf{x}) = f(\mathbf{x}) \wedge \overline{\mathrm{der}_{x_i} f(\mathbf{x})} \, ,$$

$$\max_{x_i} f(\mathbf{x}) = f(\mathbf{x}) \vee \mathrm{der}_{x_i} f(\mathbf{x}) \, .$$

The vectorial minimum and the vectorial maximum can also be expressed by the function $f(\mathbf{x}_0, \mathbf{x}_1)$ and vectorial derivatives:

$$\min_{\mathbf{x}_0} f(\mathbf{x}_0, \mathbf{x}_1) = f(\mathbf{x}_0, \mathbf{x}_1) \wedge \overline{\mathrm{der}_{\mathbf{x}_0} f(\mathbf{x}_0, \mathbf{x}_1)} \, .$$

$$\max_{\mathbf{x}_0} f(\mathbf{x}_0, \mathbf{x}_1) = f(\mathbf{x}_0, \mathbf{x}_1) \vee \mathrm{der}_{\mathbf{x}_0} f(\mathbf{x}_0, \mathbf{x}_1) \, .$$

All k-fold derivative operations can be expressed by means of the function $f(\mathbf{x}_0, \mathbf{x}_1)$ and all vectorial derivatives. The set $P(\mathbf{x}_0)$ indicates the power set of $\mathbf{x}_0$ in the following formulas:

$$\mathrm{der}_{\mathbf{x}_0}^k f(\mathbf{x}_0, \mathbf{x}_1) = \bigoplus_{\mathbf{x}_{0p} \in P(\mathbf{x}_0)} \mathrm{der}_{\mathbf{x}_{0p}} f(\mathbf{x}_0, \mathbf{x}_1) \, ,$$

$$\min_{\mathbf{x}_0}^k f(\mathbf{x}_0, \mathbf{x}_1) = f(\mathbf{x}_0, \mathbf{x}_1) \wedge \bigwedge_{\mathbf{x}_{0p} \in P(\mathbf{x}_0)} \overline{\mathrm{der}_{\mathbf{x}_{0p}} f(\mathbf{x}_0, \mathbf{x}_1)} \, ,$$

$$\max_{\mathbf{x}_0}^k f(\mathbf{x}_0, \mathbf{x}_1) = f(\mathbf{x}_0, \mathbf{x}_1) \vee \bigvee_{\mathbf{x}_{0p} \in P(\mathbf{x}_0)} \mathrm{der}_{\mathbf{x}_{0p}} f(\mathbf{x}_0, \mathbf{x}_1) \, ,$$

$$\Delta_{\mathbf{x}_0} f(\mathbf{x}_0, \mathbf{x}_1) = \bigvee_{\mathbf{x}_{0p} \in P(\mathbf{x}_0)} \mathrm{der}_{\mathbf{x}_{0p}} f(\mathbf{x}_0, \mathbf{x}_1) \, .$$

Based on these formulas any *derivative operation* (DEO) can be transformed into an expression that contains only the logic function and all its possible vectorial derivatives. Thus, each Boolean differential equation (7.94) can be solved in two steps:

1. The given Boolean differential equation

$$D_1(DEO_1(\mathbf{f}(\mathbf{x})), DEO_2(\mathbf{f}(\mathbf{x})), \ldots, \mathbf{f}(\mathbf{x}), \mathbf{dx}, \mathbf{p}, \mathbf{x}) =$$

$$D_2(DEO_1(\mathbf{f}(\mathbf{x})), DEO_2(\mathbf{f}(\mathbf{x})), \ldots, \mathbf{f}(\mathbf{x}), \mathbf{dx}, \mathbf{p}, \mathbf{x}) \qquad (7.94)$$

will be transformed into (7.91).
2. The simplified Boolean differential equation (7.91) is solved using Algorithm 4 *Separation of Functions*.

The same method can be applied to solve the most general Boolean differential equation that contains arbitrary differential operations and derivative operations of the function $f(\mathbf{x})$.

Using formulas introduced in Sect. 6.6 all total, partial, and k-fold differential operations can be mapped to expressions that only contain the function, its vectorial derivative operations, and the differentials of the variables. To solve such a most general Boolean differential equation we need three steps:

1. All differential operations are mapped onto the derivative operations of the Boolean differential equation (7.94).
2. The Boolean differential equation (7.94) is simplified to (7.91) as described above.
3. Using Algorithm 4 *Separation of Functions* we get all solution-functions.

7.8   Exercises

Prepare, as before, for each task of each exercise of this chapter a PRP for the XBOOLE-monitor XBM 2. Prefer a lexicographic order of the variables x_i. Use the help system of the XBOOLE-monitor XBM 2 to learn further details of the needed commands. Execute the created PRPs, verify the computed results, and explain the observed properties.

If you have not yet prepared the XBOOLE-monitor XBM 2 on your computer, you can get this XBOOLE-monitor free of charge by means of the following three steps:

1. **Download**:
   There are four versions of the XBOOLE-monitor XBM 2, two for Windows 10 or subsequent Windows systems (32 or 64 bits) and two for LINUX—Ubuntu (also 32 or 64 bits); you must download the version of the XBOOLE-monitor XBM 2 that fits your operating system.
   Authorized users of the online version of this chapter (https://doi.org/10.1007/978-3-030-88945-6_7) can download the XBOOLE-monitor XBM 2 directly from the web page
   https://link.springer.com/chapter/10.1007/978-3-030-88945-6_7
   where the links for the download of the XBOOLE-monitor XBM 2 are located in the part "Supplementary Information" (below the part "Abstract"). The headline above such a link indicates the associated zip-file of the XBOOLE-monitor XBM 2. The sizes of the zip-files have been provided behind the links and can be used to verify the download. A click on the link of the wanted version of the XBOOLE-monitor XBM 2 starts the download.
   Readers of the hardcopy of this book get access to the XBOOLE-monitor XBM 2 using the URL
   https://link.springer.com/chapter/10.1007/978-3-030-88945-6_7

to download the first two pages of this chapter. After this download, the same procedure as the authorized users of the online version of a chapter can be used to download the wanted version of the XBOOLE-monitor XBM 2.

2. **Unzip**: The XBOOLE-monitor XBM 2 must not be installed, but unzipped into an arbitrary directory of your computer. A convenient tool to unzip the downloaded zip-file is usually available as part of the operating system or can be downloaded from the Internet.

3. **Execute**:

   - Windows:
     The executable file of the two versions (32 or 64 bits) for Windows 10 (or subsequent Windows systems) of the XBOOLE-monitor XBM 2 is XBM2.exe; the other files in the expanded directory must remain unchanged. A double-click on the executable file XBM2.exe within the Explorer of Windows starts the XBOOLE-monitor XBM 2.

   - LINUX—Ubuntu:
     The unzipped folder of the XBOOLE-monitor XBM 2 contains for this operating system only the executable file XBM2-i386.AppImage for the version of 32 bits or XBM2-x86_64.AppImage for the version of 64 bits of the XBOOLE-monitor XBM 2. A double-click on the created AppImage-file within the file manager of LINUX—Ubuntu starts the XBOOLE-monitor XBM 2.

Exercise 7.1 (Incompletely Specified Functions)

An incompletely specified function (ISF) determines a set of completely specified functions which satisfies the properties of a lattice. Here we assume that the function or the ISF is given by a vector of function values in the order

$$\mathbf{x} = (x_1, x_2, x_3, x_4) : (0000), (0001), (0010), (0011), (0100), \ldots, (1110), (1111) \,.$$

Compute all functions and show the orthogonally minimized TVLs of the following lattices of functions:

(a) $F = (0\ 1\ 0\ 1\ 1\ 1\ 0\ 0\ 0\ 0\ 1\ 1\ 1\ 1\ 0\ 0)$.
(b) $F = (0\ 1\ 0\ 1\ 1\ 1\ 0\ 0\ 0\ -\ 1\ 1\ 1\ 1\ 0\ 0)$.
(c) $F = (0\ 1\ 0\ 1\ 1\ 1\ 0\ 0\ -\ -\ 1\ 1\ 1\ 1\ 0\ 0)$.
(d) $F = (-\ 1\ 0\ 1\ 1\ 1\ 0\ 0\ -\ -\ 1\ 1\ 1\ 1\ 0\ 0)$.

Exercise 7.2 (Mark-functions of a Lattice)

It is very convenient to describe a lattice (or the associated incompletely specified function) by the mark-functions $f_q(\mathbf{x})$ and $f_r(\mathbf{x})$. We explore in this exercise the lattice with the mark-functions:

$$f_q(\mathbf{x}) = x_1 \overline{x}_3 x_4 x_5 \vee \overline{x}_1 (x_2 \overline{x}_5 \vee \overline{x}_2 \overline{x}_4 x_5 \vee x_2 \overline{x}_3 x_4) \,,$$

$$f_r(\mathbf{x}) = x_1 (x_4 \overline{x}_5 \vee x_3 \overline{x}_5 \vee x_3 \overline{x}_4 \vee \overline{x}_2 x_3) \vee \overline{x}_1 \overline{x}_2 x_4 x_5 \,.$$

(a) Verify that the given mark-function $f_q(\mathbf{x})$ and $f_r(\mathbf{x})$ satisfy the condition

$$f_q(\mathbf{x}) \wedge f_r(\mathbf{x}) = 0$$

of a lattice. Show both the Karnaugh-maps and the TVLs of these mark-functions; use the first row of the m-fold view for the Karnaugh-maps and the second row for the TVLs. Assign the function $f_q(\mathbf{x})$ of the ON-set to column 1 and the function $f_r(\mathbf{x})$ of the OFF-set to column 3.

A lattice (or the associated incompletely specified function) that depends on up to ten variables can be represented as a single Karnaugh-map in a viewport of the XBOOLE-monitor XBM 2. The command `assign_qr` has been created for this task; note, the selected viewport must be switched to the view of the Karnaugh-map using the button located in its headline, otherwise only the TVL of the function $f_q(\mathbf{x})$ is shown. Show the Karnaugh-map of the given lattice in viewport (1,2) of the m-fold view and compare it with the Karnaugh-maps and the mark-functions shown to the left and to the right of this viewport.

(b) The number of different functions belonging to a lattice is equal to $2^{\|f_\varphi(\mathbf{x})\|}$, where the norm $\|f_\varphi(\mathbf{x})\|$ is equal to the number of function values 1 of $f_\varphi(\mathbf{x})$.

Compute the third mark-function f_φ of the given lattice and minimize this function using the command `obbc`. Show the TVL of $f_\varphi(\mathbf{x})$ in viewport (2,1) and the associated Karnaugh-map in viewport (2,2) of the m-fold view. Show in the first row the same Karnaugh-maps of the previous task so that the Karnaugh-maps of all three mark-functions and the Karnaugh-map of the lattice can be compared.

How many functions belong to the lattice explored in this exercise? Compute this number based on the function $f_\varphi(\mathbf{x})$ of the computed don't-care-set.

(c) The XBOOLE-monitor XBM 2 provides the desirable feature that the function values of $f_q(\mathbf{x})$ and $f_r(\mathbf{x})$ can be directly changed in the Karnaugh-map of a shown lattice. The details of this feature are described on the page

Edit a Karnaugh-Map for a Lattice of Functions

of the help system of the XBOOLE-monitor XBM 2. The mark-functions $f_q(\mathbf{x})$ and $f_r(\mathbf{x})$ of a lattice that have been determined in this way satisfy the condition $f_q(\mathbf{x}) \wedge f_r(\mathbf{x}) = 0$ of a lattice. Verify that the given mark-functions $f_q(\mathbf{x})$ and $f_r(\mathbf{x})$ can be determined by editing the Karnaugh-map of a lattice for the initial mark-functions $f_q'(\mathbf{x}) = 0$ and $f_r'(\mathbf{x}) = 0$. Solve this task as follows:

- Specify the mark-functions $f_q(\mathbf{x})$ (TVL 1, viewport (1,1)) and $f_r(\mathbf{x})$ (TVL 2, viewport (1,3)) and show additionally the Karnaugh-map of the lattice $\mathcal{L}\langle f_q(\mathbf{x}), f_r(\mathbf{x}) \rangle$ in viewport (1,2) of the m-fold view.

- Create mark-functions $f_q'(\mathbf{x})$ (TVL 3, viewport (2,1)) and $f_r'(\mathbf{x})$ (TVL 4, viewport (2,3)) and show additionally the Karnaugh-map of the lattice $\mathcal{L}'\langle f_q'(\mathbf{x}), f_r'(\mathbf{x}) \rangle$ in viewport (2,2) of the m-fold view.

- Edit the Karnaugh-map in viewport (2,2):

from 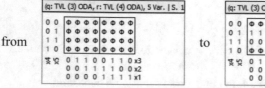 to 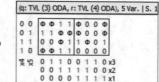 .

- Compare $f_q(\mathbf{x})$ (TVL 1) with $f_q'(\mathbf{x})$ (TVL 3) and $f_r(\mathbf{x})$ (TVL 2) with $f_r'(\mathbf{x})$ (TVL 4).

(d) Verify that the function

$$f_i(\mathbf{x}) = \overline{x}_1(x_2 \vee \overline{x}_4) \vee x_1\overline{x}_3 x_5$$

belongs to the explored lattice of functions.

Use Eq. (7.6) to compute the complement $^{\mathcal{L}}\overline{f_i(\mathbf{x})}$ related to the explored lattice. Verify that the $^{\mathcal{L}}\overline{f_i(\mathbf{x})}$ also belongs to the lattice $\mathcal{L}\langle f_q(\mathbf{x}), f_r(\mathbf{x})\rangle$ and satisfies both:

$$f_i(\mathbf{x}) \wedge^{\mathcal{L}} \overline{f_i(\mathbf{x})} = f_q(\mathbf{x}) \,,$$
$$f_i(\mathbf{x}) \vee^{\mathcal{L}} \overline{f_i(\mathbf{x})} = \overline{f_r(\mathbf{x})} \,.$$

Show the Karnaugh-maps of $f_q(\mathbf{x})$, $\mathcal{L}\langle f_q(\mathbf{x}), f_r(\mathbf{x})\rangle$, and $f_r(\mathbf{x})$ in the first row of the m-fold view, and thereunder the TVL of $f_i(\mathbf{x})$ in viewport (2,1), the Karnaugh-maps of $f_i(\mathbf{x})$ and $^{\mathcal{L}}\overline{f_i(\mathbf{x})}$ in viewports (2,2) and (2,3), respectively.

Exercise 7.3 (Exploration of a Generalized Lattice of Logic Functions)
   The aim of this exercise is the creation and analysis of a generalized lattice of logic functions. The achievement of this aim requires to solve the four task in a consecutive manner. The results of each task are therefore stored in the last line of the PRP as an $\mathtt{sdt}$-file and this $\mathtt{sdt}$-file is loaded in the first line of the subsequent PRP for further use of the already computed results.

(a) The lattice $\mathcal{L}\langle f_q(\mathbf{x}), f_r(\mathbf{x})\rangle$ is given by the mark-functions:

$$f_q(\mathbf{x}) = \overline{x}_1 x_2 \overline{x}_5 \vee x_1 \overline{x}_3 x_5 \vee x_2 \overline{x}_3 \vee \overline{x}_2 \overline{x}_4 (\overline{x}_3 \vee x_5) \,,$$
$$f_r(\mathbf{x}) = x_1 x_3 \overline{x}_5 \vee x_1 \overline{x}_2 x_4 \overline{x}_5 \vee \overline{x}_2 x_3 x_4 \vee x_1 x_2 x_3 \overline{x}_4 \vee \overline{x}_1 x_5 (\overline{x}_2 x_4 \vee x_2 x_3) \,.$$

Verify that the given mark-functions satisfy the condition of a lattice and show the Karnaugh-maps of $f_q(\mathbf{x})$, this lattice $\mathcal{L}\langle f_q(\mathbf{x}), f_r(\mathbf{x})\rangle$, and $f_r(\mathbf{x})$ in the first row of the m-fold view.
Compute the mark-function $f_\varphi(\mathbf{x})$ and evaluate this function to find the number $\mathtt{\$nf}$ of functions $f_i(\mathbf{x}) \in \mathcal{L}\langle f_q(\mathbf{x}), f_r(\mathbf{x})\rangle$.
Compute all functions that belong to the given lattice, store these functions as XBOOLE-objects starting with the object number 11, and show the Karnaugh-maps of these functions in the second row of the m-fold view and in consecutive rows (use three columns of the m-fold view to show these functions).
(b) An $\mathtt{sdt}$-file stores all TVLs and VTs of the actual XBOOLE-system, but not the variables computed in a PRP of the XBOOLE-monitor XBM 2; hence, reuse the computed TVL of the mark $f_\varphi(\mathbf{x})$ to recompute the number $\mathtt{\$nf}$ of functions $f_i(\mathbf{x}) \in \mathcal{L}\langle f_q(\mathbf{x}), f_r(\mathbf{x})\rangle$.
Compute for all functions $f_i(\mathbf{x}) \in \mathcal{L}\langle f_q(\mathbf{x}), f_r(\mathbf{x})\rangle$, which are stored as TVLs starting with the object number 11, the vectorial derivatives:

$$g_i(\mathbf{x}) = \operatorname*{der}_{(x_2, x_3, x_5)} f_i(\mathbf{x}) \,,$$

store minimized TVLs of these vectorial derivatives as XBOOLE objects starting with the object number 21, and show the Karnaugh-maps of these functions in the viewports of three columns of the m-fold view.
Verify that the set of computed vectorial derivatives satisfy the rules of a lattice. Note, all logic functions satisfy the commutative, the associative, and the absorptions laws; hence, it must only be verified, whether this set of functions is closed with regard to the two operations $\vee$ and $\wedge$.
(c) Compute the mark-function $h_q(\mathbf{x})$ and $h_r(\mathbf{x})$ of the smallest lattice that contains all functions $g_i(\mathbf{x})$ (vectorial derivatives of $f_i(\mathbf{x})$ with regard to (x_2, x_3, x_5)) computed in the previous task and stored as XBOOLE-objects (TVLs) starting with the object number 21. Compute additionally

the mark-function $h_\varphi(\mathbf{x})$ and evaluate this function to find the number \$nh of functions $h_i(\mathbf{x}) \in \mathcal{L}\langle h_q(\mathbf{x}), h_r(\mathbf{x})\rangle$.

Compute all functions $h_i(\mathbf{x}) \in \mathcal{L}\langle h_q(\mathbf{x}), h_r(\mathbf{x})\rangle$ and store minimized TVLs of these functions as XBOOLE-objects stating with the object number 101.

Show the Karnaugh-maps of $f_q(\mathbf{x})$, the lattice $\mathcal{L}\langle f_q(\mathbf{x}), f_r(\mathbf{x})\rangle$, and $f_r(\mathbf{x})$ in the first row of the m-fold view for comparison with the computed lattice $\mathcal{L}\langle h_q(\mathbf{x}), h_r(\mathbf{x})\rangle$, shown by the Karnaugh-maps of $h_q(\mathbf{x})$, the lattice $\mathcal{L}\langle h_q(\mathbf{x}), h_r(\mathbf{x})\rangle$, and $h_r(\mathbf{x})$ in the second row of the m-fold view. The functions $h_i(\mathbf{x})$ should be sequentially assigned to an arbitrary selected view due to the large number of these functions.

(d) Show for comparisons the Karnaugh-maps of $h_q(\mathbf{x})$, this lattice $\mathcal{L}\langle h_q(\mathbf{x}), h_r(\mathbf{x})\rangle$, and $h_r(\mathbf{x})$ in the first row of the m-fold view.

Compute for each function $h_i(\mathbf{x}) \in \mathcal{L}\langle h_q(\mathbf{x}), h_r(\mathbf{x})\rangle$ the vectorial derivative

$$h_i'(\mathbf{x}) = \underset{(x_2, x_3, x_5)}{\text{der}} \ h_i(\mathbf{x})$$

and verify that $h_i(\mathbf{x})$ is equal to one function $g_j(\mathbf{x})$ if $h_i'(\mathbf{x}) = 0$. How many function $h_i(\mathbf{x})$ satisfy $\text{der}_{(x_2, x_3, x_5)} h_i(\mathbf{x}) = 0$? Show the Karnaugh-maps of functions $h_i(\mathbf{x})$ that satisfy $h_i'(\mathbf{x}) = 0$ in three columns of the second row of the m-fold view and in subsequent rows. Compare these Karnaugh-maps with the Karnaugh-maps of the functions $g_i(\mathbf{x})$ computed and shown in Task (b) of this exercise. Conclude a specification of the lattice that contains the functions $g_i(\mathbf{x}) = \text{der}_{(x_2, x_3, x_5)} f_i(\mathbf{x})$ of all $f_i(\mathbf{x}) \in \mathcal{L}\langle f_q(\mathbf{x}), f_r(\mathbf{x})\rangle$.

Exercise 7.4 (Solution of Logic Equations with Regard to Variables)

(a) The following three functions are given:

$$f_1(\mathbf{x}, y_1) = x_1 \overline{x}_2 \overline{x}_4 y_1 \vee x_3 y_1 (x_2 \vee x_4) \vee \overline{x}_3 \overline{y}_1 \vee \overline{x}_1 \overline{x}_2 \overline{x}_4 \overline{y}_1 \ ,$$

$$f_2(\mathbf{x}, y_1) = x_1 \overline{x}_2 x_4 \overline{y}_1 \vee x_1 \overline{x}_3 \overline{y}_1 \vee x_2 x_3 y_1 (\overline{x}_1 \vee x_2 \vee \overline{x}_4)$$
$$\vee \overline{x}_1 \overline{x}_2 (x_4 y_1 \vee \overline{x}_4 \overline{y}_1) \vee x_2 \overline{x}_3 \overline{y}_1 \vee x_1 x_3 \overline{x}_4 y_1 \ ,$$

$$f_3(\mathbf{x}, y_1, y_2) = x_1 \overline{y}_2 (\overline{x}_2 \overline{y}_1 \vee \overline{x}_3 \overline{x}_4) \vee \overline{x}_3 (\overline{y}_1 \vee x_4 y_2)$$
$$\vee \overline{y}_1 (x_2 \overline{x}_3 y_2 \vee x_1 \overline{x}_2 \overline{x}_4) \vee x_3 y_1 (\overline{x}_1 \vee x_2 \vee x_4)$$
$$\vee \overline{x}_1 \overline{x}_4 \overline{y}_2 (\overline{x}_2 \vee y_1) \vee x_2 y_2 (\overline{x}_4 \overline{y}_1 \vee x_1 x_4) \vee \overline{x}_2 \overline{y}_1 \overline{y}_2 \ .$$

Show the Karnaugh-maps of these three functions in the m-fold view and verify whether the associated homogenous restrictive equations $f_i(\mathbf{x}, \mathbf{y}) = 0$ can be solved with regard to the variables y_1 and y_2 that occur in these expressions. Verify furthermore whether the functions $y_j = g_j(\mathbf{x})$ are uniquely specified in the case that the explored restrictive equation can be solved with regard to y_j.

(b) Compute the uniquely specified function $y_j = g_j(\mathbf{x})$ or the mark-functions $g_{jq}(\mathbf{x})$ and $g_{jr}(\mathbf{x})$ of the solution-lattice of functions $y_{jk} = g_{jk}(\mathbf{x})$ for all restrictive equations specified in Task (a) of Exercise 7.4 that are solvable with regard to $\mathbf{y}$. Verify that all computed solution-functions satisfy the associated equation. Show Karnaugh-maps of the functions $f_i(\mathbf{x}, \mathbf{y})$ for which the restrictive equations are solvable and the associated solution-functions.

(c) The following three functions are given:

$$f_1(\mathbf{x}, y_1) = (x_1\overline{x}_2x_4 \vee \overline{x}_3)\overline{y}_1 \vee \overline{x}_1\overline{x}_2(\overline{x}_4\overline{y}_1 \vee x_3y_1) \vee x_3y_1(x_2 \vee x_1\overline{x}_4) \,,$$

$$f_2(\mathbf{x}, y_1) = x_1(\overline{x}_2x_4\overline{y}_1 \vee \overline{x}_3\overline{y}_1) \vee x_2x_3\overline{y}_1(x_1 \vee \overline{x}_4)$$

$$\vee \overline{x}_1\overline{x}_2x_4y_1 \vee x_2\overline{x}_3\overline{y}_1 \vee x_1x_3\overline{x}_4y_1 \,,$$

$$f_3(\mathbf{x}, y_1, y_2) = x_1y_2(x_3y_1 \vee \overline{x}_3\overline{y}_1) \vee \overline{x}_1\overline{x}_3\overline{x}_4y_2(\overline{x}_2 \vee y_2)$$

$$\vee \overline{x}_1\overline{x}_2x_4y_1\overline{y}_2 \vee \overline{x}_1\overline{y}_1\overline{y}_2(x_2x_4 \vee x_3\overline{x}_4) \vee x_1x_2\overline{x}_4\overline{y}_1y_2 \,.$$

Show the Karnaugh-maps of these three functions in the m-fold view and verify whether the associated homogenous characteristic equations $f_i(\mathbf{x}, \mathbf{y}) = 1$ can be solved with regard to the variables y_1 and y_2 that occur in these expressions. Verify furthermore whether the functions $y_j = g_j(\mathbf{x})$ are uniquely specified in the case that the explored characteristic equation can be solved with regard to y_j.

(d) Compute the uniquely specified function $y_j = g_j(\mathbf{x})$ or the mark-functions $g_{jq}(\mathbf{x})$ and $g_{jr}(\mathbf{x})$ of the solution-lattice of functions $y_{jk} = g_{jk}(\mathbf{x})$ for all characteristic equations specified in Task (c) of Exercise 7.4 that are solvable with regard to $\mathbf{y}$. Verify that all computed solution-functions satisfy the associated equation. Show Karnaugh-maps of the functions $f_i(\mathbf{x}, \mathbf{y})$ for which the characteristic equations are solvable and the lattices or uniquely specified solution-functions. Store all functions of a solution-lattice in a sequence of XBOOLE-objects so that all solution-functions can easily be assigned to an arbitrary viewport.

Exercise 7.5 (Boolean Differential Equations)

This exercise aims to amplify the knowledge about the solution of Boolean differential equations. For that reason we restrict to Boolean differential equations that determine logic function $f_i(x_1, x_2)$. The approach to solve Boolean differential equations for functions that depend on more than two variables remains unchanged, only the size of the intermediate and solution-TVLs increases for a larger number of variables.

Show for all tasks the TVLs of the solution $SLS(\mathbf{u})$ or $SLS(\mathbf{x}, \mathbf{u})$ of the logic equation associated to the given Boolean differential equation, the transformed sets $SLS(\mathbf{v})$ or $SLS(\mathbf{x}, \mathbf{v})$, the intermediate TVLs of the separation process, and the final set of solution-functions $S(\mathbf{v})$. The Boolean differential equations of the first three tasks describe classes of logic functions and can be solved using Algorithm 3 (see page 361) and the fourth task has a set of logic functions as solution that can be computed using Algorithm 4 (see page 365). Usc Eq. (7.87) to provide the expressions of the computed functions, determined by vectors of function values.

(a) Solve the Boolean differential equation

$$\operatorname*{der}_{x_1} f(\mathbf{x}) \wedge \operatorname*{der}_{x_2} f(\mathbf{x}) = 1$$

which describes all functions $f_i(x_1, x_2)$ that are linear with regard to both x_1 and x_2.

(b) Solve the Boolean differential equation

$$\operatorname*{der}_{x_1} f(\mathbf{x}) \wedge \operatorname*{der}_{x_2} f(\mathbf{x}) = 0$$

which describes all functions $f_i(x_1, x_2)$ that do not depend at least on one of the variables x_1 or x_2.

(c) Solve the Boolean differential equation

$$\operatorname*{der}_{(x_1, x_2)}^2 f(\mathbf{x}) \vee \operatorname*{der}_{x_1} f(\mathbf{x}) \operatorname*{der}_{x_2} f(\mathbf{x}) = 0$$

which describes all functions $f_i(x_1, x_2)$ that depend on both of the variables x_1 and x_2.

(d) Solve the Boolean differential equation

$$\overline{x}_1 f(\mathbf{x}) \operatorname*{der}_{x_1} f(\mathbf{x}) \vee \overline{x}_2 f(\mathbf{x}) \operatorname*{der}_{x_2} f(\mathbf{x}) = 0$$

which describes all functions $f_i(x_1, x_2)$ that are monotonously increasing with regard to x_1 and x_2.

7.9   Solutions

Solution 7.1 (Incompletely Specified Functions)

(a) The given vector consists of 16 of function values 0 or 1; hence, this vector describes a single completely specified logic function of four variables because $2^4 = 16$. This single function is the border case of a lattice of logic functions.

Figure 7.18 shows the PRP that computes the completely specified function given by a vector of function values and also the computed TVL of this function.

The main computation steps are realized in two nested `for`-loops. The outer `for`-loop in lines 33–51 iterates over the 16 function values of the vector defined in lines 15–20 and append the conjunctions (TVL 14) for which the function is equal to 1 to solution-TVL 20 using the command `uni` in line 48. The condition of the command `if` in line 35 is satisfied when a function value 1 is specified in the position determined by the column specified in TVL 2. This column is rotated by one position to the right using the command `cco` in line 50 controlled by VTs 3 and 4. The binary vector (x_1, x_2, x_3, x_4) that belongs to a function value is determined by the value of the iteration variable `$i` of the outer `for`-loop. The used approach to convert this integer into the binary vector (TVL 14) is as follows:

- The command `copy` in line 37 copies of the repeated used basic vector $\mathbf{x} = (0000)$ (TVL 10) to TVL 14.
- The command `set` in line 38 assigns the value of `$i` to the variable `$j` because `$i` is needed to control the outer `for`-loop and `$j` will be changed while converting into the associated binary vector.
- The command `cel` in line 43 changes the bit of TVL 14 in the position of TVL 11 from 0 to 1 in the case that the last bit of `$j` is equal to 1.
- Rotating TVL 11 by one bit to the left using the command `cco` in line 45 controlled by VTs 12 and 13 as well as the division of `$j` by 2 in the inner `for`-loop in lines 39–47 facilitates that the conditional change described in the previous item computes the binary vector belonging to the selected function value of position `$i`.

The command `obb` in line 52 orthogonally minimizes the computed TVL 20 before the subsequent command assigns this result to viewport $(1,1)$ of the m-fold view. The computed TVL 20 is shown in the top-right corner of Fig. 7.18.

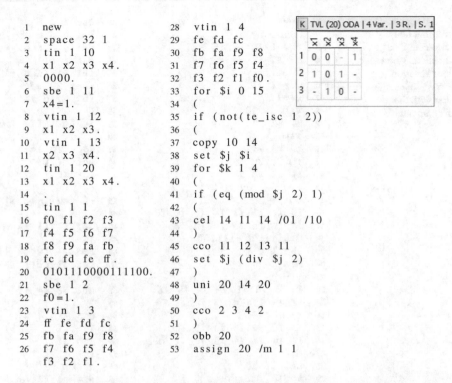

```
 1  new                      28  vtin  1  4
 2  space  32  1             29  fe  fd  fc
 3  tin  1  10               30  fb  fa  f9  f8
 4  x1  x2  x3  x4.          31  f7  f6  f5  f4
 5  0000.                    32  f3  f2  f1  f0.
 6  sbe  1  11               33  for  $i  0  15
 7  x4=1.                    34  (
 8  vtin  1  12              35  if  (not(te_isc  1  2))
 9  x1  x2  x3.              36  (
10  vtin  1  13              37  copy  10  14
11  x2  x3  x4.              38  set  $j  $i
12  tin  1  20               39  for  $k  1  4
13  x1  x2  x3  x4.          40  (
14  .                        41  if  (eq  (mod  $j  2)  1)
15  tin  1  1                42  (
16  f0  f1  f2  f3           43  cel  14  11  14  /01  /10
17  f4  f5  f6  f7           44  )
18  f8  f9  fa  fb           45  cco  11  12  13  11
19  fc  fd  fe  ff.          46  set  $j  (div  $j  2)
20  0101110000111100.        47  )
21  sbe  1  2                48  uni  20  14  20
22  f0=1.                    49  )
23  vtin  1  3               50  cco  2  3  4  2
24  ff  fe  fd  fc           51  )
25  fb  fa  f9  f8           52  obb  20
26  f7  f6  f5  f4           53  assign  20  /m  1  1
    f3  f2  f1.
```

K | TVL (20) ODA | 4 Var. | 3 R. | S. 1

	$\bar{x}_1$	$\bar{x}_2$	$\bar{x}_3$	$\bar{x}_4$
1	0	0	-	1
2	1	0	1	-
3	-	1	0	-

Fig. 7.18 Problem-program that computes the completely specified function given by a vector of function values together with the TVL of the computed function

(b) The given vector consists of 15 of function values 0 or 1 and a single dash ($-$); hence, this vector describes an incompletely specified function of four variables that can be realized by two different completely specified functions; these two functions determine a lattice of logic functions.

Figure 7.19 shows both the PRP that computes the two completely specified functions given by a vector of function values and the computed two TVLs of these functions.

The PRP of Fig. 7.19 uses the approach of the previous task to convert the vector of function values of a completely specified function twice (stored as TVL with the index $f) for the two functions of the given lattice.

Each sweep of the surrounding while-loop in lines 29–57 selects one function vector of the given lattice and excludes this completely specified function vector from the given lattice. In this way, each function is converted once.

The command stv in line 31 selects the first row of the given lattice of functions vectors. Dashes of the selected vector are excluded in the result of the command stv; this is often very welcome but here not helpful because the dashes are needed. We reconstruct the removed dash using the command uni in line 32 with an empty TVL 5 that contains all 16 variables $f_0, \ldots, f_f$; this empty TVL has easily been created using the command dif in line 27 and the given function vector.

The command cel in line 33 changes the dash of the selected vector into the value 0 so that we get the vector of a single completely specified function. The command dif in line 34 removes the selected function vector from the given lattice so that it is converted only once. The while-loop terminates when all functions of the given lattice are converted; in this case the given TVL 1 does not contain any row after the execution of this PRP.

Fig. 7.19
Problem-program that
computes two completely
specified functions given
by a vector of function
values that contain a single
dash together with the
TVLs of the computed
functions

```
 1   new
 2   space  32  1
 3   tin  1  10
 4   x1  x2  x3  x4 .
 5   0000.
 6   sbe  1  11
 7   x4 = 1 .
 8   vtin  1  12
 9   x1  x2  x3 .
10   vtin  1  13
11   x2  x3  x4 .
12   tin  1  20
13   x1  x2  x3  x4 .
14   .
15   tin  1  1
16   f0  f1  f2  f3  f4  f5  f6  f7
17   f8  f9  fa  fb  fc  fd  fe  ff .
18   010111000 - 111100.
19   sbe  1  2
20   f0 = 1 .
21   vtin  1  3
22   ff  fe  fd  fc  fb  fa  f9  f8
23   f7  f6  f5  f4  f3  f2  f1 .
24   vtin  1  4
25   fe  fd  fc  fb  fa  f9  f8
26   f7  f6  f5  f4  f3  f2  f1  f0 .
27   dif  1  1  5
28   set  $f  20
29   while  (not(te  1))
30   (
```

```
31   stv  1  1  6
32   uni  6  5  6
33   cel  6  6  /-0
34   dif  1  6  1
35   set  $f  (add  $f  1)
36   copy  20  $f
37   for  $i  0  15
38   (
39   if  (not  (te_isc  6  2))
40   (
41   copy  10  14
42   set  $j  $i
43   for  $k  1  4
44   (
45   if  (eq  (mod  $j  2)  1)
46   (
47   cel  14  11  14  /01  /10
48   )
49   cco  11  12  13  11
50   set  $j  (div  $j  2)
51   )
52   uni  $f  14  $f
53   )
54   cco  2  3  4  2
55   )
56   obb  $f
57   )
58   assign  21  /m  1  1
59   assign  22  /m  1  2
```

K	TVL (21) ODA | 4 Var. | 3 R. | S. 1			
	$\overline{x}1$	$\overline{x}2$	$\overline{x}3$	$\overline{x}4$
1	0	0	-	1
2	1	0	1	-
3	-	1	0	-

K	TVL (22) ODA | 4 Var. | 3 R. | S. 1			
	$\overline{x}1$	$\overline{x}2$	$\overline{x}3$	$\overline{x}4$
1	1	0	1	0
2	-	0	-	1
3	-	1	0	-

The computed TVLs 21 and 22 are shown below the PRP in Fig. 7.19. It can be seen that the function $f_{21}(\mathbf{x})$ is equal to 1 for eight assignments of the four variables and one additional function value 1 for $\mathbf{x} = (1001)$ belongs to the function $f_{22}(\mathbf{x})$.

(c) The given vector consists of 14 of function values 0 or 1 and two dashes $(-)$; hence, this vector describes an incompletely specified function of four variables that can be realized by four different completely specified functions; these four functions determine a lattice of logic functions.

Figure 7.20 shows both the PRP that computes the four completely specified function given by a vector of function values and the computed four TVLs of these functions. This PRP reuses the PRP of the previous task shown in Fig. 7.19; only the given vector of function values has been changed in line 18 and the additional result functions are assigned to the viewports of the second rows of the m-fold view in the last two lines of the PRP. This very easy adaption of the PRP to the task with two more resulting functions confirms the universality of the used approach.

The computed TVLs 21, 22, 23, and 24 are shown below the PRP in Fig. 7.20. It can be seen that the function $f_{21}(\mathbf{x})$ is equal to 1 for eight assignments to the four variables, the functions $f_{22}(\mathbf{x})$ and $f_{23}(\mathbf{x})$ are equal to 1 for nine such assignments, and ten function values 1 belong to the function $f_{24}(\mathbf{x})$.

Fig. 7.20
Problem-program that
computes four completely
specified functions given
by a vector of function
values that contain two
dashes together with the
TVLs of the computed
functions

```
 1  new
 2  space 32 1
 3  tin 1 10
 4  x1 x2 x3 x4.
 5  0000.
 6  sbe 1 11
 7  x4=1.
 8  vtin 1 12
 9  x1 x2 x3.
10  vtin 1 13
11  x2 x3 x4.
12  tin 1 20
13  x1 x2 x3 x4.
14  .
15  tin 1 1
16  f0 f1 f2 f3 f4 f5 f6 f7
17  f8 f9 fa fb fc fd fe ff.
18  01011100--111100.
19  sbe 1 2
20  f0=1.
21  vtin 1 3
22  ff fe fd fc fb fa f9 f8
23  f7 f6 f5 f4 f3 f2 f1.
24  vtin 1 4
25  fe fd fc fb fa f9 f8
26  f7 f6 f5 f4 f3 f2 f1 f0.
27  dif 1 1 5
28  set $f 20
29  while (not(te 1))
30  (
31  stv 1 1 6
```

```
32  uni 6 5 6
33  cel 6 6 /-0
34  dif 1 6 1
35  set $f (add $f 1)
36  copy 20 $f
37  for $i 0 15
38  (
39  if (not (te_isc 6 2))
40  (
41  copy 10 14
42  set $j $i
43  for $k 1 4
44  (
45  if (eq (mod $j 2) 1)
46  (
47  cel 14 11 14 /01 /10
48  )
49  cco 11 12 13 11
50  set $j (div $j 2)
51  )
52  uni $f 14 $f
53  )
54  cco 2 3 4 2
55  )
56  obb $f
57  )
58  assign 21 /m 1 1
59  assign 22 /m 1 2
60  assign 23 /m 2 1
61  assign 24 /m 2 2
```

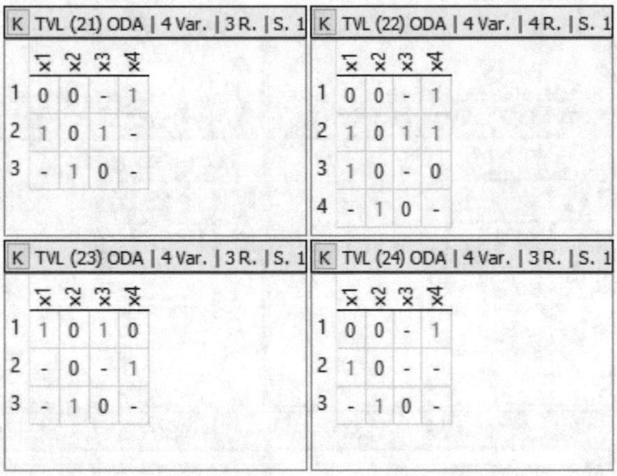

K	TVL (21) ODA	4 Var.	3 R.	S. 1
	$\overline{x}1$	$\overline{x}2$	$\overline{x}3$	$\overline{x}4$
1	0	0	-	1
2	1	0	1	-
3	-	1	0	-

K	TVL (22) ODA	4 Var.	4 R.	S. 1
	$\overline{x}1$	$\overline{x}2$	$\overline{x}3$	$\overline{x}4$
1	0	0	-	1
2	1	0	1	1
3	1	0	-	0
4	-	1	0	-

K	TVL (23) ODA	4 Var.	3 R.	S. 1
	$\overline{x}1$	$\overline{x}2$	$\overline{x}3$	$\overline{x}4$
1	1	0	1	0
2	-	0	-	1
3	-	1	0	-

K	TVL (24) ODA	4 Var.	3 R.	S. 1
	$\overline{x}1$	$\overline{x}2$	$\overline{x}3$	$\overline{x}4$
1	0	0	-	1
2	1	0	-	-
3	-	1	0	-

(d) The given vector consists of 13 of function values 0 or 1 and three dashes ($-$); hence, this vector describes an incompletely specified function of four variables that can be realized by eight different completely specified functions; these eight functions determine a lattice of logic functions.

Figure 7.21 shows both the PRP that computes all completely specified functions given by a vector of function values and shows the computed TVLs of these functions.

The PRP of Fig. 7.21 computes eight functions belonging to the vector of function values specified in line 20. This PRP reuses the universal approach of the previous tasks and increases the generality furthermore. The assignment of the computed function to a viewport has been

```
 1  new                        25  fb fa f9 f8                49  copy 10 14
 2  space 32 1                 26  f7 f6 f5 f4                50  set $j $i
 3  tin 1 10                   27  f3 f2 f1.                  51  for $k 1 4
 4  x1 x2 x3 x4.               28  vtin 1 4                   52  (
 5  0000.                      29  fe fd fc                   53  if (eq (mod $j 2) 1)
 6  sbe 1 11                   30  fb fa f9 f8                54  (
 7  x4=1.                      31  f7 f6 f5 f4                55  cel 14 11 14 /01 /10
 8  vtin 1 12                  32  f3 f2 f1 f0.               56  )
 9  x1 x2 x3.                  33  dif 1 1 5                  57  cco 11 12 13 11
10  vtin 1 13                  34  set $f 20                  58  set $j (div $j 2)
11  x2 x3 x4.                  35  set $r 1                   59  )
12  tin 1 20                   36  set $c 1                   60  uni $f 14 $f
13  x1 x2 x3 x4.               37  while (not(te 1))          61  )
14  .                          38  (                          62  cco 2 3 4 2
15  tin 1 1                    39  stv 1 1 6                  63  )
16  f0 f1 f2 f3                40  uni 6 5 6                  64  obb $f
17  f4 f5 f6 f7                41  cel 6 6 /-0                65  assign $f /m $r $c
18  f8 f9 fa fb                42  dif 1 6 1                  66  set $c (add $c 1)
19  fc fd fe ff.               43  set $f (add $f 1)          67  if (gt $c 3)
20  -1011100--111100.          44  copy 20 $f                 68  (
21  sbe 1 2                    45  for $i 0 15                69  set $r (add $r 1)
22  f0=1.                      46  (                          70  set $c 1
23  vtin 1 3                   47  if (not(te_isc 6 2))       71  )
24  ff fe fd fc                48  (                          72  )
```

K TVL (21) ODA | 4 Var. | 3 R. | S. 1

	x_1	x_2	x_3	x_4
1	0	0	-	1
2	1	0	1	-
3	-	1	0	-

K TVL (22) ODA | 4 Var. | 4 R. | S. 1

	x_1	x_2	x_3	x_4
1	1	0	1	0
2	-	0	1	1
3	1	1	0	-
4	0	-	0	-

K TVL (23) ODA | 4 Var. | 4 R. | S. 1

	x_1	x_2	x_3	x_4
1	0	0	-	1
2	1	0	1	1
3	1	0	-	0
4	-	1	0	-

K TVL (24) ODA | 4 Var. | 4 R. | S. 1

	x_1	x_2	x_3	x_4
1	1	0	-	0
2	-	0	1	1
3	1	1	0	-
4	0	-	0	-

K TVL (25) ODA | 4 Var. | 3 R. | S. 1

	x_1	x_2	x_3	x_4
1	1	0	1	0
2	-	0	-	1
3	-	1	0	-

K TVL (26) ODA | 4 Var. | 5 R. | S. 1

	x_1	x_2	x_3	x_4
1	1	0	1	0
2	-	0	1	1
3	1	1	0	0
4	1	-	0	1
5	0	-	0	-

K TVL (27) ODA | 4 Var. | 3 R. | S. 1

	x_1	x_2	x_3	x_4
1	0	0	-	1
2	1	0	-	-
3	-	1	0	-

K TVL (28) ODA | 4 Var. | 3 R. | S. 1

	x_1	x_2	x_3	x_4
1	-	0	1	1
2	1	0	1	0
3	-	-	0	-

Fig. 7.21 Universal problem-program that computes eight completely specified functions given by a vector of function values that contain three dashes together with the TVLs of the computed functions

```
 1   new                              9   x1&(x4&/x5+x3&/x5+x3&/x4+
 2   space 32 1                       10  /x2&x3)+/x1&/x2&x4&x5.
 3   avar 1                           11  set $l (te_isc 1 2)
 4   x1 x2 x3 x4 x5.                  12  assign 1 /m 1 1
 5   sbe 1 1                          13  assign 1 /m 2 1
 6   x1&/x3&x4&x5+/x1&(x2&/x5+        14  assign 2 /m 1 3
 7   /x2&/x4&x5+x2&/x3&x4).           15  assign 2 /m 2 3
 8   sbe 1 2                          16  assign_qr 1 2 /m 1 2
```

```
┌ T TVL (1) ODA │ 5 Var. │ 4 R. │ S. 1 ┐┌ (q: TVL (1) ODA, r: TVL (2) ODA), 5 Var. │ S. 1 ┐┌ T TVL (2) ODA │ 5 Var. │ 5 R. │ S. 1 ┐

 0 0   0 0 1 1 0 0 0 0                    0 0   Φ Φ 1 1 Φ 0 0 Φ                      0 0   0 0 0 0 0 1 1 0
 0 1   1 1 0 0 0 0 0 0                    0 1   1 1 Φ Φ Φ 0 0 Φ                      0 1   0 0 0 0 0 1 1 0
 1 1   0 0 0 1 1 0 0 1                    1 1   0 0 Φ 1 1 Φ 0 1                      1 1   1 1 0 0 0 0 1 0
 1 0   0 0 1 1 0 0 0 0                    1 0   Φ Φ 1 1 0 0 0 0                      1 0   0 0 0 0 1 1 1 1

 x4 x5  0 1 1 0 0 1 1 0 x3               x4 x5  0 1 1 0 0 1 1 0 x3                  x4 x5  0 1 1 0 0 1 1 0 x3
        0 0 1 1 1 1 0 0 x2                      0 0 1 1 1 1 0 0 x2                         0 0 1 1 1 1 0 0 x2
        0 0 0 0 1 1 1 1 x1                      0 0 0 0 1 1 1 1 x1                         0 0 0 0 1 1 1 1 x1
```

```
┌ K TVL (1) ODA │ 5 Var. │ 4 R. │ S. 1 ┐   -      ┌ K TVL (2) ODA │ 5 Var. │ 5 R. │ S. 1 ┐

        x1 x2 x3 x4 │ x5                                x1 x2 x3 x4 │ x5
   1    0  1  -  -  │ 0                             1   0  0  0  1  │ 1
   2    0  0  -  0  │ 1                             2   1  -  1  0  │ 0
   3    1  0  0  1  │ 1                             3   1  -  1  0  │ 1
   4    -  1  0  1  │ 1                             4   -  0  1  1  │ 1
                                                    5   1  -  -  1  │ 0
```

Fig. 7.22 Problem-program that determines the mark-functions $f_q(\mathbf{x})$ and $f_r(\mathbf{x})$ of a lattice, verifies that these functions satisfy the condition of a lattice together with the associated Karnaugh-maps and TVLs

moved to line 65 into the body of the while-loop. The integer variable $r and $c are initialize to values 1 before the while-loop starts; these integer variables determine the viewport used to show the computed TVL. The column of the m-fold view used for assignment of the next computed TVL is determined by the variable $c and will be incremented in line 66 at the end of each sweep of the while-loop. After the assignment of (arbitrarily chosen) three TVLs to one row of the m-fold view the variables $r and $c are adjusted in lines 69 and 70 to the left viewport of the next row in the m-fold view.

The computed TVLs 21 to 28 are shown below the PRP in Fig. 7.21. It can be seen that the function $f_{21}(\mathbf{x})$ is equal to 1 for eight assignments of the four variables, the functions $f_{22}(\mathbf{x})$, $f_{23}(\mathbf{x})$, and $f_{25}(\mathbf{x})$ are equal to 1 for nine, the functions $f_{24}(\mathbf{x})$, $f_{26}(\mathbf{x})$, and $f_{27}(\mathbf{x})$ are equal to 1 for ten, and $f_{28}(\mathbf{x})$ are equal to 1 for eleven such assignments.

Solution 7.2 (Mark-functions of a Lattice)

(a) Figure 7.22 shows both the PRP that verifies whether the given mark-functions $f_q(\mathbf{x})$ and $f_r(\mathbf{x})$ satisfy the condition of a lattice of logic functions and the associated Karnaugh-maps and TVLs. The mark-functions $f_q(\mathbf{x})$ (TVL 1) and $f_r(\mathbf{x})$ (TVL 2) of the lattice have been determined using two commands sbe. The result true of the command te_isc in line 11 is assigned to the variable $l and confirms that the given mark-functions specify a lattice of logic functions.

```
 1   new                          17   if (not(te 4))
 2   space 32 1                   18   (
 3   avar 1                       19   while (not(te 5))
 4   x1 x2 x3 x4 x5.              20   (
 5   sbe 1 1                      21   stv 5 1 7
 6   x1&/x3&x4&x5+/x1&(x2&/x5+    22   uni 7 6 7
 7   /x2&/x4&x5+x2&/x3&x4).       23   cel 7 7 /-0
 8   sbe 1 2                      24   dif 5 7 5
 9   x1&(x4&/x5+x3&/x5+x3&/x4+    25   set $nf (mul $nf 2)
10   /x2&x3)+/x1&/x2&x4&x5.       26   )
11   uni 1 2 3                    27   )
12   cpl 3 4                      28   assign 1 /m 1 1
13   obbc 4 4                     29   assign_qr 1 2 /m 1 2
14   copy 4 5                     30   assign 2 /m 1 3
15   dif 5 5 6                    31   assign 4 /m 2 1
16   set $nf 1                    32   assign 4 /m 2 2
```

T | TVL (1) ODA | 5 Var. | 4 R. | S. 1

0 0	0 0 1 1	0 0 0 0
0 1	1 1 0 0	0 0 0 0
1 1	0 0 0 1	1 0 0 1
1 0	0 0 1 1	0 0 0 0

```
x4         0 1 1 0 0 1 1 0  x3
x5         0 0 1 1 1 1 0 0  x2
           0 0 0 0 1 1 1 1  x1
```

(q: TVL (1) ODA, r: TVL (2) ODA), 5 Var. | S. 1

0 0	Φ Φ 1 1	Φ 0 0 Φ
0 1	1 1 Φ Φ	Φ 0 0 Φ
1 1	0 0 Φ 1	1 Φ 0 1
1 0	Φ Φ 1 1	0 0 0 0

```
x4         0 1 1 0 0 1 1 0  x3
x5         0 0 1 1 1 1 0 0  x2
           0 0 0 0 1 1 1 1  x1
```

T | TVL (2) ODA | 5 Var. | 5 R. | S. 1

0 0	0 0 0 0	0 1 1 0
0 1	0 0 0 0	0 1 1 0
1 1	1 1 0 0	0 0 1 0
1 0	0 0 0 0	1 1 1 1

```
x4         0 1 1 0 0 1 1 0  x3
x5         0 0 1 1 1 1 0 0  x2
           0 0 0 0 1 1 1 1  x1
```

K | TVL (4) ODA | 5 Var. | 4 R. | S. 1

	x_1	x_2	x_3	x_4	x_5
1	1	-	0	0	-
2	0	1	-	0	1
3	-	1	1	1	1
4	0	0	-	-	0

T | TVL (4) ODA | 5 Var. | 4 R. | S. 1

0 0	1 1 0 0	1 0 0 1
0 1	0 0 1 1	1 0 0 1
1 1	0 0 1 0	0 1 0 0
1 0	1 1 0 0	0 0 0 0

```
x4         0 1 1 0 0 1 1 0  x3
x5         0 0 1 1 1 1 0 0  x2
           0 0 0 0 1 1 1 1  x1
```

Integer variables

Name	Value
$nf	4096

Fig. 7.23 Problem-program that computes the mark-function $f_\varphi(\mathbf{x})$ and the number of functions $nf belonging to the explored lattice together with the associated Karnaugh-maps, the TVL of $f_\varphi(\mathbf{x})$, and the computed number $nf

The Karnaugh-maps and TVLs of the given mark-functions are assigned to the viewports as requested. The command `assign_qr` in line 16 uses both TVL 1 of $f_q(\mathbf{x})$ and TVL 2 of $f_r(\mathbf{x})$ to create the Karnaugh-map of the lattice in viewport (1,2) of the m-fold view; values 1 indicate in this Karnaugh-map the assignments where $f_q(\mathbf{x}) = 1$, values 0 are shown for assignments where $f_r(\mathbf{x}) = 1$, and symbols φ are assigned to the remaining fields where $f_\varphi(\mathbf{x}) = 1$.

(b) Figure 7.23 shows in the upper part the PRP that computes the third mark-function $f_\varphi(\mathbf{x})$ and evaluates this function to compute the number of functions belonging to the explored lattice. The details of the explored lattice are represented by Karnaugh-maps in the first row of the m-fold view. Both the TVL and the Karnaugh-map of the computed mark-function $f_\varphi(\mathbf{x})$ are shown in the second row of the m-fold view below the PRP in Fig. 7.23.

A formula to compute the mark-function $f_\varphi(\mathbf{x})$ can be derived from their properties (7.1) to (7.4). The disjunctions in (7.4) can be replaced by antivalence operations due to the orthogonality of the three mark-functions $f_q(\mathbf{x})$, $f_r(\mathbf{x})$, and $f_\varphi(\mathbf{x})$ (see (7.1), (7.2), and (7.3)) so that we can apply the following equivalent transformations:

$$f_q(\mathbf{x}) \vee f_r(\mathbf{x}) \vee f_\varphi(\mathbf{x}) = 1$$
$$f_q(\mathbf{x}) \oplus f_r(\mathbf{x}) \oplus f_\varphi(\mathbf{x}) = 1$$
$$f_\varphi(\mathbf{x}) = 1 \oplus f_q(\mathbf{x}) \oplus f_r(\mathbf{x})$$
$$f_\varphi(\mathbf{x}) = \overline{f_q(\mathbf{x}) \oplus f_r(\mathbf{x})}$$
$$f_\varphi(\mathbf{x}) = \overline{f_q(\mathbf{x}) \vee f_r(\mathbf{x})} \ .$$

The minimized mark-function $f_\varphi(\mathbf{x})$ is computed in lines 11–13 based on this equation.

A lattice contains only a single function for the special case of $f_\varphi(\mathbf{x}) = 0$. Each assignment of $\mathbf{x}$ with $f_\varphi(\mathbf{x}) = 1$ doubles the number of functions of the lattice. These properties are used in lines 14–27 to compute the number of functions belonging to the explored lattices. This approach changes the TVL of $f_\varphi(\mathbf{x})$ so that first TVL 4 is copied to TVL 5. The next command dif creates the needed empty TVL 6 that depends on all five variables of $f_\varphi(\mathbf{x})$. The command set in line 16 initializes the variable \$nf with the value 1.

If $f_\varphi(\mathbf{x}) \neq 0$ (checked in line 17) a multiplication of \$nf with 2 is computed for each binary vector belonging to TVL 5 within the while-loop in lines 19–26. The first three commands in this while-loop compute one binary vector (TVL 7) that belongs to the first ternary vector of TVL 5. The command dif in line 24 removes this binary vector from TVL 5 so that each binary vector causes exactly one multiplication of \$nf with 2 in the last line of the while-loop.

Due to the 12 binary vectors of $f_\varphi(\mathbf{x}) = 1$ (see TVL 4 in viewports (2,1) and (2,2)) \$nf=4096 functions belong to the explored lattice. This number of functions of the lattice of five variables confirms the advantage of the use of two mark-functions instead of the enumeration of all these functions.

(c) Figure 7.24 shows in the upper part the PRP that uses the given mark-functions $f_q(\mathbf{x})$ and $f_r(\mathbf{x})$ to compute the initial mark-functions $f_q'(\mathbf{x}) = 0$ and $f_r'(\mathbf{x}) = 0$ and assigns all the mark-functions and the Karnaugh-maps of the two lattices $\mathcal{L}\langle f_q, f_r\rangle$ and $\mathcal{L}\langle f_q', f_r'\rangle$ to the requested viewports of the m-fold view. The m-fold view that is created after the execution of the PRP and the editing of the Karnaugh-map of the lattice $\mathcal{L}\langle f_q', f_r'\rangle$ in viewport (2,2) is shown at the bottom of Fig. 7.24.

The given lattice has been prepared as in the previous two tasks. The commands dif prepare the lattice $\mathcal{L}\langle f_q'(\mathbf{x}) = 0, f_r'(\mathbf{x}) = 0\rangle$. The last three commands assign the mark-functions and the lattice $\mathcal{L}'$ to the second row of the m-fold view.

It is very easy to edit the Karnaugh-map of a lattice. A left-click on the top left element of the Karnaugh-map selects this element observably by the gray shaded background. Pressing the key N this selection is moved to the next element on the right-hand side and at the end of the row to the first element on the next row. The keys 0 or 1 assign the correspondent value to the selected element (the key 2 would assign the value ϕ, but this is not needed due to the predefined values ϕ). Note, a change of a value in the Karnaugh-map of the lattice implicitly adjusts the associated mark-function and its display in a viewport. A left-click on the region of the Gray codes finishes the selection of an element in Karnaugh-map so that the gray shaded background disappears.

The comparison of the Karnaugh-map of TVLs 1 and 3 of the ON-set-functions as well as TVLs 2 and 4 of the OFF-set-functions confirm that the mark-functions of a lattice can be determined by editing its Karnaugh-map. Alternatively, the commands te_syd 1 3 and te_syd 2 4,

```
 1   new                              10   / x2&x3 ) + / x1 &/ x2&x4&x5 .
 2   space  32  1                     11   assign  1  /m 1 1
 3   avar 1                           12   assign_qr  1  2  /m 1 2
 4   x1  x2  x3  x4  x5 .             13   assign  2  /m 1 3
 5   sbe  1  1                        14   dif  1  1  3
 6   x1&/x3&x4&x5 +/x1&(x2&/x5+       15   dif  1  1  4
 7   /x2&/x4&x5+x2&/x3&x4 ).          16   assign  3  /m 2 1
 8   sbe  1  2                        17   assign_qr  3  4  /m 2 2
 9   x1&(x4&/x5+x3&/x5+x3&/x4+        18   assign  4  /m 2 3
```

T	TVL (1) ODA | 5 Var. | 4 R. | S. 1	(q: TVL (1) ODA, r: TVL (2) ODA), 5 Var. | S. 1	T	TVL (2) ODA | 5 Var. | 5 R. | S. 1
	0 0	0 0 1 1	0 0 0 0		0 0	Φ Φ 1 1	Φ 0 0 Φ		0 0	0 0 0 0	0 1 1 0
	0 1	1 1 0 0	0 0 0 0		0 1	1 1 Φ Φ	Φ 0 0 Φ		0 1	0 0 0 0	0 1 1 0
	1 1	0 0 0 1	1 0 0 1		1 1	0 0 Φ 1	1 Φ 0 1		1 1	1 1 0 0	0 0 1 0
	1 0	0 0 1 1	0 0 0 0		1 0	Φ Φ 1 1	0 0 0 0		1 0	0 0 0 0	1 1 1 1
x4	0 1 1 0 0 1 1 0 x3	0 1 1 0 0 1 1 0 x3	x4	0 1 1 0 0 1 1 0 x3
x5	0 0 1 1 1 1 0 0 x2	0 0 1 1 1 1 0 0 x2	x5	0 0 1 1 1 1 0 0 x2
	0 0 0 0 1 1 1 1 x1	0 0 0 0 1 1 1 1 x1		0 0 0 0 1 1 1 1 x1

T	TVL (3) ODA | 5 Var. | 4 R. | S. 1	(q: TVL (3) ODA, r: TVL (4) ODA), 5 Var. | S. 1	T	TVL (4) ODA | 5 Var. | 5 R. | S. 1
	0 0	0 0 1 1	0 0 0 0		0 0	Φ Φ 1 1	Φ 0 0 Φ		0 0	0 0 0 0	0 1 1 0
	0 1	1 1 0 0	0 0 0 0		0 1	1 1 Φ Φ	Φ 0 0 Φ		0 1	0 0 0 0	0 1 1 0
	1 1	0 0 0 1	1 0 0 1		1 1	0 0 Φ 1	1 Φ 0 1		1 1	1 1 0 0	0 0 1 0
	1 0	0 0 1 1	0 0 0 0		1 0	Φ Φ 1 1	0 0 0 0		1 0	0 0 0 0	1 1 1 1
x4	0 1 1 0 0 1 1 0 x3	0 1 1 0 0 1 1 0 x3	x4	0 1 1 0 0 1 1 0 x3
x5	0 0 1 1 1 1 0 0 x2	0 0 1 1 1 1 0 0 x2	x5	0 0 1 1 1 1 0 0 x2
	0 0 0 0 1 1 1 1 x1	0 0 0 0 1 1 1 1 x1		0 0 0 0 1 1 1 1 x1

Fig. 7.24 Problem-program that realizes the visualization of the given lattice $\mathcal{L}\langle f_q(\mathbf{x}), f_r(\mathbf{x})\rangle$ and prepared lattice $\mathcal{L}'\langle f_q'(\mathbf{x}) = 0, f_r'(\mathbf{x}) = 0\rangle$ together with the m-fold view after editing $\mathcal{L}'$

typed in the console, confirm by the results $\texttt{true}$ (shown in the protocol window) this useful feature of the XBOOLE-monitor XBM 2.

(d) Figure 7.25 shows in the upper part the PRP that verifies whether the function $f_i(\mathbf{x})$ belongs to the explored lattice, computes the complement $^{\mathcal{L}}\overline{f_i(\mathbf{x})}$, verifies that this complement function also belongs to the lattice $\mathcal{L}\langle f_q(\mathbf{x}), f_r(\mathbf{x})\rangle$, and satisfies the rules of a lattice related complement. The viewports of the m-fold view show below this PRP the given lattice for comparisons, the TVL and the Karnaugh-map of the simple function $f_i(\mathbf{x})$, and the computed complement $^{\mathcal{L}}\overline{f_i(\mathbf{x})}$ related to the lattice $\mathcal{L}\langle f_q(\mathbf{x}), f_r(\mathbf{x})\rangle$.

The two results $\texttt{true}$ of the commands $\texttt{te\_dif}$ and $\texttt{te\_isc}$ in lines 13 and 14 confirm that the function $f_i(\mathbf{x})$ (TVL 3) belongs to the given lattice.

The sequence of the commands $\texttt{cpl}$, $\texttt{dif}$, and $\texttt{uni}$ in lines 15–17 compute the lattice related complement $^{\mathcal{L}}\overline{f_i(\mathbf{x})}$ (TVL 6) of the given function $f_i(\mathbf{x})$ based on (7.6). The two results $\texttt{true}$ of the commands $\texttt{te\_dif}$ and $\texttt{te\_isc}$ in lines 18 and 19 confirm that the complement function $^{\mathcal{L}}\overline{f_i(\mathbf{x})}$ (TVL 6) also belongs to the given lattice.

The two results $\texttt{true}$ of the command $\texttt{te\_syd}$ in lines 21 and 24 confirm that $f_i(\mathbf{x})$ and $^{\mathcal{L}}\overline{f_i(\mathbf{x})}$ satisfy the conditions of lattice related complements:

$$f_i(\mathbf{x}) \wedge {}^{\mathcal{L}}\overline{f_i(\mathbf{x})} = f_q(\mathbf{x}) ,$$

```
 1   new                          16   dif  4 2 5
 2   space 32 1                    17   uni  5 1 6
 3   avar 1                        18   set $nfgefq (te_dif 1 6)
 4   x1 x2 x3 x4 x5.               19   set $nflefr (te_isc 2 6)
 5   sbe 1 1                       20   isc  3 6 7
 6   x1&/x3&x4&x5+/x1&(x2&/x5+     21   set $fanfefq (te_syd 7 1)
 7   /x2&/x4&x5+x2&/x3&x4).        22   uni  3 6 8
 8   sbe 1 2                       23   cpl  2 9
 9   x1&(x4&/x5+x3&/x5+x3&/x4+     24   set $fonfenfr (te_syd 8 9)
10   /x2&x3)+/x1&/x2&x4&x5.        25   assign   1 /m 1 1
11   sbe 1 3                       26   assign_qr 1 2 /m 1 2
12   /x1&(x2+/x4)+x1&/x3&x5.       27   assign   2 /m 1 3
13   set $fgefq (te_dif 1 3)       28   assign   3 /m 2 1
14   set $flenfr (te_isc 2 3)      29   assign   3 /m 2 2
15   cpl 3 4                       30   assign   6 /m 2 3
```

```
T  TVL (1) ODA | 5 Var. | 4 R. | S. 1        (q: TVL (1) ODA, r: TVL (2) ODA), 5 Var. | S. 1      T  TVL (2) ODA | 5 Var. | 5 R. | S. 1

0 0   0 0 1 1   0 0 0 0                        0 0   Φ Φ 1 1   Φ 0 0 Φ                              0 0   0 0 0 0   0 1 1 0
0 1   1 1 0 0   0 0 0 0                        0 1   1 1 Φ Φ   Φ 0 0 Φ                              0 1   0 0 0 0   0 1 1 0
1 1   0 0 0 1   1 0 0 1                        1 1   0 0 Φ 1   1 Φ 0 1                              1 1   1 1 0 0   0 0 1 0
1 0   0 0 1 1   0 0 0 0                        1 0   Φ Φ 1 1   0 0 0 0                              1 0   0 0 0 0   1 1 1 1

x4 x5  0 1 1 0 0 1 1 0  x3                     x4 x5  0 1 1 0 0 1 1 0  x3                           x4 x5  0 1 1 0 0 1 1 0  x3
       0 0 1 1 1 1 0 0  x2                            0 0 1 1 1 1 0 0  x2                                  0 0 1 1 1 1 0 0  x2
       0 0 0 0 1 1 1 1  x1                            0 0 0 0 1 1 1 1  x1                                  0 0 0 0 1 1 1 1  x1
```

```
K  TVL (3) ODA | 5 Var. | 3 R. | S. 1        T  TVL (3) ODA | 5 Var. | 3 R. | S. 1              T  TVL (6) ODA | 5 Var. | 8 R. | S. 1

       x̄1 x̄2 x̄3 x̄4  x̄5                      0 0   1 1 1 1   0 0 0 0                             0 0   0 0 1 1   1 0 0 1
1      1  -  0  -   1                          0 1   1 1 1 1   1 0 0 1                             0 1   1 1 0 0   0 0 0 0
2      0  0  -  0   -                          1 1   0 0 1 1   1 0 0 1                             1 1   0 0 0 1   1 1 0 1
3      0  1  -  -   -                          1 0   0 0 1 1   0 0 0 0                             1 0   1 1 1 1   0 0 0 0

                                              x4 x5  0 1 1 0 0 1 1 0  x3                           x4 x5  0 1 1 0 0 1 1 0  x3
                                                     0 0 1 1 1 1 0 0  x2                                  0 0 1 1 1 1 0 0  x2
                                                     0 0 0 0 1 1 1 1  x1                                  0 0 0 0 1 1 1 1  x1
```

Fig. 7.25 Problem-program that verifies that both the given function $f_i(\mathbf{x})$ and the lattice related complement $^{\mathcal{L}}\overline{f_i(\mathbf{x})}$ belong to the explored lattice together with the m-fold view that shows all details

$$f_i(\mathbf{x}) \vee^{\mathcal{L}} \overline{f_i(\mathbf{x})} = \overline{f_r(\mathbf{x})}.$$

Both the TVL and the Karnaugh-map of $f_i(\mathbf{x})$ in viewports (2,1) and (2,2) show that this function of the lattice is simpler than the associated mark-functions. The comparison of the Karnaugh-maps of $f_i(\mathbf{x})$ (TVL 3 in viewport (2,2)) and $^{\mathcal{L}}f_i(\mathbf{x})$ (TVL 6 in viewport (2,3)) are equal to each other in the regions where $f_q(\mathbf{x}) = 1$ or $f_r(\mathbf{x}) = 1$, but they are complements for assignments where $f_\varphi(\mathbf{x}) = 1$.

Solution 7.3 (Exploration of a Generalized Lattice of Logic Functions)

(a) Figure 7.26 shows the PRP that verifies whether the given mark-functions $f_q(\mathbf{x})$ and $f_r(\mathbf{x})$ satisfy the condition of a lattice of logic functions computes the third mark-function $f_\varphi(\mathbf{x})$, generates all

Fig. 7.26
Problem-program that
computes all functions
belonging to a given lattice
and shows the
Karnaugh-maps of the
lattice and the computed
functions

```
 1   new
 2   space 32 1
 3   avar 1
 4   x1 x2 x3 x4 x5.
 5   sbe 1 1
 6   /x1&x2&/x5+x1&/x3&x5+
 7   x2&/x3+/x2&/x4&(/x3+x5).
 8   sbe 1 2
 9   x1&x3&/x5+x1&/x2&x4&/x5+
10   /x2&x3&x4+x1&x2&x3&/x4+
11   /x1&x5&(/x2&x4+x2&x3).
12   set $l (te_isc 1 2)
13   assign 1 /m 1 1
14   assign_qr 1 2 /m 1 2
15   assign 2 /m 1 3
16   uni 1 2 3
17   cpl 3 4
18   copy 4 5
19   dif 5 5 6
20   dif 5 5 8
21   set $nf 1
22   while (not(te 5))
23   (
24   stv 5 1 7
25   uni 7 6 7
26   cel 7 7 /-0
27   con 8 7 8
28   dif 5 7 5
29   set $nf (mul $nf 2)
```

```
30   )
31   set $r 2
32   set $c 1
33   for $i 0 (sub $nf 1)
34   (
35   set $fi (add $i 11)
36   copy 1 $fi
37   set $j $i
38   set $row 1
39   while (gt $j 0)
40   (
41   if (eq (mod $j 2) 1)
42   (
43   stv 8 $row 9
44   uni $fi 9 $fi
45   )
46   set $j (div $j 2)
47   set $row (add $row 1)
48   )
49   obbc $fi $fi
50   assign $fi /m $r $c
51   set $c (add $c 1)
52   if (eq $c 4)
53   (
54   set $c 1
55   set $r (add $r 1)
56   )
57   )
58   sts "e7_3_a"
```

functions $f_i(\mathbf{x})$ that belong to this lattice, and shows the Karnaugh-maps of the lattice as well as all functions belonging to this lattice.

The command te_isc in line 12 verifies whether the mark-functions $f_q(\mathbf{x})$ (TVL 1) and $f_r(\mathbf{x})$ (TVL 2) satisfy the condition of a lattice. The result $l=true confirms that these mark-functions specify a lattice. The Karnaugh-maps of the mark-functions of this lattice as well as the lattice itself are assigned to the viewports of the first row of the m-fold view in the next three lines of the PRP.

The used approach to generate the functions $f_i(\mathbf{x})$ belonging to the given lattice utilizes the mark-function $f_\varphi(\mathbf{x})$ represented as a list of binary vectors. The 1-bits of an integer counter of the index i of the wanted functions $f_i(\mathbf{x})$ in the range of 0 to $nf-1 indicate which binary vectors of $f_\varphi(\mathbf{x})$ must be appended to the TVL of $f_q(\mathbf{x})$ to get the TVL of $f_i(\mathbf{x})$.

The commands uni and cpl in lines 16 and 17 compute the TVL of the mark-function $f_\varphi(\mathbf{x})$ (TVL 4). TVL 4 is in this special case already a BVL; however, the lines 18–30 of the PRP split TVL 4 of $f_\varphi(\mathbf{x})$ into BVL 8 of this functions, for generality of the used approach, and computes in parallel the number $nf of functions $f_i(\mathbf{x})$ belonging to the lattice $\mathcal{L}\langle f_q, f_r\rangle$. In each sweep of the while-loop in lines 22–30 one binary vector (TVL 7) is selected and removed from the copied TVL 5 of $f_\varphi(\mathbf{x})$, this binary vector is appended to the BVL 8, and the number of function $nf (initialized with the values 1 before the while-loop) is multiplied by the value 2.

Each sweep of the for-loop in lines 33–57 generates one function $f_i(\mathbf{x})$ belonging to the given lattice, stores its minimized TVL as XBOOLE-object starting with the number 11, and assigns the computed function to consecutive viewports of the m-fold view starting in the second row and using three columns of this view. The TVL of each function $f_i(\mathbf{x})$ is initialized with the TVL of

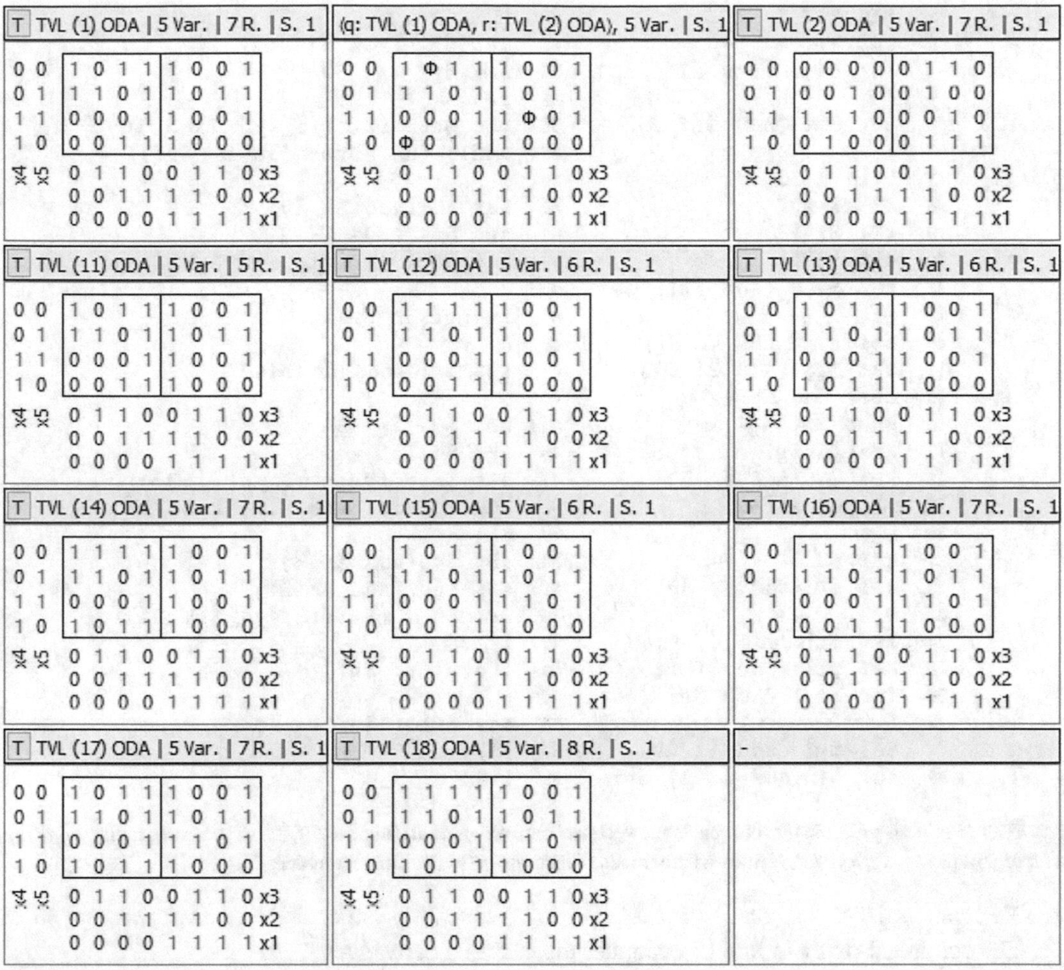

T TVL (1) ODA | 5 Var. | 7 R. | S. 1

```
0 0   1 0 1 1 | 1 0 0 1
0 1   1 1 0 1 | 1 0 1 1
1 1   0 0 0 1 | 1 0 0 1
1 0   0 0 1 1 | 1 0 0 0
x4 x5 0 1 1 0 0 1 1 0 x3
      0 0 1 1 1 1 0 0 x2
      0 0 0 0 1 1 1 1 x1
```

⟨q: TVL (1) ODA, r: TVL (2) ODA⟩, 5 Var. | S. 1

```
0 0   1 Φ 1 1 | 1 0 0 1
0 1   1 1 0 1 | 1 0 1 1
1 1   0 0 0 1 | 1 Φ 0 1
1 0   Φ 0 1 1 | 1 0 0 0
x4 x5 0 1 1 0 0 1 1 0 x3
      0 0 1 1 1 1 0 0 x2
      0 0 0 0 1 1 1 1 x1
```

T TVL (2) ODA | 5 Var. | 7 R. | S. 1

```
0 0   0 0 0 0 | 0 1 1 0
0 1   0 0 1 0 | 0 1 0 0
1 1   1 1 1 0 | 0 0 1 0
1 0   0 1 0 0 | 0 1 1 1
x4 x5 0 1 1 0 0 1 1 0 x3
      0 0 1 1 1 1 0 0 x2
      0 0 0 0 1 1 1 1 x1
```

T TVL (11) ODA | 5 Var. | 5 R. | S. 1

```
0 0   1 0 1 1 | 1 0 0 1
0 1   1 1 0 1 | 1 0 1 1
1 1   0 0 0 1 | 1 0 0 1
1 0   0 0 1 1 | 1 0 0 0
x4 x5 0 1 1 0 0 1 1 0 x3
      0 0 1 1 1 1 0 0 x2
      0 0 0 0 1 1 1 1 x1
```

T TVL (12) ODA | 5 Var. | 6 R. | S. 1

```
0 0   1 1 1 1 | 1 0 0 1
0 1   1 1 0 1 | 1 0 1 1
1 1   0 0 0 1 | 1 0 0 1
1 0   0 0 1 1 | 1 0 0 0
x4 x5 0 1 1 0 0 1 1 0 x3
      0 0 1 1 1 1 0 0 x2
      0 0 0 0 1 1 1 1 x1
```

T TVL (13) ODA | 5 Var. | 6 R. | S. 1

```
0 0   1 0 1 1 | 1 0 0 1
0 1   1 1 0 1 | 1 0 1 1
1 1   0 0 0 1 | 1 0 0 1
1 0   1 0 1 1 | 1 0 0 0
x4 x5 0 1 1 0 0 1 1 0 x3
      0 0 1 1 1 1 0 0 x2
      0 0 0 0 1 1 1 1 x1
```

T TVL (14) ODA | 5 Var. | 7 R. | S. 1

```
0 0   1 1 1 1 | 1 0 0 1
0 1   1 1 0 1 | 1 0 1 1
1 1   0 0 0 1 | 1 0 0 1
1 0   1 0 1 1 | 1 0 0 0
x4 x5 0 1 1 0 0 1 1 0 x3
      0 0 1 1 1 1 0 0 x2
      0 0 0 0 1 1 1 1 x1
```

T TVL (15) ODA | 5 Var. | 6 R. | S. 1

```
0 0   1 0 1 1 | 1 0 0 1
0 1   1 1 0 1 | 1 0 1 1
1 1   0 0 0 1 | 1 1 0 1
1 0   0 0 1 1 | 1 0 0 0
x4 x5 0 1 1 0 0 1 1 0 x3
      0 0 1 1 1 1 0 0 x2
      0 0 0 0 1 1 1 1 x1
```

T TVL (16) ODA | 5 Var. | 7 R. | S. 1

```
0 0   1 1 1 1 | 1 0 0 1
0 1   1 1 0 1 | 1 0 1 1
1 1   0 0 0 1 | 1 1 0 1
1 0   0 0 1 1 | 1 0 0 0
x4 x5 0 1 1 0 0 1 1 0 x3
      0 0 1 1 1 1 0 0 x2
      0 0 0 0 1 1 1 1 x1
```

T TVL (17) ODA | 5 Var. | 7 R. | S. 1

```
0 0   1 0 1 1 | 1 0 0 1
0 1   1 1 0 1 | 1 0 1 1
1 1   0 0 0 1 | 1 1 0 1
1 0   1 0 1 1 | 1 0 0 0
x4 x5 0 1 1 0 0 1 1 0 x3
      0 0 1 1 1 1 0 0 x2
      0 0 0 0 1 1 1 1 x1
```

T TVL (18) ODA | 5 Var. | 8 R. | S. 1

```
0 0   1 1 1 1 | 1 0 0 1
0 1   1 1 0 1 | 1 0 1 1
1 1   0 0 0 1 | 1 1 0 1
1 0   1 0 1 1 | 1 0 0 0
x4 x5 0 1 1 0 0 1 1 0 x3
      0 0 1 1 1 1 0 0 x2
      0 0 0 0 1 1 1 1 x1
```

-

Fig. 7.27 Results of the PRP shown in Fig. 7.26

$f_q(\mathbf{x})$ (TVL 1) in line 36 and extended with the binary vectors determined by the bits of integer j in the nested while-loop in lines 39–48.

The integer variables r and c determine the viewport used to show the function $f_i(\mathbf{x})$ computed in the current sweep of the for-loop. The command sts in line 58 stores the XBOOLE-system for the use in the next task.

Figure 7.27 shows in the m-fold view the Karnaugh-maps of $f_q(\mathbf{x})$ (viewport (1,1)), the lattice $\mathcal{L}\langle f_q, f_r\rangle$ (viewport (1,2)), $f_r(\mathbf{x})$ (viewport (1,3)), and in the rows 2 to 4 all eight functions $f_i(\mathbf{x})$ belonging to this lattice.

(b) Figure 7.28 shows the PRP that computes for all $f_i(\mathbf{x}) \in \mathcal{L}\langle f_q, f_r\rangle$ the vectorial derivatives $g_i(\mathbf{x}) = \mathrm{der}_{(x_2,x_3,x_5)} f_i(\mathbf{x})$ and verifies whether these functions $g_i(\mathbf{x})$ determine a lattice.

The command lds in the first line of the PRP loads all stored TVLs and VTs, but not the variables used in a PRP. Therefore, the number nf of functions of the lattice $\mathcal{L}\langle f_q, f_r\rangle$ is recomputed based on the number of binary vectors stored in the BVL 8 of $f_\varphi(\mathbf{x})$ in lines 2–6.

```
 1   lds "e7_3_a"              30   (
 2   set $nf 1                 31   set $gj (add $j 20)
 3   for $i 1 (ntv 8)          32   isc $gi $gj 30
 4   (                         33   set $k 1
 5   set $nf (mul $nf 2)       34   set $nea (not(te_syd 21 30))
 6   )                         35   while (and $nea (le $k $nf))
 7   vtin 1 20                 36   (
 8   x2 x3 x5.                 37   set $k (add $k 1)
 9   set $r 1                  38   set $gk (add $k 20)
10   set $c 1                  39   set $nea (not(te_syd $gk 30))
11   for $i 0 (sub $nf 1)      40   )
12   (                         41   if (gt $k $nf)
13   set $fi (add 11 $i)       42   (
14   set $gi (add 21 $i)       43   set $closedand false
15   derv $fi 20 $gi           44   )
16   obbc $gi $gi              45   uni $gi $gj 31
17   assign $gi /m $r $c       46   set $k 1
18   set $c (add $c 1)         47   set $neo (not(te_syd 21 31))
19   if (eq $c 4)              48   while (and $neo (le $k $nf))
20   (                         49   (
21   set $c 1                  50   set $k (add $k 1)
22   set $r (add $r 1)         51   set $gk (add $k 20)
23   ))                        52   set $neo (not(te_syd $gk 31))
24   set $closedand true       53   )
25   set $closedor true        54   if (gt $k $nf)
26   for $i 1 (sub $nf 1)      55   (
27   (                         56   set $closedor false
28   set $gi (add $i 20)       57   )))
29   for $j (add $i 1) $nf     58   sts "e7_3_b"
```

Fig. 7.28 Problem-program that computes the vectorial derivatives of all functions $f_i(\mathbf{x})$ of the given lattice $\mathcal{L}\langle f_q, f_r\rangle$ with regard to (x_2, x_3, x_5) and verifies whether these functions $g_i(\mathbf{x})$ determine a lattice $\mathcal{L}\langle g_q, g_r\rangle$

The command `derv` in line 15 computes the vectorial derivatives

$$g_i(\mathbf{x}) = \underset{(x_2, x_3, x_5)}{\mathrm{der}} f_i(\mathbf{x})$$

for all functions $f_i(\mathbf{x}) \in \mathcal{L}\langle f_q, f_r\rangle$ within the for-loop in lines 11–23 and stores the results (minimized by the subsequent command `obbc`) as sequence of XBOOLE-objects beginning with the number 21. The command `assign` in line 17 assigns within the same for-loop the computed vectorial derivatives to the viewports of the m-fold view beginning in the first row.

The computed set of vectorial derivatives determines a lattice when this set of functions is closed with regard to the operations $\wedge$ and $\vee$. This check is realized by two nested for-loops and two embedded while-loops; the outer for-loops select each pair of functions $(g_i(\mathbf{x}), g_j(\mathbf{x}))$ of the computed set of vectorial derivatives.

The command `isc` in line 32 computes the conjunction of the selected pair of vectorial derivatives and the commands `te_syd` in lines 34 or 39 compute the result `true` when either $g_1(\mathbf{x})$ or $g_k(\mathbf{x})$ is equal to the conjunction of the evaluated pair of functions. The while-loop in lines 35–40 terminates when a function $g_k(\mathbf{x})$ of the set of vectorial derivatives is equal to the function $g_i(\mathbf{x}) \wedge g_j(\mathbf{x})$ or this set does not contain such a function. An index k that is greater than nf (that is equal to the number of functions in the set $\{g_i(\mathbf{x}) | g_i(\mathbf{x}) = \mathrm{der}_{(x_2, x_3, x_5)} f_i(\mathbf{x}), 1 \le i \le nf\}$) indicates after the execution of the while-loop that no function with the required property has

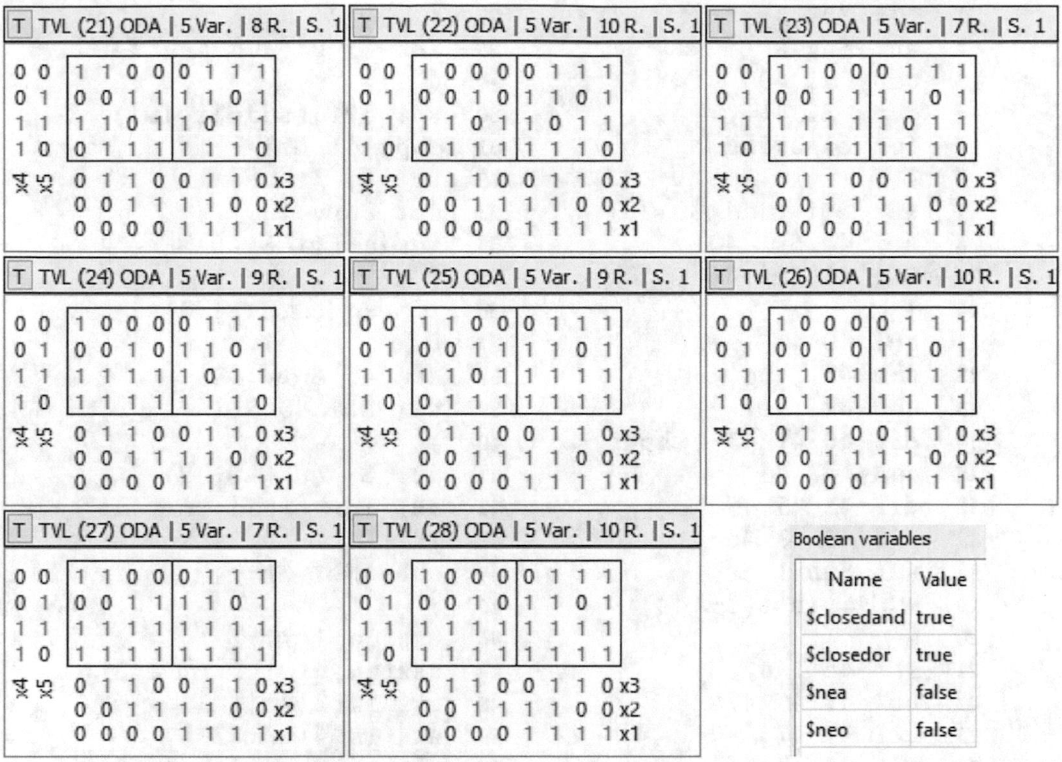

TVL (21) ODA | 5 Var. | 8 R. | S. 1

	1 1 0 0	0 1 1 1
0 0	1 1 0 0	0 1 1 1
0 1	0 0 1 1	1 1 0 1
1 1	1 1 0 1	1 0 1 1
1 0	0 1 1 1	1 1 1 0

x4 x5
```
0 1 1 0 0 1 1 0 x3
0 0 1 1 1 1 0 0 x2
0 0 0 0 1 1 1 1 x1
```

TVL (22) ODA | 5 Var. | 10 R. | S. 1

0 0	1 0 0 0	0 1 1 1
0 1	0 0 1 0	1 1 0 1
1 1	1 1 0 1	1 0 1 1
1 0	0 1 1 1	1 1 1 0

x4 x5
```
0 1 1 0 0 1 1 0 x3
0 0 1 1 1 1 0 0 x2
0 0 0 0 1 1 1 1 x1
```

TVL (23) ODA | 5 Var. | 7 R. | S. 1

0 0	1 1 0 0	0 1 1 1
0 1	0 0 1 1	1 1 0 1
1 1	1 1 1 1	1 0 1 1
1 0	1 1 1 1	1 1 1 0

x4 x5
```
0 1 1 0 0 1 1 0 x3
0 0 1 1 1 1 0 0 x2
0 0 0 0 1 1 1 1 x1
```

TVL (24) ODA | 5 Var. | 9 R. | S. 1

0 0	1 0 0 0	0 1 1 1
0 1	0 0 1 0	1 1 0 1
1 1	1 1 1 1	1 0 1 1
1 0	1 1 1 1	1 1 1 0

x4 x5
```
0 1 1 0 0 1 1 0 x3
0 0 1 1 1 1 0 0 x2
0 0 0 0 1 1 1 1 x1
```

TVL (25) ODA | 5 Var. | 9 R. | S. 1

0 0	1 1 0 0	0 1 1 1
0 1	0 0 1 1	1 1 0 1
1 1	1 1 0 1	1 1 1 1
1 0	0 1 1 1	1 1 1 1

x4 x5
```
0 1 1 0 0 1 1 0 x3
0 0 1 1 1 1 0 0 x2
0 0 0 0 1 1 1 1 x1
```

TVL (26) ODA | 5 Var. | 10 R. | S. 1

0 0	1 0 0 0	0 1 1 1
0 1	0 0 1 0	1 1 0 1
1 1	1 1 0 1	1 1 1 1
1 0	0 1 1 1	1 1 1 1

x4 x5
```
0 1 1 0 0 1 1 0 x3
0 0 1 1 1 1 0 0 x2
0 0 0 0 1 1 1 1 x1
```

TVL (27) ODA | 5 Var. | 7 R. | S. 1

0 0	1 1 0 0	0 1 1 1
0 1	0 0 1 1	1 1 0 1
1 1	1 1 1 1	1 1 1 1
1 0	1 1 1 1	1 1 1 1

x4 x5
```
0 1 1 0 0 1 1 0 x3
0 0 1 1 1 1 0 0 x2
0 0 0 0 1 1 1 1 x1
```

TVL (28) ODA | 5 Var. | 10 R. | S. 1

0 0	1 0 0 0	0 1 1 1
0 1	0 0 1 0	1 1 0 1
1 1	1 1 1 1	1 1 1 1
1 0	1 1 1 1	1 1 1 1

x4 x5
```
0 1 1 0 0 1 1 0 x3
0 0 1 1 1 1 0 0 x2
0 0 0 0 1 1 1 1 x1
```

Boolean variables

Name	Value
$closedand	true
$closedor	true
$nea	false
$neo	false

Fig. 7.29 Results of the PRP shown in Fig. 7.28

been found; in this case, the variable $closedand (which is initialized with the value true before the two nested for-loops) is changed to the value false in line 43.

The check whether the set of function $g_i(\mathbf{x})$ is closed with regard to the operation $\vee$ is realized in analogue manner using the command uni in line 45 and the subsequent commands belonging to the while-loop and the command if.

The command sts in line 52 stores the XBOOLE-system for the use in the next task.

Figure 7.29 shows the Karnaugh-maps of the eight computed vectorial derivatives and the resulting values of the Boolean variables. The values true of both $closedand and $closedor confirm that the computed set of vectorial derivatives $g_i(\mathbf{x})$ is closed with regard to the operations $\wedge$ and $\vee$; hence, these logic functions satisfy the conditions of a lattice.

(c) Figure 7.30 shows in the upper part the PRP that computes the mark-functions $h_q(\mathbf{x})$ and $h_r(\mathbf{x})$ as well as all functions $h_i(\mathbf{x})$ of the smallest lattice $\mathcal{L}\langle h_q(\mathbf{x}), h_r(\mathbf{x})\rangle$ that contains all functions $g_i(\mathbf{x})$ computed in the previous task. The details of both the basically given lattice $\mathcal{L}\langle f_q, f_r\rangle$ and the computed lattice $\mathcal{L}\langle h_q, h_r\rangle$ are shown below this PRP in two rows of the m-fold view.

The smallest function that is larger than or equal to all eight functions $g_i(\mathbf{x})$ can be computed as union of the associated eight TVLs stored as XBOOLE-objects 21 to 28. Vice versa, the largest function that is smaller than or equal to all eight functions $g_i(\mathbf{x})$ can be computed as intersection of these eight TVLs. This computation is realized by the commands in lines 2–10. The subsequent command cpl computes the mark-function $h_r(\mathbf{x})$ using the smallest function that is larger than or equal to all eight functions $g_i(\mathbf{x})$. The next two commands obb compute minimized TVLs of the mark-functions $h_q(\mathbf{x})$ (TVL 41) and $h_r(\mathbf{x})$ (TVL 42).

```
 1  lds  "e7_3_b"                27  )
 2  set  $ng 8                   28  for $i 0 (sub $nh 1)
 3  copy 21 40                   29  (
 4  copy 21 41                   30  set $hi (add $i 101)
 5  for $i 2 $ng                 31  copy 41 $hi
 6  (                            32  set $j $i
 7  set $gi (add $i 20)          33  set $row 1
 8  uni 40 $gi 40                34  while (gt $j 0)
 9  isc 41 $gi 41                35  (
10  )                            36  if (eq (mod $j 2) 1)
11  cpl 40 42                    37  (
12  obb 41 ; hq                  38  stv 46 $row 48
13  obb 42 ; hr                  39  uni $hi 48 $hi
14  dif 40 41 43 ; hphi          40  )
15  copy 43 44                   41  set $j (div $j 2)
16  dif 43 43 45                 42  set $row (add $row 1)
17  dif 43 43 46                 43  )
18  set $nh 1                    44  obbc $hi $hi
19  while (not(te 44))           45  )
20  (                            46  assign 1 /m 1 1
21  stv 44 1 47                  47  assign_qr 1 2 /m 1 2
22  uni 47 45 47                 48  assign 2 /m 1 3
23  cel 47 47 /-0                49  assign 41 /m 2 1
24  dif 44 47 44                 50  assign_qr 41 42 /m 2 2
25  con 46 47 46                 51  assign 42 /m 2 3
26  set $nh (mul $nh 2)          52  sts "e7_3_c"
```

T TVL (1) ODA | 5 Var. | 7 R. | S. 1

```
0 0  | 1 0 1 1 | 1 0 0 1
0 1  | 1 1 0 1 | 1 0 1 1
1 1  | 0 0 0 1 | 1 0 0 1
1 0  | 0 0 1 1 | 1 0 0 0
x4
x5   0 1 1 0 0 1 1 0 x3
     0 0 1 1 1 1 0 0 x2
     0 0 0 0 1 1 1 1 x1
```

(q: TVL (1) ODA, r: TVL (2) ODA), 5 Var. | S. 1

```
0 0  | 1 Φ 1 1 | 1 0 0 1
0 1  | 1 1 0 1 | 1 0 1 1
1 1  | 0 0 0 1 | 1 Φ 0 1
1 0  | Φ 0 1 1 | 1 0 0 0
x4
x5   0 1 1 0 0 1 1 0 x3
     0 0 1 1 1 1 0 0 x2
     0 0 0 0 1 1 1 1 x1
```

T TVL (2) ODA | 5 Var. | 7 R. | S. 1

```
0 0  | 0 0 0 0 | 0 1 1 0
0 1  | 0 0 1 0 | 0 1 0 0
1 1  | 1 1 1 0 | 0 0 1 0
1 0  | 0 1 0 0 | 0 1 1 1
x4
x5   0 1 1 0 0 1 1 0 x3
     0 0 1 1 1 1 0 0 x2
     0 0 0 0 1 1 1 1 x1
```

T TVL (41) ODA | 5 Var. | 8 R. | S. 1

```
0 0  | 1 0 0 0 | 0 1 1 1
0 1  | 0 0 1 0 | 1 1 0 1
1 1  | 1 1 0 1 | 1 0 1 1
1 0  | 0 1 1 1 | 1 1 1 0
x4
x5   0 1 1 0 0 1 1 0 x3
     0 0 1 1 1 1 0 0 x2
     0 0 0 0 1 1 1 1 x1
```

(q: TVL (41) ODA, r: TVL (42) ODA), 5 Var. | S. 1

```
0 0  | 1 Φ 0 0 | 0 1 1 1
0 1  | 0 0 1 Φ | 1 1 0 1
1 1  | 1 1 Φ 1 | 1 Φ 1 1
1 0  | Φ 1 1 1 | 1 1 1 Φ
x4
x5   0 1 1 0 0 1 1 0 x3
     0 0 1 1 1 1 0 0 x2
     0 0 0 0 1 1 1 1 x1
```

T TVL (42) ODA | 5 Var. | 4 R. | S. 1

```
0 0  | 0 0 1 1 | 1 0 0 0
0 1  | 1 1 0 0 | 0 0 1 0
1 1  | 0 0 0 0 | 0 0 0 0
1 0  | 0 0 0 0 | 0 0 0 0
x4
x5   0 1 1 0 0 1 1 0 x3
     0 0 1 1 1 1 0 0 x2
     0 0 0 0 1 1 1 1 x1
```

Fig. 7.30 Problem-program that computes the smallest lattice $\mathcal{L}\langle h_q, h_r \rangle$ that contains all functions $g_i(\mathbf{x})$ and shows for comparison the details of the lattices $\mathcal{L}\langle f_q, f_r \rangle$ and $\mathcal{L}\langle h_q, h_r \rangle$

```
1   lds "e7_3_c"                    20  (
2   assign 41 /m 1 1                21  set $c 1
3   assign_qr 41 42 /m 1 2          22  set $r (add $r 1)
4   assign 42 /m 1 3                23  )
5   set $ng 8                       24  set $nderv0h (add $nderv0h 1)
6   set $nh 64                      25  set $j 1
7   set $derv0g true                26  set $gj 21
8   set $nderv0h 0                  27  set $negh (not(te_syd $gj $hi))
9   set $r 2                        28  while (and $negh (le $j $ng))
10  set $c 1                        29  (
11  for $i 1 $nh                    30  set $j (add $j 1)
12  (                               31  set $gj (add $j 20)
13  set $hi (add $i 100)            32  set $negh (not(te_syd $gj $hi))
14  derv $hi 20 50                  33  )
15  if (te 50)                      34  if (gt $j $ng)
16  (                               35  (
17  assign $hi /m $r $c             36  set $derv0g false
18  set $c (add $c 1)               37  )))
19  if (eq $c 4)                    38  sts "e7_3_d"
```

Fig. 7.31 Problem-program that verifies that the condition $g_j(\mathbf{x}) = \mathrm{der}_{(x_2,x_3,x_5)}\, h_i(\mathbf{x})$ determines all functions of the generalized lattice $\mathcal{L}\langle g_q, g_r\rangle$ as subset of the lattice $\mathcal{L}\langle h_q, h_r\rangle$

The computation of all functions $h_i(\mathbf{x})$ of the lattice $\mathcal{L}\langle h_q, h_r\rangle$ reuses the approach that has been used to compute all function $f_i(\mathbf{x})$ of the lattice $\mathcal{L}\langle f_q, f_r\rangle$. The command dif in line 14 computes the TVL of the mark-function $h_\varphi(\mathbf{x})$ for which the BVL 46 is generated in lines 15–27. In parallel the number $nh of functions of lattice $\mathcal{L}\langle h_q, h_r\rangle$ has been computed.

The commands in the for-loop in lines 28–45 compute minimized TVLs of the 64 functions $h_i(\mathbf{x})$ belonging to the lattice $\mathcal{L}\langle h_q, h_r\rangle$ and store the TVLs of these functions as XBOOLE-objects 101 to 164 so that these TVLs can be displayed after the execution of the PRP.

The commands assign and assign_qr in lines 46–51 assign the lattices $\mathcal{L}\langle f_q, f_r\rangle$ and $\mathcal{L}\langle h_q, h_r\rangle$ and the associated mark-functions in two rows of the m-fold view. This part of the m-fold view has been inserted in Fig. 7.30 below the PRP. The comparison to these lattices shows that each symbol ϕ in the lattice $\mathcal{L}\langle f_q, f_r\rangle$ causes two symbols ϕ in the lattice $\mathcal{L}\langle h_q, h_r\rangle$ in a distance determined by the change of (x_2, x_3, x_5); hence, the number of function values 1 of the don't-care set is increased from $\|f_\varphi(bx)\| = 3$ to $\|h_\varphi(bx)\| = 6$, so that $2^3 = 8$ functions $f_i(\mathbf{x})$ and the same number of functions $g_i(\mathbf{x})$ exist, but the number of functions $h_i(\mathbf{x})$ is increased to $2^6 = 64$.

The command sts in line 52 stores the XBOOLE-system for the use in the next task.

(d) Figure 7.31 shows the PRP that verifies the condition that determines the generalized lattice $\mathcal{L}\langle g_q, g_r\rangle$ as sublattice of the lattice $\mathcal{L}\langle h_q, h_r\rangle$.

The smallest lattice $\mathcal{L}\langle h_q, h_r\rangle$ of 64 functions $h_i(\mathbf{x})$ that contains all eight vectorial derivatives $g_i(\mathbf{x})$ is known from the previous task. Furthermore, it is known as result of last but one task that these eight functions $g_i(\mathbf{x})$ determine also a lattice with a special property; therefore $\mathcal{L}\langle g_q, g_r, \cdot\rangle$ is denoted by *generalized lattice*. This task verifies the condition that selects the functions $g_i(\mathbf{x})$ as subset of the functions $h_i(\mathbf{x}) \in \mathcal{L}\langle h_q, h_r\rangle$. The functions $g_i(\mathbf{x})$ are the results of vectorial derivatives of the functions $f_i(\mathbf{x}) \in \mathcal{L}\langle f_q, f_r\rangle$ with regard to $\mathbf{x}_0 = (x_2, x_3, x_5)$; hence, $\mathrm{der}_{\mathbf{x}_0}\, g_i(\mathbf{x}) = 0$ and $\mathrm{der}_{\mathbf{x}_0}\, h_i(\mathbf{x}) = 0$ can be used as condition to select the functions $g_i(\mathbf{x})$ out of the lattice $\mathcal{L}\langle h_q, h_r\rangle$.

The commands in lines 2–4 assigns the details of the lattice $\mathcal{L}\langle h_q, h_r\rangle$ to the first row of the m-fold view. The lattice $\mathcal{L}\langle h_q, h_r\rangle$ and the generalized lattice $\mathcal{L}\langle g_q, g_r, \cdot\rangle$ have the same mark-functions:

$$g_q(\mathbf{x}) = h_q(\mathbf{x}), \quad \text{(TVL 41)}; \qquad g_r(\mathbf{x}) = h_r(\mathbf{x}), \quad \text{(TVL 42)}.$$

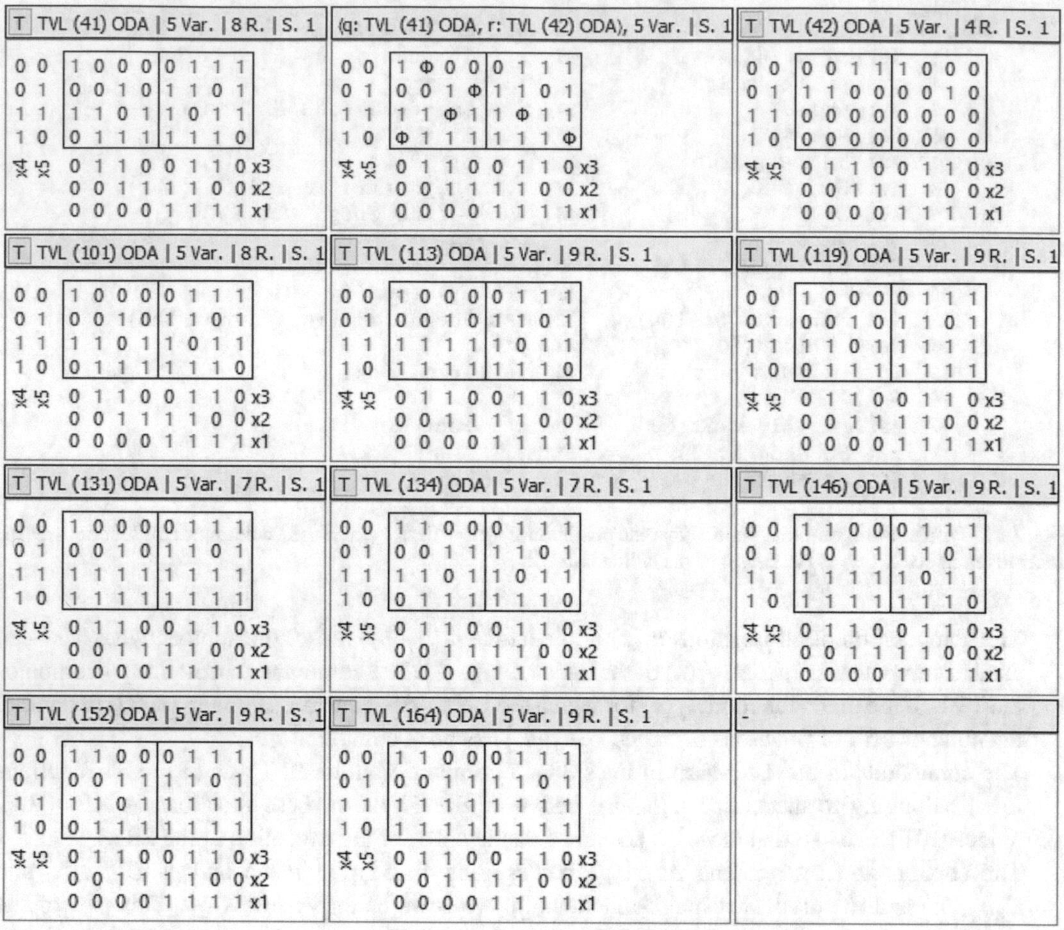

```
┌─────────────────────────────┐  ┌────────────────────────────────────┐  ┌─────────────────────────────┐
│ T  TVL (41) ODA │5 Var.│8 R.│S. 1│ (q: TVL (41) ODA, r: TVL (42) ODA), 5 Var.│S. 1│ T  TVL (42) ODA │5 Var.│4 R.│S. 1│

  0 0 │1 0 0 0│0 1 1 1            0 0 │1 Φ 0 0│0 1 1 1              0 0 │0 0 1 1│1 0 0 0
  0 1 │0 0 1 0│1 1 0 1            0 1 │0 0 1 Φ│1 1 0 1              0 1 │1 1 0 0│0 0 1 0
  1 1 │1 1 0 1│1 0 1 1            1 1 │1 1 Φ 1│1 Φ 1 1              1 1 │0 0 0 0│0 0 0 0
  1 0 │0 1 1 1│1 1 1 0            1 0 │Φ 1 1 1│1 1 1 Φ              1 0 │0 0 0 0│0 0 0 0

  x4 x5  0 1 1 0 0 1 1 0 x3       x4 x5  0 1 1 0 0 1 1 0 x3         x4 x5  0 1 1 0 0 1 1 0 x3
         0 0 1 1 1 1 0 0 x2              0 0 1 1 1 1 0 0 x2                0 0 1 1 1 1 0 0 x2
         0 0 0 0 1 1 1 1 x1              0 0 0 0 1 1 1 1 x1                0 0 0 0 1 1 1 1 x1
```

```
┌─────────────────────────────┐  ┌─────────────────────────────┐  ┌─────────────────────────────┐
│ T  TVL (101) ODA │5 Var.│8 R.│S. 1│ T  TVL (113) ODA │5 Var.│9 R.│S. 1│ T  TVL (119) ODA │5 Var.│9 R.│S. 1│

  0 0 │1 0 0 0│0 1 1 1            0 0 │1 0 0 0│0 1 1 1              0 0 │1 0 0 0│0 1 1 1
  0 1 │0 0 1 0│1 1 0 1            0 1 │0 0 1 0│1 1 0 1              0 1 │0 0 1 0│1 1 0 1
  1 1 │1 1 0 1│1 0 1 1            1 1 │1 1 1 1│1 0 1 1              1 1 │1 1 0 1│1 1 1 1
  1 0 │0 1 1 1│1 1 1 0            1 0 │1 1 1 1│1 1 1 0              1 0 │0 1 1 1│1 1 1 1

  x4 x5  0 1 1 0 0 1 1 0 x3       x4 x5  0 1 1 0 0 1 1 0 x3         x4 x5  0 1 1 0 0 1 1 0 x3
         0 0 1 1 1 1 0 0 x2              0 0 1 1 1 1 0 0 x2                0 0 1 1 1 1 0 0 x2
         0 0 0 0 1 1 1 1 x1              0 0 0 0 1 1 1 1 x1                0 0 0 0 1 1 1 1 x1
```

```
┌─────────────────────────────┐  ┌─────────────────────────────┐  ┌─────────────────────────────┐
│ T  TVL (131) ODA │5 Var.│7 R.│S. 1│ T  TVL (134) ODA │5 Var.│7 R.│S. 1│ T  TVL (146) ODA │5 Var.│9 R.│S. 1│

  0 0 │1 0 0 0│0 1 1 1            0 0 │1 1 0 0│0 1 1 1              0 0 │1 1 0 0│0 1 1 1
  0 1 │0 0 1 0│1 1 0 1            0 1 │0 0 1 1│1 1 0 1              0 1 │0 0 1 1│1 1 0 1
  1 1 │1 1 1 1│1 1 1 1            1 1 │1 1 0 1│1 0 1 1              1 1 │1 1 1 1│1 0 1 1
  1 0 │1 1 1 1│1 1 1 1            1 0 │0 1 1 1│1 1 1 0              1 0 │1 1 1 1│1 1 1 0

  x4 x5  0 1 1 0 0 1 1 0 x3       x4 x5  0 1 1 0 0 1 1 0 x3         x4 x5  0 1 1 0 0 1 1 0 x3
         0 0 1 1 1 1 0 0 x2              0 0 1 1 1 1 0 0 x2                0 0 1 1 1 1 0 0 x2
         0 0 0 0 1 1 1 1 x1              0 0 0 0 1 1 1 1 x1                0 0 0 0 1 1 1 1 x1
```

```
┌─────────────────────────────┐  ┌─────────────────────────────┐  ┌─────────────┐
│ T  TVL (152) ODA │5 Var.│9 R.│S. 1│ T  TVL (164) ODA │5 Var.│9 R.│S. 1│      -      │

  0 0 │1 1 0 0│0 1 1 1            0 0 │1 1 0 0│0 1 1 1
  0 1 │0 0 1 1│1 1 0 1            0 1 │0 0 1 1│1 1 0 1
  1 1 │1 1 0 1│1 1 1 1            1 1 │1 1 1 1│1 1 1 1
  1 0 │0 1 1 1│1 1 1 1            1 0 │1 1 1 1│1 1 1 1

  x4 x5  0 1 1 0 0 1 1 0 x3       x4 x5  0 1 1 0 0 1 1 0 x3
         0 0 1 1 1 1 0 0 x2              0 0 1 1 1 1 0 0 x2
         0 0 0 0 1 1 1 1 x1              0 0 0 0 1 1 1 1 x1
```

Fig. 7.32 Results of the PRP shown in Fig. 7.31

The main `for`-loop in lines 11–37 computes for each function $h_i(\mathbf{x})$ the vectorial derivative with regard to $\mathbf{x}_0 = (x_2, x_3, x_5)$ (line 14) and checks in the case that $\mathrm{der}_{\mathbf{x}_0} h_i(\mathbf{x}) = 0$ (condition of the command `if` in line 15) whether the evaluated function $h_i(\mathbf{x})$ is equal to one of the functions $g_i(\mathbf{x})$; the `true`-initialized variable `$derv0g` is changed to the value `false` in line 36 if such a function $g_i(\mathbf{x})$ has not been found in the `while`-loop in lines 28–33.

Functions $h_i(\mathbf{x})$ which satisfy $\mathrm{der}_{\mathbf{x}_0} h_i(\mathbf{x}) = 0$ are assigned to the viewports of the m-fold view beginning in the second row. Figure 7.32 shows that eight such functions exist. The final value `$derv0g=true` confirms that each function $h_i(\mathbf{x})$ with $\mathrm{der}_{\mathbf{x}_0} h_i(\mathbf{x}) = 0$ is equal to one of the functions $g_i(\mathbf{x})$ computed as vectorial derivatives of a function $f_i(\mathbf{x}) \in \mathcal{L}\langle f_q, f_r \rangle$.

The comparison of the Karnaugh-maps of Figs. 7.29 and 7.32 confirms that all function $h_i(\mathbf{x})$ with $\mathrm{der}_{\mathbf{x}_0} h_i(\mathbf{x}) = 0$ are identical with exactly one function $g_i(\mathbf{x})$; the object numbers of identical pairs of TVLs $(g_i(\mathbf{x}), h_j(\mathbf{x}))$ are (21, 134), (22, 101), (23, 146), (24, 113), (25, 152), (26, 119), (27, 164), and (28, 131). The integer variable `$nderv0h` counts the number of function $h_i(\mathbf{x})$ that satisfy $\mathrm{der}_{\mathbf{x}_0} h_i(\mathbf{x}) = 0$ and confirms that eight such functions exist.

The specification of the generalized lattice that contains the eight functions $g_i(\mathbf{x}) = \mathrm{der}_{(x_2, x_3, x_5)} f_i(\mathbf{x})$ of all $f_i(\mathbf{x}) \in \mathcal{L}_f \langle f_q(\mathbf{x}), f_r(\mathbf{x}) \rangle$ is

$$\mathcal{L}\left\langle g_q(\mathbf{x}), g_r(\mathbf{x}), \operatorname*{der}_{(x_2,x_3,x_5)} g(\mathbf{x}) \right\rangle ,$$

where

$$
\text{ODA}(g_q) =
\begin{array}{ccccc}
x_1 & x_2 & x_3 & x_4 & x_5 \\
\hline
0 & 0 & 0 & 0 & 0 \\
1 & 0 & 0 & 0 & - \\
0 & 1 & 1 & 0 & 1 \\
1 & - & 1 & 0 & 0 \\
1 & 1 & - & 0 & 1 \\
- & 1 & - & 1 & 0 \\
- & 0 & 1 & 1 & - \\
- & - & 0 & 1 & 1 \\
\end{array},
\qquad
\text{ODA}(g_r) =
\begin{array}{ccccc}
x_1 & x_2 & x_3 & x_4 & x_5 \\
\hline
0 & 0 & 0 & 0 & 1 \\
- & 1 & 0 & 0 & 0 \\
0 & 1 & 1 & 0 & 0 \\
- & 0 & 1 & 0 & 1 \\
\end{array}.
$$

and all functions $g_i(\mathbf{x})$ of this generalized lattice satisfy $\operatorname{der}_{(x_2,x_3,x_5)} g_i(\mathbf{x}) = 0$.

Solution 7.4 (Solution of Logic Equations with Regard to Variables)

(a) Figure 7.33 shows in the upper part the PRP that verifies whether the three restrictive equations $f_i(\mathbf{x}, \mathbf{y}) = 0$ can be solved can be solved with regard to the occurring variables y_j and determines the type of the solution (unique function or lattice of functions); the Karnaugh-maps of the given three functions are shown below this PRP.

The three functions are specified in lines 5–21 and assigned by the three subsequent commands to viewports of the first row of the m-fold view. The commands vtin in lines 25–30 create the needed VTs of the variables y_j.

Solution functions $y_j = g_j(\mathbf{x})$ of a given restrictive equation $f_i(\mathbf{x}, \mathbf{y}) = 0$ with regard to all $y_j \in \mathbf{y}$ exist if

$$\operatorname*{min}_{\mathbf{y}}{}^k f_i(\mathbf{x}, \mathbf{y}) = 0 .$$

The commands te_mink in lines 31–33 check this property for the three given functions. The results $f10s=false, $f20s=true, and $f30s=true confirm that equation $f_1(\mathbf{x}, y_1) = 0$ cannot be solved with regard to y_1 because $f_1(1, 0, 0, 0, y_1) = 1$; however, the other two explored equations are solvable with regard to all occurring variables y_j.

In the case that a logic equation is solvable with regard to all variables y_j, it can be separately checked whether the function $y_j = g_j(\mathbf{x})$ is uniquely specified or can be chosen out of a lattice of functions. These checks are realized in the last part of this PRP. The condition of the command if in line 34 is not satisfied; hence, lines 35–38 are not executed.

A restrictive equation $f_i(\mathbf{x}, y_1) = 0$ that is solvable with regard to y_1 uniquely determines the function $y_1 = g_1(\mathbf{x})$ if

$$\operatorname*{der}_{y_1} f_i(\mathbf{x}, y_1) = 1 .$$

The commands derk and te_cpl in lines 41 and 42 check this condition for the equation $f_2(\mathbf{x}, y_1) = 0$. The result $f20usy1=true confirms that the restrictive equation $f_2(\mathbf{x}, y_1) = 0$ is uniquely solvable with regard to y_1.

It must be checked separately for each variable y_j whether the associated solution-function of a restrictive equation $f_i(\mathbf{x}, \mathbf{y}) = 0$ is uniquely specified or can be chosen out of a lattice of functions. A k-fold minimum of $f_i(\mathbf{x}, \mathbf{y})$ with regard to $(\mathbf{y} \setminus y_j)$ creates a function $f_i'(\mathbf{x}, y_j)$ for which the derivative with regard to y_j is equal to 1 in the case that the solution-function $y_j = g_j(\mathbf{x})$ is uniquely specified. The two sequences of commands mink, derk and te_cpl in lines 46–51 check this condition for equation $f_3(\mathbf{x}, y_1, y_2) = 0$ first for y_1 and thereafter for y_2.

```
1    new                          27   vtin  1 5
2    space 32 1                   28   y2.
3    avar 1                       29   vtin  1 6
4    x1 x2 x3 x4 y1 y2.           30   y1 y2.
5    sbe 1 1                      31   set $f10s (te_mink 1 4)
6    x1&/x2&/x4&y1+               32   set $f20s (te_mink 2 4)
7    x3&y1&(x2+x4)+               33   set $f30s (te_mink 3 6)
8    /x3&/y1+/x1&/x2&/x4&/y1.     34   if (te_mink 1 4)
9    sbe 1 2                      35   (
10   x1&/x2&x4&/y1+x1&/x3&/y1+    36   derk 1 4 10
11   x2&x3&y1&(/x1+x2+/x4)+       37   set $f10usy1 (te_cpl 10)
12   /x1&/x2&(x4&y1+/x4&/y1)+     38   )
13   x2&/x3&/y1+x1&x3&/x4&y1.     39   if (te_mink 2 4)
14   sbe 1 3                      40   (
15   x1&/y2&(/x2&/y1+/x3&/X4)+    41   derk 2 4 20
16   /x3&(/y1+x4&y2)+             42   set $f20usy1 (te_cpl 20)
17   /y1&(x2&/x3&y2+x1&/x2&/x4)+  43   )
18   x3&y1&(/x1+x2+x4)+           44   if (te_mink 3 6)
19   /x1&/x4&/y2&(/x2+y1)+        45   (
20   x2&y2&(/x4&/y1+x1&x4)+       46   mink 3 5 30
21   /x2&/y1&/y2.                 47   derk 30 4 31
22   assign 1 /m 1 1              48   set $f30usy1 (te_cpl 31)
23   assign 2 /m 1 2              49   mink 3 4 32
24   assign 3 /m 1 3              50   derk 32 5 33
25   vtin  1 4                    51   set $f30usy2 (te_cpl 33)
26   y1.                          52   )
```

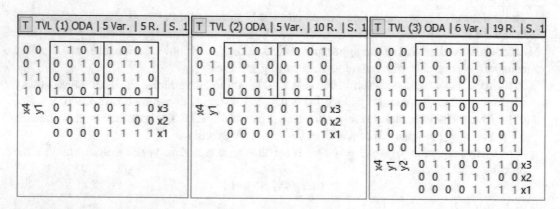

Fig. 7.33 Problem-program that verifies whether and with which expected results three restrictive equations $f_i(\mathbf{x}, \mathbf{y}) = 0$ can be solved with regard to the occurring variables y_j together with the associated Karnaugh-maps of the given function $f_i(\mathbf{x}, \mathbf{y})$

The result $f30usy1=true confirms that the restrictive equation $f_3(\mathbf{x}, y_1, y_2) = 0$ is uniquely solvable with regard to y_1, and the result $f30usy1=false means that functions $y_2 = g_{2k}(\mathbf{x})$ can be chosen out of the lattice of functions.

(b) Figure 7.34 shows the PRP that solves the equations $f_2(\mathbf{x}, y_1) = 0$ and $f_3(\mathbf{x}, y_1, y_2) = 0$ with regard to the occurring variables y_j and verifies the computed solutions.

```
 1   new                                    46   dif 46 46 51
 2   space 32 1                             47   dif 46 46 52
 3   avar 1                                 48   set $ngy2 1
 4   x1 x2 x3 x4 y1 y2.                     49   while (not(te 50))
 5   sbe 1 2                                50   (
 6   x1&/x2&x4&/y1+x1&/x3&/y1+              51   stv 50 1 53
 7   x2&x3&y1&(/x1+x2+/x4)+                 52   uni 53 51 53
 8   /x1&/x2&(x4&y1+/x4&/y1)+               53   cel 53 53 /-0
 9   x2&/x3&/y1+x1&x3&/x4&y1.               54   dif 50 53 50
10   sbe 1 3                                55   con 52 53 52
11   x1&/y2&(/x2&/y1+/x3&/X4)+              56   set $ngy2 (mul $ngy2 2)
12   /x3&(/y1+x4&y2)+                       57   )
13   /y1&(x2&/x3&y2+x1&/x2&/x4)+            58   set $c 1
14   x3&y1&(/x1+x2+x4)+                     59   set $r 3
15   /x1&/x4&/y2&(/x2+y1)+                  60   set $sf30 true
16   x2&y2&(/x4&/y1+x1&x4)+                 61   for $i 0 (sub $ngy2 1)
17   /x2&/y1&/y2.                           62   (
18   assign 2 /m 1 1                        63   set $gy2i (add $i 61)
19   assign 3 /m 2 1                        64   copy 42 $gy2i
20   sbe 1 4                                65   set $j $i
21   y1=1.                                  66   set $row 1
22   cpl 4 5                                67   while (gt $j 0)
23   sbe 1 6                                68   (
24   y2=1.                                  69   if (eq (mod $j 2) 1)
25   cpl 6 7                                70   (
26   isc 2 5 20                             71   stv 52 $row 54
27   maxk 20 4 21  ; g1 off2=0              72   uni $gy2i 54 $gy2i
28   syd 4 21 22                            73   )
29   syd 2 22 23                            74   set $j (div $j 2)
30   assign 21 /m 1 2                       75   set $row (add $row 1)
31   assign 23 /m 1 3                       76   )
32   mink 3 6 30                            77   assign $gy2i /m $r $c
33   isc 30 5 31                            78   set $c (add $c 1)
34   maxk 31 4 32  ; g1 off3=0              79   if (gt $c 3)
35   assign 32 /m 2 2                       80   (
36   syd 4 32 33                            81   set $c 1
37   mink 3 4 40                            82   set $r (add $r 1)
38   isc 40 7 41                            83   )
39   maxk 41 6 42  ; g2q off3=0             84   syd 6 $gy2i 55
40   isc 40 6 43                            85   uni 55 33 70
41   maxk 43 6 44  ; g2r off3=0             86   if (not (te_dif 3 70))
42   assign_qr 42 44 /m 2 3                 87   (
43   uni 42 44 45                           88   set $sf30 false
44   cpl 45 46  ; g2phi off3=0              89   )
45   copy 46 50                             90   )
```

Fig. 7.34 Problem-program that solves two restrictive equations with regard to the occurring variables y_j and verifies the computed solutions

The PRP in Fig. 7.34 prepares until line 25 the functions $f_2(\mathbf{x}, y_1)$ and $f_3(\mathbf{x}, y_1, y_2)$ for which the restrictive equations are solvable with regard to y_j, assigns the Karnaugh-maps of these functions to viewports (1,1) and (2,1) of the m-fold view, and creates the TVLs of y_1, $\overline{y}_1$, y_2, and $\overline{y}_2$ needed in the solution-procedure.

Equation $f_2(\mathbf{x}, y_1) = 0$ is uniquely solvable with regard to y_1; hence, the solution-function $y_1 = g_1(\mathbf{x})$ can be computed by

T TVL (2) ODA | 5 Var. | 10 R. | S. 1

```
0 0   1 1 0 1 1 0 0 1
0 1   0 0 1 0 0 1 1 0
1 1   1 1 1 0 0 1 0 0
1 0   0 0 0 1 1 0 1 1
x4 y1   0 1 1 0 0 1 1 0 x3
        0 0 1 1 1 1 0 0 x2
        0 0 0 0 1 1 1 1 x1
```

T TVL (21) ODA | 4 Var. | 6 R. | S. 1

```
0 0   1 1 1 1
0 1   0 1 1 1
1 1   0 0 0 1
1 0   1 0 0 0
x3 x4   0 1 1 0 x2
        0 0 1 1 x1
```

T TVL (23) ODA | 5 Var. | 0 R. | S. 1

```
0 0   0 0 0 0 0 0 0 0
0 1   0 0 0 0 0 0 0 0
1 1   0 0 0 0 0 0 0 0
1 0   0 0 0 0 0 0 0 0
x4 y1   0 1 1 0 0 1 1 0 x3
        0 0 1 1 1 1 0 0 x2
        0 0 0 0 1 1 1 1 x1
```

T TVL (3) ODA | 6 Var. | 19 R. | S. 1

```
0 0 0   1 1 0 1 1 0 1 1
0 0 1   1 0 1 1 1 1 1 1
0 1 1   0 1 1 0 0 1 0 0
0 1 0   1 1 1 1 1 1 0 1
1 1 0   0 1 1 0 0 1 1 0
1 1 1   1 1 1 1 1 1 1 1
1 0 1   1 0 0 1 1 1 0 1
1 0 0   1 1 0 1 1 0 1 1
x4 y1 y2   0 1 1 0 0 1 1 0 x3
           0 0 1 1 1 1 0 0 x2
           0 0 0 0 1 1 1 1 x1
```

T TVL (32) ODA | 4 Var. | 7 R. | S. 1

```
0 0   1 1 1 1
0 1   1 1 1 1
1 1   0 0 0 0
1 0   0 0 0 1
x3 x4   0 1 1 0 x2
        0 0 1 1 x1
```

(q: TVL (42) ODA, r: TVL (44) ODA), 4 Var. | S. 1

```
0 0   1 1 1 1
0 1   0 0 0 0
1 1   1 Φ 0 1
1 0   1 0 0 Φ
x3 x4   0 1 1 0 x2
        0 0 1 1 x1
```

T TVL (61) ODA | 4 Var. | 7 R. | S. 1

```
0 0   1 1 1 1
0 1   0 0 0 0
1 1   1 0 0 1
1 0   1 0 0 0
x3 x4   0 1 1 0 x2
        0 0 1 1 x1
```

T TVL (62) ODA | 4 Var. | 8 R. | S. 1

```
0 0   1 1 1 1
0 1   0 0 0 0
1 1   1 1 0 1
1 0   1 0 0 0
x3 x4   0 1 1 0 x2
        0 0 1 1 x1
```

T TVL (63) ODA | 4 Var. | 8 R. | S. 1

```
0 0   1 1 1 1
0 1   0 0 0 0
1 1   1 0 0 1
1 0   1 0 0 1
x3 x4   0 1 1 0 x2
        0 0 1 1 x1
```

T TVL (64) ODA | 4 Var. | 9 R. | S. 1

```
0 0   1 1 1 1
0 1   0 0 0 0
1 1   1 1 0 1
1 0   1 0 0 1
x3 x4   0 1 1 0 x2
        0 0 1 1 x1
```

-

-

Fig. 7.35 Results of the PRP shown in Fig. 7.34

$$g_1(\mathbf{x}) = \max_{y_1}(\overline{y}_1 \wedge f_2(\mathbf{x}, y_1)) \,.$$

The commands isc and maxk in lines 26 and 27 computes the solution-function $g_1(\mathbf{x})$ as TVL 21.

Equation $y_1 = g_1(\mathbf{x})$ can be transformed into $y_1 \oplus g_1(\mathbf{x}) = 0$; the function $f_2'(\mathbf{x}, y_1) = y_1 \oplus g_1(\mathbf{x})$ (computed by the command syd in line 28 as TVL 22) must be equal to the given function $f_2(\mathbf{x}, y_1)$. The command syd in line 29 checks this condition; the empty solution-set (TVL 23 shown in viewport (1,3) of Fig. 7.35) confirms that the solution-function

$$g_1(\mathbf{x}) = \overline{x}_1\overline{x}_2\overline{x}_4 \vee x_2\overline{x}_3 \vee x_1\overline{x}_2x_4 \vee \overline{x}_3\overline{x}_4$$

(TVL 21 shown in viewport (1,2)) has been computed correctly.

Equation $f_3(\mathbf{x}, y_1, y_2) = 0$ is uniquely solvable with regard to y_1; hence, the solution-function $y_1 = g_1(\mathbf{x})$ can be computed by

$$g_1(\mathbf{x}) = \max_{y_1}(\overline{y}_1 \wedge \min_{y_2} f_3(\mathbf{x}, y_1, y_2)) \ .$$

The commands in lines 32–35 use this formula to compute $g_1(\mathbf{x})$ as TVL 32 and to show the Karnaugh-map of this result in viewport (2,2) of the m-fold view (see Fig. 7.35).

The solution-function $y_2 = g_2(\mathbf{x})$ of the given equation $f_3(\mathbf{x}, y_1, y_2) = 0$ can be chosen out of a lattice with the mark-functions:

$$g_{2q}(\mathbf{x}) = \max_{y_2}(\overline{y}_2 \wedge \min_{y_1} f_3(\mathbf{x}, y_1, y_2)) \ ,$$

$$g_{2r}(\mathbf{x}) = \max_{y_2}(y_2 \wedge \min_{y_1} f_3(\mathbf{x}, y_1, y_2)) \ .$$

The commands in lines 37–41 use these formulas to compute the mark-functions $g_{2q}(\mathbf{x})$ (TVL 42) and $g_{2r}(\mathbf{x})$ (TVL 44) and the command `assign_qr` in line 42 assigns the Karnaugh-map of this lattice to viewport (2,3) of the m-fold view shown in Fig. 7.35.

There are several systems of equations

$$y_1 = g_1(\mathbf{x})$$

$$y_2 = g_{2i}(\mathbf{x})$$

that solve the given restrictive equation $f_3(\mathbf{x}, y_1, y_2) = 0$ with regard to y_1 and y_2 due to the lattice of solution-functions $g_{2i}(\mathbf{x})$. Each of these systems of equations can be transformed into a single restrictive equation

$$(y_1 \oplus g_1(\mathbf{x})) \vee (y_2 \oplus g_{2i}(\mathbf{x})) = 0 \ .$$

The left-hand side of these equations determines functions

$$f'_{3i}(\mathbf{x}, y_1, y_2) = (y_1 \oplus g_1(\mathbf{x})) \vee (y_2 \oplus g_{2i}(\mathbf{x}))$$

for which

$$f_3(\mathbf{x}, y_1, y_2) \le f'_{3i}(\mathbf{x}, y_1, y_2)$$

confirms that the pair of functions $(g_1(\mathbf{x}), g_{2i}(\mathbf{x}))$ is a correct solution of the explored restrictive equation $f_3(\mathbf{x}, y_1, y_2) = 0$. The final part of this PRP realizes this verification.

The functions $g_{2i}(\mathbf{x})$ are computed using the mark-function $g_{\varphi 2}(\mathbf{x})$ (computed as TVL 46 in lines 43 and 44) and transformed into the BVL 52 in lines 45–57 reusing the approach of Fig. 7.30. The commands of the `for`-loop in lines 61–90 generate the functions $g_{2i}(\mathbf{x})$ within the inner `while`-loop, store these functions as TVLs beginning at object number 61, assigns these functions as Karnaugh-maps to the viewports beginning in the third row, and compute the functions $f'_{3i}(\mathbf{x}, y_1, y_2)$ (TVL 70) so that the command `te_dif` in line 86 can verify the condition of the correct solution. The computed result `$sf30=true` confirms that each of the four functions $g_{2i}(\mathbf{x})$ (TVLs 61, ..., 64) shown in the viewports beginning in the third row of the m-fold view (see Fig. 7.35) are in combination with the function $g_1(\mathbf{x})$ (TVL 32 shown in viewport (2, 2) of the m-fold view) solutions of the restrictive equation $f_3(\mathbf{x}, y_1, y_2) = 0$ with regard to y_1 and y_2.

Three of the four solution-functions $g_{2i}(\mathbf{x})$ depend on all four variables, but the function $g_{23}(\mathbf{x})$ is independent of x_1; hence, the simplest pair of solution-functions of the restrictive equation $f_3(\mathbf{x}, \mathbf{y}) = 0$ is

$$y_1 = g_1(\mathbf{x}) = x_1 \overline{x}_2 \overline{x}_4 \vee \overline{x}_3 \,,$$

$$y_2 = g_2(\mathbf{x}) = \overline{x}_2 x_3 \vee \overline{x}_3 \overline{x}_4 \,.$$

(c) Figure 7.36 shows in the upper part the PRP that verifies whether the three given characteristic equations $f_i(\mathbf{x}, \mathbf{y}) = 1$ can be solved with regard to the occurring variables y_j and determines the type of the solution (unique function or lattice of functions); the Karnaugh-maps of the given three functions are shown below this PRP.

The three given functions are specified in lines 5–19 and shown by the commands `assign` in the three subsequent lines. The commands `vtin` in lines 23–28 create the needed VTs of the variables y_j.

Solution functions $y_j = g_j(\mathbf{x})$ of a given characteristic equation $f_i(\mathbf{x}, \mathbf{y}) = 1$ with regard to all $y_j \in \mathbf{y}$ exist if

$$\max_{\mathbf{y}}^k f_i(\mathbf{x}, \mathbf{y}) = 1 \,.$$

There are many different TVLs of a function that is equal 1, but only one TVL in ODA-form represents a function that is equal to 0. Therefore, two subsequent commands `maxk` and `te_cpl` in lines 29–34 are used to check this property for each of the three given functions. The results $f11s=true$, $f21s=false$, and $f31s=true$ confirm that equation $f_2(\mathbf{x}, y_1) = 1$ cannot be solved with regard to y_1 because $f_2(0, 0, 0, 0, y_1) = 0$, $f_2(0, 0, 1, 0, y_1) = 0$, and $f_2(0, 1, 1, 1, y_1) = 0$; however, the other two explored equations are solvable with regard to all occurring variables y_j.

In the case that a logic equation is solvable with regard to all variables y_j, it can be separately checked whether the function $y_j = g_j(\mathbf{x})$ is uniquely specified or can be chosen out of a lattice of functions. These checks are realized in this last part of the PRP.

A characteristic equation $f_i(\mathbf{x}, y_1) = 1$ that is solvable with regard to y_1 uniquely determines the function $y_j = g_j(\mathbf{x})$ if

$$\text{der}_{y_1} f_i(\mathbf{x}, y_1) = 1 \,.$$

The commands `derk` and `te_cpl` in lines 37 and 38 check this condition for equation $f_1(\mathbf{x}, y_1) = 1$. The result $f11usy1=false$ indicates that the functions $y_1 = g_1(\mathbf{x})$ can be chosen out of the lattice of functions.

The condition of the command `if` in line 40 is not satisfied; hence, lines 41–44 are not executed. It must be checked separately for each variable y_j whether the associated solution-function of the characteristic equation $f_i(\mathbf{x}, \mathbf{y}) = 1$ is uniquely specified or can be chosen out of a lattice of functions. A k-fold maximum of $f_i(\mathbf{x}, \mathbf{y})$ with regard to $(\mathbf{y} \setminus y_j)$ creates a function $f_i'(\mathbf{x}, y_j)$ for which the derivative with regard to y_j is equal to 1 in the case that the solution-function $y_j = g_j(\mathbf{x})$ is uniquely specified. The two sequences of commands `maxk`, `derk`, and `te_cpl` in lines 47–52 check this condition for equation $f_3(\mathbf{x}, y_1, y_2) = 1$ first for y_1 and thereafter for y_2. The result $f31usy1=false$ indicates that the functions $y_1 = g_1(\mathbf{x})$ can be chosen out of the lattice of functions and the result $f31usy2=true$ confirms that the characteristic equation $f_3(\mathbf{x}, y_1, y_2) = 1$ is uniquely solvable with regard to y_2.

(d) Figure 7.37 shows the PRP that solves the equations $f_1(\mathbf{x}, y_1) = 1$ and $f_3(\mathbf{x}, y_1, y_2) = 1$ with regard to the occurring variables y_j and verifies the computed solutions.

The PRP in Fig. 7.37 prepares until line 22 the functions $f_1(\mathbf{x}, y_1)$ and $f_3(\mathbf{x}, y_1, y_2)$ for which the characteristic equations are solvable with regard to y_j, assigns the Karnaugh-maps of these functions to viewports (1,1) and (2,1) of the m-fold view, and creates the TVLs of y_1, $\overline{y}_1$, y_2, and $\overline{y}_2$ needed to solve these equations with regard to all occurring variables y_j.

```
 1   new                              27   vtin 1 6
 2   space 32 1                       28   y1 y2.
 3   avar 1                           29   maxk 1 4 7
 4   x1 x2 x3 x4 y1 y2.               30   set $f11s (te_cpl 7)
 5   sbe 1 1                          31   maxk 2 4 8
 6   (x1&/x2&x4+/x3)&/y1+             32   set $f21s (te_cpl 8)
 7   /x1&/x2&(/x4&/y1+x3&y1)+         33   maxk 3 6 9
 8   x3&y1&(x2+x1&/x4).               34   set $f31s (te_cpl 9)
 9   sbe 1 2                          35   if (te_cpl 7)
10   x1&(/x2&x4&/y1+/x3&/y1)+         36   (
11   x2&x3&/y1&(x1+/x4)+              37   derk 1 4 10
12   /x1&/x2&x4&y1+                   38   set $f11usy1 (te_cpl 10)
13   x2&/x3&/y1+x1&x3&/x4&y1.         39   )
14   sbe 1 3                          40   if (te_cpl 8)
15   x1&y2&(x3&y1+/x3&/y1)+           41   (
16   /x1&/x3&/x4&y2&(/x2+y2)+         42   derk 2 4 20
17   /x1&/x2&x4&y1&/y2+               43   set $f21usy1 (te_cpl 20)
18   /x1&/y1&/y2&(x2&x4+x3&/x4)+      44   )
19   x1&x2&/x4&y1&y2.                 45   if (te_cpl 9)
20   assign 1 /m 1 1                  46   (
21   assign 2 /m 1 2                  47   maxk 3 5 30
22   assign 3 /m 1 3                  48   derk 30 4 31
23   vtin 1 4                         49   set $f31usy1 (te_cpl 31)
24   y1.                              50   maxk 3 4 32
25   vtin 1 5                         51   derk 32 5 33
26   y2.                              52   set $f31usy2 (te_cpl 33)
                                      53   )
```

```
T  TVL (1) ODA | 5 Var. | 8 R. | S. 1        T  TVL (2) ODA | 5 Var. | 9 R. | S. 1

0 0   1 1 0 1   1 0 0 1                       0 0   0 0 1 1   1 1 0 1
0 1   0 1 1 0   0 1 1 0                       0 1   0 0 0 0   0 1 1 0
1 1   0 1 1 0   0 1 0 0                       1 1   1 1 0 0   0 0 0 0
1 0   1 0 0 1   1 0 1 1                       1 0   0 0 0 1   1 1 1 1
x4 y1   0 1 1 0 0 1 1 0 x3                    x4 y1   0 1 1 0 0 1 1 0 x3
        0 0 1 1 1 1 0 0 x2                            0 0 1 1 1 1 0 0 x2
        0 0 0 0 1 1 1 1 x1                            0 0 0 0 1 1 1 1 x1


T  TVL (3) ODA | 6 Var. | 8 R. | S. 1

0 0 0   0 1 1 0   0 0 0 0
0 0 1   1 0 0 1   1 1 0 1
0 1 1   1 0 0 1   0 1 1 0
0 1 0   0 0 0 0   0 0 0 0
1 1 0   1 1 0 0   0 0 0 0
1 1 1   0 0 0 0   0 1 1 0
1 0 1   0 0 0 0   1 0 0 1
1 0 0   0 0 1 1   0 0 0 0
x4 y1 y2   0 1 1 0 0 1 1 0 x3
           0 0 1 1 1 1 0 0 x2
           0 0 0 0 1 1 1 1 x1
```

Fig. 7.36 Problem-program that verifies whether three characteristic equations $f_i(\mathbf{x}, \mathbf{y}) = 1$ can be solved with regard to the occurring variables y_j and identifies the types of the existing solutions together with the Karnaugh-maps of the given functions $f_i(\mathbf{x}, \mathbf{y})$

The solution-functions $y_1 = g_1(\mathbf{x})$ of the characteristic equation $f_1(\mathbf{x}, y_1) = 1$ can be chosen out of a lattice. The mark-functions $g_{1q}(\mathbf{x})$ (TVL 12) and $g_{1r}(\mathbf{x})$ (TVL 15) are computed based on (7.57) and (7.58) by two sequences of command isc, maxk, and cpl in lines 23–28. The subsequent commands uni and cpl compute the mark-function $g_{1\varphi}(\mathbf{x})$ (TVL 17), and the

```
 1   new                                       48   maxk 3 6 40
 2   space 32 1                                49   isc 40 5 41
 3   avar 1                                    50   maxk 41 4 42
 4   x1 x2 x3 x4 y1 y2.                        51   cpl 42 43 ; g1q off3=1
 5   sbe 1 1                                   52   isc 40 4 44
 6   (x1&/x2&x4+/x3)&/y1+                      53   maxk 44 4 45
 7   /x1&/x2&(/x4&/y1+x3&y1)+                   54   cpl 45 46 ; g1r off3=1
 8   x3&y1&(x2+x1&/x4).                        55   assign_qr 43 46 /m 2 2
 9   sbe 1 3                                   56   uni 43 46 47
10   x1&y2&(x3&y1+/x3&/y1)+                     57   cpl 47 48 ; g1phi off3=1
11   /x1&/x3&/x4&y2&(/x2+y2)+                   58   copy 48 50
12   /x1&/x2&x4&y1&/y2+                         59   dif 46 46 51
13   /x1&/y1&/y2&(x2&x4+x3&/x4)+                60   dif 46 46 52
14   x1&x2&/x4&/y1&y2.                          61   set $ngy1 1
15   assign 1 /m 1 1                           62   while (not(te 50))
16   assign 3 /m 2 1                           63   (
17   sbe 1 4                                   64   stv 50 1 53
18   y1=1.                                     65   uni 53 51 53
19   cpl 4 5                                   66   cel 53 53 /-0
20   sbe 1 6                                   67   dif 50 53 50
21   y2=1.                                     68   con 52 53 52
22   cpl 6 7                                   69   set $ngy1 (mul $ngy1 2)
23   isc 1 5 10                                70   )
24   maxk 10 4 11                              71   set $sf31 true
25   cpl 11 12 ; g1q off1=1                    72   for $i 0 (sub $ngy1 1)
26   isc 1 4 13                                73   (
27   maxk 13 4 14                              74   set $gy1i (add $i 61)
28   cpl 14 15 ; g1r off1=1                    75   copy 43 $gy1i
29   uni 12 15 16                              76   set $j $i
30   cpl 16 17 ; g1phi off1=1                  77   set $row 1
31   assign_qr 12 15 /m 1 2                    78   while (gt $j 0)
32   copy 12 21                                79   (
33   copy 14 22                                80   if (eq (mod $j 2) 1)
34   set $sf11 true                            81   (
35   for $i 21 22                              82   stv 52 $row 54
36   (                                         83   uni $gy1i 54 $gy1i
37   csd 4 $i 23                               84   )
38   if (not (te_dif 23 1))                    85   set $j (div $j 2)
39   (                                         86   set $row (add $row  1)
40   set $sf11 false                           87   )
41   ))                                        88   csd 4 $gy1i 55
42   maxk 3 4 30                               89   isc 55 34 70
43   isc 30 7 31                               90   if (not (te_dif 70 3))
44   maxk 31 6 32                              91   (
45   cpl 32 33 ; g2 off3=1                     92   set $sf31 false
46   assign 33 /m 2 3                          93   ))
47   csd 6 33 34
```

Fig. 7.37 Problem-program that solves two characteristic equations with regard to the occurring variables y_j and verifies the computed solutions

command `assign_qr` in line 31 assigns the Karnaugh-map of the computed lattice to viewport (1,2) of the m-fold view shown in Fig. 7.38.

The mark-function $g_{1\varphi}(\mathbf{x})$ is equal to 1 only for the assignment $\mathbf{x} = (0, 0, 1, 0)$; hence, this lattice contains only two functions: $g_{1q}(\mathbf{x})$ (TVL 12) and $\overline{g_{1r}(\mathbf{x})}$ (TVL 14), which are copied as TVLs 21 and 22 for the evaluation of the correctness in lines 34–41.

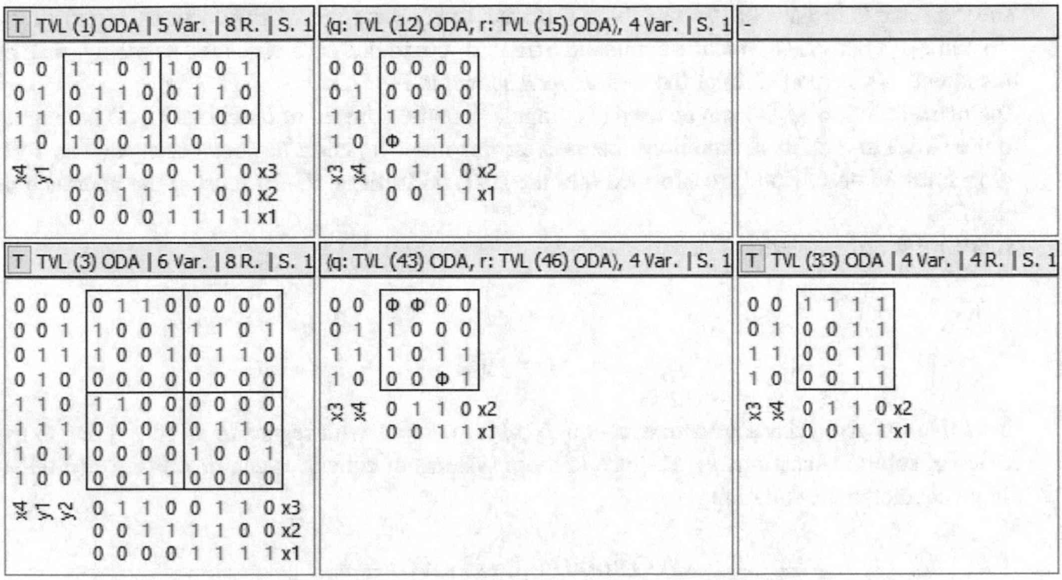

T TVL (1) ODA | 5 Var. | 8 R. | S. 1	(q: TVL (12) ODA, r: TVL (15) ODA), 4 Var. | S. 1	-

0 0 │ 1 1 0 1 │ 1 0 0 1
 0 1 │ 0 1 1 0 │ 0 1 1 0
 1 1 │ 0 1 1 0 │ 0 1 0 0
 1 0 │ 1 0 0 1 │ 1 0 1 1
 x4 x5 0 1 1 0 0 1 1 0 x3
     0 0 1 1 1 1 0 0 x2
     0 0 0 0 1 1 1 1 x1
	0 0 │ 0 0 0 0
	 0 1 │ 0 0 0 0
	 1 1 │ 1 1 1 0
	 1 0 │ Φ 1 1 1
	 x3 x4 0 1 1 0 x2
	     0 0 1 1 x1	

T TVL (3) ODA | 6 Var. | 8 R. | S. 1	(q: TVL (43) ODA, r: TVL (46) ODA), 4 Var. | S. 1	T TVL (33) ODA | 4 Var. | 4 R. | S. 1

0 0 0 │ 0 1 1 0 │ 0 0 0 0
 0 0 1 │ 1 0 0 1 │ 1 1 0 1
 0 1 1 │ 1 0 0 1 │ 0 1 1 0
 0 1 0 │ 0 0 0 0 │ 0 0 0 0
 1 1 0 │ 1 1 0 0 │ 0 0 0 0
 1 1 1 │ 0 0 0 0 │ 0 1 1 0
 1 0 1 │ 0 0 0 0 │ 1 0 0 1
 1 0 0 │ 0 0 1 1 │ 0 0 0 0
 x4 x5 x2 0 1 1 0 0 1 1 0 x3
     0 0 1 1 1 1 0 0 x2
     0 0 0 0 1 1 1 1 x1
	0 0 │ Φ Φ 0 0
	 0 1 │ 1 0 0 0
	 1 1 │ 1 0 1 1
	 1 0 │ 0 0 Φ 1
	 x3 x4 0 1 1 0 x2
	     0 0 1 1 x1
		0 0 │ 1 1 1 1
		 0 1 │ 0 0 1 1
		 1 1 │ 0 0 1 1
		 1 0 │ 0 0 1 1
		 x3 x4 0 1 1 0 x2
		     0 0 1 1 x1

Fig. 7.38 Results of the PRP shown in Fig. 7.37

Each equation $y_1 = g_1(\mathbf{x})$ can be transformed into $y_1 \odot g_1(\mathbf{x}) = 1$; the function $f_1'(\mathbf{x}, y_1) = y_1 \odot g_1(\mathbf{x})$ (computed by the command csd in line 37 as TVL 23) must be larger than or equal to the given function $f_1(\mathbf{x}, y_1)$; hence, the condition for a valid solution-function $y_1 = g_1(\mathbf{x})$ is

$$f_1'(\mathbf{x}, y_1) \wedge \overline{f_1(\mathbf{x}, y_1)} = 0 .$$

The command te_dif in line 38 checks this condition; the result $sf11=true confirms that the two solution-functions:

$$y_1 = g_{11}(\mathbf{x}) = (x_1 \vee x_2)x_3\overline{x}_4 \vee (\overline{x}_1 \vee x_2)x_3 x_4 , \text{ and}$$

$$y_1 = g_{12}(\mathbf{x}) = (\overline{x}_1 \vee x_2)x_3 \vee x_3\overline{x}_4$$

stored as TVLs 21 and 22 have been computed correctly.
Equation $f_3(\mathbf{x}, y_1, y_2) = 1$ is uniquely solvable with regard to y_2; hence, the solution-function $y_2 = g_2(\mathbf{x})$ can be computed by

$$g_2(\mathbf{x}) = \overline{\max_{y_2}(\overline{y}_2 \wedge \max_{y_1} f_3(\mathbf{x}, y_1, y_2))} .$$

The four commands in lines 42–45 compute the solution-function $g_2(\mathbf{x})$ as TVL 33. The command assign in lines 46 assigns the Karnaugh-map of $g_2(\mathbf{x})$ to viewport (2,3) of the m-fold view shown in Fig. 7.38.
The solution-functions $y_1 = g_1(\mathbf{x})$ of the characteristic equation $f_3(\mathbf{x}, y_1, y_2) = 1$ can be chosen out of a lattice with the mark-functions:

$$g_{1q}(\mathbf{x}) = \overline{\max_{y_1}(\overline{y}_1 \wedge \max_{y_2} f_3(\mathbf{x}, y_1, y_2))} ,$$

$$g_{1r}(\mathbf{x}) = \overline{\max_{y_1}(y_1 \wedge \max_{y_2} f_3(\mathbf{x}, y_1, y_2))} .$$

The commands in lines 48–54 use these formulas to compute the mark-functions $g_{1q}(\mathbf{x})$ (TVL 43) and $g_{1r}(\mathbf{x})$ (TVL 46) and the command `assign_qr` in line 55 assigns the Karnaugh-map of this lattice to viewport (2,2) of the m-fold view shown in Fig. 7.38.

The mark-function $g_{\varphi 1}(\mathbf{x})$ can be used to compute both the number of functions $g_{1i}(\mathbf{x})$ belonging to the lattice and all these functions themselves; this mark-function has been computed as TVL 48 in lines 56 and 57 and transformed into the BVL 53 in lines 58–70 reusing the approach of Fig. 7.30.

There are several systems of equations

$$y_1 = g_{1i}(\mathbf{x})$$
$$y_2 = g_2(\mathbf{x})$$

that solve the given characteristic equation $f_3(\mathbf{x}, y_1, y_2) = 1$ with regard to y_1 and y_2 due to the lattice of solution-functions $g_1(\mathbf{x})$. Each of these systems of equations can be transformed into a single characteristic equation

$$(y_1 \odot g_{1i}(\mathbf{x})) \wedge (y_2 \odot g_2(\mathbf{x})) = 1 \ .$$

The left-hand sides of these equations determine functions

$$f'_{3i}(\mathbf{x}, y_1, y_2) = (y_1 \odot g_{1i}(\mathbf{x})) \wedge (y_2 \odot g_2(\mathbf{x}))$$

for which

$$f'_{3i}(\mathbf{x}, y_1, y_2) \le f_3(\mathbf{x}, y_1, y_2)$$

confirm that the pair of functions $(g_{1i}(\mathbf{x}), g_2(\mathbf{x}))$ is a correct solution of the explored characteristic equation $f_3(\mathbf{x}, y_1, y_2) = 1$. The final part of this PRP realizes this verification for all eight pairs of solution-functions.

The commands of the `for`-loop in lines 72–93 generate the functions $g_{1i}(\mathbf{x})$ within the inner `while`-loop, store these function as TVLs beginning with the object number 61, and compute the functions $f'_{3i}(\mathbf{x}, y_1, y_2)$ (TVL 70) so that the command `te_dif` in line 90 can verify the condition of the correct solution. The computed result `$sf31=true` confirms that each of the eight functions $g_{1i}(\mathbf{x})$ specified by the mark-functions $g_{q1}(\mathbf{x})$ (TVL 43) and $g_{r1}(\mathbf{x})$ (TVL 46) and stored as TVLs 61, ..., 68 are in combination with the function $g_2(\mathbf{x})$ (TVL 33) solutions of the characteristic equation $f_3(\mathbf{x}, y_1, y_2) = 1$ with regard to y_1 and y_2. The Karnaugh-map of the lattice $\mathcal{L}_{g_1} \langle g_{1q}(\mathbf{x}), g_{1r}(\mathbf{x}) \rangle$ is shown in viewport (2,2) and the Karnaugh-map of the function $g_2(\mathbf{x})$ in viewport (2,3) of the m-fold view of Fig. 7.38.

The simplest pair of solution-functions of $f_3(\mathbf{x}, \mathbf{y}) = 1$ is

$$y_1 = g_1(\mathbf{x}) = x_1 x_3 \vee \overline{x}_1 \overline{x}_2 x_4 \ ,$$
$$y_2 = g_2(\mathbf{x}) = x_1 \vee \overline{x}_3 \overline{x}_4 \ .$$

Solution 7.5 (Boolean Differential Equations)

(a) Figure 7.39 shows both the PRP that solves the Boolean differential equation (BDE)

$$\operatorname*{der}_{x_1} f(\mathbf{x}) \wedge \operatorname*{der}_{x_2} f(\mathbf{x}) = 1$$

1	new	14	u0 u1 u2 u3 .
2	space 32 1	15	vtin 1 4
3	avar 1	16	v0 v2 .
4	v0 v1 v2 v3	17	vtin 1 5
5	u0 u1 u2 u3 .	18	v1 v3 .
6	sbe 1 1	19	vtin 1 6
7	u1&u2=1 .	20	v0 v1 .
8	sbe 1 2	21	vtin 1 7
9	v0=u0 ,	22	v2 v3 .
10	v1=u0#u1 ,	23	isc 1 2 10
11	v2=u0#u2 ,	24	maxk 10 3 11
12	v3=u0#u3 .	25	obb 11 11
13	vtin 1 3	26	cco 11 4 5 12

27	isc 11 12 13
28	obb 13 13
29	cco 13 6 7 14
30	isc 13 14 15
31	obb 15 15
32	assign 1 /m 1 1
33	assign 11 /m 1 2
34	assign 12 /m 1 3
35	assign 13 /m 2 1
36	assign 14 /m 2 2
37	assign 15 /m 2 3

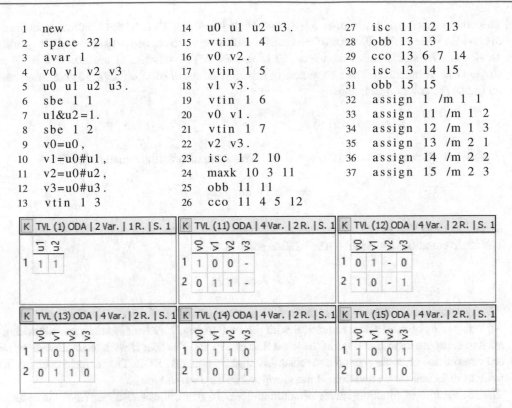

K | TVL (1) ODA | 2 Var. | 1 R. | S. 1

K	u_0	u_1
1	1	1

K | TVL (11) ODA | 4 Var. | 2 R. | S. 1

K	v_0	v_1	v_2	v_3
1	1	0	0	-
2	0	1	1	-

K | TVL (12) ODA | 4 Var. | 2 R. | S. 1

K	v_0	v_1	v_2	v_3
1	0	1	-	0
2	1	0	-	1

K | TVL (13) ODA | 4 Var. | 2 R. | S. 1

K	v_0	v_1	v_2	v_3
1	1	0	0	1
2	0	1	1	0

K | TVL (14) ODA | 4 Var. | 2 R. | S. 1

K	v_0	v_1	v_2	v_3
1	0	1	1	0
2	1	0	0	1

K | TVL (15) ODA | 4 Var. | 2 R. | S. 1

K	v_0	v_1	v_2	v_3
1	1	0	0	1
2	0	1	1	0

Fig. 7.39 Problem-program that solves the BDE that specifies all functions $f_i(x_1, x_2)$ that are linear with regard to the two variables x_1 and x_2 together with the associated TVLs

and the TVLs of the solution-steps and the computed class of solution-functions.

The logic equation associated to the BDE to solve is

$$u_1 \wedge u_2 = 1 \; ;$$

this equation is solved using the command sbe in lines 6 and 7. The system of equations and the VTs provided in lines 8–22 are needed for the subsequent transformation and separation of classes of solution-functions.

The commands isc and maxk in lines 23 and 24 transform $S(\mathbf{u})$ (TVL 1) to $S(\mathbf{v})$ (TVL 11) based on (7.85); $S(\mathbf{v})$ is subsequently minimized using the command obb.

The first sweep of the for-loop of Algorithm 3 is realized in lines 26–28. The command cco in line 26 exchanges the column of v_0 with the column of v_1 and the column of v_2 with the column of v_3 according to Table 7.10; this exchange is determined by VTs 4 and 5. The subsequent intersection excludes function vectors that do not belong to the class of solution-functions.

The second sweep of the for-loop of Algorithm 3 is realized in lines 29–31. The command cco in line 29 exchanges the column of v_0 with the column of v_2 and the column of v_1 with the column of v_3 according to Table 7.10; this exchange is determined by VTs 6 and 7. The subsequent intersection excludes function vectors that do not belong to the class of solution-functions. This second sweep does not change the set of solution-functions because the class of all solutions-functions is already determined by the first sweep for this BDE to solve.

The commands $\texttt{assign}$ in lines 32–37 assign all important TVL of the solution-procedure to the m-fold view so that all steps of the transformation and subsequent separation of the solution-classes can be studied after the execution of this PRP. The solution of the explored BDE is a single class of functions. The two vectors of function values of solution-TVL 15 determine based on (7.90) the solution-functions:

$$f_1(x_1, x_2) = x_1 \odot x_2 , \qquad\qquad\qquad f_2(x_1, x_2) = x_1 \oplus x_2 .$$

(b) Figure 7.40 shows both the PRP that solves the Boolean differential equation (BDE)

$$\operatorname*{der}_{x_1} f(\mathbf{x}) \wedge \operatorname*{der}_{x_2} f(\mathbf{x}) = 0$$

as well as the TVLs of the solution-steps and the computed classes of solution-functions. The logic equation associated to the BDE to solve is

$$u_1 \wedge u_2 = 0 ;$$

this equation is solved using the command $\texttt{sbe}$ in lines 6 and 7 of the PRP in Fig. 7.40. All other lines of this PRP are identical with the PRP of Fig. 7.39 because the same algorithm is used to compute logic functions of the same Boolean space. The four classes of solution-functions are determined after the second sweep for this slightly changed BDE. The realized exchanges of pairs of columns are well visible in the viewports of the m-fold view.

The six vectors of function values of solution-TVL 15 determine based on (7.90) the solution-functions:

$$f_1(x_1, x_2) = x_2 , \qquad\quad f_2(x_1, x_2) = \overline{x}_1 , \qquad\quad f_3(x_1, x_2) = x_1 ,$$
$$f_4(x_1, x_2) = 0 , \qquad\quad f_5(x_1, x_2) = \overline{x}_2 , \qquad\quad f_6(x_1, x_2) = 1 .$$

(c) Figure 7.41 shows both the PRP that solves the Boolean differential equation (BDE)

$$\operatorname*{der}_{(x_1,x_2)}^2 f(\mathbf{x}) \vee \operatorname*{der}_{x_1} f(\mathbf{x}) \operatorname*{der}_{x_2} f(\mathbf{x}) = 0$$

as well as the TVLs of the solution-steps and the computed classes of solution-functions. Algorithm 3 is able to solve Boolean differential equations in which the function $f(\mathbf{x})$ and all its vectorial derivative operations can occur.

The two-fold derivative in the BDE to solve can be replaced by vectorial derivative operations using the rule

$$\operatorname*{der}_{(x_1,x_2)}^2 f(x_1, x_2) = \operatorname*{der}_{x_1} f(x_1, x_2) \oplus \operatorname*{der}_{x_2} f(x_1, x_2) \oplus \operatorname*{der}_{(x_1,x_2)} f(x_1, x_2) ;$$

hence, the modified BDE to solve is

$$\left(\operatorname*{der}_{x_1} f(x_1, x_2) \oplus \operatorname*{der}_{x_2} f(x_1, x_2) \oplus \operatorname*{der}_{(x_1,x_2)} f(x_1, x_2) \right) \vee \operatorname*{der}_{x_1} f(\mathbf{x}) \operatorname*{der}_{x_2} f(\mathbf{x}) = 1 .$$

1	new	14	u0 u1 u2 u3.	27	isc 11 12 13
2	space 32 1	15	vtin 1 4	28	obb 13 13
3	avar 1	16	v0 v2.	29	cco 13 6 7 14
4	v0 v1 v2 v3	17	vtin 1 5	30	isc 13 14 15
5	u0 u1 u2 u3.	18	v1 v3.	31	obb 15 15
6	sbe 1 1	19	vtin 1 6	32	assign 1 /m 1 1
7	u1&u2=0.	20	v0 v1.	33	assign 11 /m 1 2
8	sbe 1 2	21	vtin 1 7	34	assign 12 /m 1 3
9	v0=u0,	22	v2 v3.	35	assign 13 /m 2 1
10	v1=u0#u1,	23	isc 1 2 10	36	assign 14 /m 2 2
11	v2=u0#u2,	24	maxk 10 3 11	37	assign 15 /m 2 3
12	v3=u0#u3.	25	obb 11 11		
13	vtin 1 3	26	cco 11 4 5 12		

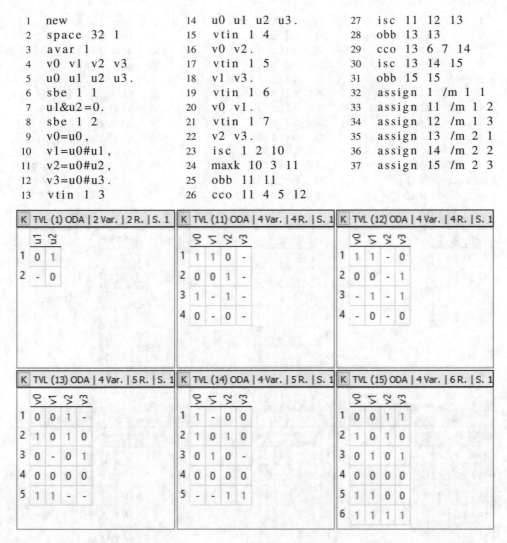

TVL (1) ODA | 2 Var. | 2 R. | S. 1

K	u_1	u_2
1	0	1
2	-	0

TVL (11) ODA | 4 Var. | 4 R. | S. 1

K	v_0	v_1	v_2	v_3
1	1	1	0	-
2	0	0	1	-
3	1	-	1	-
4	0	-	0	-

TVL (12) ODA | 4 Var. | 4 R. | S. 1

K	v_0	v_1	v_2	v_3
1	1	1	-	0
2	0	0	-	1
3	-	1	-	1
4	-	0	-	0

TVL (13) ODA | 4 Var. | 5 R. | S. 1

K	v_0	v_1	v_2	v_3
1	0	0	1	-
2	1	0	1	0
3	0	-	0	1
4	0	0	0	0
5	1	1	-	-

TVL (14) ODA | 4 Var. | 5 R. | S. 1

K	v_0	v_1	v_2	v_3
1	1	-	0	0
2	1	0	1	0
3	0	1	0	-
4	0	0	0	0
5	-	-	1	1

TVL (15) ODA | 4 Var. | 6 R. | S. 1

K	v_0	v_1	v_2	v_3
1	0	0	1	1
2	1	0	1	0
3	0	1	0	1
4	0	0	0	0
5	1	1	0	0
6	1	1	1	1

Fig. 7.40 Problem-program that solves the BDE that specifies all functions $f_i(x_1, x_2)$ that do not depend at least on one of the variables x_1 or x_2 together with the associated TVLs

The logic equation associated to this modified BDE to solve is

$$(u_1 \oplus u_2 \oplus u_3) \vee u_1 u_2 = 0 \, ;$$

this equation is solved using the command sbe in lines 6–8 of the PRP in Fig. 7.41.

All other lines of this PRP are identical with the PRP of Fig. 7.39 because the same algorithm is used to compute logic functions of the same Boolean space. The three classes of solution-functions are determined already after the transformation step; hence, the two sweeps of the separation algorithm do not change the known set of solution-functions.

The eight ternary vectors of function values of solution-TVL 15 determine based on (7.90) ten solution-functions:

1 new	14 vtin 1 3	27 cco 11 4 5 12
2 space 32 1	15 u0 u1 u2 u3 .	28 isc 11 12 13
3 avar 1	16 vtin 1 4	29 obb 13 13
4 v0 v1 v2 v3	17 v0 v2 .	30 cco 13 6 7 14
5 u0 u1 u2 u3 .	18 vtin 1 5	31 isc 13 14 15
6 sbe 1 1	19 v1 v3 .	32 obb 15 15
7 (u1#u2#u3)+	20 vtin 1 6	33 assign 1 /m 1 1
8 u1&u2=1 .	21 v0 v1 .	34 assign 11 /m 1 2
9 sbe 1 2	22 vtin 1 7	35 assign 12 /m 1 3
10 v0=u0 ,	23 v2 v3 .	36 assign 13 /m 2 1
11 v1=u0#u1 ,	24 isc 1 2 10	37 assign 14 /m 2 2
12 v2=u0#u2 ,	25 maxk 10 3 11	38 assign 15 /m 2 3
13 v3=u0#u3 .	26 obb 11 11	

K TVL (1) ODA | 3 Var. | 4 R. | S. 1

K	u_1	u_2	u_3
1	0	0	1
2	1	0	0
3	1	1	1
4	-	1	0

K TVL (11) ODA | 4 Var. | 8 R. | S. 1

K	v_0	v_1	v_2	v_3
1	1	0	1	1
2	0	1	0	0
3	0	1	1	1
4	1	0	0	0
5	1	1	0	1
6	-	0	0	1
7	-	1	1	0
8	0	0	1	0

K TVL (12) ODA | 4 Var. | 8 R. | S. 1

K	v_0	v_1	v_2	v_3
1	0	1	1	1
2	1	0	0	0
3	1	0	1	1
4	0	1	0	0
5	1	1	1	0
6	0	-	1	0
7	1	-	0	1
8	0	0	0	1

K TVL (13) ODA | 4 Var. | 8 R. | S. 1

K	v_0	v_1	v_2	v_3
1	0	1	1	1
2	1	0	0	0
3	1	1	0	1
4	1	0	-	1
5	0	0	0	1
6	1	1	1	0
7	0	1	-	0
8	0	0	1	0

K TVL (14) ODA | 4 Var. | 8 R. | S. 1

K	v_0	v_1	v_2	v_3
1	1	1	0	1
2	0	0	1	0
3	0	1	1	1
4	-	1	1	0
5	0	1	0	0
6	1	0	1	1
7	-	0	0	1
8	1	0	0	0

K TVL (15) ODA | 4 Var. | 8 R. | S. 1

K	v_0	v_1	v_2	v_3
1	1	1	0	1
2	1	0	1	1
3	1	0	0	-
4	0	0	0	1
5	1	1	1	0
6	0	1	1	-
7	0	1	0	0
8	0	0	1	0

Fig. 7.41 Problem-program that solves the BDE that specifies all functions $f_i(x_1, x_2)$ that depend on variables x_1 and x_2 together with the associated TVLs

$$f_1(x_1, x_2) = x_1 \vee \overline{x}_2 , \qquad f_2(x_1, x_2) = \overline{x}_1 \vee x_2 , \qquad f_3(x_1, x_2) = \overline{x}_1 \overline{x}_2 ,$$

$$f_4(x_1, x_2) = x_1 \odot x_2 , \qquad f_5(x_1, x_2) = x_1 x_2 , \qquad f_6(x_1, x_2) = \overline{x}_1 \vee \overline{x}_2 ,$$

$$f_7(x_1, x_2) = x_1 \oplus x_2 , \qquad f_8(x_1, x_2) = x_1 \vee x_2 , \qquad f_9(x_1, x_2) = x_1 \overline{x}_2 ,$$

$$f_{10}(x_1, x_2) = \overline{x}_1 x_2 .$$

```
 1   new                  20   vtin 1 5            39   isc 14 17 18
 2   space 32 1           21   v1 v3.              40   obb 18 21
 3   avar 1               22   vtin 1 6            41   cpl 9 22
 4   x1 x2                23   v0 v1.              42   isc 21 22 23
 5   v0 v1 v2 v3          24   vtin 1 7            43   maxk 23 9 24
 6   u0 u1 u2 u3.         25   v2 v3.              44   isc 21 9 25
 7   sbe 1 1              26   sbe 1 8             45   maxk 25 9 26
 8   /x1&u0&u1+           27   x1=1.               46   cco 26 6 7 27
 9   /x2&u0&u2=0.         28   sbe 1 9             47   isc 24 27 28
10   obb 1 1              29   x2=1.               48   obb 28 31
11   sbe 1 2              30   isc 1 2 10          49   assign 1 /m 1 1
12   v0=u0,               31   maxk 10 3 11        50   assign 11 /m 1 2
13   v1=u0#u1,            32   obb 11 11           51   assign 14 /m 2 1
14   v2=u0#u2,            33   cpl 8 12            52   assign 17 /m 2 2
15   v3=u0#u3.            34   isc 11 12 13        53   assign 21 /m 2 3
16   vtin 1 3             35   maxk 13 8 14        54   assign 24 /m 3 1
17   u0 u1 u2 u3.         36   isc 11 8 15         55   assign 27 /m 3 2
18   vtin 1 4             37   maxk 15 8 16        56   assign 31 /m 3 3
19   v0 v2.               38   cco 16 4 5 17
```

Fig. 7.42 Problem-program that solves the BDE that specifies all functions $f_i(x_1, x_2)$ that are monotonously increasing with regard to x_1 and x_2 together with the associated TVLs

(d) Figure 7.42 shows the PRP that solves the Boolean differential equation (BDE)

$$\overline{x}_1 f(\mathbf{x}) \operatorname*{der}_{x_1} f(\mathbf{x}) \vee \overline{x}_2 f(\mathbf{x}) \operatorname*{der}_{x_2} f(\mathbf{x}) = 0 .$$

Figure 7.43 shows the most important TVLs created by the PRP of Fig. 7.42.
The Boolean differential equation (BDE) of this task contains the variables x_1 and x_2; hence, Algorithm 4 must be used to solve this BDE. The logic equation associated to the BDE to solve is

$$\overline{x}_1 u_0 u_1 \vee \overline{x}_2 u_0 u_2 = 0 ;$$

this equation is solved using the command sbe in lines 7–9. The solution-set $S(\mathbf{x}, \mathbf{u})$ of this equation is subsequently minimized using the command obb. The system of equations, the VTs, and the TVLs of x_1 and x_2 provided in lines 11–29 are needed for the subsequent transformation and separation of the solution-functions.

The commands isc and maxk in lines 30 and 31 transform $S(\mathbf{x}, \mathbf{u})$ (TVL 1) to $S(\mathbf{x}, \mathbf{v})$ (TVL 11) based on (7.85); $S(\mathbf{x}, \mathbf{v})$ is subsequently minimized using the command obb.

The first sweep of the for-loop of Algorithm 4 is realized in lines 33–40, where first the subfunction $S(x_1 = 0, x_2, \mathbf{v})$ (TVL 14) and thereafter the subfunction $S(x_1 = 1, x_2, \mathbf{v})$ (TVL 16) are computed. The command cco in line 38 exchanges the column of v_0 with the column of v_1 and the column of v_2 with the column of v_3 within the subfunction $S(x_1 = 1, x_2, \mathbf{v})$ (TVL 16) according to Table 7.10; this exchange is determined by VTs 4 and 5. The subsequent intersection excludes function vectors that do not belong to the set of solution-functions. The result $S(x_2, \mathbf{v})$ of the first sweep of Algorithm 4 does not depend on x_1 anymore; it is stored as minimized TVL 21.

The second sweep of the for-loop of Algorithm 4 is realized in lines 41–48. The result function of the first sweep $S(x_2, \mathbf{v})$ is split into the subfunctions $S(x_2 = 0, \mathbf{v})$ (TVL 24) and $S(x_2 = 1, \mathbf{v})$ (TVL 26). The command cco in line 46 exchanges the column of v_0 with the column of v_2 and the column of v_1 with the column of v_3 within the subfunction $S(x_2 = 1, \mathbf{v})$ (TVL 26) according

7 Sets, Lattices, and Classes of Logic Functions

K	TVL (1) ODA | 5 Var. | 6 R. | S. 1				
	x_1	x_2	u_0	u_1	u_2
1	1	-	1	1	0
2	-	1	1	0	1
3	-	-	0	-	1
4	1	1	1	1	1
5	-	-	-	0	0
6	-	-	0	1	0

K	TVL (11) ODA | 6 Var. | 5 R. | S. 1					
	x_1	x_2	v_0	v_1	v_2	v_3
1	1	-	1	0	1	-
2	-	1	1	1	0	-
3	1	1	1	0	0	-
4	-	-	1	1	1	-
5	-	-	0	-	-	-

-

K	TVL (14) ODA | 5 Var. | 3 R. | S. 1				
	x_2	v_0	v_1	v_2	v_3
1	1	1	1	0	-
2	-	1	1	1	-
3	-	0	-	-	-

K	TVL (17) ODA | 5 Var. | 5 R. | S. 1				
	x_2	v_0	v_1	v_2	v_3
1	-	0	1	-	1
2	1	1	1	-	0
3	1	0	1	-	0
4	-	1	1	-	1
5	-	-	0	-	-

K	TVL (21) ODA | 5 Var. | 6 R. | S. 1				
	x_2	v_0	v_1	v_2	v_3
1	1	1	1	0	-
2	1	1	1	1	0
3	-	1	1	1	1
4	-	0	1	-	1
5	1	0	1	-	0
6	-	0	0	-	-

K	TVL (24) ODA | 4 Var. | 3 R. | S. 1			
	v_0	v_1	v_2	v_3
1	1	1	1	1
2	0	1	-	1
3	0	0	-	-

K	TVL (27) ODA | 4 Var. | 6 R. | S. 1			
	v_0	v_1	v_2	v_3
1	0	-	1	1
2	1	0	1	1
3	1	1	1	1
4	-	1	0	1
5	-	0	0	1
6	-	-	0	0

K	TVL (31) ODA | 4 Var. | 4 R. | S. 1			
	v_0	v_1	v_2	v_3
1	-	1	1	1
2	0	0	1	1
3	0	-	0	1
4	0	0	0	0

Fig. 7.43 Results of the PRP shown in Fig. 7.42

to Table 7.10; this exchange is determined by VTs 6 and 7. The subsequent intersection excludes function vectors that do not belong to the set of solution-functions.

The commands `assign` in lines 49–56 assign all important TVLs of the solution-procedure to the m-fold view so that all steps of the transformation and subsequent separation of the solution-functions can be studied after the execution of this PRP. Figure 7.43 shows this m-fold view with the TVLs, the solution $S(\mathbf{x}, \mathbf{u})$ of the associated equation as TVL 1, the transformed function $S(\mathbf{x}, \mathbf{v})$ as TVL 11, the intermediate results of the first sweep in the second row, and both the intermediate results of the second sweep and the final set of function vectors in the third row. The four ternary vectors of TVL 31 determine based on (7.90) six monotonously increasing functions:

$$f_1(x_1, x_2) = x_1 \vee x_2 \,, \qquad f_2(x_1, x_2) = 1 \,, \qquad f_3(x_1, x_2) = x_2 \,,$$

$$f_4(x_1, x_2) = x_1 x_2 \,, \qquad f_5(x_1, x_2) = x_1 \,, \qquad f_6(x_1, x_2) = 0 \,.$$

7.10   Supplementary Exercises

Exercise 7.6 (Sets of Logic Functions)

A set of logic function can be described by a list of binary vectors, where each vector specifies 2^n function values of one function. A ternary vector of $2^n - d$ function values 0 or 1 and d dashes ($-$) combines 2^d such binary vectors of a lattice of logic functions; hence, a list ternary vectors with 2^n columns can be used for a compact specification of each set of logic functions. A binary vector in such a list represents the special case of a lattice that contains only a single function.

Here we assume that each lattice of functions as part of the set is given by a vector of function values in the order

$$\mathbf{x} = (x_1, x_2, x_3, x_4) : (0000), (0001), (0010), (0011), (0100), \dots, (1110), (1111) \,.$$

Compute all functions of the following sets and show the orthogonally minimized TVLs. Compute and show the restricted binary vectors of functions values for comparison. How many functions belong to these sets?

(a)

	f_0	f_1	f_2	f_3	f_4	f_5	f_6	f_7	f_8	f_9	f_a	f_b	f_c	f_d	f_e	f_f
	0	1	0	1	0	1	0	1	0	0	1	1	0	1	0	1
$\mathrm{ODA}(SF_1) =$	1	1	1	1	0	0	0	0	1	1	0	1	0	0	0	1
	1	0	1	0	0	1	0	1	1	0	1	0	0	0	0	1

(b)

	f_0	f_1	f_2	f_3	f_4	f_5	f_6	f_7	f_8	f_9	f_a	f_b	f_c	f_d	f_e	f_f
	0	1	0	1	0	1	0	1	0	$-$	1	1	0	1	0	1
$\mathrm{ODA}(SF_2) =$	1	1	1	1	0	0	0	$-$	1	1	0	1	0	0	0	1
	1	0	1	0	0	1	0	1	1	0	1	0	0	$-$	0	1

$;$

(c)

	f_0	f_1	f_2	f_3	f_4	f_5	f_6	f_7	f_8	f_9	f_a	f_b	f_c	f_d	f_e	f_f
	0	1	0	1	0	1	0	$-$	0	1	1	0	1	1	$-$	1
$\mathrm{ODA}(SF_3) =$	1	1	1	1	0	0	0	$-$	1	1	$-$	1	0	0	0	1
	1	0	1	0	0	1	0	1	1	0	1	0	0	$-$	0	1

$;$

(d)

	f_0	f_1	f_2	f_3	f_4	f_5	f_6	f_7	f_8	f_9	f_a	f_b	f_c	f_d	f_e	f_f
	0	1	0	1	0	1	0	$-$	0	1	1	0	1	$-$	1	$-$
$\mathrm{ODA}(SF_4) =$	1	1	1	1	0	0	0	0	1	1	$-$	1	0	0	0	1
	1	0	1	0	0	1	0	1	1	0	1	0	0	$-$	0	1

$.$

Exercise 7.7 (Lattices of Logic Functions Specified by Their Mark-Functions)

(a) A lattice of logic function is specified by the mark-functions:

$$f_q(\mathbf{x}) = \overline{x}_2 \overline{x}_3 x_4 x_5 \vee \overline{x}_1 (x_2 \overline{x}_5 \vee \overline{x}_3 \overline{x}_4 x_6 \vee x_2 \overline{x}_3 x_4) \,,$$

$$f_r(\mathbf{x}) = x_1 (x_4 \overline{x}_5 \vee x_3 \overline{x}_6 \vee x_3 \overline{x}_4 \vee \overline{x}_2 x_3) \vee \overline{x}_1 \overline{x}_2 x_3 x_5 \,.$$

Verify that these mark-functions satisfy the condition of a lattice:

$$f_q(\mathbf{x}) f_r(\mathbf{x}) = 0 \,.$$

Check whether the functions:

$$f_1(\mathbf{x}) = \overline{x}_1 x_2 \vee \overline{x}_1 \overline{x}_4 \overline{x}_5 \vee \overline{x}_1 \overline{x}_3 x_5 \vee \overline{x}_3 x_4 x_5 \overline{x}_6 \,,$$

$$f_2(\mathbf{x}) = \overline{x}_1 x_2 \vee \overline{x}_4 \overline{x}_5 x_6 \vee \overline{x}_1 \overline{x}_3 x_5 \vee \overline{x}_3 x_4 x_5 \,,$$

$$f_3(\mathbf{x}) = \overline{x}_1 x_2 \vee \overline{x}_3 \overline{x}_4 x_6 \vee \overline{x}_1 \overline{x}_3 x_5 \vee \overline{x}_3 x_4 x_5$$

belong to this lattice and state the reason if such a check fails. Assign the Karnaugh-maps of $f_q(\mathbf{x})$, $\mathcal{L}\langle f_q(\mathbf{x}), f_r(\mathbf{x})\rangle$, and $f_r(\mathbf{x})$ to the viewports of the first row of the m-fold view. Assign additionally the Karnaugh-maps of three functions $f_i(\mathbf{x})$ to the viewports of the second row of the m-fold view so that the belonging to the lattice can visually be verified.

(b) Compute the mark-functions of the smallest lattice that contains the functions:

$$f_1(\mathbf{x}) = \overline{x}_1 x_2 \vee \overline{x}_3 \overline{x}_4 x_6 \vee \overline{x}_1 \overline{x}_3 x_5 \vee \overline{x}_3 x_4 x_5 \,,$$

$$f_2(\mathbf{x}) = \overline{x}_1 x_2 \vee \overline{x}_2 \overline{x}_4 x_6 \vee \overline{x}_1 \overline{x}_3 x_5 \vee \overline{x}_3 x_4 x_5 x_6$$

and verify that these mark-functions satisfy the condition of a lattice. Show the Karnaugh-maps of $f_1(\mathbf{x})$ and $f_2(\mathbf{x})$ in the viewports of the first row of the m-fold view and the Karnaugh-maps of $f_q(\mathbf{x})$, $\mathcal{L}\langle f_q(\mathbf{x}), f_r(\mathbf{x})\rangle$, and $f_r(\mathbf{x})$ of the computed lattice in the second row of this view.

(c) Logic functions that belong to two different lattices specify a smaller lattice. Compute the largest lattice $\mathcal{L}_3\langle f_{3q}(\mathbf{x}), f_{3r}(\mathbf{x})\rangle$ that contains all functions that belong to both the lattice $\mathcal{L}_1\langle f_{1q}(\mathbf{x}), f_{1r}(\mathbf{x})\rangle$:

$$f_{q1}(\mathbf{x}) = \overline{x}_1 x_2 x_3 \vee \overline{x}_1 \overline{x}_3 x_5 \vee \overline{x}_1 \overline{x}_3 x_4 x_5 x_6 \,,$$

$$f_{r1}(\mathbf{x}) = x_1 x_2 x_3 \vee x_4 \overline{x}_5 (x_1 \vee \overline{x}_2) \vee \overline{x}_2 x_3 x_4 \vee x_1 \overline{x}_4 \overline{x}_5 \overline{x}_6 \vee \overline{x}_2 \overline{x}_6 (x_1 x_3 \vee \overline{x}_5)$$

and the lattice $\mathcal{L}_2\langle f_{q2}(\mathbf{x}), f_{r2}(\mathbf{x})\rangle$:

$$f_{q2}(\mathbf{x}) = \overline{x}_1 x_2 x_3 x_6 \vee \overline{x}_1 \overline{x}_4 x_5 \vee x_1 x_2 \overline{x}_3 x_5 \,,$$

$$f_{r2}(\mathbf{x}) = x_1 \overline{x}_2 x_4 \vee \overline{x}_3 \overline{x}_5 x_6 \vee x_1 \overline{x}_2 \overline{x}_3 \overline{x}_4 x_5 \vee \overline{x}_2 x_4 \overline{x}_5 x_6 \,.$$

Verify that all three lattices satisfy the condition of a lattice. Show the three lattices in the rows 1, 2, and 3 of the m-fold view such that the Karnaugh-maps of the mark-functions $f_{iq}(\mathbf{x})$, the lattices $\mathcal{L}_i\langle f_{iq}(\mathbf{x}), f_{ir}(\mathbf{x})\rangle$, and the mark-functions $f_{ir}(\mathbf{x})$ are displayed side by side. Compare these Karnaugh-maps of the given two lattices and the computed intersection lattice.

(d) There are lattices of logic function which do not commonly contain any function. Verify that the given lattices $\mathcal{L}_1\langle f_{q1}(\mathbf{x}), f_{r1}(\mathbf{x})\rangle$:

$$f_{1q}(\mathbf{x}) = \overline{x}_1 x_2 x_3 \vee \overline{x}_1 \overline{x}_4 x_5 \vee \overline{x}_1 \overline{x}_2 \overline{x}_3 \vee \overline{x}_2 x_5 x_6 \,,$$

$$f_{1r}(\mathbf{x}) = x_1 x_2 x_3 \vee x_4 \overline{x}_5 (x_1 \vee \overline{x}_2 x_3) \vee \overline{x}_2 x_3 x_4 \overline{x}_5 \vee x_1 \overline{x}_4 \overline{x}_6$$

and $\mathcal{L}_2\langle f_{q2}(\mathbf{x}), f_{r2}(\mathbf{x})\rangle$:

$$f_{2q}(\mathbf{x}) = \overline{x}_1 x_2 x_3 x_6 \vee \overline{x}_1 \overline{x}_4 x_5 \vee x_1 x_2 \overline{x}_3 x_5 \,,$$

$$f_{2r}(\mathbf{x}) = x_1 \overline{x}_2 x_4 \vee \overline{x}_3 \overline{x}_5 x_6 \vee x_1 \overline{x}_2 \overline{x}_3 \overline{x}_4 x_5$$

do not contain any common function.

Try to compute a lattice $\mathcal{L}_3 \langle f_{q3}(\mathbf{x}), f_{r3}(\mathbf{x}) \rangle$ of the common functions of these two lattices and prove that this lattice is empty.

However, for arbitrary two lattices of logic functions exists always a smallest lattice that contains all functions of the explored two lattices and, with the exception of some special cases, additional functions. Compute the smallest lattice $\mathcal{L}_4 \langle f_{4q}(\mathbf{x}), f_{4r}(\mathbf{x}) \rangle$ that contains all functions of the given two lattices $\mathcal{L}_1 \langle f_{1q}(\mathbf{x}), f_{1r}(\mathbf{x}) \rangle$ and $\mathcal{L}_2 \langle f_{2q}(\mathbf{x}), f_{2r}(\mathbf{x}) \rangle$.

Show the Karnaugh-maps of the four lattices as before in the upper four rows of the m-fold view. Note, the computed function $f_{3q}(\mathbf{x})$ and $f_{3r}(\mathbf{x})$ do not satisfy the condition of a lattice; hence, the these functions are no mark-functions of a lattice.

Exercise 7.8 (Uniquely Stored Linearly Independent Directions of Change)

This exercise explores the properties of functions computed by successive vectorial derivatives which also occur when the vectorial minimum or vectorial maximum is used in such a sequence operations. These properties are utilized to store generalized lattices of logic functions using the two mark-functions $f_q(\mathbf{x})$ and $f_r(\mathbf{x})$ together with the appropriate independence matrix. Note, for real applications the XBOOLE-library and the operations of a programming language are preferred to establish and manipulate the independence matrix; however, a TVL in D-form that consists of n binary vectors for logic functions of n variables and XBOOLE-operations for sets of variables can be used to solve the respective tasks using a PRP of the XBOOLE-monitor XBM 2.

(a) Use the function:

$$f(\mathbf{x}) = \overline{x}_1 \overline{x}_2 \overline{x}_4 \vee \overline{x}_1 x_2 x_5 \vee x_1 \overline{x}_3 x_4 \vee x_2 x_3 \overline{x}_4 x_5 \vee \overline{x}_2 x_3 \overline{x}_4 \overline{x}_5$$

to compute

$$f_1(x) = \operatorname*{der}_{(x_2, x_4, x_5)} \left(\operatorname*{der}_{(x_1, x_2, x_3)} f(\mathbf{x}) \right) \quad \text{and}$$

$$f_2(x) = \operatorname*{der}_{(x_1, x_2, x_3)} \left(\operatorname*{der}_{(x_2, x_4, x_5)} f(\mathbf{x}) \right),$$

where the same vectorial derivatives are executed in different orders. Verify that the same result $f_1(\mathbf{x}) = f_2(\mathbf{x})$ will be computed independent of the used order of vectorial derivatives. Verify furthermore, that both $f_1(\mathbf{x})$ and $f_2(\mathbf{x})$ do not depend on the simultaneous change of either (x_1, x_2, x_3) or (x_2, x_4, x_5). Show and compare the Karnaugh-maps of $f(\mathbf{x})$ the vectorial derivatives with regard to one direction of change $\operatorname{der}_{(x_1, x_2, x_3)} f(\mathbf{x})$ and $\operatorname{der}_{(x_2, x_4, x_5)} f(\mathbf{x})$ as well as the results $f_1(x)$ and $f_2(x)$.

(b) A logic function of n variables that does not depend on the change with regard to k linearly independent directions of change is independent of $2^k - 1$ directions of change. Verify this property for the result function of the previous task.

Solve this task as follows:

- Reuse function $f(\mathbf{x})$ of the previous task for this explorations and show the associated Karnaugh-map.
- Prepare a TVL in D-form that specifies in two binary vectors the direction of change (x_1, x_2, x_3) by (11100) and (x_2, x_4, x_5) by (01011).

- Compute $\text{der}_{(x_1,x_2,x_3)} f(\mathbf{x})$ and $g(\mathbf{x}) = \text{der}_{(x_2,x_4,x_5)} \left(\text{der}_{(x_1,x_2,x_3)} f(\mathbf{x})\right)$ and show the associated Karnaugh-maps for comparison.
- Due to $\forall \mathbf{x} : \text{der}_{(x_1,x_2,x_3)} g(\mathbf{x}) = 0$, $\forall \mathbf{x} : \text{der}_{(x_2,x_4,x_5)} g(\mathbf{x}) = 0$, and $2^2 - 1 = 3$ there must be one more direction of change $\mathbf{x}_0$ with $\forall \mathbf{x} : \text{der}_{\mathbf{x}_0} g(\mathbf{x}) = 0$; find this direction of change and append the associated binary vector to the TVL that stores the directions of change the function $g(\mathbf{x})$ is not depending on and show this TVL.
- Compute $\text{der}_{\mathbf{x}_0} \left(\text{der}_{(x_1,x_2,x_3)} f(\mathbf{x})\right)$ and show the used directions of change as well as the Karnaugh-map of the result.
- Compute $\text{der}_{\mathbf{x}_0} \left(\text{der}_{(x_2,x_4,x_5)} f(\mathbf{x})\right)$ and show the used directions of change as well as the Karnaugh-map of the result.
- Check whether the function $g(\mathbf{x})$ and the last two results are identical functions.

(c) The result of the previous task is a TVL that stores three directions of change, where each of these direction of change is determined by the other two linearly independent directions of change. Assume that the vectorial derivatives of a lattice of functions with regard to each of these three directions of change have already been computed, the used directions of change have been stored as one of three independence matrices (here TVLs in D-form are used):

$$
\text{D}(im_1) = \begin{array}{ccccc} x_1 & x_2 & x_3 & x_4 & x_5 \\ \hline 1 & 1 & 1 & 0 & 0 \\ 0 & 0 & 0 & 0 & 0 \\ 0 & 0 & 0 & 0 & 0 \\ 0 & 0 & 0 & 0 & 0 \\ 0 & 0 & 0 & 0 & 0 \end{array} \quad
\text{D}(im_2) = \begin{array}{ccccc} x_1 & x_2 & x_3 & x_4 & x_5 \\ \hline 0 & 0 & 0 & 0 & 0 \\ 0 & 1 & 0 & 1 & 1 \\ 0 & 0 & 0 & 0 & 0 \\ 0 & 0 & 0 & 0 & 0 \\ 0 & 0 & 0 & 0 & 0 \end{array} \quad
\text{D}(im_3) = \begin{array}{ccccc} x_1 & x_2 & x_3 & x_4 & x_5 \\ \hline 1 & 0 & 1 & 1 & 1 \\ 0 & 0 & 0 & 0 & 0 \\ 0 & 0 & 0 & 0 & 0 \\ 0 & 0 & 0 & 0 & 0 \\ 0 & 0 & 0 & 0 & 0 \end{array}
$$

For each of these directions of change the two other directions of change can be used for the subsequent vectorial derivative. Provide these possible directions of change as binary vectors in TVLs in D-form:

$$
\text{D}(sd_1) = \begin{array}{ccccc} x_1 & x_2 & x_3 & x_4 & x_5 \\ \hline 1 & 0 & 1 & 1 & 1 \\ 0 & 1 & 0 & 1 & 1 \end{array}, \quad
\text{D}(sd_2) = \begin{array}{ccccc} x_1 & x_2 & x_3 & x_4 & x_5 \\ \hline 1 & 1 & 1 & 0 & 0 \\ 1 & 0 & 1 & 1 & 1 \end{array}, \quad
\text{D}(sd_3) = \begin{array}{ccccc} x_1 & x_2 & x_3 & x_4 & x_5 \\ \hline 1 & 1 & 1 & 0 & 0 \\ 0 & 1 & 0 & 1 & 1 \end{array}.
$$

Algorithm 1 of page 308 determines the minimal independent direction of change for a given independence matrix and the direction of change of the next vectorial derivative operation. Apply this algorithm for each pair of one independence matrix and one of the associated second directions of change.

Show the results of these six combinations in six rows of the m-fold view such that column 1 shows the used independence matrix, column 2 the selected second direction of change, and column 3 the computed unique vector s_{min}. Verify that each independence matrix determines a unique minimal direction of change s_{min}.

(d) Extend the PRP of the previous task such that the new independence matrix $\text{IDM}(g) = \text{UM}(\text{IDM}(f), \mathbf{x}_0)$ is computed using Algorithm 2 of page 309. Show the six computed independence matrices in column 4 of the associated row for each of the six explored cases. Verify that these six independence matrices are identical for all pairs of the given independence matrices and the associated different directions of change.

Exercise 7.9 (Applications for Derivative Operations of a Generalized Lattice)
A generalized lattice is given by its mark-functions:

$$f_q(\mathbf{x}) = \overline{x}_1(x_2 \oplus x_5) \lor x_1\overline{x}_2\overline{x}_5(x_3 \oplus x_4 \oplus x_6) \lor x_1x_2x_5\overline{(x_3 \oplus x_4 \oplus x_6)}$$

$$\lor\, x_1x_3\overline{(x_2 \oplus x_4 \oplus x_6)} \lor (x_1 \oplus x_2 \oplus x_5)\overline{(x_4 \oplus x_5 \oplus x_6)}\,,$$

$$f_r(\mathbf{x}) = \overline{x}_1\overline{x}_3(x_2 \oplus x_5) \lor \overline{x}_1\overline{x}_2\overline{x}_5(x_4 \oplus x_6) \lor \overline{x}_1x_2x_5\overline{(x_4 \oplus x_6)}$$

$$\lor\, x_1x_2\overline{x}_5(x_3 \oplus x_4 \oplus x_6) \lor x_1\overline{x}_2x_5\overline{(x_3 \oplus x_4 \oplus x_6)}\,,$$

and the associated independence matrix:

$$\mathrm{IDM}(f) = \begin{array}{c|cccccc} {}_i\diagdown{}^{j} & 1 & 2 & 3 & 4 & 5 & 6 \\ \hline 1 & 0 & 0 & 0 & 0 & 0 & 0 \\ 2 & 0 & 1 & 0 & 0 & 1 & 1 \\ 3 & 0 & 0 & 0 & 0 & 0 & 0 \\ 4 & 0 & 0 & 0 & 1 & 0 & 1 \\ 5 & 0 & 0 & 0 & 0 & 0 & 0 \\ 6 & 0 & 0 & 0 & 0 & 0 & 0 \end{array}$$

(a) How many function belong to the given generalized lattice? Does this lattice contain a function for which the restrictive equation can be solved with regard to x_3. Such functions satisfy $\forall \mathbf{x} : \min_{x_3} f(\mathbf{x}) = 0$; hence, the mark-function $f_q^{\min_{x_3}}(\mathbf{x}_1)$ must be equal to 0. Compute the independence matrix of the single minimum of the given generalized lattice separately to shorten the PRP.
Show the given generalized lattice, the computed generalized lattice of

$$\min_{x_3} \mathcal{L}\langle f_q(\mathbf{x}), f_r(\mathbf{x}), f^{id}(\mathbf{x})\rangle\,,$$

and the results of the two questions.

(b) Does the given generalized lattice contain a function for which the restrictive equation can be solved with regard to $\mathbf{x}_0 = (x_2, x_3)$. Such functions satisfy $\forall \mathbf{x} : \min^k_{\mathbf{x}_0} f(\mathbf{x}_0, \mathbf{x}_1) = 0$; hence, $f_q^{\min^k_{\mathbf{x}_0}}(\mathbf{x}_1)$ must be equal to 0. Compute the independence matrix of the k-fold minimum of the given generalized lattice separately to shorten the PRP.
Show the given generalized lattice, the computed generalized lattice of

$$\min^k_{(x_2,x_3)} \mathcal{L}\langle f_q(\mathbf{x}), f_r(\mathbf{x}), f^{id}(\mathbf{x})\rangle\,,$$

and the answer to the question.

(c) Does the given generalized lattice contain a function for which the characteristic equation can be solved with regard to x_2. Such functions satisfy $\forall \mathbf{x} : \max_{x_2} f(x_2, \mathbf{x}_1) = 1$; hence, $f_r^{\max_{x_2}}(\mathbf{x}_1)$ must be equal to 0. Compute the independence matrix of the single maximum of the given generalized lattice separately to shorten the PRP.
Show the given generalized lattice, the computed generalized lattice of

$$\max_{x_2} \mathcal{L}\langle f_q(\mathbf{x}), f_r(\mathbf{x}), f^{id}(\mathbf{x})\rangle\,,$$

and the answer to the question.

(d) Verify whether the given generalized lattice contains symmetric functions with regard to $\mathbf{x}_0 = (x_3, x_6)$. Such functions satisfy $(x_3 \oplus x_6) \, \mathrm{der}_{(x_3,x_6)} f(\mathbf{x}) = 0$; hence, $(x_3 \oplus x_6) f_q^{\mathrm{der}_{\mathbf{x}_0}}(\mathbf{x}_0, \mathbf{x}_1)$, where $\mathbf{x}_0 = (x_3, x_6)$, must be equal to 0. Take into account that solution-functions must satisfy the independence rules specified by $\mathrm{IDM}(f)$; hence, the assignment of a value 0 or 1 to a don't-care requires the use of the same value for subsets of don't-cares. Compute the independence matrix of the vectorial derivative separately to shorten the PRP. Show the given generalized lattice, the computed generalized lattice of

$$\mathop{\mathrm{der}}_{(x_3,x_6)} \mathcal{L}\langle f_q(\mathbf{x}), f_r(\mathbf{x}), f^{id}(\mathbf{x}) \rangle \,,$$

and the result of the check for the wanted symmetric functions. Verify the correctness of the computed result.

References

1. B. Steinbach, Generalized lattices of Boolean functions utilized for derivative operations, in *Materiały Konferencyjne KNWS'13*. KNWS. Łagów, Poland, June 2013, pp. 1–17. https://doi.org/10.13140/2.1.1874.3680
2. B. Steinbach, Solution of Boolean differential equations and their application for binary systems. Original title: Lösung binärer Differentialgleichungen und ihre Anwendung auf binäre Systeme (in German). Ph.D. Thesis. Technische Hochschule Karl-Marx-Stadt; now: University of Technology Chemnitz, 1981
3. B. Steinbach, C. Posthoff, *Boolean Differential Equations* (Morgan & Claypool Publishers, San Rafael, 2013). https://doi.org/10.2200/S00511ED1V01Y201305DCS042. ISBN: 978-1-6270-5241-2
4. B. Steinbach, C. Posthoff, Derivative operations for lattices of Boolean functions, in *Proceedings Reed-Muller Workshop 2013* (RM. Toyama, 2013), pp. 110–119. https://doi.org/10.13140/2.1.2398.6568

Part III
Applications

Logic, Arithmetic, and Special Functions

8

Abstract

In this chapter we explore applications of logic functions and equations that are not directly related to the design of hardware. The applications in propositional logic are surely the oldest and reach back to the developments of science in ancient Greece. Binary arithmetic is widely used in computer hardware, control systems, or other electronic devices. In this chapter we will give an outline of binary Mathematics before hardware-related problems will be discussed in more detail. Coding is a unique mapping of arbitrary objects to certain code words. Binary codes can be seen as bridges between many real-world areas and logic functions. The code words can be a subset of a given finite set so that changes of one or more bits can be detected or even corrected. The introduced *Specific Normal Form*, which uniquely expresses a logic function, differs significantly from other known normal forms and has several remarkable properties. This new normal form is used to find the most complex logic functions and to classify bent functions. Bent functions are logic functions having the largest distance to all linear functions that is a useful property for cryptographic applications.

Supplementary Information The online version of this chapter (https://doi.org/10.1007/978-3-030-88945-6_8) contains supplementary material which is available for authorized users. Please, follow the link belonging to the version of the XBOOLE-monitor XBM 2 that fits best for your operating system. This XBOOLE-monitor is needed to solve all tasks of this chapter. Instructions for starting the downloaded XBOOLE-monitor XBM 2 are given at the beginning of Section 'Exercises' in this chapter.

XBOOLE-monitor XBM 2 for Windows 10
32 bits
https://doi.org/10.1007/978-3-030-88945-6_8_MOESM1_ESM.zip (15,091 KB)

64 bits
https://doi.org/10.1007/978-3-030-88945-6_8_MOESM2_ESM.zip (14,973 KB)

XBOOLE-monitor XBM 2 for Linux Ubuntu
32 bits
https://doi.org/10.1007/978-3-030-88945-6_8_MOESM3_ESM.zip (29,522 KB)

64 bits
https://doi.org/10.1007/978-3-030-88945-6_8_MOESM4_ESM.zip (28,422 KB)

8.1   Propositional Logic

The basic construction element to be considered is the *proposition*, an expression in a language that can be assigned (without any doubt) exactly one of the values *true* or *false*. The sentence

"Brussels is the capital of Belgium."

is an example for such a proposition. In spite of its simplicity, it needs still some additional knowledge and definitions. The reader (or speaker) must know what the meaning of *Brussels* and *Belgium* is, and there must also be an accepted or acceptable definition of a "capital." Hence, the formalization of sentences very often needs careful consideration.

Another example could be the phrase

"17 is a prime number."

This is a true statement whereas the statement

"27 is a prime number"

is false because 27 can be divided by 9. Again additional knowledge is required: what means "a divisor of a (natural) number," why are 1 and 17 not considered as divisors, etc.

But in principle the considerations can start with the assumption that each proposition will be either *true* or *false*. Any additional information or knowledge that is necessary to ensure this assumption must be clearly stated.

Three very natural operations with propositions are the combinations of two propositions by **and** and **or** or the negation by **not**:

- *"Brussels is the capital of Belgium, **and** Paris is the capital of France."*
- *"Brussels is the capital of Belgium **or** France."*
- *"Brussels is **not** the capital of France."*

The second statement is again not totally precise. It abbreviates the use of two full statements by daily natural language: *"Brussels is the capital of Belgium, **or** Brussels is the capital of France."* would be an exact combination of two propositions. The **or** also shows some problems. In the daily language very often no difference is made between **or** and **either ... or**, *exclusive-or*. The concepts of countries and capitals clearly indicate that Brussels will not be the capital of both countries; hence, the two propositions cannot be true at the same time, and from this it follows that the meaning of this **or** is an **exclusive-or**. The statement that *"a given car is fast **or** cheap"* does not exclude the existence of both properties; hence, we have used the *inclusive-or* in this statement.

G. Boole himself based the idempotence of "and" on this understanding of propositions. He said that the truth of a proposition will not change when the proposition is used twice; hence, $x \wedge x = x$, and he identified $\wedge$ with the multiplication and stated now: the equations

$$x^2 = x ,$$

$$x^2 - x = 0 , \text{ and}$$

$$x(x - 1) = 0$$

have the solutions 0 and 1. *Any calculus that considers the truth of propositions can have two values only.*

All kinds of applications of "and," "or" (inclusive or exclusive), and "not" can now be based on the definition of *conjunction*, *disjunction*, *antivalence*, and *negation*. We emphasize once again that

Table 8.1 Definition of
the implication

x	y	$x \to y$
0	0	1
0	1	1
1	0	0
1	1	1

the identification of 0 and *false* and 1 and *true*, respectively, must always be observed. This is also directly reflected in *set theory*.

Let be given two sets S_1 and S_2 with the assumption that these two sets are subsets of a *universal set S* ($S_1 \subseteq S$, $S_2 \subseteq S$) with the following definitions:

$$S_1 = \{x \in S \mid p_1(x) = \text{true}\}, \quad S_2 = \{x \in S \mid p_2(x) = \text{true}\}.$$

Then it can be defined:

the set $\quad S_1 \cap S_2 \qquad\qquad$ by $\quad p_1(x) \wedge p_2(x) = \text{true}$,

the set $\quad S_1 \cup S_2 \qquad\qquad$ by $\quad p_1(x) \vee p_2(x) = \text{true}$,

the set $\quad S_1 \vartriangle S_2 \qquad\qquad$ by $\quad p_1(x) \oplus p_2(x) = \text{true}$,

the set $\quad S_1 \overline{\vartriangle} S_2 \qquad\qquad$ by $\quad p_1(x) \odot p_2(x) = \text{true}$,

the set $\quad \overline{S_1} \qquad\qquad\qquad$ by $\quad \overline{p_1(x)} = \text{true}$, i.e., $p_1(x) = \text{false}$,

the set $\quad \overline{S_2} \qquad\qquad\qquad$ by $\quad \overline{p_2(x)} = \text{true}$, i.e., $p_2(x) = \text{false}$.

The symbol $\vartriangle$ expresses the symmetric difference, and the symbol $\overline{\vartriangle}$ is used for the complement of the symmetric difference.

Very important in all fields of science, but especially in Mathematics, is the definition of the *implication*. There is a long and interesting discussion in the history of science about the definition of a "good" implication; the existing convention and definition obviously have been the most appropriate approach.

At the beginning there are very often difficulties to deal with the first two lines of Table 8.1, and these two cases should be carefully kept in mind. Let us look at the following statement:

"If it is raining, then the ground will be wet."

If the *assumption* (the part behind the *if*) is true, then the *conclusion* (the part behind *then*) must also be true (corresponding to line 4 of the definition of the implication) so that the whole statement with structure of an implication becomes true. The statement will still be logically true: if it is not raining, then the implication is not really applicable because the assumption is not satisfied. In this case it is not important whether the conclusion is false (the ground is dry) or whether the conclusion is true (the ground could be wet for another reason). The *subset relation* expresses the same understanding; let S_1 and S_2 be defined as above by propositions $p_1(x)$ and $p_2(x)$; then the subset relation will be described by the implication:

$$S_1 \subseteq S_2 \quad \equiv \quad (p_1(x) \to p_2(x)) = 1.$$

All the elements of S_1 are elements of S_2 (line 4), there could be elements outside of S_1, but elements of S_2 (line 2) and even elements outside of S_1 and S_2 (line 1). In these three cases the implication is equal to true, and there is no counterexample or contradiction to the subset relation. The only case that is forbidden (the implication is equal to false) is line 3—an element of S_1 that is not an element of S_2 cannot be found. It follows directly from these considerations that the empty set $\emptyset$ is a subset of any other set: $\emptyset \subseteq S$. $p_1(x)$ is always false, this defines the empty set, the implication is true, and, therefore, the relation holds.

It can also be seen from the definition of the implication that $(x \to y) = 1$ is equivalent to the relation $\leq$. This relation holds for the three cases $0 \leq 0, 0 \leq 1, 1 \leq 1$, and just for these three cases the implication is equal to 1.

The implication is most important and applicable in those cases where the truth of the assumption is *sufficient* for the truth of the conclusion. This has to be carefully explored and stated; from a false assumption everything can be derived—it is the best (as mentioned above) to consider the implication as not applicable. There is no information available whether the conclusion is true or false. It is relatively easy to avoid many confusions with regard to the implication by using the rule

$$(p \to q) \quad \equiv \quad (\overline{p} \vee q) \,.$$

In this context the *resolution theorem* is quite understandable. We had

$$(x \to y)(\overline{x} \to z) \to (y \vee z) \,.$$

Since for any proposition x either x or $\overline{x}$ is true, one of the assumptions must be true; hence, at least one variable of the conclusion must also be true.

The *equivalence* expresses the property that two propositions are either true or false at the same time. This can be expressed by

$$(p \odot q) \quad \equiv \quad (p \to q) \wedge (q \to p) \,.$$

The truth of p is sufficient for the truth of q and vice versa (as expressed by *equivalence*). In Mathematics the expression "p if and only if q" is used very often to express the equivalence of p and q.

The concepts of *tautology* and *contradiction* play an important role in applications of the propositional calculus. A combined proposition (consisting of several elementary propositions and connected to each other by different constructions) is a *tautology* (a *logical law*) if it is always true (independent of the truth of the elementary parts), and it is a *contradiction* if it is always false. The functions corresponding to a tautology or a contradiction are $1(\mathbf{x})$ and $0(\mathbf{x})$, resp. Finding logical laws is one of the main aims of propositional logic. The equivalence

$$(p \to q) \quad \equiv \quad (\overline{q} \to \overline{p})$$

is one of these laws and deserves special attention. It is the foundation of *proofs by contradiction*. The simplest application is the use of the right-hand side of the equivalence when the truth value of the left-hand side is known. It was already mentioned that there is no doubt about the validity of the following rule:

"If it is raining, then the ground will be wet."

When this rule is considered to be true and the observation is made that "*the ground is not wet,*" then it can be concluded that "*it is not raining.*"

Very often this equivalence will be used in the following way: the implication $p \rightarrow q$ has to be proven; in order to do this, the negated conclusion $\overline{q}$ will be taken as the new assumption, and it has to be shown that the negated previous assumption $\overline{p}$ can be derived.

Example 8.1 We explore the truth of the following sentence:

"If n is a prime number, $n > 2$, then n is an odd number."

We have the following propositions:

$$p_1 : \quad n \text{ is a prime number};$$

$$p_2 : \quad n > 2;$$

$$p : \quad p_1 \text{ and } p_2, \text{ i.e., } p_1 p_2;$$

$$q : \quad n \text{ is an odd number}.$$

Hence, the formula $p_1 p_2 \rightarrow q$ expresses the logical content of the example. Now it is assumed that n is an even number, $\overline{q} = $ true, $q = $ false. Then n can be equal to 2, which results in $\overline{p_2} = $ true, $p_2 = $ false, and, hence, $\overline{p} = $ true, $p = $ false. If $n = 4, 6, 8, \ldots$, then n can be divided by 2, $n = k_1 * 2$; hence, $\overline{p_1} = $ true, $p_1 = $ false and $\overline{p} = $ true, $p = $ false again. Other cases are not possible; hence, every time we have

$$\overline{q} \quad \rightarrow \quad \overline{p_1 p_2} \, .$$

And it should be noted that we used "odd number" as the complement of "even number." This shows that for $p \rightarrow q$ the conclusion q is a *necessary condition* for p because when q is false, then p will also be false (for a true implication $0 \rightarrow 0 = 1$). However, as before, when q is true, then p still can be true or false ($0 \rightarrow 1 = 1$, $1 \rightarrow 1 = 1$); hence, further considerations are necessary.

These concepts must be applied very carefully and errors are always possible. And it is not easy to accept that a sentence like

"If $3 < 5$, then $10 < 100$."

is really acceptable and true. The condition $3 < 5$ is true, and the conclusion $10 < 100$ is true as well; hence, the whole implication is true; however, there is no reasonable connection between the left-hand side and the right-hand side of the implication. And to accept the truth of the sentences

"If $3 > 5$, then $10 < 100$."

or

"If $3 > 5$, then $10 > 100$."

is even more difficult. As has been said before, false assumptions do not need special attention, and anything can be derived. In any case, the following equivalences can be used for the simplification of expressions containing the implication and constants:

$$(0 \rightarrow x) \equiv 1 \, , \quad (1 \rightarrow x) \equiv x \, , \quad (x \rightarrow 0) \equiv \overline{x} \, , \quad (x \rightarrow 1) \equiv 1 \, .$$

The introduction of *quantifiers* brings us to the border of propositional logic. They are used to build the so-called *predicate calculus* (predicate logic). Let us have a look at the predicate $p(x)$ with the meaning "x is a prime number." It will be intuitively clear that for any natural number $x > 1$ the answer will be "yes" or "no," and the statement $p(x)$ will be true or false (independent of the fact whether we know it or not): $p(10) = $ false, $p(11) = $ true, $p(12) = $ false, $p(13) = $ true, $\ldots$ However, since the set of natural numbers is an *infinite set*, it is not possible to write general statements about natural numbers as a statement in propositional logic. Only when the set to be considered is a *finite set*, then the expression that represents quantifiers can be written, at least in principle, as an expression of propositional logic. Two quantifiers will be used most frequently:

$$\forall x : (x \in S \to p(x)) \qquad \text{for all } x: \qquad \text{if } x \in S, \text{ then } p(x),$$
$$\exists x : (x \in S \to p(x)) \qquad \text{it exists an } x: \qquad \text{if } x \in S, \text{ then } p(x).$$

The first quantifier has the meaning that the property $p(x)$ holds for all elements of the set S, and the second expresses the fact that there is *at least* one $x \in S$ with the property $p(x)$. When the set S is a *finite* set, then the quantifiers can be expressed by logic expressions, $S = \{x_1, \ldots, x_n\}$:

$$\forall x : (x \in S \to p(x)) \quad \equiv \quad p(x_1) \wedge p(x_2) \wedge \cdots \wedge p(x_n),$$
$$\exists x : (x \in S \to p(x)) \quad \equiv \quad p(x_1) \vee p(x_2) \vee \cdots \vee p(x_n).$$

Hence, these two quantifiers can be considered as a direct generalization of propositional expressions to infinite sets of individual elements. This can also be seen by the consideration of the negation: $\neg \exists x : (x \in S \to p(x))$ has the meaning that no $x \in S$ can be found with $p(x) = \text{true}$. For a finite set this is equivalent to

$$\overline{p(x_1)} \wedge \overline{p(x_2)} \wedge \cdots \wedge \overline{p(x_n)} = \overline{p(x_1) \vee \cdots \vee p(x_n)} = \text{true}.$$

If there is no x with the property $p(x)$, then the negation must hold for all the elements to be considered. In the same way, we get that $\neg \forall x : (x \in S \to p(x))$ can be expressed by

$$\overline{p(x_1)} \vee \overline{p(x_2)} \vee \cdots \vee \overline{p(x_n)} = \overline{p(x_1) \wedge \cdots \wedge p(x_n)} = \text{true}.$$

For a property p that does not hold for all the elements of a set S, at least one counterexample must be found.

The generalization of De Morgan's laws can be given as follows:

$$\neg \forall x : p(x) \quad \equiv \quad \exists x : \overline{p(x)}, \quad x \in S,$$
$$\neg \exists x : p(x) \quad \equiv \quad \forall x : \overline{p(x)}, \quad x \in S.$$

The logical approach is very well-known in Mathematics. The starting point for building a given theory is very often a set of *axioms*. Axioms are assumed to be true without further consideration.

Example 8.2 An ordered pair $(S, \circ)$ is a *group* if S is a set, $\circ$ is a binary operation $S \times S \to S$ such that the following three axioms are satisfied:

1. $\forall x \forall y \forall z \in S: \quad (x \circ y) \circ z = x \circ (y \circ z) \qquad$ (*Associative Law*).
2. $\forall x \in S: \quad$ there is a *unit element* $e \in S$ with $e \circ x = x \circ e = x$.
3. $\forall x \in S: \quad$ there is an *inverse element* x^{-1} with $x \circ x^{-1} = x^{-1} \circ x = e$.

The *Theory of Groups* consists of all theorems that can be derived from these axioms supported by new definitions of interest, new concepts to be introduced, etc. The building of group theory very often starts with the following.

Theorem 8.1 *For any group $G = (S, \circ)$, there is one and only one unit element.*

Proof The plain logic structure of this theorem can be written as follows:

"*If* $G = (S, \circ)$ is a group, *then* there is one and only one unit element."

The second axiom requires the existence of a unit element in the sense of "at least one." We use the proof by contradiction and assume that there are at least two such elements e_1, e_2. Then, according to axiom 2, we have $\forall x \in S: e_1 \circ x = x \circ e_1 = x$ and $\forall x \in S: e_2 \circ x = x \circ e_2 = x$. These two equations must hold for e_1 and e_2 as well. We use $x = e_2$ in the first equation:

$$e_1 \circ e_2 = e_2 \circ e_1 = e_2 .$$

For $x = e_1$, the second equation results in

$$e_2 \circ e_1 = e_1 \circ e_2 = e_1 .$$

Hence, $e_1 \circ e_2 = e_1$ and $e_1 \circ e_2 = e_2$, and this results in $e_1 = e_2$, since an operation $\circ$ is a function, i.e., it is a unique mapping from $S \times S$ into S.   □

In this way, we found a theorem that is based on two assumptions:

1. The result of an operation is always uniquely determined (according to the definition of an operation).
2. There is a neutral element in a group (according to axiom 2 of groups).

Theorem 8.1 is now applicable to each group. For the set of integers with $+$ as the group operation and 0 as the unit element it has the meaning that there is one and only one 0. In other groups (e.g., the set of rational numbers without 0 and the multiplication as operation) it will have another meaning (the rational number 1 is the only neutral element in this group), but it is still based on the same theorem of group theory.

Sometimes the set of axioms will be extended, sometimes this set can be reduced, axioms can be redundant and can be replaced by other axioms, even by axioms with the opposite meaning, other axiom systems might be possible for a given theory, sometimes they could be more or less convenient, more or less "elegant," and so on. Many interesting problems have to be solved; however, the truth of the axioms to be considered will always be taken as given.

A famous example is the axiomatic system of *Euclidean* and *Non-Euclidean Geometries*. One axiom of the Euclidean Geometry requires that, for a given straight line g and a point P that is not on this straight line, there exists a second straight line g' so that P is on g', g and g' do not intersect, and they are *parallel* to each other. This axiom goes back to the famous Greek mathematician *Euclid*, and the geometry with this axiom is denoted by *Euclidean Geometry*.

This axiom has been changed to other forms completely contradicting this axiom ("No parallel straight line exists." or "An infinite number of straight lines exists."), and the resulting *Non-Euclidean Geometries* play an important role in modern Physics.

For any application of Group Theory, the set S and the operation $\circ$ have to be determined, and it must be shown that the three axioms are satisfied. We can take, for instance, $S = \mathbb{Z}$ (the set of integers), $+$ as the operation $\circ$, then the first axiom will be satisfied, the number 0 can be used as the neutral element e, and for each integer i we have the integer $(-i)$ with $i + (-i) = 0$. After it has been shown that the three axioms introduced in Example 8.2 have been satisfied, and all the theorems, concepts, and definitions of Group Theory are applicable to $G = (\mathbb{Z}, +)$. The set $G = (\mathbb{Z}, +)$ is a *model* of Group Theory, or (in daily language) it is a *group*. Because of the application of $+$, the group is very often denoted by *additive group*.

Very well-known is the *modus ponens*, which is based on the fact that

$$(p \land (p \to q)) \to q$$

is a tautology. This expresses again the property of the implication that the truth of p (the truth of the assumption) is *sufficient* for the truth of q (the truth of the conclusion) if there is a true implication. For a given true rule $p \to q$ it must be shown that the assumption is true; then the conclusion q is also true.

Another nice name *constructive dilemma* has been given to the following way of reasoning:

$$((p \to q) \land (r \to s) \land (p \lor r)) \to (q \lor s).$$

It can be verified that this expression represents a tautology as well. Hence, the modus ponens can be applied. The left-hand side $((p \to q) \land (r \to s) \land (p \lor r))$ will be true when the two implications $(p \to q)$ and $(r \to s)$ are true, and the disjunction of the two assumptions p and r, i.e., $p \lor r$, must also be true. We use the modus ponens once again: if p is true, then $p \to q$ results in $q = $ true; if r is true, then $s = $ true, and even $p = $ true, $r = $ true can happen, and then $q = $ true and $s = $ true. In each case the disjunction $q \lor s$ will be true.

The name "dilemma" indicates that we only know that $p \lor r$ is true, but not which one (either p or r or both of them). Hence, the conclusion can only be that $q \lor s = $ true.

Many rules like this "dilemma" exist, and all of them can be verified and used in the same way:

- Careful definition of the elementary propositions
- The replacement of elementary propositions by logic variables
- The modeling of complicated constructions that are given in (tricky) natural language by means of logic expressions (formulas)
- Exploration of these expressions by means of logic functions and equations

This approach was, among others, one of the original intentions of *George Boole* when he started his research in this field. It might be interesting that G. Boole himself used *conjunction*, *negation*, and *antivalence*. Later on the *disjunction* replaced the antivalence for a long time. Nowadays we are using Boolean Algebras and Boolean Rings according to our convenience.

Based on the given examples and explanations, we will show that the classical approach of propositional logic can be fully handled by means of logic functions as they have been introduced before.

The main concept are tautologies of the form $x_1 \land \cdots \land x_n \to y$, corresponding to the equation $(x_1 \land \cdots \land x_n \to y) = 1$. According to the modus ponens, the truth of $x_1, \ldots, x_n$ results in the truth of y. y is a tautological consequence of the set $X = \{x_1, \ldots, x_n\}$ of assumptions. Very often, the following transformations will be used:

$$(x_1 \land \cdots \land x_n \to y) = 1 ,$$

$$(\overline{(x_1 \land \cdots \land x_n)} \lor y) = 1 ,$$

$$\overline{(\overline{(x_1 \land \cdots \land x_n)} \lor y)} = 0 ,$$

$$x_1 \land \cdots \land x_n \land \overline{y} = 0 .$$

This means that (based on identical transformation steps) instead of proving that the implication is a tautology, the negation of the conclusion can be added to the assumptions, and this will result in a contradiction. The following example has already been used.

"If n is a prime number and $n > 2$, then n is an odd number."

Hence, we can set

$$x_1 : \quad n \text{ is a prime number;}$$

$$x_2 : \quad n > 2;$$

$$y : \quad n \text{ is an odd number;}$$

$$\overline{y} : \quad n \text{ is an even number.}$$

Note 8.1 We used the assumption that "an even number" has the meaning "not an odd number."

The expression $x_1 \wedge x_2 \wedge \overline{y}$ cannot be satisfied:

"n is an even number"

means that $n \in \{2, 4, 6, 8, \ldots\}$. For $n = 2$ the assumption x_2 is false. For any other n ($n = 4, 6, 8, \ldots$) n is divisible by 2 so that the assumption x_1 is false. Other possibilities do not exist; hence, $x_1 \wedge x_2 \wedge \overline{y}$ is a contradiction.

We used $X = \{x_1, \ldots, x_n\}$ in the sense of a *finite set* of assumptions. The *Compactness Theorem* for propositional logic shows that it is sufficient to consider such finite sets of assumptions.

Especially in Mathematics, the approach is as follows:

1. There is a set of *axioms* and a set of *assumptions* that are given or taken as true (as valid or as satisfied).
2. A given set of rules can be used to create new (additional) true propositions that will be added to the set of true axioms and assumptions.
3. A proof of a special proposition starts with the axioms and assumptions, and the proposition must be the result of a sequence of rule applications.

One possibility is the following system P that starts with one axiom and four rules:

- A1: $x_1 \vee \overline{x}_1$ holds for all formulas x_1.
- R1: $x_1 \vee (x_2 \vee x_3) \rightarrow (x_1 \vee x_2) \vee x_3$.
- R2: $x_1 \vee x_1 \rightarrow x_1$.
- R3: $x_1 \rightarrow (x_1 \vee x_2)$.
- R4: $(x_1 \vee x_2)(\overline{x}_1 \vee x_3) \rightarrow (x_2 \vee x_3)$.

The formulas of the system P can be built by means of the negation and the conjunction, and x_1, x_2, x_3 represent any formulas. $\wedge, \rightarrow, \odot$ can be introduced by means of

$$x_1 \wedge x_2 \equiv \overline{(\overline{x}_1 \vee \overline{x}_2)},$$

$$x_1 \rightarrow x_2 \equiv \overline{x}_1 \vee x_2,$$

$$x_1 \odot x_2 \equiv (x_1 \rightarrow x_2) \wedge (x_2 \rightarrow x_1).$$

We know all these rules as equivalences in the context of logic functions (that already has been considered in detail). A1, R1, R2, R3, and R4 are equal to $1(\mathbf{x})$ with an appropriate vector $\mathbf{x}$.

For convenience, many other rules can be introduced, but in principle, these four rules are completely sufficient to find all possible proofs. The most important result of these considerations is the fact that any theorem T that has been proven is the conclusion of a tautology

$$x_1 \wedge \cdots \wedge x_n \to T$$

with the assumptions $x_1, \ldots, x_n$. In principle, the application of truth tables would be completely sufficient to prove all the theorems of a given theory. However, the stepwise proofs of intermediate conclusions give very often a deeper insight into the theory to be built, the reason for the definition of new concepts, the contribution of different assumptions, and so on. These intentions also explain some requirements that good logic systems (axioms and rules) should satisfy:

1. The system should be *consistent*: there is no formula F such that F and $\overline{F}$ can be derived.
2. The system is *complete* if for every formula F, either F or $\overline{F}$ can be derived.
3. The system is *decidable* if there is an algorithm that decides whether a given formula can be proven.

The third item requires naturally an understanding (a definition) of an algorithm.

The system P specified above is consistent, not complete, and not decidable. The search for "good" systems of axioms and rules played a big role in the history of Mathematics. Additional requirements (simplicity, understandability, efficiency) also play an important role.

The following *deduction theorem* shows that the role of assumptions and conclusions can change in a given sense.

Theorem 8.2

$$(x_1 \wedge \cdots \wedge x_n \wedge x) \to y \equiv (x_1 \wedge \cdots \wedge x_n) \to (x \to y) .$$

Proof The understanding of logic functions makes theorems like this very easy to understand and to prove:

$$(x_1 \wedge \cdots \wedge x_n \wedge x) \to y \equiv \overline{(x_1 \ldots x_n x)} \vee y$$

$$\equiv \overline{x}_1 \vee \cdots \vee \overline{x}_n \vee \overline{x} \vee y$$

$$\equiv \overline{(x_1 \ldots x_n)} \vee (x \to y)$$

$$\equiv (x_1 \ldots x_n) \to (x \to y) . \qquad \square$$

We can conclude our considerations of propositional logic by citing some more logic systems:

- Principia Mathematica *(Russell, Whitehead, Hilbert, Ackermann)* (*negation and disjunction* due to $(x_1 \to x_2) \equiv (\overline{x}_1 \vee x_2)$):

  1. $(x_1 \vee x_1) \to x_1$.
  2. $x_1 \to (x_1 \vee x_2)$.
  3. $(x_1 \to x_2) \to [(x_1 \vee x_3) \to (x_2 \vee x_3)]$.

- *Rosser* (*negation and conjunction*):

  1. $x_1 \to (x_1 \wedge x_1)$.
  2. $(x_1 \wedge x_2) \to x_1$.
  3. $(x_1 \to x_2) \to \left[\overline{(x_2 \wedge x_3)} \to \overline{(x_1 \wedge x_3)} \right]$.

- System for *negation and implication*:

  1. $x_1 \rightarrow (x_2 \rightarrow x_1)$.
  2. $[x_1 \rightarrow (x_2 \rightarrow x_3)] \rightarrow [(x_1 \rightarrow x_2) \rightarrow (x_1 \rightarrow x_3)]$.
  3. $\overline{x}_1 \rightarrow (x_1 \rightarrow x_2)$.
  4. $(\overline{x}_1 \rightarrow x_1) \rightarrow x_1$.

All these systems can be used instead of the system P and have given advantages and disadvantages. At the end, however, they will build the same theory. There are many more relations between *Computer*, *Science*, and *Logic*, like *Automated Theorem Proving* or *Logic Programming*; they are not only based on *Propositional Logic*, but also on the *Predicate Calculus* and, hence, beyond the scope of this book.

8.2 Binary Arithmetic

This topic is closely related to the development of computer hardware or other electronic devices, control systems, etc., but it goes back at least to G. W. Leibniz (1646–1716). We will give an outline of binary Mathematics before the hardware-related problems will be discussed in more detail. One basic concept is the binary representation of numbers.

Theorem 8.3 *Each natural number n can be represented by a sum of powers of 2, using 0 and 1 as coefficients (digits).*

Proof The decimal system is using the representation of numbers by means of powers of 10:

$$123 = 1 * 10^2 + 2 * 10^1 + 3 * 10^0 ,$$

$$84\,905 = 8 * 10^4 + 4 * 10^3 + 9 * 10^2 + 0 * 10^1 + 5 * 10^0 .$$

The use of powers of 2 follows the same idea:

$$123 = 1 * 64 + 1 * 32 + 1 * 16 + 1 * 8 + 1 * 2 + 1 * 1$$

$$123 = 1 * 2^6 + 1 * 2^5 + 1 * 2^4 + 1 * 2^3 + 0 * 2^2 + 1 * 2^1 + 1 * 2^0$$

$$= 111\,1011_2 .$$

The index 2 emphasizes the binary system:

$$84\,905 - 1\,0100\,1011\,1010\,1001_2 . \qquad \square$$

It can be seen that the binary numbers are "much longer" (at least three times), but in principle they can be used in the same way like the decimal numbers.

It should be understood that we are not dealing with logic functions as yet, right now the two symbols are digits in the system of *dual numbers*.

The binary representation of a *natural number* (that is given in the decimal system) can be generated by the following steps:

1. The number n will be divided by 2 as long as possible.
2. The remainder of each division will be taken as the binary digit.

Example 8.3

$$
\begin{aligned}
1\,521 : 2 &= 760 & &\text{remainder } 1 * 2^0 \\
760 : 2 &= 380 & &\text{remainder } 0 * 2^1 \\
380 : 2 &= 190 & &\text{remainder } 0 * 2^2 \\
190 : 2 &= 95 & &\text{remainder } 0 * 2^3 \\
95 : 2 &= 47 & &\text{remainder } 1 * 2^4 \\
47 : 2 &= 23 & &\text{remainder } 1 * 2^5 \\
23 : 2 &= 11 & &\text{remainder } 1 * 2^6 \\
11 : 2 &= 5 & &\text{remainder } 1 * 2^7 \\
5 : 2 &= 2 & &\text{remainder } 1 * 2^8 \\
2 : 2 &= 1 & &\text{remainder } 0 * 2^9 \\
1 : 2 &= 0 & &\text{remainder } 1 * 2^{10}
\end{aligned}
$$

$$1\,521 = 1*2^{10} + 0*2^9 + 1*2^8 + 1*2^7 + 1*2^6 + 1*2^5 + 1*2^4 + 0*2^3 + 0*2^2 + 0*2^1 + 1*2^0 .$$

This can be seen quite easily when the sequence of divisions is represented as a sequence of multiplications:

$$
\begin{aligned}
1\,521 &= 2*760 + 2^0 * \underline{1} \\
&= 2*(2*380 + 0) + 1 \\
&= 2^2 * 380 + 2^1 * \underline{0} + 2^0 * \underline{1} \\
&= 2^2 * (2*190 + 0) + 2^1 * 0 + 2^0 * 1 \\
&= 2^3 * 190 + 2^2 * \underline{0} + 2^1 * \underline{0} + 2^0 * \underline{1}
\end{aligned}
$$

$$\vdots$$

The transformation of a binary number into a decimal number is possible by direct calculation:

$$1\,0110\,0101_2 = 1*2^8 + 0*2^7 + 1*2^6 + 1*2^5 + 0*2^4 + 0*2^3 + 1*2^2 + 0*2^1 + 1*2^0 = 357 .$$

Odd numbers have the digit 1 in the rightmost position, even numbers the digit 0. The use of binary numbers created the name *bit* as an abbreviation of *binary digit*. But the name "bit" is now also used when hardware and logic functions are under consideration.

The *octal* and the *hexadecimal* systems are used as well. These two systems can easily be derived from the binary representation. In order to get the *octal representation*, the binary representation of a number has to be split into groups of three bits, starting from the right and going to the left; by adding some leading zeroes on the left-hand side (if necessary), it can be assumed that the number of bits is a multiple of 3. Thereafter, the value represented by three bits will be calculated, and it will always be a value in the range of 0 to 7:

$$0 = 0 * 2^2 + 0 * 2^1 + 0 * 2^0 \,,$$
$$7 = 1 * 2^2 + 1 * 2^1 + 1 * 2^0 \,.$$

Example 8.4

$$1111011_2 = 001\ 111\ 011_2 = 173_8 \,,$$
$$101100101_2 = 101\ 100\ 101_2 = 545_8 \,.$$

The conversion of an octal number into a binary number is also very easy. Each octal digit between 0 and 7 will be represented by three bits, and the bits of different digits are always concatenated according to the sequence of the digits:

$$35\,207_8 = \underbrace{011}_{3}\ \underbrace{101}_{5}\ \underbrace{010}_{2}\ \underbrace{000}_{0}\ \underbrace{111}_{7}\ _2 \,.$$

The direct conversion of a decimal number into an octal number can use a continuous division of the decimal number by 8 (as long as possible).

$$231 = 28 * 8 + 7 \,, \qquad 28 = 3 * 8 + 4 \,;$$
$$231 = (3 * 8 + 4) * 8 + 7$$
$$= 3 * 8^2 + 4 * 8 + 7$$
$$= 347_8 \,.$$

Computer memories are (logically) built from *bytes*. One byte is a group of 8 bits. In order to make good use of these bits, the byte will be split into groups of 4 bits, and these 4 bits can represent the numbers between 0 and 15:

$$0000 = 0 * 2^3 + 0 * 2^2 + 0 * 2^1 + 0 * 2^0 = 0 \,,$$
$$1111 = 1 * 2^3 + 1 * 2^2 + 1 * 2^1 + 1 * 2^0 = 15 \,.$$

In order to consider these values as digits, the characters A, B, C, D, E, F are used for the decimal values 10, 11, 12, 13, 14, and 15, resp. This system is denoted by *hexadecimal system*.indexNumber system!hexadecimal

Example 8.5 $0101|1010|1111|0011_2$ corresponds to

$$5AF3_{16} = 5 * 16^3 + A * 16^2 + F * 16^1 + 3 * 16^0$$
$$= 5 * 4\,096 + 10 * 256 + 15 * 16 + 3$$
$$= 20\,480 + 2\,560 + 240 + 3$$
$$= 23\,283 \,.$$

Table 8.2 Lexicographic order of vectors with five bits and associated decimal number d

x_4	x_3	x_2	x_1	x_0	d	x_4	x_3	x_2	x_1	x_0	d	x_4	x_3	x_2	x_1	x_0	d	x_4	x_3	x_2	x_1	x_0	d
0	0	0	0	0	0	0	1	0	0	0	8	1	0	0	0	0	16	1	1	0	0	0	24
0	0	0	0	1	1	0	1	0	0	1	9	1	0	0	0	1	17	1	1	0	0	1	25
0	0	0	1	0	2	0	1	0	1	0	10	1	0	0	1	0	18	1	1	0	1	0	26
0	0	0	1	1	3	0	1	0	1	1	11	1	0	0	1	1	19	1	1	0	1	1	27
0	0	1	0	0	4	0	1	1	0	0	12	1	0	1	0	0	20	1	1	1	0	0	28
0	0	1	0	1	5	0	1	1	0	1	13	1	0	1	0	1	21	1	1	1	0	1	29
0	0	1	1	0	6	0	1	1	1	0	14	1	0	1	1	0	22	1	1	1	1	0	30
0	0	1	1	1	7	0	1	1	1	1	15	1	0	1	1	1	23	1	1	1	1	1	31

The bits are combined into groups of 4 bits (starting from the right), and each group will be replaced by the corresponding hexadecimal digit between 0 and F. The conversion to a decimal number now takes the hexadecimal digits between 0 and F and the appropriate powers of 16 into consideration.

It is a characteristic that the *same bit vector* can be used with base 2:

$$2^{14} + 2^{12} + 2^{11} + 2^9 + 2^7 + 2^6 + 2^5 + 2^4 + 2^1 + 2^0 = 23\,283\,.$$

This naturally holds because $16 = 2^4$; hence, $16^n = (2^4)^n = 2^{4n}$.

The octal and hexadecimal arithmetic follow the same principles as the decimal or the binary arithmetic—some training might be required, but there are no special problems.

It should be kept in mind that an enumeration of the binary numbers can be done by using the lexicographic order of the bit vectors, i.e., by using the *binary code*. In the rightmost column of the coding table, the combination 01 is repeated as often as necessary. The last but one column uses a duplication, i.e., 0011, followed by 00001111 for the next column (the third from the right), and the duplication will take place as long as necessary (Table 8.2).

As next the relations between the binary arithmetic operations and logic functions will be established. The *addition* of two bits will be performed as follows:

$$0 + 0 = 0\,, \quad 0 + 1 = 1\,, \quad 1 + 0 = 1\,.$$

In these cases there is no *carry* to the next position. For $1 + 1$, however, we get a carry to the next higher position (i.e., to the next position on the left-hand side):

$$1 + 1 = 0\,, \quad \text{carry} = 1\,.$$

This can be seen as follows:

$$\text{for any } i \geq 0: \quad 1 * 2^i + 1 * 2^i = (1 + 1) * 2^i = 2 * 2^i = 1 * 2^{i+1} + 0 * 2^i\,.$$

For the position i, the result is equal to 0; in the next higher position, an additional value 1 has to be considered. And this can go up to the leftmost position. It can always be assumed that the binary numbers to be added have the same length, and additional leading zeroes can be added on the left-hand side of the numbers.

Table 8.3 Carry and sum at position 0

x_0	y_0	Carry(x_0, y_0)	Sum(x_0, y_0)
0	0	0	0
0	1	0	1
1	0	0	1
1	1	1	0

Table 8.4 Carry and sum at position i

x_i	y_i	c_{i-1}	Carry$_i(x_i, y_i, c_{i-1})$	Sum$_i(x_i, y_i, c_{i-1})$
0	0	0	0	0
0	0	1	0	1
0	1	0	0	1
0	1	1	1	0
1	0	0	0	1
1	0	1	1	0
1	1	0	1	0
1	1	1	1	1

Example 8.6

$$
\begin{array}{r}
010\ 1101 \\
+\ 100\ 0011 \\
\hline
111\ 0000
\end{array}
$$

The carry bits can be combined into the vector $\mathbf{c} = (000\,1111)$. In order to represent this calculation by means of logic functions (or in order to prepare a hardware-based electronic implementation), we observe the fact that the carry and the sum in one position depend on two parameters (the two bits to be added) and can be expressed by Table 8.3.

It can be seen from Table 8.3 that

$$\text{carry}(x_0, y_0) = x_0 \wedge y_0 \, , \quad \text{sum}(x_0, y_0) = x_0 \oplus y_0 \, .$$

However, these two functions can be used only for the rightmost (lowest) position because it can be assumed that the carry for the lowest position is equal to 0. For any position $i > 0$, the carry that comes from position $i - 1$ has to be considered.

Hence, x_i, y_i, c_{i-1} are the *input-values* for position i (c_{i-1} being produced at position $i - 1$), and carry$_i(x_i, y_i, c_{i-1})$ and sum$_i(x_i, y_i, c_{i-1})$ are the *output-values* at position i. The values of Table 8.4 result in

$$c_i = \text{carry}_i(x_{i+1}, y_{i+1}, c_{i-1}) = x_i y_i \vee x_i c_{i-1} \vee y_i c_{i-1} \, ,$$

$$s_i = \text{sum}_i(x_i, y_i, c_i) = x_i \oplus y_i \oplus c_{i-1} \, ,$$

$$c_{-1} = 0 \, ,$$

$$i = 0, \dots, n \, .$$

Hence, we get successively (from the right to the left):

$$c_0 = x_0 y_0 \, , \qquad\qquad\qquad s_0 = x_0 \oplus y_0 \, ,$$

$$c_1 = x_1 y_1 \vee x_1 c_0 \vee y_1 c_0 \, , \qquad s_1 = x_1 \oplus y_1 \oplus c_0 \, ,$$

Fig. 8.1 The overflow of a
sum

$$
\begin{array}{r}
1011 \\
+\ 1100 \\
\hline
1\ 0111
\end{array}
$$

overflow from position 3
into position 4

$$
\begin{array}{r}
1000\ 1000\ 1000\ 1000 \\
+\ 1000\ 0000\ 1000\ 0000 \\
\hline
1\ 0000\ 1001\ 0000\ 1000
\end{array}
$$

overflow in a register
of 16 bits

Fig. 8.2 Multiplication
$13 * 13 = 169$ using binary
numbers

$$
\begin{array}{r}
1101 * 1101 \\
\hline
1101 \\
1101 \\
0000 \\
1101 \\
\hline
10101001
\end{array}
$$

Fig. 8.3 Multiplication
$13 * 13 = 169$ realized by
the addition of two
intermediate binary
numbers

$$
\begin{array}{r}
1101 * 1101 \\
\hline
1101 \\
1101 \\
\hline
100111 \\
0000 \\
\hline
1001110 \\
1101 \\
\hline
10101001
\end{array}
$$

$$c_2 = x_2 y_2 \vee x_2 c_1 \vee y_2 c_1 , \qquad\qquad s_2 = x_2 \oplus y_2 \oplus c_1 .$$

$$\vdots \qquad\qquad\qquad\qquad\qquad \vdots$$

The carry generated in position 0 will be denoted by c_0 and used as input in position 1, the carry from position 1 will be used in position 2, etc. When a carry bit equal to 1 is generated in the highest position, then the sum is one bit longer than the arguments. In relation to computer hardware, this is denoted by *overflow* when a given register size (16, 32, 64, or more bits) is not sufficient to store the result.

The two numbers in Fig. 8.1 have the bit positions $(x_{15}, \ldots, x_0)$ and $(y_{15}, \ldots, y_0)$; however, the addition results in $s_{16} = 1$.

The multiplication of two binary numbers can be implemented in the same way as for the decimal numbers; however, since we have only the values 0 and 1, we get either 0 $(x * 0 = 0)$ or x $(x * 1 = x)$.

For every value 1 in the second argument, the first vector is taken, and the value 0 in the second vector results in the vector 0000. Moving from the left to the right in the second vector shifts the vector to be taken by one position to the right (as for the decimal system). Finally, the addition of the intermediate vectors has to be implemented. Here it can happen that in one column more than three values 1 appear (Fig. 8.2). The easiest way to take this into consideration is (among others) the immediate addition of two arising vectors as shown in Fig. 8.3:

In general, the multiplication of two numbers $x = (x_n x_{n-1} \ldots x_1 x_0)$ and $y = (y_m y_{m-1} \ldots y_1 y_0)$ can be considered in the following way:

$$(x_n * 2^n + x_{n-1} * 2^{n-1} + \cdots + x_0 * 2^0) * (y_m * 2^m + y_{m-1} * 2^{m-1} + \cdots + y_0 * 2^0)$$

$$= 2^{n+m}(x_n * y_m) + 2^{n+m-1}(x_n * y_{m-1} + x_{n-1} * y_m) + \ldots$$

$$+ 2^1(x_1 * y_0 + y_1 * x_0) + 2^0(x_0 * y_0) .$$

Table 8.5 Function table
of both the multiplication
($*$) and the conjunction ($\wedge$)

x_i	y_i	$x_i * y_i$	$x_i \wedge y_i$
0	0	0	0
0	1	0	0
1	0	0	0
1	1	1	1

Fig. 8.4 Subtraction of
binary numbers

$$
\begin{array}{rl}
x & = 1\ 1\ 1\ 0 \\
y & = 1\ 1\ 0\ 0 \\
\hline
x - y & = 0\ 0\ 1\ 0
\end{array}
\qquad
\begin{array}{l}
0 - 0 = 0 \\
1 - 0 = 1 \\
1 - 1 = 0
\end{array}
$$

Fig. 8.5 Subtraction of
binary numbers: $0 - 1$ in
position 2^i using the value
1 of position 2^{i+1}

$$
\begin{array}{ccc}
2^{i+1} & 2^i & 2^{i-1} \\
\hline
1 & 0 & \ldots \\
-\ 0 & 1 & \ldots \\
\hline
0 & 1 & \ldots
\end{array}
$$

Fig. 8.6 Subtraction of
binary numbers: $0 - 1$ in
position 2^i using the value
1 of position 2^{i+2}

$$
\begin{array}{cccc}
2^{i+2} & 2^{i+1} & 2^i & 2^{i-1} \\
\hline
1 & 0 & 0 & \ldots \\
-\ 0 & 0 & 1 & \ldots \\
\hline
0 & 1 & 1 & \ldots
\end{array}
$$

It can already be seen that the multiplication of binary digits can also be transformed into logic functions, especially, when a fixed value for $n = m$ is used, for instance $n = 16$, $n = 32$, or $n = 64$. Table 8.5 shows that the multiplication can be implemented by $\wedge$.

The addition of binary numbers has been realized as before. The development of circuits for high-performance addition and multiplication is an exciting area at the border between logic functions and electronic circuits.

Now the *subtraction* will be considered. We assume $x > y$ and want to compute $x - y$. Figure 8.4 shows an example.

What has to be done if $0 - 1$ occurs in a column, say in column i? If there is a 1 in column $i + 1$, then it can be used that

$$1 * 2^{i+1} = 2 * 2^i = (1 + 1) * 2^i \ .$$

Hence, the situation $1 * 2^{i+1} + 0 * 2^i - 1 * 2^i$ can be replaced by (see Fig. 8.5)

$$(1 + 1) * 2^i - 1 * 2^i = 1 * 2^i + 1 * 2^i - 1 * 2^i = 1 * 2^i = 0 * 2^{i+1} + 1 * 2^i \ .$$

If there is another 0 in position 2^{i+1}, then we go to position 2^{i+2} and repeat the same procedure: $1 * 2^{i+2} = 4 * 2^i$, and $4 * 2^i - 1 * 2^i = 3 * 2^i$, which results in Fig. 8.6.

Since it has been assumed that $x > y$, this procedure always will be successful. If $x < y$, then the subtraction needs negative numbers. The sign of a number is also a binary information, $+$ and—are used to indicate the sign, and $+$ is mostly omitted. This means that one bit has to be used to store this information. The problem can be explained by using four bits. Table 8.6 shows a possible coding of the eight numbers $0, \ldots, 7$ in the left-hand part and the seven negative numbers $-1, \ldots, -7$ in the right-hand part.

When the four bits (sign bit and x_2, x_1, x_0) of the number $+1, \ldots, +7$ on the left-hand part and $-1, \ldots, -7$ on the right-hand part of Table 8.6 are taken as one number (i.e., the sign bit gets the meaning of a normal numerical bit), then the table shows the remarkable property that the sum of the two numbers in one line is always equal to 16:

Table 8.6 The two's
complement representation
of negative numbers

Number	Sign	x_2	x_1	x_0		Number	Sign	x_2	x_1	x_0
0	0	0	0	0						
1	0	0	0	1		-1	1	1	1	1
2	0	0	1	0		-2	1	1	1	0
3	0	0	1	1		-3	1	1	0	1
4	0	1	0	0		-4	1	1	0	0
5	0	1	0	1		-5	1	0	1	1
6	0	1	1	0		-6	1	0	1	0
7	0	1	1	1		-7	1	0	0	1

$$0001 + 1111 = 1 + 15 = 16 = 1\,0000\,,$$
$$0010 + 1110 = 2 + 14 = 16 = 1\,0000\,,$$
$$0011 + 1101 = 3 + 13 = 16 = 1\,0000\,,$$
$$0100 + 1100 = 4 + 12 = 16 = 1\,0000\,,$$
$$0101 + 1011 = 5 + 11 = 16 = 1\,0000\,,$$
$$0110 + 1010 = 6 + 10 = 16 = 1\,0000\,,$$
$$0111 + 1001 = 7 + \;\;9 = 16 = 1\,0000\,.$$

If only four bits of the result are taken into consideration, then the result of the addition is always equal to $0000 = 0$, and this expresses fully the property of negative numbers $x + (-x) = 0$. Therefore, we get the following understanding of the construction of negative binary numbers of any range:

1. We consider $x = (x_{n-1}x_{n-2}\ldots x_1x_0)$.
2. $x_n = 0$ indicates positive numbers, and $x_n = 1$ indicates negative numbers.
3. For $x = (0\ x_{n-1}x_{n-2}\ldots x_1x_0)$, we find $y = -x = (1\ y_{n-1}y_{n-2}\ldots y_1y_0)$ so that $x + y = 2^{n+1}$, i.e., $y = 2^{n+1} - x$.

The practical construction is very related to the following algorithm that expresses exactly this procedure:

1. Take the binary representation of x, $x = (0x_{n-1}x_{n-2}\ldots x_1x_0)$.
2. Find $\overline{x} = (1\overline{x}_{n-1}\overline{x}_{n-2}\ldots\overline{x}_1\overline{x}_0)$.
3. Get $y = -x = \overline{x} + 1$ by adding 1 in the rightmost position.

Example 8.7

$$5_{10} = 0101_2\,,$$

$$\overline{0101}_2 = 1010_2\,,$$

$$1010_2 + 0001_2 = 1011_2\,,$$

$$\text{signed } 1011_2 = -5_{10}\,,$$

$$\text{unsigned } 1011_2 = 11_{10}\,,$$

$$11_{10} + 5_{10} = 16_{10}\,,$$

$$16_{10} = 1\,0000_2\,.$$

It is quite natural to represent 0 by 0000; for $x = 0000$, we get $\overline{x} = 1111$, $\overline{x} + 1 = 1\,0000$, i.e., the 0 is transformed into itself. The sign bit is equal 0 for $x = 0$.

Table 8.7 The two's complement representation of $-128, \ldots, -1, 0, 1, \ldots, 127$

Number	x_7	x_6	x_5	x_4	x_3	x_2	x_1	x_0	Number	x_7	x_6	x_5	x_4	x_3	x_2	x_1	x_0
0	0	0	0	0	0	0	0	0									
1	0	0	0	0	0	0	0	1	-1	1	1	1	1	1	1	1	1
2	0	0	0	0	0	0	1	0	-2	1	1	1	1	1	1	1	0
3	0	0	0	0	0	0	1	1	-3	1	1	1	1	1	1	0	1
4	0	0	0	0	0	1	0	0	-4	1	1	1	1	1	1	0	0
				$\vdots$									$\vdots$				
124	0	1	1	1	1	1	0	0	-124	1	0	0	0	0	1	0	0
125	0	1	1	1	1	1	0	1	-125	1	0	0	0	0	0	1	1
126	0	1	1	1	1	1	1	0	-126	1	0	0	0	0	0	1	0
127	0	1	1	1	1	1	1	1	-127	1	0	0	0	0	0	0	1
									-128	1	0	0	0	0	0	0	0

Table 8.6 indicates still another approach. If only the bits $(x_2 x_1 x_0)$ are considered, then we have in the lower part the binary coding of $1, 2, 3, 4, 5, 6, 7$ on the left-hand side, and on the right-hand side we find the coding of the same numbers, now in the order $7, 6, 5, 4, 3, 2, 1$. The sum of the two numbers is always equal to $2^3 = 8$. In this representation, the number x and the number $-x$ complement each other to the power 2^{n+1} when the bits $x_n x_{n-1} \ldots x_2 x_1 x_0$ are used. This representation of negative numbers uses the *Two's Complement*.

When 8 bits will be used, Table 8.7 can be constructed in the same way.

Here the sum of the two numbers is equal to $2^8 = 256$. We can represent the numbers from -128 to 127. The Two's Complement has the big advantage that the sign is already included into the representation of the numbers. The subtraction can be directly implemented by adding these two numbers.

Example 8.8

$$7_{10} - 5_{10} = 7_{10} + (-5)_{10} = 0111_2 + 1011_2 = 1\ 0010_2 = 2_{10} \ .$$

The bit in the leftmost position of the result will simply be cut off.

$$3_{10} - 6_{10} = \quad 3_{10} + (-6)_{10} = 0011_2 + 1010_2 = 0\ 1101_2 = -3_{10} \ ,$$
$$-2_{10} - 2_{10} = (-2)_{10} + (-2)_{10} = 1110_2 + 1110_2 = 1\ 1100_2 = -4_{10} \ .$$

However, the possibility of overflow errors exists again when the available range of numbers is too small.

$$-6_{10} - 4_{10} = (-6)_{10} + (-4)_{10} = 1010_2 + 1100_2 = 1\ 0110_2 = 6_{10} \neq -10_{10} \ .$$

Only the numbers up to -8 can be represented, and -10 is outside of this range. In this case the computer hardware should indicate an error. However, at present the use of 64 bits is quite common.

The *division* will be understood as a sequence of subtractions in the same way as in the decimal system. Figure 8.7 shows two examples of divisions of binary numbers; in case **a** an integer result has been found, but in case **b** we get the remainder 2_{10}.

$$\begin{array}{r} 1\,1\,0\,1\,1 : 1\,1 = 1\,0\,0\,1 \\ -\,1\,1 \\ \hline 0\,0 \\ \hline 0\,0\,1 \\ \hline 0\,0\,1\,1 \\ -\,1\,1 \\ \hline 0 \end{array}$$

a

$$\begin{array}{r} 1\,1\,1\,0\,1 : 1\,1 = 1\,0\,0\,1 \text{ remainder } 10 \\ -\,1\,1 \\ \hline 0\,1 \\ \hline 0\,1\,0 \\ \hline 1\,0\,1 \\ -\,0\,1\,1 \\ \hline 1\,0 \end{array}$$

b

Fig. 8.7 Division of binary numbers: (**a**) $27 : 3 = 9$ without remainder, (**b**) $29 : 3 = 9$ remainder 2

Table 8.8 Binary fractions

Fraction	Binary representation	Decimal value
1/2	0.1	0.5
1/4	0.01	0.25
1/8	0.001	0.125
1/16	0.0001	0.0625
1/32	0.00001	0.03125
1/64	0.000001	0.015625
1/128	0.0000001	0.0078125
1/256	0.00000001	0.00390625
1/512	0.000000001	0.001953125

These short examples will be sufficient to understand these operations. Decimal fractions can be considered as well. The positions behind the dot get successively the values $2^{-1}, 2^{-2}, 2^{-3}, \ldots$, and all the problems of the decimal system (approximation of real numbers, rounding errors, and many more) exist also for binary numbers. Table 8.8 shows the first nine binary fractions together with their decimal equivalent.

Any decimal fraction must now be approximated by means of these binary fractions.

$$0.35 \approx 0.25 + 0.0625 + 0.03125 + 0.00390625 + 0.001953125 = 0.349609375$$

$$= 0.01_2 + 0.0001_2 + 0.00001_2 + 0.00000001_2 + 0.000000001_2$$

$$= 0.010110011_2 \,.$$

The index 2 indicates that this representation has the base 2. The arithmetic operations with these binary fractions follow the principles that have been discussed before and that are used for the normal decimal numbers in the same way. The implementations of these operations by means of hard- or software hide these details anyway.

8.3 Coding

Let $A = \{a_1, \ldots, a_n\}$ and $B = \{b_1, \ldots, b_m\}$ be two *alphabets* (finite sets of different symbols or characters). A *word* of length k is an element of A^k (or B^k). In this sense there is no difference between words of length k and vectors with k components; only the application areas will be quite different. For words the outer brackets are mostly omitted.

Special cases are A^1, all words of length 1, and $A^0 = \{\lambda\}$ where λ is the word of length 0, i.e., the *empty word*.

Very often an inductive definition of words will be used for a given alphabet $A = \{a_1, \ldots, a_n\}$:

Table 8.9 The ASCII code

Decimal interval	Hexadecimal interval	Character group
0–31	0-1F	Control characters
32–39	20-27	SPACE, !, ", #, $, %, &, '
40–47	28-2F	(,), *, +, ,, -, ., /
48–57	30-39	0, 1, …, 9
58–64	3A-40	:, ;, <, =, >, ?, @
65–90	41-5A	A, B, …, Y, Z
91–96	5B-60	[, \,], ^, _, `
97–122	61-7A	a, b, …, y, z
123–127	7B-7F	{, |, }, ~, DEL

1. λ is a word.
2. $a_1, \ldots, a_n$ are words.
3. If $u = u_0 \ldots u_k$ and $v = v_0 \ldots v_l$ are words, then also

$$w = u \| v = u_0 \ldots u_k v_0 \ldots v_l \, .$$

The operation sign $\|$ denotes the *concatenation* of words. It is an associative operation, i.e., $(u\|v)\|w = u\|(v\|w)$, and λ is a neutral element, i.e., $\lambda\|u = u\|\lambda = u$. We use $A^* = A^0 \cup A^1 \cup A^2 \cup \ldots$ for the set of all words over a given alphabet.

Example 8.9 For $B = \{0, 1\}$, we get

$$B^* = \{\lambda, 0, 1, 00, 01, 10, 11, 000, 001, 010, 011, 100, 101, 110, 111, \ldots\} \, .$$

B^* is an infinite countable set.

Note 8.2 We consider only finite words (words with a finite length k, and k is a natural number). The consideration of infinite sequences as an extension of these concepts deserves special attention.

The general concept of a *code* can be described by means of the following definition.

Definition 8.1 Let $R \subseteq A^*$, φ a unique mapping from R into B^*, for given alphabets A, B. Then φ is a coding of the set R. The set $C \subseteq B^*$ that contains all the elements $\varphi(r)$ with $r \in R$ is the code of the set R. The elements of C are the *code words*. If $v = \varphi(w)$, $v \in C$, $w \in R$, then v is the *code* of w. If $B = \{0,1\}$, then φ and C are denoted by *binary codes*.

Example 8.10 One of the first codes is the **A**merican **S**tandard **C**ode for **I**nformation **I**nterchange (ASCII). Its main intention was the representation of the characters of the keyboard within the computer, in a binary format. This code uses 7 bits, i.e., the range of numbers between 0 and 127, and the leftmost bit of a byte is free so that it can be used as *parity bit*.

The parity bit will be set in such a way that the overall number of bits in the byte is an odd number. When during the transmission of data one bit changes its value (it will be denoted by *single-bit-error* or a *one-bit-error*), then a *parity check* on the receiver side will find out that the transmitted byte is not correct. Table 8.9 gives an idea of this code.

Table 8.10 Examples of
the Unicode

Unicode	Symbol	Explanation
03E0	ꜱ	Greek letter sampi
0429	Щ	Cyrillic capital letter shcha
04BC	Ҽ	Cyrillic capital letter abkhasian che
05D8	ט	Hebrew letter tet
060F	؏	Arabic sign misra
0F16	༖	Tibetan logotype sing lhag rtags
1254	ቔ	Ethiopic syllable qhee
14D9	ᓙ	Canadian syllabics y-cree loo
189A	ᢚ	Mongolian letter manchu ali gali gha
4F86	來	CJK unified ideograph
A98B	ꦋ	Javanese letter nga lelet raswadi

Any pressing of a key (sometimes of two or even three keys) on the keyboard "translates" the character on the key (that might be also visible on the screen of the computer) into the corresponding byte content. The byte content can be easily determined by means of the hexadecimal values:

$$\text{'A'} = 41_{16} = 100\ 0001\ ,\quad \ldots,\quad \text{'Z'} = 5A_{16} = 101\ 1010\ .$$

When the leftmost bit is used as parity bit for an odd parity, then it has to be set to 1 for "A" as well as for "Z":

$$\text{'A'} = B1_{16} = 1100\ 0001\ ,\quad \ldots,\quad \text{'Z'} = CA_{16} = 1101\ 1010\ .$$

The binary coding of the character "A" (without the parity bit) contains two bits equal to 1, and four bits equal to 1 occur in the binary code of the character "Z."

An expression like $\text{'A'} = 41_{16} = 100\ 0001$ needs some attention. The "A" on the left-hand side means the character "A," an A in a hexadecimal number means the hexadecimal digit A (expressing the decimal number 10), and 100 0001 is the bitwise representation of the code word.

The set of code words can be extended by using the leftmost bit of a byte as a code bit as well. Then a bit number 9 would be required as a parity bit. This increases the number of code words to 256, and this larger code will allow the coding of symbols existing in different languages (Spanish, French, German), such as ä, ö, ü, â, á, ñ, etc. This code is very often denoted by *extended ASCII code*.

The past 25 years showed an enormous development in this area. Nowadays the *Unicode* is an international standard that can be used for all existing symbols, languages, currencies, even smileys, etc. (see Table 8.10). Its development and the recent state can be seen at

http://unicode.org/versions/Unicode12.1.0/.

It comprises now 137,929 characters. Unicode's success at unifying character sets has led to its widespread and predominant use in the internationalization and localization of computer software. The standard has been implemented in many recent technologies, including modern operating systems, XML, Java (and other programming languages), and the .NET Framework.

It is the best to go to

http://www.unicode.org/standard/WhatIsUnicode.html

where the following characteristic of the Unicode is given:

Table 8.11 Examples of the RGB color codes in hexadecimal format

Hexadecimal coding	Name of the color
7F FF 00	Chartreuse
8F BC 8F	Darkseagreen
AD FF 2F	Greenyellow
B2 22 22	Firebrick
B8 86 0B	Darkgoldenrod
BC 8F 8F	Rosybrown
C0 C0 C0	Silver
DD A0 DD	Plum
E9 96 7A	Darksalmon
EE E8 AA	Palegoldenrod
FF 8C 00	Darkorange
FF DA B9	Peachpuff
FF E4 B5	Moccasin
FF F0 F5	Lavenderblush
FF FA FA	Snow
FF FF 00	Yellow

"The Unicode Standard provides a unique number for every character, no matter what platform, device, application or language. It has been adopted by all modern software providers and now allows data to be transported through many different platforms, devices and applications without corruption. Support of Unicode forms the foundation for the representation of languages and symbols in all major operating systems, search engines, browsers, laptops, and smart phones—plus the Internet and World Wide Web (URLs, HTML, XML, CSS, JSON, etc.). Supporting Unicode is the best way to implement ISO/IEC 10646."

But the coding of information does not only relate to characters of written texts. For any image point on a computer screen, the color must also be encoded. This is normally done by means of three bytes. Each byte is assigned one color in the order of *red–green–blue* (RGB). Each byte can represent 256 different values and indicates the degree in which the respective color participates in a mixture that results in the final color according to the underlying physical principles. It can be seen that the description of colored images is rather memory-intensive. Each image point requires three bytes to define the color value of this point. Table 8.11 gives an idea of some coding together with the very nice names that are used for these colors.

The values of Table 8.11 can be understood as follows:

1. The first two digits (the first byte) relate to *red*.
2. The next two digits relate to *green*.
3. The last two digits indicate *blue*.

The value 00 is the smallest value and equals to 0%, and the value FF is the largest value and indicates 100%. Hence, FFFF00 has the meaning to take 100% of the possible *red* scale, 100% of the possible *green* scale, mix the two colors, and get *yellow*. Any other number between 00 and FF can be related to these border values. B5 for instance is equal to the decimal number 181, and this is approximately 71% of 255. In this way,

$$256 * 256 * 256 = 2^{24} = 16\,777\,216$$

different colors can be defined, and human eyes will not be able to see a difference between many of them.

Not necessarily each code word must have the same length. A good example is the old *Morse code*. This is, however, not a binary code. It needs three pieces of information, a *short* element ·, a *long* element −, and a *space* between the two other symbols. The character "e" is represented by a

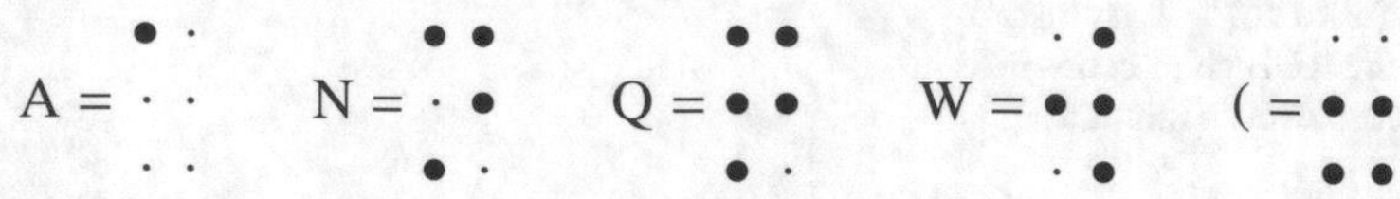

Fig. 8.8 Examples of the Braille code

$$A = \overset{\bullet\ \cdot}{\underset{\cdot\ \cdot}{\cdot\ \cdot}} \qquad N = \overset{\bullet\ \bullet}{\underset{\bullet\ \cdot}{\cdot\ \bullet}} \qquad Q = \overset{\bullet\ \bullet}{\underset{\bullet\ \cdot}{\bullet\ \bullet}} \qquad W = \overset{\cdot\ \bullet}{\underset{\cdot\ \bullet}{\bullet\ \bullet}} \qquad (= \overset{\cdot\ \cdot}{\underset{\bullet\ \bullet}{\bullet\ \bullet}}$$

single ·, the character "9" by $- - - - \cdot$, etc. The length of the code word depends on the frequency of the character in texts of the English language. The oftener the character occurs, the shorter is its code word.

A very famous binary code is the *Braille code* (invented by Louis Braille (1809–1852) for blind people). This code uses six bits in an arrangement of a 3×2-array that results in 64 possibilities. Figure 8.8 shows some examples of the Braille code.

This code is also an example in which code words are not always related to computer applications, and not necessarily the code words have to be written and read in a linear order.

A code is a *block code* when all code words have the same length (see, for instance, ASCII). A *binary block code* C with words of length k is a subset of $\mathbb{B}^k$ that can be described by its characteristic function:

$$\varphi(\mathbf{x}) = \begin{cases} 1 \text{ for } \mathbf{x} \in C \\ 0 \text{ for } \mathbf{x} \notin C \ . \end{cases}$$

Definition 8.2 The distance $d(C)$ of the binary block code C is given by

$$d(C) = \min_{(\mathbf{x},\mathbf{y}) \in C \times C} h(\mathbf{x}, \mathbf{y}) \ .$$

An (n, d) code is a code with binary words of length n and the code distance d.

Note 8.3 The function $h(\mathbf{x}, \mathbf{y})$ is the Hamming metric $h(\mathbf{x}, \mathbf{y}) = ||\mathbf{x} \oplus \mathbf{y}||$. The minimum has to be taken over all possible pairs $(\mathbf{x}, \mathbf{y}) \in C \times C$. Two words of such a code differ in at least $d(C)$ positions, i.e., at least $d(C)$ bits have to change from one code word to another.

Example 8.11 Let $n = 4$, $d = 2$, and let $\mathbf{x} = (0000)$ be a code word, $C_0 = \{(0000)\}$. Then the next code words must have two positions with a value 1; hence, $C_2 = \{(0011), (0101), (0110), (1001), (1010), (1100)\}$. All the code words in C_2 have a distance of 2 or 4 to each other. The only vector that is still available is (1111); hence, we set $C_4 = \{(1111)\}$ and get

$$C = C_0 \cup C_2 \cup C_4 \ .$$

All the code words have an even number of 1's (as indicated by the index), and only 50% of the vectors of $\mathbb{B}^4$ can be used as code words. The complement of this set with regard to $\mathbb{B}^4$, i.e.,

$$\mathbb{B}^4 \setminus C = \{(0001), (0010), (0100), (1000), (0111), (1011), (1101), (1110)\}$$

is also a code with the code distance 2. The interesting property of these codes is the fact that the change of one single bit transforms a code word into a word that is not an element of the code, which can be detected by a simple check.

Using such a code allows the *detection of single-bit-errors*. It could also be seen in the example that for a given n and a given d there is a maximum number $m(n, d)$ of possible code words.

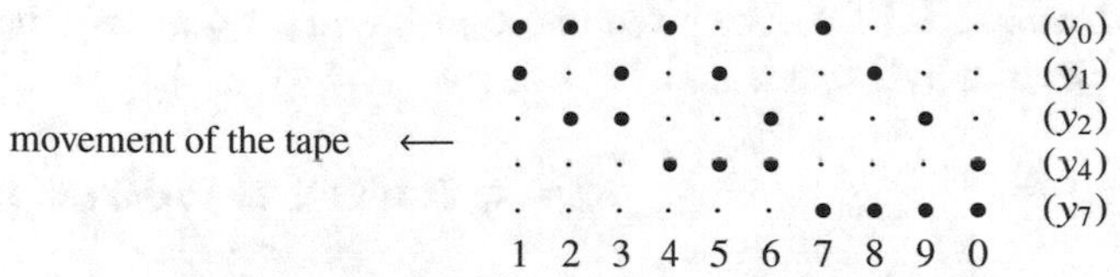

Fig. 8.9 2-out-of-5 code
on a tape

movement of the tape ⟵

(y_0)
(y_1)
(y_2)
(y_4)
(y_7)

1 2 3 4 5 6 7 8 9 0

Two famous codes that have been widely used in older "days" originated from *punched cards* and *punched tapes*. Punched cards are even older than computers. They have been invented by *H. Hollerith* (1860–1929) to accelerate the evaluation of a census in 1890. The code for punched tapes is a 2-out-of-5 code (see Fig. 8.9); each code word has 5 positions (vertically represented), two of them must be set equal to 1, three of them are equal to 0, which means that there are 10 different code words. The code words extend across the paper tape, perpendicular to the movement of the tape.

This code is still "living" as a bar code that is widely used in supermarkets, etc., to indicate items and prices in an electronically readable format. There is a *two-out-of-five bar code* with narrow and wide bars, a *three-out-of-nine bar code*, but this one is no longer a pure binary code: there are always three wide bars, but in addition to narrow bars there are also narrow spaces and wide spaces to be considered. This solution was necessary to increase the number of characters and the possibilities of error checking.

Many other codes of this type are available, but they differ only in the ways of using these bars and spaces mainly for reasons of history, readability, printability, or to be adapted to a special application area (like postal codes, library codes, etc.). The existing electronic equipment will always be able (at least in a "normal" environment) to convert these codes into a pure binary representation.

We still can explore some other important properties of binary codes.

Definition 8.3 A block code C is an *equidistant code* when the Hamming distance between two different code words is always the same: $h(\mathbf{c}_1, \mathbf{c}_2) = const.$ for any $\mathbf{c}_1, \mathbf{c}_2 \in C$.

We already have seen that all the code words of a given code can have the same weight k, $0 \le k \le n$. Different values for k result in different numbers of available code words. $k = 0$ and $k = n$ define only one single code word, $k = 1$ and $k = n - 1$ allow n different code words (one single 1 or one single 0 on n different positions), and $k \approx \frac{n}{2}$ allows the largest number of code words in such a code. The maximum number of code words in an (n, d) code with $||\mathbf{c}|| = k$ is represented by $m(n, k, d)$.

Hence, it can be seen that the set $\mathbb{B}^n$ together with many metric properties is a fundamental structure for the construction of binary codes with desirable properties.

The concepts of *error-detection* and *error-correction* have already been mentioned. These considerations have to be based on an *error-model*. Very broadly accepted are the following assumptions:

- The transmission of the code words does not change the length of the code word. This is very real when the bits are transmitted in parallel.
- Some bits within a code word can change their values from 0 to 1 or from 1 to 0; the number of bit changes in a code word has a value between 0 and n, 0 is the error-free case, and n negates the whole code word.

If $\mathbf{x}$ is the vector that has been sent and $\mathbf{y}$ is the vector that has been received, then $||\mathbf{x} \oplus \mathbf{y}||$ is the number of errors that occurred during the transmission. The possibility of a *parity check* has been mentioned before.

Example 8.12 Let be given $(x_0, x_1, x_2, x_3, x_4, x_5, x_6, x_7)$; then a parity bit can be added to these eight bits of a byte by means of

$$x_8 = x_0 \oplus x_1 \oplus x_2 \oplus x_3 \oplus x_4 \oplus x_5 \oplus x_6 \oplus x_7 \, .$$

$x_8 = 1$ for an odd number of 1's in the byte, and $x_8 = 0$ for an even number of 1's in the byte.

In each case the parity bit is set in such a way that the extended byte has an even number of 1's (*even parity*). If an odd parity is required, then

$$x_8 = x_0 \oplus x_1 \oplus x_2 \oplus x_3 \oplus x_4 \oplus x_5 \oplus x_6 \oplus x_7 \oplus 1$$

has to be used. In any case, the error in one single bit changes the parity, and on the receiver side a parity error will be indicated after the check of the parity. However, the information is not sufficient to find out the position that has been changed. Any single-bit-error changes the parity. Even worse: two-bit errors will not be detected because the parity will not change.

But the general principle of error-detection can already be seen—*define a property of the code word that can change by transmission errors and check the property on the receiver side*. The price for the detection of these errors is "affordable," and the technical resources have to be increased by 12.5% (one more bit, error-detection circuits). The 2-out-of-5 code offers similar possibilities. Error-correction follows similar principles. Generally, we can assume that a "small" environment can be defined for the code words—small changes of the code word (possibly the change of one or two bits) leave the received word in this environment, and the original code word can be re-established on the receiver side. It is not difficult to understand the following.

Theorem 8.4 *If d is the minimum distance between code words of a code C, then C can correct any $(d-1)/2$ or fewer errors. In order to correct t or fewer errors, a minimum distance of $d = 2t + 1$ is sufficient.*

For an even number d, the value of $(d-1)/2$ must be rounded down. Codes like the *Bose–Chaudhuri–Hocquenghem code*, however, require means of Higher Algebra (polynomials over Galois fields) and are mostly the topic of special courses. Another possibility, however, the so-called *Hamming* codes, can be based on the concepts of $\mathbb{B}^n$. In order to do this, $(\mathbb{B}^n, \oplus)$ will be defined as a *linear vector space*.

As an illustration we will use the following array:

$$H = \begin{pmatrix} 1\,0\,0\,0\,0\,1\,1 \\ 0\,1\,0\,0\,1\,0\,1 \\ 0\,0\,1\,0\,1\,1\,0 \\ 0\,0\,0\,1\,1\,1\,1 \end{pmatrix} . \tag{8.1}$$

As a first operation, we need the combination of a binary constant with a vector. For a given value $\lambda \in \mathbb{B}$ and a vector $\mathbf{x} = (x_1, \ldots, x_n) \in \mathbb{B}^n$, the product $\lambda \mathbf{x}$ is defined as $(\lambda x_1, \ldots, \lambda x_n)$. As in many other areas, the "multiplication sign" (here $\wedge$) has been omitted. This operation is rather trivial. Since $\lambda \in \mathbb{B}$, only two values have to be considered: $\lambda = 0$ results in the vector $\mathbf{0}$, and $\lambda = 1$ results in the vector $\mathbf{x}$.

Definition 8.4 For given binary vectors $\mathbf{v}_1, \mathbf{v}_2, \ldots, \mathbf{v}_k, \lambda_i \in \mathbb{B}, i = 1, \ldots, k$, the expression

$$\lambda_1 \mathbf{v}_1 \oplus \cdots \oplus \lambda_k \mathbf{v}_k$$

is a *linear combination* of the vectors $\mathbf{v}_1, \mathbf{v}_2, \ldots, \mathbf{v}_k$.

For any vector $(\lambda_1, \ldots, \lambda_k) \in \mathbb{B}^k$, some of the given vectors $\mathbf{v}_i$ are selected ($\lambda_i = 1$) and combined by the linear operation $\oplus$. The border case is $(\lambda_1, \ldots, \lambda_k) = \mathbf{0}$, and then the result is the vector $\mathbf{0}$; the combination of some of the given vectors by $\oplus$ is again a vector $\in \mathbb{B}^n$. When we are using

$$\mathbf{v}_1 = (1000011), \qquad \mathbf{v}_2 = (0100101), \qquad \mathbf{v}_3 = (0010110), \qquad \mathbf{v}_4 = (0001111),$$

then we can see, for instance, that

$$\mathbf{v}_1 \oplus \mathbf{v}_3 = (1010101),$$

$$\mathbf{v}_1 \oplus \mathbf{v}_3 \oplus \mathbf{v}_4 = (1011010).$$

In the first case, there is $\lambda_1 = 1, \lambda_2 = 0, \lambda_3 = 1$, and $\lambda_4 = 0$, and in the second $\lambda_1 = 1, \lambda_2 = 0, \lambda_3 = 1, \lambda_4 = 1$, etc.

The two operations $\wedge$ and $\oplus$ can now be used to define the *scalar product* of two vectors; in the same way as in *linear Algebra*, one operation is used as the multiplication, and the second operation as the addition.

Definition 8.5 For two vectors $\mathbf{x} = (x_1, \ldots, x_n)$ and $\mathbf{y} = (y_1, \ldots, y_n)$, the value

$$c = (x_1 \wedge y_1) \oplus \ldots \oplus (x_n \wedge y_n) = x_1 y_1 \oplus \ldots \oplus x_n y_n$$

is the *scalar product* of the vectors $\mathbf{x}$ and $\mathbf{y}$.

Using this scalar product allows the definition of the product of binary matrices. This is done in the well-known way: calculate the scalar product of the rows of the first matrix with the columns of the second matrix. As a very special example, we multiply one vector with a matrix.

$$(0011) \begin{pmatrix} 1 & 0 & 0 & 0 & 0 & 1 & 1 \\ 0 & 1 & 0 & 0 & 1 & 0 & 1 \\ 0 & 0 & 1 & 0 & 1 & 1 & 0 \\ 0 & 0 & 0 & 1 & 1 & 1 & 1 \end{pmatrix} = (0011001).$$

The vector (0011) will be multiplied successively with the first, the second, , the last column of the matrix H (8.1). The matrix H has been built in such a way that the first four components of the resulting vector repeat the given vector, and the last three components combine always three-out-of-four components of the original vector. This will allow the correction of some errors.

Definition 8.6 The vectors $\mathbf{v}_1, \mathbf{v}_2, \ldots, \mathbf{v}_k$ are *linearly independent* if each linear combination with a vector $(\lambda_1, \ldots, \lambda_k) \neq \mathbf{0}$ is not equal to $\mathbf{0}$; otherwise, they are *linearly dependent*.

Table 8.12 Linear
combinations of the
vectors $\mathbf{v}_i$ of the Hamming
matrix h

Linear combination	Result
$\mathbf{v}_1 \oplus \mathbf{v}_1$	(0000000)
$\mathbf{v}_1$	(1000011)
$\mathbf{v}_2$	(0100101)
$\mathbf{v}_3$	(0010110)
$\mathbf{v}_4$	(0001111)
$\mathbf{v}_1 \oplus \mathbf{v}_2$	(1100110)
$\mathbf{v}_1 \oplus \mathbf{v}_3$	(1010101)
$\mathbf{v}_1 \oplus \mathbf{v}_4$	(1001100)
$\mathbf{v}_2 \oplus \mathbf{v}_3$	(0110011)
$\mathbf{v}_2 \oplus \mathbf{v}_4$	(0101010)
$\mathbf{v}_3 \oplus \mathbf{v}_4$	(0011001)
$\mathbf{v}_1 \oplus \mathbf{v}_2 \oplus \mathbf{v}_3$	(1110000)
$\mathbf{v}_1 \oplus \mathbf{v}_2 \oplus \mathbf{v}_4$	(1101001)
$\mathbf{v}_1 \oplus \mathbf{v}_3 \oplus \mathbf{v}_4$	(1011010)
$\mathbf{v}_2 \oplus \mathbf{v}_3 \oplus \mathbf{v}_4$	(0111100)
$\mathbf{v}_1 \oplus \mathbf{v}_2 \oplus \mathbf{v}_3 \oplus \mathbf{v}_4$	(1111111)

Example 8.13 The three vectors (0110), (1011), (1101) are linearly dependent because (0110) $\oplus$ (1011) $\oplus$ (1101) $=$ (0000).

The vectors $\mathbf{v}_1, \mathbf{v}_2, \mathbf{v}_3, \mathbf{v}_4$ corresponding to the rows of the matrix H are linearly independent: all possible combinations of two, three, or four vectors by $\oplus$ are not equal to $\mathbf{0}$.

Definition 8.7 Let be given a set $L \subseteq \mathbb{B}^n$ so that with every two vectors $\mathbf{v}_1, \mathbf{v}_2 \in L$, there is also $\mathbf{v}_1 \oplus \mathbf{v}_2 \in L$. If the maximum number of linearly independent vectors in L is equal to k, then L has the dimension k.

Note 8.4 Normally in linear Algebra it is required that with $\mathbf{v}_1, \mathbf{v}_2 \in L$ also $\lambda_1 \mathbf{v}_1 \oplus \lambda_2 \mathbf{v}_2 \in L$; however, since we have here the additional assumption that $\lambda_1, \lambda_2 \in \mathbb{B}$, the general requirement is equal to the requirement above.

When $\mathbf{v}_1 = \mathbf{v}_2 = \mathbf{v}$ is used, then it can be seen that $\mathbf{0}$ must be an element of every linear vector space because of $\mathbf{v} \oplus \mathbf{v} = \mathbf{0}$.

Very often the other way can be taken: it is known that there is a set K of k linearly independent vectors. Then the set of all possible linear combinations of elements of K will be added to K. The resulting set is a *linear vector space*.

Example 8.14 We use the four vectors $\mathbf{v}_i$ of the Hamming matrix and calculate all linear combinations. Table 8.12 shows the resulting set of 16 vectors; these vectors establish a linear vector space of dimension k.

An *additive group* is another algebraic structure that can be considered.

Definition 8.8 The set $G \subseteq \mathbb{B}^n$ is an additive group when G is closed under $\oplus$.

This means that for any two vectors $\mathbf{v}_1, \mathbf{v}_2 \in G$ also $\mathbf{v}_1 \oplus \mathbf{v}_2 \in G$. Especially, $\mathbf{0}$ must be $\in G$ (because of $\mathbf{v}_1 \oplus \mathbf{v}_1 = \mathbf{0}$). In our case the two definitions result in the same set of elements: each *group* is a *linear vector space*.

Definition 8.9 If the code $C \subseteq \mathbb{B}^n$ is a group, then it is a *linear code* (a group code). When the linear code $C \in \mathbb{B}^n$ has the dimension k, then it is an (n, k) code. A binary linear code correcting one error is a Hamming code.

The set of 16 code words represented in Table 8.12 is a $(7, 4)$ code. The concept of *error-correction* has already been mentioned. This can be based on the following matrix where the *Hamming distance* of the different code words has been represented. The different code words are indicated by the numbers 1 to 16, in the order as they appear in the table. The entry d_{ij} in this matrix indicates the Hamming distance between code word i and code word j.

$$D = \begin{pmatrix}
0 & 3 & 3 & 3 & 4 & 4 & 4 & 3 & 4 & 3 & 3 & 3 & 4 & 4 & 4 & 7 \\
  & 0 & 4 & 4 & 3 & 3 & 3 & 4 & 3 & 4 & 4 & 4 & 3 & 3 & 7 & 4 \\
  &   & 0 & 4 & 3 & 3 & 3 & 4 & 3 & 4 & 4 & 4 & 3 & 7 & 3 & 4 \\
  &   &   & 0 & 3 & 3 & 3 & 4 & 3 & 4 & 4 & 4 & 7 & 3 & 3 & 4 \\
  &   &   &   & 0 & 4 & 4 & 3 & 4 & 3 & 3 & 7 & 4 & 4 & 4 & 3 \\
  &   &   &   &   & 0 & 4 & 3 & 4 & 3 & 7 & 3 & 4 & 3 & 4 & 3 \\
  &   &   &   &   &   & 0 & 3 & 4 & 7 & 3 & 3 & 3 & 4 & 4 & 3 \\
  &   &   &   &   &   &   & 0 & 7 & 4 & 4 & 4 & 3 & 3 & 3 & 4 \\
  &   &   &   &   &   &   &   & 0 & 4 & 3 & 3 & 4 & 4 & 4 & 3 \\
  &   &   &   &   &   &   &   &   & 0 & 4 & 4 & 3 & 3 & 3 & 4 \\
  &   &   &   &   &   &   &   &   &   & 0 & 4 & 3 & 3 & 3 & 4 \\
  &   &   &   &   &   &   &   &   &   &   & 0 & 3 & 3 & 3 & 4 \\
  &   &   &   &   &   &   &   &   &   &   &   & 0 & 4 & 4 & 3 \\
  &   &   &   &   &   &   &   &   &   &   &   &   & 0 & 4 & 3 \\
  &   &   &   &   &   &   &   &   &   &   &   &   &   & 0 & 3 \\
  &   &   &   &   &   &   &   &   &   &   &   &   &   &   & 0
\end{pmatrix}.$$

It can be seen that the minimum distance between two code words is ≥ 3. When one bit changes its value during the transmission of the code word, the received word is no longer a code word: in order to be transformed into another code word, three bits must change its value. The word on the receiver side will be compared with all code words, for one of these words the distance between the received word and the code word will be equal to 1, and the received word will be replaced by the code word—*the error has been corrected*.

And this approach can be generalized: for the correction of two errors, for instance, a code has to be used where the minimum distance between code words is equal to 5, two changed bits still allow to search for a code word with the distance $d = 2$, all the other code words have still at least a distance of 3 to the transmitted word, etc. A number of i errors need a code with the minimum distance of $2i + 1$.

The matrix H (8.1) has a special property: it has been built in such a way that the first four bits of a code word repeat the original four bits of the word before the encoding. Hence, the decoding is quite simple: after the transmission, find out whether the word has to be corrected. If so, then take the first four bits of the corrected word, and if not, then take the first four bits right away.

Sometimes the code given by the matrix H (8.1) is also denoted by $(7, 16, 3)$ code. Generally, a (n, M, d) code has:

Table 8.13 Recursive calculation of Reed–Muller codes

$\mathcal{RM}(0,0) = \{0, 1\}$
$\mathcal{RM}(0,1) = \{00, 11\}$
$\mathcal{RM}(1,1) = \{00, 01, 10, 11\}$
$\mathcal{RM}(0,2) = \{0000, 1111\}$
$\mathcal{RM}(1,2) = \mathcal{RM}(1,1) * \mathcal{RM}(0,1)$
$\quad = \{00, 01, 10, 11\} * \{00, 11\}$
$\quad = \{00\|00 \oplus 00, 00\|00 \oplus 11, 01\|01 \oplus 00, 01\|01 \oplus 11,$
$\qquad 10\|10 \oplus 00, 10\|10 \oplus 11, 11\|11 \oplus 00, 11\|11 \oplus 11\}$
$\quad = \{0000, 0011, 0101, 0110, 1010, 1001, 1111, 1100\}$
$\mathcal{RM}(2,2) = \{0000, 0001, 0010, 0011, 0100, 0101, 0110, 0111,$
$\qquad 1000, 1001, 1010, 1011, 1100, 1101, 1110, 1111\}$

- Code words of length n
- M code words
- A minimum distance of d

Sometimes it will be desirable to build a new code by means of the existing codes. This can be done by means of the *Plotkin construction*. For two linear binary codes C_1 and C_2 of the same length, we consider the *concatenation* of these two codes in the following sense:

$$C_1 * C_2 = \{\mathbf{u}\|\mathbf{u} \oplus \mathbf{v} \mid \mathbf{u} \in C_1, \mathbf{v} \in C_2\}.$$

The sign $\|$ indicates the concatenation of two code words; hence, every code word $\mathbf{u} \in C_1$ will be concatenated with the "sum" of this code word and any other code word $\mathbf{v} \in C_2$, i.e., with $\mathbf{u} \oplus \mathbf{v}$.

Theorem 8.5 *If C_1 is a linear (n, k_1, d_1) code, C_2 a linear (n, k_2, d_2) code, then $C_1 * C_2$ is a linear $(n, k_1 + k_2, \min(2d_1, d_2))$ code.*

A very interesting code family can be summarized as *Reed–Muller codes* $\mathcal{RM}(r, n)$. These codes have code words with the length 2^n, an order r, $0 \le r \le n$, and can be built using the following recursive definition:

1. $\mathcal{RM}(0, n) = \{\mathbf{0}, \mathbf{1}\}$.
2. $\mathcal{RM}(n, n) = \mathbb{B}^{2^n}$.
3. $\mathcal{RM}(r, n) = \mathcal{RM}(r, n - 1) * \mathcal{RM}(r - 1, n - 1)$.

These codes have the remarkable property that:

- $\mathcal{RM}(r, n)$ has $2^{f(r,n)}$ code words; the value of this function can be calculated as follows:

$$f(r, n) = \sum_{i=0}^{r} \binom{n}{i}.$$

- The minimum distance of this code is 2^{n-r}.

This recursive definition needs the calculation of many codes before a reasonable size of the code can be reached. Table 8.13 shows how Reed–Muller codes can be calculated starting with $r = 0$ and $n = 0$.

In order to find an easier way of constructing such a code, another approach can be used that can be considered as one of the first applications of Boolean differential equations. Since the code words have a length 2^n, each code word can be considered as the vector of a logic function depending on n variables. Reed–Muller codes $\mathcal{RM}(n, n)$ are all 2^{2^n} functions of n variables, which are equal to the set $\mathbb{B}^{2^n}$. Reed–Muller codes $\mathcal{RM}(r, n)$ with $r < n$ are subsets of all functions of n variables. Such subsets are the solutions of Boolean differential equations.

The Reed–Muller code $\mathcal{RM}(1, 2)$ is the solution-set of the Boolean differential equation:

$$\operatorname*{der}_{(x_1, x_2)}{}^{2} f(x_1, x_2) = 0 . \tag{8.2}$$

This solution-set contains eight function vectors belonging to five classes. The comparison with Table 8.13 confirms the following solution-classes:

- The vector 0000 describes the single function $f(x_1, x_2) = 0(x_1, x_2)$.
- The vectors {0101, 1010} describe a class of the two functions:

$$f(x_1, x_2) = x_1 \quad \text{and} \quad f(x_1, x_2) = \overline{x}_1 .$$

- The vectors {0011, 1100} describe a class of the two functions:

$$f(x_1, x_2) = x_2 \quad \text{and} \quad f(x_1, x_2) = \overline{x}_2 .$$

- The vectors {0110, 1001} describe a class of the two functions:

$$f(x_1, x_2) = x_1 \oplus x_2 \quad \text{and} \quad f(x_1, x_2) = \overline{x}_1 \oplus x_2 .$$

- The vector 1111 describes the single function $f(x_1, x_2) = 1(x_1, x_2)$.

Generalized, the Reed–Muller code $\mathcal{RM}(1, 3)$ is the solution-set of the Boolean differential equation of function $f(\mathbf{x}) = f(x_1, x_2, x_3)$:

$$\operatorname*{der}_{(x_1, x_2)}{}^{2} f(\mathbf{x}) \vee \operatorname*{der}_{(x_1, x_3)}{}^{2} f(\mathbf{x}) \vee \operatorname*{der}_{(x_2, x_3)}{}^{2} f(\mathbf{x}) = 0 . \tag{8.3}$$

The solution-set of this Boolean differential equation contains 16 function vectors of the length 8:

$$
\begin{aligned}
S(\mathbf{v}) = \mathcal{RM}(1, 3) = \{\ & 0000\,0000, \quad 1111\,1111, \\
& 0000\,1111, \quad 1111\,0000, \\
& 0011\,0011, \quad 1100\,1100, \\
& 0101\,0101, \quad 1010\,1010, \\
& 0011\,1100, \quad 1100\,0011, \\
& 0101\,1010, \quad 1010\,0101, \\
& 0110\,0110, \quad 1001\,1001, \\
& 0110\,1001, \quad 1001\,0110\ \} .
\end{aligned}
$$

Here we have two classes of one function and 7 classes of two functions.

Hence, it can be seen that the understanding of the code words as vectors of logic functions and the specification of an appropriate Boolean differential equation opens a direct approach to generate a wanted Reed–Muller code.

Now we consider a special set of functions using the property that each function of n variables can be represented in a unique way by its antivalence polynomial. Let $F(r, n)$ be the set of all functions of n variables, the antivalence polynomial of which uses only conjunctions with at most r variables.

Example 8.15 The function $f(\mathbf{x}) = 1 \oplus x_1 x_2 \oplus x_3 x_4$ is an element of $F(2, 4)$. It depends on four variables, and the conjunctions contain zero or two variables.

It is easy to see that we can find a recursive definition of such a set that is very similar to the recursive definition of the Reed–Muller codes. Here we use the representation of the function $f(\mathbf{x}) = f(x_1, \ldots, x_n) = f(x_i, \mathbf{x}_1)$ with the single derivative:

$$f(x_1, \ldots, x_n) = x_i \operatorname*{der}_{x_i} f(x_i, \mathbf{x}_1) \oplus f(x_i = 0, \mathbf{x}_1) \,,$$

known as positive Davio expansion. If $f(x_i, \mathbf{x}_1)$ is an element of $F(r, n)$, then $\operatorname{der}_{x_i} f(x_i, \mathbf{x}_1)$ must be in $F(r - 1, n - 1)$ because the conjunction with x_i adds one variable to all the conjunctions of $\operatorname{der}_{x_i} f(x_i, \mathbf{x}_1)$. For $f(x_i = 0, \mathbf{x}_1)$ the length of the conjunctions can still be equal to r; hence, $f(x_i = 0, \mathbf{x}_1) \in F(r, n - 1)$. The vector of the function $f(x_i = 0, \mathbf{x}_1) \in F(r, n)$ can be considered as the concatenation of a vector of $F(r - 1, n - 1)$ and a vector of $F(r, n - 1)$, and this is exactly the recursive schema that has been used for the definition of the Reed–Muller codes:

$$F(r, n) = F(r - 1, n - 1) * F(r, n - 1) \,.$$

The generator matrix of such a code does not need any recursion steps. The definition of $F(r, n)$ can be used directly.

Example 8.16 We want to find $F(2, 4)$. The conjunctions to be used consist of at most two variables (out of 4). Hence, we get a basic set of 11 vectors; since $2^4 = 16$, the length of the code words is equal to 16:

$$
\begin{aligned}
1 &: 1\,1\,1\,1\,1\,1\,1\,1\,1\,1\,1\,1\,1\,1\,1\,1 \,, \\
x_1 &: 0\,1\,0\,1\,0\,1\,0\,1\,0\,1\,0\,1\,0\,1\,0\,1 \,, \\
x_2 &: 0\,0\,1\,1\,0\,0\,1\,1\,0\,0\,1\,1\,0\,0\,1\,1 \,, \\
x_3 &: 0\,0\,0\,0\,1\,1\,1\,1\,0\,0\,0\,0\,1\,1\,1\,1 \,, \\
x_4 &: 0\,0\,0\,0\,0\,0\,0\,0\,1\,1\,1\,1\,1\,1\,1\,1 \,, \\
x_1 x_2 &: 0\,0\,0\,1\,0\,0\,0\,1\,0\,0\,0\,1\,0\,0\,0\,1 \,, \\
x_1 x_3 &: 0\,0\,0\,0\,0\,1\,0\,1\,0\,0\,0\,0\,0\,1\,0\,1 \,, \\
x_1 x_4 &: 0\,0\,0\,0\,0\,0\,0\,0\,0\,1\,0\,1\,0\,1\,0\,1 \,, \\
x_2 x_3 &: 0\,0\,0\,0\,0\,0\,1\,1\,0\,0\,0\,0\,0\,0\,1\,1 \,, \\
x_2 x_4 &: 0\,0\,0\,0\,0\,0\,0\,0\,0\,0\,1\,1\,0\,0\,1\,1 \,, \\
x_3 x_4 &: 0\,0\,0\,0\,0\,0\,0\,0\,0\,0\,0\,0\,1\,1\,1\,1 \,.
\end{aligned}
$$

The linear combinations of these 11 vectors are the code words of $F(2, 4)$. The minimum distance is equal to $2^{4-2} = 2^2 = 4$.

8.4 Specific Normal Form

Normal forms express logic functions in a unique manner.

The disjunctive normal form (DNF) and the antivalence normal form (ANF) use all elementary conjunctions where the function is equal to 1 to get a unique formula of the given function. Due to the orthogonality of all pairs of different elementary conjunctions, both the operation $\vee$ in the DNF and $\oplus$ in the ANF can be used to connect the elementary conjunctions in such a normal form.

Similarly, the conjunctive normal form (CNF) and the equivalence normal form (ENF) use all elementary disjunctions to get a unique formula of the given function. In this case the complements of these disjunctions indicate arguments where the function is equal to 0. Again, due to the orthogonality of all pairs of different elementary disjunctions, both the operation $\wedge$ in the CNF and $\odot$ in the ENF can be used to connect the elementary disjunctions in such a normal form.

The ANF or the ENF can be used to generate fixed polarity polynomials as further normal forms. Each variable occurs in these normal forms either negated or non-negated.

A common property of all these well-known normal forms is that 2^n coefficients of a basic formula determine the normal form of the chosen type for a given logic function. A new normal form that does not satisfy this property has been suggested for the first time in [5]. Due to the special properties of this new normal form [2,4,9], the name *Special Normal Form* (SNF) or *Specific Normal Form* (SNF) has been assigned to this unique representation of a logic function.

Here we explain how the SNF of any logic function can be generated. The definition of the SNF is based on this generation procedure. In the next two sections we utilize the SNF to find some properties of certain logic functions. The procedure to generate the SNF(f) uses an arbitrary *Exclusive-OR Sum of Products* (ESOP) of the function $f(\mathbf{x})$ as basic representation.

An algebraic property of the operation $\oplus$ and the Boolean variable x is visible in the following formulas:

$$x = \overline{x} \oplus 1 , \tag{8.4}$$

$$\overline{x} = 1 \oplus x , \tag{8.5}$$

$$1 = x \oplus \overline{x} . \tag{8.6}$$

These three formulas show that each element of the set $\{x, \overline{x}, 1\}$ has isomorphic properties. For each variable in the support of the logic function $f(\mathbf{x})$, exactly one left-hand side element of (8.4), (8.5), or (8.6) is included in each conjunction of an ESOP of the function $f(\mathbf{x})$. An application of these formulas from the left to the right doubles the number of conjunctions and is denoted by *expansion*. The reverse application of these formulas from the right to the left halves the number of conjunctions and is denoted by *compaction*.

A second important property of the operation $\oplus$ for a logic function $f(\mathbf{x})$ and a conjunction $C(\mathbf{x})$ is shown in the following formulas:

$$f(\mathbf{x}) = f(\mathbf{x}) \oplus 0(\mathbf{x}) , \tag{8.7}$$

$$0(\mathbf{x}) = C(\mathbf{x}) \oplus C(\mathbf{x}) , \tag{8.8}$$

$$f(\mathbf{x}) = f(\mathbf{x}) \oplus C(\mathbf{x}) \oplus C(\mathbf{x}) . \tag{8.9}$$

From these formulas follows that two identical conjunctions can be added to or removed from any ESOP without changing the represented function. The SNF can be defined using two simple algorithms based on the properties mentioned above.

Algorithm 5 ($\mathrm{Exp}(f)$) expands each of the given k conjunctions of n variables to 2^n conjunctions. In this way, we get a distribution of the information of the given conjunctions similar to the creation of a hologram of an object. The function $\texttt{expand()}$ in line 3 of this algorithm expands the conjunction C_j with regard to the variable x_i into the conjunctions C_{n1} and C_{n2} based on the fitting formula (8.4), (8.5), or (8.6).

Algorithm 5 $\mathrm{Exp}(f)$

Input : any ESOP of a logic function $f(\mathbf{x})$
Output : complete expansion of the logic function $f(\mathbf{x})$ with regard to all variables of its support

---

1: **for all** variables V_i of the support of $f(\mathbf{x})$ **do**
2:   **for all** conjunctions $C_j(\mathbf{x})$ of $f(\mathbf{x})$ **do**
3:     $\langle C_{n1}(\mathbf{x}), C_{n2}(\mathbf{x}) \rangle \leftarrow \texttt{expand}(C_j(\mathbf{x}), x_i)$
4:     replace $C_j(\mathbf{x})$ by $\langle C_{n1}(\mathbf{x}), C_{n2}(\mathbf{x}) \rangle$
5:   **end for**
6: **end for**

Algorithm 6 ($\mathrm{R}(f)$) reduces the given ESOP using (8.9) from the right to the left such that no conjunction occurs more than once in the final result.

Algorithm 6 $\mathrm{R}(f)$

Input : any ESOP of a logic function $f(\mathbf{x})$ containing n conjunctions $C_j(\mathbf{x})$
Output : reduced ESOP of $f(\mathbf{x})$ containing no conjunction more than once

---

1: **for** $i \leftarrow 0$ to $n-2$ **do**
2:   **for** $j \leftarrow i+1$ to $n-1$ **do**
3:     **if** $C_i(\mathbf{x}) = C_j(\mathbf{x})$ **then**
4:       $C_i(\mathbf{x}) \leftarrow C_{n-1}(\mathbf{x})$
5:       $C_j(\mathbf{x}) \leftarrow C_{n-2}(\mathbf{x})$
6:       $n \leftarrow n-2$
7:       $j \leftarrow i$
8:     **end if**
9:   **end for**
10: **end for**

Using the algorithms $\mathrm{Exp}(f)$ and $\mathrm{R}(f)$, it is possible to create a special ESOP having a number of remarkable properties that are specified and proven in [5].

Definition 8.10 ($\mathrm{SNF}(f)$) Take any ESOP of a logic function $f(\mathbf{x})$. The resulting ESOP of

$$\mathrm{SNF}(f) = \mathrm{R}(\mathrm{Exp}(f)) \tag{8.10}$$

is denoted by *Specific Normal Form* (SNF) of the logic function $f(\mathbf{x})$.

There are 2^{3^n} different ESOPs of n variables that contain no conjunction more than once, but only 2^{2^n} different functions. It has been proven in [5] that there are $2^{3^n}/2^{2^n}$ different ESOPs for each function of n variables containing each conjunction at most once. It was a surprising observation that one of these ESOPs of each function is a unique representation of the function when the procedure of (8.10) has been applied. The proof that the SNF(f) created by (8.10) is a unique representation for each given function $f(\mathbf{x})$ is given in [5]; hence, the SNF(f) is really a normal form and the term *Specific Normal Form* can be used.

8.5   Most Complex Logic Functions

A comprehensive exploration of the complexity of logic functions based on the SNF(f) and the associated extended adjacency graph has been done in [3]. It has been found that the number of conjunctions of the SNF(f) is an appropriate measure of the complexity of logic functions.

Tables 8.14 and 8.15 show the result of the evaluation of all logic functions $f(\mathbf{x})$ of one to four variables with regard to the number of ternary vectors (NTV) needed to represent both the SNF(f) (column on the left-hand side) and the exact minimal ESOP (middle column); the column on the right-hand side shows the number of logic functions having these properties. In general, it can be seen that the larger the NTV of a minimal ESOP the larger the NTV of the associated SFN. In few cases neighbored numbers of ternary vectors of minimal ESOPs belong to the same NTV of the associated SFN, but the largest NTV of a minimal ESOP occurs for all evaluated functions to such SNFs that need the largest NTV for their representation.

From these complete enumerations additional information about the most complex functions can be discovered:

1. The maximal number of ternary vectors of the most complex SNF(f) of n variables is

$$\mathrm{NTV}_{max}^{\mathrm{SNF}}(n) = 2 * 3^{n-1} .$$

2. The maximal number of the most complex logic functions of n variables is

$$\mathrm{NTV}_{max}^{\mathrm{LF}}(n) = 3 * 2^{n-1} .$$

These two formulas generalize the information of the last rows in Tables 8.14 and 8.15.

As a result of a detailed evaluation of all most complex functions, Algorithm 7 could be developed [3] that can be used to calculate the vector of function values of each most complex logic function in a recursive manner.

The required base cases are given by the most complex functions of one variable. The last row in the left-hand table of Table 8.14 shows that there are three most complex functions of one variable:

$$\mathrm{mcf}_0(1) = 1_{16} = (01) , \qquad f_{\mathrm{mcf}_0}(x_1) = x_1 ,$$

$$\mathrm{mcf}_1(1) = 2_{16} = (10) , \qquad f_{\mathrm{mcf}_1}(x_1) = \overline{x}_1 ,$$

$$\mathrm{mcf}_2(1) = 3_{16} = (11) , \qquad f_{\mathrm{mcf}_2}(x_1) = 1(x_1) .$$

These base cases are implemented in line 4 of Algorithm 7.

Table 8.14 Complexity classes of all exact minimal ESOPs of $\mathbb{B}^1$ and $\mathbb{B}^2$

All $2^{2^1} = 4$ functions of $\mathbb{B}^1$		
SNF	Min ESOP	Logic functions
0	0	1
2	1	3

All $2^{2^2} = 16$ functions of $\mathbb{B}^2$		
SNF	Min ESOP	Logic functions
0	0	1
4	1	9
6	2	6

Table 8.15 Complexity classes of all exact minimal ESOPs of $\mathbb{B}^3$ and $\mathbb{B}^4$

All $2^{2^3} = 256$ functions of $\mathbb{B}^3$		
SNF	Min ESOP	Logic functions
0	0	1
8	1	27
12	2	54
14	2	108
16	3	54
18	3	12

All $2^{2^4} = 65,536$ functions of $\mathbb{B}^4$		
SNF	Min ESOP	Logic functions
0	0	1
16	1	81
24	2	324
28	2	1296
30	2	648
32	3	648
34	3	3888
36	3	6624
36	4	108
38	3	7776
40	3	2592
40	4	6642
42	3	216
42	4	14,256
44	4	12,636
46	4	3888
46	5	1296
48	5	1944
50	5	648
54	6	24

Algorithm 7 Most complex functions: MCF(n, s)

Input : n: number of variables in the function, $n \geq 1$
Input : s: selection index of the most complex function, $0 \leq s < 3 * 2^{n-1}$
Output : function vector mcf_s of the most complex function, defined by n and s

---

1: **if** n = 1 **then**
2:      $mcf_s \leftarrow s + 1$
3: **else**
4:      $remainder \leftarrow s \bmod 3$
5:      $selection \leftarrow s/3$
6:      $direction \leftarrow selection \bmod 2$
7:      $base \leftarrow (selection/2) * 3$
8:      $len2 \leftarrow 2^{n-1}$
9:      **if** $direction = 0$ **then**
10:        $mcf_s \leftarrow$ [MCF$(n - 1, base + ((remainder + 1) \bmod 3))] * 2^{len2}$
                       $+ MCF(n - 1, base + ((remainder + 2) \bmod 3))$
11:    **else**
12:        $mcf_s \leftarrow$ [MCF$(n - 1, base + ((remainder + 3 - 1) \bmod 3))] * 2^{len2}$
                       $+ MCF(n - 1, base + ((remainder + 3 - 2) \bmod 3))$
13:      **end if**
14: **end if**

Algorithm 7 recursively calculates each most complex function. Lines 4 to 8 of Algorithm 7 prepare the recursion. The Boolean value *direction* decides about the direction in which basic functions from a lower level of recursion have to be combined. The multiplications by 2^{len2} in lines 10 and 12 move the associated function vector by its size bits so that the most outer additions in these

Table 8.16 Most complex logic functions represented by hexadecimal function vectors

$n = 1$	$n = 2$	$n = 3$	$n = 4$
3 MCF of one variable	6 MCF of two variables	12 MCF of three variables	24 MCF of four variables
mcf[0] = 1	mcf[0] = B	mcf[0] = D6	mcf[0] = 6BBD
mcf[1] = 2	mcf[1] = D	mcf[1] = 6B	mcf[1] = BDD6
mcf[2] = 3	mcf[2] = 6	mcf[2] = BD	mcf[2] = D66B
	mcf[3] = E	mcf[3] = 6D	mcf[3] = BD6B
	mcf[4] = 7	mcf[4] = B6	mcf[4] = D6BD
	mcf[5] = 9	mcf[5] = DB	mcf[5] = 6BD6
		mcf[6] = 79	mcf[6] = B6DB
		mcf[7] = 9E	mcf[7] = DB6D
		mcf[8] = E7	mcf[8] = 6DB6
		mcf[9] = 97	mcf[9] = DBB6
		mcf[10] = E9	mcf[10] = 6DDB
		mcf[11] = 7E	mcf[11] = B66D
			mcf[12] = 9EE7
			mcf[13] = E779
			mcf[14] = 799E
			mcf[15] = E79E
			mcf[16] = 79E7
			mcf[17] = 9E79
			mcf[18] = E97E
			mcf[19] = 7E97
			mcf[20] = 97E9
			mcf[21] = 7EE9
			mcf[22] = 977E
			mcf[23] = E997

lines combine the basic functions. Of course, the operations $3 - 1$ and $3 - 2$ in line 12 of the algorithm can be substituted by the values 2 and 1, respectively. The used style emphasizes the inverse direction in comparison to line 10 of Algorithm 7. Two recursive calls of the function $\text{MCF}(n, s)$ either in line 10 or in line 12 generate the required basic function.

Table 8.16 enumerates all most complex logic functions depending on the number of variables $n = 1, \ldots, 4$. The content of this table was calculated using Algorithm 7; each most complex logic function for an arbitrary number n of variables can be calculated very quickly using this algorithm.

8.6 Bent Functions

The most important application of bent functions is related to the field of cryptography. The aim of cryptosystems is the transmission of a given plain message such that the transferred message contains the given message in an encrypted form that cannot be understood by anyone else. Both the encryption and decryption are controlled by logic functions. The properties of these functions strongly influence the required efforts to decrypt a message without knowing the key.

The use of linear functions for the encryption preserves the probability of how often a certain character appears in the cipher-text. Hence, simple statistical methods can be successfully used in an attack against the corresponding cryptosystems. Taking functions with the largest distance to all linear functions within a cryptosystem hedges these attacks.

The name "bent" of a set of special functions has been introduced by O. S. Rothaus in 1976 in his pioneering paper *On "Bent" Functions* [1]. In this chapter he studied logic functions $P(\mathbf{x})$ that are defined by a mapping from the Galois field $GF(2^n)$ to $GF(2)$ and noticed that a small fraction of these functions has the property that all Fourier coefficients of $(-1)^{P(\mathbf{x})}$ are equal to ± 1. He called functions with this property *bent*.

On the first page of this paper Rothaus introduced the *Fourier coefficients* as follows:

"Now let ω be the real number -1. Then $\omega^{P(\mathbf{x})}$ is a well-defined real function on V_n, and by the character theory for abelian groups we may write:

$$\omega^{P(\mathbf{x})} = \frac{1}{2^{n/2}} \sum_{\lambda \in V_n} c(\lambda) \omega^{<\lambda, \mathbf{x}>},$$

where the $c(\lambda)$, the *Fourier coefficients* of $\omega^{P(\mathbf{x})}$, are given by

$$c(\lambda) = \frac{1}{2^{n/2}} \sum_{\mathbf{x} \in V_n} \omega^{P(\mathbf{x})} \omega^{<\lambda, \mathbf{x}>}.$$

In words, $2^{n/2} c(\lambda)$ is the number of zeros minus the number of ones of the function $P(\mathbf{x}) + <\lambda, \mathbf{x}>$.
From Parseval's equation, we know that

$$\sum_{\lambda \in V_n} c^2(\lambda) = 2^n.$$

We call $P(\mathbf{x})$ a *bent* function if all Fourier coefficients of $\omega^{P(\mathbf{x})}$ are equal to ± 1."

Based on this definition O. S. Rothaus found some important propositions:

- The Fourier transform of a bent function is a bent function.
- $P(\mathbf{x})$ is bent if and only if $\omega^{P(\mathbf{x}+\mathbf{y})}$ is a Hadamard matrix.
- If $P(\mathbf{x})$ is a bent function on V_n, then n is even, $n = 2k$; the degree of $P(\mathbf{x})$ is at most k, except in the case $k = 1$.
- If $P(\mathbf{x})$ is a bent function on V_{2k} of degree k, $k > 3$, then $P(\mathbf{x})$ is irreducible.

Alternatively, it is possible to define the bent function with regard to the set of all linear functions (Definition 4.17 on page 153, the Hamming distance $h(f, g)$, and the nonlinearity $\mathrm{NL}(f)$.

Definition 8.11 (Hamming Distance $h(f, g)$) The *Hamming distance* $h(f, g)$ between two logic functions, $f(\mathbf{x})$ and $g(\mathbf{x})$, is the number of positions (argument vectors) with different values. The function $f(\mathbf{x}) \oplus g(\mathbf{x})$ is equal to 1 for these argument vectors.

Example 8.17 We determine the Hamming distance $h(f, g)$ between the linear function

$$f(\mathbf{x}) = x_1 \oplus x_2 \oplus x_3 \oplus x_4$$

and the bent function

$$g(\mathbf{x}) = x_1 x_2 \oplus x_3 x_4.$$

Figure 8.10 shows the Karnaugh-maps of $f(\mathbf{x})$, $g(\mathbf{x})$, $f(\mathbf{x}) \oplus g(\mathbf{x})$.
The Hamming distance $h(f, g) = 6$ due to the six values 1 in the Karnaugh-map of $f(\mathbf{x}) \oplus g(\mathbf{x})$.

Definition 8.12 (Nonlinearity $\mathrm{NL}(f)$) The *nonlinearity* $\mathrm{NL}(f)$ of a logic function $f(\mathbf{x})$ is the minimum of all Hamming distances between this function and all linear functions.

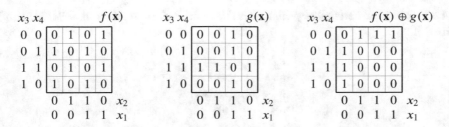

Fig. 8.10 Karnaugh-maps of the linear function $f(\mathbf{x})$, the bent function $g(\mathbf{x})$, and the function $f(\mathbf{x}) \oplus g(\mathbf{x})$ that helps to determine the Hamming distance $h(f, g)$

As introduced above [1], Rothaus has considered the algebraic structure of *Galois fields* for the definition of bent functions. The set $\mathbb{B} = \{0, 1\}$ together with $\wedge$ as multiplication and $\oplus$ as addition satisfies the axioms of a Galois field $GF(\mathbb{B})$ as well as $\mathbb{B}^n$ with the same operations (indicated by $GF(\mathbb{B}^n)$).

Functions from $GF(\mathbb{B}^n)$ into $GF(\mathbb{B})$ allow the definition of a *Fourier transformation*, and for bent functions all Fourier coefficients had to be equal to ± 1. It could be shown that such functions exist only if n is even. In this case the set of bent functions is equal to the set of functions with maximal nonlinearity. Therefore, it is common to define bent functions only when n is even.

Definition 8.13 (Bent Function) Let $f(\mathbf{x})$ be a logic function of n variables, where n is even. $f_b(\mathbf{x})$ is a *bent function* if its nonlinearity $NL(f)$ is as large as possible.

This means that after the calculation of the nonlinearity of each nonlinear function the maximum of all these values has to be found, and all nonlinear functions with this maximal nonlinearity are the bent functions $f_b(x_1, \ldots, x_n)$.

Boolean Differential Equations of Bent Functions

It is known from Sect. 7.7 that the solution of a Boolean differential equation is a set of logic functions. Bent functions are logic functions with the special *bent* property. Hence, Boolean differential equations can be used to specify the set of bent functions of a given even number n of variables.

As a result of our research, we found Boolean differential equations that have all or a subset of bent functions as solution. We published these Boolean differential equations in [7, 8] and demonstrated in [6] their solution using a previous version of the XBOOLE-monitor.

Theorem 8.6 *All bent functions of two variables are the solution of the Boolean differential equation:*

$$\operatorname*{der}_{(x_1, x_2)}^2 f(x_1, x_2) = 1 . \tag{8.11}$$

It is an interesting observation that the Boolean differential equation (8.2) of the set $\mathcal{RM}(1, 2)$ (the set of all linear functions of $\mathbb{B}^2$) differs from the Boolean differential equation (8.11) of all bent functions of two variables (all nonlinear functions of $\mathbb{B}^2$) only on the right-hand side that is equal to 0 for (8.2) and equal to 1 for (8.11). Table 8.17 shows the solution-set of the Boolean differential equation (8.11) that consists of two classes of nonlinear functions.

Theorem 8.7 *All bent functions of four variables are the solution of the Boolean differential equation:*

$$\operatorname*{der}_{(x_1,x_2)}{}^{2} f(\mathbf{x}) \operatorname*{der}_{(x_3,x_4)}{}^{2} f(\mathbf{x})$$

$$\oplus \operatorname*{der}_{(x_1,x_3)}{}^{2} f(\mathbf{x}) \operatorname*{der}_{(x_2,x_4)}{}^{2} f(\mathbf{x})$$

$$\oplus \operatorname*{der}_{(x_1,x_4)}{}^{2} f(\mathbf{x}) \operatorname*{der}_{(x_2,x_3)}{}^{2} f(\mathbf{x}) = 1 . \tag{8.12}$$

The solution-set of the Boolean differential equation (8.12) contains exactly all 896 bent functions of four variables that satisfy Definition 8.13.

The comparison of (8.11) and (8.12) shows a principle that can be used to specify Boolean differential equations for bent functions of more than four variables. This principle can be recursively used to specify Boolean differential equations for all even numbers of variables. However, the created Boolean differential equations have only a subset of all bent functions as solution. The construction principle is as follows:

- Prepare $n - 1$ two-fold derivatives of the function $f(x_1, \ldots, x_n)$ with regard to the two variables x_1 and x_i, $i = 2, \ldots, n$.
- Combine these two-fold derivatives by $\wedge$ with the expression of the Boolean differential equation of bent functions for $n - 2$ variables that are not used in the two-fold derivative.
- Combine all these expressions by $\oplus$ on the left-hand side of the Boolean differential equation that is equal to 1.
- All solutions of these Boolean differential equations are bent functions; however, if $n > 4$, then only a subset of bent functions satisfied these Boolean differential equations.

Theorem 8.8 applies this construction principle for bent functions of six variables:

Theorem 8.8 *The solution of the Boolean differential equation:*

$$\operatorname*{der}_{(x_1,x_2)}{}^{2} f \left(\operatorname*{der}_{(x_3,x_4)}{}^{2} f \operatorname*{der}_{(x_5,x_6)}{}^{2} f \oplus \operatorname*{der}_{(x_3,x_5)}{}^{2} f \operatorname*{der}_{(x_4,x_6)}{}^{2} f \oplus \operatorname*{der}_{(x_3,x_6)}{}^{2} f \operatorname*{der}_{(x_4,x_5)}{}^{2} f \right)$$

$$\oplus \operatorname*{der}_{(x_1,x_3)}{}^{2} f \left(\operatorname*{der}_{(x_2,x_4)}{}^{2} f \operatorname*{der}_{(x_5,x_6)}{}^{2} f \oplus \operatorname*{der}_{(x_2,x_5)}{}^{2} f \operatorname*{der}_{(x_4,x_6)}{}^{2} f \oplus \operatorname*{der}_{(x_2,x_6)}{}^{2} f \operatorname*{der}_{(x_4,x_5)}{}^{2} f \right)$$

$$\oplus \operatorname*{der}_{(x_1,x_4)}{}^{2} f \left(\operatorname*{der}_{(x_2,x_3)}{}^{2} f \operatorname*{der}_{(x_5,x_6)}{}^{2} f \oplus \operatorname*{der}_{(x_2,x_5)}{}^{2} f \operatorname*{der}_{(x_3,x_6)}{}^{2} f \oplus \operatorname*{der}_{(x_2,x_6)}{}^{2} f \operatorname*{der}_{(x_3,x_5)}{}^{2} f \right)$$

$$\oplus \operatorname*{der}_{(x_1,x_5)}{}^{2} f \left(\operatorname*{der}_{(x_2,x_3)}{}^{2} f \operatorname*{der}_{(x_4,x_6)}{}^{2} f \oplus \operatorname*{der}_{(x_2,x_4)}{}^{2} f \operatorname*{der}_{(x_3,x_6)}{}^{2} f \oplus \operatorname*{der}_{(x_2,x_6)}{}^{2} f \operatorname*{der}_{(x_3,x_4)}{}^{2} f \right)$$

$$\oplus \operatorname*{der}_{(x_1,x_6)}{}^{2} f \left(\operatorname*{der}_{(x_2,x_3)}{}^{2} f \operatorname*{der}_{(x_4,x_5)}{}^{2} f \oplus \operatorname*{der}_{(x_2,x_4)}{}^{2} f \operatorname*{der}_{(x_3,x_5)}{}^{2} f \oplus \operatorname*{der}_{(x_2,x_5)}{}^{2} f \operatorname*{der}_{(x_3,x_4)}{}^{2} f \right) = 1 \tag{8.13}$$

is a subset of bent functions $f = f(x_1, \ldots, x_6)$.

Table 8.17 All eight bent functions of two variables

Class	Solution-vector	Bent function
1	1000	$f_{b1}(x_1, x_2) = \overline{x}_1\overline{x}_2$
1	0100	$f_{b2}(x_1, x_2) = x_1\overline{x}_2$
1	0010	$f_{b3}(x_1, x_2) = \overline{x}_1 x_2$
1	0001	$f_{b4}(x_1, x_2) = x_1 x_2$
2	0111	$f_{b5}(x_1, x_2) = x_1 \vee x_2$
2	1011	$f_{b6}(x_1, x_2) = \overline{x}_1 \vee x_2$
2	1101	$f_{b7}(x_1, x_2) = x_1 \vee \overline{x}_2$
2	1110	$f_{b8}(x_1, x_2) = \overline{x}_1 \vee \overline{x}_2$

Table 8.18 Distribution of the eight bent functions over SNF(f)-classes for all 16 logic functions of two variables

NTV in the		Number of	
SNF	Minimal ESOP	All functions	Bent functions
0	0	1	0
4	1	9	4
6	2	6	4

Classes of Bent Functions Identified by the Specific Normal Form SNF(f)

The number of cubes of the *Specific Normal Form* SNF(f) is a simple measure of the complexity of a logic function $f(\mathbf{x})$. Table 8.18 shows how the bent functions of two variables are distributed regarding the NTV of the SNF(f).

Table 8.18 reveals that the bent functions can be distinguished regarding their complexity. The four bent functions with NTV(SNF(f)) = 4 can be expressed by a single conjunction in ESOP expression, e.g.,

$$f(x_1, x_2) = x_1 x_2 .$$

The bent functions with NTV(SNF(f)) = 6 need one conjunction more in the ESOP expression, e.g.,

$$f(x_1, x_2) = x_1 \vee x_2 = x_1 \oplus x_2 \oplus x_1 x_2 = x_1 \oplus \overline{x}_1 x_2 .$$

These bent functions belong to the most complex logic functions in $\mathbb{B}^2$.

These basic observations motivated us to explore the known bent functions of four variables more in detail. Table 8.19 enumerates the analog results of Table 8.18 for all logic functions of four variables.

Table 8.19 shows again that bent function of four variables is distributed over several SNF-classes. Consequently, the bent functions in the Boolean space $\mathbb{B}^4$ have again different complexities. Contrary to the Boolean space $\mathbb{B}^2$, no bent function of four variables belongs to the class of the most complex logic functions over $\mathbb{B}^4$.

An interesting observation of this first experiment is that for all bent functions $f_b(\mathbf{x})$ of four variables the number of cubes in the SNF(f_b) modulo 4 is equal to 2. A more detailed analysis is necessary to detect further properties of bent functions.

It is known that the weight (the number of function values 1) of bent functions of n variables is equal to

$$2^{n-1} \pm 2^{\frac{n}{2}-1} .$$

Therefore, we distinguish the bent functions of four variables between the possible weights of $2^3 - 2^1 = 6$ and $2^3 + 2^1 = 10$. Table 8.20 reveals that for each bent function of the weight 6 exists an associated bent function of the weight 10. These pairs of bent functions are complements of each other. The complement of a bent function of the weight 6 requires an EXOR-operation with a constant

Table 8.19 Distribution of the 896 bent functions into classes of the SNF(f) for all 65,536 logic functions of four variables

NTV in the		Number of	
SNF	Minimal ESOP	All functions	Bent functions
0	0	1	0
16	1	81	0
24	2	324	0
28	2	1296	0
30	2	648	48
32	3	648	0
34	3	3888	240
36	3	6624	0
36	4	108	0
38	3	7776	384
40	3	2592	0
40	4	6642	0
42	3	216	0
42	4	14,256	192
44	4	12,636	0
46	4	3888	0
46	5	1296	16
48	5	1944	0
50	5	648	16
54	6	24	0

Table 8.20 Distribution of the 896 bent functions over the SNF(f)-classes of $\mathbb{B}^4$ distinguished by the weights 6 and 10

NTV in the		Number of bent functions	
SNF	minimal ESOP	of weight 6	of weight 10
30	2	$48 = 3 * 16$	0
34	3	$192 = 12 * 16$	$48 = 3 * 16$
38	3	$192 = 12 * 16$	$192 = 12 * 16$
42	4	0	$192 = 12 * 16$
46	5	$16 = 1 * 16$	0
50	5	0	$16 = 1 * 16$

1 that leads to four additional ternary vectors in the SNF of the bent function of weight 10. It should be mentioned that in some cases the minimal ESOPs of a bent function $f_b(\mathbf{x})$ and their complement $\overline{f_b(\mathbf{x})}$ contain the same number of terms.

For each NTV(SNF(f)) and each weight of the bent function, we identified a number of classes of $2^4 = 16$ bent functions (see Table 8.20). All 16 bent functions of such a class can be generated by replacing x_i with $\overline{x}_i$ for all combinations of i, $i = 1, \ldots, 4$.

The expressions of the bent functions have a unique pattern for each value of NTV(SNF(f_b)) and each weight $w(f_b)$ of the bent function. The following expressions show the positive polarity ESOPs of an example of a bent function f_{bi} for the different values of NTV(SNF(f_{bi})).

$$\text{NTV}(\text{SNF}(f_{b1})) = 30 \qquad w(f_{b1}) = 6 \qquad f_{b1}(\mathbf{x}) = x_1 x_2 \oplus x_3 x_4 \,,$$

$$\text{NTV}(\text{SNF}(f_{b2})) = 34 \qquad w(f_{b2}) = 6 \qquad f_{b2}(\mathbf{x}) = x_1 x_2 \oplus x_3 x_4 \oplus x_1 x_3 \,,$$

$$\text{NTV}(\text{SNF}(f_{b3})) = 38 \qquad w(f_{b3}) = 6 \qquad f_{b3}(\mathbf{x}) = x_1 x_2 \oplus x_3 x_4 \oplus x_1 x_3 \oplus x_1 x_4 \,,$$

$$\text{NTV}(\text{SNF}(f_{b4})) = 46 \qquad w(f_{b4}) = 6 \qquad f_{b4}(\mathbf{x}) = x_1 x_2 \oplus x_1 x_3 \oplus x_1 x_4$$

$$\oplus x_2 x_3 \oplus x_2 x_4 \oplus x_3 x_4 \oplus 1 \,.$$

The complement $\overline{f_b(\mathbf{x})}$ of a bent function $f_b(\mathbf{x})$ of four variables with the weight $w(f_b) = 6$ is also a bent function but has a weight $w(\overline{f_b(\mathbf{x})}) = 10$, and the number of ternary vectors of their SNFs satisfies

$$\text{NTV}(\text{SNF}(\overline{f_b(\mathbf{x})})) = \text{NTV}(\text{SNF}(f_b(\mathbf{x}))) + 4$$

as can be seen in the following examples:

$$\text{NTV}(\text{SNF}(\overline{f}_{b1})) = 34 \qquad w(\overline{f}_{b1}) = 10 \qquad \overline{f_{b1}(\mathbf{x})} = x_1 x_2 \oplus x_3 x_4 \oplus 1 \,,$$

$$\text{NTV}(\text{SNF}(\overline{f}_{b2})) = 38 \qquad w(\overline{f}_{b2}) = 10 \qquad \overline{f_{b2}(\mathbf{x})} = x_1 x_2 \oplus x_3 x_4 \oplus x_1 x_3 \oplus 1 \,,$$

$$\text{NTV}(\text{SNF}(\overline{f}_{b3})) = 42 \qquad w(\overline{f}_{b3}) = 10 \qquad \overline{f_{b3}(\mathbf{x})} = x_1 x_2 \oplus x_3 x_4 \oplus x_1 x_3$$
$$\oplus x_1 x_4 \oplus 1 \,,$$

$$\text{NTV}(\text{SNF}(\overline{f}_{b4})) = 50 \qquad w(\overline{f}_{b4}) = 10 \qquad \overline{f_{b4}(\mathbf{x})} = x_1 x_2 \oplus x_1 x_3 \oplus x_1 x_4$$
$$\oplus x_2 x_3 \oplus x_2 x_4 \oplus x_3 x_4 \,.$$

8.7 Exercises

Prepare again for each task of each exercise of this chapter a PRP for the XBOOLE-monitor XBM 2. Prefer a lexicographic order of the variables, but use the usual order of variables in case of exercises regarding arithmetic operations that starts on the left-hand side with the variable x_{15} of the most significant bit (in the case that 16 bits are used) and ends on the right-hand side with the variable x_0 of the least significant bit. Use the help system of the XBOOLE-monitor XBM 2 to extend your knowledge about the details of used commands. Execute the created PRPs, verify the computed results, and explain the observed properties.

If you have not yet prepared the XBOOLE-monitor XBM 2 on your computer, you can get this XBOOLE-monitor free of charge by means of the following three steps:

1. **Download**:

There are four versions of the XBOOLE-monitor XBM 2, two for Windows 10 or subsequent Windows systems (32 or 64 bits) and two for LINUX—Ubuntu (also 32 or 64 bits); you must download the version of the XBOOLE-monitor XBM 2 that fits to your operating system. Authorized users of the online version of this chapter (https://doi.org/10.1007/978-3-030-88945-6_8) can download the XBOOLE-monitor XBM 2 directly from the web page

https://link.springer.com/chapter/10.1007/978-3-030-88945-6_8

where the links for the download of the XBOOLE-monitor XBM 2 are located in the part "Supplementary Information" (below the part "Abstract"). The headline above such a link indicates the associated zip-file of the XBOOLE-monitor XBM 2. The sizes of the zip-files have been provided behind the links and can be used to verify the download. A click on the link of the wanted version of the XBOOLE-monitor XBM 2 starts the download.

Readers of the hardcopy of this book get access to the XBOOLE-monitor XBM 2 using the URL

https://link.springer.com/chapter/10.1007/978-3-030-88945-6_8

to download the first two pages of this chapter. After this download, the same procedure as the authorized users of the online version of a chapter can be used to download the wanted version of the XBOOLE-monitor XBM 2.

2. **Unzip**: The XBOOLE-monitor XBM 2 must not be installed but unzipped into an arbitrary directory of your computer. A convenient tool for unziping the downloaded zip-file is usually available as part of the operating system or can be downloaded from the Internet.

3. **Execute**:

- Windows:
  The executable file of the two versions (32 or 64 bits) for Windows 10 (or subsequent Windows systems) of the XBOOLE-monitor XBM 2 is XBM2.exe; the other files in the expanded directory must remain unchanged. A double-click on the executable file XBM2.exe within the Explorer of Windows starts the XBOOLE-monitor XBM 2.
- LINUX—Ubuntu:
  The unzipped folder of the XBOOLE-monitor XBM 2 contains for this operating system only the executable file XBM2-i386.AppImage for the version of 32 bits or XBM2-x86_64.AppImage for the version of 64 bits of the XBOOLE-monitor XBM 2. A double-click on the created AppImage-file within the file manager of LINUX—Ubuntu starts the XBOOLE-monitor XBM 2.

Exercise 8.1 (Propositional Logic)

(a) Verify that the following inference rules are always valid:

$$(x \to y) \equiv (\overline{x} \vee y),$$
$$(x \to y) \equiv \overline{(x \wedge \overline{y})}.$$

(b) Verify that the following paradoxes of implications are tautologies:

$$x \to (x \to y) \equiv 1,$$
$$(x \to y) \vee (y \to x) \equiv 1.$$

(c) Verify the truth of the following rules:

$$x \vee \overline{x} \equiv 1,$$
$$((x \vee y) \wedge (\overline{x} \vee z)) \to (y \vee z) \equiv 1.$$

(d) Verify the following implications:

$$(x \wedge (x \to y)) \to y,$$
$$(\overline{y} \wedge (x \to y)) \to \overline{x}.$$

Fig. 8.11 Sequence of
four code words encoded
using the 2-out-of-5 code

	y_7	y_4	y_2	y_1	y_0
	0	0	0	1	1
ODA(scw) =	1	0	0	0	1
	1	1	0	0	0
	0	1	0	0	1

Fig. 8.12 Four data words
to be encoded using the
Hamming code $H(7, 4)$

	x_1	x_2	x_3	x_4
	0	1	1	0
ODA(dw) =	1	1	0	1
	0	0	1	1
	1	0	0	1

Exercise 8.2 (Binary Arithmetic)

(a) Convert the positive integers 13 and 12345 into a binary numbers of 16 bits, store these binary vectors as two rows of a TVL, and show this TVL in the 1-fold view. Convert thereafter these two binary vectors back into the integer results $r1 and $r2.

(b) Extend the PRP of the previous task such that the binary representations of the integer values $a=13 and $b=12345 are added, the binary vector of the computed sum is connected to the TVL of the given two binary vectors, and the TVL of these three binary vectors is shown in the 1-fold view. Convert thereafter the binary vectors of the computed sum back into the integer number $s.

(c) Convert the integers 12, −12, 19283, and −19283 into binary numbers of 16 bits using the two's complement, store these binary vectors as four rows of a TVL, and show this TVL in the 1-fold view. Convert thereafter these four binary vectors back into the integer results $r1, ..., $r4.

(d) Modify the PRP of the previous task such that difference $a − b$ of $a=12 and $b=19283 represented as binary vectors will be computed, the result vector is connected to the TVL of the given two binary vectors, and the TVL of these three binary vectors is shown in the 1-fold view. Convert thereafter the binary vector of the computed difference back into the integer number $d.

Exercise 8.3 (Coding)

(a) Decode the sequence of code words of the TVL scw of Fig. 8.11.
    Prepare for the decoding procedure a coding table where the index of the row specifies the data word (integer) of the code word of this row; start with row number one and use row with the index ten for the data word 0. Store the result as the integer $w that consists of the decoded digits where the left digit belongs to the code word of the first row of the TVL that stores the given sequence of code words (see Fig. 8.11).

(b) Determine a system of logic equations that describes the coding table of the Hamming code $H(7, 4)$; seven bits in a code word of this code encode four data bits. The bits in the code word with an index of 2^k, $k = 0, 1, 2$, i.e., y_1, y_2, and y_4, are parity bits. The value of k also determines the bit of the binary encoded index of the variables of a code word that are evaluated to get an even parity. The remaining bits of the code word store the bits of the data word.
    Solve the created system of equations and show this coding table.

(c) Use the coding table of the Hamming code $H(7, 4)$ (as created in the previous task) to encode the four data word $\mathbf{x} = (x_1, x_2, x_3, x_4)$, which are specified in the TVL of Fig. 8.12, store the associated code words in the same order in a TVL, and show this TVL to the right of the given TVL of data words.
    Use the same coding table of the Hamming code $H(7, 4)$ to decode the five code words $\mathbf{y} = (y_1, y_2, y_3, y_4, y_5, y_6, y_7)$ of the TVL of Fig. 8.13, store the associated data words in the same

Fig. 8.13 Five code
words to be decoded using
the Hamming code $H(7, 4)$

$$
\mathrm{ODA}(cw) = \begin{array}{ccccccc}
y_1 & y_2 & y_3 & y_4 & y_5 & y_6 & y_7 \\
\hline
1 & 1 & 1 & 0 & 0 & 0 & 0 \\
0 & 0 & 0 & 0 & 1 & 1 & 1 \\
0 & 0 & 1 & 0 & 1 & 1 & 0 \\
0 & 1 & 0 & 1 & 0 & 1 & 0 \\
1 & 0 & 1 & 0 & 0 & 1 & 0
\end{array}
$$

Fig. 8.14 TVL in A-form
used as example for a
complete expansion

$$
\mathrm{A}(f) = \begin{array}{ccc}
x_1 & x_2 & x_3 \\
\hline
0 & 1 & - \\
- & 0 & 1
\end{array}
$$

Fig. 8.15 TVL in A-form
used as example for
removing all pairs of
identical conjunctions

$$
\mathrm{A}(f) = \begin{array}{ccccc}
x_1 & x_2 & x_3 & x_4 & x_5 \\
\hline
- & 0 & 1 & - & 1 \\
0 & - & 0 & 1 & - \\
1 & 1 & - & 0 & 0 \\
0 & - & 0 & 1 & - \\
1 & 0 & 0 & - & 1 \\
- & 0 & 0 & 1 & - \\
1 & 1 & - & 0 & 0 \\
- & 1 & 1 & - & - \\
0 & 0 & - & 1 & -
\end{array}
$$

$$
\mathrm{A}(f) = \begin{array}{ccc}
x_1 & x_2 & x_3 \\
\hline
0 & - & 1 \\
1 & 0 & 1
\end{array}
$$

Fig. 8.16 TVL in A-form used as example to compute its *Specific Normal Form*

order in a TVL, and show this TVL to the right of the given TVL of code words. Display a vector
of four dashes in the case that a given vector **y** does not belong to the valid code words of the
Hamming code $H(7, 4)$.

(d) The code words of the Reed–Muller code $\mathcal{RM}(1, 2)$ are the solutions of the Boolean differential
equation (8.2) on page 443. Remember that

$$
\operatorname*{der}_{(x_1,x_2)}^{2} f(x_1, x_2) = \operatorname*{der}_{x_1} f(x_1, x_2) \oplus \operatorname*{der}_{x_2} f(x_1, x_2) \oplus \operatorname*{der}_{(x_1,x_2)} f(x_1, x_2) .
$$

Solve the Boolean differential equation (8.2) and show the solution-TVL of the associated logic
equation $D(\mathbf{u}) = 0$ and the TVL of solution-vectors **v**.

Exercise 8.4 (Specific Normal Form)

(a) Expand the all ternary vectors of the TVL of Fig. 8.14 in all positions using Algorithm 5 Exp(f)
of page 446. Show the given TVL and the TVL of the result side by side in the m-fold view and
thereunder the associated Karnaugh-maps. Verify that the expanded antivalence expression also
describes the given function.

(b) Use Algorithm 6 R(f) of page 446 to remove all pairs of identical ternary vectors from the TVL
of Fig. 8.15. Show the given TVL and the TVL of the result side by side in the m-fold view and
thereunder the associated Karnaugh-maps. Verify that the described function remains unchanged
when pairs of identical conjunctions are deleted.

(c) Compute the *Specific Normal Form* (SFN) of the logic function $f(\mathbf{x})$ specified in Fig. 8.16
using Eq. (8.10) of Definition 8.10. Show the TVL of the given function $f(\mathbf{x})$, the TVL of the
completely expanded function $f(\mathbf{x})$, and the TVL of SNF(f) side by side in the m-fold view.

$$A(f_1) = \frac{\begin{array}{ccc} x_1 & x_2 & x_3 \end{array}}{\begin{array}{ccc} 0 & 1 & - \\ - & 1 & 0 \end{array}} \qquad A(f_2) = \frac{\begin{array}{ccc} x_1 & x_2 & x_3 \end{array}}{\begin{array}{ccc} 0 & 0 & - \\ 0 & 1 & 0 \end{array}} \qquad A(f_3) = \frac{\begin{array}{ccc} x_1 & x_2 & x_3 \end{array}}{\begin{array}{ccc} 0 & 1 & - \\ 1 & - & 0 \end{array}}$$

Fig. 8.17 Three TVLs in A-form used to compare the number of conjunctions in their *Specific Normal Forms*

(d) Demonstrate that the number of columns with different values of two ternary vectors but not the values itself determine the number of conjunctions in the SNF(f_i); use the three TVLs of Fig. 8.17 for this check. Show the given three TVLs in the first row of the m-fold view and the associated *Specific Normal Forms* thereunder.

Exercise 8.5 (Most Complex Functions)

(a) A logic function $f(x) : \mathbb{B}^1 \to \mathbb{B}^1$ needs zero, one, or two conjunctions for their representation as disjunctive form. The complexity of these simple functions can be determined by the ternary vectors in the associated ODA-form. Compute for all $2^{2^1} = 4$ functions of $\mathbb{B}^1$ the ODA-form minimized using the command obbc and the associated vector of function values. Create TVLs that represent the sets of vectors of function values determined by the same number of conjunctions in the associated minimized ODA-form. Conclude from these TVLs how many and which functions belong to the set of most complex functions of one variable.

(b) Verify whether the simple approach of the previous task can be used to determine the most complex functions $f(\mathbf{x}) : \mathbb{B}^2 \to \mathbb{B}^1$. Compute therefore TVLs that represent the sets of vectors of function values determined by the same number of conjunctions in the associated minimized ODA-form. Compare the set of vectors of function values with the largest number of conjunctions in the associated minimized ODA-form with the known most complex functions of Table 8.16 for $n = 2$.

(c) A minimized ODA-form of a logic function can require more conjunctions than the simplest non-orthogonal representation of this function. Alternatively to the approach of the last two tasks, the complexity of a function is uniquely determined by the number of conjunctions in its specific normal form. Compute the SNF(f) of all $2^{2^3} = 256$ functions $f(\mathbf{x}) : \mathbb{B}^3 \to \mathbb{B}^1$ and create the TVL of the set of vectors of function values of the most complex functions. Compare this set of functions with the known most complex functions of Table 8.16 for $n = 3$. Note, this approach requires three nested loops in the PRP; hence, a very large number of commands must be executed (about half a million) and written into the protocol, which needs some minutes. This task can be seen as a border case for the use of the XBOOLE-monitor XBM 2; a program that uses the XBOOLE-library computes the same result much faster.

(d) Compute the sets of vectors of function values that represent the most complex functions belonging to the Boolean spaces $\mathbb{B}^1$, $\mathbb{B}^2$, $\mathbb{B}^3$, and $\mathbb{B}^4$. Use for this purpose the approach of Algorithm 7 of page 448, but change the recursive procedure into an iterative approach. Compare these four sets of vectors of function values with the known most complex functions of Table 8.16.

8.8   Solutions

Solution 8.1 (Propositional Logic)

(a) Figure 8.18 shows both the PRP that verifies the given two inference rules and the values of the Boolean variables that confirm that these rules are valid for all truth values of x and y.

```
1   new                        6   (x>y)=(/x+y).
2   space 32 1                 7   sbe  1 2
3   avar 1                     8   (x>y)=/(x&/y).
4   x y.                       9   set $ir1 (te_cpl 1)
5   sbe 1 1                   10   set $ir2 (te_cpl 2)
```

Boolean variables	
Name	Value
$ir1	true
$ir2	true

Fig. 8.18 Problem-program that verifies two inference rules together with the computed results

```
1   new                        6   x>(y>x)=1.
2   space 32 1                 7   sbe  1 2
3   avar 1                     8   (x>y)+(y>x)=1.
4   x y.                       9   set $pt1 (te_cpl 1)
5   sbe 1 1                   10   set $pt2 (te_cpl 2)
```

Boolean variables	
Name	Value
$pt1	true
$pt2	true

Fig. 8.19 Problem-program that verifies two paradoxes of implications and shows the computed Boolean values that confirm the truth of these paradoxes

```
1   new                        6   x+/x=1.
2   space 32 1                 7   sbe  1 2
3   avar 1                     8   (x+y)&(/x+z)>(y+z)=1.
4   x y z.                     9   set $r1 (te_cpl 1)
5   sbe 1 1                   10   set $r2 (te_cpl 2)
```

Boolean variables	
Name	Value
$r1	true
$r2	true

Fig. 8.20 Problem-program that verifies two rules and shows the computed Boolean values that confirm the truth of these tautologies

All operations of the given inference rules, i.e., the negation, the conjunction, the disjunction, the implication, and the equivalence, are available and can be used in expressions of logic equations; hence, two commands `sbe` compute all assignments to the variables x and y for which these rules are satisfied. The result of a command `te_cpl` is equal to the value `true` if the complement of the evaluated set is empty.

The computed values `true` of the variables `$ir1` and `$ir2` confirm that the given two inference rules are valid for all possible combinations of the values of x and y.

(b) Figure 8.19 shows both the PRP that verifies the given two paradoxes of implications are tautologies and the computed Boolean values confirm this verification.

The same approach as in the previous task has been used to verify whether the given two paradoxes of implications are true. The result `$pt1=true` confirms that a true proposition x is implied by any proposition y, and the second result `$pt2=true` confirms that for any two propositions, one implies the other.

(c) Figure 8.20 shows both the PRP that verifies the given two tautologies and the computed Boolean values that confirm this verification.

The first tautology is the *law of excluded middle*. The left-hand side of the implication in the second rule is equal to 1 if $y \vee z$ is equal to 1; exactly this expression is given on the right-hand side of this implication so that we also have a tautology.

(d) Figure 8.21 shows both the PRP that verifies the given two derivation rules and the computed Boolean values that confirm this verification.

The first rule is the *modus ponens*, and the second rule is the *modus tollens*. The results `$i1=true` and `$i2=true` confirm that these two implications are satisfied for all possible combinations of the values of x and y.

					Boolean variables	
1	new	6	(x&(x>y)) > y .		Name	Value
2	space 32 1	7	sbe 1 2		$i1	true
3	avar 1	8	(/y&(x>y))>/x .			
4	x y z .	9	set $i1 (te_cpl 1)		$i2	true
5	sbe 1 1	10	set $i2 (te_cpl 2)			

Fig. 8.21 Problem-program that verifies two implications and shows the computed Boolean values that confirm the truth of these derivation rules

1	new		24	(
2	space 32 1		25	if (eq (mod $n 2) 1)
3	tin 1 1		26	(cel 11 4 11 /01)
4	x15 x14 x13 x12 x11 x10 x9 x8		27	set $n (div $n 2)
5	x7 x6 x5 x4 x3 x2 x1 x0.		28	cco 4 2 3 4
6	0000000000000000.		29	)
7	vtin 1 2		30	con 10 11 10
8	x15 x14 x13 x12 x11 x10 x9 x8		31	set $r 0
9	x7 x6 x5 x4 x3 x2 x1.		32	set $vb 1
10	vtin 1 3		33	for $i 0 15
11	x14 x13 x12 x11 x10 x9 x8		34	(
12	x7 x6 x5 x4 x3 x2 x1 x0.		35	if (not (te_isc 11 4))
13	sbe 1 4		36	(set $r (add $r $vb))
14	x0=1.		37	set $vb (mul $vb 2)
15	dif 1 1 10		38	cco 4 2 3 4
16	for $case 1 2		39	)
17	(		40	if (eq $case 1)
18	if (eq $case 1)		41	(set $r1 $r)
19	(set $n 13)		42	if (eq $case 2)
20	if (eq $case 2)		43	(set $r2 $r)
21	(set $n 12345)		44	)
22	copy 1 11		45	assign 10 /1 1 1
23	for $i 0 15			

TVL (10) D	16 Var.	2 R.	S. 1														
	x15	x14	x13	x12	x11	x10	x9	x8	x7	x6	x5	x4	x3	x2	x1	x0
1	0	0	0	0	0	0	0	0	0	0	0	0	1	1	0	1
2	0	0	1	1	0	0	0	0	0	0	1	1	1	0	0	1

Fig. 8.22 Problem-program that converts two integer numbers into binary vector and vice versa together with the TVL of the computed binary vectors

Solution 8.2 (Binary Arithmetic)

(a) Figure 8.22 shows the PRP that converts first the integer numbers 13 and 12345 into binary vector and thereafter these binary vectors back into integer numbers; below this PRP the computed binary vectors are shown.

The for-loop in lines 23–29 converts an integer into a binary vector stored as TVL 11. This TVL is initialized with a vector of 16 binary values 0 in line 22. The operation mod in line 25 identifies whether the evaluated integer is an odd number; the command cel in line 26 changes in this case the associated bit of TVL 11 determined by the only variable of TVL 4 from 0 to 1. The evaluation of the next higher bit of the integer number is prepared by its division by 2. The command cco in line 28 rotates, controlled by VTs 2 and 3, the value 1 of TVL 4 by one bit to the right so that the needed position for a possible change of the binary vector is specified in the next sweep of the for-loop. The command con in line 30 connects the computed binary vector to the resulting TVL 11.

The for-loop in lines 33–39 converts the binary vector of TVL 11 back to the integer result $r that is initialized by the value 0 in line 31. The variable $vb carries the value belonging to the evaluated bit of the binary vector evaluated in the recent sweep of the for-loop; this variable has been initialized with $2^0 = 1$. The command te_isc evaluates the bit of the binary vector of TVL 11 indicated by the only variable of TVL 4. The command add in line 36 adds the power of two, belonging to the evaluated bit, to the result value $r. The adjustment of $vb is realized by a multiplication with the value 2. The command cco in line 38 rotates as before, controlled by VTs 2 and 3, the value 1 of TVL 4 by one bit to the right so that the next position of binary vector is specified for the evaluation in the next sweep of the for-loop.

The outer for-loop in lines 16–44 organizes together with the command if at the beginning and end that the main part of the conversions is used twice. The two binary vectors of TVL 10 as well as the result values $r1=13 and $r1=12345 confirm that PRP of Fig. 8.22 converts the given integers correctly in both directions.

(b) Figure 8.23 shows both the PRP that computes the sum of two integers represented as binary vectors and the binary vectors of this addition. The given integers $a=13 and $b=12345 are first converted into the binary vectors used to compute the binary vector of the sum that is converted back to an integer at the end of this PRP.

The approach of the previous task is reused to convert the integer $a=13 into a binary vector in lines 24–35. The created binary vector is connected to TVL 10 for the display of the results and stored as TVL 21 for its use in the addition procedure. Analogously, the integer $b=12345 is converted into a binary vector in lines 36–47 that is stored as TVL 22.

The addition of the two binary vectors stored as TVLs 21 and 22 is realized in the for-loop in lines 61–81; in each sweep of this loop the bit s_i of the sum is computed using the bits a_i and b_i as well as the carry of the previous position. This computation starts with the least significant bit and uses the value 0 as initial carry of the previous position.

The command isc in line 77 realizes the addition of a_i, b_i, and the carry of the previous position c_{i-1}; the results are the binary values of the sum s_i and the carry c_i. The solution of the system of equations solved in lines 17–19 comprises the logic functions $s = f(a, b, c_0)$ and $c = g(a, b, c_0)$. The commands in lines 63–76 prepare the single ternary vector of TVL 35 with the binary values of a_i and b_i of the recent position i, the carry of the previous position c_{i-1}, and dashes for the function values of s_i and c_i.

The command if and all therein used commands in lines 78 and 79 insert the computed value s_i into the solution-vector of TVL 23. The command cco in line 80 rotates, controlled by VTs 2 and 3, the value 1 of TVL 4 by one bit to the right so that next position of binary vector is specified for the evaluation in the next sweep of the for-loop.

The computed TVL 23 of the sum is appended to TVL 10 behind this for-loop, and the subsequent command assign in line 83 shows the binary vectors of the two summands and the computed sum in the 1-fold view.

The last part of this PRP reuses the approach of the previous task to convert the binary vector of the computed sum back into the integer variable $s. The shown value $s=12358 confirms that the binary addition of $a=13 and $b=12345 has been computed correctly.

```
 1   new                                   47   copy  11  22  ;  b
 2   space  32  1                          48   vtin  1  30
 3   tin  1  1                             49   a .
 4   x15  x14  x13  x12  x11  x10  x9      50   vtin  1  31
 5   x8  x7  x6  x5  x4  x3  x2  x1  x0.   51   b .
 6   0000000000000000.                     52   vtin  1  32
 7   vtin  1  2                            53   c0 .
 8   x15  x14  x13  x12  x11  x10  x9      54   vtin  1  33
 9   x8  x7  x6  x5  x4  x3  x2  x1.       55   s .
10   vtin  1  3                            56   vtin  1  34
11   x14  x13  x12  x11  x10  x9  x8       57   c .
12   x7  x6  x5  x4  x3  x2  x1  x0.       58   sv_get  33  36
13   sbe  1  4                             59   copy  7  35
14   x0=1.                                 60   copy  1  23
15   avar  1                               61   for  $i  0  15
16   c  s  a  b  c0.                       62   (
17   sbe  1  5                             63   cco  35  34  32  35  ;  c -> c0
18   s=a#b#c0 ,                            64   cel  35  30  35  /10  /-0  ;  a=0
19   c=a&b+a&c0+b&c0 .                     65   cel  35  31  35  /10  /-0  ;  b=0
20   tin  1  7                             66   if  (not  (te_isc  21  4))
21   c  s  a  b  c0.                       67   (
22   00000.                                68   cel  35  30  35  /01  ;  ai
23   dif  1  1  10                         69   )
24   set  $a  13                           70   if  (not  (te_isc  22  4))
25   set  $n  $a                           71   (
26   copy  1  11                           72   cel  35  31  35  /01  ;  bi
27   for  $i  0  15                        73   )
28   (                                     74   cel  35  33  35  /0-  /1-  ;  s=-
29   if  (eq  (mod  $n  2)  1)             75   cel  35  34  35  /0-  /1-  ;  c=-
30   (cel  11  4  11  /01)                 76   orth  35  35
31   set  $n  (div  $n  2)                 77   isc  35  5  35  ;  addition
32   cco  4  2  3  4                       78   if  (not  (te_isc  35  36))
33   )                                     79   (cel  23  4  23  /01)
34   con  10  11  10                       80   cco  4  2  3  4
35   copy  11  21  ;  a                    81   )
36   set  $b  12345                        82   con  10  23  10
37   set  $n  $b                           83   assign  10  /1  1  1
38   copy  1  11                           84   set  $s  0
39   for  $i  0  15                        85   set  $vb  1
40   (                                     86   for  $i  0  15
41   if  (eq  (mod  $n  2)  1)             87   (
42   (cel  11  4  11  /01)                 88   if  (not  (te_isc  23  4))
43   set  $n  (div  $n  2)                 89   (set  $s  (add  $s  $vb))
44   cco  4  2  3  4                       90   set  $vb  (mul  $vb  2)
45   )                                     91   cco  4  2  3  4
46   con  10  11  10                       92   )
```

TVL (10) D	16 Var.	3 R.	S. 1														
	x15	x14	x13	x12	x11	x10	x9	x8	x7	x6	x5	x4	x3	x2	x1	x0
1	0	0	0	0	0	0	0	0	0	0	0	0	1	1	0	1
2	0	0	1	1	0	0	0	0	0	0	1	1	1	0	0	1
3	0	0	1	1	0	0	0	0	0	1	0	0	0	1	1	0

Fig. 8.23 Problem-program that adds two integers using their created representation as binary vectors and converts the computed sum to an integer value together with the binary results

(c) Figure 8.24 shows a PRP that converts the integers 12, −12, 19283, and −19283 into binary vectors encoded by the two's complement and thereafter these binary vectors back into integer numbers.

The PRP of Fig. 8.24 is able to convert positive and negative integers into the associated binary vectors encoded in the two's complement and realizes also the inverse transformation. The same approach as used in the previous two tasks is reused to convert the positive numbers in both directions. The used procedure to convert a negative integer into the binary vector consists of three steps:

(1) Replace the negative number by the corresponding positive number realized by a multiplication with −1.
(2) Convert the positive number reusing the approach of the previous two tasks.
(3) Compute the two's complement using a complement of the created binary vector followed by an increment.

These three steps are used in reverse order to convert the binary vector of a negative number back into the corresponding integer:

(1) Compute the two's complement using a complement of the created binary vector followed by an increment.
(2) Convert the binary vector of the positive number reusing the approach of the previous two tasks.
(3) Replace the positive number by the corresponding negative number realized by a multiplication with −1.

The positive or negative number is given as value of the variable $n. The command lt in line 28 checks whether a negative number is given; if this is the case, the negative number is replaced by the associated positive number in line 30 and the type of the number is stored as Boolean value of variable $lt0.

The commands in lines 33–40 convert the positive integer $n into the binary vector stored as TVL 11.

The Boolean variable $lt0 is equal to true if a primarily negative number is given; in this case the two's complement of the binary vector of TVL 11 is computed in lines 42–61. The command cel in line 43 exchanged the values 0 and 1 in all positions of the binary vector of TVL 1.

The commands in lines 44–60 realize an increment of the binary vector. Starting with the least significant bit, a value 1 is changed to 0 as long as a value 1 is detected; the first detected value 0 is changed into a value 1 and further changes are disabled by the change of the control variable $reddy to the value true. The binary vector of the completely converted integer is appended to TVL 10 in line 62.

The value of the most significant bit (here x_{15}) is equal to 1 for the two's complement of a negative number. This property is checked in line 63. If TVL 11 contains a binary vector of a negative number, this property is stored as $lt0=true in line 65, the complement is computed using the command cel in line 66, and the increment is computed as before in lines 67–82. In this way, we get a binary vector of the positive number as TVL 11 and the associated sign information as value of $lt0.

The binary vector of the positive number is converted into the integer variable $nr in lines 83–91 reusing the approach of the previous two tasks. The command mul in line 93 changes $nr from a positive to the associated negative value in the case that $lt0=true.

The outer for-loop organizes together with the command if in the first and last part of its body that the commands in lines 27–93 are reused to convert the four given integers in both directions.

```
1   new                                    52   set $ready true
2   space 32 1                             53   ))
3   tin 1 1                                54   if (not $ready)
4   x15 x14 x13 x12 x11 x10 x9 x8          55   (
5   x7 x6 x5 x4 x3 x2 x1 x0.               56   if (not (te_isc 11 4))
6   0000000000000000.                      57   (cel 11 4 11 /10)
7   vtin 1 2                               58   )
8   x15 x14 x13 x12 x11 x10 x9 x8          59   cco 4 2 3 4
9   x7 x6 x5 x4 x3 x2 x1.                  60   )
10  vtin 1 3                               61   )
11  x14 x13 x12 x11 x10 x9 x8             62   con 10 11 10
12  x7 x6 x5 x4 x3 x2 x1 x0.              63   if (te_isc 11 5)
13  sbe 1 4                                64   (
14  x0=1.                                  65   set $lt0 true
15  sbe 1 5                                66   cel 11 1 11 /01 /10
16  x15=0.                                 67   set $ready false
17  dif 1 1 10                             68   for $i 0 15
18  for $case 1 4                          69   (
19  (if (eq $case 1)                       70   if (not $ready)
20  (set $n 12)                            71   (
21  if (eq $case 2)                        72   if (te_isc 11 4)
22  (set $n -12)                           73   (
23  if (eq $case 3)                        74   cel 11 4 11 /01
24  (set $n 19283)                         75   set $ready true
25  if (eq $case 4)                        76   ))
26  (set $n -19283)                        77   if (not $ready)
27  set $lt0 false                         78   (
28  if (lt $n 0)                           79   if (not (te_isc 11 4))
29  (                                      80   (cel 11 4 11 /10))
30  set $n (mul $n -1)                     81   cco 4 2 3 4
31  set $lt0 true                          82   ))
32  )                                      83   set $nr 0
33  copy 1 11                              84   set $vb 1
34  for $i 0 15                            85   for $i 0 15
35  (                                      86   (
36  if (eq (mod $n 2) 1)                   87   if (not (te_isc 11 4))
37  (cel 11 4 11 /01)                      88   (set $nr (add $nr $vb))
38  set $n (div $n 2)                      89   set $vb (mul $vb 2)
39  cco 4 2 3 4                            90   cco 4 2 3 4
40  )                                      91   )
41  if $lt0                                92   if $lt0
42  (                                      93   (set $nr (mul $nr -1))
43  cel 11 1 11 /01 /10                    94   if (eq $case 1)
44  set $ready false                       95   (set $nr1 $nr)
45  for $i 0 15                            96   if (eq $case 2)
46  (                                      97   (set $nr2 $nr)
47  if (not $ready)                        98   if (eq $case 3)
48  (                                      99   (set $nr3 $nr)
49  if (te_isc 11 4)                       100  if (eq $case 4)
50  (                                      101  (set $nr4 $nr)
51  cel 11 4 11 /01                        102  assign 10 /1 1 1
```

Fig. 8.24 Problem-program that converts two positive and two negative integer numbers into binary vector and vice versa

TVL (10) D | 16 Var. | 4 R. | S. 1															
x_{15}	x_{14}	x_{13}	x_{12}	x_{11}	x_{10}	x_9	x_8	x_7	x_6	x_5	x_4	x_3	x_2	x_1	x_0
0	0	0	0	0	0	0	0	0	0	0	0	1	1	0	0
1	1	1	1	1	1	1	1	1	1	1	1	0	1	0	0
0	1	0	0	1	0	1	1	0	1	0	1	0	0	1	1
1	0	1	1	0	1	0	0	1	0	1	0	1	1	0	1

Integer variables

Name	Value
$case	4
$i	15
$n	0
$nr	-19283
$nr1	12
$nr2	-12
$nr3	19283
$nr4	-19283
$vb	65536

Fig. 8.25 Binary vectors encoded by the two's complement belonging to the four integer values converted into the problem-program of Fig. 8.24 and the results of all integer variables

Figure 8.25 shows the four binary vectors computed as a result of the given integer values of the RPR shown in Fig. 8.24. It can be seen that the value 1 of the bit x_{15} indicates negative values. The numbers $nr1=12, $nr2=-12, $nr3=19283, and $nr4=-19283 stored as result of the transformation of the binary vectors to integers confirm that the PRP solves the specified task as expected.

(d) Figure 8.26 shows both the PRP that computes the difference $d = a - b$ of the two integers $a = 12$ and $b = 19283$ using their representation as two's complement encoded binary vectors and the details of the results.

The computation of the difference of binary vectors is realized by the addition of the binary vectors of a and $-b$; hence, components of the previous tasks can be reused.

The commands in lines 29–40 store the integer 12 as variable $a, convert this integer into the binary vector of TVL 11, append the created binary vector to TVL 10 for its visualization, and store it additionally as TVL 21 for the evaluation in the bitwise addition.

The two's complement is needed for subtrahend b; hence, the integer 19283 is stored as variable $b and also converted into the binary vector of TVL 11 in lines 41–50. However, the created binary vector of the positive number must be subsequently transformed into its two's complement; this is done in lines 51–70 where at the end the two's complement of b is appended to TVL 10 for its visualization and additionally stored as TVL 22 for the evaluation in the bitwise addition.

The commands in lines 71–100 realize the bitwise addition of the two 16-bit binary vectors of a and $-b$ stored as TVLs 21 and 22. The subsequent command con appends the computed difference vector of TVL 23 to TVL 10, and the command assign assigns this TVL to the 1-fold view.

The final part of this PRP converts the computed binary vector of the difference back into the integer variable $d. The binary vector of the computed difference stored in TVL 23 is copied to TVL 24 that is used for transformation into the integer number; hence, the sign information of the TVL 23 remains unchanged. The commands in lines 104–124 replace the two's complement of a negative number into the binary vector of the associated positive number. The commands in

```
 1  new                              46  if (eq (mod $n 2) 1)          91  if (not (te_isc 22 4))
 2  space 32 1                       47  (cel 11 4 11 /01)             92  (cel 35 31 35 /01)
 3  tin 1 1                          48  set $n (div $n 2)             93  cel 35 33 35 /0- /1-
 4  x15 x14 x13 x12 x11              49  cco 4 2 3 4                   94  cel 35 34 35 /0- /1-
 5  x10 x9 x8 x7 x6 x5               50  )                             95  orth 35 35
 6  x4 x3 x2 x1 x0.                  51  cel 11 1 11 /01 /10           96  isc 35 5 35 ; addition
 7  0000000000000000.                52  set $ready false              97  if (not(te_isc 35 36))
 8  vtin 1 2                         53  for $i 0 15                   98  (cel 23 4 23 /01)
 9  x15 x14 x13 x12 x11              54  (                             99  cco 4 2 3 4
10  x10 x9 x8 x7 x6 x5               55  if (not $ready)              100  )
11  x4 x3 x2 x1.                     56  (                            101  con 10 23 10
12  vtin 1 3                         57  if (te_isc 11 4)             102  assign 10 /1 1 1
13  x14 x13 x12 x11 x10              58  (                            103  copy 23 24
14  x9 x8 x7 x6 x5 x4                59  cel 11 4 11 /01              104  if (te_isc 23 6)
15  x3 x2 x1 x0.                     60  set $ready true             105  (
16  sbe 1 4                          61  ))                           106  cel 24 1 24 /01 /10
17  x0 = 1.                          62  if (not $ready)              107  set $ready false
18  avar 1                           63  (                            108  for $i 0 15
19  c s a b c0.                      64  if (not(te_isc 11 4))        109  (
20  sbe 1 5                          65  (cel 11 4 11 /10)            110  if (not $ready)
21  s=a#b#c0 ,                       66  )                            111  (
22  c=a&b+a&c0+b&c0 .                67  cco 4 2 3 4                  112  if (te_isc 24 4)
23  sbe 1 6                          68  )                            113  (
24  x15 = 0.                         69  con 10 11 10                 114  cel 24 4 24 /01
25  tin 1 7                          70  copy 11 22 ; -b              115  set $ready true
26  c s a b c0.                      71  vtin 1 30                    116  ))
27  00000.                           72  a.                           117  if (not $ready)
28  dif 1 1 10                       73  vtin 1 31                    118  (
29  set $a 12                        74  b.                           119  if (not(te_isc 11 4))
30  set $n $a                        75  vtin 1 32                    120  (cel 24 4 24 /10)
31  copy 1 11                        76  c0.                          121  )
32  for $i 0 15                      77  vtin 1 33                    122  cco 4 2 3 4
33  (                                78  s.                           123  )
34  if (eq (mod $n 2) 1)             79  vtin 1 34                    124  )
35  (cel 11 4 11 /01)                80  c.                           125  set $d 0
36  set $n (div $n 2)                81  sv_get 33 36                 126  set $vb 1
37  cco 4 2 3 4                      82  copy 7 35                    127  for $i 0 15
38  )                                83  copy 1 23                    128  (
39  con 10 11 10                     84  for $i 0 15                  129  if (not(te_isc 24 4))
40  copy 11 21 ; a                   85  (                            130  (set $d (add $d $vb))
41  set $b 19283                     86  cco 35 34 32 35              131  set $vb (mul $vb 2)
42  set $n $b                        87  cel 35 30 35 /10 /-0         132  cco 4 2 3 4
43  copy 1 11                        88  cel 35 31 35 /10 /-0         133  )
44  for $i 0 15                      89  if (not (te_isc 21 4))       134  if (te_isc 23 6)
45  (                                90  (cel 35 30 35 /01)           135  (set $d (mul $d -1))
```

TVL (10) D | 16 Var. | 3 R. | S. 1															
	$x15$	$x14$	$x13$	$x12$	$x11$	$x10$	$x9$	$x8$	$x7$	$x6$	$x5$	$x4$	$x3$	$x2$	$x1$	$x0$
1	0	0	0	0	0	0	0	0	0	0	0	0	1	1	0	0
2	1	0	1	1	0	1	0	0	1	0	1	0	1	1	0	1
3	1	0	1	1	0	1	0	0	1	0	1	1	1	0	0	1

Boolean variables

Name	Value
$ready	true

Integer variables

Name	Value
$a	12
$b	19283
$d	-19271
$i	15
$n	0
$vb	65536

Fig. 8.26 Problem-program that subtracts two integers using their binary vectors encoded as two's complements and converts the computed difference into an integer value together with the values of all control variables

```
 1   new                        14   01001                    27   (
 2   space  32  1               15   01010                    28   stv  2  $j  4
 3   tin  1  1                  16   01100                    29   if  (te_syd  3  4)
 4   y7  y4  y2  y1  y0.        17   10001                    30   (
 5   00011                      18   10010                    31   set  $d  (mod  $j  10)
 6   10001                      19   10100                    32   set  $w  (add  $w  $d)
 7   11000                      20   11000.                   33   )
 8   01001.                     21   set  $w  0               34   )
 9   tin  1  2                  22   for  $i  1  4            35   )
10   y7  y4  y2  y1  y0.        23   (                        36   assign  1  /m  1  1
11   00011                      24   stv  1  $i  3            37   assign  2  /m  1  2
12   00101                      25   set  $w  (mul  $w  10)
13   00110                      26   for  $j  1  10
```

K	TVL (1) ODA | 5 Var. | 4 R. | S. 1				
	y7	y4	y2	y1	y0
1	0	0	0	1	1
2	1	0	0	0	1
3	1	1	0	0	0
4	0	1	0	0	1

K	TVL (2) ODA | 5 Var. | 10 R. | S. 1				
	y7	y4	y2	y1	y0
1	0	0	0	1	1
2	0	0	1	0	1
3	0	0	1	1	0
4	0	1	0	0	1
5	0	1	0	1	0
6	0	1	1	0	0
7	1	0	0	0	1
8	1	0	0	1	0
9	1	0	1	0	0
10	1	1	0	0	0

Integer variables

Name	Value
$d	4
$i	4
$j	10
$w	1704

Fig. 8.27 Problem-program that decodes the sequence of 2-out-of-5 encoded code words of TVL 1 into the single integer $w together with table of code words and the computed result

lines 125–133 convert the positive binary vector into the positive integer number of the difference. The last two commands of this PRP change this positive number into the corresponding negative number if the computed difference is smaller than zero (indicated by the value of x_{15} in the binary vector of TVL 23).

Below the PRP shows TVL 10 the binary vectors of a, $-b$, and $d = a - b$. The highlighted value $d=-19271 confirms that the difference $12 - 19283$ has been computed correctly.

Solution 8.3 (Coding)

(a) Figure 8.27 shows the PRP that decodes the sequence of 2-out-of-5 encoded code words given as TVL 1 into the integer $w, the prepared table of code words stored as TVL 2, and all used integer variables with the highlighted result $w=1704.

The given sequence of code words has been specified as TVL 1. Thereafter, the table of code words of the 2-out-of-5 code is defined according to Fig. 8.9 as TVL 2; the indices of the rows 1 to 9 are the source symbols corresponding to the code words of these rows, and the code of the digit 0 is stored in the row with the index 10.

One code word of the given sequence is decoded in the inner for-loop in lines 26–34. The code word of TVL 2 belonging to the index $j of this loop is selected by the command stv in line 28

		x1	x2	x3	x4	y1	y2	y3	y4	y5	y6	y7
TVL (1) ODA | 11 Var. | 16 R. | S. 1												
1		1	1	1	1	1	1	1	1	1	1	1
2		0	1	1	1	0	0	0	1	1	1	1
3		1	0	1	1	0	1	1	0	0	1	1
4		0	0	1	1	1	0	0	0	0	1	1
5		1	1	0	1	1	0	1	0	1	0	1
6		0	1	0	1	0	1	0	0	1	0	1
7		1	0	0	1	0	0	1	1	0	0	1
8		0	0	0	1	1	1	0	1	0	0	1
9		1	1	1	0	0	0	1	0	1	1	0
10		0	1	1	0	1	1	0	0	1	1	0
11		1	0	1	0	1	0	1	1	0	1	0
12		0	0	1	0	0	1	0	1	0	1	0
13		1	1	0	0	0	1	1	1	1	0	0
14		0	1	0	0	1	0	0	1	1	0	0
15		1	0	0	0	1	1	1	0	0	0	0
16		0	0	0	0	0	0	0	0	0	0	0

```
1   new
2   space 32 1
3   avar 1
4   x1 x2 x3 x4
5   y1 y2 y3 y4 y5 y6 y7 .
6   sbe 1 1
7   y1=x1#x2#x4 ,
8   y2=x1#x3#x4 ,
9   y3=x1 ,
10  y4=x2#x3#x4 ,
11  y5=x2 ,
12  y6=x3 ,
13  y7=x4 .
14  assign 1 /1
```

Fig. 8.28 Problem-program that creates the coding table of the Hamming code $H(7,4)$ and the result of its execution

and compared by the command te_syd with the code word that must be decoded in the actual sweep of this for-loop. The command mod in line 31 maps the index $j=10$ to the needed digit 0, and the command add in line 32 inserts the decoded digit d to the last digit of the resulting integer w.

The command stv in line 24 in the body of the outer for-loop in lines 22–35 selects in each sweep one code word of the given TVL 1 for its decoding in this sweep. The command mul in line 25 moves the zero-initialized integer variable w by one decimal digit to the left so that the decoded digit can be inserted in line 32.

The highlighted value $w=1704$ in Fig. 8.27 shows the decoded result.

(b) Figure 8.28 shows the PRP that solves a system of logic equations that specifies the coding table of the Hamming code $H(7,4)$. The code words $\mathbf{y}$ are created for all 16 data words $\mathbf{x}$. The computed coding table confirms that the values of three tuples have an even parity:

- (y_1, y_3, y_5, y_7), where index bit 0 is equal to 1
- (y_2, y_3, y_6, y_7), where index bit 1 is equal to 1
- (y_4, y_5, y_6, y_7), where index bit 2 is equal to 1.

(c) Figure 8.29 shows both the PRP that uses the same coding table of the Hamming code $H(7,4)$ for the encoding of the four given data words and the decoding of the five given potential code words, and the TVLs of the corresponding binary vectors.

1 new	18 1101	35 con 11 4 11
2 space 32 1	19 0011	36)
3 avar 1	20 1001.	37 ctin 1 21 /0
4 x1 x2 x3 x4	21 tin 1 20	38 for $i 1 5
5 y1 y2 y3 y4	22 y1 y2 y3 y4	39 (
6 y5 y6 y7 .	23 y5 y6 y7 .	40 stv 20 $i 2
7 sbe 1 1	24 1110000	41 isc 2 1 3
8 y1=x1#x2#x4 ,	25 0000111	42 maxk 3 20 4
9 y2=x1#x3#x4 ,	26 0010110	43 if (te 4)
10 y3=x1 ,	27 0101010	44 (cpl 4 4)
11 y4=x2#x3#x4 ,	28 1010010.	45 con 21 4 21
12 y5=x2 ,	29 ctin 1 11 /0	46)
13 y6=x3 ,	30 for $i 1 4	47 assign 10 /m 1 1
14 y7=x4 .	31 (	48 assign 11 /m 1 2
15 tin 1 10	32 stv 10 $i 2	49 assign 20 /m 2 1
16 x1 x2 x3 x4 .	33 isc 2 1 3	50 assign 21 /m 2 2
17 0110	34 maxk 3 10 4	

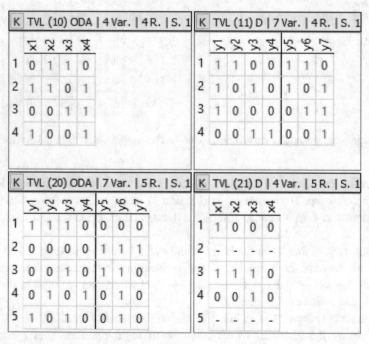

Fig. 8.29 Problem-program that encodes four data words into the associated code words and thereafter decodes five potential code words into the associated data words based on the Hamming code $H(7, 4)$ together with the results of its execution

The coding table of the Hamming code $H(7, 4)$, the given data words, and potential code words are determined in the first part of this PRP.

The coding table contains for each of the possible 16 data words $\mathbf{x}$ the associated code word $\mathbf{y}$; hence, each data word can be encoded. The for-loop in lines 30–36 realizes the encoding of the four given data words. The command stv in line 32 selects the data word to be encoded. The encoding is done by the command `isc` in line 33, and the command `maxk` in line 34 removes the used data bits from the code word. Finally, the command `con` appends in line 35 the computed code word to TVL 11 that has been prepared as an empty TVL before the execution of this loop. Only a subset of binary vectors **y** are valid code words of the Hamming code $H(7, 4)$; hence, it can be detected whether a received vector **y** has been changed into a vector not belonging to the valid code words. It is known that change of any single bit can be corrected using the information of the other bit of **y**. Here we only distinguish between valid and not valid code words; a vector of dashes as decoded data word **x** indicates that the given vector **y** is no valid code word.

The `for`-loop in lines 38–46 realizes the decoding of the five given potential code words. The same coding table (TVL 1) as used to encode the data words **x** as well as basically the same few steps of computation can also be used to decode the potential code words **y**; this emphasizes the universality of this approach.

The command `stv` in line 40 selects the potential code word to be decoded. The decoding is done by the command `isc` in line 41, and the command `maxk` in line 42 removes the used **y** bits to get the bits **x** of the data word.

An empty TVL as a result of the intersection remains empty as result of the k-fold maximum; the command `te` in line 43 detects such an empty TVL that indicates that the evaluated vector **y** is not a valid code word; the conditionally executed command `cpl` in line 44 transforms the empty TVL of **x** into a TVL with dashes in all positions of **x**. Finally, the command `con` appends in line 45 the computed data word to TVL 21 that has been prepared as an empty TVL before the execution of this loop.

The four commands `assign` assign side by side the TVLs of the encoding in the first row of the m-fold view and the TVLs of the decoding in the second row. Dash vectors of the rows 2 and 5 in TVL 21 indicate that the associated vectors of TVL 20 are no valid code words.

(d) Figure 8.30 shows both the PRP that solves based on Algorithm 3 of page 361 the Boolean differential equation (8.2) of the Reed–Muller code $\mathcal{RM}(1, 2)$ and the TVLs of the solution-set of the associated logic equation $D(\mathbf{u}) = 0$ as well as the computed eight code vectors of $\mathcal{RM}(1, 2)$.

The Boolean differential equation to solve

$$\operatorname*{der}_{(x_1, x_2)}{}^2 f(x_1, x_2) = 0$$

can be transformed into

$$\operatorname*{der}_{x_1} f(x_1, x_2) \oplus \operatorname*{der}_{x_2} f(x_1, x_2) \oplus \operatorname*{der}_{(x_1, x_2)} f(x_1, x_2) = 0 \,.$$

The associated logic equation is solved in lines 6 and 7. The commands `isc` in line 23 and `maxk` in line 24 transform the intermediate set of local solutions $SLS(\mathbf{u})$ into $S(\mathbf{v})$ using TVL 2 and VT 3. The two pairs of commands `cco` and `isc` in lines 25–28 realize the two sweeps of the `for`-loop in Algorithm 3 controlled by the appropriate prepared VTs.

TVL 8 shows the computed eight code words of the Reed–Muller code $\mathcal{RM}(1, 2)$.

1 new	11 v2=u0#u2,	21 vtin 1 7
2 space 32 1	12 v3=u0#u3.	22 v2 v3.
3 avar 1	13 vtin 1 3	23 isc 1 2 8
4 v0 v1 v2 v3	14 u0 u1 u2 u3.	24 maxk 8 3 8
5 u0 u1 u2 u3.	15 vtin 1 4	25 cco 8 4 5 9
6 sbe 1 1	16 v0 v2.	26 isc 8 9 8
7 u1#u2#u3=0.	17 vtin 1 5	27 cco 8 6 7 9
8 sbe 1 2	18 v1 v3.	28 isc 8 9 8
9 v0=u0,	19 vtin 1 6	29 assign 1 /m 1 1
10 v1=u0#u1,	20 v0 v1.	30 assign 8 /m 1 2

K	TVL (1) ODA	3 Var.	4 R.	S. 1
	u1	u2	u3
1	0	1	1
2	1	1	0
3	1	0	1
4	0	0	0

K	TVL (8) ODA	4 Var.	8 R.	S. 1
	v0	v1	v2	v3
1	0	0	1	1
2	1	1	0	0
3	1	0	0	1
4	0	1	1	0
5	0	1	0	1
6	1	0	1	0
7	1	1	1	1
8	0	0	0	0

Fig. 8.30 Problem-program that solves the Boolean differential equation of the Reed–Muller code $\mathcal{RM}(1, 2)$ together with the solutions of the associated logic equation and the final solution-set

Solution 8.4 (Specific Normal Form)

(a) Figure 8.31 shows both the PRP that implements Algorithm 5 to realize the expansion of the given two ternary vectors in all three positions and the requested representations of the results.

Algorithm 5 can be very efficiently implemented using few commands of the XBOOLE-monitor XBM 2; the short while-loop in lines 8–13 realizes all steps of this algorithm.

The command copy in line 7 creates a copy of the given TVL so that the given TVL 1 remains unchanged. The command sv_next controls the while-loop and provides for each sweep one of the three variables of the copied TVL 10 as TVL 11. The two commands cel in lines 10 and 11 change each of the possible values 0, 1, or—once in an arbitrarily chosen other value and thereafter remaining other values; hence, an explicit check of the existing value is not necessary. The command con in line 12 appends the two differently changed TVLs 11 and 12 to TVL 10 used in the next sweep of the while-loop or representing the result.

The command te_syd in line 18 checks whether the given and expanded TVLs describe the same function. This command requires orthogonal TVLs as input that are prepared in the two preceding command orth. The result $equal=true confirms that TVLs 1 and 10 describe the same function. The Karnaugh-maps belonging to these two TVLs also confirm this property. Each of the two given ternary vectors of three variables has been expanded into $2^3 = 8$ ternary vectors.

```
 1   new
 2   space 32 1
 3   tin 1 1 /a
 4   x1 x2 x3.
 5   01-
 6   -01.
 7   copy 1 10
 8   while (sv_next 10 11 11)
 9   (
10   cel 10 11 12 /01 /1- /-0
11   cel 10 11 13 /0- /10 /-1
12   con 12 13 10
13   )
14   assign 1 /m 1 1
15   assign 10 /m 1 2
16   orth 1 30
17   orth 10 31
18   set $equal (te_syd 30 31)
19   assign 1 /m 1 1
20   assign 10 /m 2 1
21   assign 1 /m 2 1
22   assign 10 /m 2 2
```

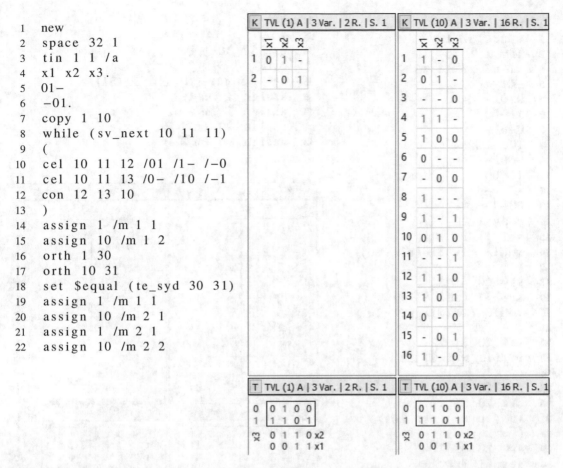

K	TVL (1) A	3 Var.	2 R.	S. 1	
	$\bar{x}_1$	$\bar{x}_2$	$\bar{x}_3$
1	0	1	-
2	-	0	1

K	TVL (10) A	3 Var.	16 R.	S. 1	
	$\bar{x}_1$	$\bar{x}_2$	$\bar{x}_3$
1	1	-	0
2	0	1	-
3	-	-	0
4	1	1	-
5	1	0	0
6	0	-	-
7	-	0	0
8	1	-	-
9	1	-	1
10	0	1	0
11	-	-	1
12	1	1	0
13	1	0	1
14	0	-	0
15	-	0	1
16	1	-	0

T TVL (1) A | 3 Var. | 2 R. | S. 1

```
0   0 1 0 0
1   1 1 0 1
x3  0 1 1 0 x2
    0 0 1 1 x1
```

T TVL (10) A | 3 Var. | 16 R. | S. 1

```
0   0 1 0 0
1   1 1 0 1
x3  0 1 1 0 x2
    0 0 1 1 x1
```

Fig. 8.31 Problem-program that completely expands the given TVL based on Algorithm 5 together with the given and expanded TVLs

(b) Figure 8.32 shows the PRP that removes all pairs of equal ternary vectors and verifies whether this simplification changes the represented function. The comparison of the given TVL 1 and the simplified TVL 20 shows that two pairs of identical ternary vectors $\{(2, 4), (3, 7)\}$ have been removed. Identical Karnaugh-maps of these functions as well as the result $equal=true confirm that removing of these pairs of identical ternary vectors does not change the represented function.

The command copy in line 14 creates a copy of the given TVL so that the given TVL 1 remains unchanged. The outer while-loop in lines 16–35 determines the index i (initialized by the value 1) of the first ternary vector to compare, and the inner while-loop in lines 20–33 specifies the index j (initialized by the value $i + 1$) of the second ternary vector used for comparison. The command lt in line 16 terminates the outer while-loop when the last but one ternary vector has been evaluated, and the command le in line 20 terminates the inner while-loop when last ternary vector has been compared.

The command stv in lines 18 and 22 selects the ternary vectors that are compared by the command te_syd. The command dtv in lines 25 and 26 deletes the detected two identical ternary vectors; note, the second vector is first deleted because the indices of all subsequent TVs are reduced by 1 as a consequence of the removed vector. The commands stv and set in

```
 1  new                            34  set $i (add $i 1)
 2  space 32 1                     35  )
 3  tin 1 1 /a                     36  orth 1 30
 4  x1 x2 x3 x4 x5.                37  orth 20 31
 5  -01-1                          38  set $equal (te_syd 30 31)
 6  0-01-                          39  assign 1 /m 1 1
 7  11-00                          40  assign 1 /m 2 1
 8  0-01-                          41  assign 20 /m 1 2
 9  100-1                          42  assign 20 /m 2 2
10  -001-
11  11-00
12  -11--
13  00-1-.
14  copy 1 20
15  set $i 1
16  while ( lt $i (ntv 20))
17  (
18  stv 20 $i 21
19  set $j (add $i 1)
20  while ( le $j (ntv 20))
21  (
22  stv 20 $j 22
23  if (te_syd 21 22)
24  (
25  dtv 20 $j
26  dtv 20 $i
27  if ( lt $i (ntv 20))
28  (
29  stv 20 $i 21
30  set $j $i
31  ))
32  set $j (add $j 1)
33  )
```

K | TVL (1) A | 5 Var. | 9 R. | S. 1

	x1	x2	x3	x4	x5
1	-	0	1	-	1
2	0	-	0	1	-
3	1	1	-	0	0
4	0	-	0	1	-
5	1	0	0	-	1
6	-	0	0	1	-
7	1	1	-	0	0
8	-	1	1	-	-
9	0	0	-	1	-

K | TVL (20) A | 5 Var. | 5 R. | S. 1

	x1	x2	x3	x4	x5
1	-	0	1	-	1
2	1	0	0	-	1
3	-	0	0	1	-
4	-	1	1	-	-
5	0	0	-	1	-

T | TVL (1) A | 5 Var. | 9 R. | S. 1

```
0 0    0 0 1 0 0 0 1 0 0
0 1    0 1 1 0 0 1 1 1
1 1    0 0 1 0 0 1 1 0
1 0    0 1 1 0 0 1 0 1
x4 x5  0 1 1 0 0 1 1 0  x3
       0 0 1 1 1 1 0 0  x2
       0 0 0 0 1 1 1 1  x1
```

T | TVL (20) A | 5 Var. | 5 R. | S. 1

```
0 0    0 0 1 0 0 1 0 0
0 1    0 1 1 0 0 1 1 1
1 1    0 0 1 0 0 1 1 0
1 0    0 1 1 0 0 1 0 1
x4 x5  0 1 1 0 0 1 1 0  x3
       0 0 1 1 1 1 0 0  x2
       0 0 0 0 1 1 1 1  x1
```

Fig. 8.32 Problem-program that realizes Algorithm 6 to remove all pairs of equal ternary vectors from the given TVL in A-form together with the given and reduced TVLs

lines 29 and 30 prepare the comparison of pairs of ternary vectors after removing two identical ternary vectors.

(c) Figure 8.33 shows both the PRP that computes the SNF(f) and the associated TVLs. This PRP reuses the main parts of the PRPs explained in the last two tasks; the complete expansion of the given TVL is computed in lines 7–13, and subsequently, Algorithm 6 has been implemented in lines 14–35.

(d) Figure 8.34 shows the PRP in which the functions $f_1(\mathbf{x})$ (TVL 1), $f_2(\mathbf{x})$ (TVL 2), and $f_3(\mathbf{x})$ (TVL 3) are determined and the associated specific normal norms SNF(f_1) (TVL 4), SNF(f_2) (TVL 5), and SNF(f_3) (TVL 6) are computed within the body of the for-loop in lines 15–50. The last two commands assign of this loop assign the TVLs of the evaluated function and the associated SNF as requested to the viewports of the m-fold view.

Figure 8.35 confirms that the number of columns with different ternary values but not the values itself determines the number of ternary vectors of an SNF; the two ternary vectors of TVLs 1 and

```
1   new                              20   while ( le $j (ntv 20))
2   space 32 1                       21   (
3   tin 1 1 /a                       22   stv 20 $j 22
4   x1 x2 x3 .                       23   if ( te_syd 21 22 )
5   0-1                              24   (
6   101 .                            25   dtv 20 $j
7   copy 1 10                        26   dtv 20 $i
8   while ( sv_next 10 11 11 )       27   if ( lt $i (ntv 20))
9   (                                28   (
10  cel 10 11 12 /01 /1- /-0         29   stv 20 $i 21
11  cel 10 11 13 /0- /10 /-1         30   set $j $i
12  con 12 13 10                     31   ))
13  )                                32   set $j (add $j 1)
14  copy 10 20                       33   )
15  set $i 1                         34   set $i (add $i 1)
16  while ( lt $i (ntv 20))          35   )
17  (                                36   assign 1 /m 1 1
18  stv 20 $i 21                     37   assign 10 /m 1 2
19  set $j (add $i 1)                38   assign 20 /m 1 3
```

K	TVL (1) A | 3 Var. | 2 R. | S. 1		
	x1	x2	x3
1	0	-	1
2	1	0	1

K	TVL (10) A | 3 Var. | 16 R. | S. 1		
	x1	x2	x3
1	1	0	-
2	-	1	-
3	-	0	-
4	0	1	-
5	1	1	-
6	-	-	-
7	-	1	-
8	0	-	-
9	1	0	0
10	-	1	0
11	-	0	0
12	0	1	0
13	1	1	0
14	-	-	0
15	-	1	0
16	0	-	0

K	TVL (20) A | 3 Var. | 12 R. | S. 1		
	x1	x2	x3
1	1	0	-
2	-	0	-
3	0	1	-
4	1	1	-
5	-	-	-
6	0	-	-
7	1	0	0
8	-	0	0
9	0	1	0
10	1	1	0
11	-	-	0
12	0	-	0

Fig. 8.33 Problem-program that computes the SNF(f) together the TVLs of the given function $f(\mathbf{x})$, the completely expanded function, and the SNF(f)

```
 1   new                                26   while (lt $i (ntv 20))
 2   space 32 1                         27   (
 3   tin 1 1 /a                         28   stv 20 $i 21
 4   x1 x2 x3.                          29   set $j (add $i 1)
 5   01-                                30   while (le $j (ntv 20))
 6   -10.                               31   (
 7   tin 1 2 /a                         32   stv 20 $j 22
 8   x1 x2 x3.                          33   if (te_syd 21 22)
 9   00-                                34   (
10   010.                               35   dtv 20 $j
11   tin 1 3 /a                         36   dtv 20 $i
12   x1 x2 x3.                          37   if (lt $i (ntv 20))
13   01-                                38   (
14   1-0.                               39   stv 20 $i 21
15   for $f 1 3                         40   set $j $i
16   (                                  41   ))
17   copy $f 10                         42   set $j (add $j 1)
18   while (sv_next 10 11 11)           43   )
19   (                                  44   set $i (add $i 1)
20   cel 10 11 12 /01 /1- /-0           45   )
21   cel 10 11 13 /0- /10 /-1           46   set $snf (add $f 3)
22   con 12 13 10                       47   copy 20 $snf
23   )                                  48   assign $f /m 1 $f
24   copy 10 20                         49   assign $snf /m 2 $f
25   set $i 1                           50   )
```

Fig. 8.34 Problem-program that computes the Specific Normal Forms of three functions

2 are different in two columns and the associated SNFs contain in each case 12 ternary vectors. The three different columns of the two ternary vectors of TVL 3 cause 14 ternary vectors in the SNF because only one pair of equal ternary vectors could be deleted.

Solution 8.5 (Most Complex Functions)

(a) Figure 8.36 shows the PRP that computes minimized ODA-forms of all logic functions of one variable and takes the number of conjunctions of these TVLs as measure of the complexity. The associated vectors of function values are stored as TVLs 20 (no conjunction in the ODA-form), 21 (one conjunction in the ODA-form), and 22 (two conjunctions in the ODA-form, such functions do not exist); these TVLs are shown below the PRP in Fig. 8.36.

The outer for-loop in lines 22–45 iterates over the four functions of $\mathbb{B}^1$. The inner for-loop in lines 27–37 iterates over the two function values and creates both the ODA-form (TVL 11, using the commands stv and uni) and the vector of function values (TVL 12, using the command cel) of the logic function determined by the index $i of the outer for-loop.

The command obbc in line 38 minimizes the number of rows in TVL 11, and the subsequent commands until the end of the outer for-loop append the computed vector of function values to the TVL that represents functions with an associated number of conjunctions in the minimal ODA-form (TVL 20, 21, or 22).

The three computed TVLs shown in Fig. 8.36 reveal that the function $f_0(x_1) = 0$ does not need any ternary vector for its representation and the other three functions $f_1(x_1) = \overline{x}_1$, $f_2(x_1) = x_1$, and $f_3(x_1) = 1$ can be represented by one ternary vector with the values 0, 1, or $-$, respectively. There is no function of one variable that needs two ternary vectors in their minimized ODA-form;

K	TVL (1) A | 3 Var. | 2R. | S. 1		
	$\overline{x}_1$	x_2	x_3
1	0	1	-
2	-	1	0

K	TVL (2) A | 3 Var. | 2R. | S. 1		
	$\overline{x}_1$	x_2	x_3
1	0	0	-
2	0	1	0

K	TVL (3) A | 3 Var. | 2R. | S. 1		
	$\overline{x}_1$	x_2	x_3
1	0	1	-
2	1	-	0

K	TVL (4) A | 3 Var. | 12R. | S. 1		
	$\overline{x}_1$	x_2	x_3
1	1	-	0
2	0	-	1
3	-	-	0
4	1	0	0
5	0	0	1
6	-	0	0
7	0	-	-
8	-	-	1
9	1	-	-
10	0	0	-
11	-	0	1
12	1	0	-

K	TVL (5) A | 3 Var. | 12R. | S. 1		
	$\overline{x}_1$	x_2	x_3
1	1	1	0
2	-	1	0
3	1	-	0
4	1	0	1
5	-	-	0
6	-	0	1
7	1	1	1
8	1	-	-
9	-	1	1
10	-	-	-
11	1	0	-
12	-	0	-

K	TVL (6) A | 3 Var. | 14R. | S. 1		
	$\overline{x}_1$	x_2	x_3
1	1	-	0
2	-	-	0
3	0	0	1
4	1	0	0
5	-	1	1
6	-	0	0
7	0	1	1
8	1	-	1
9	-	0	-
10	-	-	1
11	0	0	-
12	1	0	1
13	-	1	-
14	0	1	-

Fig. 8.35 Results of the PRP shown in Fig. 8.34

hence, the most complex functions of $\mathbb{B}^1$ are the three functions: $f_1(x_1) = \overline{x}_1$, $f_2(x_1) = x_1$, and $f_3(x_1) = 1$.

(b) Figure 8.37 shows the PRP that computes minimized ODA-forms of all logic functions of two variables and takes the number of conjunctions of these TVLs as measure of the complexity.

The problem-program in Fig. 8.37 uses the same approach as explained in the solution of the previous task. Only the initial TVLs and VTs as well as the intervals of the two for-loops have been extended to satisfy the properties of $\mathbb{B}^2$.

The computed TVLs shown at the bottom of Fig. 8.37 confirm that no function of two variables needs more than two conjunctions in its minimized ODA-form. The set of vectors of function values for the most complex functions of two variables coincides with the known most complex functions of Table 8.16 for $n = 2$; hence, the number of ternary vectors of a minimized ODA-form of a logic function of two variables can be used to determine its complexity.

(c) Figure 8.38 shows both the PRP that computes all most complex functions $f_i(\mathbf{x})$ of three variables based on the number of conjunctions of $\text{SNF}(f_i)$ and the 12 vectors of function values $(f_7, f_6, f_5, f_4, f_3, f_2, f_1, f_0)$ of the most complex functions $f_i(x_1, x_2, x_3)$.

The outer for-loop in lines 27–79 iterates over the $2^{2^3} = 256$ functions of $\mathbb{B}^3$. The inner for-loop in lines 32–42 iterates over the eight function values and creates as intermediate result

```
 1   new                          25   copy  1  11
 2   space  32  1                 26   copy  2  12
 3   tin  1  1                    27   for  $j  0  1
 4   x1 .                         28   (
 5   .                            29   if  (eq  (mod  $f  2)  1)
 6   tin  1  2                    30   (
 7   f1  f0 .                     31   stv  6  (add  $j  1)  7
 8   00 .                         32   uni  11  7  11
 9   sbe  1  3                    33   cel  12  3  12  /01
10   f0 = 1 .                     34   )
11   vtin  1  4                   35   cco  3  4  5  3
12   f1 .                         36   set  $f  (div  $f  2)
13   vtin  1  5                   37   )
14   f0 .                         38   obbc  11  11
15   tin  1  6  /d                39   if  (eq  (ntv  11)  0)
16   x1 .                         40   (con  20  12  20)
17   0                            41   if  (eq  (ntv  11)  1)
18   1 .                          42   (con  21  12  21)
19   dif  2  2  20                43   if  (eq  (ntv  11)  2)
20   dif  2  2  21                44   (con  22  12  22)
21   dif  2  2  22                45   )
22   for  $i  0  3                46   assign  20  /m  1  1
23   (                            47   assign  21  /m  1  2
24   set  $f  $i                  48   assign  22  /m  1  3
```

K	TVL (20) D | 2 Var. | 1 R. | S. 1	
	강	우
1	0	0

K	TVL (21) D | 2 Var. | 3 R. | S. 1	
	강	우
1	0	1
2	1	0
3	1	1

K	TVL (22) ODA | 2 Var. | 0 R. | S. 1	
	강	우

Fig. 8.36 Problem-program that explores the complexity of all functions of one variable and creates associated sets of vectors of function values together with the computed TVL of these sets

both the ODA-form (TVL 11) and the vector of function values (TVL 12) of the logic function determined by the index $i of the outer for-loop.

The ODA-form of the function $f_i(\mathbf{x})$ to explore (TVL 11) must be transformed into its specific normal form $\text{SNF}(f_i)$. The form of TVL 11 of this intermediate function is changed into an A-form of TVL 110 that is completely expanded by the commands in the while-loop in lines 44–49 based on Algorithm 5.

The second step to compute the specific normal form $\text{SNF}(f_i)$ consists in removing all pairs of equal ternary vectors of the expanded TVL 110 based on Algorithm 6. The completely expanded TVL 110 is copied to TVL 120 for which Algorithm 6 ($R(f)$ that removes all pairs of equal ternary vectors) has been implemented in two nested while-loops in lines 52–71. The two commands dtv in lines 61 and 62 delete the identified pair of equal ternary vectors starting with the vector of the higher index $jsnf.

The number $nmax has been initialized with the value 0 and stores later the maximal number of ternary vectors of the TVL of an already computed $\text{SNF}(f_i)$. The command con in line 73 appends the associated vector of function values to the set of solutions in the case that $nmax is

1 new	26 set $f $i
2 space 32 1	27 copy 1 11
3 tin 1 1	28 copy 2 12
4 x1 x2.	29 for $j 0 3
5 .	30 (
6 tin 1 2	31 if (eq (mod $f 2) 1)
7 f3 f2 f1 f0.	32 (
8 0000.	33 stv 6 (add $j 1) 7
9 sbe 1 3	34 uni 11 7 11
10 f0 = 1.	35 cel 12 3 12 /01
11 vtin 1 4	36)
12 f3 f2 f1.	37 cco 3 4 5 3
13 vtin 1 5	38 set $f (div $f 2)
14 f2 f1 f0.	39)
15 tin 1 6 /d	40 obbc 11 11
16 x1 x2.	41 if (eq (ntv 11) 0)
17 00	42 (con 20 12 20)
18 10	43 if (eq (ntv 11) 1)
19 01	44 (con 21 12 21)
20 11.	45 if (eq (ntv 11) 2)
21 dif 2 2 20	46 (con 22 12 22)
22 dif 2 2 21	47)
23 dif 2 2 22	48 assign 20 /m 1 1
24 for $i 0 15	49 assign 21 /m 1 2
25 (	50 assign 22 /m 1 3

K	TVL (20) D	4 Var.	1 R.	S. 1
	f3	f2	f1	f0
1	0	0	0	0

K	TVL (21) D	4 Var.	9 R.	S. 1
	f3	f2	f1	f0
1	0	0	0	1
2	0	0	1	0
3	0	0	1	1
4	0	1	0	0
5	0	1	0	1
6	1	0	0	0
7	1	0	1	0
8	1	1	0	0
9	1	1	1	1

K	TVL (22) D	4 Var.	6 R.	S. 1
	f3	f2	f1	f0
1	0	1	1	0
2	0	1	1	1
3	1	0	0	1
4	1	0	1	1
5	1	1	0	1
6	1	1	1	0

Fig. 8.37 Problem program that explores the complexity of all functions of two variables and creates associated sets of vectors of function value together with the computed TVL of these sets

equal to the number of ternary vectors of the SNF(f_i) (TVL 120). The command copy in line 76 replaces the vectors of function values of the so far found largest specific normal forms (TVL 100) by TVL 12 of the new larger SNF(f_i), and the commands set and ntv in line 77 store the new larger number of ternary vectors of an analyzed SNF(f_i) as new value of $nmax.

```
 1   new
 2   space 32 1
 3   tin 1 1
 4   x1 x2 x3.
 5   .
 6   tin 1 2
 7   f7 f6 f5 f4 f3 f2 f1 f0.
 8   00000000.
 9   sbe 1 3
10   f0=1.
11   vtin 1 4
12   f7 f6 f5 f4 f3 f2 f1.
13   vtin 1 5
14   f6 f5 f4 f3 f2 f1 f0.
15   tin 1 6 /d
16   x1 x2 x3.
17   000
18   100
19   010
20   110
21   001
22   101
23   011
24   111.
25   dif 2 2 100
26   set $nmax 0
27   for $i 0 255
28   (
29   set $f $i
30   copy 1 11
31   copy 2 12
32   for $j 0 7
33   (
34   if (eq (mod $f 2) 1)
35   (
36   stv 6 (add $j 1) 7
37   uni 11 7 11
38   cel 12 3 12 /01
39   )
40   cco 3 4 5 3
41   set $f (div $f 2)
42   )
43   sform 11 /a 110
44   while (sv_next 110 111 111)
45   (
46   cel 110 111 112 /01 /1- /-0
47   cel 110 111 113 /0- /10 /-1
48   con 112 113 110
49   )
50   copy 110 120
```

```
51   set $isnf 1
52   while (lt $isnf (ntv 120))
53   (
54   stv 120 $isnf 121
55   set $jsnf (add $isnf 1)
56   while (le $jsnf (ntv 120))
57   (
58   stv 120 $jsnf 122
59   if (te_syd 121 122)
60   (
61   dtv 120 $jsnf
62   dtv 120 $isnf
63   if (lt $isnf (ntv 120))
64   (
65   stv 120 $isnf 121
66   set $jsnf $isnf
67   ))
68   set $jsnf (add $jsnf 1)
69   )
70   set $isnf (add $isnf 1)
71   )
72   if (eq (ntv 120) $nmax)
73   (con 100 12 100)
74   if (gt (ntv 120) $nmax)
75   (
76   copy 12 100
77   set $nmax (ntv 120)
78   )
79   )
80   assign 100 /1
```

K	TVL (100) D	8 Var.	12 R.	S. 1					
	f7	f6	f5	f4	f3	f2	f1	f0
1	0	1	1	0	1	0	1	1
2	0	1	1	0	1	1	0	1
3	0	1	1	1	1	0	0	1
4	0	1	1	1	1	1	1	0
5	1	0	0	1	0	1	1	1
6	1	0	0	1	1	1	1	0
7	1	0	1	1	0	1	1	0
8	1	0	1	1	1	1	0	1
9	1	1	0	1	0	1	1	0
10	1	1	0	1	1	0	1	1
11	1	1	1	0	0	1	1	1
12	1	1	1	0	1	0	0	1

Fig. 8.38 Problem-program that computes the most complex functions of three variables based on the number of conjunctions in their specific normal forms together with the set of computed solution-functions

The TVL in Fig. 8.38 shows the 12 vectors of function values for which the $\mathrm{SNF}(f_i)$ has the largest number of ternary vectors: $\$nmax=18$. Exactly these 12 functions have been specified in Table 8.16 as most complex functions of $n = 3$ variables.

(d) Figure 8.39 shows a PRP that computes the vectors of function values of all most complex functions depending on one to four variables using an iterative version of Algorithm 7 ($\mathrm{MCF}(n, s)$ on page 448).

The outer for-loop in lines 38–112 iterates over the Boolean spaces $\mathbb{B}^1$ to $\mathbb{B}^4$ so that the most complex functions for these spaces are iteratively computed. The number $\$smax$ of most complex functions is equal to $3 * 2^{n-1}$, where n is the dimension of the Boolean space $\mathbb{B}^n$. The value of $\$smax$ is computed in lines 40–46. The subsequent command sub decrements this value so that the index of the computed most complex functions is in the range of $0, \ldots, smax - 1$ as used in Table 8.16. The inner for-loop in lines 49–110 computes the $\$smax$ most complex functions of the Boolean space determined by the index $\$n$ of the outer for-loop.

Algorithm 7 specifies a special simple approach to compute the most complex functions for the initial case of $\mathbb{B}^1$. The commands in lines 51–61 realize this approach by converting the integer $\$f$ into an associated binary vector that is appended to TVL 11 of the most complex functions of $\mathbb{B}^1$ (see Fig. 8.40).

Algorithm 7 distinguishes between two cases to compute a most complex function of $\mathbb{B}^n, n > 1$, based on the known most complex function of $\mathbb{B}^{n-1}$; case $\$d=0$ is realized in the body of the command if in lines 69–88 and case $\$d=1$ in the body of the command if in lines 89–108, respectively. The command stv selects the determined vectors of function values of the most complex function of the next smaller Boolean space, the command cco moves the selected vectors to the higher valued parts of the vectors to compute (this realizes the multiplication with 2^{len2} of Algorithm 7, and the command isc combines the two parts of vectors of function values of the new most complex function. The command con appends the computed vectors to the TVL of most complex functions associated to the Boolean space determined by the index $\$n$ of the outer for-loop.

Figure 8.40 shows the vectors of function values of all most complex functions of $\mathbb{B}^1$ (TVL 11), $\mathbb{B}^2$ (TVL 12), $\mathbb{B}^3$ (TVL 13), and $\mathbb{B}^4$ (TVL 14). The comparison of Fig. 8.40 and Table 8.16 confirms that the PRP of Fig. 8.39 computes the most complex functions of $\mathbb{B}^1$ to $\mathbb{B}^4$ correctly. This PRP can be extended to compute the most complex functions of larger Boolean spaces.

8.9   Supplementary Exercises

Exercise 8.6 (Logic Systems)
   Verify that:

(a) The logic system P specified on page 421 by the axiom A1 and the rules R1–R4
(b) The logic system *Principia Mathematica* specified on page 422 by three rules
(c) The logic system of *Rosser* specified on page 422 also by three rules
(d) The logic system for *negation and implication* specified on page 423 by four rules

are tautologies.

Exercise 8.7 (Binary Addition, Subtraction, Multiplication, and Division)
   The following four pairs of integers are given:

$$a_1 = 27, \qquad\qquad\qquad b_1 = 7,$$

```
 1   new                                      57   set $f (div $f 2)
 2   space 32 1                               58   if (eq (mod $f 2) 1)
 3   avar 1                                   59   (cel 20 2 20 /01)
 4   f15 f14 f13 f12 f11 f10 f9 f8            60   con $mcf 20 $mcf
 5   f7 f6 f5 f4 f3 f2 f1 f0 .                61   )
 6   vtin 1 1                                 62   if (gt $n 1)
 7   f0 .                                     63   (
 8   vtin 1 2                                 64   set $r (mod $s 3)
 9   f1 .                                     65   set $sel (div $s 3)
10   vtin 1 3                                 66   set $d (mod $sel 2)
11   f1 f0 .                                  67   set $b (div $sel 2)
12   vtin 1 4                                 68   set $b (mul $b 3)
13   f3 f2 .                                  69   if (eq $d 0)
14   vtin 1 5                                 70   (
15   f3 f2 f1 f0 .                            71   set $i (add $r 1)
16   vtin 1 6                                 72   set $i (mod $i 3)
17   f7 f6 f5 f4 .                            73   set $i (add $i $b)
18   vtin 1 7                                 74   set $i (add $i 1)
19   f7 f6 f5 f4 f3 f2 f1 f0 .                75   set $st (add $n 9)
20   vtin 1 8                                 76   set $hv (mul $n 2)
21   f15 f14 f13 f12 f11 f10 f9 f8 .          77   set $lv (sub $hv 1)
22   tin 1 9                                  78   stv $st $i 21
23   f1 f0 .                                  79   cco 21 $lv $hv 22
24   00 .                                     80   set $i (add $r 2)
25   tin 1 11                                 81   set $i (mod $i 3)
26   f1 f0 .                                  82   set $i (add $i $b)
27   .                                        83   set $i (add $i 1)
28   tin 1 12                                 84   set $st (add $n 9)
29   f3 f2 f1 f0 .                            85   stv $st $i 23
30   .                                        86   isc 22 23 24
31   tin 1 13                                 87   con $mcf 24 $mcf
32   f7 f6 f5 f4 f3 f2 f1 f0 .                88   )
33   .                                        89   if (eq $d 1)
34   tin 1 14                                 90   (
35   f15 f14 f13 f12 f11 f10 f9 f8            91   set $i (add $r 2)
36   f7 f6 f5 f4 f3 f2 f1 f0 .                92   set $i (mod $i 3)
37   .                                        93   set $i (add $i $b)
38   for $n 1 4                               94   set $i (add $i 1)
39   (                                        95   set $st (add $n 9)
40   set $smax 1                              96   set $hv (mul $n 2)
41   if (gt $n 1)                             97   set $lv (sub $hv 1)
42   (                                        98   stv $st $i 21
43   for $h 2 $n                              99   cco 21 $lv $hv 22
44   (set $smax (mul $smax 2))               100   set $i (add $r 1)
45   )                                       101   set $i (mod $i 3)
46   set $smax (mul $smax 3)                 102   set $i (add $i $b)
47   set $smax (sub $smax 1)                 103   set $i (add $i 1)
48   set $mcf (add $n 10)                    104   set $st (add $n 9)
49   for $s 0 $smax                          105   stv $st $i 23
50   (                                       106   isc 22 23 24
51   if (eq $n 1)                            107   con $mcf 24 $mcf
52   (                                       108   )
53   copy 9 20                               109   )
54   set $f (add $s 1)                       110   )
55   if (eq (mod $f 2) 1)                    111   assign $mcf /m $n 1
56   (cel 20 1 20 /01)                       112   )
```

Fig. 8.39 Problem-program that computes the sets of vectors of function values that represent the most complex functions belonging to the Boolean spaces $\mathbb{B}^1$ to $\mathbb{B}^4$

K	TVL (11) D	2 Var.	3 R.	S. 1	
		f1	f0
1		0	1
2		1	0
3		1	1

K	TVL (12) D	4 Var.	6 R.	S. 1			
	f3	f2	f1	f0
1	1	0	1	1
2	1	1	0	1
3	0	1	1	0
4	1	1	1	0
5	0	1	1	1
6	1	0	0	1

K	TVL (13) D	8 Var.	12 R.	S. 1							
	f7	f6	f5	f4	f3	f2	f1	f0
1	1	1	0	1	0	1	1	0
2	0	1	1	0	1	0	1	1
3	1	0	1	1	1	1	0	1
4	0	1	1	0	1	1	0	1
5	1	0	1	1	0	1	1	0
6	1	1	0	1	1	0	1	1
7	0	1	1	1	1	0	0	1
8	1	0	0	1	1	1	1	0
9	1	1	1	0	0	1	1	1
10	1	0	0	1	0	1	1	1
11	1	1	1	0	1	0	0	1
12	0	1	1	1	1	1	1	0

	TVL (14) D	16 Var.	24 R.	S. 1													
	f15	f14	f13	f12	f11	f10	f9	f8	f7	f6	f5	f4	f3	f2	f1	f0
1	0	1	1	0	1	0	1	1	1	0	1	1	1	1	0	1
2	1	0	1	1	1	1	0	1	1	1	0	1	0	1	1	0
3	1	1	0	1	0	1	1	0	0	1	1	0	1	0	1	1
4	1	0	1	1	1	1	0	1	0	1	1	0	1	0	1	1
5	1	1	0	1	0	1	1	0	1	0	1	1	1	1	0	1
6	0	1	1	0	1	0	1	1	1	1	0	1	0	1	1	0
7	1	0	1	1	0	1	1	0	1	1	0	1	1	0	1	1
8	1	1	0	1	1	0	1	1	0	1	1	0	1	1	0	1
9	0	1	1	0	1	1	0	1	1	0	1	1	0	1	1	0
10	1	1	0	1	1	0	1	1	1	0	1	1	0	1	1	0
11	0	1	1	0	1	1	0	1	1	1	0	1	1	0	1	1
12	1	0	1	1	0	1	1	0	0	1	1	0	1	1	0	1
13	1	0	0	1	1	1	1	0	1	1	1	0	0	1	1	1
14	1	1	1	0	0	1	1	1	0	1	1	1	1	0	0	1
15	0	1	1	1	1	0	0	1	1	0	0	1	1	1	1	0
16	1	1	1	0	0	1	1	1	1	0	0	1	1	1	1	0
17	0	1	1	1	1	0	0	1	1	1	1	0	0	1	1	1
18	1	0	0	1	1	1	1	0	0	1	1	1	1	0	0	1
19	1	1	1	0	1	0	0	1	0	1	1	1	1	1	1	0
20	0	1	1	1	1	1	1	0	1	0	0	1	0	1	1	1
21	1	0	0	1	0	1	1	1	1	1	1	0	1	0	0	1
22	0	1	1	1	1	1	1	0	1	1	1	0	1	0	0	1
23	1	0	0	1	0	1	1	1	0	1	1	1	1	1	1	0
24	1	1	1	0	1	0	0	1	1	0	0	1	0	1	1	1

Fig. 8.40 Results of the PRP shown in Fig. 8.39

$$a_2 = -6, \qquad\qquad b_2 = 4,$$
$$a_3 = 83, \qquad\qquad b_3 = -15,$$
$$a_4 = -66, \qquad\qquad b_4 = -10$$

Convert these integers into 16-bit binary vectors and use these vectors to compute:

(a) The sum s_i, such that $s_i = a_i + b_i$, $i = 1, \ldots, 4$
(b) The difference d_i, such that $d_i = a_i - b_i$, $i = 1, \ldots, 4$
(c) The product p_i, such that $p_i = a_i \cdot b_i$, $i = 1, \ldots, 4$
(d) The quotient q_i and the remainder r_i, such that $a_i = q_i \cdot b_i + r_i$, $i = 1, \ldots, 4$, $|r_i| < |b_i|$, and the remainder r_i has the same sign as the dividend a_i

Fig. 8.41 Four data words to be encoded using the Hamming code $H(15, 11)$

	x_1	x_2	x_3	x_4	x_5	x_6	x_7	x_8	x_9	x_{10}	x_{11}
	0	1	1	0	0	0	1	1	1	0	0
$\mathrm{ODA}(dw) =$	1	1	1	0	1	0	0	1	0	0	1
	1	0	0	1	1	1	1	1	1	0	0
	1	0	0	1	0	1	0	1	0	1	0

Fig. 8.42 Five received code words to be decoded using the Hamming code $H(15, 11)$

	y_1	y_2	y_3	y_4	y_5	y_6	y_7	y_8	y_9	y_{10}	y_{11}	y_{12}	y_{13}	y_{14}	y_{15}
	1	0	1	1	0	0	1	1	1	1	1	1	1	0	0
	1	0	1	1	0	1	0	1	1	1	0	0	1	1	0
$\mathrm{ODA}(cw) =$	0	1	1	0	0	0	1	1	1	1	0	0	0	1	0
	0	1	0	0	0	0	0	0	0	1	1	1	0	0	1
	0	0	1	1	0	0	1	1	0	1	0	1	0	0	0

Show for each of the four tasks four TVLs that contain the 16-bit binary vectors of the integers of a_i, b_i, and the results. Convert the computed binary vectors into integers and verify the results.

Exercise 8.8 (Coding)

(a) Create a table that can be used for both the encoding and the decoding between five-bit binary vectors of the dual code and the Gray code. Use a system of logic equations that specifies these mappings and show this table ordered by the vectors of the dual code.

(b) Create a table that can be used for both the encoding and the decoding between the binary-coded decimal (BCD) code and the Aiken code for the digits 0–9. The Aiken code is a complementary binary-coded decimal (BCD) code. The weights of the bits of the BCD code are 8-4-2-1 and of the Aiken code 2-4-2-1. The code words of the Aiken code of the pairs of digits $(0, 9)$, $(1, 8)$, $(2, 7)$, $(3, 6)$, and $(4, 5)$ are complements to each other due to these weights. Use a system of logic equations that specifies the direct mapping for the digits 0–4 and compute complemented vectors for the digits 5–9. Show the encoding–decoding table ordered by the BCD code. Verify that the created coding table satisfied the properties of the Aiken code.

(c) The rate, i.e., the ratio between the number of data bits and the total number of bits of a Hamming code, becomes better (closer to the value 1) when larger Hamming codes are used: for $H(7, 4)$ the rate is $4/7 \approx 0.571$ and for $H(15, 11)$ we have $11/15 \approx 0.733$; hence, larger Hamming codes are preferred.

Use a system of logic equations to create a table that can be used for both the encoding and the decoding between a 11-bit data word and the 15-bit code word of the Hamming code $H(15, 11)$. Use the created coding table to encode the four data words $\mathbf{x} = (x_1, x_2, \ldots, x_{11})$ that are specified in the TVL of Fig. 8.41, store the associated code words in the same order in a TVL, and show this TVL to the right of the given TVL of data words.

The Hamming code has the property that a single-bit-error can be corrected. Assume that the TVL of Fig. 8.42 shows five received code words $\mathbf{y} = (y_1, y_2, \ldots, y_{15})$, which can be real code words of the Hamming code $H(15, 11)$ or changed in maximal one bit position. Decode the five received code words in three steps:

a. Determine for each received code word maximal one bit that must be changed to get a real code word; compute and show a TVL that corresponds to the TVL of the received code words and contains binary vectors where a value 1 indicates a received incorrect bit.

b. Correct the received code words into real code words using the information computed in the previous step; compute and show a TVL that contains the corrected code word.

c. Decode the corrected code words using the coding table of the Hamming code $H(15, 11)$; compute and show a TVL that contains the decoded data words.

Table 8.21 Associations of the vectorial derivatives to the model variables u_k

u_0	u_1	u_2	u_3	u_4	u_5	u_6	u_7
f	$\operatorname{der}_{x_1} f$	$\operatorname{der}_{x_2} f$	$\operatorname{der}_{(x_1,x_2)} f$	$\operatorname{der}_{x_3} f$	$\operatorname{der}_{(x_1,x_3)} f$	$\operatorname{der}_{(x_2,x_3)} f$	$\operatorname{der}_{(x_1,x_2,x_3)} f$

(d) The code words of the Reed–Muller code $\mathcal{RM}(1, 3)$ are the solutions of the Boolean differential equation (8.3) on page 443. Remember that

$$\operatorname*{der}_{(x_i,x_j)}{}^2 f(x_i, x_j, \mathbf{x}_1) = \operatorname*{der}_{x_i} f(x_i, x_j, \mathbf{x}_1) \oplus \operatorname*{der}_{x_j} f(x_i, x_j, \mathbf{x}_1) \oplus \operatorname*{der}_{(x_i,x_j)} f(x_i, x_j, \mathbf{x}_1) .$$

The associations of the vectorial derivatives to the model variables u_k for the Boolean space $\mathbb{B}^3$ are shown in Table 8.21.

Solve the Boolean differential equation (8.3) and show the solution-TVL of the associated logic equation $D(\mathbf{u}) = 0$ as well as the TVL of solution-vectors $\mathbf{v}$. How many code words belong to the Reed–Muller code $\mathcal{RM}(1, 3)$?

Exercise 8.9 (Bent Functions)

(a) Compute all bent functions of two variables based on Definition 8.13.
It is convenient to compute the required Hamming distance $h(f, g)$ using the commands sv_syd and sv_size; hence, both the linear functions and the functions to evaluate should be represented by vectors of function values.
Create the $2^{2+1} = 8$ linear functions of $\mathbb{B}^2$ using Eq. (4.11) of Definition 4.17 specified for two variables

$$f_l(x_1, x_2) = a_0 \oplus a_1 x_1 \oplus a_2 x_2$$

and the vectors of function values of $f_0(x_1, x_2) = 1(x_1, x_2)$, $f_1(x_1, x_2) = x_1$, and $f_2(x_1, x_2) = x_2$.
Compute for each nonlinear function the Hamming distances to all linear functions, select from these distances the smallest value, and assign the evaluated function to the set of bent functions if no other nonlinear function has a larger value as smallest Hamming distance to all linear functions. Show the computed sets of vectors of function values of all linear functions and all bent functions.

(b) The set of all bent functions of two variables is the solution of the Boolean differential equation (8.11). Solve this Boolean differential equation and show the set of solutions of the associated logic equation and the vectors of function values of the computed bent functions. How many bent functions of two variables exist?

(c) The set of all bent functions of four variables is the solution of the Boolean differential equation (8.12). Solve this Boolean differential equation (BDE) and show the set of solutions of the associated logic equation and the vectors of function values of the computed bent functions. How many bent functions of four variables exist?
The transformation of the set of local solutions $SLS(\mathbf{u})$ to the initial solution-set $S(\mathbf{v})$ can be realized using a transformation-matrix $T(\mathbf{u}, \mathbf{v})$ that can be created as solution of a system of logic equations. Use this approach as step in the procedure to solve the BDE (8.12). How many vectors belong to the TVL $T(\mathbf{u}, \mathbf{v})$?

(d) It is known that weight of bent functions of four variables is either equal to $8 - 2 = 6$ or equal to $8 + 2 = 10$. Each of these two subsets of bent functions contains functions with different complexities. The most complex bent functions of these two subsets are the solution of the Boolean differential equation:

$$\operatorname*{der}_{(x_1,x_2)}{}^2 f(\mathbf{x}) \operatorname*{der}_{(x_1,x_3)}{}^2 f(\mathbf{x}) \operatorname*{der}_{(x_1,x_4)}{}^2 f(\mathbf{x}) \operatorname*{der}_{(x_2,x_3)}{}^2 f(\mathbf{x}) \operatorname*{der}_{(x_2,x_4)}{}^2 f(\mathbf{x}) \operatorname*{der}_{(x_3,x_4)}{}^2 f(\mathbf{x}) = 1 \, . \tag{8.14}$$

Compute the set of most complex bent functions of four variables by solving the Boolean differential equation (8.14). How many most complex bent functions of four variables exist? Avoid the creation of the large transformation-matrix $T(\mathbf{u}, \mathbf{v})$ and utilize instead for the needed transformation from $SLS(\mathbf{u})$ into $S(\mathbf{v})$ that

$$v_0 = u_0 \, ,$$

$$\text{if} \quad u_0 = 0 \, , \qquad\qquad \text{then} \quad v_i = u_i, \; i = 1, \dots, 15 \, , \text{ and}$$

$$\text{if} \quad u_0 = 1 \, , \qquad\qquad \text{then} \quad v_i = \overline{u}_i, \; i = 1, \dots, 15 \, .$$

Which commands of the XBOOLE-monitor XBM 2 can be used for this transformation?

References

1. O.S. Rothaus. On "Bent" functions. J. Comb. Theory (A) **20**, 300–305 (1976). https://doi.org/10.1016/0097-3165(76)90024-8
2. B. Steinbach, Adjacency graph of the SNF as source of information, in *Boolean Problems, Proceedings of the 7th International Workshops on Boolean Problems*, ed. by B. Steinbach. IWSBP 7. Freiberg, Germany: Freiberg University of Mining and Technology, Sept. 2006, pp. 19–28. ISBN: 978-3-86012-287-7
3. B. Steinbach, Most complex Boolean functions, in *Proceedings Reed-Muller 2007* (RM. Oslo, 2007), pp. 13–23
4. B. Steinbach, A. De Vos, The shape of the SNF as a source of information, in *Boolean Problems, Proceedings of the 8th International Workshops on Boolean Problems*, ed. by B. Steinbach. IWSBP 8. Freiberg, Germany: Freiberg University of Mining and Technology, Sept. 2008, pp. 127–136. ISBN: 978-3-86012-346-1
5. B. Steinbach, A. Mishchenko, SNF: A Special Normal Form for ESOPs, in *Proceedings of the 5th International Workshop on Application of the Reed-Muller Expansion in Circuit Design* (RM, Starkville, 2001), pp. 66–81
6. B. Steinbach, C. Posthoff, *Boolean Differential Equations* (Morgan & Claypool Publishers, San Rafael, 2013). ISBN: 978-1-6270-5241-2. https://doi.org/10.2200/S00511ED1V01Y201305DCS042
7. B. Steinbach, C. Posthoff, Classes of Bent functions identified by specific normal forms and generated using Boolean differential equations. Facta Universitatis (Nis). Electrical Eng. **24**(3), 357–383 (2011). https://doi.org/10.2298/FUEE1103357S
8. B. Steinbach, C. Posthoff, Classification and generation of Bent functions, in *Proceedings Reed-Muller 2011 Workshop*. RM. Tuusula, Finland, May 2011, pp. 81–91
9. B. Steinbach, V. Yanchurkin, M. Lukac. On SNF optimization: a functional comparison of methods, in *Proceedings of the 6th International Symposium on Representations and Methodology of Future Computing Technology* (RM, Trier, 2003), pp. 11–18

SAT-Problems

9

Abstract

Many finite discrete problems can be modeled as satisfiability (SAT) problems. There are different types of SAT-problems. Most common are CD-SAT-formulas which consist of conjunctions of disjunctions of Boolean variables and this expression is equal to 1. Such disjunctions are also called clauses. New CDC-SAT-formulas are conjunctions of disjunctions of conjunctions of Boolean variables and allow a more compact specification of the problem. Besides special algorithms for SAT-problems (SAT-solvers) ternary vector lists are an appropriate data structure to express and solve all such problems. SAT-problems belong to the class of NP-complete problems. The modeling and solution of many SAT-problems will be explored. Studied problem classes are placement problems, covering problems, path problems, and coloring problems. Some of the selected examples have their root many centuries ago, but also very recent research results will be presented. Hints for efficient solutions using a single central processing unit (CPU), several CPU-cores of a multi-processor, or even the huge number of cores of a graphical processing unit (GPU) will be given.

Supplementary Information The online version of this chapter (https://doi.org/10.1007/978-3-030-88945-6_9) contains supplementary material which is available for authorized users. Please, follow the link belonging to the version of the XBOOLE-monitor XBM 2 that fits best for your operating system. This XBOOLE-monitor is needed to solve all tasks of this chapter. Instructions for starting the downloaded XBOOLE-monitor XBM 2 are given at the beginning of Section 'Exercises' in this chapter.

XBOOLE-monitor XBM 2 for Windows 10
32 bits
https://doi.org/10.1007/978-3-030-88945-6_9_MOESM1_ESM.zip (15,091 KB)

64 bits
https://doi.org/10.1007/978-3-030-88945-6_9_MOESM2_ESM.zip (14,973 KB)

XBOOLE-monitor XBM 2 for Linux Ubuntu
32 bits
https://doi.org/10.1007/978-3-030-88945-6_9_MOESM3_ESM.zip (29,522 KB)

64 bits
https://doi.org/10.1007/978-3-030-88945-6_9_MOESM4_ESM.zip (28,422 KB)

9.1   Specification and Solution Methods

In Chap. 5 the basic methods for the solution of logic equations have been presented. It could be seen that *ternary vectors* and *lists of ternary vectors* were the main data structure to be used. Logic operations have been transferred to operations with sets of binary vectors. Because of its importance we address the *SAT-problem*, by separate presentations using SAT as an abbreviation for *satisfiability*.

SAT-Problems have their roots in *Artificial Intelligence* where many rules $(x \rightarrow y) = 1$ have to be processed, but these problems are very important for the discussion of the complexity of algorithms as well. The most popular and important definition of a SAT-problem is as follows.

Definition 9.1 Let $\mathbf{x} = (x_1, x_2, \ldots, x_n)$, $D_1, \ldots, D_m$ disjunctions of literals of some variables of the vector $\mathbf{x}$ and, finally $D_1 D_2 \ldots D_m = 1$ a logic equation. Then the finding of solutions of this equation is denoted by *SAT-problem*.

This definition means that m disjunctions of negated and non-negated variables are given, and the conjunctive form consisting of these disjunctions must be equal to 1.

Note 9.1 There is still a more general possibility how SAT-problems can be defined: given any correct formula (expression)—is there a vector $\mathbf{x} \in \mathbb{B}^n$ that satisfies the formula, i.e., is the function represented by this formula not equal to 0? However, the transformation of this formula into a conjunctive form transforms this most general problem into the SAT-problem defined above. The term "*satisfiable*" always relates to the problem of finding a vector $\mathbf{x}$ with $f(\mathbf{x}) = 1$, where $f(\mathbf{x})$ is the function defined by the expression.

Example 9.1 An example of a SAT-problem is

$$(x_1 \vee x_3 \vee \overline{x}_5)(x_1 \vee \overline{x}_2 \vee \overline{x}_3)(x_2 \vee \overline{x}_4 \vee \overline{x}_5) = 1 .$$

The left-hand side of this equation is a conjunctive form of a function $f(\mathbf{x})$.

The possible previous shape of rules is no longer visible: The disjunction $(x_1 \vee x_3 \vee \overline{x}_5)$ can come from different rules:

$$x_5 \rightarrow x_1 \vee x_3 ,$$
$$\overline{x}_1 \overline{x}_3 \rightarrow \overline{x}_5 ,$$
$$\overline{x}_1 x_5 \rightarrow x_3 .$$

The application of the rule $a \rightarrow b = \overline{a} \vee b$ to each application always gives the same result. The reverse transformation, however, is unique.

Why is this problem important and difficult? The conjunctive form is equal to 1 if and only if each disjunction is equal to 1. For n variables we have 2^n different $\mathbf{x}$-vectors. Each vector can be a solution; hence, each vector must be checked. In the worst case this has to be done for all $\mathbf{x}$-vectors; for m disjunctions $m \times 2^n$ tests have to be performed. And the number of possible disjunctions is also growing exponentially.

We remember the concept of a *ternary vector* in order to represent disjunctions: We use 0 for a negated variable, 1 for a non-negated variable, and $-$ in the case that a variable does not occur in a given disjunction. This results in 3^n different ternary vectors, or omitting the "empty" disjunction $(- - \cdots -)$, in $3^n - 1$ different disjunctions. Hence, a rough estimate of the upper bound of the complexity of the SAT-problem results in $m = 3^n - 1$ tests for 2^n vectors (i.e., in a maximum of $(3^n - 1) \times 2^n$ tests), and this is quite challenging for "larger" values of n.

Note 9.2 By using the complement of the equation, we get the equivalent SAT-problem that for any reason is not so popular like the first one:

$$\overline{f(\mathbf{x})} = \overline{x}_1 \overline{x}_3 x_5 \vee \overline{x}_1 x_2 x_3 \vee \overline{x}_2 x_4 x_5 = 0 \,.$$

Here each conjunction must be equal to 0, and again all the vectors of the whole $\mathbb{B}^n$ are solution-candidates. These two problems will be handled later on using the same means.

There are several variations of this problem:

- 3-SAT: Each disjunction contains exactly three literals.
- k-SAT: Each disjunction contains exactly k literals.
- Monotonous 3-SAT: The conjunctive form of $f(\mathbf{x})$ can be split into two parts $f_1(\mathbf{x})$ and $f_2(\mathbf{x})$, and $f_1(\mathbf{x})$ contains only non-negated variables, $f_2(\mathbf{x})$ contains only negated variables.
- 1-satisfiability: Is there at most one vector that satisfies a given conjunctive form?
- #SAT: How many vectors satisfy a given conjunctive form?
- Maximum clauses: Is there a conjunctive form that represents the same function like the given conjunctive form of $f(\mathbf{x})$ and has at most k clauses?
- MAXSAT: Let be given a conjunctive form that is not satisfiable. What is the maximum number of clauses of this conjunctive form that still can be satisfied?
- CDC-SAT: Very often the constraints are known just from the beginning of the solution of a problem. In this case, they can sometimes be combined and included into a SAT-problem, but the brackets do not contain single literals, they contain *conjunctions* which are combined by $\vee$, and these disjunctions are combined by $\wedge$:

$$f(x_1, \ldots, x_n) = (C_{i_1} \vee \cdots \vee C_{i_n})(C_{k_1} \vee \cdots \vee C_{k_m}) \ldots (C_{l_1} \vee \cdots \vee C_{l_p}) \,.$$

Hence, the CDC-SAT-formula is a conjunction (C) of disjunctions (D) of conjunctions (C) that contain a different number of literals of the variables $x_1, \ldots, x_n$.

The SAT-problem relates to the whole area of modeling by *constraints*. We understand each disjunction of variables (also denoted by *clause*) as the description of a set of binary vectors that will satisfy this disjunction. The intersection of two of these sets will reduce the number of solutions. At the end only those vectors will be a solution that have not been excluded by a constraint. When the conjunctive form is not satisfiable, then there are too many constraints—and the question of the MAXSAT-problem is therefore to keep as many clauses as possible and still to have some solutions.

Many problems can be expressed as a SAT-problem. Therefore, there has been a comprehensive research in this field [1], and special SAT-solvers are provided to solve these problems. The approach that uses (the parallel processing of) ternary vectors solves many of these problems simultaneously. It avoids many transformations and special considerations and allows a broad range of efficient solutions.

We will start with some discussions of possible approaches based on TVLs. We are using the following example equation:

$$(x_1 \vee \overline{x}_3 \vee x_5)(\overline{x}_2 \vee x_3 \vee x_4)(x_1 \vee x_2 \vee \overline{x}_5)(\overline{x}_3 \vee \overline{x}_4 \vee \overline{x}_5) = 1 . \qquad (9.1)$$

It has already a special format: All the disjunctions contain three literals; it is an example of 3-SAT.

Determine All Solutions

In order to find all solutions, we use the orthogonal coding for each bracket and combine these lists for the single brackets by means of the intersection.

x_1 x_2 x_3 x_4 x_5		x_1 x_2 x_3 x_4 x_5		x_1 x_2 x_3 x_4 x_5		x_1 x_2 x_3 x_4 x_5
1 − − − −		− 0 − − −		1 − − − −		− − 0 − −
0 − 0 − −	$\cap$	− 1 1 − −	$\cap$	0 1 − − −	$\cap$	− − 1 0 −
0 − 1 − 1		− 1 0 1 −		0 0 − − 0		− − 1 1 0

	x_1 x_2 x_3 x_4 x_5		x_1 x_2 x_3 x_4 x_5		x_1 x_2 x_3 x_4 x_5		x_1 x_2 x_3 x_4 x_5
			1 − 0 − −		1 0 0 − −		1 0 0 − −
x_1 x_2 x_3 x_4 x_5			1 − 1 0 −		1 0 1 0 −		1 − 1 0 −
1 0 − − −			1 − 1 1 0		1 0 1 1 0		1 − 1 1 0 .
1 1 1 − −			0 1 0 − −		1 1 1 0 −		0 0 0 − 0
= 1 1 0 1 −	$\cap$		0 1 1 0 −	=	1 1 1 1 0	=	− 1 0 1 −
0 0 0 − −			0 1 1 1 0		1 1 0 1 −		0 1 1 0 1
0 1 0 1 −			0 0 0 − 0		0 0 0 − 0		
0 0 1 − 1			0 0 1 0 0		0 1 0 1 −		
0 1 1 − 1			0 0 1 1 0		0 1 1 0 1		

It is interesting to see that there are only nine ternary vectors after the sequence of intersections. The orthogonality considerably reduces the number of solution-vectors. Three pairs of these nine ternary vectors can be combined, so that only six ternary vectors represent all 17 solutions.

Determine Some Solutions Fast

This approach implements a *depth-first search* with *backtracking*. In each list only one vector will be used; the intermediate results will be stored. It is tried to reach the last list with a non-empty intersection. As soon as the intersection will be empty, the next vector of the same list will be used. If it is the last vector in a list, then we go one list back and use the next vector in this previous list. In principle, the sequence of actions is as follows:

case 1 :	$((1.1 \cap 2.1) \cap 3.1) \cap 4.1$
case 2 :	$((1.1 \cap 2.1) \cap 3.1) \cap 4.2$
case 3 :	$((1.1 \cap 2.1) \cap 3.1) \cap 4.3$
case 4 :	$((1.1 \cap 2.1) \cap 3.2) \cap 4.1$
case 5 :	$((1.1 \cap 2.1) \cap 3.2) \cap 4.2$
$\vdots$	$\vdots$
case 81 :	$((1.3 \cap 2.3) \cap 3.3) \cap 4.3 .$

Always when a vector of the last list is used and the intersection is not empty, then we found a (ternary) solution-vector. However, as soon as the intersection is empty, the next vector can be taken. The chances to find a solution fast are relatively high, especially when the vectors with many $-$ are at the beginning of the list. Some steps of our example:

$$\frac{\begin{array}{ccccc} x_1 & x_2 & x_3 & x_4 & x_5 \\ 1 & - & - & - & - \end{array}}{1.1} \cap \frac{\begin{array}{ccccc} x_1 & x_2 & x_3 & x_4 & x_5 \\ - & 0 & - & - & - \end{array}}{2.1} \cap \frac{\begin{array}{ccccc} x_1 & x_2 & x_3 & x_4 & x_5 \\ 1 & - & - & - & - \end{array}}{3.1} \cap \frac{\begin{array}{ccccc} x_1 & x_2 & x_3 & x_4 & x_5 \\ - & - & 0 & - & - \end{array}}{4.1}$$

$$= \frac{\begin{array}{ccccc} x_1 & x_2 & x_3 & x_4 & x_5 \\ 1 & 0 & - & - & - \end{array}}{1.1 \cap 2.1} \cap \frac{\begin{array}{ccccc} x_1 & x_2 & x_3 & x_4 & x_5 \\ 1 & - & - & - & - \end{array}}{3.1} \cap \frac{\begin{array}{ccccc} x_1 & x_2 & x_3 & x_4 & x_5 \\ - & - & 0 & - & - \end{array}}{4.1}$$

$$= \frac{\begin{array}{ccccc} x_1 & x_2 & x_3 & x_4 & x_5 \\ 1 & 0 & - & - & - \end{array}}{1.1 \cap 2.1 \cap 3.1} \cap \frac{\begin{array}{ccccc} x_1 & x_2 & x_3 & x_4 & x_5 \\ - & - & 0 & - & - \end{array}}{4.1}$$

$$= \frac{\begin{array}{ccccc} x_1 & x_2 & x_3 & x_4 & x_5 \\ 1 & 0 & 0 & - & - \end{array}}{1.1 \cap 2.1 \cap 3.1 \cap 4.1}.$$

Already after three intersections the first solution-vector has been found. It is also a big advantage to store all the intermediate results from the left to the right.

$$\frac{\begin{array}{ccccc} x_1 & x_2 & x_3 & x_4 & x_5 \\ 1 & 0 & - & - & - \end{array}}{}$$

is the intersection $((1.1 \cap 2.1) \cap 3.1)$ and can be used with all the vectors of the last list:

$$\frac{\begin{array}{ccccc} x_1 & x_2 & x_3 & x_4 & x_5 \\ 1 & 0 & - & - & - \end{array}}{1.1 \cap 2.1 \cap 3.1} \cap \frac{\begin{array}{ccccc} x_1 & x_2 & x_3 & x_4 & x_5 \\ - & - & 0 & - & - \end{array}}{4.1} = \frac{\begin{array}{ccccc} x_1 & x_2 & x_3 & x_4 & x_5 \\ 1 & 0 & 0 & - & - \end{array}}{1.1 \cap 2.1 \cap 3.1 \cap 4.1},$$

$$\frac{\begin{array}{ccccc} x_1 & x_2 & x_3 & x_4 & x_5 \\ 1 & 0 & - & - & - \end{array}}{1.1 \cap 2.1 \cap 3.1} \cap \frac{\begin{array}{ccccc} x_1 & x_2 & x_3 & x_4 & x_5 \\ - & - & 1 & 0 & - \end{array}}{4.2} = \frac{\begin{array}{ccccc} x_1 & x_2 & x_3 & x_4 & x_5 \\ 1 & 0 & 1 & 0 & - \end{array}}{1.1 \cap 2.1 \cap 3.1 \cap 4.2},$$

$$\frac{\begin{array}{ccccc} x_1 & x_2 & x_3 & x_4 & x_5 \\ 1 & 0 & - & - & - \end{array}}{1.1 \cap 2.1 \cap 3.1} \cap \frac{\begin{array}{ccccc} x_1 & x_2 & x_3 & x_4 & x_5 \\ - & - & 1 & 1 & 0 \end{array}}{4.3} = \frac{\begin{array}{ccccc} x_1 & x_2 & x_3 & x_4 & x_5 \\ 1 & 0 & 1 & 1 & 0 \end{array}}{1.1 \cap 2.1 \cap 3.1 \cap 4.3}.$$

Then we go back to $(1.1 \cap 2.1)$ which is also equal to

$$\frac{\begin{array}{ccccc} x_1 & x_2 & x_3 & x_4 & x_5 \\ 1 & 0 & - & - & - \end{array}}{}$$

and take the next vectors of the third list, etc.:

$$\frac{\dfrac{x_1\ x_2\ x_3\ x_4\ x_5}{1\ \ 0\ -\ -\ -}}{1.1 \cap 2.1} \cap \frac{\dfrac{x_1\ x_2\ x_3\ x_4\ x_5}{0\ \ 1\ -\ -\ -}}{3.2} = \frac{x_1\ x_2\ x_3\ x_4\ x_5}{1.1 \cap 2.1 \cap 3.2}\ ,$$

$$\frac{\dfrac{x_1\ x_2\ x_3\ x_4\ x_5}{1\ \ 0\ -\ -\ -}}{1.1 \cap 2.1} \cap \frac{\dfrac{x_1\ x_2\ x_3\ x_4\ x_5}{0\ \ 0\ -\ -\ 0}}{3.3} = \frac{x_1\ x_2\ x_3\ x_4\ x_5}{1.1 \cap 2.1 \cap 3.3}\ .$$

Due to the empty lists after two intersections, in these two cases no vector of the fourth list must be used.

Thereafter, we must go back to the second list, the second vector in this list has to be taken, and so on. After executing all possible intersections, the results of the two methods will naturally be the same.

The Use of the Complement

All so far explored approaches directly solved the SAT-problem $C(f(\mathbf{x})) = 1$. The solution-set remains unchanged when both sides of this equation are negated. Using the negation with regard to De Morgan on the left-hand side leads to the equivalent equation $D(\overline{f(\mathbf{x})}) = 0$. This approach has the welcome property that the ternary vector of a conjunction of $D(\overline{f(\mathbf{x})})$ also describes a set of binary vectors that satisfy $D(\overline{f(\mathbf{x})}) = 1$; hence, this set belongs to the complement of the needed solution. Obviously, the complement of the wanted solution-set can very easily be constructed using the negation with regard to De Morgan, and the subsequent calculation of the associated complement generates the final solution. For our example (9.1), four vectors describe the function:

$$D(\overline{f(\mathbf{x})}) = \frac{\begin{array}{ccccc} x_1 & x_2 & x_3 & x_4 & x_5 \end{array}}{\begin{array}{ccccc} 0 & - & 1 & - & 0 \\ - & 1 & 0 & 0 & - \\ 0 & 0 & - & - & 1 \\ - & - & 1 & 1 & 1 \end{array}}\ .$$

The orthogonalization of this TVL results in the solution-set:

$$S(D(\overline{f(\mathbf{x})}) = 1) = S(C(f(\mathbf{x})) = 0) = \frac{\begin{array}{ccccc} x_1 & x_2 & x_3 & x_4 & x_5 \end{array}}{\begin{array}{ccccc} 0 & - & 1 & - & 0 \\ - & 1 & 0 & 0 & - \\ 0 & 0 & - & - & 1 \\ - & 1 & 1 & 1 & 1 \\ 1 & 0 & 1 & 1 & 1 \end{array}}\ .$$

By counting the positions with $-$, we see that 15 binary vectors satisfy both $D(\overline{f(\mathbf{x})}) = 1$ and $C(f(\mathbf{x})) = 0$; hence, $C(f(\mathbf{x})) = 1$ has $32 - 15 = 17$ solutions.

The simplest way to find these 17 solutions consists in generating all vectors from (00000) to (11111) and testing for each vector whether it is orthogonal to all the vectors of the solution-set $S(C(f) = 0)$ or not. (00000) is orthogonal to all the vectors; hence, it is a solution of $C(f(\mathbf{x})) = 1$. This algorithm is "not very intelligent"; however, the simplicity of the test and the parallel processing justify this approach.

The most elegant way, however, consists in the direct calculation of the complement set. This can be done by orthogonalization of the TVL of the Boolean space $\mathbb{B}^5$ $D(1(x_1, x_2, x_3, x_4, x_5))$ with regard to all ternary vectors of $D(\overline{f(\mathbf{x})})$:

$$
\frac{x_1\ x_2\ x_3\ x_4\ x_5}{-\ -\ -\ -\ -}\ \Big\backslash\ \frac{x_1\ x_2\ x_3\ x_4\ x_5}{0\ -\ 1\ -\ 0}\ =\
\begin{array}{ccccc}
x_1 & x_2 & x_3 & x_4 & x_5 \\
\hline
1 & - & - & - & - \\
0 & - & 0 & - & - \\
0 & - & 1 & - & 1
\end{array}\ ,
$$

$$
\begin{array}{ccccc}
x_1 & x_2 & x_3 & x_4 & x_5 \\
\hline
1 & - & - & - & - \\
0 & - & 0 & - & - \\
0 & - & 1 & - & 1
\end{array}
\ \Big\backslash\ \frac{x_1\ x_2\ x_3\ x_4\ x_5}{-\ 1\ 0\ 0\ -}\ =\
\begin{array}{ccccc}
x_1 & x_2 & x_3 & x_4 & x_5 \\
\hline
1 & 0 & - & - & - \\
1 & 1 & 1 & - & - \\
1 & 1 & 0 & 1 & - \\
0 & 0 & 0 & - & - \\
0 & 1 & 0 & 1 & - \\
0 & - & 1 & - & 1
\end{array}\ ,
$$

$$
\begin{array}{ccccc}
x_1 & x_2 & x_3 & x_4 & x_5 \\
\hline
1 & 0 & - & - & - \\
1 & 1 & 1 & - & - \\
1 & 1 & 0 & 1 & - \\
0 & 0 & 0 & - & - \\
0 & 1 & 0 & 1 & - \\
0 & - & 1 & - & 1
\end{array}
\ \Big\backslash\ \frac{x_1\ x_2\ x_3\ x_4\ x_5}{0\ 0\ -\ -\ 1}\ =\
\begin{array}{ccccc}
x_1 & x_2 & x_3 & x_4 & x_5 \\
\hline
1 & 0 & - & - & - \\
1 & 1 & 1 & - & - \\
1 & 1 & 0 & 1 & - \\
0 & 0 & 0 & - & 0 \\
0 & 1 & 0 & 1 & - \\
0 & 1 & 1 & - & 1
\end{array}\ ,
$$

$$
\begin{array}{ccccc}
x_1 & x_2 & x_3 & x_4 & x_5 \\
\hline
1 & 0 & - & - & - \\
1 & 1 & 1 & - & - \\
1 & 1 & 0 & 1 & - \\
0 & 0 & 0 & - & 0 \\
0 & 1 & 0 & 1 & - \\
0 & 1 & 1 & - & 1
\end{array}
\ \Big\backslash\ \frac{x_1\ x_2\ x_3\ x_4\ x_5}{-\ -\ 1\ 1\ 1}\ =\
\begin{array}{ccccc}
x_1 & x_2 & x_3 & x_4 & x_5 \\
\hline
1 & 0 & 0 & - & - \\
1 & 0 & 1 & 0 & - \\
1 & 0 & 1 & 1 & 0 \\
1 & 1 & 1 & 0 & - \\
1 & 1 & 1 & 1 & 0 \\
1 & 1 & 0 & 1 & - \\
0 & 0 & 0 & - & 0 \\
0 & 1 & 0 & 1 & - \\
0 & 1 & 1 & 0 & 1
\end{array}\ .
$$

The nine ternary vectors of the resulting TVL describe the 17 solutions of the SAT-problem (9.1). Three pairs of these nine ternary vectors can be combined (2 + 4, 3 + 5, and 6 + 8), so that only six ternary vectors represent all 17 solutions:

$$
S(C(f(\mathbf{x})) = 1) =
\begin{array}{ccccc}
x_1 & x_2 & x_3 & x_4 & x_5 \\
\hline
1 & 0 & 0 & - & - \\
1 & - & 1 & 0 & - \\
1 & - & 1 & 1 & 0 \\
- & 1 & 0 & 1 & - \\
0 & 0 & 0 & - & 0 \\
0 & 1 & 1 & 0 & 1
\end{array}\ .
$$

Table 9.1 Verification of the resolution law

x	y	z	$x \to y$	$x \vee z$	$(x \to y)(x \vee z)$	$(y \vee z)$	$(x \to y)(x \vee z) \to (y \vee z)$
0	0	0	1	0	0	0	1
0	0	1	1	1	1	1	1
0	1	0	1	0	0	1	1
0	1	1	1	1	1	1	1
1	0	0	0	1	0	0	1
1	0	1	0	1	0	1	1
1	1	0	1	1	1	1	1
1	1	1	1	1	1	1	1

The Resolution Principle

Many algorithms have been developed in order to show that an expression in conjunctive form is not satisfiable (i.e., that it represents the function $0(\mathbf{x})$) using the so-called *resolution principle*. The simplest form of this principle can be expressed as follows.

Theorem 9.1 *It holds for all* $x, y, z \in \mathbb{B}$ *that*

$$(x \to y)(x \vee z) \to (y \vee z) \equiv 1(x, y, z) . \tag{9.2}$$

The formula on the left-hand side of (9.2) represents the function that is constant equal to 1; it is a *logical law* or a *tautology* (see the applications in logic). The importance of this law for the satisfiability problem can be seen step by step:

1. According to the definition of the implication, we have

$$x \to y = \overline{x} \vee y .$$

2. We verify the required identities as shown in Table 9.1.
3. It can be seen that

$$(x \to y)(x \vee z) = (\overline{x} \vee y)(x \vee z) \le (y \vee z) .$$

This relation is fundamental for proving that an expression is not satisfiable. Resolution-based algorithms are looking for two clauses with a complementary variable (like x and $\overline{x}$). The consideration of these two clauses can be replaced by the exploration of one clause. The literals x and $\overline{x}$ simply can be "omitted," the other variables can be combined into one clause, the resulting clause $y \vee z$ is denoted by *resolvent* of the two other clauses, indicated by R. We have, for instance, $R(\overline{x} \vee y, x \vee z) = y \vee z$. Progress can be made when other clauses can be found by means of which also the variable y and the variable z can be deleted, and finally the *empty* clause will appear not containing any variable and therefore representing the function $0(x, y, z)$. Due to the $\le$-relation all the other intermediate formulas and finally the leftmost formula must also represent $0(x, y, z)$.

Example 9.2 Let

$$f(x, y, z) = (x \vee z)(\overline{x} \vee y)(\overline{x} \vee \overline{y})(\overline{y} \vee \overline{z})(x \vee y) ,$$

and the SAT-problem $f(x, y, z) = 1$ has to be solved. The five clauses of $f(x, y, z)$ can be combined as follows:

$$R(x \vee z, \overline{x} \vee y) = y \vee z,$$

$$R(x \vee z, \overline{x} \vee \overline{y}) = \overline{y} \vee z,$$

$$R(\overline{x} \vee y, \overline{y} \vee \overline{z}) = \overline{x} \vee \overline{z},$$

$$R(\overline{y} \vee \overline{z}, x \vee y) = x \vee \overline{z}.$$

These four new clauses will be added to the original five clauses and can be handled in the same way. Because of the idempotence of the conjunction ($x \wedge x = x$), the clauses can be used several times. By adding these four clauses to $f(x, y, z)$, we get a function $f'(x, y, z)$:

$$f'(x, y, z) = f(x, y, z) \wedge (y \vee z)(\overline{y} \vee z)(\overline{x} \vee \overline{z})(x \vee \overline{z}),$$

and it holds that

$$f(x, y, z) \leq f'(x, y, z).$$

We continue as before and get:

$$R(y \vee z, \overline{y} \vee z) = z,$$

$$R(\overline{x} \vee \overline{z}, x \vee \overline{z}) = \overline{z},$$

so that the function $f''(x, y, z)$ can be created:

$$f''(x, y, z) = f'(x, y, z) \wedge (z)(\overline{z}),$$

and this means that

$$f(x, y, z) \leq f'(x, y, z) \leq f''(x, y, z).$$

In the final step we get:

$$R(z, \overline{z}) = 0,$$

and create the function

$$f'''(x, y, z) = f''(x, y, z) \wedge (0)$$

i.e.,

$$f(x, y, z) \leq f'(x, y, z) \leq f''(x, y, z) \leq f'''(x, y, z) = 0(x, y, z).$$

Hence, the original function $f(x, y, z)$ has to be less or equal to $0(x, y, z)$, and this is naturally equivalent to the fact that $f(x, y, z)$ itself is equal to $0(x, y, z)$; hence, the SAT-problem $f(x, y, z) = 1$ is not satisfiable.

It should be mentioned that we enumerated in this example only a subset of resolvents needed to show that the explored SAT-problem is not satisfiable. The additionally existing resolvents do not change the result but facilitate other ways to find the same result.

A huge amount of algorithms have been designed and implemented for this problem considering strategies and heuristic rules for the selection of variables and clauses to be considered in order to achieve rather "small" search trees.

The use of (orthogonal) ternary vectors and especially the parallel implementation of the intersection makes all these strategies obsolete (for propositional logic). The direct intersection can be applied, and automatically (without any special searching) the resolution principle is observed. This can easily be seen by the following.

Example 9.3 We consider the equation

$$(\overline{x}_1 \vee x_2 \vee x_4)(x_1 \vee x_3 \vee \overline{x}_5) = 1 .$$

We see that the resolution would be possible (using x_1 and $\overline{x}_1$); however, the parallel intersection of the solution-vectors considers the resolution possibility automatically:

x_1	x_2	x_3	x_4	x_5		x_1	x_2	x_3	x_4	x_5		x_1	x_2	x_3	x_4	x_5
0	–	–	–	–		1	–	–	–	–		0	–	1	–	–
1	1	–	–	–	$\cap$	0	–	1	–	–	=	0	–	0	–	0
1	0	–	1	–		0	–	0	–	0		1	1	–	–	–
												1	0	–	1	–

The intersections of the pairs of ternary vectors $(1, 1)$, $(2, 2)$, $(2, 3)$, $(3, 2)$, and $(3, 3)$ are empty and therefore omitted; the intersection of the other four pairs of ternary vectors $(1, 2)$, $(1, 3)$, $(2, 1)$, and $(3, 1)$ are included into the resulting TVL. Special considerations are not necessary, the *intersection* of ternary vectors and the *orthogonality* fully replace the *resolution principle*.

Example 9.4 The equation $(x \vee z)(\overline{x} \vee y)(\overline{x} \vee \overline{y})(\overline{y} \vee \overline{z})(x \vee y) = 1$ from above results in the empty set after four matrix intersections (always the two leftmost TVLs are used):

x	y	z		x	y	z		x	y	z		x	y	z		x	y	z
1	–	–	$\cap$	0	–	–	$\cap$	0	–	–	$\cap$	–	0	–	$\cap$	1	–	–
0	–	1		1	1	–		1	0	–		–	1	0		0	1	–

		x	y	z		x	y	z		x	y	z		x	y	z
=		1	1	–	$\cap$	0	–	–	$\cap$	–	0	–	$\cap$	1	–	–
		0	–	1		1	0	–		–	1	0		0	1	–

		x	y	z		x	y	z		x	y	z
=		0	–	1	$\cap$	–	0	–	$\cap$	1	–	–
						–	1	0		0	1	–

		x	y	z		x	y	z
=		0	0	1	$\cap$	1	–	–
						0	1	–

		x	y	z
=				

The MAXSAT-Problem

This is also a good place to remember the MAXSAT-problem. We saw here that the equation $(x \vee z)(\overline{x} \vee y)(\overline{x} \vee \overline{y})(\overline{y} \vee \overline{z})(x \vee y) = 1$ from above is unsatisfiable. But the successive application of the intersection showed even more. After the intersection of the first four clauses we still had a solution-vector $(x, y, z) = (001)$; hence, only one clause has to be eliminated, and the MAXSAT-problem is already solved, without any further consideration.

It might still be interesting to find out whether the last clause must be eliminated, or whether it also could be another one. This can be explored in the same way, and we could use intermediate results: In order to find out whether the fourth clause can be eliminated, we can use the existing intersection of the first three clauses and intersect with the fifth clause, and so on. Generally, when we reach an empty intersection, then we have at least a lower bound for MAXSAT, and further tests can try to increase this lower bound, but the search does not have to start from the scratch.

HORN-Clauses

We want to mention another set of conjunctive forms that was considered by many algorithms, the so-called *HORN-formulas*. A disjunction is a HORN-*clause* if it contains at most one non-negated variable: $x_1 \vee \overline{x}_2 \vee \overline{x}_3$, $\overline{x}_1 \vee \overline{x}_2$, or x_3 alone are HORN-clauses, $x \vee y \vee \overline{z}$ is not a HORN-clause. The benefit of a HORN-formula is that it can be verified in polynomial time whether this formula can be satisfied. When we are using ternary vectors as before, then there is no need for a special consideration of HORN-clauses. The candidate vectors for the solution of the SAT-problem can be built as before.

The Use of Prolog as SAT-solver

Prolog (from "programming in logic") is a programming language that was developed in the early 1970s by the French computer scientist Alain Colmerauer and enables declarative programming. It is considered the most important logic programming language. The main structures of Prolog are facts and rules.

SWI Prolog [14] provides the library `clpb` (Constraint Logic Programming over Boolean Variables) that defines among other Boolean expressions (*Expr*), logic operations (e.g., $\sim$ NOT, + OR, * AND), and the interface predicates `sat(+Expr)` and `labeling(+Vs)`. The statement

$$\texttt{use\_module(library(clpb)).}$$

activates this library so that subsequently defined SAT-problems can easily be determined and solved.

Figure 9.1 shows the declaration of the SAT-equation (9.1) and all computed solutions using the SWI Prolog inference engine. The first solution is shown in line 3, immediately after the input of the SAT-problem in lines 1 and 2 and pressing the `enter`-button. Pressing the `tab`-button shows in each case the next solution until the dot (.) at the end of the line indicates that no further solutions exist and the prompt (2 ?-) for the next command is shown. The same 17 solutions as in the approaches using TVLs have been computed using Prolog.

9.2   Complexity and NP-Completeness

The SAT-problem discussed before has a very special meaning in the general Theory of Complexity—it is the so-called **NP**-*complete problem*. We will not deal with this problem here, we will only give the basic definitions so that the reader gets an understanding of this concept.

One basic model for discussing complexity issues is the *Turing machine*. There are many different types of these machines; the basic model has a non-finite memory tape (memory cells) for writing,

```
1    1 ?- sat((X1+ ~X3+X5)*(~X2+X3+X4)*(X1+X2+ ~X5)*(~X3+ ~X4+ ~X5)),
2    labeling([X1,X2,X3,X4,X5]).
3    X1 = X3, X3 = X5, X5 = X2, X2 = X4, X4 = 0 ;
4    X1 = X3, X3 = X5, X5 = X2, X2 = 0,
5    X4 = 1 ;
6    X1 = X3, X3 = X5, X5 = 0,                          22   X1 = X5, X5 = X2, X2 = X4, X4 = 1,
7    X2 = X4, X4 = 1 ;                                  23   X3 = 0 ;
8    X1 = X3, X3 = 0,                                   24   X1 = X3, X3 = 1,
9    X5 = X2, X2 = X4, X4 = 1 ;                         25   X5 = X2, X2 = X4, X4 = 0 ;
10   X1 = X4, X4 = 0,                                   26   X1 = X3, X3 = X4, X4 = 1,
11   X3 = X5, X5 = X2, X2 = 1 ;                         27   X5 = X2, X2 = 0 ;
12   X1 = 1,                                            28   X1 = X3, X3 = X2, X2 = 1,
13   X3 = X5, X5 = X2, X2 = X4, X4 = 0 ;                29   X5 = X4, X4 = 0 ;
14   X1 = X4, X4 = 1,                                   30   X1 = X3, X3 = X2, X2 = X4, X4 = 1,
15   X3 = X5, X5 = X2, X2 = 0 ;                         31   X5 = 0 ;
16   X1 = X2, X2 = X4, X4 = 1,                          32   X1 = X3, X3 = X5, X5 = 1,
17   X3 = X5, X5 = 0 ;                                  33   X2 = X4, X4 = 0 ;
18   X1 = X5, X5 = 1,                                   34   X1 = X3, X3 = X5, X5 = X2, X2 = 1,
19   X3 = X2, X2 = X4, X4 = 0 ;                         35   X4 = 0.
20   X1 = X5, X5 = X4, X4 = 1,                          36
21   X3 = X2, X2 = 0 ;                                  37   2 ?-
```

Fig. 9.1 Solving the SAT-equation (9.1) using SWI Prolog: specification of the problem and all computed 17 solutions

storing, and reading information; there is only a sequential access to the cells. The machine is controlled by a stored program; at the beginning of the working the program is available as well as the input-data. The number of steps required for the solution of a problem is considered as the *time complexity*; the memory space required for the input, intermediate results, and the output is considered as the *memory complexity*. Quite naturally these two parameters depend on the size of the input; hence, it is possible or desirable to express these parameters as a function of the size of the input. If X is a given input, then we use $|X|$ for the size of the input.

In our area this size can be, for instance, the number of conjunctions or disjunctions used to represent a function, or the number of characters to represent a logic formula and so on. The set of all possible problems of a given category is indicated by Π; $P_i(X) \in \Pi$ indicates a special problem of this category with the input X.

Definition 9.2 **P** is the set of problems that can be solved in polynomial time by a deterministic Turing machine:

$\Pi \in \mathbf{P}$ if and only if each problem $P_i(X) \in \Pi$ can be solved by a deterministic Turing machine in at most k steps, and k can be calculated by means of a polynomial of $|X|$.

Hence, we have for such a problem $P_i(X)$ of Π:

$$k = a_n|X|^n + a_{n-1}|X|^{n-1} + \ldots + a_1|X| + a_0 . \tag{9.3}$$

Problems that can be solved in polynomial time are considered as computationally feasible. The next definition uses the concept of a *non-deterministic Turing machine*. The difference between a deterministic and a non-deterministic machine consists in the ability of the non-deterministic machine to make in some situations a "good" guess. If we want to know, for instance, whether a conjunctive form is satisfiable, then the machine guesses a satisfying vector **x** (if it exists), and the algorithm (program) has to verify only that the vector really satisfies the form, which can be done in polynomial time.

Definition 9.3 **NP** is the set of problems that can be solved in polynomial time by a non-deterministic Turing machine: $\Pi \in$ **NP** if and only if each problem $P_i(X) \in \Pi$ can be solved by a non-deterministic Turing machine in at most k steps, and k can be expressed by a polynomial of $|X|$.

This ability of a "good guess" has in practice to be replaced by considerable computational efforts, for instance by checking all binary vectors $\mathbf{x} \in \mathbb{B}^n$. Up to now it is only known that (trivially)

$$\mathbf{P} \subseteq \mathbf{NP} \, .$$

The question whether

$$\mathbf{P} = \mathbf{NP} \quad \text{or} \quad \mathbf{P} \neq \mathbf{NP}$$

is still an unsolved problem; hence, it is an important research topic. At this point in time it will not be surprising that the satisfiability problem for conjunctive forms is in this class **NP**.

Very often we find the situation that one problem of a given category can be transformed into a problem of another category, mostly in the hope that there are good problem-solving methods, algorithms, or programs available. Hence, the next definition is very understandable.

Definition 9.4 The problem class Π_1 can be polynomially reduced to the problem class Π_2 if there is a transformation of any problem of the class Π_1 into a problem of the class Π_2 that can be calculated by a deterministic Turing machine in polynomial time.

Here we have a representation of a problem of the class Π_1 as an input for a deterministic Turing machine, and the output of this machine is a representation of a problem of the class Π_2 that has been written on the tape. This transformation step is important because if a polynomial algorithm for one problem of the second class has been found, then there is also a polynomial algorithm for the problem of the first class: the polynomial transformation plus the polynomial solution plus the polynomial transformation of the results still can be described by a polynomial.

The final definition now introduces the concept of an **NP**-complete problem.

Definition 9.5 The problem class $\Pi \in$ **NP** is an **NP**-complete problem if and only if each class $\Pi_i \in$ **NP** can be polynomially reduced to Π.

And we will state the most important result that relates to logic functions and goes back to *S. A. Cook* [2].

Theorem 9.2

1. *The satisfiability problem for logic formulas is* **NP**-*complete.*
2. *The satisfiability problem for conjunctive forms is* **NP**-*complete.*
3. *The satisfiability problem for $k-SAT$ is* **NP**-*complete.*

The class $k-$SAT contains all conjunctive forms where each disjunction has exactly k literals, $k \geq 3$. At present many **NP**-complete problems are known, several of them are problems related to logic functions. The construction of a polynomial algorithm for one of these problems would mean that

$$\mathbf{P} = \mathbf{NP}$$

Fig. 9.2 Four ternary vectors lists to describe the assignment of the queens to the four columns of a 4×4 chessboard

$$D(QC_1) =$$

a1	a2	a3	a4	b1	b2	b3	b4	c1	c2	c3	c4	d1	d2	d3	d4
1	0	0	0	0	0	–	–	0	–	0	–	0	–	–	0
0	1	0	0	0	0	0	–	–	0	–	0	–	0	–	–
0	0	1	0	–	0	0	0	0	–	0	–	–	–	0	–
0	0	0	1	–	–	0	0	–	0	–	0	0	–	–	0

$$D(QC_2) =$$

a1	a2	a3	a4	b1	b2	b3	b4	c1	c2	c3	c4	d1	d2	d3	d4
0	0	–	–	1	0	0	0	0	0	–	–	0	–	0	–
0	0	0	–	0	1	0	0	0	0	0	–	–	0	–	0
–	0	0	0	0	0	1	0	–	0	0	0	0	–	0	–
–	–	0	0	0	0	0	1	–	–	0	0	–	0	–	0

$$D(QC_3) =$$

a1	a2	a3	a4	b1	b2	b3	b4	c1	c2	c3	c4	d1	d2	d3	d4
0	–	0	–	0	0	–	–	1	0	0	0	0	0	–	–
–	0	–	0	0	0	0	–	0	1	0	0	0	0	0	–
0	–	0	–	–	0	0	0	0	0	1	0	–	0	0	0
–	0	–	0	–	–	0	0	0	0	0	1	–	–	0	0

$$D(QC_4) =$$

a1	a2	a3	a4	b1	b2	b3	b4	c1	c2	c3	c4	d1	d2	d3	d4
0	–	–	0	0	–	0	–	0	0	–	–	1	0	0	0
–	0	–	–	–	0	–	0	0	0	0	–	0	1	0	0
–	–	0	–	0	–	0	–	–	0	0	0	0	0	1	0
0	–	–	0	–	0	–	0	–	–	0	0	0	0	0	1

because the existing polynomial-time transformations would result in a polynomial algorithm for all these problems.

The next sections will show, however, that the transformation of many problems into a SAT-problem and their solution by means of TVLs is a very powerful approach, which allows us the solution of extremely complex problems. The statement that problems of exponential complexity cannot be solved in a reasonable time loses its generality and its importance. It is possible to solve problems that are far beyond human possibilities, and the understanding of the solution is also not possible. There must be a firm belief that the program, the computer, and the algorithm work correctly.

9.3   Placement Problems

The Queens Problem

The first step to solve a problem is its specification by an adequate model. We take as example the modeling of a well-known combinatorial problem, the placement of 8 queens on a chessboard that do not attack each other. It is traditionally solved by *searching* and goes back to 1845 (*Bezzel*). *C. F. Gauss* found 72 solutions and *Franz Nauck* found all 92 solutions later in 1850. We study this problem to introduce the concept of Boolean equations into the solution of combinatorial problems.

We demonstrate a solution for a board 4×4. Figure 9.2 represents the four complete TVLs where each of them describes the conditions of one column of a $4 \times 4 -$ chessboard. A value 1 in these TVLs specifies the position of a queen and the values 0 in the same row the positions attacked by this queen and therefore forbidden for other queens.

Fig. 9.3 All solutions of four queens on a $4 \times 4-$ chessboard which do not attack each other

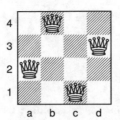 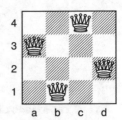

$\mathrm{D}(QC_1) =$

a1	a2	a3	a4	a5	a6	a7	a8	b1	b2	b3	b4	b5	...	c1	c2	c3	c4	c5	...	d1	d2	d3	d4	d5	...
1	0	0	0	0	0	0	0	0	0	−	−	−	...	0	−	0	−	−	...	0	−	−	0	−	...
0	1	0	0	0	0	0	0	0	0	0	−	−	...	−	0	−	0	−	...	−	0	−	−	0	...
0	0	1	0	0	0	0	0	−	0	0	0	−	...	0	−	0	−	0	...	−	−	0	−	−	...
0	0	0	1	0	0	0	0	−	−	0	0	0	...	−	0	−	0	−	...	0	−	−	0	−	...
0	0	0	0	1	0	0	0	−	−	−	0	0	...	−	−	0	−	0	...	−	0	−	−	0	...
0	0	0	0	0	1	0	0	−	−	−	−	0	...	−	−	−	0	−	...	−	−	0	−	−	...
0	0	0	0	0	0	1	0	−	−	−	−	−	...	−	−	−	−	0	...	−	−	−	0	−	...
0	0	0	0	0	0	0	1	−	−	−	−	−	...	−	−	−	−	−	...	−	−	−	−	0	...

Fig. 9.4 Ternary vectors to describe the assignment of a queen to one field of the first column and all resulting restrictions for other fields of the board

The four rows of one of these TVLs are disjunctive connected conjunctions and express that only one queen can be assigned to the selected column. Exactly one queen must be assigned to each of the four columns; hence, the intersection of the four TVLs delivers all solutions:

$$\mathrm{D}(SQ_{44}) = \bigwedge_{i=1}^{4} \mathrm{D}(QC_i) \,. \tag{9.4}$$

Due to the disjunctive forms of the four functions QC_i, $i = 1, \ldots, 4$, (9.4) is the left-hand side of a CDC-SAT-problem that has only two solutions:

	a1	a2	a3	a4	b1	b2	b3	b4	c1	c2	c3	c4	d1	d2	d3	d4
$\mathrm{D}(SQ_{44}) =$	0	1	0	0	0	0	0	1	1	0	0	0	0	0	1	0
	0	0	1	0	1	0	0	0	0	0	0	1	0	1	0	0

which are shown in Fig. 9.3.

For a good understanding of the model you can always go back to Fig. 9.2 which is the result of the procedure described below. It can be seen in this table that any placement of a queen on a $4 \times 4-$ chessboard leaves only six fields blank within the twelve remaining fields of the other columns of the chessboard.

We transform the problem (now for a board 8×8) into the area of logic equations in the following way (see Fig. 9.4). We use rules and find the equation by the following steps:

1. The variables $a1, a2, \ldots, h8$ are used as names for the 64 fields of the chessboard, as usual. They get, however, a binary meaning in the following way:

$$(a1 = 1) \quad \equiv \quad \text{there is a queen on field } a1,$$

$$(a1 = 0) \quad \equiv \quad \text{there is no queen on field } a1,$$

$$\vdots$$

2. The knowledge of the placing possibilities can be expressed as follows:
   In each column and each row precisely one queen can be placed; at most one queen can occur in each diagonal.

   - In consideration of the first vertical we write the respective possibilities:

   $$a1 \vee a2 \vee a3 \vee a4 \vee a5 \vee a6 \vee a7 \vee a8 = 1 \,. \tag{9.5}$$

   For the other verticals we get analogous equations, altogether eight such equations.
   - For the placement on the field $a1$ we get seven constraints for the first vertical:

   $$a1 \rightarrow \overline{a2} = 1 \,, \quad \ldots, \quad a1 \rightarrow \overline{a8} = 1 \,.$$

   - We transform the implications and combine the constraints into one equation:

   $$(\overline{a1} \vee \overline{a2}) \wedge \quad \ldots \quad \wedge (\overline{a1} \vee \overline{a8}) = 1 \,.$$

   - The same constraints must be written for the fields of the first row and the fields of the diagonal from $a1$ to $h8$.

   $$(\overline{a1} \vee \overline{b1}) \wedge \quad \ldots \quad \wedge (\overline{a1} \vee \overline{h1}) = 1 \,,$$

   $$(\overline{a1} \vee \overline{b2}) \wedge \quad \ldots \quad \wedge (\overline{a1} \vee \overline{h8}) = 1 \,.$$

   - Using the conjunction, all the single constraints can be included into one equation:

   $$(\overline{a1} \vee \overline{a2}) \ldots (\overline{a1} \vee \overline{a8})(\overline{a1} \vee \overline{b1}) \ldots (\overline{a1} \vee \overline{h1})(\overline{a1} \vee \overline{b2}) \ldots (\overline{a1} \vee \overline{h8}) = 1$$

   that can be transformed from the conjunctive form into a disjunctive form:

   $$\overline{a1} \vee \overline{a2} \wedge \cdots \wedge \overline{a8} \wedge \overline{b1} \wedge \cdots \wedge \overline{h1} \wedge \overline{b2} \wedge \cdots \wedge \overline{h8} = 1 \,. \tag{9.6}$$

   Now the effects of a queen on the field $a1$ have been fully considered.
   - This procedure has to be repeated for the other seven fields $a2, \ldots, a8$.
   - The system of the *requirement* Eq. (9.5) and the eight equations of the type (9.6) for the associated *restrictions* of the first column are included into one TVL in disjunctive form $D(QC_1)$.

3. We determine the constraints for the other columns in the same way. For each column we get one TVL: $D(QC_2), \ldots, D(QC_8)$.
4. Now we have the constraints for all placements and the requirement for all columns. The solution of the CDC-SAT-equation:

$$D(SQ_{88}) = \bigwedge_{i=1}^{8} D(QC_i) \tag{9.7}$$

gives a complete solution of the problem.

Naturally this procedure looks lengthy and cumbersome. The use of *ternary vectors as the appropriate data structure* allows us to bypass all these steps.

Fig. 9.5 Eight queens on a chessboard which do not attack each other

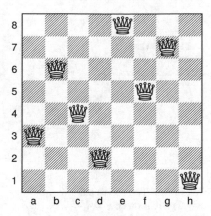

We use vectors of 64 components corresponding to the 64 fields of the board (see Fig. 9.4).

The setting of a queen on the field $a1$ can be seen by the value 1 for this field, but the *additional effect* of this setting is also included into this vector: It is not allowed to set a queen on other fields of the same column and the first row and the respective diagonal from $a1$ to $h8$; by $-$ it is indicated that there is no knowledge with regard to this field. So the eight vectors for the column (a) are considered as an $\vee$ of these possibilities; hence, we get the disjunctive form $D(QC_1)$ as shown in Fig. 9.4.

In this way, we get a very interesting extension of the SAT-problem. No longer disjunctions of single literals are considered, but *disjunctions of conjunctions*—do not forget that each ternary vector represents a conjunction. So we reach the following equation that can be built directly, without all the intermediate steps:

$$(C_{a1} \vee C_{a2} \vee C_{a3} \vee C_{a4} \vee C_{a5} \vee C_{a6} \vee C_{a7} \vee C_{a8})$$
$$\wedge (C_{b1} \vee C_{b2} \vee C_{b3} \vee C_{b4} \vee C_{b5} \vee C_{b6} \vee C_{b7} \vee C_{b8})$$
$$\dots$$
$$\wedge (C_{h1} \vee C_{h2} \vee C_{h3} \vee C_{h4} \vee C_{h5} \vee C_{h6} \vee C_{h7} \vee C_{h8}) = 1 \; .$$

The intersection of the eight TVLs $D(QC_1)$, $\dots$, $D(QC_8)$ completely solves the problem! *Search procedures are no longer necessary to find the final result*, the solution of a *Boolean equation* (in the form of a CDC-SAT-formula) delivers all solutions. Figure 9.5 shows one of the 92 solutions of the chessboard 8×8.

It is not difficult to extend this methodology to smaller or larger board sizes (see above for $n = 4$). At present, the problem (as far as we know) has been solved for $n = 27$ (TU Dresden, Germany). The number of

$$234, 907, 967, 154, 122, 528$$

possible solutions has been published, after 1 year of work time.

There must be a very good reason to spend these efforts onto the solution of such a problem. And it must be understood that there is no evidence that the result is correct. Only a successful repetition can support the result. Similar problems show the same complexity for their solution, but the results might be easier to understand.

The search procedures have some difficulties when one or more queens have been set already. The problem is now: *Is there a solution that considers some given fixed points?* In this case the models which have been based on ternary vectors, however, are working even faster because the whole TVL for the columns of this setting degenerates to a single vector.

Fig. 9.6 Nine queens and
one pawn or ten queens
and two pawns on a
chessboard which do not
attack each other

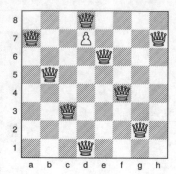 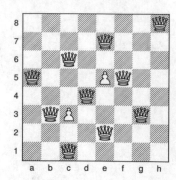

Table 9.2 Two pawns and
$n + 2$ queens on an $n \times n-$
chessboard

Rows	Columns	Queens	Solutions	Time in ms
6	6	8	0	140
7	7	9	4	640
8	8	10	44	2421
9	9	11	280	8515
10	10	12	1208	26,359
11	11	13	11,432	76,359
12	12	14	96,476	308,375

Example 9.5 This example shows how to explore the hypothesis that on a chessboard 8×8 the placement of one pawn allows $8 + 1 = 9$ queens and the addition of two pawns allows $8 + 2 = 10$ queens. Additionally, the placements of the pawns should be found.

It is easy to understand that an appropriate position of a pawn interrupts the vertical, horizontal, and the tow diagonal action lines of a queen. Hence, the column and the line of the pawn position can allow the placement of two queens. The way of thinking remains as before; however, the possibilities and the constraints change a bit (see Fig. 9.6):

- The fields of the pawns are not available for the queens.
- Some of the constraints will be shorter.

For the column c in the right chessboard of Fig. 9.6 we have now

$$(c1 \, \overline{c2} \vee \overline{c1} \, c2)$$

$$\wedge \, (c4 \, \overline{c5} \, \overline{c6} \, \overline{c7} \, \overline{c8} \vee \overline{c4} \, c5 \, \overline{c6} \, \overline{c7} \, \overline{c8} \vee$$

$$\overline{c4} \, \overline{c5} \, c6 \, \overline{c7} \, \overline{c8} \vee \overline{c4} \, \overline{c5} \, \overline{c6} \, c7 \, \overline{c8} \vee \overline{c4} \, \overline{c5} \, \overline{c6} \, \overline{c7} \, c8) = 1 \,.$$

We achieved a complete solution of the problem by arranging a single pawn as well as all pairs of pawns on the columns from b to g and on the rows from 2 to 7. Figure 9.6 shows one solution with a pawn on $d7$ and a solution with pawns on $c3$ and $e5$. There are 44 solutions of ten queens and two pawns on a $8 \times 8 -$ chessboard. Table 9.2 enumerates the number of solution and the time for their calculation for two pawns and $n + 2$ queens on chessboards of several sizes.

We used the programming system XBOOLE [6]. The time naturally depends on the respective hardware.

Table 9.3 Results and conjectures given by Schwarzkopf in [5]

Condition	Number of pieces	Number of solutions
$n = m = 1$	1	1
$n = m > 1$	$2n - 2$	2^n
$n < m, n$ odd	$n + m - 1$	1?
$n < m, n$ even, m odd	$n + m - 1$	1?
$n < m, n, m$ even	$n + m - 2$	?

The Bishops Problem

In this subsection we will show how the solution of SAT-problems can be used to build theories in other areas.

Schwarzkopf published in [5] a well-defined problem which had not been solved until then:

Let be given a chessboard with m rows and n columns. Arrange a number of *bishops* on the board not attacking each other, but attacking all empty fields on the board.

A problem like this very often appears in different shapes:

1. Find the minimum number of pieces required to attack all the empty fields.
2. Find the maximum number of pieces that can be arranged without attacking each other.
3. How many different positions (solutions) exist for a given number of pieces (bishops)?

Table 9.3 shows the starting point.

The case $m = n = 1$ is trivial. A single bishop is placed on the single field of the board. Surprisingly the number of solutions for a quadratic board $n \times n$, $n > 1$, is equal to 2^n. It is a remarkable achievement that Schwarzkopf found this result. He also detected that the solution numbers depend on whether the numbers of rows and columns are odd or even. Other conjectures, however, could not be confirmed; by using the solutions of Boolean equations it could be seen that corrections and extensions are necessary.

We assign a variable x_{ij} to each field of the board. The index i indicates the row, and the index j the column of the board, respectively:

$$x_{ij} = \begin{cases} 1 & \text{if a bishop is on the field } (i, j), \\ 0 & \text{otherwise}. \end{cases}$$

A bishop can attack other pieces on at most two diagonal lines. Hence, no other bishops are allowed on these diagonals. For each field a *restriction* can be specified as part of the model.

We assume as example a $4 \times 4 -$ chessboard (see Fig. 9.7), the counting of rows and columns will be used as usually, from the bottom up and from the left to the right. The condition for a bishop on the field $(3, 2)$ can be expressed by the following equation:

$$C_{32} = x_{32} \wedge \overline{x}_{14} \wedge \overline{x}_{21} \wedge \overline{x}_{23} \wedge \overline{x}_{41} \wedge \overline{x}_{43} = 1 \,.$$

The second type of rules expresses *requirements*. One bishop must be placed on each diagonal, any field on the diagonal is allowed:

$$C_{12} \vee C_{23} \vee C_{34} = 1 \,.$$

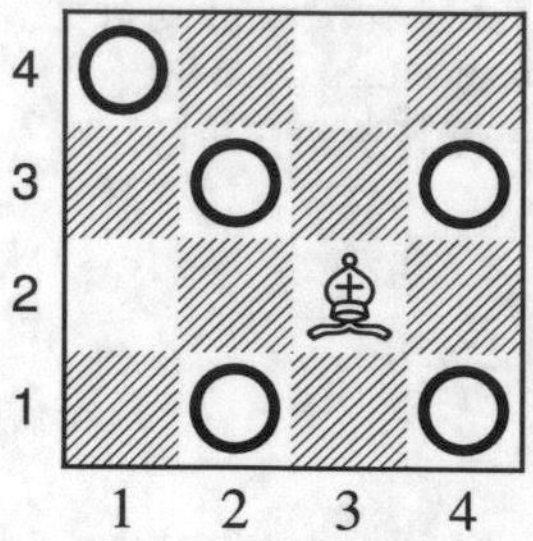

Fig. 9.7 The bishop on field (3,2) attacks on the 4 × 4 chessboard five fields marked by circles

Table 9.4 Numbers of solutions for larger chessboards

m	n	8	9	10	11	12	13	14	15	16	17	18
8	1	1										
8	2	4										
8	3	25	30	9								
8	4			144								
8	5				289	170	25					
8	6					324						
8	7					1024	64	1				
8	8							256				
9	1		1									
9	2	1	2	1								
9	3		40	64	16							
9	4			169	130	25						
9	5				775	750	75					
9	6					1681	656	64				
9	7						2646	972	27			
9	8							4096	128	1		
9	9								3456	512		
10	1			1								
10	2			4								
10	3			64	80	25						
10	4					484						
10	5					1764	1008	144				
10	6							3844				
10	7							9801	2574	169		
10	8									2916		
10	9									16,384	256	1
10	10											1024

Such an equation must be created for each diagonal. The four fields in the corners of the chessboard are on one diagonal only, all the other fields are on two diagonals.

From the approach for the placements of queens follows that for any values of rows and columns the respective equations can be built and solved.

In comparison with the initial table we can see much more information of the solutions of the problem. These values can now be studied and extended to a mathematical description of the results depending on m and n (which is beyond the scope of the SAT-problem itself).

Table 9.4 proves for the chessboard $n < m$, n odd, and the number of Bishops $n + m - 1$ that the conjecture of Schwarzkopf is wrong: e.g., for $n = 3$, $m = 8$ there are nine solutions with $3 + 8 - 1 = 10$ Bishops and not only one solution. The other conjecture of Schwarzkopf is also

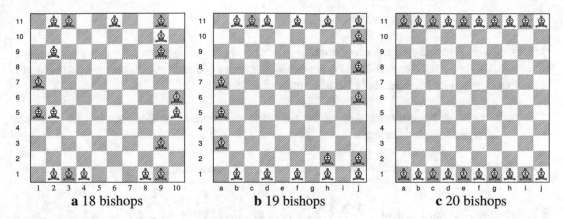

a 18 bishops   **b** 19 bishops   **c** 20 bishops

Fig. 9.8 Selected solutions of the Bishop Problem on the chessboard 10×11

Table 9.5 Fibonacci numbers F_n and the newly defined **Bishop numbers** B_n

Recurrence relation								
for $n > 2$	$n = 1$	$n = 2$	$n = 3$	$n = 4$	$n = 5$	$n = 6$	$n = 7$	$n = 8$
Fibonacci numbers								
$F_n = F_{n-2} + F_{n-1}$	1	1	2	3	5	8	13	21
Bishop numbers								
$B_n = B_{n-2} + 2 * B_{n-1}$	1	1	3	7	17	41	99	239

Table 9.6 Number of Bishops for the case $m - n = 3, m > 4, n > 1$

Number of bishops	Number of solutions
$2 * m - 6$	$B_n{}^2$
$2 * m - 5$	$2 * B_n * F_n$
$2 * m - 4$	$F_n{}^2$

wrong: for $n = 4$, $m = 9$ (where n is even and m is odd) there are 25 solutions with $4 + 9 - 1 = 12$ Bishops and not only one solution. This conjecture is satisfied only in some case: e.g., $n = 2$, $m = 9$ or $n = 8$, $m = 9$.

Three solutions for a given board size (10×11 in Fig. 9.8) can result in very nice different images.

Figure 9.8 shows a selection of solutions of the Bishop Problem on the chessboard 10×11: **a** one of 65,536 positionings with 18 bishops; **b** one of 512 positionings with 19 bishops; **c** the single positioning with 20 bishops.

The evaluation of these results leads us to interesting insights [7]. Most surprisingly we found a relation of the number of solutions to the well-known *Fibonacci numbers* F_n. To express all solution numbers of the Bishop problem we discovered a similar sequence of numbers and gave them the name *Bishop numbers* B_n. Table 9.5 show the details of these two series of numbers.

Using both the *Fibonacci numbers* and *Bishop numbers* we found the short formulas of Table 9.6 to express the number of bishops for certain sizes of the chessboard.

Further details about the Bishop problem for chessboards until a size of 16×16 are given in [7].

Latin Squares and Sudoku

Latin rectangles and *Latin squares* have a long history and go back to ancient times. L. Euler (1707–1783) was the most famous mathematician who contributed to the enormous theory that exists now. At present the game *Sudoku* is the most popular example for a Latin square.

We start with the following definition.

Definition 9.6 A rectangular matrix of dimension $m \times n$, $m \leq n$, each row of which is a permutation of the elements of a set S consisting of n elements, and in the columns each element occurs at most once. For $m = n$ a Latin rectangle is a Latin square of order n.

In order to make the understanding of the definition as simple as possible we use the basic set $S = \{1, \ldots, n\}$ and small values of n.

Example 9.6 Let $n = 5$. Then the following representation shows Latin rectangles:

$$R_1 \;=\; \begin{vmatrix} 1 & 3 & 2 & 5 & 4 \\ 5 & 2 & 3 & 4 & 1 \end{vmatrix} \quad \text{or} \quad R_2 \;=\; \begin{vmatrix} 1 & 3 & 2 & 5 & 4 \\ 5 & 2 & 3 & 4 & 1 \\ 3 & 1 & 4 & 2 & 5 \end{vmatrix}.$$

As a special case we have Latin squares with $m = n$:

$$R_3 \;=\; \begin{vmatrix} 1 & 2 & 3 & 4 & 5 \\ 5 & 1 & 2 & 3 & 4 \\ 4 & 5 & 1 & 2 & 3 \\ 3 & 4 & 5 & 1 & 2 \\ 2 & 3 & 4 & 5 & 1 \end{vmatrix}.$$

One of the problems arising for Latin squares is the completion of a square where some values already have been set. Sudoku is an example for this problem. It is a rather popular problem and a special case of a Latin square; in addition to the requirement that each row and each column contains the numbers $1, \ldots, 9$, some boxes of the size 3×3 must be filled by these numbers as well.

Logic variables x_{ijk} can be used to express the information about the field of the row i and the column j as follows:

$$x_{ijk} = \begin{cases} 1 & \text{if the number } k \text{ is assigned to the field } (i, j)\,, \\ 0 & \text{if the number } k \text{ is prohibited on the field } (i, j)\,. \end{cases}$$

The requirements and constraints can be stated by one single conjunction C_{ijk} for each number on each field:

$$
\begin{aligned}
C_{111} =\;\; & x_{111}\overline{x}_{112}\overline{x}_{113}\overline{x}_{114}\overline{x}_{115}\overline{x}_{116}\overline{x}_{117}\overline{x}_{118}\overline{x}_{119} && \text{only 1 on the field} \\
\wedge\;\; & \overline{x}_{121}\overline{x}_{131}\overline{x}_{141}\overline{x}_{151}\overline{x}_{161}\overline{x}_{171}\overline{x}_{181}\overline{x}_{191} && \text{no other 1 vertically} \\
\wedge\;\; & \overline{x}_{211}\overline{x}_{311}\overline{x}_{411}\overline{x}_{511}\overline{x}_{611}\overline{x}_{711}\overline{x}_{811}\overline{x}_{911} && \text{no other 1 horizontally} \\
\wedge\;\; & \overline{x}_{221}\overline{x}_{231}\overline{x}_{321}\overline{x}_{331} && \text{no other 1 in the box.}
\end{aligned}
$$

This conjunction describes completely the setting of the number 1 on the field $(1, 1)$ and *all the consequences*; hence, for each field we need nine conjunctions for the nine different possible numbers. These conjunctions have the advantage that they are orthogonal to each other, for one field exactly one variable can be equal to 1. The Boolean equation

$$C_{111} \vee C_{112} \vee C_{113} \vee C_{114} \vee C_{115} \vee C_{116} \vee C_{117} \vee C_{118} \vee C_{119} = 1$$

describes the nine possibilities which exist for the field $(1, 1)$ and *all the consequences* resulting from a given setting. The translation into ternary vectors results in a TVL with nine rows and at all we get again a CDC-SAT-problem to solve.

In principle, 729 such conjunctions are possible; however, each given setting (such as $x_{195} = 1$, i.e., the number 5 is assigned to the field $(1, 9)$) selects one out of nine conjunctions. Now it is only necessary to write a tiny program for the input of some given values and the transformation into a conjunctive form of a Boolean equation. A comparison of several approaches for this problem has been published in [12].

9.4 Covering Problems

Set Covers

Given a set of elements $\mathcal{U} = \{1, 2, \ldots, m\}$ and a set $\mathcal{S} = \{S_1, \ldots, S_n\}$ of subsets of $\mathcal{U}$ with $S_1 \cup S_2 \cup \ldots \cup S_n = \mathcal{U}$. Find the smallest subset $\mathcal{C} \subseteq \mathcal{S}$ of sets the union of which is equal to $\mathcal{U}$.

Example 9.7 Let $\mathcal{U} = \{1, 2, 3, 4, 5\}$ and $S = \{\{1, 2, 3\}, \{2, 4\}, \{3, 4\}, \{2, 4, 5\}\}$. We can cover all of the elements with the two sets $\{\{1, 2, 3\}, \{2, 4, 5\}\}$ because

$$\{1, 2, 3\} \cup \{2, 4, 5\} = \mathcal{U} .$$

It is easy to find an appropriate SAT-equation. For each element of $\mathcal{U}$, we write down the sets that contain this element:

$$\mathcal{C} = \underbrace{S_1}_{1} \underbrace{(S_1 \vee S_2 \vee S_4)}_{2} \underbrace{(S_1 \vee S_3)}_{3} \underbrace{(S_2 \vee S_3 \vee S_4)}_{4} \underbrace{S_4}_{5} = S_1 S_4 = 1 . \tag{9.8}$$

This equation shows for each element the subsets that can be taken for its inclusion. This can also be expressed very clearly by using the *characteristic functions* of the subsets (Table 9.7a).

The columns *with exactly one value 1 must be included*, in this tiny example the sets S_1 and S_4 must be used, and this is already sufficient (Table 9.7b). It cannot be avoided that the element 2 is covered by two sets.

It is not even necessary to transform equation (9.8) into a proper SAT-format, the symbols of the sets can be used as in (9.8). The definition of logic variables could be done as follows:

$$x_i = 1 \quad \text{if and only if the explored element is covered by the set } S_i .$$

This results in the equation

$$\underbrace{x_1}_{1} \underbrace{(x_1 \vee x_2 \vee x_4)}_{2} \underbrace{(x_1 \vee x_3)}_{3} \underbrace{(x_2 \vee x_3 \vee x_4)}_{4} \underbrace{x_4}_{5} = 1 .$$

Table 9.7 Set cover: **(a)** characteristic functions, **(b)** minimal cover

a

χ	1	2	3	4	5
$\chi(S_1)$	1	1	1	0	0
$\chi(S_2)$	0	1	0	1	0
$\chi(S_3)$	0	0	1	1	0
$\chi(S_4)$	0	1	0	1	1

b

χ	1	2	3	4	5
$\chi(S_1)$	1	1	1	0	0
$\chi(S_4)$	0	1	0	1	1

Table 9.8 Exact set cover

χ	1 2 3 4 5 6 7
$\chi(S_1)$	1 0 0 1 0 0 1
$\chi(S_2)$	1 0 0 1 0 0 0
$\chi(S_3)$	0 0 0 1 1 0 1
$\chi(S_4)$	0 0 1 0 1 1 0
$\chi(S_5)$	0 1 1 0 0 1 1
$\chi(S_6)$	0 1 0 0 0 0 1

The solution-set of this equation consists of four binary vectors represented by the single ternary vector $(x_1, x_2, x_3, x_4) = (1 - -1)$. It is a property of this problem that all the variables are non-negated. It is a *unate covering problem*. The assignment of values 1 to all variables is a trivial solution, but solutions with a minimal number of values 1 are wanted. Replacing the two dashes in the solution-vector by values 0 leads to the same minimal set cover $(x_1, x_2, x_3, x_4) = (1001)$ as shown in Table 9.7b.

Sometimes it is required that in the final solution any element is covered only by one set. A solution of this problem will be denoted by *"exact set cover."* Our previous example shows that a solution of an exact set cover does not always exist.

We use the example given in Table 9.8. For the element 1, we can either use the set S_1 or the set S_2 expressed by $(x_1\overline{x}_2 \vee \overline{x}_1 x_2)$. The same step will be done for the other elements, which finally results in the following equation:

$$\underbrace{(x_1\overline{x}_2 \vee \overline{x}_1 x_2)}_{1} \underbrace{(x_5\overline{x}_6 \vee \overline{x}_5 x_6)}_{2} \underbrace{(x_4\overline{x}_5 \vee \overline{x}_4 x_5)}_{3}$$

$$\wedge \underbrace{(x_1\overline{x}_2\overline{x}_3 \vee \overline{x}_1 x_2\overline{x}_3 \vee \overline{x}_1\overline{x}_2 x_3)}_{4} \underbrace{(x_3\overline{x}_4 \vee \overline{x}_3 x_4)}_{6} \underbrace{(x_4\overline{x}_5 \vee \overline{x}_4 x_5)}_{6}$$

$$\wedge \underbrace{(x_1\overline{x}_3\overline{x}_5\overline{x}_6 \vee \overline{x}_1 x_3\overline{x}_5\overline{x}_6 \vee \overline{x}_1\overline{x}_3 x_5\overline{x}_6 \vee \overline{x}_1\overline{x}_3\overline{x}_5 x_6)}_{7} = 1 .$$

The single solution-vector $(x_1, x_2, x_3, x_4, x_5, x_6) = (010101)$ determines the set of sets $\{S_2, S_4, S_6\}$ as the only existing exact set cover:

$$\{1, 4\} \cup \{3, 5, 6\} \cup \{2, 7\} = \{1, 2, 3, 4, 5, 6, 7\} .$$

Vertex Covers

A *vertex cover* (also referred to as *node cover*) of a graph is a set of vertices such that each edge of the graph is incident to at least one vertex of the set. It is easy to find the logic model. For each edge from vertex i to vertex j the vertex i **or** the vertex j **or** both of them must be an element of the *vertex cover*. Figure 9.9 shows an example of a graph, the associated CD-SAT-formula, and the solution. This is an example for a 2-SAT-problem—each bracket contains two variables which indicate the vertices that are connected by the selected edge. In this way, we get three sets of vertex covers:

- $\{1, 3\}$.
- $\{2, 3\}$.
- $\{1, 2, 4, 5, 6\}$.

Fig. 9.9 Example of a
vertex cover: (**a**) graph, (**b**)
SAT-equation, (**c**) minimal
solutions

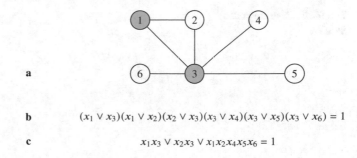

b $(x_1 \lor x_3)(x_1 \lor x_2)(x_2 \lor x_3)(x_3 \lor x_4)(x_3 \lor x_5)(x_3 \lor x_6) = 1$

c $x_1 x_3 \lor x_2 x_3 \lor x_1 x_2 x_4 x_5 x_6 = 1$

The first two sets have the smallest number of elements.

The complement of a vertex cover is the *set of independent vertices*. No vertex of this set is connected to another vertex of the set:

- $\overline{\{1, 3\}} = \{2, 4, 5, 6\}$.
- $\overline{\{2, 3\}} = \{1, 4, 5, 6\}$.
- $\overline{\{1, 2, 4, 5, 6\}} = \{3\}$.

Unate Covering Problems

Problems of the vertex cover belong to the class of *unate covering problems* (UCP). An important application of UCP consists in selecting a minimal set of prime conjunctions in circuit design. This class of problems has two special properties:

1. The clauses of the SAT-formula only contain non-negated variables.
2. The wanted solution consists of minimal sets; that means no element can be removed from a solution-set without losing the satisfiability.

The first property can be seen in Fig. 9.9b. Due to the second property, SAT-solvers are not well suitable to solve such problems. The assignment of the value 1 to all variables satisfies the given equation but does in most cases not belong to the wanted minimal solution. Hence, the utilization of a SAT-solver requires the calculation of all solutions followed by a sieving with regard to minimal solutions.

A basic method to solve a UCP consists in the application of the distributive, the idempotent, and the absorption rule. In [9] we found that the absorption applied to each intermediate result reduces the needed time by a factor of about 10^3. We used this improved algorithm as basis for later comparisons.

A more powerful approach to solve UCP [11] utilizes the XBOOLE-operations NDM (*negation with regard to De Morgan*), CPL (*complement*), OBBC (*orthogonal block-building with changes*), and CEL (*exchange elements of selected columns*). This approach reduces the time to solve a UCP by a factor of about 10^5 on a single CPU-core. Instead of the calculation of the complement as a whole, several difference operations can be executed on the available cores on the CPU. Using four CPU-cores, where one of them operates as an intelligent master, an additional speedup factor of more than 350 was measured [8].

By means of special algorithms on a GPU we were able to speed up the solution time by the remarkable factor of about 10^{11} [10]. Further optimization of this GPU-approach additionally reduced the solution time by a factor of 3 [13].

Fig. 9.10 Edge cover of a
graph

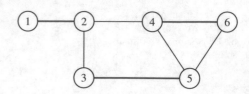

Edge Covers

An *edge cover* of a graph is a set of edges such that every vertex of the graph is incident to at least
one edge of the set. Find the minimum edge cover!

It is again not very difficult to find an appropriate model. For each vertex we combine the edges
that are connected to the selected vertex by $\vee$. We use the graph of Fig. 9.10 and get the following
CD-SAT-formula:

$$x_{12}(x_{12} \vee x_{23} \vee x_{24})(x_{23} \vee x_{35})(x_{24} \vee x_{45} \vee x_{46})(x_{35} \vee x_{45} \vee x_{56})(x_{46} \vee x_{56}) = 1 \ .$$

The solution for the edge cover: Solve the CD-SAT-problem and take the solution with a minimum
number of values 1. The three red colored edges of Fig. 9.10 belong to the edge cover of this graph.

The calculation of minimal edge covers belongs as the calculation of minimal vertex covers to the
unate covering problems.

Cliques

A *clique* is a complete subgraph; each vertex of a clique is connected by edges with all other vertices
of this subgraph.

Example 9.8 We reuse the graph of Fig. 9.10 as example and look for triangles. The three sides of a
triangle must be edges of the graph.

A schematic solution uses a very simple model: We select any three vertices and check whether
the three required edges exist in the given graph. This leads to one equation for each possible clique:

$$x_{12}x_{13}\overline{x}_{14}\overline{x}_{15}\overline{x}_{16}x_{23}\overline{x}_{24}\overline{x}_{25}\overline{x}_{26}\overline{x}_{34}\overline{x}_{35}\overline{x}_{36}\overline{x}_{45}\overline{x}_{46}\overline{x}_{56} = 1$$

$$x_{12}\overline{x}_{13}x_{14}\overline{x}_{15}\overline{x}_{16}\overline{x}_{23}x_{24}\overline{x}_{25}\overline{x}_{26}\overline{x}_{34}\overline{x}_{35}\overline{x}_{36}\overline{x}_{45}\overline{x}_{46}\overline{x}_{56} = 1$$

$$x_{12}\overline{x}_{13}\overline{x}_{14}x_{15}\overline{x}_{16}\overline{x}_{23}\overline{x}_{24}x_{25}\overline{x}_{26}\overline{x}_{34}\overline{x}_{35}\overline{x}_{36}\overline{x}_{45}\overline{x}_{46}\overline{x}_{56} = 1$$

$$\vdots$$

$$\overline{x}_{12}\overline{x}_{13}\overline{x}_{14}\overline{x}_{15}\overline{x}_{16}\overline{x}_{23}\overline{x}_{24}\overline{x}_{25}\overline{x}_{26}\overline{x}_{34}\overline{x}_{35}\overline{x}_{36}x_{45}x_{46}x_{56} = 1 \ . \tag{9.9}$$

The number of triangles is growing very fast (here $\binom{6}{3} = 20$); however, non-existing edges reduce
the size of the problem. In this example only the solution-set of the last equation of (9.9) is not empty.
Hence, 4 - 5 - 6 is a complete subgraph with three nodes (Fig. 9.11).

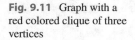

Fig. 9.11 Graph with a red colored clique of three vertices

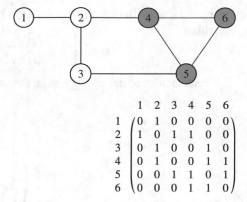

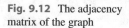

Fig. 9.12 The adjacency matrix of the graph

$$
\begin{array}{c}
\ \ 1\ \ 2\ \ 3\ \ 4\ \ 5\ \ 6\\
\begin{array}{c}1\\2\\3\\4\\5\\6\end{array}
\left(
\begin{array}{cccccc}
0 & 1 & 0 & 0 & 0 & 0\\
1 & 0 & 1 & 1 & 0 & 0\\
0 & 1 & 0 & 0 & 1 & 0\\
0 & 1 & 0 & 0 & 1 & 1\\
0 & 0 & 1 & 1 & 0 & 1\\
0 & 0 & 0 & 1 & 1 & 0
\end{array}
\right)
\end{array}
$$

Figure 9.12 shows the adjacency matrix of the graph that also can be considered as a logic function and used to detect existing cliques of a certain size as follows:

- Which columns and rows have to be deleted to get a submatrix with values 1 only (except the main diagonal)?
- Here we delete the rows and columns 1 - 2 - 3.

Larger cliques can be built from triangles—a complete subgraph consisting of the nodes 1 - 2 - 3 - 4 must be built from four triangles 1 - 2 - 3, 1 - 2 - 4, 1 - 3 - 4, and 2 - 3 - 4. This property (also for more than 4 vertices) can be efficiently included into Boolean models.

9.5 Path Problems

Hamiltonian Paths

We will use another very old problem to define a next combinatorial concept.

Let be given a graph with some vertices and edges. Is it possible to find a sequence of edges (a path) such that *each* vertex will be used once and only once? Such a path will be denoted by *Hamiltonian Path*. It is denoted by *Hamiltonian Cycle* when the last step will go back to the starting point. Several questions can be stated for a given graph:

- Is there a path that all nodes will be used?
- If so, how many possibilities exist?
- If not, find Hamiltonian subgraphs.

The problem of using all vertices in a graph goes back to the nineth century A.D..:

- Is it possible that a knight tour on a chessboard uses every field on the board precisely once?

If the answer is "yes," then we have an example for a Hamiltonian cycle. It is known that there are solutions for board sizes $n \geq 5$.

You can find wonderful patterns for very different board sizes and a comprehensive theory for this field. The Internet even shows videos where the knight is moving directly on the screen.

The example shows, as we have already seen before, the transformation into a graph and into a Boolean equation.

We show again the transformation into a Boolean equation. The edges can be expressed by Boolean variables:

$$x_{ij} = \begin{cases} 1 & \text{if the edge is used from node } i \text{ to node } j, \\ 0 & \text{otherwise} . \end{cases} \tag{9.10}$$

To make this definition applicable, the vertices (the fields on the chessboard) get a single number:

$$\begin{array}{ccccc} 1 & 2 & 3 & 4 & 5 \\ 5 \begin{pmatrix} 5 & 10 & 15 & 20 & 25 \\ 4 & 9 & 14 & 19 & 24 \\ 3 & 8 & 13 & 18 & 23 \\ 2 & 7 & 12 & 17 & 22 \\ 1 & 6 & 11 & 16 & 21 \end{pmatrix} . \end{array}$$

The definition of a Hamiltonian cycle implies the following rules:

1. An edge from node i to vertex j, i.e., $x_{ij} = 1$, prohibits that the reverse edge is used, i.e., $x_{ji} = 0$.
2. An edge from node i to vertex j, i.e., $x_{ij} = 1$, prohibits all edges to other destination vertices k, i.e., $x_{ik} = 0$.
3. An edge from node i to vertex j, i.e., $x_{ij} = 1$, prohibits all edges from other source vertices k, i.e., $x_{kj} = 0$.

These rules are now applied to each field of the chessboard. The easiest step is the application to the corner fields. For the vertex 1, we get, for instance:

$$(x_{12,1}\overline{x}_{1,12}\overline{x}_{12,3}\overline{x}_{12,9}\overline{x}_{12,19}\overline{x}_{12,23}\overline{x}_{12,21}\overline{x}_{8,1} \vee x_{8,1}\overline{x}_{1,8}\overline{x}_{8,5}\overline{x}_{8,15}\overline{x}_{8,19}\overline{x}_{8,17}\overline{x}_{8,11}\overline{x}_{12,1}) = 1 .$$

The path of the knight can use the field $a1$ only in two ways: $c2 - a1 - b3$ ($12 - 1 - 8$) or in reverse order $b3 - a1 - c2$ ($8 - 1 - 12$). The number of arriving edges and therefore the size of the DC-clauses increases for fields in middle of the board.

A (small) program for the definition of the graph will be helpful. Furthermore, it is not necessary to generate the equation itself. A direct generation of the respective TVL will be the best.

The equations for all fields will be combined into one single equation which is again a CDC-SAT-equation (conjunctions of disjunctions consisting of conjunctions again). It ensures that each vertex will be used once and only once as start vertex of an edge, and all solutions will be found simultaneously. Figure 9.13 shows one solution of a knight tour on a chessboard 8×8.

The CDC-SAT-formula must specify for each vertex a DC-clause that contains for each incoming edge a conjunction that satisfies the three rules introduced above. These clauses are sufficient to determine all potential Hamiltonian cycles because all vertices are in this case both a source and a destination of an edge. The DC-clauses that describe potential Hamiltonian paths must additionally contain conjunctions that determine each possible end vertex of the Hamiltonian path by conjunctions of incoming edges for which no outgoing edge exists.

The three rules of a *Hamiltonian Cycle* introduced above are only necessary conditions because not all edges of the given graph must belong to this path, so that several separate cycles can use each vertex once and only once. This gap can be closed by an analysis of the computed potential *Hamiltonian Cycles* whether all vertices of these subgraphs are reachable from an arbitrary chosen vertex of the evaluated graph.

Fig. 9.13 One solution for the knights problem on a chessboard 8×8

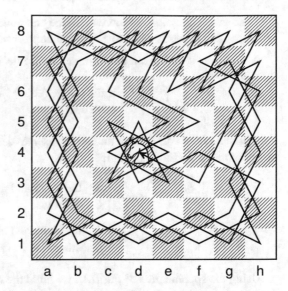

An easy approach for this analysis is the computation of the reachable graph. For this purpose the graph of a potential *Hamiltonian Cycle* must be extended by all transitive edges which can be computed in $\lceil \log_2 n \rceil$ steps, where n is the number of vertices of the original graph. $\lceil \log_2 n \rceil$ determines the smallest integer that is larger than $\log_2 n$. An analyzed graph of edges that satisfy the all rules of a *Hamiltonian Path* specified above is really:

- A *Hamiltonian Path* when the computed path begins in exactly one vertex and can reach all other vertices starting in this vertex.
- A *Hamiltonian Cycle* when the derived reachable graph contains edges from an arbitrary chosen vertex to all vertices of this graph.

Eulerian Paths

An *Eulerian Path* in a graph has the property that *each* edge will be used once and only once. This problem has been explored by Leonhard Euler (1707–1783) looking at the bridges in Königsberg (now: Kaliningrad, Russia), see Fig. 9.14.

He found out that such a path is only possible when the number of edges attached to one node is an *even number*; two nodes can be attached by an odd number of edges; these nodes must be used as start and end node of the path. If the number of edges attached to a node is an odd number for more than two nodes, then you get stuck in such a node. Therefore, the desired trip is not available since all four nodes have an odd number of edges.

An *Eulerian Path* satisfies two necessary conditions:

- Each edge between two nodes must be used in exactly one of two possible directions.
- The number of arriving edge must not differ from the number of leaving edges by more than one in each node.

Additionally an Eulerian Path must satisfy the sufficient conditions:

- The graph contains not more than one start node of the Eulerian Path, where the number of used outgoing edges is equal to 1 plus the number of used incoming edges.

- The graph contains not more than one end node of the Eulerian Path, where the number of used incoming edges is equal to 1 plus the number of used outgoing edges.

A way over a bridge can be expressed by Boolean variables:

$$x_{iYZ} = \begin{cases} 1 & \text{if the bridge (edge) } i \text{ is used from node } Y \text{ to node } Z \,, \\ 0 & \text{otherwise} \,. \end{cases}$$

Additional model variables s_Y and e_Y are used to determine the property that a node is a start node or an end node of an Eulerian Path:

$$s_Y = \begin{cases} 1 & \text{the Eulerian Path starts at the node } Y \,, \\ 0 & \text{otherwise} \,, \end{cases}$$

$$e_Y = \begin{cases} 1 & \text{the Eulerian Path ends at the node } Y \,, \\ 0 & \text{otherwise} \,. \end{cases}$$

Using the specification of Fig. 9.14 we build the respective Boolean equation that is again a CDC-SAT-formula:

$$(x_{1AB}\overline{x}_{1BA} \vee \overline{x}_{1AB}x_{1BA})(x_{2AB}\overline{x}_{2BA} \vee \overline{x}_{2AB}x_{2BA})(x_{3AD}\overline{x}_{3DA} \vee \overline{x}_{3AD}x_{3DA})$$

$$\wedge (x_{4BC}\overline{x}_{4CB} \vee \overline{x}_{4BC}x_{4CB})(x_{5BC}\overline{x}_{5CB} \vee \overline{x}_{5BC}x_{5CB})(x_{6BD}\overline{x}_{6DB} \vee \overline{x}_{6BD}x_{6DB})$$

$$\wedge (x_{7CD}\overline{x}_{7DC} \vee \overline{x}_{7CD}x_{7DC})$$

$$\wedge (x_{1BA}x_{2AB}x_{3AD}s_a \vee x_{2BA}x_{1AB}x_{3AD}s_a \vee x_{3DA}x_{1AB}x_{2AB}s_a$$

$$\vee x_{1BA}x_{2BA}x_{3AD}e_a \vee x_{1BA}x_{3DA}x_{2AB}e_a \vee x_{2BA}x_{3DA}x_{1AB}e_a)$$

$$\wedge (x_{1AB}x_{2AB}x_{4BC}x_{5BC}x_{6BD}s_b \vee x_{1AB}x_{4CB}x_{2BA}x_{5BC}x_{6BD}s_b$$

$$\vee x_{1AB}x_{5CB}x_{2BA}x_{4BC}x_{6BD}s_b \vee x_{1AB}x_{6DB}x_{2BA}x_{4BC}x_{5BC}s_b$$

$$\vee x_{2AB}x_{4CB}x_{1BA}x_{5BC}x_{6BD}s_b \vee x_{2AB}x_{5CB}x_{1BA}x_{4BC}x_{6BD}s_b$$

$$\vee x_{2AB}x_{6DB}x_{1BA}x_{4BC}x_{5BC}s_b \vee x_{4CB}x_{5CB}x_{1BA}x_{2BA}x_{6BD}s_b$$

$$\vee x_{4CB}x_{6DB}x_{1BA}x_{2BA}x_{5BC}s_b \vee x_{5CB}x_{6DB}x_{1BA}x_{2BA}x_{4BC}s_b$$

$$\vee x_{1AB}x_{2AB}x_{4CB}x_{5BC}x_{6BD}e_b \vee x_{1AB}x_{2AB}x_{5CB}x_{4BC}x_{6BD}e_b$$

$$\vee x_{1AB}x_{2AB}x_{6DB}x_{4BC}x_{5BC}e_b \vee x_{1AB}x_{4CB}x_{5CB}x_{2BA}x_{6BD}e_b$$

$$\vee x_{1AB}x_{4CB}x_{6DB}x_{2BA}x_{5BC}e_b \vee x_{1AB}x_{5CB}x_{6DB}x_{2BA}x_{4BC}e_b$$

$$\vee x_{2AB}x_{4CB}x_{5CB}x_{1BA}x_{6BD}e_b \vee x_{2AB}x_{4CB}x_{6DB}x_{1BA}x_{5BC}e_b$$

$$\vee x_{2AB}x_{5CB}x_{6DB}x_{1BA}x_{4BC}e_b \vee x_{4CB}x_{5CB}x_{6DB}x_{1BA}x_{2BA}e_b)$$

$$\wedge (x_{4BC}x_{5CB}x_{7CD}s_c \vee x_{5BC}x_{4CB}x_{7CD}s_c \vee x_{7DC}x_{4CB}x_{5CB}s_c$$

$$\vee x_{4BC}x_{5BC}x_{7CD}e_c \vee x_{4BC}x_{7DC}x_{5CB}e_c \vee x_{5BC}x_{7DC}x_{4CB}e_c)$$

$$\wedge (x_{3AD}x_{6DB}x_{7DC}s_d \vee x_{6BD}x_{3DA}x_{7DC}s_d \vee x_{7CD}x_{3DA}x_{6DB}s_d$$

$$\vee x_{3AD}x_{6BD}x_{7DC}e_d \vee x_{3AD}x_{7CD}x_{6DB}e_d \vee x_{6BD}x_{7CD}x_{3DA}e_d)$$

$$\wedge (\overline{s}_b\overline{s}_c\overline{s}_d \vee \overline{s}_a\overline{s}_c\overline{s}_d \vee \overline{s}_a\overline{s}_b\overline{s}_d \vee \overline{s}_a\overline{s}_b\overline{s}_c)$$

$$\wedge (\overline{e}_b\overline{e}_c\overline{e}_d \vee \overline{e}_a\overline{e}_c\overline{e}_d \vee \overline{e}_a\overline{e}_b\overline{e}_d \vee \overline{e}_a\overline{e}_b\overline{e}_c) = 1 \,.$$

Fig. 9.14 The bridges in Koenigsberg

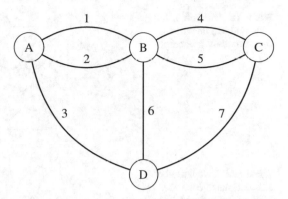

The first seven disjunctions determine that each edge is used in exactly one direction. The next four disjunctions specify all possibilities to use the edges attached to one of the nodes. The last two disjunctions express the sufficient conditions that at most one start node and at most one end node can occur in an Eulerian Path. The solution of this equation is the empty set; hence, no Eulerian Path exists for the graph of Fig. 9.14.

The special property of Eulerian paths that each edge of the graph must be used in exactly one direction facilitates a simplification of the CDC-SAT-equation. A single logic variable can be used instead of two variables for the two directions of an edge; e.g., $x_1 = x_{1AB}$ describes the use of the edge 1 from A to B and $\overline{x}_1 = x_{1BA}$ the reverse direction where the edge 1 is used from B to A. The seven disjunctions that restrict to one of these two directions can be avoided. The number of logic variables to model the edges used in the Eulerian path of the graph shown in Fig. 9.14 can be reduced in this way from 14 to seven.

Similar equations can be built for any graphs. All solution-vectors show the existing Eulerian paths.

9.6   Coloring Problems

Vertex Coloring—Birkhoff's Diamond

This is an example of coloring problems. The vertices of the graph are supposed to be colored by four colors such that connected vertices have different colors. The constraints of this problem can be described using Boolean variables $x_{v,c}$, where the index v indicates the number of the vertex and the second index c defines the color which will be used for the respective node: 1 = red, 2 = green, 3 = blue, or 4 = yellow.

We use vertex 6 as an example for the requirements of the model:

$$(x_{6,1} \vee x_{6,2} \vee x_{6,3} \vee x_{6,4}) = 1 \,. \tag{9.11}$$

This equation describes that one of the four colors must be assigned to the vertex 6. Now this requirement can be extended by the restrictions:

$$x_{6,1}\overline{x}_{6,2}\overline{x}_{6,3}\overline{x}_{6,4}\overline{x}_{5,1}\overline{x}_{7,1}\overline{x}_{1,1} \vee x_{6,2}\overline{x}_{6,1}\overline{x}_{6,3}\overline{x}_{6,4}\overline{x}_{5,2}\overline{x}_{7,2}\overline{x}_{1,2}$$

$$\vee \, x_{6,3}\overline{x}_{6,1}\overline{x}_{6,2}\overline{x}_{6,4}\overline{x}_{5,3}\overline{x}_{7,3}\overline{x}_{1,3} \vee x_{6,4}\overline{x}_{6,1}\overline{x}_{6,2}\overline{x}_{6,3}\overline{x}_{5,4}\overline{x}_{7,4}\overline{x}_{1,4} = 1 \,. \tag{9.12}$$

Fig. 9.15 Birkhoff's
Diamond: (**a**) uncolored,
(**b**) colored by four colors

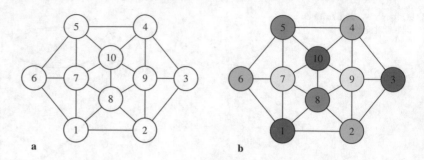

a   b

Table 9.9 Calculation of
all solutions to color the
Birkhoff's Diamond using
3, 4, or 5 colors

Number of				Time in
Nodes	Colors	Variables	Solutions	seconds
10	3	30	0	0.00
10	4	40	576	0.00
10	5	50	40,800	0.02

Each vertex must be colored by exactly one of the four colors. If, e.g., the vertex 6 is colored in red ($x_{6,1} = 1$), then this vertex cannot be green ($\overline{x}_{6,2} = 1$), blue ($\overline{x}_{6,3} = 1$), and yellow ($\overline{x}_{6,4} = 1$). Further restrictions result from the adjacent vertices. Figure 9.15a shows that the vertex 6 is connected with the vertices 1, 5, and 7. The last three literals in the first conjunction of (9.12) describe that the assignment of the color 1 (red) to the vertex 6 prohibits the assignment of the same color to the adjacent vertices 1, 5, and 7. The other three conjunctions of (9.12) describe analog restrictions for the other three colors. There are ten such disjunctions of inner conjunctions connected by the outer conjunctions in this CDC-SAT-model; two of them contain seven, four of them eight, and the remaining four of them nine variables in the inner conjunctions.

The ternary vectors for the conjunctions can be built directly—it is not necessary to write down all the disjunctions of inner conjunctions that describe the problem. The size of the search space is equal to $4^{10} = 1,048,576$, i.e., four colors for ten vertices. The whole CDC-SAT-problem is modeled by only ten disjunctions and the solution requires their intersection. Figure 9.15b shows one of the 576 colorings of the ten vertices of Birkhoff's Diamond. Table 9.9 shows our experimental results [4] solving this CDC-SAT-problem for 3, 4, and 5 colors.

This graph coloring can also be modeled as CD-SAT-problem that consists of ten clauses for the requirements of the ten vertices like (9.11), $10 * (3 + 2 + 1) = 60$ clauses for the color restrictions of the ten vertices, e.g., $(\overline{x}_{1,1} \vee \overline{x}_{1,2})$, and $21 * 4 = 84$ clauses for the color restrictions of the 21 edges, e.g., $(\overline{x}_{1,1} \vee \overline{x}_{2,1})$. Hence, the CD-SAT-problem consists of 154 clauses.

This example is part of an old and very famous problem:

- Is it possible to color the countries on a map by means of four colors in such a way that neighbored countries get different colors?

Birkhoff's Diamond is already a model of such a problem. We consider the vertices of the graph as the different countries, and the edges represent the neighborship relation: The country 6 has the countries 5, 7, 1 as neighbors, the country 7 has five neighbors 1, 5, 6, 8, 10, and so on (see Fig. 9.16).

The problem has been formulated in 1852 by *Francis Guthrie*. The mathematician *Arthur Cayley* published the problem in 1878. Since these days many efforts have been spent on the solution of the problem.

Fig. 9.16 One solution for
the coloring: **a** Birkhoff's
Diamond as a colored
graph, **b** Birkhoff's
Diamond as a colored map

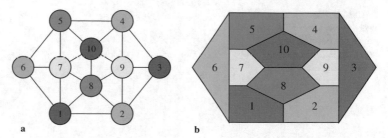

Fig. 9.17 Complete graph
with five vertices and no
monochromatic triangle

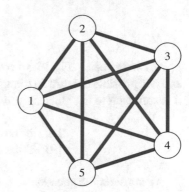

A hot discussion started when the two scientists *Kenneth Appel* and *Wolfgang Haken* of the
University of Illinois 1976 explored 1936 (later 1476) critical cases and used a computer program
to check all these cases. Since those days a hot discussion is conducted whether such computer-based
proofs are legitimate or not.

In this regard the use of logic equations is even more constructive: A given map will be translated
into a graph, and the respective Boolean equation will be solved. The answer will be not only a simple
yes, but many real colorings will be offered.

Edge Coloring—Ramsey's Theorem

Another famous problem field is dealing with the so-called *Ramsey numbers*. Again the available
theoretical investigations fill many journals and books, but because of the complexity of the problem
not so many solutions are available. The origin of the problem is *Dirichlet's pigeonhole principle*. It
is very easy to understand and was explicitly stated in 1834. *If n objects are distributed to m sets*
(n, m > 0) and n > m, then at least one set must get two elements. We use an application of this
principle.

Theorem 9.3 *There exists a least positive integer R(r, s) for which every blue-red edge coloring of a*
complete graph on R(r, s) vertices contains a blue clique on r vertices or a red clique on s vertices.
This number is denoted by Ramsey number.

The two arguments *r* and *s* indicate that we are using two colors. If the number of vertices is large
enough, then there must be a triangle where all the edges are colored in red or in blue.

Figure 9.17 shows that $R(3, 3) > 5$ because it shows a complete graph with five vertices without
a monochromatic triangle. The next possibility is $R(3, 3) = 6$.

Fig. 9.18 Complete graph with six vertices, 19 of 20 edges are colored such that no triangle is completely red or completely blue

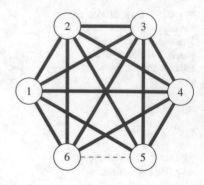

As can be seen in Fig. 9.18 a complete graph of six vertices has 15 edges between different vertices, and it is possible to build 20 triangles. Since two values red and blue must be taken into consideration, it is sufficient to use Boolean variables for the color of the edges:

$$x_{ij} = \begin{cases} 1 & \text{if the color of the edge from vertex } i \text{ to vertex } j \text{ is red ,} \\ 0 & \text{if the color of the edge from vertex } i \text{ to vertex } j \text{ is blue .} \end{cases}$$

We consider the equation

$$(x_{12} \vee x_{13} \vee x_{23})(\overline{x}_{12} \vee \overline{x}_{13} \vee \overline{x}_{23}) = 1$$

for the triangle $1 - 2 - 3$ as example.

$x_{12} = 1$ ensures that the edge $1 - 2$ is red, $\overline{x}_{23} = 0$ sets the color of the edge $2 - 3$ to blue, etc.; hence, the solution of this equation is all colorings of the triangle $1 - 2 - 3$ which are not completely red or completely blue.

Again the use of ternary vectors is very helpful. The consideration of one triangle excludes two of the possible eight colorings.

$$
\begin{array}{ccc}
\begin{array}{ccc}
x_{12} & x_{13} & x_{23} \\
\hline
1 & - & - \\
0 & 1 & - \\
0 & 0 & 1 \\
\end{array}
& \cap
\begin{array}{ccc}
x_{12} & x_{13} & x_{23} \\
\hline
0 & - & - \\
1 & 0 & - \\
1 & 1 & 0 \\
\end{array}
& =
\begin{array}{ccc}
x_{12} & x_{13} & x_{23} \\
\hline
1 & 0 & - \\
1 & 1 & 0 \\
0 & 1 & - \\
0 & 0 & 1 \\
\end{array}
\end{array}
\qquad (9.13)
$$

Therefore, it is appropriate to use the TVL on the right-hand side of Eq. (9.13) for each triangle. If this is applied to the 20 possible triangles of the graph with six vertices followed by the intersection of the TVLs, then it can be seen that the equation does not have any solutions anymore, and this proves that $R(3, 3)$ is equal to 6. Figure 9.18 shows that both a red edge and a blue edge cause a monochromatic triangle (red: e.g., $1 - 5 - 6$; blue: e.g., $2 - 5 - 6$).

It is also very interesting to transfer the problems of Ramsey numbers into grid problems. Figure 9.19 shows the transformation of the successful coloring of Fig. 9.17 of the edges of a complete graph of five nodes.

Each edge will now be used as a vertex of a graph: The node 24 represents, for instance, the edge from vertex 2 to vertex 4. The color of the vertex is equal to the color of the edge. Now we use the elements of one triangle and indicate the respective graph. Figure 9.19 shows the triangles $12 - 13 - 23$, $12 - 14 - 24$, $23 - 24 - 34$, $24 - 25 - 45$, and $34 - 35 - 45$. The other triangles

Fig. 9.19 $R(3, 3)$ as a grid problem

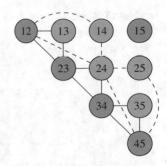

Table 9.10 Some known Ramsey numbers or intervals

Red\Blue	3	4	5	6	7	8	9	10
3		6	9	14	18	23	28	36	40-42
4		9	18	25	36-41	49-61	58-84	73-115	92-149
5			14	25	43-49	58-87	80-143	101-216	126-316	144-442
⋮	⋮	⋮	⋮	⋮	⋮	⋮	⋮	⋮

have been omitted due to the readability of the grid. The main diagonal has been omitted as well as the elements below the main diagonal because the edges of the triangles are not directed and the representation would be symmetric. The nodes must be colored such that no monochromatic triangles will be found! The respective Boolean equations can be built as before.

Now the problem can be extended by using not only triangles, but structures with more than three vertices. Table 9.10 enumerates some known Ramsey numbers. It can be seen, for instance, that $R(3, 6) = 18$. This means that a complete graph with 18 vertices is necessary to generate a coloring with two colors which has either a monochromatic triangle or a monochromatic hexagon. The table also shows that only a small set of Ramsey numbers is known. This depends naturally on the complexity of the problem. The number of triangles and hexagons is growing very fast, and a large number of Boolean variables must be introduced. The next chapter will show how to proceed for such extreme complexities.

It is also possible to use more than two colors.

Example 9.9 It has been found, for instance, that $R(3, 3, 3) = 17$. This number of vertices is necessary to enforce a monochromatic triangle in one of three colors a, b, or c. There are $\binom{17}{3} = 680$ different triangles, and their edges can be colored with one of three colors. For each triangle exist 27 different colorings.

Let us take three vertices $1 - 2 - 3$ and three colors a, b, and c. All non-monochromatic colorings of this triangle are solutions of the equation

$$(x_{12a}\overline{x}_{12b}\overline{x}_{12c} \vee \overline{x}_{12a}x_{12b}\overline{x}_{12c} \vee \overline{x}_{12a}\overline{x}_{12b}x_{12c})$$
$$\wedge (x_{13a}\overline{x}_{13b}\overline{x}_{13c} \vee \overline{x}_{13a}x_{13b}\overline{x}_{13c} \vee \overline{x}_{13a}\overline{x}_{13b}x_{13c})$$
$$\wedge (x_{23a}\overline{x}_{23b}\overline{x}_{23c} \vee \overline{x}_{23a}x_{23b}\overline{x}_{23c} \vee \overline{x}_{23a}\overline{x}_{23b}x_{23c})$$
$$\wedge (\overline{x}_{12a} \vee \overline{x}_{13a} \vee \overline{x}_{23a})(\overline{x}_{12b} \vee \overline{x}_{13b} \vee \overline{x}_{23b})(\overline{x}_{12c} \vee \overline{x}_{13c} \vee \overline{x}_{23c}) = 1 .$$

The conjunction of 680 such equations will confirm the result.

Table 9.11 Some
Pythagorean triples

(3, 4, 5)	(5, 12, 13)	(8, 15, 17)	(7, 24, 25)	(20, 21, 29)
(12, 35, 37)	(9, 40, 41)	(28, 45, 53)	(11, 60, 61)	(16, 63, 65)
(33, 56, 65)	(48, 55, 73)	(13, 84, 85)	(36, 77, 85)	(39, 80, 89)
(65, 72, 97)	(20, 99, 101)	(60, 91, 109)	(15, 112, 113)	(44, 117, 125)
(88, 105, 137)	(17, 144, 145)	(24, 143, 145)	(51, 140, 149)	(85, 132, 157)
(119, 120, 169)	(52, 165, 173)	(19, 180, 181)	(57, 176, 185)	(104, 153, 185)
(95, 168, 193)	(28, 195, 197)	(84, 187, 205)	(133, 156, 205)	(21, 220, 221)
(140, 171, 221)	(60, 221, 229)	(105, 208, 233)	(120, 209, 241)	(32, 255, 257)
(23, 264, 265)	(96, 247, 265)	(69, 260, 269)	(115, 252, 277)	(160, 231, 281)

Pythagorean Triples

The graph and the vertices to be colored are not always defined by complete graphs and the definition of the numbers of colors (see, e.g., [3]). In this publication a problem has been solved that takes as basic structure the graph which is defined by Pythagorean triples.

Definition 9.7 Three numbers (a, b, c) are a Pythagorean triple if they satisfy the equation

$$a^2 + b^2 = c^2 . \tag{9.14}$$

Such a triple (a, b, c) is now considered as a triangle with the corners a, b, and c. The colors blue and red are used to color the nodes. It is now required to find colorings that avoid monochromatic triangles of Pythagorean triples. The respective Ramsey number of this graph defines the border for monochromatic numbers.

Very many different triangles will appear, and some of them will have common vertices (see, for instance, $(3, 4, 5)$ and $(5, 12, 13)$). It is an easy programming exercise to write a program that generates as many triples as desired. Table 9.11 enumerates some Pythagorean triples.

The translation into a Boolean equation is easy. Again for each triple the right-hand side of Eq. (9.13) and the intersection of the resulting lists will be used. The use of ternary vectors also can again make full use of parallelization (as mentioned before).

Why is 7825 the Ramsey number of this coloring problem? The coloring of Fig. 9.20 hardly can be checked by human beings. This is a remarkable difference to the next chapter: There the problem is also very complex, but the result still can be checked by (patient) human beings.

The solution in Fig. 9.20 shows 7825 as the border case. This is understandable because the additional use of 7825 produces seven more Pythagorean triples (see Table 9.12). The pairs (625,7800), (1584,7663), and (2191,7512) in Table 9.12 have been colored already in blue. This would require the red color for 7825. However, the pair (5180,5865) already uses only red, which forces 7825 to blue. This conflict confirms 7825 as the border of the problem.

It is quite clear that these considerations fully depend on the given solution up to 7824. There are, however, many other colorings, and one of them could possibly allow the inclusion of 7825. Who knows? And this is now causing hot discussions among scientists about what to do in such a situation.

This shows also a big difference to the solution of the grid problem in the next chapter. The calculations are also very difficult and require many different mathematical considerations and an extremely skilled programmer. The final result, however, still can be checked by patient human beings.

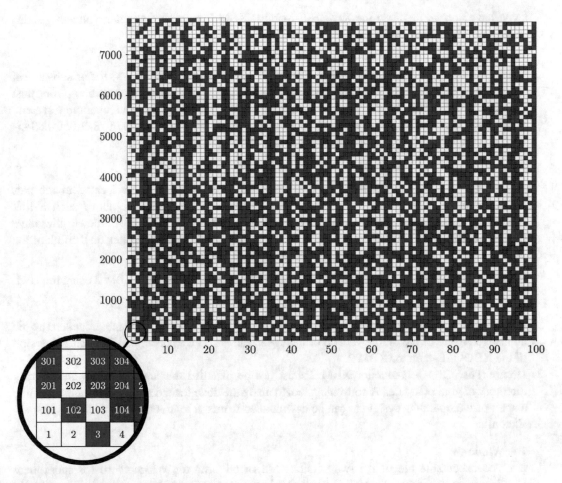

Fig. 9.20 The solution of coloring Pythagorean triples (source: [3])

Table 9.12 Pythagorean triples with $c = 7825$

a	b	c
625	7800	7825
1584	7663	7825
2191	7512	7825
2784	7313	7825
4180	6615	7825
4695	6260	7825
5180	5865	7825

9.7 Exercises

Prepare again for each task of each exercise of this chapter a PRP for the XBOOLE-monitor XBM 2. Prefer a lexicographic order of the variables and use the help system of the XBOOLE-monitor XBM 2 to extend your knowledge about the details of needed commands. Execute the created PRPs, verify the computed results, and explain the observed properties.

If you have not yet prepared the XBOOLE-monitor XBM 2 on your computer, you can get this XBOOLE-monitor free of charge by means of the following three steps:

1. **Download**:
   There are four versions of the XBOOLE-monitor XBM 2, two for Windows 10 or subsequent Windows systems (32 or 64 bits) and two for LINUX - Ubuntu (also 32 or 64 bits); you must download the version of the XBOOLE-monitor XBM 2 that fits to your operating system. Authorized users of the online version of this chapter (https://doi.org/10.1007/978-3-030-88945-6_9) can download the XBOOLE-monitor XBM 2 directly from the web page

   https://link.springer.com/chapter/10.1007/978-3-030-88945-6_9

   where the links for the download of the XBOOLE-monitor XBM 2 are located in the part "Supplementary Information" (below the part "Abstract"). The headline above such a link indicates the associated zip-file of the XBOOLE-monitor XBM 2. The sizes of the zip-files have been provided behind the links and can be used to verify the download. A click on the link of the wanted version of the XBOOLE-monitor XBM 2 starts the download.

   Readers of the hardcopy of this book get access to the XBOOLE-monitor XBM 2 using the URL

   https://link.springer.com/chapter/10.1007/978-3-030-88945-6_9

   to download the first two pages of this chapter. After this download, the same procedure as the authorized users of the online version of a chapter can be used to download the wanted version of the XBOOLE-monitor XBM 2.
2. **Unzip**: The XBOOLE-monitor XBM 2 must not be installed, but unzipped into an arbitrary directory of your computer. A convenient tool to unzip the downloaded zip-file is usually available as part of the operating system or can be downloaded from the Internet.
3. **Execute**:

   - Windows:
     The executable file of the two versions (32 or 64 bits) for Windows 10 (or subsequent Windows systems) of the XBOOLE-monitor XBM 2 is XBM2.exe; the other files in the expanded directory must remain unchanged. A double-click on the executable file XBM2.exe within the Explorer of Windows starts the XBOOLE-monitor XBM 2.
   - LINUX - Ubuntu:
     The unzipped folder of the XBOOLE-monitor XBM 2 contains for this operating system only the executable file XBM2-i386.AppImage for the version of 32 bits or XBM2-x86_64.AppImage for the version of 64 bits of the XBOOLE-monitor XBM 2. A double-click on the created AppImage-file within the file manager of LINUX - Ubuntu starts the XBOOLE-monitor XBM 2.

Exercise 9.1 (Solution Methods for SAT-Problems)
   The aim of this exercise is the exploration of several methods to find solutions of the logic equation:

$$(x_3 \vee x_6 \vee x_8)(x_1 \vee \overline{x}_5 \vee \overline{x}_6)(\overline{x}_4 \vee x_7 \vee x_8)(x_2 \vee x_4 \vee x_6) \wedge$$
$$(x_1 \vee \overline{x}_3 \vee \overline{x}_7)(x_4 \vee x_5 \vee \overline{x}_7)(\overline{x}_1 \vee x_6 \vee \overline{x}_8)(\overline{x}_2 \vee x_5 \vee \overline{x}_6)(\overline{x}_1 \vee \overline{x}_5 \vee \overline{x}_6) = 1$$

that specifies a 3-SAT-problem of nine clauses and eight variables.

(a) Compute all solutions of the specified 3-SAT-problem by splitting the given equation into nine homogenous characteristic equations determined by the clauses, solving these nine equations, and computing the intersections of these nine partial solution-sets. Use the command obbc to minimize the final solution-set and show the TVLs of the solutions before and after this minimization. How many solution-vectors exist and how many ternary vectors express these solutions in the minimized TVL. Hint: In the menu "Extras" exist the item "Information about an Object"; there you can select the object number of the solution-TVL and get the value "Rho" that is ratio between the number of binary vectors represented by the selected TVL of n variables and all 2^n possible binary vectors.

(b) Compute all solutions of the specified 3-SAT-problem using the negation with regard to De Morgan and the complement. Specify in this case the 3-SAT-problem by the associated TVL in conjunctive form. Use the command obbc to minimize the final solution-set and show the TVLs of the solutions before and after this minimization. How many solution-vectors exist and how many ternary vectors express these solutions in the minimized TVL. Verify your solution by comparison with solution-set computed using the command sbe. Show the prepared TVL in K-form of the 3-SAT-problem, the TVL in ODA-form of complement of the solution-set computed with the command ndm, the TVL in ODA-form of the solution-set computed with the command cpl, and the TVL of the solution-set minimized using the command obbc.

(c) Compute one ternary vector that determines some solutions of the specified 3-SAT-problem using the intersection of single ternary vectors in a depth-first search with backtracking. How many intersections between two ternary vectors are needed to compute the requested solution-vector? Show the computed solution-vectors in the 1-fold view and for the analysis of the implemented approach the intermediate TVLs in the m-fold view after finishing this computation.

(d) Compute one ternary vector that determines some solutions of the specified 3-SAT-problem using the difference between two single ternary vectors in a depth-first search with backtracking. How many difference operations between two ternary vectors are need to compute the requested solution-vector? Show the computed solution-vectors in the 1-fold view and for the analysis of the implemented approach the intermediate TVLs in the m-fold view after finishing this computation.

Exercise 9.2 (Placement Problems)
The aim of this exercise is the exploration of several methods to solve the placement problem for rocks on a chess board of the size 4×4. Use the name of variables $a1, \ldots, d4$ as usual on a chess board. The letters $a, \ldots, d$ indicate the columns on the board from the left to the right and the numbers $1, \ldots, 4$ the rows in bottom up order. A value 1 of such a logic variable means that a rock is located on the associated field. A rock can move and threaten other pieces only in the same row or the same column of the board. Compute all placements with a maximal number of rocks that do not threaten each other. How many solutions exist for this SAT-problem? Show chess boards with the calculated solutions besides on the solution-TVL.

(a) Specify a CD-SAT-formula for this problem and solve it using the command sbe. How many clauses and literals are needed in this SAT-formula?

(b) Specify a CDC-SAT-formula for this problem and solve it using the command obc. How many DC-clauses and literals are needed in this CDC-SAT-formula?

(c) Express the DC-clauses of this problem by TVLs in ODA-form and compute the solution of this SAT-problem by intersections. How many intersections must be computed?

(d) The DC-clauses of this problem are very similar because each of these clauses is determined by the possible rocks on one column of the same size. Determine the CDC-SAT-problem for the maximal number of rocks on a chess board of the size 4×4 using a single TVL that specifies the DC-clause for one column, generate the other DC-clauses using commands cco, and solve the problem as before by intersections.

9.8   Solutions

Solution 9.1 (Solution Methods for SAT-Problems)

(a) Figure 9.21 shows both the PRP that computes all solutions of the given 3-SAT-problem and the TVLs in ODA-form of the solution-sets before and after their minimization using the command obbc.

The commands sbe in lines 6–23 solve the nine homogenous characteristic equations determined by the clauses of the given 3-SAT-problem. The command copy in line 24 prepares the root TVL for the subsequent eight intersection with the TVLs of the other partial solutions in the for-loop in lines 25–28. The command obbc reduces the number of ternary vectors in the solution-TVL from 15 to 11.

The values of Rho of the two solution-TVLs 10 and 11 are equal to 0.171875; hence, due to $n = 8$ variables of these TVLs there are $2^8 * 0.171875 = 44$ binary solution-vectors of the given 3-SAT-problem which are represented by 11 ternary vectors of solution-TVL 11.

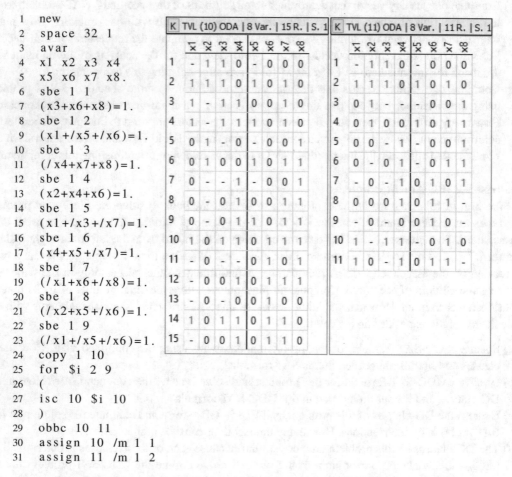

```
1    new
2    space  32  1
3    avar
4    x1  x2  x3  x4
5    x5  x6  x7  x8.
6    sbe  1  1
7    (x3+x6+x8)=1.
8    sbe  1  2
9    (x1+/x5+/x6)=1.
10   sbe  1  3
11   (/x4+x7+x8)=1.
12   sbe  1  4
13   (x2+x4+x6)=1.
14   sbe  1  5
15   (x1+/x3+/x7)=1.
16   sbe  1  6
17   (x4+x5+/x7)=1.
18   sbe  1  7
19   (/x1+x6+/x8)=1.
20   sbe  1  8
21   (/x2+x5+/x6)=1.
22   sbe  1  9
23   (/x1+/x5+/x6)=1.
24   copy  1  10
25   for  $i  2  9
26   (
27   isc  10  $i  10
28   )
29   obbc  10  11
30   assign  10  /m  1  1
31   assign  11  /m  1  2
```

K	TVL (10) ODA	8 Var.	15 R.	S. 1					
	x1	x2	x3	x4	x5	x6	x7	x8
1	-	1	1	0	-	0	0	0
2	1	1	1	0	1	0	1	0
3	1	-	1	1	0	0	1	0
4	1	-	1	1	1	0	1	0
5	0	1	-	0	-	0	0	1
6	0	1	0	0	1	0	1	1
7	0	-	-	1	-	0	0	1
8	0	-	0	1	0	0	1	1
9	0	-	0	1	1	0	1	1
10	1	0	1	1	0	1	1	1
11	-	0	-	-	0	1	0	1
12	-	0	0	1	0	1	1	1
13	-	0	-	0	0	1	0	0
14	1	0	1	1	0	1	1	0
15	-	0	0	1	0	1	1	0

K	TVL (11) ODA	8 Var.	11 R.	S. 1					
	x1	x2	x3	x4	x5	x6	x7	x8
1	-	1	1	0	-	0	0	0
2	1	1	1	0	1	0	1	0
3	0	1	-	-	-	0	0	1
4	0	1	0	0	1	0	1	1
5	0	0	-	1	-	0	0	1
6	0	-	0	1	-	0	1	1
7	-	0	-	1	0	1	0	1
8	0	0	0	1	0	1	1	-
9	-	0	-	0	0	1	0	-
10	1	-	1	1	-	0	1	0
11	1	0	-	1	0	1	1	-

Fig. 9.21 Problem-program that computes all solutions of the given 3-SAT-problem using intersections between the partial solution-sets together with the computed results

```
1   new
2   space 32 1
3   tin 1 1 /k
4   x1 x2 x3 x4
5   x5 x6 x7 x8 .
6   --1--1-1
7   1---00--
8   ---0--11
9   -1-1-1--
10  1-0---0-
11  ---11-0-
12  0-----1-0
13  -0--10--
14  0---00-- .
15  ndm 1 2
16  cpl 2 3
17  obbc 3 4
18  sbe 1 5
19  (x3+x6+x8)&
20  (x1+/x5+/x6)&
21  (/x4+x7+x8)&
22  (x2+x4+x6)&
23  (x1+/x3+/x7)&
24  (x4+x5+/x7)&
25  (/x1+x6+/x8)&
26  (/x2+x5+/x6)&
27  (/x1+/x5+/x6)
28  =1.
29  te_syd 3 5
30  te_syd 4 5
31  assign 1 /4 1 1
32  assign 2 /4 1 2
33  assign 3 /4 2 1
34  assign 4 /4 2 2
```

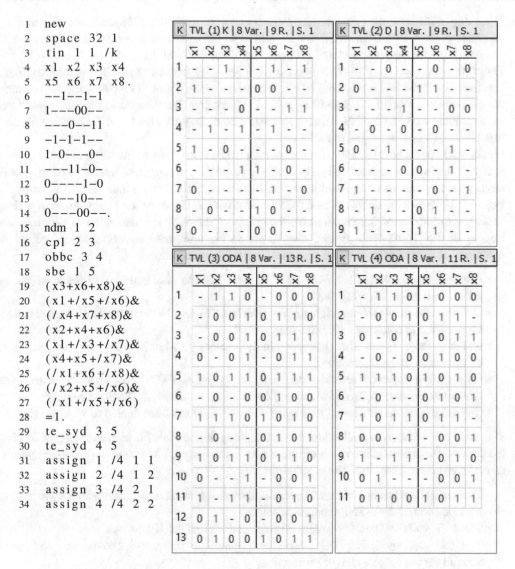

TVL (1) K | 8 Var. | 9 R. | S. 1

K	x1	x2	x3	x4	x5	x6	x7	x8
1	-	-	1	-	-	1	-	1
2	1	-	-	-	0	0	-	-
3	-	-	-	0	-	-	1	1
4	-	1	-	1	-	1	-	-
5	1	-	0	-	-	-	0	-
6	-	-	-	1	1	-	0	-
7	0	-	-	-	-	1	-	0
8	-	0	-	-	1	0	-	-
9	0	-	-	-	0	0	-	-

TVL (2) D | 8 Var. | 9 R. | S. 1

K	x1	x2	x3	x4	x5	x6	x7	x8
1	-	-	0	-	-	0	-	0
2	0	-	-	-	1	1	-	-
3	-	-	-	1	-	-	0	0
4	-	0	-	0	-	0	-	-
5	0	-	1	-	-	-	1	-
6	-	-	-	0	0	-	1	-
7	1	-	-	-	-	0	-	1
8	-	1	-	-	0	1	-	-
9	1	-	-	-	1	1	-	-

TVL (3) ODA | 8 Var. | 13 R. | S. 1

K	x1	x2	x3	x4	x5	x6	x7	x8
1	-	1	1	0	-	0	0	0
2	-	0	0	1	0	1	1	0
3	-	0	0	1	0	1	1	1
4	0	-	0	1	-	0	1	1
5	1	0	1	1	0	1	1	1
6	-	0	-	0	0	1	0	0
7	1	1	1	0	1	0	1	0
8	-	0	-	-	0	1	0	1
9	1	0	1	1	0	1	1	0
10	0	-	-	1	-	0	0	1
11	1	-	1	1	-	0	1	0
12	0	1	-	0	-	0	0	1
13	0	1	0	0	1	0	1	1

TVL (4) ODA | 8 Var. | 11 R. | S. 1

K	x1	x2	x3	x4	x5	x6	x7	x8
1	-	1	1	0	-	0	0	0
2	-	0	0	1	0	1	1	-
3	0	-	0	1	-	0	1	1
4	-	0	-	0	0	1	0	0
5	1	1	1	0	1	0	1	0
6	-	0	-	-	0	1	0	1
7	1	0	1	1	0	1	1	-
8	0	0	-	1	-	0	0	1
9	1	-	1	1	-	0	1	0
10	0	1	-	-	-	0	0	1
11	0	1	0	0	1	0	1	1

Fig. 9.22 Problem-program that computes all solutions of the given 3-SAT-problem using the complement of the set of non-solutions together with the computed results

(b) Figure 9.22 shows the PRP that computes all solutions of the given 3-SAT-problem using the very efficient method based on the complement.

The 3-SAT-problem has been specified by TVL 1 in conjunctive form as shown as left TVL in the upper row of Fig. 9.22. The command ndm computes the set of non-solutions simply by exchanging the value 0 and 1 in TVL 1 together with the change of the K-form into the D-form; the result is shown in the upper right corner of Fig. 9.22. The command cpl in line 16 is the main step of this approach that computes the complement of the non-solutions of the given 3-SAT-problem so that we get the wanted set of solutions as TVL 3 (left TVL in the lower row of Fig. 9.22). The command obbc reduces the number of ternary vectors of the solution-TVL from 13 to 11.

The value of *Rho* for the two solution-TVLs 3 and 4 is equal to 0.171875; hence, due to $n = 8$ variables of these TVLs there are $2^8 * 0.171875 = 44$ binary solution-vectors of the given 3-SAT-problem which are represented by 11 ternary vectors of solution-TVL 11.

The command sbe in line 18–28 solves the given 3-SAT-problem directly; hence, this is a simpler approach to solve a SAT-problem. The solution values true (shown in the protocol as solution of the commands te_syd 3 5 and te_syd 4 5) confirm that the solution-sets computed with the command cpl 2 3 (TVL 3), the command obbc 3 4 (TVL 4), and the command sbe in line 18–28 (TVL 5) are equal to each other.

(c) Figure 9.23 shows both the PRP that computes a single ternary vector as solution of the given 3-SAT-problem of nine clauses and eight variables based on a modicum of intersections of selected ternary vectors and the computed ternary vector that determines four binary solution-vectors.

Figure 9.24 shows the details of intermediate results after finishing the computation of a single ternary solution-vector. The rows of the viewports of Fig. 9.24 are associated to the nine given clauses of the 3-SAT-problem and row 10 shows the computed ternary solution-vector. The meanings of the columns of the viewports of Fig. 9.24 are as follows:

- Column one shows the solution-sets (TVLs 1–9) of the nine equation of the given clauses solved in lines 5–22 of Fig. 9.23.
- Column two shows the remaining ternary vectors (TVLs 11–19) of the associated clauses so far not used to compute a solution-vector.
- Column three shows the selected ternary vectors (TVLs 21–29) used for the recent computation of the intersection.
- Column four shows the results of the intersection (TVLs 31–39) of the selected ternary vector (shown in viewport (i,3)) and the so far known intermediate result shown in viewport $(i-1, 4)$ (in case of row one the ternary vector of viewport (i,3) has been copied to viewport (i,4)).

The main computation is realized in the while-loop in lines 29–78. In each sweep of this loop the TVL of the intermediate result of one clause is computed. The *design by contract* approach is used for this while-loop. The preconditions are:

- The variable $i indicates the clause that has to be evaluated in the actual sweep of the while-loop of the 3-SAT-problem to solve.
- The TVL with the index i contains all conjunctions that satisfy the clause i.
- The TVL with the index $10 + i$ contains conjunctions that satisfy the clause i and are not used so far to compute the global solution.
- The value true of the variable $unknown indicates that it is unknown whether a solution of the given 3-SAT-problem exists; hence, the commands of the body of the while-loop must be executed.
- TVL 10 determines the function $1(x_1, x_2, x_3, x_4, x_5, x_6, x_7, x_8)$;

And the postconditions are:

- The value false of the variable $unknown indicates that no further computations are needed; in this case the value false of the variable $solution states that no solution exists for this 3-SAT-problem and the value true of the variable $solution indicates that one ternary solution-vector of the 3-SAT-problem has been computed and stored as TVL 40.
- The TVL with the index $20 + i$ contains the selected conjunction that has been stored in before as first ternary vector in the TVL with the index $10 + i$.
- The first ternary vector in the TVL with the index $10 + i$ is deleted.
- TVL 21 is copied as TVL 31 in the case of $i=1.
- The TVL with the index $30 + i$ is computed as intersection of TVL $29 + i$ (the last computed partial solution-vector) and $20 + i$ (the selected vector that solves the clause i).

```
 1  new                                  48  if (gt $i 1)
 2  space 32 1                           49  (
 3  avar                                 50  if (not (te (add $i 10)))
 4  x1 x2 x3 x4 x5 x6 x7 x8.             51  (
 5  sbe 1 1                              52  if (le $i 9)
 6  (x3+x6+x8)=1.                        53  (
 7  sbe 1 2                              54  stv (add $i 10) 1 (add $i 20)
 8  (x1+/x5+/x6)=1.                      55  dtv (add $i 10) 1 (add $i 10)
 9  sbe 1 3                              56  set $r (add $i 30)
10  (/x4+x7+x8)=1.                       57  isc (add $i 29) (add $i 20) $r
11  sbe 1 4                              58  if (not (te (add $i 30)))
12  (x2+x4+x6)=1.                        59  (
13  sbe 1 5                              60  if (eq $i 9)
14  (x1+/x3+/x7)=1.                      61  (
15  sbe 1 6                              62  set $solution true
16  (x4+x5+/x7)=1.                       63  set $unknown false
17  sbe 1 7                              64  isc 39 10 40
18  (/x1+x6+/x8)=1.                      65  assign 40 /1
19  sbe 1 8                              66  )
20  (/x2+x5+/x6)=1.                      67  if (lt $i 9)
21  sbe 1 9                              68  (
22  (/x1+/x5+/x6)=1.                     69  set $i (add $i 1)
23  tin 1 10                             70  copy $i (add $i 10)
24  x1 x2 x3 x4 x5 x6 x7 x8.             71  )
25  --------.                            72  )
26  copy 1 11                            73  )
27  set $i 1                             74  )
28  set $unknown true                    75  if (te (add $i 10))
29  while ($unknown)                     76  (set $i (sub $i 1))
30  (                                    77  )
31  if (eq $i 1)                         78  )
32  (                                    79  for $j 1 9
33  if (not (te 11))                     80  (
34  (                                    81  assign $j /m $j 1
35  stv 11 1 21                          82  assign (add $j 10) /m $j 2
36  dtv 11 1 11                          83  assign (add $j 20) /m $j 3
37  copy 21 31                           84  assign (add $j 30) /m $j 4
38  copy 2 12                            85  )
39  set $i 2                             86  if $solution
40  )                                    87  (assign 40 /m 10 4)
41  if (te 11)
42  (
43  set $solution false
44  set $unknown false
45  set $i 1
46  )
47  )
```

K	TVL (40) ODA | 8 Var. | 1 R. | S. 1						
	x1	x2	x3	x4	x5	x6	x7	x8
1	-	1	1	0	-	0	0	0

Fig. 9.23 Problem-program that computes a single ternary vector as solution of a 3-SAT-problem using intersections together with the computed solution-vector

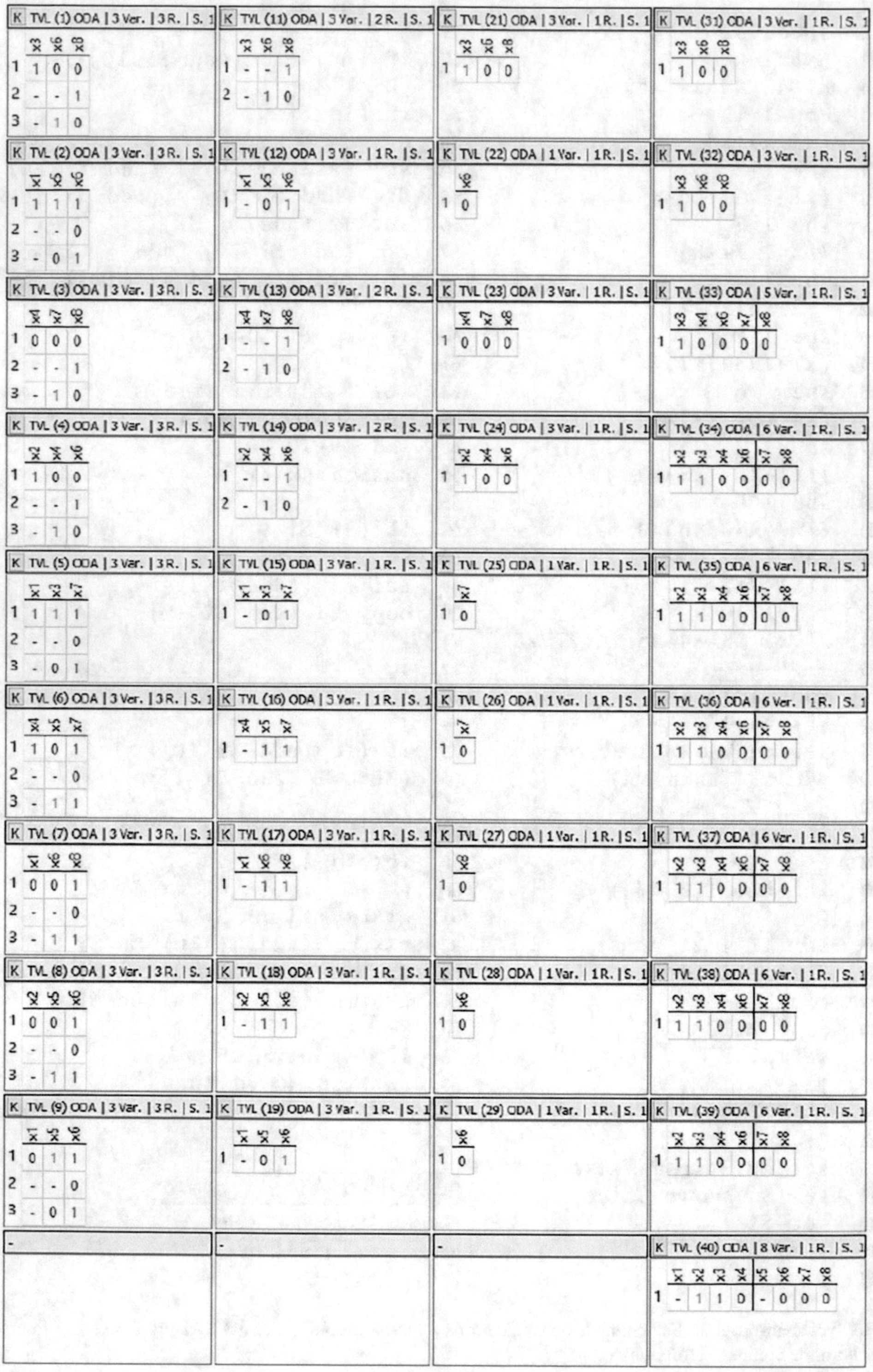

Fig. 9.24 Intermediate results to compute a single ternary vector as solution of a 3-SAT-problem using intersections

- The value of the variable $i is incremented when $i < 9$ and the computed TVL with the index $30 + i$ is not empty.
- The value of the variable $i remains unchanged when the computed TVL with the index $30 + i$ is empty and the TVL of the remaining vectors with the index $10 + i$ is not empty.
- The value of the variable $i is decremented when both the computed TVL with the index $30 + i$ and the TVL of the remaining vectors with the index $10 + i$ are empty (backtracking).
- The value of the variable $unknown is set to `false` and the value of the variable $solution is set to `true` when the TVL with the index $30 + i$ is not empty and $i = 9$ (the wanted solution-vector has been computed).
- The TVL with the index $i + 1$ is copied as TVL with the index $i + 11$ when the computed TVL with the index $30 + i$ is not empty and $i < 9$ (precondition for the incremented value of i).
- The computed solution-vector is extended to all eight variables.
- The values of the variables $unknown and $solution are set to `false` when $i = 1$ and the TVL with the index $10 + i$ is empty (there exists no solution for the given 3-SAT-problem).

The commands before line 29 are needed to satisfy the precondition of the `while`-loop.
The first clause must be handled in a special manner; this is done in the body of the `if`-command in lines 31–47. The selected ternary vector (TVL 21) is copied as TVL 31 instead of an intersection with $1(\mathbf{x})$. The second peculiarity for the case $i = 1$ is that an empty TVL 11 indicates that no solution of the 3-SAT-problem exists; the commands in lines 41–46 realize the associated postconditions, and the command `set $i 1` prevents the execution of the body of the subsequent command `if`.
Almost all computations for the clauses $1 < i \leq 9$ can be executed in the same manner; this is done in the body of the `if`-command in lines 48–77. Here it must be distinguished between the cases that one more ternary vector as solution of the clause i exists (this case is handled in lines 50–74) or that the TVL of the remaining vectors that solve the clause i is empty (this case in handled in lines 75 and 76).
The main action for $1 < i \leq 9$ and at least one remaining solution-vector in the TVL with the index $i + 10$ is the intersection (in line 57) of the ternary vector (selected in line 54 and thereafter deleted in line 55) and the so-far computed partial solution of the TVL with the index $i + 29$. The result of this intersection determines the further steps of the computation. A solution-vector of the given 3-SAT-problem has been found in the case that the result of the intersection is not empty and $i = 9$; the commands in lines 60–66 realize the associated postconditions. The preconditions for the evaluation of the next clause are prepared in lines 67–71 in the case that the result of the intersection is not empty and $i < 9$.
A computed solution-vector is immediately shown in the 1-fold view due to the command `assign` in line 65. The commands in lines 79–87 assign the final TVLs of all intermediate results so that the used approach can be explored in detail.
The evaluation of the protocol confirms that 14 intersections between single ternary vectors have been executed to compute the wanted solution-vector of the 3-SAT-problem of nine clauses. One additional intersection has been used to expand the found solution-vector to a ternary vector of all eight variables.

(d) Figure 9.25 shows both the PRP that computes a single ternary vector as solution of the given 3-SAT-problem of nine clauses and eight variables based on a modicum of difference operations of selected ternary vectors as well as the computed ternary vector that determines four binary solution-vectors.

```
 1   new                                  36   if  (not  (te  (add  $i  20)))
 2   space  32  1                         37   (
 3   tin  1  1  /k                        38   if  (eq  $i  9)
 4   x1  x2  x3  x4  x5  x6  x7  x8 .      39   (
 5   --1--1-1                             40   stv  29  1  39
 6   1---00--                             41   dtv  29  1  29
 7   ---0--11                             42   set  $solution  true
 8   -1-1-1--                             43   set  $unknown  false
 9   1-0---0-                             44   isc  39  10  40
10   ---11-0-                             45   assign  40  /1
11   0----1-0                             46   )
12   -0--10--                             47   if  (lt  $i  9)
13   0---00--.                            48   (
14   tin  1  10                           49   stv  (add  $i  20)  1  (add  $i  30)
15   x1  x2  x3  x4  x5  x6  x7  x8 .      50   dtv  (add  $i  20)  1  (add  $i  20)
16   --------.                            51   stv  1  (add  $i  1)  (add  $i  11)
17   stv  1  1  11                        52   ndm  (add  $i  11)  (add  $i  11)
18   ndm  11  11                          53   set  $r  (add  $i  21)
19   cpl  11  21                          54   dif  (add  $i  30)  (add  $i  11)  $r
20   set  $i  1                           55   set  $i  (add  $i  1)
21   set  $unknown  true                  56   ))))
22   while  ($unknown)                    57   for  $j  1  9
23   (                                    58   (
24   if  (and  (eq  $i  1)  (te  21))     59   assign  (add  $j  10)  /m  $j  1
25   (                                    60   assign  (add  $j  20)  /m  $j  2
26   set  $solution  false                61   assign  (add  $j  30)  /m  $j  3
27   set  $unknown  false                 62   )
28   set  $i  0                           63   if  $solution
29   )                                    64   (assign  40  /m  10  3)
30   if  (ge  $i  1)
31   (
32   if  (te  (add  $i  20))
33   (
34   set  $i  (sub  $i  1)
35   )
```

K	TVL (40) ODA	8 Var.	1 R.	S. 1
	$x1$ $x2$ $x3$ $x4$	$x5$ $x6$ $x7$ $x8$		
1	- 1 1 0	- 0 0 0		

Fig. 9.25 Problem-program that computes a single ternary vector as solution of a 3-SAT-problem using difference operations together with the computed solution-vector

Figure 9.26 shows the details of intermediate results after finishing the computation of a single ternary solution-vector. The rows of the viewports of Fig. 9.26 are associated to the nine given clauses of the 3-SAT-problem and row 10 shows the computed ternary solution-vector. The meanings of the columns of the viewports of Fig. 9.26 are as follows:

- Column one shows the ternary vectors (TVLs 11–19) that do not satisfy the nine associated clauses determined by the rows of TVL 1.
- Column two shows the remaining ternary vectors (TVLs 21–29) of the associated clauses so far not used to compute a solution-vector.

K TVL (11) ODA | 3 Var. | 1 R. | S. 1

	x_3	x_6	x_8
1	0	0	0

K TVL (12) ODA | 3 Var. | 1 R. | S. 1

	x_1	x_5	x_6
1	0	1	1

K TVL (13) ODA | 3 Var. | 1 R. | S. 1

	x_4	x_7	x_8
1	1	0	0

K TVL (14) ODA | 3 Var. | 1 R. | S. 1

	x_2	x_4	x_6
1	0	0	0

K TVL (15) ODA | 3 Var. | 1 R. | S. 1

	x_1	x_3	x_7
1	0	1	1

K TVL (16) ODA | 3 Var. | 1 R. | S. 1

	x_4	x_5	x_7
1	0	0	1

K TVL (17) ODA | 3 Var. | 1 R. | S. 1

	x_1	x_6	x_8
1	1	0	1

K TVL (18) ODA | 3 Var. | 1 R. | S. 1

	x_2	x_5	x_6
1	1	0	1

K TVL (19) ODA | 3 Var. | 1 R. | S. 1

	x_1	x_5	x_6
1	1	1	1

K TVL (21) ODA | 3 Var. | 2 R. | S. 1

	x_3	x_6	x_8
1	-	-	1
2	-	1	0

K TVL (22) ODA | 5 Var. | 0 R. | S. 1

	x_1	x_3	x_5	x_6	x_8

K TVL (23) ODA | 5 Var. | 1 R. | S. 1

	x_3	x_4	x_6	x_7	x_8
1	1	-	0	1	0

K TVL (24) ODA | 6 Var. | 0 R. | S. 1

	x_2	x_3	x_4	x_6	x_7	x_8

K TVL (25) ODA | 7 Var. | 0 R. | S. 1

	x_1	x_2	x_3	x_4	x_6	x_7	x_8

K TVL (26) ODA | 7 Var. | 0 R. | S. 1

	x_2	x_3	x_4	x_5	x_6	x_7	x_8

K TVL (27) ODA | 7 Var. | 0 R. | S. 1

	x_1	x_2	x_3	x_4	x_6	x_7	x_8

K TVL (28) ODA | 7 Var. | 0 R. | S. 1

	x_2	x_3	x_4	x_5	x_6	x_7	x_8

K TVL (29) ODA | 8 Var. | 1 R. | S. 1

	x_1	x_2	x_3	x_4	x_5	x_6	x_7	x_8
1	-	1	1	0	-	0	0	0

K TVL (31) ODA | 3 Var. | 1 R. | S. 1

	x_3	x_6	x_8
1	1	0	0

K TVL (32) ODA | 3 Var. | 1 R. | S. 1

	x_3	x_6	x_8
1	1	0	0

K TVL (33) ODA | 5 Var. | 1 R. | S. 1

	x_3	x_4	x_6	x_7	x_8
1	1	0	0	0	0

K TVL (34) ODA | 6 Var. | 1 R. | S. 1

	x_2	x_3	x_4	x_6	x_7	x_8
1	1	1	0	0	0	0

K TVL (35) ODA | 6 Var. | 1 R. | S. 1

	x_2	x_3	x_4	x_6	x_7	x_8
1	1	1	0	0	0	0

K TVL (36) ODA | 6 Var. | 1 R. | S. 1

	x_2	x_3	x_4	x_6	x_7	x_8
1	1	1	0	0	0	0

K TVL (37) ODA | 6 Var. | 1 R. | S. 1

	x_2	x_3	x_4	x_6	x_7	x_8
1	1	1	0	0	0	0

K TVL (38) ODA | 6 Var. | 1 R. | S. 1

	x_2	x_3	x_4	x_6	x_7	x_8
1	1	1	0	0	0	0

K TVL (39) ODA | 6 Var. | 1 R. | S. 1

	x_2	x_3	x_4	x_6	x_7	x_8
1	1	1	0	0	0	0

K TVL (40) ODA | 8 Var. | 1 R. | S. 1

	x_1	x_2	x_3	x_4	x_5	x_6	x_7	x_8
1	-	1	1	0	-	0	0	0

Fig. 9.26 Intermediate results to compute a single ternary vector as solution of a 3-SAT-problem using difference operations

- Column three shows the selected ternary vectors (TVLs 31–39) used as minuend of the difference with the vector of non-solutions of the next clause (for $i < 9$).

The main computation is realized in the while-loop in lines 22–56 as in the previous task. In each sweep of this loop the TVL of the intermediate result of one clause is computed. The *design by contract* approach is also used for this while-loop. The preconditions are:

- The variable $i indicates the clause that has to be evaluated in the actual sweep of the while-loop of the 3-SAT-problem to solve.
- The ternary vector with the index i of TVL 1 in conjunctive form determines the clause i.
- The TVL with the index $10 + i$ contains the conjunction that determines all non-solutions of the clause i.
- The TVL with the index $20 + i$ contains conjunctions which satisfy the clauses $1, \ldots, i$.
- The value true of the variable $unknown indicates that it is unknown whether a solution of the given 3-SAT-problem exists; hence, the commands of the body of the while-loop must be executed in this case.
- TVL 10 determines the function $1(x_1, x_2, x_3, x_4, x_5, x_6, x_7, x_8)$.

And the postconditions are:

- The value false of the variable $unknown indicates that no further computations are needed; in this case the value false of the variable $solution states that no solution exists for this 3-SAT-problem and the value true of the variable $solution indicates that one ternary solution-vector of the 3-SAT-problem has been computed and stored as TVL 40.
- The first ternary vector of the TVL with the index $20 + i$ is selected as TVL with the index $30 + i$ as partial solution and thereafter deleted in the TVL with the index $20 + i$.
- In the cases $1 \leq i < 9$, the ternary vector of the non-solution of the next clause $(i + 1)$ is computed and stored as TVL with the index $11 + i$.
- In the cases $1 \leq i < 9$, the difference between the selected ternary vector (TVL with the index $30 + i$) and the non-solution of the next clause (TVL with the index $11 + i$) is computed and stored as TVL with the index $21 + i$.
- The value of the variable $i is incremented when $i < 9$ and the computed TVL with the index $20 + i$ is not empty.
- The value of the variable $i is decremented when the TVL with the index $20 + i$ is empty (backtracking).
- The value of the variable $unknown is set to false and the value of the variable $solution is set to true when the TVL with the index $20 + i$ is not empty and $i = 9$ (the wanted solution-vector has been computed).
- The computed solution-vector is extended to all eight variables.
- The values of the variables $unknown and $solution are set to false when $i = 1$ and the TVL with the index $20 + i$ is empty (there exists no solution for the given 3-SAT-problem).

The commands before line 22 are needed to satisfy the precondition of the while-loop. The commands stv and ndm in lines 17 and 18 compute the non-solutions of the first clause and the command cpl in lines 19 computes the associated solutions.

An empty TVL 21 indicates in the case $i = 1$ that no solution of the 3-SAT-problem exists; the commands in lines 24–29 realize the associated postconditions and the command set $i 0 prevents the execution of the body of the subsequent command if. The value 0 is used because the same computations must be executed for $i = 1$ and $i > 1$.

An empty TVL with the index $20+i$ indicates that no solution of the given 3-SAT-problem exists for the previously selected vectors; hence, the decrement of the variable $\$i$ in line 34 leads to a backtracking step in this case.

The difference between the selected solution-vector that satisfies the clauses $1, \ldots, i$ and the ternary vector of the non-solutions of the clause $i + 1$ (for $i < 9$) is computed in the sweep of the while-loop with the value i; hence, a solution-vector of the given 3-SAT-problem is found when the TVL with the index $20 + i$ is not empty and $i = 9$. The commands in lines 38–46 realize the respective postconditions.

The main computations are executed in the case that $i < 9$ and the TVL with the index $20 + i$ is not empty. The commands stv and dtv in lines 49 and 50 select the first vector of the TVL with the index $20 + i$ as partial solution and the commands stv and ndm in lines 51 and 52 compute the non-solutions of the next clause. The difference between these two ternary vectors in line 54 is the main step of this approach. The increment in the subsequent command prepares the evaluation of the next clause in the next sweep of the while-loop.

A computed solution-vector is immediately shown in the 1-fold view due to the command assign in line 45. The commands in lines 57–87 assign the final TVLs of all intermediate results so that the used approach can be explored in detail.

The evaluation of the protocol confirms that eight difference operations between single ternary vectors have been executed to compute the wanted solution-vector of the 3-SAT-problem of nine clauses. A single intersection in line 44 expands the computed solution-vector to a ternary vector of all eight variables.

Solution 9.2 (Placement Problems)

(a) Figure 9.27 shows the PRP that solves the CD-SAT-problem that determines all placements of a maximal number of rocks on a chess board of the size 4×4 which do not threaten each other. The clauses of four variables define the requirements that at least one rock must be placed on a column $a, \ldots, d$. The subsequent sets of clauses of two variables specify the restrictions that an assigned rock prohibits other rock in the same column as well as the same row.

The specification of this CD-SAT-problem consists of 100 clauses and 208 literals. This SAT-problem is a special form of a logic equation; hence, the command sbe has been used to solve this problem.

Figure 9.28 shows the solution-TVL of the SAT-problem specified and solved in the PRP of Fig. 9.27 and well ordered the chess boards of the $4! = 24$ solutions.

(b) Figure 9.29 shows the PRP that solves the CDC-SAT-problem that determines all placements of a maximal number of rocks on a chess board of the size 4×4 which do not threaten each other. Each DC-clause specify both the requirements and the associated restrictions for the placement of a rock in one of the four columns of the chess board.

The specification of this CDC-SAT-problem consists of four DC-clauses and 112 literals at all. This CDC-SAT-problem is also a special form of a logic equation; hence, the command sbe has been used to solve this problem.

The same 24 solution-vectors as shown in Fig. 9.28 are computed as solution of this CDC-SAT-problem.

(c) Figure 9.30 shows the PRP that solves the CDC-SAT-problem that determines all placements of a maximal number of rocks on a chess board of the size 4×4 which do not threaten each other. The four DC-clauses of the previous task are here specified by TVLs 1–4. The use of these TVL is very convenient because

```
1    new
2    space 32 1
3    avar
4    a1  a2  a3  a4  b1  b2  b3  b4  c1  c2  c3  c4  d1  d2  d3  d4.
5    sbe 1 1
6    (a1+a2+a3+a4)&
7    (/a1+/a2)&(/a1+/a3)&(/a1+/a4)&(/a1+/b1)&(/a1+/c1)&(/a1+/d1)&
8    (/a2+/a1)&(/a2+/a3)&(/a2+/a4)&(/a2+/b2)&(/a2+/c2)&(/a2+/d2)&
9    (/a3+/a1)&(/a3+/a2)&(/a3+/a4)&(/a3+/b3)&(/a3+/c3)&(/a3+/d3)&
10   (/a4+/a1)&(/a4+/a2)&(/a4+/a3)&(/a4+/b4)&(/a4+/c4)&(/a4+/d4)&
11   (b1+b2+b3+b4)&
12   (/b1+/b2)&(/b1+/b3)&(/b1+/b4)&(/b1+/a1)&(/b1+/c1)&(/b1+/d1)&
13   (/b2+/b1)&(/b2+/b3)&(/b2+/b4)&(/b2+/a2)&(/b2+/c2)&(/b2+/d2)&
14   (/b3+/b1)&(/b3+/b2)&(/b3+/b4)&(/b3+/a3)&(/b3+/c3)&(/b3+/d3)&
15   (/b4+/b1)&(/b4+/b2)&(/b4+/b3)&(/b4+/a4)&(/b4+/c4)&(/b4+/d4)&
16   (c1+c2+c3+c4)&
17   (/c1+/c2)&(/c1+/c3)&(/c1+/c4)&(/c1+/a1)&(/c1+/b1)&(/c1+/d1)&
18   (/c2+/c1)&(/c2+/c3)&(/c2+/c4)&(/c2+/a2)&(/c2+/b2)&(/c2+/d2)&
19   (/c3+/c1)&(/c3+/c2)&(/c3+/c4)&(/c3+/a3)&(/c3+/b3)&(/c3+/d3)&
20   (/c4+/c1)&(/c4+/c2)&(/c4+/c3)&(/c4+/a4)&(/c4+/b4)&(/c4+/d4)&
21   (d1+d2+d3+d4)&
22   (/d1+/d2)&(/d1+/d3)&(/d1+/d4)&(/d1+/a1)&(/d1+/b1)&(/d1+/c1)&
23   (/d2+/d1)&(/d2+/d3)&(/d2+/d4)&(/d2+/a2)&(/d2+/b2)&(/d2+/c2)&
24   (/d3+/d1)&(/d3+/d2)&(/d3+/d4)&(/d3+/a3)&(/d3+/b3)&(/d3+/c3)&
25   (/d4+/d1)&(/d4+/d2)&(/d4+/d3)&(/d4+/a4)&(/d4+/b4)&(/d4+/c4)
26   =1.
27   assign 1 /1
```

Fig. 9.27 Problem-program that specifies and solves the SAT-problem to compute all placements of a maximal number of rocks on a chess board of the size 4×4 which do not threaten each other

- The properties of the modeled problem become better visible.
- The effort to type the four conjunctions of a DC-clause as ternary vectors is smaller than their typing as logic expression.

Three commands isc in the for-loop at end of the PRP of Fig. 9.30 solve this CDC-SAT-problem very fast.

The same 24 solution-vectors as shown in Fig. 9.28 (however in a different order) are computed as solution of this CDC-SAT-problem. The order of the binary vectors of this solution has been used to arrange the chess boards that show the 24 solutions in Fig. 9.28.

(d) Figure 9.31 shows the PRP that solves the CDC-SAT-problem that determines all placements of a maximal number of rocks on a chess board of the size 4×4 which do not threaten each other. The TVL of the first DC-clause (the specification of the rules for the rocks in column a) and four VTs of the variables that specify the fields of the four columns are sufficient as specification of the CDC-SAT-problem due to the similarity of the four DC-clause.

The command copy in line 18 of Fig. 9.31 uses the given TVL 1 to prepare the initial TVL of the solution. The command cco in line 21 of the subsequent for-loop creates the temporary TVL 6 of the next needed DC-clause determined by the used VTs, and the command isc restricts the solution-set (TVL 5) by this temporary TVL.

The same 24 solution-vectors as shown in Fig. 9.28 (in the same order of the previous task) are computed as solution of this CDC-SAT-problem.

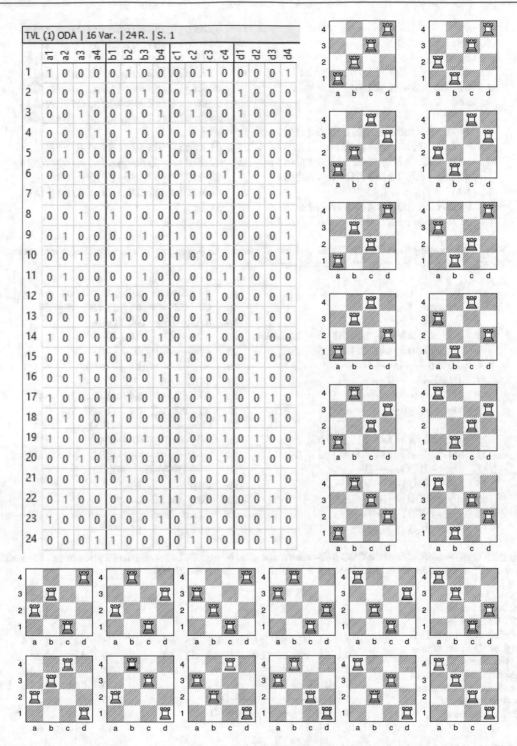

TVL (1) ODA	16 Var.	24 R.	S. 1														
	a1	a2	a3	a4	b1	b2	b3	b4	c1	c2	c3	c4	d1	d2	d3	d4
1	1	0	0	0	0	1	0	0	0	0	1	0	0	0	0	1
2	0	0	0	1	0	0	1	0	0	1	0	0	1	0	0	0
3	0	0	1	0	0	0	0	1	0	1	0	0	1	0	0	0
4	0	0	0	1	0	1	0	0	0	0	1	0	1	0	0	0
5	0	1	0	0	0	0	0	1	0	0	1	0	1	0	0	0
6	0	0	1	0	0	1	0	0	0	0	0	1	1	0	0	0
7	1	0	0	0	0	0	1	0	0	1	0	0	0	0	0	1
8	0	0	1	0	1	0	0	0	0	1	0	0	0	0	0	1
9	0	1	0	0	0	0	1	0	1	0	0	0	0	0	0	1
10	0	0	1	0	0	1	0	0	1	0	0	0	0	0	0	1
11	0	1	0	0	0	0	1	0	0	0	0	1	1	0	0	0
12	0	1	0	0	1	0	0	0	0	0	1	0	0	0	0	1
13	0	0	0	1	1	0	0	0	0	0	1	0	0	1	0	0
14	1	0	0	0	0	0	0	1	0	0	1	0	0	1	0	0
15	0	0	0	1	0	0	1	0	1	0	0	0	0	1	0	0
16	0	0	1	0	0	0	0	1	1	0	0	0	0	1	0	0
17	1	0	0	0	0	1	0	0	0	0	0	1	0	0	1	0
18	0	1	0	0	1	0	0	0	0	0	0	1	0	0	1	0
19	1	0	0	0	0	0	1	0	0	0	0	1	0	1	0	0
20	0	0	1	0	1	0	0	0	0	0	0	1	0	1	0	0
21	0	0	0	1	0	1	0	0	1	0	0	0	0	0	1	0
22	0	1	0	0	0	0	0	1	1	0	0	0	0	0	1	0
23	1	0	0	0	0	0	0	1	0	1	0	0	0	0	1	0
24	0	0	0	1	1	0	0	0	0	1	0	0	0	0	1	0

Fig. 9.28 All placements of a maximal number of rocks on a chess board of the size 4 × 4 computed by the PRP of Fig. 9.27

```
 1   new                                    15   b3&/b1&/b2&/b4&/a3&/c3&/d3+
 2   space 32 1                             16   b4&/b1&/b2&/b3&/a4&/c4&/d4
 3   avar                                   17   )&(
 4   a1 a2 a3 a4 b1 b2 b3 b4                 18   c1&/c2&/c3&/c4&/a1&/b1&/d1+
 5   c1 c2 c3 c4 d1 d2 d3 d4 .              19   c2&/c1&/c3&/c4&/a2&/b2&/d2+
 6   sbe 1 1                                20   c3&/c1&/c2&/c4&/a3&/b3&/d3+
 7   (                                      21   c4&/c1&/c2&/c3&/a4&/b4&/d4
 8   a1&/a2&/a3&/a4&/b1&/c1&/d1+            22   )&(
 9   a2&/a1&/a3&/a4&/b2&/c2&/d2+            23   d1&/d2&/d3&/d4&/a1&/b1&/c1+
10   a3&/a1&/a2&/a4&/b3&/c3&/d3+            24   d2&/d1&/d3&/d4&/a2&/b2&/c2+
11   a4&/a1&/a2&/a3&/b4&/c4&/d4            25   d3&/d1&/d2&/d4&/a3&/b3&/c3+
12   )&(                                    26   d4&/d1&/d2&/d3&/a4&/b4&/c4
13   b1&/b2&/b3&/b4&/a1&/c1&/d1+            27   )=1 .
14   b2&/b1&/b3&/b4&/a2&/c2&/d2+            28   assign 1 /1
```

Fig. 9.29 Problem-program of the CDC-SAT-problem to compute all placements of a maximal number of rocks on a chess board of the size 4×4 which do not threaten each other

```
 1   new                                    18   a1 a2 a3 a4 b1 b2 b3 b4
 2   space 32 1                             19   c1 c2 c3 c4 d1 d2 d3 d4 .
 3   tin 1 1                                20   0---0---10000---
 4   a1 a2 a3 a4 b1 b2 b3 b4                 21   -0---0--0100-0--
 5   c1 c2 c3 c4 d1 d2 d3 d4 .              22   --0---0-0010--0-
 6   10000---0---0---                       23   ---0---00001---0.
 7   0100-0---0---0--                       24   tin 1 4
 8   0010--0---0---0-                       25   a1 a2 a3 a4 b1 b2 b3 b4
 9   0001---0---0---0.                      26   c1 c2 c3 c4 d1 d2 d3 d4 .
10   tin 1 2                                27   0---0---0---1000
11   a1 a2 a3 a4 b1 b2 b3 b4                 28   -0---0---0--0100
12   c1 c2 c3 c4 d1 d2 d3 d4 .              29   --0---0---0-0010
13   0---10000---0---                       30   ---0---0---00001.
14   -0--0100-0---0--                       31   copy 1 5
15   --0-0010--0---0-                       32   for $i 2 4
16   ---00001---0---0.                      33   (isc 5 $i 5)
17   tin 1 3                                34   assign 5 /1
```

Fig. 9.30 Problem-program of the CDC-SAT-problem specified by four TVLs to compute all placements of a maximal number of rocks on a chess board of the size 4×4 which do not threaten each other

Fig. 9.31
Problem-program of the
CDC-SAT-problem
specified by a single TVL
and four VTs to compute
all placements of a
maximal number of rocks
on a chess board of the size
4×4 which do not
threaten each other

```
 1   new                        13   b1 b2 b3 b4 .
 2   space 32 1                 14   vtin 1 13
 3   tin 1 1                    15   c1 c2 c3 c4 .
 4   a1 a2 a3 a4 b1 b2 b3 b4    16   vtin 1 14
 5   c1 c2 c3 c4 d1 d2 d3 d4 . 17   d1 d2 d3 d4 .
 6   10000---0---0---           18   copy 1 5
 7   0100-0---0---0--           19   for $i 12 14
 8   0010--0---0---0-           20   (
 9   0001---0---0---0.          21   cco 1 11 $i 6
10   vtin 1 11                  22   isc 5 6 5
11   a1 a2 a3 a4 .              23   )
12   vtin 1 12                  24   assign 5 /1
```

9.9 Supplementary Exercises

Exercise 9.3 (Covering Problems)

(a) Compute all minimal assignments of knights on a chessboard of the size 4×4, such that these knights threaten (cover) all $4^2 = 16$ fields. A value 1 of a logic variable $a1, \ldots, d4$ indicates that a knight is located on this field on the board. What is the minimal number of knights that threaten (cover) all 16 fields? How many minimal covers of knights exist for a chessboard of the size 4×4?

One movement of a knight consists of either two fields horizontally followed by one field vertically or two fields vertically followed by one field horizontally; hence, one knight threatens (covers) eight fields on the board when it is surrounded by at least two fields in each direction.

Determine a SAT-equation in which each clause specifies all positions of knights that threaten (cover) the selected field. This SAT-equation belongs to the unate covering problems; hence, the solution consists of both all minimal covers and all covers specified by a minimal cover and additional knights. The minimal covers can be selected by counting the values 1 in the solution-vectors. This can be done using the commands `cel`, `stv`, and `sv_size`; the command `cel` is used to replace the values 0 of by dashes, the command `stv` omits the dashes in the selected ternary vector, and the command `sv_size` counts the remaining values 1 of the selected ternary vector. The TVL of the solution-set minimized using the command `obbc` strongly reduces the effort for this evaluation.

(b) This and the next two tasks are related to the strategy board game denoted by *nine-men morris* also known as mill or several other names. Figure 9.32 shows the board of this game; this graph has 24 vertices labeled by $v_{01}, \ldots, v_{24}$ and 32 edges $e_{01}, \ldots, e_{32}$. The pieces (the men) must be located at the corners or at the connection point of the edges.

What is the minimal number of pieces $npmin that must be located on the vertices such that all edges are adjacent with at least one of the located pieces? How many such minimal placements exist? Compute all these minimal placements.

It is convenient to specify the needed 32 clauses as ternary vectors of a TVL in K-form that depends on 24 variables. The solution of this vertex cover problem can be computed using the commands `ndm` and `cpl` which transform the specified TVL of the function $f_{vc}(\mathbf{v})$ from the conjunctive form into the disjunctive form. The selection of the solutions with a minimal number of values 1 can be realized in the same way as used in the previous task.

(c) Compute all exact minimal vertex covers in which each edge of the nine-men morris board is adjacent with exactly one of the placed pieces. How many such exact minimal vertex covers exist? Are the computed exact minimal vertex covers of the graph defined by the nine-men morris board a subset of the minimal vertex covers computed in the previous task?

The specification of all exact minimal vertex covers of a given graph leads to a CDC-SAT-formula. The clause $(v_{01} \vee v_{02})$ that describes the condition of a cover of the edge e_{01} must be replaced by the DC-clause $(v_{01} v_{02} \vee v_{01} v_{02})$; the clauses of the other edges must be transformed in the same way. This transformation can be separately realized for each clause using the commands `stv`, `ndm`, `cpl`, and `cel`. The wanted solution of the CDC-SAT-problem is the result of intersections of all these generated TVLs. Hence, it is convenient to reuse the TVL in K-form of the previous task as basic description and realize both the required transformations and the intersection operations in the body of a `for`-loop.

Note, all exact minimal vertex covers are computed much faster than all minimal vertex covers because the selection of vectors with a minimal number of values 1 is not needed.

Fig. 9.32 Board of the
game nine-men morris
with labeled 24 vertices
and 32 edges

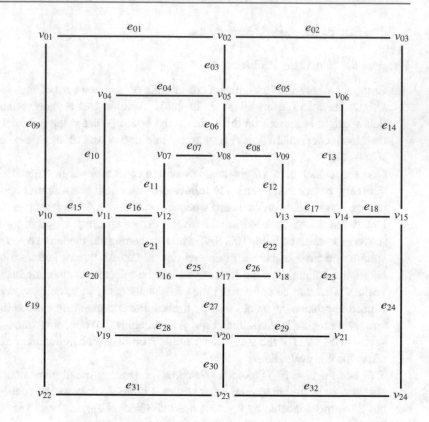

(d) Compute all exact minimal edge covers in which each vertex of the nine-men morris board is
adjacent with exactly one of the selected edges. How many such exact minimal edge covers
exist?

The specification of all exact minimal edge covers of the given graph leads to a CDC-SAT-formula
with 24 DC-clauses depending at all on the 32 variables of the edges. The effort to solve this
SAT-problem can also be reduced when the TVL in K-form of the associated minimal edge cover
problem (not exact) is used as basic description. The transformation of the clauses of the vertices
into DC-clauses can be realized as before separately for each clause using the commands stv,
ndm, cpl, and cel because this approach is not restricted to two variables in the clause. We get,
e.g., for the vertex v_{02} from the clause $(e_{01} \vee e_{02} \vee e_{03})$ the DC-clause $(e_{01}\bar{e}_{02}\bar{e}_{03} \vee \bar{e}_{01}e_{02}\bar{e}_{03} \vee \bar{e}_{01}\bar{e}_{02}e_{03})$.

Exercise 9.4 (Path and Coloring Problems)

This exercise explores several properties of the graph of eight vertices and 13 edges shown in
Fig. 9.33.

(a) Compute all Hamiltonian paths and Hamiltonian cycles of the graph shown in Fig. 9.33. Use
variables x_{ij} that are equal to 1 if an edge from vertex i to vertex j is used in the Hamiltonian
path and 0 otherwise. How many Hamiltonian paths exist in this graph? How many Hamiltonian
cycles exist in this graph?

Determine a CDC-SAT-formula that describes the necessary conditions for Hamiltonian paths.
The solution of this CDC-SAT-formula are sets of directed edges that specify either a potential
Hamiltonian path or a potential Hamiltonian cycle. Transform these sets of edges determined

Fig. 9.33 Explored graph of eight vertices and 13 edges

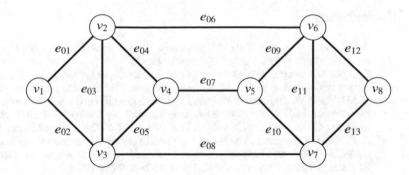

by the binary vector of the variables x_{ij} of each computed potential Hamiltonian path or cycle into a graph that specifies each directed edge by the index of a start vertex binary encoded by the variables (s_2, s_1, s_0) and the index of the end vertex also binary encoded by the variables (sn_2, sn_1, sn_0) as follows: $v_1 \rightarrow (s_2, s_1, s_0) = (001)$, $v_2 \rightarrow (s_2, s_1, s_0) = (010)$, $\dots v_7 \rightarrow (s_2, s_1, s_0) = (111)$, and $v_8 \rightarrow (s_2, s_1, s_0) = (000)$. Use another Boolean space for this model of the potential Hamiltonian graph. Verify for each computed directed graph by means of the associated reachable graph whether it describes a complete sequence of edges of a real Hamiltonian path or a single cycle of a real Hamiltonian cycle.

Note, there are at all more than 1000 potential Hamiltonian paths and cycles; hence, the created PRP needs in comparison to the most other exercises a longer time to solve this task. Adding the integer 2000 as parameter `iob` of the `while`-loop, in which the transformations and evaluations of the potential Hamiltonian paths and cycles are executed, avoids the dialog that can be used to break this loop after 1000 sweeps.

Show the TVL of the potential Hamiltonian paths or cycles in viewport (1,1), the TVL of real Hamiltonian paths in viewport (2,1), and TVL of real Hamiltonian cycles in viewport (3,1) of the m-fold view.

(b) Compute all Eulerian paths of the graph shown in Fig. 9.33. Use variables x_{ij} that are equal to 1 if an edge from vertex i to vertex j is used in the Eulerian path and 0 otherwise. How many Eulerian paths exist in this graph? Are there Eulerian cycles? Indicate the start vertex and the end vertex of the Eulerian paths which are no Eulerian cycles. Show the TVL of the computed Eulerian paths in the 1-fold view.

(c) What is the largest number of colors for which no vertex coloring of the graph, shown in Fig. 9.33, exists. Determine and solve an associated CDC-SAT-equation that verifies your assumption. Specify a second CDC-SAT-equation that has all minimal vertex colorings of the graph, shown in Fig. 9.33, as solution. How many minimal vertex colorings exist for this graph?

Use variables v_{ic} that are equal to 1 if the vertex with the index i is colored with the color number c. Show the TVLs of these two coloring problems in the m-fold view.

(d) Alternatively to the coloring of the vertices of a graph, it is possible to assign different colors to the edges of a graph. An edge coloring of a graph is an assignment of different colors to the edges of a graph so that no two edges adjacent with the same vertex have the same color.

Compute all minimal edge colorings of the graph, shown in Fig. 9.33. Use variables e_{ijc} that are equal to 1 if the edge with the two digits index ij is colored with the color number c and 0 otherwise. What is the smallest possible number of colors of an edge coloring of the graph, shown in Fig. 9.33? How many variables are needed to specify the CDC-SAT-equation that determines all minimal edge colorings of the explored graph? How many minimal edge coloring exist for this graph? Show the TVL of all minimal edge colorings in the 1-fold view.

References

1. A. Biere et al., eds., *Handbook of Satisfiability*, vol. 185. Frontiers in Artificial Intelligence and Applications (IOS Press, New York, 2009). ISBN: 978-1-58603-929-5
2. S. Cook, The complexity of theorem-proving procedures, in *Proceedings of the Third Annual ACMSymposium on Theory of Computing*. STOC (ACM, Shaker Heights, 1971), pp. 151–158. https://doi.org/10.1145/800157.805047
3. M.J.H. Heule, O. Kullmann, V.W. Marek, Solving and verifying the Boolean pythagorean triples problem via cube-and-conquer, in *Theory and Applications of Satisfiability Testing—SAT 2016*, ed. by N. Creignou, D. Le Berre (Springer, Cham, 2016), pp. 228–245. ISBN: 978-3-319-40970-2. https://doi.org/10.1007/978-3-319-40970-2_15
4. C. Posthoff, B. Steinbach, The solution of discrete constraint problems using Boolean models, in *Proceedings of the 2nd International Conference on Agents and Artificial Intelligence*, ed. by J. Filipe, A. Fred, B. Sharp. ICAART 2 (Valencia, Spain, 2010), pp. 487–493. ISBN: 978-989-674-021-4
5. B. Schwarzkopf, Without cover on a rectangular chessboard, in *Feenschach* (1990). German title: "Ohne Deckung auf dem Rechteck", pp. 272–275
6. B. Steinbach, XBOOLE—a toolbox for modelling, simulation, and analysis of large digital systems. Syst. Anal. Model. Simul. **9**(4), 297–312 (1992). ISSN: 0232-9298
7. B. Steinbach, C. Posthoff, New results based on Boolean Models, in *Boolean Problems, Proceedings of the 9th International Workshops on Boolean Problems*, ed. by B. Steinbach. IWSBP 9 (Freiberg University of Mining and Technology, Freiberg, 2010), pp. 29–36. ISBN: 978-3-86012-404-8
8. B. Steinbach, C. Posthoff, Parallel solution of covering problems super-linear speedup on a small set of cores, in *International Journal on Computing, Global Science and Technology Forum*. GSTF 1.2 (2011), pp. 113–122. ISSN: 2010-2283
9. B. Steinbach, C. Posthoff, Improvements of the construction of exact minimal covers of Boolean functions, in *Computer Aided Systems Theory – EUROCAST 2011. Lecture Notes in Computer Science*, ed. by R. Moreno-Díaz, F. Pichler, A. Quesada-Arencibia, vol. 6928, (Springer, Berlin, Heidelberg. ISBN 978-3-642-27578-4, 2012). https://doi.org/10.1007/978-3-642-27579-1_35
10. B. Steinbach, C. Posthoff, Sources and obstacles for parallelization—a comprehensive exploration of the unate covering problem using both CPU and GPU, in *GPU Computing with Applications in Digital Logic*, ed. by J. Astola et al. TICSP 62 (Tampere International Center for Signal Processing, Tampere, 2012), pp. 63–96. ISBN: 978-952-15-2920-7. https://doi.org/10.13140/2.1.4266.4320
11. B. Steinbach, C. Posthoff, Fast calculation of exact minimal unate coverings on both the CPU and the GPU, in *Proceedings of the 14th International Conference on Computer Aided Systems Theory—EUROCAST 2013—Part II*, ed. by A. Quesada-Arencibia et al. LNCS 8112 (Springer, Las Palmas, 2013), pp. 234–241. ISBN: 978-0-99245-180-6. https://doi.org/10.1007/978-3-642-53862-9_30
12. B. Steinbach, C. Posthoff, Multiple-valued problem solvers—comparison of several approaches, in *Proceedings of the IEEE 44th International Symposium on Multiple-Valued Logic*. ISMVL 44. Bremen, Germany (2014), pp. 25–31. https://doi.org/10.1109/ISMVL.2014.13
13. B. Steinbach, C. Posthoff, Evaluation and optimization of GPU based unate covering algorithms, in *Computer Aided Systems Theory—EUROCAST 2015*, ed. by A. Quesada-Arencibia et al. LNCS 9520 (Springer, Las Palmas, 2015), pp. 617–624. ISBN: 978-3-319-27340-2. https://doi.org/10.1007/978-3-319-27340-2_76
14. *SWI Prolog—Reference Manuel*. Updated for version 8.2.4 (2021). https://www.swi-prolog.org/pldoc/doc_for?object=manual

Extremely Complex Problems

10

Abstract

The problem of rectangle-free assignments of four colors to a grid will be explored in this chapter. This problem can also be considered as a problem of Ramsey numbers. Nether theoretical approaches of mathematicians nor programs running on the largest computers were able to solve this task for the grids 17×17, 17×18, 18×17, 18×18, 12×21, and 21×12 before we solved these problems. The reason for that is the extreme complexity of $5.23 * 10^{151}$ for the grid 12×21 and even $1.17 * 10^{195}$ for the grid 18×18. We will show that it is not sufficient to use logic equations; many other mathematical concepts also have to be used in order to solve the problem using a usual PC. We started this research because we wanted to know how far the power of logic equations and of ternary vectors will reach to solve a problem of this extreme complexity. The successful solution of such a complex problem can be taken as the borderline for the solution of similar problems. The detailed steps of the solution for the grids 18×18 and 12×21 are completely different; however, the general approach of a very deep analysis and utilization of all properties of the problem guided us to the successful solutions.

Supplementary Information The online version of this chapter (https://doi.org/10.1007/978-3-030-88945-6_10) contains supplementary material which is available for authorized users. Please, follow the link belonging to the version of the XBOOLE-monitor XBM 2 that fits best for your operating system. This XBOOLE-monitor is needed to solve all tasks of this chapter. Instructions for starting the downloaded XBOOLE-monitor XBM 2 are given at the beginning of Section 'Exercises' in this chapter.

XBOOLE-monitor XBM 2 for Windows 10
32 bits
https://doi.org/10.1007/978-3-030-88945-6_10_MOESM1_ESM.zip (15,091 KB)

64 bits
https://doi.org/10.1007/978-3-030-88945-6_10_MOESM2_ESM.zip (14,973 KB)

XBOOLE-monitor XBM 2 for Linux Ubuntu
32 bits
https://doi.org/10.1007/978-3-030-88945-6_10_MOESM3_ESM.zip (29,522 KB)

64 bits
https://doi.org/10.1007/978-3-030-88945-6_10_MOESM4_ESM.zip (28,422 KB)

© Springer Nature Switzerland AG 2022
B. Steinbach, C. Posthoff, *Logic Functions and Equations*,
https://doi.org/10.1007/978-3-030-88945-6_10

10.1   Rectangle-Free Four-Colored

Definition 10.1 A bipartite graph is a graph, whose set of vertices (nodes) are decomposed into two disjoint sets such that no two vertices within the same set are connected.

Figure 10.1 shows an example of a bipartite graph 6 × 6.

Definition 10.1 can be extended to a *complete bipartite graph*: Each vertex of one set is connected with each vertex of the second set (see Fig. 10.2a). An alternative representation of a bipartite graph is a grid:

- Each vertex of the first set is assigned to a row of the grid.
- Each vertex of the second set is assigned to a column of the grid.
- The color of the edge between the vertex Ai and Bj of the bipartite graph is assigned to the grid element of the row i and the column j.

Figure 10.2b shows the completely colored grid that represents the bipartite graph of Fig. 10.2a.

The problem to solve is:

> Find a coloring of the edges of a bipartite graph that does not contain a
> *monochromatic four-sided figure*.

We met the problem in the format of the grid for which a complete coloring must be found such that the same color does not occur in the four intersection points specified by each pair of rows and each pair of columns; hence, each four-sided figure will be represented as a *rectangle*. Here the existence of *Ramsey numbers* can be anticipated: When the number of nodes is growing, then it might occur that four colors are not sufficient for avoiding monochromatic rectangles.

Figure 10.3 emphasizes the rectangle that represents the sequence of edges (A1,B2), (B2,A3), (A3,B1), (B1,A1). Figure 10.3a repeats the coloring of the graph shown in Fig. 10.2; it can be checked that this four-colored grid does not contain a monochromatic rectangle. A counterexample is shown in Fig. 10.3b where all four corners of the emphasized rectangle are red-colored.

When we started our research four problems remained open:

- Can a grid 17 × 17 be colored by four colors without monochromatic rectangles?

Fig. 10.1 Some colored edges of a bipartite graph with six vertices in each partition

Fig. 10.2 Completely colored bipartite graph 3 × 3: (**a**) graph, (**b**) grid

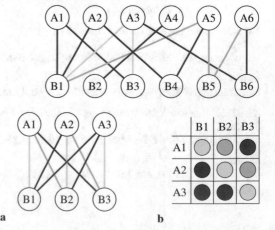

Fig. 10.3 The rectangle of the rows A1, A3 and the columns B1, B2: (**a**) rectangle-free four-colored grid, (**b**) a gird that contains a monochromatic (red) rectangle

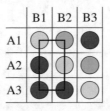 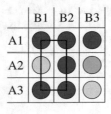

a   b

- Can a grid 17×18 be colored by four colors without monochromatic rectangles?
- Can a grid 18×18 be colored by four colors without monochromatic rectangles?
- Can a grid 12×21 be colored by four colors without monochromatic rectangles?

The grids 18×17 and 21×12 must not be considered due to the symmetry with regard to the problems 17×18 and 12×21. The central point of our efforts was the problem 18×18—a solution of this problem would have been a solution of the problems 18×17 and 17×17 as well.

10.2 Discussion of the Complexity

As a first step it is recommended to get an understanding of the complexity of the problem.

Number of Vertices
The first complexity measure is the *number of vertices*. This can easily be calculated as the product of the number of rows and columns. Any combination (n, m) with $n, m \geq 2$ are possible for the size of the grid. For the problems to be considered we have

$$\text{grid } 17 \times 17: \quad 289 \text{ nodes}, \qquad\qquad \text{grid } 17 \times 18: \quad 306 \text{ nodes},$$
$$\text{grid } 18 \times 18: \quad 324 \text{ nodes}, \qquad\qquad \text{grid } 12 \times 21: \quad 252 \text{ nodes}.$$

Number of Rectangles
The *number of rectangles* can be calculated by selecting two rows and two columns. Their intersection defines four corner points of a rectangle. The selection of the rows 2 and 6 as well as the use of the columns 4 and 7, for instance, determine the four corner points (2,4), (2,7), (6,4), and (6,7). The number of selections of two out of n elements can be calculated by using the *binomial coefficient*:

$$\binom{n}{2} = \frac{n * (n - 1)}{2}.$$

Hence, for the problems we get the following number of pairs of rows or columns:

$$\binom{12}{2} = 66, \qquad \binom{17}{2} = 136, \qquad \binom{18}{2} = 153, \qquad \binom{21}{2} = 210.$$

Table 10.1 shows how many rectangles must be evaluated for the explored grids.

Table 10.1 Numbers of rectangles in the explored grids

| Grid of the size | Number of pairs of | | Number of |
Rows × Columns	Rows	Columns	rectangles
12 × 21	66	210	13,860
17 × 17	136	136	18,496
17 × 18	136	153	20,808
18 × 18	153	153	23,409

Table 10.2 The complexity of the unsolved grid problems

Grid	Number of nodes	Number of rectangles	Number of colorings
12 × 21	252	13,860	$4^{252} = 2^{504} \approx 5.2374 * 10^{151}$
17 × 17	289	18,496	$4^{289} = 2^{578} \approx 9.8932 * 10^{173}$
17 × 18	306	20,808	$4^{306} = 2^{612} \approx 1.6996 * 10^{184}$
18 × 18	324	23,409	$4^{324} = 2^{648} \approx 1.1680 * 10^{195}$

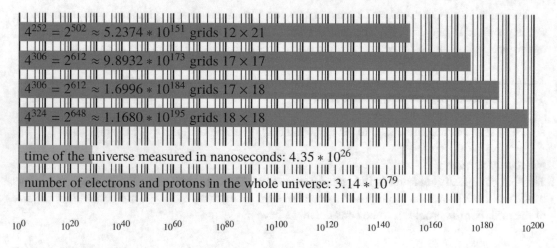

$4^{252} = 2^{502} \approx 5.2374 * 10^{151}$ grids 12 × 21

$4^{306} = 2^{612} \approx 9.8932 * 10^{173}$ grids 17 × 17

$4^{306} = 2^{612} \approx 1.6996 * 10^{184}$ grids 17 × 18

$4^{324} = 2^{648} \approx 1.1680 * 10^{195}$ grids 18 × 18

time of the universe measured in nanoseconds: $4.35 * 10^{26}$

number of electrons and protons in the whole universe: $3.14 * 10^{79}$

10^0 10^{20} 10^{40} 10^{60} 10^{80} 10^{100} 10^{120} 10^{140} 10^{160} 10^{180} 10^{200}

Fig. 10.4 Number of different four-colored grids drawn in a logarithmic scale and comparisons in time and space

Number of Colorings

Since we know that four colors are available for each node and the colors can be used arbitrarily, the powers $4^{\text{number of nodes}}$ define the size of the search space. Table 10.2 summarizes the results.

The size of the set of possible colorings is the crucial point. They are far beyond good and evil. These numbers do not even have a name in natural language. To get a certain perception of these extremely large numbers we tried to find known large numbers of real physical occurrences:

- The time between the big bang and today has been determined in [5] to be $13.831 \pm 0.038 * 10^9$ years; assuming that a computer is able to evaluate one four-colored grid within a single nanosecond, then this computer could check $4.36 * 10^{26}$ color patterns in this very long period of time.

- As comparison in the space we use the number of elementary particles in the whole universe; the astro-physicist Arthur Eddington determined already in 1938 that there are $1.57 * 10^{79}$ protons and the same number of electrons in the whole universe [1]; hence, a uniform distribution of all color patterns of the grid 18 × 18 would assign $3.72 * 10^{115}$ to each of these huge number of elementary particles.

Figure 10.4 illustrates these relationships and confirms that we are faced with a really extremely complex problem.

In many problems of this kind the number of solution-candidates gets smaller and smaller when the search gets deeper and deeper. However, if we have a solution of the problem 17×17, then it might be a candidate for the solution of the problem 18×18. The given solution, however, represents

$$17! * 17! * 4! \approx 3.0363 * 10^{30}$$

solutions because any permutation of rows, columns, and colors is a solution as well. Which one might be used for the next step?

10.3 Basic Approaches and Results

The Creation of a Boolean Model
The creation of a Boolean model consists of three steps:

- The introduction of Boolean variables and a correct definition of their meaning.
- The formalization of the needed requirements.
- The description of the constraints.

We will start with the introduction of Boolean variables:

$$y_{ijr} = \begin{cases} 1 & \text{if the color of the node } (i, j) \text{ is red}, \\ 0 & \text{otherwise}. \end{cases}$$

$$y_{ijg} = \begin{cases} 1 & \text{if the color of the node } (i, j) \text{ is green}, \\ 0 & \text{otherwise}. \end{cases}$$

$$y_{ijb} = \begin{cases} 1 & \text{if the color of the node } (i, j) \text{ is blue}, \\ 0 & \text{otherwise}. \end{cases}$$

$$y_{ijy} = \begin{cases} 1 & \text{if the color of the node } (i, j) \text{ is yellow}, \\ 0 & \text{otherwise}. \end{cases}$$

It is self-evident that i and j always will have values between 1 and 18 for the grid of the size 18×18. As a test, it can be checked that these variables allow us a unique description of each coloring of the grids. In this way, each node requires four Boolean variables.

Note 10.1 We are using here y_{ij} for a moment and return to x_{ij} after an important optimization step.

The problem requires the coloring of the nodes by one of four colors. This requirement can be formalized for each field (i, j) by

$$y_{ijr} \vee y_{ijg} \vee y_{ijb} \vee y_{ijy} = 1. \tag{10.1}$$

For each node a coloring must be given. Here is the possibility to include immediately the constraint that each node can get only one color. This constraint is also self-evident. Hence,

$$y_{ijr}\overline{y}_{ijg}\overline{y}_{ijb}\overline{y}_{ijy} \vee \overline{y}_{ijr}y_{ijg}\overline{y}_{ijb}\overline{y}_{ijy}$$

$$\vee \overline{y}_{ijr}\overline{y}_{ijg}y_{ijb}\overline{y}_{ijy} \vee \overline{y}_{ijr}\overline{y}_{ijg}\overline{y}_{ijb}y_{ijy} = 1. \tag{10.2}$$

Table 10.3 Mapping of the four colors of the node (i, j) to the two Boolean variables x_{ija} and x_{ijb}

Color		x_{ija}	x_{ijb}
Red	●	0	0
Green	◐	1	0
Blue	●	0	1
Yellow	○	1	1

Such an equation is again necessary for each node.

As constraint, we have the request that all monochromatic rectangles are forbidden. For a good readability we use the corner points from above: (2,4), (2,7), (6,4), and (6,7). However, such an equation must be used for each rectangle which results in 23,409 equations! For the example we get

$$y_{24r} \, y_{27r} \, y_{64r} \, y_{67r} \lor y_{24g} \, y_{27g} \, y_{64g} \, y_{67g}$$

$$\lor y_{24b} \, y_{27b} \, y_{64b} \, y_{67b} \lor y_{24y} \, y_{27y} \, y_{64y} \, y_{67y} = 0 \,. \tag{10.3}$$

These equations can be transformed into the typical SAT-format:

$$(\overline{y}_{24r} \lor \overline{y}_{27r} \lor \overline{y}_{64r} \lor \overline{y}_{67r}) \land (\overline{y}_{24g} \lor \overline{y}_{27g} \lor \overline{y}_{64g} \lor \overline{y}_{67g})$$

$$\land (\overline{y}_{24b} \lor \overline{y}_{27b} \lor \overline{y}_{64b} \lor \overline{y}_{67b}) \land (\overline{y}_{24y} \lor \overline{y}_{27y} \lor \overline{y}_{64y} \lor \overline{y}_{67y}) = 1.$$

These steps complete a correct model of the problem. The solution of this system of equations can now start. Before this will be tried, we see a possibility for the optimization of the model.

An Optimized Boolean Model

Due to the high number of variables, we are using the possibility of a binary coding of some properties of the model. The four colors have been represented by the four index values r, g, b, y. These four values can be encoded by two Boolean values. Table 10.3 shows the used mapping: The color of any node (i, j) can now be represented by two Boolean variables x_{ija} and x_{ijb}.

The first positive effect: Eq. (10.2) are no longer necessary because any assignment of values to the two variables of a node immediately indicates one color and excludes the other colors.

The equations for the rectangles get another shape as well: We go back and transform Eq. (10.3). The conjunction of the left-hand side of

$$x_{24a} x_{24b} x_{27a} x_{27b} x_{64a} x_{64b} x_{67a} x_{67b} = 0 \,, \tag{10.4}$$

for instance, describes the situation that the color of the four nodes of the given rectangle is yellow, an example for a forbidden monochromatic rectangle; hence, a restrictive equation with the value 0 on the right-hand side or, equivalently,

$$\overline{x}_{24a} \lor \overline{x}_{24b} \lor \overline{x}_{27a} \lor \overline{x}_{27b} \lor \overline{x}_{64a} \lor \overline{x}_{64b} \lor \overline{x}_{67a} \lor \overline{x}_{67b} = 1$$

avoids this situation.

We repeat this transformation for the other colors and finish at the end with one equation for excluding a monochromatic rectangle:

$$(\overline{x}_{24a} \vee \overline{x}_{24b} \vee \overline{x}_{27a} \vee \overline{x}_{27b} \vee \overline{x}_{64a} \vee \overline{x}_{64b} \vee \overline{x}_{67a} \vee \overline{x}_{67b})$$

$$\wedge (x_{24a} \vee \overline{x}_{24b} \vee x_{27a} \vee \overline{x}_{27b} \vee x_{64a} \vee \overline{x}_{64b} \vee x_{67a} \vee \overline{x}_{67b})$$

$$\wedge (\overline{x}_{24a} \vee x_{24b} \vee \overline{x}_{27a} \vee x_{27b} \vee \overline{x}_{64a} \vee x_{64b} \vee \overline{x}_{67a} \vee x_{67b})$$

$$\wedge (x_{24a} \vee x_{24b} \vee x_{27a} \vee x_{27b} \vee x_{64a} \vee x_{64b} \vee x_{67a} \vee x_{67b}) = 1 \ . \tag{10.5}$$

Hence, only two binary variables are necessary for the description of the coloring of one node. The equations still must be written for each rectangle. Symbolically we combine the equations beginning with the smallest rectangle in the left upper corner up to the smallest rectangle in the right lower corner:

$$(\overline{x}_{1,1a} \vee \overline{x}_{1,1b} \vee \overline{x}_{1,2a} \vee \overline{x}_{1,2b} \vee \overline{x}_{2,1a} \vee \overline{x}_{2,1b} \vee \overline{x}_{2,2a} \vee \overline{x}_{2,2b})$$

$$\wedge (x_{1,1a} \vee \overline{x}_{1,1b} \vee x_{1,2a} \vee \overline{x}_{1,2b} \vee x_{2,1a} \vee \overline{x}_{2,1b} \vee x_{2,2a} \vee \overline{x}_{2,2b})$$

$$\wedge (\overline{x}_{1,1a} \vee x_{1,1b} \vee \overline{x}_{1,2a} \vee x_{1,2b} \vee \overline{x}_{2,1a} \vee x_{2,1b} \vee \overline{x}_{2,2a} \vee x_{2,2b})$$

$$\wedge (x_{1,1a} \vee x_{1,1b} \vee x_{1,2a} \vee x_{1,2b} \vee x_{2,1a} \vee x_{2,1b} \vee x_{2,2a} \vee x_{2,2b}) = 1$$

$$\vdots$$

$$(\overline{x}_{17,17a} \vee \overline{x}_{17,17b} \vee \overline{x}_{17,18a} \vee \overline{x}_{17,18b} \vee \overline{x}_{18,17a} \vee \overline{x}_{18,17b} \vee \overline{x}_{18,18a} \vee \overline{x}_{18,18b})$$

$$\wedge (x_{17,17a} \vee \overline{x}_{17,17b} \vee x_{17,18a} \vee \overline{x}_{17,18b} \vee x_{18,17a} \vee \overline{x}_{18,17b} \vee x_{18,18a} \vee \overline{x}_{18,18b})$$

$$\wedge (\overline{x}_{17,17a} \vee x_{17,17b} \vee \overline{x}_{17,18a} \vee x_{17,18b} \vee \overline{x}_{18,17a} \vee x_{18,17b} \vee \overline{x}_{18,18a} \vee x_{18,18b})$$

$$\wedge (x_{17,17a} \vee x_{17,17b} \vee x_{17,18a} \vee x_{17,18b} \vee x_{18,17a} \vee x_{18,17b} \vee x_{18,18a} \vee x_{18,18b}) = 1 \ . \tag{10.6}$$

If we would use four lines for each rectangle and assume 40 lines on a page the resulting equation of the rectangle-free four-colored grid 18×18 would fill a book with 2341 pages!

It is always useful to start with small examples. The smallest rectangle is the rectangle 2×2. This is still a very easy job for XBOOLE. We build the four TVLs using adjusted Eq. (10.4) and get 252 solutions using the intersection.

However, it is also possible to write down the TVL $ODA(f_{mcr})$ of the four monochromatic rectangles:

$$ODA(f_{mcr}) = \begin{array}{cccccccc} x_{1,1,a} & x_{1,1,b} & x_{1,2,a} & x_{1,2,b} & x_{2,1,a} & x_{2,1,b} & x_{2,2,a} & x_{2,2,b} \\ \hline 0 & 0 & 0 & 0 & 0 & 0 & 0 & 0 \\ 0 & 1 & 0 & 1 & 0 & 1 & 0 & 1 \\ 1 & 0 & 1 & 0 & 1 & 0 & 1 & 0 \\ 1 & 1 & 1 & 1 & 1 & 1 & 1 & 1 \end{array} \ . \tag{10.7}$$

Now we use the complement for this TVL, which is available in XBOOLE, and get the 252 solutions as well. We continue with this procedure, Table 10.4 shows the results up to seven rows.

It can be seen that the share of monochromatic rectangles gets larger and larger, and it can be expected that at a given point of the increased size of the grid 100% for the share of monochromatic rectangles will be seen, which means that the existence of monochromatic rectangles cannot be avoided anymore.

And it can be seen as well that *the solution of the respective equations works.*

Table 10.4 Solutions of
the Boolean equations for
four-colored grids up to m
rows and two columns

Rows	Columns	Variables	Ternary vectors	Not mono-chromatic	Mono-chromatic	Percentage
2	2	8	24	252	4	1.56%
3	2	12	304	3912	184	4.49%
4	2	16	3416	59,928	5608	8.56%
5	2	20	36,736	906,912	141,664	13.51%
6	2	24	387,840	13,571,712	3,205,504	19.11%
7	2	28	4,061,824	201,014,784	67,420,672	25.12%

Table 10.5 Selected solutions using a fixed uniform distribution of the colors in the upper row and in the left column

Rows	Columns	Variables	TV	Solutions	Memory in kB	Time in sec
2	2	8	1	4	1	0.000
2	12	48	6912	2,361,960	1365	0.023
3	2	12	1	16	1	0.002
3	8	48	4,616,388	136,603,152	424,679	7.844
4	2	16	1	64	1	0.002
4	5	40	674,264	12,870,096	113,135	1.000
5	2	20	1	256	1	0.002
5	4	40	573,508	12,870,096	133,010	0.824
6	2	24	4	960	2	0.003
6	3	36	15,928	797,544	11,367	0.020
7	2	28	16	3600	4	0.004
7	3	42	183,152	10,493,136	95,565	0.314
8	2	32	64	13,500	14	0.005
8	3	48	2,152,819	136,603,152	910,656	4.457
9	2	36	256	50,625	52	0.007
10	2	40	768	182,250	153	0.008
15	2	60	147,456	104,162,436	29,090	0.386
19	2	76	6,553,600	14,999,390,784	1,292,802	21.378

Utilization of Permutations of Colors, Rows, and Columns

The limit in the previous approach was the required memory of about 800 Megabytes to represent
the correct color patterns of the grid $G_{7,2}$, which could be calculated in less than 5 s. To break the
limitation set by memory requirements we used some heuristic properties of the problem:

1. Knowing one single correct solution of the four-color problem, $4! = 24$ permutations of this
   solution with regard to the four colors are also correct solutions.
2. Each permutation of rows and columns of a given correct solution pattern creates also another
   pattern that is a correct solution.
3. A nearly uniform distribution of the colors in both the rows and columns is given for the largest
   number of correct four-colored grids.

Hence, in [10] we confined to the calculations of solutions with a *fixed uniform distribution of the
colors in the upper row and in the left column*. We restricted in this experiment the calculation to
12 columns of the grid and 2 Gigabytes of available memory. Table 10.5 shows some results of this
restricted calculation given in [10].

Figure 10.5 depicts the four selected four-colored grids of the first row in Table 10.5. These grids
have the fixed sequence of colors (red, green) in the upper row and the left column. For the remaining
field in the grid each of the four colors is used once.

Fig. 10.5 Grids $G_{2,2}$ with fixed colors in the upper row and in the left column

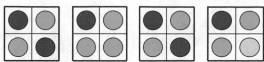

Using the same memory size, the number of Boolean variables could be enlarged from 28 for $G_{7,2}$ to 76 for $G_{19,2}$. That means, by utilizing properties of the four-color problem mentioned above, we have solved problems that are $2^{48} = 281,474,976,710,656$ times larger than before, but again the available memory size restricts the solution of the four-color problem for larger grids.

Solution of Larger Problems in a Restricted Size of Memory
The fundamental algorithm to solve the grid problem for four colors is quite easy: The intersection of the solutions of (10.5) for each rectangle must be calculated iteratively. The solution of (10.5) can be computed as complement of TVL (10.7) that is created for the particular rectangle. The sequence of the two operations complement (CPL in XBOOLE) and intersection (ISC in XBOOLE) can be combined into the single more efficient operation difference (DIF in XBOOLE). The limitation of this algorithm is the huge number of ternary vectors in intermediate TVLs.

The remaining task (RT) after each successfully executed difference operation is completely specified by the TVL of the intermediate solutions (IS) and the set of TVLs of the type (10.7) for the remaining rectangles. An additional recursive approach facilitates the solution of significantly larger problems. The key idea of this approach is the following:

- When the number of ternary vectors in the TVL IS exceeds a predefined limit (`SplitLimit`) the *remaining task* RT is split into two subtasks: RT1 and RT2 by dividing the ternary vectors of IS into the TVLs IS1 and IS2.

- RT2 is stored on a stack and the computation continues with the solution of RT1.

- The split of the recent RT into RT1 and RT2 will be performed whenever the TVL IS exceeds a predefined limit after execution of a difference operation.

- If the difference leads to an empty TVL there is no solution within the evaluated subspace and the computation continues with the topmost RT stored on the stack.

- If the result of TVL IS and the last TVL in the set of rectangles is not empty at least one solution pattern has been found so that the computation can be terminated.

Table 10.6 summarizes the results of this approach for grids of 12 rows and `SplitLimit=` 400. This recursive approach facilitates the solution of larger grids within a limited memory space. Solutions for four-colored grids which are modeled with up to 384 Boolean variables were found instead of 76 variables in the second (already improved) approach. This means that the explained recursive approach for the problem of rectangle-free four-color grids allows the solution of problems which are $2^{308} = 5.214812 * 10^{92}$ times larger than before. However, now the needed time restricts the solution for even larger grids. An approach to reduce the required time is given by parallel computing [9].

Applying SAT-Solvers
The system of logic equations (10.6) can be transformed into a single SAT-equation; hence, SAT-solvers can be used to find at least one solution of the four-colored grid problem. The power of SAT-solvers [3] has improved over the last decades. We tried to solve the four-color grid problem using the best SAT-solvers from the SAT-competitions of the last years.

Table 10.6 Recursive solution using a stack and the `SplitLimit`= 400

Number of					Maximal	Time in
Rows	Columns	Variables	TV	Correct solutions	stack size	seconds
12	2	48	337	6620	3	0.011
12	3	72	147	2423	22	0.029
12	4	96	319	7386	30	0.056
12	5	120	236	1188	47	0.095
12	6	144	181	1040	61	0.147
12	7	168	231	627	69	0.216
12	8	192	109	413	81	0.287
12	9	216	72	227	79	0.398
12	10	240	34	103	87	0.516
12	11	264	40	109	88	0.645
12	12	288	112	293	82	0.806
12	13	312	51	81	81	1.054
12	14	336	82	1415	97	1.290
12	15	360	1	1	80	2.361
12	16	384	2	3	81	426.903

Table 10.7 Time to solve quadratic four-colored grids using different SAT-solvers

Number of			Time in minutes:seconds			
Rows	Columns	Variables	Clasp	Lingeling	Plingeling	Precosat
12	12	288	0:00.196	0:00.900	0:00.990	0:00.368
13	13	338	0:00.326	0:01.335	0:04.642	0:00.578
14	14	392	0:00.559	0:03.940	0:02.073	0:00.578
15	15	450	46:30.716	54:02.304	73:05.210	120:51.739

Table 10.7 shows the required time to find the first solution for quadratic four-colored grids $G_{12,12}$, $G_{13,13}$, $G_{14,14}$, and $G_{15,15}$ using the SAT-solvers *clasp* [4], *lingeling* [2], *plingeling* [2], and *precosat* [2]. Figure 10.6 shows the rectangle-free four-colored grid $G_{15,15}$ found by *clasp* after 46.5 min.

From the utilization of the SAT-solvers we learned that

1. SAT-solvers are powerful tools that are able to solve the tasks of rectangle-free four-colored grids up to $G_{15,15}$.
2. It was not possible to calculate a rectangle-free four-colored grid larger than $G_{15,15}$.

The reasons for the second statement are in the first place that the search space for the four-colored grid $G_{16,16}$ is $2^{62} = 4.61 * 10^{18}$ times larger than the search space for the four-colored grid $G_{15,15}$; hence, it can be expected that it takes approximately 6 million times the age of the Earth to find a rectangle-free four-colored grid $G_{16,16}$. The second reason is that the fraction of rectangle-free four-colored grids is even further reduced for the larger grid.

10.4 Construction of Four-Colored Grids $G_{16,16}$ and $G_{16,20}$

The found solution for $G_{15,15}$ of Fig. 10.6 gave us a hint how larger four-colored grids may be constructed. The evaluation of the rows and columns over the whole grid shows that the colors are nearly uniformly distributed. However, single colors dominate in the sub-grids.

We utilized this observation of the solution for the grid $G_{15,15}$ for the construction of a solution of the grid $G_{16,16}$. Due to the 16 positions in the rows and columns of $G_{16,16}$ and the four allowed colors sub-grids $G_{4,4}$ are taken into account. From our complete solutions for small grids, we know

Fig. 10.6 Rectangle-free four-colored grid 15×15 found by the SAT-solver *clasp*

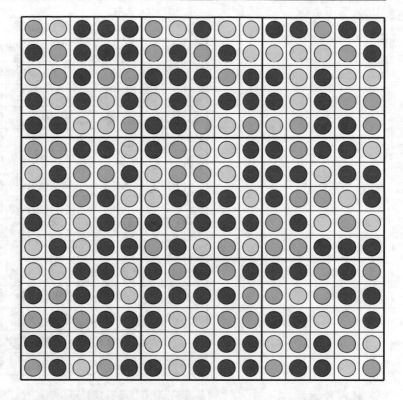

that maximal nine positions can be occupied by a single color without violation of the rectangle-free condition.

Such maximal four-colored grids $G_{4,4}$ cannot be repeated in rows or columns, but in diagonal order. In this way, a $G_{8,8}$ grid can be dominated by two colors. In such a grid $G_{8,8}$ each of the two red-dominated sub-grids $G_{4,4}$ can be extended by a single position of the color green. Vice versa, each of the two green-dominated sub-grids $G_{4,4}$ can be extended by a single position of the color red.

It remain $8*8 - 4*9 - 4*1 = 24$ positions which can be filled up with the colors blue and yellow, respectively. This assignment is realized such that

- Each of the colors blue and yellow occurs only in once in both a row and column of the red-dominated sub-grids $G_{4,4}$.
- The colors blue and yellow are exchanged in the green-dominated sub-grids $G_{4,4}$.

We assigned the grids $G_{8,8}$ dominated by the colors red and green to the top left and bottom right region of the grid $G_{16,16}$. Analog grids $G_{8,8}$ dominated by the colors blue and yellow can be built and assigned to the top right and bottom left region. Figure 10.7 shows the constructed four-colored grid $G_{16,16}$.

In the sub-grids $G_{4,4}$ of Fig. 10.7 each pair of rows and each pair of columns are covered by the dominating color. Hence, in these ranges each color is only allowed in a single position. Cyclically shifted combinations of all four colors allow the extension of the four-colored grid $G_{16,16}$ of Fig. 10.7 to the rectangle-free four-colored grid $G_{16,20}$ as shown in Fig. 10.8.

Unfortunately, it is not possible to build a rectangle-free four-colored grid $G_{17,17}$ that contains the four-colored grid $G_{16,16}$ of Fig. 10.7.

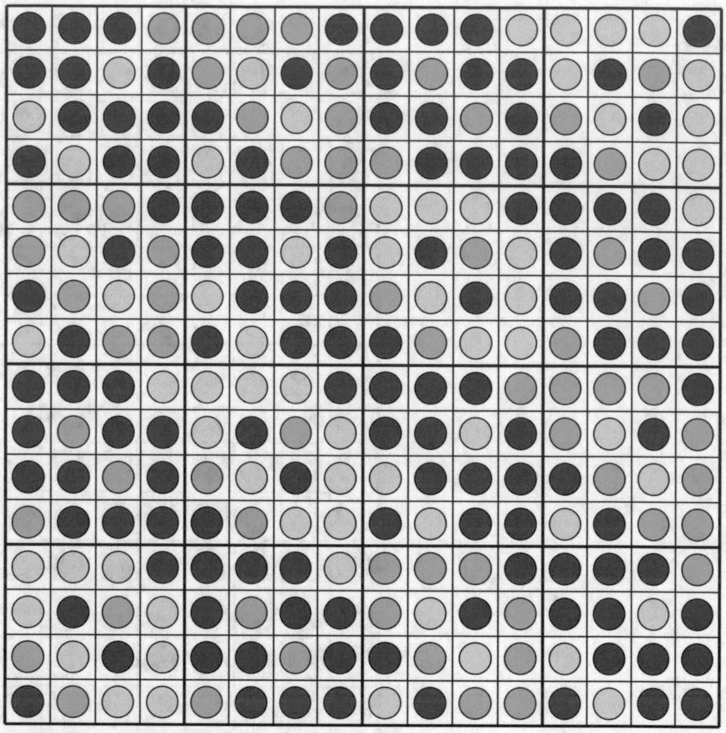

Fig. 10.7 Four-colored grid $G_{16,16}$ constructed using sub-grids $G_{4,4}$ dominated by nine assignments of the same color

Fig. 10.8 Rectangle-free four-colored grid $G_{20,16}$ constructed using maximal one-colored sub-grids $G_{4,4}$ and cyclic extensions of all four colors in the right four columns

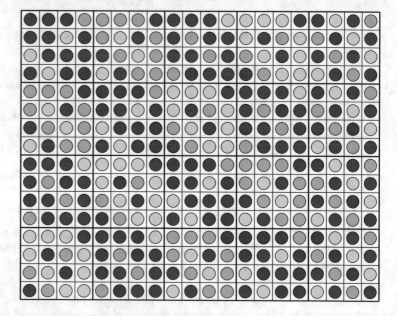

10.5   Solution of All Four-Colored Grids up to 18×18

Restriction to One Color

Due to the high complexity, we segment the solution-process. The first step is limited to one single color. Due to the *pigeonhole principle*, at least one fourth of the grid positions must be covered by the first color without contradiction to the color restrictions. When such a partial solution is known, the same fill-up step must be executed taking into account the already fixed positions of the grid. This procedure must be repeated for all four colors. The advantage of this approach is that a single Boolean variable describes whether the color is assigned to a grid position or not. Such a restriction to one half of the needed Boolean variables drastically reduces the search space from

$$2^{2*18*18} \approx 1.17 * 10^{195} \quad \text{to} \quad 2^{18*18} \approx 3,42 * 10^{97} \, .$$

We can go back to the equations using one variable to express the assignment of the selected color. For instance, the rectangle $(1,1)$, $(1,2)$, $(2,1)$, $(2,2)$ requires the constraint

$$x_{11}x_{12}x_{21}x_{22} = 0$$

that four values equal to 1 are forbidden, which can be transformed into

$$\overline{x}_{11} \vee \overline{x}_{12} \vee \overline{x}_{21} \vee \overline{x}_{22} = 1 \, .$$

We built, as done before, these constraints for all rectangles and solved the resulting system of equations using a SAT-solver. However, the disadvantage of this approach is that setting all variables to 0 satisfies all equations: there is no conflict when the selected color is not assigned to any position of the grid. However, we needed a solution with 81 values 1 which is equal to 25% of the grid 18×18.

Soft-Computing with Utilization of Permutations
We developed a quite complicated algorithm based on a *soft-computing* approach that utilizes permutation classes [6]. For the sake of space, we must exclude the details of this approach. Using XBOOLE, we implemented this algorithm that allowed us to find within few seconds solutions with maximal rectangle-free assignments of values 1 to grids even larger than 18×18. Figure 10.9 shows a rectangle-free assignment of 81 values 1 to the 324 grid positions of $G_{18,18}$ calculated using the approach presented in [6].

However, our effort to fill up the rectangle-free one-colored grid $G_{18,18}$ of Fig. 10.9 with the second color on again 81 grid positions failed. This results from the fact that the freedom for the choice of the positions is restricted by the assignments of the first color. We learned from this approach that it is not enough to know a correct coloring for one color, these assignments must not constrain the assignment of the other colors.

Cyclic Color Assignments
The fewest restrictions for the coloring of a grid by the four colors are given when the number of assignments to the grid positions is equal for all four colors. For square grids $G_{m,m}$ with an even number of rows and columns, quadruples of all grid positions can be chosen which contain all four colors.

There are several possibilities for the selection of quadruples. One of them is the cyclic rotation of a chosen grid position by 90 degrees around the center of the grid.

Figure 10.10 illustrates this possibility for the grid $G_{4,4}$. The quadruples are labeled by the letters r, s, t, and u. The attached index specifies the element of the quadruple.

In addition to the color restriction for the chosen single color we can require that this color occurs exactly once in each quadruple. This property can be expressed by two additional rules. For the corners of the grid of Fig. 10.10, for instance, we model as the first rule the requirement:

Fig. 10.9 Rectangle-free grid $G_{18,18}$ for one single color

1	1	1	0	0	0	0	0	0	1	1	0	0	0	0	0	0	0
1	0	0	1	1	0	0	0	0	0	0	0	1	1	0	0	0	0
1	0	0	0	0	1	1	0	0	0	0	0	0	1	1	0	0	0
0	1	0	1	0	0	0	1	0	0	0	0	0	1	0	1	0	0
0	1	0	0	0	1	0	0	1	0	0	1	0	0	0	0	1	0
0	0	1	0	1	0	0	0	1	0	0	0	0	0	1	1	0	0
0	0	1	0	0	0	1	1	0	0	0	0	1	0	0	0	1	0
0	0	0	1	0	0	1	0	1	1	0	0	0	0	0	0	0	1
0	0	0	0	1	1	0	1	0	0	1	0	0	0	0	0	0	1
1	0	0	0	0	0	0	0	0	0	0	0	0	0	0	1	1	1
0	1	0	0	0	0	0	0	0	0	0	0	1	0	1	0	0	1
0	0	1	0	0	0	0	0	0	0	0	1	0	1	0	0	0	1
0	0	0	1	0	0	0	0	0	0	1	0	0	0	1	0	1	0
0	0	0	0	1	0	0	0	0	1	0	0	0	1	0	0	1	0
0	0	0	0	0	1	0	0	0	1	0	0	1	0	0	1	0	0
0	0	0	0	0	0	1	0	0	0	1	1	0	0	0	1	0	0
0	0	0	0	0	0	0	1	0	1	0	1	0	0	1	0	0	0
0	0	0	0	0	0	0	0	1	0	1	0	1	1	0	0	0	0

Fig. 10.10 Cyclic quadruples for a grid $G_{4,4}$

r_1	s_1	t_1	r_2
t_4	u_1	u_2	s_2
s_4	u_4	u_3	t_2
r_4	t_3	s_3	r_3

$$r_1 \vee r_2 \vee r_3 \vee r_4 = 1 \,,$$

such that at least one variable r_i must be equal to 1. As the second rule, the additional restriction

$$(r_1 r_2) \vee (r_1 r_3) \vee (r_1 r_4) \vee (r_2 r_3) \vee (r_2 r_4) \vee (r_3 r_4) = 0$$

prohibits that more than one variable r_i is equal to 1. Hence, we get 7 additional clauses

$$(r_1 \vee r_2 \vee r_3 \vee r_4) \wedge (\overline{r}_1 \vee \overline{r}_2) \wedge (\overline{r}_1 \vee \overline{r}_3) \wedge (\overline{r}_1 \vee \overline{r}_4) \wedge (\overline{r}_2 \vee \overline{r}_3) \wedge (\overline{r}_2 \vee \overline{r}_4) \wedge (\overline{r}_3 \vee \overline{r}_4) = 1 \quad (10.8)$$

for each quadruple in the SAT-equation.

A SAT-formula has been constructed that ensures, as before, that the result is rectangle-free, including now adjusted equations (10.8) for each cyclic quadruple. The solution of such a SAT-formula for a square grid of even numbers of rows and columns must assign exactly one fourth of the variables to the value 1. Such a solution can be used rotated by 90 degrees for the second color, rotated by 180 degrees for the third color, and rotated by 270 degrees for the fourth color without any contradiction. We generated a file for this SAT-equation which depends on 324 variables; the file contains

$$\binom{18}{2} * \binom{18}{2} + \frac{18^2}{4} * 7 = 153 * 153 + \frac{324 * 7}{4} = 23,976$$

clauses for the grid $G_{18,18}$. We tried to find a solution using the SAT-solver *clasp*. The first cyclically reusable solution for the grid $G_{18,18}$ has been found after 2 days, 10 h, 58 min, and 21.503 s. Figure 10.11a shows this solution for the first color of the grid $G_{18,18}$.

Using the core solution of Fig. 10.11a we have constructed the rectangle-free four-colored grid $G_{18,18}$ of Fig. 10.11b by three times rotating around the grid center by 90 degrees each and assigning the next color [7].

Many other rectangle-free four-colored grids can be created from the solution in Fig. 10.11b by permutations of rows, columns, and colors. Several valid rectangle-free four-colored grids $G_{17,18}$ can be derived from the four-colored grid $G_{18,18}$ by removing any single row, and by removing any single column we get rectangle-free four-colored grids $G_{18,17}$. Obviously, several so far unknown rectangle-free four-colored grids $G_{17,17}$ can be selected from the four-colored grid of Fig. 10.11b by removing both any single row and any single column.

It should be mentioned that the approach of cyclically reusable assignments of a single color can be applied to rectangle-free four-colored square grids of an odd number of rows and columns, too. The central position must be colored with an arbitrary chosen color, e.g., the color red; hence, the variable of this central position must occur as single variable in a clause of the SAT-equation. Figure 10.12 shows the principle of the quadruple assignment in this case.

10.6 Solution of Four-Colored Grids with the Size 12 × 21

Analysis of the Properties
Since the grid $G_{12,21}$ can be transformed into a grid $G_{21,12}$ by turning the grid to the right by 90 degrees, only the grid $G_{12,21}$ will be explored. This grid contains $12 * 21 = 252$ nodes in 12 rows and 21 columns. Due to the four colors, $252/4 = 63$ nodes must be rectangle-freely colored with one color. This part of the solution-process needs many considerations that are borrowed from Combinatorics.

An obvious solution could be the uniform allocation to the columns. In this case $63/21 = 3$ nodes must get the same color. The use of all four colors would completely fill the column.

After the consideration of the columns also the rows must be considered. A uniform distribution of the four colors is not possible since each row has 21 elements—21 is not divisible by four with a remainder equal to 0. A relatively uniform allocation can be achieved when three colors get five and one color gets six positions ($3 * 5 + 6 = 21$). Such a uniform allocation of the colors to the rows and columns creates smaller grids with 12 rows and six columns which have one row completely filled with one color, and in each of these columns two nodes must be colored in different rows with this color. It has been proven that such a coloring cannot exist in a grid $G_{12,21}$.

These considerations lead to further conclusions. The six columns with the same color in a selected row require that two of these columns contain only two nodes of this color (see Fig. 10.13: columns four, five, and, six). In order to compensate the missing third colored node three times, it is necessary to have three columns with four nodes of the selected color in the rectangle-free four-colored grid $G_{12,21}$ (see Fig. 10.13: one, two, and, three). These three columns produce the fewest restrictions if the colorings do not overlap in the rows. The required 63 colorings with one color can only be

Fig. 10.11 Rectangle-free grid $G_{18,18}$: (**a**) for one single color that can be conflict-free reused for the other colors after a rotation of 90 degrees each, (**b**) completely merged for four colors

0	0	1	1	0	0	1	0	0	0	0	0	0	0	1	0	1	0
0	0	0	0	0	0	1	0	0	0	1	1	0	0	0	1	0	1
0	0	0	0	1	0	0	0	0	1	0	1	0	1	0	0	1	0
0	1	0	1	0	0	0	0	1	0	0	1	0	0	0	0	0	0
1	0	0	1	0	1	0	0	0	0	0	0	0	1	0	0	0	1
1	1	0	0	0	0	0	1	0	0	0	0	0	0	0	1	1	0
1	0	0	0	1	0	1	0	1	0	0	0	0	0	0	0	0	0
0	0	0	0	1	1	0	0	0	0	0	0	0	0	1	1	0	0
0	1	0	0	0	1	1	0	0	1	0	0	1	0	0	0	0	0
0	1	1	0	1	0	0	0	0	0	0	0	0	0	0	0	0	1
0	0	0	0	0	0	0	1	1	1	0	0	0	0	1	0	0	1
0	0	0	1	1	0	0	1	0	0	1	0	1	0	0	0	0	0
1	0	0	0	0	0	0	0	0	0	0	1	1	0	1	0	0	0
0	0	0	0	0	1	0	0	1	0	1	0	0	0	0	0	1	0
0	0	1	0	0	0	0	0	1	0	0	0	1	1	0	1	0	0
0	0	1	0	0	1	0	1	0	0	0	1	0	0	0	0	0	0
0	1	0	0	0	0	0	0	0	0	1	0	0	1	1	0	0	0
1	0	1	0	0	0	0	0	0	1	1	0	0	0	0	0	0	0

a

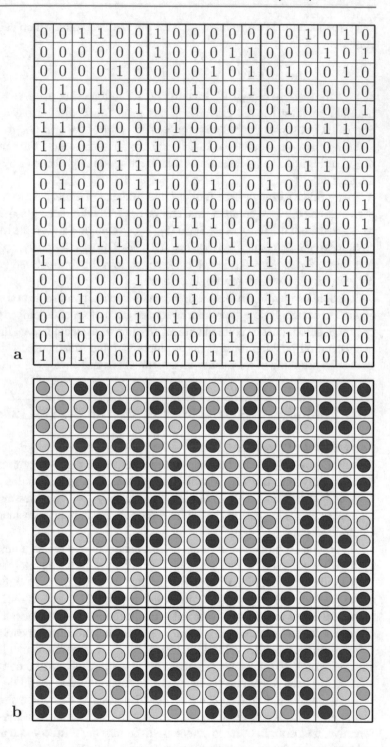

b

achieved if the color appears in three columns four times, in three columns two times, and three times in the remaining 15 columns.

By exchanging some columns, it is possible to build the stair pattern that can be seen in the upper eight rows of the columns seven to 21 in Fig. 10.13. Colorings with the patterns of several *Latin*

Fig. 10.12 Cyclic quadruple in a grid $G_{5,5}$

r_1	s_1	t_1	u_1	r_2
u_4	v_1	w_1	v_2	s_2
t_4	w_4	x_1	w_2	t_2
s_4	v_4	w_3	v_3	u_2
r_4	u_3	t_3	s_3	r_3

	1			4			7			10			14			18				
1	0	0	1	1	0	1	1	1	0	0	0	0	0	0	0	0	0	0	0	0
1	0	0	0	0	0	0	0	0	1	1	1	1	0	0	0	0	0	0	0	0
1	0	0	0	0	0	0	0	0	0	0	0	0	1	1	1	1	0	0	0	0
1	0	0	0	0	0	0	0	0	0	0	0	0	0	0	0	0	1	1	1	1
0	1	0	1	0	1	0	0	0	1	0	0	0	1	0	0	0	1	0	0	0
0	1	0	0	0	0	1	0	0	0	1	0	0	0	1	0	0	0	1	0	0
0	1	0	0	0	0	0	1	0	0	0	1	0	0	0	1	0	0	0	1	0
0	1	0	0	0	0	0	0	1	0	0	0	1	0	0	0	1	0	0	0	1
0	0	1	0	1	1	0	0	0	0	1	0	0	0	0	1	0	0	0	0	1
0	0	1	0	0	0	1	0	0	1	0	0	0	0	0	0	1	0	0	1	0
0	0	1	0	0	0	0	0	1	0	0	1	0	1	0	0	0	0	1	0	0
0	0	1	0	0	0	0	1	0	0	0	0	1	0	1	0	0	1	0	0	0

Fig. 10.13 The rectangle-free grid $G_{12,21}$ with one color

squares below the staggered arrangement facilitate the complete rectangle-free colorings of the grid for a single color.

Necessary Assignment of Colors to the Grid 12×6

The next problem is the conflict-free combination of the assignment of Fig. 10.13 for all four colors. It must be taken into account that columns with two, three, or four positions with one color must be combined to full colorings of the 12 grid elements in each column. This can be achieved only when in one column two grid elements are colored with one color and four grid elements with another color. We demonstrate the possibilities to combine the columns with two and four colors in a grid $G_{12,6}$. This construction is the key to the solution for the rectangle-free four-colored grid 12×21.

Since there are $4 * 3 = 12$ columns with two colors, two of these columns must overlap each other in the grid $G_{12,6}$. This is possible for the first three colors by means of a cyclic overlap in one column. Columns with grid elements of the fourth color are used for the columns two, four, and six. Figure 10.14a shows in the upper part the selected assignment of the two-elements-coloring to the columns and below the resulting assignments of two times two colors in the grid $G_{12,6}$ as required, without overlap in the rows.

We know from Fig. 10.13 that columns with two or four positions of the same color do not overlap. There are

$$\binom{8}{4}^6 = 70^6 = 1.117\,649 * 10^{11} \tag{10.9}$$

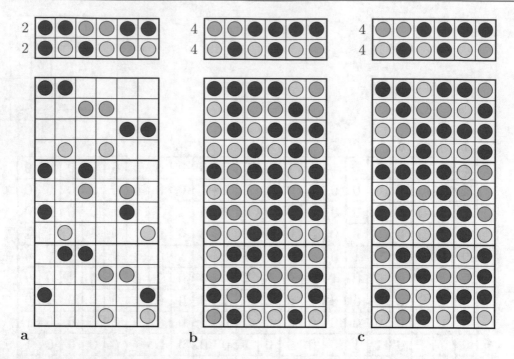

Fig. 10.14 Construction of the left six columns of the rectangle-free grid $G_{12,21}$: (**a**) entangled assignment of two colors in two positions, (**b**) first extension of (**a**) with the other two colors in four positions, (**c**) second extension of (**a**) with the other two colors in four positions

possibilities to place four nodes with one of the remaining two colors into each of the six columns of Fig. 10.14a. We showed by a long sequence of combinatorial considerations, the construction and solution of a SAT-equation that only two out of more than 10^{11} colorings are rectangle-free [8]. Other rectangle-free four-colored sub-grids $G_{12,6}$ can be generated from each of these two color patterns by permutations of rows, columns, and/or colors.

The Solution: Restricted SAT-Problem
The SAT-equation for $m = 12$ rows and $n = 21$ columns depends on $12 * 21 * 2 = 504$ variables and needs 55,440 clauses.

Even after several weeks of continuous working, the SAT-solver *clasp-2.0.0-st-win32* could not find a solution. Knowing the rectangle-free partial grids $G_{12,6}$ of Fig. 10.14b and c, we experimented in the following way:

1. The SAT-problem with a given solution for the partial grid $G_{12,6}$ needs 504 variables (360 of them are not known) and $55,440 + 144 = 55,584$ clauses; it could not be solved by a SAT-solver over several weeks.
2. From the analysis we know that each of the four colors must occur three times in the columns seven to 21; hence, we added for all $\binom{12}{4}$ forbidden combinations of positions in these 15 columns and each of the four colors an additional clause to the SAT-equation: This more precise SAT-equation consists of $55,584 + 29,700 = 85,284$ clauses and could not be solved over several weeks.
3. In this step clauses have been added that additionally determine the first color in the columns seven to 21 according to Fig. 10.13; the SAT-equation contains with this extension 85,374 clauses

and depends on 504 Boolean variables, but the free variables are restricted to $15 * 9 * 2 = 270$; the SAT-solver needed less than one second to find out that there is no solution.

What to do now? Is there still a possibility to find a rectangle-free solution? We went back to Fig. 10.13. The staggered assignment of the colors was necessary in the two upper groups of four rows. However, several possibilities exist for the lowest four rows which can be based on *Latin squares*.

Therefore, we removed the clauses for the assignment of the first color according to Fig. 10.13 for the rows eight to 12 and the columns seven to 21. This adjustment increases the number of free variables by 30 to 300, but the number of clauses decreases by 30 to 85,344. The SAT-solver *clasp-2.0.0-st-win32* found the first solution of a rectangle-free four-colored grid $G_{12,21}$ using the partial grid $G_{12,6}$ of Fig. 10.14b after 343.305 s and for the partial grid $G_{12,6}$ of Fig. 10.14c after 577.235 s.

Figure 10.15 shows the two solutions. It can be seen that the partial grids of Fig. 10.14b and c have been directly included into the solutions of Fig. 10.15.

To get an answer to the question why no solution exits for our third SAT-equation with 270 free variables and 85,374 clauses, we tried to adjust the found solution of Fig. 10.15a for the red color to the rectangle-free grid $G_{12,21}$ with one color of Fig. 10.13. The color patterns in the upper eight rows and the right 15 columns can be reconstructed by permutations of rows. The color pattern in the left six columns can be thereafter reconstructed by permutations of columns. Figure 10.16 shows the result of these permutations based on the grid of Fig. 10.15a.

The comparison of Figs. 10.13 and 10.16 reveals that different *Latin squares* in the three lower right sub-grids $G_{4,4}$ are used. This confirms our assumption that not each combination of Latin squares of one color can be completed with assignments of the other three colors to a rectangle-free completely four-colored grid $G_{12,21}$.

The turning of the rectangle-free four-colored grids of Fig. 10.15 by 90 degrees produces rectangle-free four-colored grids $G_{21,12}$.

Number of Solutions and Their Classification

The *SAT-Solver clasp-2.0.0-st-win32* facilitates the calculation of all solutions. This calculation took 453.880 s for the partial grid $G_{12,6}$ of the Fig. 10.14b and 745.700 s for the partial grid of the Fig. 10.14c. There are 38,926 different solutions for each partial grid. These solutions can be classified according to the following criteria:

1. Which of the four existing Latin squares will be used for the filling of the rows nine to 12 and the columns seven to 21 (with regard to Fig. 10.13).
2. Permutation of the three neighbored sub-grids $G_{4,4}$ in the rows nine to 12 and the columns 10 to 21.
3. Permutation of the three rows 10 to 12.

According to these three items we get $4 * 3! * 3! = 4 * 6 * 6 = 144$ permutation classes.

We assigned the 38,926 solutions for each of the two partial grids $G_{12,6}$ to these permutation classes and found out that (in both cases) 100 of these classes are free of solutions. The number of solutions for the other classes is in the range of 2 to 15,720 for each partial grid; the detailed distribution of all solutions to the remaining 44 classes is given in [8] for the two partial grids $G_{12,6}$ of Fig. 10.14b and c.

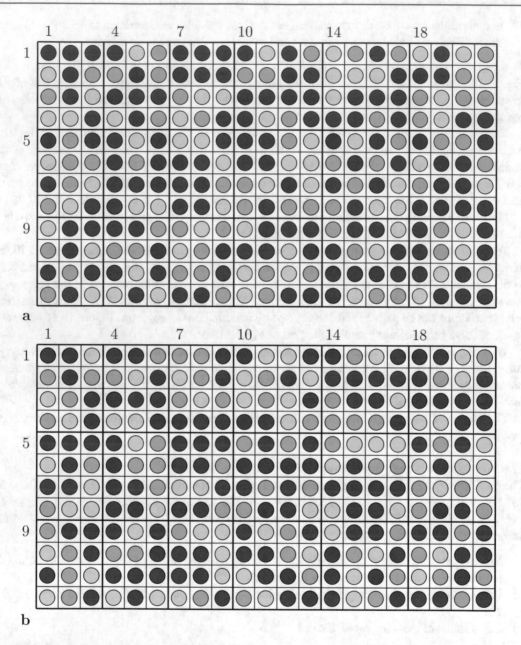

Fig. 10.15 Rectangle-free four-color grids $G_{12,21}$: (**a**) extension of the partial grid $G_{12,6}$ from Fig. 10.14b; (**b**) extension of the partial grid $G_{12,6}$ from Fig. 10.14c

10.7 Exercises

Prepare for each task of this chapter a PRP for the XBOOLE-monitor XBM 2. Prefer a lexicographic order of the variables and use the help system of the XBOOLE-monitor XBM 2 to extend your knowledge about the details of needed commands. Consider alternative solution-methods and follow

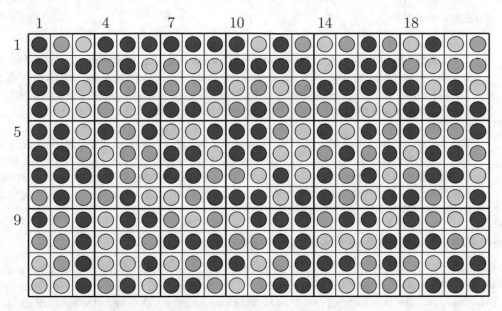

Fig. 10.16 Rectangle-free four-colored grid $G_{12,21}$ adjusted to the upper 12 rows and the left six columns of Fig. 10.13 using permutations of row and columns starting with the grid of Fig. 10.15a

the hints, given in the task, to find an efficient approach. Execute the created PRPs, verify the computed results, and explain the observed properties.

If you have not yet prepared the XBOOLE-monitor XBM 2 on your computer, you can get this XBOOLE-monitor free of charge by means of the following three steps:

1. **Download**:
   There are four versions of the XBOOLE-monitor XBM 2, two for Windows 10 or subsequent Windows systems (32 or 64 bits), and two for LINUX - Ubuntu (also 32 or 64 bits); you must download the version of the XBOOLE-monitor XBM 2 that fits your operating system. Authorized users of the online version of this chapter (https://doi.org/10.1007/978-3-030-88945-6_10) can download the XBOOLE-monitor XBM 2 directly from the web page

   https://link.springer.com/chapter/10.1007/978-3-030-88945-6_10

   where the links for the download of the XBOOLE-monitor XBM 2 are located in the part "Supplementary Information" (below the part "Abstract"). The headline above such a link indicates the associated zip-file of the XBOOLE-monitor XBM 2. The sizes of the zip-files have been provided behind the links and can be used to verify the download. A click on the link of the wanted version of the XBOOLE-monitor XBM 2 starts the download.

   Readers of the hardcopy of this book get access to the XBOOLE-monitor XBM 2 using the URL

   https://link.springer.com/chapter/10.1007/978-3-030-88945-6_10

   to download the first two pages of this chapter. After this download, the same procedure as the authorized users of the online version of a chapter can be used to download the wanted version of the XBOOLE-monitor XBM 2.

2. **Unzip**: The XBOOLE-monitor XBM 2 must not be installed, but unzipped into an arbitrary directory of your computer. A convenient tool to unzip the downloaded zip-file is usually available as part of the operating system or can be downloaded from the Internet.

3. **Execute**:

- Windows:

  The executable file of the two versions (32 or 64 bits) for Windows 10 (or subsequent Windows systems) of the XBOOLE-monitor XBM 2 is XBM2.exe; the other files in the expanded directory must remain unchanged. A double-click on the executable file XBM2.exe within the Explorer of Windows starts the XBOOLE-monitor XBM 2.

- LINUX–Ubuntu:

  The unzipped folder of the XBOOLE-monitor XBM 2 contains for this operating system only the executable file XBM2-i386.AppImage for the version of 32 bits or XBM2-x86_64.AppImage for the version of 64 bits of the XBOOLE-monitor XBM 2. A double-click on the created AppImage-file within the file manager of LINUX–Ubuntu starts the XBOOLE-monitor XBM 2.

Exercise 10.1 (Rectangle-Free Four-Colored Grids)

(a) Compute all rectangle-free four-colored grids of the size 3×3. Solve this task by a single restrictive equation, where the conjunctions of the function on the left-hand side specify the prohibited rectangles colored by one of the four colors. Use the encoding of the four colors by two Boolean variables as specified in Table 10.3. How many ternary vectors belong to the solution-TVL of this equation? Minimize this number of ternary vectors using the command obb. How many ternary vectors describe the same set of solutions in the minimized TVL?

Compute the number of different rectangle-free assignments of the four colors to the grid of the size 3×3. Utilize for this task the property that the command stv excludes the dashes in the representation of the selected ternary vector. What is the benefit of this command in the context of this task? How many different colorings with four colors exist at all for the grids of the size 3×3? What is the percentage of rectangle-free assignments on all assignments of the four colors to the grid of the size 3×3?

Note, the main command sbe and obb are executed in less than one second, but the computation of several thousand XBOOLE-objects requires about one minute to compute the number of the already known solutions; this task can be solved much faster when the functions of the XBOOLE-library are directly used.

(b) Compute all rectangle-free four-colored grids of the size 3×4. The single additional column of this grid would require a logic function in the restrictive equation that contains twice as much conjunctions as in the previous task and increases the number of variables from 18 to 24; hence, a more efficient approach should be used to solve this task.

TVL (10.7) describes the four monochromatic rectangles determined by the upper two rows and the left two columns. The body of this TVL can be reused for all other rectangles when the associated variables are assigned to the columns of this TVL. The substitution of the variables can be realized by means of commands cco controlled by the VTs determined by the rows and columns of the grid. Take into account that the result of the command cco depends on used order of the columns. Commands dif are convenient to exclude all these uniquely colored rectangles from the solution-set.

Show the generated TVLs for the 18 rectangles in the m-fold view so that the correct assignments of the variables can be checked. Show the computed solution-set in the 1-fold view. How many ternary vectors belong to this solution-TVL? We forgo the computation of the number of solutions due to the very large number of ternary vectors; this number should be computed using the functions of the XBOOLE-library.

(c) Verify whether the four-colored grid of the size 18×18, shown in Fig. 10.11b, does not contain any rectangle that is completely colored by any of the four colors.

It would be possible to determine $18 \cdot 18 \cdot 4 = 1296$ logic variables that determine whether the selected color has been assigned to the associated field of the grid and to verify the $23,409 \cdot 4 = 93,636$ monochromatic rectangles. The encoding of the four colors by two variables for each field would restrict the number of variables to $1296/2 = 648$, but requires more complicated rules for the check. Hence, a more efficient approach to solve this task is desirable.

Commands for sets of variables can be used to solve this task. Commands stv are convenient to create the required TVLs of sets of variables based on a TVL in D-form for each color where values 1 indicate the position of the assigned color and dashes are omitted in the generated TVLs. Use as basic information four TVLs in D-form for the four colors that determine the assignments of associated color of the four-colored grid of Fig. 10.11b.

Which checks are additionally necessary to verify that the TVLs defined for the four colors of the grid determine a complete unique coloring?

(d) Verify whether the four-colored grid of the size 21×12, shown in Fig. 10.16, does not contain any rectangle that is completely colored by one of the four colors. Reuse the approach of the previous task to solve this task.

10.8 Solutions

Solution 10.1 (Rectangle-Free Four-Colored Grids)

(a) Figure 10.17 shows both the PRP that computes all rectangle-free four-colored grids of the size 3×3, the last ten rows of the solution-TVL of the logic equation consisting of $ntv1=11,696$ ternary vectors, the last ten rows of the minimized solution-TVL consisting of $ntv2=5258$ ternary vectors, and the values of the computed integer variables, where $ns=228,984$ is the computed number of differently four-colored rectangle-free grids of the size 3×3.

$3 \cdot 3 \cdot 2 = 18$ variables are needed to compute all rectangle-free four-colored grids of the size 3×3 using the encoding of Table 10.3; these 18 variables are specified in the command avar to get a well ordered representation of the solution.

Each conjunction in the logic equation determines one monochromatic rectangle for one of the four colors and one of the nine rectangles. The parentheses in this restrictive equation have no influence to the computed result; but they emphasize the four restrictions belonging to one of the nine rectangles.

The command sv_size in line 68 counts the number of variables of the solution-TVL; this number is stored as integer variable $nv. The number of solutions $ns is initialized with the value 0 and computed in the outer for-loop in lines 69–83. The for-loop has been implemented in the XBOOLE-monitor XBM 2 with the feature that the user can break this loop after 1000 sweeps. This limit can be changed using the last parameter in the command for; this limit has been changed to the value 6.000 for the for-loop in line 69, due to the 5258 ternary vectors of the minimized TVL, which must be evaluated.

The command stv in line 71 selects in each sweep of the for-loop one ternary vector of the minimized solution-TVL and excludes the dashes in the created XBOOLE-object; hence, the command sv_size in line 72 provides the number of values 0 and 1 of the selected ternary vector. The command sub in line 73 computes the number of dashes of the selected ternary vector as difference of the values of all variables $nv and all values 0 and 1 of the selected ternary vector.

```
1   new
2   space 32 1
3   avar 1
4   x11a x11b x12a x12b x13a x13b
5   x21a x21b x22a x22b x23a x23b
6   x31a x31b x32a x32b x33a x33b.
7   sbe 1 1
8   (
9   /x11a&/x11b&/x12a&/x12b&
10  /x21a&/x21b&/x22a&/x22b+
11  x11a&/x11b&x12a&/x12b&x21a&/x21b&x22a&/x22b+
12  /x11a&x11b&/x12a&x12b&/x21a&x21b&/x22a&x22b+
13  x11a&x11b&x12a&x12b&x21a&x21b&x22a&x22b)
14  +(
15  /x11a&/x11b&/x13a&/x13b&
16  /x21a&/x21b&/x23a&/x23b+
17  x11a&/x11b&x13a&/x13b&x21a&/x21b&x23a&/x23b+
18  /x11a&x11b&/x13a&x13b&/x21a&x21b&/x23a&x23b+
19  x11a&x11b&x13a&x13b&x21a&x21b&x23a&x23b)
20  +(
21  /x12a&/x12b&/x13a&/x13b&
22  /x22a&/x22b&/x23a&/x23b+
23  x12a&/x12b&x13a&/x13b&x22a&/x22b&x23a&/x23b+
24  /x12a&x12b&/x13a&x13b&/x22a&x22b&/x23a&x23b+
25  x12a&x12b&x13a&x13b&x22a&x22b&x23a&x23b)
26  +(
27  /x11a&/x11b&/x12a&/x12b&
28  /x31a&/x31b&/x32a&/x32b+
29  x11a&/x11b&x12a&/x12b&x31a&/x31b&x32a&/x32b+
30  /x11a&x11b&/x12a&x12b&/x31a&x31b&/x32a&x32b+
31  x11a&x11b&x12a&x12b&x31a&x31b&x32a&x32b)
32  +(
33  /x11a&/x11b&/x13a&/x13b&
34  /x31a&/x31b&/x33a&/x33b+
35  x11a&/x11b&x13a&/x13b&x31a&/x31b&x33a&/x33b+
36  /x11a&x11b&/x13a&x13b&/x31a&x31b&/x33a&x33b+
37  x11a&x11b&x13a&x13b&x31a&x31b&x33a&x33b)
38  +(
39  /x12a&/x12b&/x13a&/x13b&
40  /x32a&/x32b&/x33a&/x33b+
41  x12a&/x12b&x13a&/x13b&x32a&/x32b&x33a&/x33b+
42  /x12a&x12b&/x13a&x13b&/x32a&x32b&/x33a&x33b+
43  x12a&x12b&x13a&x13b&x32a&x32b&x33a&x33b)
44  +(
45  /x21a&/x21b&/x22a&/x22b&
46  /x31a&/x31b&/x32a&/x32b+
47  x21a&/x21b&x22a&/x22b&x31a&/x31b&x32a&/x32b+
48  /x21a&x21b&/x22a&x22b&/x31a&x31b&/x32a&x32b+
49  x21a&x21b&x22a&x22b&x31a&x31b&x32a&x32b)
50  +(
51  /x21a&/x21b&/x23a&/x23b&
52  /x31a&/x31b&/x33a&/x33b+
53  x21a&/x21b&x23a&/x23b&x31a&/x31b&x33a&/x33b+
54  /x21a&x21b&/x23a&x23b&/x31a&x31b&/x33a&x33b+
55  x21a&x21b&x23a&x23b&x31a&x31b&x33a&x33b)
56  +(
57  /x22a&/x22b&/x23a&/x23b&
58  /x32a&/x32b&/x33a&/x33b+
59  x22a&/x22b&x23a&/x23b&x32a&/x32b&x33a&/x33b+
60  /x22a&x22b&/x23a&x23b&/x32a&x32b&/x33a&x33b+
61  x22a&x22b&x23a&x23b&x32a&x32b&x33a&x33b)=0.
62  assign 1 /m 1 1
63  set $ntv1 (ntv 1)
64  obb 1 2
65  assign 2 /m 2 1
66  set $ntv2 (ntv 2)
67  set $ns 0
68  set $nv (sv_size 1)
69  for $i 1 (ntv 2) 6000
70  (
71  stv 2 $i 3
72  set $n01 (sv_size 3)
73  set $nd (sub $nv $n01)
74  set $nbv 1
75  if (gt $nd 0)
76  (
77  for $j 1 $nd
78  (
79  set $nbv (mul $nbv 2)
80  )
81  )
82  set $ns (add $ns $nbv)
83  )
```

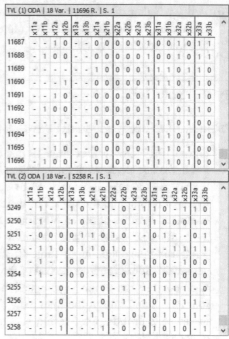

Integer variables	
Name	Value
$i	5258
$j	9
$n01	9
$nbv	512
$nd	9
$ns	228984
$ntv1	11696
$ntv2	5258
$nv	18

Fig. 10.17 Problem-program that computes all rectangle-free four-colored grids of the size 3 × 3 together with the computed results

1	new	33	x24a x24b.
2	space 32 1	34	vtin 1 31
3	avar 1	35	x31a x31b.
4	x11a x11b x12a x12b	36	vtin 1 32
5	x13a x13b x14a x14b	37	x32a x32b.
6	x21a x21b x22a x22b	38	vtin 1 33
7	x23a x23b x24a x24b	39	x33a x33b.
8	x31a x31b x32a x32b	40	vtin 1 34
9	x33a x33b x34a x34b.	41	x34a x34b.
10	ctin 1 1 /1	42	set $i 101
11	tin 1 101	43	for $r1 1 2
12	x11a x11b x12a x12b	44	(
13	x21a x21b x22a x22b.	45	for $r2 (add $r1 1) 3
14	00000000	46	(
15	10101010	47	for $c1 1 3
16	01010101	48	(
17	11111111.	49	for $c2 (add $c1 1) 4
18	vtin 1 11	50	(
19	x11a x11b.	51	cco 101 22 (add $c2 (mul $r2 10)) $i
20	vtin 1 12	52	cco $i 12 (add $c2 (mul $r1 10)) $i
21	x12a x12b.	53	cco $i 21 (add $c1 (mul $r2 10)) $i
22	vtin 1 13	54	cco $i 11 (add $c1 (mul $r1 10)) $i
23	x13a x13b.	55	dif 1 $i 1
24	vtin 1 14	56	set $rmfv (add (mod (sub $i 101) 6) 1)
25	x14a x14b.	57	set $cmfv (add (div (sub $i 101) 6) 1)
26	vtin 1 21	58	assign $i /m $rmfv $cmfv
27	x21a x21b.	59	set $i (add $i 1)
28	vtin 1 22	60	)
29	x22a x22b.	61	)
30	vtin 1 23	62	)
31	x23a x23b.	63	)
32	vtin 1 24	64	assign 1 /1

Fig. 10.18 Problem-program that computes all rectangle-free four-colored grids of the size 3×4

The inner for-loop in lines 77–80 computes the value $nbv as 2 to the power of the number of dashes using $nd. The command if in line 75 avoids the execution of the inner for-loop in the case that the selected ternary vector does not contain any dash. The command add in line 82 computes the sum of the binary vectors of the previous rows and the ternary vector evaluated in the recent sweep.

There are $2^{18} = 262,144$ different four-colored grids of the size 3×3 and $ns=228,984$ of these colorings are rectangle-free, which is equal to 87.35%.

(b) Figure 10.18 shows the PRP that generates 18 TVLs of all monochromatic rectangles of four colors of the grid of the size 3×4 and removes these rectangles from the set of all possible colorings of four colors of this grid using commands dif.

Figure 10.19 shows the generated 18 TVLs used to solve this task.

The extension of the grid of the size 3×3 by one column to the size 3×4 doubles the number of conjunctions; hence, it is convenient to generate TVLs that represent these conjunctions and compute the wanted solution-set by excluding all these monochromatic rectangles from the set of all possible colorings of four colors.

A second useful property is that the four conjunctions have the same structure for all rectangles, but are specified by different variables; hence, commands cco can be used to adjust a single

TVL (101) ODA | 8 Var. | 4 R. | S. 1

K	x11a	x11b	x12a	x12b	x21a	x21b	x22a	x22b
1	0	0	0	0	0	0	0	0
2	1	0	1	0	1	0	1	0
3	0	1	0	1	0	1	0	1
4	1	1	1	1	1	1	1	1

TVL (102) ODA | 8 Var. | 4 R. | S. 1

K	x11a	x11b	x13a	x13b	x21a	x21b	x23a	x23b
1	0	0	0	0	0	0	0	0
2	1	0	1	0	1	0	1	0
3	0	1	0	1	0	1	0	1
4	1	1	1	1	1	1	1	1

TVL (103) ODA | 8 Var. | 4 R. | S. 1

K	x11a	x11b	x14a	x14b	x21a	x21b	x24a	x24b
1	0	0	0	0	0	0	0	0
2	1	0	1	0	1	0	1	0
3	0	1	0	1	0	1	0	1
4	1	1	1	1	1	1	1	1

TVL (104) ODA | 8 Var. | 4 R. | S. 1

K	x12a	x12b	x13a	x13b	x22a	x22b	x23a	x23b
1	0	0	0	0	0	0	0	0
2	1	0	1	0	1	0	1	0
3	0	1	0	1	0	1	0	1
4	1	1	1	1	1	1	1	1

TVL (105) ODA | 8 Var. | 4 R. | S. 1

K	x12a	x12b	x14a	x14b	x22a	x22b	x24a	x24b
1	0	0	0	0	0	0	0	0
2	1	0	1	0	1	0	1	0
3	0	1	0	1	0	1	0	1
4	1	1	1	1	1	1	1	1

TVL (106) ODA | 8 Var. | 4 R. | S. 1

K	x13a	x13b	x14a	x14b	x23a	x23b	x24a	x24b
1	0	0	0	0	0	0	0	0
2	1	0	1	0	1	0	1	0
3	0	1	0	1	0	1	0	1
4	1	1	1	1	1	1	1	1

TVL (107) ODA | 8 Var. | 4 R. | S. 1

K	x11a	x11b	x12a	x12b	x31a	x31b	x32a	x32b
1	0	0	0	0	0	0	0	0
2	1	0	1	0	1	0	1	0
3	0	1	0	1	0	1	0	1
4	1	1	1	1	1	1	1	1

TVL (108) ODA | 8 Var. | 4 R. | S. 1

K	x11a	x11b	x13a	x13b	x31a	x31b	x33a	x33b
1	0	0	0	0	0	0	0	0
2	1	0	1	0	1	0	1	0
3	0	1	0	1	0	1	0	1
4	1	1	1	1	1	1	1	1

TVL (109) ODA | 8 Var. | 4 R. | S. 1

K	x11a	x11b	x14a	x14b	x31a	x31b	x34a	x34b
1	0	0	0	0	0	0	0	0
2	1	0	1	0	1	0	1	0
3	0	1	0	1	0	1	0	1
4	1	1	1	1	1	1	1	1

TVL (110) ODA | 8 Var. | 4 R. | S. 1

K	x12a	x12b	x13a	x13b	x32a	x32b	x33a	x33b
1	0	0	0	0	0	0	0	0
2	1	0	1	0	1	0	1	0
3	0	1	0	1	0	1	0	1
4	1	1	1	1	1	1	1	1

TVL (111) ODA | 8 Var. | 4 R. | S. 1

K	x12a	x12b	x14a	x14b	x32a	x32b	x34a	x34b
1	0	0	0	0	0	0	0	0
2	1	0	1	0	1	0	1	0
3	0	1	0	1	0	1	0	1
4	1	1	1	1	1	1	1	1

TVL (112) ODA | 8 Var. | 4 R. | S. 1

K	x13a	x13b	x14a	x14b	x33a	x33b	x34a	x34b
1	0	0	0	0	0	0	0	0
2	1	0	1	0	1	0	1	0
3	0	1	0	1	0	1	0	1
4	1	1	1	1	1	1	1	1

TVL (113) ODA | 8 Var. | 4 R. | S. 1

K	x21a	x21b	x22a	x22b	x31a	x31b	x32a	x32b
1	0	0	0	0	0	0	0	0
2	1	0	1	0	1	0	1	0
3	0	1	0	1	0	1	0	1
4	1	1	1	1	1	1	1	1

TVL (114) ODA | 8 Var. | 4 R. | S. 1

K	x21a	x21b	x23a	x23b	x31a	x31b	x33a	x33b
1	0	0	0	0	0	0	0	0
2	1	0	1	0	1	0	1	0
3	0	1	0	1	0	1	0	1
4	1	1	1	1	1	1	1	1

TVL (115) ODA | 8 Var. | 4 R. | S. 1

K	x21a	x21b	x24a	x24b	x31a	x31b	x34a	x34b
1	0	0	0	0	0	0	0	0
2	1	0	1	0	1	0	1	0
3	0	1	0	1	0	1	0	1
4	1	1	1	1	1	1	1	1

TVL (116) ODA | 8 Var. | 4 R. | S. 1

K	x22a	x22b	x23a	x23b	x32a	x32b	x33a	x33b
1	0	0	0	0	0	0	0	0
2	1	0	1	0	1	0	1	0
3	0	1	0	1	0	1	0	1
4	1	1	1	1	1	1	1	1

TVL (117) ODA | 8 Var. | 4 R. | S. 1

K	x22a	x22b	x24a	x24b	x32a	x32b	x34a	x34b
1	0	0	0	0	0	0	0	0
2	1	0	1	0	1	0	1	0
3	0	1	0	1	0	1	0	1
4	1	1	1	1	1	1	1	1

TVL (118) ODA | 8 Var. | 4 R. | S. 1

K	x23a	x23b	x24a	x24b	x33a	x33b	x34a	x34b
1	0	0	0	0	0	0	0	0
2	1	0	1	0	1	0	1	0
3	0	1	0	1	0	1	0	1
4	1	1	1	1	1	1	1	1

Fig. 10.19 Generated TVLs of all monochromatic rectangles of four colors of the grid 3×4

prepared TVL that determines the four monochromatic rectangles for one pair of rows and one pair of columns to the other rectangles. Equation (10.7) specifies this TVL for one rectangle and lines 11–17 define this TVL in the PRP of Fig. 10.18 for the upper two rows and the left two columns.

TVL (1) ODA	24 Var.	342816 R.	S. 1																					
	x11a	x11b	x12a	x12b	x13a	x13b	x14a	x14b	x21a	x21b	x22a	x22b	x23a	x23b	x24a	x24b	x31a	x31b	x32a	x32b	x33a	x33b	x34a	x34b
342812	-	0	1	1	-	0	0	1	1	1	1	1	1	1	1	1	-	0	-	0	1	1	-	0
342813	-	0	1	1	-	0	0	1	1	1	1	1	1	1	1	1	-	0	0	1	1	1	-	0
342814	-	0	-	0	-	0	1	1	1	1	1	1	1	1	1	1	-	0	0	1	1	1	-	0
342815	0	1	-	0	-	0	1	1	1	1	1	1	1	1	1	1	-	0	-	0	1	1	-	0
342816	0	1	-	0	-	0	1	1	1	1	1	1	1	1	1	1	-	0	0	1	1	1	-	0

Fig. 10.20 Last five of 342,816 ternary vector of the computed TVL that describes all rectangle-free four-colored grids of the size 3×4

The VTs in lines 18–41 define the variables that encode the color of row r and column c using the index $i = r \cdot 10^1 + c \cdot 10^0$. These indices are both well understandable by human beings and easy to compute in a PRP. The VTs are used in the commands cco in lines 51–54 to adjust the TVL of monochromatic rectangles to the needed pair of rows and pair of columns.

The commands cco adjust first the variables of the field (r_2, c_2) followed by the fields (r_2, c_1), (r_1, c_2), and (r_1, c_1) so that no destination field coincides with an unused source field. The main operation is the command dif in line 55 that excludes specified monochromatic rectangles from the solution-set. The adjusted TVL is needed only in this command and could be removed immediately; the commands in lines 56–59 are inserted to store and to assign the generated 18 intermediate TVLs to the viewports of the m-fold view (see Fig. 10.19) to enable a check after running the PRP of Fig. 10.18.

The TVL of the solution-set of all rectangle-free four-colored grids of the size 3×4 is shown in the 1-fold view. This TVL depends on 24 variables and consists of 342,816 ternary vectors. Figure 10.20 shows only the last part of the computed solution-set due to this large number of ternary vectors. The generation of the 18 intermediate TVLs and the computation of the 342,816 ternary vector of the solution-set have been executed in the XBOOLE-monitor XBM 2 in less than one second.

(c) Figure 10.21 shows the first part of the PRP that verifies whether the four-colored grid of the size 18×18, shown in Fig. 10.11b, does not contain any rectangle colored completely by one of the four colors. Values 1 in the four TVLs, specified in lines 4–97, determine that the color attached to the TVL has been assigned to the associated field. The colors have been assigned to the TVLs as follows:

- TVL 19: red.
- TVL 39: green.
- TVL 59: blue.
- TVL 79: yellow.

The numbers of these TVL have been chosen such that the 18 TVLs are usable for the TVLs that indicate the sets of variables of the 18 rows of the grid. The assignment of these four TVLs to the viewports of the 4-fold view facilitates the comparison of these TVLs after the execution of this PRP in the XBOOLE-monitor XBM 2 with the four-colored grid of Fig. 10.11b.

Fig. 10.21 First part of the problem-program that verifies a four-colored grid of the size 18×18

```
 1  new
 2  space 32 1
 3  ; red
 4  tin 1 19 /d
 5  c1 c2 c3 c4 c5 c6
 6  c7 c8 c9 c10 c11 c12
 7  c13 c14 c15 c16 c17 c18.
 8  --11--1--------1-1-
 9  ------1---11---1-1
10  ----1----1-1-1--1-
11  -1-1----1--1-----
12  1--1-1--------1---1
13  11-----1--------11-
14  1---1-1-1---------
15  ----11---------11--
16  -1---11--1--1-----
17  -11-1------------1
18  -------111----1--1
19  ---11--1--1-1-----
20  1-----------11-1---
21  -----1--1-1------1-
22  --1-----1---11-1--
23  --1--1-1--1-------
24  -1--------1--11---
25  1-1------11--------.
26  assign 19 /4 1 1
27  ; green
28  tin 1 39 /d
29  c1 c2 c3 c4 c5 c6
30  c7 c8 c9 c10 c11 c12
31  c13 c14 c15 c16 c17 c18.
32  1----1-----111----
33  -1-------11--1-1---
34  1-11----1---------1
35  ------1-------11--1
36  ------1-1-11----1--
37  --1-1----11--1----
38  ----------1-1----11
39  --1---11----1-----
40  ---11--1---1--1--
41  1-------1-1------1-
42  11--1-1----------1-
43  --1--1--------111-
44  ---1-11--1--------
45  -1-1----------1-1--
46  -1---1-1--1------1
47  ---1------1-1---1-
48  ----1--------1--1-1
49  -------11-----1--1-.
50  assign 39 /4 1 2
```

```
51  ; blue
52  tin 1 59 /d
53  c1 c2 c3 c4 c5 c6
54  c7 c8 c9 c10 c11 c12
55  c13 c14 c15 c16 c17 c18.
56  -------11-------1-1
57  ---11--1---------1-
58  ------1---1-1--1--
59  --1-11---1------1--
60  -1-----1-1--1-----
61  ---1-11-----------1
62  -----1-1--1--11---
63  1--1----111-------
64  1------------1-11-
65  -----1--1--11--1-
66  --11--------11----
67  ----------1-1-1---1
68  -11--------1-----11
69  1---1-------1-1--1
70  -------1--1-----1-1-
71  -1--1-1-1-----1----
72  1-1---11---1------
73  -1-1--------1--11--.
74  assign 59 /4 2 1
75  ; yellow
76  tin 1 79 /d
77  c1 c2 c3 c4 c5 c6
78  c7 c8 c9 c10 c11 c12
79  c13 c14 c15 c16 c17 c18.
80  -1--1----11-------
81  1-1--1--------1----
82  -1---1-1-------1---
83  1------1--1-1-1---1-
84  --1-1----------1-1-
85  --------1--11-1--
86  -111--------1--1--
87  -1----------1-1--11
88  --1------1-1-------1
89  ---1--1--1---1--11---
90  -----1-----11---1--
91  11----1-1---------
92  ----1--11-----1-1--
93  --1---11-1-1-------
94  1--11------1------
95  1--------1----11-1
96  ---1-1-1--11-------1-
97  ----111-----1----1.
98  assign 79 /4 2 2
```

Figure 10.22 shows the second part of the PRP that verifies whether the four-colored grid of the size 18×18, shown in Fig. 10.11b, does not contain any rectangle colored completely by one of the four colors.

The utilization of XBOOLE-operations for sets of variables requires that the 18 rows of the given four TVLs are transformed into separate TVLs; each of the created TVLs represents the set of column-variables c_i which are indicated by values 1 in the associated row of one of the given TVLs. The command stv in line 107 solves this task controlled by the two surrounded for-loops in lines 100–108. The indices 1 to 18 are used to represent the 18 sets of column-variables c_i of red colors determined in the rows of TVL 19. The offset of the indices of TVLs of the sets of column-variables c_i of the green (blue, or yellow) colors is equal to 20 (40, or 60).

```
 99   ; split                      127   (
100   for $row 1 18                 128   for $color1 1 3
101   (                             129   (
102   for $color 1 4                130   set $c1m1 (sub $color1 1)
103   (                             131   set $svcr1 (add $row (mul $c1m1 20))
104   set $tc (sub (mul $color 20) 1)  132   for $color2 (add $color1 1) 4
105   set $cm1 (sub $color 1)       133   (
106   set $svcr (add $row (mul $cm1 20))  134   set $c2m1 (sub $color2 1)
107   stv $tc $row $svcr            135   set $svcr2 (add $row (mul $c2m1 20))
108   ))                            136   sv_isc $svcr1 $svcr2 100
109   ; cover                       137   if (gt (sv_size 100) 0)
110   set $complete true            138   (
111   for $row 1 18                 139   set $sc false
112   (                             140   ))))
113   ctin 1 100 /0                 141   ; rectangle free
114   for $color 1 4                142   set $rf true
115   (                             143   for $color 1 4
116   set $cm1 (sub $color 1)       144   (
117   set $svcr (add $row (mul $cm1 20))  145   set $ci (mul (sub $color 1) 20)
118   sv_uni 100 $svcr 100          146   for $row1 (add $ci 1) (add $ci 17)
119   )                             147   (
120   if (ne (sv_size 100) 18)      148   for $row2 (add $row1 1) (add $ci 18)
121   (                             149   (
122   set $complete false           150   sv_isc $row1 $row2 100
123   ))                            151   if (gt (sv_size 100) 1)
124   ; single color                152   (
125   set $sc true                  153   set $rffalse
126   for $row 1 18                 154   ))))
```

Fig. 10.22 Second part of the problem-program that verifies a four-colored grid of the size 18×18

Firstly, it must be checked whether the given four TVLs determine an assignment of one color to each field of the grid 18×18. This check for a *cover* is realized in lines 110–123. The outer for-loop in this part iterates over the 18 sets of column-variables belonging to the 18 rows and the inner for-loop over the four colors. The command sv_uni in line 118 computes the union of the sets of column-variables of the four colors for selected row. The number of variables of such a computed set must be equal to 18 (due to the 18 columns). The command sv_size in line 120 computes this number of variables in this set and the Boolean variable $complete (that have been initialized with the value true) is changed to the value false if not all 18 column-variables belong to the computed set of the selected row. The final value $complete=true confirms that at least one color has been assigned to each field in the four given TVLs.

The second necessary condition is that each field is colored by not more than one color. This check whether only a *single color* has been assigned to each field is realized in lines 125–140. The outer for-loop in this part iterates again over the 18 sets of column-variables belonging to the 18 rows. The inner two nested for-loops iterates over the $\binom{4}{2} = 6$ pairs of different colors. The command sv_isc in line 136 computes for the chosen case a set of column-variables of different colors belonging to the same field. These sets must be empty; this property is checked by the commands in line 137. The Boolean variable $sc has been initialized with the value true in line 125; this value is changed to false in line 139 if more than one color has been assigned to one field. The final value $sc=true confirms that no field with more than one assigned color exists in the four given TVLs. The successful check of these two rules confirms that the four given TVLs determine a disjoint cover of the four-colored grid of the size 18×18.

It remains the check whether this grid does not contain any rectangle that is completely colored by one of the four colors. Due to the prepared 4×18 sets of column-variables only $4 \cdot \binom{18}{2} = 612$ pairs of these sets must be evaluated. A monochromatic rectangle exists when the intersection of two

```
 1  new
 2  space 32 1
 3  ; red
 4  tin 1 19 /d
 5  c1 c2 c3 c4 c5 c6 c7 c8 c9
 6  c10 c11 c12 c13 c14 c15
 7  c16 c17 c18 c19 c20 c21 .
 8  1--11-111--------------
 9  1---------1111---------
10  1--------------1111----
11  1-----------------1111
12  -1-1-1---1---1---1---
13  -1-----1---1---1---1--
14  -1-----1---1---1---1-
15  -1-------1---1---1---1
16  --1-11----1----1-----1
17  --1----1----1------11---
18  --1-----11----1----1-
19  --1----1----11----1--.
20  assign 19 /4 1 1
21  ; green
22  tin 1 39 /d
23  c1 c2 c3 c4 c5 c6 c7 c8 c9
24  c10 c11 c12 c13 c14 c15
25  c16 c17 c18 c19 c20 c21 .
26  -1-----------1-1-1---1
27  ---1--1-----------1-11
28  ----1-11--1-1---------
29  ---1------1-111-------
30  ----1------1-----1-11-
31  --1-11--------1-1----1
32  ----1---11----1--1---
33  1-11---1-------1---1--
34  -1----1-1----1----1--
35  11---1---11---------1-
36  -1-----1---1---1-1---
37  ---1----1-1----11----.
38  assign 39 /4 1 2

39  ; blue
40  tin 1 59 /d
41  c1 c2 c3 c4 c5 c6 c7 c8 c9
42  c10 c11 c12 c13 c14 c15
43  c16 c17 c18 c19 c20 c21 .
44  -----1---1-1---1--1--
45  -11-1----------111----
46  -1-1-1--1---------1-1-
47  -----111--1---1------
48  1--------1-1----1----1
49  1------1-1-------1--1-
50  1-11--1------1----1--
51  ----1----11--1---1---
52  1----------11-1--1---
53  ----1--11---1------1--
54  -----1------11--1---1
55  ----1-1----1-------11.
56  assign 59 /4 2 1
57  ; yellow
58  tin 1 79 /d
59  c1 c2 c3 c4 c5 c6 c7 c8 c9
60  c10 c11 c12 c13 c14 c15
61  c16 c17 c18 c19 c20 c21 .
62  --1--------1--1---1-1-
63  -----1-11----1----1--
64  --1------1-1-------1-1
65  -11-1---1-------11----
66  --1---11----1-1------
67  ---1----1--11----1---
68  -----1----1-1---1---1
69  -----11----1---1---1-
70  ---1---1-1-------1--1-
71  ---1----------111----1
72  1--11-1---1--------1--
73  11---1---1----1--1---.
74  assign 79 /4 2 2
```

Fig. 10.23 First part of the problem-program that verifies a four-colored grid of the size 12×21

of these sets of the same color for different rows contain more than one column-variables; hence, the command sv_isc in line 150 computes again the set to evaluate, but here a single column-variable of the same color is permitted and more than one column-variables in the set computed in line 150 indicate a monochromatic rectangle. The commands in line 151 verify whether the analyzed sets of two rows of one color contain a monochromatic rectangle. The Boolean variable $rf has been initialized with the value true in line 142; this value is changed to false in line 153 if a monochromatic rectangle has been detected. The final value $rf=true confirms that no monochromatic rectangle exists in the four-colored grid of the size 18×18 determined by the four given TVLs.

```
75   ; split                                    103  (
76   for $row 1 12                              104  for $color1 1 3
77   (                                          105  (
78   for $color 1 4                             106  set $c1m1 (sub $color1 1)
79   (                                          107  set $svcr1 (add $row (mul $c1m1 20))
80   set $tc (sub (mul $color 20) 1)            108  for $color2 (add $color1 1) 4
81   set $cm1 (sub $color 1)                    109  (
82   set $svcr (add $row (mul $cm1 20))         110  set $c2m1 (sub $color2 1)
83   stv $tc $row $svcr                         111  set $svcr2 (add $row (mul $c2m1 20))
84   ))                                         112  sv_isc $svcr1 $svcr2 100
85   ; cover                                    113  if (gt (sv_size 100) 0)
86   set $complete true                         114  (
87   for $row 1 12                              115  set $sc false
88   (                                          116  ))))
89   ctin 1 100 /0                              117  ; rectangle free
90   for $color 1 4                             118  set $rf true
91   (                                          119  for $color 1 4
92   set $cm1 (sub $color 1)                    120  (
93   set $svcr (add $row (mul $cm1 20))         121  set $ci (mul (sub $color 1) 20)
94   sv_uni 100 $svcr 100                       122  for $row1 (add $ci 1) (add $ci 11)
95   )                                          123  (
96   if (ne (sv_size 100) 21)                   124  for $row2 (add $row1 1) (add $ci 12)
97   (                                          125  (
98   set $complete false                        126  sv_isc $row1 $row2 100
99   ))                                         127  if (gt (sv_size 100) 1)
100  ; single color                             128  (
101  set $sc true                               129  set $rf false
102  for $row 1 12                              130  ))))
```

Fig. 10.24 Second part of the problem-program that verifies a four-colored grid of the size 12×21

Boolean variables	
Name	Value
$complete	true
$rf	true
$sc	true

Fig. 10.25 Results of the verification whether the four-colored grid of the size 12×21 shown in Fig. 10.16 contain any rectangle that is completely colored by one of the four colors

(d) Figure 10.23 shows the first part of the PRP that verifies whether the four-colored grid of the size 12×21, shown in Fig. 10.16, does not contain any rectangle that is completely colored by one of the four colors.

Figure 10.24 shows the second part of the PRP that verifies whether the four-colored grid of the size 12×21 specified in the first part of this PRP (see Fig. 10.23) does not contain any rectangle that is completely colored by one of the four colors.

The solution of this task uses the approach that has been explained in detail for the previous task for the gird 18×18. The four TVLs of the four colors are specified in Fig. 10.23 based on the requested grid 12×21 of Fig. 10.16. In the second part of this PRP (see Fig. 10.24) the number of columns has been changed from 18 to 21 and the number of rows from 18 to 12.

The three computed variables of Fig. 10.25 confirm that the four-colored grid of the size 12×21, shown in Fig. 10.16, does not contain any rectangle that is completely colored by one of the four colors.

10.9   Supplementary Exercises

Exercise 10.2 (Tasks Needing Many Logic Variables for Their Solution)

(a) The safe in a hotel room can be opened by a key number consisting of four decimal digits. It is known that the last digit of this key is equal to nine and that the result of a multiplication of this key number by a single decimal digit that is larger than one results in a decimal number of also four digits with the property that the same digits as in the key number appear in reverse order. Suffice these propositions for a unique reconstruction of key number? If yes, compute the key number using the BCD code (Binary-Coded Decimal) and verify your result. How many logic variables describe this problem? Show the TVL of the computed solution that contains all used logic variables. Convert the found solution back to the decimal representation of the key number. Figure 10.26 shows a schema of the multiplication where the decimal digits a, b, c, d of the key number and the unknown factor f are encoded by four binary variables each. The result of each multiplication of two decimal digits requires again two decimal digits, encoded by variables x_{ij} in Fig. 10.26. The result of the addition of two decimal digits and a binary carry c_{ai} can be represented by one decimal digit and a carry for the next higher decimal value.

Here are some hints to solve this task. The following restrictions follow from the propositions:

- Due to $d = 9$: $d_0 = d_3 = 1$ and $d_1 = d_2 = 0$.
- Due to four digits in the result: $c_{a4} = e_0 = e_1 = e_2 = e_3 = 0$.
- Due to $f > 1$: $f_1 \vee f_2 \vee f_3 = 1$.

A TVL in D-form of the 10 binary encoded values of one digit is useful for both encoding and decoding of a decimal digit. A TVL that describes the multiplication of two BCD digits can be generated and used to compute the elementary multiplications. Similarly, a TVL that describes the addition of two BCD digits and a carry can be generated and used to compute the sum of two BCD digits x_i. Commands cco and VTs of the variables of single BCD digits can be used to map and reuse these TVLs for the needed partial operations. The peculiarities of the addition of the lowest and highest decimal digit can be directly expressed by systems of logic equations.

- due to four digits in the result: $c_{a4} = e_0 = e_1 = e_2 = e_3 = 0$; and
- due to $f > 1$: $f_1 \vee f_2 \vee f_3 = 1$.

f_3	f_2	f_1	f_0	$*$	a_3	a_2	a_1	a_0	b_3	b_2	b_1	b_0	c_3	c_2	c_1	c_0	d_3	d_2	d_1	d_0
													x_{23}	x_{22}	x_{21}	x_{20}	x_{13}	x_{12}	x_{11}	x_{10}
									x_{43}	x_{42}	x_{41}	x_{40}	x_{33}	x_{32}	x_{31}	x_{30}				
					x_{63}	x_{62}	x_{61}	x_{60}	x_{53}	x_{52}	x_{51}	x_{50}								
	x_{83}	x_{82}	x_{81}	x_{80}	x_{73}	x_{72}	x_{71}	x_{70}												
						c_{a4}					c_{a3}				c_{a2}				c_{a1}	
	e_3	e_2	e_1	e_0	d_3	d_2	d_1	d_0	c_3	c_2	c_1	c_0	b_3	b_2	b_1	b_0	a_3	a_2	a_1	a_0

Fig. 10.26 Schema that determines the computation of the mirrored key number

(b) Five men were sitting at a bar side by side. They can be distinguished by colors of their hair as well as the worn garments and trousers:

- The five different colors of their hair are blond (cb), black (ck), gray (cg), red (cr), and white (cw).
- The five different garments are a jacket (gj), a pullover (gp), a shirt (gs), a cardigan (gc), and a vest (gv).
- The five different trousers are cord trousers (tc), jeans (tj), shorts (ts), leather trousers (tl), and cotton trousers (tn).

A detective has to find out in which order these five men were sitting on the bar. Especially, he should answer the following questions:

- On which bar stool was sitting the man with gray hair?
- On which bar stool was sitting the man with the vest?
- On which bar stool was sitting the man with the shorts?

He asked the people at the bar and got the following answers:

a. The man with the shirt was not sitting next to the man with the vest.
b. The man with the red hair was sitting next to the man with gray hair.
c. The man with the leather trousers was not sitting next to the man with black hair.
d. The man with the vest was not sitting next to the man with blond hair.
e. The man with the vest wore the cotton trousers.
f. The man with the blond hair was sitting next to the man with black hair.
g. The man with the jeans was not sitting next to the man with cardigan.
h. The man with the jeans was not sitting next to the man with white hair.
i. The man with the gray hair does not wore the leather trousers.
j. The man with the jeans was sitting next to the man with the shirt.
k. The man with the leather trousers was sitting on the fourth bar stool.
l. The man with the jacket wore the cord trousers.
m. The man with the cardigan was sitting next to the man with white hair.
n. The man with the pullover was sitting next to man with shirt.
o. The man with the jacket has blond hair.
p. The man with the jacket was not sitting on the third bar stool.
q. The man with the cardigan was sitting next to the man with the vest.

We can define five logic variables for each of these properties. For instance, the man with the blond hair could be sitting on one of the five bar stools; hence, we define the logic variables cb_1, cb_2, cb_3, cb_4, and cb_5 where $cb_i = 1$ means that the man with the blond hair was sitting on the bar stool with the number i. In this way, we get $3 \cdot 5 \cdot 5 = 75$ logic variables. Knowing that the solution belongs to the

$$2^{75} = 37,778,931,862,957,161,709,568$$

possible binary vectors of 75 variables, it is necessary to utilize as much as possible restrictions for a subset of variables before additional variables are inserted into the solution-procedure. Besides the 17 propositions restrictions can be derived from the basic knowledge because each of the 15 characteristics of five men exists exactly once. All conditions of this task can be described by logic equations. It is more convenient to specify the conditions directly by TVLs and compute the global solution using intersections. Commands cco and appropriate VTs facilitate the reuse of intermediate results about basic properties.

(c) This and the next task are related to the popular game *Sudoku*. The aim of this game is to fill the missing digits 1 to 9 into the fields of a matrix of the size 9×9 such that the digits 1 to 9 occur exactly once in:

- Each row.
- Each column.
- Each sub-square of the size 3×3 defined by the rows and columns $(1, 2, 3)$, $(4, 5, 6)$, and $(7, 8, 9)$.

Predefined digits 1 to 9 in some of these fields (the key) determine that only one assignment of the digits 1 to 9 to the remaining fields satisfies the rules of Sudoku.

Logic variables x_{rcv} can be used to describe the assignments of the values to the fields of Sudoku as follows:

$$x_{rcv} = \begin{cases} 1 \Leftrightarrow \text{ the field of row } r \text{ and column } c \text{ carries the value } v\ , \\ 0 \Leftrightarrow \text{ the value of the field } (r, c) \text{ cannot be equal to } v\ . \end{cases}$$

Due to the nine rows, the nine columns, and the nine values $9 \cdot 9 \cdot 9 = 729$ such variables are needed to describe the assignments of a Sudoku. Basically, it is more convenient to solve problems of such a large number of variables directly using the XBOOLE-library. Accepting additional time caused by the wrapping software of the XBOOLE-monitor XBM 2 and the restricted possibilities to express the program sequence, it is possible to solve such large problems using the XBOOLE-monitor XBM 2.

Define a VT of the 729 variables x_{rcv} ordered by their indices to facilitate the access to the variables based on integer numbers.

A Sudoku can be solved using a CDC-SAT-formula. Generate a TVL of 729 rows and 729 columns where each ternary vector specifies an inner conjunction of this SAT-formula that describes the assignment of one of the nine digits to one field ($x_{rcv} = 1$) and all the consequences (28 negated variables specified by same value in the associated row, column, and sub-square, as well as the other values in the same field). Organize the PRP such that the values 1 occur in the main diagonal of this TVL.

Both the VT and the generated TVL can be used to solve the Sudoku of an arbitrary key. Store therefore the result of this task as sdt-file at the end of the PRP.

(d) Solve the Sudoku with the key shown in Fig. 10.27. Load the sdt-file stored in the previous task and use the defined VT and the generated TVL to solve the Sudoku with the key shown in Fig. 10.27. How many variables x_{rcv} are already determined by the key? Verify that exactly one of the values 0 or 1 has been assigned to all 729 variables x_{rcv} in the final solution. What is the largest number of ternary vectors that occurs as solution of an intersection with a DC-clause? This number indicates the largest number of partial assignments used in parallel as intermediate solution. Verify that a computed result satisfies the rules of a Sudoku.

Here are some hints to solve this task. Each DC-clause to solve the Sudoku is determined by nine ternary vectors of the TVL generated in the previous task; the first index i_{DC1} of such an interval of ternary vectors satisfies $i \mod 9 = i_{DC1}$, where i is the index of the row in the generated TVL that describes the 729 constrains of any Sudoku.

The DC-clauses can be applied in any order to compute the unique final result; however, the used order strongly influences the number of intermediate ternary vectors. The intersection of ten such clauses for fields with two unknown values would result in $2^{10} = 1024$ ternary vectors

Fig. 10.27 Key of a Sudoku that is hard to solve

	5		7					
8				4				
		4					6	5
1							8	
	6		4				1	2
				1	3			
4						7		8
9	8					2		
7		2		3				

of up to 729 variables. When the number of unknown values in these ten field is equal to five (near the middle of all nine values) then already $5^{10} = 9,765,625$ such ternary vectors would be computed. That shows that DC-clauses with a small number of unknown values should be preferred.

The assignment of a digit to one field restricts the possible values in the fields of the same row, the same column, and the associated sub-square. Due to these reduced numbers of unknown values, it is necessary to compute the numbers of unknown values for the not determined fields after each intersection with a DC-clause.

The intersection of the constrains belonging to the digits of the key results in a single ternary vector that determines already many variable $x_{rcv} = 0$ of fields (r, c) which do not belong to the key; hence, the computation of the solution should be started with the intersections of ternary vectors belonging to the key.

The DC-clause of each field must be applied only once and can be avoided for the fields of the key. A binary vector of variables f_{rc} can be used to control this helpful restriction. This binary vector can be initialized with values 1 for variables f_{rc} that indicate the key and extended by $f_{rc} = 1$ for the field (r, c) where the DC-clause has been applied. The solution-procedure terminates when all 81 variables f_{rc} of this vector are equal to 1. This binary vector can be managed in a second Boolean space of 81 variables.

References

1. J.D. Barrow, *The Constants of Nature: The Numbers That Encode the Deepest Secrets of the Universe* (Knopf Doubleday Publishing Group, 2009). ISBN:978-0-3754-2221-8
2. A. Biere, *Lingeling, Plingeling, PicoSAT and PrecoSAT at SAT Race 2010*. Tech. rep. 1. Linz, Austria, Aug. 2010
3. A. Biere et al. (eds.), *Handbook of Satisfiability*, vol. 185. Frontiers in Artificial Intelligence and Applications (IOS Press, Feb. 2009). ISBN:978-1-58603-929-5
4. M. Gebser et al., clasp: A conflict-driven answer set solver, in *International Conference on Logic Programming and Nonmonotonic Reasoning (LPNMR 2007)*, eds. by C. Baral, G. Brewka, J. Schlipf. LNCS, vol. 4483. (Springer, Berlin, Heidelberg, Germany, 2007), pp. 260–265. ISBN:978-3-540-72199-4. https://doi.org/10.1007/978-3-540-72200-7_23

5. Planck Collaboration, Planck 2013 results. I. Overview of products and scientific results. Astron. Astrophys. **571**(11), 1–44 (2014). https://doi.org/10.1051/0004-6361/201321529

6. B. Steinbach, C. Posthoff, Utilization of permutation classes for solving extremely complex 4-colorable rectangle-free grids, in *Proceedings of the IEEE 2012 International Conference on Systems and Informatics.* (ICSAI, Yantai, China, May 2012), pp. 2361–2370. ISBN:978-1-4673-0197-8. https://doi.org/10.13140/2.1.2365.8887

7. B. Steinbach, C. Posthoff, Highly complex 4-colored rectanglefree grids – Solution unsolved multiple-valued problems. J. Multiple Valued Logic Soft Comput. **MVLSC 24**(1-4), 369–404 (2014). ISSN:1542-3980

8. B. Steinbach, C. Posthoff, The last unsolved four-colored rectangle-free grid: The solution of extremely complex multiple-valued problems. J. Multiple Valued Logic Soft Comput. **MVLSC 25**(4-5), 461–490 (2015). ISSN:1542-3980

9. B. Steinbach, W. Wessely, C. Posthoff, Several approaches to parallel computing in the Boolean domain, in *First International Conference on Parallel, Distributed and Grid Computing*, eds. by P. Chaudhuri et al. (PDGC 1. Solan, H.P., India, Oct. 2010), pp. 6–11. https://doi.org/10.1109/PDGC.2010.5679611

10. B. Steinbach, C. Posthoff, W. Wessely, Approaches to shift the complexity limitations of Boolean problems, in *Computer - Aided Design of Discrete Devices - CAD DD 2010, Proceedings of the Seventh International Conference* (CAD DD 7, Minsk, Belarus, Nov. 2010), pp. 84–91. ISBN:978-985-6744-63-4

Combinational Circuits

11

Abstract

The analysis, synthesis, and test of combinational circuits is a major field of applications of logic functions and equations. We introduce models that can be used to describe either the behavior or the structure of several realizations of combinational circuits. Based on these models, we provide the methods for several analysis tasks. There are two main approaches for the synthesis of combinational circuits: covering and decomposition methods. Covering methods are widely used and suitable for circuits of a small number of variables due to technological restrictions. We give an overview about these methods and demonstrate their application by means of synthesis examples. Decomposition methods facilitate the synthesis of multilevel circuits for larger numbers of variables, but their theory is more complicated. We give an overview of different decomposition methods and explain the newest results with regard to strong, weak, and vectorial bi-decompositions for both single logic functions and lattices of logic functions. We explain new possibilities of bi-decompositions utilizing the extensions of the Boolean Differential Calculus, provided in this book. The test of combinational circuits is needed to discard circuits that do not show the expected behavior. We explain methods to calculate the needed test-pattern using both

Supplementary Information The online version of this chapter (https://doi.org/10.1007/978-3-030-88945-6_11) contains supplementary material which is available for authorized users. Please, follow the link belonging to the version of the XBOOLE-monitor XBM 2 that fits best for your operating system. This XBOOLE-monitor is needed to solve all tasks of this chapter. Instructions for starting the downloaded XBOOLE-monitor XBM 2 are given at the beginning of Section 'Exercises' in this chapter.

XBOOLE-monitor XBM 2 for Windows 10
32 bits
https://doi.org/10.1007/978-3-030-88945-6_11_MOESM1_ESM.zip (15,091 KB)

64 bits
https://doi.org/10.1007/978-3-030-88945-6_11_MOESM2_ESM.zip (14,973 KB)

XBOOLE-monitor XBM 2 for Linux Ubuntu
32 bits
https://doi.org/10.1007/978-3-030-88945-6_11_MOESM3_ESM.zip (29,522 KB)

64 bits
https://doi.org/10.1007/978-3-030-88945-6_11_MOESM4_ESM.zip (28,422 KB)

© Springer Nature Switzerland AG 2022
B. Steinbach, C. Posthoff, *Logic Functions and Equations*,
https://doi.org/10.1007/978-3-030-88945-6_11

the traditional network-model of the *sensible path* and the more powerful network-model of the *sensible point* for internal signals and internal branches. Using this new model, test-patterns for all non-redundant gate-connection in the circuit can be computed.

11.1   The Circuit Model

Combinational circuits realize logic functions by means of *switching elements*. There must be a physical carrier for the values 0 and 1. Low and high air pressure represent, for instance, these values in pneumatic circuits. If the air is substituted by liquid, again low and high pressure represent these values in hydraulic circuits. Both pneumatic and hydraulic combinational circuits are used in control systems, but mostly the *electrical voltage* is the *physical carrier* for the logic values 0 and 1. A threshold divides the maximum voltage used in electronic circuits into two intervals. Usually, the interval of lower voltage is associated to the logic value 0, and voltage values above the fixed threshold represent the logic value 1.

A *combinational circuit* consists of switching elements that realize logic operations. These elements are denoted by *gates*. The voltages at the inputs of a gate determine the voltage at its output. A wire connects the output of one gate with at least one input of another gate or with the output of the whole circuit. The output of a gate determines the voltage of the wire and controls the voltage of the connected inputs of other gates. Both the gates and the wired connections between them determine the behavior of the combinational circuit. The outputs of a given circuit structure depend therefore only on the values (expressed by voltages) at the inputs of the circuit.

Several technical elements, such as electrical contacts, electronic tubes, or transistors, can be used to switch the electrical voltage. At the beginning of the technological development, electrical contacts (relays) have been used to realize electrical combinational circuits. At present, electrical contacts are applied especially to switch very large voltages and currents. The most frequently used switching elements of combinational circuits are transistors. A very comprehensive technological research and development in the last decades made it possible to implement today more than 10^{10} transistors on a single chip.

In order to extend the understanding of how the switching processes are realized in practice, we consider the structure and the behavior of logic gates on the electrical level. There are several types of transistors. Two of them, the n-channel MOSFET and the p-channel MOSFET, are commonly used in the technology denoted by CMOS. MOSFET is the abbreviation of *Metal Oxide Semiconductor Field Effect Transistor*, and CMOS abbreviates *Complementary Metal Oxide Semiconductor*. Each transistor has three connections: *source–gate–drain*.

Note 11.1   Gate means in the context of a MOSFET the control connection of this transistor. The gate is insulated from *source* and *drain* and controls the current between them by the gate–source voltage. This voltage controls whether the drain is insulated from the source or not.

If the voltage at the gate is high, the negative charges of the n-channel transistor fill the channel between source and drain and create a temporary connection between them. Under the same conditions, the positive charges of the p-channel transistor are displaced from the channel between source and drain so that these connections are insulated from each other.

Vice versa, if the voltage at the gate is low, the negative charges of the n-channel transistor are displaced from the channel between source and drain so that now these connections are insulated

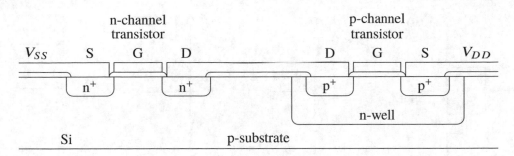

Fig. 11.1 Physical structure of a NOT-gate in CMOS-technology

from each other, and the positive charges of the p-channel transistor fill the channel between source and drain and create a temporary connection between them. Thus, MOSFETs can simply be modeled by switches. A high voltage at the gates switches the n-channel MOSFET *ON* and the p-channel MOSFET *OFF*, and, vice versa, if a low voltage controls the gates, then the n-channel MOSFET is switched *OFF* and the p-channel MOSFET *ON*, respectively. In CMOS-technology, pairs of these complementary MOSFETs are series-connected between the high circuit voltage V_{DD} and the low circuit voltage V_{SS}, and their gates are controlled by the same voltage. Figure 11.1 shows the physical realization of such a pair of transistor in a cross-section through the silicon chip.

Figure 11.2a shows the electrical structure of such a pair of transistors. If there is a constant gate voltage (low or high), one of these transistors is switched *OFF* so that no current flows. The power consumption results from the multiplication of current and voltage and is under this condition almost equal to zero. For that reason, the CMOS-technology is preferred in comparison to technologies based on one of these transistor types. However, if the gate voltage switches from low to high or vice versa, then the charge of the connection line of both transistors moves from or to the voltage lines of the circuit. Consequently, power consumption happens during the switching process.

The power consumption increasingly influences the synthesis of integrated circuits because a MOSFET can switch more than 10^9 times per second and more than 10^{10} MOSFETs can be integrated on one chip.

Figure 11.2 summarizes information of such a circuit from several points of view. Figure 11.2a shows the electrical circuit structure. The MOSFETs can be modeled by switching contacts, and the voltages by logic values. Figure 11.2b depicts the position of these contacts for the input $x = 0$. The value 1 is connected by the upper contact to the output $y = 1$, and the lower contact is open. As can be seen in Fig. 11.2c: if the input $x = 1$, then the value 0 is connected by the lower contact to the output $y = 0$, and the upper contact is open. In both cases the negated value of the input x appears as the output y so that this electrical circuit realizes a logic NOT-gate. Sometimes a NOT-gate is denoted by *inverter*. The function table in Fig. 11.2d summarizes the behavior represented in Fig. 11.2b and c. This logic function can be expressed by the logic function $y = f(x) = \overline{x}$ or the symbol shown in Fig. 11.2d.

Note 11.2 Generally, the bubble at the output of the NOT-gate symbol means that the logic value is negated at this position of a logic diagram.

Each pair of MOSFETs is controlled by one input. If several such pairs of MOSFETs are combined, then logic gates depending on more than one input can be built. For each input-pattern of the circuit, the structure of a gate must satisfy two conditions:

1. the voltage V_{DD} must be insulated from the voltage V_{SS};
2. either a low or a high voltage appears at the output.

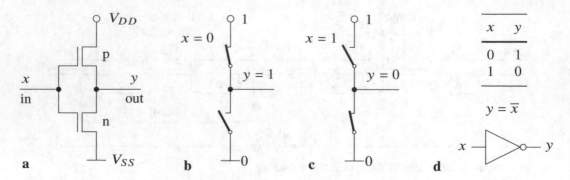

Fig. 11.2 NOT-gate: (**a**) structure in CMOS-technology, (**b**) switching model for $x = 0$, (**c**) switching model for $x = 1$, and (**d**) realized logic function and symbol used in logic diagrams

These two conditions are satisfied if either all p-channel MOSFETs are switched in parallel and the associated n-channel MOSFETs are series-connected or vice versa.

As can be seen in Fig. 11.3a, a NAND-gate of two inputs is built if the p-channel MOSFETs are switched in parallel and the n-channel MOSFETs are series-connected. The term NAND abbreviates NOT-AND. Figure 11.3b, ..., e describes the positions of the switches in the associated switching model. These four configurations confirm that both conditions mentioned above are satisfied for all possible input-patterns. The function table in Fig. 11.3f summarizes the behavior for all input-patterns explored in the previous four sub-figures. The function table describes the logic function $y = f(\mathbf{x}) = \overline{x_1 \wedge x_2}$. The logic symbol of a NAND-gate is also shown in Fig. 11.3f.

Note 11.3 The term NAND associates a combination of a NOT-gate and an AND-gate where the output of the AND-gate controls the input of the NOT-gate.

The four MOSFETs of Fig. 11.3a directly realize this combination of two logic operations. The CMOS-technology requires a combination of a NAND-gate and a NOT-gate in order to get an AND-gate.

Similarly, other logic gates can be realized using pairs of complementary MOSFETs. An electrical circuit structure behaves as a logic NOR-gate if the p-channel transistors are series-connected and the associated n-channel transistors are connected in parallel, respectively.

The behavior of a combinational circuit with the output y and the inputs $\mathbf{x}$ can be expressed by the logic function $y = f(\mathbf{x})$. This combinational circuit can be described more in detail by a structure of several gates and their connections by wires. The behavior of a gate is given by a logic function that is determined by the internal structure. Each input-pattern causes an associated value at the output of the gate. Such a vector of dependent values is denoted by *phase*. The list of all possible phases completely describes the logic behavior of a gate. Table 11.1 summarizes the three equivalent representations of several gates:

- the *symbol*,
- the *logic function*, and
- the *list of phases*.

The time is not included into the model on this logic level; real logic gates, however, need some nanoseconds to create the output-value depending on the input-values. The delay of a gate depends on the relation between length and width of the electrical gate area of the MOSFETs and the number

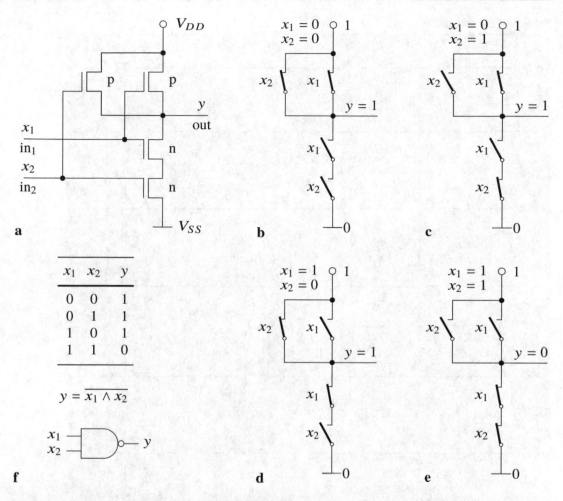

Fig. 11.3 NAND-gate: (**a**) structure in CMOS-technology, (**b**) switching model for the case $x_1 = 0$, $x_2 = 0$, (**c**) switching model for the case $x_1 = 0$, $x_2 = 1$, (**d**) switching model for the case $x_1 = 1$, $x_2 = 0$, (**e**) switching model for the case $x_1 = 1$, $x_2 = 1$, and (**f**) realized logic function and symbol

of inputs. Additionally, the delay of the gate is influenced by the output environment. A longer wire at the gate-output and additionally connected gates increase the delay.

A gate can have more than two inputs. Theoretically the number of inputs is not restricted, but there is a practical limit for the number of inputs. The reason for this limitation is the *threshold voltage* of the MOSFET. If the required number of gate-inputs is larger than the number of available inputs, the gate can be substituted by a cascade of gates.

A logic (wiring) diagram describes the structure of a combinational circuit graphically. The connection lines in such a diagram are labeled by logic variables. Alternatively, a system of logic equations can be used. Each equation describes the behavior of one gate. Variables with the same name express the connections. A third approach to describe the structure of a combinational circuit is a set of local lists of phases. Each list describes one gate. The connections are modeled by the variable names of the columns. This approach is suited for computer calculations. Figure 11.4 shows these three types of structure descriptions of the same combinational circuit.

Table 11.1 Combinational gates

Symbol	Name	Logic function	List of phases
	NOT	$o = \bar{i}$	$i\ \ o$ 0 1 1 0
	AND	$o = i_1 \wedge i_2$	$i_1\ \ i_2\ \ o$ 0 – 0 1 0 0 1 1 1
	OR	$o = i_1 \vee i_2$	$i_1\ \ i_2\ \ o$ 0 0 0 0 1 1 1 – 1
	EXOR	$o = i_1 \oplus i_2$	$i_1\ \ i_2\ \ o$ 0 0 0 0 1 1 1 0 1 1 1 0
	NAND	$o = \overline{i_1 \wedge i_2}$	$i_1\ \ i_2\ \ o$ 0 – 1 1 0 1 1 1 0
	NOR	$o = \overline{i_1 \vee i_2}$	$i_1\ \ i_2\ \ o$ 0 0 1 0 1 0 1 – 0
	MUX	$o = \bar{i}_0 \wedge i_1 \vee i_0 \wedge i_2$	$i_0\ \ i_1\ \ i_2\ \ o$ 0 0 – 0 0 1 – 1 1 – 0 0 1 – 1 1
	DEMUX	$o_1 = \bar{i}_0 \wedge i_1$ $o_2 = i_0 \wedge i_1$	$i_0\ \ i_1\ \ o_1\ \ o_2$ 0 0 0 0 0 1 1 0 1 0 0 0 1 1 0 1

The gates of a combinational circuit can be ordered in levels. The gates of the first level are controlled by the inputs of the circuit only. The outputs of first-level gates together with the inputs of the circuit control the gates of the second level, and so on. The combinational circuit in Fig. 11.4a has three levels. A multilevel circuit is the most general case.

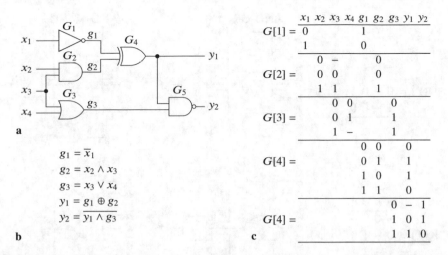

$$g_1 = \overline{x}_1$$
$$g_2 = x_2 \wedge x_3$$
$$g_3 = x_3 \vee x_4$$
$$y_1 = g_1 \oplus g_2$$
$$y_2 = \overline{y_1 \wedge g_3}$$

	x_1	x_2	x_3	x_4	g_1	g_2	g_3	y_1	y_2
$G[1] =$	0				1				
	1				0				
$G[2] =$		0	–			0			
		0	0			0			
		1	1			1			
$G[3] =$			0	0			0		
			0	1			1		
			1	–			1		
$G[4] =$					0	0		0	
					0	1		1	
					1	0		1	
					1	1		0	
$G[4] =$							0	–	1
							1	0	1
							1	1	0

a **b** **c**

Fig. 11.4 Structure of a combinational circuit: (**a**) logic diagram, (**b**) system of logic equations, and (**c**) set of local lists of phases

Each logic function can be realized by a two-level combinational circuit if two conditions are satisfied:

1. both the values of the logic variables and their complements are available at the same time; and
2. the number of gate-inputs will not be restricted.

Two-level combinational circuits can directly be created using the four basic forms of logic functions. From a logic function in disjunctive form , we get a two-level circuit where the outputs of AND-gates of the first level are connected to the inputs of an OR-gate at the second level. The same logic function will be realized if all gates in this structure are substituted by NAND-gates. This change of the circuit structure can be realized based on a two-step transformation. In the first step the disjunctive form of the logic function is negated twice, which does not change the function. The application of De Morgan's law to inner complement converts the $\vee$ of the disjunctive form into $\wedge$, and the existing conjunctions are negated. The outer negation together with the newly created $\wedge$-operations can be combined by a NAND-gate as well.

Example 11.1 The logic function in disjunctive form (11.1) is to be transformed into a NAND–NAND-structure (11.2):

$$f(a,b,c,d) = \overline{b}\,c \vee \overline{a}\,\overline{c}\,d \vee a\,\overline{c}\,\overline{d} \tag{11.1}$$

$$= \overline{\overline{\overline{b}\,c \vee \overline{a}\,\overline{c}\,d \vee a\,\overline{c}\,\overline{d}}}$$

$$= \overline{\overline{\overline{b}\,c} \wedge \overline{\overline{a}\,\overline{c}\,d} \wedge \overline{a\,\overline{c}\,\overline{d}}}\,. \tag{11.2}$$

Figure 11.5a shows the TVL of this logic function in disjunctive form (DF), the associated two-level AND–OR-circuit structure, and the alternative two-level NAND–NAND-circuit.

Note 11.4 The inputs of the first level gates are the same in both the two-level AND–OR- and the two-level NAND–NAND-circuits.

Example 11.2 The logic function (11.1) can also be expressed by the conjunctive form (11.3) and is transformed in two steps into a NOR–NOR-structure (11.4):

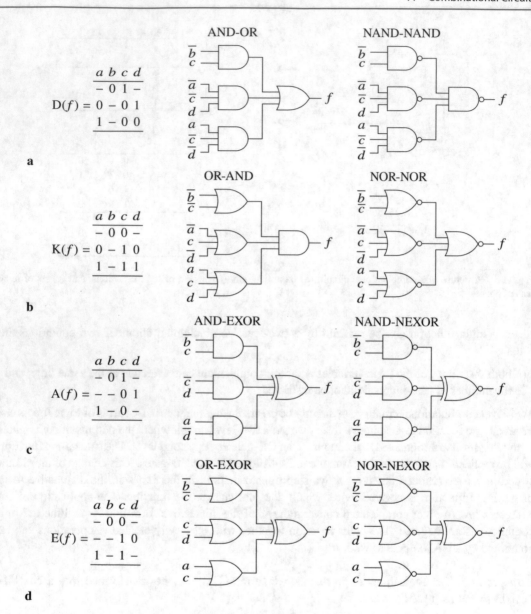

Fig. 11.5 Alternative circuit structures of two-level combinational circuits of the logic function $f = f(a, b, c, d) = \bar{b}c \vee \bar{c}(a \oplus d)$ based on: (**a**) the disjunctive form (DF), (**b**) the conjunctive form (CF), (**c**) the antivalence form (AF), and (**d**) the equivalence form (EF)

$$f(a, b, c, d) = (\bar{b} \vee \bar{c})(\bar{a} \vee c \vee \bar{d})(a \vee c \vee d) \tag{11.3}$$

$$= \overline{\overline{(\bar{b} \vee \bar{c})(\bar{a} \vee c \vee \bar{d})(a \vee c \vee d)}} \tag{11.4}$$

$$= \overline{\overline{(\bar{b} \vee \bar{c})} \vee \overline{(\bar{a} \vee c \vee \bar{d})} \vee \overline{(a \vee c \vee d)}}.$$

Figure 11.5b shows the TVL of the function (11.3) in conjunctive form (CF), the associated two-level OR–AND-circuit structure, and the alternative two-level NOR–NOR-circuit.

Conjunctions of literals are connected by EXOR-operations in an antivalence form. Consequently, this form of a logic function can be directly mapped onto a two-level AND–EXOR-circuit. The formal result of the transformation by two negations and the partial application of De Morgan's law results in NAND-expressions that are combined by equivalence operations. The equivalence of three inputs shows the same result as the antivalence. Due to the outer negation, a NEXOR-gate of three inputs is used at the second level.

Note 11.5 The result of the antivalence is the complement of equivalence for an even number of terms: $a \oplus b = \overline{a \odot b}$. However, the same function is created if an odd number of terms is connected by either antivalence or equivalence operations: $a \oplus b \oplus c = a \odot b \odot c$. Due to these properties, the outputs of the NAND-gates in the transformed AF must be connected either by an EXOR-gate or a NEXOR-gate. The NEXOR-gate works like an EXOR-gate and an additional NOT-gate on the output. A slightly changed transformation procedure can be used to get NAND-expressions for the first level. Each conjunction C is replaced by the equivalent expression $\overline{C} \oplus 1$. Afterward we remove pairs of 1s. A single remaining constant 1 causes a NEXOR-gate at the second level; otherwise, an EXOR-gate is used.

Example 11.3 Due to the orthogonality of (11.1), an associated orthogonal antivalence form (11.5) can be created by replacing OR-operations by EXOR-operations. Formula (11.5) can be simplified to the non-orthogonal antivalence form (11.6). The result of the transformation (11.7) contains a global negation and can therefore be realized by a NAND–NEXOR-circuit:

$$f(a, b, c, d) = \overline{b}\,c \oplus \overline{a}\,\overline{c}\,d \oplus a\,\overline{c}\,\overline{d} \tag{11.5}$$

$$= \overline{b}\,c \oplus \overline{c}\,d \odot a\,\overline{c} \tag{11.6}$$

$$= \overline{\overline{b}\,c} \oplus 1 \oplus \overline{\overline{c}\,d} \oplus 1 \oplus \overline{a\,\overline{c}} \oplus 1$$

$$= \overline{\overline{\overline{b}\,c} \oplus \overline{\overline{c}\,d} \oplus \overline{a\,\overline{c}}}. \tag{11.7}$$

Figure 11.5c shows the TVL of the antivalence form (AF) of the function (11.6), the associated two-level AND–EXOR-circuit structure, and the alternative two-level NAND–NEXOR-circuit.

The last two-level combinational circuits are created from the equivalence form (EF) of the logic function. The disjunctions of the equivalence form are realized by the OR-gates of the first level. The equivalence operation can be realized by an EXOR-gate or a NEXOR-gate. In order to decide which of these gates must be used, we express the equivalence $D_1 \odot D_2$ by the the equivalent expression $D_1 \oplus D_2 \oplus 1$ and remove pairs of 1's because of $1 \oplus 1 = 0$. If there is an odd number of disjunctions and consequently an even number of $\odot$-operations in the equivalence form, all 1's are removed, and an EXOR-gate must be used at the second level. Otherwise there is an even number of disjunctions in the equivalence form, and the single remaining constant 1 can be expressed by the NOT-operation that is a part of the NEXOR-gate. The transformation of the disjunctions into NOR-gates is based on $D = \overline{D} \oplus 1$.

Example 11.4 The logic function (11.1) can alternatively be expressed by both the conjunctive form (11.3) and the equivalence form (11.8). Due to the orthogonality of (11.3), Formula (11.8) can be created by replacing all $\wedge$-operations of (11.3) by $\odot$-operations. The result of further simplifications is the non-orthogonal equivalence form (11.9). Due to the two $\odot$-operations in the equivalence form

(11.9), the OR–EXOR-structure (11.10) can be created. The local transformation of the disjunctions into NOR-expressions leads finally to an alternative two-level NOR–NEXOR-circuit specified by (11.11):

$$f(a, b, c, d) = (\overline{b} \vee \overline{c}) \odot (\overline{a} \vee c \vee \overline{d}) \odot (a \vee c \vee d) \tag{11.8}$$

$$= (\overline{b} \vee \overline{c}) \odot (c \vee \overline{d}) \odot (a \vee c) \tag{11.9}$$

$$= (\overline{b} \vee \overline{c}) \oplus 1 \oplus (c \vee \overline{d}) \oplus 1 \oplus (a \vee c)$$

$$= (\overline{b} \vee \overline{c}) \oplus (c \vee \overline{d}) \oplus (a \vee c) \tag{11.10}$$

$$= \overline{(\overline{b} \vee \overline{c})} \oplus 1 \oplus \overline{(c \vee \overline{d})} \oplus 1 \oplus \overline{(a \vee c)} \oplus 1$$

$$= \overline{\overline{(\overline{b} \vee \overline{c})} \oplus \overline{(c \vee \overline{d})} \oplus \overline{(a \vee c)}}. \tag{11.11}$$

Figure 11.5d shows the TVL of the function (11.9) in equivalence form (EF), the associated two-level OR–EXOR circuit structure, and the alternative two-level NOR–NEXOR-circuit.

Note 11.6 An even number of disjunctions in the equivalence form leads to an OR–NEXOR- or NOR–EXOR-structure of the two-level combinational circuit.

As can be seen in Fig. 11.5, each row of the TVLs results in one gate at the first level, and the number of literals in a row determines the number of inputs of the corresponding gate. A value 1 in the TVL corresponds to an associated variable at the gate-input, and a value 0 to a negated variable, respectively. The number of rows and the number of 1s and 0s determine directly the size of the circuit. Thus, non-orthogonal minimized TVLs in the basic forms of logic functions are preferred to describe the circuit structure, and these TVLs are called *structure-TVLs*.

The gates of the first level can be reused for several different logic functions. *Programmable logic arrays* (PLAs) use this possibility. A PLA consists of an AND-matrix as the first level and an OR-matrix as the second level of a set of logic functions. Connections in the AND-matrix determine the non-negated or negated variables of the conjunctions. Connections in the OR-matrix select the conjunctions of the logic functions. Figure 11.6a shows an example of a PLA-structure. The realized logic functions are listed in Fig. 11.6b.

The structure of a PLA can be modeled by a TVL denoted by *structure-TVL of a PLA* and indicated by S_{PLA}. This TVL consists of columns that express the literals of the conjunctions and the outputs (logic functions). There is a direct relation between the AND-lines of the PLA and the rows of the TVL. The elements in a row of the structure-TVL of a PLA have the following meanings:

- 0 for a literal: connection to the negated line of the variable,
- 1 for a literal: connection to the non-negated line of the variable,
- − for a literal: no connection to the variable,
- 0 for a function: no connection, and
- 1 for a function: connection to the OR-gate of the function.

Figure 11.6c depicts the structure-TVL of the PLA of Fig. 11.6a.

Logic functions and equations are convenient to describe combinational circuits based on gates introduced in Table 11.1; however, properties of such circuits besides their logic behavior remain out of the scope. Such additional properties are the time interval a logic gate needs to switch between the output-values 0 and 1 caused by a change on at least one input or the energy consumed by a logic gate for such a change. One of the greatest challenges in the technological progress of integrated circuits

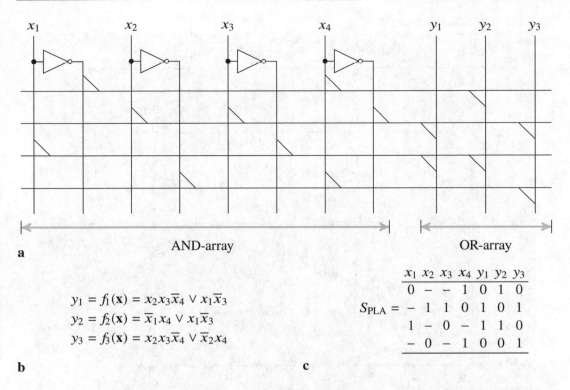

$$y_1 = f_1(\mathbf{x}) = x_2 x_3 \overline{x}_4 \vee x_1 \overline{x}_3$$
$$y_2 = f_2(\mathbf{x}) = \overline{x}_1 x_4 \vee x_1 \overline{x}_3$$
$$y_3 = f_3(\mathbf{x}) = x_2 x_3 \overline{x}_4 \vee \overline{x}_2 x_4$$

$$S_{\text{PLA}} = \begin{array}{ccccccc} x_1 & x_2 & x_3 & x_4 & y_1 & y_2 & y_3 \\ \hline 0 & - & - & 1 & 0 & 1 & 0 \\ - & 1 & 1 & 0 & 1 & 0 & 1 \\ 1 & - & 0 & - & 1 & 1 & 0 \\ - & 0 & - & 1 & 0 & 0 & 1 \\ \hline \end{array}$$

b   **c**

Fig. 11.6 Programmable logic array: (**a**) structure, (**b**) set of logic functions, and (**c**) structure-TVL of the PLA

is the very high temperature of the circuits caused by the change of electrical energy into thermal energy.

One way forward to low-power computing consists in the use of reversible circuits. The gates of Table 11.1 (except the NOT-gate) are irreversible gates. We take the AND-gate as example; it has two input-bits and only one output-bit. Landauer combined the theory of information, the thermodynamics, and the statistical physics and found that the loss of one bit of information causes an energy of $k_B T \ln 2$, where k_B is the Boltzmann constant and T the absolute temperature (see [7]).

Motivated by Landauer's principle and the tight relationship to the quantum computing (which is out of the scope of this book), reversible circuits are increasingly studied with the aim of low-power computing. The number of outputs of a reversible circuit is equal to the number of inputs, and the output-patterns can only be a permutation of the input-patterns. A sequence of reversible gates assigned to the lines of the reversible circuit contributes with their local permutations to the global permutation between the patterns of the inputs and outputs.

Table 11.2 shows often used reversible gates that are characterized by changing only a single target line. Toffoli-gates can be extended to more than two control lines.

Sometimes negated variables (represented by empty circles as connections to the control lines) are used to control CNOT- or Toffoli-gates. Also OR-controlled Toffoli-gates have been suggested where the connections to the control lines are represented by small triangles. The first three gates of Table 11.2 are sufficient to realize each reversible function. The Fredkin-gate is also a complete basis for all reversible functions. There are several other reversible gates (see [5]).

Figure 11.7 shows that the structure of a *reversible combinational circuit* can be described using the same three approaches as introduced for non-reversible combinational circuit. The logic diagram of Fig. 11.7a describes the structure of a reversible combinational circuit graphically. Each line

Table 11.2 Reversible gates

Symbol	Name	Logic function	List of phases
i —⊕— o	NOT	$o = \bar{i}$	i o 0 1 1 0
i_1 —•— o_1 i_2 —⊕— o_2	CNOT (controlled NOT) (Feynman)	$o_1 = i_1$ $o_2 = i_1 \oplus i_2$	i_1 i_2 o_1 o_2 0 0 0 0 0 1 0 1 1 0 1 1 1 1 1 0
i_1 —•— o_1 i_2 —•— o_2 i_3 —⊕— o_3	Toffoli	$o_1 = i_1$ $o_2 = i_2$ $o_3 = i_1 i_2 \oplus i_3$	i_1 i_2 i_3 o_1 o_2 o_3 0 0 0 0 0 0 0 0 1 0 0 1 0 1 0 0 1 0 0 1 1 0 1 1 1 0 0 1 0 0 1 0 1 1 0 1 1 1 0 1 1 1 1 1 1 1 1 0
i_1 —✕— o_1 i_2 —✕— o_2	SWAP	$o_1 = i_2$ $o_2 = i_1$	i_1 i_2 o_1 o_2 0 0 0 0 0 1 1 0 1 0 0 1 1 1 1 1
i_1 —•— o_1 i_2 —✕— o_2 i_3 —✕— o_3	Fredkin (controlled SWAP)	$o_1 = i_1$ $o_2 = \bar{i}_1 i_2 \oplus i_1 i_3$ $o_3 = \bar{i}_1 i_3 \oplus i_1 i_2$	i_1 i_2 i_3 o_1 o_2 o_3 0 0 0 0 0 0 0 0 1 0 0 1 0 1 0 0 1 0 0 1 1 0 1 1 1 0 0 1 0 0 1 0 1 1 1 0 1 1 0 1 0 1 1 1 1 1 1 1

between an input x_i and an output y_i is a target line for one of the three gates in this example. Due to the variables g_{ji} attached to each line between two gates, each gate can be specified using only its neighborhood variables. Alternatively to the logic diagram, a system of logic equations (see Fig. 11.7b) or a set of local lists of phases (see Fig. 11.7c) can be used to determine the structure of a reversible combinational circuit.

Not only gates but also contacts of relays can be used to realize a logic function. Such switching contacts can also be implemented by the different types of transistors. Table 11.3 shows the basic model of the two types of contacts. The variable x represents in a logic expression the associated make contact and the variable $\bar{x}$ a break contact labeled by x.

The series connection of such contacts realizes a logic AND-function and the parallel connection a logic OR-function. The combination of these two types of connections leads to circuit structures shown in Fig. 11.8 that directly express a disjunctive form or a conjunctive form of a logic function.

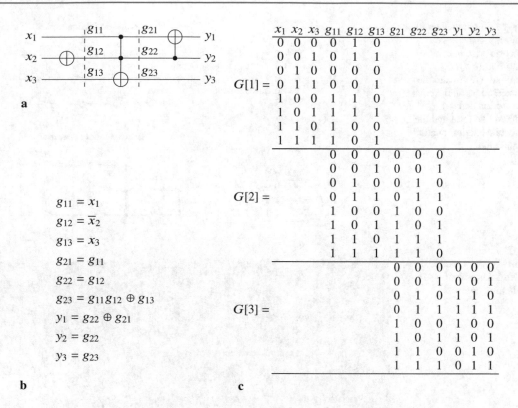

Fig. 11.7 Structure of a reversible combinational circuit: (**a**) logic diagram, (**b**) system of logic equations, and (**c**) set of local lists of phases

Table 11.3 Basic model of switching contacts

Type	Symbol	Logic function	Not activated behavior
Make contact	──o ╱ ── y $\;x$	$y = f(x) = x$	This contact is normally open
Break contact	──o ╲ ── y $\;x$	$y = f(x) = \overline{x}$	This contact is normally closed

The use of contacts is not restricted to the two types of circuit structures shown in Fig. 11.8. Arbitrary contact circuits can be modeled by means of the *contact potential model*. In this model not only the logic values that control the positions of the contacts but also the logic values on the connection lines of the contacts are modeled by logic variables. Table 11.4 shows the details of the contact potential model. A given circuit where make and break contacts are arbitrarily connected can be modeled by a set of logic equations of the two types given in Table 11.4.

The contact potential model facilitates the specification of contact circuits that determine the value of the output only for a subset of input-pattern; for other input-patterns, the output can be equal to both 0 or 1. Several of such circuits can control the same *bus-wire*.

Above we used logic functions and equations as model of the structure of combinational circuits. Logic functions and equations are also convenient to model the behavior of logic combinational circuits.

Each input-pattern **x** causes one of the values 0 or 1 at the output y of the combinational circuit. Thus, a logic function $f(\mathbf{x})$ is a suitable model of the behavior, and Eq. (11.12) describes additionally that the output-wire labeled by the variable y is the carrier of the function value:

Fig. 11.8 Contact circuits that realize the logic function $f(a, b, c, d) = \bar{b} c \vee \bar{c} (a \oplus d)$: (**a**) disjunctive form and the associated parallel–series connection and (**b**) conjunctive form and the associated series–parallel connection

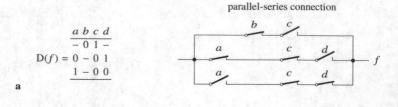

$$D(f) = \begin{array}{cccc} a & b & c & d \\ \hline - & 0 & 1 & - \\ 0 & - & 0 & 1 \\ 1 & - & 0 & 0 \\ \hline \end{array}$$

a

parallel-series connection

$$K(f) = \begin{array}{cccc} a & b & c & d \\ \hline - & 0 & 0 & - \\ 0 & - & 1 & 0 \\ 1 & - & 1 & 1 \\ \hline \end{array}$$

b

series-parallel connection

Table 11.4 Contact potential model of switching contacts

Type	Symbol	Logic equation	List of phases
Make contact	$p_0 \;\text{—}\overset{x}{\text{—}}\; p_1$	$\bar{x} \vee x(p_0 \odot p_1) = 1$	$F(p_0, x, p_1) = \begin{array}{ccc} p_0 & x & p_1 \\ \hline - & 0 & - \\ 0 & 1 & 0 \\ 1 & 1 & 1 \\ \hline \end{array}$
Break contact	$p_0 \;\text{—}\overset{x}{\text{—}}\; p_1$	$x \vee \bar{x}(p_0 \odot p_1) = 1$	$F(p_0, x, p_1) = \begin{array}{ccc} p_0 & x & p_1 \\ \hline - & 1 & - \\ 0 & 0 & 0 \\ 1 & 0 & 1 \\ \hline \end{array}$

$$y = f(\mathbf{x}) \,. \tag{11.12}$$

Equation (11.12) separates the output-variable y on the left-hand side from the input-variables in the function expression on the right-hand side of the equation; hence, Eq. (11.12) is denoted by *explicit system-equation*.

Note 11.7 Each logic function can occur on the right-hand side of (11.12). The solution-set of (11.12) describes all observable phases of the associated combinational circuit.

The *implicit system-equation* (11.13) of a combinational circuit originates from the explicit system-equation (11.12) by applying the equivalence with $f(\mathbf{x})$ on both sides:

$$y \odot f(\mathbf{x}) = f(\mathbf{x}) \odot f(\mathbf{x}) \,,$$
$$F(\mathbf{x}, y) = 1 \,. \tag{11.13}$$

Note 11.8 Not each system-function $F(\mathbf{x}, y)$ describes the behavior of a combinational circuit. A system-function $F(\mathbf{x}, y)$ can determine the allowed behavior of a combinational circuit that is a superset of the implemented behavior. Any partial behavior can also be expressed by a system-function $F(\mathbf{x}, y)$.

Sometimes not all 2^n patterns can occur at the input of the combinational circuit. It is not necessary to define the behavior for these input-patterns. Such input-patterns without defined function values are denoted by *don't-cares* $DC \subseteq \mathbb{B}^n$. An *incompletely specified function* (ISF) $f^C(\mathbf{x})$ maps the remaining care set $C = \mathbb{B}^n \setminus DC$ into $\mathbb{B}$. If there are $|DC|$ don't-care patterns, then $2^{|DC|}$ logic functions $f(\mathbf{x}) \in F^C(\mathbf{x})$ exist. We call the set of logic functions $F^C(\mathbf{x})$ (11.14) the *characteristic function set* of the incompletely specified function $f^C(\mathbf{x})$:

$$F^C(\mathbf{x}) = \left\{ f(\mathbf{x}) \mid \forall \mathbf{c} \in C \quad f(\mathbf{c}) = f^C(\mathbf{c}) \right\} . \tag{11.14}$$

This set of logic functions has the properties of a lattice and can be determined by two of the following *mark-functions*:

- $f_q(\mathbf{x})$: ON-set-function, $f(\mathbf{x})$ must be equal to 1,
- $f_r(\mathbf{x})$: OFF-set-function, $f(\mathbf{x})$ must be equal to 0,
- $f_\varphi(\mathbf{x})$: don't-care-function, any value is allowed for $f(\mathbf{x})$.

These functions are mutually disjoint:

$$f_q(\mathbf{x}) \wedge f_r(\mathbf{x}) = 0 , \tag{11.15}$$

$$f_q(\mathbf{x}) \wedge f_\varphi(\mathbf{x}) = 0 , \tag{11.16}$$

$$f_\varphi(\mathbf{x}) \wedge f_r(\mathbf{x}) = 0 , \tag{11.17}$$

and cover the whole Boolean space:

$$f_q(\mathbf{x}) \vee f_r(\mathbf{x}) \vee f_\varphi(\mathbf{x}) = 1 . \tag{11.18}$$

If two of the mark-functions are known, then the third function can be derived:

$$f_q(\mathbf{x}) = \overline{f_r(\mathbf{x}) \vee f_\varphi(\mathbf{x})} , \tag{11.19}$$

$$f_r(\mathbf{x}) = \overline{f_q(\mathbf{x}) \vee f_\varphi(\mathbf{x})} , \tag{11.20}$$

$$f_\varphi(\mathbf{x}) = \overline{f_q(\mathbf{x}) \vee f_r(\mathbf{x})} . \tag{11.21}$$

The mark-functions and the system-function $F(\mathbf{x}, y)$ represent the same information. Therefore, they can be transformed into each other:

$$F(\mathbf{x}, y) = (y \odot f_q(\mathbf{x})) \vee (y \oplus f_r(\mathbf{x})) , \tag{11.22}$$

$$F(\mathbf{x}, y) = f_\varphi(\mathbf{x}) \vee (y \oplus f_r(\mathbf{x})) , \tag{11.23}$$

$$F(\mathbf{x}, y) = (y \odot f_q(\mathbf{x})) \vee f_\varphi(\mathbf{x}) , \tag{11.24}$$

$$f_\varphi(\mathbf{x}) = \min_y F(\mathbf{x}, y) , \tag{11.25}$$

$$f_q(\mathbf{x}) = \max_y (y \wedge F(\mathbf{x}) \wedge \overline{\min_y F(\mathbf{x}, y)}) , \tag{11.26}$$

$$f_r(\mathbf{x}) = \max_y (\overline{y} \wedge F(\mathbf{x}) \wedge \overline{\min_y F(\mathbf{x}, y)}) . \tag{11.27}$$

Let us finally consider a combinational circuit of n inputs $\mathbf{x}$ and k outputs $\mathbf{y}$. As a generalization of (11.12), we get a set of explicit system-equations (11.28):

$$y_1 = f_1(\mathbf{x}) \,,$$
$$y_2 = f_2(\mathbf{x}) \,,$$
$$\vdots \quad \vdots$$
$$y_k = f_k(\mathbf{x}) \,. \tag{11.28}$$

This system of equations can be transformed into a single implicit system-equation (11.29):

$$y_1 \odot f_1(\mathbf{x}) = 1 \,,$$
$$y_2 \odot f_2(\mathbf{x}) = 1 \,,$$
$$\vdots \quad \vdots$$
$$y_k \odot f_k(\mathbf{x}) = 1 \,,$$

$$(y_1 \odot f_1(\mathbf{x})) \wedge (y_2 \odot f_2(\mathbf{x})) \wedge \cdots \wedge (y_k \odot f_k(\mathbf{x})) = 1 \,,$$
$$F(\mathbf{x}, \mathbf{y}) = 1 \,. \tag{11.29}$$

The *implicit system-function* $F(\mathbf{x}, \mathbf{y})$ in (11.29) describes the realized behavior of the modeled combinational circuit. Again an implicit system-function $F(\mathbf{x}, \mathbf{y})$ can define all requirements and the degrees of freedom of a required combinational circuit. In extension to the implicit system-function $F(\mathbf{x}, y)$ of a circuit with one output, the implicit system-function $F(\mathbf{x}, \mathbf{y})$ of a circuit with k outputs can describe additional relations between the output-functions.

Similar to the outputs $\mathbf{y}$, an implicit system-function can include internal variables of the combinational circuit.

11.2   Analysis

The main analysis task is the calculation of the behavior of a given circuit structure. There are two methods to solve this problem. The first method is the *substitution-method* and uses the logic equations of the gates. This method can find the logic function of a single output or the implicit system-function of the whole combinational circuit.

We start with the calculation of the logic function belonging to a single output. The starting point is the equation of the selected output of the circuit. Internal variables in this equation are substituted by the right-hand expression of the associated equations of other gates. This procedure is repeated until only input-literals occur on the right-hand side of the equation. In a next step a homogenous characteristic equation like (11.13) is created. In a final step the expression is simplified to an orthogonal disjunctive form. The benefit of this representation is that each ternary vector associated to a conjunction of $F(\mathbf{x}, y)$ determines a subset of solutions of Eq. (11.13). Consequently, a representation of $F(\mathbf{x}, y)$ as TVL in an orthogonal disjunctive form contains all phases of the combinational circuit and describes the input–output-behavior completely.

Example 11.5 We calculate the behavior of the output y_2 of the combinational circuit of Fig. 11.4a that is determined by the logic equations of Fig. 11.4b (which are repeated here):

$$g_1 = \overline{x}_1 , \tag{11.30}$$

$$g_2 = x_2 \wedge x_3 , \tag{11.31}$$

$$g_3 = x_3 \vee x_4 , \tag{11.32}$$

$$y_1 = g_1 \oplus g_2 , \tag{11.33}$$

$$y_2 = \overline{y_1 \wedge g_3} . \tag{11.34}$$

The substitution of (11.33) and (11.32) into (11.34) results in (11.35). In the next iteration (11.30) and (11.31) are substituted in (11.35) so that we get (11.36):

$$y_2 = \overline{(g_1 \oplus g_2) \wedge (x_3 \vee x_4)} , \tag{11.35}$$

$$y_2 = \overline{(\overline{x}_1 \oplus x_2 \wedge x_3) \wedge (x_3 \vee x_4)} . \tag{11.36}$$

The result of an equivalence operation with the right-hand expression on both sides of (11.36) is the homogenous characteristic equation (11.37):

$$y_2 \odot \overline{(\overline{x}_1 \oplus x_2 x_3)(x_3 \vee x_4)} = 1 . \tag{11.37}$$

The left-hand expression of (11.37) represents the implicit system-function $F_2(\mathbf{x}, y_2)$. This expression must be simplified and transformed into an orthogonal disjunctive form. By applying the law $a \odot b = ab \vee \overline{a}\,\overline{b}$, we get (11.38) from (11.37):

$$y_2 \overline{(\overline{x}_1 \oplus x_2 x_3)(x_3 \vee x_4)} \vee \overline{y}_2 (\overline{x}_1 \oplus x_2 x_3)(x_3 \vee x_4) = 1 . \tag{11.38}$$

As next step, we use De Morgan's law to remove the big complement from (11.38) and get (11.39)

$$y_2((x_1 \odot (\overline{x}_2 \vee \overline{x}_3)) \vee \overline{x}_3 \overline{x}_4) \vee \overline{y}_2 (\overline{x}_1 \oplus x_2 x_3)(x_3 \vee x_4) = 1 . \tag{11.39}$$

Both the equivalence and the antivalence operation in (11.39) have been replaced by equivalent disjunctive forms in (11.40):

$$y_2(x_1 \overline{x}_2 \vee x_1 \overline{x}_3 \vee \overline{x}_1 x_2 x_3 \vee \overline{x}_3 \overline{x}_4)$$
$$\vee \overline{y}_2 (\overline{x}_1 \overline{x}_2 \vee \overline{x}_1 \overline{x}_3 \vee x_1 x_2 x_3)(x_3 \vee x_4) = 1 . \tag{11.40}$$

In the next step we extend the term $\overline{x}_3 \overline{x}_4$ by $x_1 \vee \overline{x}_1$ and apply distributive law to the last two expressions in parentheses of (11.40):

$$y_2(x_1 \overline{x}_2 \vee x_1 \overline{x}_3 \vee \overline{x}_1 x_2 x_3 \vee x_1 \overline{x}_3 \overline{x}_4 \vee \overline{x}_1 \overline{x}_3 \overline{x}_4)$$
$$\vee \overline{y}_2 (\overline{x}_1 \overline{x}_2 x_3 \vee x_1 x_2 x_3 \vee \overline{x}_1 \overline{x}_2 x_4 \vee \overline{x}_1 \overline{x}_3 x_4 \vee x_1 x_2 x_3 x_4) = 1 . \tag{11.41}$$

From (11.41) to (11.42), the expression is simplified by two absorptions. The term $x_1 \overline{x}_3 \overline{x}_4$ will be absorbed by $x_1 \overline{x}_3$ and the last term by the term $x_1 x_2 x_3$, respectively.

$$y_2(x_1\overline{x}_2 \lor x_1\overline{x}_3 \lor \overline{x}_1 x_2 x_3 \lor \overline{x}_1 \overline{x}_3 \overline{x}_4)$$

$$\lor \overline{y}_2(\overline{x}_1 \overline{x}_2 x_3 \lor x_1 x_2 x_3 \lor \overline{x}_1 \overline{x}_2 x_4 \lor \overline{x}_1 \overline{x}_3 x_4) = 1 . \tag{11.42}$$

Once more two non-orthogonal terms are extended to find an orthogonal expression. From (11.42) to (11.43), the term $x_1\overline{x}_3$ has been extended by $x_2 \lor \overline{x}_2$ and the term $\overline{x}_1\overline{x}_2 x_4$ by $x_3 \lor \overline{x}_3$:

$$y_2(x_1\overline{x}_2 \lor x_1 x_2 \overline{x}_3 \lor x_1\overline{x}_2\overline{x}_3 \lor \overline{x}_1 x_2 x_3 \lor \overline{x}_1 \overline{x}_3 \overline{x}_4)$$

$$\lor \overline{y}_2(\overline{x}_1 \overline{x}_2 x_3 \lor x_1 x_2 x_3 \lor \overline{x}_1 \overline{x}_2 x_3 x_4 \lor \overline{x}_1 \overline{x}_2 \overline{x}_3 x_4 \lor \overline{x}_1 \overline{x}_3 x_4) = 1 . \tag{11.43}$$

Three absorptions and two applications of the distributive law lead to the final orthogonal disjunctive expression of $F_2(\mathbf{x}, y_2)$ on the left-hand side of (11.44). In detail, the term $x_1\overline{x}_2\overline{x}_3$ is absorbed by $x_1\overline{x}_2$, the term $\overline{x}_1\overline{x}_2 x_3 x_4$ is absorbed by $\overline{x}_1\overline{x}_2 x_3$, and the term $\overline{x}_1\overline{x}_2\overline{x}_3 x_4$ is absorbed by $\overline{x}_1\overline{x}_3 x_4$:

$$x_1\overline{x}_2 y_2 \lor x_1 x_2 \overline{x}_3 y_2 \lor \overline{x}_1 x_2 x_3 y_2 \lor \overline{x}_1 \overline{x}_3 \overline{x}_4 y_2$$

$$\lor \overline{x}_1 \overline{x}_2 x_3 \overline{y}_2 \lor x_1 x_2 x_3 \overline{y}_2 \lor \overline{x}_1 \overline{x}_3 x_4 \overline{y}_2 = 1 . \tag{11.44}$$

Figure 11.9 shows the calculated implicit system-function $F_2(\mathbf{x}, y_2)$ as an orthogonal TVL in disjunctive form.

The same TVL is the solution of each equation (11.37), ..., (11.44) and describes all phases observable on all inputs and the output y_2 of the combinational circuit of Fig. 11.4a.

Note 11.9 Each of the 16 input-patterns occurs together with the output-value of y_2 in exactly one row of the list of phases in Fig. 11.9.

The expression on the left-hand side of Eq. (11.44) describes the system-function $F_2(\mathbf{x}, y_2)$ (11.46). Similarly, the system-function $F_1(\mathbf{x}, y_1)$ (11.45) can be calculated:

$$F_1(\mathbf{x}, y_1) = \overline{x}_1\overline{x}_2 y_1 \lor \overline{x}_1 x_2 \overline{x}_3 y_1 \lor x_1 x_2 x_3 y_1$$

$$\lor x_1\overline{x}_2\overline{y}_1 \lor x_1 x_2 \overline{x}_3 \overline{y}_1 \lor \overline{x}_1 x_2 x_3 \overline{y}_1 , \tag{11.45}$$

$$F_2(\mathbf{x}, y_2) = x_1\overline{x}_2 y_2 \lor x_1 x_2 \overline{x}_3 y_2 \lor \overline{x}_1 x_2 x_3 y_2 \lor \overline{x}_1 \overline{x}_3 \overline{x}_4 y_2$$

$$\lor \overline{x}_1 \overline{x}_2 x_3 \overline{y}_2 \lor x_1 x_2 x_3 \overline{y}_2 \lor \overline{x}_1 \overline{x}_3 x_4 \overline{y}_2 . \tag{11.46}$$

The system-function (11.29) describes the behavior of the entire combinational circuit. It can be calculated by (11.47), based on the partial system-functions $F_i(\mathbf{x}, y_i)$ of all single outputs:

Fig. 11.9 Behavior of the circuit of Fig. 11.4a with regard to the output y_2, described by the system-function $F_2(\mathbf{x}, y_2)$ as the solution of Eq. (11.44)

$$\mathrm{ODA}(F_2(\mathbf{x}, y_2)) = $$

x_1	x_2	x_3	x_4	y_2
1	0	–	–	1
1	1	0	–	1
0	1	1	–	1
0	–	0	0	1
0	0	1	–	0
1	1	1	–	0
0	–	0	1	0

$$F(\mathbf{x}, \mathbf{y}) = \bigwedge_{i=1}^{k} F_i(\mathbf{x}, y_i) \,. \tag{11.47}$$

Example 11.6 The system-function $F(\mathbf{x}, \mathbf{y})$ of the combinational circuit shown in Fig. 11.4a is calculated by means of (11.47) and the given partial system of functions (11.45) and (11.46). The final result is again represented as an orthogonal disjunctive form:

$$
\begin{aligned}
F(\mathbf{x}, \mathbf{y}) = \quad & (\overline{x}_1 \overline{x}_2 y_1 \vee \overline{x}_1 x_2 \overline{x}_3 y_1 \vee x_1 x_2 x_3 y_1 \\
& \vee x_1 \overline{x}_2 \overline{y}_1 \vee x_1 x_2 \overline{x}_3 \overline{y}_1 \vee \overline{x}_1 x_2 x_3 \overline{y}_1) \\
& \wedge (x_1 \overline{x}_2 y_2 \vee x_1 x_2 \overline{x}_3 y_2 \vee \overline{x}_1 x_2 x_3 y_2 \vee \overline{x}_1 \overline{x}_3 \overline{x}_4 y_2 \\
& \vee \overline{x}_1 \overline{x}_2 x_3 \overline{y}_2 \vee x_1 x_2 x_3 \overline{y}_2 \vee \overline{x}_1 \overline{x}_3 x_4 \overline{y}_2) \,, \\[2mm]
F(\mathbf{x}, \mathbf{y}) = \quad & \overline{x}_1 \overline{x}_2 \overline{x}_3 \overline{x}_4 y_1 y_2 \vee \overline{x}_1 \overline{x}_2 x_3 y_1 \overline{y}_2 \vee \overline{x}_1 \overline{x}_2 \overline{x}_3 x_4 \ y_1 \overline{y}_2 \\
& \vee \overline{x}_1 x_2 \overline{x}_3 \overline{x}_4 y_1 y_2 \vee \overline{x}_1 x_2 \overline{x}_3 x_4 y_1 \overline{y}_2 \vee x_1 x_2 x_3 y_1 \overline{y}_2 \\
& \vee x_1 \overline{x}_2 \overline{y}_1 y_2 \vee x_1 x_2 \overline{x}_3 \overline{y}_1 y_2 \vee \overline{x}_1 x_2 x_3 \overline{y}_1 y_2 \,, \\[2mm]
F(\mathbf{x}, \mathbf{y}) = \quad & \overline{x}_1 \overline{x}_3 \overline{x}_4 y_1 y_2 \vee \overline{x}_1 \overline{x}_2 x_3 y_1 \overline{y}_2 \vee \overline{x}_1 \overline{x}_3 x_4 \ y_1 \overline{y}_2 \\
& \vee x_1 x_2 x_3 y_1 \overline{y}_2 \vee x_1 \overline{x}_2 \overline{y}_1 y_2 \vee x_1 x_2 \overline{x}_3 \overline{y}_1 y_2 \vee \overline{x}_1 x_2 x_3 \overline{y}_1 y_2 \,.
\end{aligned}
$$

Figure 11.10 shows the system-function $F(\mathbf{x}, \mathbf{y})$ as TVL in orthogonal disjunctive form. The seven rows of this TVL contain all 16 phases observable at the circuit of Fig. 11.4a. The same TVL is the result of the system-equation $F(\mathbf{x}, \mathbf{y}) = 1$.

Alternatively, the behavior of a combinational circuit can be calculated using the *equation-method*. The main difference to the substitution-method is that the internal variables are not removed. There are two possibilities to apply the equation-method. First the set of logic equations of the gates can be transformed into a single equation (11.29) that contains additionally the variables $\mathbf{g}$ of the internal wires. The solution of the system-equation $F(\mathbf{x}, \mathbf{g}, \mathbf{y}) = 1$ is the list of phases $(\mathbf{x}, \mathbf{g}, \mathbf{y})$ that completely describes the behavior of the circuit.

In the second version of the equation-method each equation is separately solved and the set of solutions is calculated by intersections of these solution-sets. The result is the same list of phases $(\mathbf{x}, \mathbf{g}, \mathbf{y})$ as calculated by means of the first version of the equation-method.

Example 11.7 We calculate the behavior of the whole combinational circuit of Fig. 11.4a using the second version of the equation-method. The solutions of the logic equations of the gates are already

Fig. 11.10 Behavior of the circuit of Fig. 11.4a

$$
ODA(F(\mathbf{x}, y_1, y_2)) =
\begin{array}{cccc|cc}
x_1 & x_2 & x_3 & x_4 & y_1 & y_2 \\
\hline
0 & - & 0 & 0 & 1 & 1 \\
0 & 0 & 1 & - & 1 & 0 \\
0 & - & 0 & 1 & 1 & 0 \\
1 & 1 & 1 & - & 1 & 0 \\
1 & 0 & - & - & 0 & 1 \\
1 & 1 & 0 & - & 0 & 1 \\
0 & 1 & 1 & - & 0 & 1 \\
\end{array}
$$

Fig. 11.11 Intermediate results of the equation-method applied to the combinational circuit of Fig. 11.4a

$$\bigcap_{i=1}^{2} G[i] =$$

x_1	x_2	x_3	x_4	g_1	g_2	g_3	y_1	y_2
0	0	–		1	0			
0	1	0		1	0			
0	1	1		1	1			
1	0	–		0	0			
1	1	0		0	0			
1	1	1		0	1			

$$\bigcap_{i=1}^{3} G[i] =$$

x_1	x_2	x_3	x_4	g_1	g_2	g_3	y_1	y_2
0	0	0	0	1	0	0		
0	0	0	1	1	0	1		
0	0	1	–	1	0	1		
0	1	0	0	1	0	0		
0	1	0	1	1	0	1		
0	1	1	–	1	1	1		
1	0	0	0	0	0	0		
1	0	0	1	0	0	1		
1	0	1	–	0	0	1		
1	1	0	0	0	0	0		
1	1	0	1	0	0	1		
1	1	1	–	0	1	1		

$$\bigcap_{i=1}^{4} G[i] =$$

x_1	x_2	x_3	x_4	g_1	g_2	g_3	y_1	y_2
0	0	0	0	1	0	0	1	
0	0	0	1	1	0	1	1	
0	0	1	–	1	0	1	1	
0	1	0	0	1	0	0	1	
0	1	0	1	1	0	1	1	
0	1	1	–	1	1	1	0	
1	0	0	0	0	0	0	0	
1	0	0	1	0	0	1	0	
1	0	1	–	0	0	1	0	
1	1	0	0	0	0	0	0	
1	1	0	1	0	0	1	0	
1	1	1	–	0	1	1	1	

given in Fig. 11.4c. Figure 11.11 shows the intermediate results of the intersection of the local lists of phases $G[i]$, $i = 1, 2, 3, 4$.

We calculate the intersection of $G[1]$ and $G[2]$ in the first step. There is no common column in these TVLs; hence, their intersection combines each row of $G[1]$ with all three rows of $G[2]$, and we get six resulting vectors.

In the next step we calculate the intersection of $\bigcap_{i=1}^{2} G[i]$ and $G[3]$. These TVLs overlap in column x_3. Since $x_3 = -$ in the first row of $\bigcap_{i=1}^{2} G[i]$, this row can be combined with each row of $G[3]$. The value $x_3 = 0$ in the second row of $\bigcap_{i=1}^{2} G[i]$ selects the first two rows of $G[3]$ as the result of the intersection. In the third row of $\bigcap_{i=1}^{2} G[i]$ we have $x_3 = 1$ so that only the combination

Fig. 11.12 Final results ODA($F(\mathbf{x}, \mathbf{g}, \mathbf{y})$) of the equation-method applied to the combinational circuit of Fig. 11.4a

$$F(\mathbf{x}, \mathbf{g}, \mathbf{y}) = \bigcap_{i=1}^{5} G[i] =$$

x_1	x_2	x_3	x_4	g_1	g_2	g_3	y_1	y_2
0	0	0	0	1	0	0	1	1
0	0	0	1	1	0	1	1	0
0	0	1	–	1	0	1	1	0
0	1	0	0	1	0	0	1	1
0	1	0	1	1	0	1	1	0
0	1	1	–	1	1	1	0	1
1	0	0	0	0	0	0	0	1
1	0	0	1	0	0	1	0	1
1	0	1	–	0	0	1	0	1
1	1	0	0	0	0	0	0	1
1	1	0	1	0	0	1	0	1
1	1	1	–	0	1	1	1	0

$=$

x_1	x_2	x_3	x_4	g_1	g_2	g_3	y_1	y_2
0	0	1	–	1	0	1	1	0
0	–	0	0	1	0	0	1	1
0	–	0	1	1	0	1	1	0
0	1	1	–	1	1	1	0	1
1	0	1	–	0	0	1	0	1
1	–	0	0	0	0	0	0	1
1	–	0	1	0	0	1	0	1
1	1	1	–	0	1	1	1	0

with the last row of $G[3]$ can be used as part of the intersection. The other rows of $\bigcap_{i=1}^{2} G[i]$ are considered analogously.

In the third step we calculate the intersection of $\bigcap_{i=1}^{3} G[i]$ and $G[4]$. These TVLs overlap in the columns g_1 and g_2. These two TVLs possess only values 0 or 1 in the columns g_1 and g_2; hence, no row of $\bigcap_{i=1}^{3} G[i]$ can be combined with more than one row of $G[4]$. Since all combinations of the values 0 and 1 occur in the columns g_1 and g_2 of $G[4]$, all rows of $\bigcap_{i=1}^{3} G[i]$ are part of the result $\bigcap_{i=1}^{4} G[i]$ that includes the additional column y_1 from $G[4]$.

The final result of the intersections $\bigcap_{i=1}^{5} G[i]$ is shown in Fig. 11.12. This TVL also describes the implicit system-function $F(\mathbf{x}, \mathbf{g}, \mathbf{y})$ in an orthogonal disjunctive form. Each phase contains the values of the internal wires and the outputs for given input-patterns. The pairs of rows (1,4), (2,5), (7,10), and (8,11) differ only in the value of x_2; hence, each of these pairs can be merged into a single ternary vector so that only eight rows of the TVL ODA($F(\mathbf{x}, \mathbf{g}, \mathbf{y})$) are needed to express the complete behavior of the combinational circuit of Fig. 11.4.

The *input–output-system-function* $F(\mathbf{x}, \mathbf{y})$ can now easily be calculated using the *complete system-function* $F(\mathbf{x}, \mathbf{g}, \mathbf{y})$. A condition for an input–output-phase is the existence of values of the internal wires that conform to the input–output-values. The k-fold maximum checks this property and removes the variables of the internal wires, for $\mathbf{g} = (g_1, \ldots, g_k)$:

$$F(\mathbf{x}, \mathbf{y}) = \max_{\mathbf{g}}^{k} F(\mathbf{x}, \mathbf{g}, \mathbf{y}) \ . \tag{11.48}$$

Example 11.8 The input–output-system-function $F(\mathbf{x}, \mathbf{y})$ has to be calculated as an orthogonal minimized disjunctive form of the given complete system-function $F(\mathbf{x}, \mathbf{g}, \mathbf{y})$. The associated combinational circuit is shown in Fig. 11.4. This problem can be solved in two steps. The first step applies Formula (11.48), and the second step realizes an orthogonal minimization. The results of both steps are shown in Fig. 11.13.

A comparison of Figs. 11.10 and 11.13 confirms that both methods find the same description of the behavior.

Note 11.10 The order of the rows in the TVLs can change because the commutative law of the excluded middle holds.

The behavior of a PLA can be calculated using the associated structure-TVL $S_{\text{PLA}}(\mathbf{x}, \mathbf{y})$. The values 0 and 1 in the columns $\mathbf{y}$ indicate the existence of connections to the conjunctions of the AND-matrix. A connection affects the behavior, but a missing connection does not give any information about the

Fig. 11.13 Input–output-system-function $\mathrm{ODA}(F(\mathbf{x},\mathbf{y}))$ of the combinational circuit of Fig. 11.4a

$$
\mathrm{ODA}(F(\mathbf{x},\mathbf{y})) =
\begin{array}{cccccc}
x_1 & x_2 & x_3 & x_4 & y_1 & y_2 \\
\hline
0 & 0 & 0 & 0 & 1 & 1 \\
0 & 0 & 0 & 1 & 1 & 0 \\
0 & 0 & 1 & - & 1 & 0 \\
0 & 1 & 0 & 0 & 1 & 1 \\
0 & 1 & 0 & 1 & 1 & 0 \\
0 & 1 & 1 & - & 0 & 1 \\
1 & 0 & 0 & 0 & 0 & 1 \\
1 & 0 & 0 & 1 & 0 & 1 \\
1 & 0 & 1 & - & 0 & 1 \\
1 & 1 & 0 & 0 & 0 & 1 \\
1 & 1 & 0 & 1 & 0 & 1 \\
1 & 1 & 1 & - & 1 & 0 \\
\end{array}
=
\begin{array}{cccccc}
x_1 & x_2 & x_3 & x_4 & y_1 & y_2 \\
\hline
0 & 0 & 1 & - & 1 & 0 \\
0 & - & 0 & 0 & 1 & 1 \\
0 & - & 0 & 1 & 1 & 0 \\
0 & 1 & 1 & - & 0 & 1 \\
1 & 1 & 0 & - & 0 & 1 \\
1 & 1 & 1 & - & 1 & 0 \\
1 & 0 & - & - & 0 & 1 \\
\end{array}
$$

Fig. 11.14 System-function $F(\mathbf{x},\mathbf{y})$ of the PLA of Fig. 11.6a

$$
\mathrm{ODA}(F(\mathbf{x},\mathbf{y})) =
\begin{array}{ccccccc}
x_1 & x_2 & x_3 & x_4 & y_1 & y_2 & y_3 \\
\hline
- & 1 & 1 & 0 & 1 & 0 & 1 \\
1 & 0 & 0 & 1 & 1 & 1 & 1 \\
1 & 1 & 0 & 1 & 1 & 1 & 0 \\
1 & - & 0 & 0 & 1 & 1 & 0 \\
- & 0 & 1 & 0 & 0 & 0 & 0 \\
0 & - & 0 & 0 & 0 & 0 & 0 \\
0 & 0 & - & 1 & 0 & 1 & 1 \\
0 & 1 & - & 1 & 0 & 1 & 0 \\
1 & 0 & 1 & 1 & 0 & 0 & 1 \\
1 & 1 & 1 & 1 & 0 & 0 & 0 \\
\end{array}
$$

behavior of the output-function. Formula (11.49) uses this property in order to calculate the behavior $F(\mathbf{x},\mathbf{y})$ of the PLA-structure $S_{\mathrm{PLA}}(\mathbf{x},\mathbf{y})$:

$$
F(\mathbf{x},\mathbf{y}) = \bigwedge_{i=1}^{k}\left(y_i \odot \max_{\mathbf{y}}^{k}(y_i \wedge S_{\mathrm{PLA}}(\mathbf{x},\mathbf{y})) \right). \tag{11.49}
$$

For each output, the used conjunctions are separately selected. After removing the unnecessary partial structure information by means of the k-fold maximum, we get the function of the selected output. The system-function $F(\mathbf{x}, y_i)$ is created by applying (11.13). Finally, the input–output-system-function $F(\mathbf{x},\mathbf{y})$ can be calculated based on (11.29) that is used in (11.49).

Example 11.9 Using (11.49), we calculate the system-function $F(\mathbf{x},\mathbf{y})$ of the PLA of Fig. 11.6a. Figure 11.14 shows the system-function $F(\mathbf{x},\mathbf{y})$ that describes the behavior of the PLA.

The comparison of the structure-TVL in Fig. 11.6c and the associated behavioral description in Fig. 11.14 shows obvious differences. While each vector $\mathbf{y}$ in the structure-TVL contains both the values 0 and 1 as indicator of a connection, in the behavioral TVL there appear vectors $\mathbf{y}$ where all values are equal to 0 or all values are equal to 1. The second behavioral vector of Fig. 11.14 expresses, for instance, that all three outputs y_i are equal to 1 for the assignment $(x_1, x_2, x_3, x_4) = (1, 0, 0, 1)$ to the inputs of the PLA. This behavior is realized in the third row of the PLA for the outputs y_1 and y_2, but the fourth row of the PLA creates $y_3 = 1$ for this input-pattern. The last vector of Fig. 11.14 is an example for the case that all three outputs y_i are equal to 0. The comparison with the PLA-structure shows that none of the four conjunctions covers the input-pattern $(x_1, x_2, x_3, x_4) = (1, 1, 1, 1)$.

The methods explained for non-reversible combinational circuits can be reused for reversible combinational circuits because the their structures are also specified by systems of logic equations or sets of local lists of phases. The only peculiarity is that the number of functions observable at the outputs y_i and the number of intermediate values g_{ji} at the outputs of the reversible gates j are equal to the number of inputs x_i.

The values of at least two variables of each pair of rows of a list of phases of a reversible gate or a whole reversible circuit are different; hence, it is not possible to merge any pair of such binary vectors into a ternary vector so that 2^k binary vectors describe the behavior of a reversible circuit of k inputs. The variables of intermediate functions g_{ji} show for each input-pattern the associated values behind the reversible gate j; fortunately, these variables do not increase the number of binary vectors that describe the behavior of the reversible circuit. The *input–output-system-function* $F(\mathbf{x}, \mathbf{y})$ can also be computed using the *complete system-function* $F(\mathbf{x}, \mathbf{g}, \mathbf{y})$ and a k-fold maximum as specified in Eq. (11.48).

The solution-set of a system of k logic equations that describes a single reversible gate together with unchanged wires is the associated local list of phases; hence, the intersection of such local lists of phases of the reversible circuit avoids some intermediate solution-steps. This can be seen in Fig. 11.7 where the following mappings exist:

- Equations 1, 2, and 3 have the solution-set $G[1]$;
- Equations 4, 5, and 6 have the solution-set $G[2]$; and
- Equations 7, 8, and 9 have the solution-set $G[3]$.

The number of binary vectors of the intermediate results remain restricted to 2^k when the local list of phases, used in an intersection, belongs to a reversible gate that is a neighbor of the so far computed part of the circuit; the direction (from left to right or vice versa) in which the local lists of phases are used in the intersections has no influence on the number of rows in the intermediate and final TVL.

Example 11.10 We compute the behavior of the reversible combinational circuit of Fig. 11.7a using the local lists of phases shown in Fig. 11.7c. Figure 11.15 shows the steps to compute the complete system-function $F(\mathbf{x}, \mathbf{g}, \mathbf{y})$ that describes the complete behavior of this reversible combinational circuit.

Alternatively, the intersection of the nine solution-sets of the nine equations of Fig. 11.7b can be computed and has also the complete system-function $F(\mathbf{x}, \mathbf{g}, \mathbf{y})$ as result.

Particularly easy is the computation of the behavior of two-level gate structures as shown in Fig. 11.5 and the two types of contact circuits of Fig. 11.8. The structure of these circuits is modeled by TVLs in the four non-orthogonal forms: $D(f)$, $K(f)$, $A(f)$, and $E(f)$. Each of these circuits represents a single logic function. The associated behavior can be expressed by an orthogonal TVL in ODA-form; each ternary vector of ODA(f) describes input-patterns for which the output of the circuit is equal to 1. Hence, the analysis task for all these circuits can be reduced to the transformation of a given non-orthogonal TVL to the associated orthogonal TVL in ODA form.

Figure 11.16 shows a scheme of the needed transformation steps. Starting points are the TVLs in the white blocks. Depending on the given circuit structure, one of these TVLs can be created as shown in Figs. 11.5 and 11.8. As the first step, the given non-orthogonal TVL will be orthogonalized

Fig. 11.15 Behavior of the reversible combinational circuit of Fig. 11.7a: steps to compute the complete system-function $F(\mathbf{x}, \mathbf{g}, \mathbf{y})$ and the restriction to the input–output-system-function $F(\mathbf{x}, \mathbf{y})$

$$\bigcap_{i=1}^{2} G[i] =$$

x_1	x_2	x_3	g_{11}	g_{12}	g_{13}	g_{21}	g_{22}	g_{23}	y_1	y_2	y_3
0	0	0	0	1	0	0	1	0			
0	0	1	0	1	1	0	1	1			
0	1	0	0	0	0	0	0	0			
0	1	1	0	0	1	0	0	1			
1	0	0	1	1	0	1	1	1			
1	0	1	1	1	1	1	1	0			
1	1	0	1	0	0	1	0	0			
1	1	1	1	0	1	1	0	1			

$$F(\mathbf{x}, \mathbf{g}, \mathbf{y}) = \bigcap_{i=1}^{3} G[i] =$$

x_1	x_2	x_3	g_{11}	g_{12}	g_{13}	g_{21}	g_{22}	g_{23}	y_1	y_2	y_3
0	0	0	0	1	0	0	1	0	1	1	0
0	0	1	0	1	1	0	1	1	1	1	1
0	1	0	0	0	0	0	0	0	0	0	0
0	1	1	0	0	1	0	0	1	0	0	1
1	0	0	1	1	0	1	1	1	0	1	1
1	0	1	1	1	1	1	1	0	0	1	0
1	1	0	1	0	0	1	0	0	1	0	0
1	1	1	1	0	1	1	0	1	1	0	1

$$F(\mathbf{x}, \mathbf{y}) = \max{}_{\mathbf{g}}{}^{k} F(\mathbf{x}, \mathbf{g}, \mathbf{y}) =$$

x_1	x_2	x_3	y_1	y_2	y_3
0	0	0	1	1	0
0	0	1	1	1	1
0	1	0	0	0	0
0	1	1	0	0	1
1	0	0	0	1	1
1	0	1	0	1	0
1	1	0	1	0	0
1	1	1	1	0	1

using Algorithm ORTH (red blocks in Fig. 11.16) that takes the given form of the TVL into account for both the realized transformation and the form of the result. In case of a given TVL $D(f)$ or $K(f)$ the wanted behavioral description $ODA(f)$ (the green block in Fig. 11.16) is already reached.

In the case of the other two basic forms the intermediate orthogonal TVL, which can be used as a conjunctive or equivalence form ($OKE(f)$), will be transformed into the wanted TVL $ODA(f)$ in two further steps:

1. the negation with regard to the law of De Morgan (red block NDM in Fig. 11.16) creates a TVL in ODA-form of the negated function;
2. the complement operation (red block CPL in Fig. 11.16) computes the wanted behavioral description $ODA(f)$.

The names in the red blocks of Fig. 11.16 are the function names provided in the XBOOLE-library [13]; the commands of the XBOOLE-monitor XBM 2 [17] replace the characters of these names by lower-case characters.

Example 11.11 The behavior of the two-level NOR–NOR circuit of Fig. 11.5b is wanted. The associated TVL $K(f)$ is already given in Fig. 11.5b. Figure 11.17 demonstrates the three steps for the calculation of the behavior of this NOR–NOR-circuit. In the first step the orthogonal TVL $OKE(f)$ is created; in this special case, all pairs of ternary vectors of the given TVL $K(f)$ are already orthogonal to each other; hence, only the form has been changed. In the second step the negation with regard to De Morgan is executed by exchanging the values 0 and 1 as well as the assignment of the new

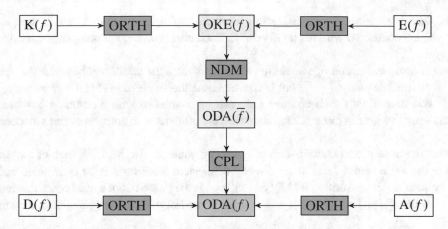

Fig. 11.16 Scheme to calculate the behavior of all types of two-level circuits

$$
K(f) = \begin{array}{cccc} a & b & c & d \\ \hline - & 0 & 0 & - \\ 0 & - & 1 & 0 \\ 1 & - & 1 & 1 \end{array} \longrightarrow \boxed{ORTH} \longrightarrow OKE(f) = \begin{array}{cccc} a & b & c & d \\ \hline - & 0 & 0 & - \\ 0 & - & 1 & 0 \\ 1 & - & 1 & 1 \end{array}
$$

$$
OKE(f) = \begin{array}{cccc} a & b & c & d \\ \hline - & 0 & 0 & - \\ 0 & - & 1 & 0 \\ 1 & - & 1 & 1 \end{array} \longrightarrow \boxed{NDM} \longrightarrow ODA(\overline{f}) = \begin{array}{cccc} a & b & c & d \\ \hline - & 1 & 1 & - \\ 1 & - & 0 & 1 \\ 0 & - & 0 & 0 \end{array}
$$

$$
ODA(\overline{f}) = \begin{array}{cccc} a & b & c & d \\ \hline - & 1 & 1 & - \\ 1 & - & 0 & 1 \\ 0 & - & 0 & 0 \end{array} \longrightarrow \boxed{CPL} \longrightarrow ODA(f) = \begin{array}{cccc} a & b & c & d \\ \hline - & 0 & 1 & - \\ 0 & - & 0 & 1 \\ 1 & - & 0 & 0 \end{array}
$$

Fig. 11.17 Calculation of the behavior of the NOR–NOR-circuits of Fig. 11.5b

specification of the TVL $ODA(\overline{f})$. In the final step the complement is calculated in order to get the TVL $ODA(f)$ from the given TVL $ODA(\overline{f})$. It can be seen that the output of this circuit is equal to 1 for eight input-patterns summarized by the three orthogonal ternary vectors of the calculated behavior.

The analysis method of two-level circuits can also be used for *circuits of switching contacts* with the structures of a parallel-series connection or a series–parallel connection as introduced in Fig. 11.8. For instance, the same transformation steps of Example 11.11 also calculate the behavior of the series–parallel connection of Fig. 11.8b because this structure of switches is described by the same TVL $K(f)$ as the explored in NOR–NOR-circuit.

More flexible circuit structures of switching contacts can be analyzed by means of the *contact potential model*. Both make and break contacts can be connected on several points of different potentials. One or more such potential points are fixed to the logic value 1, and at least one other potential point is connected with the logic value 0. Each other connection point can be used as output of the circuit.

Such an unrestricted circuit of switching contacts provides the additional behavior that for certain input-patterns a chosen output potential is connected neither with the value 1 nor with the value 0. More than one output with such an open state can be connected with a common bus-wire, which extends the use of switching circuits. Bus structures are often used in computers and other controlling systems.

However, the unrestricted combination of switching contacts also bears the risk of a *short-circuit fault* where the logic values 1 and 0 are directly connected. Such fault must be avoided and can be detected by an analysis. Assuming that $F_i(p_{0i}, x_j, p_{1i})$ is the phase list of a used contact as introduced in Table 11.4 and $F_i(p_{1i}) = 1$ and $F_i(p_{0i}) = 0$ the phase lists of the fixed potentials, then the phase list

$$F_g(\mathbf{x}, \mathbf{p}) = \bigwedge_{i=1}^{k} F_i(p_{0i}, x_j, p_{1i}) \tag{11.50}$$

describes the *conflict-free global behavior* of the circuit; no phase of a short cut belongs to this phase list. Hence, all input-patterns of *short-circuit faults* $F_{scf}(\mathbf{x})$ can be calculated by:

$$F_{scf}(\mathbf{x}) = \overline{\max_{\mathbf{p}}^{k} F_g(\mathbf{x}, \mathbf{p})} \ . \tag{11.51}$$

Only circuits of switching contacts that satisfy the condition

$$F_{scf}(\mathbf{x}) \equiv 0$$

can really be used. Hence, an analysis regarding possible short-circuit faults is necessary for all freely configured circuits of switching contacts.

Example 11.12 It is the aim of this example to verify whether the circuit of Fig. 11.18 has short-circuit faults or not.

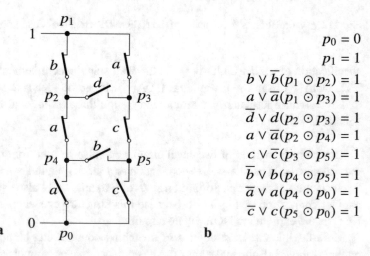

Fig. 11.18 Circuit of switching contacts containing a short-circuit fault: (**a**) circuit structure and (**b**) associated system of logic equations

$$p_0 = 0$$
$$p_1 = 1$$
$$b \vee \overline{b}(p_1 \odot p_2) = 1$$
$$a \vee \overline{a}(p_1 \odot p_3) = 1$$
$$\overline{d} \vee d(p_2 \odot p_3) = 1$$
$$a \vee \overline{a}(p_2 \odot p_4) = 1$$
$$c \vee \overline{c}(p_3 \odot p_5) = 1$$
$$\overline{b} \vee b(p_4 \odot p_5) = 1$$
$$\overline{a} \vee a(p_4 \odot p_0) = 1$$
$$\overline{c} \vee c(p_5 \odot p_0) = 1$$

As solution of the system of equations of Fig. 11.18b, we get the global list of phases:

$$\mathrm{ODA}(F_g(a,b,c,d,\mathbf{p})) = \begin{array}{cccccccccc} a & b & c & d & p_0 & p_1 & p_2 & p_3 & p_3 & p_5 \\ \hline 0 & - & 0 & - & 0 & 1 & 1 & 1 & 1 & 1 \\ 1 & 0 & 0 & - & 0 & 1 & 1 & 1 & 0 & 1 \\ 0 & 0 & 1 & - & 0 & 1 & 1 & 1 & 1 & 0 \\ 1 & - & 1 & - & 0 & 1 & 1 & 1 & 0 & 0 \\ - & 1 & 1 & 0 & 0 & 1 & 0 & 1 & 0 & 0 \\ 1 & - & - & 0 & 0 & 1 & 1 & 0 & 0 & 0 \\ 1 & 1 & - & - & 0 & 1 & 0 & 0 & 0 & 0 \\ \hline \end{array}$$

which is used in (11.51) so that we get

$$F_{scf}(a,b,c,d) = \overline{a}bcd .$$

For this input-pattern, the short-circuit path is

$$1 - p_1 - a - p_3 - d - p_2 - a - p_4 - b - p_5 - c - p_0 - 0 .$$

Due to this short-circuit fault, the circuit of Fig. 11.18a cannot be used when all input-patterns are possible.

Found short-circuit faults can be used as hint to modify the circuit such that the no short-circuit fault remains. The short-circuit fault in the circuit of Fig. 11.18a can be, for instance, eliminated by the replacement of the make contact d by a series of the two make contacts a and d. A repeated analysis of the modified circuit regarding short-circuit faults should be executed.

For the circuit of switching contacts without a short-circuit fault and the used output p_o can be analyzed which input-patterns cause an *open output* (not connected with 0 as well as not connected with 1); both the output-values 0 and 1 are possible for such input-patterns:

$$F_{open}(\mathbf{x}) = \min_{p_o} \left(\max_{\mathbf{p} \backslash p_o}{}^{k-1} F_g(\mathbf{x}, \mathbf{p}) \right) . \tag{11.52}$$

The phase list $F_{det}(\mathbf{x}, p_o)$ describes in this case the input–output-phases for the *input-patterns* $\mathbf{x}$, *which determine the value of the output* either to 0 or to 1:

$$F_{det}(\mathbf{x}, p_o) = \overline{F_{open}(\mathbf{x})} \wedge \max_{\mathbf{p} \backslash p_o}{}^{k-1} F_g(\mathbf{x}, \mathbf{p}) . \tag{11.53}$$

Example 11.13 We modified the circuit of Fig. 11.18 by inserting the additional make contact a between the potential p_2 and the new potential p_6 as shown in Fig. 11.19.

The global list of phases of the circuit of Fig. 11.19 is

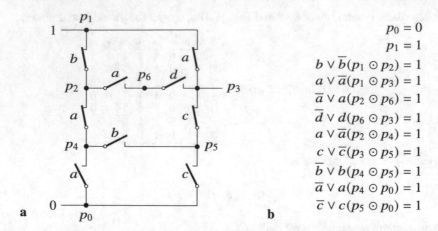

Fig. 11.19 Circuit of switching contacts containing no short-circuit fault: (a) circuit structure and (b) associated system of logic equations

$$
\mathrm{ODA}(F_g(a,b,c,d,\mathbf{p})) =
\begin{array}{cccc|ccccccc}
a & b & c & d & p_0 & p_1 & p_2 & p_3 & p_3 & p_5 & p_6 \\
\hline
0 & - & 0 & 0 & 0 & 1 & 1 & 1 & 1 & 1 & - \\
0 & - & 0 & 1 & 0 & 1 & 1 & 1 & 1 & 1 & 1 \\
1 & 0 & 0 & - & 0 & 1 & 1 & 1 & 0 & 1 & 1 \\
0 & 0 & 1 & 0 & 0 & 1 & 1 & 1 & 1 & 0 & - \\
0 & 0 & 1 & 1 & 0 & 1 & 1 & 1 & 1 & 0 & 1 \\
1 & - & 1 & - & 0 & 1 & 1 & 1 & 0 & 0 & 1 \\
- & 1 & 1 & 0 & 0 & 1 & 0 & 1 & 0 & 0 & 0 \\
0 & 1 & 1 & - & 0 & 1 & 0 & 1 & 0 & 0 & 1 \\
1 & - & - & 0 & 0 & 1 & 1 & 0 & 0 & 0 & 1 \\
1 & 1 & - & - & 0 & 1 & 0 & 0 & 0 & 0 & 0 \\
\end{array}
.
$$

The solution of (11.51)

$$ F_{scf}(a,b,c,d) = 0 $$

confirms that the circuit of Fig. 11.19a does not have any short-circuit fault. After this check, we analyze the behavior at the chosen output p_3. Using (11.52), we get

$$
\mathrm{ODA}(F_{open}(a,b,c,d)) =
\begin{array}{cccc}
a & b & c & d \\
\hline
1 & 0 & - & 0 \\
1 & 1 & 1 & - \\
\end{array}
$$

so that the output p_3 is open for four of the 16 input-patterns. By means of (11.53), the determined input–output-phases have been computed:

$$
\mathrm{ODA}(F_{det}(a,b,c,d)) =
\begin{array}{ccccc}
a & b & c & d & p_3 \\
\hline
1 & 0 & - & 1 & 1 \\
0 & - & - & - & 1 \\
1 & 1 & 0 & - & 0 \\
\end{array}
.
$$

Hence, for ten input-patterns the output p_3 is equal to 1, and for the remaining two input-patterns the value 0 is determined for this output.

The complete system-function $F(\mathbf{x}, \mathbf{g}, \mathbf{y})$ is an excellent behavioral model of a combinational circuit. Many special analysis problems can be solved by means of this function. The full understanding of the solution-process of logic equations and the application of Boolean derivatives cooperate very elegantly and efficiently.

In a *forward simulation* the outputs $\mathbf{y}$ must be calculated for a given input-pattern $in(\mathbf{x})$. This analysis problem is solved by the following formula:

$$out(\mathbf{y}) = \max_{\mathbf{x}, \mathbf{g}}^{n+k} (in(\mathbf{x}) \wedge F(\mathbf{x}, \mathbf{g}, \mathbf{y})) .$$

Classical *backward simulations* of combinational circuits are more complicated than forward simulations because several input-patterns can cause the known output-pattern. Based on the complete system-function $F(\mathbf{x}, \mathbf{g}, \mathbf{y})$, the backward simulation is as simple as the forward simulation. In a backward simulation the inputs $\mathbf{x}$ must be calculated for a given output-pattern $out(\mathbf{y})$. This analysis problem is solved by the following equation:

$$in(\mathbf{x}) = \max_{\mathbf{g}, \mathbf{y}}^{k+m} (out(\mathbf{y}) \wedge F(\mathbf{x}, \mathbf{g}, \mathbf{y})) . \tag{11.54}$$

A further analytical problem is the detection of input–output-patterns $inout(\mathbf{x}, \mathbf{y})$ that occur together with a given pattern observable at the internal wires $internal(\mathbf{g})$. The formula

$$inout(\mathbf{x}, \mathbf{y}) = \max_{\mathbf{g}}^{k} (internal(\mathbf{g}) \wedge F(\mathbf{x}, \mathbf{g}, \mathbf{y}))$$

solves this problem again by using the complete system-function $F(\mathbf{x}, \mathbf{g}, \mathbf{y})$.

Example 11.14 We assume that $internal(g_1, g_2, g_3)$ is equal to $(1, 1, 1)$ for the combinational circuit of Fig. 11.4a with the complete system-function $F(\mathbf{x}, \mathbf{g}, \mathbf{y})$ of Fig. 11.12, then the restricted input–output-behavior is

$$inout(x_1, x_2, x_3, x_4, y_1, y_2) = (0, 1, 1, -, 0, 1) .$$

Large digital systems consist of several parts where one part may be a combinational circuit that controls another part of the system. If the considered combinational circuit with m outputs cannot create all 2^m output-patterns, then there is some freedom in the synthesis of the controlled system part. A further aim of the analysis of a combinational circuit is therefore the calculation of the don't-care-function $f_\varphi(\mathbf{y})$ of the circuit-outputs. Based on the system-function $F(\mathbf{x}, \mathbf{y})$, this task is solved as follows:

$$f_\varphi(\mathbf{y}) = \overline{\max_{\mathbf{x}}^{n} F(\mathbf{x}, \mathbf{y})} . \tag{11.55}$$

Example 11.15 The don't-care-function $f_\varphi(\mathbf{y})$ of the outputs of the PLA of Fig. 11.6a with the system-function of Fig. 11.14 has to be calculated. Using (11.55), we get the result $\varphi(\mathbf{y}) = y_1 \overline{y}_2 \overline{y}_3$. It is not directly visible in the structure of the PLA of Fig. 11.6a, but in the associated system-function $F(\mathbf{x}, \mathbf{y})$ of Fig. 11.14 shows that all patterns except

$$(y_1, y_2, y_3) = (1, 0, 0)$$

occur as possible output-patterns of this PLA.

It is a property of combinational circuits that each output-function uniquely depends on the inputs; hence, the logic function $f_i(\mathbf{x})$ of the output y_i can be computed using the system-function $F(\mathbf{x}, \mathbf{y})$:

$$f_i(\mathbf{x}) = \max_{\mathbf{y}}^m (y_i \wedge F(\mathbf{x}, \mathbf{y})) \ .$$

Each function $y = f(\mathbf{x})$ can be analyzed with regard to its dynamic properties. The single derivation (6.30) with regard to the input x_i determines all patterns of the other inputs for which the change of the input x_i changes the value of the output $y = f(\mathbf{x})$. If Eq. (11.56) holds for all input-patterns, then the function $f(\mathbf{x})$ is independent of the input x_i:

$$\det_{x_i} f(\mathbf{x}) = 0 \ . \tag{11.56}$$

Note 11.11 The function $f(\mathbf{x})$ is known in this analysis.

Consequently, the single derivative of $f(\mathbf{x})$ in the Boolean differential equation (11.56) can be calculated so that a logic equation arises and has to be solved.

Similarly, all the other derivative operations can be applied in order to analyze a special dynamic behavior of the output-function $f(\mathbf{x})$.

In the case that several inputs change their values at the same time the behavior can be determined by means of vectorial derivative operations. The vectorial minimum (6.9) determines, for instance, the patterns of the inputs $\mathbf{x}_1$ for which the output-value remains constant equal to 1 if all values of the inputs $\mathbf{x}_0$ change their values at the same time.

The influence of all patterns of the k inputs $\mathbf{x}_0$ to the output of the function $y = f(\mathbf{x}) = f(\mathbf{x}_0, \mathbf{x}_1)$ can be analyzed by means of k-fold derivative operations. The solution of Equation

$$\max_{\mathbf{x}_0}^k f(\mathbf{x}_0, \mathbf{x}_1) = 0$$

determines the patterns of the inputs $\mathbf{x}_1$ for which the output-value remains constant equal to 0 for all input-patterns $\mathbf{x}_0$.

Models that use logic functions do not explicitly consider the time. For that reason, it can be supposed that several variables change their values at the same time. Physically (i.e., in the sense of electronics) it is not possible to change two logic values exactly at the same moment. A small time difference can be observed between the points in time where different variables change their values if a very fine time-scale is used.

The simultaneous change of two logic values can therefore be modeled using three time-steps. In the first time-step both variables have their starting value. In the second time-step one of the two variables has changed the value. Both variables have changed their values in the third time-step.

If the values of two variables change and the function has the same function value in the first and third time-steps, no change of the function value happens in the logic model. If there is the opposite function value at the second time-step, an additional short impulse appears at the output. Such a disturbing additional impulse is denoted by *glitch*. A *hazard* is the possibility that a glitch can occur. A hazard causes, for instance, an error if the number of changes of the output is counted. There are three conditions for such a *function-hazard*:

1. two input-variables change their values;
2. the function has the same value before and after these changes; and
3. there are different function values in the associated subspace.

Fig. 11.20 Types of function-hazards

shapes of the output-signals

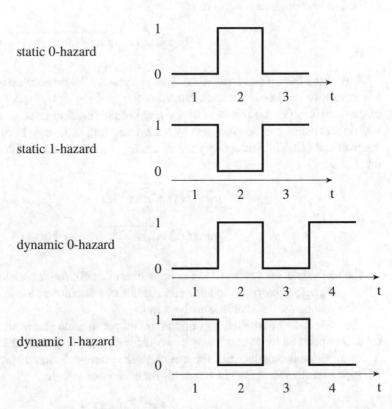

Equation (11.57) considers these conditions, and the solutions are the patterns where a hazard can occur when the variables (x_i, x_j) change their values, for $\mathbf{x} = (x_i, x_j, \mathbf{x}_1)$:

$$\overline{\underset{(x_i,x_j)}{\text{der}} \ f(x_i, x_j, \mathbf{x}_1)} \wedge \Delta_{(x_i,x_j)} f(x_i, x_j, \mathbf{x}_1) = 1 . \tag{11.57}$$

There are different types of hazards. First we distinguish between *static* and *dynamic* hazards. Static hazards have been introduced above. Their characteristic feature is that the function has the same value before and after two input-variables changed their values. A dynamic hazard requires the change of three input-variables; it is a concatenation of a static hazard and a change of the function value caused by the change of the third input-variable. A second distinction of both types of hazards is the function value before all changes. Figure 11.20 shows the shapes of the output-signals as a function of the time for the classified hazards.

The solutions of Eq. (11.58) describe all static 0-hazards of the function $f(x_i, x_j, \mathbf{x}_1)$ with regard to the change of the variables (x_i, x_j). The complement of the vectorial maximum determines that the values of $f(x_i, x_j, \mathbf{x}_1)$ are equal to 0 before and after these changes. There must be at least one intermediate function value 1 if the 2-fold maximum is equal to 1:

$$\overline{\underset{(x_i,x_j)}{\max} \ f(x_i, x_j, \mathbf{x}_1)} \wedge \underset{(x_i,x_j)}{\max}{}^2 f(x_i, x_j, \mathbf{x}_1) = 1 . \tag{11.58}$$

All static 1-hazards of $f(x_i, x_j, \mathbf{x}_1)$ with regard to the change of variables (x_i, x_j) can be calculated using Eq. (11.59). The vectorial minimum determines that values of $f(x_i, x_j, \mathbf{x}_1)$ are equal to 1 before and after these changes. There must be at least one intermediate function value 0 if the complement of the 2-fold minimum is equal to 1:

$$\min_{(x_i,x_j)} f(x_i, x_j, \mathbf{x_1}) \wedge \overline{\min^2_{(x_i,x_j)} f(x_i, x_j, \mathbf{x_1})} = 1 \ . \tag{11.59}$$

A dynamic hazard can occur when three variables of the function change their values. The dynamic hazard can be modeled by a static hazard of the function $f(x_i, x_j, x_k, \mathbf{x_1})$ with regard to the change of the variables (x_i, x_j) followed by a change of the function value caused by the change of x_k. The additional change can be expressed by a single derivative. Using (11.58), the dynamic 0-hazards are the result of (11.60). Analogously the dynamic 1-hazards are calculated by means of (11.61), based on (11.59):

$$\overline{\max_{(x_i,x_j)} f(\mathbf{x})} \wedge \max^2_{(x_i,x_j)} f(\mathbf{x}) \wedge \det_{x_k} f(\mathbf{x}) = 1 \ , \tag{11.60}$$

$$\min_{(x_i,x_j)} f(\mathbf{x}) \wedge \overline{\min^2_{(x_i,x_j)} f(\mathbf{x})} \wedge \det_{x_k} f(\mathbf{x}) = 1 \ . \tag{11.61}$$

The analysis is not restricted to the properties of realized combinational circuits. A given system-function $F_g(\mathbf{x}, \mathbf{y})$ describes the allowed behavior of a circuit that has to be synthesized. This system-function can be the object of the analysis too.

The system-function $F_g(\mathbf{x}, \mathbf{y})$ summarizes all permissible phases. A combinational circuit for the behavior specified by $F_g(\mathbf{x}, \mathbf{y})$ only exists if for each input-pattern at least one phase is determined. A combinational circuit exists for a *realizable* system-function $F_g(\mathbf{x}, \mathbf{y})$; a given system-function $F_g(\mathbf{x}, \mathbf{y})$ is realizable if Eq. (11.62) holds for each input-pattern:

$$\max_{\mathbf{y}}^k F_g(\mathbf{x}, \mathbf{y}) = 1 \ . \tag{11.62}$$

A realizable system-function $F_g(\mathbf{x}, \mathbf{y})$ can permit several functions for each output. The cost of the combinational circuit strongly depends on the number of inputs of the circuit. An important analysis problem is therefore to check whether the system-function $F_g(\mathbf{x}, \mathbf{y})$ contains for each output $\mathbf{y}$ a function that does not depend on the input x_i. If Formula (11.63) holds for all $(\mathbf{x} \setminus x_i)$, this complex property has been successfully checked. Phases that do not depend on x_i occur in $F_g(\mathbf{x}, \mathbf{y})$ with both the values $x_i = 0$ and $x_i = 1$. The single minimum of $F_g(\mathbf{x}, \mathbf{y})$ with regard to x_i selects these phases. This restricted system-function must be realizable:

$$\max_{\mathbf{y}}^k \left[\min_{x_i} F_g(\mathbf{x}, \mathbf{y})) \right] = 1 \ . \tag{11.63}$$

Some analysis problems take only one single output-function into account. The given system-function $F_{gi}(\mathbf{x}, y_i)$ can be extracted from $F_g(\mathbf{x}, \mathbf{y})$ by (11.64):

$$F_{gi}(\mathbf{x}, y_i) = \max_{\mathbf{y} \setminus y_i}^{k-1} F_g(\mathbf{x}, \mathbf{y}) \ . \tag{11.64}$$

It can be analyzed whether an output-function $y_i = f_i(\mathbf{x})$ is uniquely defined by a system-function $F_{gi}(\mathbf{x}, y_i)$. If the following equation:

$$\det_{y_i} F_{gi}(\mathbf{x}, y_i) = 1 \ . \tag{11.65}$$

is satisfied, the system-function $F_{gi}(\mathbf{x}, y_i)$ determines exactly one output-value y_i for each input-pattern $\mathbf{x}$.

There are many other analytical problems that can be expressed by logic equations or Boolean differential equations. The above examples should help to find for each analytical problem a convenient solution.

11.3 Covering Methods for Synthesis

The goal of logic synthesis consists in transforming the behavioral description given by a logic function into the structure of a combinational circuit. This task is more difficult than the reverse analytical problem. The reason for that is that a given combinational circuit realizes exactly one logic function, but there are many different combinational circuits for the same logic function. The result of the synthesis can be selected by means of additional conditions such as the delay, the power consumption, or the required chip area.

There are two classes of methods to synthesize a combinational circuit. The first class summarizes the *covering methods*. An AND-gate of n inputs creates for one input-vector the value 1 at the output. This value 1 covers one value 1 of a logic function of n inputs if the output of this AND-gate is connected to an input of an OR-gate of the function. If an AND-gate of an AND–OR-circuit is controlled only by $n - k$ inputs, this AND-gate covers 2^k values 1 of a logic function that depends on n inputs. The objective of the covering method in the synthesis of an AND–OR-circuit consists in covering all values 1 of the logic function by a small number of AND-gates that have a small number of inputs. This covering method can be realized based on the disjunctive form. As shown in Fig. 11.5, the disjunctive form of a logic function $f(\mathbf{x})$ can be mapped to both an AND–OR-circuit and a NAND–NAND-circuit. Alternatively, a parallel-series connection as shown in Fig. 11.8 can be used to realize the synthesized disjunctive form of the logic function $f(\mathbf{x})$.

The aim of another covering method are two-level circuits that have OR–AND- or NOR–NOR-structures. This covering method minimizes the associated conjunctive form of the logic function. Using the duality between a disjunctive form and a conjunctive form, we can omit the detailed study of the covering method for circuits in OR–AND- or NOR–NOR-structures. The required minimized conjunctive form of a logic function can be created in three steps:

1. express the function $\overline{f(\mathbf{x})}$ in a disjunctive form;
2. minimize this disjunctive form;
3. transform the minimized disjunctive form of the function $\overline{f(\mathbf{x})}$ into a minimized function $f(\mathbf{x})$ by using the law of De Morgan.

As mentioned above, the smaller the number of literals in a conjunction of a disjunctive form, the more function values 1 will be covered. The basic principles to minimize logic functions have already been explained in Sect. 4.4. Here we will show some more specialized concepts and especially some more detailed examples.

$x \vee \overline{x} = 1$ is the basic law to minimize a disjunctive form. We express a conjunction of $|\mathbf{x}_0|$ literals by $C_0(\mathbf{x}_0)$ and assume that there are two conjunctions $C_1(\mathbf{x}_0, x_i)$ and $C_2(\mathbf{x}_0, x_i)$, defined by (11.66) and (11.67) in the disjunctive form of $f(\mathbf{x})$:

$$C_1(\mathbf{x}_0, x_i) = x_i \wedge C_0(\mathbf{x}_0) \,, \tag{11.66}$$

$$C_2(\mathbf{x}_0, x_i) = \overline{x}_i \wedge C_0(\mathbf{x}_0) \,. \tag{11.67}$$

These two conjunctions can be replaced by the single conjunction (11.68) that depends on one variable less:

$$C_1(\mathbf{x}_0, x_i) \vee C_2(\mathbf{x}_0, x_i) = x_i \wedge C_0(\mathbf{x}_0) \vee \overline{x}_i \wedge C_0(\mathbf{x}_0)$$

$$= (x_i \vee \overline{x}_i) \wedge C_0(\mathbf{x}_0)$$

$$= 1 \wedge C_0(\mathbf{x}_0)$$

$$= C_0(\mathbf{x}_0) . \tag{11.68}$$

Note 11.12 Both the conjunction $C_1(\mathbf{x}_0, x_i)$ and $C_2(\mathbf{x}_0, x_i)$ are covered by the conjunction $C_0(\mathbf{x}_0)$:

$$C_1(\mathbf{x}_0, x_i) \vee C_0(\mathbf{x}_0) = C_0(\mathbf{x}_0) ,$$

$$C_2(\mathbf{x}_0, x_i) \vee C_0(\mathbf{x}_0) = C_0(\mathbf{x}_0) .$$

Since $C_0(\mathbf{x}_0)$ does not contain the variable x_i, the conjunctions $C_1(\mathbf{x}_0, x_i)$ and $C_2(\mathbf{x}_0, x_i)$, which contain additionally either x_i or $\overline{x}_i$, cannot be part of a minimal disjunctive form.

Vice versa, the question arises to determine the conjunctions that occur in a minimal disjunctive form of $f(\mathbf{x})$. Conjunctions that can be part of minimal disjunctive forms of $f(\mathbf{x})$ are denoted by *prime conjunctions*. A *prime conjunction* $PC_i(\mathbf{x}_0)$ of a logic function $f(\mathbf{x})$ is defined by two properties:

1. the prime conjunction $PC_i(\mathbf{x}_0)$ covers only function values 1 of $f(\mathbf{x})$;
2. there is no other conjunction $C_j(\mathbf{x}_1)$ of $f(\mathbf{x})$ that completely covers the prime conjunction $PC_i(\mathbf{x}_0)$.

These two properties can be transformed into the following condition that can be checked by an algorithm:

- the condition $PC_i(\mathbf{x}_0) \leq f(\mathbf{x})$ holds for $PC_i(\mathbf{x}_0)$;
- each omission of one literal in this conjunction causes that this property is not satisfied anymore.

Example 11.16 The logic function $f(a, b, c)$ is given by the disjunctive form

$$f(a, b, c) = \overline{b}c \vee ab \vee a\overline{b}\overline{c} . \tag{11.69}$$

Figure 11.21 illustrates for this function how prime conjunctions can be constructed using the idea given above.

Each Karnaugh-map of Fig. 11.21 depicts the function (11.69). The three conjunctions are visualized by different colors. Equally colored fields of a Karnaugh-map indicate that the associated function value 1 is covered by the same conjunction. Red colored fields emphasize a conflict between the extended conjunction and a given function value 0. Whether the first conjunction $\overline{b}c$ is a prime conjunction is analyzed by means of the upper three Karnaugh-maps. Figure 11.21a1 emphasizes this yellow colored conjunction by a frame. Each element 0 or 1 in the ternary vector of the conjunction is a literal that can be removed. We begin systematically from the left and remove first the literal $\overline{b}$. The remaining conjunction is c that covers the whole lower row in the Karnaugh-map of Fig. 11.21a2. This conjunction covers the red colored value $f(0, 1, 1) = 0$; hence, c is not acceptable as a conjunction of $f(a, b, c)$. Next we remove the literal c and get the conjunction $\overline{b}$ that is emphasized by a frame in Fig. 11.21a3. This conjunction doubles the covered area to the whole left and right columns of the Karnaugh-map and contains three values 1, but also the red colored value $f(0, 0, 0) = 0$; hence, $\overline{b}$ is also not acceptable as conjunction of $f(a, b, c)$. It can be concluded that the conjunction $\overline{b}c$ is a prime conjunction of $f(a, b, c)$.

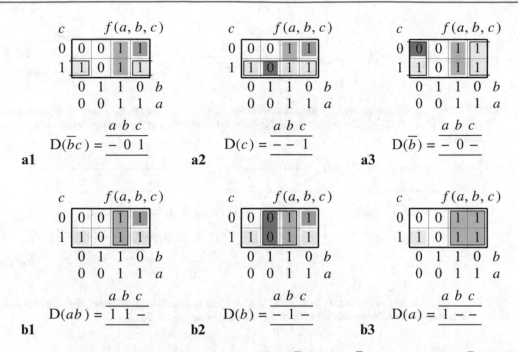

Fig. 11.21 The Karnaugh-maps of the function $f(a, b, c) = \bar{b}c \vee ab \vee a\bar{b}\bar{c}$: (**a1**, **a2**, **a3**) check $\bar{b}c$ for prime conjunctions and (**b1**, **b2**, **b3**) check ab for prime conjunctions

Using the same method, the second conjunction ab must be checked. All details of this check are visualized in the lower row of Fig. 11.21 where the green colored fields in the Karnaugh-map of Fig. 11.21b1 indicate the analyzed conjunction in the third column. Removing the literal a, we get the conjunction b that covers the two columns in the middle of the Karnaugh-map in Fig. 11.21b2. Due to the red colored fields $f(0, 1, 0) = f(0, 1, 1) = 0$, the conjunction b covers two zeros of $f(a, b, c)$. Consequently, the conjunction b is not acceptable as a conjunction of $f(a, b, c)$. Finally, we have to remove the literal b and get the conjunction a. This conjunction covers four values 1 on the right-hand side of the Karnaugh-map, as can be seen by the green filled frame in Fig. 11.21b3. Since no value 0 of $f(a, b, c)$ is covered by b, it follows that the analyzed conjunction ab is no prime conjunction of $f(a, b, c)$.

The conjunction a (consisting only of one single variable) is a prime conjunction of $f(a, b, c)$ since $a \leq f(a, b, c)$, and $f(a, b, c)$ is not constant equal to 1; hence, a cannot be omitted and replaced by $1(a, b, c)$. Using the same method, we find the prime conjunction a from the last given conjunction $a\bar{b}\bar{c}$ again.

The two prime conjunctions in the example of Fig. 11.21 are necessary to cover the function. Thus, the minimal disjunctive form is

$$f(a, b, c) = \bar{b}c \vee a \, . \tag{11.70}$$

The two two-level gate structures as well as the parallel-series connection of switching contacts associated to this disjunctive form are shown in Fig. 11.22.

The logic operations NOT, AND, and OR can also be realized by a reversible circuit using ancilla lines. Fig. 11.23 shows reversible circuits of these three logic operations and their use to realize the minimal disjunctive form (11.70). Repeated gates (see, e.g., line a in Fig. 11.23d) restore the previous function of this line so that it can be reused; this approach can be used to reduce the number of ancilla lines.

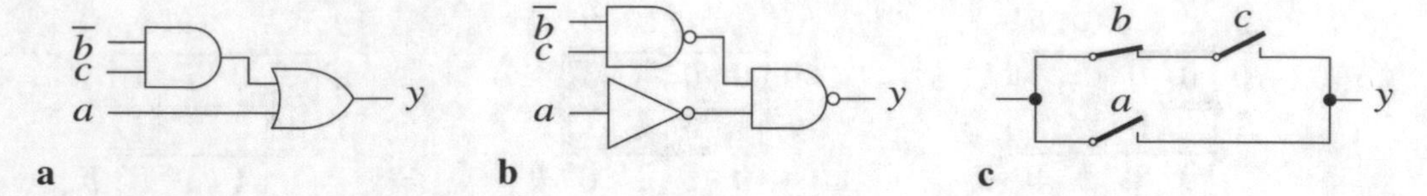

Fig. 11.22 Structures of the synthesized two-level combinational circuits associated to the minimal disjunctive form $y = f(a, b, c) = \overline{b}\,c \vee a$: (**a**) AND–OR-circuit, (**b**) NAND–NAND-circuit, and (**c**) parallel–series connection of switching contacts

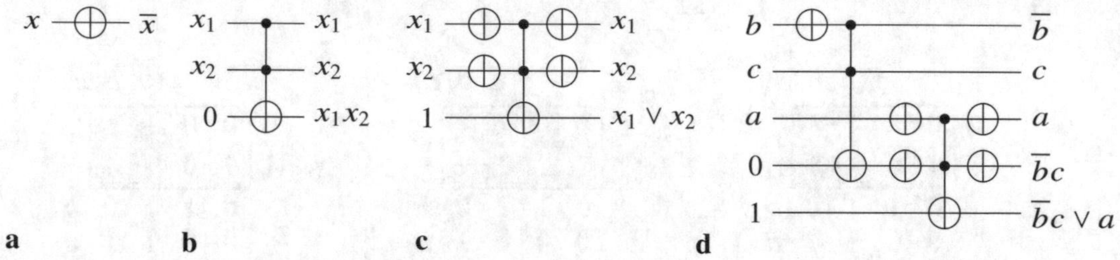

Fig. 11.23 Reversible circuits of elementary logic operations and their application: (**a**) NOT ($\overline{x}$), (**b**) AND ($x_1 \wedge x_2$), (**c**) OR ($x_1 \vee x_2$), and (**d**) reversible circuit of the function $f(a, b, c) = \overline{b}c \vee a$ (11.70)

The most general method—the application of the consensus, together with the absorption law—can be applied as soon as a disjunctive form for the function $f(\mathbf{x})$ to be considered is known. This is especially the case when all vectors $\mathbf{x} \in \mathbb{B}^n$ with $f(\mathbf{x}) = 1$ are known; see Sect. 4.4. Alternatively, the classical Quine–McCluskey algorithm can be used in order to calculate all prime conjunctions of a logic function.

Generally, not all prime conjunctions are necessary to cover the logic function completely. A further subtask in the synthesis of two-level circuits by covering methods consists in the selection of a set of prime conjunctions that cover the function without any redundant prime conjunction. The general procedure has been explained in Sect. 4.4; hence, the following example should be understandable.

Example 11.17 We look for all minimal disjunctive forms of the logic function $f(a, b, c, d)$ that is used in the Karnaugh-maps of Fig. 11.24. Eight prime conjunctions exist for this function. Each of these prime conjunctions is:

- enumerated by a model variable p_i;
- expressed by a ternary vector; and
- a colored block in the Karnaugh-map located above this vector in Fig. 11.24.

For a better understanding, the disjunctions of the cover function $cov_f(\mathbf{p})$ are labeled by the selection vector (a, b, c, d):

$$
\begin{aligned}
cov_f(\mathbf{p}) = \; & \underbrace{(p_2 \vee p_3)}_{0010} \wedge \underbrace{(p_2 \vee p_5)}_{0011} \wedge \underbrace{(p_4)}_{0101} \wedge \underbrace{(p_3 \vee p_6)}_{0110} \wedge \underbrace{(p_1)}_{1000} \wedge \underbrace{(p_5 \vee p_7)}_{1011} \\
& \wedge \underbrace{(p_1 \vee p_8)}_{1100} \wedge \underbrace{(p_4 \vee p_8)}_{1101} \wedge \underbrace{(p_6 \vee p_8)}_{1110} \wedge \underbrace{(p_7 \vee p_8)}_{1111} .
\end{aligned}
$$

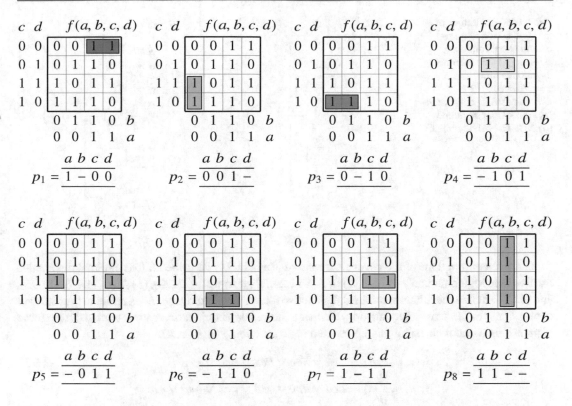

Fig. 11.24 Prime conjunctions of the logic function $f(a, b, c, d)$ expressed by ternary vectors and emphasized in associated Karnaugh-maps

We simplify this cover function by two absorptions and re-order the disjunctions for easy applications of the distributive law:

$$cov_f(\mathbf{p}) = (p_1)(p_4)(p_2 \vee p_3)(p_3 \vee p_6)(p_2 \vee p_5)(p_5 \vee p_7)$$
$$(p_6 \vee p_8)(p_7 \vee p_8).$$

The application of the distributive law, in combination with the absorption law, results in the disjunctive form of the cover function:

$$cov_f(\mathbf{p}) = (p_1)(p_4)(p_2 p_6 \vee p_3)(p_2 p_7 \vee p_5)(p_6 p_7 \vee p_8)$$
$$= (p_1 p_4)(p_2 p_6 p_7 \vee p_2 p_5 p_6 \vee p_2 p_3 p_7 \vee p_3 p_5)(p_6 p_7 \vee p_8)$$
$$= (p_1 p_4)(p_2 p_6 p_7 \vee p_2 p_6 p_7 p_8 \vee p_2 p_5 p_6 p_7 \vee p_2 p_5 p_6 p_8$$
$$\vee p_2 p_3 p_6 p_7 \vee p_2 p_3 p_7 p_8 \vee p_3 p_5 p_6 p_7 \vee p_3 p_5 p_8)$$
$$= (p_1 p_4)(p_2 p_6 p_7 \vee p_2 p_5 p_6 p_8 \vee p_2 p_3 p_7 p_8 \vee p_3 p_5 p_6 p_7 \vee p_3 p_5 p_8)$$
$$= p_1 p_2 p_4 p_6 p_7 \vee p_1 p_2 p_4 p_5 p_6 p_8 \vee p_1 p_2 p_3 p_4 p_7 p_8 \vee$$
$$p_1 p_3 p_4 p_5 p_6 p_7 \vee p_1 p_3 p_4 p_5 p_8. \tag{11.71}$$

Fig. 11.25 Structures of
the synthesized two-level
combinational circuits
associated to the minimal
disjunctive form
$y = f_5(a, b, c, d) =$
$a\overline{c}\overline{d} \vee \overline{a}cd \vee b\overline{c}d \vee \overline{b}cd \vee ab$:
(**a**) AND–OR-circuit and
(**b**) NAND–NAND-circuit

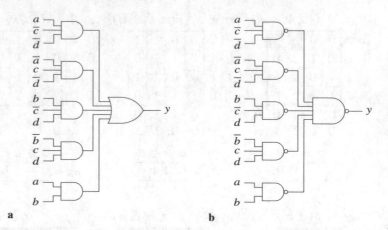

The five conjunctions of the cover function $cov_f(\mathbf{p})$ (11.71) describe all five minimal disjunctive forms of the logic function $f(a, b, c, d)$ of Fig. 11.24. The formulas (11.72), (11.73), (11.74), (11.75), and (11.76) define the different minimal disjunctive forms. The introduced index refers to the different expressions for the same logic function. Neither a literal nor a conjunction can be removed from these expressions without changing the represented logic function $f(a, b, c, d)$:

$$f_1(a, b, c, d) = a\overline{c}\overline{d} \vee \overline{a}\overline{b}c \vee b\overline{c}d \vee bc\overline{d} \vee acd \,, \tag{11.72}$$

$$f_2(a, b, c, d) = a\overline{c}\overline{d} \vee \overline{a}\overline{b}c \vee b\overline{c}d \vee \overline{b}cd \vee bc\overline{d} \vee ab \,, \tag{11.73}$$

$$f_3(a, b, c, d) = a\overline{c}\overline{d} \vee \overline{a}\overline{b}c \vee \overline{a}c\overline{d} \vee b\overline{c}d \vee acd \vee ab \,, \tag{11.74}$$

$$f_4(a, b, c, d) = a\overline{c}\overline{d} \vee \overline{a}c\overline{d} \vee b\overline{c}d \vee \overline{b}cd \vee bc\overline{d} \vee acd \,, \tag{11.75}$$

$$f_5(a, b, c, d) = a\overline{c}\overline{d} \vee \overline{a}c\overline{d} \vee b\overline{c}d \vee \overline{b}cd \vee ab \,. \tag{11.76}$$

The expressions of $f_2(a, b, c, d)$, $f_3(a, b, c, d)$, and $f_4(a, b, c, d)$ need six conjunctions, but $f_1(a, b, c, d)$ or $f_5(a, b, c, d)$ can be realized by five AND-gates in the first level. The expression $f_5(a, b, c, d)$ needs one literal less than the expression for $f_1(a, b, c, d)$. Thus, we decided to use the disjunctive form $f_5(a, b, c, d)$ for an AND–OR-circuit. Figure 11.25 shows both the AND–OR-circuit and the NAND–NAND-circuit of the synthesized function of the selected expression $f_5(a, b, c, d)$.

There are five different minimal expressions for the given function specified by the Karnaugh-maps of Fig. 11.24, but all of them contain the prime conjunctions labeled by p_1 and p_4. Such prime conjunctions are denoted by *essential prime conjunctions*. It is easier to select a minimal disjunctive form out of all prime conjunctions by means of heuristic approaches when the essential prime conjunctions are known. A prime conjunction PC_i is an essential prime conjunction if the following inequality holds:

$$PC_i \wedge \overline{\bigvee_{j=1,\, j\neq i}^{k} PC_j} \neq 0 \,. \tag{11.77}$$

Let us apply (11.77) to the previous example. In the case of $PC_1 = a\overline{c}\overline{d}$ we get $a\overline{b}\overline{c}\overline{d} \neq 0$. It can be seen in Fig. 11.24 that this elementary conjunction is only covered by PC_1. For $PC_4 = b\overline{c}d$, the application of (11.77) results in $\overline{a}b\overline{c}d \neq 0$. For all the other prime conjunctions of Example 11.17, the left-hand side of (11.77) is equal to 0. Consequently, these prime conjunctions are no essential prime conjunctions.

It is easy to generalize the covering method for incompletely specified functions. In this case the prime conjunctions can cover additionally the don't-care-function $f_\varphi(\mathbf{x})$. This degree of freedom can lead to a smaller number of literals in the prime conjunctions. A *prime conjunction* $PC_i(\mathbf{x}_0)$ of an incompletely specified function with the ON-set-function $f_q(\mathbf{x})$ and the don't-care-function $f_\varphi(\mathbf{x})$ is defined by three properties:

1. the prime conjunction $PC_i(\mathbf{x}_0)$ covers only function values 1 of the logic function $f_q(\mathbf{x}) \vee f_\varphi(\mathbf{x})$;
2. there is no other conjunction $C_j(\mathbf{x}_1)$ of $f_q(\mathbf{x}) \vee f_\varphi(\mathbf{x})$ that covers the prime conjunction $PC_i(\mathbf{x}_0)$ completely;
3. the prime conjunction $PC_i(\mathbf{x}_0)$ covers at least one function value 1 of $f_q(\mathbf{x})$.

A minimal complete cover of an incompletely specified function requires only that all function values for $\mathbf{c}_j$ with $f_q(\mathbf{c}_j) = 1$ are covered by at least one prime conjunction. Thus, the cover function can be restricted to the following $cov_q(\mathbf{p})$:

$$cov_q(\mathbf{p}) = \bigwedge_{\{j \,|\, f_q(\mathbf{c}_j)=1\}} \bigvee_{\{i \,|\, PC_i(\mathbf{c}_j)=1\}} p_i \; . \tag{11.78}$$

Covering methods are very useful if their results can be mapped directly to two-level circuits. If the required number of inputs is larger than the available number of inputs of a logic gate, then cascades of gates must be built and the optimality will be lost. In this case it is better to use *decomposition methods* for the synthesis the combinational circuit.

11.4 Decomposition Methods for Synthesis

In opposition to the covering methods (which cover the logic function by a small number of prime conjunctions), the *decomposition methods* try to decompose a given logic function into two or more simpler subfunctions. This problem is significantly more difficult because neither the subfunctions nor the function that combines the created subfunctions to the logic function to be decomposed is known.

Separation of a Single Variable

The simplest approach of a decomposition of a logic function $f(\mathbf{x}_0, x_i)$ is the separation of a single variable x_i using an AND-, an OR-, or an EXOR-gate. The remaining function $g(\mathbf{x}_0)$ of such a separation depends on one variable less than the given function and is therefore simpler. Unfortunately, the separation of a variable is not possible for all functions. A given logic function must satisfy a certain condition to separate a selected variable using one of the gates mentioned above. Next we show the circuit structure, the property of the circuit, the condition that a variable can be separated, and the formula to compute the decomposition-function $g(\mathbf{x}_0)$ for all three types of gate as well as the separation of both x_i and $\overline{x}_i$.

Figure 11.26 shows the circuit structure in which the variable x_i is separated using an AND-gate.

The variable x_i can be separated from the function $f(\mathbf{x}_0, x_i)$ using an AND-gate if this function can be represented as

$$f(\mathbf{x}_0, x_i) = x_i \wedge g(\mathbf{x}_0) \; ; \tag{11.79}$$

Fig. 11.26 Circuit where x_i is separated using an AND-gate

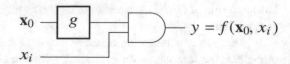

this is possible if Condition

$$\max_{x_i} (\overline{x}_i \wedge f(\mathbf{x}_0, x_i)) = 0 \tag{11.80}$$

is satisfied. The fitness of Condition (11.80) for the separation of x_i using an AND-gate can easily be verified by the substitution of the given property (11.79) into (11.80):

$$\max_{x_i} (\overline{x}_i \wedge f(\mathbf{x}_0, x_i)) = \max_{x_i} (\overline{x}_i \wedge x_i \wedge g(\mathbf{x}_0)) = \max_{x_i} (0) = 0 \, .$$

The decomposition-function $g(\mathbf{x}_0)$ of the circuit shown in Fig. 11.26 for the separation of x_i using an AND-gate can be computed as follows:

$$g(\mathbf{x}_0) = \max_{x_i} (x_i \wedge f(\mathbf{x}_0, x_i)) \, . \tag{11.81}$$

The substitution of (11.79) into (11.81), the utilization of the property that $g(\mathbf{x}_0)$ does not depend on x_i, and several steps of simplifications confirm that Eq. (11.81) correctly determines the decomposition-function $g(\mathbf{x}_0)$ of a separation of x_i using an AND-gate:

$$\max_{x_i} (x_i \wedge f(\mathbf{x}_0, x_i)) = \max_{x_i} (x_i \wedge x_i \wedge g(\mathbf{x}_0)) = \max_{x_i} (g(\mathbf{x}_0) \wedge x_i)$$

$$= g(\mathbf{x}_0) \wedge \max_{x_i} (x_i) = g(\mathbf{x}_0) \wedge 1 = g(\mathbf{x}_0) \, .$$

While $f(\mathbf{x}_0, x_i = 0)$ must be equal to 0 for the separation of x_i using an AND-gate, the dual property, $f(\mathbf{x}_0, x_i = 1) = 1$, must be satisfied for the separation of x_i using an OR-gate. Figure 11.27 shows the circuit structure in which the variable x_i is separated using an OR-gate.

The variable x_i can be separated from the function $f(\mathbf{x}_0, x_i)$ using an OR-gate if this function can be represented as

$$f(\mathbf{x}_0, x_i) = x_i \vee g(\mathbf{x}_0) \, ; \tag{11.82}$$

this is possible if Condition

$$\min_{x_i} (\overline{x}_i \vee f(\mathbf{x}_0, x_i)) = 1 \tag{11.83}$$

is satisfied. The fitness of Condition (11.83) for the separation of x_i using an OR-gate can easily be verified by the substitution of the given property (11.82) into (11.83):

$$\min_{x_i} (\overline{x}_i \vee f(\mathbf{x}_0, x_i)) = \min_{x_i} (\overline{x}_i \vee x_i \vee g(\mathbf{x}_0)) = \min_{x_i} (1) = 1 \, .$$

The decomposition-function $g(\mathbf{x}_0)$ of the circuit shown in Fig. 11.27 for the separation of x_i using an OR-gate can be computed as follows:

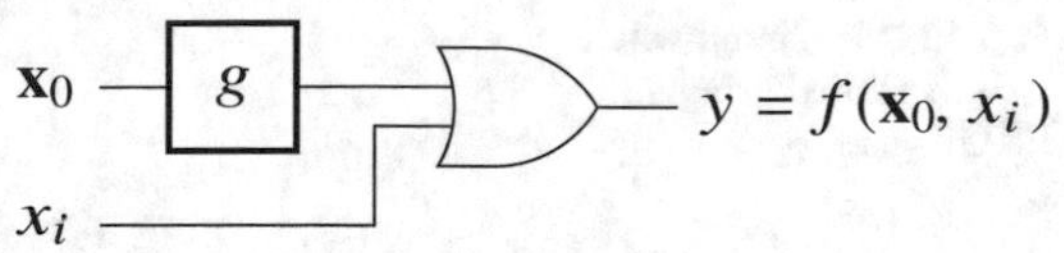

Fig. 11.27 Circuit where x_i is separated using an OR-gate

$$g(\mathbf{x}_0) = \max_{x_i} (\overline{x}_i \wedge f(\mathbf{x}_0, x_i)) \ . \tag{11.84}$$

The substitution of (11.82) into (11.84), the utilization of the property that $g(\mathbf{x}_0)$ does not depend on x_i, and several steps of simplifications confirm that Eq. (11.84) correctly determines the decomposition-function $g(\mathbf{x}_0)$ of a separation of x_i using an OR-gate:

$$\max_{x_i} (\overline{x}_i \wedge f(\mathbf{x}_0, x_i)) = \max_{x_i} (\overline{x}_i \wedge (x_i \vee g(\mathbf{x}_0))) = \max_{x_i} (0 \vee \overline{x}_i \wedge g(\mathbf{x}_0))$$

$$= g(\mathbf{x}_0) \wedge \max_{x_i} (\overline{x}_i) = g(\mathbf{x}_0) \wedge 1 = g(\mathbf{x}_0) \ .$$

The third gate that can be used to separate the variable x_i from a logic function $f(\mathbf{x}_0, x_i)$ is the EXOR-gate; Fig. 11.28 shows the associated circuit structure.

The variable x_i can be separated from the function $f(\mathbf{x}_0, x_i)$ using an EXOR-gate if this function can be represented as

$$f(\mathbf{x}_0, x_i) = x_i \oplus g(\mathbf{x}_0) \ ; \tag{11.85}$$

hence, such a function must be linear with regard to x_i and the Condition to be satisfied is

$$\operatorname*{der}_{x_i} f(\mathbf{x}_0, x_i) = 1 \ . \tag{11.86}$$

The fitness of Condition (11.86) for the separation of x_i using an EXOR-gate can easily be verified by the substitution of the given property (11.85) into (11.86):

$$\operatorname*{der}_{x_i} f(\mathbf{x}_0, x_i) = \operatorname*{der}_{x_i} (x_i \oplus g(\mathbf{x}_0)) = \operatorname*{der}_{x_i} x_i \oplus \operatorname*{der}_{x_i} g(\mathbf{x}_0) = 1 \oplus 0 = 1 \ .$$

The decomposition-function $g(\mathbf{x}_0)$ of the circuit shown in Fig. 11.28 for the separation of x_i using an EXOR-gate can be computed as follows:

$$g(\mathbf{x}_0) = \max_{x_i} (\overline{x}_i \wedge f(\mathbf{x}_0, x_i)) \ . \tag{11.87}$$

The substitution of (11.85) into (11.87), the utilization of the property that $g(\mathbf{x}_0)$ does not depend on x_i, and several steps of simplifications confirm that Eq. (11.87) correctly determines the decomposition-function $g(\mathbf{x}_0)$ of a separation of x_i using an EXOR-gate:

$$\max_{x_i} (\overline{x}_i \wedge f(\mathbf{x}_0, x_i)) = \max_{x_i} (\overline{x}_i \wedge (x_i \oplus g(\mathbf{x}_0))) = \max_{x_i} (0 \oplus \overline{x}_i \wedge g(\mathbf{x}_0))$$

$$= g(\mathbf{x}_0) \wedge \max_{x_i} (\overline{x}_i) = g(\mathbf{x}_0) \wedge 1 = g(\mathbf{x}_0) \ .$$

Fig. 11.28 Circuit where x_i is separated using an EXOR-gate

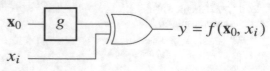

Fig. 11.29 Circuit where $\overline{x}_i$ is separated using an AND-gate

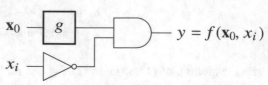

If not any of the above explained separations of the variable x_i is possible the separation of the negated variable $\overline{x}_i$ can be explored.

Figure 11.29 shows the circuit structure in which the variable $\overline{x}_i$ is separated using an AND-gate.

The variable $\overline{x}_i$ can be separated from the function $f(\mathbf{x}_0, x_i)$ using an AND-gate if this function can be represented as

$$f(\mathbf{x}_0, x_i) = \overline{x}_i \wedge g(\mathbf{x}_0) ; \tag{11.88}$$

this is possible if Condition

$$\max_{x_i} (x_i \wedge f(\mathbf{x}_0, x_i)) = 0 \tag{11.89}$$

is satisfied. The fitness of Condition (11.89) for the separation of $\overline{x}_i$ using an AND-gate can easily be verified by the substitution of the given property (11.88) into (11.89):

$$\max_{x_i} (x_i \wedge f(\mathbf{x}_0, x_i)) = \max_{x_i} (x_i \wedge \overline{x}_i \wedge g(\mathbf{x}_0)) = \max_{x_i} (0) = 0 .$$

The decomposition-function $g(\mathbf{x}_0)$ of the circuit shown in Fig. 11.29 for the separation of $\overline{x}_i$ using an AND-gate can be computed as follows:

$$g(\mathbf{x}_0) = \max_{x_i} (\overline{x}_i \wedge f(\mathbf{x}_0, x_i)) . \tag{11.90}$$

The substitution of (11.88) into (11.90), the utilization of the property that $g(\mathbf{x}_0)$ does not depend on x_i, and several steps of simplifications confirm that Eq. (11.81) correctly determines the decomposition-function $g(\mathbf{x}_0)$ of a separation of $\overline{x}_i$ using an AND-gate:

$$\max_{x_i} (\overline{x}_i \wedge f(\mathbf{x}_0, x_i)) = \max_{x_i} (\overline{x}_i \wedge \overline{x}_i \wedge g(\mathbf{x}_0)) = \max_{x_i} (g(\mathbf{x}_0) \wedge \overline{x}_i)$$

$$= g(\mathbf{x}_0) \wedge \max_{x_i} (\overline{x}_i) = g(\mathbf{x}_0) \wedge 1 = g(\mathbf{x}_0) .$$

The use of an OR-gate instead of an AND-gate in the separation of the negated variable $\overline{x}_i$ from the function $f(\mathbf{x}_0, x_i)$ leads to dual properties. Figure 11.30 shows the circuit structure in which the variable $\overline{x}_i$ is separated using an OR-gate.

The variable $\overline{x}_i$ can be separated from the function $f(\mathbf{x}_0, x_i)$ using an OR-gate if this function can be represented as

Fig. 11.30 Circuit where $\overline{x}_i$ is separated using an OR-gate

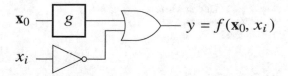

$$f(\mathbf{x}_0, x_i) = \overline{x}_i \vee g(\mathbf{x}_0) \; ; \tag{11.91}$$

this is possible if Condition

$$\min_{x_i} (x_i \vee f(\mathbf{x}_0, x_i)) = 1 \tag{11.92}$$

is satisfied. The fitness of Condition (11.92) for the separation of $\overline{x}_i$ using an OR-gate can easily be verified by the substitution of the given property (11.91) into (11.92):

$$\min_{x_i} (x_i \vee f(\mathbf{x}_0, x_i)) = \min_{x_i} (x_i \vee \overline{x}_i \vee g(\mathbf{x}_0)) = \min_{x_i} (1) = 1 \; .$$

The decomposition-function $g(\mathbf{x}_0)$ of the circuit shown in Fig. 11.30 for the separation of $\overline{x}_i$ using an OR-gate can be computed as follows:

$$g(\mathbf{x}_0) = \max_{x_i} (x_i \wedge f(\mathbf{x}_0, x_i)) \; . \tag{11.93}$$

The substitution of (11.91) into (11.93), the utilization of the property that $g(\mathbf{x}_0)$ does not depend on x_i, and several steps of simplifications confirm that Eq. (11.93) correctly determines the decomposition-function $g(\mathbf{x}_0)$ of a separation of $\overline{x}_i$ using an OR-gate:

$$\max_{x_i} (x_i \wedge f(\mathbf{x}_0, x_i)) = \max_{x_i} (x_i \wedge (\overline{x}_i \vee g(\mathbf{x}_0))) = \max_{x_i} (0 \vee \overline{x}_i \wedge g(\mathbf{x}_0))$$

$$= g(\mathbf{x}_0) \wedge \max_{\overline{x}_i} (x_i) = g(\mathbf{x}_0) \wedge 1 = g(\mathbf{x}_0) \; .$$

The separation of the negated variable $\overline{x}_i$ from the logic function $f(\mathbf{x}_0, x_i)$ is possible if the separation of the non-negated variable x_i can be separated from the same function. This follows from the rule: $a \oplus b = \overline{a} \oplus \overline{b}$. For completeness we provide the details for this last case of separation of a single variable from the function $f(\mathbf{x}_0, x_i)$.

Figure 11.31 shows the circuit structure in which the negated variable $\overline{x}_i$ is separated from the function $f(\mathbf{x}_0, x_i)$ using an EXOR-gate.

The variable $\overline{x}_i$ can be separated from the function $f(\mathbf{x}_0, x_i)$ using an EXOR-gate if this function can be represented as

$$f(\mathbf{x}_0, x_i) = \overline{x}_i \oplus g(\mathbf{x}_0) \; ; \tag{11.94}$$

hence, such a function must be linear with regard to $\overline{x}_i$ and the Condition to be satisfied is

$$\mathop{\mathrm{der}}_{x_i} f(\mathbf{x}_0, x_i) = 1 \; . \tag{11.95}$$

Fig. 11.31 Circuit where
$\overline{x}_i$ is separated using an
EXOR-gate

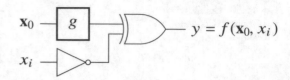

The fitness of Condition (11.95) for the separation of $\overline{x}_i$ using an EXOR-gate can easily be verified by the substitution of the given property (11.94) into (11.95):

$$\operatorname*{der}_{x_i} f(\mathbf{x}_0, x_i) = \operatorname*{der}_{x_i} (\overline{x}_i \oplus g(\mathbf{x}_0)) = \operatorname*{der}_{x_i} \overline{x}_i \oplus \operatorname*{der}_{x_i} g(\mathbf{x}_0) = 1 \oplus 0 = 1 \ .$$

The decomposition-function $g(\mathbf{x}_0)$ of the circuit shown in Fig. 11.31 for the separation of $\overline{x}_i$ using an EXOR-gate can be computed as follows:

$$g(\mathbf{x}_0) = \max_{x_i} (x_i \wedge f(\mathbf{x}_0, x_i)) \ . \tag{11.96}$$

The substitution of (11.94) into (11.96), the utilization of the property that $g(\mathbf{x}_0)$ does not depend on x_i, and several steps of simplifications confirm that Eq. (11.96) correctly determines the decomposition-function $g(\mathbf{x}_0)$ of a separation of $\overline{x}_i$ using an EXOR-gate:

$$\max_{x_i} (x_i \wedge f(\mathbf{x}_0, x_i)) = \max_{x_i} (x_i \wedge (\overline{x}_i \oplus g(\mathbf{x}_0))) = \max_{x_i} (0 \oplus x_i \wedge g(\mathbf{x}_0))$$

$$= g(\mathbf{x}_0) \wedge \max_{x_i} (x_i) = g(\mathbf{x}_0) \wedge 1 = g(\mathbf{x}_0) \ .$$

Taking 2^n, where n is the number of variables a logic function $f(\mathbf{x}_0, x_i)$ is depending on, as a simple complexity measure, the complexity of the decomposition-function $g(\mathbf{x}_0)$, created as result of the separation of a single variable from the function $f(\mathbf{x}_0, x_i)$, is equal to 2^{n-1}, i.e., equal to one half of the complexity of basic function $f(\mathbf{x}_0, x_i)$. Due to this significant simplification, the separation of a single variable should be used. Unfortunately, the larger the number of variables the function $f(\mathbf{x}_0, x_i)$ is depending on, the smaller is the percentage of functions for which a separation of a single variable exists. Therefore further decomposition methods have been explored.

Pipe Decompositions
Approximately at the same time, *Ashenhurst* [1] and *Povarov* [10] suggested similar approaches to decompose a logic function. The Ashenhurst-decomposition (11.97) is a non-disjoint decomposition of the logic function $f(\mathbf{x}_0, \mathbf{x}_1, \mathbf{x}_2)$ into the decomposition-functions $g(\mathbf{x}_0, \mathbf{x}_2)$ and $h(g(\mathbf{x}_0, \mathbf{x}_2), \mathbf{x}_1, \mathbf{x}_2)$:

$$y = f(\mathbf{x}_0, \mathbf{x}_1, \mathbf{x}_2) = h(g(\mathbf{x}_0, \mathbf{x}_2), \mathbf{x}_1, \mathbf{x}_2) \ . \tag{11.97}$$

The benefit of an Ashenhurst-decomposition increases when the number of variables in the common set of variables $\mathbf{x}_2$ decreases. Unfortunately, the number of functions for which an Ashenhurst-decomposition exists is smaller for smaller sets of variables $\mathbf{x}_2$.

Povarov suggested the disjoint decomposition (11.98) that has less degrees of freedom to select the decomposition-functions $g(\mathbf{x}_0)$ and $h(g, \mathbf{x}_1)$, but the most simple circuit structure:

$$y = f(\mathbf{x}_0, \mathbf{x}_1) = h(g(\mathbf{x}_0), \mathbf{x}_1) \ . \tag{11.98}$$

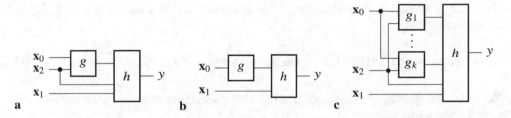

Fig. 11.32 Pipe decompositions of a logic function $f(\mathbf{x})$, suggested by: (**a**) Ashenhurst, (**b**) Povarov, and (**c**) Curtis

However, the method of Ashenhurst to find the functions $g(\mathbf{x}_0, \mathbf{x}_2)$ and $h(g, \mathbf{x}_1, \mathbf{x}_2)$ by means of large tables was not practicable. The function $h(g, \mathbf{x}_1, \mathbf{x}_2)$ is simpler than the given logic function $f(\mathbf{x}_0, \mathbf{x}_1, \mathbf{x}_2)$ if the set of variables $\mathbf{x}_0$ contains at least two variables.

Several years later, *Curtis* [4] published a more general decomposition method (11.99) together with a practicable algorithm. Curtis replaced the single decomposition-function $g(\mathbf{x}_0, \mathbf{x}_2)$ by a set of decomposition-functions $g_1(\mathbf{x}_0, \mathbf{x}_2), \ldots, g_k(\mathbf{x}_0, \mathbf{x}_2)$:

$$y = f(\mathbf{x}_0, \mathbf{x}_1, \mathbf{x}_2) = h(g_1(\mathbf{x}_0, \mathbf{x}_2), \ldots, g_k(\mathbf{x}_0, \mathbf{x}_2), \mathbf{x}_1, \mathbf{x}_2) \,. \tag{11.99}$$

The function $h(g_1, \ldots, g_k, \mathbf{x}_1, \mathbf{x}_2)$ of the Curtis-decomposition is simpler than the given logic function $f(\mathbf{x}_0, \mathbf{x}_1, \mathbf{x}_2)$ if the set of variables $\mathbf{x}_0$ contains at least $k + 1$ variables.

Figure 11.32 shows the structures of these three decompositions and emphasizes both the uniform basic structure and the modifications in the number of functions g and their dependencies on the input-variables. Generally, the functions h are controlled by functions g in all three decomposition structures. For that reason, we denote all these approaches by *pipe decomposition*.

It is a property of the given function $y = f(\mathbf{x})$ whether one of these decompositions exists for a chosen assignment of the variables $\mathbf{x}$ to the subsets of variables $\mathbf{x}_0$, $\mathbf{x}_1$, and optionally $\mathbf{x}_2$. One method to verify whether an Ashenhurst-decomposition for chosen subsets of variables $\mathbf{x}_0$ and $\mathbf{x}_1$ utilizes the general structure of the function $h(g, \mathbf{x}_1, \mathbf{x}_2)$:

$$h(g, \mathbf{x}_1, \mathbf{x}_2) = h_1(\mathbf{x}_1, \mathbf{x}_2) \vee g \wedge h_2(\mathbf{x}_1, \mathbf{x}_2) \vee \overline{g} \wedge h_3(\mathbf{x}_1, \mathbf{x}_2) \,, \tag{11.100}$$

where g is the output-variable of the decomposition-function $g(\mathbf{x}_0)$.

The function $h_1(\mathbf{x}_1, \mathbf{x}_2)$ is determined by

$$h_1(\mathbf{x}_1, \mathbf{x}_2) = \min_{\mathbf{x}_0}^k f(\mathbf{x}_0, \mathbf{x}_1, \mathbf{x}_2) \tag{11.101}$$

if an Ashenhurst-decomposition of $y = f(\mathbf{x}_0, \mathbf{x}_1, \mathbf{x}_2)$ with regard to k variables $\mathbf{x}_0$ and l variables $\mathbf{x}_1$ exists. The decomposition-function $h_2(\mathbf{x}_1, \mathbf{x}_2)$ is not uniquely defined; one possible function $h_2(\mathbf{x}_1, \mathbf{x}_2)$ is

$$h_2(\mathbf{x}_1, \mathbf{x}_2) = \max_{\mathbf{x}_0}^k \left(\bigwedge_{x_i \in \mathbf{x}_0} x_i \wedge (f(\mathbf{x}_0, \mathbf{x}_1, \mathbf{x}_2) \oplus h_1(\mathbf{x}_1, \mathbf{x}_2)) \right) . \tag{11.102}$$

The function $h_3(\mathbf{x}_1, \mathbf{x}_2)$ is determined by the given function $f(\mathbf{x}_0, \mathbf{x}_1, \mathbf{x}_2)$, the uniquely specified function $h_1(\mathbf{x}_1, \mathbf{x}_2)$, the chosen function $h_2(\mathbf{x}_1, \mathbf{x}_2)$, and the used distribution of the variables:

$$h_3(\mathbf{x}_1, \mathbf{x}_2) = \overline{h_1(\mathbf{x}_1, \mathbf{x}_2)} \wedge \overline{h_2(\mathbf{x}_1, \mathbf{x}_2)} \wedge \max_{\mathbf{x}_0}^k f(\mathbf{x}_0, \mathbf{x}_1, \mathbf{x}_2) \,. \tag{11.103}$$

The potential decomposition-function $g(\mathbf{x}_0, \mathbf{x}_2)$ is determined by the chosen functions $h_2(\mathbf{x}_1, \mathbf{x}_2)$ and $h_3(\mathbf{x}_1, \mathbf{x}_2)$ as follows:

$$g(\mathbf{x}_0, \mathbf{x}_2) = \max_{\mathbf{x}_1}^l \left(h_2(\mathbf{x}_1, \mathbf{x}_2) \wedge f(\mathbf{x}_0, \mathbf{x}_1, \mathbf{x}_2) \vee h_3(\mathbf{x}_1, \mathbf{x}_2) \wedge \overline{f(\mathbf{x}_0, \mathbf{x}_1, \mathbf{x}_2)} \right) . \qquad (11.104)$$

The composed function $f'(\mathbf{x}_0, \mathbf{x}_1, \mathbf{x}_2)$ of a potential Ashenhurst-decomposition can built using all these decomposition-functions:

$$f'(\mathbf{x}_0, \mathbf{x}_1, \mathbf{x}_2) = h_1(\mathbf{x}_1, \mathbf{x}_2) \vee g(\mathbf{x}_0, \mathbf{x}_2)h_2(\mathbf{x}_1, \mathbf{x}_2) \vee \overline{g(\mathbf{x}_0, \mathbf{x}_2)}h_3(\mathbf{x}_1, \mathbf{x}_2) . \qquad (11.105)$$

If Condition

$$f(\mathbf{x}_0, \mathbf{x}_1, \mathbf{x}_2) \equiv f'(\mathbf{x}_0, \mathbf{x}_1, \mathbf{x}_2) \qquad (11.106)$$

is satisfied, an Ashenhurst-decomposition of the logic function $y = f(\mathbf{x}_0, \mathbf{x}_1, \mathbf{x}_2)$ with regard to the dedicated subsets of variables $\mathbf{x}_0$ and $\mathbf{x}_1$ really exists.

It is necessary to know the subsets of variables $\mathbf{x}_0$ and $\mathbf{x}_1$ to evaluate the rules to compute the functions $g(\mathbf{x}_0, \mathbf{x}_2)$ and $h(g, \mathbf{x}_1, \mathbf{x}_2)$ of an Ashenhurst-decomposition; however, these subsets are initially unknown, and there are many possibilities to split the set of all variables into the three sets of variables for which the rules given above can be checked.

A more practical approach is the computation of an initial Ashenhurst-decomposition, where the set of variables $\mathbf{x}_0$ contains two variables, a single variable x_i is assigned to the set of variables $\mathbf{x}_1$, and all other variables belong to the remaining set of variables $\mathbf{x}_2$. The set of variables $\mathbf{x}_0$ of the function $g(\mathbf{x}_0, \mathbf{x}_2)$ must contain at least two variables; otherwise, the number of variables of the function $h(g, \mathbf{x}_1, \mathbf{x}_2)$ is not smaller than the number of variables of the given function $f(\mathbf{x}_0, \mathbf{x}_1, \mathbf{x}_2)$. Knowing such an initial Ashenhurst-decomposition, a selected variable of the set of variables $\mathbf{x}_2$ can be moved preferably to the set of variables $\mathbf{x}_0$, but alternatively also to the set of variables $\mathbf{x}_1$, to check whether such an improved Ashenhurst-decomposition or even a Povarov-decomposition exists.

The restriction of the set of variables $\mathbf{x}_1$ to a single variable x_i facilitates the use of Condition

$$\Delta_{\mathbf{x}_0} \left(\det_{x_i} f(\mathbf{x}_0, x_i, \mathbf{x}_2) \right) \wedge \min_{x_i} \left(\Delta_{\mathbf{x}_0} f(\mathbf{x}_0, x_i, \mathbf{x}_2) \right) = 0 \qquad (11.107)$$

for a direct check whether an Ashenhurst-decomposition

$$f(\mathbf{x}_0, x_i, \mathbf{x}_2) = h(g(\mathbf{x}_0, \mathbf{x}_2), x_i, \mathbf{x}_2) \qquad (11.108)$$

exists. The two subfunctions $f(\mathbf{x}_0, x_i = 0, \mathbf{x}_2)$ and $f(\mathbf{x}_0, x_i = 1, \mathbf{x}_2)$ determine for each space $\mathbf{x}_2 = const.$ whether such an initial Ashenhurst-decomposition exists:

1. $f(\mathbf{x}_0, x_i = 0, \mathbf{x}_2) = f(\mathbf{x}_0, x_i = 1, \mathbf{x}_2)$: the derivative with regard to x_i is constant equal to 0 so that the first Δ-operation of (11.107) is equal to 0;
2. $f(\mathbf{x}_0, x_i = 0, \mathbf{x}_2) = \overline{f(\mathbf{x}_0, x_i = 1, \mathbf{x}_2)}$: the derivative with regard to x_i is constant equal to 1 so that the first Δ-operation of (11.107) is also equal to 0;
3. $f(\mathbf{x}_0, x_i = 0, \mathbf{x}_2)$ is constant equal to 0 or constant equal to 1 so that the second Δ-operation with regard to $\mathbf{x}_0$ is equal to 0 and the subsequent minimum with regard to x_i is equal to 0; and
4. $f(\mathbf{x}_0, x_i = 1, \mathbf{x}_2)$ is constant equal to 0 or constant equal to 1 so that the second Δ-operation with regard to $\mathbf{x}_0$ is equal to 0 and the subsequent minimum with regard to x_i is equal to 0.

Condition (11.107) is satisfied for all these four cases for which an Ashenhurst-decomposition (11.108) exists. For all other functions $f(\mathbf{x}_0, x_i, \mathbf{x}_2)$, no Ashenhurst-decomposition exists because at least for one pattern of $\mathbf{x}_2$ the derivative with regard to x_i is not constant so that the first part of (11.107) is equal to 1 and both $f(\mathbf{x}_0, x_i = 0, \mathbf{x}_2)$ and $f(\mathbf{x}_0, x_i = 1, \mathbf{x}_2)$ are not constant functions so that the second part of (11.107) is also equal to 1; hence, Condition (11.107) is not satisfied.

The synthesis method introduced for the Ashenhurst-decomposition can be adopted for both the Povarov-decomposition and the Curtis-decomposition:

- a Povarov-decomposition is an Ashenhurst-decomposition for the special case that the set of variables $\mathbf{x}_2$ is an empty set; hence, all the rules given above can also be used to verify whether a Povarov-decomposition exists and to determine the associated decomposition-functions $g(\mathbf{x}_0)$ and $h(g(\mathbf{x}_0), \mathbf{x}_1)$;
- additionally to the split into $2^{|\mathbf{x}_2|}$ decompositions for all different patterns of the commonly used variables $\mathbf{x}_2 = const.$, in the case of the Curtis-decomposition the k decomposition-functions $g_i(\mathbf{x}_0, \mathbf{x}_2)$ must be taken into account.

Alternatively, BDDs [13], SAT-solvers [6], and further techniques have recently been utilized for these types of decompositions.

Example 11.18 It is our first aim to verify whether a Povarov-decomposition exists for the given function:

$$f(\mathbf{x}) = x_1 x_2 x_4 \vee \overline{x}_1 \overline{x}_3 \overline{x}_5 \vee \overline{x}_2 \overline{x}_3 \overline{x}_5 \vee x_3 x_4$$

with regard to the dedicated sets of variables

$$\mathbf{x}_0 = (x_1, x_2, x_3) \quad \text{and} \quad \mathbf{x}_1 = (x_4, x_5) .$$

Figure 11.33 shows the Karnaugh-maps of the given function, all potential decomposition-functions, and the composed function $f'(\mathbf{x}_0, \mathbf{x}_1)$. All these function have been calculated using the formulas given above.

It can be seen that Condition (11.106) is satisfied. Hence, in the second step the really usable decomposition-functions are determined by:

$$y = h(g, \mathbf{x}_1) = x_4 \overline{x}_5 \vee g \wedge x_4 x_5 \vee \overline{g} \wedge \overline{x}_4 \overline{x}_5 ,$$

$$g = g(\mathbf{x}_0) = x_1 x_2 \vee x_3 .$$

The green colored block in the Karnaugh-map of $f'(\mathbf{x}_0, \mathbf{x}_1)$ emphasizes the chosen decomposition-function $g(\mathbf{x}_0)$ that is selected by the function $h_2(\mathbf{x}_1)$, and the yellow colored block indicates the associated function $\overline{g(\mathbf{x}_0)}$ selected by the function $h_3(\mathbf{x}_1)$.

Bi-Decompositions of Logic Functions

A completely different approach is the *bi-decomposition*, first suggested in [3] and [2]. The new idea bases on the observation that the final output-gate in a circuit structure combines the subfunctions of its inputs to the logic function at the output. This final gate is in many cases a two-input gate. The condition for a bi-decomposition is that the two subfunctions that control the gate depend on fewer variables than the output-function. The recursive application of such a decomposition terminates in logic functions depending on a single variable. Originally this type of decomposition was denoted by *grouping*, due to three groups of variables that influence the decomposition. The first group of variables is a part of the support of the first subfunction, the second group of variables is a part of

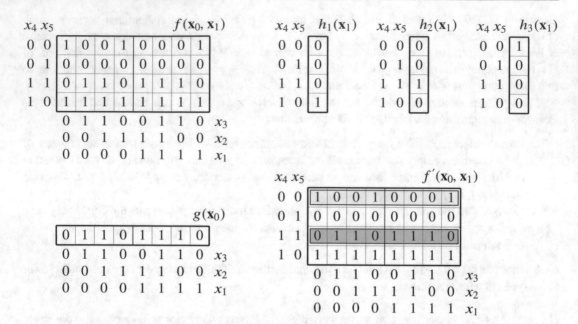

Fig. 11.33 Karnaugh-maps of all decomposition-functions of a Povarov-decomposition of the function $f(\mathbf{x}_0, \mathbf{x}_1)$ with regard to the dedicated sets of variables $\mathbf{x}_0 = (x_1, x_2, x_3)$ and $\mathbf{x}_1 = (x_4, x_5)$

the support of the second subfunction, and the third group of variables controls both subfunctions. In the last years the term *bi-decomposition* is preferred for this type of decomposition because the given logic function is split into exactly two subfunctions on the same circuit level.

A group of German researchers (first at the University of Chemnitz and later at the University of Freiberg) extended the knowledge about the theory and application of the bi-decomposition. Collaborations with research groups from the Portland State University (Oregon, USA) [9] and the Belorussian Academy of Sciences intensified these improvements [23]. Also a group in Japan [11] studied this type of decomposition, but their special method prefers disjoint bi-decomposition or restricts to small common sets. A comprehensive introduction into the most important results of the bi-decomposition has been published in [16]. The use of decision diagrams for bi-decompositions has been explored in [13].

Experimental comparisons [9] confirm that the bi-decomposition is an excellent method to synthesize multilevel circuits. It can be used for both completely specified logic functions and incompletely specified logic functions. During the synthesis process, incompletely specified decomposition-functions are created even if a completely specified logic function is given. The optimal utilization of functional properties for the bi-decomposition leads to circuit structures with a small number of levels and, thus, a very small delay. The area of the circuits synthesized by the method of bi-decomposition is implicitly minimized, especially if many EXOR-bi-decompositions occur. Due to these and further desirable properties, the basics of the bi-decomposition are introduced in the following.

Generally, the bi-decomposition is determined by a two-input logic gate. The logic function to be decomposed is created by the function of a gate that uses the logic functions of its two inputs. The results of a bi-decomposition step are one logic gate of the multilevel circuit and two logic functions that control the inputs of this gate. In contrast, the pipe decomposition does not determine a gate.

The bi-decomposition takes advantage of the properties of two-input logic gates AND, OR, or EXOR. All these gates create a more complex logic function at the output using the two simpler logic functions of the inputs. As special case, the inputs can be logic variables.

Several questions arise around the bi-decomposition:

- Which logic gates mentioned above should be taken to decompose a given logic function?
- How can the decomposition-functions be calculated?
- Which variables the decomposition-functions are depending on?
- How the bi-decomposition can be used to synthesize a complete multilevel circuit?

We answer these and other questions first for the case of a given completely specified logic function.

The key idea of the bi-decomposition is to decompose a given logic function into two simpler functions. The connection of these two simpler functions by an AND-, an OR-, or an EXOR-gate creates the given function. It is not trivial to measure the complexity of logic functions, but generally a logic function depending on fewer variables is simpler than a logic function depending on more variables. A recursive procedure of decompositions terminates with the specification of the complete structure of the circuit when the number of independent variables of the logic functions is reduced at least by one in each step.

Definition 11.1 (Strong π-Bi-Decompositions for Logic Functions) A bi-decomposition of the logic function $f(\mathbf{x}_a, \mathbf{x}_b, \mathbf{x}_c)$ with regard to the *decomposition-operator* $\pi(g, h)$ and the *dedicated sets* of variables $\mathbf{x}_a$ and $\mathbf{x}_b$ is a pair of functions $\langle g(\mathbf{x}_a, \mathbf{x}_c), h(\mathbf{x}_b, \mathbf{x}_c)\rangle$ with

$$f(\mathbf{x}_a, \mathbf{x}_b, \mathbf{x}_c) = \pi\left(g(\mathbf{x}_a, \mathbf{x}_c), h(\mathbf{x}_b, \mathbf{x}_c)\right) . \tag{11.109}$$

The sets of variables $\mathbf{x}_a$ and $\mathbf{x}_b$ must not be empty. The set of variables $\mathbf{x}_c$ is the *common set*. If the set of variables $\mathbf{x}_c$ is empty, the bi-decomposition is denoted by *disjoint*. The logic functions $g(\mathbf{x}_a, \mathbf{x}_c)$ and $h(\mathbf{x}_b, \mathbf{x}_c)$ are denoted by *decomposition-functions*. The *decomposition-operator* $\pi(g, h)$ can be realized by an OR-, an AND-, or an EXOR-gate.

We abbreviate the specification "bi-decomposition with regard to the decomposition operator $\pi(g, h)$" by the term π-*bi-decomposition*. The operator $\pi(g, h)$ can be realized by an OR-gate, an AND-gate, or an EXOR-gate. The associated circuit structures of these three types of bi-decompositions are shown in Fig. 11.34.

We start with the disjoint OR-bi-decomposition

$$f(\mathbf{x}_a, \mathbf{x}_b) = g(\mathbf{x}_a) \vee h(\mathbf{x}_b) \tag{11.110}$$

because it is easy to understand. An example of this bi-decomposition is shown in Fig. 11.35. The dedicated sets of variables $\mathbf{x}_a = (a, b)$ and $\mathbf{x}_b = (c, d)$ are selected in such a way that the minterms of $\mathbf{x}_a$ define the columns of the Karnaugh-map of $f(a, b, c, d)$ and the minterms of $\mathbf{x}_b$ the rows, respectively. In order to calculate the OR-operation between $g(a, b)$ and $h(c, d)$, these two functions must be transformed into the common Boolean space $\mathbb{B}^4$ that expands each value 1 of $g(a, b)$ into a column of four values 1 and each value 1 of $h(c, d)$ into a row of four values 1. In order to create an OR-bi-decomposable function $f(a, b, c, d)$, we take as example $g(a, b) = \bar{a}b$ and $h(c, d) = c \odot d$. These two functions and the result of the OR-operation $f(a, b, c, d) = g(a, b) \vee h(c, d)$ are displayed in the Karnaugh-maps of Fig. 11.35. There are two types of columns in this Karnaugh-map of $f(a, b, c, d)$: one contains some values 0, and the other one contains only values 1. The function $g(a, b)$ must be equal to 0 for a minterm (a, b) where at least one value 0 occurs in the associated

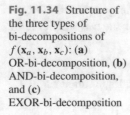

Fig. 11.34 Structure of the three types of bi-decompositions of $f(\mathbf{x}_a, \mathbf{x}_b, \mathbf{x}_c)$: (**a**) OR-bi-decomposition, (**b**) AND-bi-decomposition, and (**c**) EXOR-bi-decomposition

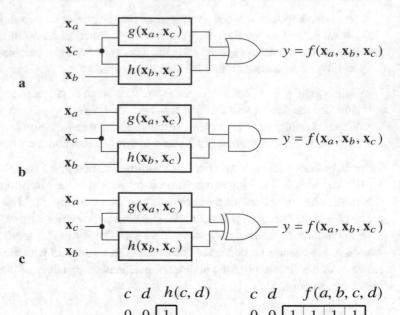

Fig. 11.35 Disjoint OR-bi-decomposition of $f(a, b, c, d) = \overline{a}b \vee (c \odot d)$ into $g(a, b)$ and $h(c, d)$

column of $f(a, b, c, d)$ because the value 1 dominates in the OR-operation. The function $g(a, b)$ can be equal to 1 for a minterm (a, b) if the associated column of $f(a, b, c, d)$ does not contain any value 1. Corresponding conditions are valid for the function $h(c, d)$ and the rows of $f(a, b, c, d)$.

An OR-bi-decomposition for the function $f(a, b, c, d)$ of Fig. 11.35 exists because all values 1 are covered by the values 1 of the functions $g(a, b)$ and $h(c, d)$:

$$f(a, b, c, d) = (\overline{a}b) \vee (cd \vee \overline{c}\overline{d}) .$$

Assume that the function $f(a, b, c, d)$ has been changed to $f'(a, b, c, d)$ by switching the value $f(0, 0, 0, 1) = 0$ (emphasized by the green box in Fig. 11.35) into the value $f'(0, 0, 0, 1) = 1$. The Karnaugh-map of $f'(a, b, c, d)$ contains the values 0 (emphasized by yellow boxes in Fig. 11.35) in the first column at $f'(0, 0, 1, 0) = 0$ and in the second row at $f'(1, 1, 0, 1) = 0$. These two values 0 require that $g'(0, 0) = 0$ and $h'(0, 1) = 0$ in an explored OR-bi-decomposition of $f'(a, b, c, d)$. Consequently, the result of the OR-operation is $f'(0, 0, 0, 1) = g'(0, 0) \vee h'(0, 1) = 0$; however, this is a contradiction to the assumption $f'(0, 0, 0, 1) = 1$. This example proves that a disjoint OR-bi-decomposition for given dedicated sets $\mathbf{x}_a$ and $\mathbf{x}_b$ does not exist for each logic function.

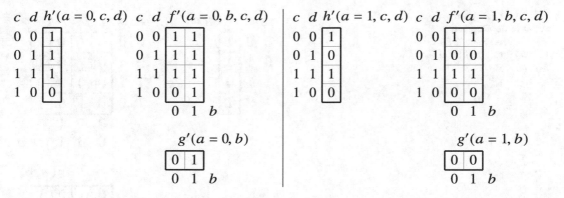

Fig. 11.36 Non-disjoint OR-bi-decomposition of the function $f'(a, b, c, d)$ into $g'(a, b)$ and $h'(a, c, d)$

A disjoint bi-decomposition creates the simplest decomposition-functions but imposes the strongest condition onto the function to be decomposed. More OR-bi-decompositions exist in the non-disjoint case of the dedicated sets of variables used in the decomposition-functions. By moving variables from the dedicated sets $\mathbf{x}_a$ or $\mathbf{x}_b$ into the common set $\mathbf{x}_c$, the function $f(\mathbf{x}_a, \mathbf{x}_b, \mathbf{x}_c)$ can be divided into $2^{|\mathbf{x}_c|}$ subfunctions $f(\mathbf{x}_a, \mathbf{x}_b, \mathbf{x}_c = const.)$. Each of these subfunctions can be checked separately for the OR-bi-decomposition because the two decomposition-functions $g(\mathbf{x}_a, \mathbf{x}_c)$ and $h(\mathbf{x}_b, \mathbf{x}_c)$ depend on the common set $\mathbf{x}_c$.

It has been shown that for the function $f'(a, b, c, d)$ and the dedicated sets of variables $\mathbf{x}_a = (a, b)$ and $\mathbf{x}_b = (c, d)$ no disjoint OR-bi-decomposition exists. If the variable a is moved into the common set $\mathbf{x}_c = (a)$, the dedicated sets for a possible OR-bi-decomposition are $\mathbf{x}_a = (b)$ and $\mathbf{x}_b = (c, d)$. Figure 11.36 shows that this non-disjoint OR-bi-decomposition exists. The analysis of the non-disjoint OR-bi-decomposition is shown in Fig. 11.36 on the left-hand side in the Boolean subspace defined by $a = 0$ and on the right-hand side for $a = 1$, respectively.

Note 11.13 In contrast to Fig. 11.35, the function $h'(a = 0, c, d) = \bar{c} \vee d$ is different from $h'(a = 1, c, d) = c \odot d$ so that the decomposition-functions $g'(a, b)$ and $h'(a, c, d)$ can be created.

The number of variables in the set $\mathbf{x}_c$ is not restricted. The chances to find an OR-bi-decomposition increase with the number of variables in the common set $\mathbf{x}_c$. Theorem 11.1 shows the condition that a function $f(\mathbf{x}_a, \mathbf{x}_b, \mathbf{x}_c)$ is OR-bi-decomposable. This is the case if the projections in both direction $\mathbf{x}_a$ and $\mathbf{x}_b$ of the values 0 expressed by $\overline{f(\mathbf{x}_a, \mathbf{x}_b, \mathbf{x}_c)}$ do not overlap with values 1 expressed by $f(\mathbf{x}_a, \mathbf{x}_b, \mathbf{x}_c)$.

Theorem 11.1 (Strong OR-Bi-Decomposition for a Logic Function) *The logic function $f(\mathbf{x}_a, \mathbf{x}_b, \mathbf{x}_c)$ is OR-bi-decomposable with regard to the dedicated sets $\mathbf{x}_a$ and $\mathbf{x}_b$ if and only if*

$$f(\mathbf{x}_a, \mathbf{x}_b, \mathbf{x}_c) \wedge \max_{\mathbf{x}_a}^k \overline{f(\mathbf{x}_a, \mathbf{x}_b, \mathbf{x}_c)} \wedge \max_{\mathbf{x}_b}^k \overline{f(\mathbf{x}_a, \mathbf{x}_b, \mathbf{x}_c)} = 0 \, . \tag{11.111}$$

If a logic function $f(\mathbf{x}_a, \mathbf{x}_b, \mathbf{x}_c)$ satisfies Condition (11.111) of an OR-bi-decomposition, then admissible decomposition-functions $g(\mathbf{x}_a, \mathbf{x}_c)$ and $h(\mathbf{x}_b, \mathbf{x}_c)$ must be calculated in order to use this bi-decomposition in the circuit. Generally, several pairs of decomposition-functions can exist. The decomposition-functions must meet certain conditions. A value 0 of $f(\mathbf{x}_a, \mathbf{x}_b, \mathbf{x}_c)$ in a subspace defined by $\mathbf{x}_a = const.$ and $\mathbf{x}_c = const.$ requires a value 0 for the selected minterm of $g(\mathbf{x}_a, \mathbf{x}_c)$.

Fig. 11.37 Disjoint AND-bi-decomposition of $f(a,b,c,d) = (a \oplus b) \wedge (\overline{c} \vee \overline{d})$ into $g(a,b)$ and $h(c,d)$

c	d	h(c,d)
0	0	1
0	1	1
1	1	0
1	0	1

c	d	f(a,b,c,d)			
0	0	0	1	0	1
0	1	0	1	0	1
1	1	0	0	0	0
1	0	0	1	0	1

0	1	1	0	b
0	0	1	1	a

$g(a,b)$

0	1	0	1

0	1	1	0	b
0	0	1	1	a

This describes directly the property of the k-fold minimum. An equivalent condition is valid for decomposition-functions $h(\mathbf{x}_b, \mathbf{x}_c)$ with regard to the subspace defined by $\mathbf{x}_b = const.$ and $\mathbf{x}_c = const.$ Therefore, the decomposition-functions of the OR-bi-decomposition can be calculated by:

$$g(\mathbf{x}_a, \mathbf{x}_c) = \min_{\mathbf{x}_b}^{k} f(\mathbf{x}_a, \mathbf{x}_b, \mathbf{x}_c) \,, \tag{11.112}$$

$$h(\mathbf{x}_b, \mathbf{x}_c) = \min_{\mathbf{x}_a}^{k} f(\mathbf{x}_a, \mathbf{x}_b, \mathbf{x}_c) \,. \tag{11.113}$$

Now we are going to explore the disjoint AND-bi-decomposition

$$f(\mathbf{x}_a, \mathbf{x}_b) = g(\mathbf{x}_a) \wedge h(\mathbf{x}_b) \,. \tag{11.114}$$

Figure 11.37 shows an example of the AND-bi-decomposition of the function $f(a,b,c,d)$ with regard to the dedicated sets $\mathbf{x}_a = (a,b)$ and $\mathbf{x}_b = (c,d)$. The AND-operation of the AND-bi-decomposition requires values 1 in the two functions $g(a,b)$ and $h(c,d)$ in order to create a value 1 in the function $f(a,b,c,d)$; hence, an AND-bi-decomposition does not exist if the projections of the values 1 given by $f(\mathbf{x}_a, \mathbf{x}_b)$ in both direction $\mathbf{x}_a$ and $\mathbf{x}_b$ overlap with values 0 expressed by $\overline{f(\mathbf{x}_a, \mathbf{x}_b)}$. The change of the value $f(0,1,0,1) = 1$ into a value $f(0,1,0,1) = 0$ (emphasized by the green box in Fig. 11.37) would cause such a contradiction due to the values $f(1,0,0,1) = 1$ and $f(0,1,1,0) = 1$ highlighted by yellow boxes in Fig. 11.37.

The generalization to the non-disjoint case is similar to the OR-bi-decomposition. Theorem 11.2 describes the condition that must be satisfied if a logic function is AND-bi-decomposable.

Theorem 11.2 (Strong AND-Bi-Decomposition for a Logic Function) *The logic function* $f(\mathbf{x}_a, \mathbf{x}_b, \mathbf{x}_c)$ *is AND-bi-decomposable with regard to the dedicated sets* $\mathbf{x}_a$ *and* $\mathbf{x}_b$ *if and only if*

$$\overline{f(\mathbf{x}_a, \mathbf{x}_b, \mathbf{x}_c)} \wedge \max_{\mathbf{x}_a}^{k} f(\mathbf{x}_a, \mathbf{x}_b, \mathbf{x}_c) \wedge \max_{\mathbf{x}_b}^{k} f(\mathbf{x}_a, \mathbf{x}_b, \mathbf{x}_c) = 0 \,. \tag{11.115}$$

Again there are several pairs of decomposition-functions $g(\mathbf{x}_a, \mathbf{x}_c)$ and $h(\mathbf{x}_b, \mathbf{x}_c)$ of an AND-bi-decomposition. Here, a value 1 in a suitable subspace requires a value 1 for the associated minterm of the decomposition-functions. This describes directly the property of the k-fold maximum, so that the decomposition-functions of the AND-bi-decomposition can be calculated by:

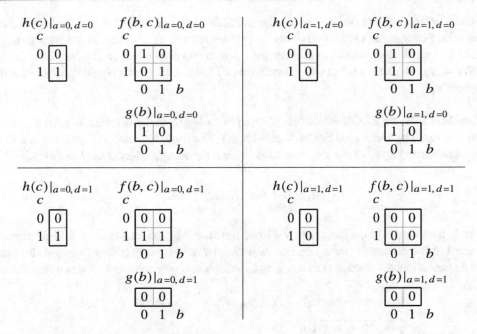

Fig. 11.38 Non-disjoint EXOR-bi-decomposition of $f(a, b, c, d)$ into $g(a, b, d)$ and $h(a, c, d)$ with regard to the dedicated sets $\mathbf{x}_a = (b)$, $\mathbf{x}_b = (c)$, and the common set $\mathbf{x}_c = (a, d)$

$$g(\mathbf{x}_a, \mathbf{x}_c) = \max_{\mathbf{x}_b}^k f(\mathbf{x}_a, \mathbf{x}_b, \mathbf{x}_c) , \qquad (11.116)$$

$$h(\mathbf{x}_b, \mathbf{x}_c) = \max_{\mathbf{x}_a}^k f(\mathbf{x}_a, \mathbf{x}_b, \mathbf{x}_c) . \qquad (11.117)$$

Quite different properties facilitate the testing for possible EXOR-bi-decompositions. It is easier to start with non-disjoint EXOR-bi-decompositions where each dedicated set $\mathbf{x}_a$ and $\mathbf{x}_b$ contains only one single variable. Figure 11.38 shows an example of a non-disjoint EXOR-bi-decomposition of $f(a, b, c, d)$ with regard to the dedicated sets $\mathbf{x}_a = (b)$ and $\mathbf{x}_b = (c)$.

The Karnaugh-maps of $f(b, c)$ in Fig. 11.38 show four of the eight logic functions of two variables that can be decomposed by an EXOR-gate into the decomposition-functions $g(b)$ and $h(c)$. These eight functions are the results of EXOR-operations between all pairs of the four logic functions depending on a single variable each; partially this can be seen in Fig. 11.38. The following functions of two variables are EXOR-bi-decomposable:

$$f(b, c) = 0 , \qquad f(b, c) = 1 , \qquad f(b, c) = b , \qquad f(b, c) = \overline{b} ,$$

$$f(b, c) = c , \qquad f(b, c) = \overline{c} , \qquad f(b, c) = b \oplus c , \qquad f(b, c) = \overline{b \oplus c} .$$

Exactly these eight of all 16 logic functions of two variables have an even number of function values 1, a property indicated by the complement of the 2-fold derivative as stated in Theorem 11.3.

Theorem 11.3 (Strong EXOR-Bi-Decomposition for a Logic Function with Regard to the Single Variables x_a and x_b) *The logic function $f(x_a, x_b, \mathbf{x}_c)$ is EXOR-bi-decomposable with regard to the single variables x_a and x_b if and only if*

$$\underset{(x_a, x_b)}{\mathrm{der}}^2 f(x_a, x_b, \mathbf{x}_c) = 0 . \qquad (11.118)$$

As can be seen in Fig. 11.38, the rows of the Karnaugh-maps of $f(b, c)$ contain either the function $g(b)$ or $\overline{g(b)}$. Because each row covers only two values of the function, the derivative with regard to the variable b must be constant in each subspace defined by $a = const.$ and $d = const.$ Based on this observation, Theorem 11.3 can be generalized to Theorem 11.4 for a single variable x_a and a set of variables $\mathbf{x}_b$.

Theorem 11.4 (Strong EXOR-Bi-Decomposition for a Logic Function with Regard to the Single Variable x_a and the Dedicated Set of Variables $\mathbf{x}_b$) *The logic function $f(x_a, \mathbf{x}_b, \mathbf{x}_c)$ is EXOR-bi-decomposable with regard to the single variable x_a and the set of variables $\mathbf{x}_b$ if and only if*

$$\Delta_{\mathbf{x}_b} \left(\operatorname*{der}_{x_a} f(x_a, \mathbf{x}_b, \mathbf{x}_c) \right) = 0 . \tag{11.119}$$

There is no formula to decide whether a logic function $f(\mathbf{x}_a, \mathbf{x}_b, \mathbf{x}_c)$ is EXOR-bi-decomposable with regard to the dedicated sets $\mathbf{x}_a$ and $\mathbf{x}_b$; however, by means of an indirect method, it is easy to answer this question. As discussed above, in each subspace $\mathbf{x}_c = const.$ only the two functions

$$g(\mathbf{x}_a, \mathbf{x}_c = const.) = f(\mathbf{x}_a, \mathbf{x}_b = const., \mathbf{x}_c = const.) \quad \text{and}$$

$$\overline{g(\mathbf{x}_a, \mathbf{x}_c = const.)}$$

are possible to realize an EXOR-bi-decomposition. One of these two functions must always be the complement of the other one. Theorem 11.5 takes advantage of this property and delivers, if possible, an admissible pair of decomposition-functions for the EXOR-bi-decomposition.

Theorem 11.5 (Strong EXOR-Bi-Decomposition for a Logic Function with Regard to the Dedicated Sets $\mathbf{x}_a$ and $\mathbf{x}_b$) *Let (11.120) be the potential decomposition-function $g(\mathbf{x}_a, \mathbf{x}_c)$ of the logic function $f(\mathbf{x}_a, \mathbf{x}_b, \mathbf{x}_c)$:*

$$g(\mathbf{x}_a, \mathbf{x}_c) = f(\mathbf{x}_a, \mathbf{x}_b = (0, 0, \ldots, 0), \mathbf{x}_c) , \tag{11.120}$$

then $f(\mathbf{x}_a, \mathbf{x}_b, \mathbf{x}_c)$ is EXOR-bi-decomposable with regard to the dedicated sets $\mathbf{x}_a$ and $\mathbf{x}_b$ if and only if

$$\Delta_{\mathbf{x}_a} (f(\mathbf{x}_a, \mathbf{x}_b, \mathbf{x}_c) \oplus f(\mathbf{x}_a, \mathbf{x}_b = (0, 0, \ldots, 0), \mathbf{x}_c)) = 0 . \tag{11.121}$$

If (11.121) holds, then (11.120) and

$$h(\mathbf{x}_b, \mathbf{x}_c) = \max_{\mathbf{x}_a}^k (f(\mathbf{x}_a, \mathbf{x}_b, \mathbf{x}_c) \oplus f(\mathbf{x}_a, \mathbf{x}_b = (0, 0, \ldots, 0), \mathbf{x}_c)) \tag{11.122}$$

is an admissible pair of decomposition-functions of the EXOR-bi-decomposition of $f(\mathbf{x}_a, \mathbf{x}_b, \mathbf{x}_c)$ with regard to the dedicated sets $\mathbf{x}_a$ and $\mathbf{x}_b$.

Example 11.19 Figure 11.39 shows how Theorem 11.5 can be applied to look for a disjoint EXOR-bi-decomposition where each of the dedicated sets contains more than one variable. We try to find an EXOR-bi-decomposition of the logic function

$$f(a, b, c, d) = a\,c \vee a\,\overline{d} \vee \overline{b}\,c \vee \overline{b}\,\overline{d} \vee \overline{a}\,b\,\overline{c}\,d$$

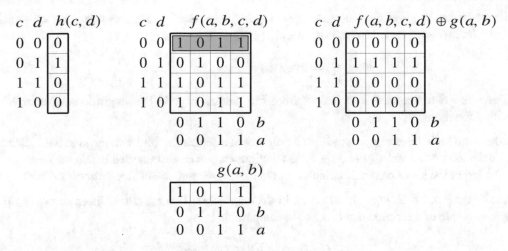

Fig. 11.39 Disjoint EXOR-bi-decomposition of the logic function $f(a, b, c, d)$ with regard to the dedicated sets $\mathbf{x}_a = (a, b)$ and $\mathbf{x}_b = (c, d)$ into the decomposition-functions $g(a, b)$ and $h(c, d)$

with regard to the dedicated sets $\mathbf{x}_a = (a, b)$ and $\mathbf{x}_b = (c, d)$. Using (11.120), the potential decomposition-function $g(a, b)$ is taken from the first row (emphasized by the green box) of the Karnaugh-map of $f(a, b, c, d)$ in Fig. 11.39. The result of the EXOR-operation of (11.121) is represented in the Karnaugh-map on the right-hand side of Fig. 11.39. Only the second row contains four values 1; hence, all subspaces of the rows contain constant functions so that the result of the Δ-operation in (11.121) is equal to 0 and $g(a, b) = a \vee \overline{b}$ is a decomposition-function. The decomposition-function $h(c, d) = \overline{c}d$ is calculated by the 2-fold maximum of $f(a, b, c, d) \oplus g(a, b)$ with regard to (a, b).

A slightly changed approach to compute the decomposition-functions of the EXOR-bi-decomposition uses the single variable x_a and the maximally extended set of variables $\mathbf{x}_b$ that satisfy Condition (11.119). The decomposition-function $g(x_a, \mathbf{x}_c)$ can be computed by:

$$g(x_a, \mathbf{x}_c) = f(x_a, \mathbf{x}_b = (0, 0, \ldots, 0), \mathbf{x}_c) \tag{11.123}$$

and the associated decomposition-function $h(\mathbf{x}_b, \mathbf{x}_c)$ by:

$$h(\mathbf{x}_b, \mathbf{x}_c) = \max_{x_a}(f(x_a, \mathbf{x}_b, \mathbf{x}_c) \oplus f(x_a, \mathbf{x}_b = (0, 0, \ldots, 0), \mathbf{x}_c)) . \tag{11.124}$$

The decomposition-function $h(\mathbf{x}_b, \mathbf{x}_c)$ depends on all variables of the set $\mathbf{x}_b$; however, it is possible that this function does not depend on all variables of the set $\mathbf{x}_c$. Hence, Condition (6.36) can be used to check whether $h(\mathbf{x}_b, \mathbf{x}_c)$ does not depend on $x_i \in \mathbf{x}_c$ and in the case the simplified function $h(\mathbf{x}_b, \mathbf{x}_c \setminus x_i)$ can be computed by:

$$h(\mathbf{x}_b, \mathbf{x}_c') = h(\mathbf{x}_b, \mathbf{x}_c \setminus x_i) = \max_{x_i} h(\mathbf{x}_b, \mathbf{x}_c) . \tag{11.125}$$

A reduced set $\mathbf{x}_c \setminus x_i$ implicitly increases the set $\mathbf{x}_a$ by the variable x_i. The function $g(x_a, \mathbf{x}_c)$ remains unchanged; however, the role of the variable x_i is changed:

$$g(\mathbf{x}_a, \mathbf{x}'_c) = g(x_a, \mathbf{x}_c) = g(x_a \cup x_i, \mathbf{x}_c \setminus x_i) \,. \tag{11.126}$$

There are two important conclusions from our studies of π-bi-decompositions of completely specified functions:

1. there are logic functions possessing the property of a selected π-bi-decomposition with regard to dedicated sets $\mathbf{x}_a$ and $\mathbf{x}_b$, but other logic functions cannot be decomposed in this manner;
2. the larger the set of common variables $\mathbf{x}_c$, the more bi-decomposable logic functions exist.

Unfortunately, not all logic functions are bi-decomposable. This has already been shown in [3]. A simple example of a function that is not bi-decomposable is

$$f(a, b, c) = \overline{b}\,\overline{c} \vee \overline{a}\,\overline{c} \vee \overline{a}\,\overline{b} \,,$$

visualized in the center of Fig. 11.41 on page 635. This is a symmetric function. Each selection of a pair of variables for the dedicated sets $\mathbf{x}_a$ and $\mathbf{x}_b$ divides the Boolean space $\mathbb{B}^3$ into two subspaces $\mathbb{B}^2$, characterized by the value of the remaining variable. The values 1 of $f(a, b, c)$ are distributed between these subspaces such that one subspace contains one value 1 and the other one three values 1, respectively. Thus, one subfunction of $f(a, b, c)$ allows only an AND-bi-decomposition and the other one only an OR-bi-decomposition. There is no common π-bi-decomposition.

In order to overcome this gap, Le [8] suggested the *weak bi-decomposition*. For a clear differentiation, we use the name *strong bi-decompositions* for the three types of π-bi-decompositions explored above.

Definition 11.2 (Weak π-Bi-Decompositions for Logic Functions) A weak bi-decomposition of the logic function $f(\mathbf{x}_a, \mathbf{x}_c)$ with regard to the *decomposition-operator* $\pi(g, h)$ and the dedicated set $\mathbf{x}_a \neq \emptyset$ is a pair of functions $\langle g(\mathbf{x}_a, \mathbf{x}_c), h(\mathbf{x}_c) \rangle$ with

$$f(\mathbf{x}_a, \mathbf{x}_c) = \pi \left(g(\mathbf{x}_a, \mathbf{x}_c), h(\mathbf{x}_c) \right), \tag{11.127}$$

where $g(\mathbf{x}_a, \mathbf{x}_c) \in \mathcal{L} \langle g_q(\mathbf{x}_a, \mathbf{x}_c), r_q(\mathbf{x}_a, \mathbf{x}_c) \rangle$ and

$$g_q(\mathbf{x}_a, \mathbf{x}_c) < f(\mathbf{x}_a, \mathbf{x}_c) \qquad \text{and} \qquad g_r(\mathbf{x}_a, \mathbf{x}_c) = \overline{f(\mathbf{x}_a, \mathbf{x}_c)} \,, \tag{11.128}$$

or

$$g_q(\mathbf{x}_a, \mathbf{x}_c) = f(\mathbf{x}_a, \mathbf{x}_c) \qquad \text{and} \qquad g_r(\mathbf{x}_a, \mathbf{x}_c) < \overline{f(\mathbf{x}_a, \mathbf{x}_c)} \,. \tag{11.129}$$

The set $\mathbf{x}_c$ is the common set. The functions $g(\mathbf{x}_a, \mathbf{x}_c)$ and $h(\mathbf{x}_c)$ are denoted by decomposition-functions. The *decomposition-operator* $\pi(g, h)$ can be realized by an OR- or an AND-gate.

The function $h(\mathbf{x}_c)$ of a weak bi-decomposition is simpler than the given function $f(\mathbf{x}_a, \mathbf{x}_c)$ due to the smaller number of variables. The function $g(\mathbf{x}_a, \mathbf{x}_c)$ of a weak bi-decomposition can be chosen out of the lattice $\mathcal{L} \langle g_q(\mathbf{x}_a, \mathbf{x}_c), r_q(\mathbf{x}_a, \mathbf{x}_c) \rangle$; if one of the conditions (11.128) or (11.129) is satisfied, at least one of the possible functions $g(\mathbf{x}_a, \mathbf{x}_c)$ is also simpler than the given function $f(\mathbf{x}_a, \mathbf{x}_c)$.

Using an arbitrary function $h(\mathbf{x})$, a function $g(\mathbf{x}_a, \mathbf{x}_c)$ can be calculated by:

Fig. 11.40 Structure of the two types of weak bi-decompositions of $f(\mathbf{x}_a, \mathbf{x}_c)$ with regard to $\mathbf{x}_a$: (**a**) weak OR-bi-decomposition and (**b**) weak AND-bi-decomposition

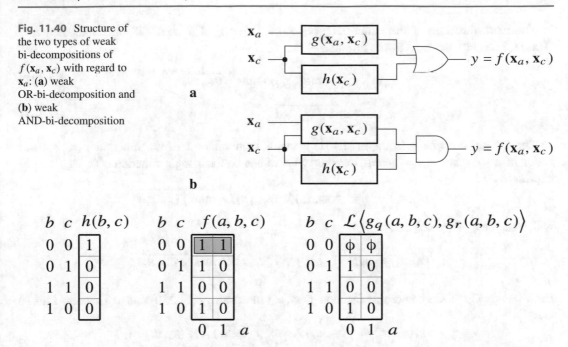

Fig. 11.41 Weak OR-bi-decomposition of $f(a, b, c) = \overline{b}\,\overline{c} \vee \overline{a}\,\overline{c} \vee \overline{a}\,\overline{b}$, with regard to the dedicated set $\mathbf{x}_a = (a)$ into the decomposition-lattice $\mathcal{L}\langle g_q, g_r \rangle$ and the decomposition-function $h(b, c)$

$$g(\mathbf{x}_a, \mathbf{x}_c) = f(\mathbf{x}_a, \mathbf{x}_c) \oplus h(\mathbf{x}_c) .$$

This function $g(\mathbf{x}_a, \mathbf{x}_c)$ satisfies together with the used function $h(\mathbf{x}_c)$ the construction rule (11.127) of a weak bi-decomposition:

$$f(\mathbf{x}_a, \mathbf{x}_c) = g(\mathbf{x}_a, \mathbf{x}_c) \oplus h(\mathbf{x}_c)$$

for $\pi(g, h) = \text{EXOR}(g, h)$. However, in the case of an EXOR-gate neither the condition (11.128) nor (11.129) can be satisfied so that even more complex functions $g(\mathbf{x}_a, \mathbf{x}_c)$ can be caused by the chosen function $h(\mathbf{x}_c)$. Therefore, we avoid the weak EXOR-bi-decomposition.

The comparison of Fig. 11.34 (on page 628) and Fig. 11.40 shows that the difference between the weak π-bi-decomposition and the strong π-bi-decomposition is that there is no dedicated set $\mathbf{x}_b$ in the weak π-bi-decomposition of $f(\mathbf{x}_a, \mathbf{x}_c)$.

An example of a weak OR-bi-decomposition with regard to the dedicated set $\mathbf{x}_a = (a)$ and the common set $\mathbf{x}_c = (b, c)$ is shown in Fig. 11.41. The value 1 of $h(b = 0, c = 0)$ at the input of the OR-gate causes two values 1 at the output for $f(a, b = 0, c = 0)$ without any constraint for the function $g(a, b, c)$ in the subspace $(b = 0, c = 0)$. Consequently, $g_\varphi(a, b, c) = \overline{b}\,\overline{c}$, and Condition (11.128) becomes true. As specified in Theorem 11.6, the condition for a weak OR-bi-decomposition is that at least one subspace $\mathbf{x}_c = const.$ contains only values 1.

Theorem 11.6 (Weak OR-Bi-Decomposition for a Logic Function) *The function $f(\mathbf{x}_a, \mathbf{x}_c)$ is weakly OR-bi-decomposable with regard to the dedicated set $\mathbf{x}_a$ if and only if*

$$\min_{\mathbf{x}_a}^k f(\mathbf{x}_a, \mathbf{x}_c) \neq 0 . \tag{11.130}$$

The mark-functions of the lattice $\mathcal{L}\langle g_q(\mathbf{x}_a, \mathbf{x}_c), r_q(\mathbf{x}_a, \mathbf{x}_c)\rangle$ of a weak OR-bi-decomposition of $f(\mathbf{x}_a, \mathbf{x}_c)$ can be calculated by the formulas:

$$g_q(\mathbf{x}_a, \mathbf{x}_c) = f(\mathbf{x}_a, \mathbf{x}_c) \wedge \max_{\mathbf{x}_a}^k \overline{f(\mathbf{x}_a, \mathbf{x}_c)}, \tag{11.131}$$

$$g_r(\mathbf{x}_a, \mathbf{x}_c) = \overline{f(\mathbf{x}_a, \mathbf{x}_c)}. \tag{11.132}$$

The ON-set-function $g_q(\mathbf{x}_a, \mathbf{x}_c)$ (11.131) contains such values 1 of the function $f(\mathbf{x}_a, \mathbf{x}_c)$ that occur in a subspace $\mathbf{x}_c = const.$ together with values 0. Each logic function $f(\mathbf{x}_a, \mathbf{x}_c)$ can be expressed by

$$f(\mathbf{x}_a, \mathbf{x}_c) = \min_{\mathbf{x}_a}^k f(\mathbf{x}_a, \mathbf{x}_c) \vee f(\mathbf{x}_a, \mathbf{x}_c) \wedge \overline{\min_{\mathbf{x}_a}^k f(\mathbf{x}_a, \mathbf{x}_c)},$$

which can be transformed into

$$f(\mathbf{x}_a, \mathbf{x}_c) = \min_{\mathbf{x}_a}^k f(\mathbf{x}_a, \mathbf{x}_c) \vee f(\mathbf{x}_a, \mathbf{x}_c) \wedge \max_{\mathbf{x}_a}^k \overline{f(\mathbf{x}_a, \mathbf{x}_c)}.$$

Each weakly OR-bi-decomposable function $f(\mathbf{x}_a, \mathbf{x}_c)$ must satisfy (11.130) due to Theorem 11.6, so that

$$g_q(\mathbf{x}_a, \mathbf{x}_c) = f(\mathbf{x}_a, \mathbf{x}_c) \wedge \max_{\mathbf{x}_a}^k \overline{f(\mathbf{x}_a, \mathbf{x}_c)} < f(\mathbf{x}_a, \mathbf{x}_c)$$

holds and the decomposition-lattice $\mathcal{L}\langle g_q, g_r\rangle$ specified by (11.131) and (11.132) satisfies Condition (11.128) of a weak OR-bi-decomposition.

The benefit of the weak OR-bi-decomposition is visualized in Fig. 11.41 and will be recognized in the next level of the π-bi-decomposition. The lattice $\mathcal{L}\langle g_q, g_r\rangle$ contains the function $g(a, b, c) = \overline{a}(\overline{b} \vee \overline{c})$ for which the disjoint AND-bi-decomposition into $g'(a) = \overline{a}$ and $h'(b, c) = \overline{b} \wedge c$ exists. Using the selected decomposition-function $g(\mathbf{x}_a, \mathbf{x}_c)$, the decomposition-function $h(\mathbf{x}_c)$ of a weak OR-bi-decomposition of $f(\mathbf{x}_a, \mathbf{x}_c)$ can be calculated by:

$$h(\mathbf{x}_c) = \max_{\mathbf{x}_a}^k \left(f(\mathbf{x}_a, \mathbf{x}_c) \wedge \overline{g(\mathbf{x}_a, \mathbf{x}_c)} \right). \tag{11.133}$$

Figure 11.42 shows an example of a weak AND-bi-decomposition of the function $f(\mathbf{x}_a, \mathbf{x}_c)$ with regard to the dedicated set $\mathbf{x}_a = (a, b)$ and the common set $\mathbf{x}_c = (c, d)$. It can be checked that there is no strong π-bi-decomposition of the function $f(a, b, c, d)$ specified in Fig. 11.42. The value 0 of $h(c = 0, d = 1)$ at an input of an AND-gate causes four values 0 at the output $f(a, b, c = 0, d = 1)$ without any constraint for the decomposition-function $g(a, b, c = 0, d = 1)$. Consequently, $g_\varphi(a, b, c, d) = \overline{c}d$, and Condition (11.129) becomes true. As specified in Theorem 11.7, the condition for a weak AND-bi-decomposition is that at least one subspace $\mathbf{x}_c = const.$ contains only values 0.

Theorem 11.7 (Weak AND-Bi-Decomposition for a Logic Function) *The function* $f(\mathbf{x}_a, \mathbf{x}_c)$ *is weakly AND-bi-decomposable with regard to the dedicated set* $\mathbf{x}_a$ *if and only if*

$$\min_{\mathbf{x}_a}^k \overline{f(\mathbf{x}_a, \mathbf{x}_c)} \neq 0. \tag{11.134}$$

The mark-functions of the decomposition-lattice $\mathcal{L}\langle g_q(\mathbf{x}_a, \mathbf{x}_c), r_q(\mathbf{x}_a, \mathbf{x}_c)\rangle$ of a weak AND-bi-decomposition of $f(\mathbf{x}_a, \mathbf{x}_c)$ can be calculated by the formulas:

$$g_q(\mathbf{x}_a, \mathbf{x}_c) = f(\mathbf{x}_a, \mathbf{x}_c), \tag{11.135}$$

c	d	h(c, d)
0	0	1
0	1	0
1	1	1
1	0	1

c	d	$f(a, b, c, d)$			
0	0	1	0	1	1
0	1	0	0	0	0
1	1	1	0	1	1
1	0	1	0	1	0
		0	1	1	0	b
		0	0	1	1	a

c	d	$\mathcal{L}\langle g_q(a, b, c, d), g_r(a, b, c, d)\rangle$			
0	0	1	0	1	1
0	1	ϕ	ϕ	ϕ	ϕ
1	1	1	0	1	1
1	0	1	0	1	0
		0	1	1	0	b
		0	0	1	1	a

Fig. 11.42 Weak AND-bi-decomposition of $f(a, b, c, d) = (a \vee \overline{b})(c \odot d) \vee (a \odot b)c\overline{d}$ with regard to the dedicated set $\mathbf{x}_a = (a, b)$ into the lattice $\mathcal{L}\langle g_q, g_r\rangle$ and the function $h(c, d)$

$$g_r(\mathbf{x}_a, \mathbf{x}_c) = \overline{f(\mathbf{x}_a, \mathbf{x}_c)} \wedge \max_{\mathbf{x}_a}^k f(\mathbf{x}_a, \mathbf{x}_c) . \tag{11.136}$$

The OFF-set-function $g_r(\mathbf{x}_a, \mathbf{x}_c)$ (11.135) is equal to 1 for such values 0 of the function $f(\mathbf{x}_a, \mathbf{x}_c)$ that occur in a subspace $\mathbf{x}_c = const.$ together with values 1. The complement $\overline{f(\mathbf{x}_a, \mathbf{x}_c)}$ of each logic function $f(\mathbf{x}_a, \mathbf{x}_c)$ can be expressed by

$$\overline{f(\mathbf{x}_a, \mathbf{x}_c)} = \min_{\mathbf{x}_a}^k f(\mathbf{x}_a, \mathbf{x}_c) \vee \overline{f(\mathbf{x}_a, \mathbf{x}_c)} \wedge \overline{\min_{\mathbf{x}_a}^k f(\mathbf{x}_a, \mathbf{x}_c)} ,$$

which can be transformed into

$$\overline{f(\mathbf{x}_a, \mathbf{x}_c)} = \min_{\mathbf{x}_a}^k f(\mathbf{x}_a, \mathbf{x}_c) \vee \overline{f(\mathbf{x}_a, \mathbf{x}_c)} \wedge \max_{\mathbf{x}_a}^k f(\mathbf{x}_a, \mathbf{x}_c) .$$

Each weakly AND-bi-decomposable function $f(\mathbf{x}_a, \mathbf{x}_c)$ must satisfy (11.134) due to Theorem 11.7, so that

$$g_r(\mathbf{x}_a, \mathbf{x}_c) = \overline{f(\mathbf{x}_a, \mathbf{x}_c)} \wedge \max_{\mathbf{x}_a}^k f(\mathbf{x}_a, \mathbf{x}_c) < \overline{f(\mathbf{x}_a, \mathbf{x}_c)}$$

holds and the lattice $\mathcal{L}\langle g_q, g_r\rangle$ specified by (11.135) and (11.136) satisfies Condition (11.129) of a weak AND-bi-decomposition.

Using the selected function $g(\mathbf{x}_a, \mathbf{x}_c)$, the decomposition-function $h(\mathbf{x}_c)$ of a weak AND-bi-decomposition of $f(\mathbf{x}_a, \mathbf{x}_c)$ can be calculated by:

$$h(\mathbf{x}_c) = \min_{\mathbf{x}_a}^k \left(f(\mathbf{x}_a, \mathbf{x}_c) \vee \overline{g(\mathbf{x}_a, \mathbf{x}_c)} \right) . \tag{11.137}$$

The benefit of the weak AND-bi-decomposition of $f(a, b, c, d)$ with regard to $\mathbf{x}_a = (a, b)$ is emphasized in Fig. 11.42. The resulting lattice $\mathcal{L}\langle g_q, g_r\rangle$ contains a function $g'(a, b, c, d)$ that is again weak AND-bi-decomposable, but with regard to $\mathbf{x}_a = (c, d)$. After these two weak AND-bi-decompositions, we get a lattice that contains a function $g''(a, b, c, d)$ for which a disjoint OR-bi-decomposition with regard to the dedicated sets $\mathbf{x}_a = (a, b)$ and $\mathbf{x}_b = (c, d)$ exists. Figure 11.43 shows the complete structure of the circuit synthesized by this sequence of two weak and one strong bi-decompositions. Weak π-bi-decompositions have the benefit that they can be used for logic functions for which no strong π-bi-decomposition exists. The drawback of weak π-bi-decompositions is that the decomposition-function $g(\mathbf{x}_a, \mathbf{x}_c)$ depends on all variables of the given function $f(\mathbf{x}_a, \mathbf{x}_c)$ and only the decomposition-function $h(\mathbf{x}_c)$ depends on less variables. This causes signal paths of different lengths in the circuit. Figure 11.43 shows, for instance, that the path from

Fig. 11.43 Circuit of the function of Fig. 11.42 synthesized by two consecutive weak AND-bi-decompositions and a final disjoint OR-bi-decomposition

Fig. 11.44 Structure and conditions of the three types of vectorial bi-decompositions of $f(\mathbf{x})$: (**a**) vectorial OR-bi-decomposition, (**b**) vectorial AND-bi-decomposition, (**c**) vectorial EXOR-bi-decomposition, and (**d**) conditions of all vectorial bi-decompositions

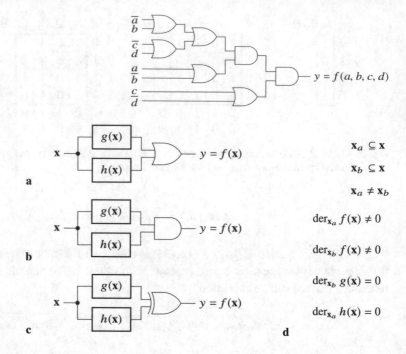

the input $\overline{c}$ to the output y contains four gates, but the path from the input c to the output y only two gates. Hence, weak π-bi-decompositions cause significantly different delays.

Due to this observation, the question has arisen whether other π-bi-decompositions exist for functions that are not strongly bi-decomposable. The key to answer this question is the fundamental property of each bi-decomposition that the decomposition-functions must be simpler than the given function. This property will be achieved by a smaller number of variables the decomposition-function is depending on. A logic function does not depend on a variable x_i if its single derivative is equal to 0:

$$\operatorname{der}_{x_i} f(x_i, \mathbf{x}_1) = 0 .$$

Hence, from a more general point of view, a derivative that is equal 0 indicates a simplification. We know that the single derivative is a special case of the vectorial derivatives. For that reason, we explored *vectorial π-bi-decompositions* in [18]. As shown in Fig. 11.44, the decomposition-functions $g(\mathbf{x})$ and $h(\mathbf{x})$ of all three types of vectorial bi-decompositions depend on the same variables as the given function $f(\mathbf{x})$, but all these decomposition-functions are simpler than the given function due to the constraints specified in Fig. 11.44d.

Definition 11.3 (Vectorial π-Bi-Decompositions for Logic Functions) A vectorial π-bi-decomposition:

$$f(\mathbf{x}) = \pi(g(\mathbf{x}), h(\mathbf{x})) \tag{11.138}$$

decomposes a given logic function $f(\mathbf{x})$ with regard to the dedicated subsets of variables $\mathbf{x}_a \subseteq \mathbf{x}$ and $\mathbf{x}_b \subseteq \mathbf{x}$, $\mathbf{x}_a \neq \mathbf{x}_b$, into the decomposition-functions $g(\mathbf{x})$ and $h(\mathbf{x})$ that satisfy

$$\operatorname{der}_{\mathbf{x}_b} g(\mathbf{x}) = 0 , \tag{11.139}$$

$$\operatorname{der}_{\mathbf{x}_a} h(\mathbf{x}) = 0 . \tag{11.140}$$

The *decomposition-operator* $\pi(g, h)$ can be realized by an OR-, an AND-, or an EXOR-gate.

Based on this common definition, the following theorems specify for all three types of vectorial bi-decompositions both the condition and formulas of possible decomposition-functions. We provide simple examples that explore each type of these vectorial bi-decompositions. At the first glance, it seems that the vectorial bi-decompositions do not contribute to a simplification because the decomposition-functions depend on the same variables as the given function. We demonstrate in our examples the benefit of the used vectorial bi-decomposition for the subsequent bi-decomposition and compare the synthesized circuit with utilization of only weak and strong bi-decompositions. The proofs of the following theorem are given in [18].

Theorem 11.8 (Vectorial OR-Bi-Decomposition for a Logic Function) *A vectorial OR-bi-decomposition of the logic function* $f(\mathbf{x})$ *with regard to the dedicated sets* $\mathbf{x}_a \subseteq \mathbf{x}$ *and* $\mathbf{x}_b \subseteq \mathbf{x}$, $\mathbf{x}_a \neq \mathbf{x}_b$ *exists if and only if*

$$f(\mathbf{x}) \wedge \max_{\mathbf{x}_a} \overline{f(\mathbf{x})} \wedge \max_{\mathbf{x}_b} \overline{f(\mathbf{x})} = 0 . \tag{11.141}$$

Possible decomposition-functions are

$$g(\mathbf{x}) = \min_{\mathbf{x}_b} f(\mathbf{x}) , \tag{11.142}$$

$$h(\mathbf{x}) = \min_{\mathbf{x}_a} f(\mathbf{x}) . \tag{11.143}$$

Example 11.20 (Vectorial OR-Bi-decomposition) We are going to synthesize the symmetric function

$$f(a, b, c, d) = \overline{a}bcd \vee a\overline{b}cd \vee ab\overline{c}d \vee abc\overline{d} \tag{11.144}$$

for which no strong bi-decomposition exists. However, Condition (11.141) of the vectorial OR-bi-decomposition is satisfied for the function (11.144) and the dedicated sets $\mathbf{x}_a = (a, d)$ and $\mathbf{x}_b = (b, c)$. Figure 11.45 shows in the upper row the Karnaugh-maps of the three subfunctions of Condition (11.141) and at the bottom the decomposition-functions $g(a, b, c, d)$ and $h(a, b, c, d)$ calculated by (11.142) and (11.143), respectively. Due to (11.142), the decomposition-function $g(a, b, c, d)$ satisfies Condition (11.139); and due to (11.143) Condition (11.140) is satisfied by $h(a, b, c, d)$.

In the decomposition-function $g(a, b, c, d)$ can be expressed by

$$g(a, b, c, d) = ad \wedge (b \oplus c) ,$$

where the possible disjoint strong AND-bi-decomposition becomes directly visible. For the decomposition-function

$$h(a, b, c, d) = bc \wedge (a \oplus d) ,$$

there exists also a disjoint strong AND-bi-decomposition. Figure 11.46a shows the circuit structure that uses the explored vectorial OR-bi-decomposition and Fig. 11.46b the best known circuit structure synthesized by using only weak and strong bi-decompositions.

The comparison of the circuit structures of Fig. 11.46 reveals the benefit of the vectorial OR-bi-decomposition; the longest path is reduced from four to three gates and one gate less is needed.

Dual properties determine the vectorial AND-bi-Decomposition.

Theorem 11.9 (Vectorial AND-Bi-Decomposition for a Logic Function) *A vectorial AND-bi-decomposition of the logic function* $f(\mathbf{x})$ *with regard to the dedicated sets* $\mathbf{x}_a \subseteq \mathbf{x}$ *and* $\mathbf{x}_b \subseteq \mathbf{x}$,

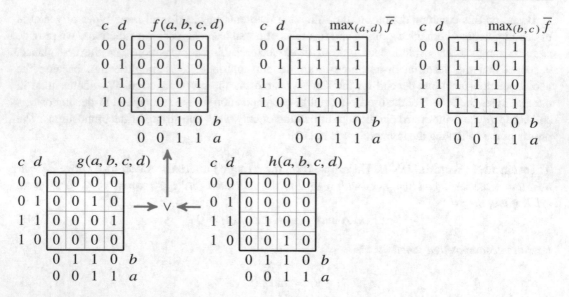

Fig. 11.45 Karnaugh-maps of the vectorial OR-bi-decomposition of the function $f(a, b, c, d)$ (11.144) with regard to the dedicated sets $\mathbf{x}_a = (a, d)$ and $\mathbf{x}_b = (c, d)$

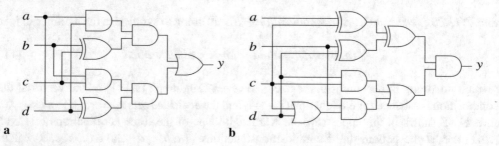

Fig. 11.46 Circuit structure of the function $y = f(a, b, c, d)$ (11.144): (**a**) synthesized using the explored vectorial OR-bi-decomposition for the output-gate and (**b**) synthesized using only weak and strong bi-decompositions

$\mathbf{x}_a \neq \mathbf{x}_b$ *exists if and only if*

$$\overline{f(\mathbf{x})} \wedge \max_{\mathbf{x}_a} f(\mathbf{x}) \wedge \max_{\mathbf{x}_b} f(\mathbf{x}) = 0 . \tag{11.145}$$

Possible decomposition-functions are

$$g(\mathbf{x}) = \max_{\mathbf{x}_b} f(\mathbf{x}) , \tag{11.146}$$

$$h(\mathbf{x}) = \max_{\mathbf{x}_a} f(\mathbf{x}) . \tag{11.147}$$

Example 11.21 (Vectorial AND-Bi-decomposition) We take another symmetric function

$$f(a, b, c, d) = \overline{a}\,\overline{b} \vee \overline{a}\,\overline{c} \vee \overline{a}\,\overline{d} \vee \overline{b}\,\overline{c} \vee \overline{b}\,\overline{d} \vee \overline{c}\,\overline{d} \vee abcd \tag{11.148}$$

for which no strong bi-decomposition exists. However, Condition (11.145) of the vectorial AND-bi-decomposition is satisfied for the function (11.148) and the dedicated sets $\mathbf{x}_a = (a, d)$ and $\mathbf{x}_b = (b, c)$. Figure 11.47 shows in the upper row on the left-hand side the Karnaugh-map of the given function

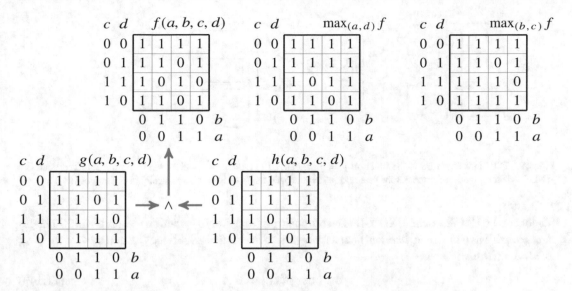

Fig. 11.47 Karnaugh-maps of the vectorial AND-bi-decomposition of the function (11.148) with regard to the dedicated sets $\mathbf{x}_a = (a, d)$ and $\mathbf{x}_b = (c, d)$

$f(a, b, c, d)$ and to the right of this the other two subfunctions of Condition (11.145). The conjunction of the complement of the function values in the left Karnaugh-map and the function values of the right two Karnaugh-maps results in a function that is equal to 0 so that Condition (11.145) of the vectorial AND-bi-decomposition is satisfied. The Karnaugh-maps in the second row of Fig. 11.47 show the decomposition-functions $g(a, b, c, d)$ and $h(a, b, c, d)$ calculated by (11.146) and (11.147), respectively. The decomposition-function $g(a, b, c, d)$ satisfies Condition (11.139) due to (11.146), and the decomposition-function $h(a, b, c, d)$ satisfies Condition (11.140) due to (11.147).

The decomposition-function $g(a, b, c, d)$ can be expressed by

$$g(a, b, c, d) = \overline{(a \wedge d)} \vee \overline{(b \oplus c)} \,,$$

where the possible disjoint strong OR-bi-decomposition becomes directly visible. For the decomposition-function

$$h(a, b, c, d) = \overline{(b \wedge c)} \vee \overline{(a \oplus d)}$$

exists also a disjoint strong OR-bi-decomposition. Figure 11.48a shows the circuit structure that uses the explored vectorial AND-bi-decomposition, and Fig. 11.48b the best known circuit structure synthesized by using only weak and strong bi-decompositions.

As can be seen in Fig. 11.48, the benefit of the vectorial AND-bi-decomposition is also a reduced length of the longest path from four to three gates and a one gate less than the best circuit synthesized using only weak and strong bi-decompositions.

The vectorial EXOR-bi-decomposition is a generalization of the strong EXOR-bi-decomposition with regard to a single variable in each of the dedicated sets. Instead of single derivative operations with regard to these two variables, vectorial derivative operations with regard to the dedicated subsets of variables $\mathbf{x}_a$ and $\mathbf{x}_b$ are used.

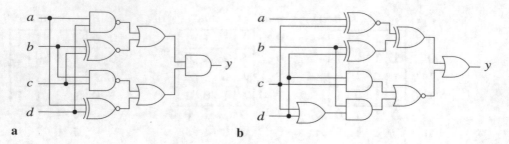

Fig. 11.48 Circuit structure of the function $y = f(a, b, c, d)$ (11.148): (**a**) synthesized using the explored vectorial AND-bi-decomposition for the output-gate and (**b**) synthesized using only weak and strong bi-decompositions

Theorem 11.10 (Vectorial EXOR-Bi-Decomposition for a Logic Function) *A vectorial EXOR-bi-decomposition of the logic function $f(\mathbf{x})$ with regard to the dedicated sets $\mathbf{x}_a \subseteq \mathbf{x}$ and $\mathbf{x}_b \subseteq \mathbf{x}$, $\mathbf{x}_a \neq \mathbf{x}_b$ exists if and only if*

$$\operatorname*{der}_{\mathbf{x}_b}\left(\operatorname*{der}_{\mathbf{x}_a} f(\mathbf{x})\right) = 0 . \tag{11.149}$$

Possible decomposition-functions are

$$g(\mathbf{x}) = \operatorname*{der}_{\mathbf{x}_b}(x_{bi} \wedge f(\mathbf{x})) , \tag{11.150}$$

$$h(\mathbf{x}) = f(\mathbf{x}) \oplus g(\mathbf{x}) , \tag{11.151}$$

where

$$x_{bi} \in \mathbf{x}_b . \tag{11.152}$$

It follows from (11.150) that Condition (11.139) is satisfied, but it is not obvious that Condition (11.140) holds for the specified decomposition-function $h(\mathbf{x})$ (11.151). The associated proof is given in [18]; however, it is also an interesting exercise for the reader.

Example 11.22 An analysis with regard to all pairs of variables confirms that for the function

$$f(a, b, c, d) = a\overline{d} \vee b\overline{d} \vee \overline{a}d \vee bc \vee c\overline{d} \tag{11.153}$$

neither a strong AND-bi-decomposition nor a strong EXOR-bi-decomposition exists. The existence of a strong OR-bi-decomposition of this function with regard to the dedicated sets $\mathbf{x}_a = (a)$ and $\mathbf{x}_b = (b, c)$ is an additional challenge for the comparison with the vectorial EXOR-bi-decomposition of this function.

Figure 11.49 shows in the upper row on the left-hand side the Karnaugh-map of the given function $f(a, b, c, d)$ and to the right of this the results of the two consecutive vectorial derivatives of f with regard to the sets $\mathbf{x}_a = (a, d)$ and $\mathbf{x}_b = (b, c)$, which confirm that a vectorial EXOR-bi-decomposition for the explored function and the used dedicated sets exists. The use of both $x_{bi} = b$ and $x_{bi} = c$ in (11.150) results in the Karnaugh-maps in the second row of Fig. 11.49, which shows the decomposition-functions

$$g(a, b, c, d) = \overline{a} \vee \overline{d} \vee \overline{(b \oplus c)} = \overline{(a \wedge d) \wedge (b \oplus c)}$$

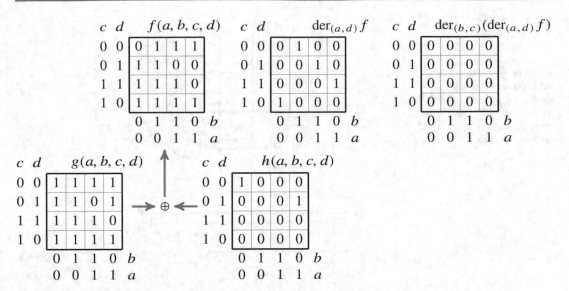

Fig. 11.49 Karnaugh-maps of the vectorial EXOR-bi-decomposition of the function (11.153) with regard to the dedicated sets $\mathbf{x}_a = (a, d)$ and $\mathbf{x}_b = (b, c)$

and

$$h(a, b, c, d) = \overline{b} \wedge \overline{c} \wedge \overline{(a \oplus d)} = \overline{(b \vee c) \vee (a \oplus d)}$$

calculated by (11.150) and (11.151), respectively. Due to $f = g \oplus h = \overline{g} \oplus \overline{h}$, the decomposition-functions $g(a, b, c, d)$ and $h(a, b, c, d)$ can be replaced by the functions $\overline{g(a, b, c, d)}$ and $\overline{h(a, b, c, d)}$ in each EXOR-bi-decomposition. We utilize this rule in this example. Figure 11.50 reveals that the circuit of the function $y = f(a, b, c, d)$ synthesized using the vectorial EXOR-bi-decomposition needs the same number of gates as the circuit that uses a strong OR-bi-decomposition for the output gate; however, the maximal path length contains in case of the used strong OR-bi-decomposition five gates, but for the utilized vectorial EXOR-bi-decomposition only three gates. Hence, utilizing the vectorial EXOR-bi-decomposition instead of the common use of only strong and weak bi-decomposition facilitates a shorter delay of the circuit without an increased the number of gates.

An interesting and very useful property of all vectorial bi-decompositions that contain two variables in the dedicated sets has recently been published in [14]. The theoretical basis is the relationship between the vectorial and k-fold derivative shown in Theorem 11.11.

Theorem 11.11 (Relation Between the a Vectorial Bi-Decomposition and an EXOR-Bi-Decomposition) *Let $f(\mathbf{x}_0, \mathbf{x}_1) = f(x_i, x_j, \mathbf{x}_1)$ be a logic function. If*

$$\operatorname*{der}_{(x_i, x_j)} f(\mathbf{x}_0, \mathbf{x}_1) = 0 \tag{11.154}$$

is satisfied, then it also holds

$$\operatorname*{der}_{(x_i, x_j)}^{2} f(\mathbf{x}_0, \mathbf{x}_1) = 0 . \tag{11.155}$$

Because (11.154) is the condition of the decomposition-function of a vectorial bi-decomposition where the dedicated set contains exactly two variables and (11.155) is the condition of a strong

Fig. 11.50 Circuit structure of the function $y = f(a, b, c, d)$ (11.153): (**a**) synthesized using the explored vectorial EXOR-bi-decomposition for the output-gate and (**b**) synthesized using only weak and strong bi-decompositions

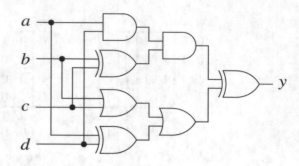

a

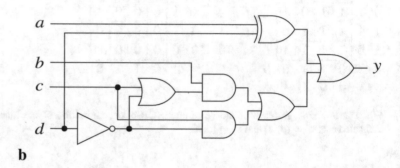

b

EXOR-bi-decomposition, each decomposition-function of vectorial bi-decomposition that does not depend on the simultaneous change of exactly two variables is strongly EXOR-bi-decomposable with regard to these two variables. No additional checks are needed in such a case.

Example 11.23 The decomposition-function $g(a, b, c, d)$ of the vectorial OR-bi-decomposition shown in Fig. 11.45 can be expressed by:

$$g(a, b, c, d) = (a \wedge d) \wedge (b \oplus c) = (b \wedge (a \wedge d)) \oplus (c \wedge (a \wedge d)),$$

where the second expression emphasizes the strong EXOR-bi-decomposition of the function $g(a, b, c, d)$ with regard to the dedicated sets $\mathbf{x}_a = (b)$ and $\mathbf{x}_b = (c)$. The parentheses in this expression indicate an associated circuit structure using only gates with two inputs.

It is a peculiarity of this example that the functions $g'(a, b, d) = abd$ and $h'(a, c, d) = acd$ contain the same subfunctions $g''(a, d) = ad$. Therefore, the utilization of the distributive law transforms $(b \wedge (a \wedge d)) \oplus (c \wedge (a \wedge d)))$ back to $(a \wedge d) \wedge (b \oplus c)$ so that the number of needed gates to realize this decomposition-function can be reduced from four to three.

Bi-Decompositions for Lattices of Logic Functions

Disjoint bi-decompositions having approximately the same number of variables in the dedicated sets are preferable because such bi-decompositions lead to small and fast circuits. However, it is a property of the given function whether a certain bi-decomposition and chosen variables for the dedicated sets exists. Hence, the use of an arbitrary function of a lattice of logic function increases the chance to synthesize a smaller and faster circuit in comparison with the synthesis of a single function. This is one reason why we explore bi-decompositions for lattices of logic functions.

Even if a circuit structure must be synthesized for single function, a weak bi-decomposition creates a lattice of logic functions for the decomposition-function $g(\mathbf{x}_a, \mathbf{x}_c)$ so that an approach for the further bi-decomposition based on a lattice of logic functions is needed.

Basically there are two possibilities for the bi-decomposition of a lattice of logic functions:

1. successively select each logic function of the lattice, check all types of bi-decompositions for all possible dedicated sets, and use the best found bi-decomposition;
2. commonly check all functions of the lattice for all types of bi-decompositions for all possible dedicated sets and use the best found bi-decomposition.

The effort to find the best bi-decomposition is obviously lesser for the second approach; hence, we explore this approach in the following. The effort of the check for the different assignments of the variables to the dedicated sets remains despite the common analysis of all functions of a lattice. A comprehensive analysis of all conditions of bi-decomposition led to the very useful result:

> If a strong or weak bi-decomposition is not possible for certain dedicated sets of variables, then this bi-decomposition is also not possible when a variable is moved from the common set to one of the dedicated sets.

Hence, it is a convenient strategy to start the analysis for a bi-decomposition where each dedicated set contains a single variable and alternately increase one of these sets by a variable of the previous common set. A *compact* bi-decomposition is found when no further variable can be moved from the common set $\mathbf{x}_c$ to one of the dedicated sets $\mathbf{x}_a$ or $\mathbf{x}_b$ without the loss of the property of the explored bi-decomposition. Compact bi-decompositions should be preferred because they facilitate the smallest circuits with the shortest delay.

The circuit structure of a bi-decomposition analyzed for a lattice of logic functions is the same as for a single function; hence, we must not repeat these structures. However, we are going to provide conditions that check whether a lattice contains a function for which the explored bi-decomposition exists.

If a lattice of logic functions contains at least one function for which a checked bi-decomposition exists, usually several pairs of decomposition-functions can be used. The set of possible functions satisfies for each decomposition-function the properties of a lattice. However, the used decomposition-function of one of these lattices restricts the lattice of the other decomposition-function. Therefore,

- we provide possible mark-functions of the lattice $\mathcal{L}\langle g_q(\mathbf{x}_a, \mathbf{x}_c), g_r(\mathbf{x}_a, \mathbf{x}_c)\rangle$;
- we expect that the synthesis of this lattice will be executed next so that the realized function $g(\mathbf{x}_a, \mathbf{x}_c)$ is known;
- we provide possible mark-functions of the lattice $\mathcal{L}\langle h_q(\mathbf{x}_b, \mathbf{x}_c), h_r(\mathbf{x}_b, \mathbf{x}_c)\rangle$ that can be used together with the realized function $g(\mathbf{x}_a, \mathbf{x}_c)$;
- based on this lattice, the decomposition-function $h(\mathbf{x}_b, \mathbf{x}_c)$ can be synthesized.

When the lattices of the decomposition-functions provided in this section and the procedure mentioned above are used, the synthesized circuit structures have the welcome property that they are completely testable with regard to all single stuck-at faults on all inputs and outputs of the gates and of the whole circuit. More details about this welcome property have been published in [22].

Fig. 11.51 Disjoint OR-bi-decomposition of $f(a,b,c,d) \in \mathcal{L}\langle f_q, f_r\rangle$ with regard to $\mathbf{x}_a = (a,b)$ and $\mathbf{x}_b = (c,d)$

c	d	$\mathcal{L}\langle h_q, h_r\rangle$
0	0	φ
0	1	0
1	1	1
1	0	0

c	d	$\mathcal{L}\langle f_q, f_r\rangle$			
0	0	φ	φ	φ	φ
0	1	0	φ	1	φ
1	1	1	φ	1	1
1	0	0	φ	1	0
		0	1	1	0	b
		0	0	1	1	a

$\mathcal{L}\langle g_q, g_r\rangle$

0	φ	1	0
0	1	1	0	b
0	0	1	1	a

Example 11.24 Figure 11.51 shows an example of a disjoint OR-bi-decomposition of the lattice $\mathcal{L}\langle f_q(\mathbf{x}_a, \mathbf{x}_b), f_r(\mathbf{x}_a, \mathbf{x}_b)\rangle$ into the lattices

$$\mathcal{L}\langle g_q(\mathbf{x}_a), g_r(\mathbf{x}_a)\rangle \quad \text{and} \quad \mathcal{L}\langle h_q(\mathbf{x}_b), h_r(\mathbf{x}_b)\rangle$$

with regard to the dedicated sets $\mathbf{x}_a = (a,b)$ and $\mathbf{x}_b = (c,d)$.

The given lattice $\mathcal{L}\langle f_q(a,b,c,d), f_r(a,b,c,d)\rangle$ contains eight don't-cares, labeled by φ in the Karnaugh-map, so that $2^8 = 256$ logic functions have to be checked for an OR-bi-decomposition. All of them have the same function values in the care set $f_\varphi(a,b,c,d) = 0$. A function value 0 of all functions of the given lattice requires appropriate function values 0 in the decomposition-functions due to the OR-operation of the decomposition. As can be seen in Fig. 11.51, the values 0 of the given lattice $\mathcal{L}\langle f_q(a,b,c,d), f_r(a,b,c,d)\rangle$ (emphasized by green boxes) are projected to the bottom into the lattices $\mathcal{L}\langle g_q(a,b), g_r(a,b)\rangle$ and to the left into the lattices $\mathcal{L}\langle h_q(c,d), h_r(c,d)\rangle$. We assume that the don't-care in the yellow box of Fig. 11.51 is changed into $f(1,0,0,1) = 1$. In this case an OR-bi-decomposition with regard to the dedicated sets $\mathbf{x}_a = (a,b)$ and $\mathbf{x}_b = (c,d)$ is not possible due to the necessary values 0 of $g(a = 1, b = 0)$ and $h(c = 0, d = 1)$. This shows that a larger don't-care-function $f_\varphi(a,b,c,d)$ improves the chance to find an OR-bi-decomposition, and secondly, the don't-care-function $f_\varphi(a,b,c,d)$ does not restrict the property of OR-bi-decomposition of the given lattice $\mathcal{L}\langle f_q(a,b,c,d), f_r(a,b,c,d)\rangle$. Thus, by means of the mark-functions $f_q(a,b,c,d)$ and $f_r(a,b,c,d)$, it is possible to decide whether this lattice contains a logic function $f(a,b,c,d)$ that is OR-bi-decomposable.

The explored disjoint OR-bi-decomposition of a lattice of logic functions can be generalized to the non-disjoint OR-bi-decomposition in the same way as in the case of a completely specified logic function.

The general condition for a non-disjoint OR-bi-decomposition of the lattice $\mathcal{L}\langle f_q(\mathbf{x}_a, \mathbf{x}_b, \mathbf{x}_c), f_r(\mathbf{x}_a, \mathbf{x}_b, \mathbf{x}_c)\rangle$ is given in Theorem 11.12 that also describes the disjoint case for an empty common set $\mathbf{x}_c$. An OR-bi-decomposition of the lattice $\mathcal{L}\langle f_q(\mathbf{x}_a, \mathbf{x}_b, \mathbf{x}_c), f_r(\mathbf{x}_a, \mathbf{x}_b, \mathbf{x}_c)\rangle$ exists if the projections in both directions $\mathbf{x}_a$ and $\mathbf{x}_b$ of the values 0 expressed by the OFF-set-function $f_r(\mathbf{x}_a, \mathbf{x}_b, \mathbf{x}_c)$ do not overlap with values 1 expressed by the ON-set-function $f_q(\mathbf{x}_a, \mathbf{x}_b, \mathbf{x}_c)$.

Theorem 11.12 (Strong OR-Bi-Decomposition for a Lattice of Logic Functions) *The lattice* $\mathcal{L}\langle f_q(\mathbf{x}_a, \mathbf{x}_b, \mathbf{x}_c), f_r(\mathbf{x}_a, \mathbf{x}_b, \mathbf{x}_c)\rangle$ *contains at least one logic function* $f(\mathbf{x}_a, \mathbf{x}_b, \mathbf{x}_c)$ *that is OR-bi-decomposable with regard to the dedicated sets* $\mathbf{x}_a$ *and* $\mathbf{x}_b$ *if and only if*

$$f_q(\mathbf{x}_a, \mathbf{x}_b, \mathbf{x}_c) \wedge \max_{\mathbf{x}_a}^k f_r(\mathbf{x}_a, \mathbf{x}_b, \mathbf{x}_c) \wedge \max_{\mathbf{x}_b}^k f_r(\mathbf{x}_a, \mathbf{x}_b, \mathbf{x}_c) = 0 . \qquad (11.156)$$

Figure 11.51 shows that the decomposition-functions of the OR-bi-decomposition of a lattice can also be chosen out of lattices. Formulas (11.157) and (11.158) specify necessary and sufficient conditions to calculate the don't-care-functions $g_\varphi(\mathbf{x}_a, \mathbf{x}_c)$ and $h_\varphi(\mathbf{x}_b, \mathbf{x}_c)$ independent of each other for all different types of strong bi-decompositions depending on the don't-care-function $f_\varphi(\mathbf{x}_a, \mathbf{x}_b, \mathbf{x}_c)$:

$$g_\varphi(\mathbf{x}_a, \mathbf{x}_c) = \min_{\mathbf{x}_b}^k f_\varphi(\mathbf{x}_a, \mathbf{x}_b, \mathbf{x}_c) , \qquad (11.157)$$

$$h_\varphi(\mathbf{x}_b, \mathbf{x}_c) = \min_{\mathbf{x}_a}^k f_\varphi(\mathbf{x}_a, \mathbf{x}_b, \mathbf{x}_c) . \qquad (11.158)$$

Larger don't-care-functions are possible, but such don't-care-functions depend on each other. The chosen values 1 of the decomposition-function $g(\mathbf{x}_a, \mathbf{x}_c)$ can extend the don't-care set of the decomposition-function $h(\mathbf{x}_b, \mathbf{x}_c)$. A reverse consideration results in a rule that determines the ON-set-function of the decomposition-lattice $\mathcal{L}\langle g_q(\mathbf{x}_a, \mathbf{x}_c), g_r(\mathbf{x}_a, \mathbf{x}_c)\rangle$.

The ON-set-function $g_q(\mathbf{x}_a, \mathbf{x}_c)$ of an OR-bi-decomposition must cover only such values 1 of the given lattice (determined by $f_q(\mathbf{x}_a, \mathbf{x}_b, \mathbf{x}_c)$) that overlap with $\max_{\mathbf{x}_a}^k f_r(\mathbf{x}_a, \mathbf{x}_b, \mathbf{x}_c)$ because these values 1 cannot be covered by $h_q(\mathbf{x}_b, \mathbf{x}_c)$. Due to the OR-operation, all values 0 must be mapped from the lattice $\mathcal{L}\langle f_q, f_r\rangle$ into the lattice $\mathcal{L}\langle g_q, g_r\rangle$. The mark-functions for $g_q(\mathbf{x}_a, \mathbf{x}_c)$ and $g_r(\mathbf{x}_a, \mathbf{x}_c)$ of an OR-bi-decomposition of a given lattice $\mathcal{L}\langle f_q, f_r\rangle$ can be calculated by:

$$g_q(\mathbf{x}_a, \mathbf{x}_c) = \max_{\mathbf{x}_b}^k (f_q(\mathbf{x}_a, \mathbf{x}_b, \mathbf{x}_c) \wedge \max_{\mathbf{x}_a}^k f_r(\mathbf{x}_a, \mathbf{x}_b, \mathbf{x}_c)) , \qquad (11.159)$$

$$g_r(\mathbf{x}_a, \mathbf{x}_c) = \max_{\mathbf{x}_b}^k f_r(\mathbf{x}_a, \mathbf{x}_b, \mathbf{x}_c) . \qquad (11.160)$$

Formulas (11.159) and (11.160) describe the largest lattice $\mathcal{L}\langle g_q, g_r\rangle$ of all feasible decomposition-functions $g(\mathbf{x}_a, \mathbf{x}_c)$. The realized decomposition-function $g(\mathbf{x}_a, \mathbf{x}_c)$ influences the lattice of the decomposition-functions $h(\mathbf{x}_b, \mathbf{x}_c)$ as follows: the term $f_q(\mathbf{x}_a, \mathbf{x}_b, \mathbf{x}_c) \wedge \overline{g(\mathbf{x}_a, \mathbf{x}_c)}$ of (11.161) specifies all values 1 of the given lattice $\mathcal{L}\langle f_q, f_r\rangle$ covered by values 0 of the function $g(\mathbf{x}_a, \mathbf{x}_c)$; hence, these values 1 must be assigned to $h_q(\mathbf{x}_b, \mathbf{x}_c)$:

$$h_q(\mathbf{x}_b, \mathbf{x}_c) = \max_{\mathbf{x}_a}^k \left(f_q(\mathbf{x}_a, \mathbf{x}_b, \mathbf{x}_c) \wedge \overline{g(\mathbf{x}_a, \mathbf{x}_c)} \right) , \qquad (11.161)$$

$$h_r(\mathbf{x}_b, \mathbf{x}_c) = \max_{\mathbf{x}_a}^k f_r(\mathbf{x}_a, \mathbf{x}_b, \mathbf{x}_c) . \qquad (11.162)$$

Dual properties hold for AND-bi-decompositions of lattices of logic functions.

Example 11.25 Figure 11.52 shows an example of a non-disjoint AND-bi-decomposition of the lattice $\mathcal{L}\langle f_q(\mathbf{x}_a, \mathbf{x}_b, \mathbf{x}_c), f_r(\mathbf{x}_a, \mathbf{x}_b, \mathbf{x}_c)\rangle$ into the lattices

$$\mathcal{L}\langle g_q(\mathbf{x}_a, \mathbf{x}_c), g_r(\mathbf{x}_a, \mathbf{x}_c)\rangle \quad \text{and} \quad \mathcal{L}\langle h_q(\mathbf{x}_b, \mathbf{x}_c), h_r(\mathbf{x}_b, \mathbf{x}_c)\rangle$$

with regard to the dedicated sets $\mathbf{x}_a = (b, c)$ and $\mathbf{x}_b = (d, e)$; hence, the common set is $\mathbf{x}_c = (a)$.

Fig. 11.52 Non-disjoint
AND-bi-decomposition of
$f(a, b, c, d, e) \in \mathcal{L}\langle f_q, f_r\rangle$
with regard to
$\mathbf{x}_a = (b, c)$ and $\mathbf{x}_b = (d, e)$

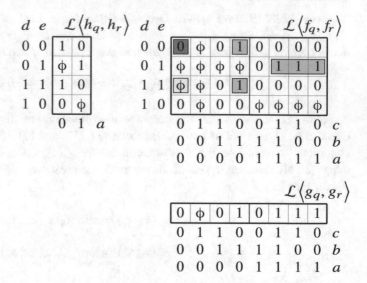

The given lattice $\mathcal{L}\langle f_q(a, b, c, d, e), f_r(a, b, c, d, e)\rangle$ contains 12 don't-cares, labeled by ϕ in the Karnaugh-map, so that $2^{12} = 4096$ logic functions have to be checked for an AND-bi-decomposition. All of them have the same function values in the care set $f_\varphi(a, b, c, d, e) = 0$. Due to the common variable a, the subfunctions $f(a = 0)$ and $f(a = 1)$ can be separately analyzed for an AND-bi-decomposition. A function value 1 of all functions of the given lattice requires appropriate function values 1 in the decomposition-functions due to the AND-operation of the decomposition. As can be seen in Fig. 11.52, the values 1 from the given lattice $\mathcal{L}\langle f_q(a, b, c, d, e), f_r(a, b, c, d, e)\rangle$ (emphasized by green boxes) are projected to the bottom into the lattice $\mathcal{L}\langle g_q(a, b, c), g_r(a, b, c)\rangle$ and to the left into the lattice $\mathcal{L}\langle h_q(a, d, e), h_r(a, d, e)\rangle$ where different results are feasible for the subspaces $a = 0$ and $a = 1$. Let us assume that the don't-care in the yellow box of Fig. 11.52 is changed into $f(0, 0, 0, 1, 1) = 1$. This change would require the change of $g(a = 0, b = 0, c = 0)$ from 0 to 1 so that an AND-bi-decomposition with regard to the dedicated sets $\mathbf{x}_a = (b, c)$ and $\mathbf{x}_b = (d, e)$ is not possible due to $f(0, 0, 0, 0, 0) = 0$ (emphasized by a red box in Fig. 11.52) and the necessary values 1 of $g(a = 0, b = 0, c = 0)$ and $h(a = 0, d = 0, e = 0)$. This shows that a larger don't-care-function $f_\varphi(a, b, c, d, e)$ also improves the chance to find an AND-bi-decomposition.

This example demonstrates that it is possible to decide by means of the mark-functions $f_q(a, b, c, d, e)$ and $f_r(a, b, c, d, e)$ whether this lattice contains a logic function $f(a, b, c, d, e)$ that is AND-bi-decomposable.

The general condition for a non-disjoint AND-bi-decomposition of the lattice $\mathcal{L}\langle f_q(\mathbf{x}_a, \mathbf{x}_b, \mathbf{x}_c), f_r(\mathbf{x}_a, \mathbf{x}_b, \mathbf{x}_c)\rangle$ is given in Theorem 11.13 that also describes the disjoint case for an empty common set $\mathbf{x}_c$. An AND-bi-decomposition of the lattice $\mathcal{L}\langle f_q(\mathbf{x}_a, \mathbf{x}_b, \mathbf{x}_c), f_r(\mathbf{x}_a, \mathbf{x}_b, \mathbf{x}_c)\rangle$ exists if the projections in both direction $\mathbf{x}_a$ and $\mathbf{x}_b$ of the values 1 expressed by the ON-set-function $f_q(\mathbf{x}_a, \mathbf{x}_b, \mathbf{x}_c)$ do not overlap with values 0 expressed by the OFF-set-function $f_r(\mathbf{x}_a, \mathbf{x}_b, \mathbf{x}_c)$.

Theorem 11.13 (Strong AND-Bi-Decomposition for a Lattice of Logic Functions) *The lattice* $\mathcal{L}\langle f_q(\mathbf{x}_a, \mathbf{x}_b, \mathbf{x}_c), f_r(\mathbf{x}_a, \mathbf{x}_b, \mathbf{x}_c)\rangle$ *contains at least one logic function* $f(\mathbf{x}_a, \mathbf{x}_b, \mathbf{x}_c)$ *that is AND-bi-decomposable with regard to the dedicated sets* $\mathbf{x}_a$ *and* $\mathbf{x}_b$ *if and only if*

$$f_r(\mathbf{x}_a, \mathbf{x}_b, \mathbf{x}_c) \wedge \max_{\mathbf{x}_a}^{k} f_q(\mathbf{x}_a, \mathbf{x}_b, \mathbf{x}_c) \wedge \max_{\mathbf{x}_b}^{k} f_q(\mathbf{x}_a, \mathbf{x}_b, \mathbf{x}_c) = 0 . \tag{11.163}$$

The OFF-set-function $g_r(\mathbf{x}_a, \mathbf{x}_c)$ of an AND-bi-decomposition must cover only such values 0 of the given lattice (determined by $f_r(\mathbf{x}_a, \mathbf{x}_b, \mathbf{x}_c)$) that overlap with $\max_{\mathbf{x}_a}{}^k f_q(\mathbf{x}_a, \mathbf{x}_b, \mathbf{x}_c)$ because these values 0 cannot be contributed by $h_r(\mathbf{x}_b, \mathbf{x}_c)$. Due to the AND-operation, all values 1 must be mapped from the lattice $\mathcal{L}\langle f_q, f_r\rangle$ into the lattice $\mathcal{L}\langle g_q, g_r\rangle$. The mark-functions $g_q(\mathbf{x}_a, \mathbf{x}_c)$ and $g_r(\mathbf{x}_a, \mathbf{x}_c)$ of an AND-bi-decomposition of a given lattice $\mathcal{L}\langle f_q, f_r\rangle$ can be calculated by:

$$g_q(\mathbf{x}_a, \mathbf{x}_c) = \max_{\mathbf{x}_b}{}^k f_q(\mathbf{x}_a, \mathbf{x}_b, \mathbf{x}_c) \,, \tag{11.164}$$

$$g_r(\mathbf{x}_a, \mathbf{x}_c) = \max_{\mathbf{x}_b}{}^k (f_r(\mathbf{x}_a, \mathbf{x}_b, \mathbf{x}_c) \wedge \max_{\mathbf{x}_a}{}^k f_q(\mathbf{x}_a, \mathbf{x}_b, \mathbf{x}_c)) \,. \tag{11.165}$$

Formulas (11.164) and (11.165) describe the largest lattice $\mathcal{L}\langle g_q, g_r\rangle$ of all feasible decomposition-functions $g(\mathbf{x}_a, \mathbf{x}_c)$. The realized decomposition-function $g(\mathbf{x}_a, \mathbf{x}_c)$ influences the lattice of the decomposition-functions $h(\mathbf{x}_b, \mathbf{x}_c)$ as follows: the term $f_r(\mathbf{x}_a, \mathbf{x}_b, \mathbf{x}_c) \wedge g(\mathbf{x}_a, \mathbf{x}_c)$ of (11.167) specifies all values 0 of the given lattice $\mathcal{L}\langle f_q, f_r\rangle$ covered by values 1 of $g(\mathbf{x}_a, \mathbf{x}_c)$; hence, these values 0 must belong to $h_r(\mathbf{x}_b, \mathbf{x}_c)$:

$$h_q(\mathbf{x}_b, \mathbf{x}_c) = \max_{\mathbf{x}_a}{}^k f_q(\mathbf{x}_a, \mathbf{x}_b, \mathbf{x}_c) \,, \tag{11.166}$$

$$h_r(\mathbf{x}_b, \mathbf{x}_c) = \max_{\mathbf{x}_a}{}^k (f_r(\mathbf{x}_a, \mathbf{x}_b, \mathbf{x}_c) \wedge g(\mathbf{x}_a, \mathbf{x}_c)) \,. \tag{11.167}$$

The operations $\vee$ and $\wedge$ specify a Boolean Algebra, but the operation $\oplus$ determines the properties of a Boolean Ring. Due to these different algebraic structures and the tight relationship between a lattice of logic functions and a Boolean Algebra, the existence of both a compact OR- and a compact AND-bi-decomposition can be detected in a given lattice of logic functions using (11.156) or (11.163). Until the year 2016, only a generalization of (11.119) was known so that it could be merely verified whether an EXOR-bi-decomposition of a given lattice of logic functions with regard to a single variable x_a and a set of variables $\mathbf{x}_b$ exists. Recently we found a basic solution for this problem [20]; hence, now it is possible to determine maximal dedicated sets $\mathbf{x}_a$ and $\mathbf{x}_b$ for each type of a strong bi-decomposition of a lattice of logic functions. Here we explain this basic solution and extend it to a more general one.

As generalization of Theorem 11.4, a maximal dedicated set $\mathbf{x}_b$ of a lattice of logic function of a strong EXOR-bi-decomposition can be found; however, this generalization restricts the dedicated set $\mathbf{x}_a$ to a single variable x_a.

Theorem 11.14 (Strong EXOR-Bi-Decomposition for a Lattice of Logic Functions with Regard to the Single Variable x_a and Dedicated Set $\mathbf{x}_b$) *The lattice $\mathcal{L}\langle f_q(x_a, \mathbf{x}_b, \mathbf{x}_c), f_r(x_a, \mathbf{x}_b, \mathbf{x}_c)\rangle$ contains at least one logic function $f(x_a, \mathbf{x}_b, \mathbf{x}_c)$ that is EXOR-bi-decomposable with regard to the single variable x_a and the dedicated set of variables $\mathbf{x}_b$ if and only if*

$$\max_{\mathbf{x}_b}{}^k f_q^{\mathrm{der}_{x_a}}(\mathbf{x}_b, \mathbf{x}_c) \wedge f_r^{\mathrm{der}_{x_a}}(\mathbf{x}_b, \mathbf{x}_c) = 0 \,. \tag{11.168}$$

Equation (11.168) uses the mark-functions

$$f_q^{\mathrm{der}_{x_a}}(\mathbf{x}_b, \mathbf{x}_c) \quad \text{and} \quad f_r^{\mathrm{der}_{x_a}}(\mathbf{x}_b, \mathbf{x}_c)$$

Fig. 11.53 Non-disjoint
EXOR-bi-decomposition
of $f(a, b, c, d) \in \mathcal{L}\langle f_q, f_r\rangle$
with regard to $x_a = (b)$ and
$\mathbf{x}_b = (c, d)$

c	d	$\mathcal{L}\langle h_q, h_r\rangle$	
0	0	0	1
0	1	1	0
1	1	ϕ	1
1	0	1	ϕ

$\qquad\qquad$ 0 $\quad$ 1 $\quad a$

c	d	$\mathcal{L}\langle f_q, f_r\rangle$			
0	0	0	0	0	1
0	1	1	ϕ	1	0
1	1	ϕ	ϕ	ϕ	1
1	0	1	1	ϕ	ϕ

$\qquad$ 0 $\quad$ 1 $\quad$ 1 $\quad$ 0 $\quad b$
$\qquad$ 0 $\quad$ 0 $\quad$ 1 $\quad$ 1 $\quad a$

$g(a, b)$

0	0	1	0

$\qquad$ 0 $\quad$ 1 $\quad$ 1 $\quad$ 0 $\quad b$
$\qquad$ 0 $\quad$ 0 $\quad$ 1 $\quad$ 1 $\quad a$

of the single derivative of the given lattice with regard to the variable x_a (see Theorem 7.2 on page 315 et seqq.). The given lattice of logic functions $\mathcal{L}\langle f_q(x_a, \mathbf{x}_b, \mathbf{x}_c), f_r(x_a, \mathbf{x}_b, \mathbf{x}_c)\rangle$ contains at least one logic function $f(x_a, \mathbf{x}_b, \mathbf{x}_c)$ that is EXOR-bi-decomposable with regard to the variable x_a and the dedicated set $\mathbf{x}_b$, if a change of $f(x_a, \mathbf{x}_b, \mathbf{x}_c) \in \mathcal{L}\langle f_q, f_r\rangle$ with regard to x_a (expressed by $f_q^{\mathrm{der}_{x_a}}(\mathbf{x}_b, \mathbf{x}_c)$) does not meet a pair of values that remains unchanged for the change of x_a (expressed by $f_r^{\mathrm{der}_{x_a}}(\mathbf{x}_b, \mathbf{x}_c)$) in the same subspace ($\mathbf{x}_b, \mathbf{x}_c = const.$).

Example 11.26 Figure 11.53 shows an example of a non-disjoint EXOR-bi-decomposition of a given lattice $\mathcal{L}\langle f_q(a, b, c, d), f_r(a, b, c, d)\rangle$ into the function $g(a, b)$ and the lattice $\mathcal{L}\langle h_q(a, c, d), h_r(a, c, d)\rangle$ with regard to the variable $x_a = (b)$ and the dedicated set $\mathbf{x}_b = (c, d)$ so that the common variable is $x_c = (a)$.

The given lattice $\mathcal{L}\langle f_q(a, b, c, d), f_r(a, b, c, d)\rangle$ contains six don't-cares, labeled by ϕ in the Karnaugh-map, so that $2^6 = 64$ logic functions have to be checked for the EXOR-bi-decomposition. All of them have the same function values in the care set $f_\varphi(a, b, c, d) = 0$. Due to the common variable $x_c = (a)$, two subspaces must be taken into account: $a = 0$ and $a = 1$. For $a = 0$, equal function values of all $f(a, b, c, d) \in \mathcal{L}\langle f_q, f_r\rangle$ are determined for $f(a = 0, b, c = 0, d = 0)$ and $f(a = 0, b, c = 1, d = 0)$; hence, the single derivative $\mathrm{der}_b\, g(a = 0, b)$ must be equal to 0. If the don't-care in the left yellow box of Fig. 11.53 is changed into $f(0, 1, 0, 1) = 0$, the explored non-disjoint EXOR-bi-decomposition does not exist due to the different function values for $f(a = 0, b, c = 0, d = 1)$ and the equal function values, e.g., for $f(a = 0, b, c = 0, d = 0)$.

For $a = 1$, different function values of all $f(a, b, c, d) \in \mathcal{L}\langle f_q, f_r\rangle$ are determined for $f(a = 1, b, c = 0, d = 0)$ and $f(a = 1, b, c = 0, d = 1)$; hence, the single derivative $\mathrm{der}_b\, g(a = 1, b)$ must be equal to 1. If the don't-care in the right yellow box of Fig. 11.53 is changed into $f(1, 1, 1, 1) = 1$, the explored non-disjoint EXOR-bi-decomposition also does not exist due to the equal function values for $f(a = 1, b, c = 1, d = 1)$ and the different function values, e.g., for $f(a = 1, b, c = 0, d = 0)$.

Hence, by means of the mark-functions $f_q(a, b, c, d)$ and $f_r(a, b, c, d)$, it is possible to decide whether this lattice contains a logic function $f(a, b, c, d)$ that is EXOR-bi-decomposable, but (11.168) restricts the dedicated set $\mathbf{x}_a$ to a single variable x_a that is equal to (b) in the explored example.

The given lattice $\mathcal{L}\langle f_q, f_r\rangle$ of a strong EXOR-bi-decomposition with regard to a single variable x_a and a dedicated set of variables $\mathbf{x}_b$ determines not the decomposition-function $g(x_a, \mathbf{x}_c)$, but only the values of the derivative $\text{der}_{x_a} g(x_a, \mathbf{x}_c)$ for each subspace $\mathbf{x}_c = const$. Hence, there are four possible functions $g(a, b)$ that can be used as decomposition-function of the examples explored in Fig. 11.53:

$$g_1(a, b) = ab \oplus 0 , \quad g_2(a, b) = ab \oplus a , \quad g_3(a, b) = ab \oplus \overline{a} , \quad g_4(a, b) = ab \oplus 1 .$$

Using one of these decomposition-functions $g(a, b)$, the associated lattice of the decomposition-functions $h(a, c, d) \in \mathcal{L} \langle h_q, h_r \rangle$ can be determined.

The antivalence of one function with each function of a given lattice of functions results again in a lattice of functions. This approach has been used to calculate the lattice of decomposition-functions $h(\mathbf{x}_b, \mathbf{x}_c)$ for a selected decomposition-function $g(x_a, \mathbf{x}_c)$:

$$g(x_a, \mathbf{x}_c) = x_a \wedge \max_{\mathbf{x}_b}{}^k f_q^{\text{der}_{x_a}} (\mathbf{x}_b, \mathbf{x}_c) , \tag{11.169}$$

$$h_q(\mathbf{x}_b, \mathbf{x}_c) = \max_{x_a} \left(\overline{g(x_a, \mathbf{x}_c)} f_q(x_a, \mathbf{x}_b, \mathbf{x}_c) \vee g(x_a, \mathbf{x}_c) f_r(x_a, \mathbf{x}_b, \mathbf{x}_c) \right) , \tag{11.170}$$

$$h_r(\mathbf{x}_b, \mathbf{x}_c) = \max_{x_a} \left(\overline{g(x_a, \mathbf{x}_c)} f_r(x_a, \mathbf{x}_b, \mathbf{x}_c) \vee g(x_a, \mathbf{x}_c) f_q(x_a, \mathbf{x}_b, \mathbf{x}_c) \right) . \tag{11.171}$$

Figure 11.53 shows the decomposition-functions of the EXOR-bi-decomposition calculated by (11.169), (11.170), and (11.171).

A compact bi-decomposition requires as much as possible variables in the dedicated sets $\mathbf{x}_a$ and $\mathbf{x}_b$; hence, the number of commonly used variables $\mathbf{x}_c$ is as small as possible and the decomposition-functions $g(\mathbf{x}_a, \mathbf{x}_c)$ and $h(\mathbf{x}_b, \mathbf{x}_c)$ will be the simplest in case of a desirable compact bi-decomposition.

Condition (11.168) allows us to find the maximal dedicated set $\mathbf{x}_b$ of an EXOR-bi-decomposition, but it unfortunately restricts to a single variable x_a instead of a dedicated set $\mathbf{x}_a$. As initial situation we can calculate the decomposition-function $g(x_a, \mathbf{x}_c)$ using (11.169) and the lattice of decomposition-functions $h(\mathbf{x}_b, \mathbf{x}_c) \in \mathcal{L} \langle h_q(\mathbf{x}_b, \mathbf{x}_c), h_r(\mathbf{x}_b, \mathbf{x}_c) \rangle$ using (11.170) and (11.171). The set of all variables $\mathbf{x}$ is distributed to the disjoint sets $\mathbf{x}_a = x_a$, $\mathbf{x}_b$, and $\mathbf{x}_c$.

We assume that $\mathbf{x}_b$ contains as much as possible variables because it can be verified by (11.168) that no other variable can be added to the dedicated set $\mathbf{x}_b$ without the loosing of the property of an EXOR-bi-decomposition of the given lattice. A found EXOR-bi-decomposition is not compact if at least one variable can be moved from $\mathbf{x}_c$ to $\mathbf{x}_a$. Moving a variable x_i from $\mathbf{x}_c$ to $\mathbf{x}_a$ does not change the set of variables the function $g(\mathbf{x}_a, \mathbf{x}_c)$ is depending on; however, it reduces the support of the function $h(\mathbf{x}_b, \mathbf{x}_c)$. We split the set of variables $\mathbf{x}_c$ into x_i and $\mathbf{x}_{c0}$ to demonstrate the reassignment of x_i from $\mathbf{x}_c$ to $\mathbf{x}_a$. For the exploration of the properties of this reassignment, we assume furthermore that $\mathbf{x}_{c0} = \emptyset$.

The reassignment of the variable x_i to $\mathbf{x}_a$ requires the change of the function $h(\mathbf{x}_b, x_i)$ to $h'(\mathbf{x}_b)$. First we explore the case that all functions $h(\mathbf{x}_b, x_i)$ of the decomposition-lattice $\mathcal{L} \langle h_q, h_r \rangle$ linearly depend on the variable x_i; this assumption has been used in [20]. In the context of the EXOR-bi-decomposition, the following transformation steps show the key idea to solve the problem:

$$g(\mathbf{x}_{a0}, x_i) \oplus h(\mathbf{x}_b, x_i) = g(\mathbf{x}_{a0}, x_i) \oplus (x_i \oplus h'(\mathbf{x}_b)) \tag{11.172}$$

$$= (g(\mathbf{x}_{a0}, x_i) \oplus x_i) \oplus h'(\mathbf{x}_b) \tag{11.173}$$

$$= g'(\mathbf{x}_{a0}, x_i) \oplus h'(\mathbf{x}_b) \tag{11.174}$$

$$= g'(\mathbf{x}_a) \oplus h'(\mathbf{x}_b) . \tag{11.175}$$

- The left-hand side of (11.172) emphasizes the variable x_i as the only element of the given set of variables $\mathbf{x}_c$.
- The right-hand side of (11.172) requires that the function $h(\mathbf{x}_b, x_i)$ is linear in x_i. This property enables or prohibits the whole transformation.
- The step from (11.172) to (11.173) moves the variable x_i to the other decomposition-function of the EXOR-bi-decomposition.
- The step from (11.173) to (11.174) includes the x_i into the new decomposition-function $g'(\mathbf{x}_{a0}, x_i)$. This transformation is possible without any restrictions.
- The step from (11.174) to (11.175) emphasizes that x_i does not belong anymore to the common set $\mathbf{x}_c$ because $h'(\mathbf{x}_b)$ does not depend on x_i; hence, x_i extends the dedicated set of variables $\mathbf{x}_{a0}$ to $\mathbf{x}_a = (\mathbf{x}_a 0, x_i)$.

The only condition for all transformations from (11.172) to (11.175) is that the lattice $\mathcal{L}\langle h_q(\mathbf{x}_b, \mathbf{x}_c), h_r(\mathbf{x}_b, \mathbf{x}_c)\rangle$ contains at least one function that satisfies

$$\operatorname*{der}_{x_i} h(\mathbf{x}_b, x_i) = 1 \ . \tag{11.176}$$

Theorem 11.15 (Lattice Containing a Function that is Linear with Regard to x_i) *A lattice $\mathcal{L}\langle h_q(\mathbf{x}_b, x_i), h_r(\mathbf{x}_b, x_i)\rangle$ of logic functions contains at least one function $h(\mathbf{x}_b, x_i)$ that can be represented by:*

$$h(\mathbf{x}_b, x_i) = x_i \oplus h'(\mathbf{x}_b) \tag{11.177}$$

if Condition

$$h_r^{\operatorname{der}_{x_i}}(\mathbf{x}_b) = 0 \tag{11.178}$$

is satisfied.

The proof of Theorem 11.15 is given in [20].

Corollary 11.1 *If the common set $\mathbf{x}_c$ of an EXOR-bi-decomposition contains only the variable x_i and Condition (11.178) is satisfied, then this variable can be moved to the dedicated set $\mathbf{x}_a$ using the transformation of $g(\mathbf{x}_{a0}, x_i)$ to $g'(\mathbf{x}_a)$:*

$$g'(\mathbf{x}_a) = x_i \oplus g(\mathbf{x}_{a0}, x_i) \tag{11.179}$$

and the associated new lattice $\mathcal{L}\langle h_q'(\mathbf{x}_b), h_r'(\mathbf{x}_b)\rangle$ is adjusted by

$$h_q'(\mathbf{x}_b) = \max_{x_i}\left(\overline{x}_i\, h_q(\mathbf{x}_b, x_i) \vee x_i\, h_r(\mathbf{x}_b, x_i)\right) , \tag{11.180}$$

$$h_r'(\mathbf{x}_b) = \max_{x_i}\left(\overline{x}_i\, h_r(\mathbf{x}_b, x_i) \vee x_i\, h_q(\mathbf{x}_b, x_i)\right) . \tag{11.181}$$

Example 11.27 We use the results of Example 11.26 shown in Fig. 11.53 as starting point. The lattice of decomposition-functions $h(a, c, d)$ of this basic EXOR-bi-decomposition with regard to the single variable $x_a = (b)$ and the dedicated set $\mathbf{x}_b = (c, d)$ has the mark-functions:

$$h_q(a, c, d) = a \oplus c \oplus d \ ,$$

$$h_r(a, c, d) = \overline{c}(a \oplus \overline{d}) \ .$$

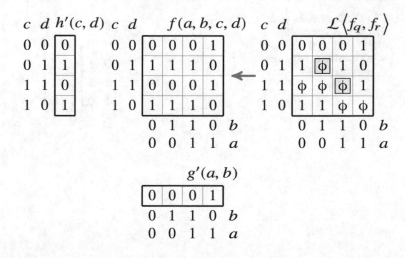

Fig. 11.54 Disjoint EXOR-bi-decomposition of $f(a, b, c, d) \in \mathcal{L}\langle f_q, f_r \rangle$ with regard to $\mathbf{x}_a = (a, b)$ and $\mathbf{x}_b = (c, d)$

There is only one variable $x_i = (a)$ in the common set $\mathbf{x}_c$; hence, we check Condition (11.178) for this variable using (7.25)

$$h_r^{\mathrm{der}_a}(c, d) = \min_a h_q(a, c, d) \vee \min_a h_r(a, c, d)$$

$$= (0 \oplus c \oplus d) \wedge (1 \oplus c \oplus d) \vee (\overline{c}(0 \oplus \overline{d})) \wedge (\overline{c}(1 \oplus \overline{d}))$$

$$= (c \oplus d)\overline{(c \oplus d)} \vee \overline{c}\,\overline{d}\,d$$

$$= 0.$$

Condition (11.178) is satisfied, and the rules of Corollary 11.1 can be used to transform the decomposition-function $g(a, b)$ into $g'(a, b)$ and adjust the associated lattice of $h(a, c, d)$ to $h'(c, d)$. Figure 11.54 shows the resulting disjoint EXOR-bi-decomposition that exists for one function of the given lattice so that the lattice of $h'(c, d)$ is also restricted to a single function.

There is a second possibility to extend the single variable x_a to a dedicated set $\mathbf{x}_a$ of an EXOR-bi-decomposition for a given lattice of logic functions. Even if all functions of the given lattice depend on all variables that occur in the mark-functions $f_q(\mathbf{x}_a, \mathbf{x}_b, x_i)$ and $f_r(\mathbf{x}_a, \mathbf{x}_b, x_i)$, the antivalence of this lattice with the decomposition-function $g(\mathbf{x}_a, x_i)$ can determine a lattice $\mathcal{L}\langle h_q(\mathbf{x}_b, x_i), h_r(\mathbf{x}_b, x_i) \rangle$ that contains at least one function $h(\mathbf{x}_b, x_i)$ that does not depend on x_i.

Example 11.28 Figure 11.55 shows an example of a non-disjoint EXOR-bi-decomposition of the lattice $\mathcal{L}\langle f_q(a, b, c, d), f_r(a, b, c, d) \rangle$ into the function $g(a, b)$ and the lattice $\mathcal{L}\langle h_q(a, c, d), h_r(a, c, d) \rangle$ with regard to the dedicated variable $x_a = (b)$ and the dedicated set $\mathbf{x}_b = (c, d)$ so that the common variable is $x_c = x_i = (a)$. In comparison to Example 11.26 (see Fig. 11.53) only the function values in the subspace $(a = 1, c = 0)$ are replaced by their complements.

Due to the different values of $f(a = 0, b = 1, c = 0, d = 0) = 0$ and $f(a = 1, b = 1, c = 0, d = 0) = 1$, all functions of the given lattice $\mathcal{L}\langle f_q, f_r \rangle$ depend on the common variable a. However, the resulting lattice $\mathcal{L}\langle h_q, h_r \rangle$ contains the function $h(c, d) = c \vee d$ that is independent of the common variable a of this bi-decomposition. Hence, the lattice $\mathcal{L}\langle h_q, h_r \rangle$ can directly be used to select a decomposition-function $h(b, c)$ that does not depend on common variable a.

Fig. 11.55 Non-disjoint EXOR-bi-decomposition of $f(a, b, c, d) \in \mathcal{L}\langle f_q, f_r\rangle$ with regard to $x_a = (b)$ and $\mathbf{x}_b = (c, d)$

c d	$\mathcal{L}\langle h_q, h_r\rangle$		c d	$\mathcal{L}\langle f_q, f_r\rangle$			
0 0	0	0	0 0	0	0	1	0
0 1	1	1	0 1	1	φ	0	1
1 1	φ	1	1 1	φ	φ	φ	1
1 0	1	φ	1 0	1	1	φ	φ
	0	1		0	1	1	0
		a		0	0	1	1

$0\ 1\quad a$

$0\ 1\ 1\ 0\quad b$
$0\ 0\ 1\ 1\quad a$

$g(a, b)$

0	0	1	0
0	1	1	0	b
0	0	1	1	a

The only condition for this second possibility to restrict the common set of an EXOR-bi-decomposition is that the lattice $\mathcal{L}\langle h_q(\mathbf{x}_b, x_i), h_r(\mathbf{x}_b, x_i)\rangle$ contains at least one function that satisfies

$$\operatorname*{der}_{x_i} h(\mathbf{x}_b, x_i) = 0 \,. \tag{11.182}$$

Theorem 11.16 generalizes the finding of Example 11.28.

Theorem 11.16 (Lattice Containing a Function that does not Depend on x_i) *A lattice $\mathcal{L}\langle h_q(\mathbf{x}_b, x_i), h_r(\mathbf{x}_b, x_i)\rangle$ of logic functions contains at least one function $h(\mathbf{x}_b)$ that does not depend on x_i if Condition*

$$h_q^{\operatorname{der}_{x_i}}(\mathbf{x}_b) = 0 \tag{11.183}$$

is satisfied.

Corollary 11.2 *If the common set $\mathbf{x}_c$ of an EXOR-bi-decomposition contains only the variable x_i and Condition (11.183) is satisfied, then this variable can be moved to the dedicated set $\mathbf{x}_a$ using the unchanged decomposition-function*

$$g(\mathbf{x}_a) = g(\mathbf{x}_{a0}, x_i) \tag{11.184}$$

and the associated new lattice $\mathcal{L}\langle h'_q(\mathbf{x}_b), h'_r(\mathbf{x}_b)\rangle$ is adjusted by:

$$h'_q(\mathbf{x}_b) = \max_{x_i}\left(h_q(\mathbf{x}_b, x_i)\right) \,, \tag{11.185}$$

$$h'_r(\mathbf{x}_b) = \max_{x_i}\left(h_r(\mathbf{x}_b, x_i)\right) \,. \tag{11.186}$$

Example 11.29 Here, we use the results of Example 11.28 shown in Fig. 11.55 as starting point. The lattice of decomposition-functions $h(a, c, d)$ of this basic EXOR-bi-decomposition with regard to the single variable $x_a = (b)$ and the dedicated set $\mathbf{x}_b = (c, d)$ has the mark-functions:

$$h_q(a, c, d) = \overline{c}d \vee c(a \oplus \overline{d}) \,,$$

$$h_r(a, c, d) = \overline{c}\,\overline{d} \,.$$

Fig. 11.56 Disjoint
EXOR-bi-decomposition
of $f(a, b, c, d) \in \mathcal{L}\langle f_q, f_r\rangle$
with regard to $\mathbf{x}_a = (a, b)$
and $\mathbf{x}_b = (c, d)$

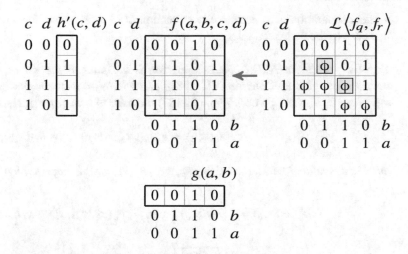

There is only one variable $x_i = (a)$ in the common set $\mathbf{x}_c$; hence, we check Condition (11.183) for this variable using (7.24)

$$h_q^{\mathrm{der}_a}(c, d) = \max_a h_q(a, c, d) \wedge \max_a h_r(a, c, d)$$

$$= \left((\overline{c}d \vee c(0 \oplus \overline{d})) \vee (\overline{c}d \vee c(1 \oplus \overline{d}))\right) \wedge (\overline{c}d \vee \overline{c}d)$$

$$= (\overline{c}d \vee c\overline{d} \vee \overline{c}d \vee cd) \wedge \overline{c}d$$

$$= 0 \, .$$

Condition (11.183) is satisfied, and the rules of Corollary 11.2 can be used to adjust the associated lattice of $h(a, c, d)$ to $h'(c, d)$. Figure 11.56 shows the resulting disjoint EXOR-bi-decomposition that exists for one function of the given lattice so that the lattice of $h'(c, d)$ is also restricted to a single function.

Now we generalize the found results of a single variable x_i in the common set to a common set $\mathbf{x}_c$ that contains more than one variable. The variable $x_i \in \mathbf{x}_c$ can be moved into the dedicated set $\mathbf{x}_a$ of an EXOR-bi-decomposition if for each subspace $\mathbf{x}_{c_0} = const.$ either Condition (11.178) of Theorem 11.15 or Condition (11.183) of Theorem 11.16 is satisfied.

Theorem 11.17 (Strong EXOR-Bi-Decomposition for a Lattice of Logic Functions with Regard to the Dedicated Sets $\mathbf{x}_a$ and $\mathbf{x}_b$) *The variable $x_i \in \mathbf{x}_c = (\mathbf{x}_{c0}, x_i)$ of the lattice of decomposition-functions $h(\mathbf{x}_b, \mathbf{x}_c) \in \mathcal{L}\langle h_q, h_r\rangle$ of an EXOR-bi-decomposition can be moved from the common set $\mathbf{x}_c$ to the dedicated set $\mathbf{x}_a$ if and only if Condition*

$$\max_{\mathbf{x}_b}^k \left(h_q^{\mathrm{der}_{x_i}}(\mathbf{x}_b, \mathbf{x}_{c0})\right) \wedge \max_{\mathbf{x}_b}^k \left(h_r^{\mathrm{der}_{x_i}}(\mathbf{x}_b, \mathbf{x}_{c0})\right) = 0 \tag{11.187}$$

is satisfied.

Proof Equation (11.187) holds if for each subspace $\mathbf{x}_{c_0} = const.$ either Condition (11.178) or Condition (11.183) is satisfied. Vice versa, if Condition (11.187) is not satisfied, then there is at least one subspace $\mathbf{x}_{c_0} = const.$ where the lattice of $\mathcal{L}\langle h_q(\mathbf{x}_b, \mathbf{x}_c), h_r(\mathbf{x}_b, \mathbf{x}_c)\rangle$ contains neither after a linear separation of x_i nor directly a function $h(\mathbf{x}_b, \mathbf{x}_{c_0})$; hence, the variable x_i cannot be moved in this case to the dedicated set $\mathbf{x}_a$.   □

Corollary 11.3 *An EXOR-bi-decomposition is compact, if the set of variables* $\mathbf{x}_b$ *is as large as possible (verified by Condition (11.168)), and within an iterative procedure all variables* x_i *of the initial set* $\mathbf{x}_c$ *that satisfy Condition (11.187) are used to transform* $g(\mathbf{x}_{a0}, \mathbf{x}_{c0}, x_i)$ *to* $g'(\mathbf{x}_a, \mathbf{x}_{c0})$ *by:*

$$g'_q(\mathbf{x}_a, \mathbf{x}_{c0}) = g(\mathbf{x}_{a0}, \mathbf{x}_{c0}, x_i) \oplus (x_i \wedge h_m(\mathbf{x}_{c0})) \tag{11.188}$$

and the associated new lattice $\mathcal{L}\langle h'_q(\mathbf{x}_b, \mathbf{x}_{c0}), h'_r(\mathbf{x}_b, \mathbf{x}_{c0})\rangle$ *is adjusted by:*

$$h'_q(\mathbf{x}_b, \mathbf{x}_{c0}) = h_m(\mathbf{x}_{c0}) \left(\max_{x_i} \left(\overline{x}_i \, h_q(\mathbf{x}_b, \mathbf{x}_{c0}, x_i) \vee x_i \, h_r(\mathbf{x}_b, \mathbf{x}_{c0}, x_i) \right) \right)$$

$$\vee \, \overline{h_m(\mathbf{x}_{c0})} \left(\max_{x_i} h_q(\mathbf{x}_b, \mathbf{x}_{c0}, x_i) \right) , \tag{11.189}$$

$$h'_r(\mathbf{x}_b, \mathbf{x}_{c0}) = h_m(\mathbf{x}_{c0}) \left(\max_{x_i} \left(\overline{x}_i \, h_r(\mathbf{x}_b, \mathbf{x}_{c0}, x_i) \vee x_i \, h_q(\mathbf{x}_b, \mathbf{x}_{c0}, x_i) \right) \right)$$

$$\vee \, \overline{h_m(\mathbf{x}_{c0})} \left(\max_{x_i} h_r(\mathbf{x}_b, \mathbf{x}_{c0}, x_i) \right) , \tag{11.190}$$

where

$$h_m(\mathbf{x}_{c0}) = \max_{\mathbf{x}_b}^k \left(h_r^{\text{der}_{x_i}}(\mathbf{x}_b, \mathbf{x}_{c0}) \right)$$

determines subspaces $\mathbf{x}_{c0} = const.$ *for which* x_i *must be linearly separated from the lattice* $\mathcal{L}\langle h_q(\mathbf{x}_b, \mathbf{x}_c), h_r(\mathbf{x}_b, \mathbf{x}_c)\rangle$.

Using the theoretical background provided above, a compact EXOR-bi-decomposition of a lattice of logic function can be computed in three main steps:

1. check for an EXOR-bi-decomposition of the given lattice using Condition (11.168) and all pairs of single variables the given lattice is depending on (break if no EXOR-bi-decomposition for any pair of variables exists);
2. extend the known dedicated variable x_b to the largest dedicated set $\mathbf{x}_b$ that satisfies Condition (11.168) by checking all variables of the common set as an additional candidate of $\mathbf{x}_b$;
3. extend the known dedicated variable x_a to the largest dedicated set $\mathbf{x}_a$ using all variables x_i that satisfy Condition (11.187) and adjust for each successfully checked $x_i \in \mathbf{x}_c$ the decomposition-function $g(\mathbf{x}_{a0}, \mathbf{x}_{c0}, x_i)$

$$\text{from} \quad g(\mathbf{x}_{a0}, \mathbf{x}_{c0}, x_i) \quad \text{to} \quad g'(\mathbf{x}_a, \mathbf{x}_{c0})$$

using (11.188) and the lattice of the decomposition-function $h(\mathbf{x}_b, \mathbf{x}_{c0}, x_i)$

$$\text{from} \quad \mathcal{L}\langle h_q(\mathbf{x}_b, \mathbf{x}_{c0}, x_i), h_r(\mathbf{x}_b, \mathbf{x}_{c0}, x_i)\rangle \quad \text{to} \quad \mathcal{L}\langle h'_q(\mathbf{x}_b, \mathbf{x}_{c0}), h'_r(\mathbf{x}_b, \mathbf{x}_{c0})\rangle$$

using (11.189) and (11.190).

A lattice of logic functions increases the chance that at least for one function of the lattice a strong bi-decomposition exists. Nevertheless, it is possible that no function of a given lattice is strongly bi-decomposable. In such a case it is helpful that weak bi-decompositions can also be directly detected using the mark-functions of the given lattice. The decrease of the requirements for weak bi-decompositions reveals when Definition 11.2 for a single given logic function is compared with Definition 11.4 for a lattice of functions from which an arbitrary function can be chosen for the weak bi-decomposition.

Definition 11.4 (Weak π-Bi-Decompositions for Lattices of Logic Functions) A weak bi-decomposition of the lattice of logic function

$$f(\mathbf{x}_a, \mathbf{x}_c) \in \mathcal{L}\langle f_q(\mathbf{x}_a, \mathbf{x}_c), f_r(\mathbf{x}_a, \mathbf{x}_c)\rangle$$

with regard to the *decomposition-operator* $\pi(g, h)$ and the dedicated set $\mathbf{x}_a \neq \emptyset$ is a pair of lattices of logic functions

$$\langle g(\mathbf{x}_a, \mathbf{x}_c) \in \mathcal{L}\langle g_q(\mathbf{x}_a, \mathbf{x}_c), g_r(\mathbf{x}_a, \mathbf{x}_c)\rangle, h(\mathbf{x}_c) \in \mathcal{L}\langle h_q(\mathbf{x}_c), h_r(\mathbf{x}_c)\rangle\rangle$$

with

$$f(\mathbf{x}_a, \mathbf{x}_c) = \pi\,(g(\mathbf{x}_a, \mathbf{x}_c), h(\mathbf{x}_c)) , \qquad (11.191)$$

where

$$g_q(\mathbf{x}_a, \mathbf{x}_c) < f_q(\mathbf{x}_a, \mathbf{x}_c) \qquad \text{and} \qquad g_r(\mathbf{x}_a, \mathbf{x}_c) = f_r(\mathbf{x}_a, \mathbf{x}_c) , \qquad (11.192)$$

or

$$g_q(\mathbf{x}_a, \mathbf{x}_c) = f_q(\mathbf{x}_a, \mathbf{x}_c) \qquad \text{and} \qquad g_r(\mathbf{x}_a, \mathbf{x}_c) < f_r(\mathbf{x}_a, \mathbf{x}_c) . \qquad (11.193)$$

The set $\mathbf{x}_c$ is the common set. The functions $g(\mathbf{x}_a, \mathbf{x}_c)$ and $h(\mathbf{x}_c)$ are denoted by decomposition-functions. The *decomposition-operator* $\pi(g, h)$ can be realized by an OR- or an AND-gate.

A weak OR-bi-decomposition of the lattice $\mathcal{L}\langle f_q(\mathbf{x}_a, \mathbf{x}_c), f_r(\mathbf{x}_a, \mathbf{x}_c)\rangle$ with regard to the dedicated set $\mathbf{x}_a$ exists if there is a subspace $\mathbf{x}_c = const.$ that contains at least one value 1 but no value 0 because in this case the value $h(\mathbf{x}_c = const.) = 1$ facilitates due to the used OR-gate and increased number of don't-cares of the lattice $\mathcal{L}\langle g_q(\mathbf{x}_a, \mathbf{x}_c), g_r(\mathbf{x}_a, \mathbf{x}_c)\rangle$.

Theorem 11.18 (Weak OR-Bi-Decomposition for a Lattice of Logic Functions) *The lattice* $\mathcal{L}\langle f_q(\mathbf{x}_a, \mathbf{x}_c), f_r(\mathbf{x}_a, \mathbf{x}_c)\rangle$ *contains at least one logic function* $f(\mathbf{x}_a, \mathbf{x}_c)$ *that is weakly OR-bi-decomposable with regard to the dedicated set* $\mathbf{x}_a$ *if and only if*

$$f_q(\mathbf{x}_a, \mathbf{x}_c) \wedge \overline{\max_{\mathbf{x}_a}^k f_r(\mathbf{x}_a, \mathbf{x}_c)} \neq 0 . \qquad (11.194)$$

The mark-functions of the lattice $\mathcal{L}\langle g_q(\mathbf{x}_a, \mathbf{x}_c), g_r(\mathbf{x}_a, \mathbf{x}_c)\rangle$ of a weak OR-bi-decomposition of the lattice $\mathcal{L}\langle f_q(\mathbf{x}_a, \mathbf{x}_c), f_r(\mathbf{x}_a, \mathbf{x}_c)\rangle$ can be calculated by:

$$g_q(\mathbf{x}_a, \mathbf{x}_c) = f_q(\mathbf{x}_a, \mathbf{x}_c) \wedge \overline{\max_{\mathbf{x}_a}^k f_r(\mathbf{x}_a, \mathbf{x}_c)} , \qquad (11.195)$$

$$g_r(\mathbf{x}_a, \mathbf{x}_c) = f_r(\mathbf{x}_a, \mathbf{x}_c) . \qquad (11.196)$$

The ON-set-function $g_q(\mathbf{x}_a, \mathbf{x}_c)$ in (11.195) contains such values 1 of the ON-set-function $f_q(\mathbf{x}_a, \mathbf{x}_c)$ that occur in a subspace $\mathbf{x}_c = const.$ together with values 0 of the OFF-set-function $f_r(\mathbf{x}_a, \mathbf{x}_c)$.

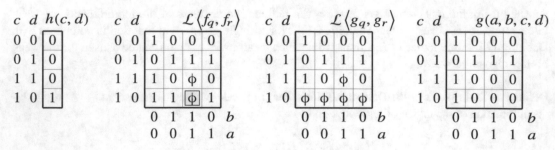

Fig. 11.57 Weak OR-bi-decomposition of the lattice $\mathcal{L}\langle f_q, f_r\rangle$ with regard to the dedicated set $\mathbf{x}_a = (a, b)$ into the lattice $\mathcal{L}\langle g_q, g_r\rangle$ and the function $h(c, d)$

If Condition (11.194) is satisfied, then the ON-set-function $g_q(\mathbf{x}_a, \mathbf{x}_c)$ is smaller than $f_q(\mathbf{x}_a, \mathbf{x}_c)$ so that (11.192) holds for (11.195) and (11.196). Using the selected decomposition-function $g(\mathbf{x}_a, \mathbf{x}_c)$, the mark-functions of $\mathcal{L}\langle h_q(\mathbf{x}_c), h_r(\mathbf{x}_c)\rangle$ of a weak OR-bi-decomposition of $\mathcal{L}\langle f_q(\mathbf{x}_a, \mathbf{x}_c), f_r(\mathbf{x}_a, \mathbf{x}_c)\rangle$ can be calculated by:

$$h_q(\mathbf{x}_c) = \max_{\mathbf{x}_a}^{k}\left(f_q(\mathbf{x}_a, \mathbf{x}_c) \wedge \overline{g(\mathbf{x}_a, \mathbf{x}_c)}\right), \tag{11.197}$$

$$h_r(\mathbf{x}_c) = \max_{\mathbf{x}_a}^{k} f_r(\mathbf{x}_a, \mathbf{x}_c). \tag{11.198}$$

Example 11.30 There is no function $f(a, b, c, d) \in \mathcal{L}\langle f_q, f_r\rangle$ shown in Fig. 11.57 for which a strong bi-decomposition exists. However, Condition (11.194) of a weak OR-bi-decomposition with regard to $\mathbf{x}_a = (a, b)$ holds because the lower row of the Karnaugh-map of $\mathcal{L}\langle f_q, f_r\rangle$ contains values 1 but no value 0.

The yellow box reveals the usefulness of a given lattice instead of a single function; in case of $f(1, 1, 1, 0) = 0$, the weak OR-bi-decomposition is not possible with regard to $\mathbf{x}_a = (a, b)$, but only with regard to each single variable. The benefit of the weak OR-bi-decomposition is that the number of don't-cares is increased due to $g_q(a, b, c, d) < f_q(a, b, c, d)$; the lattice of $f(a, b, c, d)$ contains $2^2 = 4$ functions, but this number is increased to $2^5 = 32$ for the lattice of $g(a, b, c, d)$. For one of these 32 functions $g(a, b, c, d)$, there exists in this example a strong disjoint EXOR-bi-decomposition with regard to $\mathbf{x}_a = (a, b)$ and $\mathbf{x}_b = (c, d)$ into $g'(a, b) = \overline{a}\overline{b}$ and $h'(c, d) = \overline{c}d$, as can be seen for the chosen function $g(a, b, c, d)$ in the right Karnaugh-map of Fig. 11.57.

The resulting lattice $h(\mathbf{x}_c) \in \mathcal{L}\langle h_q(\mathbf{x}_c), h_r(\mathbf{x}_c)\rangle$ contains in this example only the single function $h(c, d) = c\overline{d}$.

Dual properties hold for the weak AND-bi-decomposition of a lattice of logic functions $f(\mathbf{x}_a, \mathbf{x}_c) \in \mathcal{L}\langle f_q(\mathbf{x}_a, \mathbf{x}_c), f_r(\mathbf{x}_a, \mathbf{x}_c)\rangle$.

A weak AND-bi-decomposition of the lattice $\mathcal{L}\langle f_q(\mathbf{x}_a, \mathbf{x}_c), f_r(\mathbf{x}_a, \mathbf{x}_c)\rangle$ with regard to the dedicated set $\mathbf{x}_a$ exists if there is a subspace $\mathbf{x}_c = const.$ that contains at least one value 0 but no value 1 because in this case the value $h(\mathbf{x}_c = const.) = 0$ facilitates due to the used AND-gate and increased number of don't-cares of the lattice $\mathcal{L}\langle g_q(\mathbf{x}_a, \mathbf{x}_c), g_r(\mathbf{x}_a, \mathbf{x}_c)\rangle$.

Theorem 11.19 (Weak AND-Bi-Decomposition for a Lattice of Logic Functions) *The lattice* $\mathcal{L}\langle f_q(\mathbf{x}_a, \mathbf{x}_c), f_r(\mathbf{x}_a, \mathbf{x}_c)\rangle$ *contains at least one logic function* $f(\mathbf{x}_a, \mathbf{x}_c)$ *that is weakly AND-bi-decomposable with regard to the dedicated set* $\mathbf{x}_a$ *if and only if*

$$f_r(\mathbf{x}_a, \mathbf{x}_c) \wedge \overline{\max_{\mathbf{x}_a}^{k} f_q(\mathbf{x}_a, \mathbf{x}_c)} \neq 0. \tag{11.199}$$

The mark-functions of the lattice $\mathcal{L}\langle g_q(\mathbf{x}_a, \mathbf{x}_c), g_r(\mathbf{x}_a, \mathbf{x}_c)\rangle$ of a weak AND-bi-decomposition of the lattice $\mathcal{L}\langle f_q(\mathbf{x}_a, \mathbf{x}_c), f_r(\mathbf{x}_a, \mathbf{x}_c)\rangle$ can be calculated by:

$$g_q(\mathbf{x}_a, \mathbf{x}_c) = f_q(\mathbf{x}_a, \mathbf{x}_c) \,, \tag{11.200}$$

$$g_r(\mathbf{x}_a, \mathbf{x}_c) = f_r(\mathbf{x}_a, \mathbf{x}_c) \wedge \max_{\mathbf{x}_a}^k f_q(\mathbf{x}_a, \mathbf{x}_c) \,. \tag{11.201}$$

The OFF-set-function $g_r(\mathbf{x}_a, \mathbf{x}_c)$ in (11.201) contains such values 0 of the OFF-set-function $f_r(\mathbf{x}_a, \mathbf{x}_c)$ that occur in a subspace $\mathbf{x}_c = const.$ together with values 1 of the ON-set-function $f_q(\mathbf{x}_a, \mathbf{x}_c)$.

If Condition (11.199) is satisfied, then the OFF-set-function $g_r(\mathbf{x}_a, \mathbf{x}_c)$ is smaller than $f_r(\mathbf{x}_a, \mathbf{x}_c)$ so that (11.193) holds for (11.200) and (11.201). Using the selected decomposition-function $g(\mathbf{x}_a, \mathbf{x}_c)$, the mark-functions of $\mathcal{L}\langle h_q(\mathbf{x}_c), h_r(\mathbf{x}_c)\rangle$ of a weak AND-bi-decomposition of $\mathcal{L}\langle f_q(\mathbf{x}_a, \mathbf{x}_c), f_r(\mathbf{x}_a, \mathbf{x}_c)\rangle$ can be calculated by:

$$h_q(\mathbf{x}_c) = \max_{\mathbf{x}_a}^k f_q(\mathbf{x}_a, \mathbf{x}_c) \,, \tag{11.202}$$

$$h_r(\mathbf{x}_c) = \max_{\mathbf{x}_a}^k (f_r(\mathbf{x}_a, \mathbf{x}_c) \wedge g(\mathbf{x}_a, \mathbf{x}_c)) \,. \tag{11.203}$$

Example 11.31 The lattice $\mathcal{L}\langle f_q, f_r\rangle$ shown in Fig. 11.58 contains no function $f(a, b, c, d)$ for which a strong bi-decomposition exists. However, Condition (11.199) of a weak AND-bi-decomposition with regard to $\mathbf{x}_a = (a, b)$ holds because the row $(c, d) = (1, 1)$ of the Karnaugh-map of $\mathcal{L}\langle f_q, f_r\rangle$ contains values 0 but no value 1.

The usefulness of a given lattice instead of a single function is emphasized by the yellow box in the Karnaugh-map of Fig. 11.58; in case of $f(0, 1, 1, 1) = 1$, the weak AND-bi-decomposition is not possible with regard to $\mathbf{x}_a = (a, b)$, but only with regard to each single variable. The benefit of the weak AND-bi-decomposition is that the number of don't-cares is increased due to $g_r(a, b, c, d) < f_r(a, b, c, d)$; the lattice of $f(a, b, c, d)$ contains $2^3 = 8$ functions, but this number is increased to $2^5 = 32$ for the lattice of $g(a, b, c, d)$. Unfortunately, the extended lattice of functions $g(a, b, c, d) \in \mathcal{L}\langle g_q, g_r\rangle$ also does not contain a function for which a strong bi-decomposition exists; however, a consecutive weak OR-bi-decomposition of the lattice $\mathcal{L}\langle g_q, g_r\rangle$ with regard to $\mathbf{x}_a = (c, d)$ into $\mathcal{L}\langle g'_q, g'_r\rangle$ (see the right Karnaugh-map of Fig. 11.58) facilitates a strong AND-bi-decomposition into

$$g''(a, b) = a \vee \overline{b} \quad \text{and} \quad h''(c, d) = \overline{c}\,\overline{d} \,.$$

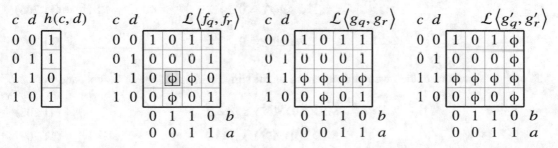

Fig. 11.58 Weak AND-bi-decomposition of the lattice $\mathcal{L}\langle f_q, f_r\rangle$ with regard to the dedicated set $\mathbf{x}_a = (a, b)$ into the lattice $\mathcal{L}\langle g_q, g_r\rangle$ and the function $h(c, d)$

The resulting lattice $h(\mathbf{x}_c) \in \mathcal{L}\langle h_q(\mathbf{x}_c), h_r(\mathbf{x}_c)\rangle$ contains in this example only the single function

$$h(c, d) = \overline{c} \vee \overline{d}$$

due to the chosen decomposition-function

$$g(a, b, c) = (a\overline{b}) \vee ((a \vee \overline{b}) \wedge \overline{c}\overline{d}) \, .$$

The last two examples have shown that bi-decompositions are possible for lattices that do not facilitate any strong bi-decomposition. The increased don't-care set of the lattice of the decomposition-function g contributes to a consecutive strong bi-decomposition. However, as shown in Example 11.31, more than one weak bi-decompositions can be required to get a lattice for which a strong bi-decomposition exists. An alternative to a weak bi-decomposition can be a vectorial bi-decomposition. Recently we found that it is also possible to detect all three types of vectorial bi-decompositions directly based on the mark-functions of a given lattice [19, 21].

The variables a lattice is depending on are implicitly known for strong and weak bi-decompositions. Due to our aim to emphasize the main properties of these bi-decompositions, we avoided the explicit presentation of the independence matrix and the associated independence function in these cases. However, due to the independence of the simultaneous of several variables in case of the vectorial bi-decomposition, we use both the independence matrix and the associated independence function (see Chap. 7, especially Sects. 7.2 and 7.3) to specify the vectorial bi-decompositions of lattices of logic functions precisely.

Definition 11.5 (Vectorial π-Bi-Decompositions for Lattices of Logic Functions) A lattice

$$\mathcal{L}\langle f_q(\mathbf{x}), f_r(\mathbf{x}), f^{id}(\mathbf{x})\rangle \tag{11.204}$$

contains at least one function $f(\mathbf{x})$ for which exists a vectorial OR-, a vectorial AND-, or a vectorial EXOR-bi-decomposition with regard to the subsets of variables $\mathbf{x}_a \subseteq \mathbf{x}$ and $\mathbf{x}_b \subseteq \mathbf{x}$, $\mathbf{x}_a \neq \mathbf{x}_b$, if

1. $f(\mathbf{x})$ can be expressed by

$$f(\mathbf{x}) = \pi(g(\mathbf{x}), h(\mathbf{x})) \, , \tag{11.205}$$

   where the decomposition-operator $\pi(g, h)$ can be realized by an OR-, an AND-, or an EXOR-gate,
2. the decomposition-functions $g(\mathbf{x})$ and $h(\mathbf{x})$ satisfy

$$\operatorname*{der}_{\mathbf{x}_b} g(\mathbf{x}) = 0 \, , \tag{11.206}$$

$$\operatorname*{der}_{\mathbf{x}_a} h(\mathbf{x}) = 0 \, , \tag{11.207}$$

3. and the functions of the given lattice depend on the simultaneous change of both $\mathbf{x}_a$ and $\mathbf{x}_b$:

$$\mathrm{MIDC}(\mathrm{IDM}(f), \mathbf{x}_a) \neq \mathbf{0} \, , \tag{11.208}$$

$$\mathrm{MIDC}(\mathrm{IDM}(f), \mathbf{x}_b) \neq \mathbf{0} \, . \tag{11.209}$$

Theorem 11.20 (Vectorial OR-Bi-Decomposition for a Lattice of Logic Functions) *A lattice defined by* (11.204) *contains at least one logic function* $f(\mathbf{x})$ *for which exists a vectorial OR-decomposition with regard to the dedicated sets* $\mathbf{x}_a \subseteq \mathbf{x}$ *and* $\mathbf{x}_b \subseteq \mathbf{x}$, $\mathbf{x}_a \neq \mathbf{x}_b$, *that satisfy* (11.208) *and* (11.209) *if and only if*

$$f_q(\mathbf{x}) \wedge \max_{\mathbf{x}_a} f_r(\mathbf{x}) \wedge \max_{\mathbf{x}_b} f_r(\mathbf{x}) = 0 . \tag{11.210}$$

The proof of Theorem 11.20 is given in [21].

Possible decomposition-functions $g(\mathbf{x})$ can be chosen from the lattice

$$\mathcal{L}\langle g_q(\mathbf{x}), g_r(\mathbf{x}), g^{id}(\mathbf{x})\rangle$$

with

$$g_q(\mathbf{x}) = \max_{\mathbf{x}_b}(f_q(\mathbf{x}) \wedge \max_{\mathbf{x}_a} f_r(\mathbf{x})) , \tag{11.211}$$

$$g_r(\mathbf{x}) = \max_{\mathbf{x}_b} f_r(\mathbf{x}) , \tag{11.212}$$

$$g^{id}(\mathbf{x}) = f^{id}(\mathbf{x}) \vee \operatorname*{der}_{\mathbf{x}_b} f(\mathbf{x}) , \tag{11.213}$$

so that

$$\text{IDM}(g) = \text{UM}(\text{IDM}(f), \mathbf{x}_b) . \tag{11.214}$$

Assuming that the function $g(\mathbf{x})$ has been selected from this lattice to realize the circuit, possible decomposition-functions $h(\mathbf{x})$ can be chosen from the lattice $\mathcal{L}\langle h_q(\mathbf{x}), h_r(\mathbf{x}), h^{id}(\mathbf{x})\rangle$ with

$$h_q(\mathbf{x}) = \max_{\mathbf{x}_a}(f_q(\mathbf{x}) \wedge \overline{g(\mathbf{x})}) , \tag{11.215}$$

$$h_r(\mathbf{x}) = \max_{\mathbf{x}_a} f_r(\mathbf{x}) , \tag{11.216}$$

$$h^{id}(\mathbf{x}) = f^{id}(\mathbf{x}) \vee \operatorname*{der}_{\mathbf{x}_a} f(\mathbf{x}) , \tag{11.217}$$

so that

$$\text{IDM}(h) = \text{UM}(\text{IDM}(f), \mathbf{x}_a) . \tag{11.218}$$

Example 11.32 For no function $f(a, b, c, d, e) \in \mathcal{L}\langle f_q, f_r\rangle$ defined in Fig. 11.59 exists a strong bi-decomposition. The empty independence matrix IDM(f) indicates that this lattice depends on all $2^5 - 1 = 31$ directions of change.

Condition (11.210) of a vectorial OR-bi-decomposition is satisfied for the given lattice $\mathcal{L}\langle f_q, f_r\rangle$ with regard to the dedicated sets $\mathbf{x}_a = (a, e)$ and $\mathbf{x}_b = (a, c, e)$ so that the lattice $\mathcal{L}\langle g_q, g_r\rangle$ could be calculated using (11.211) and (11.212). This lattice and the associated independent matrix are shown in Fig. 11.59. For the chosen function $g(a, b, c, d, e) = \overline{b}\,\overline{d} \wedge (a\,c\,e \vee \overline{a}\,\overline{c}\,\overline{e})$ exists a strong AND-bi-decomposition with regard to $\mathbf{x}_a = (b, d)$ and $\mathbf{x}_b = (a, c, e)$. Using this decomposition-function, the lattice $\mathcal{L}\langle h_q, h_r\rangle$ of the vectorial OR-bi-decomposition has been computed using (11.215) and (11.216). This lattice contains only one function, but this function does not depend on the simultaneous change of the variables a and e; hence, due to Theorem 11.11, a strong EXOR-bi-decomposition exists for this function $h(a, b, c, d, e)$. Figure 11.59 also shows the lattice of the decomposition-function $h(a, b, c, d, e)$ and the associated independence matrix.

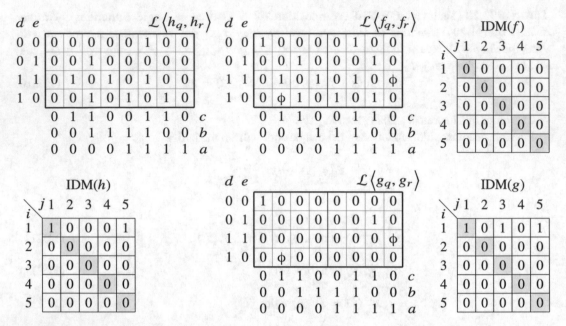

Fig. 11.59 Vectorial OR-bi-decomposition of $f(a, b, c, d, e) \in \mathcal{L}\langle f_q, f_r \rangle$ with regard to $\mathbf{x}_a = (a, e)$ and $\mathbf{x}_b = (a, c, e)$

The vectorial AND-bi-decomposition of a lattice holds dual properties in comparison to the vectorial OR-bi-decomposition.

Theorem 11.21 (Vectorial AND-Bi-Decomposition for a Lattice of Logic Functions) *A lattice defined by* (11.204) *contains at least one logic function* $f(\mathbf{x})$ *for which exists a vectorial AND-decomposition with regard to the dedicated sets* $\mathbf{x}_a \subseteq \mathbf{x}$ *and* $\mathbf{x}_b \subseteq \mathbf{x}$, $\mathbf{x}_a \neq \mathbf{x}_b$, *that satisfy* (11.208) *and* (11.209) *if and only if*

$$f_r(\mathbf{x}) \wedge \max_{\mathbf{x}_a} f_q(\mathbf{x}) \wedge \max_{\mathbf{x}_b} f_q(\mathbf{x}) = 0 . \tag{11.219}$$

The proof of Theorem 11.21 can be carried out in the same manner as for Theorem 11.20 due to the duality between the Boolean Algebras $(\mathbb{B}^n, \vee, \wedge, \bar{\ }, \mathbf{0}, \mathbf{1})$ and $(\mathbb{B}^n, \wedge, \vee, \bar{\ }, \mathbf{1}, \mathbf{0})$.

Possible decomposition-functions $g(\mathbf{x})$ can be chosen from the lattice

$$\mathcal{L}\langle g_q(\mathbf{x}), g_r(\mathbf{x}), g^{id}(\mathbf{x}) \rangle$$

with

$$g_q(\mathbf{x}) = \max_{\mathbf{x}_b} f_q(\mathbf{x}) , \tag{11.220}$$

$$g_r(\mathbf{x}) = \max_{\mathbf{x}_b} (f_r(\mathbf{x}) \wedge \max_{\mathbf{x}_a} f_q(\mathbf{x})) , \tag{11.221}$$

$$g^{id}(\mathbf{x}) = f^{id}(\mathbf{x}) \vee \operatorname*{der}_{\mathbf{x}_b} f(\mathbf{x}) , \tag{11.222}$$

so that

$$\mathrm{IDM}(g) = \mathrm{UM}(\mathrm{IDM}(f), \mathbf{x}_b) . \tag{11.223}$$

Assuming that the function $g(\mathbf{x})$ has been selected from this lattice to realize the circuit, possible decomposition-functions $h(\mathbf{x})$ can be chosen from the lattice $\mathcal{L}\langle h_q(\mathbf{x}), h_r(\mathbf{x}), h^{id}(\mathbf{x})\rangle$ with

$$h_q(\mathbf{x}) = \max_{\mathbf{x}_a} f_q(\mathbf{x}) , \tag{11.224}$$

$$h_r(\mathbf{x}) = \max_{\mathbf{x}_a}(f_r(\mathbf{x}) \wedge g(\mathbf{x})) , \tag{11.225}$$

$$h^{id}(\mathbf{x}) = f^{id}(\mathbf{x}) \vee \operatorname*{der}_{\mathbf{x}_a} f(\mathbf{x}) , \tag{11.226}$$

so that

$$\mathrm{IDM}(h) = \mathrm{UM}(\mathrm{IDM}(f), \mathbf{x}_a) . \tag{11.227}$$

Example 11.33 Figure 11.59 shows the Karnaugh-map of the given lattice $\mathcal{L}\langle f_q, f_r\rangle$. For no function $f(a, b, c, d, e)$ of this lattice exists a strong bi-decomposition. The empty independence matrix $\mathrm{IDM}(f)$ indicates that this lattice depends on all $2^5 - 1 = 31$ directions of change.

Condition (11.219) of a vectorial AND-bi-decomposition is satisfied for the given lattice $\mathcal{L}\langle f_q, f_r\rangle$ with regard to the dedicated sets $\mathbf{x}_a = (b, d)$ and $\mathbf{x}_b = (b, d, e)$ so that the lattice $\mathcal{L}\langle g_q, g_r\rangle$ could be calculated using (11.220) and (11.221). This lattice and the associated independent matrix are shown in Fig. 11.60. For the chosen function $g(a, b, c, d, e) = (a \vee c) \vee (b \vee d \vee e)(\overline{b} \vee \overline{d} \vee \overline{e})$ exists a disjoint OR-bi-decomposition with regard to $\mathbf{x}_a = (a, c)$ and $\mathbf{x}_b = (b, d, e)$. Using this decomposition-function, the lattice $\mathcal{L}\langle h_q, h_r\rangle$ of the vectorial AND-bi-decomposition has been computed using (11.224) and (11.225). This lattice contains only one function, but this function does not depend on the simultaneous change of the variables b and d; hence, due to Theorem 11.11, a strong EXOR-bi-decomposition for the function $h(a, b, c, d, e)$ exists. The decomposition-function $h(a, b, c, d, e)$ and the associated independence matrix are also shown in Fig. 11.60.

The vectorial EXOR-bi-decomposition can also be generalized for a lattice of logic functions.

Theorem 11.22 (Vectorial EXOR-Bi-Decomposition for a Lattice of Logic Functions) *A lattice defined by (11.204) contains at least one logic function $f(\mathbf{x})$ for which exists a vectorial EXOR-decomposition with regard to the dedicated sets $\mathbf{x}_a \subseteq \mathbf{x}$ and $\mathbf{x}_b \subseteq \mathbf{x}$, $\mathbf{x}_a \neq \mathbf{x}_b$, that satisfy (11.208) and (11.209) if and only if*

$$\max_{\mathbf{x}_b} f_q^{\mathrm{der}_{\mathbf{x}_a}}(\mathbf{x}) \wedge f_r^{\mathrm{der}_{\mathbf{x}_a}}(\mathbf{x}) = 0 . \tag{11.228}$$

Condition (11.228) generalizes the basic rule (11.118) of an EXOR-bi-decomposition of a single function with regard to single variables in the dedicated sets to the EXOR-bi-decomposition for a lattice of logic function with regard the simultaneous change of the variables specified by the dedicated sets. The proof of Theorem 11.22 is given in [21].

The vectorial EXOR-bi-decomposition evaluates quadruples of function values such that an even number of function values must occur in each of these quadruples; hence, don't-cares facilitate that this condition can be satisfied. The calculation of the decomposition-functions requires that each single don't-care within a quadruple of functions values must be replaced by a function value such that the rule of an even number of values 1 within a quadruple is satisfied.

One possibility of this adjustment consists in the following procedure:

1. function $f_\varphi(\mathbf{x})$ of the don't-care set is determined using the other two mark-functions of the given lattice:

$$f_\varphi(\mathbf{x}) = \overline{f_q(\mathbf{x}) \vee f_r(\mathbf{x})} ; \tag{11.229}$$

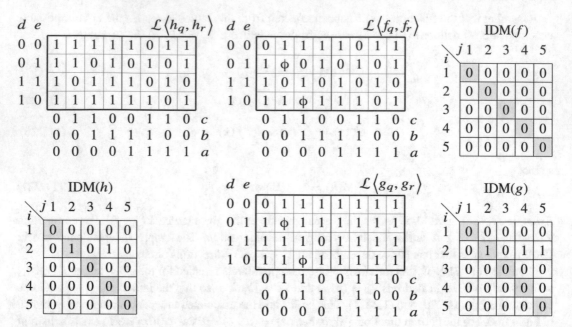

Fig. 11.60 Vectorial AND-bi-decomposition of $f(a, b, c, d, e) \in \mathcal{L}\langle f_q, f_r\rangle$ with regard to $\mathbf{x}_a = (b, d)$ and $\mathbf{x}_b = (b, d, e)$

2. the ON-set-function $f_q(\mathbf{x})$ is extended to the completely specified function $f_{cs}(\mathbf{x})$ replacing all don't-cares by values 1:

$$f_{cs}(\mathbf{x}) = f_q(\mathbf{x}) \vee f_\varphi(\mathbf{x}) \; ; \tag{11.230}$$

3. an adjustment function $f_a(\mathbf{x})$ is calculated that specifies the detected conflicts if all don't-cares are replaced by values 1; this function utilizes the check for vectorial EXOR-bi-decompositions of the function $f_{cs}(\mathbf{x})$:

$$f_a(\mathbf{x}) = \operatorname*{der}_{\mathbf{x}_b} \left(\operatorname*{der}_{\mathbf{x}_a} f_{cs}(\mathbf{x}) \right) \; ; \tag{11.231}$$

4. the function $f_{odd}(\mathbf{x})$ determines quadruples of function values in which an odd number of don't-cares occur:

$$f_{odd}(\mathbf{x}) = \operatorname*{der}_{\mathbf{x}_b}(\operatorname*{der}_{\mathbf{x}_a} f_\varphi(\mathbf{x})) \; ; \tag{11.232}$$

5. the function $f_{e2}(\mathbf{x})$ determines quadruples of function values that contain exactly two don't-cares:

$$f_{e2}(\mathbf{x}) = \overline{f_{odd}(\mathbf{x})} \wedge \overline{\min_{\mathbf{x}_b}(\min_{\mathbf{x}_a} f_\varphi(\mathbf{x}))} \wedge \max_{\mathbf{x}_b}(\max_{\mathbf{x}_a} f_\varphi(\mathbf{x})) \; ; \tag{11.233}$$

6. exactly one don't-care within each quadruple of $f_{e2}(\mathbf{x})$ can be determined by:

$$f_{s1}(\mathbf{x}) = f_\varphi(\mathbf{x}) f_{e2}(\mathbf{x}) \wedge (x_{ai} \overline{\operatorname*{der}_{\mathbf{x}_a} f_\varphi(\mathbf{x})} \vee x_{bj} \overline{\operatorname*{der}_{\mathbf{x}_b} f_\varphi(\mathbf{x})} \vee x_{ai} \overline{\operatorname*{der}_{(\mathbf{x}_a \triangle \mathbf{x}_b)} f_\varphi(\mathbf{x})}) \; , \tag{11.234}$$

where $x_{ai} \in \mathbf{x}_a$ and $x_{bj} \in \mathbf{x}_b$;

7. the mark-functions of the adjusted lattice $\mathcal{L}\langle f_q'(\mathbf{x}), f_r'(\mathbf{x})\rangle$ are determined by:

$$f_q'(\mathbf{x}) = f_q(\mathbf{x}) \vee f_\varphi(\mathbf{x}) \overline{f_a(\mathbf{x})} f_{odd}(\mathbf{x}) \vee f_a(\mathbf{x}) f_{s1}(\mathbf{x}) \; , \tag{11.235}$$

$$f_r'(\mathbf{x}) = f_r(\mathbf{x}) \vee f_\varphi(\mathbf{x}) f_a(\mathbf{x}) f_{odd}(\mathbf{x}) \vee \overline{f_a(\mathbf{x})} f_{s1}(\mathbf{x}) \vee f_\varphi(\mathbf{x}) f_{e2}(\mathbf{x}) \overline{f_{s1}(\mathbf{x})} \; . \tag{11.236}$$

Possible decomposition-functions $g(\mathbf{x})$ can be chosen from the lattice

$$\mathcal{L}\langle g_q(\mathbf{x}), g_r(\mathbf{x}), g^{id}(\mathbf{x})\rangle$$

with

$$g_q(\mathbf{x}) = \max_{\mathbf{x}_b} \left(x_{bj} \wedge f_q'(\mathbf{x}) \right), \tag{11.237}$$

$$g_r(\mathbf{x}) = \max_{\mathbf{x}_b} \left(x_{bj} \wedge f_r'(\mathbf{x}) \right), \tag{11.238}$$

$$g^{id}(\mathbf{x}) = f^{id}(\mathbf{x}) \vee \operatorname*{der}_{\mathbf{x}_b} f(\mathbf{x}), \tag{11.239}$$

so that

$$\mathrm{IDM}(g) = \mathrm{UM}(\mathrm{IDM}(f), \mathbf{x}_b). \tag{11.240}$$

Assuming that the function $g(\mathbf{x})$ has been selected from this lattice to realize the circuit, the possible decomposition-functions $h(\mathbf{x})$ can be chosen from the lattice $\mathcal{L}\langle h_q(\mathbf{x}), h_r(\mathbf{x}), h^{id}(\mathbf{x})\rangle$ with

$$h_q(\mathbf{x}) = \max_{\mathbf{x}_a} \left(\left(f_q'(\mathbf{x}) \wedge \overline{g(\mathbf{x})} \right) \vee \left(f_r'(\mathbf{x}) \wedge g(\mathbf{x}) \right) \right), \tag{11.241}$$

$$h_r(\mathbf{x}) = \max_{\mathbf{x}_a} \left(\left(f_r'(\mathbf{x}) \wedge \overline{g(\mathbf{x})} \right) \vee \left(f_q'(\mathbf{x}) \wedge g(\mathbf{x}) \right) \right), \tag{11.242}$$

$$h^{id}(\mathbf{x}) = f^{id}(\mathbf{x}) \vee \operatorname*{der}_{\mathbf{x}_a} f(\mathbf{x}), \tag{11.243}$$

so that

$$\mathrm{IDM}(h) = \mathrm{UM}(\mathrm{IDM}(f), \mathbf{x}_a). \tag{11.244}$$

Example 11.34 For no function $f(a, b, c, d, e) \in \mathcal{L}\langle f_q, f_r\rangle$ defined in Fig. 11.61 exists a strong bi-decomposition. The empty independence matrix $\mathrm{IDM}(f)$ indicates that this lattice depends on all $2^5 - 1 = 31$ directions of change.

Condition (11.228) of a vectorial EXOR-bi-decomposition is satisfied for the given lattice $\mathcal{L}\langle f_q, f_r\rangle$ with regard to the dedicated sets $\mathbf{x}_a = (b, e)$ and $\mathbf{x}_b = (c, d)$ so that next the adjusted lattice $\mathcal{L}\langle f_q', f_r'\rangle$ can be computed. The value $f_a(0, 0, 1, 1, 0) = 1$ indicates that the assumption of the value 1 for this don't-care was wrong; hence, $f_r'(0, 0, 1, 1, 0) = 1$. The value $f_a(1, 0, 0, 0, 1) = 0$ indicates the opposite case that the assumption of the value 1 for this don't-care was correct; hence, $f_q'(1, 0, 0, 0, 1) = 1$. The adjusted lattice $\mathcal{L}\langle f_q', f_r'\rangle$ contains in this example only a single function because the two don't-cares belong to different quadruples and must therefore be replaced by fixed values. The decomposition-lattice $\mathcal{L}\langle g_q, g_r\rangle$ has been calculated using $x_{bj} = c$ in (11.237) and (11.238); this lattice contains only a single function. Figure 11.61 shows the computed lattice $\mathcal{L}\langle g_q, g_r\rangle$ and the associated independent matrix. Alternatively the use of $x_{bj} = d$ leads to another correct pair of decomposition-functions $g(a, b, c, d, e)$ and $h(a, b, c, d, e)$. Due to the two variables in the dedicated set $\mathbf{x}_b = (c, d)$, it follows from Theorem 11.11 that a strong EXOR-bi-decomposition with regard to $x_a = c$ and $x_b = d$ exists for the chosen function

$$g(a, b, c, d, e) = c(b\overline{e} \vee \overline{a}be) \oplus (db\overline{e} \vee \overline{d}\overline{a}be).$$

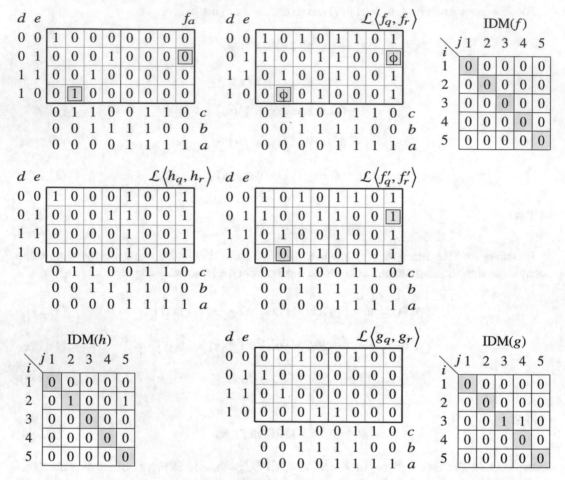

Fig. 11.61 Vectorial EXOR-bi-decomposition of $f(a, b, c, d, e) \in \mathcal{L}\langle f_q, f_r\rangle$ with regard to $\mathbf{x}_a = (b, e)$ and $\mathbf{x}_b = (c, d)$

Using this decomposition-function, the lattice $\mathcal{L}\langle h_q, h_r\rangle$ of the vectorial EXOR-bi-decomposition has been computed using (11.241) and (11.242). This lattice contains only one function, but this function does not depend on the simultaneous change of the variables b and e. Figure 11.61 also shows the lattice of the decomposition-function $h(a, b, c, d, e)$ and the associated independence matrix. Possible further decompositions of the function $h(a, b, c, d, e)$ can be seen in the following expression:

$$h(a, b, c, d, e) = \overline{c}(a \vee \overline{d}(b \odot e)) .$$

Completeness of the Bi-Decomposition

The recursive application of all three types of strong π-bi-decompositions together with weak OR-bi-decompositions and weak AND-bi-decompositions facilitates a complete multilevel synthesis of each logic function. The proof of Theorem 11.23 is more comprehensible than the explanation by an example; hence, we use this representation here.

Theorem 11.23 (Completeness of the π-Bi-Decomposition)

If the lattice $\mathcal{L}\langle f_q(x_a, \mathbf{x}_c), f_r(x_a, \mathbf{x}_c)\rangle$ contains only logic functions $f(x_a, \mathbf{x}_c)$ that are neither weakly OR-bi-decomposable nor weakly AND-bi-decomposable with regard to a single variable x_a, then this lattice $\mathcal{L}\langle f_q(x_a, \mathbf{x}_c), f_r(x_a, \mathbf{x}_c)\rangle$ contains at least one logic function $f(x_a, \mathbf{x}_b) = f(x_a, \mathbf{x}_c)$ for which a disjoint EXOR-bi-decomposition with regard to the single variable x_a and the dedicated set $\mathbf{x}_b$ exists. The variables $\mathbf{x}_c$ of the weak bi-decompositions are used as variables $\mathbf{x}_b$ of the disjoint EXOR-bi-decomposition.

Proof If the lattice $\mathcal{L}\langle f_q(x_a, \mathbf{x}_c), f_r(x_a, \mathbf{x}_c)\rangle$ does not contain a function $f(x_a, \mathbf{x}_c)$ that is weakly OR-bi-decomposable with regard to the single variable x_a, then (11.245) follows from (11.194) and can be equivalently transformed into (11.246) and (11.247):

$$f_q(x_a, \mathbf{x}_c) \wedge \overline{\max_{x_a} f_r(x_a, \mathbf{x}_c)} = 0 \,, \tag{11.245}$$

$$f_q(x_a, \mathbf{x}_c) \wedge \min_{x_a} \overline{f_r(x_a, \mathbf{x}_c)} = 0 \,, \tag{11.246}$$

$$f_q(x_a, \mathbf{x}_c) \wedge \min_{x_a} \left(f_q(x_a, \mathbf{x}_c) \vee f_\varphi(x_a, \mathbf{x}_c) \right) = 0 \,. \tag{11.247}$$

By removing $f_\varphi(x_a, \mathbf{x}_c)$ from the left-hand side of (11.247), we get Eq. (11.248). Due to (6.42) on page 244, Eq. (11.249) is equivalent to Eq. (11.248):

$$f_q(x_a, \mathbf{x}_c) \wedge \min_{x_a} f_q(x_a, \mathbf{x}_c) = 0 \,, \tag{11.248}$$

$$\min_{x_a} f_q(x_a, \mathbf{x}_c) = 0 \,. \tag{11.249}$$

If the lattice $\mathcal{L}\langle f_q(x_a, \mathbf{x}_c), f_r(x_a, \mathbf{x}_c)\rangle$ does not contain a function $f(x_a, \mathbf{x}_c)$ that is weakly AND-bi-decomposable with regard to the single variable x_a, then (11.250) follows from (11.199) and can be equivalently transformed into (11.251) and (11.252):

$$f_r(x_a, \mathbf{x}_c) \wedge \overline{\max_{x_a} f_q(x_a, \mathbf{x}_c)} = 0 \,, \tag{11.250}$$

$$f_r(x_a, \mathbf{x}_c) \wedge \min_{x_a} \overline{f_q(x_a, \mathbf{x}_c)} = 0 \,, \tag{11.251}$$

$$f_r(x_a, \mathbf{x}_c) \wedge \min_{x_a} \left(f_r(x_a, \mathbf{x}_c) \vee f_\varphi(x_a, \mathbf{x}_c) \right) = 0 \,. \tag{11.252}$$

By removing $f_\varphi(x_a, \mathbf{x}_c)$ from the left-hand side of (11.252), we get Eq. (11.253). Due to (6.42) on page 244, Eq. (11.254) is equivalent to Eq. (11.253):

$$f_r(x_a, \mathbf{x}_c) \wedge \min_{x_a} f_r(x_a, \mathbf{x}_c)) = 0 \,, \tag{11.253}$$

$$\min_{x_a} f_r(x_a, \mathbf{x}_c)) = 0 \,. \tag{11.254}$$

The substitution of (11.249) and (11.254) into the OFF-set-function (7.25) of the single derivative of the lattice $\mathcal{L}\langle f_q(x_a, \mathbf{x}_c), f_r(x_a, \mathbf{x}_c)\rangle$ with regard to x_a results in (11.255):

$$f_r^{\mathrm{der}_{x_a}}(\mathbf{x}_c) = \min_{x_a} f_q(x_a, \mathbf{x}_c) \vee \min_{x_a} f_r(x_a, \mathbf{x}_c) = 0 \vee 0 = 0 \,. \tag{11.255}$$

The final substitution of (11.255) into Condition (11.168) shows that, for the dedicated set $\mathbf{x}_b = \mathbf{x}_c$, Condition (11.256) of the disjoint EXOR-bi-decomposition with regard to x_a and the dedicated set $\mathbf{x}_b$ holds:

$$\max_{\mathbf{x}_b}{}^k f_q^{\mathrm{der}_{x_a}}(\mathbf{x}_b) \wedge f_r^{\mathrm{der}_{x_a}}(\mathbf{x}_b) = \max_{\mathbf{x}_b}{}^k f_q^{\mathrm{der}_{x_a}}(\mathbf{x}_b) \wedge 0 = 0 \ . \tag{11.256}$$

□

It can be checked, based on the logic equations introduced above, whether there is a π-bi-decomposition of a single logic function or a lattice of logic functions with regard to certain dedicated sets of variables. For each logic function exists at least one of the strong OR-, AND-, or EXOR-bi-decompositions or either a weak OR- or a weak AND-bi-decomposition. The vectorial OR-, AND-, or EXOR-bi-decompositions additionally facilitate to the synthesis of circuits with nearly uniform short paths between the inputs and outputs. The decomposition-functions are less complex than the logic function to be decomposed and can be selected from lattice of logic functions, even when a completely specified function is given. These lattices are specified by a pair of logic mark-functions and the independence functions that provides additional hints about the independence of certain directions of change. Based on the introduced bi-decompositions, a recursive algorithm can be used to synthesize a multilevel combinational circuit. In [16] such an algorithm has been suggested that uses, in addition to the theory of bi-decomposition, approved heuristics for the selection of the decomposition type and the dedicated sets of variables. Experimental results in [9] combine this synthesis approach with a technology mapping and outperformed other decomposition methods.

11.5   Test

Network-Model: Sensible Path

The aim of the test of combinational circuits is to prove whether their behavior is equal to the logic function to be realized. A complete check requires the comparison of the output-value of the combinational circuit with the function value for each of the 2^n input-patterns. This is a very time-consuming task because in each step the behavior for only one input-pattern can be checked. Assuming that the test of one input-pattern requires one millisecond and the combinational circuit has 20 inputs, it takes more than 15 min to test all the different (more than one million) input-patterns.

Therefore, the behavioral test is often replaced by a structural test in order to reduce the required time and the proportional costs. The structural test is based on an *error-model*. There is an error in the circuit if the logic value at a certain wire is different from the correct value for the given input. The combinational circuit consists of logic gates and wires. There are many different reasons for a logic error. Since the wires connect the logic gates, it is sufficient to assume that the inputs and outputs of the combinational circuit and the inputs and outputs of the logic gates are the sources of errors. The results of detailed studies show that almost all technical faults can be detected using the *stuck-at error-model*. The assumption of this model is that each technical fault results in a constant value at a certain gate-connection.

There are two types of stuck-at errors. A *stuck-at-zero-error* (SA0) expresses the case that a technical fault causes a constant value equal to 0. Otherwise, if the technical fault causes a value constant equal to 1, the stuck-at error-model indicates a *stuck-at-one-error* (SA1).

Note 11.14 There are further error-models like *stuck-open* or *stuck-closed*. The required tests for these error-models can be created by pairs of stuck-at test-patterns. For that reason, we confine here to the stuck-at error-model.

Fig. 11.62
Network-model: *sensible*
path

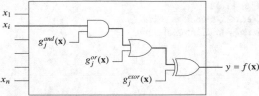

As mentioned above, stuck-at errors may occur in a combinational circuit at the inputs and outputs of the gates or of the whole circuit. Of course, there can be several stuck-at errors at the same time in the circuit. The number of all possible combinations of single stuck-at errors forming multiple errors is huge. Fortunately, it could be shown that almost all multiple stuck-at errors can be detected by a set of test-patterns for all single stuck-at errors. Therefore, we restrict ourselves to the *single stuck-at error-model* that requires test-patterns for $2 * (n_i + n_o + n_{gc})$ stuck-at errors, where n_i is the number of inputs of the combinational circuit, n_o is the number of its outputs, and n_{gc} is the number of input- and output-connections of all logic gates.

For technical reasons, usually only the inputs and the outputs, but not the internal wires of the circuit, are available. In order to test an internal gate-connection, it is necessary to find a logic path from the inputs through considered gate-connections to the outputs of the circuit. Widely used is the network-model of a *sensible path*, which means that there is a path from one input through considered gate-connections to one output of the circuit. Figure 11.62 visualizes a sensible path from the input x_i through several gates to the output $y = f(\mathbf{x})$.

Such a sensible path can be a physical path that satisfies several conditions:

1. if the sensible path goes through an AND-gate, the values of the other inputs of this gate must be constant equal to 1;
2. if the sensible path goes through an OR-gate, the values of the other inputs of this gate must be constant equal to 0;
3. if the sensible path goes through an EXOR-gate, the values of the other inputs of this gate must be constant.

There is no stuck-at error on the sensible path, if these conditions become true and a change of the input-value x_i of the sensible path causes the change of its output $f(\mathbf{x})$. Formula (11.257) emphasizes that there are very strong constraints for a sensible path:

$$\bigwedge_j \left(g_j^{and}(\mathbf{x}) \, \overline{\operatorname*{der}_{x_i} g_j^{and}(\mathbf{x})} \right) \wedge \bigwedge_j \left(\overline{g_j^{or}(\mathbf{x})} \, \overline{\operatorname*{der}_{x_i} g_j^{or}(\mathbf{x})} \right)$$

$$\wedge \bigwedge_j \left(\overline{\operatorname*{der}_{x_i} g_j^{exor}(\mathbf{x})} \right) \wedge \operatorname*{der}_{x_i} f(\mathbf{x}) = 1 \, . \tag{11.257}$$

The functions $g_j^{and}(\mathbf{x})$ control all AND-gates of the sensible path, without the inputs of the sensible path itself. The required constant values equal to 1 guarantee that the change of the value x_i changes the outputs of these AND-gates. The functions $g_j^{or}(\mathbf{x})$ control all OR-gates of the sensible path, without the inputs of the sensible path itself. The output of an OR-gate on the sensible path depends only on its input on the sensible path if its other inputs are constant equal to 0. The functions $g_j^{exor}(\mathbf{x})$ control all EXOR-gates of the sensible path, without the inputs of the sensible path itself. The test condition of EXOR-gates is weaker because the output of an EXOR-gate on the sensible path depends

Fig. 11.63
Network-model: *sensible*
point—local internal signal

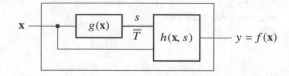

on its input of the sensible path for each constant value on its other inputs not belonging to the sensible path.

The left-hand side of Eq. (11.257) is a conjunction of many logic functions and derivatives that restrict the possible set of solutions. For that reason, not each physical path is a sensible path. It can be shown that, in many cases, there is no sensible path that goes through a testable internal gate-input. In such cases a sensible path algorithm wastes a lot of time for the analysis of all physical paths without success. To overcome this gap, the network-model of a *sensible point* has been suggested in [15].

Network-Model: Sensible Point—Local Internal Signal
It seems to be a contradiction that, for an input of an internal gate, both an SA0- and an SA1-test exist, but no sensible path. The reason for this phenomenon is that Condition (11.257) is sufficient, but not necessary. The sensible point model allows to specify test conditions that are necessary and sufficient as well. Figure 11.63 shows the simplest form of this network-model where the sensible point is emphasized by a thick red line and labeled by the logic variable s.

There are several types of the network-model *sensible point*. As shown in Fig. 11.63, the simplest type of the sensible point model is denoted by *local internal signal* and restricts the sensible point to a connection from the gate-output $s = g(\mathbf{x})$ to a single gate-input s of $h(\mathbf{x}, s)$ through the wire s. We introduce the logic model variable T that determines the type of the stuck-at error. The abbreviation *SAT-error* means both the SA0-error and SA1-error:

$$T = \begin{cases} 0, & \text{if a stuck-at-0-error (SA0) is considered}, \\ 1, & \text{if a stuck-at-1-error (SA1) is considered}. \end{cases} \tag{11.258}$$

It is not possible to distinguish between SAT-errors that occur at the gate-output $s = g(\mathbf{x})$, at the connection wire s, or at the gate-input s of $h(\mathbf{x}, s)$. Generally, there are three necessary conditions for a test-pattern to detect an SAT-error at the sensible point s of Fig. 11.63:

1. It is necessary to control the circuit such that the function value $s = g(\mathbf{x})$ is the complement of the stuck-at-T-error to be checked at the sensible point s. Only in this case, an SAT-error can cause a change of the circuit-output. The solutions of Eq. (11.259) facilitate an effect of an SAT-error. Therefore, we denote Condition (11.259) by *error-controllability*. Since the values of T must be the complement of s, the term $\overline{T}$ is associated to the sensible point s in Fig. 11.63:

$$g(\mathbf{x}) \oplus T = 1. \tag{11.259}$$

2. We assume that only the output of the circuit is directly observable. The effect of the internal error at the sensible point s must influence the function value at the circuit-output. Thus, a further necessary condition for test-patterns of the sensible point s is the *error-observability*. The value at the sensible point s is observable by the value at the circuit-output for all solution-vectors $\mathbf{x}$ of:

$$\operatorname*{der}_{s} h(\mathbf{x}, s) = 1. \tag{11.260}$$

3. It is necessary to compare the observed output-value y with the expected function value of $f(\mathbf{x})$ in order to evaluate whether there is an SAT-error at the sensible point s or not. The expected value y is determined by the input-pattern $\mathbf{x}$ in the solution of Eq. (11.261). The solution of Eq. (11.261) facilitates the *error-evaluation*:

$$f(\mathbf{x}) \odot y = 1 . \tag{11.261}$$

The necessary conditions (11.259), (11.260), and (11.261) can be combined with the necessary and sufficient condition (11.262). The solution of Eq. (11.262) consists of test-patterns $(\mathbf{x}, y, T)$. Each of these patterns detects an SAT-error if the circuit does not show the output-value y controlled by the input-pattern $\mathbf{x}$:

$$(g(\mathbf{x}) \oplus T) \wedge \operatorname*{der}_{s} h(\mathbf{x}, s) \wedge (f(\mathbf{x}) \odot y) = 1 . \tag{11.262}$$

The functions $g(\mathbf{x})$, $h(\mathbf{x}, s)$, and $f(\mathbf{x})$ are required in order to solve Eq. (11.262). These functions can be calculated as explained in Sect. 11.2 (Analysis) of this chapter. One SA0- and one SA1-test-pattern can be selected from the solution-set of (11.262) to check for these errors at the sensible point s.

The amount of calculations can be reduced. We calculate the required system-behavior $F^R(\mathbf{x}, y)$ using the equation-method and the lists of phases of the gates. According to the *network-model* of the *sensible point* for a *local internal signal*, we substitute $\overline{T}$ for s in the list of phases of the output-gate of the sub-circuit $s = g(\mathbf{x})$. The intersection of the modified lists of phases of all gates followed by a k-fold maximum with regard to the internal variables without T and s leads to a so-called list of error-phases $F^E(\mathbf{x}, s, T, y)$. It was shown in [15] that Eq. (11.262) can be transformed into Eq. (11.263) that uses the *list of required phases* $F^R(\mathbf{x}, y)$ and the *list of error-phases* $F^E(\mathbf{x}, s, T, y)$:

$$F^R(\mathbf{x}, y) \wedge \operatorname*{der}_{s} F^E(\mathbf{x}, s, T, y) = 1 . \tag{11.263}$$

If the solution-set of (11.262) or (11.263) is empty, then there is no input-vector $\mathbf{x}$ that detects an SAT-error at the sensible point. In such a case the output-gate of the sub-circuit $g(\mathbf{x})$ cannot influence the behavior of the circuit; hence, this gate is redundant. It can be removed from the circuit. Consequently, other gates that do not control any gate can be removed, recursively. The circuit is finally simplified by removing those gate-inputs that are not necessary.

Network-Model: Sensible Point—Local Internal Branch
The network-model of Fig. 11.63 is restricted to a sensible point where the gate-output $s = g(\mathbf{x})$ controls only one gate-input of the sub-circuit $h(\mathbf{x}, s)$. Generally, there can be a branch of the wire at the sensible point s. In that case several gates of the sub-circuit $h(\mathbf{x}, s)$ are controlled by the gate-output $s = g(\mathbf{x})$. This more general case is considered in the type *local internal branch* of the network-model *sensible point*. Figure 11.64 shows the assumptions of this generalized model.

There are two extensions in the network-model *sensible point* of the type *local internal signal* in Fig. 11.63 to the type *local internal branch* in Fig. 11.64. First, the logic cut at the sensible point s introduces the independent model variables s_i at the gate-inputs of the sub-circuit $h(\mathbf{x}, s)$ that are originally connected to the branched wire s. Both the required behavior and the behavior in the case that an error occurs on a selected gate-input s_i can be modeled by means of these independent variables. Second, the circuit can have m outputs $y_i = f_i(\mathbf{x})$. An error can be detected if a deviation from the required behavior occurs at least at one output.

The information about both the required behavior and the error-behavior caused by all gate-connections of the sensible point s of the combinational circuit is summarized in the *list of error-phases* $F^E(\mathbf{x}, \mathbf{y}, \mathbf{s}, T)$. This list of error-phases can be calculated using modified lists of phases. Only

Fig. 11.64
Network-model: *sensible point*—local internal branch

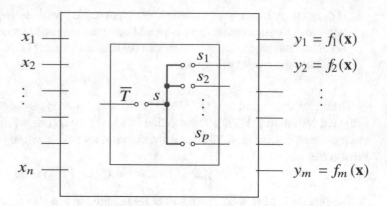

such lists of phases are modified that are directly controlled by the value of the sensible point. In the modified lists of phases the model variables s_i are used. Additionally the function

$$F_T(g_i, T) = g_i \oplus T \tag{11.264}$$

models the controllability-condition of the sensible point s located on the output g_i of the gate G_i. Figure 11.65 explains the calculation of the list of error-phases for a very simple example. The internal variable g_1 of the circuit in Fig. 11.65a is selected as the sensible point s. The branch of this wire is emphasized by thick red lines.

The required behavior is determined by the set of local lists of phases shown in Fig. 11.65b. Each of these lists of phases describes the behavior of the logic gate by the system-function $G_i(\mathbf{x}, \mathbf{g}, \mathbf{y})$.

Note 11.15 Not all logic variables occur in such a function of a selected gate.

The system-function $F^R(\mathbf{x}, \mathbf{y})$ of the required behavior can be calculated by (11.265):

$$F^R(\mathbf{x}, \mathbf{y}) = \max_{\mathbf{g}}^m \left(\bigwedge_{i=1}^{l} G_i(\mathbf{x}, \mathbf{g}, \mathbf{y}) \right). \tag{11.265}$$

Simple modifications of the functions $G_i(\mathbf{x}, \mathbf{g}, \mathbf{y})$ around the selected internal branch result in functions $GM_i(\mathbf{x}, \mathbf{g}, \mathbf{y}, \mathbf{s})$ and facilitate the specification of all possible stuck-at-T-errors at the gate-connections of the sensible point. The modified lists of phases of the circuit of Fig. 11.65a are given in Fig. 11.65c. The comparison with Fig. 11.65b shows that only the functions GM_3 and GM_4 are modified. The inputs of the associated two gates are connected to the sensible point. The input-variable g_1 is substituted by the unique variables s_1 and s_2, respectively. This modification models a logic cut at the sensible point. The function $F_T(g_1, T)$ is equal to 1 if the controllability-condition of the sensible point is satisfied. The *system-error-function* $F^E(\mathbf{x}, \mathbf{y}, \mathbf{s}, T)$ describes both the error-behavior and the required behavior and can be calculated by (11.266). This formula assumes that the sensible point s is located at the internal branch g_i:

$$F^E(\mathbf{x}, \mathbf{y}, \mathbf{s}, T) = \max_{\mathbf{g}}^k \left(F_T(g_i, T) \wedge \bigwedge_{i=1}^{l} GM_i(\mathbf{x}, \mathbf{g}, \mathbf{y}, \mathbf{s}) \right). \tag{11.266}$$

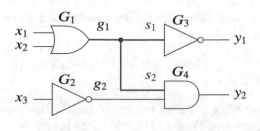

a

b

	x_1	x_2	x_3	g_1	g_2	y_1	y_2	s_1	s_2	T
$G_1 =$	1	–		1						
	0	1		1						
	0	0		0						
$G_2 =$			0		1					
			1		0					
$G_3 =$						0		1		
						1		0		
$G_4 =$					0		0		–	
					1		0		0	
					1		1		1	

c

	x_1	x_2	x_3	g_1	g_2	y_1	y_2	s_1	s_2	T
$GM_1 =$	1	–		1						
	0	1		1						
	0	0		0						
$GM_2 =$			0		1					
			1		0					
$GM_3 =$						1		0		
						0		1		
$GM_4 =$							–		0	0
							0		0	1
							1		1	1
$F_T =$				0						1
				1						0

d

$F^R =$

x_1	x_2	x_3	y_1	y_2
1	–	0	0	1
1	–	1	0	0
0	1	0	0	1
0	1	1	0	0
0	0	–	1	0

e

$F^E =$

x_1	x_2	x_3	y_1	y_2	s_1	s_2	T
1	–	0	1	1	0	1	0
1	–	0	0	1	1	1	0
1	0	0	1	0	0	0	0
1	0	1	1	0	0	–	0
1	0	0	0	0	1	0	0
1	0	1	0	0	1	–	0
0	1	0	1	1	0	1	0
0	1	0	0	1	1	1	0
–	1	0	1	0	0	0	0
–	1	1	1	0	0	–	0
–	1	0	0	0	1	0	0
–	1	1	0	0	1	–	0
0	0	0	1	1	0	1	1
0	0	0	0	1	1	1	1
0	0	0	1	0	0	0	1
0	0	1	1	0	0	–	1
0	0	0	0	0	1	0	1
0	0	1	0	0	1	–	1

Fig. 11.65 Preparing the basic data for the calculation of the test-patterns of the sensible point g_1 where a branch exists: (**a**) scheme of the combinational circuit, (**b**) set of local lists of phases, (**c**) set of modified local lists of phases, (**d**) list of required phases, and (**e**) list of error-phases

The *branch-function* $F_B(\mathbf{s}, T)$ describes the required behavior of the branch of the sensible point:

$$F_B(\mathbf{s}, T) = T \bigwedge_{j=1}^{p} \overline{s}_j \vee \overline{T} \bigwedge_{j=1}^{p} s_j \ . \tag{11.267}$$

The *required behavior* $F^R(\mathbf{x}, \mathbf{f})$ can be calculated using the system-error-function $F^E(\mathbf{x}, \mathbf{f}, \mathbf{s}, T)$ (11.266) and the branch-function $F_B(\mathbf{s}, T)$ (11.267) by means of:

$$F^R(\mathbf{x}, \mathbf{y}) = \max_{(\mathbf{s}, T)}^{m} \left(F^E(\mathbf{x}, \mathbf{y}, \mathbf{s}, T) \wedge F_B(\mathbf{s}, T) \right) \ . \tag{11.268}$$

The place where an error may occur at the sensible point can be the output of the gate that controls the sensible point or each of the controlled gate-inputs. We introduce the logic variable *place* that specifies one of these places:

$$place = \begin{cases} T, & \text{if the error place is the source of the branch } s, \\ s_j, & \text{the error place is the sink } s_j \text{ of the branch } s. \end{cases} \tag{11.269}$$

The maximum of $F_B(\mathbf{s}, T)$ with regard to s_j enables a possible error-behavior at the gate-input s_j because s_j lost the dependency on T. It is not obvious that the maximum of $F_B(\mathbf{s}, T)$ with regard to T enables a possible error-behavior at the gate that controls the branch of the sensible point. All the variables s_j carry in this case the same value, either the value 0 or the value 1; hence, one of these vectors models a stuck-at error at the gate that controls the branch of the sensible point.

Generally, there are different sets of test-patterns for each connection of a gate with the branching wire of the sensible point. All these sets of test-patterns result from Eq. (11.270) where the variable *place*, defined in (11.269), selects the place at the sensible point for which the calculated test-patterns are valid:

$$\text{der}_{\mathbf{s}}^{p} \left(F^E(\mathbf{x}, \mathbf{y}, \mathbf{s}, T) \wedge \max_{place} F_B(\mathbf{s}, T) \right) \wedge F^R(\mathbf{x}, \mathbf{y}) = 1 \ . \tag{11.270}$$

If the solution of (11.270) is empty for a certain place of the sensible point, there is definitely no possibility to check this place of the combinational circuit by any pattern of circuit inputs and observation of circuit-outputs. The structure of the synthesized combinational circuit determines its test properties.

The synthesis method of the previous section creates multilevel circuits that are testable for stuck-at faults at each gate-connection. The equations to calculate the mark-function of the lattices of the decomposition-functions satisfy a theorem proved in [8]. This theorem states that there is a completely stuck-at testable circuit for each lattice of logic functions, and no test-pattern from the don't-care-function of this lattice is necessary.

The synthesis method for multilevel circuits from the previous section can be extended such that the test-patterns are calculated in parallel to the synthesis process; see [22]. An important benefit of this method is that only known logic functions of the synthesis process are needed to calculate the test-patterns. Each physical path of such a particular multilevel combinational circuit is completely testable for stuck-at errors. All test-patterns of SAT-errors of a path from the input x_i to the output of the synthesized function are the solution of:

$$(x_i \oplus T) \wedge \bigwedge_{j} tc_j(\mathbf{x}) = 1 \ , \tag{11.271}$$

where the *test condition* $tc_j(\mathbf{x})$ of the decomposition step j is defined by:

$$
tc_j(\mathbf{x}) = \begin{cases}
\max_{\mathbf{x}_a}{}^k f_r(\mathbf{x}_a, \mathbf{x}_b, \mathbf{x}_c) & \begin{array}{l} \text{for the } g\text{-branch of} \\ \text{an OR-bi-decomposition or} \\ \text{a weak OR-bi-decomposition ,} \end{array} \\[1em]
\max_{\mathbf{x}_a}{}^k f_q(\mathbf{x}_a, \mathbf{x}_b, \mathbf{x}_c) & \begin{array}{l} \text{for the } g\text{-branch of} \\ \text{an AND-bi-decomposition or} \\ \text{a weak AND-bi-decomposition ,} \end{array} \\[1em]
\overline{g(\mathbf{x}_a, \mathbf{x}_c)} & \begin{array}{l} \text{for the } h\text{-branch of} \\ \text{an OR-bi-decomposition or} \\ \text{a weak OR-bi-decomposition ,} \end{array} \\[1em]
g(\mathbf{x}_a, \mathbf{x}_c) & \begin{array}{l} \text{for the } h\text{-branch of} \\ \text{an AND-bi-decomposition or} \\ \text{a weak AND-bi-decomposition ,} \end{array} \\[1em]
1 & \begin{array}{l} \text{for each branch of} \\ \text{an EXOR-bi-decomposition .} \end{array}
\end{cases}
\tag{11.272}
$$

The solution of Eq. (11.271) contains all test-patterns that can detect both stuck-at-0 and stuck-at-1 errors at each gate connection on the evaluated path. There is at least one $SA0$- and one $SA1$-test-pattern in the solution of Eq. (11.271) although the solution-set is restricted by many test-conditions for the AND-gates and the OR-gates on the path.

Note 11.16 All physical paths satisfy the conditions of a sensible path if the synthesis rules for the multilevel circuit from above are applied.

11.6 Exercises

Prepare for each task of each exercise of this chapter a PRP for the XBOOLE-monitor XBM 2. Prefer a lexicographic order of variables and use the help system of the XBOOLE-monitor XBM 2 to extend your knowledge about the details of needed commands. Follow the hints, given in the tasks, to store the PRPs of circuit models for later execution or execute the PRPs immediately.

If you have not yet prepared the XBOOLE-monitor XBM 2 on your computer, you can get this XBOOLE-monitor free of charge by means of the following three steps:

1. **Download**:
   There are four versions of the XBOOLE-monitor XBM 2, two for Windows 10 or subsequent Windows systems (32 or 64 bits) and two for LINUX—Ubuntu (also 32 or 64 bits); you must download the version of the XBOOLE-monitor XBM 2 that fits to your operating system.
   Authorized users of the online version of this chapter (https://doi.org/10.1007/978-3-030-88945-6_11) can download the XBOOLE-monitor XBM 2 directly from the web page
   https://link.springer.com/chapter/10.1007/978-3-030-88945-6_11
   where the links for the download of the XBOOLE-monitor XBM 2 are located in the part "Supplementary Information" (below the part "Abstract"). The headline above such a link indicates the associated zip-file of the XBOOLE-monitor XBM 2. The sizes of the zip-files have been provided behind the links and can be used to verify the download. A click on the link of the wanted version of the XBOOLE-monitor XBM 2 starts the download.

Readers of the hardcopy of this book get access to the XBOOLE-monitor XBM 2 using the URL
https://link.springer.com/chapter/10.1007/978-3-030-88945-6_11
to download the first two pages of this chapter. After this download, the same procedure as the authorized users of the online version of a chapter can be used to download the wanted version of the XBOOLE-monitor XBM 2.

2. **Unzip**: The XBOOLE-monitor XBM 2 must not be installed but must be unzipped into an arbitrary directory of your computer. A convenient tool for unzipping the downloaded zip-file is usually available as part of the operating system or can be downloaded from the Internet.

3. **Execute**:

   - Windows:
     The executable file of the two versions (32 or 64 bits) for Windows 10 (or subsequent Windows systems) of the XBOOLE-monitor XBM 2 is XBM2.exe; the other files in the expanded directory must remain unchanged. A double-click on the executable file XBM2.exe within the Explorer of Windows starts the XBOOLE-monitor XBM 2.
   - LINUX—Ubuntu:
     The unzipped folder of the XBOOLE-monitor XBM 2 contains for this operating system only the executable file XBM2-i386.AppImage for the version of 32 bits or XBM2-x86_64.AppImage for the version of 64 bits of the XBOOLE-monitor XBM 2. A double-click on the created AppImage-file within the file manager of LINUX—Ubuntu starts the XBOOLE-monitor XBM 2.

Exercise 11.1 (Structural Models of a Combinational Circuit)

The structure of a combinational circuit is shown in Fig. 11.66. Prepare PRPs that specify the structure of the circuit shown in Fig. 11.66 using:

(a) a system of logic equations,
(b) a set of local lists of phases,
(c) two TVLs in D-form of the functions $y_1 = f_1(\mathbf{x})$ and $y_2 = f_2(\mathbf{x})$, and
(d) a structure-TVL of an equivalent PLA.

Store the four created PRPs for later use.

Exercise 11.2 (Behavioral Models of Combinational Circuits)

(a) The behavior of a combinational circuit is given by the following system of logic equations:

$$y_1 = (((x_1 x_4 (\overline{x}_5 \vee x_6)) \oplus (\overline{x}_1 \overline{x}_4 \vee (x_2 \oplus x_3))) \vee (x_1 \overline{x}_2 \overline{x}_3 \vee \overline{x}_1 x_2 \overline{x}_3 \vee \overline{x}_1 \overline{x}_2 x_3))$$

$$\oplus (x_3 ((x_2 \vee \overline{x}_4) \oplus x_1)),$$

Fig. 11.66 Structure of a combinational circuit that realizes two functions $y_i = f_i(\mathbf{x})$

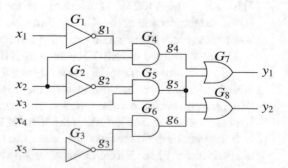

$$y_2 = ((\overline{x}_3 x_4 (x_1 \oplus x_6)) \oplus (x_1 x_2 \vee x_5)) \vee (\overline{x}_1 x_2 \vee x_2 \overline{x}_3 \vee \overline{x}_3 x_4 \vee x_4 \overline{x}_5 \vee \overline{x}_5 x_6) .$$

Transform this system of equations into the equivalent system-function $F(\mathbf{x}, \mathbf{y})$ and minimize the associated TVL in ODA-form.

The characteristic equation of this system-function must be uniquely solvable with regard to y_1 and y_2 due to the given explicit system of equations. Verify this property.

Show the computed and minimized TVL of $F(\mathbf{x}, \mathbf{y})$ and store the XBOOLE-system for the use in the next task.

(b) Load the XBOOLE-system of the previous task and split the system-function $F(\mathbf{x}, \mathbf{y})$ into two function-TVLs that specify the behavior of $y_1 = f_1(\mathbf{x})$ and $y_2 = f_2(\mathbf{x})$. Show the computed and minimized TVLs of $f_1(\mathbf{x})$ and $f_2(\mathbf{x})$ in the m-fold view.

(c) A system-function describes the allowed behavior of a circuit; hence, it can also specify lattices of functions. Besides this welcome property to choose a convenient function out of such a lattice, the realization of a system-function $F(\mathbf{x}, y)$ by a combinatorial circuit requires that the equation $F(\mathbf{x}, y) = 1$ can be solved with regard to the variable y.

Verify whether the equation $F(\mathbf{x}, y) = 1$ of the system-function

$$F(\mathbf{x}, y) = ((x_1 \vee x_2 \overline{x}_3 \vee (\overline{x}_3 \oplus x_5)) y) \oplus (x_2 \overline{x}_3 \overline{y} \vee x_1 x_6)) \vee \overline{x}_2 (x_3 \oplus x_5) \overline{y}$$

$$\vee x_1 \overline{x}_2 x_6 (\overline{x}_3 \oplus x_5) y \vee (x_2 \overline{x}_4 x_6 y \oplus x_3 \overline{x}_5 \overline{x}_6 \overline{y}) \vee x_2 x_4 x_6$$

is solvable with regard to y. Verify furthermore in case of a solvable equation whether the function $y = f(\mathbf{x})$ is uniquely specified or can be selected out of a lattice. Compute either the uniquely specified function $y = f(\mathbf{x})$ or the lattice of functions belonging to the incompletely specified function. Show the Karnaugh-map associated to the function $f(\mathbf{x})$ or the computed lattice $\mathcal{L}\langle f_q, f_r \rangle$. Show for comparison the Karnaugh-map of the system-function $F(\mathbf{x}, y)$ in the m-fold view using the same assignments of the variables $\mathbf{x}$ to the horizontal and vertical Gray codes; use the help system of the XBOOLE-monitor XBM 2 to learn how the distribution of variables of a Karnaugh-map can be changed.

(d) Figure 11.67 shows the Karnaugh-map of a lattice of functions (an incompletely specified function). Determine this lattice by the mark-functions $f_q(\mathbf{x})$ and $f_r(\mathbf{x})$ and show for comparison the Karnaugh-map of this lattice.

Compute the third mark-function $f_\varphi(\mathbf{x})$ and verify that the three mark-functions are pairwise disjoint and cover the whole Boolean space.

Fig. 11.67
Karnaugh-map of a lattice
of functions that specifies
the admissible behavior of
a combinational circuit

$x_4\,x_5$								$\mathcal{L}\langle f_q, f_r \rangle$
0 0	ϕ	ϕ	1	1	ϕ	ϕ	0	ϕ
0 1	1	0	0	1	1	0	0	1
1 1	1	1	ϕ	0	0	ϕ	0	1
1 0	1	1	ϕ	0	0	0	0	1

	0	1	1	0	0	1	1	0	x_3
	0	0	1	1	1	1	0	0	x_2
	0	0	0	0	1	1	1	1	x_1

Compute the system-function $F(\mathbf{x}, y)$ of this lattice using each pair of mark-functions and verify that the computed system-functions are equal to each other despite differently combined ternary vectors. Show the three computed system-functions in the second row of the m-fold view.

Exercise 11.3 (Analysis: Compute the Behavior of a Combinational Circuit)

(a) Compute the behavior of the combinational circuit based on the structure model (system of logic equations of the gates) specified in the PRP prepared in Exercise 11.1 (a). Extend therefore this PRP such that both the global list of phases and the input–output list of phases (the system-function $F(\mathbf{x}, \mathbf{y})$) are computed as minimized TVL in ODA-form. Show these two TVLs in the m-fold view.

(b) Compute the behavior of the combinational circuit based on the structure model (set of local lists of phases of the gates) specified in the PRP prepared in Exercise 11.1 (b). Extend therefore this PRP such that both the global list of phases and input–output list of phases (the system-function $F(\mathbf{x}, \mathbf{y})$) are computed as minimized TVL in ODA-form. Show these two TVLs in the m-fold view.

(c) Compute the behavior of the combinational circuit based on the structure model (TVLs in D-form of the functions $y_1 = f_1(\mathbf{x})$ and $y_2 = f_2(\mathbf{x})$) specified in the PRP prepared in Exercise 11.1 (c). Extend therefore this PRP such that the input–output lists of phases are separately computed as TVLs in ODA-form for the two outputs y_1 and y_2. Combine these two behavioral descriptions into the minimized input–output list of phases for the two outputs. Show these three TVLs in the m-fold view.

(d) Compute the behavior of the combinational circuit based on the structure model (structure-TVL of the equivalent PLA) specified in the PRP prepared in Exercise 11.1 (d). Extend therefore this PRP such that the input–output lists of phases is computed based on Eq. (11.49). Show both the structure-TVL in D-form and the computed system-function in ODA-form in the m-fold view. Note that the single ternary vectors of the structure-TVL carry the structure information and must therefore be separately computed.

Exercise 11.4 (Partial Behavior Filtered out of Lists of Phases)

The structure of a combinational circuit with ten inputs is shown in Fig. 11.68; hence, $2^{10} = 1024$ phases describe its behavior.

(a) Solve the system of logic equations that describes the structure of this combinational circuit to get the global list of phases as behavioral description. How many ternary vectors belong to this solution-set? Use the command obbc to minimize the computed global list of phases. How many ternary vectors belong to the minimized solution-set?

Define a VT of the internal variables g_i and compute the input–output list of phases and use the command obbc to minimize this list. How many ternary vectors belong to the minimized input–output list of phases?

Organize the PRP such that

- the minimized global list of phases is stored as object number 1;
- the minimized input–output list of phases is stored as object number 2; and
- the VT of the internal variables is stored as object number 3.

Use the command sts to store the XBOOLE-system of these general behavioral descriptions as an sdt-file for the use in the next three tasks of this exercise.

Fig. 11.68 Structure of a combinational circuit that realizes five functions $y_i = f_i(\mathbf{x})$

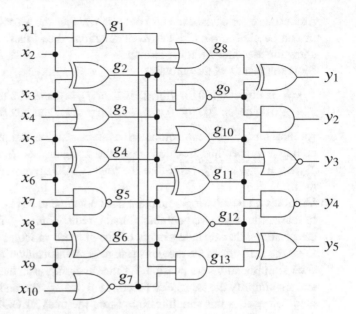

What is the global behavior determined by the inputs $x_1 = x_2 = x_3 = x_4 = 1$ and $x_7 = x_8 = x_9 = x_{10} = 0$? Show additionally the minimize representation of the input–output phases for the same restriction of the inputs.

(b) What is the global behavior determined by the inputs $y_1 = y_2 = y_4 = y_5 = 1$ and $y_3 = 0$? Show additionally the minimize representation of the input–output-phases for the same restriction of the outputs.

(c) What is the global behavior determined by the internal variables $g_2 = g_{13} = 1$ and $g_3 = g_{12} = 0$? Show additionally the minimize representation of the input–output-phases for the same restriction of the internal variables.

(d) Which patterns can occur at the outputs of the combinational circuit shown in Fig. 11.68? Compute and show the minimize representation of patterns that cannot occur at the outputs of the combinational circuit. What can be concluded from these results?

Exercise 11.5 (Simple Covering Method to Synthesize a Combinational Circuit)

The following system-function describes the permitted behavior of a combinational circuit.

$$F(\mathbf{x}, \mathbf{y}) = \overline{x}_1 x_2 \overline{x}_4 \overline{x}_5 \overline{x}_6 \overline{y}_1 \overline{y}_2 \vee (\overline{x}_1 \overline{x}_2 x_3 \overline{x}_4 x_6 \vee x_1 x_2 \overline{x}_6 \vee \overline{x}_2 x_3 x_4 \overline{x}_6) \overline{y}_1 y_2$$

$$\vee (x_1 x_2 (x_5 \vee \overline{x}_6) \vee x_1 \overline{x}_2 (x_3 x_6 \vee \overline{x}_4) \vee \overline{x}_3 \overline{x}_4 (x_1 \vee x_6)$$

$$\vee \overline{x}_2 \overline{x}_6 (\overline{x}_3 x_4 \vee \overline{x}_4 \overline{x}_5) \vee x_2 x_4 x_6) y_1 \overline{y}_2$$

$$\vee (\overline{x}_1 \overline{x}_2 (\overline{x}_3 \vee \overline{x}_4 \overline{x}_6) \vee x_1 \overline{x}_2 x_4 (\overline{x}_3 \vee x_6) \vee x_2 (x_1 x_3 x_4 \vee \overline{x}_4 x_6)$$

$$\vee x_1 \overline{x}_5 x_6 (x_4 \vee x_3) \vee x_1 \overline{x}_3 \overline{x}_4 \overline{x}_6 \vee x_4 x_6) y_1 y_2 .$$

The aim of this exercise is the synthesis of this circuit using a simple covering method that computes a minimal disjunctive form for each of the two outputs y_1 and y_2.

(a) Show the Karnaugh-map of $F(\mathbf{x}, \mathbf{y})$ and verify whether at least one output-pattern is permitted for each input-pattern, so that a combinational circuit can realize the given behavior. Verify furthermore for each of the two outputs whether the associated function is completely specified or can be chosen out of a lattice of functions (i.e., this function is incompletely specified) and show the associated Karnaugh-maps.

Store the results of the outputs:

- y_1 as objects: 100 for $f_{1q}(\mathbf{x})$, 101 for $f_{1r}(\mathbf{x})$, and 102 for $f_{1\varphi}(\mathbf{x})$; and
- y_2 as objects: 200 for $f_{2q}(\mathbf{x})$, 201 for $f_{2r}(\mathbf{x})$, and 202 for $f_{2\varphi}(\mathbf{x})$,

so that the subsequent computation steps can be uniquely executed. Use for a computed completely specified function $f(\mathbf{x})$ the assignments $f_{iq}(\mathbf{x}) = f(\mathbf{x})$, $f_{ir}(\mathbf{x}) = \overline{f(\mathbf{x})}$, and $f_{i\varphi}(\mathbf{x}) = \overline{f_{iq}(\mathbf{x}) \vee f_{ir}(\mathbf{x})}$. Store the final XBOOLE-system as sdt-file for the use in the next task.

(b) Variables on which a logic function does not depend can be excluded and hence reduce the effort to realize the associated combinational circuit. Similarly, functions out of a lattice of functions are preferred that are independent of one or more variables x_i.

Load the sdt-file of the previous task to get the specifications of the two outputs. Verify whether such simplifications are possible for the function, resp., the lattice of functions of the outputs y_1 and y_2, simplify the associated functions $f_{iq}(\mathbf{x})$, $f_{ir}(\mathbf{x})$, and $f_{i\varphi}(\mathbf{x})$, and show the two Karnaugh-maps of y_i after this simplification. Store the final XBOOLE-system as sdt-file for the use in the next task.

(c) Compute sufficient sets of prime conjunctions needed to cover the functions of the two outputs y_1 and y_2. Load therefore the sdt-file of the previous task, use the command obbc to minimize the two functions $f_{1q}(\mathbf{x})$ and $f_{2q}(\mathbf{x})$, and expand each of the resulting ternary vectors such that it represents a prime conjunction of the associated function, resp., lattice of functions.

Store the set of prime conjunctions of $f_{1q}(\mathbf{x})$ ($f_{2q}(\mathbf{x})$) as TVL in D-form using the object number 120 (220) and show for comparison the TVLs 100, 120, 200, and 220 in the m-fold view. Store the final XBOOLE-system as sdt-file for the use in the next task.

(d) Compute two minimal sets of prime conjunctions that cover the functions of the two outputs y_1 and y_2. Load therefore the sdt-file of the previous task, remove as much as possible ternary vectors from the two sets of prime conjunctions such that the remaining ternary vectors cover the associated function $f_{iq}(\mathbf{x})$.

Store the computed minimal set of prime conjunctions of $f_{1q}(\mathbf{x})$ ($f_{2q}(\mathbf{x})$) as TVL in D-form using the object number 121 (221) and show for comparison the TVLs 120, 121, 220, and 221 in the m-fold view. Draw the circuit structure that realizes the computed minimal disjunctive forms using AND- and OR-gates up to three inputs and NOT-gates.

Exercise 11.6 (Synthesis by Separation of a Single Variable or a Function)

The aim of this exercise is the complete synthesis of a combinational circuit that realizes the logic function:

$$f(\mathbf{x}) = \overline{x}_1 x_5 \overline{x}_9 (\overline{x}_4 \vee \overline{x}_6 x_7) \vee x_5 \overline{x}_8 \vee x_1 x_5 x_9 (x_2 x_3 \overline{x}_4 \vee \overline{x}_4 \overline{x}_6 \vee \overline{x}_7)$$

$$\vee x_2 x_4 x_5 x_6 (x_1 \overline{x}_3 \overline{x}_9 \vee \overline{x}_1 x_3 x_7) \vee \overline{x}_1 x_5 x_6 x_7 x_9 (\overline{x}_2 \vee \overline{x}_3)$$

$$\vee x_4 x_5 (\overline{x}_1 \overline{x}_6 x_7 \vee x_1 \overline{x}_7 x_8 \vee x_1 \overline{x}_2 x_6 \overline{x}_9) \ .$$

It has been checked that this function depends on all nine variables.

(a) Verify whether a variable x_i or a negated variable $\overline{x}_i$ can be separated from the given function $f(x_i, \mathbf{x}_0)$ using an AND-gate. Break the search for such a separation when a variable has been found that can be separated using an AND-gate. Store the TVL of this variable as object number 10, and the associated TVL of the decomposition-function $g(\mathbf{x}_0)$ as object number 11; store the TVL of the given function as object 11 in the case that no variable could be separated using an AND-gate. Show the minimized TVLs of the given function $f(x_i, \mathbf{x}_0)$, the computed decomposition-function $g(\mathbf{x}_0)$, and the separated variable. Store the final XBOOLE-system as sdt-file for the use in the next task.

(b) Load the sdt-file of the previous task and verify whether a variable x_i or a negated variable $\bar{x}_i$ can be separated from the function $f_1(x_i, \mathbf{x}_0)$ given as TVL 11 using an OR-gate. Break the search for such a separation when a variable has been found that can be separated using an OR-gate. Store the TVL of this variable as object number 20, and the associated TVL of the decomposition-function $g_1(\mathbf{x}_0)$ as object number 21; store the TVL of the loaded function $f_1(x_i, \mathbf{x}_0)$ (TVL 11) as object 21 in the case that no variable could be separated using an OR-gate. Show the minimized TVLs of the given function $f_1(x_i, \mathbf{x}_0)$, the computed decomposition-function $g_1(\mathbf{x}_0)$, and the separated variable. Store the final XBOOLE-system as sdt-file for the use in the next task.

(c) Load the sdt-file of the previous task and verify whether a variable x_i can be separated from the function $f_2(x_i, \mathbf{x}_0)$ given as TVL 21 using an EXOR-gate. Break the search for such a separation when a variable has been found that can be separated using an EXOR-gate. Store the TVL of this variable as object number 30, and the associated TVL of the decomposition-function $g_2(\mathbf{x}_0)$ as object number 31; store the TVL of the loaded function $f_2(x_i, \mathbf{x}_0)$ (TVL 21) as object 31 in the case that no variable could be separated using an EXOR-gate. Show the minimized TVLs of the given function $f_2(x_i, \mathbf{x}_0)$, the computed decomposition-function $g_2(\mathbf{x}_0)$, and the separated variable. Store the final XBOOLE-system as sdt-file for the use in the next task.

(d) Load the sdt-file of the previous task and verify whether an Ashenhurst-decomposition exists for the function $f_3(\mathbf{x})$ given as TVL 31 and try to increase the subsets of variables $\mathbf{x}_0$ and $\mathbf{x}_1$ to get finally a Povarov-decomposition. Compute the decomposition-functions $g(\mathbf{x}_0)$ and $h(g, \mathbf{x}_1)$ and check whether these two functions determine the given function $f_3(\mathbf{x}) = h(g(\mathbf{x}_0), \mathbf{x}_1)$. Show the Karnaugh-map of the function to decompose $f_3(\mathbf{x})$ (TVL 31), the subsets of variables $\mathbf{x}_0$ and $\mathbf{x}_1$ of the computed initial Ashenhurst-decomposition, the extended subsets of variables $\mathbf{x}_0$ and $\mathbf{x}_1$ of the final Povarov-decomposition, the Karnaugh-maps of the decomposition-functions $g(\mathbf{x}_0)$ and $h(g, \mathbf{x}_1)$, and for comparison in a neighbor viewport of $f_3(\mathbf{x})$ the Karnaugh-map of $h(g(\mathbf{x}_0), \mathbf{x}_1)$. Compute the function $f(\mathbf{x})$ based on all separations of this exercise and verify that this reconstructed function is equal to the given function stored as TVL 1. Draw the circuit structure based on all decomposition-results.

Exercise 11.7 (Synthesis Using Several Types of Strong Bi-decompositions)

The aim of this exercise is the complete synthesis of a combinational circuit with the output y that realizes the logic function:

$$f(\mathbf{x}) = x_1 x_2 (\bar{x}_4 \vee \bar{x}_5) \vee \bar{x}_1 \bar{x}_3 x_5 \bar{x}_6 (\bar{x}_2 \vee \bar{x}_4) \vee \bar{x}_1 x_2 \bar{x}_3 \bar{x}_5 \bar{x}_6 .$$

It has been checked that this function depends on all six variables.

(a) Determine the dedicated sets of variables $\mathbf{x}_a$ and $\mathbf{x}_b$ of an AND-bi-decomposition and compute the decomposition-functions $g(\mathbf{x}_a, \mathbf{x}_c)$ and $h(\mathbf{x}_b, \mathbf{x}_c)$ such that

$$f(\mathbf{x}_a, \mathbf{x}_b, \mathbf{x}_c) = g(\mathbf{x}_a, \mathbf{x}_c) \wedge h(\mathbf{x}_b, \mathbf{x}_c) .$$

Find first an initial pair of variables usable for the dedicated sets of variables $\mathbf{x}_a$ and $\mathbf{x}_b$ of the explored AND-bi-decomposition and extend these sets thereafter as much as possible so that a compact AND-bi-decomposition is determined. Compute the decomposition-function $g(\mathbf{x}_a, \mathbf{x}_c)$ as TVL 20 and the decomposition-function $h(\mathbf{x}_b, \mathbf{x}_c)$ as TVL 21. Show the TVL of the function to decompose, the found pair of initial variables of both dedicated sets, both maximally extended dedicated sets of variables $\mathbf{x}_a$ and $\mathbf{x}_b$, and the Karnaugh-maps of the computed decomposition-functions $g(\mathbf{x}_a, \mathbf{x}_c)$ and $h(\mathbf{x}_b, \mathbf{x}_c)$. Store the final XBOOLE-system as `sdt`-file for the use in the next two tasks.

(b) Load the `sdt`-file of the previous task to get the specification of the function $f_1(\mathbf{x}) = g(\mathbf{x}_a, \mathbf{x}_c)$ (TVL 20), identify the dedicated sets of variables $\mathbf{x}_a$ and $\mathbf{x}_b$ of an EXOR-bi-decomposition, and compute the decomposition-functions $g_1(\mathbf{x}_a, \mathbf{x}_c)$ and $h_1(\mathbf{x}_b, \mathbf{x}_c)$ so that

$$f_1(\mathbf{x}_a, \mathbf{x}_b, \mathbf{x}_c) = g_1(\mathbf{x}_a, \mathbf{x}_c) \oplus h_1(\mathbf{x}_b, \mathbf{x}_c) \ .$$

Find first an initial pair of variables usable for the dedicated sets of variables $\mathbf{x}_a$ and $\mathbf{x}_b$ of the explored EXOR-bi-decomposition and extend the set $\mathbf{x}_b$ thereafter as much as possible. Compute the decomposition-functions $g_1(\mathbf{x}_a, \mathbf{x}_c)$ and the decomposition-functions $h_1(\mathbf{x}_b, \mathbf{x}_c)$. Verify whether the decomposition-functions $h_1(\mathbf{x}_b, \mathbf{x}_c)$ depend on all variables of the common set of variables $\mathbf{x}_c$ and exclude all independent variables $x_i \in \mathbf{x}_c$ of the function $h_1(\mathbf{x}_b, \mathbf{x}_c)$ so that a compact EXOR-bi-decomposition is computed. Show the TVL of the function to decompose $f_1(\mathbf{x})$, the determined pair of initial variables of both dedicated sets, the two maximally extended dedicated sets of variables $\mathbf{x}_a$ and $\mathbf{x}_b$, and the Karnaugh-maps of the computed decomposition-functions $g_1(\mathbf{x}_a, \mathbf{x}_c)$ and $h_1(\mathbf{x}_b, \mathbf{x}_c)$.

(c) Load the `sdt`-file of Task (a) of this exercise to get the specification of the function $f_2(\mathbf{x}) = h(\mathbf{x}_a, \mathbf{x}_c)$ (TVL 21), determine the dedicated sets of variables $\mathbf{x}_a$ and $\mathbf{x}_b$ of an OR-bi-decomposition, and compute the decomposition-functions $g_2(\mathbf{x}_a, \mathbf{x}_c)$ and $h_2(\mathbf{x}_b, \mathbf{x}_c)$ so that

$$f_2(\mathbf{x}_a, \mathbf{x}_b, \mathbf{x}_c) = g_2(\mathbf{x}_a, \mathbf{x}_c) \vee h_2(\mathbf{x}_b, \mathbf{x}_c) \ .$$

Find first an initial pair of variables usable for the dedicated sets of variables $\mathbf{x}_a$ and $\mathbf{x}_b$ of the explored OR-bi-decomposition and extend these sets thereafter as much as possible so that a compact OR-bi-decomposition is determined. Compute the decomposition-functions $g_2(\mathbf{x}_a, \mathbf{x}_c)$ and $h_2(\mathbf{x}_b, \mathbf{x}_c)$. Show the TVL of the function to decompose, the found pair of initial variables of both dedicated sets, both maximally extended dedicated sets of variables $\mathbf{x}_a$ and $\mathbf{x}_b$, and the Karnaugh-maps of the computed decomposition-functions $g_2(\mathbf{x}_a, \mathbf{x}_c)$ and $h_2(\mathbf{x}_b, \mathbf{x}_c)$.

(d) Decomposition-functions that depend at most on two variables can directly be realized by a single gate so that no further decompositions are needed. The only decomposition-function of the second level that depends on more than two variables is $f_3(\mathbf{x}) = h_1(x_1, x_3, x_6)$. Specify $f_3(\mathbf{x})$ as TVL taken from Task (b), determine the dedicated sets of variables $\mathbf{x}_a$ and $\mathbf{x}_b$ of an AND-bi-decomposition, and compute the decomposition-functions $g_3(\mathbf{x}_a, \mathbf{x}_c)$ and $h_3(\mathbf{x}_b, \mathbf{x}_c)$ so that

$$f_3(\mathbf{x}_a, \mathbf{x}_b, \mathbf{x}_c) = g_3(\mathbf{x}_a, \mathbf{x}_c) \wedge h_3(\mathbf{x}_b, \mathbf{x}_c) \ .$$

Find first an initial pair of variables usable for the dedicated sets of variables $\mathbf{x}_a$ and $\mathbf{x}_b$ of the explored AND-bi-decomposition and extend these sets thereafter as much as possible so that a compact AND-bi-decomposition is determined. Compute the decomposition-functions $g_3(\mathbf{x}_a, \mathbf{x}_c)$ and $h_3(\mathbf{x}_b, \mathbf{x}_c)$. Show the TVL of the function to decompose, the found pair of initial variables of both dedicated sets, both maximally extended dedicated sets of variables $\mathbf{x}_a$ and $\mathbf{x}_b$, and the

Karnaugh-maps of the computed decomposition-functions $g_3(\mathbf{x}_a, \mathbf{x}_c)$ and $h_3(\mathbf{x}_b, \mathbf{x}_c)$. Draw the circuit structure of the combinational circuit synthesized in this exercise using four strong bi-decompositions and indicate the computed decomposition functions.

Exercise 11.8 (Synthesis Using Weak and Strong Bi-Decompositions)

The synthesis of each combinational function can be completely realized when additionally to the strong AND-, OR-, and EXOR-bi-decompositions also weak AND- and OR-bi-decompositions are used (see Theorem 11.23). We use in this exercise the function:

$$f(\mathbf{x}) = \overline{x}_1 \overline{x}_2 \overline{x}_4 (\overline{x}_3 \vee x_5) \vee x_1 x_2 \overline{x}_3 (x_4 \vee \overline{x}_5) \vee \overline{x}_4 x_5 (x_1 x_2 x_3 \vee \overline{x}_1 \overline{x}_3)$$

$$\vee x_4 (x_1 \overline{x}_2 \overline{x}_5 \vee \overline{x}_1 x_2 x_3) \vee x_1 \overline{x}_2 x_3 (x_4 \vee \overline{x}_5) \vee x_2 x_3 \overline{x}_4 \overline{x}_5$$

to explore the relations between the two weak bi-decomposition and the strong EXOR-bi-decomposition for a complete synthesis. The chosen function is one of the very rare functions of five variables for which no strong bi-decomposition exists.

(a) Verify that no strong bi-decomposition for the given function and any pair of variables exists. Show the Karnaugh-map of the function to decompose and if exist the pairs of dedicated variables of the first found OR-, AND-, and EXOR-bi-decompositions.

(b) At least one weak bi-decomposition is possible when no strong EXOR-bi-decomposition exists. Verify whether a weak OR-bi-decomposition of the given function with regard to a single variable x_a exists, compute in case of a successful check the decomposition-lattice $\mathcal{L}\langle g_{1q}(x_a, \mathbf{x}_c), g_{1r}(x_a, \mathbf{x}_c)\rangle$, and show the Karnaugh-map of this lattice and the TVL of the chosen variable x_a. Store the TVLs of the chosen variable x_a as object 10, the function $g_{1q}(x_a, \mathbf{x}_c)$ as object 11, and the function $g_{1r}(x_a, \mathbf{x}_c)$ as object 12 in the case that a weak OR-bi-decomposition exists or the given function $f(\mathbf{x})$ as object 11 and $\overline{f(\mathbf{x})}$ as object 12 otherwise.

The extension of the single given function to the lattice of functions increases the possibilities for bi-decompositions. Verify whether the computed lattice contains a least one function for which a strong bi-decomposition exists. Store the final XBOOLE-system as sdt-file for the use in the next task.

(c) In the case that decomposition-lattice $\mathcal{L}\langle g_{1q}(x_a, \mathbf{x}_c), g_{1r}(x_a, \mathbf{x}_c)\rangle$ of a weak OR-bi-decomposition does not contain a function for which a strong EXOR-bi-decomposition exists, a weak AND-bi-decomposition exists for this lattice with regard to the same variables x_a.

Load the XBOOLE-system stored at the end of the previous task, verify this property, and compute the decomposition-lattice $\mathcal{L}\langle g_{2q}(x_a, \mathbf{x}_c), g_{2r}(x_a, \mathbf{x}_c)\rangle$ of the weak AND-bi-decomposition. Show the Karnaugh-maps of the given lattice of the OR-bi-decomposition $\mathcal{L}\langle g_{1q}(x_a, \mathbf{x}_c), g_{1r}(x_a, \mathbf{x}_c)\rangle$ and the computed lattice of the weak AND-bi-decomposition $\mathcal{L}\langle g_{2q}(x_a, \mathbf{x}_c), g_{2r}(x_a, \mathbf{x}_c)\rangle$ as well as the TVL of the chosen variable x_a. Store the TVLs of the chosen variable x_a as object 20, the function $g_{2q}(x_a, \mathbf{x}_c)$ as object 21, and the function $g_{2r}(x_a, \mathbf{x}_c)$ as object 22. Store the final XBOOLE-system as sdt-file for the use in the next task.

(d) After a weak OR-bi-decomposition and a weak AND-bi-decomposition with regard to the same variable x_a, an EXOR-bi-decomposition with regard to this variable and at least one other variable exists.

Load the XBOOLE-system stored at the end of the previous task and determine a variable x_b so that the lattice $\mathcal{L}\langle g_{2q}(x_a, \mathbf{x}_c), g_{2r}(x_a, \mathbf{x}_c)\rangle$ contains a least one function for which a strong EXOR-bi-decomposition with regard to the pair of variables x_a and x_b exists. Increase the dedicated set of variables $\mathbf{x}_b$ as much as possible to get a compact strong EXOR-bi-decomposition and compute the associated decomposition-function $g_3(x_a)$.

Extend the PRP to compute the lattices $\mathcal{L}\langle h_{3q}(\mathbf{x}_b), h_{3r}(\mathbf{x}_b)\rangle$ of the EXOR-bi-decomposition (the set of common variables $\mathbf{x}_c$ is empty due to the disjoint EXOR-bi-decomposition), show the associated Karnaugh-map, choose a simple decomposition-function $h_3(\mathbf{x}_b)$, and show the Karnaugh-map of this function, too. Repeat this procedure to determine the decomposition-function $h_2(\mathbf{x}_c)$ of the weak AND-bi-decomposition and the decomposition-function $h_1(\mathbf{x}_c)$ of the weak OR-bi-decomposition.

Compute finally the function of the synthesized circuit, show the associated Karnaugh-map, and check that this function is equal to the given function of this exercise. Draw the synthesized circuit using gates of up to two inputs and indicate the three used bi-decompositions.

Exercise 11.9 (Test-Patterns of a Combinational Circuit)

The aim of this exercise is the computation of test-pattern for selected sensible paths and sensible points at a local internal signal or a local internal branch of the combinational circuit shown in Fig. 11.66.

(a) Compute for each of the three paths between the input x_2 and one of the outputs y_1 or y_2 based on Condition (11.257) all test-patterns that can be used to detect both the SA0- and the SA1-errors on all wires of these paths. Show these three sets of test-patterns.

(b) The network-model of the sensible point facilitates to distinguish between test-patterns that detect either SA0-errors or SA1-errors and implicitly inserts the expected output-values into the test-patterns. This network-model uses the system-function $F^R(\mathbf{x}, \mathbf{y})$ that describes the required behavior of the combinational circuit. Compute this system-function based on Eq. (11.265) and store the created XBOOLE-system as sdt-file for the computation of test-pattern in the next two tasks.

(c) The aim of this task is the computation of all test-patterns for the local internal signal g_2 at the output of the gate G_2. Load therefore the sdt-file from the previous, generate the model of modified gates and $F_T(g_i, T)$ (11.264) to compute the list of error-phase $F^E(\mathbf{x}, \mathbf{y}, \mathbf{s}, T)$ based on Eq. (11.266), and compute the wanted test-patterns using (11.263). Show the list of error-phase $F^E(\mathbf{x}, \mathbf{y}, s, T)$ and the computed set of test-patterns for the local internal signal g_2.

(d) Different test-patterns can exist at a connection of gates with a local internal branch. Compute in this task separately the test-patterns for the inputs of the gates G_7 and G_8 that are controlled by the output of the gate G_5 and the test-patterns for this output. Load therefore the sdt-file from Task (b) of this exercise, generate the model of modified gates as before to compute the list of error-phase $F^E(\mathbf{x}, \mathbf{y}, \mathbf{s}, T)$ based on Eq. (11.266), and compute and show the requested three sets of test-patterns of the local internal branch g_5 using (11.270).

11.7   Solutions

Solution 11.1 (Structural Models of a Combinational Circuit)

(a) Figure 11.69 shows the PRP that uses a system of logic equations to describe the structure of the combinational circuit of Fig. 11.66. Each equation of this system specifies the type of a used gate. Identical variables indicate the connections between the input, the gates, and the outputs. Alternatively, eight logic equations can used be to describe each gate separately.

(b) Figure 11.70 shows the PRP that uses eight local lists of phases to describe the structure of the combinational circuit of Fig. 11.66. Each local lists of phases specifies the type of a used gate based on Table 11.1. Identical variables indicate the connections between the input, the gates, and the outputs.

```
1    new                          7    y1  y2 .              13   g5=g2&x3 ,
2    space  32  1                 8    sbe  1  1             14   g6=x4&g3 ,
3    avar  1                      9    g1 =/x1 ,             15   y1=g4+g5 ,
4    x1  x2  x3  x4  x5          10    g2 =/x2 ,             16   y2=g5+g6 .
5    g1  g2  g3                  11    g3 =/x5 ,
6    g4  g5  g6                  12    g4=g1&x2 ,
```

Fig. 11.69 Problem-program that specifies the structure of the combinational circuit of Fig. 11.66 using a system of logic equations

```
1    new                        16   tin  1  3            31   x4  g3  g6 .
2    space  32  1               17   x5  g3 .             32   0-0
3    avar  1                    18   01                   33   100
4    x1  x2  x3  x4  x5         19   10 .                 34   111 .
5    g1  g2  g3                 20   tin  1  4            35   tin  1  7
6    g4  g5  g6                 21   g1  x2  g4 .         36   g4  g5  y1 .
7    y1  y2 .                   22   0-0                  37   000
8    tin  1  1                  23   100                  38   011
9    x1  g1 .                   24   111 .                39   1-1 .
10   01                         25   tin  1  5            40   tin  1  8
11   10 .                       26   g2  x3  g5 .         41   g5  g6  y2 .
12   tin  1  2                  27   0-0                  42   000
13   x2  g2 .                   28   100                  43   011
14   01                         29   111 .                44   1-1 .
15   10 .                       30   tin  1  6
```

Fig. 11.70 Problem-program that specifies the structure of the combinational circuit of Fig. 11.66 using a set of local lists of phases

Fig. 11.71
Problem-program that
specifies the structure of
the combinational circuit
of Fig. 11.66 using two
TVLs in D-form

```
1    new                          8    x1  x2  x3 .
2    space  32  1                 9    01-
3    avar  1                     10    -01 .
4    x1  x2  x3  x4  x5          11    tin  1  2  /d
5    g1  g2  g3  g4  g5  g6      12    x2  x3  x4  x5 .
6    y1  y2 .                    13    01--
7    tin  1  1  /d               14    --10 .
```

(c) Figure 11.71 shows the PRP that describes the structure of the combinational circuit of Fig. 11.66 in a very compact manner. These two TVLs specify the AND–OR-structure (indicated by the D-form) for the two functions $y_i = f_i(\mathbf{x})$. Each ternary vector represents an AND-gate controlled by the inputs. A value 0 of such a vector expresses that a NOT-gate exists in the path to the associated input.

(d) Figure 11.72 shows the PRP that represents the combinational circuit of Fig. 11.66 using the structure-TVL of an equivalent PLA. The rules for such a TVL are explained in the beginning on page 588. Note that a value 0 in the columns of y_1 or y_2 does not determine a function value 0 but expresses only that the conjunction of this row is not used in the associated function.

Solution 11.2 (Behavioral Models of Combinational Circuits)

(a) Figure 11.73 shows on the right-hand side the PRP that computes the minimized orthogonal system-function $F(\mathbf{x}, \mathbf{y})$ and the computed TVL of this system-function. The PRP verifies additionally that the equation $F(\mathbf{x}, \mathbf{y}) = 1$ is uniquely solvable with regard to y_1 and y_2.

```
1   new                          7   tin 1 1 /d
2   space 32 1                   8   x1 x2 x3 x4 x5 y1 y2.
3   avar 1                       9   01---10
4   x1 x2 x3 x4 x5              10   -01--11
5   g1 g2 g3 g4 g5 g6          11   ---1001.
6   y1 y2.
```

Fig. 11.72 Problem-program that specifies the structure of the combinational circuit of Fig. 11.66 using the structure-TVL of an equivalent PLA

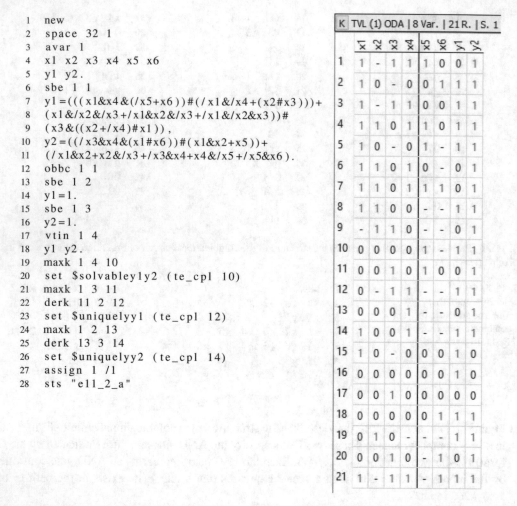

```
1   new
2   space 32 1
3   avar 1
4   x1 x2 x3 x4 x5 x6
5   y1 y2.
6   sbe 1 1
7   y1=(((x1&x4&(/x5+x6))#(/x1&/x4+(x2#x3)))+
8   (x1&/x2&/x3+/x1&x2&/x3+/x1&/x2&x3))#
9   (x3&((x2+/x4)#x1)),
10  y2=(((/x3&x4&(x1#x6))#(x1&x2+x5))+
11  (/x1&x2+x2&/x3+/x3&x4+x4&/x5+/x5&x6).
12  obbc 1 1
13  sbe 1 2
14  y1=1.
15  sbe 1 3
16  y2=1.
17  vtin 1 4
18  y1 y2.
19  maxk 1 4 10
20  set $solvabley1y2 (te_cpl 10)
21  maxk 1 3 11
22  derk 11 2 12
23  set $uniquelyy1 (te_cpl 12)
24  maxk 1 2 13
25  derk 13 3 14
26  set $uniquelyy2 (te_cpl 14)
27  assign 1 /1
28  sts "e11_2_a"
```

K	x1	x2	x3	x4	x5	x6	y1	y2
1	1	-	1	1	1	0	0	1
2	1	0	-	0	0	1	1	1
3	1	-	1	1	0	0	1	1
4	1	1	0	1	1	0	1	1
5	1	0	-	0	1	-	1	1
6	1	1	0	1	0	-	0	1
7	1	1	0	1	1	1	0	1
8	1	1	0	0	-	-	1	1
9	-	1	1	0	-	-	0	1
10	0	0	0	0	1	-	1	1
11	0	0	1	0	1	0	0	1
12	0	-	1	1	-	-	1	1
13	0	0	0	1	-	-	0	1
14	1	0	0	1	-	-	1	1
15	1	0	-	0	0	0	1	0
16	0	0	0	0	0	0	1	0
17	0	0	1	0	0	0	0	0
18	0	0	0	0	0	1	1	1
19	0	1	0	-	-	-	1	1
20	0	0	1	0	-	1	0	1
21	1	-	1	1	-	1	1	1

Table header: K | TVL (1) ODA | 8 Var. | 21 R. | S. 1

Fig. 11.73 Problem-program that computes the system-function $F(\mathbf{x}, \mathbf{y})$ and verifies that the equation $F(\mathbf{x}, \mathbf{y}) = 1$ is uniquely solvable with regard to y_1 and y_2 together with the TVL of $F(\mathbf{x}, \mathbf{y})$

The transformation of the given system of logic equations into the equivalent system-function $F(\mathbf{x}, \mathbf{y})$ has been realized by the command sbe; the solution of this system of two equations is the global list of phase represented by a TVL in ODA-form that describes the input–output-behavior of the associated combinational circuit. This TVL is also a representation of the wanted system-function $F(\mathbf{x}, \mathbf{y})$. The command obbc in line 12 reduces the number of orthogonal ternary vectors from 62 to 21.

```
1    lds  "e11_2_a"
2    isc  1 2 20
3    maxk 20 4 21
4    obbc 21 22 ; f1
5    isc  1 3 30
6    maxk 30 4 31
7    obbc 31 32 ; f2
8    assign 22 /m 1 1
9    assign 32 /m 1 2
```

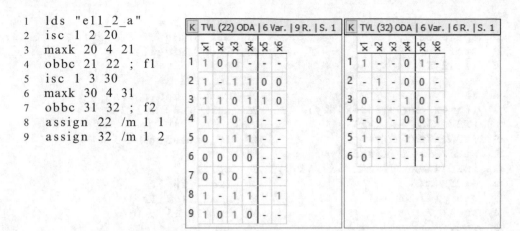

K	TVL (22) ODA | 6 Var. | 9 R. | S. 1							K	TVL (32) ODA | 6 Var. | 6 R. | S. 1					
	x_1	x_2	x_3	x_4	x_5	x_6			x_1	x_2	x_3	x_4	x_5	x_6
1	1	0	0	-	-	-		1	1	-	-	0	1	-
2	1	-	1	1	0	0		2	-	1	-	0	0	-
3	1	1	0	1	1	0		3	0	-	-	1	0	-
4	1	1	0	0	-	-		4	-	0	-	0	0	1
5	0	-	1	1	-	-		5	1	-	-	1	-	-
6	0	0	0	0	-	-		6	0	-	-	-	1	-
7	0	1	0	-	-	-								
8	1	-	1	1	-	1								
9	1	0	1	0	-	-								

Fig. 11.74 Problem-program that computes the minimized function-TVLs $y_1 = f_1(\mathbf{x})$ and $y_2 = f_2(\mathbf{x})$ together with the representation of these two TVLs

Equation (7.56) has been used in lines 19 and 20 to verify whether the equation $F(\mathbf{x}, \mathbf{y}) = 1$ is solvable with regard to y_1 and y_2; the result value $solvabley1y2=true confirms this property.

Condition (7.59) has been used twice; the result $uniquelyy1=true (computed in lines 21–23) confirms as expected that $F(\mathbf{x}, \mathbf{y}) = 1$ is uniquely solvable with regard to y_1, and the result $uniquelyy2=true (computed in lines 24–26) confirms that this property is also satisfied for y_2.

The command sts in the last line stores the complete XBOOLE-system as file "e11_2_a.sdt" for the use in the next task.

(b) Figure 11.74 shows both the PRP that computes and minimizes the function-TVLs $y_1 = f_1(\mathbf{x})$ and $y_2 = f_2(\mathbf{x})$ belonging to the given system-function $F(\mathbf{x}, \mathbf{y})$ and on the right-hand side the TVLs of these two functions.

The command lds in the first line loads the file "e11_2_a.sdt" so that the XBOOLE-system of the previous task is available and all XBOOLE-objects (TVLs and VTs) can be reused with unchanged object numbers.

Knowing that $F(\mathbf{x}, \mathbf{y}) = 1$ is uniquely solvable with regard to y_1 and y_2, Eq. (7.60) could be used to compute these two function-TVLs.

(c) Figure 11.75 shows both the PRP that checks whether the given system-function $F(\mathbf{x}, y)$ can be realized by a combinational circuit and the Karnaugh-maps of the given system-function $F(\mathbf{x}, y)$ and the computed lattice $\mathcal{L}\langle f_q, f_r \rangle$ that partially determines the behavior of a possible combinational circuit.

The PRP verifies as first step whether the equation $F(\mathbf{x}, y) = 1$ of the given system-function $F(\mathbf{x}, y)$ is solvable with regard to y using Eq. (7.56) in lines 16 and 17; the result $solvabley=true confirms that the given system-function $F(\mathbf{x}, y)$ satisfies this necessary property; otherwise, the command if in line 18 would avoid the execution of all following commands.

Condition (7.59) has been used next to verify whether given system-function $F(\mathbf{x}, y)$ determines a uniquely specified function $y = f(\mathbf{x})$; the operation maxk of (7.59) has been avoided because only a single variable y occurs in $F(\mathbf{x}, y)$. The result of the single derivative in line 20 is not equal to 1 for all $\mathbf{x}$ so that the result $uniquef=false indicates that $F(\mathbf{x}, y) = 1$ is not uniquely

```
1   new                               20   derk 1 2 4
2   space 32 1                        21   set $uniquef (te_cpl 4)
3   avar 1                            22   if ($uniquef)
4   x1 x2 x3 x4 x5 x6                 23   (
5   y .                               24   isc 1 2 5
6   sbe 1 1                           25   maxk 5 2 6 ; f(x)
7   ((x1+x2&/x3+(/x3#x5)&y)#          26   assign 6 /m 1 2
8   (x2&/x3&/y+x1&x6))+               27   )
9   /x2&(x3#x5)&/y+                   28   if (not $uniquef)
10  x1&/x2&x6&(/x3#x5)&y+             29   (
11  (x2&/x4&x6&y#x3&/x5&/x6&/y)       30   cpl 2 10
12  +x2&x4&x6 .                       31   isc 1 10 11
13  assign 1 /m 1 1                   32   maxk 11 2 12
14  sbe 1 2                           33   cpl 12 13 ; fq(x)
15  y=1 .                             34   isc 1 2 14
16  maxk 1 2 3                        35   maxk 14 2 15
17  set $solvabley (te_cpl 3)         36   cpl 15 16 ; fr(x)
18  if ($solvabley)                   37   assign_qr 13 16 /m 1 2
19  (                                 38   ))
```

```
┌─────────────────────────────────────────┐ ┌──────────────────────────────────────────────┐
│ T  TVL (1) ODA | 7 Var. | 22 R. | S. 1    │ │ (q: TVL (14) ODA, r: TVL (18) ODA), 6 Var. |S.1│
├─────────────────────────────────────────┤ ├──────────────────────────────────────────────┤
│ 0 0 0  0 0 1 1 1 1 0 0 0 0 1 1 1 1 1 1    │ │ 0 0  1 1 0 0 0 0 1 1 1 1 Φ Φ Φ Φ Φ Φ           │
│ 0 0 1  1 1 0 0 0 0 1 1 1 1 1 1 1 1 1 1    │ │ 0 1  1 1 0 0 1 Φ Φ 1 1 Φ Φ 1 0 0 1 1           │
│ 0 1 1  1 1 0 0 1 1 1 1 1 1 1 1 0 0 1 1    │ │ 1 1  0 0 1 1 1 Φ Φ 1 1 Φ Φ 1 1 1 0 0           │
│ 0 1 0  0 0 1 1 0 1 1 0 0 1 1 0 1 1 0 0    │ │ 1 0  0 0 1 1 1 1 1 1 1 1 Φ Φ Φ Φ Φ Φ           │
│ 1 1 0  1 1 0 0 0 1 1 0 0 1 1 0 0 0 1 1    │ │                                                │
│ 1 1 1  0 0 1 1 1 1 1 1 1 1 1 1 1 1 0 0    │ │ x5 x6                                          │
│ 1 0 1  0 0 1 1 1 1 1 1 1 1 1 1 1 1 1 1    │ │      0 1 1 0 0 1 1 0 0 1 1 0 0 1 1 0 x4        │
│ 1 0 0  1 1 0 0 0 0 0 0 0 0 1 1 1 1 1 1    │ │      0 0 1 1 1 1 0 0 0 0 1 1 1 1 0 0 x3        │
│                                           │ │      0 0 0 0 1 1 1 1 1 1 1 1 0 0 0 0 x2        │
│ x5 x6 y                                   │ │      0 0 0 0 0 0 0 0 1 1 1 1 1 1 1 1 x1        │
│       0 1 1 0 0 1 1 0 0 1 1 0 0 1 1 0 x4  │ │                                                │
│       0 0 1 1 1 1 0 0 0 0 1 1 1 1 0 0 x3  │ │                                                │
│       0 0 0 0 1 1 1 1 1 1 1 1 0 0 0 0 x2  │ │                                                │
│       0 0 0 0 0 0 0 0 1 1 1 1 1 1 1 1 x1  │ │                                                │
└─────────────────────────────────────────┘ └──────────────────────────────────────────────┘
```

Fig. 11.75 Problem-program that computes either the uniquely specified function $y = f(\mathbf{x})$ or the mark-functions $f_q(\mathbf{x})$ and $f_r(\mathbf{x})$ of the lattice $\mathcal{L}\langle f_q, f_r\rangle$ together with the Karnaugh-maps of the given system-function $F(\mathbf{x}, y)$ and the computed lattice

solvable with regard to y. Due to this result, the computation and visualization of the uniquely specified function $y = f(\mathbf{x})$ based on Eq. (7.60) in lines 24–26 is skipped.

Knowing that the function of the combinational circuit can be chosen out of a lattice $\mathcal{L}\langle f_q, f_r\rangle$, the mark-function $f_q(\mathbf{x})$ is computed in lines 30–33 based on Eq. (7.57) and the computation of the associated mark-function $f_r(\mathbf{x})$ follows in lines 34–36 based on Eq. (7.58). The command assign_qr in line 37 assigns the two mark-functions $f_q(\mathbf{x})$ and $f_r(\mathbf{x})$ of the computed lattice to a single Karnaugh-map shown in viewport (1,2) of the m-fold view. The variable x_4 has been moved from the vertical to the horizontal Gray code for an easier comparison with the Karnaugh-map of the given system-function $F(\mathbf{x}, y)$ shown in viewport (1,1) of the m-fold view.

The evaluation of these two Karnaugh-maps confirms the correctness of the solution of this task. Four pairs $(F(\mathbf{x} = const., y = 0), F(\mathbf{x} = const., y = 1))$ of function values are possible at all:

- the pair (00) does not appear in the Karnaugh-map of $F(\mathbf{x}, y)$; this pattern would prohibit each function value of $f(\mathbf{x})$ (the result $solvabley=true confirms this necessary property);

- the pair (01) in the Karnaugh-map of $F(\mathbf{x}, y)$ determines that $f(\mathbf{x}) = 1$ in the associated field of $\mathbf{x}$ in the Karnaugh-map of $\mathcal{L}\langle f_q, f_r \rangle$; for these patterns of $\mathbf{x}$, we get $f_q(\mathbf{x}) = 1$ and $f_r(\mathbf{x}) = 0$;
- the pair (10) in the Karnaugh-map of $F(\mathbf{x}, y)$ determines that $f(\mathbf{x}) = 0$ in the associated field of $\mathbf{x}$ in the Karnaugh-map of $\mathcal{L}\langle f_q, f_r \rangle$; for these patterns of $\mathbf{x}$, we get $f_q(\mathbf{x}) = 0$ and $f_r(\mathbf{x}) = 1$; or
- the pair (11) in the Karnaugh-map of $F(\mathbf{x}, y)$ determines that both $f(\mathbf{x}) = 0$ and $f(\mathbf{x}) = 1$ are acceptable function values of $f(\mathbf{x})$ indicated by the symbol ϕ in the associated field of $\mathbf{x}$ in the Karnaugh-map of $\mathcal{L}\langle f_q, f_r \rangle$; for these patterns of $\mathbf{x}$, we get $f_q(\mathbf{x}) = 1$ and $f_r(\mathbf{x}) = 1$.

(d) Figure 11.76 shows the PRP that specifies the mark-functions $f_q(\mathbf{x})$ of the ON-set and $f_r(\mathbf{x})$ of the OFF-set as TVLs, computes the third mark-function $f_\varphi(\mathbf{x})$ of the don't-care set, verifies that these mark-functions are a disjoint cover of the Boolean space $\mathbb{B}^{|\mathbf{x}|}$, computes the system-function $F(\mathbf{x}, y)$ using each pair of mark-functions, and checks that these three system-functions are equal to each other despite of their different representations as TVLs in ODA-form. Figure 11.76 shows furthermore the Karnaugh-map specified by the mark-functions $f_q(\mathbf{x})$ and $f_r(\mathbf{x})$, the TVLs of these two functions, and the TVLs of the system-function $F(\mathbf{x}, y)$ computed using the three different pairs of mark-functions.

The mark-functions $f_q(\mathbf{x})$ and $f_r(\mathbf{x})$ have been specified as TVLs 1 and 2 (see viewports (1,2) and (1,3) of the m-fold view) based on the given Karnaugh-map of the lattice $\mathcal{L}\langle f_q, f_r \rangle$. The Karnaugh-map in viewport (1,1) is created based on these two mark-functions, and the comparison with the given Karnaugh-map confirms that functions $f_q(\mathbf{x})$ and $f_r(\mathbf{x})$ are correctly specified.

The mark-function $f_\varphi(\mathbf{x})$ of the don't-care set is computed based on (11.21) in lines 19 and 20. The result `true` of the three subsequent commands `te_isc` confirms that the three mark-functions are pairwise disjoint. One more property of the three mark-function is that they completely cover the Boolean space. Condition (11.18) of this property is verified in lines 24–26; the result `$cover=true` confirms that this property is also satisfied.

The system-function $F(\mathbf{x}, y)$ can be computed based on each of the three pairs of mark-functions. The system-function $F(\mathbf{x}, y)$ is computed in lines 29–32 as TVL 23 using Eq. (11.22). Next, the system-function $F(\mathbf{x}, y)$ is computed in lines 33–35 as TVL 33 using Eq. (11.24). Finally, the system-function $F(\mathbf{x}, y)$ is computed in lines 36–38 as TVL 43 using Eq. (11.23). The three computed system-functions are shown in the second row of the m-fold view. The results `true` of the final three commands `te_syd` confirm that the computed three system-functions are equal to each other despite their different representations.

Solution 11.3 (Analysis: Compute the Behavior of a Combinational Circuit)

(a) Figure 11.77 shows the PRP that computes the global list of phases and the reduced input–output list of phases (representing the system-function $F(\mathbf{x}, y)$) together the view of these two TVLs. The global list of phases is a comprehensive behavioral description; it contains for each input-pattern of the combinational circuit, shown in Fig. 11.66, the values at the outputs of all gates that include the outputs of the circuit. The solution of the system of equations in line 8–16 is this global list of phases represented as a TVL in ODA-form. The $2^5 = 32$ phases are represented by 18 ternary vectors; the values of $g_1, \ldots, g_6$ prevent that the command `obbc` in line 20 reduces the number of ternary vectors even more.

```
 1  new                                    24  uni 1 2 10
 2  space 32 1                             25  uni 10 3 10
 3  avar 1                                 26  set $cover ( te_cpl 10)
 4  x1 x2 x3 x4 x5                         27  sbe 1 4
 5  y.                                     28  y=1.
 6  tin 1 1                                29  csd 1 4 20
 7  x1 x2 x3 x4 x5.                        30  syd 2 4 21
 8  01-00                                  31  uni 20 21 22
 9  00-1-                                  32  obbc 22 23 ; F: fq, fr
10  -001-                                  33  csd 1 4 30
11  --001. ; fq                            34  uni 30 3 31
12  tin 1 2                                35  obbc 31 33 ; F: fq, fphi
13  x1 x2 x3 x4 x5.                        36  syd 2 4 40
14  101--                                  37  uni 40 3 41
15  -101-                                  38  obbc 41 43 ; F: fr, fphi
16  --101                                  39  assign 1 /m 1 2
17  11-10. ; fr                            40  assign 2 /m 1 3
18  assign_qr 1 2 /m 1 1                   41  assign 23 /m 2 1
19  uni 1 2 3                              42  assign 33 /m 2 2
20  cpl 3 3 ; fphi                         43  assign 43 /m 2 3
21  set $qrdisjoint ( te_isc 1 2)          44  set $equalf1f2 ( te_syd 23 33)
22  set $qphidisjoint ( te_isc 1 3)        45  set $equalf1f3 ( te_syd 23 43)
23  set $rphidisjoint ( te_isc 2 3)        46  set $equalf2f3 ( te_syd 33 43)
```

(q: TVL (1) ODA, r: TVL (2) ODA), 5 Var. | S. 1

```
0 0 | Φ Φ 1 1 | Φ Φ 0 Φ
0 1 | 1 0 0 1 | 1 0 0 1
1 1 | 1 1 Φ 0 | 0 Φ 0 1
1 0 | 1 1 Φ 0 | 0 0 0 1
x4
x5   0 1 1 0 0 1 1 0  x3
     0 0 1 1 1 1 0 0  x2
     0 0 0 0 1 1 1 1  x1
```

K TVL (1) ODA | 5 Var. | 4 R. | S. 1

	x1	x2	x3	x4	x5
1	0	1	-	0	0
2	0	0	-	1	-
3	1	0	0	1	-
4	-	-	0	0	1

K TVL (2) ODA | 5 Var. | 5 R. | S. 1

	x1	x2	x3	x4	x5
1	1	0	1	-	-
2	-	1	0	1	-
3	0	0	1	0	1
4	1	1	1	1	0
5	-	1	1	0	1

K TVL (23) ODA | 6 Var. | 12 R. | S. 1

	x1	x2	x3	x4	x5	y
1	1	-	0	0	0	0
2	1	1	0	1	-	0
3	0	0	-	0	0	0
4	1	-	1	-	-	0
5	0	-	1	0	1	0
6	0	-	1	1	1	1
7	-	0	0	1	-	1
8	0	1	-	1	-	0
9	-	-	0	0	-	1
10	0	-	1	-	0	1
11	1	1	1	1	1	1
12	1	1	1	0	0	1

K TVL (33) ODA | 6 Var. | 15 R. | S. 1

	x1	x2	x3	x4	x5	y
1	-	1	-	0	0	1
2	0	0	-	1	-	1
3	1	0	0	1	-	1
4	-	-	0	0	1	1
5	1	1	1	1	0	0
6	1	0	1	1	-	0
7	1	-	1	0	-	0
8	0	-	1	0	1	0
9	1	1	0	1	1	0
10	0	1	-	1	-	0
11	1	1	0	-	0	0
12	0	1	1	1	-	1
13	1	1	1	1	1	-
14	0	0	1	0	0	-
15	-	0	0	0	0	-

K TVL (43) ODA | 6 Var. | 13 R. | S. 1

	x1	x2	x3	x4	x5	y
1	1	0	1	1	-	0
2	-	1	-	1	-	0
3	-	-	1	0	1	0
4	-	0	0	1	1	1
5	0	0	1	0	0	1
6	-	1	-	0	0	1
7	0	0	1	1	1	1
8	-	-	0	0	1	1
9	1	-	-	0	0	0
10	0	-	1	1	0	1
11	-	1	1	1	1	1
12	0	0	-	0	0	0
13	-	0	0	-	0	1

Fig. 11.76 Problem-program that computes the system-function $F(\mathbf{x}, y)$ using each pair of mark-functions and compares the system-functions with each other together with the computed results

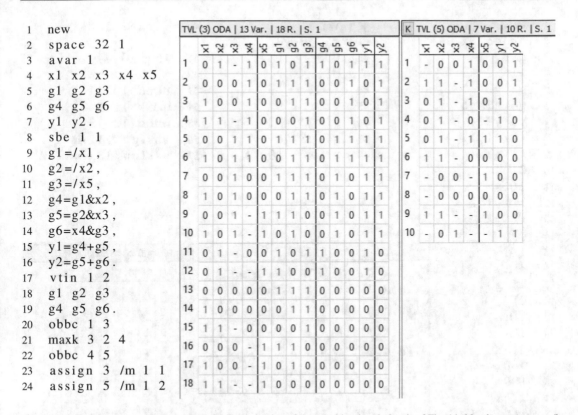

```
 1   new
 2   space  32  1
 3   avar  1
 4   x1  x2  x3  x4  x5
 5   g1  g2  g3
 6   g4  g5  g6
 7   y1  y2 .
 8   sbe  1  1
 9   g1 =/ x1 ,
10   g2 =/ x2 ,
11   g3 =/ x5 ,
12   g4 = g1 & x2 ,
13   g5 = g2 & x3 ,
14   g6 = x4 & g3 ,
15   y1 = g4 + g5 ,
16   y2 = g5 + g6 .
17   vtin  1  2
18   g1  g2  g3
19   g4  g5  g6 .
20   obbc  1  3
21   maxk  3  2  4
22   obbc  4  5
23   assign  3  /m  1  1
24   assign  5  /m  1  2
```

TVL (3) ODA | 13 Var. | 18 R. | S. 1

	x1	x2	x3	x4	x5	g1	g2	g3	g4	g5	g6	y1	y2
1	0	1	-	1	0	1	0	1	1	1	0	1	1
2	0	0	0	1	0	1	1	1	0	0	1	0	1
3	1	0	0	1	0	0	1	1	0	0	1	0	1
4	1	1	-	1	0	0	0	1	0	0	1	0	1
5	0	0	1	1	0	1	1	1	0	1	1	1	1
6	1	0	1	1	0	0	1	1	0	1	1	1	1
7	0	0	1	0	0	1	1	1	0	1	0	1	1
8	1	0	1	0	0	0	1	1	0	1	0	1	1
9	0	0	1	-	1	1	1	0	0	1	0	1	1
10	1	0	1	-	1	0	1	0	0	1	0	1	1
11	0	1	-	0	0	1	0	1	1	0	0	1	0
12	0	1	-	-	1	1	0	0	1	0	0	1	0
13	0	0	0	0	0	1	1	1	0	0	0	0	0
14	1	0	0	0	0	0	1	1	0	0	0	0	0
15	1	1	-	0	0	0	0	1	0	0	0	0	0
16	0	0	0	-	1	1	1	0	0	0	0	0	0
17	1	0	0	-	1	0	1	0	0	0	0	0	0
18	1	1	-	-	1	0	0	0	0	0	0	0	0

K TVL (5) ODA | 7 Var. | 10 R. | S. 1

	x1	x2	x3	x4	x5	y1	y2
1	-	0	0	1	0	0	1
2	1	1	-	1	0	0	1
3	0	1	-	1	0	1	1
4	0	1	-	0	-	1	0
5	0	1	-	1	1	1	0
6	1	1	-	0	0	0	0
7	-	0	0	-	1	0	0
8	-	0	0	0	0	0	0
9	1	1	-	-	1	0	0
10	-	0	1	-	-	1	1

Fig. 11.77 Problem-program that computes the behavior of the combinational circuit of Fig. 11.66 using a system of logic equations together with the two computed lists of phases

The command maxk in line 21 excludes the variables $g_1, \ldots, g_6$ (determined as VT 2) from the global list of phases so that the input–output-behavior remains. The command obbc in line 22 reduced the number of ternary vectors of the input–output list of phase from 18 to 10.

(b) Figure 11.78 shows both the PRP that computes the same two lists of phases as in the previous task, but based on the local lists of phase of the eight gates, and again the two computed TVLs.

The command isc in the body of the for-loop in lines 49–52 computes the global list of phases; their number of ternary vectors could be reduced from 24 to 18 by the command obbc in line 53. The command maxk in line 54 excludes again the variables $g_1, \ldots, g_6$ (determined as VT 9) from the global list of phases so that the input–output-behavior remains. The command obbc in line 55 could reduce the number of ternary vectors of the input–output list of phase also from 18 to 10.

(c) Figure 11.79 shows the PRP that computes three lists of phases together with these computed TVLs.

The two TVLs in D-form determine the structure of the combinational circuit of Fig. 11.66 more compactly but avoid the explicit specification of variables g_i of the internal gates. Each ternary vector implicitly describes an AND-gate, and values 0 in these vectors determine NOT-gates.

The solution of an equations $y_i = f_i(\mathbf{x})$ is the separate input–output lists of phases of the output y_i. The command csd in line 19 solves the equation $y_1 = f_1(\mathbf{x})$ but requires an orthogonal TVL of the function $f_1(\mathbf{x})$ that is computed as before by the command orth in line 18.

```
1    new                    40   tin  1  8             49   for  $i  1  8
2    space  32  1           41   g5  g6  y2.           50   (
3    avar  1                42   000                   51   isc  10  $i  10
4    x1  x2  x3  x4  x5      43   011                   52   )
5    g1  g2  g3             44   1-1.                  53   obbc  10  11
6    g4  g5  g6             45   vtin  1  9             54   maxk  11  9  12
7    y1  y2.                46   g1  g2  g3             55   obbc  12  13
8    tin  1  1              47   g4  g5  g6.           56   assign  11  /m  1  1
9    x1  g1.                48   ctin  1  10  /1       57   assign  13  /m  1  2
10   01
11   10.
12   tin  1  2
13   x2  g2.
14   01
15   10.
16   tin  1  3
17   x5  g3.
18   01
19   10.
20   tin  1  4
21   g1  x2  g4.
22   0-0
23   100
24   111.
25   tin  1  5
26   g2  x3  g5.
27   0-0
28   100
29   111.
30   tin  1  6
31   x4  g3  g6.
32   0-0
33   100
34   111.
35   tin  1  7
36   g4  g5  y1.
37   000
38   011
39   1-1.
```

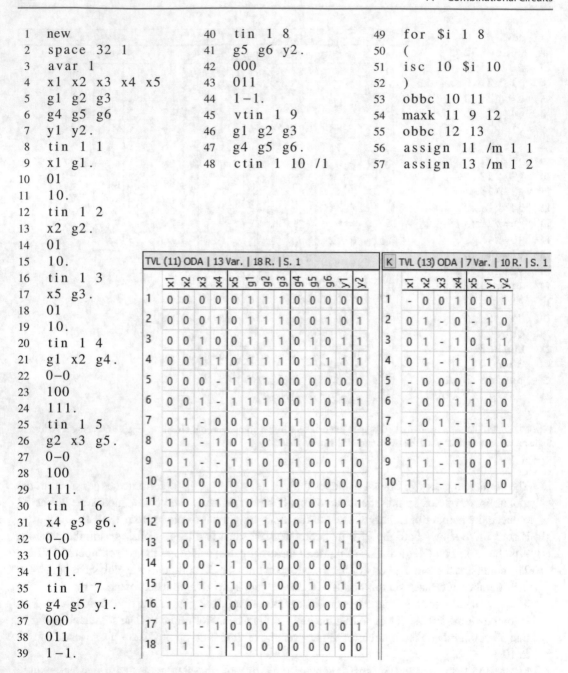

Fig. 11.78 Problem-program that computes the behavior of the combinational circuit of Fig. 11.66 using a set of local lists of phases together with the two computed lists of phases

Similarly, the phase list of the input–output-behavior of the second output is computed as solution of the equation $y_2 = f_2(\mathbf{x})$ using the command orth in line 20 and the command csd in line 21. These two lists of phase are merged using the command isc in line 22. The command obbc in line 23 reduces the number of ternary vectors in the input–output list of phase of both outputs from 16 to 10.

```
 1   new                    10   tin  1  2  /d            19   csd  3  5  6
 2   space  32  1           11   x2  x3  x4  x5 .         20   orth  2  7
 3   avar  1                12   01--                     21   csd  4  7  8
 4   x1  x2  x3  x4  x5     13   --10.                    22   isc  6  8  9
 5   y1  y2 .               14   sbe  1  3                23   obbc  9  10
 6   tin  1  1  /d          15   y1=1.                    24   assign  6  /m  1  1
 7   x1  x2  x3 .           16   sbe  1  4                25   assign  8  /m  1  2
 8   01-                    17   y2=1.                    26   assign  10  /m  1  3
 9   -01.                   18   orth  1  5
```

K	TVL (6) ODA	4 Var.	4 R.	S. 1
	x1	x2	x3	y1
1	0	1	-	1
2	-	0	1	1
3	1	1	-	0
4	-	0	0	0

K	TVL (8) ODA	5 Var.	7 R.	S. 1	
	x2	x3	x4	x5	y2
1	0	1	0	0	1
2	0	1	-	1	1
3	-	-	1	0	1
4	1	1	0	0	0
5	-	0	0	0	0
6	1	1	-	1	0
7	-	0	-	1	0

K	TVL (10) ODA	7 Var.	10 R.	S. 1			
	x1	x2	x3	x4	x5	y1	y2
1	0	1	-	1	0	1	1
2	0	1	-	0	0	1	0
3	0	1	-	-	1	1	0
4	1	1	-	1	0	0	1
5	1	1	-	0	0	0	0
6	1	1	-	-	1	0	0
7	-	0	0	1	0	0	1
8	-	0	0	0	0	0	0
9	-	0	0	-	1	0	0
10	-	0	1	-	-	1	1

Fig. 11.79 Problem-program that computes the behavior of the combinational circuit of Fig. 11.66 using two TVLs in D-form together with two lists of phases that describe the behavior of the two outputs separately and the combined list of phases

(d) Figure 11.80 shows the PRP that computes the input–output list of phases based on the given structure-TVL of an equivalent PLA and the TVLs of the structural and behavioral description of the combinational circuit of Fig. 11.66 together with the requested results.

The structure-TVL of a PLA is the most compact representation of a combinational circuit that realizes several functions using an AND–OR-structure. However, their transformation into the behavioral description of the input–output list of phases requires that the ternary vectors must be separately evaluated for each output. The two nested for-loops in lines 16–28 solve this task according Eq. (11.49). The inner for-loop iterates over the ternary vectors of structure-TVL of the PLA and computes the function-TVL 10 for the output determined by the outer for-loop.

The command csd in line 26 computes the input–output list of phases for the output y_i determined by the outer for-loop, and the command isc summarizes these two separate list of phases into a common list of phases of both outputs. The command obbc in line 29 reduces finally the number of ternary vectors in the input–output list of phase of both outputs from 16 to 10.

```
 1  new                    22  isc 11 $i 12        27  isc 20 21 20
 2  space 32 1             23  maxk 12 4 13         28  )
 3  tin 1 1 /d             24  uni 10 13 10         29  obbc 20 22
 4  x1 x2 x3 x4 x5         25  )                    30  assign 1 /m 1 1
 5  y1 y2 .                26  csd 10 $i 21         31  assign 22 /m 1 2
 6  01---10
 7  -01--11
 8  ---1001.
 9  sbe 1 2
10  y1=1.
11  sbe 1 3
12  y2=1.
13  vtin 1 4
14  y1 y2 .
15  ctin 1 20 /1
16  for $i 2 3
17  (
18  ctin 1 10 /0
19  for $j 1 (ntv 1)
20  (
21  stv 1 $j 11
```

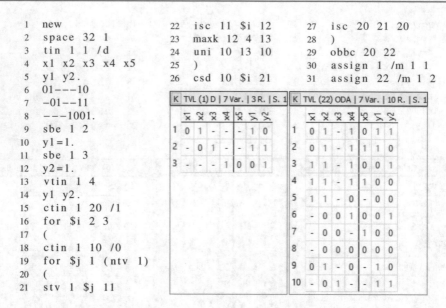

TVL (1) D | 7 Var. | 3 R. | S. 1

K	x_1	x_2	x_3	x_4	x_5	y_1	y_2
1	0	1	-	-	-	1	0
2	-	0	1	-	-	1	1
3	-	-	-	1	0	0	1

TVL (22) ODA | 7 Var. | 10 R. | S. 1

K	x_1	x_2	x_3	x_4	x_5	y_1	y_2
1	0	1	-	1	0	1	1
2	0	1	-	1	1	1	0
3	1	1	-	1	0	0	1
4	1	1	-	1	1	0	0
5	1	1	-	0	-	0	0
6	-	0	0	1	0	0	1
7	-	0	0	-	1	0	0
8	-	0	0	0	0	0	0
9	0	1	-	0	-	1	0
10	-	0	1	-	-	1	1

Fig. 11.80 Problem-program that computes the behavior of the combinational circuit of Fig. 11.66 using a structure-TVL of an equivalent PLA together with the given structure-TVL and the computed list of phases

Solution 11.4 (Partial Behavior Filtered out of Lists of Phases)

(a) Figure 11.81 shows both the PRP that computes the minimized global and input–output list of phases of the given combinational circuit and selects a partial behavior determined by given input-values and the computed results.

The command sbe in lines 10–28 computes the global list of phases that represents the 1024 phases with 432 ternary vectors; the command obbc in line 29 reduces this number of ternary vectors to 225 and stores the result as XBOOLE-object 1.

The commands maxk and obbc in lines 34 and 35 compute the minimized input–output list of phases using VT 3; the result is stored as XBOOLE-object 2. This object expresses the 1024 phases with 68 ternary vectors. The command sts in line 36 stores the computed behavioral descriptions according to the requested protocol; this command finishes the general part of the behavioral analysis.

The two lists of phases can be displayed in an arbitrary viewport of the XBOOLE-monitor XBM 2. It can be cumbersome to find some wanted phases within such slightly larger lists. Commands of the XBOOLE-monitor XBM 2 can be used to filter these information out of the given lists of phases.

The ternary vector of TVL 10 determines four input-patterns for which the global behavior is requested. The command isc in line 41 filters the associated four phases out of the global list of phases, and the subsequent command assigns the result to viewport (1,1) of the m-fold view.

The commands maxk and obbc in lines 43 and 44 exclude the internal variables g_i and minimize the result to a single ternary vector (shown in viewport (2,1) of the m-fold view). This ternary vector emphasizes that the values of x_5 and x_6 have for the given values of the other inputs no influence on the values of the outputs.

(b) Figure 11.82 shows both the PRP that loads the behavioral description of the combinational circuit of Fig. 11.68 computed in Exercise 11.4 (a) using the command lds in line 1 and selects a partial behavior determined by the output-values $(y_1, y_2, y_3, y_4, y_5) = (11011)$ and the computed results.

1	new	16	g6=x8#x9 ,	31	g1 g2 g3 g4 g5
2	space 32 1	17	g7 =/(x9+x10) ,	32	g6 g7 g8 g9 g10
3	avar 1	18	g8=g1+g7+g2 ,	33	g11 g12 g13 .
4	x1 x2 x3 x4 x5	19	g9 =/(g2&g3) ,	34	maxk 1 3 5
5	x6 x7 x8 x9 x10	20	g10=g3+g4 ,	35	obbc 5 2
6	g1 g2 g3 g4 g5	21	g11=g4#g5 ,	36	sts "e11_4_a"
7	g6 g7 g8 g9 g10	22	g12 =/(g5+g6) ,	37	tin 1 10
8	g11 g12 g13	23	g13=g6&g2&g7 ,	38	x1 x2 x3 x4 x5
9	y1 y2 y3 y4 y5 .	24	y1=g8#g9 ,	39	x6 x7 x8 x9 x10 .
10	sbe 1 4	25	y2=g9&g10 ,	40	1111--0000 .
11	g1=x1&x2 ,	26	y3 =/(g10+g11) ,	41	isc 1 10 11
12	g2=x2#x3 ,	27	y4 =/(g11&g12) ,	42	assign 11 /m 1 1
13	g3=x3+x4+x5 ,	28	y5=g12#g13 .	43	maxk 11 3 12
14	g4=x5#x6 ,	29	obbc 4 1	44	obbc 12 13
15	g5 =/(x6&x7&x8) ,	30	vtin 1 3	45	assign 13 /m 2 1

TVL (11) ODA | 28 Var. | 4R. | S. 1

	x_1	x_2	x_3	x_4	x_5	x_6	x_7	x_8	x_9	x_{10}	g_1	g_2	g_3	g_4	g_5	g_6	g_7	g_8	g_9	g_{10}	g_{11}	g_{12}	g_{13}	y_1	y_2	y_3	y_4	y_5
1	1	1	1	1	0	1	0	0	0	0	1	0	1	1	1	0	1	1	1	1	0	0	0	0	1	0	1	0
2	1	1	1	1	1	0	0	0	0	0	1	0	1	1	1	0	1	1	1	1	0	0	0	0	1	0	1	0
3	1	1	1	1	1	1	0	0	0	0	1	0	1	0	1	0	1	1	1	1	1	0	0	0	1	0	1	0
4	1	1	1	1	0	0	0	0	0	0	1	0	1	0	1	0	1	1	1	1	1	0	0	0	1	0	1	0

TVL (13) ODA | 15 Var. | 1R. | S. 1

	x_1	x_2	x_3	x_4	x_5	x_6	x_7	x_8	x_9	x_{10}	y_1	y_2	y_3	y_4	y_5
1	1	1	1	1	1	-	-	0	0	0	0	1	0	1	0

Fig. 11.81 Problem-program that computes the behavior of the combinational circuit of Fig. 11.68 and selects a partial behavior determined by given input-values together with the two computed partial lists of phases

The filtering procedure is the same as in the previous task; only the specification of TVL 10 has been changed. There are 12 global phases represented by two ternary vectors for which the output-values are equal to the specification of TVL 10. The filtered global phases confirm that all internal variables have a constant value for the output-values $\mathbf{y} = (11011)$. Both the filtered global phases and input–output-phases confirm that the values of the inputs x_4 and x_{10} have no influence on the explored output-pattern.

(c) Figure 11.83 shows both the PRP that also reuses the behavioral description of the combinational circuit of Fig. 11.68 computed in Exercise 11.4 (a) and selects partial behavior determined by the values $g_2 = 1$, $g_3 = 0$, $g_{12} = 0$, and $g_{13} = 1$ of internal variables and the computed results.

Using the same filtering procedure as before, we get eight global phases represented by six ternary vectors for which the internal values are equal to the specification of TVL 10. Two different output-patterns occur for the specified internal values. The filtered six ternary vectors of global phases have been minimized to two ternary vectors of eight input–output-phases. It can be seen

```
1  lds "e11_4_a"          4  11011.          7  maxk 11 3 12
2  tin 1 10               5  isc 10 1 11     8  obbc 12 13
3  y1 y2 y3 y4 y5.        6  assign 11 /m 1 1  9  assign 13 /m 2 1
```

TVL (11) ODA | 28 Var. | 2 R. | S. 1

	x1	x2	x3	x4	x5	x6	x7	x8	x9	x10	g1	g2	g3	g4	g5	g6	g7	g8	g9	g10	g11	g12	g13	y1	y2	y3	y4	y5
1	0	1	1	-	1	1	1	1	1	-	0	0	1	0	0	0	0	0	1	1	0	1	0	1	1	0	1	1
2	-	0	0	-	1	1	1	1	1	-	0	0	1	0	0	0	0	0	1	1	0	1	0	1	1	0	1	1

TVL (13) ODA | 15 Var. | 2 R. | S. 1

	x1	x2	x3	x4	x5	x6	x7	x8	x9	x10	y1	y2	y3	y4	y5
1	0	1	1	-	1	1	1	1	1	-	1	1	0	1	1
2	-	0	0	-	1	1	1	1	1	-	1	1	0	1	1

Fig. 11.82 Problem-program that selects a partial behavior determined by given output-values of the combinational circuit of Fig. 11.68 together with the two computed partial lists of phases

```
1  lds "e11_4_a"          4  1001.           7  maxk 11 3 12
2  tin 1 10               5  isc 10 1 11     8  obbc 12 13
3  g2 g3 g12 g13.         6  assign 11 /m 1 1  9  assign 13 /m 2 1
```

TVL (11) ODA | 28 Var. | 6 R. | S. 1

	x1	x2	x3	x4	x5	x6	x7	x8	x9	x10	g1	g2	g3	g4	g5	g6	g7	g8	g9	g10	g11	g12	g13	y1	y2	y3	y4	y5
1	0	1	0	0	0	1	0	1	0	0	0	1	0	1	1	1	1	1	1	1	0	0	1	0	1	0	1	1
2	1	1	0	0	0	1	0	1	0	0	1	1	0	1	1	1	1	1	1	1	0	0	1	0	1	0	1	1
3	0	1	0	0	0	1	1	1	0	0	0	1	0	1	0	1	1	1	1	1	1	0	1	0	1	0	1	1
4	1	1	0	0	0	1	1	1	0	0	1	1	0	1	0	1	1	1	1	1	1	0	1	0	1	0	1	1
5	0	1	0	0	0	0	-	1	0	0	0	1	0	0	1	1	1	1	1	0	1	0	1	0	0	0	1	1
6	1	1	0	0	0	0	-	1	0	0	1	1	0	0	1	1	1	1	1	0	1	0	1	0	0	0	1	1

TVL (13) ODA | 15 Var. | 2 R. | S. 1

	x1	x2	x3	x4	x5	x6	x7	x8	x9	x10	y1	y2	y3	y4	y5
1	-	1	0	0	0	0	-	1	0	0	0	0	0	1	1
2	-	1	0	0	0	1	-	1	0	0	0	1	0	1	1

Fig. 11.83 Problem-program that selects a partial behavior determined by the values of internal variables g_2, g_3, g_{12}, and g_{13} of the combinational circuit of Fig. 11.68 together with the two computed partial lists of phases

that the specified internal values occur only for constant values of $x_2 = x_8 = 1$ and $x_3 = x_4 = x_5 = x_9 = x_{10} = 0$.

```
1    lds  "e11_4_a"
2    vtin 1 10
3    x1 x2 x3 x4 x5
4    x6 x7 x8 x9 x10.
5    maxk 2 10 11
6    obbc 11 12
7    assign 12 /m 1 1
8    cpl 12 13
9    obbc 13 14
10   assign 14 /m 1 2
```

K	TVL (12) ODA	5 Var.	3 R.	S. 1
	y_1 y_2 y_3 y_4	y_5		
1	0 1 0 0	1		
2	1 - 0 0	1		
3	- - 0 1	-		

K	TVL (14) ODA	5 Var.	3 R.	S. 1
	y_1 y_2 y_3 y_4	y_5		
1	0 0 0 0	1		
2	- - 0 0	0		
3	- - 1 -	-		

Fig. 11.84 Problem-program that computes the occurring output-patterns and their complement of the combinational circuit of Fig. 11.68 together with these two sets of patterns

(d) Figure 11.84 shows both the PRP that filters the occurring output-patterns and their complement using the input–output list of phases of the combinational circuit of Fig. 11.68 and these two sets of patterns.

Here we do not select rows but columns from the input–output list of phases. The command maxk in line 5 excludes the columns of the input-variables from the input–output list of phases, and the command obbc transforms the computed 11 binary vectors into three ternary vectors shown in viewport (1,1) of the m-fold view. A noteworthy observation of this set of possible output-patterns is that the output y_3 is constant equal to 0; this information can be used to simplify the circuit structure by removing the NOR-gate of this output.

The computed complement of the possible output-pattern confirms this observation. The set of not occurring output-patterns (TVL 14 in viewport (1,2) of the m-fold view) shows additionally that at least one of the outputs (y_1, y_2, y_4) or (y_4, y_5) must be equal to 1.

Solution 11.5 (Simple Covering Method to Synthesize a Combinational Circuit)

(a) Figure 11.85 shows the PRP that verifies whether the equation $F(\mathbf{x}, \mathbf{y}) = 1$ of the given system-function $F(\mathbf{x}, \mathbf{y})$ is solvable with regard to the two output-variables y_1 and y_2. If $F(\mathbf{x}, \mathbf{y})$ satisfies this property, this PRP verifies furthermore whether these functions of are completely or incompletely specified and computes either the completely specified function or the lattice from which a function can be chosen. Figure 11.86 shows the computed Karnaugh-map of the evaluated system-function $F(\mathbf{x}, \mathbf{y})$, the Karnaugh-maps that describe the permitted behavior separately for each output, and the results of two-level evaluation.

The command sbe in lines 6–21 solves the system-equation $F(\mathbf{x}, \mathbf{y}) = 1$. The four subsequent commands minimize the representation of the computed TVL and prepare the data needed for their evaluation. The command assign in line 29 shows this function in the 1-fold view where the visualization as a Karnaugh-map has been chosen and the variables have been associated to the two boarders such that it can easily compared with the Karnaugh-maps of the two outputs (see the help system of the XBOOLE-monitor XBM 2 for such changes).

The commands maxk and te_cpl in lines 31 and 32 utilize Eq. (7.56) to verify whether the given system-function can be realized by a combinational circuit. The result \$complete=true indicates that the given system-function satisfies this condition; for each input-pattern exist is at least one output-pattern permitted.

The commands maxk, derk, and te_cpl in lines 35–37 utilize Eq. (7.59) to verify whether the function $y_1 = f_1(\mathbf{x})$ is uniquely determined by the given system-function; the result \$csy1=true (completely specified function $y_1 = f_1(\mathbf{x})$) confirms this property. Consequently,

```
 1   new                                    42   cpl 100 101 ; /y1
 2   space 32 1                             43   )
 3   avar 1                                 44   if (not (te_cpl 7))
 4   x1 x2 x3 x4 x5 x6                      45   (
 5   y1 y2.                                 46   set $csy1 false
 6   sbe 1 1                                47   cpl 2 11
 7   /x1&x2&/x4&/x5&/x6&/y1&/y2+            48   isc 1 11 12
 8   (/x1&/x2&x3&/x4&x6+/x1&x2&/x6+         49   maxk 12 4 13
 9   /x2&x3&x4&/x6)&/y1&y2+                 50   cpl 13 100 ; y1q
10   (x1&x2&(x5+/x6)+                       51   isc 1 2 14
11   x1&/x2&(x3&x6+/x4)+                    52   maxk 14 4 15
12   /x3&/x4&(x1+x6)+                       53   cpl 15 101 ; y1r
13   /x2&/x6&(/x3&x4+/x4&/x5)+              54   )
14   x2&x4&x6)&y1&/y2+                      55   uni 100 101 102
15   (/x1&/x2&(/x3+/x4&/x6)+                56   cpl 102 102 ; y1phi
16   x1&/x2&x4&(/x3+x6)+                    57   assign_qr 100 101 /m 1 1 ; y1
17   x2&(x1&x3&x4+/x4&x6)+                  58   maxk 1 2 8
18   x1&/x5&x6&(x4+x3)+                     59   derk 8 3 9
19   x1&/x3&/x4&/x6+                        60   if (te_cpl 9)
20   x4&x6)&y1&y2                           61   (
21   =1.                                    62   set $csy2 true
22   obbc 1 1                               63   isc 1 3 20
23   sbe 1 2                                64   maxk 20 4 200 ; y2
24   y1=1.                                  65   cpl 200 201 ; /y2
25   sbe 1 3                                66   )
26   y2=1.                                  67   if (not (te_cpl 9))
27   vtin 1 4                               68   (
28   y1 y2.                                 69   set $csy2 false
29   assign 1 /1                            70   cpl 3 21
30   set $complete false                    71   isc 1 21 22
31   maxk 1 4 5                             72   maxk 22 4 23
32   if (te_cpl 5)                          73   cpl 23 200 ; y2q
33   (                                      74   isc 1 3 24
34   set $complete true                     75   maxk 24 4 25
35   maxk 1 3 6                             76   cpl 25 201 ; y2r
36   derk 6 2 7                             77   )
37   if (te_cpl 7)                          78   uni 200 201 202
38   (                                      79   cpl 202 202 ; y2phi
39   set $csy1 true                         80   assign_qr 200 201 /m 2 1 ; y2
40   isc 1 2 10                             81   )
41   maxk 10 4 100 ; y1                     82   sts "e11_5_a"
```

Fig. 11.85 Problem-program that verifies whether system-function of Exercise 11.5 can be realized by a combinational circuit and computes the completely specified function or the lattice of permitted functions for the two outputs y_1 and y_2

the commands isc and maxk in lines 40 and 41 are used to compute the function $y_1 = f_1(\mathbf{x}) = f_{1q}(\mathbf{x})$ (stored as requested as TVL 100) based on Eq. (7.60). The command cpl in line 42 computes the associated function $f_{1r}(\mathbf{x})$ as TVL 101.

The commands in lines 44–54 would compute the mark-functions of a lattice for the output y_1; however, these commands will not be executed because the function $y_1 = f_1(\mathbf{x})$ is completely specified so that the if-condition in line 44 has the value false.

The function $f_{1\varphi}(\mathbf{x})$ is determined by the mark-functions $f_{1q}(\mathbf{x})$ (TVL 100) and $f_{1r}(\mathbf{x})$ (TVL 101) so that the commands uni and cpl in lines 55 and 56 can be used to compute $f_{1\varphi}(\mathbf{x})$ (TVL 102) for both a completely or an incompletely specified function for the output y_1. The command assign_qr in line 57 facilitates the visualization of the Karnaugh-map for the output y_1 for the two possible results; this Karnaugh-map (see Fig. 11.86) contains only values 0 and 1 (no ϕ) and specifies therefore the computed completely specified function $y_1 = f_1(\mathbf{x})$.

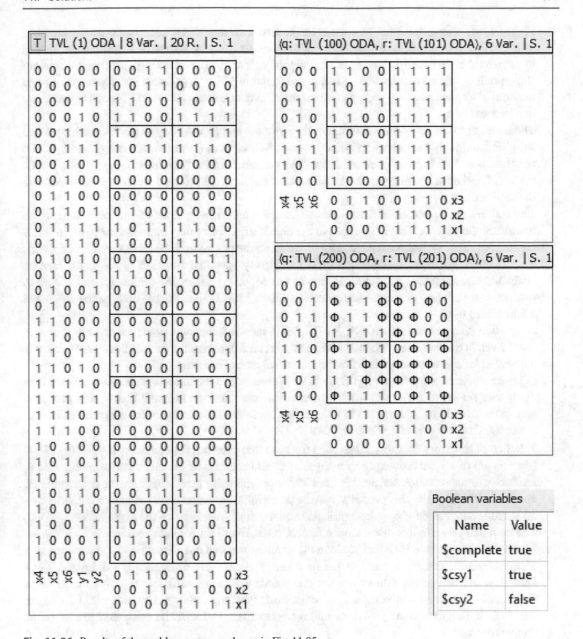

Fig. 11.86 Results of the problem-program shown in Fig. 11.85

The commands in lines 58–80 repeat the program part in lines 35–57 adjusted to the output y_2. The result $csy2=false$ (the function $y_2 = f_2(\mathbf{x})$ is not completely specified) confirms that this function can be chosen out of a lattice of functions. The commands in lines 70–76 compute the mark-functions $f_{2q}(\mathbf{x})$ and $f_{2r}(\mathbf{x})$ based on Eqs. (7.57) and (7.58). The computed lattice of functions that can be used for the output y_2 is shown in viewport (2,1) of the m–fold view; the 32 elements ϕ of this Karnaugh-map indicate that $2^{32} = 4.294.967.296$ functions are usable for the output y_2.

The command sts in line 82 stores the final XBOOLE-system for later use.

(b) Figure 11.87 shows the PRP that loads in the first line the XBOOLE-system of the previous task and excludes variables either from the function or from the lattice of functions not needed to express the permitted behavior at the outputs y_1 and y_2; the simplified Karnaugh-maps are shown below this PRP. It can be seen that the function $y_1 = f_1(\mathbf{x})$ does not depend on x_5 and the lattice of functions $\mathcal{L}\langle f_{2q}(\mathbf{x}), f_{2r}(\mathbf{x})\rangle$ of the output y_2 contains $2^8 = 256$ functions which are independent of x_2.

The index $\$i$ of the for-loop determines the output y_i so that the body of this loop analyzes and simplifies uniquely the behavioral descriptions of all outputs (here the outputs y_1 and y_2). The commands set in lines 4–8 determine the object numbers of the functions $f_{iq}(\mathbf{x})$, $f_{ir}(\mathbf{x})$, $f_{i\varphi}(\mathbf{x})$, the set of variables occurring in the specification of y_i, and the selected variable used to evaluate the dependency.

The commands sv_next in lines 13 and 27 require an unchanged set of variables from which iteratively the next variable is assigned to the object $\$selvar$; the command sv_get in line 9 creates this set of variables as object $\$svyi$ based on the not minimized set of variables of $f_{iq}(\mathbf{x})$. The function $f_{i\varphi}(\mathbf{x}) = 0$ indicates that a completely specified function $y_i = f_i(\mathbf{x})$ must be evaluated; the command te in line 11 checks this property; hence, the evaluation with regard to independent variables of a completely specified function is realized in the body of the command if in lines 11–20.

A function does not depend on x_i if Eq. (11.56) is satisfied; the commands if and te_derk in line 15 verify this property, and the conditional executed commands maxk in lines 17–19 exclude a detected independent variable from the associated mark-functions.

A lattice of functions $y_i = \mathcal{L}\langle f_{iq}(\mathbf{x}), f_{ir}(\mathbf{x})\rangle$ is given when the associated mark-function of the don't-care set satisfies $f_{i\varphi}(\mathbf{x}) \neq 0$. The commands not and te in line 22 check this property; hence, the evaluation with regard to independent variables of a lattice of functions is realized in the body of the command if in lines 22–37.

A lattice of functions contains at least one function that does not depend on x_i if $f_q^{der_{x_i}}(\mathbf{x})$ (7.24) (the ON-set of all derivatives with regard to x_i of all functions of the given lattice) is equal to 0; the two commands maxk and isc in lines 29–31 compute this mark-function, the commands if and te in line 32 verify this property, and the conditional executed commands maxk in lines 34–36 exclude a detected independent variable from the mark-functions of the given lattice.

The command sts in line 40 stores the final XBOOLE-system for later use.

(c) Figure 11.88 shows the PRP that loads in the first line the XBOOLE-system of the previous task and expands the ternary vectors of the functions $f_{iq}(\mathbf{x})$ of the orthogonally minimized TVLs such that sets of prime conjunctions are computed, which are sufficient to cover functions of the outputs y_1 and y_2; on the right-hand side of this PRP, the orthogonally minimized TVLs are shown in the upper row of viewports and thereunder the TVLs of the computed sets of prime conjunctions.

A conjunction $C(x_i, \mathbf{x}_0)$ that satisfies $C(x_i, \mathbf{x}_0) \wedge f(\mathbf{x}) = C(x_i, \mathbf{x}_0)$ (i.e., this conjunction is part of a disjunctive form of this function) can be replaced by the conjunction $C(\mathbf{x}_0)$ if $C(x_i, \mathbf{x}_0)$ holds

$$\max_{x_i} C(x_i, \mathbf{x}_0) \wedge \overline{f(\mathbf{x})} = 0 \,. \tag{11.273}$$

Patterns with $f_\varphi(\mathbf{x}) = 1$ can additionally be used as part of such an expanded conjunction in the case of a lattice of functions; hence, a conjunction $C(x_i, \mathbf{x}_0)$ that satisfies $C(x_i, \mathbf{x}_0) \wedge f_q(\mathbf{x}) = C(x_i, \mathbf{x}_0)$ (i.e., this conjunction is part of a disjunctive form of this mark-function $f_q(\mathbf{x})$ of a lattice of functions) can be replaced by the conjunction $C(\mathbf{x}_0)$ if $C(x_i, \mathbf{x}_0)$ holds

$$\max_{x_i} C(x_i, \mathbf{x}_0) \wedge f_r(\mathbf{x}) = 0 \,. \tag{11.274}$$

```
1   lds "e11_5_a"
2   for $i 1 2
3   (
4   set $yiq (mul $i 100)
5   set $yir (add 1 (mul $i 100))
6   set $yiphi (add 2 (mul $i 100))
7   set $svyi (add 10 (mul $i 100))
8   set $selvar (add 11 (mul $i 100))
9   sv_get $yiq $svyi
10  ; yi=fi(x) completly specified function
11  if (te $yiphi)
12  (
13  while (sv_next $svyi $selvar $selvar)
14  (
15  if (te_derk $yiq $selvar)
16  (
17  maxk $yiq $selvar $yiq
18  maxk $yir $selvar $yir
19  maxk $yiphi $selvar $yiphi
20  )))
21  ; yi=fi(x) incompletly specified function
22  if (not (te $yiphi))
23  (
24  set $maxkq (add 12 (mul $i 100))
25  set $maxkr (add 13 (mul $i 100))
26  set $derklq (add 14 (mul $i 100))
27  while (sv_next $svyi $selvar $selvar)
28  (
29  maxk $yiq $selvar $maxkq
30  maxk $yir $selvar $maxkr
31  isc $maxkq $maxkr $derklq ; q of derivative of the lattice
32  if (te $derklq)
33  (
34  maxk $yiq $selvar $yiq
35  maxk $yir $selvar $yir
36  maxk $yiphi $selvar $yiphi
37  )))
38  assign_qr $yiq $yir /m 1 $i
39  )
40  sts "e11_5_b"
```

(q: TVL (100) ODA, r: TVL (101) ODA), 5 Var. | S. 1	(q: TVL (200) ODA, r: TVL (201) ODA), 5 Var. | S. 1

```
0 0   1 1 0 0 | 1 1 1 1           0 0   Φ 1 1 Φ | 0 1 0 Φ
0 1   1 0 1 1 | 1 1 1 1           0 1   Φ 1 1 1 | 1 Φ 1 0
1 1   1 1 1 1 | 1 1 1 1           1 1   Φ 1 1 1 | 0 Φ 1 0
1 0   1 0 0 0 | 1 1 0 1           1 0   1 1 1 1 | 0 1 0 Φ

x4 x6   0 1 1 0 0 1 1 0 x3         x5 x6   0 1 1 0 0 1 1 0 x4
        0 0 1 1 1 1 0 0 x2                 0 0 1 1 1 1 0 0 x3
        0 0 0 0 1 1 1 1 x1                 0 0 0 0 1 1 1 1 x1
```

Fig. 11.87 Problem-program that excludes variables from the function or the lattice of functions not needed to realize the permitted behavior at the outputs y_1 and y_2 together with the two computed Karnaugh-maps

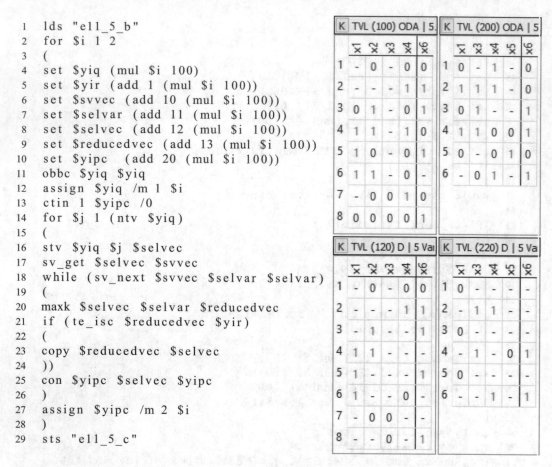

```
1    lds "ell_5_b"
2    for $i 1 2
3    (
4    set $yiq (mul $i 100)
5    set $yir (add 1 (mul $i 100))
6    set $svvec (add 10 (mul $i 100))
7    set $selvar (add 11 (mul $i 100))
8    set $selvec (add 12 (mul $i 100))
9    set $reducedvec (add 13 (mul $i 100))
10   set $yipc  (add 20 (mul $i 100))
11   obbc $yiq $yiq
12   assign $yiq /m 1 $i
13   ctin 1 $yipc /0
14   for $j 1 (ntv $yiq)
15   (
16   stv $yiq $j $selvec
17   sv_get $selvec $svvec
18   while (sv_next $svvec $selvar $selvar)
19   (
20   maxk $selvec $selvar $reducedvec
21   if (te_isc $reducedvec $yir)
22   (
23   copy $reducedvec $selvec
24   ))
25   con $yipc $selvec $yipc
26   )
27   assign $yipc /m 2 $i
28   )
29   sts "ell_5_c"
```

K	TVL (100) ODA | 5				
	x_1	x_2	x_3	x_4	x_6
1	-	0	-	0	0
2	-	-	-	1	1
3	0	1	-	0	1
4	1	1	-	1	0
5	1	0	-	0	1
6	1	1	-	0	-
7	-	0	0	1	0
8	0	0	0	0	1

K	TVL (200) ODA | 5				
	x_1	x_3	x_4	x_5	x_6
1	0	-	1	-	0
2	1	1	1	-	0
3	0	1	-	-	1
4	1	1	0	0	1
5	0	-	0	1	0
6	-	0	1	-	1

K	TVL (120) D | 5 Var				
	x_1	x_2	x_3	x_4	x_6
1	-	0	-	0	0
2	-	-	-	1	1
3	-	1	-	-	1
4	1	1	-	-	-
5	1	-	-	-	1
6	1	-	-	0	-
7	-	0	0	-	-
8	-	-	0	-	1

K	TVL (220) D | 5 Va				
	x_1	x_3	x_4	x_5	x_6
1	0	-	-	-	-
2	-	1	1	-	-
3	0	-	-	-	-
4	-	1	-	0	1
5	0	-	-	-	-
6	-	-	1	-	1

Fig. 11.88 Problem-program that computes two sufficient sets of prime conjunctions that cover the functions of the outputs y_1 and y_2 together with the minimized orthogonal TVLs and associated TVLs of prime conjunctions

Determining a completely specified function $f(\mathbf{x})$ as special case of a lattice $\mathcal{L}\{f_q(\mathbf{x}) = f(\mathbf{x})$, $f_r(\mathbf{x}) = \overline{f(\mathbf{x})})\}$ with $f_\varphi(\mathbf{x}) = 0$ facilitates the unique use of (11.274) for both completely and incompletely specified functions. The check of (11.274) for all variables and the associated substitution of $C(x_i, \mathbf{x}_0)$ by $C(\mathbf{x}_0)$ generates a prime conjunction for each conjunction belonging to $f_{iq}(\mathbf{x})$, $i = 1, 2$. These expansions to prime conjunctions are realized in the body of the outer for-loop in lines 4–27 for the outputs y_1 and y_2.

The commands set in lines 4–10 determine the object numbers of the functions $f_{iq}(\mathbf{x})$, $f_{ir}(\mathbf{x})$, the set of variables occurring in the specification of y_i, the selected variable, the selected ternary vector, the reduced ternary vector that describes finally a prime conjunction, and the sufficient set of prime conjunctions to realize the function of the output y_i.

Each disjunctive form of the mark-function $f_{iq}(\mathbf{x})$ can be used as source of the expansion to a sufficient set of prime conjunctions. The command obbc in line 11 minimizes the given mark-function $f_{iq}(\mathbf{x})$; a comparison with the results (TVLs 100 and 200) of the previous task shows that the number of ternary vectors has been reduced in case of:

- $f_{1q}(\mathbf{x})$ from 11 to eight; and
- $f_{2q}(\mathbf{x})$ from 12 to six,

so that less ternary vectors must be expanded to prime conjunctions. The subsequent command `assign` shows the orthogonally minimized TVL in the viewport of the first row of the m-fold view.

The inner `for`-loop in lines 14–26 iterates over all ternary vectors of the mark-function $f_{iq}(\mathbf{x})$. The command `stv` in line 16 selects the ternary vector j as object `$selvec` to be transformed into a prime conjunction.

The command `sv_next` in line 18 requires an unchanged set of variables from which iteratively the next variable is assigned to the object `$selvar`; the command `sv_get` in line 17 creates this set of variables as object `$svvec` based on the variables belonging to the object `$selvec`. The commands `maxk` and `te_isc` in lines 20 and 21 realize Condition (11.274); if this condition is satisfied, the reduced vector replaces in line 23 the object `$selvec` for possible further expansions.

The command `con` in line 25 appends the created prime conjunction to the set of prime conjunctions of the output y_i that have been initialized as empty TVL in line 13.

The computed sufficient sets of prime conjunctions to realize the functions of the outputs y_1 and y_2 are assigned to the second row of the m-fold view using the command `assign` in line 27. The command `sts` in line 29 stores the final XBOOLE-system for later use.

(d) Figure 11.89 shows the PRP that loads in the first line the XBOOLE-system of the previous task and removes redundant prime conjunctions with the result of minimal disjunctive forms that cover the functions of the outputs y_1 and y_2; on the right-hand side of this PRP, the TVLs of the given sufficient sets of prime conjunctions are shown in the upper row of viewports and thereunder TVLs of the computed minimal disjunctive forms.

A prime conjunction is redundant and can therefore be removed if the remaining prime conjunctions completely cover the function $f_{iq}(\mathbf{x})$ that is equal to $f_i(\mathbf{x})$ in case of a completely specified function; hence, the same procedure can be used to compute the wanted minimal disjunctive forms for the outputs y_1 and y_2 in the body of the `for`-loop.

The commands `set` in lines 4–8 determine the object numbers of the functions $f_{iq}(\mathbf{x})$, $f_{ir}(\mathbf{x})$, the TVL in which one ternary vector has been removed, the sufficient set of prime conjunctions to realize the function of the output y_i, and the TVL of the minimal disjunctive form of the output y_i. The command `assign` in line 9 shows the given set of prime conjunctions belonging to the output y_i in the first row of the m-fold view for an easy comparison with the associated minimal disjunctive form that will be shown later thereunder.

The command `copy` in line 10 copies the set of prime conjunctions that is sufficient to cover the function $f_{iq}(\mathbf{x})$ to the TVL `$yipcmin` from which the redundant prime conjunctions will be removed. The commands `dtv` and `te_dif` are the main operations to exclude redundant prime conjunctions from the TVL `$yipcmin`; the command `dtv` in line 14 creates the TVL `$reducedtvl` in which the prime conjunction with the index j of the TVL `$yipcmin` has been deleted, and the command `te_dif` in line 15 verifies whether the reduced set of prime conjunction `$reducedtvl` covers the functions $f_{iq}(\mathbf{x})$. The command `copy` in line 17 replaces the TVL `$yipcmin` by the reduced set of prime conjunction `$reducedtvl` if this condition is satisfied.

The check with regard to redundancy must be executed for all prime conjunctions of the TVL `$yipcmin`. The number of ternary vectors of this TVL is changed when a redundant prime conjunction has been removed. A `for`-loop that iterates over the ternary vectors of the TVL `$yipcmin` cannot be used to determine the index j of the ternary vector to evaluate because the index j can refer to a ternary vector that does not exist due to a removed redundant prime conjunction. This problem has been solved in the PRP of Fig. 11.89 by a `while`-loop in lines 12–20 in which the prime conjunctions are evaluated in reverse order starting with the largest index j.

```
1    lds "e11_5_c"
2    for $i 1 2
3    (
4    set $yiq (mul $i 100)
5    set $yir (add 1 (mul $i 100))
6    set $reducedtvl (add 14 (mul $i 100))
7    set $yipc   (add 20 (mul $i 100))
8    set $yipcmin (add 21 (mul $i 100))
9    assign $yipc /m 1 $i
10   copy $yipc $yipcmin
11   set $j (ntv $yipc)
12   while (ge $j 1)
13   (
14   dtv $yipcmin $j $reducedtvl
15   if (te_dif $yiq $reducedtvl)
16   (
17   copy $reducedtvl $yipcmin
18   )
19   set $j (sub $j 1)
20   )
21   assign $yipcmin /m 2 $i
22   assign_qr $yiq $yir /4 1 $i
23   assign $yipcmin /4 2 $i
24   )
```

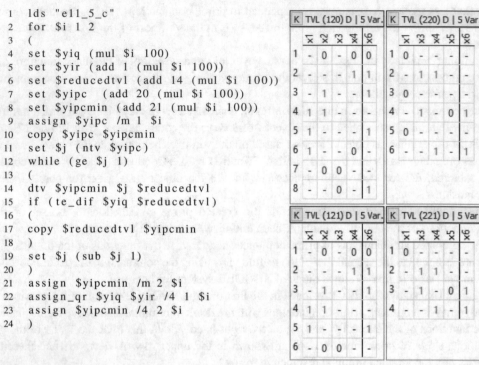

K	TVL (120) D	5 Var.				K	TVL (220) D	5 Var.			
	x1	x2	x3	x4	x6		x1	x3	x4	x5	x6
1	-	0	-	0	0	1	0	-	-	-	-
2	-	-	-	1	1	2	-	1	1	-	-
3	-	1	-	-	1	3	0	-	-	-	-
4	1	1	-	-	-	4	-	1	-	0	1
5	1	-	-	-	1	5	0	-	-	-	-
6	1	-	-	0	-	6	-	-	1	-	1
7	-	0	0	-	-						
8	-	-	0	-	1						

K	TVL (121) D	5 Var.				K	TVL (221) D	5 Var.			
	x1	x2	x3	x4	x6		x1	x3	x4	x5	x6
1	-	0	-	0	0	1	0	-	-	-	-
2	-	-	-	1	1	2	-	1	1	-	-
3	-	1	-	-	1	3	-	1	-	0	1
4	1	1	-	-	-	4	-	-	1	-	1
5	1	-	-	-	1						
6	-	0	0	-	-						

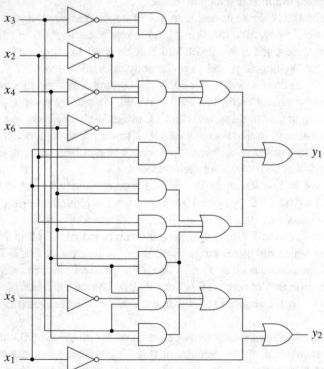

Fig. 11.89 Problem-program that restricts the given two sufficient sets of prime conjunctions to TVLs of minimal disjunctive forms that cover the functions of the outputs y_1 and y_2 together with given and computed TVLs as well as the computed combinational circuit

(q: TVL (100) ODA, r: TVL (101) ODA), 5 Var.	S. 1	(q: TVL (200) ODA, r: TVL (201) ODA), 5 Var.	S. 1

```
0 0  1 1 0 0 1 1 1 1        0 0  Φ 1 1 Φ 0 1 0 Φ
0 1  1 0 1 1 1 1 1 1        0 1  Φ 1 1 1 1 Φ 1 0
1 1  1 1 1 1 1 1 1 1        1 1  Φ 1 1 1 0 Φ 1 0
1 0  1 0 0 0 1 1 0 1        1 0  1 1 1 1 0 1 0 Φ

x4 x6  0 1 1 0 0 1 1 0 x3    x5 x6  0 1 1 0 0 1 1 0 x4
       0 0 1 1 1 1 0 0 x2           0 0 1 1 1 1 0 0 x3
       0 0 0 0 1 1 1 1 x1           0 0 0 0 1 1 1 1 x1
```

T TVL (121) D	5 Var.	6 R.	S. 1	T TVL (221) D	5 Var.	4 R.	S. 1

```
0 0  1 1 0 0 1 1 1 1        0 0  1 1 1 1 0 1 0 0
0 1  1 0 1 1 1 1 1 1        0 1  1 1 1 1 1 1 1 0
1 1  1 1 1 1 1 1 1 1        1 1  1 1 1 1 0 1 1 0
1 0  1 0 0 0 1 1 0 1        1 0  1 1 1 1 0 1 0 0

x4 x6  0 1 1 0 0 1 1 0 x3    x5 x6  0 1 1 0 0 1 1 0 x4
       0 0 1 1 1 1 0 0 x2           0 0 1 1 1 1 0 0 x3
       0 0 0 0 1 1 1 1 x1           0 0 0 0 1 1 1 1 x1
```

Fig. 11.90 Results of the problem-program shown in Fig. 11.89 shown in the 4-fold view

The command `assign` in line 21 shows the computed minimal disjunctive form in the viewport below the associated sufficient set of prime conjunctions. It can be seen that two redundant ternary vectors have been removed for each output y_j and that the prime conjunction $x_4 x_6$ can be used for both y_1 and y_2. These two TVLs have been used to draw the associated combinational circuit shown below the PRP in Fig. 11.89.

The last two commands show the Karnaugh-maps of the used orthogonal mark-functions and the computed minimal disjunctive forms for the outputs y_1 and y_2 (see Fig. 11.90) in the 4-fold view. These Karnaugh-maps confirm that the completely specified function $y_1 = f_1(\mathbf{x})$ is exactly covered by the six prime conjunctions of TVL 121 and out of the lattice of functions shown in viewport (1,2) a permitted function $y_2 = f_2(\mathbf{x})$ covers the mark-function $f_{2q}(\mathbf{x})$ and six of the eight don't-cares.

Solution 11.6 (Synthesis by Separation of a Single Variable or a Function)

(a) Figure 11.91 shows both the PRP that transforms the expression of the given function $f(\mathbf{x})$ into a minimized TVL, copies this TVL as object 11 for the case that no variable can be separated by an AND-gate (line 12), and iteratively verifies whether x_i or $\overline{x}_i$ can be separated from $f(\mathbf{x})$ using an AND-gate as well as the computed results.

The check whether x_i or $\overline{x}_i$ can be separated from $f(x_i, \mathbf{x}_0)$ using an AND-gate is implemented in the body of the `while`-loop in lines 14–43. The Boolean variable $found is used to break this loop when either x_i or $\overline{x}_i$ can be separated from $f(\mathbf{x})$ using an AND-gate.

The commands `isc` and `te_maxk` in lines 21 and 22 realize Condition (11.80) to check whether the non-negated variable x_i can be separated from $f(x_i, \mathbf{x}_0)$ using an AND-gate. In the case that this condition is satisfied, the TVL of the separated variable x_i is stored as object 10 in line 24, the minimized TVL of the decomposition-function $g(\mathbf{x}_0)$ is computed based on (11.81) in lines 25–27, and value of the variable $found is changed to the value `true` to avoid further checks for separations of a variable from the given function $f(\mathbf{x})$.

```
1   new
2   space 32 1
3   avar
4   x1 x2 x3 x4 x5 x6 x7 x8 x9.
5   sbe 1 1
6   /x1&x5&/x9&(/x4+/x6&x7)+
7   x5&/x8+x1&x5&x9&(x2&x3&/x4+/x4&/x6+/x7)+
8   x2&x4&x5&x6&(x1&/x3&/x9+/x1&x3&x7)+
9   /x1&x5&x6&x7&x9&(/x2+/x3)+
10  x4&x5&(/x1&/x6&x7+x1&/x7&x8+x1&/x2&x6&/x9).
11  obbc 1 1
12  copy 1 11 ; g=f
13  set $found false
14  while (and (not $found) (sv_next 1 2 2))
15  (
16  sv_get 2 3 ; xi
17  cpl 3 4 ; /xi
18  ; check the
19  ; separation of
20  ; xi using AND
21  isc 1 4 5
22  if (te_maxk 5 3)
23  (
24  copy 3 10
25  isc 1 3 11
26  maxk 11 3 11
27  obbc 11 11
28  set $found true
29  )
30  ; check the
31  ; separation of
32  ; /xi using AND
33  if (not $found)
34  (
35  isc 1 3 5
36  if (te_maxk 5 3)
37  (
38  copy 4 10
39  isc 1 4 11
40  maxk 11 3 11
41  obbc 11 11
42  set $found true
43  )))
44  assign 1 /m 1 1
45  assign 10 /m 2 1
46  assign 11 /m 1 2
47  sts "e11_6_a"
```

K TVL (1) ODA | 9 Var. | 12 R. | S. 1

K	x1	x2	x3	x4	x5	x6	x7	x8	x9
1	1	0	1	1	1	1	1	1	0
2	0	-	-	1	1	0	1	1	-
3	0	1	0	-	1	1	1	1	1
4	0	0	-	-	1	1	1	1	1
5	1	-	0	1	1	1	1	1	0
6	0	1	1	1	1	1	1	1	-
7	1	-	-	1	1	-	0	1	-
8	1	-	-	0	1	-	0	1	1
9	1	-	-	0	1	0	1	1	1
10	1	1	1	0	1	1	1	1	1
11	-	-	-	-	1	-	-	0	-
12	0	-	-	0	1	-	-	1	0

K TVL (11) ODA | 8 Var. | 12 R. | S. 1

K	x1	x2	x3	x4	x6	x7	x8	x9
1	1	0	-	1	1	1	1	0
2	0	-	-	1	0	1	1	-
3	0	-	0	-	1	1	1	1
4	0	0	1	-	1	1	1	1
5	1	1	0	1	1	1	1	0
6	0	1	1	1	1	1	1	-
7	1	-	-	1	-	0	1	0
8	1	-	-	-	-	0	1	1
9	1	-	-	0	0	1	1	1
10	1	1	1	0	1	1	1	1
11	-	-	-	-	-	-	0	-
12	0	-	-	0	-	-	1	0

K TVL (10) ODA | 1 Var. | 1 R. | S. 1   -

K	x5
1	1

Fig. 11.91 Problem-program that verifies whether a variable or a negated variable can be separated using an AND-gate together with the representation of the TVLs of the function to decompose, the decomposition-function, and the separated variable

The check whether the negated variable $\overline{x}_i$ can be separated from $f(x_i, \mathbf{x}_0)$ using an AND-gate is only executed in the body of the command if in line 33 if the separation of the variable x_i from $f(x_i, \mathbf{x}_0)$ using an AND-gate is not possible. The commands isc and te_maxk in lines 35 and 36 realize Condition (11.89) to check whether the negated variable $\overline{x}_i$ can be separated from $f(x_i, \mathbf{x}_0)$ using an AND-gate. In the case that this condition is satisfied, the TVL of the separated negated variable $\overline{x}_i$ is stored as object 10 in line 38, the minimized TVL of the decomposition-function $g(\mathbf{x}_0)$ is computed based on (11.90) in lines 39–41, and the value of the variable $found is changed to the value true to avoid further checks for separations of a variable from the given function $f(x_i, \mathbf{x}_0)$.

The three commands assign behind the while-loop show the TVLs of given function $f(x_i, \mathbf{x}_0)$ in viewport (1,1), the computed decomposition-function $g(\mathbf{x}_0)$ in viewport (1,2), and the separated variable $x_i = x_5$ in viewport (2,1) of the m-fold view. The possibility to separate the variable $x_i = x_5$ using an AND-gate is already identifiable by the values 1 in all rows of the column x_5 of TVL 1 of the given function $f(x_5, \mathbf{x}_0)$. The command sts in line 47 stores the final XBOOLE-system for further synthesis-steps.

(b) Figure 11.92 shows both the PRP that loads the XBOOLE-system stored in the previous task in which the function to decompose is stored as object number 11, copies this TVL as object 21 for the case that no variable can be separated by an OR-gate (line 2), and iteratively verifies whether x_i or $\overline{x}_i$ can be separated from $f_1(\mathbf{x})$ using an OR-gate as well as the computed results.

The check whether x_i or $\overline{x}_i$ can be separated from $f_1(x_i, \mathbf{x}_0)$ using an OR-gate is implemented in the body of the while-loop in lines 4–35. The Boolean variable $found is used to break this loop when either x_i or $\overline{x}_i$ can be separated from $f(\mathbf{x})$ using an OR-gate.

The commands uni, mink, and te_cpl in lines 11, 12, and 13 realize Condition (11.83) to check whether the non-negated variable x_i can be separated from $f_1(x_i, \mathbf{x}_0)$ using an OR-gate. In the case that this condition is satisfied, the TVL of the separated variable x_i is stored as object 20 in line 15, the minimized TVL of the decomposition-function $g(\mathbf{x}_0)$ is computed based on (11.84) in lines 16–18, and value of the variable $found is changed to the value true to avoid further checks for separations of a variable from the given function $f_1(x_i, \mathbf{x}_0)$.

The check whether the negated variable $\overline{x}_i$ can be separated from $f_1(x_i, \mathbf{x}_0)$ using an OR-gate is only executed in the body of the command if in line 24 if the separation of the variable x_i from $f_1(x_i, \mathbf{x}_0)$ using an OR-gate is not possible. The commands uni, mink, and te_cpl in lines 26, 27, and 28 realize Condition (11.92) to check whether the negated variable $\overline{x}_i$ can be separated from $f_1(x_i, \mathbf{x}_0)$ using an OR-gate. In the case that this condition is satisfied, the TVL of the separated negated variable $\overline{x}_i$ is stored as object 20 in line 30, the minimized TVL of the decomposition-function $g_1(\mathbf{x}_0)$ is computed based on (11.93) in lines 31–33, and the value of the variable $found is changed to the value true to avoid further checks for separations of a variable from the given function $f_1(x_i, \mathbf{x}_0)$.

The three commands assign behind the while-loop show the TVLs of given function $f_1(x_i, \mathbf{x}_0)$ in viewport (1,1), the computed decomposition-function $g_1(\mathbf{x}_0)$ in viewport (1,2), and the separated negated variable $\overline{x}_8$ in viewport (2,1) of the m-fold view. The possibility to separate the variable $\overline{x}_8$ using an OR-gate is already identifiable by row 11 of TVL 11 of the given function $f_1(x_8, \mathbf{x}_0)$ that contains only in the column of x_8 a value 0 and in all other columns dashes. The command sts in line 39 stores the final XBOOLE-system for further synthesis-steps.

(c) It is known that either both x_i and $\overline{x}_i$ can be separated from $f(x_i, \mathbf{x}_0)$ using an EXOR-gate or neither x_i nor $\overline{x}_i$. The PRP of Fig. 11.93 verifies therefore only whether x_i can be separated from $f(x_i, \mathbf{x}_0)$ using an EXOR-gate. Figure 11.93 shows also the computed results of this separation using an EXOR-gate.

```
1   lds "e11_6_a"
2   copy 11 21 ; g1=f1
3   set $found false
4   while (and (not $found) (sv_next 11 12 12))
5   (
6   sv_get 12 13 ; xi
7   cpl 13 14 ; /xi
8   ; check the
9   ; separation of
10  ; xi using OR
11  uni 11 14 15
12  mink 15 13 16
13  if (te_cpl 16)
14  (
15  copy 13 20
16  isc 11 14 21
17  maxk 21 13 21
18  obbc 21 21
19  set $found true
20  )
21  ; check the
22  ; separation of
23  ; /xi using OR
24  if (not $found)
25  (
26  uni 11 13 15
27  mink 15 13 16
28  if (te_cpl 16)
29  (
30  copy 14 20
31  isc 11 13 21
32  maxk 21 13 21
33  obbc 21 21
34  set $found true
35  )))
36  assign 11 /m 1 1
37  assign 20 /m 2 1
38  assign 21 /m 1 2
39  sts "e11_6_b"
```

K	TVL (11) ODA	8 Var.	12 R.	S. 1					K	TVL (21) ODA	7 Var.	11 R.	S. 1			
	x_1	x_2	x_3	x_4	x_6	x_7	x_8	x_9		x_1	x_2	x_3	x_4	x_6	x_7	x_9
1	1	0	-	1	1	1	1	0	1	1	0	1	1	1	1	0
2	0	-	-	1	0	1	1	-	2	0	-	-	1	0	1	-
3	0	-	0	-	1	1	1	1	3	0	1	0	-	1	1	1
4	0	0	1	-	1	1	1	1	4	0	0	-	-	1	1	1
5	1	1	0	1	1	1	1	0	5	1	-	0	1	1	1	0
6	0	1	1	1	1	1	1	-	6	0	1	1	1	1	1	-
7	1	-	-	1	-	0	1	0	7	1	-	-	1	-	0	-
8	1	-	-	-	-	0	1	1	8	1	-	-	0	1	0	1
9	1	-	-	0	0	1	1	1	9	1	-	-	0	0	-	1
10	1	1	1	0	1	1	1	1	10	1	1	1	0	1	1	1
11	-	-	-	-	-	-	0	-	11	0	-	-	0	-	-	0
12	0	-	-	0	-	-	1	0								

K	TVL (20) ODA	1 Var.	1 R.	S. 1	-
	x_8	
1	0	

Fig. 11.92 Problem-program that verifies whether a variable or a negated variable can be separated using an OR-gate together with the representation of the TVLs of the function to decompose, the decomposition-function, and the separated variable

The command `lds` in the first line loads the XBOOLE-system stored in the previous task in which the function to decompose is stored as object number 21, and the command `copy` in the second line copies this TVL as object 31 for the case that no variable can be separated by an EXOR-gate.

The check whether x_i can be separated from $f_2(x_i, \mathbf{x}_0)$ using an EXOR-gate is implemented in the body of the `while`-loop in lines 4–19. The Boolean variable $found is used to break this loop when the variable x_i can be separated from $f(\mathbf{x})$ using an EXOR-gate.

```
1   lds "e11_6_b"
2   copy 21 31 ; g2=f2
3   set $found false
4   while (and (not $found) (sv_next 21 22 22))
5   (
6   sv_get 22 23 ; xi
7   cpl 23 24 ; /xi
8   ; check the
9   ; separation of
10  ; xi using EXOR
11  derk 21 23 25
12  if (te_cpl 25)
13  (
14  copy 23 30
15  isc 21 24 31
16  maxk 31 23 31
17  obbc 31 31
18  set $found true
19  ))
20  assign 21 /m 1 1
21  assign 30 /m 2 1
22  assign 31 /m 1 2
23  sts "e11_6_c"
```

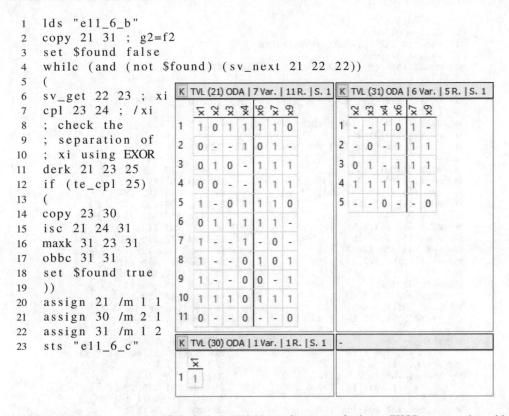

K TVL (21) ODA | 7 Var. | 11 R. | S. 1

	x_1	x_2	x_3	x_4	x_6	x_7	x_9
1	1	0	1	1	1	1	0
2	0	-	-	1	0	1	-
3	0	1	0	-	1	1	1
4	0	0	-	-	1	1	1
5	1	-	0	1	1	1	0
6	0	1	1	1	1	1	-
7	1	-	-	1	-	0	-
8	1	-	-	0	1	0	1
9	1	-	-	0	0	-	1
10	1	1	1	0	1	1	1
11	0	-	-	0	-	-	0

K TVL (31) ODA | 6 Var. | 5 R. | S. 1

	x_2	x_3	x_4	x_6	x_7	x_9
1	-	-	1	0	1	-
2	-	0	-	1	1	1
3	0	1	-	1	1	1
4	1	1	1	1	1	-
5	-	-	0	-	-	0

K TVL (30) ODA | 1 Var. | 1 R. | S. 1 | -

	x_1
1	1

Fig. 11.93 Problem-program that verifies whether a variable can be separated using an EXOR-gate together with the representation of the TVLs of the function to decompose, the decomposition-function, and the separated variable

The commands `derk` and `te_cpl` in lines 11 and 12 realize Condition (11.86) to check whether the variable x_i can be separated from $f_2(x_i, \mathbf{x}_0)$ using an EXOR-gate. In the case that this condition is satisfied, the TVL of the separated variable x_i is stored as object 30 in line 14, the minimized TVL of the decomposition-function $g(\mathbf{x}_0)$ is computed based on (11.87) in lines 15–17, and the value of the variable `$found` is changed to the value `true` to avoid further checks for separations of a variable from the given function $f_2(x_i, \mathbf{x}_0)$.

The three commands `assign` behind the `while`-loop show the TVLs of given function $f_2(x_i, \mathbf{x}_0)$ in viewport (1,1), the computed decomposition-function $g_2(\mathbf{x}_0)$ in viewport (1,2), and the separated variable x_1 in viewport (2,1) of the m-fold view. It is not obviously visible in TVL 21 of the given function $f_2(x_i, \mathbf{x}_0)$ that the variable x_1 can be separated from this function using an EXOR-gate. The command `sts` in line 23 stores the final XBOOLE-system for further synthesis-steps.

(d) Figure 11.94 shows the PRP that loads the XBOOLE-system stored in the previous task in which the function to decompose is stored as object number 31, computes an initial Ashenhurst-decomposition, extend this non-disjoint decomposition to a disjoint Pavarov-decomposition, and verifies the computed results.

The three nested `while`-loops in lines 5–29 organize iteratively all pairs of two variables that are assigned to the set of variables $\mathbf{x}_0$ (TVL 103) and a single variable is chosen as variable x_i out of the remaining variables (TVL 105) so that in the inner body of these loops Condition (11.107) of an Ashenhurst-decomposition can be checked. The result `true` of the command `te_isc` in

```
 1   lds "e11_6_c"
 2   assign 31 /m 1 1
 3   set $is false ; initial solution
 4   sv_get 31 101
 5   while (and (not $is) (sv_next 31 100 100))
 6   (
 7   sv_dif 101 100 101
 8   while (and (not $is) (sv_next 101 102 102))
 9   (
10   sv_uni 100 102 103 ; x0
11   sv_dif 31 103 104
12   while (and (not $is) (sv_next 104 105 105))
13   (
14   mink 31 103 110
15   maxk 31 103 111
16   syd 110 111 112 ; delta f,x0
17   mink 112 105 113
18   derk 31 105 114
19   mink 114 103 115
20   maxk 114 103 116
21   syd 115 116 117
22   if (te_isc 113 117)
23   (
24   set $is true
25   copy 103 120 ; x0
26   copy 105 121 ; x1
27   sv_dif 104 105 122 ; x2
28   )
29   )))
30   copy 121 106 ; x1
31   assign 103 /m 2 1 ; x0
32   assign 106 /m 2 2 ; x1
33   while (sv_next 122 123 123)
34   (
35   sv_uni 120 123 124 ; x0
36   mink 31 124 130 ; h1
37   sv_get 124 131 ; x0
38   syd 31 130 132
39   isc 131 132 133
40   maxk 133 131 134 ; h2
41   maxk 31 124 135
42   dif 135 130 136
43   dif 136 134 137 ; h3
44   isc 31 134 138
45   dif 137 31 139
46   uni 138 139 140
47   maxk 140 121 141 ; g(x0,x2)
48   isc 141 134 142
49   dif 137 141 143
50   uni 142 143 144
51   uni 144 130 145 ; f'
52   if (te_syd 31 145)
53   (copy 124 120)
54   if (not (te_syd 31 145))
55   (
56   sv_uni 121 123 125 ; x1
57   mink 31 120 130 ; h1
58   sv_get 120 131 ; x0
59   syd 31 130 132
60   isc 131 132 133
```

```
61   maxk 133 131 134 ; h2
62   maxk 31 120 135
63   dif 135 130 136
64   dif 136 134 137 ; h3
65   isc 31 134 138
66   dif 137 31 139
67   uni 138 139 140
68   maxk 140 125 141 ; g(x0,x2)
69   isc 141 134 142
70   dif 137 141 143
71   uni 142 143 144
72   uni 144 130 145 ; f'
73   if (te_syd 31 145)
74   (copy 125 121)
75   ))
76   assign 120 /m 3 1 ; x0
77   assign 121 /m 3 2 ; x1
78   sbe 1 150
79   g.
80   mink 31 120 130 ; h1
81   sv_get 120 131 ; x0
82   syd 31 130 132
83   isc 131 132 133
84   maxk 133 131 134 ; h2
85   maxk 31 120 135
86   dif 135 130 136
87   dif 136 134 137 ; h3
88   isc 31 134 138
89   dif 137 31 139
90   uni 138 139 140
91   maxk 140 121 141 ; g(x0)
92   isc 150 134 151
93   dif 137 150 152
94   uni 151 152 153
95   uni 153 130 154 ; h(g,x1)
96   assign 141 /m 4 1 ; g
97   assign 154 /m 4 2 ; h
98   csd 141 150 160
99   isc 160 154 161
100  maxk 161 150 162
101  assign 162 /m 1 2
102  set $equal false
103  syd 162 30 170
104  uni 170 20 171
105  isc 171 10 172
106  if (te_syd 172 1)
107  (set $equal true)
```

Fig. 11.94 Problem-program that computes a Povarov-decomposition of the function $f_2(\mathbf{x})$ and verifies whether the completely synthesized circuit realizes the initially given function $f(\mathbf{x})$

line 22 indicates that this condition is satisfied and an Ashenhurst-decomposition with regard to $\mathbf{x}_0$ (TVL 103) and x_i (TVL 105) exists; in this case, the value of the variable $\$is$ (initial solution) is changed to `true` in line 24 so that the three nested `while`-loops are terminated, the subsequent two commands `copy` store the found initial sets of variables as x_0 (TVL 120) and x_1 (TVL 121) for later extensions, and the command `sv_dif` in line 27 computes the set of variables as x_2 (TVL 122) that is later used as source of variables of potential extensions of the sets x_0 and x_1. The set x_1 has been copied in line 30 from TVL 121 to TVL 106 because x_i (TVL 105) has been already changed to the next variable in the last executed command `sv_next` of the inner `while`-loop in line 12. The subsequent two commands `assign` show the found dedicated sets of variables x_0 and x_1 of the initial Ashenhurst-decomposition in viewports (2,1) and (2,2) of the m-fold view of Fig. 11.95.

The aim of the `while`-loop in lines 33–75 is the extension of dedicated sets of variables x_0 and x_1 with permitted variables of the set x_2. A chosen variable (TVL 123) is firstly inserted into the temporary set x_0 (TVL 124 in line 35) of the decomposition-function $g(x_0, x_2)$. The commands in lines 36–51 compute the function $f'(x_0, x_1, x_2)$ (11.105) based on the computed potential decomposition-functions $g(x_0, x_2)$ and the three subfunction $h_i(x_1, x_2)$ using Eqs. (11.101) to (11.104). The assignment of the chosen variable to the dedicated set x_0 is permitted when Condition (11.106) checked in line 52 is satisfied; in this case, the temporary set x_0 (TVL 124) is permanently stored as set x_0 (TVL 120).

Otherwise, the chosen variable is inserted into the temporary set x_1 (TVL 125 in line 56), and the check with regard to an Ashenhurst-decomposition for the changed temporary sets of variables x_0 and x_1 is repeated in lines 57–72. The assignment of the chosen variable to the dedicated set x_1 is permitted when Condition (11.106) checked in line 73 is satisfied; in this case, the temporary set x_1 (TVL 125) is permanently stored as set x_1 (TVL 121).

The two commands `assign` in lines 76 and 77 show the final extended dedicated sets of variables x_0 and x_1 of the decomposition in viewports (3,1) and (3,2) of the m-fold view of Fig. 11.95. All six variables of $f(x)$ could be assigned to the dedicated sets of variables x_0 and x_1; hence, a disjoint Povarov-decomposition has been found.

The decomposition-functions $g(x_0)$ and $h(g, x_0)$ are computed in lines 78–95 using the final extended dedicated sets of variables x_0 and x_1 and Eqs. (11.100) to (11.104). The two commands `assign` in lines 96 and 97 show the Karnaugh-maps of the computed decomposition-functions $g(x_0)$ and $h(g, x_0)$ of the Povarov-decomposition in viewports (4,1) and (4,2) of the m-fold view of Fig. 11.95.

The realized function $f_2'(x_0, x_1)$ (TVL 162) of the Povarov-decomposition is computed in lines 98–100. The Karnaugh-map of $f_2'(x_0, x_1)$ (TVL 162) is assigned by the subsequent command to viewport (1,2) so that this function can easily be compared with the given function $f_2(x_0, x_1)$ (TVL 31) shown in viewport (1,1) of the m-fold view. The distribution of the variables has been changed in these two Karnaugh-maps such that the dedicated sets of variables x_0 and x_1 are assigned to the two boarders so that the function $g(x_0)$ is shown in the row $(x_4, x_7, x_9) = (110)$ where the function $h_2(x_1)$ is equal to 1, the function $\overline{g(x_0)}$ is shown in the row $(x_4, x_7, x_9) = (011)$ where the function $h_3(x_1)$ is equal to 1, and the three rows that are constant equal to 1 indicate the function $h_1(x_1)$.

The commands in lines 103–105 compute the complete realized function $f(x)$ (TVL 172), and the result `$equal=true` of the comparison with the given function $f(x)$ (TVL 1) confirms the correct solution of this exercise.

Figure 11.95 shows below the computed Karnaugh-maps and dedicated sets of the initial Ashenhurst-decomposition and final Povarov-decomposition of the circuit structure synthesized by several types of separations.

T	TVL (31) ODA	6 Var.	5 R.	S. 1

```
0 0 0   1 1 1 1   1 1 1 1
0 0 1   0 0 0 0   0 0 0 0
0 1 1   0 1 1 0   0 0 1 0
0 1 0   1 1 1 1   1 1 1 1
1 1 0   1 0 0 1   1 1 0 1
1 1 1   1 1 1 1   1 1 1 1
1 0 1   0 0 0 0   0 0 0 0
1 0 0   0 0 0 0   0 0 0 0
x4 x7 x9  0 1 1 0  0 1 1 0  x6
          0 0 1 1  1 1 0 0  x3
          0 0 0 0  1 1 1 1  x2
```

T	TVL (162) ODA	6 Var.	12 R.	S. 1

```
0 0 0   1 1 1 1   1 1 1 1
0 0 1   0 0 0 0   0 0 0 0
0 1 1   0 1 1 0   0 0 1 0
0 1 0   1 1 1 1   1 1 1 1
1 1 0   1 0 0 1   1 1 0 1
1 1 1   1 1 1 1   1 1 1 1
1 0 1   0 0 0 0   0 0 0 0
1 0 0   0 0 0 0   0 0 0 0
x4 x7 x9  0 1 1 0  0 1 1 0  x6
          0 0 1 1  1 1 0 0  x3
          0 0 0 0  1 1 1 1  x2
```

K	TVL (103) ODA	2 Var.	0 R.	S. 1

```
x̄2 x̄3
```

K	TVL (106) ODA	1 Var.	0 R.	S. 1

```
x̄4
```

K	TVL (120) ODA	3 Var.	0 R.	S. 1

```
x̄2 x̄3 x̄6
```

K	TVL (121) ODA	3 Var.	0 R.	S. 1

```
x̄4 x̄7 x̄9
```

T	TVL (141) ODA	3 Var.	2 R.	S. 1

```
0   1 1 1 1
1   0 0 1 0
x6  0 1 1 0 x3
    0 0 1 1 x2
```

T	TVL (154) ODA	4 Var.	4 R.	S. 1

```
0 0   1 1 0 0
0 1   1 1 1 0
1 1   0 0 1 0
1 0   0 1 1 0
x9 x5  0 1 1 0 x7
       0 0 1 1 x4
```

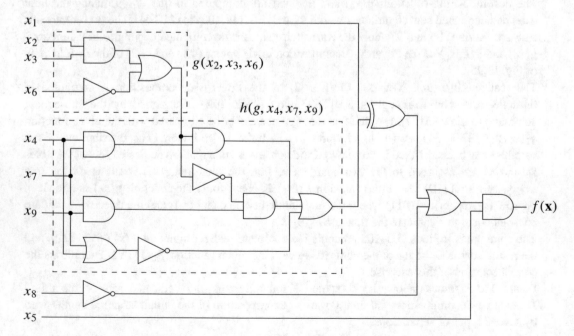

Fig. 11.95 Solution of the PRP shown in Fig. 11.94 together with the circuit synthesized by several types of separations in Exercise 11.6

```
 1   new
 2   space 32 1
 3   avar 1
 4   x1 x2 x3 x4 x5 x6 .
 5   sbe 1 1
 6   x1&x2 &(/x4+/x5)+/x1&/x3&x5&/x6 &((/x2+/x4)+/x1&x2&/x3&/x5&/x6 .
 7   assign 1 /m 1 1
 8   set $is false ; initial solution
 9   sv_get 1 2 ; usable for xb
10   while (and (not $is) (sv_next 1 3 3))
11   (
12   sv_dif 2 3 2
13   if (ge (sv_size 2) 1)
14   (
15   while (and (not $is) (sv_next 2 4 4))
16   (
17   maxk 1 3 5                        39   dif 16 1 18
18   maxk 1 4 6                        40   if (te_isc 17 18)
19   dif 5 1 7                         41   (
20   if (te_isc 6 7)                   42   copy 14 10 ; xa
21   (                                 43   )
22   set $is true                      44   if (not (te_isc 17 18))
23   copy 3 10 ; xa                    45   (
24   copy 4 11 ; xb                    46   sv_uni 11 13 15
25   ))))                              47   maxk 1 10 16
26   copy 10 3                         48   maxk 1 15 17
27   copy 11 4                         49   dif 16 1 18
28   assign 3 /m 2 1                   50   if (te_isc 17 18)
29   assign 4 /m 2 2                   51   (
30   sv_dif 1 10 12                    52   copy 15 11 ; xb
31   sv_dif 12 11 12                   53   ))))
32   if (ge (sv_size 12) 1)            54   assign 10 /m 3 1
33   (                                 55   assign 11 /m 3 2
34   while (sv_next 12 13 13)          56   maxk 1 11 20
35   (                                 57   maxk 1 10 21
36   sv_uni 10 13 14                   58   assign 20 /m 4 1
37   maxk 1 14 16                      59   assign 21 /m 4 2
38   maxk 1 11 17                      60   sts "ell_7_a"
```

Fig. 11.96 Problem-program that computes a compact AND-bi-decomposition

Solution 11.7 (Synthesis Using Several Types of Strong Bi-Decompositions)

(a) Figure 11.96 shows the PRP that computes the two decomposition-functions $g(\mathbf{x}_a, \mathbf{x}_c)$ and $h(\mathbf{x}_b, \mathbf{x}_c)$ of an AND-bi-decomposition of the function $f(\mathbf{x})$ specified as object 1; the associated solutions are shown in Fig. 11.97.

The command sbe in lines 5 and 6 determines the function $f(\mathbf{x})$ to be decomposed using an AND-gate, and the subsequent command assign shows the TVL of this function in viewport $(1,1)$ of the m-fold view. The two nested while-loops in lines 10–25 select iteratively all pairs of variables of $\mathbf{x}$ as objects 3 (x_a) and 4 (x_b). The commands in lines 12 and 13 exclude the selected variable x_a from the set of all variables $\mathbf{x}$ and verify whether the remaining set contains at least one variable so that the variable x_b can be chosen from the set of variables stored as object 2. The commands in lines 17–20 implement Condition (11.115) to check whether an AND-bi-decomposition of $f(\mathbf{x})$ with regard to the variables x_a and x_b exists. In the case that this condition is satisfied, the value of the variable $is (initial solution, initialized with the value false) is

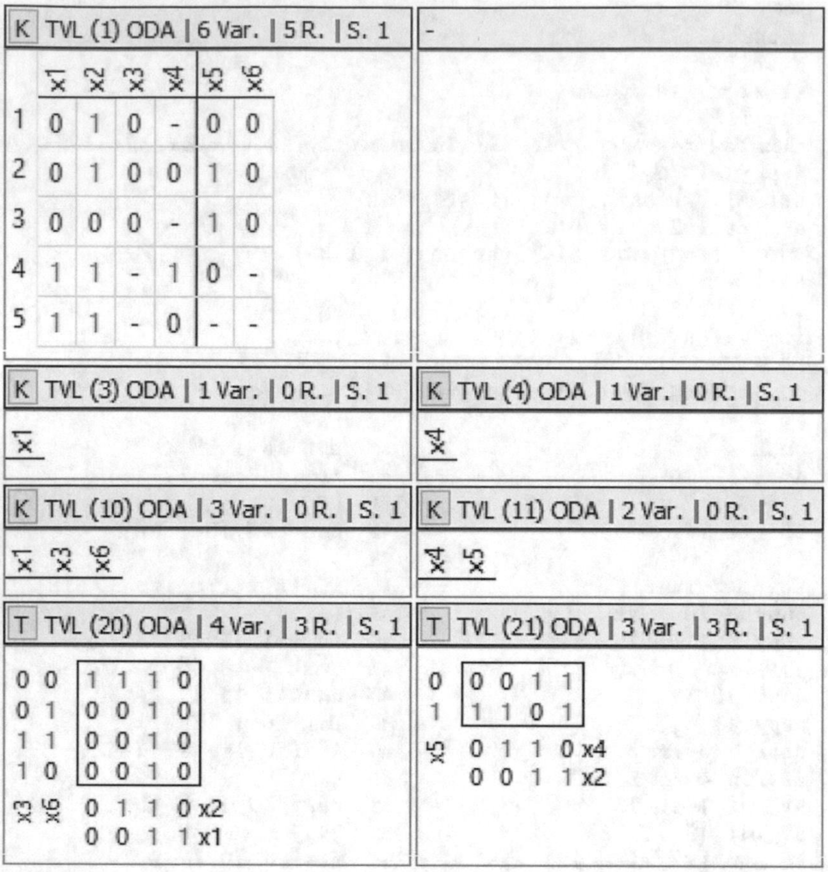

Fig. 11.97 Solution of the PRP shown in Fig. 11.96 of the AND-bi-decomposition computed as Exercise 11.7 (a)

changed to `true` and the TVLs of the variables x_a and x_b are copied to the objects 10 and 11 for later expansions.

The value `true` of variable `$is` terminates the two nested `while`-loops in lines 10–25 if an initial AND-bi-decomposition of $f(\mathbf{x})$ has been found. The commands `copy` in lines 26 and 27 restore the variables x_a and x_b of the initially found AND-bi-decomposition, and the subsequent two commands `assign` show these variables in the second row of the m-fold view.

The two commands `sv_dif` in lines 30 and 31 prepare the set of variables that can be used to extend one of the dedicated sets $\mathbf{x}_a$ or $\mathbf{x}_b$. In the case that this set is not empty (checked in line 32) the `while`-loop in lines 34–53 selects in each sweep one of these variables for an optional extension of one of the dedicated sets of variables of the AND-bi-decomposition of $f(\mathbf{x})$.

The command `sv_uni` in line 36 inserts the selected variable into the temporary set $\mathbf{x}_a$ (TVL 14), and the subsequent four commands implement Condition (11.115) to check whether an AND-bi-decomposition of $f(\mathbf{x})$ with regard to the extended set $\mathbf{x}_a$ and the last known set $\mathbf{x}_b$ exists. The command `copy` in line 42 permanently includes the selected variable into the dedicated set $\mathbf{x}_a$ in the case that this condition is satisfied; otherwise, the command `sv_uni` in line 46 inserts the selected variable into the temporary set $\mathbf{x}_b$ (TVL 15) and the subsequent four commands implement Condition (11.115) to check whether an AND-bi-decomposition of $f(\mathbf{x})$ with regard to the last known set $\mathbf{x}_a$ and the extended set $\mathbf{x}_b$ exists. The command `copy` in line 52 permanently includes the selected variable into the dedicated set $\mathbf{x}_b$ in the case that this condition is satisfied. The two commands `assign` in lines 54 and 55 show the extended dedicated sets of variables $\mathbf{x}_a$ and $\mathbf{x}_b$ in the third row of the m-fold view.

The two commands `maxk` in lines 56 and 57 compute the decomposition-functions $g(\mathbf{x}_a, \mathbf{x}_c)$ and $h(\mathbf{x}_b, \mathbf{x}_c)$ based on (11.116) and (11.117) as required as objects 20 and 21. The subsequent two commands `assign` show the Karnaugh-maps of these decomposition-functions in the fourth row of the m-fold view. The common set $\mathbf{x}_2$ of this compact AND-bi-decomposition contains only the variable x_2. The command `sts` in line 60 stores the final XBOOLE-system for the further decompositions in the next two tasks.

(b) Figure 11.98 shows the PRP that computes the two decomposition-functions $g_1(\mathbf{x}_a, \mathbf{x}_c)$ and $h_1(\mathbf{x}_b, \mathbf{x}_c)$ of an EXOR-bi-decomposition of the function $f_1(\mathbf{x})$ (TVL 20, assigned to viewport (1,1)) computed as decomposition-function $g(\mathbf{x})$ of the AND-bi-decomposition in the previous task. The associated solutions are shown in Fig. 11.99.

A necessary condition for a compact EXOR-bi-decomposition is that the functions to decompose are EXOR-bi-decomposable with regard to two single variables x_a and x_b. The check for such an initial EXOR-bi-decomposition is realized in the PRP of Fig. 11.98 by the command `te_derk` in line 13 that implements Condition (11.118). The two surrounding while-loops in lines 5–18 select similar to the previous task all pairs of variables x_a and x_b, and the command `sv_uni` in line 12 combines these variables to a set used to determine the 2-fold derivative in line 13. If a pair of variables x_a and x_b has been found, for which an initial EXOR-bi-decomposition exists, the variable `$is` is changed to the value `true` in line 15 to terminate the two nested while-loops in lines 5–18, the TVLs of successfully evaluated variables are saved in lines 16 and 17 because the variables x_a and x_b have been changed (or even deleted) by the commands `sv_next` in the conditions of the two while-loops. The TVLs of the variables x_a and x_b for which the initial EXOR-bi-decomposition exists are assigned to the viewports of the second row of the m-fold view.

The two commands `copy` in lines 21 and 22 prepare new TVLs of the dedicated sets of variables used for potential extensions; this preserves TVLs 41 and 42 of the initially found variables x_a and x_b. The two commands `sv_dif` in lines 23 and 24 compute the set of variables that can be used to extend the initial dedicated sets of variables. Potential extensions are checked and executed in the body of the command `if` in lines 25–52.

Condition (11.119) can be used to check whether an EXOR-bi-decomposition with regard to a single variable x_a and a set of variables $\mathbf{x}_b$ exists; however, each of the variables of initial EXOR-bi-decomposition can be chosen as variable x_a. The Boolean variables `$ex1` (extension of $\mathbf{x}_1$) and `$ex2` (extension of $\mathbf{x}_2$) are initialized by values `true` and indicate that the associated set of variables (TVL 51 in case of `$ex1=true` or TVL 52 in case of `$ex2=true`) takes the role of the set $\mathbf{x}_b$ for possible extensions.

The while-loop in lines 29–52 selects iteratively each variable of the set $\mathbf{x} \setminus (\{x_a\} \cup \{x_b\})$ for potential extensions of $\mathbf{x}_b$. The commands in lines 33–37 implement Condition (11.119); these commands are executed when the variable `$ex1=true`. In the case that this condition is satisfied, the command `sv_uni` in line 39 permanently extends the set of variables $\mathbf{x}_b$ stored as TVL 51, and the subsequent change of `$ex2` to the value `false` excludes the other dedicated set (TVL 52) from possible extensions.

As long as `$ex2=true`, the commands in lines 44–48 are executed to check Condition (11.119) using the extension of TVL 52. In the case that this condition is satisfied, the command `sv_uni` in line 50 permanently extends the set of variables $\mathbf{x}_b$ stored as TVL 52, and the subsequent change of `$ex1` to the value `false` excludes the other dedicated set (TVL 51) from possible extensions.

```
 1  lds "e11_7_a"
 2  assign 20 /m 1 1
 3  set $is false ; initial solution
 4  sv_get 20 30 ; usable for xb
 5  while (and (not $is) (sv_next 20 31 31))
 6  (
 7  sv_dif 30 31 30
 8  if (ge (sv_size 30) 1)
 9  (
10  while (and (not $is) (sv_next 30 32 32))
11  (
12  sv_uni 31 32 33                 48  if (te_syd 46 47)
13  if (te_derk 20 33)              49  (
14  (                               50  sv_uni 52 43 52
15  set $is true                    51  set $ex1 false
16  copy 31 41                      52  ))))
17  copy 32 42                      53  if ($ex1)
18  ))))                            54  (
19  assign 41 /m 2 1                55  copy 52 53 ; xa
20  assign 42 /m 2 2                56  copy 51 54 ; xb
21  copy 41 51                      57  )
22  copy 42 52                      58  if ($ex2)
23  sv_dif 20 41 40                 59  (
24  sv_dif 40 42 40                 60  copy 51 53 ; xa
25  if (ge (sv_size 40) 1)          61  copy 52 54 ; xb
26  (                               62  )
27  set $ex1 true                   63  assign 53 /m 3 1 ; xa
28  set $ex2 true                   64  assign 54 /m 3 2 ; xb
29  while (sv_next 40 43 43)        65  sv_get 54 55 ; xb
30  (                               66  cel 55 55 55 /01 /10
31  if ($ex1)                       67  isc 20 55 56
32  (                               68  maxk 56 54 57 ; g
33  sv_uni 51 43 44                 69  syd 20 57 58
34  derk 20 42 45                   70  maxk 58 53 59 ; h
35  mink 45 44 46                   71  sv_dif 20 53 60
36  maxk 45 44 47                   72  sv_dif 60 54 60
37  if (te_syd 46 47)               73  if (ge (sv_size 60) 1)
38  (                               74  (
39  sv_uni 51 43 51                 75  while (sv_next 60 61 61)
40  set $ex2 false                  76  (
41  ))                              77  if (te_derk 59 61)
42  if ($ex2)                       78  (
43  (                               79  maxk 59 61 59
44  sv_uni 52 43 44                 80  sv_uni 51 61 51
45  derk 20 41 45                   81  )))
46  mink 45 44 46                   82  assign 57 /m 4 1
47  maxk 45 44 47                   83  assign 59 /m 4 2
```

Fig. 11.98 Problem-program that computes a compact EXOR-bi-decomposition

The final assignment of the optionally expanded set of variables $\mathbf{x}_b$ and single variable x_a is realized in lines 53–62 based on the values of the Boolean variables $ex1 and $ex2. The two subsequent commands assign show these dedicated sets in the third row of the m-fold view.

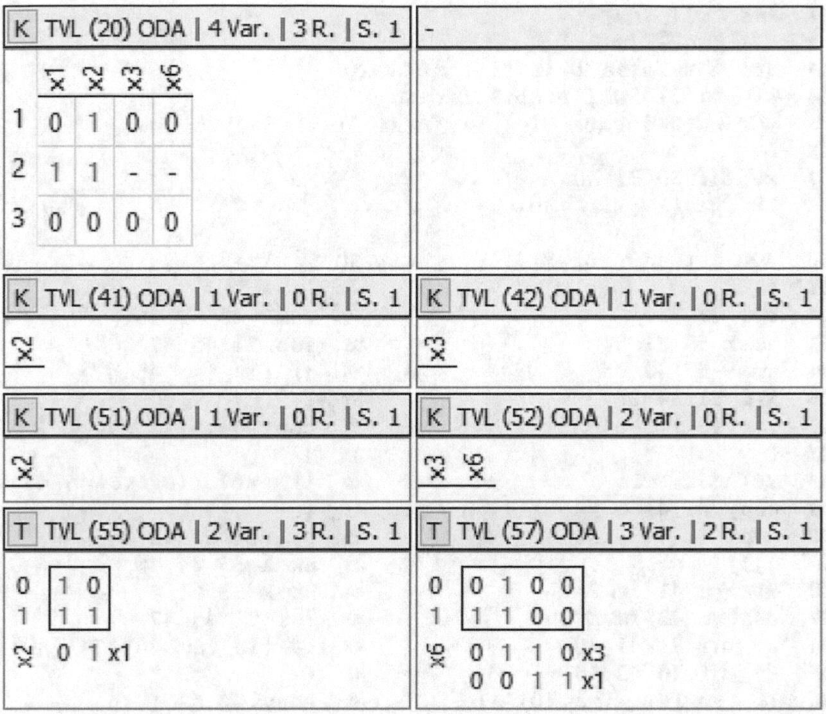

| K | TVL (20) ODA | 4 Var. | 3 R. | S. 1 | - |

	x_1	x_2	x_3	x_6
1	0	1	0	0
2	1	1	-	-
3	0	0	0	0

| K | TVL (41) ODA | 1 Var. | 0 R. | S. 1 |

x_2

| K | TVL (42) ODA | 1 Var. | 0 R. | S. 1 |

x_3

| K | TVL (51) ODA | 1 Var. | 0 R. | S. 1 |

x_2

| K | TVL (52) ODA | 2 Var. | 0 R. | S. 1 |

x_3 x_6

| T | TVL (55) ODA | 2 Var. | 3 R. | S. 1 |

```
0 | 1 0 |
1 | 1 1 |
x2  0 1 x1
```

| T | TVL (57) ODA | 3 Var. | 2 R. | S. 1 |

```
0 | 0 1 0 0 |
1 | 1 1 0 0 |
x6  0 1 1 0 x3
    0 0 1 1 x1
```

Fig. 11.99 Solution of the PRP shown in Fig. 11.98 of the EXOR-bi-decomposition computed as Exercise 11.7 (b)

The commands in lines 65–70 compute the decomposition-functions $g(x_a, \mathbf{x}_c)$ and $h(\mathbf{x}_b, \mathbf{x}_c)$ using Eqs. (11.123) and (11.124). The set of variables $\mathbf{x}_c$ is computed by the commands sv_dif in lines 71 and 72.

In the case that the set of variables $\mathbf{x}_c$ is not empty (checked in line 73), it is checked whether the function $h(\mathbf{x}_b, \mathbf{x}_c)$ does not depend on $x_i \in \mathbf{x}_c$ controlled by the while-loop in lines 75–81. If the function $h(\mathbf{x}_b, \mathbf{x}_c)$ does not depend on $x_i \in \mathbf{x}_c$, this variable is removed from the TVL of $h(\mathbf{x}_b, \mathbf{x}_c)$ using the command maxk in line 79 and added to the dedicated set $\mathbf{x}_a$ using the command sv_uni in line 80. The subsequent two commands assign show the Karnaugh-maps of the computed decomposition-functions $g(x_a, \mathbf{x}_c)$ and $h(\mathbf{x}_b, \mathbf{x}_c)$. It can be seen that $h(\mathbf{x}_b, \mathbf{x}_c)$ depends on the only variable $x_1 \in \mathbf{x}_c$ so that the dedicated set $\mathbf{x}_a$ of this EXOR-bi-decomposition contains also only the single variables x_2.

(c) Figure 11.100 shows the PRP that computes the decomposition-functions $g_2(\mathbf{x}_a, \mathbf{x}_c)$ and $h_2(\mathbf{x}_b, \mathbf{x}_c)$ of an OR-bi-decomposition of the function $f_2(\mathbf{x})$ (TVL 21, assigned to viewport $(1,1)$) computed as decomposition-function $h(\mathbf{x})$ of the AND-bi-decomposition of the first task of this exercise. The associated solutions are shown in Fig. 11.101.

The two nested while-loops in lines 5–21 select iteratively all pairs of variables of $\mathbf{x}$ as objects 31 (x_a) and 32 (x_b). The commands in lines 7 and 8 exclude the selected variable x_a from the set of all variables $\mathbf{x}$ and verify whether the remaining set contains at least one variable so that the variable x_b can be chosen from the set of variables stored as object 30. The commands in lines 12–16 implement Condition (11.111) to check whether an OR-bi-decomposition of $f_2(\mathbf{x})$ with regard to the variables x_a and x_b exists. In the case that this condition is satisfied, the value of the variable $is (initial solution, initialized with the value false) is changed to true and the TVLs of the variables x_a and x_b are copied to the objects 41 and 42 for later expansions. The

```
1   lds "ell_7_a"
2   assign 21 /m 1 1
3   set $is false ; initial solution
4   sv_get 21 30 ; usable for xb
5   while (and (not $is) (sv_next 21 31 31))
6   (
7   sv_dif 30 31 30
8   if (ge (sv_size 30) 1)
9   (
10  while (and (not $is) (sv_next 30 32 32))
11  (
12  cpl 21 33                              34  maxk 33 52 46
13  maxk 33 31 34                          35  isc 21 45 47
14  maxk 33 32 35                          36  if (te_isc 46 47)
15  isc 21 34 36                           37  (
16  if (te_isc 35 36)                      38  copy 44 51 ; xa
17  (                                      39  )
18  set $is true                           40  if (not (te_isc 46 47))
19  copy 31 41 ; xa                        41  (
20  copy 32 42 ; xb                        42  sv_uni 52 43 44
21  ))))                                   43  maxk 33 44 45
22  assign 41 /m 2 1                       44  maxk 33 51 46
23  assign 42 /m 2 2                       45  isc 21 45 47
24  sv_dif 21 41 40                        46  if (te_isc 46 47)
25  sv_dif 40 42 40                        47  (
26  if (ge (sv_size 40) 1)                 48  copy 44 52 ; xb
27  (                                      49  ))))
28  copy 41 51 ; xa                        50  assign 51 /m 3 1
29  copy 42 52 ; xb                        51  assign 52 /m 3 2
30  while (sv_next 40 43 43)               52  mink 21 52 48
31  (                                      53  mink 21 51 49
32  sv_uni 51 43 44                        54  assign 48 /m 4 1
33  maxk 33 44 45                          55  assign 49 /m 4 2
```

Fig. 11.100 Problem-program that computes a compact OR-bi-decomposition

value `true` of variable `$is` terminates the two nested `while`-loops in lines 5–21 if an initial OR-bi-decomposition of $f_2(\mathbf{x})$ has been found. The two commands `assign` in lines 22 and 23 show the found variables of the initial OR-bi-decomposition in the second row of the m-fold view.

The two commands `sv_dif` in lines 24 and 25 compute the set of variables that can be used to extend the initial dedicated sets of variables. Potential extensions are checked and executed in the body of the command `if` in lines 26–49. The two commands `copy` in lines 28 and 29 prepare new TVLs of the dedicated sets of variables used for potential extensions; this preserves TVLs 41 and 42 of the initially found variables x_a and x_b. The `while`-loop in lines 30–49 selects in each sweep one of the variables of TVL 40 for an optional extension of one of the dedicated sets of variables of the OR-bi-decomposition of $f_2(\mathbf{x})$.

The command `sv_uni` in line 32 inserts the selected variable into the temporary set $\mathbf{x}_a$ (TVL 44), and the subsequent four commands implement Condition (11.111) to check whether an OR-bi-decomposition of $f_2(\mathbf{x})$ with regard to the extended set $\mathbf{x}_a$ and the last known set $\mathbf{x}_b$ exists.

The command `copy` in line 38 permanently includes the selected variable into the dedicated set $\mathbf{x}_a$ in the case that this condition is satisfied; otherwise, the command `sv_uni` in line 42 inserts the selected variable into the temporary set $\mathbf{x}_b$ (TVL 44), and the subsequent four commands

Fig. 11.101 Solution of the PRP shown in Fig. 11.100 of the OR-bi-decomposition computed as Exercise 11.7 (c)

K	TVL (21) ODA | 3 Var. | 3 R. | S. 1	-

	x_2	x_4	x_5
1	1	-	0
2	1	0	1
3	0	-	1

K	TVL (41) ODA | 1 Var. | 0 R. | S. 1	K	TVL (42) ODA | 1 Var. | 0 R. | S. 1

x_2 x_4

K	TVL (51) ODA | 1 Var. | 0 R. | S. 1	K	TVL (52) ODA | 1 Var. | 0 R. | S. 1

x_2 x_4

T	TVL (48) ODA | 2 Var. | 2 R. | S. 1	T	TVL (49) ODA | 2 Var. | 1 R. | S. 1

	0	1
0	0	1
1	1	0
x_5	0	1 x2

	0	1
0	0	0
1	1	0
x_5	0	1 x4

implement Condition (11.111) to check whether an OR-bi-decomposition of $f_2(\mathbf{x})$ with regard to the last known set $\mathbf{x}_a$ and the extended set $\mathbf{x}_b$ exists. The command copy in line 48 permanently includes the selected variable into the dedicated set $\mathbf{x}_b$ in the case that this condition is satisfied. The two commands assign in lines 50 and 51 show the extended dedicated sets of variables $\mathbf{x}_a$ and $\mathbf{x}_b$ in the third row of the m-fold view. It can be seen that an extension neither of $\mathbf{x}_b$ nor of $\mathbf{x}_a$ is possible for this OR-bi-decomposition.

The two commands mink in lines 52 and 53 compute the decomposition-functions $g_2(\mathbf{x}_a, \mathbf{x}_c)$ and $h_2(\mathbf{x}_b, \mathbf{x}_c)$ based on (11.112) and (11.113), and subsequent two commands assign show the Karnaugh-maps of these decomposition-functions in the fourth row of the m-fold view.

(d) Figure 11.102 shows the PRP that computes the decomposition-functions $g_3(\mathbf{x}_a, \mathbf{x}_c)$ and $h_3(\mathbf{x}_b, \mathbf{x}_c)$ of an AND-bi-decomposition of the function $f_3(\mathbf{x})$ computed as decomposition-function $h(\mathbf{x}_b, \mathbf{x}_c)$ (TVL 57) of the EXOR-bi-decomposition shown in Fig. 11.98 and specified as TVL 1 (function $f_3(\mathbf{x})$, assigned to viewport (1,1)). The associated solutions are shown in Fig. 11.103.

An AND-bi-decomposition of the function $f(\mathbf{x})$ has already been implemented in the PRP of Fig. 11.96. Hence, this PRP can be reused in an adjusted version to decompose the function $f_3(\mathbf{x})$ computed as decomposition-function $h(\mathbf{x}_b, \mathbf{x}_c)$ (TVL 57) of the EXOR-bi-decomposition shown in Fig. 11.98. The Karnaugh-map of this function is shown in viewport (4,2) of Fig. 11.99. The commands avar and sbe in lines 3–6 of Fig. 11.96 have been replaced in the PRP of Fig. 11.102 by the command tin in lines 3–6 that determines the function $f_3(\mathbf{x})$. Furthermore, the command sts in line 60 of Fig. 11.96 has been removed in the adjusted version of the PRP for the AND-bi-decomposition of the function $f_3(\mathbf{x})$ because the computed results are not needed in other PRPs. All other commands of the PRP of Fig. 11.96 remain unchanged in the adjusted version of this PRP in Fig. 11.102; hence, the details of the PRP in Fig. 11.102 have been already explained in the solution of Task (a) of this exercise and are therefore not repeated.

The results in Fig. 11.103 show that a disjoint AND-bi-decomposition with regard to the single variable x_1 in the dedicated set $\mathbf{x}_a$ and the two variables x_3 and x_6 in the associated set $\mathbf{x}_b$ has been

```
 1  new
 2  space 32 1
 3  tin 1 1
 4  x1 x3 x7.
 5  010
 6  0-1.
 7  assign 1 /m 1 1
 8  set $is false ; initial solution
 9  sv_get 1 2 ; usable for xb
10  while (and (not $is) (sv_next 1 3 3))
11  (
12  sv_dif 2 3 2
13  if (ge (sv_size 2) 1)
14  (
15  while (and (not $is) (sv_next 2 4 4))
16  (
17  maxk 1 3 5
18  maxk 1 4 6                        39  dif 16 1 18
19  dif 5 1 7                         40  if (te_isc 17 18)
20  if (te_isc 6 7)                   41  (
21  (                                 42  copy 14 10
22  set $is true                      43  )
23  copy 3 10                         44  if (not (te_isc 17 18))
24  copy 4 11                         45  (
25  ))))                              46  sv_uni 11 13 15
26  copy 10 3                         47  maxk 1 10 16
27  copy 11 4                         48  maxk 1 15 17
28  assign 3 /m 2 1                   49  dif 16 1 18
29  assign 4 /m 2 2                   50  if (te_isc 17 18)
30  sv_dif 1 10 12                    51  (
31  sv_dif 12 11 12                   52  copy 15 11
32  if ( ge ( sv_size 12) 1)          53  ))))
33  (                                 54  assign 10 /m 3 1
34  while (sv_next 12 13 13)          55  assign 11 /m 3 2
35  (                                 56  maxk 1 11 20
36  sv_uni 10 13 14                   57  maxk 1 10 21
37  maxk 1 14 16                      58  assign 20 /m 4 1
38  maxk 1 11 17                      59  assign 21 /m 4 2
```

Fig. 11.102 Problem-program that computes a compact AND-bi-decomposition

computed. The exploration regarding a separation of $\overline{x}_1$ from the function $f_3(\mathbf{x})$ would determine the same result.

The circuit structure of the synthesized circuit in Fig. 11.103 reveals the advantages of the used strong bi-decompositions. At most three gates belong to any path between an input and the output and only ten gates with not more than two inputs are needed to realize the given function $f(\mathbf{x})$ that depends on six variables.

Solution 11.8 (Synthesis Using Weak and Strong Bi-Decompositions)

(a) Figure 11.104 shows both the PRP that verifies whether a strong bi-decomposition for the given function exists and requested results.

K	TVL (1) ODA | 3 Var. | 2 R. | S. 1	-

	$\overline{x}_1$	x_3	x_6
1	0	1	0
2	0	-	1

K	TVL (3) ODA | 1 Var. | 0 R. | S. 1	K	TVL (4) ODA | 1 Var. | 0 R. | S. 1
	$\overline{x}_1$		x_3

K	TVL (10) ODA | 1 Var. | 0 R. | S. 1	K	TVL (11) ODA | 2 Var. | 0 R. | S. 1
	$\overline{x}_1$		x_3 x_6

T	TVL (20) ODA | 1 Var. | 1 R. | S. 1	T	TVL (21) ODA | 2 Var. | 2 R. | S. 1

TVL (20):

```
 1 0
 0 1 x1
```

TVL (21):

```
      0   0 1
      1   1 1
     x6   0 1 x3
```

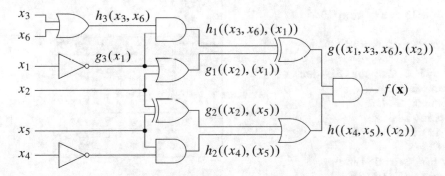

Fig. 11.103 Solution of the PRP shown in Fig. 11.102 of the AND-bi-decomposition computed as Exercise 11.7 (d) and the complete circuit synthesized by the four bi-decompositions of this exercise

The command sbe in lines 5–11 determines the given function $f(\mathbf{x})$, and the subsequent command assign is used to show the associated Karnaugh-map in viewport (1,1) of the m-fold view.

The two nested while-loops in lines 17–62 prepare all pairs of variables of the given function as TVLs 3 and 4. The Boolean variables $orbd, $andbd, and $exorbd are initialized with the value false and control the evaluation of the three types of strong bi-decompositions; the value of such a variable is changed to true when the associated strong bi decomposition with regard to the provided pair of dedicated variables exists. The commands if in lines 24, 38, and 51 exclude the check for the associated type of a detected bi-decomposition with regard to further pairs of dedicated variables.

Condition (11.111) of a strong OR-bi-decomposition of the given function (TVL 1) with regard to the selected pair of dedicated variables (TVLs 3 and 4) is implemented by the commands in lines 26–30. In the case that this condition is satisfied the value of the variable $orbd is changed to true and the current TVLs of the dedicated variables are stored as TVLs 10 and 11 and shown in the first row of the m-fold view.

```
 1   new                                          41   maxk  1  4  7
 2   space  32  1                                 42   dif  6  1  8
 3   avar  1                                      43   if  (te_isc  8  7)
 4   x1  x2  x3  x4  x5.                           44   (
 5   sbe  1  1                                     45   set  $andbd  true
 6   /x1&/x2&/x4&(/x3+x5)+                         46   copy  3  12
 7   x1&x2&/x3&(x4+/x5)+                           47   copy  4  13
 8   /x4&x5&(x1&x2&x3+/x1&/x3)+                    48   assign  12  /m  2  2
 9   x4&(x1&/x2&/x5+/x1&x2&x3)+                    49   assign  13  /m  2  3
10   x1&/x2&x3&(x4+/x5)+                           50   ))
11   x2&x3&/x4&/x5.                                51   if  (not  $exorbd)
12   assign  1  /m  1  1                           52   (
13   set  $orbd  false                            53   sv_uni  3  4  6  ;  exor  bi−dec  f
14   set  $andbd  false                           54   if  (te_derk  1  6)
15   set  $exorbd  false                          55   (
16   sv_get  1  5  ;  usable  for  xb             56   set  $exorbd  true
17   while  (sv_next  1  3  3)                     57   copy  3  14
18   (                                             58   copy  4  15
19   sv_dif  5  3  5                               59   assign  14  /m  3  2
20   if  (ge  (sv_size  5)  1)                     60   assign  15  /m  3  3
21   (                                             61   ))
22   while  (sv_next  5  4  4)                     62   )))
23   (
24   if  (not  $orbd)
25   (
26   cpl  1  2  ;  or  bi−dec  f
27   maxk  2  3  6
28   maxk  2  4  7
29   isc  1  6  8
30   if  (te_isc  8  7)
31   (
32   set  $orbd  true
33   copy  3  10
34   copy  4  11
35   assign  10  /m  1  2
36   assign  11  /m  1  3
37   ))
38   if  (not  $andbd)
39   (
40   maxk  1  3  6  ;  and  bi−dec  f
```

T	TVL (1) ODA	5 Var.	11 R.	S. 1

0 0	1 0 1 0	1 1 1 0
0 1	1 1 0 1	0 1 0 0
1 1	0 0 1 0	1 0 1 0
1 0	0 0 1 0	1 0 1 1
x4 x5	0 1 1 0 0 1 1 0 x3
	0 0 1 1 1 1 0 0 x2
	0 0 0 0 1 1 1 1 x1

Boolean variables

Name	Value
$andbd	false
$exorbd	false
$orbd	false

Fig. 11.104 Problem-program that determines the function given in Exercise 11.8, shows the associated Karnaugh-map, and verifies whether a strong bi-decomposition of this function exists together with the computed results

Condition (11.115) of a strong AND-bi-decomposition of the given function (TVL 1) with regard to the selected pair of dedicated variables (TVLs 3 and 4) is implemented by the commands in lines 40–43. In the case that this condition is satisfied the value of the variable $andbd is changed to true and the current TVLs of the dedicated variables are stored as TVLs 12 and 13 and shown in the second row of the m-fold view.

Condition (11.118) of a strong EXOR-bi-decomposition of the given function (TVL 1) with regard to the selected pair of dedicated variables (TVLs 3 and 4) is implemented by the commands in lines 53 and 54. In the case that this condition is satisfied the value of the variable $\$exorbd$ is changed to `true` and the current TVLs of the dedicated variables are stored as TVLs 14 and 15 and shown in the third row of the m-fold view.

The resulting values `false` of the Boolean variables $\$orbd$, $\$andbd$, and $\$exorbd$ confirm that neither a strong OR-, a strong AND-, nor a strong EXOR-bi-decomposition for the given function $f(\mathbf{x})$ and any pair of dedicated variables exists.

(b) Figure 11.105 shows the PRP that determines and shows the given function as in the previous task, checks whether a weak OR-bi-decomposition with regard to a single variable x_a exists, computes the decomposition-lattice $\mathcal{L}\langle g_{1q}(x_a, \mathbf{x}_c), g_{1r}(x_a, \mathbf{x}_c)\rangle$ of the possible weak OR-bi-decomposition, and verifies whether this lattice contains at least one function for which a strong bi-decomposition exists; the Karnaugh-maps of the given function and the computed decomposition-lattice as well as the TVL of the dedicated variable $x_a = x_1$ of the found weak OR-bi-decomposition are shown below this PRP.

The dedicated set $\mathbf{x}_b$ is empty for each weak bi-decomposition; hence, the command `sv_next` in the single `while`-loop in lines 14–24 is used to select the variable x_a for which the possibility of a weak OR-bi-decomposition is checked. Condition (11.130) is implemented in lines 18 and 19 to check whether a weak OR-bi-decomposition of the given function with regard to the selected variable x_a exists. In the case that this condition is satisfied the control variable $\$worbd$ is changed from the initial value `false` to the value `true`, the TVL of the current dedicated variable x_a is stored as requested as object 10, and shown in viewport (1,3) of the m-fold view.

The commands `copy` and `cpl` in lines 25 and 26 transform the given function into the mark-functions of the associated lattice to satisfy the protocol requirements for the case that no weak OR-bi-decomposition of the given function exists.

The mark-function of the decomposition-lattice $\mathcal{L}\langle g_{1q}(x_a, \mathbf{x}_c), g_{1r}(x_a, \mathbf{x}_c)\rangle$ is computed using Eqs. (11.131) and (11.132) in the case that a weak OR-bi-decomposition has been detected. These equations are implemented in lines 29–31 using the commands `cpl`, `maxk`, and `isc`. The subsequent command `assign_qr` shows the Karnaugh-map of the computed lattice in viewport (1,2) of the m-fold view.

The part of the PRP in lines 34–89 verifies whether the computed decomposition-lattice $\mathcal{L}\langle g_{1q}(x_a, \mathbf{x}_c), g_{1r}(x_a, \mathbf{x}_c)\rangle$ contains at least one function for which a strong bi-decomposition exists. The structure of this part is equal to the PRP in Fig. 11.104 of the previous task. The difference to the previous PRP is that all functions of a lattice and not only a single function must be checked for all three types of strong bi-decompositions. Using Condition (11.156) of the strong OR-bi-decomposition (implemented in lines 48–51), Condition (11.163) of the strong AND-bi-decomposition (implemented in lines 61–64), and Condition (11.168) of the strong EXOR-bi-decomposition (implemented in lines 74–81) all functions of the lattice are commonly verified for the associated type of a strong bi-decomposition. TVLs of found pairs of dedicated variables are stored as objects 100 to 105 so that these objects remain unchanged when in the following task XBOOLE-objects with numbers smaller than 100 are used.

The value of the variable $\$worbd$ is equal to `true`, and the values of the other three variables $\$orbd$, $\$andbd$, and $\$exorbd$ are equal to `false` after the execution of this PRP; this confirms that the decomposition-lattice $\mathcal{L}\langle g_{1q}(x_a, \mathbf{x}_c), g_{1r}(x_a, \mathbf{x}_c)\rangle$ of the weak OR-bi-decomposition has been computed and that this lattice does not contain any function for which a strong bi-decomposition exists.

The command `sts` in line 90 stores the final XBOOLE-system that contains the mark-functions of the decomposition-lattice $\mathcal{L}\langle g_{1q}(x_a, \mathbf{x}_c), g_{1r}(x_a, \mathbf{x}_c)\rangle$ as objects 11 and 12 in an `sdt`-file for the decomposition of this lattice in the next task.

```
 1   new
 2   space 32 1
 3   avar 1
 4   x1 x2 x3 x4 x5.
 5   sbe 1 1
 6   /x1&/x2&/x4&(/x3+x5)+
 7   x1&x2&/x3&(x4+/x5)+
 8   /x4&x5&(x1&x2&x3+/x1&/x3)+
 9   x4&(x1&/x2&/x5+/x1&x2&x3)+
10   x1&/x2&x3&(x4+/x5)+
11   x2&x3&/x4&/x5.
12   assign 1 /m 1 1
13   set $worbd false
14   while (sv_next 1 3 3)
15   (
16   if (not $worbd)
17   (
18   mink 1 3 5 ; wor bi-dec f
19   if (ge (ntv 5) 0)
20   (
21   set $worbd true
22   copy 3 10
23   assign 10 /m 1 3
24   )))
25   copy 1 11 ; fq
26   cpl 1 12 ; fr
27   if ($worbd)
28   (
29   cpl 1 12 ; gr
30   maxk 12 10 7
31   isc 1 7 11 ; gq
32   assign_qr 11 12 /m 1 2
33   )
34   set $orbd false
35   set $andbd false
36   set $exorbd false
37   sv_uni 11 12 2 ; usable for xa
38   copy 2 5 ; usable for xb
39   while (sv_next 2 3 3)
40   (
41   sv_dif 5 3 5
42   if (ge (sv_size 5) 1)
43   (
44   while (sv_next 5 4 4)
45   (
```

```
46   if (not $orbd)
47   (
48   maxk 12 3 6 ; or bi-dec L
49   maxk 12 4 7
50   isc 11 6 8
51   if (te_isc 8 7)
52   (
53   set $orbd true
54   copy 3 100
55   copy 4 101
56   assign 100 /m 2 1
57   assign 101 /m 2 2
58   ))
59   if (not $andbd)
60   (
61   maxk 11 3 6 ; and bi-dec L
62   maxk 11 4 7
63   isc 12 6 8
64   if (te_isc 8 7)
65   (
66   set $andbd true
67   copy 3 102
68   copy 4 103
69   assign 102 /m 3 1
70   assign 103 /m 3 2
71   ))
72   if (not $exorbd)
73   (
74   maxk 11 3 13 ; exor bi-dec L
75   maxk 12 3 14
76   isc 13 14 15 ; fq der xa
77   mink 11 3 16
78   mink 12 3 17
79   uni 16 17 18 ; fr der xa
80   maxk 15 4 19
81   if (te_isc 19 18)
82   (
83   set $exorbd true
84   copy 3 104
85   copy 4 105
86   assign 104 /m 4 1
87   assign 105 /m 4 2
88   ))
89   )))
90   sts "e11_8_b"
```

T	TVL (1) ODA	5 Var.	11 R.	S. 1

```
0 0   1 0 1 0   1 1 1 0
0 1   1 1 0 1   0 1 0 0
1 1   0 0 1 0   1 0 1 0
1 0   0 0 1 0   1 0 1 1

x4 x5   0 1 1 0 0 1 1 0  x3
        0 0 1 1 1 1 0 0  x2
        0 0 0 0 1 1 1 1  x1
```

(q: TVL (11) ODA, r: TVL (12) ODA), 5 Var.	S. 1

```
0 0   1 0 Φ 0   1 Φ 1 0
0 1   1 1 0 1   0 1 0 0
1 1   0 0 1 0   1 0 1 0
1 0   0 0 1 0   1 0 1 1

x4 x5   0 1 1 0 0 1 1 0  x3
        0 0 1 1 1 1 0 0  x2
        0 0 0 0 1 1 1 1  x1
```

K	TVL (10) ODA	1 Var.	0 R.	S. 1

```
x̄
```

Fig. 11.105 Problem-program that computes a weak OR-bi-decomposition of the given function and verifies whether a strong bi-decomposition of the computed decomposition-lattice $\mathcal{L}\langle g_{1q}(x_a, \mathbf{x}_c), g_{1r}(x_a, \mathbf{x}_c)\rangle$ (shown in viewport (1,2)) exists

```
 1   lds  "ell_8_b"                    12   assign  20  /m 1 3
 2   assign_qr  11  12  /m 1 1         13   )
 3   copy  11  21  ;  g1q             14   if  ($wandbd)
 4   copy  12  22  ;  g1r             15   (
 5   set  $wandbd  false             16   copy  11  21  ;  g2q
 6   maxk  11  10  6  ; wand bi-dec L  17   maxk  11  10  7
 7   dif  12  6  7                    18   isc  12  7  22  ;  g2r
 8   if  (ge  (ntv  7)  0)            19   assign_qr  21  22  /m 1 2
 9   (                                20   )
10   set  $wandbd  true              21   sts  "ell_8_c"
11   copy  10  20
```

(q: TVL (11) ODA, r: TVL (12) ODA), 5 Var. | S. 1	(q: TVL (21) ODA, r: TVL (22) ODA), 5 Var. | S. 1	K TVL (20) ODA | 1 Var. | 0 R. | S. 1

Karnaugh-maps below the program headers.

Viewport (1,1):

```
0 0 | 1 0 Φ 0 | 1 Φ 1 0
0 1 | 1 1 0 1 | 0 1 0 0
1 1 | 0 0 1 0 | 1 0 1 0
1 0 | 0 0 1 0 | 1 0 1 1

        0 1 1 0 0 1 1 0 x3
        0 0 1 1 1 1 0 0 x2
        0 0 0 0 1 1 1 1 x1
```

Viewport (1,2):

```
0 0 | 1 0 Φ 0 | 1 Φ 1 0
0 1 | 1 1 0 1 | 0 1 0 0
1 1 | Φ 0 1 0 | 1 0 1 Φ
1 0 | 0 0 1 0 | 1 0 1 1

        0 1 1 0 0 1 1 0 x3
        0 0 1 1 1 1 0 0 x2
        0 0 0 0 1 1 1 1 x1
```

K TVL (20):

```
  x̄
```

Fig. 11.106 Problem-program that computes the lattice $\mathcal{L}\langle g_{2q}(x_a, \mathbf{x}_c), g_{2r}(x_a, \mathbf{x}_c)\rangle$ (see viewport (1,2)) of a weak AND-bi-decomposition of the provided lattice $\mathcal{L}\langle g_{1q}(x_a, \mathbf{x}_c), g_{1r}(x_a, \mathbf{x}_c)\rangle$ (see viewport (1,1)) with regard to the dedicated variable $x_a = x_1$

(c) Figure 11.106 shows the PRP that loads the XBOOLE-system of the previous task using the command `lds` in the first line, shows the provided lattice to decompose in viewport (1,1) of the m-fold view, verifies that a weak AND-bi-decomposition of the given lattice $\mathcal{L}\langle g_{1q}(x_a, \mathbf{x}_c), g_{1r}(x_a, \mathbf{x}_c)\rangle$ (provided as TVLs 11 and 12) with regard to the variable $x_a = x_1$ (provided as TVL 10) exists, and computes the decomposition-lattice $\mathcal{L}\langle g_{2q}(x_a, \mathbf{x}_c), g_{2r}(x_a, \mathbf{x}_c)\rangle$. The Karnaugh-maps of these two lattices and the TVL of the used dedicated variable are shown below the PRP in Fig. 11.106.

The command `assign_qr` in line 2 shows for comparison the Karnaugh-map of the provided lattice in viewport (1,1) of the m-fold view. The subsequent two commands `copy` transfer the mark-function of the given lattice to the objects 21 and 22 to satisfy the protocol requirements for the case that no weak AND-bi-decomposition of the given lattice exists.

Condition (11.199) to check whether a lattice contains at least one function for which a weak AND-bi-decomposition with regard to the dedicated variable x_a exists is implemented in lines 6–8. If this condition is satisfied (as expected), the value of the variable `$wandbd` is changed from the initial value `false` to the value `true` and the TVL of the dedicated variable $x_a = x_1$ is copied to the object 20 to satisfy the requirements of the protocol.

In the case that the existence of the weak AND-bi-decomposition with regard to the dedicated variable x_a has been confirmed, the mark-functions of the decomposition-lattice $\mathcal{L}\langle g_{2q}(x_a, \mathbf{x}_c), g_{2r}(x_a, \mathbf{x}_c)\rangle$ are computed using Eqs. (11.200) and (11.201) that are implemented in lines 16–18 using the commands `copy`, `maxk`, and `isc`. The subsequent command `assign_qr` shows the Karnaugh-map computed lattice in viewport (1,2) of the m-fold view.

The value variable `$wandbd` is equal to `true` after the execution of this PRP; this confirms that the evaluated lattice $\mathcal{L}\langle g_{1q}(x_a, \mathbf{x}_c), g_{1r}(x_a, \mathbf{x}_c)\rangle$ contains at least one function for which a weak AND-bi-decomposition exists. The command `sts` in line 21 stores the final XBOOLE-system for the decomposition of the lattice $\mathcal{L}\langle g_{2q}(x_a, \mathbf{x}_c), g_{2r}(x_a, \mathbf{x}_c)\rangle$ in the next task. TVLs of the mark-functions of this lattice are stored as objects 21 and 22.

```
 1  lds "ell_8_c"                            47  copy 7 44
 2  assign_qr 21 22 /m 1 1                   48  )))
 3  copy 21 31 ; fq                          49  assign 44 /m 3 3
 4  copy 22 32 ; fr                          50  maxk 15 44 35
 5  sv_uni 21 22 5 ; usable for xa           51  sv_get 43 36
 6  sv_uni 21 22 6 ; usable for xb           52  isc 35 36 30 ; g3 exorbd
 7  set $exorbd false                        53  assign 30 /m 1 2
 8  while (sv_next 5 3 3)                     54  cpl 30 37 ; /g3
 9  (                                        55  isc 37 21 45
10  sv_dif 6 3 6                             56  isc 30 22 46
11  if (ge (sv_size 6) 1)                    57  uni 45 46 47
12  (                                        58  maxk 47 43 31 ; h3q exor bi-dec L
13  while (sv_next 6 4 4)                    59  isc 37 22 48
14  (                                        60  isc 30 21 49
15  if (not $exorbd)                         61  uni 48 49 50
16  (                                        62  maxk 50 43 32 ; h3r exor bi-dec L
17  maxk 21 3 13 ; exor bi-dec L             63  assign_qr 31 32 /m 2 1
18  maxk 22 3 14                             64  )
19  isc 13 14 15 ; fq der xa                 65  ; h3 exor bi-decomposition
20  mink 21 3 16                             66  sbe 1 51
21  mink 22 3 17                             67  (x2&x3#/x4)&/((x2#x3)&/x5).
22  uni 16 17 18 ; fr der xa                 68  assign 51 /m 2 2 ; h3 exorbd
23  maxk 15 4 19                             69  syd 30 51 52 ; g2 wabd
24  if (te_isc 19 18)                        70  maxk 11 10 53 ; h2q wabd
25  (                                        71  isc 12 52 54
26  set $exorbd true                         72  maxk 54 10 55 ; h2r wabd
27  copy 3 33 ; xa                           73  assign_qr 53 55 /m 3 1
28  copy 4 34 ; xb                           74  ; h2 weak AND-bi-decomposition
29  assign 33 /m 1 3                         75  sbe 1 56
30  assign 34 /m 2 3                         76  x2+x3+/x4+/x5.
31  )))))                                    77  assign 56 /m 3 2 ; h2 wabd
32  if ($exorbd)                             78  isc 52 56 57 ; g1 wobd
33  (                                        79  cpl 57 58
34  copy 33 43 ; xa                          80  isc 1 58 59
35  copy 34 44 ; xb                          81  maxk 59 10 60 ; h1q wobd
36  sv_uni 21 22 6                           82  cpl 1 2
37  sv_dif 6 33 6                            83  maxk 2 10 61 ; h1r wobd
38  sv_dif 6 34 6 ; usable for xb            84  assign_qr 60 61 /m 4 1
39  if (ge (sv_size 6) 1)                    85  ; h1 weak OR-bi-decomposition
40  (                                        86  sbe 1 63
41  while (sv_next 6 4 4)                    87  x2&x3&/x4&/x5.
42  (                                        88  assign 63 /m 4 2 ; h1 wobd
43  sv_uni 44 4 7                            89  uni 57 63 64 ; f
44  maxk 15 7 19                             90  assign 64 /m 4 3
45  if (te_isc 19 18)                        91  set $equal (te_syd 1 64)
46  (
```

Fig. 11.107 Problem-program that completes the synthesis based on a weak OR-bi-decomposition, a weak AND-bi-decomposition, and a strong EXOR-bi-decomposition

(d) Figure 11.107 shows the PRP that consists of four parts; comments beginning with a semicolon indicate the three parts that have been iteratively appended using the results of the previous part. The first part of this PRP in lines 1–64 loads the sdt-file stored in the previous task that contains the mark-functions of the lattice to decompose $\mathcal{L}\langle g_{2q}(x_a, \mathbf{x}_c), g_{2r}(x_a, \mathbf{x}_c)\rangle$ as TVLs 21 and 22, shows the Karnaugh-map of this lattice in viewport (1,1) of the m-fold view (line 2), verifies whether this lattice contains at least one function for which an EXOR-bi-decomposition exists (the found pair of dedicates variables is shown in viewport (1,3) $x_a = x_1$ and (2,3) $x_b = x_2$ of the

m-fold view), expands the single variable of the initial dedicated set $\mathbf{x}_b$ of the found EXOR-bi-decomposition as much as possible, and computes the decomposition-function $g_3(x_a)$ (shown in viewport (1,2) of the m-fold view) and the associated lattice of decomposition-functions $\mathcal{L}\langle h_{3q}(x_a, \mathbf{x}_c), h_{3r}(x_a, \mathbf{x}_c)\rangle$ (shown in viewport (2,1) of the m-fold view).

The second part of this PRP in lines 65–73 determines the function $h_3(\mathbf{x}_b)$ (shown in viewport (2,2) of the m-fold view) belonging to the lattice $\mathcal{L}\langle h_{3q}(x_a, \mathbf{x}_c), h_{3r}(x_a, \mathbf{x}_c)\rangle$ and uses this function to compute the lattice $\mathcal{L}\langle h_{2q}(x_a, \mathbf{x}_c), h_{2r}(x_a, \mathbf{x}_c)\rangle$ (shown in viewport (3,1) of the m-fold view) of the weak AND-bi-decomposition.

The third part of this PRP in lines 74–84 determines the function $h_2(\mathbf{x}_c)$ (shown in viewport (3,2) of the m-fold view) belonging to the lattice $\mathcal{L}\langle h_{2q}(\mathbf{x}_c), h_{2r}(\mathbf{x}_c)\rangle$ and uses this decomposition-function to compute the lattice $\mathcal{L}\langle h_{1q}(\mathbf{x}_c), h_{1r}(\mathbf{x}_c)\rangle$ (shown in viewport (4,1) of the m-fold view) of the weak OR-bi-decomposition.

The last part of this PRP in lines 85–91 determines the function $h_1(\mathbf{x}_c)$ (shown in viewport (4,2) of the m-fold view) belonging to the lattice $\mathcal{L}\langle h_{1q}(\mathbf{x}_c), h_{1r}(\mathbf{x}_c)\rangle$ that computes the function $f(\mathbf{x})$ realized by the synthesized circuit, shows the Karnaugh-map of $f(\mathbf{x})$ in viewport (4,3) of the m-fold view for comparison with the give function of this exercise, and verifies successfully that the synthesized circuit realizes the given function $f(\mathbf{x})$.

Figure 11.108 shows the results of the PRP shown in Fig. 11.107 together with the synthesized circuit.

The first part of the PRP in Fig. 11.107 shows the complete procedure to compute a compact EXOR-bi-decomposition of the lattice determined by the mark-functions $g_{2q}(\mathbf{x})$ (TVL 21) and $g_{2r}(\mathbf{x})$ (TVL 22). The two nested while-loops in lines 8–31 prepare all pairs of variables of the given lattice as TVLs 3 and 4. The commands in lines 17–19 implement Eq. (7.24) of the function $f_q^{\mathrm{der}_{x_a}}(\mathbf{x})$, and the commands in lines 20–22 the associated mark-function $f_r^{\mathrm{der}_{x_a}}(\mathbf{x})$ based on Eq. (7.25); these mark-functions describe the lattice that results as derivative with regard to x_a (TVL 3) of the given lattice. The commands in lines 23 and 24 implement Condition (11.168) to verify whether an initial EXOR-bi-decomposition with regard to the selected variables x_a (TVL 3) and x_b (TVL 4) exists. The value of the variable $exorbd is changed from the initial value false to true if this condition is satisfied; the dedicated variables x_a and x_b are stored as objects 33 and 34 and shown in viewports (1,3) and (2,3) in this case.

The second step in the first part of the PRP in Fig. 11.107 aims to increase the dedicated set $\mathbf{x}_b$ as much as possible; the objects 43 and 44 are used to store the final dedicated sets of the EXOR-bi-decomposition. The commands in lines 36–38 determine all variables except x_a and x_b as possible candidates to increase the dedicated set $\mathbf{x}_b$. The while-loop in lines 41–48 iteratively provides the variables usable to increase the dedicated set $\mathbf{x}_b$. The command sv_uni in line 43 temporarily inserts the selected variable to the dedicated set $\mathbf{x}_b$ (TVL 7). The commands in lines 44 and 45 implement Condition (11.168) to verify whether an extended EXOR-bi-decomposition with regard to the selected variables x_a (TVL 3) and $\mathbf{x}_b$ (TVL 7) exists; TVLs 15 and 18 can be reused because the dedicated variable x_a remains unchanged. The temporary dedicated set $\mathbf{x}_b$ (TVL 7) is permanently stored as TVL 44 if this condition is satisfied. The command assign in line 49 assigns the maximally increased dedicated set $\mathbf{x}_b$ to viewport (3,3). The result of this analysis is that the lattice to decompose $\mathcal{L}\langle g_{2q}(x_a, \mathbf{x}_c), g_{2r}(x_a, \mathbf{x}_c)\rangle$ contains a function for which a disjoint EXOR-bi-decomposition with regard to $x_a = x_1$ and $\mathbf{x}_b = (x_2, x_3, x_4, x_5)$ exists.

The commands in lines 50–52 implement Eq. (11.169) to compute the decomposition-function $g_3(x_1) = x_1$; the Karnaugh-map of this decomposition-function is shown in viewport (1,2) of the m-fold view.

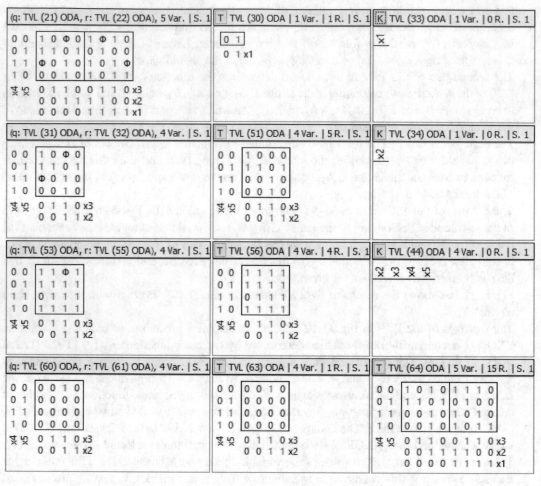

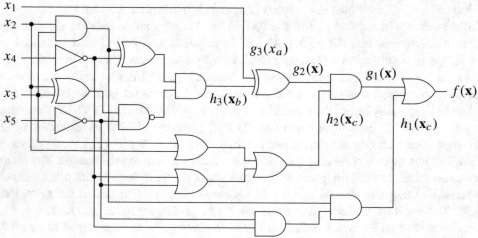

Fig. 11.108 Solutions of the PRP shown in Fig. 11.107 and the complete circuit synthesized in this exercise using a weak OR-bi-decomposition, a weak AND-bi-decomposition, and a strong EXOR-bi-decomposition

The commands in lines 54–62 compute the mark-functions of the decomposition-lattice $\mathcal{L}\langle h_{3q}(\mathbf{x}_b), h_{3r}(\mathbf{x}_b)\rangle$ of the EXOR-bi-decomposition using Eqs. (11.170) and (11.171). The subsequent command assigns the Karnaugh-map of this lattice to viewport (2,1) of the m-fold view.

This is the end of the first part of this PRP because the chosen function of the lattice $\mathcal{L}\langle h_{3q}(\mathbf{x}_b), h_{3r}(\mathbf{x}_b)\rangle$ of the EXOR-bi-decomposition is required to compute the realized function $g_2(\mathbf{x})$ at the output of the EXOR-gate.

The aim of the *second part* of this PRP is the computation of the decomposition-lattice $\mathcal{L}\langle h_{2q}(\mathbf{x}_c), h_{2r}(\mathbf{x}_c)\rangle$ of the weak AND-bi-decomposition that requires the specification of the decomposition-function $g_2(\mathbf{x})$ and the selection of the function $h_3(\mathbf{x}_b)$ out of the known decomposition-lattice.

The command sbe in lines 66 and 67 determines the realized function $h_3(\mathbf{x}_b)$ based on further bi-decompositions of the lattice $\mathcal{L}\langle h_{3q}(\mathbf{x}_b), h_{3r}(\mathbf{x}_b)\rangle$ that can be synthesized using strong bi-decompositions in an additionally supplementary task; the expression of this function reveals that a strong AND-bi-decomposition and thereafter a disjoint EXOR-bi-decomposition and a disjoint OR-bi-decomposition have been detected. The subsequent command shows the Karnaugh-map of this function in viewport (2,2) of the m-fold view.

The command syd in line 69 computes the function $g_2(\mathbf{x})$ using the decomposition-functions $g_3(x_a)$ and $h_3(\mathbf{x}_b)$ of the EXOR-bi-decomposition. The commands in lines 70–72 implement Eqs. (11.202) and (11.203) to compute the mark-functions of the decomposition-lattice $\mathcal{L}\langle h_{2q}(\mathbf{x}_c), h_{2r}(\mathbf{x}_c)\rangle$ of the weak AND-bi-decomposition, and the subsequent command shows the Karnaugh-map of this lattice in viewport (3,1).

The aim of the *third part* of this PRP is the computation of the decomposition-lattice $\mathcal{L}\langle h_{1q}(\mathbf{x}_c), h_{1r}(\mathbf{x}_c)\rangle$ of the weak OR-bi-decomposition that requires the selection of the function $h_2(\mathbf{x}_c)$ out of the known decomposition-lattice and the computation of the decomposition-function $g_1(\mathbf{x})$.

The simplest function $h_2(\mathbf{x}_c) = x_2 \vee x_3 \vee \overline{x}_4 \vee \overline{x}_5$ of the decomposition-lattice $\mathcal{L}\langle h_{2q}(\mathbf{x}_c), h_{2r}(\mathbf{x}_c)\rangle$ uses the value 1 in the position of the don't-care. This function can be realized using three OR-gates of two inputs and two NOT-gates to get the negated variables $\overline{x}_4$ and $\overline{x}_5$. The command sbe in lines 75 and 76 determines this simple function $h_2(\mathbf{x}_b)$, and the subsequent command shows the Karnaugh-map of this function in viewport (3,2) of the m-fold view.

The command isc in line 78 computes the function $g_1(\mathbf{x})$ using the decomposition-functions $g_2(x_a)$ and $h_2(\mathbf{x}_b)$ of the AND-bi-decomposition. The commands in lines 79–83 implement Eqs. (11.197) and (11.198) to compute the mark-functions of the decomposition-lattice $\mathcal{L}\langle h_{1q}(\mathbf{x}_c), h_{1r}(\mathbf{x}_c)\rangle$ of the weak OR-bi-decomposition, and the subsequent command shows the Karnaugh-map of this lattice in viewport (4,1) of the m-fold view.

The aim of the *fourth part* of this PRP is the computation of the function realized by the synthesized circuit and the comparison with the given function. The decomposition-lattice $\mathcal{L}\langle h_{1q}(\mathbf{x}_c), h_{1r}(\mathbf{x}_c)\rangle$ of the weak OR-bi-decomposition contains only the function $h_1(\mathbf{x}_c) = x_2 x_3 \overline{x}_4 \overline{x}_5$ that is determined by the command sbe in lines 86 and 87 and shown in viewport (4,2) of the m-fold view by the subsequent command assign.

The command uni in line 89 computes the function $f(\mathbf{x})$ realized by the synthesized circuit using the decomposition-functions $g_1(\mathbf{x})$ and $h_1(\mathbf{x}_c)$ of the weak OR-bi-decomposition. The command assign shows the Karnaugh-map of realized function $f(\mathbf{x})$ in viewport (4,3) of the m-fold view. The command te_syd in line 91 compares the given function $f(\mathbf{x})$ (TVL 1) with the realized function (TVL 64). The result $equal=true confirms that the synthesized circuit (shown in Fig. 11.108) realizes the given function $f(\mathbf{x})$ of this exercise.

1 new	22 ; x2–G4–G7–y1	43 cpl 4 35
2 space 32 1	23 derk 1 9 10	44 isc 20 31 36
3 avar 1	24 cpl 10 11	45 isc 36 35 37
4 x1 x2 x3 x4 x5	25 derk 5 9 12	46 isc 37 33 38
5 y1 y2 .	26 cpl 12 13	47 isc 38 34 39
6 sbe 1 1	27 derk 7 9 14	48 obb 39 39
7 / x1 .	28 cpl 5 15	49 assign 39 /m 1 2
8 sbe 1 2	29 isc 1 11 16	50 ; x2–G2–G5–G8–y2
9 / x2 .	30 isc 16 15 17	51 derk 20 9 40
10 sbe 1 3	31 isc 17 13 18	52 cpl 40 41
11 / x5 .	32 isc 18 14 19	53 derk 6 9 42
12 sbe 1 4	33 obb 19 19	54 cpl 42 43
13 / x1&x2 .	34 assign 19 /m 1 1	55 derk 8 9 44
14 sbe 1 5	35 ; x2–G2–G5–G7–y1	56 cpl 6 45
15 / x2&x3 .	36 sbe 1 20	57 isc 20 41 46
16 sbe 1 6	37 x3 .	58 isc 46 45 47
17 x4&/x5 .	38 derk 20 9 30	59 isc 47 43 48
18 uni 4 5 7	39 cpl 30 31	60 isc 48 44 49
19 uni 5 6 8	40 derk 4 9 32	61 obb 49 49
20 vtin 1 9	41 cpl 32 33	62 assign 49 /m 1 3
21 x2 .	42 derk 7 9 34	

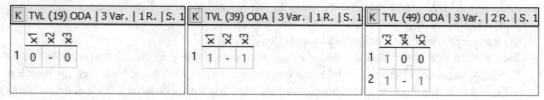

Fig. 11.109 Problem-program that computes the test-pattern (shown below this PRP) for three paths of the combinational circuit shown in Fig. 11.66 using the network-model: sensible path

Solution 11.9 (Test-Patterns of a Combinational Circuit)

(a) Figure 11.109 shows the PRP that uses the network-model *sensible path* to compute the test-patterns of three paths of the combinational circuit shown in Fig. 11.66 indicated by the comments in lines 22, 35, and 50 and below this PRP the computed test-patterns.

The commands sbe and uni in lines 6–19 determine the functions of all outputs of the gates of the explored combinational circuit. The commands derk and cpl of this PRP compute the functions that are combined by the commands isc according to Eq. (11.257) to compute the test-patterns for the selected path.

The benefit of the test-patterns computed using the network-model of the sensible path is that both SA0- and SA1-errors are detected on all wires of the chosen path when the determined inputs of a test-pattern remain unchanged and the change on the input x_2 does not cause the change of the associated output. One drawback of the network-model of the sensible path is that all controlling functions of the path and the output-function are required to compute the test-patterns. Another drawback is that Eq. (11.257) does not compute test-pattern that can detect either an SA0- or an SA1-error.

(b) Figure 11.110 shows the PRP that computes and shows the function of the required behavior $F^R(\mathbf{x}, \mathbf{y})$ of the combinational circuit shown in Fig. 11.66.

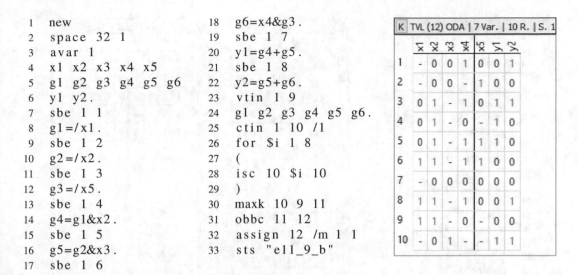

```
1   new                          18   g6=x4&g3 .
2   space 32 1                   19   sbe 1 7
3   avar 1                       20   y1=g4+g5 .
4   x1 x2 x3 x4 x5               21   sbe 1 8
5   g1 g2 g3 g4 g5 g6            22   y2=g5+g6 .
6   y1 y2 .                      23   vtin 1 9
7   sbe 1 1                      24   g1 g2 g3 g4 g5 g6 .
8   g1=/x1 .                     25   ctin 1 10 /1
9   sbe 1 2                      26   for $i 1 8
10  g2=/x2 .                     27   (
11  sbe 1 3                      28   isc 10 $i 10
12  g3=/x5 .                     29   )
13  sbe 1 4                      30   maxk 10 9 11
14  g4=g1&x2 .                   31   obbc 11 12
15  sbe 1 5                      32   assign 12 /m 1 1
16  g5=g2&x3 .                   33   sts "e11_9_b"
17  sbe 1 6
```

K	TVL (12) ODA	7 Var.	10 R.	S. 1				
	x_1	x_2	x_3	x_4	x_5	y_1	y_2
1	-	0	0	1	0	0	1
2	-	0	0	-	1	0	0
3	0	1	-	1	0	1	1
4	0	1	-	0	-	1	0
5	0	1	-	1	1	1	0
6	1	1	-	1	1	0	0
7	-	0	0	0	0	0	0
8	1	1	-	1	0	0	1
9	1	1	-	0	-	0	0
10	-	0	1	-	-	1	1

Fig. 11.110 Problem-program that computes the function $F^R(\mathbf{x}, \mathbf{y})$ of the combinational circuit shown in Fig. 11.66 using the local lists of phases of all gates together with the computed result

The commands sbe in lines 7–22 determine the local list of phase using logic equations of the gates. The object numbers of these TVLs correspond to the numbers of the gates; this facilitates modifications required to compute the list of error-phases $F^E(\mathbf{x}, \mathbf{y}, \mathbf{s}, T)$ in the next two tasks.

The repeated execution of the command isc in line 28 in the body of the for-loop in lines 26–29 computes the global list of phase of the explored circuit. The subsequent command maxk removes all internal variables g_i. The command obbc in line 31 reduces the number of ternary vectors from 18 to 10 before the TVL 12 of the minimize function $F^R(\mathbf{x}, \mathbf{y})$ is assigned to viewport (1,1) of the m-fold view. The last command sts stores the complete XBOOLE-system as file "e11_9_b.sdt" for the use in the next two tasks.

(c) Figure 11.111 shows the PRP that loads the sdt-file of the previous task, prepares the functions $GM_i(\mathbf{x}, \mathbf{y}, \mathbf{g}, \mathbf{s})$ modified to detect errors at the local internal signal g_2, defines the function $F_T(g_2, T)$, and computes the list of error-phases $F^E(\mathbf{x}, \mathbf{y}, \mathbf{s}, T)$ and thereafter the complete set of test-patterns for the chosen sensible point. The computed function $F^E(\mathbf{x}, \mathbf{y}, \mathbf{s}, T)$ and all test-patterns usable to detect SA0- and SA1-errors at the local internal signal g_2 are shown on the right-hand side of Fig. 11.111.

The first for-loop in lines 2–5 copies the eight local list of phase of the gates as source of a modified model. The signal g_2 controls on input of the gate G_5; hence, the command sbe in lines 6 and 7 modifies this local list of phases. The command sbe in lines 8 and 9 defines the function $F_T(g_2, T)$ needed to distinguish between test-patterns usable to detect SA0- or SA1-errors.

The second for-loop in lines 11–14 and the subsequent commands maxk and obbc compute and minimize the list of error-phases $F^E(\mathbf{x}, \mathbf{y}, \mathbf{s}, T)$.

Knowing the functions $F^R(\mathbf{x}, \mathbf{y})$ and $F^E(\mathbf{x}, \mathbf{y}, \mathbf{s}, T)$, only two commands (_derk and isc in lines 17 and 18) are needed to compute all test-patterns for the chosen internal signal g_2 using Eq. (11.263). The computed and subsequently minimized TVL of all test-patterns of the local internal signal g_2 are assigned to viewport (1,2) of the m-fold view.

(d) Figure 11.112 shows the PRP that loads the sdt-file stored in Task (b) of this exercise, prepares the functions $GM_i(\mathbf{x}, \mathbf{y}, \mathbf{g}, \mathbf{s})$ modified to detect errors at the local internal branch g_5, defines

```
1    lds  "e11_9_b"
2    for  $i  1  8
3    (
4    copy  $i  (add  $i  20)
5    )
6    sbe  1  25
7    g5=s&x3.
8    sbe  1  29
9    g2#T=1.
10   ctin  1  20  /1
11   for  $i  21  29
12   (
13   isc  20  $i  20
14   )
15   maxk  20  9  31
16   obbc  31  32
17   _derk  32  <s>  33
18   isc  12  33  34
19   obbc  34  34
20   assign  32  /m  1  1
21   assign  34  /m  1  2
```

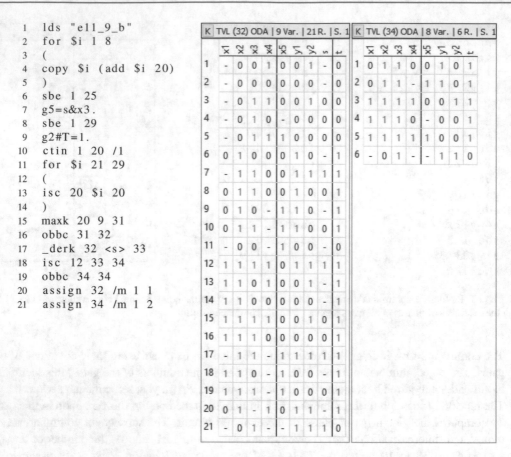

K	x_1	x_2	x_3	x_4	x_5	y_1	y_2	s	t
1	-	0	0	1	0	0	1	-	0
2	-	0	0	0	0	0	0	-	0
3	-	0	1	1	0	0	1	0	0
4	-	0	1	0	-	0	0	0	0
5	-	0	1	1	1	0	0	0	0
6	0	1	0	0	0	1	0	-	1
7	-	1	1	0	0	1	1	1	1
8	0	1	1	0	0	1	0	0	1
9	0	1	0	-	1	1	0	-	1
10	0	1	1	-	1	1	0	0	1
11	-	0	0	-	1	0	0	-	0
12	1	1	1	1	0	1	1	1	1
13	1	1	0	1	0	0	1	-	1
14	1	1	0	0	0	0	0	-	1
15	1	1	1	1	0	0	1	0	1
16	1	1	1	0	0	0	0	0	1
17	-	1	1	-	1	1	1	1	1
18	1	1	0	-	1	0	0	-	1
19	1	1	1	-	1	0	0	0	1
20	0	1	-	1	0	1	1	-	1
21	-	0	1	-	-	1	1	1	0

K	x_1	x_2	x_3	x_4	x_5	y_1	y_2	t
1	0	1	1	0	0	1	0	1
2	0	1	1	-	1	1	0	1
3	1	1	1	1	0	0	1	1
4	1	1	1	0	-	0	0	1
5	1	1	1	1	1	0	0	1
6	-	0	1	-	-	1	1	0

Fig. 11.111 Problem-program that computes all test-patterns usable to detect SA0- and SA1-errors at the local internal signal g_2 of the combinational circuit shown in Fig. 11.66 together with the list of error-phases $F^E(\mathbf{x}, \mathbf{y}, s, T)$ and the computed test-patterns

the functions $F_T(g_5, T)$ and $F_B(\mathbf{s}, T)$, and computes the list of error-phases $F^E(\mathbf{x}, \mathbf{y}, s, T)$ and thereafter the three complete sets of test-patterns for the connections of the chosen sensible branch. The TVLs of the three sets of test-patterns usable to detect SA0- and SA1-errors at the inputs g_5 of the gates G_7 and G_8 as well as the output of the gate G_5 of the local internal branch g_5 are shown below the PRP in Fig. 11.112.

The modified model is prepared similar to the previous task; due to the branch of g_5, the two commands sbe in lines 6–9 determine the modified local lists of phases of the gates G_7 and G_8. The command sbe in lines 10 and 11 defines the function $F_T(g_9, T)$ needed to distinguish between test-patterns usable to detect either SA0- or SA1-errors. The command tin in lines 14–17 determines additionally the function $F_B(\mathbf{s}, T)$ needed to distinguish between the source and two sinks s_1 and s_2 of this branch. The second for-loop in lines 19–22 and the subsequent commands maxk and obbc compute and minimize the list of error-phases $F^E(\mathbf{x}, \mathbf{y}, \mathbf{s}, T)$.

The commands in lines 25–29 implement Eq. (11.270) to compute all test-patterns of the sink s_1 (input of the gate G_7) of the branch g_5. The same equation is implemented in lines 30–34 to compute all test-patterns of the sink s_2 (input of the gate G_8) and in lines 35–39 to compute all test-patterns of the output g_5 of the gate G_5. The last three commands assign show the three computed sets of test-pattern in the viewports of the first row of the m-fold view.

```
 1  lds "e11_9_b"          15  t s1 s2 .            29  obbc 44 44
 2  for $i 1 8             16  100                  30  _maxk 14 <s2> 51
 3  (                      17  011 .                31  isc 51 32 52
 4  copy $i (add $i 20)    18  ctin 1 20 /1         32  derk 52 13 53
 5  )                      19  for $i 21 29         33  isc 53 12 54
 6  sbe 1 27               20  (                    34  obbc 54 54
 7  y1=g4+s1 .             21  isc 20 $i 20         35  _maxk 14 <t> 61
 8  sbe 1 28               22  )                    36  isc 61 32 62
 9  y2=s2+g6 .             23  maxk 20 9 31         37  derk 62 13 63
10  sbe 1 29               24  obbc 31 32           38  isc 63 12 64
11  g5#t =1 .              25  _maxk 14 <s1> 41     39  obbc 64 64
12  vtin 1 13              26  isc 41 32 42         40  assign 44 /m 1 1
13  s1 s2 .                27  derk 42 13 43        41  assign 54 /m 1 2
14  tin 1 14               28  isc 43 12 44         42  assign 64 /m 1 3
```

K TVL (44) ODA | 8 Var. | 7 R. | S. 1

K	x_1	x_2	x_3	x_4	x_5	y_1	y_2	t
1	1	1	-	1	0	0	1	1
2	-	0	0	1	0	0	1	1
3	-	0	1	-	-	1	1	0
4	1	1	-	1	1	0	0	1
5	-	0	0	1	1	0	0	1
6	-	0	0	0	-	0	0	1
7	1	1	-	0	-	0	0	1

K TVL (54) ODA | 8 Var. | 8 R. | S. 1

K	x_1	x_2	x_3	x_4	x_5	y_1	y_2	t
1	0	1	-	0	-	1	0	1
2	0	1	-	1	1	1	0	1
3	1	1	-	1	1	0	0	1
4	-	0	0	1	1	0	0	1
5	-	0	0	0	-	0	0	1
6	1	1	-	0	-	0	0	1
7	-	0	1	-	1	1	1	0
8	-	0	1	0	0	1	1	0

K TVL (64) ODA | 8 Var. | 9 R. | S. 1

K	x_1	x_2	x_3	x_4	x_5	y_1	y_2	t
1	1	1	-	1	0	0	1	1
2	-	0	0	1	0	0	1	1
3	0	1	-	0	-	1	0	1
4	0	1	-	1	1	1	0	1
5	1	1	-	1	1	0	0	1
6	-	0	0	1	1	0	0	1
7	-	0	0	0	-	0	0	1
8	1	1	-	0	-	0	0	1
9	-	0	1	-	-	1	1	0

Fig. 11.112 Problem-program that computes all test-patterns usable to detect SA0- and SA1-errors at the local internal branch g_5 of the combinational circuit shown in Fig. 11.66 together with the computed three sets test-patterns for all connections of this branch with the gates G_5, G_7, and G_8

The selection of test-patterns out of these sets facilitates not only the detection but also the localization of an error. The short PRPs to compute all test-patterns of a local internal signal or a local internal branch confirm the advantage of the network-model of the sensible point in comparison to the network-model of the sensible path.

11.8 Supplementary Exercises

Exercise 11.10 (Behavior of a Combinational Circuit Realized by Gates)

The structure of a combinational circuit is shown in Fig. 11.113. Prepare PRPs that compute the behavior of the circuit shown in Fig. 11.113 using:

(a) a system of logic equations,
(b) a set of local lists of phases,
(c) three TVLs in D-form of the functions $y_i = f_i(\mathbf{x})$, and
(d) a structure-TVL of an equivalent PLA.

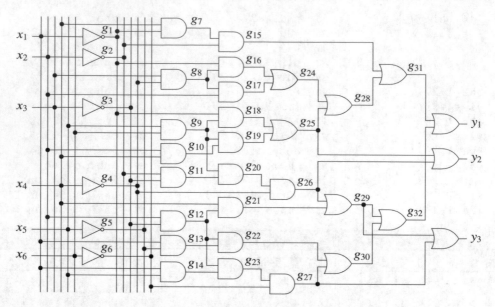

Fig. 11.113 Structure of a combinational circuit that realizes three functions $y_i = f_i(\mathbf{x})$

Fig. 11.114 Structure of
a combinational circuit that
realizes the function
$y = f(\mathbf{x})$

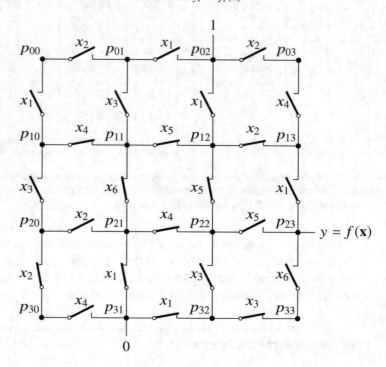

Show in each PRP the computed intermediate results and the minimized input–output list of phases $F(\mathbf{x}, \mathbf{y})$. Verify the result in the last three PRPs using the system of logic equations prepared in the first task.

Exercise 11.11 (Behavior of a Combinational Circuit Realized by Switches)

The structure of a combinational circuit realized by switches is shown in Fig. 11.114. Such a circuit facilitates input-patterns for which the *output is open* but requires a carefully synthesis to avoid *short-circuit faults*.

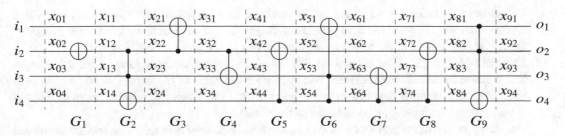

Fig. 11.115 Structure of a reversible combinational circuit of nine gates using four lines

(a) Compute the global list of phases $F(\mathbf{x}, \mathbf{p}, y)$ as well as the input–output list of phases $F(\mathbf{x}, y)$, show the associated TVLs, and store the final XBOOLE-system as sdt-file for the use in the next three tasks.

(b) Load the sdt-file of the previous task and verify whether a short-circuit fault occurs for any input-pattern; i.e., check whether $F_{scf}(\mathbf{x}) \equiv 0$.

(c) Load the sdt-file of Task (a) of this exercise and compute all input-patterns $F_{open}(\mathbf{x})$ for which the output y is open, i.e., for these input-pattern there exists neither a connection to the value 0 nor a connection to the value 1. Show the TVL of $F_{open}(\mathbf{x})$.

(d) Load the sdt-file of Task (a) of this exercise and compute all input–output-phases $F_{det}(\mathbf{x}, y)$ for the input-patterns, which determines the value of the output y either to 0 or to 1. Show the TVL of $F_{det}(\mathbf{x}, y)$.

Exercise 11.12 (Behavior of a Reversible Combinational Circuit)

The structure of a reversible combinational circuit is shown in Fig. 11.115.

(a) Describe firstly the local list of phases $F_j(\mathbf{x}_{j-1}, \mathbf{x}_j)$ of each gate G_j using a system of logic equations and show these nine local lists of phases in the m-fold view. Compute secondly the global behavior of the given reversible circuit as list of phases $F(\mathbf{x}_0, \ldots, \mathbf{x}_9)$ and show this TVL in the 1-fold view. Compute finally the input–output-behavior of this reversible circuit as list of phases $F(\mathbf{x}_0, \mathbf{x}_9)$ and show this TVL in one more viewport of the m-fold view.

(b) The number of variables used to compute the system-function $F(\mathbf{i}, \mathbf{o})$ of the input–output-behavior of the given reversible circuit can be restricted to four input-variables i_i, four output-variables, and four helping variables h_i. Determine the initial system-function $F(\mathbf{i}, \mathbf{o})$ for a reversible circuit of four lines that contains no gate. Describe successively each gate of the given reversible circuit using a system of four equations $o_i = f(h_1, h_2, h_3, h_4)$ and use the commands cco, isc, and maxk to compute the system-function $F(\mathbf{i}, \mathbf{o})$ where the variables $\mathbf{o}$ determine the outputs of the currently added gate. The system-function $F(\mathbf{i}, \mathbf{o})$ of the input–output-behavior of the complete reversible circuit is the result of this procedure when all gates are added in the given order.

(c) A TVL $S(\mathbf{h})$ in D-form can be used to determine the structural information of the given reversible circuit that uses only NOT-, controlled NOT-, and Toffoli-gates. Each gate is specified in this TVL by one ternary vector; the order of the ternary vectors corresponds with the order of the gates in the circuit. The single value 0 of such a ternary vector determines the target line of the associated gate, possible values 1 indicate the lines that control this gate, and the remaining dashes belong to lines the gate is not depending on.

Generate in an inner `for`-loop the list of phases $F(\mathbf{h}, \mathbf{o})$ belonging to the gate of a selected ternary vector of $S(\mathbf{h})$ and use the approach of the previous task to compute in an outer `for`-loop the list of phases $F(\mathbf{i}, \mathbf{o})$ of the input–output-behavior of the given reversible circuit.

(d) The gates G_5 and G_8 of the given reversible circuit are controlled NOT-gates on the same target line 2 that are controlled by the same line 4. Neither the value of the target line nor the value of the control line changes between these two gates; hence, these two gates can be removed without any change of the input–output-behavior of the circuit.

Use the approach of the previous task to compute the behavior of both the given and the simplified reversible circuit and verify whether these two circuits have the same input–output-behavior. Show the structure-TVLs $S_{given}(\mathbf{h})$ and $S_{simplified}(\mathbf{h})$ as well as the associated TVLs and of the input–output-behavior $F_{given}(\mathbf{i}, \mathbf{o})$ and $F_{simplified}(\mathbf{i}, \mathbf{o})$ in the m-fold view.

Exercise 11.13 (Synthesis Using Covering Methods)

The aim of this exercise is the synthesis of the function $y_1 = f_1(\mathbf{x})$ defined as the first function of the circuit shown in Fig. 11.113 using covering methods.

(a) Compute all prime conjunctions of $f_1(\mathbf{x})$ using the consensus rule (4.12) to create new conjunctions and the absorption rule to exclude conjunctions that are covered by another conjunction within an iterative procedure. Show the minimized TVL in ODA-form of the function $f_1(\mathbf{x})$ used as source, the newly created conjunctions after each sweep of the iterative approach, and the computed TVL in D-form of all prime conjunctions of $f_1(\mathbf{x})$. How many prime conjunctions exist for the explored function? Store the final XBOOLE-system as `sdt`-file for the use in the next task.

(b) Load the `sdt`-file of the previous task and select all essential prime conjunctions. Show the TVLs of the explored function $f_1(\mathbf{x})$, the set of all prime conjunctions (taken from the loaded `sdt`-file), and the selected essential prime conjunctions. How many essential prime conjunctions exist for the explored function? Store the final XBOOLE-system as `sdt`-file for the use in the next task.

(c) Load the `sdt`-file of the previous task and use a greedy approach to extend the set of the essential prime conjunctions using in each sweep the conjunction that covers the largest number of additional values 1 until a complete cover of the explored function $f_1(\mathbf{x})$ is reached. Show the TVLs of the explored function $f_1(\mathbf{x})$, the set of all prime conjunctions (taken from the loaded `sdt`-file), the determined essential prime conjunctions, and the computed complete cover of $f_1(\mathbf{x})$. How many conjunctions belong to the computed minimal disjunctive form of the function $f_1(\mathbf{x})$? How many gates with up to two inputs are needed to realize the circuit based on the computed minimal disjunctive form? Compare this number with the number of gates that realize the function $f_1(\mathbf{x})$ in the circuit shown in Fig. 11.113.

(d) Load the `sdt`-file of Task (a) of this exercise to get the set of all prime conjunctions and compute all exact minimal disjunctive forms of $f_1(\mathbf{x})$ (such a form consists of the smallest possible number of prime conjunction that completely cover the explored function). Generate therefore the conjunctive form of the function $cov(\mathbf{p})$ (11.78) in a second Boolean space, use the commands `cpl`, `ndm`, and `obbc` to transform this TVL into a disjunctive form; the smallest numbers of values 1 in the resulting ternary vectors determine the wanted exact minimal disjunctive forms of $f_1(\mathbf{x})$. What is the minimal number of prime conjunctions needed to express the function $f_1(\mathbf{x})$? How many minimal disjunctive forms of $f_1(\mathbf{x})$ exist? How many exact minimal disjunctive forms of $f_1(\mathbf{x})$ exist? Show the TVLs of the explored function $f_1(\mathbf{x})$, the set of all prime conjunctions (taken from the loaded `sdt`-file), the generated function $cov(\mathbf{p})$ in conjunctive form, the select vectors $\mathbf{p}$ that specify the exact minimal disjunctive forms, and the TVLs in D-form belonging to these exact minimal disjunctive forms of $f_1(\mathbf{x})$. How many gates with up to two inputs are

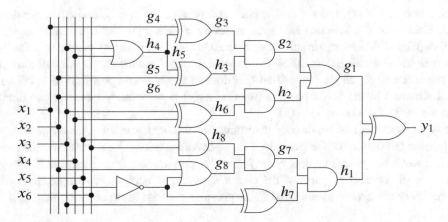

Fig. 11.116 Structure of the combinational circuit that realizes the function $y_1 = f_1(\mathbf{x})$ of Fig. 11.113 synthesized using several bi-decompositions

needed to realize the circuit based on the computed exact minimal disjunctive forms? Compare this number with the number of gates that realize the function $f_1(\mathbf{x})$ in the circuit shown in Fig. 11.113.

Exercise 11.14 (Synthesis Using Strong and Weak Bi-decompositions)

The aim of this exercise is the complete synthesis of the function $y_1 = f_1(\mathbf{x})$ defined as the first function of the circuit shown in Fig. 11.113 using bi-decompositions; this is the same function as synthesized in the previous exercise so that a direct comparison is possible.

Figure 11.116 shows circuit synthesized by strong and weak bi-decompositions in this exercise; the associated labels of the decomposition-functions are used for a clear specification of the four tasks of this exercise. This circuit structure facilitates also the verification of the computed results.

Prepare the PRP for the tasks of this exercise in a sequence of parts and use the observed results of each part for the implementation of the next part. Check for each function to decompose whether a strong bi-decomposition with regard to a selected pair of dedicated variables exists and extend thereafter the dedicated sets of variables of the detected type of bi-decomposition as much as possible. Prefer a weak OR-bi-decomposition in the case that no strong bi-decomposition exists. Use in a weak bi-decomposition only one variable x_a so that the decomposition-functions depend almost on the same numbers of variables. Take care, lattices of function can occur even though a completely specified function is given. Terminate the recursive bi-decompositions when a decomposition-function depends on two or less variables. Use for each bi-decomposition different XBOOLE-object so that previous results can be used in later steps of the synthesis. Show the main results of each bi-decomposition. Forward the XBOOLE-system from a solved task to the PRP of the next task.

(a) Compute recursively all bi-decompositions $y_1 = f_1(\mathbf{x})$ until the resulting decomposition-functions depend on two or less variables. The results of this task are the functions or the lattices for $g_1(\mathbf{x})$, $g_2(\mathbf{x})$, $g_3(\mathbf{x})$, $g_4(\mathbf{x})$, and $h_4(\mathbf{x})$.

(b) Compute recursively all bi-decompositions so that the circuit structure of the decomposition-function $g_1(\mathbf{x})$ is completely known. The results of this task are the functions or the lattices for $h_3(\mathbf{x})$, $g_5(\mathbf{x})$, $h_5(\mathbf{x})$, $h_2(\mathbf{x})$, $g_6(\mathbf{x})$, and $h_6(\mathbf{x})$. Use in these computations the selected function $g_i(\mathbf{x})$ to compute the associated decomposition-function $h_i(\mathbf{x})$.

(c) Compute recursively all bi-decompositions of the function $h_1(\mathbf{x})$ until the resulting decomposition-functions depend on two or less variables. The results of this task will be functions or lattices for $h_1(\mathbf{x})$, $g_7(\mathbf{x})$, and $g_8(\mathbf{x})$. There are several types of bi-decompositions for the function $g_7(\mathbf{x})$. Check for this function all types of strong bi-decompositions and store separately the possible pairs of dedicated variables. Prefer the dedicated set with less variables in the procedure of extending these sets. Select the bi-decomposition with the smallest set of common variables.

(d) Compute recursively all decomposition-functions of the bi-decompositions of $h_1(\mathbf{x})$ so that the circuit structure of the decomposition-function $h_1(\mathbf{x})$ is completely known; this completes the synthesis of the function $y_1 = f_1(\mathbf{x})$.

Specify all functions that are directly determined by the input-variables. Compute recursively all functions at the outputs of the gates used for bi-decompositions. Verify whether the synthesized circuit realizes the given function $y_1 = f_1(\mathbf{x})$. Compare the number of gates in the longest path as well as the number of gates of the circuit synthesized using bi-decompositions with these parameters of the circuits synthesized in the previous exercise using covering methods.

References

1. R.L. Ashenhurst, The decomposition of switching functions, in *Annals of Computation Laboratory*, vol. 29 (Harvard University, Cambridge, MA, USA, 1959), pp. 74–116
2. D. Bochmann, F. Dresig, B. Steinbach, A new approach for multilevel circuit design, in *European Conference on Design Automation*. EURODAC '91 (IEEE Computer Society Press, Amsterdam, The Netherlands, 1991), pp. 374–377. ISBN:0-8186-2130-3
3. P. Böhlau, A decomposition strategy for the logic design based on properties of the function. In German: Eine Dekompositionsstrategie für den Logikentwurf auf der Basis funktionstypischer Eigenschaften. PhD thesis. Technical University Karl-Marx-Stadt, Germany, 1987
4. H.A. Curtis, *A New Approach to the Design of Switching Circuits* (Van Nostrand, Princeton, NJ, USA, 1962)
5. R. Garipelly, P.M. Kiran, A. Kumar, A review on reversible logic gates and their implementation. Int. J. Emerg. Tech. Adv. Eng. **3**(3), 183–191 (2013). ISSN:417–423.
6. S. Khatri, K. Gulati, (eds.), *Advanced Techniques in Logic Synthesis, Optimizations and Applications* (Springer, New York, NY, USA, 2011). ISBN:978-1-4419-7517-1. https://doi.org/10.1007/978-1-4419-7518-8
7. R. Landauer, Irreversibility and heat generation in the computing process. IBM J. Res. Dev. **5**(3), 183–191 (1961). https://doi.org/10.1147/rd.53.0183
8. T. Le, Testability of combinational circuits–Theory and design. In German: Testbarkeit kombinatorischer Schaltungen—Theorie und Entwurf. PhD thesis. Technical University Karl-Marx-Stadt, Germany, 1989
9. A. Mishenko, B. Steinbach, M. Perkowski, An algorithm for bidecomposition of logic functions, in *Proceedings on the 38th Design Automation Conference*, DAC 28 (IEEE Computer Society Press, Las Vegas, NV, USA, 2001), pp. 18–22. ISBN:1-58113-297-2. https://doi.org/10.1145/378239.378353
10. G.N. Povarov, About functional decomposition of Boolean functions. Rep. Acad. Sci. USSR **DAN 94**(5) (1954). In Russian
11. T. Sasao, J. Butler, On bi-decompositions of logic functions, in *Proceedings of International Workshop on Logic Synthesis 1997*, IWLS. Lake Tahoe City, USA, 1997
12. B. Steinbach, XBOOLE—A toolbox for modelling, simulation, and analysis of large digital systems. Syst. Anal. Model. Simul. **9**(4), 297–312 (Sept. 1992). ISSN:0232-9298
13. B. Steinbach, Decomposition using decision diagrams, in *Decision Diagram Technique for Micro- and Nanoelectronic Design, Handbook*, eds. by S. Yanushkevich et al. (CRC Press, Boca Raton, London, New York, 2006), pp. 509–544. ISBN:0-8493-3424-1. https://doi.org/10.1201/9781420037586
14. B. Steinbach, Relationships between vectorial bi-decompositions and strong EXOR-Bi-decompositions, in *Proceedings of the 25th International Workshop on Post-Binary ULSI Systems*, ULSI 25. Sapporo, Hokkaido, Japan, 2016
15. B. Steinbach, R. Hilbert, Fast test patterns generation using the Boolean differential calculus, in *Errors in Finite State Machines*, eds. by D. Bochmann, R. Ubar. in German (title of the contribution) Schnelle Testsatzgenerierung gestützt auf den Booleschen Differentialkalkül, (title of the book) Fehler in Automaten (Verlag Technik, Berlin, Germany, 1989), pp. 45–90. ISBN:3-341-00683-4
16. B. Steinbach, C. Lang, Exploiting functional properties of Boolean functions for optimal multi-level design by bi-decomposition, in *Artificial Intelligence in Logic Design*, ed. by S. Yanushkevich. SECS 766 (Springer, Dordrecht, The Netherlands, 2004), pp. 159–200. ISBN:978-90-481-6583-4. https://doi.org/10.1007/978-1-4020-2075-9_6

17. B. Steinbach, C. Posthoff, *Logic Functions and Equations - Examples and Exercises* (Springer Science + Business Media B.V., 2009). ISBN:978-1-4020-9594-8. https://doi.org/10.1007/978-1-4020-9595-5

18. B. Steinbach, C. Posthoff, Vectorial bi-decompositions of logic functions, in *Proceedings of the Reed-Muller Workshop 2015*, RM. Waterloo, Canada, 2015

19. B. Steinbach, C. Posthoff, Vectorial bi-decompositions for lattices of Boolean functions, in *Proceedings of the 12th International Workshops on Boolean Problems*, ed. by B. Steinbach. IWSBP 12. (Freiberg University of Mining and Technology, Freiberg, Germany, Sept. 2016), pp. 93–104. ISBN:978-3-86012-488-8

20. B. Steinbach, C. Posthoff, Compact XOR-Bi-decomposition for generalized lattices of Boolean functions, in *Proceedings Reed-Muller Workshop 2017*, RM 13. Novi Sad, Serbia, 2017

21. B. Steinbach, C. Posthoff, Vectorial bi-decompositions for lattices of Boolean functions, in *Further Improvements in the Boolean Domain*, ed. by B. Steinbach (Cambridge Scholars Publishing, Newcastle upon Tyne, UK, Jan. 2018), pp. 175–198. ISBN:978-1-5275-0371-7

22. B. Steinbach, M. Stöckert, Design of fully testable circuits by functional decomposition and implicit test pattern generation, in *Proceedings of 12th IEEE VLSI Test Symposium*, VTS 12. Cherry Hill, New Jersey, USA, 1994

23. B. Steinbach, A. Zakrevski, Three models and some theorems on decomposition of Boolean functions, in *Proceedings of the 3rd International Workshops on Boolean Problems*, ed. by B. Steinbach. IWSBP 3 (Freiberg University of Mining and Technology, Freiberg, Germany, Sept. 1998), pp. 11–18. ISBN:3-86012-069-7

Sequential Circuits

12

Abstract

The behavior of finite state machines is determined not only by the input-values but also by the internal states. Sequential circuits are technical realizations of finite state machines. We introduce models that can be used to describe either the behavior or the structure of several realizations of both asynchronous and synchronous sequential circuits. The behavior of these two types of sequential circuits can be calculated using the same basic approach; however, there are peculiarities that must be considered. Therefore, we describe the analysis of asynchronous and synchronous sequential circuits in separate sections. The synthesis task must take these peculiarities also into account; hence, we explain synthesis methods separately for asynchronous and synchronous sequential circuits. The common synthesis task can be realized by the solution of the system-equation with regard to memory-variables and output-variables. For the synthesis of synchronous sequential circuits, we explain a very efficient approach that avoids the explicit calculation of the memory-functions but directly computes the controlling functions of the flip-flops. In all these cases non-deterministic finite state machines are utilized that contribute to the synthesis

Supplementary Information The online version of this chapter (https://doi.org/10.1007/978-3-030-88945-6_12) contains supplementary material which is available for authorized users. Please, follow the link belonging to the version of the XBOOLE-monitor XBM 2 that fits best for your operating system. This XBOOLE-monitor is needed to solve all tasks of this chapter. Instructions for starting the downloaded XBOOLE-monitor XBM 2 are given at the beginning of Section 'Exercises' in this chapter.

XBOOLE-monitor XBM 2 for Windows 10
32 bits
https://doi.org/10.1007/978-3-030-88945-6_12_MOESM1_ESM.zip (15,091 KB)

64 bits
https://doi.org/10.1007/978-3-030-88945-6_12_MOESM2_ESM.zip (14,973 KB)

XBOOLE-monitor XBM 2 for Linux Ubuntu
32 bits
https://doi.org/10.1007/978-3-030-88945-6_12_MOESM3_ESM.zip (29,522 KB)

64 bits
https://doi.org/10.1007/978-3-030-88945-6_12_MOESM4_ESM.zip (28,422 KB)

of optimized sequential circuits. Several approaches for the hardware–software co-design of finite state machines complete this chapter.

12.1   The Circuit Model

A *finite state machine* is a mathematical model of a special type of systems. One of the possible practical implementations of finite state machines are sequential circuits. We study first sequential circuits and generalize then these concepts to finite state machines.

Sequential circuits can be built by the same logic gates like combinational circuits. What is the difference between these classes of circuits? We know from the previous section that a combinational circuit maps an input-pattern to exactly one output-pattern. For that reason, combinational circuits are used, for instance, to calculate the sum of two numbers. The sum of $a = 2$ and $b = 3$ is equal to $y = 5$ at any time. An appropriate combinational circuit has four inputs and three outputs. If the values $a_1 = 1$, $a_0 = 0$, $b_1 = 1$, and $b_0 = 1$ are used as the inputs of the combinational circuit, its outputs adjust to $y_2 = 1$, $y_1 = 0$, and $y_0 = 1$, independent of the previous input-values.

The peculiarity of sequential circuits is that different output-patterns can occur for the same input-pattern. Why does the behavior change in sequential circuits although the same logic gates from combinational circuits are used? It is based on the use of connections by wires. We can order the gates in levels such that in a combinational circuit the gates of the first level are only controlled by the inputs, the gates of the second level are controlled by the inputs and the outputs of the first level, and so on. The gates of a certain level in a combinational circuit control only gates of higher levels or outputs of the circuit. In opposite to this rule, there are wires in a sequential circuit that connect the output of a higher-level gate to the input of a lower-level gate. In this way, loops of wires and logic gates are built in sequential circuits.

Example 12.1   We explore the effect of such a loop using a circuit of two NAND-gates (G_1, G_2) and two inputs (x_1, x_2) and connect the output of gate G_1 to one input of gate G_2; hence, G_2 is on the second level because the output of G_1 is an input of G_2. Additional connections from the inputs x_1 and x_2 to different inputs of the two gates as well as a connection from the output of the NAND-gate G_2 to the output of the circuit y initially result in a combinational circuit.

Now we connect the output of the NAND-gate G_2 of the second level to the so far unused input of the NAND-gate G_1 of the first level. This returning wire creates a loop in the circuit and changes the combinational behavior into a sequential behavior. Figure 12.1 depicts the structure of the sequential circuit and its behavior for several time-steps. The returning wire is emphasized by a thick red line.

The two inputs x_1 and x_2 are equal to 0 in *time-step* 0. If one input of a NAND-gate is equal to 0, then the output of the NAND-gate is equal to 1 independent of the value of the other input. Hence, the output-values of the two NAND-gates G_1 and G_2 are equal to 1. In this case the returning value 1 does not influence the behavior. The values at all points of the explored sequential circuit are shown in the upper row of Fig. 12.1.

We switch the input x_2 from 0 to 1 in *time-step* 1. The output of the NAND-gate G_1 remains equal to 1 because $x_1 = 0$. The output of the NAND-gate G_2 switches from 1 to 0 because its two inputs are equal to 1. The changed value on the returning wire does not influence the output-value of the NAND-gate G_1 due to the constant value 0 at the input x_1. The second row of Fig. 12.1 shows this partial behavior.

We switch also the input x_1 from 0 to 1 in *time-step* 2. The output-value of the NAND-gate G_1 depends now on the input-value of the returning wire and remains equal to 1 because the returning wire carries the value 0 from the output of the NAND-gate G_2. Both inputs of the NAND-gate G_2 are equal to 1, and thus the output of the NAND-gate G_2 remains unchanged, i.e., equal to 0. All information about the sequential circuit in this time-step has been summarized in the third row of Fig. 12.1.

We explore the final possible input-pattern in *time-step* 3. For that purpose, we switch the input x_2 back from 1 to 0. The input-value 0 of the input x_2 causes the output-value 1 of the NAND-gate G_2. This value 1 returns to the second input of the NAND-gate G_1. Now both inputs of the NAND-gate G_1 are equal to 1 so that consequently the output of the NAND-gate G_1 switches to 0. This change of the value does not have any influence on other values in the sequential circuit because the input x_2 is equal to 0. The adjusted values of time-step 3 are shown in the fourth row of Fig. 12.1.

Until now, we studied all four possible input-patterns of a circuit of two inputs. From the function table in the right part of Fig. 12.1, we get the logic function $y = f(\mathbf{x}) = \overline{x}_2$, and it seems that this circuit can be simplified to a single NOT-gate. The behavior in time-step 4 will show that this guess is not true.

We switch again the input x_2, now from 0 to 1 in *time-step* 4. In this case the output-value of the NAND-gate G_2 remains equal to 1 because the output-value of the NAND-gate G_1 is equal to 0. The two inputs of the NAND-gate G_1 do not change their values 1 because the output-value of the NAND-gate G_2 remains equal to 1; hence, the output of NAND-gate G_1 remains equal to 0 All logic values of this case are shown in the lower row of Fig. 12.1.

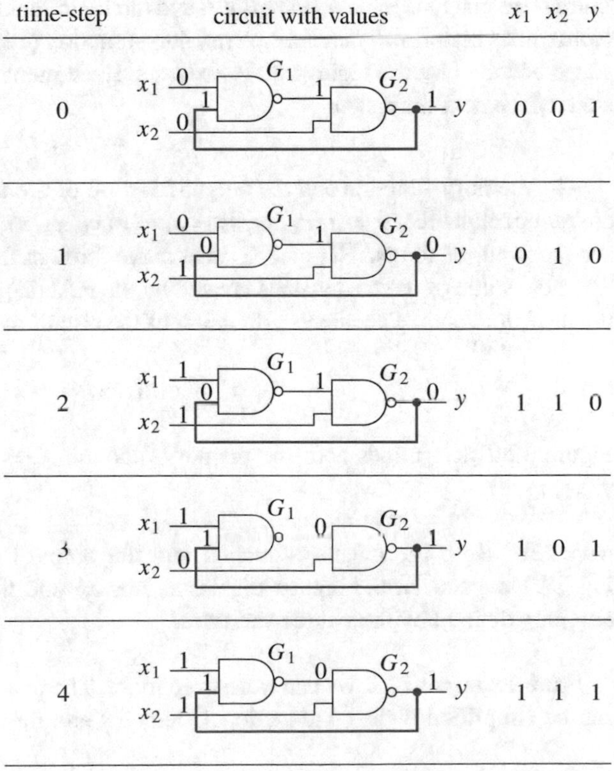

Fig. 12.1 Observable behavior of a sequential circuit

Fig. 12.2 Unique
behavior of the sequential
circuit of Fig. 12.1

x_1	x_2	s	s'	y
0	0	0	1	1
0	0	1	1	1
0	1	0	0	0
0	1	1	0	0
1	0	0	1	1
1	0	1	1	1
1	1	0	0	0
1	1	1	1	1

The comparison of time-steps 2 and 4 in Fig. 12.1 reveals the sequential behavior. The output y carries different values for identical input-patterns in these two time-steps. This observation proves that:

1. the circuit of Fig. 12.1 is no combinational circuit;
2. in a sequential circuit, different output values can occur for the same input-pattern; and
3. sequential circuits can be built from combinational circuits extended by returning wires.

The different output-values for the same input-pattern indicate that there must be an additional source of information that contributes to the logic output-function. This source cannot be outside of the circuit because all input signals are already taken into account; hence, there must be internal states that influence the behavior of the sequential circuit. If the number of such internal states is finite, then they can be realized by logic functions as well. The finite number of states motivates the term *finite state machine* (FSM). We denote the logic functions that implement the states of a finite state machine *memory-functions*, and the logic functions the create the outputs *output-functions* (sometimes also *result function*). A memory-function (12.1) depends on the input-variables $\mathbf{x}$ and the states encoded by the logic *state-variables* $\mathbf{s}$. Each memory-function defines one bit of the encoded state of the next time-step:

$$s_i' = g_i(\mathbf{x}, \mathbf{s}) \,. \tag{12.1}$$

The memory-function and the output-function of the sequential circuit explored in Example 12.1 are identical; hence, $y = f(x_1, x_2, s) = s' = g(x_1, x_2, s)$. The state-variable s carries the value of the second input of the NAND-gate G_1 that came through the returning wire in the previous time-step. The new value of the next state is created by the NAND-gate G_2 and stored in the state-variable s' of the next time-step. The memory-function of the circuit of Example 12.1 is

$$s' = g(x_1, x_2, s) = x_1 s \vee \overline{x}_2 \,.$$

Figure 12.2 determines both the memory-function $s' = g(x_1, x_2, s)$ and the output-function $y = f(x_1, x_2, s)$.

Note 12.1 Both the memory-function and the output-function of the sequential circuit shown in Fig. 12.1 depend on the input-variables x_1 and x_2 and the state-variable s. All function values are uniquely defined by these three variables.

From Example 12.1, we can learn even more. The analysis and the synthesis of sequential circuits can be simplified if the combinational elements and the memory-elements are separated from each other. For that reason, we are looking for special memory-elements. We have seen that the circuit of Fig. 12.1 can store one bit. Therefore, we can use this circuit as a memory-element.

Concentrated memory-elements are usually denoted by *flip-flops*. As can be seen in Fig. 12.1, the NAND-gates G_1 and G_2 carry opposite output-values, except for the input-pattern $(0, 0)$. This is a valuable property because without an additional NOT-gate both the direct and the inverted value of such a memory-element can be used in a subsequently controlled combinational circuit. For that reason, the input-pattern $(0, 0)$ is not allowed when the circuit of Fig. 12.1 is used as a flip-flop.

The remaining three input-patterns have special properties:

- the input-pattern $(x_1 = 1, x_2 = 0)$ sets the output of the flip-flop to the value 1 independent of the recent state; such an input-pattern is denoted by *set-condition*;
- the input-pattern $(x_1 = 0, x_2 = 1)$ realizes the reset of the output of the flip-flop to the value 0 independent of the recent state; such an input-pattern is denoted by *reset-condition*.
- the input-pattern $(x_1 = 1, x_2 = 1)$ stores the recent state of the flip-flop so that the output of the flip-flop depends on its previous output-value; such an input-pattern is denoted by *memory-condition*.

There are several types of flip-flops. The set-condition of our explored flip-flop is satisfied by $x_2 = 0$. Therefore, x_2 can be renamed into $\overline{S}$. The reset-condition of this flip-flop is satisfied by $x_1 = 0$. Therefore, x_1 can be renamed into $\overline{R}$, and the whole circuit is denoted by $\overline{R}\,\overline{S}$-flip-flop. We get a well-known RS-flip-flop when additional NOT-gates are used at each input.

If the three conditions introduced above are used to control a flip-flop, two inputs are necessary. The exclusion of the fourth input-pattern in the case of the RS-flip-flop complicates the synthesis process. How the fourth input-pattern can beneficially be used? One possibility is the reuse of one of the three other conditions. However, there is one more possible behavior that can be associated to the fourth input-pattern; this additional behavior changes the recent state of the flip-flop, and the associated input-pattern is denoted by *change-condition*. The application of this change-condition requires that the flip-flop is additionally controlled by a clock-signal. An associated input-pattern selects the behavior of the desired change of the flip-flop-output that is executed at a rising or falling edge of the clock-signal. It depends on the flip-flop structure which edge will be used.

Caused by a clock-signal all flip-flops evaluate their input-values only at fixed points in time. Such sequential circuits are denoted by *synchronous sequential circuits*. The advantages of synchronous sequential circuits are that the combinational gates that are controlled by the flip-flops have constant state-values in the period of time between clock-signals. The clock-signal changes its value in fixed periods of time. Short periods of time of the clock-signal allow high working frequencies and reduce the required total time to solve a task. Of course, the time periods of the clock-signal must be long enough so that the combinational part of the circuit can create the valid output-pattern depending on the last state- and input-values.

All four controlling conditions of a flip-flop mentioned above are realized in the widely used JK-flip-flop shown in the fourth row of Table 12.1. This flip-flop is controlled by the inputs J and K and the clock-signal c. We assume that the values of J and K are evaluated for falling edges of the clock-signal. The JK-flip-flop works as follows:

- $J = 1, K = 0$: set-condition;
- $J = 0, K = 1$: reset-condition;
- $J = 0, K = 0$: memory-condition; and
- $J = 1, K = 1$: change-condition.

Table 12.1 Clocked flip-flops

Symbol	Graph	Logic function	List of phases

D-FF — Logic function: $q' = D$

D	q	q'
0	–	0
1	–	1

T-FF — Logic function: $q' = T \oplus q$

T	q	q'
0	0	0
0	1	1
1	0	1
1	1	0

DE-FF — Logic function: $q' = \overline{E}q \vee DE$

D	E	q	q'
–	0	0	0
–	0	1	1
0	1	–	0
1	1	–	1

JK-FF — Logic function: $q' = J\overline{q} \vee \overline{K}q$

J	K	q	q'
1	–	0	1
–	0	1	1
0	–	0	0
–	1	1	0

RS-FF — Logic function: $q' = S \vee \overline{R}q$, restriction: $RS = 0$

R	S	q	q'
0	0	0	0
0	0	1	1
0	1	–	1
1	0	–	0

The behavior of this JK-flip-flop can be described in detail by a logic function that depends on the inputs J and K, the clock-variable c, and its differential dc:

$$q' = (c\, \mathrm{d}c)\,(J\,\overline{q} \vee \overline{K}\,q) \vee \overline{(c\, \mathrm{d}c)}\,q \,. \tag{12.2}$$

The term $(c\, \mathrm{d}c)$ describes the falling edge of the clock-signal. The logic function (12.2) can be simplified when the real clock-signal is substituted by the variable $c_m = (c \wedge \mathrm{d}c)$ as model of the clock-variable. This simplification is useful if a common clock must be modeled and the clock-signal itself is not controlled by other parts of the circuit:

$$q' = c_m\,(J\,\overline{q} \vee \overline{K}\,q) \vee \overline{c}_m\,q \,. \tag{12.3}$$

Equation (12.3) emphasized the typical property of each clocked flip-flop. The flip-flop stores the previous state-value if $c_m = 0$. Otherwise, in case of $c_m = 1$, the behavior of the flip-flop is controlled by its input-variables. In the context of these background knowledge, it is possible to omit the clock-variable c_m:

$$q' = J\,\overline{q} \vee \overline{K}\,q \,. \tag{12.4}$$

Equation (12.4) describes the behavior of the JK-flip-flop in the case of a clock-event. Using (12.4), it is implicitly assumed that all flip-flops in the synchronous sequential circuit are controlled by the same clock-signal that is drawn in the diagram of the circuit but not mentioned in the logic description.

The memory-condition is satisfied by the clock; hence, clocked flip-flops require only one controlling input. Using the input D of the D-flip-flop (*delay*-flip-flop), the set-condition ($D = 1$) or reset-condition ($D = 0$) can be selected. By means of the input T of the T-flip-flop (*trigger*-flip-flop), the memory-condition ($T = 0$) or the change-condition ($T = 1$) can be chosen. Clocked flip-flops of a more general behavior need at least two inputs.

Table 12.1 shows five types of clocked flip-flops that are used very often. The central input on the left-hand side of the flip-flop symbol is the clock-input. The other inputs are labeled inside of the flip-flop symbol. On the output-side of the flip-flops, both the stored value q and its complement $\overline{q}$ are available.

The flip-flops of Table 12.1 have the behavior of simple finite state machines. All of them have two stable states represented by the nodes in the associated graphs. The weights on the edges in the graph describe the conditions of the transitions. In order to describe the flip-flop behavior, an alternative to the graph is the logic function. The Boolean variable q carries the value of the recent stable state of the flip-flop. The associated variable q' describes the state-value after the next clock-event.

The lists of phases in the last column of Table 12.1 provide the solutions of the logic equations of the associated flip-flop. Each phase describes which new state-value q' is created at the point in time of the next clock-event depending on the current state-value q and the input-values of the flip-flop.

A peculiarity must be considered in the case of the RS-flip-flop (*reset-set*-flip-flop). The RS-flip-flop uses only three working conditions:

- $R = 0, S = 1$: the set-condition;
- $R = 1, S = 0$: the reset-condition; and
- $R = 0, S = 0$: the memory-condition.

The input-pattern $R = 1, S = 1$ is not permitted and must be a constraint of the output of the combinational circuit that controls the RS flip flop.

If all flip-flops are summarized in a memory-block, then two types of the general structure of finite state machines can be distinguished. As shown in Fig. 12.3, the difference consists in the use of the input-variables:

- the output-functions $\lambda(\mathbf{s})$ depend only on the state-variables $\mathbf{s}$ in the case of finite state machines of the *Moore-type*;
- in contrast, output-functions $\lambda(\mathbf{x}, \mathbf{s})$ determine the outputs in the case of finite state machines of the *Mealy-type*.

The same type of memory-functions $\delta(\mathbf{x}, \mathbf{s})$ is used in each of these two types of finite state machines.

Fig. 12.3 Basic structures of finite state machines: (**a**) Moore-type and (**b**) Mealy-type

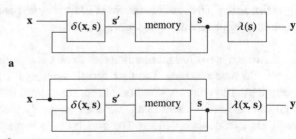

The structure of a finite state machine can be described by a system of logic equations. Taking the sub-circuits of Fig. 12.3 into account, this system of logic equations consists of the system of memory-equations $s_i' = \delta_i(\mathbf{x}, \mathbf{s})$ and the system of result-equations $y_j = \lambda_j(\mathbf{x}, \mathbf{s})$ (in the case of the most general Mealy-type). The system of logic equations (12.5) specifies a finite state machine of the Moore-type. The logic functions on the right-hand sides of these equations can be mapped into combinational circuits as shown in the previous chapter. Thus, the structural aspects of the sequential circuit are explicitly expressed by:

$$s_1' = \delta_1(\mathbf{x}, \mathbf{s}) ,$$
$$\vdots$$
$$s_l' = \delta_l(\mathbf{x}, \mathbf{s}) ,$$
$$y_1 = \lambda_1(\mathbf{s}) ,$$
$$\vdots$$
$$y_m = \lambda_m(\mathbf{s}) . \tag{12.5}$$

The system of Eq. (12.5) can be transformed into the single implicit equation (12.6), creating first a system of homogenous characteristic equations:

$$s_1' \odot \delta_1(\mathbf{x}, \mathbf{s}) = 1 ,$$
$$\vdots$$
$$y_m \odot \lambda_m(\mathbf{s}) = 1$$

and connecting thereafter the expressions of the left-hand side using conjunctions:

$$(s_1' \odot \delta_1(\mathbf{x}, \mathbf{s})) \wedge (s_2' \odot \delta_2(\mathbf{x}, \mathbf{s})) \wedge \ldots \wedge (y_m \odot \lambda_m(\mathbf{s})) = 1 ,$$
$$F(\mathbf{x}, \mathbf{s}, \mathbf{s}', \mathbf{y}) = 1 . \tag{12.6}$$

We denote (12.6) by *system-equation of a finite state machine*. In contrast to the system of logic equations (12.5), the system-equation (12.6) describes the behavior of the finite state machine. The solution of the system-equation (12.6) is a list of phases. Each phase $(\mathbf{x}, \mathbf{s}, \mathbf{s}', \mathbf{y})$ expresses explicitly the next state $\mathbf{s}'$ that will be reached in the next time-step and the outputs $\mathbf{y}$ depending on the actual state $\mathbf{s}$ and input-pattern $\mathbf{x}$.

Note 12.2 If the system-equation (12.6) is created from the system of logic equations (12.5), it describes the deterministic behavior of the associated finite state machine. That means for each input-pattern and each state exactly one state will be reached and one output-pattern will be created at the next time-step. The system-equation (12.6) is more general. It can describe finite state machines that cannot be implemented by a sequential circuit.

There are two different situations in which a sequential circuit cannot realize the specified behavior of a finite state machine. The first situation occurs if for a given input and a given state no phase is permitted. The contrary situation allows for a given input and a given state several of phases. In the first case the finite state machine is not completely specified. In the second case a *non-deterministic finite state machine* is defined. The possibility to implement one of the alternative phases for a certain input-pattern and state can be utilized for an optimized synthesis of a sequential circuit.

The above investigations have been done for finite state machines of the Moore-type. The system of logic equations (12.7) defines the structure of a finite state machine of the Mealy-type:

Fig. 12.4 Behavior of a finite state machine of Mealy-type: (**a**) list of phases and (**b**) graph

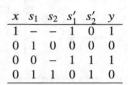

x	s_1	s_2	s_1'	s_2'	y
1	–	–	1	0	1
0	1	0	0	0	0
0	0	–	1	1	1
0	1	1	0	1	0

a

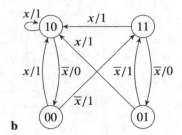

b

$$s_1' = \delta_1(\mathbf{x}, \mathbf{s}) \,,$$

$$\vdots$$

$$s_l' = \delta_l(\mathbf{x}, \mathbf{s}) \,,$$

$$y_1 = \lambda_1(\mathbf{x}, \mathbf{s}) \,,$$

$$\vdots$$

$$y_m = \lambda_m(\mathbf{x}, \mathbf{s}) \,. \tag{12.7}$$

Using the same method, it can also be transformed into a system-equation (12.6).

The formal representation of system-equation (12.6) is the same for finite state machines of the Moore-type and the Mealy-type, but the respective system-function $F(\mathbf{x}, \mathbf{s}, \mathbf{s}', \mathbf{y})$ determines one of these types. The solution of system-equation (12.6) is a list of phases. Such a list of phases can be used as behavioral specification of a finite state machine; this description is easily understandable for human beings, and the handling by computer programs is rather simple. Figure 12.4a shows a list of phases of a finite state machine of the Mealy-type.

In the states $(s_1, s_2) = (1, 0)$ and $(s_1, s_2) = (1, 1)$ the output y depends directly on the input x. A list of phases can be directly mapped onto a *graph*. Each phase represents an edge in the graph of the finite state machine. Such an edge starts at the node of the state in the recent time-step and ends at the node of the state in the next time-step. In the case of finite state machines of the Mealy-type the edge is labeled by a pair that shows the input condition and the output-pattern. The input condition selects the edge that must be used to reach the next state, and the output-pattern occurs for these input-values at the outputs. Figure 12.4b shows the graph of the finite state machine that is defined by the list of phases in Fig. 12.4a.

The outputs of a finite state machine of the Moore-type do not depend directly on the inputs; hence, in the associated graph each node is labeled by an output-pattern, and each edge is only labeled by an input-pattern.

12.2 Analysis of Asynchronous Sequential Circuits

The basic analysis task for a finite state machine is the investigation of the behavior of a given sequential circuit. There is one general approach to solve this task for all types of sequential circuits, but the details differ from asynchronous sequential circuits to synchronous sequential circuits with flip-flops. The entire behavior of a sequential circuit results from the combination of the local behavior of all switching elements in the circuit. Similar to the analysis of combinational circuits, the local behavior of each switching element can be expressed by a logic equation. These equations are combined into a system of equations. The solution of such a system of logic equations describes all details of the global behavior. Some of these details can be removed in order to get a behavioral description of the finite state machine independent of the special implementation.

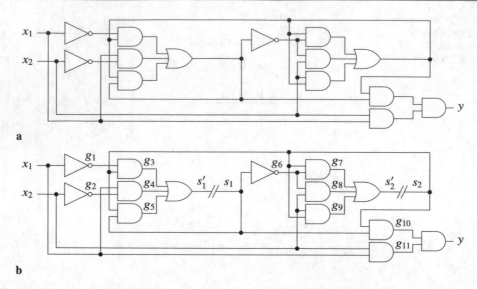

Fig. 12.5 Structure of an asynchronous sequential circuit: (**a**) original circuit and (**b**) circuit with cuts of the loops and labeled wires

Let us start with the analysis of an asynchronous sequential circuit given in Fig. 12.5a. This circuit consists of 14 logic gates and has two inputs and one output, and it is not controlled by a clock-signal. The loops in this circuit indicate that it is a sequential circuit. The absence of a clock-signal specifies it as an asynchronous sequential circuit. There are two loops in the circuit of Fig. 12.5a; the first loop is determined by the AND–OR-circuit on the left-hand side, and the second loop comprises the AND–OR-circuit on the right-hand side. Each loop realizes implicitly one memory-variable.

The first step of the analysis of an asynchronous sequential circuit is the definition of the state-variables and the points in the circuit where the memory-functions can be observed. In order to achieve this definition, we cut each loop of the sequential circuit at the output of a gate in the loop. As shown in Figure 12.5b, the outputs of these gates are labeled by the variables s_i' describing the next state and the wire behind the cut by the associated state-variable s_i.

Note 12.3 The introduced cuts transform the structure of the sequential circuit into a combinational structure and create both additional outputs of the memory-functions and associated inputs of the state-variables. All gate-outputs in the circuit that are not labeled as yet will be labeled by logic variables g_i. Figure 12.5b shows the result of this analysis step.

In the second step we express the behavior of each gate by a logic equation and combine all these equations into one system:

$$
\begin{aligned}
g_1 &= \overline{x}_1 , & g_7 &= g_6 \wedge s_2 , \\
g_2 &= \overline{x}_2 , & g_8 &= x_2 \wedge g_6 , \\
g_3 &= g_1 \wedge s_2 , & g_9 &= x_2 \wedge s_2 , \\
g_4 &= x_1 \wedge g_2 , & s_2' &= g_7 \vee g_8 \vee g_9 , \\
g_5 &= s_1 \wedge s_2 , & g_{10} &= s_1 \wedge s_2 , \\
s_1' &= g_3 \vee g_4 \vee g_5 , & g_{11} &= x_1 \wedge x_2 , \\
g_6 &= \overline{s}_1 , & y &= g_{10} \wedge g_{11} .
\end{aligned} \tag{12.8}
$$

x_1	x_2	s_1	s_2	s_1'	s_2'	y	g_1	g_2	g_3	g_4	g_5	g_6	g_7	g_8	g_9	g_{10}	g_{11}
0	0	0	0	0	0	0	1	1	0	0	0	1	0	0	0	0	0
0	0	0	1	1	1	0	1	1	1	0	0	1	1	0	0	0	0
0	0	1	0	0	0	0	1	1	0	0	0	0	0	0	0	0	0
0	0	1	1	1	0	0	1	1	1	0	1	0	0	0	0	1	0
0	1	0	0	0	1	0	1	0	0	0	0	1	0	1	0	0	0
0	1	0	1	1	1	0	1	0	1	0	0	1	1	1	1	0	0
0	1	1	0	0	0	0	1	0	0	0	0	0	0	0	0	0	0
0	1	1	1	1	1	0	1	0	1	0	1	0	0	0	1	1	0
1	0	0	0	1	0	0	0	1	0	1	0	1	0	0	0	0	0
1	0	0	1	1	1	0	0	1	0	1	0	1	1	0	0	0	0
1	0	1	0	1	0	0	0	1	0	1	0	0	0	0	0	0	0
1	0	1	1	1	0	0	0	1	0	1	1	0	0	0	0	1	0
1	1	0	0	0	1	0	0	0	0	0	0	1	0	1	0	0	1
1	1	0	1	0	1	0	0	0	0	0	0	1	1	1	1	0	1
1	1	1	0	0	0	0	0	0	0	0	0	0	0	0	0	0	1
1	1	1	1	1	1	1	0	0	0	0	1	0	0	0	1	1	1

Fig. 12.6 Detailed behavior of the asynchronous sequential circuit of Fig. 12.5, solution of (12.8) or (12.9)

Note 12.4 Equal expressions of g_5 and g_{10} reveal that these two gates realize the same function; hence, we get a simpler circuit with same behavior when the AND-gate g_{10} is removed and g_5 is reused for the output y.

The system of Eqs. (12.8) can be transformed into the single system-equation (12.9):

$$F(x_1, x_2, s_1, s_2, s_1', s_2', y, g_1, g_2, g_3, g_4, g_5, g_6, g_7, g_8, g_9, g_{10}, g_{11}) = 1 . \qquad (12.9)$$

In the third step we solve the equation-system (12.8) or the system-equation (12.9) and get 16 solution-vectors because there are four independent variables, the input-variables (x_1, x_2) and the state-variables (s_1, s_2), respectively. Figure 12.6 enumerates these 16 phases of the asynchronous sequential circuit of Fig. 12.5 ordered by the input- and state-variables.

Each phase describes the behavior of the sequential circuit at a fixed point in time determined by the logic values of the inputs (x_1, x_2) and the internal state (s_1, s_2) at all gate-outputs. Especially the value of the output y and the values of the memory-functions (s_1', s_2') of the next point in time are specified.

In order to study the behavior of the finite state machine realized by the sequential circuit of Fig. 12.5, it is not necessary to know the values of the gate-outputs labeled by g_i. The system-function $F(\mathbf{x}, \mathbf{s}, \mathbf{s}', y)$ (12.10) can be derived from the detailed system function on the left-hand side of Eq. (12.9) using the k-fold maximum:

$$F(\mathbf{x}, \mathbf{s}, \mathbf{s}', y) = \max_{\mathbf{g}}^{k} F(\mathbf{x}, \mathbf{s}, \mathbf{s}', y, \mathbf{g}) . \qquad (12.10)$$

The solution of the system-equation (12.11) describes the behavior of the analyzed finite state machine:

$$F(x_1, x_2, s_1, s_2, s_1', s_2', y) = 1 . \qquad (12.11)$$

This solution can be expressed in a more compact manner as shown in Fig. 12.7.

Fig. 12.7 Behavior of the
finite state machine
associated to the
asynchronous sequential
circuit of Fig. 12.5,
solution of (12.11)

x_1	x_2	s_1	s_2	s_1'	s_2'	y
1	1	1	1	1	1	1
1	1	0	1	0	1	0
0	1	–	1	1	1	0
–	1	0	0	0	1	0
–	1	1	0	0	0	0
–	0	1	1	1	0	0
–	0	0	1	1	1	0
1	0	–	0	1	0	0
0	0	–	0	0	0	0

Fig. 12.8 Partial behavior
restricted to the stable
states of the asynchronous
circuit of Fig. 12.5,
solution of (12.12)

stable state	x_1	x_2	s_1	s_2	s_1'	s_2'	y
S1	0	0	0	0	0	0	0
S2	1	1	0	1	0	1	0
S3	0	1	1	1	1	1	0
S4	1	1	1	1	1	1	1
S5	1	0	1	0	1	0	0

The state-change of an asynchronous finite state machine is caused by a change of at least one input-variable. The state that has been reached can cause a further change of the state. Such a sequence of state-changes stops if a *stable state* has been reached; hence, a special analysis task for an asynchronous finite state machine consists in finding the stable states for each input-pattern.

Stable states of an asynchronous finite state machine require that each state-variable s_i is equal to s_i' because in this case no changes in the circuit are stimulated for a constant input-pattern. This condition can be described by a system of logic equations that consists of the system-equation of the asynchronous finite state machine and an equation $s_i' = s_i$ for each state-variable. Equation-system (12.12) describes the partial behavior restricted to the stable states of the asynchronous finite state machine of Fig. 12.5. Figure 12.8 enumerates the solution-vectors of (12.12):

$$F(x_1, x_2, s_1, s_2, s_1', s_2', y) = 1 \, ,$$

$$s_1' = s_1 \, ,$$

$$s_2' = s_2 \, . \tag{12.12}$$

The stable states of an asynchronous finite state machine are associated to the input-patterns. Therefore, we can uniquely characterize the stable states by the values of the input-variables (x_1, x_2) and the state-variables (s_1, s_2). We use the labels from the left column of Fig. 12.8 as the names of the stable states.

As mentioned above, a stable state keeps unchanged when the input-values are not changed. There can be small time differences between the change of the input-values when more than one input changes its value. Since these single changes in a given order can cause different state-changes, the restriction that only one input changes its value at a given point in time will be used. Figure 12.7 describes in detail all changes of the states and the observable value on the output.

Figure 12.7 is also a common function table for the memory-functions

$$s_1' = \delta_1(\mathbf{x}, \mathbf{s}) \, , \qquad\qquad\qquad s_2' = \delta_2(\mathbf{x}, \mathbf{s})$$

and the output-function $y = \lambda(\mathbf{x}, \mathbf{s})$. Figure 12.9 represents these functions by Karnaugh-maps.

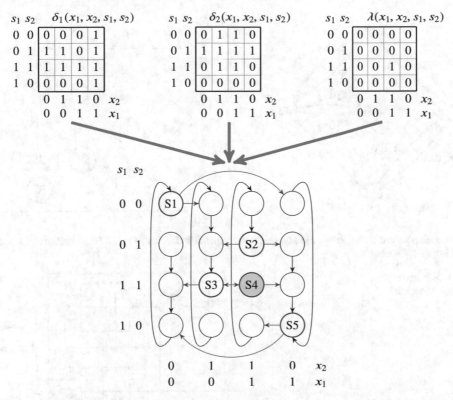

Fig. 12.9 Karnaugh-maps of the memory-function $s'_1 = \delta_1(\mathbf{x}, \mathbf{s})$, $s'_2 = \delta_2(\mathbf{x}, \mathbf{s})$ and the output function $y = \lambda(\mathbf{x}, \mathbf{s})$ merged into the detailed behavioral graph of the asynchronous sequential circuit of Fig. 12.5

The values of the three Karnaugh-maps of Fig. 12.9 are merged in this figure into the detailed behavioral graph of the asynchronous finite state machine. Stable states are indicated by red circles, labeled by the name of the state; the loops at the stable states are omitted for a better clearness. The green filled node determines the output-value $y = 1$. Blue edges start in stable states and describe the change of a single input. Black edges indicate the transition used to reach a stable state.

The behavioral description can be restricted to the stable states and the transitions between them. Both a graph and a table of the finite state machine can express this more abstract behavior. The transitions between the stable states are determined by edges in the graph or simply by the label of the next state in the table. Loops at the stable states are omitted. We used a green colored state node in the graph to indicate that the output y is equal to 1. Figure 12.10 shows these two alternatively usable behavioral descriptions for the circuit of Fig. 12.5.

The graph of Fig. 12.10a emphasizes behavioral properties of the asynchronous finite state machine of Fig. 12.5. There is a chain of stable states S2–S5–S1–S3–S4. It is possible to switch between neighborhood states of this chain in both directions. Additional transitions from S2 to S3 and from S4 to S5 are only allowed in this direction. The output is equal to 1 only in the stable state S4 in which both x_1 and x_2 are equal to 1. This finite state machine is obviously of the Mealy-type.

Note 12.5 The state ($s_1 = 1$, $s_2 = 1$) is split into the stable states S3 and S4 as can be seen in the graph of Fig. 12.9. Since these stable states are additionally determined by the inputs (x_1, x_2), it is possible to associate the output-value with the nodes of the graph of Fig. 12.10 and not as usually with the edges of the graph of the Mealy-type.

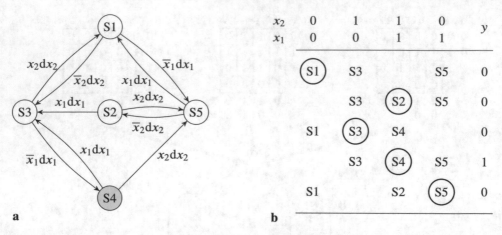

Fig. 12.10 Behavioral descriptions of the asynchronous sequential circuit of Fig. 12.5: (a) behavioral graph and (b) table of the finite state machine

Fig. 12.11 Structure of a synchronous sequential circuit

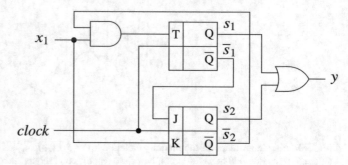

12.3 Analysis of Synchronous Sequential Circuits

The analysis of a synchronous finite state machine that uses flip-flops as memory-elements is simpler than the analysis of asynchronous finite state machines discussed above. It is not necessary to cut some wires in order to get the state-variables and gate-outputs of the memory-functions. Each flip-flop determines both the associated state-variable and memory-function.

The entire behavior of a clocked sequential circuit results from the combination of the local behavior of all switching elements, including the flip-flops. The outputs of the flip-flops represent the state-variables. The memory-functions are determined by the flip-flop-equations and the combinational functions controlling the inputs of the flip-flops. The clock-signal must not be taken into account if all flip-flops of the circuit are controlled by the same clock-signal.

We demonstrate the analysis of a clocked sequential circuit using a circuit that contains one T-flip-flop and one JK-flip-flop. Figure 12.11 depicts the sequential circuit to be analyzed.

The controlling functions of the flip-flop inputs (12.13), (12.14), and (12.15) can be taken directly from the scheme of Fig. 12.11:

$$T = x_1 \wedge \bar{s}_2 \,, \tag{12.13}$$

$$J = \bar{s}_1 \,, \tag{12.14}$$

$$K = x_1 \,. \tag{12.15}$$

Fig. 12.12 Behavior of the synchronous finite state machine belonging to the clocked sequential circuit of Fig. 12.11: (**a**) list of phases and (**b**) graph

x_1	s_1	s_2	s_1'	s_2'	y
0	0	0	0	1	0
1	0	0	1	1	0
0	0	1	0	1	1
1	0	1	0	0	1
0	1	0	1	0	1
1	1	0	0	0	1
0	1	1	1	1	1
1	1	1	1	0	1

a

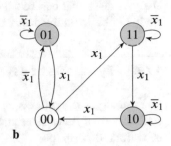

b

The memory-functions are directly associated with the corresponding flip-flop functions defined in Table 12.1. These functions must be assigned to the flip-flops of the circuit. In order to do this, the general state-variable q must be substituted by the state-variable s_i of the flip-flop in the circuit and the variable of the memory-function q' by s_i', respectively:

$$s_1' = T \oplus s_1 ,$$ (12.16)

$$s_2' = J\,\overline{s}_2 \vee \overline{K}\,s_2 .$$ (12.17)

The variables of the flip-flop inputs in (12.16) and (12.17) can be replaced by the controlling functions (12.13), (12.14), and (12.15) of the flip-flop. The resulting memory-functions (12.18) and (12.19) describe, together with the output-function (12.20) in a common system of Eqs. (12.18), (12.19), and (12.20), the behavior of the synchronous finite state machine:

$$s_1' = (x_1 \wedge \overline{s}_2) \oplus s_1 ,$$ (12.18)

$$s_2' = \overline{s}_1\overline{s}_2 \vee \overline{x}_1 s_2 ,$$ (12.19)

$$y = s_1 \vee s_2 .$$ (12.20)

Note 12.6 The output-function (12.20) directly depends only on the state-variables (s_1, s_2); hence, the synchronous finite state machine has the Moore-type.

The solution of the system of Eqs. (12.18), (12.19), and (12.20) is the list of phases that can be observed at the clocked sequential circuit of Fig. 12.11. Figure 12.12a depicts these phases.

Since all flip-flops change their states depending on the same clock-signal, all states of a synchronous finite state machine are stable states; hence, the list of phases of Fig. 12.12a can directly be visualized as a graph shown in Fig. 12.12b. Special analysis problems do not exist.

Green colored nodes indicate in Fig. 12.12b that the output is equal to 1 in these states. A general property of a synchronous finite state machine is that for each fixed input-pattern a sequence of transitions occurs that ends in a cycle. As can be seen in Fig. 12.12b, the finite state machine of Fig. 12.11 has the following behavior: the cycle $(0, 0) - (1, 1) - (1, 0) - (0, 0)$ is reached from each state if the input $x_1 = 1$. In the opposite case ($x_1 = 0$) three separate cycles of length 1 occur in the states $(0, 1)$, $(1, 0)$, and $(1, 1)$. The loop at the state $(0, 1)$ is reached from the state $(0, 0)$ if $x_1 = 0$.

The basic representation of a finite state machine for many further analysis problems is the system-equation (12.6). It is easy to calculate which transitions start from a certain state or end in this state. The check for reachability or the calculation of the shortest path between selected states are also analysis tasks that can be solved by algorithms using logic functions and equations. We suggest to study books about the theory of graphs (see, e.g., [1]) to extend the knowledge in this area.

12.4   Synthesis of Asynchronous Finite State Machines

Basically, the synthesis and the analysis of a finite state machine are reverse tasks; hence, the synthesis of a finite state machine consists of all steps to create a circuit structure from a given behavioral description.

The basic behavioral description of a finite state machine is the system-equation (12.6). The solution of this equation is a list of phases. Vice versa, the system-function $F(\mathbf{x}, \mathbf{s}, \mathbf{s}', \mathbf{y})$ can be created from the list of phases of the finite state machine. Each phase specifies a conjunction. Values 1 of the phase are mapped to non-negated variables and values 0 to negated variables. The system-function connects all conjunctions by $\vee$-operations. Due to these relations, we use the system-equation (12.6), the system-function on its left-hand side, and the list of phases as equivalent behavioral descriptions of a finite state machine. An equivalent graphical representation is the graph of the finite state machine. *Note 12.7* Any list of phases can be given to synthesize an associated sequential circuit. Some of such lists of phases describe non-deterministic finite state machines. In that case one of the covered deterministic finite state machines can be mapped onto the circuit structure.

The given list of phases can be incompletely specified. That means that at least for one input-pattern $\mathbf{x}$ and one state $\mathbf{s}$ no next state $\mathbf{s}'$ and no output $\mathbf{y}$ are defined. Such a behavior cannot occur in a sequential circuit. Therefore, the first step in the synthesis process of a finite state machine is the analysis whether it is realizable. A finite state machine is realizable if for each input-pattern $\mathbf{x}$ and each state $\mathbf{s}$ at least one pattern $(\mathbf{s}', \mathbf{y})$ is possible and defined. This can be checked by Eq. (12.21) that must satisfied for each pattern $(\mathbf{x}, \mathbf{s})$:

$$\max_{(\mathbf{s}',\mathbf{y})}^k F(\mathbf{x}, \mathbf{s}, \mathbf{s}', \mathbf{y}) = 1 . \tag{12.21}$$

If $F(\mathbf{x}, \mathbf{s}, \mathbf{s}', \mathbf{y})$ is not realizable, then the list of phases must be extended by the missing phases $F_m(\mathbf{x}, \mathbf{s})$ determined by:

$$F_m(\mathbf{x}, \mathbf{s}) = \overline{\max_{(\mathbf{s}',\mathbf{y})}^k F(\mathbf{x}, \mathbf{s}, \mathbf{s}', \mathbf{y})} .$$

The number of states of a finite state machine will not necessarily be a power of two. In the sequential circuit the states are represented by the state-variables. If there are l state-variables, then 2^l states are realized in the sequential circuit. It is necessary to define the behavior also for such states that are not required by the finite state machine to be synthesized, but necessary in the sequential circuit. These states fill up the number of required states to the next higher power of two.

It is possible to define a fixed behavior for the supplemented states. For instance, a selected state can be reached for all of them in one step. This strong definition may lead to an overhead of switching elements. On the other side, advantage can be taken from the previously not defined behavior of the supplemented states. Any transition from the supplemented states to the required states ensures the required behavior after the first time-step. The only restriction is that no loops occur over the supplemented states. If transitions from each supplemented state to all required states are permitted, a non-deterministic finite state machine has been specified.

As can be seen in Fig. 12.3, the general structure of a finite state machine consists of three blocks. Two of them only need combinational switching elements for their realization. The result-

functions $\lambda(\mathbf{s})$ of the Moore-type or $\lambda(\mathbf{x}, \mathbf{s})$ of the Mealy-type of a finite state machine are realized by combinational circuits. Generally, the system-function for the result-part can be calculated by

$$F_\lambda(\mathbf{x}, \mathbf{s}, \mathbf{y}) = \max_{\mathbf{s}'}^k F(\mathbf{x}, \mathbf{s}, \mathbf{s}', \mathbf{y}) . \tag{12.22}$$

Note 12.8 In the case of the Moore-type F_λ depends only on $(\mathbf{s}, \mathbf{y})$. The synthesis of this part can be done as explained in the previous chapter.

The sequential behavior is described by the system-function $F_\delta(\mathbf{x}, \mathbf{s}, \mathbf{s}')$ that can be calculated by

$$F_\delta(\mathbf{x}, \mathbf{s}, \mathbf{s}') = \max_{\mathbf{y}}^k F(\mathbf{x}, \mathbf{s}, \mathbf{s}', \mathbf{y}) . \tag{12.23}$$

In the case of asynchronous sequential circuits, there is no dedicated memory-block, and the sequential behavior is realized by returning wires. The remaining block of the memory-functions $\delta(\mathbf{x}, \mathbf{s})$ are combinational functions. The system-equation (12.24) must be solved with regard to $\mathbf{s}'$. This is again a synthesis task for combinational circuits as explained in the previous chapter:

$$F_\delta(\mathbf{x}, \mathbf{s}, \mathbf{s}') = 1 . \tag{12.24}$$

Peculiarities of asynchronous finite state machines are:

- the restriction to the change of a single input to be sure that the expected stable state is reached; and
- the possibility that alternative sequences of state-changes can reach the same stable state.

The second peculiarity mentioned above can be used as source to optimize the sequential circuit to be synthesized because an arbitrary sequence of non-stable states can be used, which reaches the expected stable state. In this way, we get the behavior of finite state machines that is non-deterministic with regard to the non-stable states, but deterministic with regard to the stable states.

We demonstrate the synthesis of an asynchronous finite state machine using the analysis result of Sect. 12.2.

Figure 12.13a shows the graph (repeated from Fig. 12.9) of the asynchronous finite state machine that is realized by the sequential circuit of Fig. 12.5 together with the used encoding.

The longest sequence of state transitions needs for the explored FSM three steps for $(x_1 = 0, x_2 = 0)$ from the state $(s_1 = 0, s_2 = 1)$ to the stable state S1 encoded by $(s_1 = 0, s_2 = 0)$ or for $(x_1 = 0, x_2 = 1)$ from the state $(s_1 = 1, s_2 = 0)$ to the stable state S3 encoded by $(s_1 = 1, s_2 = 1)$. Shorter sequences of such state-changes are preferred because they result in faster circuits.

In the case of a single stable state for a constant input-pattern, all possible transitions to this state can be used and direct transition should be preferred. Therefore, we redirected the transition for $(x_1 = 0, x_2 = 0)$ from the state $(s_1 = 0, s_2 = 1)$ directly to the stable state S1 and for $(x_1 = 0, x_2 = 1)$ from the state $(s_1 = 1, s_2 = 0)$ to the stable state S3. Alternative state transitions are possible in the case that the distance is larger than one to a single stable state for a constant input-pattern; this choice can be used to get a simpler circuit. This opportunity has been utilized in the non-stable states with a distance of two to the stable states for all input-patterns except $(x_1 = 1, x_2 = 1)$; Figure 12.13b emphasizes these transitions by green colored edges. This extension of the permitted behavior leads to a non-deterministic system-function of the finite state machine. The system-function associated to the non-deterministic graph of Fig. 12.13b is

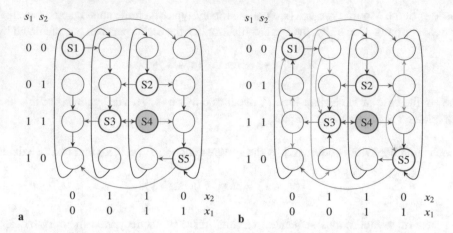

Fig. 12.13 Graph of an asynchronous finite state machine: (**a**) result of the analysis of the asynchronous sequential circuit of Fig. 12.5 and (**b**) non-deterministic modified graph used for the synthesis

$$
F(\mathbf{x}, \mathbf{s}, \mathbf{s}', y) =
\begin{array}{ccccccc}
x_1 & x_2 & s_1 & s_2 & s_1' & s_2' & y \\
\hline
0 & 0 & 1 & 1 & 0 & 1 & 0 \\
0 & 0 & - & - & 0 & 0 & 0 \\
0 & 1 & 0 & 0 & 0 & 1 & 0 & 0 \\
0 & 1 & - & - & 1 & 1 & 0 \\
- & 1 & 0 & 0 & 0 & 1 & 0 \\
1 & 1 & 0 & 1 & 0 & 1 & 0 \\
1 & 1 & 1 & 1 & 1 & 1 & 1 \\
1 & 1 & 1 & 0 & 0 & 0 & 0 \\
1 & 0 & 0 & - & 1 & 0 & 0 \\
1 & 0 & 0 & 1 & 1 & 1 & 0 \\
1 & 0 & 0 & 1 & 0 & 0 & 0 \\
- & 0 & 1 & 1 & 1 & 0 & 0 \\
1 & 0 & 1 & 0 & 1 & 0 & 0 \\
\end{array}
.
\tag{12.25}
$$

The synthesis task is now to solve of the system-equation

$$
F(\mathbf{x}, \mathbf{s}, \mathbf{s}', y) = 1
\tag{12.26}
$$

with regard to s_1', s_2', and y, where $F(\mathbf{x}, \mathbf{s}, \mathbf{s}', y)$ of (12.25) is used. The check whether the given behavior is realizable by a sequential circuit can be performed using Condition (12.21). The result of this check is that the system-function (12.25) is realizable.

Now we apply the known procedure to solve the system-equation (12.26) with regard to s_1', s_2', and y. The mark-functions of the lattice of functions $s_1' = \delta_1(\mathbf{x}, \mathbf{s})$ can be computed by:

$$
\delta_{1q}(\mathbf{x}, \mathbf{s}) = \overline{\max_{(\mathbf{s}', y)}^{k}(\overline{s}_1' \wedge F(\mathbf{x}, \mathbf{s}, \mathbf{s}', y))},
\tag{12.27}
$$

$$
\delta_{1r}(\mathbf{x}, \mathbf{s}) = \overline{\max_{(\mathbf{s}', y)}^{k}(s_1' \wedge F(\mathbf{x}, \mathbf{s}, \mathbf{s}', y))}.
\tag{12.28}
$$

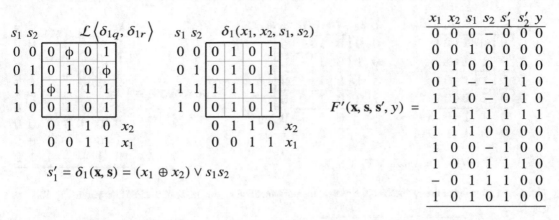

$$s_1' = \delta_1(\mathbf{x}, \mathbf{s}) = (x_1 \oplus x_2) \vee s_1 s_2$$

Fig. 12.14 Results of the synthesis of the first memory-function $s_1' = \delta_1(\mathbf{x}, \mathbf{s})$ of the FSM specified in Fig. 12.13b

Using this lattice, the synthesis of the combinational circuit of the memory-function $\delta_1(\mathbf{x}, \mathbf{s})$ is executed. The function of $s_1' = \delta_1(\mathbf{x}, \mathbf{s})$ is chosen as the result of this procedure. This function must be used to restrict the system-function so that arbitrary functions of the lattices of $s_2' = \delta_2(\mathbf{x}, \mathbf{s})$ and $y = \lambda(\mathbf{x}, \mathbf{s})$ can be chosen. The new system-function $F'(\mathbf{x}, \mathbf{s}, \mathbf{s}', y)$:

$$F'(\mathbf{x}, \mathbf{s}, \mathbf{s}', y) = F(\mathbf{x}, \mathbf{s}, \mathbf{s}', y) \wedge (s_1' \odot \delta_1(\mathbf{x}, \mathbf{s})) \tag{12.29}$$

replaces in further synthesis-steps the previous system-function $F(\mathbf{x}, \mathbf{s}, \mathbf{s}', y)$. Figure 12.14 shows the Karnaugh-maps of the calculated lattice $\mathcal{L}_1\langle\delta_{1q}, \delta_{1r}\rangle$, the chosen function $\delta_1(\mathbf{x}, \mathbf{s})$, and the restricted system-function.

It can be seen in Fig. 12.14 that the largest function of the lattice $\mathcal{L}\langle\delta_{1q}, \delta_{1r}\rangle$ has been chosen due to the smallest effort for their realization as combinational circuit. Due to this selection, the number of phases has been reduced from 22 to 18; two choices for (x_1, x_2, s_1, s_2) equal to (0100) and (1001) remain.

Now we solve the system-equation (12.26) where $F(\mathbf{x}, \mathbf{s}, \mathbf{s}', y)$ has been replaced by $F'(\mathbf{x}, \mathbf{s}, \mathbf{s}', y)$ of Fig. 12.14 with regard to s_2' to get the lattice of the second memory-function $s_2' = \delta_2(\mathbf{x}, \mathbf{s})$. Adjusted formulas to compute the associated mark-functions are

$$\delta_{2q}(\mathbf{x}, \mathbf{s}) = \overline{\max_{(\mathbf{s}', y)}^k (\overline{s_2'} \wedge F(\mathbf{x}, \mathbf{s}, \mathbf{s}', y))}, \tag{12.30}$$

$$\delta_{2r}(\mathbf{x}, \mathbf{s}) = \overline{\max_{(\mathbf{s}', y)}^k (s_2' \wedge F(\mathbf{x}, \mathbf{s}, \mathbf{s}', y))}. \tag{12.31}$$

The result of the combinational synthesis of the lattice $\mathcal{L}\langle\delta_{2q}, \delta_{2r}\rangle$ is the function $s_2' = \delta_2(\mathbf{x}, \mathbf{s})$ that restricts the permitted remaining behavior again:

$$F'(\mathbf{x}, \mathbf{s}, \mathbf{s}', y) = F(\mathbf{x}, \mathbf{s}, \mathbf{s}', y) \wedge (s_2' \odot \delta_2(\mathbf{x}, \mathbf{s})). \tag{12.32}$$

Figure 12.15 shows the Karnaugh-maps of the calculated lattice $\mathcal{L}\langle\delta_{2q}, \delta_{2r}\rangle$, the chosen function $\delta_2(\mathbf{x}, \mathbf{s})$, and the even more restricted system-function.

As can be seen in Fig. 12.15, 16 permitted phases remain after this second synthesis-step. At all $2^{2+2} = 16$, phases must be determined due to the two inputs and two memory-variables; hence, after

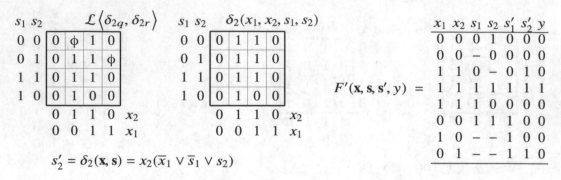

$$s_2' = \delta_2(\mathbf{x},\mathbf{s}) = x_2(\overline{x}_1 \vee \overline{s}_1 \vee s_2)$$

Fig. 12.15 Results of the synthesis of the second memory-function $s_2' = \delta_2(\mathbf{x},\mathbf{s})$ of the FSM defined in Fig. 12.13b

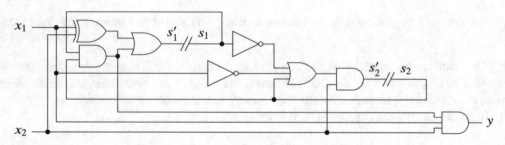

Fig. 12.16 Asynchronous sequential circuit of the FSM defined in Fig. 12.13b

the synthesis of the two memory-functions, the given non-deterministic FSM has been restricted to a deterministic FSM.

For the synthesis of the output-function $y = \lambda(\mathbf{x},\mathbf{s})$, we have to solve the system-equation (12.26) with regard to y where $F'(\mathbf{x},\mathbf{s},\mathbf{s}',y)$ of Fig. 12.15 must be used for $F(\mathbf{x},\mathbf{s},\mathbf{s}',y)$. It can be verified by

$$\operatorname*{der}_{y}\left(\operatorname*{max}_{\mathbf{s}'}^{k} F(\mathbf{x},\mathbf{s},\mathbf{s}',y)\right) = 1 \qquad (12.33)$$

that this system-function is uniquely solvable with regard to y; hence, we can compute output-function $y = \lambda(\mathbf{x},\mathbf{s})$ by:

$$\lambda(\mathbf{x},\mathbf{s}) = \operatorname*{max}_{(\mathbf{s}',y)}^{k}(y \wedge F(\mathbf{x},\mathbf{s},\mathbf{s}',y)) \qquad (12.34)$$

and get

$$y = \lambda(\mathbf{x},\mathbf{s}) = x_1 x_2 s_1 s_2 .$$

Figure 12.16 shows the synthesized asynchronous sequential circuit where the state-variables and the variables of the memory-functions are assigned to the right and left of the cuts in the loops. In the real circuit these cuts do not exist.

An analysis of the synthesized asynchronous sequential circuit confirms that this circuit has a deterministic behavior that belongs to the given non-deterministic FSM. The global list of phases of the synthesized asynchronous sequential circuit is already shown in Fig. 12.15 as system-function $F'(\mathbf{x},\mathbf{s},\mathbf{s}',y)$. Figure 12.17 depicts the graphs of the given non-deterministic FSM and the realized FSM; it confirms that the synthesized asynchronous sequential circuit realizes a permitted deterministic behavior.

At the beginning of the synthesis, it is not clear which of the alternatively usable transitions facilitate the simplest circuit. The longest sequence of state transitions could be reduced from three

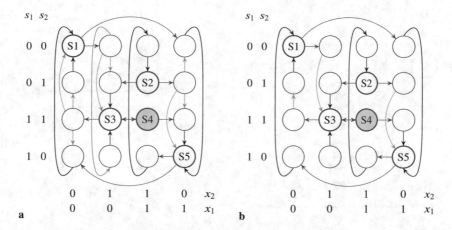

Fig. 12.17 Comparison of graphs of an asynchronous FSM: (**a**) non-deterministic modified graph used for the synthesis, result of analysis of the asynchronous circuit of Fig. 12.5, and (**b**) realized deterministic graph of the synthesized circuit

to two and the number of gates with up to three inputs has been reduced from 14 to eight as a result of executed synthesis using the same encoding of the states by the Boolean variables s_1 and s_2.

There is one more possibility to synthesize a small sequential circuit that utilizes that:

- any encoding of the stable states by Boolean variables can be used; and
- stable states for different input-patterns can be encoded by the same pattern of the state-variables.

A method suggested by David A. Huffman [3] and Samuel H. Caldwell [2] (known as Huffman–Caldwell method) specifies the following steps:

1. represent the behavior of the FSM by a table in which stable states are emphasized by circles and next states are specified for each stable state and each change of a single input-variable;
2. construct a compatibility graph in which:

   - nodes of the stable states are connected by a solid edge (pair of unconditional compatible states) if there are no different values in the associated rows of the table of the FSM neither for the states nor for the outputs;
   - nodes of the stable states are connected by a dashed edge (pair of conditional compatible states) if there are no different values in the associated rows of the table of the FSM for the states but different values for at least one output;

3. encode the stable states of complete subgraphs of the compatibility graph by the same pattern of state-variables; and
4. synthesize the sequential circuit using the chosen encoding as explained above.

Figure 12.18 shows all steps of this synthesis method for the FSM of our previous synthesis. The simplified system-function $F(\mathbf{x}, s, s', y)$ contains no choice in this case so that both the memory-function $s' = \delta(\mathbf{x}, s)$ and the output-function $y = \lambda(\mathbf{x}, s)$ are completely specified.

The compatibility graph of Fig. 12.18b contains two disjoint complete subgraphs (S1, S2, S5) and (S3, S4); hence, an encoding using a single memory variable is possible. Due to the used conditional compatibility between S3 and S4, we get an asynchronous sequential circuit of the Mealy-type. Again less gates (five instead of eight) are needed to realize the required behavior; the main reason for this improvement is the restriction to a single memory-variable.

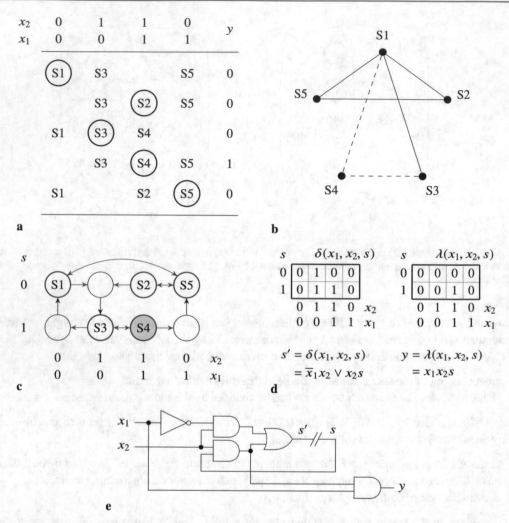

a

b

c

d

e

Fig. 12.18 Synthesis of the asynchronous FSM of Fig. 12.13a using the Huffman–Caldwell method: (**a**) table of the FSM, (**b**) compatibility graph, (**c**) simplified graph of the FSM, (**d**) Karnaugh-maps and functions, and (**e**) circuit structure

12.5   Synthesis of Synchronous Finite State Machines

The basic synthesis task is the same for both asynchronous and synchronous finite state machines; it consists in solving the system-equation (12.6) with regard to the variables $\mathbf{s}'$ of the memory-functions and the variables $\mathbf{y}$ of the output-functions. However, there are significant differences for the synthesis between these two types of finite state machines:

- the peculiarities of asynchronous finite state machines, that at each point in time only a single input-variable can be changed and does not hold for synchronous finite state machines; hence, without this restriction, the synthesis becomes easier; but
- an additional challenge for the synthesis of synchronous finite state machines is that not only the memory-functions but also the flip-flop controlling functions must be calculated.

The key problem of the synthesis of a synchronous finite state machine is that two different equations must be solved with regard to certain variables, and the first solution-function determines the second equation:

1. the system-equation (12.6) must be solved with regard to the variables s_i'; the functions $s_i' = \delta_i(\mathbf{x}, \mathbf{s})$ can be chosen in general out of lattices of functions, but not each combination of functions is permitted;
2. the chosen function $s_i' = \delta_i(\mathbf{x}, \mathbf{s})$ must be substituted in the flip-flop-equation of the used type and the resulting equation must be solved with regard to the input-variables of the flip-flop; in general, the flip-flop controlling functions can also be chosen out of lattices of functions, but again, not each combination of these functions is permitted.

It is very time consuming to

- calculate all lattices of flip-flop controlling functions;
- verify which combinations of these functions satisfy the expected behavior;
- realize for all of them by the synthesis of combinational circuits; and
- select the best circuit structure of the synchronous sequential circuit.

We suggest Algorithm 8 for the synthesis of a synchronous sequential circuit for a non-deterministic finite state machine.

The key to resolve the problems mentioned above is to merge the behavior of the given FSM and the behavior of the flip-flop associated to the memory-function of s_i' in line 4 of Algorithm 8 into the function $F_i(in1_i, in2_i, s_i, s_i')$ that describes the permitted non-deterministic behavior of the flip-flop inputs $in1$ and $in2$. After that, the known synthesis method for a lattice of functions is used (lines 5–7 for the first input of the selected flip-flop, lines 9–11 for the second input of the selected flip-flop, and later in lines 19–21 for the output i). Any function of the calculated lattices can be used for the combinational synthesis due to the restriction to the remaining conflict-free behavior in lines 8, 12, and 22. The restriction to the memory-behavior in line 2 or the behavior of the output-functions in line 17 excluded unneeded variables for the subsequent computations. The really implemented deterministic behavior is determined for the flip-flop i in line 13, for all so far computed memory-functions in line 14, for all memory-functions (together with the permitted behavior of the output-functions) in line 16, and additionally for all output-functions in line 24; hence, we get the deterministic system-function $F(\mathbf{x}, \mathbf{s}, \mathbf{s}', \mathbf{y})$ of the realized behavior in line 24.

Algorithm 8 can be used for both deterministic and non-deterministic finite state machines. Non-realizable finite state machines are excluded from the synthesis procedure by the check in line 1. The input-variables $in[i]$ of the system-function $F_i^{\text{ff}}(in1_i, in2_i, s_i, s_i')$ of the flip-flop-type will be replaced by the respective variables, such as j_i and k_i or d_i and e_i, and so on. An important benefit of Algorithm 8 is that it is not necessary to solve the system-equation with regard to the variables s_i'.

Example 12.2 We assume that the finite state machine to be synthesized is controlled by one input x and has two outputs (y_1, y_2). The outputs are equal to the state-variables: $y_1 = s_1$ and $y_2 = s_2$; hence, no effort for the output-functions is necessary, and we can concentrate our work on the sequential part of the circuit. When the input x is equal to 1, the three states $(10) \rightarrow (11) \rightarrow (01)$ must be reached in a cycle in this sequence. In the other case when x is equal to 0, the sequential circuit must be switched between the states (10) and (01). When the input x switches from 1 to 0 while the state (11) is active, the shorter cycle must be reached in the next time-step. It is not determined whether in this case the state (10) or (01) is reached as the next state. The state (00) is not required for the working behavior. The only demand for this state is that in the next time-step one of the working states (10), (11), or (01) will be reached. Two clocked DE-flip-flops and NAND-gates have to be used in the sequential circuit

---

Algorithm 8 Synthesis of synchronous sequential circuits SSSC(F)

Input : system-function $F(\mathbf{x}, \mathbf{s}, \mathbf{s}', \mathbf{y})$ that determines the permitted behavior by $F(\mathbf{x}, \mathbf{s}, \mathbf{s}', \mathbf{y}) = 1$
Input : for each $s_i' \in \mathbf{s}'$, the characteristic flip-flop function $F_i^{\mathrm{ff}}(in1_i, in2_i, s_i, s_i')$ of the used flip-flop
Output : synchronous sequential circuit that realizes one permitted behavior, if $F(\mathbf{x}, \mathbf{s}, \mathbf{s}', \mathbf{y}) = 1$ describes a realizable FSM

1: **if** $\max_{(\mathbf{s}',\mathbf{y})}{}^k F(\mathbf{x}, \mathbf{s}, \mathbf{s}', \mathbf{y}) = 1$ **then**
2:      $F_\delta(\mathbf{x}, \mathbf{s}, \mathbf{s}') \leftarrow \max_{\mathbf{y}}{}^k F(\mathbf{x}, \mathbf{s}, \mathbf{s}', \mathbf{y})$
3:      **for all** $s_i' \in \mathbf{s}'$ **do**
4:          $F_i(\mathbf{x}, \mathbf{s}, in1_i, in2_i) \leftarrow \max_{\mathbf{s}'}{}^k \left(F_\delta(\mathbf{x}, \mathbf{s}, \mathbf{s}') \wedge F_i^{\mathrm{ff}}(in1_i, in2_i, s_i, s_i') \right)$
5:          $in1_{iq}(\mathbf{x}, \mathbf{s}) \leftarrow \overline{\max_{(in1_i, in2_i)}^k \left(\overline{in1_i} \wedge F_i(\mathbf{x}, \mathbf{s}, in1_i, in2_i) \right)}$
6:          $in1_{ir}(\mathbf{x}, \mathbf{s}) \leftarrow \max_{(in1_i, in2_i)}^k \left(in1_i \wedge F_i(\mathbf{x}, \mathbf{s}, in1_i, in2_i) \right)$
7:          $in1_i(\mathbf{x}, \mathbf{s}) \leftarrow$ combinational synthesis of $\mathcal{L}\langle in1_{iq}(\mathbf{x}, \mathbf{s}), in1_{ir}(\mathbf{x}, \mathbf{s}) \rangle$
8:          $F_i(\mathbf{x}, \mathbf{s}, in1_i, in2_i) \leftarrow F_i(\mathbf{x}, \mathbf{s}, in1_i, in2_i) \wedge (in1_i \odot in1_i(\mathbf{x}, \mathbf{s}))$
9:          $in2_{iq}(\mathbf{x}, \mathbf{s}) \leftarrow \overline{\max_{(in2_i, in2_i)}^k \left(\overline{in2_i} \wedge F_i(\mathbf{x}, \mathbf{s}, in1_i, in2_i) \right)}$
10:         $in2_{ir}(\mathbf{x}, \mathbf{s}) \leftarrow \max_{(in2_i, in2_i)}^k \left(in2_i \wedge F_i(\mathbf{x}, \mathbf{s}, in1_i, in2_i) \right)$
11:         $in2_i(\mathbf{x}, \mathbf{s}) \leftarrow$ combinational synthesis of $\mathcal{L}\langle in2_{iq}(\mathbf{x}, \mathbf{s}), in2_{ir}(\mathbf{x}, \mathbf{s}) \rangle$
12:         $F_i(\mathbf{x}, \mathbf{s}, in1_i, in2_i) \leftarrow F_i(\mathbf{x}, \mathbf{s}, in1_i, in2_i) \wedge (in2_i \odot in2_i(\mathbf{x}, \mathbf{s}))$
13:         $F_{\delta i}(\mathbf{x}, \mathbf{s}, s_i') \leftarrow \max_{(in1_i, in2_i)}^k \left(F_i^{\mathrm{ff}}(in1_i, in2_i, s_i, s_i') \wedge F_i(\mathbf{x}, \mathbf{s}, in1_i, in2_i) \right)$
14:         $F_\delta(\mathbf{x}, \mathbf{s}, \mathbf{s}') \leftarrow F_\delta(\mathbf{x}, \mathbf{s}, \mathbf{s}') \wedge F_{\delta i}(\mathbf{x}, \mathbf{s}, s_i')$
15:     **end for**
16:     $F(\mathbf{x}, \mathbf{s}, \mathbf{s}', \mathbf{y}) \leftarrow F(\mathbf{x}, \mathbf{s}, \mathbf{s}', \mathbf{y}) \wedge F_\delta(\mathbf{x}, \mathbf{s}, \mathbf{s}')$
17:     $F_\lambda(\mathbf{x}, \mathbf{s}, \mathbf{y}) \leftarrow \max_{\mathbf{s}'}^k F(\mathbf{x}, \mathbf{s}, \mathbf{s}', \mathbf{y})$
18:     **for all** $y_i \in \mathbf{y}$ **do**
19:         $\lambda_{iq}(\mathbf{x}, \mathbf{s}) \leftarrow \overline{\max_{\mathbf{y}}^k \left(\overline{y_i} \wedge F_\lambda(\mathbf{x}, \mathbf{s}, \mathbf{y}) \right)}$
20:         $\lambda_{ir}(\mathbf{x}, \mathbf{s}) \leftarrow \max_{\mathbf{y}}^k \left(y_i \wedge F_\lambda(\mathbf{x}, \mathbf{s}, \mathbf{y}) \right)$
21:         $\lambda_i(\mathbf{x}, \mathbf{s}) \leftarrow$ combinational synthesis of $\mathcal{L}\langle \lambda_{iq}(\mathbf{x}, \mathbf{s}), \lambda_{ir}(\mathbf{x}, \mathbf{s}) \rangle$
22:         $F_\lambda(\mathbf{x}, \mathbf{s}, \mathbf{y}) \leftarrow F_\lambda(\mathbf{x}, \mathbf{s}, \mathbf{y}) \wedge (y_i \odot \lambda_i(\mathbf{x}, \mathbf{s}))$
23:     **end for**
24:     $F(\mathbf{x}, \mathbf{s}, \mathbf{s}', \mathbf{y}) \leftarrow F(\mathbf{x}, \mathbf{s}, \mathbf{s}', \mathbf{y}) \wedge F_\lambda(\mathbf{x}, \mathbf{s}, \mathbf{y})$
25: **else**
26:     $F(\mathbf{x}, \mathbf{s}, \mathbf{s}', \mathbf{y})$ is not realizable
27: **end if**

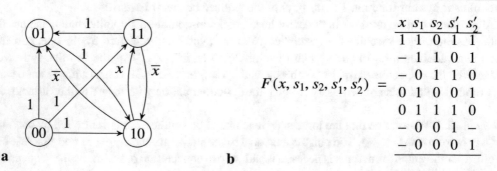

Fig. 12.19 A non-deterministic finite state machine to be synthesized: (**a**) graph in which the states are labeled by (s_1, s_2) and (**b**) system-function expressed by an orthogonal list of phases

to be synthesized. Figure 12.19a depicts the graph of the permitted behavior of the non-deterministic finite state machine.

$$F_1^{\text{ff}}(d_1, e_1, s_1, s_1') = \begin{array}{cccc} d_1 & e_1 & s_1 & s_1' \\ \hline - & 0 & 0 & 0 \\ - & 0 & 1 & 1 \\ 0 & 1 & - & 0 \\ 1 & 1 & - & 1 \\ \hline \end{array}$$

$$F_1(x, s_1, s_2, d_1, e_1) = \begin{array}{ccccc} x & s_1 & s_2 & d_1 & e_1 \\ \hline 1 & 1 & 0 & - & 0 \\ 1 & 1 & 0 & 1 & 1 \\ - & 1 & 1 & 0 & 1 \\ - & 0 & 1 & 1 & 1 \\ 0 & 1 & 0 & 0 & 1 \\ 0 & 1 & 1 & 0 & 0 \\ 0 & 1 & 1 & 1 & - \\ - & 0 & 0 & - & - \\ \hline \end{array}$$

a   b

Fig. 12.20 DE-flip-flop 1: (**a**) general behavior and (**b**) behavior adjusted to s_1' of the FSM of Fig. 12.19

$$d_{1q}(x, \mathbf{s}) = \begin{array}{ccc} x & s_1 & s_2 \\ \hline - & 0 & 1 \\ \hline \end{array}$$

$$d_{r1}(x, \mathbf{s}) = \begin{array}{ccc} x & s_1 & s_2 \\ \hline 0 & 1 & 0 \\ 1 & 1 & 1 \\ \hline \end{array}$$

$\mathcal{L}\langle d_{1q}, d_{1r} \rangle$

$s_1\,s_2$	0	1
0 0	ϕ	ϕ
0 1	1	1
1 1	ϕ	0
1 0	0	ϕ

x

$d_1(x, \mathbf{s})$

$s_1\,s_2$	0	1
0 0	1	1
0 1	1	1
1 1	0	0
1 0	0	0

x

$$d_1(x, \mathbf{s}) = \overline{s}_1$$

Fig. 12.21 Synthesis results for the input d_1 of the first DE-flip-flop

$$e_{1q}(x, \mathbf{s}) = \begin{array}{ccc} x & s_1 & s_2 \\ \hline 0 & 1 & 0 \\ 1 & 1 & 1 \\ - & 0 & 1 \\ \hline \end{array}$$

$$e_{r1}(x, \mathbf{s}) = \begin{array}{ccc} x & s_1 & s_2 \\ \hline 1 & 1 & 0 \\ \hline \end{array}$$

$\mathcal{L}\langle e_{1q}, e_{1r} \rangle$

$s_1\,s_2$	0	1
0 0	ϕ	ϕ
0 1	1	1
1 1	ϕ	1
1 0	1	0

x

$e_1(x, \mathbf{s})$

$s_1\,s_2$	0	1
0 0	1	0
0 1	1	1
1 1	1	1
1 0	1	0

x

$$e_1(x, \mathbf{s}) = x \wedge \overline{s}_2$$

Fig. 12.22 Synthesis results for the input e_1 of the first DE-flip-flop

The finite state machine of Fig. 12.19 is realizable due to the successful check of Eq. (12.21) in line 1 of Algorithm 8. This property is also clearly visible in the list of phases in Fig. 12.19b where all possible patterns (x, s_1, s_2) occur. The list of phases of Fig. 12.19b permits 13 phases, from which eight phases must be used. This list of phases specifies the system-function $F(x, s_1, s_2, s_1', s_2')$ that is already restricted to the memory behavior so that the execution of the formula in line 2 of Algorithm 8 results in $F_\delta(x, s_1, s_2, s_1', s_2') = F(x, s_1, s_2, s_1', s_2')$.

The list of phases of the DE-flip-flop is taken from Table 12.1. The variable names are adjusted to the first DE-flip-flop of the sequential circuit to be synthesized as shown in Fig. 12.20.

The controlling functions of any DE-flip-flop can be chosen out of lattices of functions; hence, the system function of the first DE-flip-flop adjusted to s_1' of the FSM that has to be synthesized comprises 19 phases from which eight phases must be used (see Fig. 12.20b). Figure 12.21 shows the results of the synthesis for the input d_1 of the first DE-flip-flop based on lines 5–7 of Algorithm 8.

Due to the chosen controlling function $d_1(x, \mathbf{s}) = \overline{s}_1$, the formula in line 8 of Algorithm 8 reduces the actual system-function $F_1(x, s_1, s_2, d_1, e_1)$ from 19 to 11 phases; hence, there remain choices for the controlling function of the input e_1 of the first DE-flip-flop. Figure 12.22 shows the synthesis results for the input e_1 of the first DE-flip-flop based on lines 9–11 of Algorithm 8.

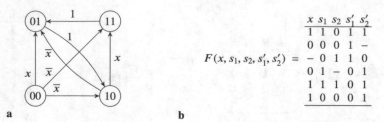

$$F(x, s_1, s_2, s'_1, s'_2) = \begin{array}{ccccc} x & s_1 & s_2 & s'_1 & s'_2 \\ \hline 1 & 1 & 0 & 1 & 1 \\ 0 & 0 & 0 & 1 & - \\ - & 0 & 1 & 1 & 0 \\ 0 & 1 & - & 0 & 1 \\ 1 & 1 & 1 & 1 & 0 \\ 1 & 0 & 0 & 0 & 1 \end{array}$$

a　　　　　　　　　　　　　　　**b**

Fig. 12.23 The remaining non-deterministic behavior of the partially synthesized FSM: (**a**) graph in which the states are labeled by (s_1, s_2) and (**b**) system-function expressed by an orthogonal list of phases

$$F_2^{\text{ff}}(d_2, e_2, s_2, s'_2) = \begin{array}{cccc} d_2 & e_2 & s_2 & s'_2 \\ \hline - & 0 & 0 & 0 \\ - & 0 & 1 & 1 \\ 0 & 1 & - & 0 \\ 1 & 1 & - & 1 \end{array}$$

$$F_2(x, s_1, s_2, d_2, e_2) = \begin{array}{ccccc} x & s_1 & s_2 & d_2 & e_2 \\ \hline 0 & 0 & 0 & 0 & 1 \\ 0 & 0 & 0 & - & 0 \\ - & 0 & 1 & 0 & 1 \\ - & 1 & 1 & - & 0 \\ - & 1 & 1 & 1 & 1 \\ - & - & 0 & 1 & 1 \end{array}$$

a　　　　　　　　　　　　　　　**b**

Fig. 12.24 DE-flip-flop 2: (**a**) general behavior and (**b**) behavior adjusted to s'_2 of the FSM of Fig. 12.19 where s'_1 is already determined by the functions $d_1(x, \mathbf{s})$ and $e_1(x, \mathbf{s})$

Fig. 12.25 Synthesis results for the input d_2 of the second DE-flip-flop

$$d_{2q}(x, \mathbf{s}) = \begin{array}{ccc} x & s_1 & s_2 \\ \hline 1 & 0 & 0 \\ - & 1 & 0 \end{array}$$

$$d_{2r}(x, \mathbf{s}) = \begin{array}{ccc} x & s_1 & s_2 \\ \hline - & 0 & 1 \end{array}$$

$$\begin{array}{c|cc} s_1\,s_2 & \mathcal{L}\langle d_{2q}, d_{2r}\rangle & \\ \hline 0\ 0 & \phi & 1 \\ 0\ 1 & 0 & 0 \\ 1\ 1 & \phi & \phi \\ 1\ 0 & 0 & 1 \\ & 0 & 1 \ \ x \end{array}$$

$$\begin{array}{c|cc} s_1\,s_2 & d_2(x, \mathbf{s}) & \\ \hline 0\ 0 & 1 & 1 \\ 0\ 1 & 0 & 0 \\ 1\ 1 & 0 & 0 \\ 1\ 0 & 1 & 1 \\ & 0 & 1 \ \ x \end{array}$$

$$d_2(x, \mathbf{s}) = \bar{s}_2$$

The execution of the formula in line 12 of Algorithm 8 restricts $F_1(x, s_1, s_2, d_1, e_1)$ to a deterministic system-function, but as a result of lines 13 and 14 there remains a non-deterministic system-function $F_\delta(x, s_1, s_2, s'_1, s'_2)$ that is only partially restricted.

Figure 12.23 shows both the graph and the system-function that remains after the synthesis of the controlling functions of the first DE-flip-flop. It can be seen that starting from the state $(0, 0)$ either the state $(1, 0)$ or the state $(1, 1)$ will be reached when the input x is equal to 0.

In the second sweep of the `for`-loop of lines 3–15 of Algorithm 8, we synthesize the controlling functions for the second DE-flip-flop. The list of phases of the DE-flip-flop is again taken from Table 12.1, and the names of variables are adjusted to the second DE-flip-flop (see Fig. 12.24a) used to realize the memory-function $s'_2 = \delta_2(x, \mathbf{s})$.

The transformation of the permitted behavior of the finite state machine to be synthesized regarding line 4 of Algorithm 8 results in the system-function $F_2(x, s_2, s_2, d_2, e_2)$ (see Fig. 12.24b) that permits 15 phases from which eight phases must be used. Figure 12.25 shows the synthesis results for the input d_2 of the second DE-flip-flop based on lines 5–7 of Algorithm 8.

Due to the chosen controlling function $d_2(x, \mathbf{s}) = \bar{s}_2$, the formula in line 8 of Algorithm 8 reduces the actual system-function F_1 from 15 to nine phases; hence, there remains a single choice for the controlling function of the input e_2 of the second DE-flip-flop. Figure 12.26 shows the synthesis results for the input e_2 of the second DE-flip-flop based on lines 9–11 of Algorithm 8.

The restrictions in lines 12–14 of Algorithm 8 result in a deterministic system-function $F_\delta(x, s_1, s_2, s'_1, s'_2)$ that is realized by the synthesized circuit. Figure 12.27 shows this synchronous sequential circuit as well as the associated graph and the realized system-function. We can skip the

$$e_{2q}(x, \mathbf{s}) = \begin{array}{ccc} x & s_1 & s_2 \\ \hline 1 & 0 & 0 \\ - & 0 & 1 \\ - & 1 & 0 \end{array}$$

$$e_{r2}(x, \mathbf{s}) = \begin{array}{ccc} x & s_1 & s_2 \\ \hline - & 1 & 1 \end{array}$$

s_1 s_2	$\mathcal{L}\langle e_{2q}, e_{2r}\rangle$	
0 0	ϕ	1
0 1	1	1
1 1	0	0
1 0	1	1
	0	1 x

s_1 s_2	$e_2(x, \mathbf{s})$	
0 0	1	1
0 1	1	1
1 1	0	0
1 0	1	1
	0	1 x

$$e_1(x, \mathbf{s}) = \overline{s_1 \wedge s_2}$$

Fig. 12.26 Synthesis results for the input e_2 of the second DE-flip-flop

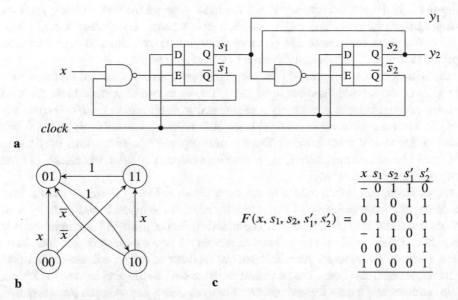

$$F(x, s_1, s_2, s_1', s_2') = \begin{array}{ccccc} x & s_1 & s_2 & s_1' & s_2' \\ \hline - & 0 & 1 & 1 & 0 \\ 1 & 1 & 0 & 1 & 1 \\ 0 & 1 & 0 & 0 & 1 \\ - & 1 & 1 & 0 & 1 \\ 0 & 0 & 0 & 1 & 1 \\ 1 & 0 & 0 & 0 & 1 \end{array}$$

Fig. 12.27 Synthesized deterministic FSM: (**a**) synchronous sequential circuit, (**b**) graph in which the states are labeled by (s_1, s_2), and (**c**) system-function expressed by an orthogonal list of phases

subsequent part of Algorithm 8 in which the combinational circuit of the output-functions will be synthesized due to the assumption that $y_1 = s_1$ and $y_2 = s_2$.

It is interesting to see that different states are reached for different values of the input x in the state $(s_1, s_2) = (0, 0)$. This result confirms that non-deterministic finite state machines provide advantages for the synthesis of optimized sequential circuits.

Note 12.9 The clock-signal has not been taken into consideration during the synthesis process, but it must be added in the diagram of the circuit.

12.6 Hardware Software Co-design

The realization of a finite state machine needs both a generally usable technical base and a possibility to specify the special required behavior. This is comparable to a computer where the real behavior is specified by the software that controls the hardware base. There are several different hardware bases to realize a finite state machine. The selection of such a hardware base influences both the synthesis efforts and the properties of the synthesized finite state machine in terms of time and space.

The simplest hardware base of a finite state machine consists of logic gates. As introduced above, returning wires change combinational circuits into asynchronous sequential circuits of finite state machines. The next stable state will be reached in the shortest possible time because all gates work in parallel. This advantage requires a very high effort for the synthesis because certain unstable state and possible inconsistencies (hazards) must be avoided.

A more comfortable hardware base of a finite state machine adds clocked flip-flops to the logic gates. If all clock-inputs of the flip-flops are connected to the same clock-signal, a synchronous sequential circuit of a finite state machine is created. The logic gates realize the controlling functions of the flip-flops. If the clock-frequency is not too high—that means that all logic gates completed their changes before the next clock-event—no hazards or unstable states must be taken into account during the synthesis process; hence, the previous disadvantages are removed. The price is a slightly longer period of time for the transition between two stable states.

The real behavior for both bases of hardware explained above is specified by the wires between the logic gates or the flip-flops and the logic gates. The wires can be included into the hardware so that the finite state machine is completely implemented in hardware. In these two cases we have to carry out the hardware synthesis as shown above. An optimal solution of this task can be found if the number of inputs and states is small. Otherwise it requires extensive efforts even to calculate a nearly optimal circuit structure; hence, alternative approaches that need less efforts to synthesize a finite state machine are welcome.

The central element of an alternative hardware base is a *read-only memory* (ROM). The address inputs **a** of the ROM are divided and partially associated to the inputs **x** and partially used for the state-variables **s** of the finite state machine. The stored content of the ROM represents logic functions at its outputs **o**. These functions can be used as memory-functions **s**$'$ and output-functions **y** of the finite state machine. The memory-functions that are available at certain outputs of the ROM control directly the inputs of D-flip-flops. The hardware structure does not depend on the real behavior of the finite state machine, and it only depends on the number of inputs and required state-variables; hence, we have a fixed hardware structure for a large class of finite state machines. An example of such a ROM-based structure of a finite state machine is shown in Fig. 12.28.

The synthesis task for such a ROM structure of a finite state machine is very easy. The memory-functions and the output-functions must be represented in a function table ordered by the decimal equivalent of the input- and state-variables. All function columns of this table can be stored directly into the ROM. The content of the ROM is a sequence of 0's and 1's and can be interpreted as software. This software influences the configuration-step during the production of the ROM. The run-time properties of such a circuit are similar to a fixed-wired sequential circuit of logic gates and flip-flops. The advantages of the ROM structure consist in the fixed hardware structure and the easy synthesis of the software to specify the required behavior of the finite state machine. The limits of the ROM approach of a finite state machine result from the available ROM-sizes.

Another slightly changed hardware structure can be handled by the same synthesis method as the previously introduced ROM-structure for a finite state machine, but it offers additional valuable features. The only change is that the ROM is substituted by a *random-access memory* (RAM). If the RAM stores the same pattern as a ROM, the implemented finite state machine has the same behavior. The difference is that the content of a RAM can change during run-time, which affects the behavior of the finite state machine. This emphasizes the software character of the behavioral description by the content of the RAM.

A possible application of such a RAM-based finite state machine is the realization of a large finite state machine using a hierarchical architecture. It is the task of the global finite state machine in a two-level architecture to select which of the local finite state machines must be active at a certain point in time. The benefit of a RAM-based finite state machine on the local level in this architecture is

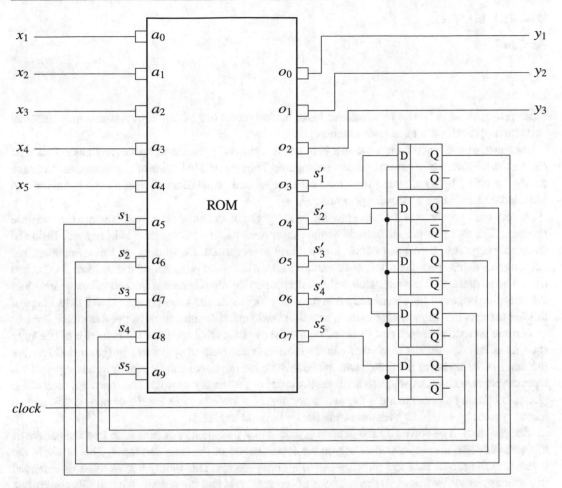

Fig. 12.28 Structure of a finite state machine based on a 1-KBit ROM configured for five inputs **x**, three outputs **y**, and 32 states, encoded by five state-variables **s**

that the same hardware can implement different behaviors of several local finite state machines. For this approach, it is only necessary to store all required RAM-patterns in a separate memory and to transfer the right pattern into the RAM in order to switch the behavior of the local finite state machine. It is obvious that the same principle can be used across several levels.

A desirable side-effect of this approach is that local finite state machines need less input- and state-variables. Consequently, the number of address variables of the RAM can be reduced.

Note 12.10 Each removed address variable reduces the number of bits to be stored in the RAM by 50%.

A further advantage of this approach is the effect that only a given part of the whole circuit is active at each point in time. This reduces the power consumption of the circuit, which restricts considerably the practical implementation of very large-scale integrated circuits.

There is a lot of programmable hardware structures that use basically the same idea to adapt a finite state machine to pre-produced hardware structures. Very popular are, for instance, the *logic cell arrays* (LCAs) or the *programmable logic devices* (PLDs). We do not describe the details of these and

Table 12.2 Binary code (a, b) of a ternary element t

t	a	b
0	0	1
1	1	1
–	0	0

other programmable hardware structures here, but we direct our attention to a completely different software approach for a finite state machine.

As stated above, each finite state machine needs a generally usable technological base. This base can be the hardware of a general-purpose computer. The required behavior of the finite state machine can be specified by a software program. There are several possibilities to specify the behavior of a finite state machine by a software program.

A very easy approach is a direct mapping of the graph of the finite state machine into the software program. For that purpose, each node of the graph is modeled by a separate subprogram. Inside of these subprograms, the values of the inputs must be evaluated. Depending on this evaluation, the subprogram determines the new output-values and calls the subprogram of the state to be reached next. The advantage of this approach is the restriction to the local behavior in each subprogram. Only the outgoing edges of the associated node must be modeled in the subprogram. The disadvantage of this approach is that the software of each new finite state machine must be developed from the scratch.

A more general approach can be based on the list of phases that specifies the behavior of the finite state machine. Such a list of phases can be used as a database of a general program that realizes the required behavior of the finite state machine. The list of phases can be stored in a compact TVL representation without any requirements on the order of the ternary vectors. The representation of the phases by ternary vectors reduces the necessary memory space to store the list of phases. We store a ternary vector by two binary vectors using the code in Table 12.2.

The problem to be solved by the general program of a finite state machine is to find the pattern of the next state and the output depending on the recent values of the state and input. The values of the inputs and the recent state are put together to a binary vector. This binary vector must be extended to a ternary vector by dashes in the columns of the next state and the output. We call this controlling vector **c**. This vector **c** of the independent variables must be compared with the ternary vectors of the list of phases $\mathbf{p}[1, \ldots, n_l]$. If **c** is orthogonal to $\mathbf{p}[i]$, then the ternary vector $\mathbf{p}[i]$ does not contain a phase matching the controlling vector **c**. Otherwise a matching phase has been found. The values of the new state and the outputs can be taken directly from the detected non-orthogonal vector $\mathbf{p}[i]$. Further comparisons of **c** with regard to the remaining vectors $\mathbf{p}[j]$, $j > i$, are not necessary and will be omitted. The logic equation for a matching phase is (12.35) that must be satisfied for a ternary vector $\mathbf{p}[i]$ of the list of phases. Practically Eq. (12.35) can be solved in parallel for all columns $k, k = 1, \ldots, n$:

$$0 = \bigvee_{k=1}^{n} (c_{ak} \oplus p[i]_{ak}) \wedge c_{bk} \wedge p[i]_{bk} , \qquad (12.35)$$

where **c** is encoded by $(\mathbf{c}_a, \mathbf{c}_b)$ and $\mathbf{p}[i]$ by $(\mathbf{p}[i]_a, \mathbf{p}[i]_b)$.

The advantage of this general approach is that no synthesis effort is necessary. The behavioral description of the finite state machine by ternary lists of phases can directly be used to implement the finite state machine by a computer. The price for the elimination of any synthesis-steps is the longer delay to switch to the next state.

The hardware–software co-design can soften this disadvantage. For that purpose, we use the hierarchical structure of finite state machines. Most of the transitions in such a structure occur in the partial finite state machines on the lowest level. For these partial finite state machines, hardware solutions should be preferred. The task of the higher-level finite state machines and especially for the top-level finite state machine consists in activating the required lower-level finite state machines. Such controlling actions do not occur very often so that a longer delay influences the general behavior only slightly.

A good combination of the short delay of the finite state machines during run-time and a minimal synthesis effort will be reached if in a hierarchical structure the lowest-level finite state machines are built by a RAM-structure and the other finite state machines in the higher levels of the hierarchy by the general software approach based on lists of phases. The change of the behavior of a finite state machine in an upper level requires only the change of a pointer to the next valid list of phases. The change of the behavior of the lowest-level finite state machine requires the download of the memory pattern to the RAM. This subtask causes an additional delay. This delay can be reduced if more than one RAM-structure is available. In this case the next needed lowest-level finite state machines can be prepared during the run-time of another finite state machine on the lowest-level of the hierarchy. The transition between two lowest-level finite state machines needs in this case only a very short activation time.

12.7 Exercises

Prepare for each task of each exercise of this chapter a PRP for the XBOOLE-monitor XBM 2. Prefer a lexicographic order of the variables and use the help system of the XBOOLE-monitor XBM 2 to extend your knowledge about the details of used commands. Note that a prime ($'$) is not allowed as a character in a name of a variable in the XBOOLE-monitor XBM 2; hence, you should use sf_i (state-value at the following point in time) instead of s_i'.

If you have not yet prepared the XBOOLE-monitor XBM 2 on your computer, you can get this XBOOLE-monitor free of charge by means of the following three steps:

1. **Download**:
   There are four versions of the XBOOLE-monitor XBM 2, two for Windows 10 or subsequent Windows systems (32 or 64 bits) and two for LINUX—Ubuntu (also 32 or 64 bits); you must download the version of the XBOOLE-monitor XBM 2 that fits to your operating system.
   Authorized users of the online version of this chapter(https://doi.org/10.1007/978-3-030-88945-6_12) can download the XBOOLE-monitor XBM 2 directly from the web page
   https://link.springer.com/chapter/10.1007/978-3-030-88945-6_12
   where the links for the download of the XBOOLE-monitor XBM 2 are located in the part "Supplementary Information" (below the part "Abstract"). The headline above such a link indicates the associated zip-file of the XBOOLE-monitor XBM 2. The sizes of the zip-files have been provided behind the links and can be used to verify the download. A click on the link of the wanted version of the XBOOLE-monitor XBM 2 starts the download.
   Readers of the hardcopy of this book get access to the XBOOLE-monitor XBM 2 using the URL
   https://link.springer.com/chapter/10.1007/978-3-030-88945-6_12
   to download the first two pages of this chapter. After this download, the same procedure as the authorized users of the online version of a chapter can be used to download the wanted version of the XBOOLE-monitor XBM 2.
2. **Unzip**: The XBOOLE-monitor XBM 2 must not be installed but must be unzipped into an arbitrary directory of your computer. A convenient tool for unzipping the downloaded zip-file is usually available as part of the operating system or can be downloaded from the Internet.

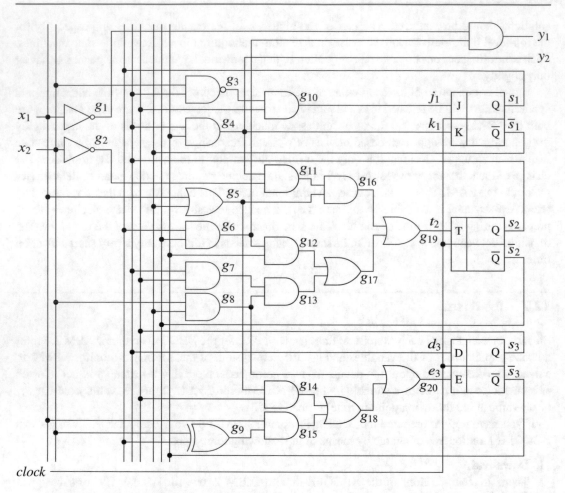

Fig. 12.29 Structure of a sequential circuit with two inputs, three flip-flops, and two outputs

3. **Execute**:

- Windows:
  The executable file of the two versions (32 or 64 bits) for Windows 10 (or subsequent Windows systems) of the XBOOLE-monitor XBM 2 is XBM2.exe; the other files in the expanded directory must remain unchanged. A double-click on the executable file XBM2.exe within the Explorer of Windows starts the XBOOLE-monitor XBM 2.
- LINUX—Ubuntu:
  The unzipped folder of the XBOOLE-monitor XBM 2 contains for this operating system only the executable file XBM2-i386.AppImage for the version of 32 bits or XBM2-x86_64.AppImage for the version of 64 bits of the XBOOLE-monitor XBM 2. A double-click on the created AppImage-file within the file manager of LINUX—Ubuntu starts the XBOOLE-monitor XBM 2.

Exercise 12.1 (Analysis of a Synchronous Sequential Circuit)
The structure of a synchronous sequential circuit is shown in Fig. 12.29.

(a) Compute the behavior of the sequential circuit shown in Fig. 12.29 using a system of logic equations of the gates and flip-flops. Show both the global list of phases (that contains the values observable on all wires of the circuit as well as the values of the state reached after the following clock-signal) and the minimized list of phases that represents the system-function $F(\mathbf{x}, \mathbf{s}, \mathbf{s}', \mathbf{y})$. Store the final XBOOLE-system for the use in all other tasks of this exercise.

(b) Load the XBOOLE-system stored in the previous task so that a comparison of the behavior of the explored sequential circuit computed using different approaches becomes possible. Compute thereafter the behavior of the sequential circuit shown in Fig. 12.29 using a set of local lists of phases of the gates and flip-flops. The flip-flops in this circuit provide both the state-variable s_i and its complement $\bar{s}_i$; extend therefore the list of phases of the used flip-flops provided in Table 12.1 by variables nsi representing $\bar{s}_i$. Verify whether the solution of a system of logic equations and the approach based on a set of local list of phases have the same system-function $F(\mathbf{x}, \mathbf{s}, \mathbf{s}', \mathbf{y})$ as solution.

(c) Initial states of a finite state machine (FSM) are states that cannot be reached from any other state of this FSM.
Load the XBOOLE-system stored in the first task of this exercise to get the system-function $F(\mathbf{x}, \mathbf{s}, \mathbf{s}', \mathbf{y})$ of the sequential circuit shown in Fig. 12.29 and compute all initial states of this circuit. How many initial states exist in this circuit? Show the TVL of the computed initial states.

(d) Final states of a finite state machine (FSM) are states from which no other state of this FSM can be reached.
Load the XBOOLE-system stored in the first task of this exercise to get the system-function $F(\mathbf{x}, \mathbf{s}, \mathbf{s}', \mathbf{y})$ of the sequential circuit shown in Fig. 12.29 and compute all final states of this circuit. How many final states exist in this circuit? Show the TVL of the computed final states.
Draw the graph of the FSM realized by the sequential circuit shown in Fig. 12.29 and verify the results of the last two tasks of this exercise. What is the type of the explored finite state machine?

Exercise 12.2 (Properties of a Finite State Machine)

Figure 12.30 specifies the behavior of a finite state machine. It can easily be checked that it is a FSM of the Moore-type.

(a) Compute all cycles of the given FSM that have a length of two edges and satisfy the additional restrictions that the input x_2 is a constant equal to 1 and that different states belong to the cycle, i.e., two repeated used loops at the same state do not belong to the solution. How many such cycles exist? Show the set of edges belonging to these cycles. Store the TVL of the given system-function as sdt-file so that it can be reused in the next tasks.

(b) Assume that the input-pattern (x_1, x_2) is a constant equal to (10), and at the outputs (y_1, y_2) the sequence of patterns $(01) \rightarrow (11) \rightarrow (10)$ has been observed. Is this information sufficient to conclude the three states belonging to these observed output-patterns? Which state will be reached after the next clock when the input-pattern remains unchanged? Express the state-variables s_i at the time-step $t = 0, \ldots, 3$ in the PRP using variables sit (e.g., s40 specifies the value of s_4 at time $t = 0$); analogously, the output-variables y_j at the time-step t are expressed in the PRP by variables yjt. Show all possible sequences of edges and the associated output-patterns that satisfy the observed output-pattern for the first three time-steps. Reuse the stored XBOOLE-system for this analysis.

(c) An edge $A \rightarrow B$ of the reachability graph connects the state A with the state B in the case that the associated finite state machine contains at least one sequence of edges such that the state B can be reached from the state A. The inputs and outputs are excluded in the reachability graph.

Fig. 12.30
System-function
$F(\mathbf{x}, \mathbf{s}, \mathbf{s}', \mathbf{y})$ of a finite
state machine expressed by
a TVL in ODA-form

	x_1	x_2	s_1	s_2	s_3	s_4	sf_1	sf_2	sf_3	sf_4	y_1	y_2
	1	–	0	0	0	0	0	0	0	1	0	1
	–	–	0	0	0	1	0	0	1	0	1	1
	–	–	0	0	1	0	0	0	1	1	0	0
	–	0	0	0	1	1	0	0	0	1	0	1
	–	1	0	0	1	1	0	1	0	0	0	1
	–	–	0	1	0	0	0	1	0	1	0	0
	–	–	0	1	0	1	0	1	1	0	0	1
	–	1	0	1	1	0	0	1	0	1	1	0
	0	0	0	1	1	0	1	0	1	0	1	0
	1	0	0	1	1	0	0	1	1	1	1	0
	–	–	0	1	1	1	1	0	0	0	0	1
ODA$(F) =$	1	–	1	0	0	0	1	0	1	0	1	1
	0	–	1	0	0	0	1	0	0	1	1	1
	–	–	1	0	0	1	0	0	0	0	1	0
	0	–	0	0	0	0	1	0	1	0	0	1
	–	–	1	0	1	0	0	1	1	0	1	1
	0	–	1	1	1	1	1	1	1	1	0	0
	1	–	1	1	1	1	0	1	1	0	0	0
	–	–	1	0	1	1	1	1	0	0	0	1
	–	–	1	1	0	0	1	1	0	1	1	1
	–	1	1	0	1	1	1	0	0	1	0	
	0	0	1	1	0	1	1	0	1	1	1	0
	1	0	1	1	0	1	1	1	1	0	1	0
	–	–	1	1	1	0	1	0	1	1	1	1

Fig. 12.31 Cyclic
sequences of output-values
in the cases of (**a**) $x = 0$
and (**b**) $x = 1$

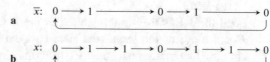

Load the TVL of the explored FSM (stored in the first task of this exercise) and compute the associated reachability graph. Show both the TVL of the given FSM restricted to the existing edges between the states and the TVL of the associated reachability graph. How many edges belong to the computed reachability graph and how many ternary vectors are needed to store this minimized TVL? Store the final XBOOLE-systems for the use in the next task.

(d) A component of a FSM is a set of states where all states of the component can be reached from each state of this component.

Load the `sdt`-file stored in the previous task and compute all components of the explored FSM. How many components exist? Show the sets of states for each computed component. Draw the graph of the explored FSM to verify the results of all four tasks of this exercise.

Exercise 12.3 (Synthesis of a Synchronous Sequential Circuit)

The aim of this exercise is the synthesis of a synchronous sequential circuit that generates the cyclic sequence of values shown in Fig. 12.31a at the output y in the case that the input x is equal to 0 or the extended cyclic sequence of values shown in Fig. 12.31b at the output y in the case that the input x is equal to 1. This circuit must be realized using DE-flip-flops. The behavior of a DE-flip-flop is equal to the D-flip-flop in the case that the enable input E is equal to 1; hence, a D-flip-flop can be used in such a special case. Forward the XBOOLE-system from one task to the following task so that VTs and previously computed TVLs can be reused.

The finite state machine requires seven states that can be encoded by three state-variables. These three state-variables specify eight states. It must be avoided that the FSM does not leave the additional state; the free choice of an edge of this additional state to an arbitrary state of the working cycle should be used to reduce the cost to implement this circuit. For the same reason, an arbitrary output-value

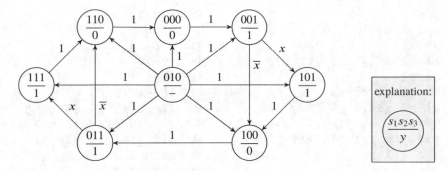

Fig. 12.32 Initial graph of the non-deterministic finite state machine to be synthesized

can be used for the additional state. Use Algorithm 8 of page 764 for the synthesis of the wanted synchronous sequential circuit. Figure 12.32 shows the graph of the non-deterministic finite state machine to synthesize.

(a) Specify the system-function $F(x, \mathbf{s}, \mathbf{sf}, y)$ of the non-deterministic finite state machine and verify whether this FSM is realizable by a sequential circuit. Compute lattices of the controlling functions $d_1(x, \mathbf{s})$ and $e_1(x, \mathbf{s})$ of the first DE-flip-flop and select simple functions of these lattices. Restrict the given non-deterministic behavior according to the chosen controlling functions. Show the Karnaugh-maps of the computed lattices and the chosen controlling functions.

(b) Compute lattices of the controlling functions $d_2(x, \mathbf{s})$ and $e_2(x, \mathbf{s})$ of the second DE-flip-flop and select simple functions of these lattices. Restrict the given non-deterministic behavior according to the chosen controlling functions. Show the Karnaugh-maps of the computed lattices and the chosen controlling functions.

(c) Compute lattices of the controlling functions $d_3(x, \mathbf{s})$ and $e_3(x, \mathbf{s})$ of the third DE-flip-flop and select simple functions of these lattices. Restrict the given non-deterministic behavior according to the chosen controlling functions. Show the Karnaugh-maps of the computed lattices and the chosen controlling functions.

(d) Compute lattice of the output-function $y(\mathbf{s})$ and select a simple function of this lattice. Restrict the given non-deterministic behavior according to the chosen output-function. Show the Karnaugh-maps of the computed lattice and the chosen output-function. Show furthermore the TVL of the realized deterministic FSM and verify that this function is covered by the given non-deterministic finite state machine and that the three memory-functions and the output-function are uniquely specified. Draw the synthesized synchronous sequential circuit and the graph of the finite state machine realized by this circuit.

12.8 Solutions

Solution 12.1 (Analysis of a Sequential Circuit)

(a) Figure 12.33 shows the PRP that computes the behavior of the given sequential circuit as well as the list of phases (the system-function $F(\mathbf{x}, \mathbf{s}, \mathbf{s'}, \mathbf{y})$) computed as finial result.

The command avar in lines 3–9 determines well-ordered all used variables. The command sbe in lines 10–40 computes the set of solution-vectors of the system of logic equations that determines the synchronous sequential circuit specified in Fig. 12.29. Due to the two input-

```
 1   new                              37   d3=s2 ,
 2   space 64 1                       38   e3=g20 ,
 3   avar 1                           39   y1=g1&g6 ,
 4   x1 x2 s1 s2 s3 sf1 sf2 sf3       40   y2=g4 .
 5   y1 y2                            41   vtin 1 2
 6   j1 k1 t2 d3 e3                   42   j1 k1 t2 d3 e3
 7   g1 g2 g3 g4 g5 g6 g7 g8          43   g1 g2 g3 g4 g5 g6 g7 g8
 8   g9 g10 g11 g12 g13 g14           44   g9 g10 g11 g12 g13 g14
 9   g15 g16 g17 g18 g19 g20 .        45   g15 g16 g17 g18 g19 g20 .
10   sbe 1 1                          46   maxk 1 2 3
11   g1 =/x1 ,                        47   obbc 3 4
12   g2 =/x2 ,                        48   assign 1 /1 1 1
13   g3 =/(x2&s2) ,                   49   assign 4 /4 1 1
14   g4=s2&/s3 ,                      50   sts "e12_1_a"
15   g5=x1+/s1 ,
16   g6=s1&/s2 ,
17   g7=s2&s3 ,
18   g8=x2&/s1 ,
19   g9=x2#s1 ,
20   g10=g3&s3 ,
21   g11=g4&g2 ,
22   g12=g6&/s3 ,
23   g13=g7&g8 ,
24   g14=g5&s3 ,
25   g15=x1&g9 ,
26   g16=g5&g11 ,
27   g17=g12+g13 ,
28   g18=g14+g15 ,
29   g19=g16+g17 ,
30   g20=g8+g18 ,
31   sf1=j1&/s1+/k1&s1 ,
32   j1=g10 ,
33   k1=g4 ,
34   sf2=t2#s2 ,
35   t2=g19 ,
36   sf3 =/e3&s3+d3&e3 ,
```

K	TVL (4) ODA	10 Var.	14 R.	S. 1						
	x_1	x_2	s_1	s_2	s_3	sf_1	sf_2	sf_3	y_1	y_2
1	1	0	1	1	0	0	0	1	0	1
2	-	1	0	1	0	0	1	1	0	1
3	-	0	0	1	0	0	0	0	0	1
4	-	1	1	1	0	0	1	0	0	1
5	0	0	1	1	0	0	1	0	0	1
6	0	-	1	0	1	1	0	1	1	0
7	0	-	1	0	0	1	1	0	1	0
8	1	-	-	0	1	1	0	0	0	0
9	1	-	1	0	0	1	1	0	0	0
10	-	1	0	1	1	0	0	1	0	0
11	-	-	1	1	1	1	1	1	0	0
12	-	0	0	1	1	1	1	1	0	0
13	0	-	0	0	1	1	0	0	0	0
14	-	-	0	0	0	0	0	0	0	0

Fig. 12.33 Problem-program that computes the behavior of the sequential circuit specified in Fig. 12.29 using a system of logic equations together with the computed system-function $F(\mathbf{x}, \mathbf{s}, \mathbf{s}', \mathbf{y})$

variables x_1 and x_2 and the three variables s_1, s_2, and s_3 that determine the states, there are $2^{2+3} = 2^5 = 32$ phases. All values of these phases are shown in Fig. 12.34.

The command maxk in line 46 excludes all those variables from the list of phases, which are not required to express the behavior of the given sequential circuit. The subsequent command obbc minimizes the number of rows in this list of phases shown in Fig. 12.33 that describes the behavior $F(\mathbf{x}, \mathbf{s}, \mathbf{s}', \mathbf{y})$ of the analyzed sequential circuit.

(b) Figure 12.35 shows the PRP that computes the behavior of the given sequential circuit using local lists of phases of all elements of this circuit.

TVL (1) ODA | 35 Var. | 32 R. | S. 1

	x1	x2	s1	s2	s3	sf1	sf2	sf3	y1	y2	j1	k1	t2	d3	e3	g1	g2	g3	g4	g5	g6	g7	g8	g9	g10	g11	g12	g13	g14	g15	g16	g17	g18	g19	g20
1	1	0	1	1	0	0	0	1	0	1	0	1	1	1	1	0	1	1	1	1	0	0	0	1	0	1	0	0	0	1	1	0	1	1	1
2	1	1	0	1	0	0	1	1	0	1	0	1	0	1	1	0	0	0	1	1	0	0	1	1	0	0	0	0	0	1	0	0	1	0	1
3	0	1	0	1	0	0	1	1	0	1	0	1	0	1	1	1	0	0	1	1	0	0	1	1	0	0	0	0	0	0	0	0	0	0	1
4	0	0	0	1	0	0	0	0	0	1	0	1	1	1	0	1	1	1	1	1	0	0	0	0	0	1	0	0	0	0	1	0	0	1	0
5	1	0	0	1	0	0	0	0	0	1	0	1	1	1	0	0	1	1	1	1	0	0	0	0	0	1	0	0	0	0	1	0	0	1	0
6	0	0	1	1	0	0	1	0	0	1	0	1	0	1	0	1	1	1	1	0	0	0	0	1	0	1	0	0	0	0	0	0	0	0	0
7	1	1	1	1	0	0	1	0	0	1	0	1	0	1	0	0	0	0	1	1	0	0	0	0	0	0	0	0	0	0	0	0	0	0	0
8	0	1	1	1	0	0	1	0	0	1	0	1	0	1	0	1	0	0	1	0	0	0	0	0	0	0	0	0	0	0	0	0	0	0	0
9	0	0	1	0	1	1	0	1	1	0	1	0	0	0	0	1	1	1	0	0	1	0	0	1	1	0	0	0	0	0	0	0	0	0	0
10	0	1	1	0	1	1	0	1	1	0	1	0	0	0	0	1	0	1	0	0	1	0	0	0	1	0	0	0	0	0	0	0	0	0	0
11	0	0	1	0	0	1	1	0	1	0	0	0	1	0	0	1	1	1	0	0	1	0	0	1	0	0	1	0	0	0	0	1	0	1	0
12	0	1	1	0	0	1	1	0	1	0	0	0	1	0	0	1	0	1	0	0	1	0	0	0	0	0	1	0	0	0	0	1	0	1	0
13	1	0	1	0	0	1	1	0	0	0	0	0	1	0	1	0	1	1	0	1	1	0	0	1	0	0	1	0	0	1	0	1	1	1	1
14	1	0	1	0	1	1	0	0	0	0	1	0	0	0	1	0	1	1	0	1	1	0	0	1	1	0	0	0	1	1	0	0	1	0	1
15	1	1	1	0	1	1	0	0	0	0	1	0	0	0	1	0	0	1	0	1	1	0	0	0	1	0	0	0	1	0	0	0	1	0	1
16	1	1	1	0	0	1	1	0	0	0	0	0	1	0	0	0	0	1	0	1	1	0	0	0	0	0	1	0	0	0	0	1	0	1	0
17	1	1	0	1	1	0	0	1	0	0	0	0	1	1	1	0	0	0	0	1	0	1	1	1	0	0	0	1	1	1	0	1	1	1	1
18	0	1	0	1	1	0	0	1	0	0	0	0	1	1	1	1	0	0	0	1	0	1	1	1	0	0	0	1	1	0	0	1	1	1	1
19	1	0	1	1	1	1	1	1	0	0	1	0	0	1	1	0	1	1	0	1	0	1	0	1	1	0	0	0	1	1	0	0	1	0	1
20	0	0	0	1	1	1	1	1	0	0	1	0	0	1	1	1	1	1	0	1	0	1	0	0	1	0	0	0	1	0	0	0	1	0	1
21	1	0	0	1	1	1	1	1	0	0	1	0	0	1	1	0	1	1	0	1	0	1	0	0	1	0	0	0	1	0	0	0	1	0	1
22	1	1	1	1	1	1	1	1	0	0	0	0	0	1	1	0	0	0	0	1	0	1	0	0	0	0	0	0	1	0	0	0	1	0	1
23	0	0	0	0	1	1	0	0	0	0	1	0	0	0	1	1	1	1	0	1	0	0	0	0	1	0	0	0	1	0	0	0	1	0	1
24	1	0	0	0	1	1	0	0	0	0	1	0	0	0	0	1	1	1	0	1	0	0	0	0	1	0	0	0	1	0	0	0	1	0	1
25	1	1	0	0	1	1	0	0	0	0	1	0	0	0	1	0	0	1	0	1	0	0	1	1	1	0	0	0	1	1	0	0	1	0	1
26	0	1	0	0	1	1	0	0	0	0	1	0	0	0	1	1	0	1	0	1	0	0	1	1	1	0	0	0	1	0	0	0	1	0	1
27	1	1	0	0	0	0	0	0	0	0	0	0	0	0	1	0	0	1	0	1	0	0	1	1	0	0	0	0	0	1	0	0	1	0	1
28	0	1	0	0	0	0	0	0	0	0	0	0	0	0	1	1	0	1	0	1	0	0	1	1	0	0	0	0	0	0	0	0	0	0	1
29	0	0	1	1	1	1	1	1	0	0	1	0	0	0	1	0	1	1	1	0	0	0	1	0	1	1	0	0	0	0	0	0	0	0	0
30	0	1	1	1	1	1	1	1	0	0	0	0	0	1	0	1	0	0	0	0	0	1	0	0	0	0	0	0	0	0	0	0	0	0	0
31	0	0	0	0	0	0	0	0	0	0	0	0	0	0	0	1	1	1	0	1	0	0	0	0	0	0	0	0	0	0	0	0	0	0	0
32	1	0	0	0	0	0	0	0	0	0	0	0	0	0	0	0	1	1	0	1	0	0	0	0	0	0	0	0	0	0	0	0	0	0	0

Fig. 12.34 Global list of phases of the sequential circuit of Fig. 12.29 computed in the PRP of Fig. 12.33

The head line of a TVL determines the used variables but does not allow negated variables. Therefore variables nsi are used to express the additional outputs $\bar{s}_i$ of the three flip-flops. The command vtin in lines 2 and 3 specifies these variables so that they can be removed from the final behavioral description. The commands tin in lines 5–152 determine the local lists of phases of all elements of the synchronous sequential circuit specified in Fig. 12.29; the head lines determine the connection-variables and the rows the behavior of the element.

Fig. 12.35
Problem-program that
computes the behavior of
the sequential circuit
specified in Fig. 12.29
using local lists of phases

```
1    lds "e12_1_a"        54   tin 1 21           109  tin 1 32
2    vtin 1 5             55   g4 g2 g11.         110  g4 y2.
3    ns1 ns2 ns3.         56   0-0               111  00
4    ctin 1 10 /1         57   100               112  11.
5    tin 1 11             58   111.              113  tin 1 33
6    x1 g1.               59   tin 1 22          114  g10 j1.
7    01                   60   g6 ns3 g12.       115  00
8    10.                  61   0-0               116  11.
9    tin 1 12             62   100               117  tin 1 34
10   x2 g2.               63   111.              118  g4 k1.
11   01                   64   tin 1 23          119  00
12   10.                  65   g7 g8 g13.        120  11.
13   tin 1 13             66   0-0               121  tin 1 35
14   x2 s2 g3.            67   100               122  g19 t2.
15   0-1                  68   111.              123  00
16   101                  69   tin 1 24          124  11.
17   110.                 70   g5 s3 g14.        125  tin 1 36
18   tin 1 14             71   0-0               126  s2 d3.
19   s2 ns3 g4.           72   100               127  00
20   0-0                  73   111.              128  11.
21   100                  74   tin 1 25          129  tin 1 37
22   111.                 75   x1 g9 g15.        130  g20 e3.
23   tin 1 15             76   0-0               131  00
24   x1 ns1 g5.           77   100               132  11.
25   1-1                  78   111.              133  tin 1 38
26   011                  79   tin 1 26          134  j1 k1 s1 ns1 sf1.
27   000.                 80   g5 g11 g16.       135  1-011
28   tin 1 16             81   0-0               136  -0101
29   s1 ns2 g6.           82   100               137  0-010
30   0-0                  83   111.              138  -1100.
31   100                  84   tin 1 27          139  tin 1 39
32   111.                 85   g12 g13 g17.      140  t2 s2 ns2 sf2.
33   tin 1 17             86   1-1               141  0010
34   s2 s3 g7.            87   011               142  0101
35   0-0                  88   000.              143  1011
36   100                  89   tin 1 28          144  1100.
37   111.                 90   g14 g15 g18.      145  tin 1 40
38   tin 1 18             91   1-1               146  d3 e3 s3 ns3 sf3.
39   x2 ns1 g8.           92   011               147  -0010
40   0-0                  93   000.              148  -0101
41   100                  94   tin 1 29          149  01010
42   111.                 95   g16 g17 g19.      150  01100
43   tin 1 19             96   1-1               151  11011
44   x2 s1 g9.            97   011               152  11101.
45   000                  98   000.              153  for $i 11 40
46   110                  99   tin 1 30          154  (
47   011                  100  g8 g18 g20.       155  isc 10 $i 10
48   101.                 101  1-1               156  )
49   tin 1 20             102  011               157  maxk 10 2 41
50   g3 s3 g10.           103  000.              158  maxk 41 5 42
51   0-0                  104  tin 1 31          159  obbc 42 43
52   100                  105  g1 g6 y1.         160  assign 10 /1 1 1
53   111.                 106  0-0               161  assign 4 /4 1 1
                          107  100               162  assign 43 /4 1 2
                          108  111.              163  set $equal (te_syd 4 43)
```

The command `ctin` in line 4 defines a logic function that is a constant equal to 1 so that the command `isc` in the body of the `for`-loop in lines 153–156 computes the global list of phases of the whole circuit. The subsequent two commands `maxk` exclude again all those variables, which are not needed to express the behavior of the given sequential circuit, from the list of phases, and the command `obbc` minimizes the computed list of phases that specifies the system-function $F(\mathbf{x}, \mathbf{s}, \mathbf{s}', \mathbf{y})$ of the analyzed synchronous sequential circuit. The result $equal=true, computed in the last line of this PRP, confirms that the same behavior as in the previous task has been computed using the approach of local lists of phases; therefore, we do not repeatedly show

```
 1   lds  "e12_1_a"              26   maxk  20 6 21
 2   vtin  1  5                  27   cco  21 6 7 22
 3   x1 x2 y1 y2.                28   cpl  22 23
 4   vtin  1  6                  29   assign  23  /m 2 1
 5   s1 s2 s3.
 6   vtin  1  7
 7   sf1 sf2 sf3.
 8   maxk  4 5 10
 9   assign  10  /m 1 1
10   syd  10 10 8
11   copy  10 11
12   ctin  1 20 /0
13   while (not (te 11))
14   (
15   stv  11 1 12
16   uni  12 8 13
17   cel  13 10 14 /-0
18   dif  11 14 11
19   cco  14 6 7 15
20   if (not (te_syd 14 15))
21   (
22   uni  20 14 20
23   )
24   )
25   assign  20  /m 1 2
```

K	TVL (10) ODA | 6 Var. | 11 R. | S. 1					
	s1	s2	s3	sf1	sf2	sf3
1	1	1	0	0	0	1
2	0	1	0	0	1	1
3	0	1	0	0	0	0
4	1	1	0	0	1	0
5	0	0	0	0	0	0
6	1	0	1	1	0	1
7	1	0	0	1	1	0
8	-	0	1	1	0	0
9	0	1	1	1	1	1
10	0	1	1	0	0	1
11	1	1	1	1	1	1

K	TVL (20) ODA | 6 Var. | 9 R. | S. 1					
	s1	s2	s3	sf1	sf2	sf3
1	1	1	0	0	0	1
2	0	1	1	0	0	1
3	0	1	1	1	1	1
4	0	0	1	1	0	0
5	1	0	1	1	0	0
6	1	0	0	1	1	0
7	1	1	0	0	1	0
8	0	1	0	0	0	0
9	0	1	0	0	1	1

K	TVL (23) ODA | 3 Var. | 1 R. | S. 1		
	s1	s2	s3
1	1	0	1

-

Fig. 12.36 Problem-program that computes the initial states of the sequential circuit specified in Fig. 12.29 together with the computed solution

the computed lists of phases in a figure, but the commands `assign` depict them in the viewports of the XBOOLE-monitor XBM 2.

(c) Figure 12.36 shows the PRP that computes the initial states of the given sequential circuit based on the system-function $F(\mathbf{x}, \mathbf{s}, \mathbf{s}', \mathbf{y})$; the computed solution (TVL 23) shows that only a single initial state exists.

Neither the input- nor the output-variables have any influence on the property whether a state of the FSM is an initial state; hence, these variables are specified in the VT 5 (in lines 2 and 3) and excluded from the loaded system-function (TVL 4) using the command `maxk` in line 8. The resulting TVL 10 is shown in viewport (1,1) of the m-fold view.

An edge that begins and ends at the same state is the only edge that may arrive in an initial state. Therefore, the commands in lines 10–25 remove all such loops on states that are from the set of edges. The commands in the `while`-loop in lines 13–24 select each edge exactly once and stores the binary vector of an edge that is no loop into TVL 20 that has been initialized as an empty TVL using the command `ctin` in line 12.

The procedure that provides each element (in this case an edge encoded by (s_1, s_2, s_3) for the start node and (sf_1, sf_2, sf_3) for the end node (the state reached at the following point in time) changes the set of these elements; hence, the command `copy` in line 11 prepares a copy (TVL 11) of the unchanged set to evaluate (TVL 10). The command `stv` in line 15 selects the first ternary vector of the set to evaluate but removes implicitly the dashes that are required in this procedure. These dashes are restored using the command `uni` in line 16 and the empty ternary vector of all variables of TVL 10 created by the command `syd` in line 10. The command `cel` in line 17 replaces all dashes by values 0 so that exactly one element of the set (one edge) has been selected. The command dif in line 18 removes this selected binary vector from the set to evaluate (TVL 11) so that an empty TVL 11 indicates that all vectors have been evaluated and the while-loop terminates.

It is necessary to distinguish between an edge that connects different vertices and an edge that begins and ends at the same vertex (a loop). The command cco in line 19 changes controlled by VTs 6 and 7 (defined in lines 4–7) the direction of the selected edge. The command te_syd in line 20 compares the given edge with the edge of the reverse direction; if these two edges are different, the command uni in line 22 inserts the set of edge without any loops (TVL 20). This intermediate result, i.e., the set of edges of the evaluated FSM without all loops, is shown in viewport (1,2) of the m-fold view.

Initial states cannot be reached by an edge (except a loop); hence, all states encoded by (sf_1, sf_2, sf_3) of TVL 20 are no initial states. The command maxk in line 26 removes the beginning vertex of the edges in TVL 20, and the subsequent command cco renames the binary vector that are no initial states from (sf_1, sf_2, sf_3) to (s_1, s_2, s_3) so that the command cpl in line 28 computes the requested initial states. TVL 23 in viewport (2,1) of the m-fold view in Fig. 12.36 shows that only the state $(s_1, s_2, s_3) = (101)$ is an initial state of the evaluated FSM belonging to the sequential circuit shown in Fig. 12.29.

(d) Figure 12.37 shows the PRP that computes the final states of the given sequential circuit based on the system-function $F(\mathbf{x}, \mathbf{s}, \mathbf{s'}, \mathbf{y})$; the computed solution (TVL 22) shows that two final states exist.

An edge that begins and ends at the same state is the only edge that may leave a final state. Therefore, the set of edges without any loop is computed in the same manner as in the previous task in lines 1–24.

Final states cannot be left by an edge (except a loop); hence, all states encoded by (s_1, s_2, s_3) of TVL 20 are no final states. The command maxk in line 26 removes the ending states of the edges in TVL 20 so that the command cpl in line 27 computes the requested finial states. TVL 22 (see Fig. 12.37) shows that the evaluated FSM belonging to the sequential circuit shown in Fig. 12.29 has two final states $(s_1, s_2, s_3) = (000, 111)$.

Figure 12.38 depicts the graph of the FSM realized by the sequential circuit shown in Fig. 12.29 and expressed by the system-function $F(\mathbf{x}, \mathbf{s}, \mathbf{s'}, \mathbf{y})$ computed as TVL 4 (see Fig. 12.33). This graph confirms that the single initial state is $(s_1, s_2, s_3) = (101)$ and the two final states are $(s_1, s_2, s_3) = (000, 111)$. The outputs of this FSM depend on both input- and state-variables; hence, it is a FSM of the Mealy-type.

```
1   lds "e12_1_a"        13   while (not (te 11))       25   assign 20 /m 1 2
2   vtin 1 5             14   (                         26   maxk 20 7 21
3   x1 x2 y1 y2.         15   stv 11 1 12               27   cpl 21 22
4   vtin 1 6             16   uni 12 8 13               28   assign 22 /m 2 1
5   s1 s2 s3.            17   cel 13 10 14 /-0
6   vtin 1 7             18   dif 11 14 11
7   sf1 sf2 sf3.         19   cco 14 6 7 15
8   maxk 4 5 10          20   if (not (te_syd 14 15))
9   assign 10 /m 1 1     21   (
10  syd 10 10 8          22   uni 20 14 20
11  copy 10 11           23   )
12  ctin 1 20 /0         24   )
```

K TVL (22) ODA | 3 Var. | 2 R. | S. 1

	s_1	s_2	s_3
1	0	0	0
2	1	1	1

Fig. 12.37 Problem-program that computes the final states of the sequential circuit specified in Fig. 12.29 together with the computed solution

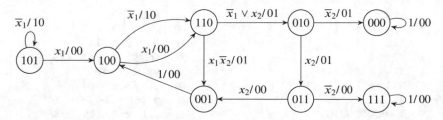

Fig. 12.38 Graph of the finite state machine realized by the sequential circuit shown in Fig. 12.29

Solution 12.2 (Properties of a Finite State Machine)

(a) Figure 12.39 shows the PRP that prepares and stores the TVL of the explored finite state machine and computes thereafter the cycles of the length two that are active in the case $x_2 = 1$.

The TVL of the explored FSM specified in Fig. 12.30 is determined using the command tin in lines 3–31. The subsequent command assign shows this TVL of 24 rows in the 1-fold view; we do not repeatedly print this TVL because it is identical with the given TVL in Fig. 12.30. The command sts in line 33 stores this TVL as sdt-file for later use.

The values of the outputs do not influence the edges of the wanted cycles; hence, the command _maxk in line 34 excludes these variables from the TVL of the FSM.

Only cycles for $x_2 = 1$ are requested; hence, the commands sbe, isc, and _maxk in lines 35–38 restrict the explored FSM to this behavior.

The VTs defined in lines 38–42 are needed to specify the states of the FSM in the two consecutive time-steps **s** and **sf**. The command cco in line 43 changes the direction of edges between pairs of states.

The command isc in line 44 selects edges for which the reverse edges exist; hence, these edges either belong to the wanted cycles or describe loops on states of this FSM. The command copy in line 45 creates a copy of this set so that the set of edges of the wanted cycles and possible loops remains unchanged. The commands in lines 46–57 remove the loops from the intermediate set of edges so that only the edges of the wanted cycles of the length two remain.

The command maxk in line 46 computes the set of states belonging to the wanted cycles or having a loop. All these states are evaluated in while-loop in lines 48–57. The command stv selects the first row of this set, the command uni expands the selected vector to all four variables of the state (using the empty TVL created by the command syd in line 47) so that possibly removed dashes are restored, and the command cel substitutes possible dashes by values 0 so that a single state has been selected. The commands cco and isc in lines 53 and 54 create the loop for the selected state so that this loop can be removed from the set of computed edges using the command dif in line 55. The command dif in line 56 removes the evaluated state from the set of states so that the while-loop terminates.

The last command of this PRP assigns the computed set of edges belonging to cycles of the length two to viewport (1,1) of the m-fold view (see Fig. 12.39). The two cycles of the length two that are active for $x_2 = 1$ in the explored FSM switch either between (0101) and (0110) or between (1101) and (1100).

(b) Figure 12.40 shows the PRP that restricts the behavior as requested to the constant input-pattern $(x_1, x_2) = (10)$ and computes the sequence of possible states fitting to a sequence of observed output-patterns $(01) \rightarrow (11) \rightarrow (10)$.

The command lds in the first line loads the XBOOLE-system containing the system-function to explore as TVL 1. The VTs in lines 2–9 determine the variables of the inputs **x**, the states **s**, the

1 new	35 sbe 1 3
2 space 64 1	36 x2=1.
3 tin 1 1	37 isc 2 3 4
4 x1 x2	38 _maxk 4 <x1 x2> 5
5 s1 s2 s3 s4	39 vtin 1 6
6 sf1 sf2 sf3 sf4	40 s1 s2 s3 s4.
7 y1 y2.	41 vtin 1 7
8 1-0000000101	42 sf1 sf2 sf3 sf4.
9 --0001001011	43 cco 5 6 7 10
10 --0010001100	44 isc 5 10 11
11 -00011000101	45 copy 11 12
12 -10011010001	46 maxk 11 7 13
13 --0100010100	47 syd 13 13 14
14 --0101011001	48 while (not (te 13))
15 -10110010110	49 (
16 000110101010	50 stv 13 1 15
17 100110011110	51 uni 15 14 16
18 --0111100001	52 cel 16 13 17 /-0
19 1-1000101011	53 cco 17 6 7 18
20 0-1000100111	54 isc 17 18 19
21 --1001000010	55 dif 12 19 12
22 0-0000101001	56 dif 13 16 13
23 --1010011011	57)
24 0-1111111100	58 assign 12 /m 1 1
25 1-1111101100	
26 --1011110001	
27 --1100110111	
28 -11101110010	
29 001101101110	
30 101101111010	
31 --1110101111.	
32 assign 1 /1 1 1	
33 sts "e12_2_a"	
34 _maxk 1 <y1 y2> 2	

K	TVL (12) ODA	8 Var.	4 R.	S. 1

	s1	s2	s3	s4	sf1	sf2	sf3	sf4
1	0	1	0	1	0	1	1	0
2	0	1	1	0	0	1	0	1
3	1	1	0	1	1	1	0	0
4	1	1	0	0	1	1	0	1

Fig. 12.39 Problem-program that computes the cycles of the length two for $x_2 = 1$ existing in the explored FSM specified in Fig. 12.30 together with the computed solution

states reached after the next following clock **sf**, and the outputs **y**. The commands in lines 10–13 restrict the given behavior as requested to the constant input-pattern $(x_1, x_2) = (10)$. The subsequent command `assign` shows this intermediate TVL in the 1-fold view.

The commands `vtin` in lines 16–28 specify the variables of the states in the first four clock-intervals $0, \ldots, 3$ as well as the output-variables in the first three clock-intervals $0, \ldots, 2$. The three subsequent commands `tin` specify the output-patterns observed in these clock-intervals.

The three commands `cco` in lines 38–40 map the restricted behavior to the initial point in time 0, and the subsequent command `isc` restricts these phases to the observed output-pattern $(y10, y20) = (01)$ at this point in time. TVL 41 in Fig. 12.40 shows that five phases satisfy these conditions.

The three commands `cco` in lines 43–45 map the restricted behavior to the time-step 1, and the subsequent command `isc` restricts these phases to the observed output-pattern $(y11, y21) = (11)$ at this point in time. The command `isc` in line 47 computes the phases

```
1   lds "e12_2_a"          19  vtin 1 12           37  10.
2   vtin 1 2               20  s12 s22 s32 s42.    38  cco 8 3 10 40
3   x1 x2.                 21  vtin 1 13           39  cco 40 4 11 40
4   vtin 1 3               22  s13 s23 s33 s43.    40  cco 40 5 20 40
5   s1 s2 s3 s4.           23  vtin 1 20           41  isc 40 30 41
6   vtin 1 4               24  y10 y20.            42  assign 41 /m 1 1
7   sf1 sf2 sf3 sf4.       25  vtin 1 21           43  cco 8 3 11 50
8   vtin 1 5               26  y11 y21.            44  cco 50 4 12 50
9   y1 y2.                 27  vtin 1 22           45  cco 50 5 21 50
10  sbe 1 6                28  y12 y22.            46  isc 50 31 51
11  x1&/x2=1.              29  tin 1 30            47  isc 51 41 52
12  isc 1 6 7              30  y10 y20.            48  assign 52 /m 2 1
13  maxk 7 2 8             31  01.                 49  cco 8 3 12 60
14  assign 8 /1            32  tin 1 31            50  cco 60 4 13 60
15  vtin 1 10              33  y11 y21.            51  cco 60 5 22 60
16  s10 s20 s30 s40.       34  11.                 52  isc 60 32 61
17  vtin 1 11              35  tin 1 32            53  isc 61 52 62
18  s11 s21 s31 s41.       36  y12 y22.            54  assign 62 /m 3 1
```

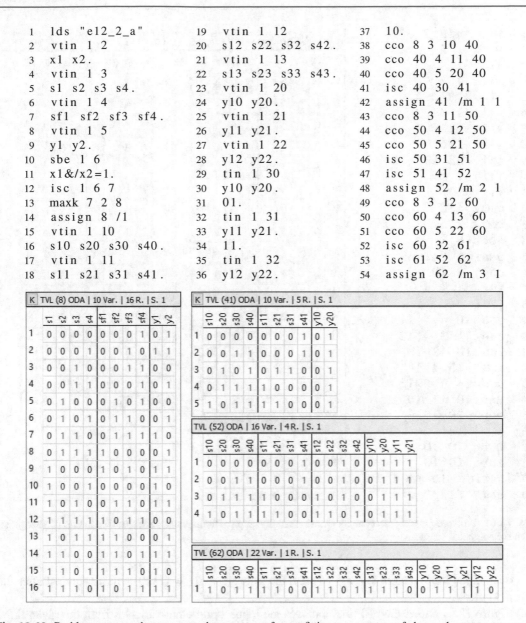

Fig. 12.40 Problem-program that computes the sequence of states fitting to a sequence of observed output-patterns of the explored FSM specified in Fig. 12.30 together with the computed solutions

over two clock-intervals that satisfy the sequence of two observed output-patterns. TVL 52 in Fig. 12.40 shows that four phases satisfy these conditions.

Analogously, the three commands cco in lines 49–51 map the restricted behavior to the time-step 2, and the subsequent command isc restricts these phases to the observed output-pattern $(y12, y22) = (10)$ at this point in time. The command isc in line 53 computes the phases over three clock-intervals that satisfy the sequence of two observed output-patterns. TVL 62 in Fig. 12.40 shows as final result that only one phase satisfies these conditions; hence, the state $\mathbf{s}[t = 2] = (1101)$ is uniquely determined by given sequence of output-patterns, and with an

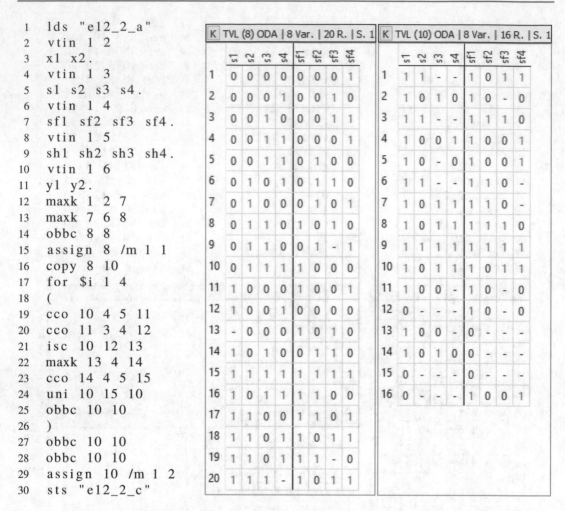

```
1   lds "e12_2_a"
2   vtin 1 2
3   x1 x2 .
4   vtin 1 3
5   s1 s2 s3 s4 .
6   vtin 1 4
7   sf1 sf2 sf3 sf4 .
8   vtin 1 5
9   sh1 sh2 sh3 sh4 .
10  vtin 1 6
11  y1 y2 .
12  maxk 1 2 7
13  maxk 7 6 8
14  obbc 8 8
15  assign 8 /m 1 1
16  copy 8 10
17  for $i 1 4
18  (
19  cco 10 4 5 11
20  cco 11 3 4 12
21  isc 10 12 13
22  maxk 13 4 14
23  cco 14 4 5 15
24  uni 10 15 10
25  obbc 10 10
26  )
27  obbc 10 10
28  obbc 10 10
29  assign 10 /m 1 2
30  sts "e12_2_c"
```

TVL (8) ODA | 8 Var. | 20 R. | S. 1

K	s1	s2	s3	s4	sf1	sf2	sf3	sf4
1	0	0	0	0	0	0	0	1
2	0	0	0	1	0	0	1	0
3	0	0	1	0	0	0	1	1
4	0	0	1	1	0	0	0	1
5	0	0	1	1	0	1	0	0
6	0	1	0	1	0	1	1	0
7	0	1	0	0	0	1	0	1
8	0	1	1	0	1	0	1	0
9	0	1	1	0	0	1	-	1
10	0	1	1	1	1	0	0	0
11	1	0	0	0	1	0	0	1
12	1	0	0	1	0	0	0	0
13	-	0	0	0	1	0	1	0
14	1	0	1	0	0	1	1	0
15	1	1	1	1	1	1	1	1
16	1	0	1	1	1	1	0	0
17	1	1	0	0	1	1	0	1
18	1	1	0	1	1	0	1	1
19	1	1	0	1	1	1	-	0
20	1	1	1	-	1	0	1	1

TVL (10) ODA | 8 Var. | 16 R. | S. 1

K	s1	s2	s3	s4	sf1	sf2	sf3	sf4
1	1	1	-	-	1	0	1	1
2	1	0	1	0	1	0	-	0
3	1	1	-	-	1	1	1	0
4	1	0	0	1	1	0	0	1
5	1	0	-	0	1	0	0	1
6	1	1	-	-	1	1	0	-
7	1	0	1	1	1	1	0	-
8	1	0	1	1	1	1	1	0
9	1	1	1	1	1	1	1	1
10	1	0	1	1	1	0	1	1
11	1	0	0	-	1	0	-	0
12	0	-	-	-	1	0	-	0
13	1	0	0	-	0	-	-	-
14	1	0	1	0	0	-	-	-
15	0	-	-	-	0	-	-	-
16	0	-	-	-	1	0	0	1

Fig. 12.41 Problem-program that computes the reachability graph of the explored FSM specified in Fig. 12.30 together with the computed solutions

unchanged input-pattern $(x_1, x_2) = (10)$, the state $\mathbf{s}[t = 3] = (1110)$ will be reached after the next clock.

(c) Figure 12.41 shows the PRP that removes the input- and output-variables from the given finite state machine (TVL 8) and extends this graph by all transitive edges to get the reachability graph (TVL 10).

The command `lds` in the first line loads the XBOOLE-system containing the system-function to explore as TVL 1. The VTs in lines 2–11 determine the variables of the inputs $\mathbf{x}$, the states $\mathbf{s}$, the states reached after the next following clock $\mathbf{sf}$, the states reached after two clocks $\mathbf{sh}$, and the outputs $\mathbf{y}$. The two commands `maxk` in lines 12 and 13 exclude the input- and output-variables from the given system-function and the command `obbc` in line 14 reduces the number of ternary vectors to represent this graph. The subsequent command `assign` shows this intermediate TVL in viewport (1,1) of the m-fold view. This TVL represents 24 edges by 20 ternary vectors. The command `copy` in line 16 prepares the TVL that will be extended to the reachability graph.

The computation of transitive edges for all pairs of consecutive edges is the key to compute the reachability graph. Figure 12.42 shows the procedure that solves this task.

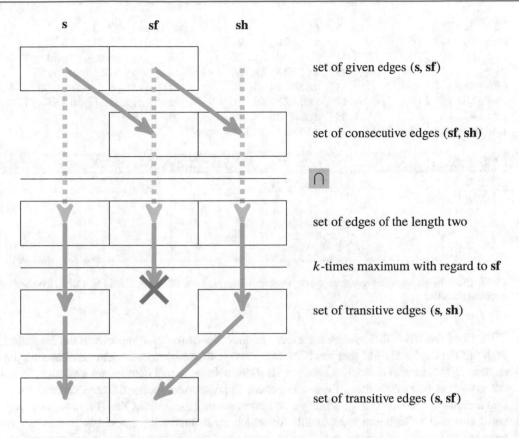

Fig. 12.42 Computation of transitive edges

The two commands `cco` in lines 19 and 20 create a new TVL containing all given edges shifted by one clock-period. The command `isc` in line 21 concatenates the given edges with the consecutive edges. The command `maxk` in line 22 removes the intermediate state of the concatenated edges, and the command `cco` in line 23 completes the computation of transitive edges of the given graph.

The command `uni` in line 24 combines the computed transitive edges with the given edges, and the subsequent command `obbc` minimizes the number of rows of the computed graph. The result of the first sweep of the `for`-loop in lines 17–26 is the set of edges that describe transitions of each state to all states that can be reached after one or two clock-steps. Using all these edges in the second sweep of this `for`-loop, we get as a result already the set of edges that describe transitions of each state to all states that can be reached after one, two, three, and four clock-steps; hence, the maximal distance of the computed edges increases exponentially. Four sweeps of the `for`-loop are sufficient to compute the reachability graph due to the four variables used to encode the states. TVL 10 in Fig. 12.41 shows the computed reachability graph minimized by two commands `obbc`; this TVL represents 142 edges with 16 ternary vectors.

The command `sts` in the last line stores the complete XBOOLE-system as file "e12_2_c.sdt" so that the reachability graph and the specified VTs can be reused in the next task.

(d) Figure 12.43 shows the PRP that reuses the reachability graph computed in the previous task to compute the components of the explored finite state machine.

```
 1   lds  "e12_2_c"          9   (                      17   uni  27 28 29
 2   cco  10 3 4 16          10  stv  20 1 22           18   dif  20 29 20
 3   isc  16 10 17           11  uni  22 21 23          19   set  $comp (add $i 30)
 4   obbc 17 18              12  cel  23 21 24 /-0      20   copy 28 $comp
 5   copy 18 20              13  maxk 24 4 25           21   assign $comp /m 1 $i
 6   syd  18 18 21           14  isc  25 20 26          22   set  $i (add $i 1)
 7   set  $i 1               15  maxk 26 3 27           23   )
 8   while (not (te 20))     16  cco  27 3 4 28
```

K	TVL (31) ODA	4 Var.	3 R.	S. 1
	s1	s2	s3	s4
1	1	1	0	-
2	1	1	1	0
3	1	0	1	1

K	TVL (32) ODA	4 Var.	3 R.	S. 1
	s1	s2	s3	s4
1	1	0	-	0
2	1	0	0	1
3	0	-	-	-

K	TVL (33) ODA	4 Var.	1 R.	S. 1
	s1	s2	s3	s4
1	1	1	1	1

Fig. 12.43 Problem-program that computes the components of the explored FSM specified in Fig. 12.30 together with the computed solutions

TVL 10 of the XBOOLE-system loaded in the first line of the PRP represents the reachability graph of the explored FSM that has been computed in the previous task. The reachability graph contains edges in both directions between all states belonging to a component and edges in only one direction from the states of one component to another component. The command cco in line 2 changes the direction of all edges so that the command isc in line 3 removes the edges between states of different components for which no reverse edge exists but preserves edges between the states belonging to the same component due to the existing reverse edges. The subsequent command obbc minimizes this TVL, and the command copy in line 5 copies the created complete component graph used to control the selection of the components.

Each sweep of the while-loop in lines 8–23 identifies all states belonging to one of the components and assigns these states to a viewport of the m-fold view. The command stv in line 10 selects the first ternary vector of the component graph. The subsequent commands uni and cel select the binary vector of a single edge replacing possible dashes of the original ternary vector by values 0. The command maxk in line 13 removes the end state of this edge so that TVL 25 specifies exactly one state of a component.

The command isc in line 14 computes all edges of the component graph that begin in the selected state; the end states of these edges are all states belonging to one component. The subsequent commands maxk and cco restrict to these end states of the selected edges and map them to the state-variables s, which finishes the computation of the states of one component.

The command uni in line 17 generates the set of all edges beginning at the states of the selected component so that the subsequent command dif removes all edges of the selected component from the component graph; hence, in the next sweep of the while-loop will be computed another component or the while-loop terminates when all components have been computed. The command copy in line 20 stores the computed set of states belonging to the same component to a TVL uniquely determined by the integer variable $comp.

Figure 12.43 shows that the explored FSM consists of three components consisting of four states, eleven states, and a single state. Figure 12.44 depicts the graph of the explored FSM. This graph can be used to verify that all computed solutions of this exercise are correct.

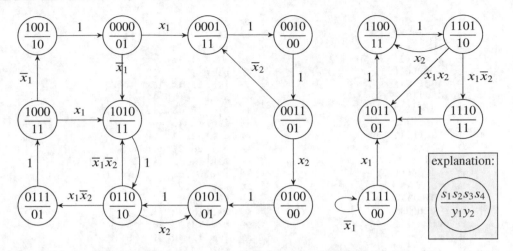

Fig. 12.44 Graph of the finite state machine specified by the system-function in Fig. 12.30

Solution 12.3 (Synthesis of a Synchronous Sequential Circuit)

(a) Figure 12.45 shows the PRP that specifies the non-deterministic system-function $F(x, \mathbf{s}, \mathbf{sf}, y)$ according to the graph of Fig. 12.32, verifies whether this system-function is realizable, and computes the lattices of the two controlling functions of the first DE-flip-flop. The chosen controlling functions partially restrict the non-deterministic behavior. It can be seen that the two computed lattices contain $2^8 = 256$ functions each; hence, very simple functions could be chosen. The command `tin` in lines 3–16 specifies the given non-deterministic FSM, and the subsequent commands `vtin` determine the input-variable x, the state-variables $\mathbf{s}$, the state-variables $\mathbf{sf}$ reached after the next following clock, and the output-variable y.

The non-deterministic system-function will be restricted due to the chosen controlling functions of the flip-flops; for these changes, a copy of the given system-function has been prepared in line 25. The subsequent command assigns this TVL to the 1-fold view.

The commands in lines 27–29 check whether the given system-function is realizable. The result `$realizable=true` confirms that the given system-function satisfies this necessary condition.

The command `maxk` in line 30 restricts the global system-function $F(x, \mathbf{s}, \mathbf{sf}, y)$ to the system-function $F_\delta(x, \mathbf{s}, \mathbf{sf})$ (TVL 8) of the memory-variables.

The commands in lines 31–50 compute the lattice $\mathcal{L}\langle d_{1q}(x, \mathbf{s}), d_{1r}(x, \mathbf{s})\rangle$ according to lines 4–6 of Algorithm 8; the associated Karnaugh-map is shown in viewport $(1,1)$ of the m-fold view. The very simple function $d_1 = s_3$ (in Algorithm 8: $in1_1(\mathbf{x}, \mathbf{s})$) has been chosen out of this lattice and assigned to viewport $(2,1)$ of the m-fold view.

The commands in lines 55–66 compute the lattice $\mathcal{L}\langle e_{1q}(x, \mathbf{s}), e_{1r}(x, \mathbf{s})\rangle$ according to lines 8–10 of Algorithm 8, the associated Karnaugh-map is shown in viewport $(1,2)$ of the m-fold view. The extremely simple function $e_1 = 1$ (in Algorithm 8: $in2_1(\mathbf{x}, \mathbf{s})$) has been chosen out of this lattice and assigned to viewport $(2,2)$ of the m-fold view. Due to this function, a D-flip-flop replaces the basically selected first DE-flip-flop.

The commands in lines 70–74 compute the system-function $F_1(x, \mathbf{s}, d_1, e_1)$ of the first DE-flip-flop, the realized behavior of the first state-variable $F_{\delta 1}(x, \mathbf{s}, sf_1)$, and the restricted global memory-function $F_\delta(x, \mathbf{s}, \mathbf{sf})$ (TVL 108) according to lines 12–14 of Algorithm 8.

```
 1  new                                    39  sbe 1 13
 2  space  32  1                           40  d1 .
 3  tin  1                                 41  cpl 13 14
 4  x  s1  s2  s3  sf1  sf2  sf3  y .       42  vtin 1 15
 5  -0000010                               43  d1 e1 .
 6  10011011                               44  isc 12 14 16
 7  00011001                               45  maxk 16 15 17
 8  -1011001                               46  cpl 17 18 ; d1q
 9  -1000110                               47  isc 12 13 19
10  10111111                               48  maxk 19 15 20
11  00111101                               49  cpl 20 21 ; d1r
12  -1111101                               50  assign_qr 18 21 /m 1 1
13  -1100000                               51  ; chosen function d1
14  -0101---                               52  sbe 1 22
15  -01000--                               53  s3 .
16  -010011-.                              54  assign 22 /m 2 1
17  vtin  1  2                             55  csd 13 22 40
18  x .                                    56  isc 12 40 41 ; F1
19  vtin  1  3                             57  sbe 1 23
20  s1  s2  s3 .                           58  e1 .
21  vtin  1  4                             59  cpl 23 24
22  sf1  sf2  sf3 .                        60  isc 41 24 26
23  vtin  1  5                             61  maxk 26 15 27
24  y .                                    62  cpl 27 28 ; e1q
25  copy  1  6                             63  isc 41 23 29
26  assign  6  /1                          64  maxk 29 15 30
27  maxk  6  4  7                          65  cpl 30 31 ; e1r
28  maxk  7  5  7                          66  assign_qr 28 31 /m 1 2
29  set $realizable (te_cpl 7)             67  ; chosen function e1
30  maxk  6  5  8  ; Fd                    68  ctin 1 32 /1
31  tin  1  10                             69  assign 32 /m 2 2
32  d1  e1  s1  sf1 .                       70  csd 23 32 42
33  -000                                   71  isc 41 42 43 ; F1
34  -011                                   72  isc 43 10 44
35  01-0                                   73  maxk 44 15 45 ; Fd1
36  11-1.                                  74  isc 8 45 108 ; Fd
37  isc  8  10  11                         75  sts "e12_3_a"
38  maxk  11  4  12  ; F1
```

(q: TVL (18) ODA, r: TVL (21) ODA), 4 Var.	S. 1	(q: TVL (28) ODA, r: TVL (31) ODA), 4 Var.	S. 1

```
0 0  Φ 0 0 Φ              0 0  Φ 1 1 Φ
0 1  1 Φ Φ 1              0 1  1 Φ Φ 1
1 1  1 Φ Φ 1              1 1  1 Φ Φ 1
1 0  Φ 0 0 Φ              1 0  Φ 1 1 Φ
s2 s3  0 1 1 0 s1         s2 s3  0 1 1 0 s1
       0 0 1 1 x                 0 0 1 1 x
```

T TVL (22) ODA	1 Var.	1 R.	S. 1	T TVL (32) ODA	0 Var.	1 R.	S. 1

```
0 1                      1
0 1 s3
```

Fig. 12.45 Problem-program that checks whether the non-deterministic FSM specified in Fig. 12.32 is realizable and computes the controlling function of the first DE-flip-flop together with the Karnaugh-maps of the computed lattices and the chosen controlling functions

The command sts in line 75 stores the XBOOLE-system so that the synthesis can be continued in the next task.

```
 1   lds  "e12_3_a"                   25   assign  122  /m 2  1
 2   tin  1  110                      26   csd  113  122  140
 3   d2  e2  s2  sf2 .                27   isc  112  140  141  ; F2
 4   -000                            28   sbe  1  123
 5   -011                            29   e2 .
 6   01-0                            30   cpl  123  124
 7   11-1.                           31   isc  141  124  126
 8   isc  108  110  111              32   maxk  126  115  127
 9   maxk  111  4  112  ; F2         33   cpl  127  128  ; e2q
10   sbe  1  113                     34   isc  141  123  129
11   d2 .                            35   maxk  129  115  130
12   cpl  113  114                   36   cpl  130  131  ; e2r
13   vtin  1  115                    37   assign_qr  128  131  /m 1  2
14   d2  e2 .                        38   ; chosen  function  e2
15   isc  112  114  116              39   sbe  1  132
16   maxk  116  115  117             40   s1&/s3 .
17   cpl  117  118  ; d2q            41   assign  132  /m 2  2
18   isc  112  113  119              42   csd  123  132  142
19   maxk  119  115  120             43   isc  141  142  143  ; F2
20   cpl  120  121  ; d2r            44   isc  143  110  144
21   assign_qr  118  121  /m 1  1    45   maxk  144  115  145  ; Fd2
22   ; chosen  function  d2          46   isc  108  145  208  ; Fd
23   sbe  1  122                     47   sts  "e12_3_b"
24   / s2 .
```

(q: TVL (118) ODA, r: TVL (121) ODA), 4 Var. | S. 1	(q: TVL (128) ODA, r: TVL (131) ODA), 4 Var. | S. 1

0 0 Φ 1 1 Φ
0 1 Φ Φ Φ Φ
1 1 Φ Φ Φ Φ
1 0 Φ 0 0 Φ
 	
ss 0 1 1 0 s1
      0 0 1 1 x
	0 0 0 1 1 0
	0 1 0 0 0 0
	1 1 0 0 0 0
	1 0 Φ 1 1 Φ
	
	ss 0 1 1 0 s1
	      0 0 1 1 x

T TVL (122) ODA | 1 Var. | 1 R. | S. 1	T TVL (132) ODA | 2 Var. | 1 R. | S. 1
1 0 0 1 s2	0 0 1 1 0 0 s 0 1 s1

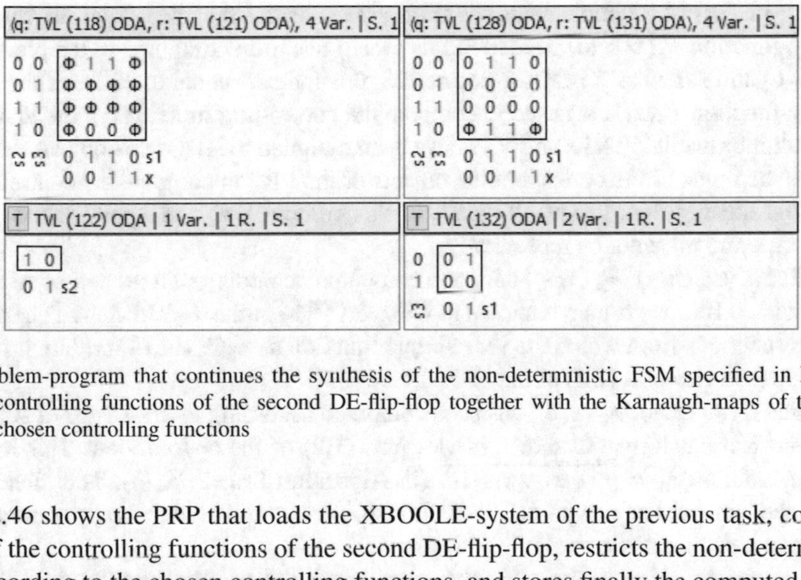

Fig. 12.46 Problem-program that continues the synthesis of the non-deterministic FSM specified in Fig. 12.32 by computing the controlling functions of the second DE-flip-flop together with the Karnaugh-maps of the computed lattices and the chosen controlling functions

(b) Figure 12.46 shows the PRP that loads the XBOOLE-system of the previous task, computes the lattices of the controlling functions of the second DE-flip-flop, restricts the non-deterministic behavior according to the chosen controlling functions, and stores finally the computed XBOOLE-system so that the synthesis can be continued in the next task.

The PRP of Fig. 12.46 implements the second sweep of the first for-loop of Algorithm 8. The memory-function $F_\delta(x, s, sf)$ needed in this sweep has been computed in the previous PRP (see Fig. 12.45) and stored as XBOOLE-object 108; this index is equal to 100 plus the index eight of memory-function $F_\delta(x, s, sf)$ used to compute the controlling functions of the first DE-flip-flop. Analogously, all other intermediate TVLs and VTs have also been assigned to XBOOLE-objects with an index of 100 plus the index of the corresponding objects of the PRP to compute the controlling functions of the first DE-flip-flop; hence, all details of this computation correspond to the explanation in the previous task and must not be repeated. These different object numbers have been used so that all intermediate results can be shown after the computation. A program that uses the XBOOLE-library would reuse the same XBOOLE-object within a `for`-loop for all flip-flops and invoke a subprogram that realizes the synthesis of the computed lattices of combinational functions.

The lattice $\mathcal{L}\langle d_{2q}(x, \mathbf{s}), d_{2r}(x, \mathbf{s})\rangle$ has been computed according to lines 4–6 of Algorithm 8, and the associated Karnaugh-map is shown in viewport (1,1) of the m-fold view. This lattice contains $2^{12} = 4096$ functions so that the very simple function $d_2 = \bar{s}_2$ (in Algorithm 8: $in1_2(\mathbf{x}, \mathbf{s})$) could be chosen out of this lattice (see viewport (2,1) of the m-fold view).

The lattice $\mathcal{L}\langle e_{2q}(x, \mathbf{s}), e_{2r}(x, \mathbf{s})\rangle$ has been computed according to lines 8–10 of Algorithm 8; the associated Karnaugh-map is shown in viewport (1,2) of the m-fold view and contains $2^2 = 4$ functions where the simplest function is $e_2 = s_1\bar{s}_3$ (in Algorithm 8: $in2_2(\mathbf{x}, \mathbf{s})$). This function has been chosen out of this lattice, assigned to viewport (2,2) of the m-fold view, and used to restrict the non-deterministic memory-function again.

The restricted memory-function $F_\delta(x, \mathbf{s}, \mathbf{sf})$ after the synthesis of the controlling functions of the second DE-flip-flop is stored as XBOOLE-object 208. The command `sts` in the last line stores the XBOOLE-system so that the synthesis can be continued in the next task.

(c) Figure 12.47 shows the PRP that loads the XBOOLE-system of the previous task, computes the lattices of the controlling functions of the third DE-flip-flop, restricts the non-deterministic behavior according to the chosen controlling functions, and stores finally the computed XBOOLE-system so that the synthesis can be continued in the next task.

The PRP of Fig. 12.47 implements the third sweep of the first `for`-loop of Algorithm 8. The memory-function $F_\delta(x, \mathbf{s}, \mathbf{sf})$ needed in this sweep has been computed in the previous PRP (see Fig. 12.46) and stored as XBOOLE-object 208; this index is equal to 200 plus the index eight of memory-function $F_\delta(x, \mathbf{s}, \mathbf{sf})$ used to compute the controlling functions of the first DE-flip-flop. Again, all intermediate TVLs and VTs have been assigned to XBOOLE-objects with an index of 200 plus the index of the corresponding objects of the PRP to compute the controlling functions of the first DE-flip-flop; hence, all details of this computation correspond to the explanation in the last task and must not be repeated.

The lattice $\mathcal{L}\langle d_{3q}(x, \mathbf{s}), d_{3r}(x, \mathbf{s})\rangle$ has been computed according to lines 4–6 of Algorithm 8, and the associated Karnaugh-map is shown in viewport (1,1) of the m-fold view. This lattice contains $2^4 = 16$ functions from which the very simple function $d_3 = \bar{s}_3$ (in Algorithm 8: $in1_3(\mathbf{x}, \mathbf{s})$) has be chosen (see viewport (2,1) of the m-fold view).

The lattice $\mathcal{L}\langle e_{3q}(x, \mathbf{s}), e_{3r}(x, \mathbf{s})\rangle$ has been computed according to lines 8–10 of Algorithm 8; the associated Karnaugh-map is shown in viewport (1,2) of the m-fold view. This lattice contains only the function $e_3 = \overline{(x\bar{s}_1s_3 \vee s_1s_2\bar{s}_3)}$ (in Algorithm 8: $in2_3(\mathbf{x}, \mathbf{s})$). This function has been determined, assigned to viewport (2,2) of the m-fold view, and used to finally restrict the memory-function to a deterministic one.

The deterministic memory-function $F_\delta(x, \mathbf{s}, \mathbf{sf})$ after the synthesis of the third DE-flip-flop is stored as XBOOLE-object 308. The command `sts` in the last line stores the XBOOLE-system so that the synthesis of the output-function can be realized in the next task.

(d) Figure 12.48 shows the PRP that loads the XBOOLE-system of the previous task, computes the lattice of the output-function, restricts the non-deterministic behavior according to this function, and verifies the system-function of the synthesized synchronous sequential circuit.

The command `lds` in line 1 loads the XBOOLE-system stored in the previous task. This XBOOLE-system provides the deterministic memory-function $F_\delta(x, \mathbf{s}, \mathbf{sf})$ as object 308. The command `isc` in line 2 restricts the global system-function (TVL 6) according to line 16 of Algorithm 8, and the subsequent command `maxk` the non-deterministic system-function $F_\lambda(x, \mathbf{s}, y)$ according to line 17 of Algorithm 8.

```
 1    lds  "e12_3_b"                    25    assign  222  /m 2 1
 2    tin  1  210                       26    csd  213  222  240
 3    d3  e3  s3  sf3 .                 27    isc  212  240  241  ;  F3
 4    −000                              28    sbe  1  223
 5    −011                              29    e3 .
 6    01−0                              30    cpl  223  224
 7    11−1 .                            31    isc  241  224  226
 8    isc  208  210  211                32    maxk  226  215  227
 9    maxk  211  4  212  ;  F3          33    cpl  227  228  ;  e3q
10    sbe  1  213                       34    isc  241  223  229
11    d3 .                             35    maxk  229  215  230
12    cpl  213  214                     36    cpl  230  231  ;  e2r
13    vtin  1  215                      37    assign_qr  228  231  /m 1 2
14    d3  e3 .                          38    ;  chosen  function  e3
15    isc  212  214  216                39    sbe  1  232
16    maxk  216  215  217               40    /( x&/s1&s3+s1&s2&/s3 ) .
17    cpl  217  218  ;  d3q             41    assign  232  /m 2 2
18    isc  212  213  219                42    csd  223  232  242
19    maxk  219  215  220               43    isc  241  242  243  ;  F3
20    cpl  220  221  ;  d3r             44    isc  243  210  244
21    assign_qr  218  221  /m 1 1       45    maxk  244  215  245  ;  Fd3
22    ;  chosen  function  d3           46    isc  208  245  308  ;  Fd
23    sbe  1  222                       47    sts  "e12_3_c"
24    /s3 .
```

(q: TVL (218) ODA, r: TVL (221) ODA), 4 Var. | S. 1	(q: TVL (228) ODA, r: TVL (231) ODA), 4 Var. | S. 1
0 0 | 1 1 1 1	0 0 | 1 1 1 1
0 1 | 0 0 0 Φ	0 1 | 1 1 1 0
1 1 | 0 0 0 Φ	1 1 | 1 1 1 0
1 0 | 1 Φ Φ 1	1 0 | 1 0 0 1
$s_2 s_3$ 0 1 1 0 s1	$s_2 s_3$ 0 1 1 0 s1
0 0 1 1 x	0 0 1 1 x

T TVL (222) ODA | 1 Var. | 1 R. | S. 1	T TVL (232) ODA | 4 Var. | 4 R. | S. 1
1 0	0 0 | 1 1 1 1
0 1 s3	0 1 | 1 1 1 0
	1 1 | 1 1 1 0
	1 0 | 1 0 0 1
	$s_2 s_3$ 0 1 1 0 s1
	0 0 1 1 x

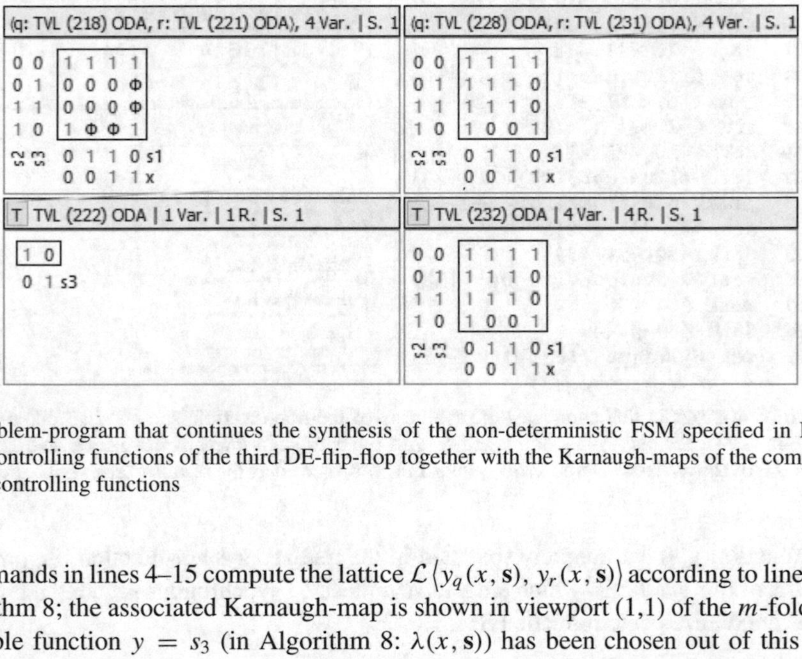

Fig. 12.47 Problem-program that continues the synthesis of the non-deterministic FSM specified in Fig. 12.32 by computing the controlling functions of the third DE-flip-flop together with the Karnaugh-maps of the computed lattices and the chosen controlling functions

The commands in lines 4–15 compute the lattice $\mathcal{L}\langle y_q(x, \mathbf{s}), y_r(x, \mathbf{s})\rangle$ according to lines 19 and 20 of Algorithm 8; the associated Karnaugh-map is shown in viewport $(1,1)$ of the m-fold view. The very simple function $y = s_3$ (in Algorithm 8: $\lambda(x, \mathbf{s})$) has been chosen out of this lattice and assigned to viewport $(2,1)$ of the m-fold view.

The commands in lines 20 and 21 compute the realized deterministic system-function $F_\lambda(x, \mathbf{s}, y)$ according to line 22 of Algorithm 8, and the command isc in line 22 computes the realized global system-function $F(x, \mathbf{s}, \mathbf{sf}, y)$ that is shown in viewport $(3,1)$ of the m-fold view. This finishes the synthesis of the synchronous sequential circuit of the non-deterministic FSM specified in Fig. 12.32.

The commands in line 24–39 verify the realized system-function. The five computed Boolean values confirm that the realized system-function is covered by the given non-deterministic system-function ($covered=true), and both all three memory-function $sf_i(x, \mathbf{s})$ and the

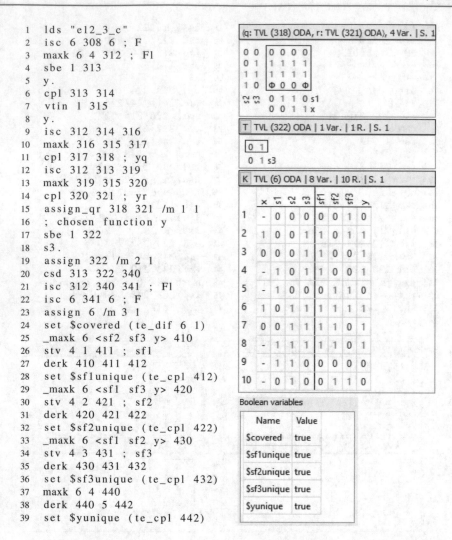

```
1   lds  "e12_3_c"
2   isc  6 308 6  ; F
3   maxk 6 4 312  ; F1
4   sbe  1 313
5   y.
6   cpl  313 314
7   vtin 1 315
8   y.
9   isc  312 314 316
10  maxk 316 315 317
11  cpl  317 318  ; yq
12  isc  312 313 319
13  maxk 319 315 320
14  cpl  320 321  ; yr
15  assign_qr 318 321 /m 1 1
16  ; chosen function y
17  sbe  1 322
18  s3.
19  assign 322 /m 2 1
20  csd  313 322 340
21  isc  312 340 341  ; F1
22  isc  6 341 6  ; F
23  assign 6 /m 3 1
24  set $covered (te_dif 6 1)
25  _maxk 6 <sf2 sf3 y> 410
26  stv  4 1 411  ; sf1
27  derk 410 411 412
28  set $sf1unique (te_cpl 412)
29  _maxk 6 <sf1 sf3 y> 420
30  stv  4 2 421  ; sf2
31  derk 420 421 422
32  set $sf2unique (te_cpl 422)
33  _maxk 6 <sf1 sf2 y> 430
34  stv  4 3 431  ; sf3
35  derk 430 431 432
36  set $sf3unique (te_cpl 432)
37  maxk 6 4 440
38  derk 440 5 442
39  set $yunique (te_cpl 442)
```

Fig. 12.48 Problem-program that completes the synthesis of the non-deterministic FSM specified in Fig. 12.32 by computing the lattice of the output-function together with the Karnaugh-maps of this lattice and the chosen output-function, the TVL of the realized deterministic system-function, as well as the results of the verification

output-function $y(\mathbf{s})$ are uniquely specified in the realized system-function. Figure 12.49 shows the results of this synthesis, which are **a** the synthesized synchronous sequential circuit and **b** the graph of the realized deterministic FSM.

12.9 Supplementary Exercises

Exercise 12.4 (Analysis of an Asynchronous Sequential Circuit)

The structure of an asynchronous sequential circuit is shown in Fig. 12.50. Suitable cut points of the three loops indicate already the memory-functions $sf_i(x, \mathbf{s})$ and state-variables $\mathbf{s}$.

Solve each of the following tasks using a PRP and forward the computed XBOOLE-system to the next task using an sdt-file.

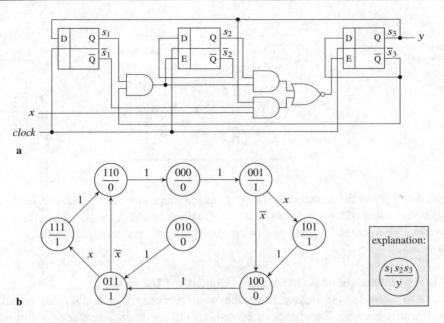

Fig. 12.49 Results of this synthesis of the non-deterministic FSM specified in Fig. 12.32: (**a**) the synthesized synchronous sequential circuit and (**b**) the graph of the realized deterministic FSM

Fig. 12.50 Structure of the asynchronous sequential circuit to analyze

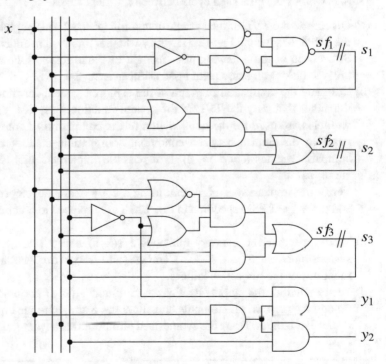

(a) Compute and show the global list of phases.

(b) Compute and show the phases of stable states.

(c) Compute and show the set of all phases that reach a stable state and the set of remaining phase. Belong the remaining phase to a sequence that reach a stable state or cause these phases a non-desirable cycle between unstable states?

Fig. 12.51 Specification
of the finite state machine
to be synthesized: (**a**) table
of the FSM and (**b**)
encoding of the states

x	0	1	y_1	y_2
Ⓐ	B	0	0	
C	Ⓑ	1	0	
Ⓒ	D	0	0	
A	Ⓓ	0	1	

a

s_2		
0	A	D
1	B	C
	0	1 s_1

b

(d) Compute and show the sequence of cyclic repeated output-patterns **y** that can be observed by changing the value of the input x several times; extend this sequence of patterns by the details of the associated phases $(x, \mathbf{s}, \mathbf{sf})$. Verify whether this cycle can surely be reached from the other stable states.

Exercise 12.5 (Synthesis of an Asynchronous Sequential Circuit)
    Figure 12.51 specifies the behavior and the used encoding of a finite state machine to be synthesized in this exercise. This behavior corresponds to the main behavior of the asynchronous sequential circuit analyzed in the previous exercise after removing the unneeded separated stable states as well as the unwanted cycle between transient states.

(a) Determine the TVL of the system-function $F(x, \mathbf{s}, \mathbf{sf}, \mathbf{y})$ (the global list of phases) that specifies the behavior given in Fig. 12.51. Verify whether this system-function is realizable.
    Solve each of the following tasks using a PRP and forward the computed XBOOLE-system of the first task to all subsequent tasks using an sdt-file.
(b) Compute the compatibility graph of the explored system-function; implement this PRP in such a manner that only the TVL of the system-function $F(\mathbf{x}, \mathbf{s}, \mathbf{sf}, \mathbf{y})$ and VTs of the four types of variables are used for the computation of the compatibility graph. Use the additional variables *cond* to distinguish between conditional compatible states (*cond* = 1) and unconditional compatible states (*cond* = 0). Is it possible implement this FSM using less than two state-variables?
    Verify the implemented algorithm using a copy of the developed PRP in which only the TVL and the VTs of FSM specified in Fig. 12.18 are replaced to compute the associated compatibility graph.
(c) Verify whether the memory-functions $sf_1(x, \mathbf{s})$ and $sf_2(x, \mathbf{s})$ are uniquely specified by the system-function $F(\mathbf{x}, \mathbf{s}, \mathbf{sf}, \mathbf{y})$. Compute and show the Karnaugh-maps of these memory-functions or the respective lattices.
(d) Verify whether the output-functions $y_1(x, \mathbf{s})$ and $y_2(x, \mathbf{s})$ are uniquely specified by the system-function $F(\mathbf{x}, \mathbf{s}, \mathbf{sf}, \mathbf{y})$. Compute and show the Karnaugh-maps of these output-functions or the respective lattices. Draw the synthesized circuit of this asynchronous FSM.

References

1. A. Bondy, M.R. Murty, *Graph Theory* (Springer, London, 2008). ISBN: 978-1-84628-969-9
2. S.H. Caldwell, *Switching Circuits and Logical Design* (Wiley, New York, 1958). ISBN: 978-0-4711-2969-1
3. D.A. Huffman, The synthesis of sequential switching circuits. J. Franklin Inst. **257**(3), 161–190 (1954). https://doi.org/10.1016/0016-0032(54)90574-8

Bibliography

1   The Concepts of XBOOLE

1. D. Bochmann, B. Steinbach, *Logic Design with XBOOLE (in German: Logikentwurf mit XBOOLE)* in German (Verlag Technik GmbH, Berlin, 1991). ISBN: 3-341-01006-8
2. F. Dresig et al., Programming with XBOOLE (in German: Programmieren mit XBOOLE), in *Series of Scientific Publications of the Chemnitz University of Technology (in German: Wissenschaftliche Schriftenreihe der Technischen Universität Chemnitz)* in German (1992), pp. 1–119. ISSN: 0863-0755
3. B. Steinbach, XBOOLE—a toolbox for modelling, simulation, and analysis of large digital systems. Syst. Anal. Model. Simul. **9**(4), 297–312 (1992). ISSN: 0232-9298
4. B. Steinbach, C. Posthoff. *EAGLE Start-up Aid - Efficient Computations with XBOOLE (in German: EAGLE Starthilfe - Effiziente Berechnungen mit XBOOLE)* (Edition am Gutenbergplatz, Leipzig, 2015). ISBN: 978-3-95922-081-1

2   The XBOOLE-Monitor XBM 2

5. C. Posthoff, B. Steinbach, *Logic Functions and Equations – Binary Models for Computer Science*, 2nd edn. (Springer, Cham, 2019)
6. B. Steinbach, C. Posthoff, Boolean Differential Calculus – Theory and Applications. J. Comput. Theor. Nanosci. **7**(6), 933–981 (2010). ISSN: 1546-1955. https://doi.org/10.1166/jctn.2010.1441
7. B. Steinbach, C. Posthoff, *Boolean Differential Calculus* (Morgan & Claypool Publishers, San Rafael, 2017). ISBN: 978-1-6270-5922-0. https://doi.org/10.2200/S00766ED1V01Y201704DCS052

3   Basic Algebraic Structures

8. G. Boole, *The Mathematical Analysis of Logic: Being an Essay Towards a Calculus of Deductive Reasoning (Classic Reprint)* (Forgotten Books, London, 2016). ISBN: 978-1-4400-6642-9
9. R.E. Bryant, Graph-based algorithms for Boolean function manipulation, in *IEEE Transaction on Computers* C-35.8 (1986), pp. 677–691. ISSN: 0018-9340. https://doi.org/10.1109/TC.1986.1676819
10. P. Clote, E. Kranakis, *Boolean Functions and Computation Models* (Springer, Berlin, 2002). ISBN: 978-3-662-04943-3. https://doi.org/10.1007/978-3-662-04943-3
11. I. Grattan-Guinness, G. Bornet, eds., *George Boole: Selected Manuscripts on Logic and its Philosophy*. Science Networks. Historical Studies 20 (Birkhäuser, Berlin, 1997). ISBN: 978-3-7643-5456-5
12. D. MacHale, *The Life and Work of George Boole: A Prelude to the Digital Age* (Cork University Press, Cork, 2014). ISBN: 978-1-78205-004-9
13. C. Meinel, T. Theobald, *Algorithms and Data Structures in VLSI Design* (Springer, Berlin, 1998). ISBN: 978-3-540-64486-6. https://doi.org/10.1007/978-3-642-58940-9
14. C. Posthoff, B. Steinbach, *Logic Functions and Equations—Binary Models for Computer Science* (Springer, The Netherlands, 2004). ISBN: 978-1-44195-261-5. https://doi.org/10.1007/978-1-4020-2938-7

© Springer Nature Switzerland AG 2022
B. Steinbach, C. Posthoff, *Logic Functions and Equations*,
https://doi.org/10.1007/978-3-030-88945-6

15. C. Posthoff, B. Steinbach, *Logic Functions and Equations—Binary Models for Computer Science, Second Edition* (Springer, Cham, 2019). ISBN: 978-3-030-02419-2. https://doi.org/10.1007/978-3-030-02420-8

16. S. Rudeanu, *Lattice Functions and Equations*. Discrete Mathematics and Theoretical Computer Science (Springer, London, 2001). ISBN: 978-1-85233-266-2. https://doi.org/10.1007/978-1-4471-0241-0

17. C.E. Shannon, A symbolic analysis of relay and switching circuits, in *Transactions of the American Institute of Electrical Engineers*, vol. 57(12) (1938), pp. 713–723. ISSN: 0096-3860. https://doi.org/10.1109/T-AIEE.1938. 5057767

18. B. Steinbach, XBOOLE—A Toolbox for Modelling, Simulation, and Analysis of Large Digital Systems, in *Systems Analysis and Modeling Simulation*, vol. 9(4) (1992), pp. 297–312. ISSN: 0232-9298

19. B. Steinbach, C. Posthoff. *Logic Functions and Equations—Examples and Exercises* (Springer, Berlin, 2009). ISBN: 978-1-4020-9594-8. https://doi.org/10.1007/978-1-4020-9595-5

20. B. Steinbach, C. Posthoff, *Boolean Differential Equations* (Morgan and Claypool Publishers, San Rafael, 2013). ISBN: 978-1-6270-5241-2. https://doi.org/10.2200/S00511ED1V01Y201305DCS042

21. B. Steinbach, C. Posthoff, *EAGLE Start-up Aid—Efficient Computations with XBOOLE (in German: EAGLE Starthilfe—Effiziente Berechnungen mit XBOOLE)* (Edition am Gutenbergplatz, Leipzig, 2015). ISBN: 978-3-95922-081-1

22. B. Steinbach, C. Posthoff, *Boolean Differential Calculus* (Morgan and Claypool Publishers, San Rafael, 2017). ISBN: 978-1-6270-5922-0. https://doi.org/10.2200/S00766ED1V01Y201704DCS052

23. B. Steinbach, M. Werner, XBOOLE-CUDA—fast Boolean operations on the GPU, in *Boolean Problems, Proceedings of the 11th International Workshops on Boolean Problems* ed. by B. Steinbach. IWSBP 11 (Freiberg University of Mining and Technology, Freiberg, 2014), pp. 75–84. ISBN: 978-3-86012-488-8

24. B. Steinbach, M. Werner, XBOOLE-CUDA—Fast Calculations on the GPU, in *Problems and New Solutions in the Boolean Domain*, ed. by B. Steinbach (Cambridge Scholars Publishing, Cambridge, 2016), pp. 117–149. ISBN: 978-1-4438-8947-6

25. I. Wegener. *Branching Programs and Binary Decision Diagrams: Theory and Applications*. Discrete Mathematics and Applications (Society for Industrial and Applied Mathematics (SIAM), Philadelphia, 2000). ISBN: 978-0-89871-458-6. https://doi.org/10.1137/1.9780898719789

4   Logic Functions

26. R.E. Bryant, Graph-based algorithms for Boolean function manipulation. IEEE Trans. Comput. **C-35**(8), 677–691 (Aug. 1986). ISSN:0018-9340. https://doi.org/10.1109/TC.1986.1676819

27. E.J. McCluskey, Minimization of Boolean functions. Bell Syst. Tech. J. **35**(6), 1417–1444 (Nov. 1956). ISSN:0005-8580. https://doi.org/10.1002/j.1538-7305.1956.tb03835.x

28. W.V. Quine, A way to simplify truth functions. Am. Math. Monthly **62**(9), 627–631 (Nov. 1955). https://doi.org/10.2307/2307285

29. W.V. Quine, The problem of simplifying truth functions. Am. Math. Monthly **59**(8), 521–531 (Oct. 1952). https://doi.org/10.2307/2308219

30. T. Sasao, M. Fujita, (eds.), *Representations of Discrete Functions* (Kluwer Academic Publishers, Boston, London, Dordrecht, 1996). ISBN:0-7923-9720-7. https://doi.org/10.1007/978-1-4613-1385-4

31. C.E. Shannon, A symbolic analysis of relay and switching circuits. Trans. Am. Inst. Electric. Eng. **57**(12), 713–723 (Dec. 1938). ISSN:0096-3860. https://doi.org/10.1109/T-AIEE.1938.5057767

32. B. Steinbach, XBOOLE–A toolbox for modelling, simulation, and analysis of large digital systems. Syst. Anal. Model. Simul. **9**(4), 297–312 (Sept. 1992). ISSN:0232-9298

33. B. Steinbach, M. Werner, XBOOLE-CUDA - Fast Boolean operations on the GPU, in *Proceedings of the 11th International Workshops on Boolean Problems*, ed. by B. Steinbach. IWSBP 11 (Freiberg University of Mining and Technology, Freiberg, Germany, Sept. 2014), pp. 75–84. ISBN:978-3-86012-488-8

34. B. Steinbach, M. Werner, XBOOLE-CUDA - Fast calculations on the GPU, in *Problems and New Solutions in the Boolean Domain*, ed. by B. Steinbach (Cambridge Scholars Publishing, Newcastle upon Tyne, UK, Apr. 2016), pp. 117–149. ISBN:978-1-4438-8947-6

35. S.N. Yanushkevich et al. (eds.), *Decision Diagram Techniques for Micro- and Nanoelectronic Design - Handbook* (CRC Press, Taylor & Francis Group, Boca Raton, London, New York, 2006). ISBN:978-0-8493-3424-5. https://doi.org/10.1201/9781420037586

5 Logic Equations

36. D. Bochmann, B. Steinbach, *Logic Design with XBOOLE* (in German: Logikentwurf mit XBOOLE) (Verlag Technik GmbH, Berlin, Germany, 1991). ISBN:3-341-01006-8
37. F. Dresig et al., Programming with XBOOLE (in German: Programmieren mit XBOOLE), in *Series of Scientific Publications of the Chemnitz University of Technology* (in German: Wissenschaftliche Schriftenreihe der Technischen Universität Chemnitz) (1992), pp. 1–119. ISSN:0863-0755
38. B. Steinbach, XBOOLE—A toolbox for modelling, simulation, and analysis of large digital systems. Syst. Anal. Model. Simul. **9**(4), 297–312 (Sept. 1992). ISSN: 0232-9298
39. B. Steinbach, C. Posthoff, *EAGLE Start-up Aid - Efficient Computations with XBOOLE* (in German: EAGLE Starthilfe - Effiziente Berechnungen mit XBOOLE) (Edition am Gutenbergplatz, Leipzig, Germany, 2015). ISBN:978-3-95922-081-1

6 Boolean Differential Calculus

40. D. Bochmann, C. Posthoff, *Binäre Dynamische Systeme* (Akademie, Oldenbourg, 1981). ISBN: 978-3-48625-071-8
41. C. Posthoff, B. Steinbach, *Logic Functions and Equations—Binary Models for Computer Science.* (Springer, The Netherlands, 2004). ISBN: 978-1-44195-261-5. https://doi.org/10.1007/978-1-4020-2938-7
42. B. Steinbach, C. Posthoff, *Logic Functions and Equations—Examples and Exercises* (Springer, Berlin, 2009). ISBN: 978-1-4020-9594-8. https://doi.org/10.1007/978-1-4020-9595-5
43. B. Steinbach, C. Posthoff, Boolean differential calculus—theory and applications. J. Comput. Theor. Nanosci. **7**(6), 933–981 (2010). ISSN: 1546-1955. https://doi.org/10.1166/jctn.2010.1441
44. B. Steinbach, C. Posthoff, *Boolean Differential Equations* (Morgan and Claypool Publishers, San Rafael, 2013). ISBN: 978-1-6270-5241-2. https://doi.org/10.2200/S00511ED1V01Y201305DCS042
45. B. Steinbach, C. Posthoff, *Boolean Differential Calculus* (Morgan and Claypool Publishers, California, 2017). ISBN: 978-1-6270-5922-0. https://doi.org/10.2200/S00766ED1V01Y201704DCS052
46. A. Thayse, *Boolean Calculus of Differences* (Springer, Berlin, 1981). ISBN: 978-3-54010-286-1

7 Sets, Lattices, and Classes of Logic Functions

47. B. Steinbach, Generalized lattices of Boolean functions utilized for derivative operations, in *Materiały Konferencyjne KNWS'13*. KNWS. Łagów, Poland, June 2013, pp. 1–17. https://doi.org/10.13140/2.1.1874.3680
48. B. Steinbach, Solution of Boolean differential equations and their application for binary systems. Original title: Lösung binärer Differentialgleichungen und ihre Anwendung auf binäre Systeme (in German). Ph.D. Thesis. Technische Hochschule Karl-Marx-Stadt; now: University of Technology Chemnitz, 1981
49. B. Steinbach, C. Posthoff, *Boolean Differential Equations* (Morgan & Claypool Publishers, San Rafael, 2013). https://doi.org/10.2200/S00511ED1V01Y201305DCS042. ISBN: 978-1-6270-5241-2
50. B. Steinbach, C. Posthoff, Derivative operations for lattices of Boolean functions, in *Proceedings Reed-Muller Workshop 2013* (RM. Toyama, 2013), pp. 110–119. https://doi.org/10.13140/2.1.2398.6568

8 Logic, Arithmetic, and Special Functions

51. O.S. Rothaus. On "Bent" functions. J. Comb. Theory (A) **20**, 300–305 (1976). https://doi.org/10.1016/0097-3165(76)90024-8
52. B. Steinbach, Adjacency graph of the SNF as source of information, in *Boolean Problems, Proceedings of the 7th International Workshops on Boolean Problems*, ed. by B. Steinbach. IWSBP 7. Freiberg, Germany: Freiberg University of Mining and Technology, Sept. 2006, pp. 19–28. ISBN: 978-3-86012-287-7
53. B. Steinbach, Most complex Boolean functions, in *Proceedings Reed-Muller 2007* (RM. Oslo, 2007), pp. 13–23
54. B. Steinbach, A. De Vos, The shape of the SNF as a source of information, in *Boolean Problems, Proceedings of the 8th International Workshops on Boolean Problems*, ed. by B. Steinbach. IWSBP 8. Freiberg, Germany: Freiberg University of Mining and Technology, Sept. 2008, pp. 127–136. ISBN: 978-3-86012-346-1

55. B. Steinbach, A. Mishchenko, SNF: A Special Normal Form for ESOPs, in *Proceedings of the 5th International Workshop on Application of the Reed-Muller Expansion in Circuit Design* (RM, Starkville, 2001), pp. 66–81

56. B. Steinbach, C. Posthoff, *Boolean Differential Equations* (Morgan & Claypool Publishers, San Rafael, 2013). ISBN: 978-1-6270-5241-2. https://doi.org/10.2200/S00511ED1V01Y201305DCS042

57. B. Steinbach, C. Posthoff, Classes of Bent functions identified by specific normal forms and generated using Boolean differential equations. Facta Universitatis (Nis). Electrical Eng. **24**(3), 357–383 (2011). https://doi.org/10.2298/FUEE1103357S

58. B. Steinbach, C. Posthoff, Classification and generation of Bent functions, in *Proceedings Reed-Muller 2011 Workshop*. RM. Tuusula, Finland, May 2011, pp. 81–91

59. B. Steinbach, V. Yanchurkin, M. Lukac. On SNF optimization: a functional comparison of methods, in *Proceedings of the 6th International Symposium on Representations and Methodology of Future Computing Technology* (RM, Trier, 2003), pp. 11–18

9 SAT-Problems

60. A. Biere et al., eds., *Handbook of Satisfiability*, vol. 185. Frontiers in Artificial Intelligence and Applications (IOS Press, New York, 2009). ISBN: 978-1-58603-929-5

61. S. Cook, The complexity of theorem-proving procedures, in *Proceedings of the Third Annual ACM Symposium on Theory of Computing*. STOC (ACM, Shaker Heights, 1971), pp. 151–158. https://doi.org/10.1145/800157.805047

62. M.J.H. Heule, O. Kullmann, V.W. Marek, Solving and verifying the Boolean pythagorean triples problem via cube-and-conquer, in *Theory and Applications of Satisfiability Testing—SAT 2016*, ed. by N. Creignou, D. Le Berre (Springer, Cham, 2016), pp. 228–245. ISBN: 978-3-319-40970-2. https://doi.org/10.1007/978-3-319-40970-2_15

63. C. Posthoff, B. Steinbach, The solution of discrete constraint problems using Boolean models, in *Proceedings of the 2nd International Conference on Agents and Artificial Intelligence*, ed. by J. Filipe, A. Fred, B. Sharp. ICAART 2 (Valencia, Spain, 2010), pp. 487–493. ISBN: 978-989-674-021-4

64. B. Schwarzkopf, Without cover on a rectangular chessboard, in *Feenschach* (1990). German title: "Ohne Deckung auf dem Rechteck", pp. 272–275

65. B. Steinbach, XBOOLE—a toolbox for modelling, simulation, and analysis of large digital systems. Syst. Anal. Model. Simul. **9**(4), 297–312 (1992). ISSN: 0232-9298

66. B. Steinbach, C. Posthoff, New results based on Boolean Models, in *Boolean Problems, Proceedings of the 9th International Workshops on Boolean Problems*, ed. by B. Steinbach. IWSBP 9 (Freiberg University of Mining and Technology, Freiberg, 2010), pp. 29–36. ISBN: 978-3-86012-404-8

67. B. Steinbach, C. Posthoff, Parallel solution of covering problems super-linear speedup on a small set of cores, in *International Journal on Computing, Global Science and Technology Forum*. GSTF 1.2 (2011), pp. 113–122. ISSN: 2010–2283

68. B. Steinbach, C. Posthoff, Improvements of the construction of exact minimal covers of Boolean functions, in *Computer Aided Systems Theory – EUROCAST 2011. Lecture Notes in Computer Science*, ed. by R. Moreno-Díaz, F. Pichler, A. Quesada-Arencibia, vol. 6928, (Springer, Berlin, Heidelberg. ISBN 978-3-642-27578-4, 2012). https://doi.org/10.1007/978-3-642-27579-1_35

69. B. Steinbach, C. Posthoff, Sources and obstacles for parallelization—a comprehensive exploration of the unate covering problem using both CPU and GPU, in *GPU Computing with Applications in Digital Logic*, ed. by J. Astola et al. TICSP 62 (Tampere International Center for Signal Processing, Tampere, 2012), pp. 63–96. ISBN: 978-952-15-2920-7. https://doi.org/10.13140/2.1.4266.4320

70. B. Steinbach, C. Posthoff, Fast calculation of exact minimal unate coverings on both the CPU and the GPU, in *Proceedings of the 14th International Conference on Computer Aided Systems Theory—EUROCAST 2013—Part II*, ed. by A. Quesada-Arencibia et al. LNCS 8112 (Springer, Las Palmas, 2013), pp. 234–241. ISBN: 978-0-99245-180-6. https://doi.org/10.1007/978-3-642-53862-9_30

71. B. Steinbach, C. Posthoff, Multiple-valued problem solvers—comparison of several approaches, in *Proceedings of the IEEE 44th International Symposium on Multiple-Valued Logic*. ISMVL 44. Bremen, Germany (2014), pp. 25–31. https://doi.org/10.1109/ISMVL.2014.13

72. B. Steinbach, C. Posthoff, Evaluation and optimization of GPU based unate covering algorithms, in *Computer Aided Systems Theory—EUROCAST 2015*, ed. by A. Quesada-Arencibia et al. LNCS 9520 (Springer, Las Palmas, 2015), pp. 617–624. ISBN: 978-3-319-27340-2. https://doi.org/10.1007/978-3-319-27340-2_76

73. *SWI Prolog—Reference Manuel*. Updated for version 8.2.4 (2021). https://www.swi-prolog.org/pldoc/doc_for?object=manual

10   Extremely Complex Problems

74. J.D. Barrow, *The Constants of Nature: The Numbers That Encode the Deepest Secrets of the Universe* (Knopf Doubleday Publishing Group, 2009). ISBN:978-0-3754-2221-8
75. A. Biere, *Lingeling, Plingeling, PicoSAT and PrecoSAT at SAT Race 2010*. Tech. rep. 1. Linz, Austria, Aug. 2010
76. A. Biere et al. (eds.), *Handbook of Satisfiability*, vol. 185. Frontiers in Artificial Intelligence and Applications (IOS Press, Feb. 2009). ISBN:978-1-58603-929-5
77. M. Gebser et al., clasp: A conflict-driven answer set solver, in *International Conference on Logic Programming and Nonmonotonic Reasoning (LPNMR 2007)*, eds. by C. Baral, G. Brewka, J. Schlipf. LNCS, vol. 4483. (Springer, Berlin, Heidelberg, Germany, 2007), pp. 260–265. ISBN:978-3-540-72199-4. https://doi.org/10.1007/978-3-540-72200-7_23
78. Planck Collaboration, Planck 2013 results. I. Overview of products and scientific results. Astron. Astrophys. **571**(11), 1–44 (2014). https://doi.org/10.1051/0004-6361/201321529
79. B. Steinbach, C. Posthoff, Utilization of permutation classes for solving extremely complex 4-colorable rectangle-free grids, in *Proceedings of the IEEE 2012 International Conference on Systems and Informatics*. (ICSAI, Yantai, China, May 2012), pp. 2361–2370. ISBN:978-1-4673-0197-8. https://doi.org/10.13140/2.1.2365.8887
80. B. Steinbach, C. Posthoff, Highly complex 4-colored rectanglefree grids – Solution unsolved multiple-valued problems. J. Multiple Valued Logic Soft Comput. **MVLSC 24**(1-4), 369–404 (2014). ISSN:1542-3980
81. B. Steinbach, C. Posthoff, The last unsolved four-colored rectangle-free grid: The solution of extremely complex multiple-valued problems. J. Multiple Valued Logic Soft Comput. **MVLSC 25**(4-5), 461–490 (2015). ISSN:1542-3980
82. B. Steinbach, W. Wessely, C. Posthoff, Several approaches to parallel computing in the Boolean domain, in *First International Conference on Parallel, Distributed and Grid Computing*, eds. by P. Chaudhuri et al. (PDGC 1. Solan, H.P., India, Oct. 2010), pp. 6–11. https://doi.org/10.1109/PDGC.2010.5679611
83. B. Steinbach, C. Posthoff, W. Wessely, Approaches to shift the complexity limitations of Boolean problems, in *Computer - Aided Design of Discrete Devices - CAD DD 2010, Proceedings of the Seventh International Conference* (CAD DD 7, Minsk, Belarus, Nov. 2010), pp. 84–91. ISBN:978-985-6744-63-4

11   Combinational Circuits

84. R.L. Ashenhurst, The decomposition of switching functions, in *Annals of Computation Laboratory*, vol. 29 (Harvard University, Cambridge, MA, USA, 1959), pp. 74–116
85. D. Bochmann, F. Dresig, B. Steinbach, A new approach for multilevel circuit design, in *European Conference on Design Automation*. EURODAC '91 (IEEE Computer Society Press, Amsterdam, The Netherlands, 1991), pp. 374–377. ISBN:0-8186-2130-3
86. P. Böhlau, A decomposition strategy for the logic design based on properties of the function. In German: Eine Dekompositionsstrategie für den Logikentwurf auf der Basis funktionstypischer Eigenschaften. PhD thesis. Technical University Karl-Marx-Stadt, Germany, 1987
87. H.A. Curtis, *A New Approach to the Design of Switching Circuits* (Van Nostrand, Princeton, NJ, USA, 1962)
88. R. Garipelly, P.M. Kiran, A. Kumar, A review on reversible logic gates and their implementation. Int. J. Emerg. Tech. Adv. Eng. **3**(3), 183–191 (2013). ISSN:417–423
89. S. Khatri, K. Gulati, (eds.), *Advanced Techniques in Logic Synthesis, Optimizations and Applications* (Springer, New York, NY, USA, 2011). ISBN:978-1-4419-7517-1. https://doi.org/10.1007/978-1-4419-7518-8
90. R. Landauer, Irreversibility and heat generation in the computing process. IBM J. Res. Dev. **5**(3), 183–191 (1961). https://doi.org/10.1147/rd.53.0183
91. T. Le, Testability of combinational circuits–Theory and design. In German: Testbarkeit kombinatorischer Schaltungen—Theorie und Entwurf. PhD thesis. Technical University Karl-Marx-Stadt, Germany, 1989
92. A. Mishchenko, B. Steinbach, M. Perkowski, An algorithm for bidecomposition of logic functions, in *Proceedings on the 38th Design Automation Conference*, DAC 28 (IEEE Computer Society Press, Las Vegas, NV, USA, 2001), pp. 18–22. ISBN:1-58113-297-2. https://doi.org/10.1145/378239.378353
93. G.N. Povarov, About functional decomposition of Boolean functions. Rep. Acad. Sci. USSR **DAN 94**(5) (1954). In Russian
94. T. Sasao, J. Butler, On bi-decompositions of logic functions, in *Proceedings of International Workshop on Logic Synthesis 1997*, IWLS. Lake Tahoe City, USA, 1997
95. B. Steinbach, XBOOLE—A toolbox for modelling, simulation, and analysis of large digital systems. Syst. Anal. Model. Simul. **9**(4), 297–312 (Sept. 1992). ISSN:0232-9298

96. B. Steinbach, Decomposition using decision diagrams, in *Decision Diagram Technique for Micro- and Nanoelectronic Design, Handbook*, eds. by S. Yanushkevich et al. (CRC Press, Boca Raton, London, New York, 2006), pp. 509–544. ISBN:0-8493-3424-1

97. B. Steinbach, Relationships between vectorial bi-decompositions and strong EXOR-Bi-decompositions, in *Proceedings of the 25th International Workshop on Post-Binary ULSI Systems*, ULSI 25. Sapporo, Hokkaido, Japan, 2016

98. B. Steinbach, R. Hilbert, Fast test patterns generation using the Boolean differential calculus, in *Errors in Finite State Machines*, eds. by D. Bochmann, R. Ubar. in German (title of the contribution) Schnelle Testsatzgenerierung gestützt auf den Booleschen Differentialkalkül, (title of the book) Fehler in Automaten (Verlag Technik, Berlin, Germany, 1989), pp. 45–90. ISBN:3-341-00683-4

99. B. Steinbach, C. Lang, Exploiting functional properties of Boolean functions for optimal multi-level design by bi-decomposition, in *Artificial Intelligence in Logic Design*, ed. by S. Yanushkevich. SECS 766 (Springer, Dordrecht, The Netherlands, 2004), pp. 159–200. ISBN:978-90-481-6583-4. https://doi.org/10.1007/978-1-4020-2075-9_6

100. B. Steinbach, C. Posthoff, *Logic Functions and Equations - Examples and Exercises* (Springer Science + Business Media B.V., 2009). ISBN:978-1-4020-9594-8. https://doi.org/10.1007/978-1-4020-9595-5

101. B. Steinbach, C. Posthoff, Vectorial bi-decompositions of logic functions, in *Proceedings of the Reed-Muller Workshop 2015*, RM. Waterloo, Canada, 2015

102. B. Steinbach, C. Posthoff, Vectorial bi-decompositions for lattices of Boolean functions, in *Proceedings of the 12th International Workshops on Boolean Problems*, ed. by B. Steinbach. IWSBP 12. (Freiberg University of Mining and Technology, Freiberg, Germany, Sept. 2016), pp. 93–104. ISBN:978-3-86012-488-8

103. B. Steinbach, C. Posthoff, Compact XOR-Bi-decomposition for generalized lattices of Boolean functions, in *Proceedings Reed-Muller Workshop 2017*, RM 13. Novi Sad, Serbia, 2017

104. B. Steinbach, C. Posthoff, Vectorial bi-decompositions for lattices of Boolean functions, in *Further Improvements in the Boolean Domain*, ed. by B. Steinbach (Cambridge Scholars Publishing, Newcastle upon Tyne, UK, Jan. 2018), pp. 175–198. ISBN:978-1-5275-0371-7

105. B. Steinbach, M. Stöckert, Design of fully testable circuits by functional decomposition and implicit test pattern generation, in *Proceedings of 12th IEEE VLSI Test Symposium*, VTS 12. Cherry Hill, New Jersey, USA, 1994

106. B. Steinbach, A. Zakrevski, Three models and some theorems on decomposition of Boolean functions, in *Proceedings of the 3rd International Workshops on Boolean Problems*, ed. by B. Steinbach. IWSBP 3 (Freiberg University of Mining and Technology, Freiberg, Germany, Sept. 1998), pp. 11–18. ISBN:3-86012-069-7

12 Sequential Circuits

107. A. Bondy, M.R. Murty, *Graph Theory* (Springer, London, UK, 2008). ISBN: 978-1-84628-969-9
108. S.H. Caldwell, *Switching Circuits and Logical Design* (Wiley, New York, USA, 1958). ISBN:978-0-4711-2969-1
109. D.A. Huffman, The synthesis of sequential switching circuits. J. Franklin Inst. **257**(3), 161–190 (1954). https://doi.org/10.1016/0016-0032(54)90574-8

Index

© Springer Nature Switzerland AG 2022
B. Steinbach, C. Posthoff, *Logic Functions and Equations*,
https://doi.org/10.1007/978-3-030-88945-6

Printed in the United States
by Baker & Taylor Publisher Services